D0161101

Sequential Logic

Analysis and Synthesis

Sequential Logic

Analysis and Synthesis

Joseph Cavanagh
Santa Clara University

CRC is an imprint of the Taylor & Francis Group,
an informa business

Published in 2007 by
CRC Press
Taylor & Francis Group
6000 Broken Sound Parkway NW, Suite 300
Boca Raton, FL 33487-2742

Printed in the United States of America on acid-free paper
10 9 8 7 6 5 4 3 2 1

International Standard Book Number-10: 0-8493-7564-9 (Hardcover)
International Standard Book Number-13: 978-0-8493-7564-4 (Hardcover)
Library of Congress Card Number 2006002621

Library of Congress Cataloging-in-Publication Data

Cavanagh, Joseph J. F.
Sequential logic : analysis and synthesis / Joseph Cavanagh.
p. cm.
Includes bibliographical references and index.
ISBN-13: 978-0-8493-7564-4 (alk. paper)
ISBN-10: 0-8493-7564-9 (alk. paper)
1. Sequential machine theory. I. Title.

QA267.5.S4C38 2006
511.3'5--dc22 2006002621

Visit the Taylor & Francis Web site at
http://www.taylorandfrancis.com

and the CRC Press Web site at
http://www.crcpress.com

To my son Brad, who urged me onward.

CONTENTS

Preface xi

Chapter 1 Review of Combinational Logic 1

1.1 Number Systems 2
1.1.1 Binary Number System 3
1.1.2 Octal Number System 4
1.1.3 Decimal Number System 4
1.1.4 Hexadecimal Number System 5
1.2 Number Representations 8
1.2.1 Sign Magnitude 8
1.2.2 Diminished-Radix Complement 9
1.2.3 Radix Complement 10
1.3 Boolean Algebra 12
1.4 Minimization Techniques 18
1.4.1 Algebraic Minimization 19
1.4.2 Karnaugh Maps 20
1.4.3 Quine-McCluskey Algorithm 28
1.5 Logic Symbols 35
1.6 Analysis of Combinational Logic 37
1.7 Synthesis of Combinational Logic 42
1.8 Multiplexers 45
1.9 Decoders 47
1.10 Encoders 53
1.11 Comparators 54
1.12 Storage Elements 56
1.12.1 *SR* Latch 57
1.12.2 *D* Flip-Flop 57
1.12.3 *JK* Flip-Flop 59
1.12.4 *T* Flip-Flop 60
1.13 Programmable Logic Devices 61
1.13.1 Programmable Read-Only Memories 61
1.13.2 Programmable Array Logic 64
1.13.3 Programmable Logic Array 67
1.14 Problems 69

Chapter 2 Analysis of Synchronous Sequential Machines 77

2.1 Sequential Circuits 78
2.1.1 Machine Alphabets 79
2.1.2 Formal Definition of a Synchronous Sequential Machine 82

2.2 Classes of Sequential Machines 89
2.2.1 Combinational Logic 89
2.2.2 Registers 93
2.2.3 Counters 98
2.2.4 Moore Machines 105
2.2.5 Mealy Machines 110
2.2.6 Asynchronous Sequential Machines 121
2.2.7 Additional Definitions for Synchronous Sequential Machines 123
2.3 Methods of Analysis 136
2.3.1 Next-State Table 136
2.3.2 Present-State Map 137
2.3.3 Next-State Map 137
2.3.4 Input Map 138
2.3.5 Output Map 140
2.3.6 Timing Diagram 141
2.3.7 State Diagram 143
2.3.8 Analysis Examples 146
2.4 Complete and Incomplete Synchronous Sequential Machines 161
2.4.1 Complete Synchronous Sequential Machines 161
2.4.2 Incomplete Synchronous Sequential Machines 162
2.5 Problems 166

Chapter 3 Synthesis of Synchronous Sequential Machines 1 181

3.1 Synthesis Procedure 182
3.1.1 Equivalent States 183
3.2 Synchronous Registers 197
3.2.1 Parallel-In, Parallel-Out Registers 197
3.2.2 Parallel-In, Serial-Out Registers 201
3.2.3 Serial-In, Parallel-Out Registers 205
3.2.4 Serial-In, Serial-Out Registers 208
3.2.5 Linear Feedback Shift Registers 212
3.2.6 Combinational Shifter 218
3.3 Synchronous Counters 223
3.3.1 Modulo-8 Counter 223
3.3.2 Modulo-10 Counter 234
3.3.3 Johnson Counter 245
3.3.4 Binary-to-Gray Code Converter 248
3.4 Moore Machines 254
3.5 Mealy Machines 277
3.6 Moore-Mealy Equivalence 298
3.6.1 Mealy-to-Moore Transformation 298
3.6.2 Moore-to-Mealy Transformation 307
3.7 Output Glitches 307

3.7.1 Glitch Elimination Using State Code Assignment 312
3.7.2 Glitch Elimination Using Storage Elements 319
3.7.3 Glitch Elimination Using Complemented Clock 323
3.7.4 Glitch Elimination Using Delayed Clock 326
3.7.5 Glitches and Output Maps 333
3.7.6 Compendium of Output Glitches 339
3.8 Problems 344

Chapter 4 Synthesis of Synchronous Sequential Machines 2 361

4.1 Multiplexers for δ Next-State Logic 361
4.1.1 Linear-Select Multiplexers 363
4.1.2 Nonlinear-Select Multiplexers 377
4.2 Decoders for λ Output Logic 400
4.3 Programmable Logic Devices 412
4.3.1 Programmable Read-Only Memory 413
4.3.2 Programmable Array Logic 421
4.3.3 Programmable Logic Array 432
4.3.4 Field-Programmable Gate Array 437
4.4 Microprocessor-Controlled Sequential Machines 448
4.4.1 General Considerations 449
4.4.2 Mealy Machine Synthesis 453
4.4.3 Machine State Augmentation 461
4.4.4 Moore and Mealy Outputs 466
4.4.5 System Architecture 467
4.4.6 Multiple Machines 477
4.5 Sequential Iterative Machines 481
4.6 Error Detection in Synchronous Sequential Machines 489
4.7 Problems 500

Chapter 5 Analysis of Asynchronous Sequential Machines 519

5.1 Introduction 520
5.2 Fundamental-Mode Model 522
5.3 Methods of Analysis 526
5.4 Hazards 553
5.4.1 Static Hazards 553
5.4.2 Dynamic Hazards 568
5.4.3 Essential Hazards 572
5.4.4 Multiple-Order Hazards 577
5.5 Oscillations 578
5.6 Races 582
5.6.1 Noncritical Races 583
5.6.2 Cycles 586

5.6.3 Critical Races 586
5.7 Problems 590

Chapter 6 Synthesis of Asynchronous Sequential Machines 607

6.1 Introduction 608
6.2 Synthesis Procedure 610
6.2.1 State Diagram 612
6.2.2 Primitive Flow Table 616
6.2.3 Equivalent States 632
6.2.4 Merger Diagram 645
6.2.5 Merged Flow Table 656
6.2.6 Excitation Maps and Equations 661
6.2.7 Output Maps and Equations 691
6.2.8 Logic Diagram 710
6.3 Synthesis Examples 716
6.3.1 Mealy Machine with Two Inputs and One Output 716
6.3.2 Mealy Machine with Two Inputs and One Output Using a Programmable Logic Array (PLA) 727
6.3.3 Moore Machine with One Input and One Output 735
6.3.4 Mealy Machine with Two Inputs and Two Outputs 741
6.3.5 Mealy Machine with Three Inputs and One Output 755
6.3.6 Mealy Machine with Two Inputs and Two Outputs 765
6.4 Problems 777

Chapter 7 Pulse-Mode Asynchronous Sequential Machines 807

7.1 Analysis Procedure 809
7.1.1 *SR* Latches as Storage Elements 810
7.1.2 *SR* Latches with *D* Flip-Flops as Storage Elements 817
7.2 Synthesis Procedure 823
7.2.1 *SR* Latches as Storage Elements 823
7.2.2 *T* Flip-Flops as Storage Elements 830
7.2.3 *SR-T* Flip-Flops as Storage Elements 836
7.2.4 *SR* Latches with *D* Flip-Flops as Storage Elements 844
7.3 Problems 850

Appendix Answers to Selected Problems 861

Index 889

PREFACE

The field of digital logic consists primarily of analysis and synthesis of combinational and sequential logic circuits, also referred to as finite-state machines. Finite-state machines are designed into every computer. They occur in the form of counters, shift registers, microprogram control sequencers, sequence detectors, and many other sequential structures. The principal characteristic of combinational logic is that the outputs are a function of the present inputs only, whereas, the outputs of sequential logic are a function of the input sequence; that is, the input history. Sequential logic, therefore, requires storage elements which indicate the present state of the machine relative to a unique sequence of inputs.

Sequential logic is partitioned into synchronous and asynchronous sequential machines. Synchronous sequential machines are controlled by a system clock which provides the triggering mechanism to cause state changes. Asynchronous sequential machines have no clocking mechanism — the machines change state upon the application of input signals. The input signals provide the means to enable the sequential machines to proceed through a prescribed sequence of states.

During the last three decades, the design of digital computers has progressed from hand-drawn logic diagrams to automated logic diagrams to hardware description languages, where the latter precludes the need for logic diagrams of any type. Although the methods for portraying the hardware functions may have changed, the necessity for rigorous analysis and synthesis remains unchanged.

The purpose of this book is to present a thorough exposition of the analysis and synthesis of both synchronous and asynchronous sequential machines. Emphasis is placed on structured and rigorous design principles that can be applied to practical applications. Each step of the analysis and synthesis procedures is clearly delineated. Each method that is presented is expounded in sufficient detail with several accompanying examples. Many analysis and synthesis examples use mixed-logic symbols which incorporate both positive- and negative-input logic gates for NAND and NOR logic, while other examples utilize only positive-input logic gates. The use of mixed logic parallels the use of these symbols in the industry.

This book is intended to be tutorial in nature, and as such, is comprehensive and self-contained. All concepts and examples are explained in detail and carried out to completion. The book provides techniques which help to develop logical thinking and lend rigor to the analysis and synthesis procedures. The design philosophies that are presented can be easily extended to apply to any digital system.

Chapter 1 reviews the analysis and synthesis of combinational logic. A knowledge of combinational logic is a prerequisite for a study of sequential logic and it is assumed that the reader has the necessary background. An overview of number systems and number representations is included. This chapter also contains a review of boolean algebra, including axioms and theorems. Minimization techniques are presented using boolean algebra, Karnaugh maps, map-entered variables, and the Quine-McCluskey algorithm. An introduction to the ANSI/IEEE Std. 91-1984 logic symbols is presented, including their correlation with the distinctive shape logic symbols of IEEE Std. 91-1962. The ANSI/IEEE Std. 91-1984 defines symbols and associated terminology which describe a circuit function with a minimal amount of reference to technical data manuals. These symbols are used only in Chapter 1; the remaining chapters use the distinctive shape symbols, since these are more widely used and more easily understood. Simple combinational logic circuits are analyzed and synthesized. There are also sections describing latches, flip-flops, and programmable logic devices.

Chapter 2 presents a formal definition of sequential machines. The machine alphabets are described and include the input alphabet, the state alphabet, and the output alphabet. Six general classes of machines are described with accompanying equations for the major functions of each machine class. A constituent part of this chapter is additional definitions for sequential machines, which include state machines, submachines, terminal states, strongly connected machines, machine homomorphism, machine isomorphism, and equivalent machines. The main objective of this chapter is to present methods of analysis for synchronous sequential machines. These techniques form the basic mechanisms for effective analysis. Each method of analysis is accompanied by appropriate examples. The final section discusses complete and incomplete synchronous sequential machines.

Chapter 3 presents the first of two chapters on the synthesis of synchronous sequential machines. The synthesis procedure is outlined, then methods are described to determine state equivalence. If equivalent states can be identified, then redundant states can be eliminated, resulting in a machine with a minimal number of logic gates. The primary focus of this chapter is on the synthesis of deterministic synchronous sequential machines in which the next state is uniquely determined by the present state and the present inputs. The synthesis of each class of machine is presented with accompanying examples. A section is included on Moore-Mealy machine equivalence which includes Moore-to-Mealy and Mealy-to-Moore transformations. A final section discusses output glitches, which are momentary erroneous outputs. Several methods are presented to nullify the effects of these spurious signals which wreak havoc in digital systems, especially if they occur on an output signal.

Chapter 4 covers alternative methods for synthesizing synchronous sequential machines, in which logic functions other than discrete logic gates are utilized for both the input and output logic. These methods include logic macros such as multiplexers for the combinational input logic and decoders for the output logic. Programmable logic devices are presented in which synchronous sequential machines are synthesized using programmable read-only memories, programmable array logic devices, programmable logic array devices, and field-programmable gate arrays. A section is included on microprocessors and illustrates their use in the synthesis of synchronous sequential machines. Microprocessors are extremely versatile, especially for sequential machines with a large number of states or for multiple machines. Iterative networks are discussed and a method is presented to convert a combinational iterative network to a functionally equivalent sequential machine. A final section provides a discourse on error detection in synchronous sequential machines.

Chapter 5 covers the analysis of asynchronous sequential machines. The fundamental-mode model is introduced and is a major consideration in both the analysis and synthesis of asynchronous sequential machines. Methods of analysis are presented which completely characterize the operation of the machine. The mechanisms include an excitation map, a next-state table, a state diagram, and a flow table. Three significant items that are of paramount concern in the analysis and synthesis of asynchronous sequential machines are discussed in detail: hazards, oscillations, and races. Hazards are the result of varying propagation delays caused by logic gates, wires, and different signal path lengths. Hazards can produce erroneous transient signals on the outputs. Oscillations occur when a change to the input vector causes the machine to oscillate between two or more unstable states. Races occur when a change to the input vector results in simultaneous changes to two or more excitation variables.

Chapter 6 presents an exhaustive discourse on the synthesis of asynchronous sequential machines. The chapter develops a systematic method for this relatively complex procedure. Each step in the synthesis procedure employs several examples which help to clarify the corresponding step. The synthesis procedure is developed as individual fragments of the complete process. Using this approach, each step in the procedure can be reviewed as a separate entity. The complete synthesis process is explicated by several worked examples. Only through dedicated perseverance can the synthesis of asynchronous sequential machines be effectively learned. To achieve this goal, several examples are presented detailing the synthesis procedure in a step-by-step process.

Chapter 7 presents analysis and synthesis procedures for pulse-mode asynchronous sequential machines. Many situations are encountered in digital engineering where the input signals occur as pulses and in which there is no periodic clock signal to synchronize the operation of the machine. A vending machine is a typical example. Due to the stringent requirements of input pulse characteristics, however, pulse-mode machines are less preferred than synchronous sequential machines. Analysis of pulse-mode asynchronous sequential machines consists of deriving a next-state table, input maps, output maps, and a state diagram for a given logic diagram in much the same manner as for synchronous sequential machines. The synthesis procedure is described using several different types of storage elements.

It is assumed that the reader has an adequate background in the analysis and synthesis of combinational logic. This book presents basic and advanced concepts in sequential machine analysis and synthesis and is designed for practicing electrical and computer engineers, graduate students who require a noncredit course in sequential logic to supplement their program of studies, and for undergraduate students in electrical and computer engineering.

I would like to express my appreciation and thanks to the following people who gave generously of their time and expertise to review the manuscript and submit comments: Professor Daniel W. Lewis, Chair, Department of Computer Engineering, Santa Clara University who supported me in all my endeavors; Dr. Geri Lamble; Steve Midford; and Ron Lewerenz. Thanks also to Nora Konopka and the staff at Taylor & Francis for their support.

Joseph Cavanagh

1.1 *Number Systems*
1.2 *Number Representations*
1.3 *Boolean Algebra*
1.4 *Minimization Techniques*
1.5 *Logic Symbols*
1.6 *Analysis*
1.7 *Synthesis*
1.8 *Multiplexers*
1.9 *Decoders*
1.10 *Encoders*
1.11 *Comparators*
1.12 *Storage Elements*
1.13 *Programmable Logic Devices*
1.14 *Problems*

1 Review of Combinational Logic

The purpose of this book is to analyze and synthesize finite-state machines; that is, synchronous and asynchronous sequential machines. Prior to presenting sequential machines in Chapter 2, however, a review of combinational logic will be covered. It is assumed that a reader who is interested in studying sequential logic has an adequate background in the analysis and synthesis of combinational logic. This chapter, therefore, presents only an overview of combinational logic. *Combinational logic* refers to logic circuits whose present output values depend only upon the present input values. A combinational circuit is a special case of a sequential circuit in which there is no storage capability. The word combinational is used interchangeably with combinatorial.

A combinational logic circuit is comprised of combinational logic elements (or logic primitives), which have one or more inputs and at least one output. The input and output variables are characterized by discrete states such that, at any instant of time, the state of each output is completely determined by the states of the inputs. A combinational circuit refers to a logic circuit which performs the same fixed mapping of inputs into outputs, regardless of the past input history and may be considered as a one-state sequential machine.

This chapter will discuss number systems and number representations, define the logical operations encountered in different logic families, and briefly review the IEEE Standard: Graphic Symbols for Logic Functions, ANSI/IEEE Std. 91-1984. The boolean axioms and theorems will be listed, without proof. Three minimization techniques will be presented: algebraic minimization using boolean algebra, Karnaugh maps, including map-entered variables, and the Quine-McCluskey algorithm.

The important, and sometimes confusing, topic of logic assertion levels will also be discussed. Logic assertion levels are especially difficult to understand by new students in the field of digital logic design. Combinational logic circuits will be analyzed and synthesized (designed) using different logic families. Chapter 1 will also briefly review the different storage elements that are used in the synthesis of synchronous and asynchronous sequential machines.

1.1 Number Systems

Digital systems contain signals that are represented by two values: 0 or 1, which describe the signal alphabet. Numerical data, or operands, are expressed in various positional number systems for each radix. This section will discuss binary, octal, decimal, and hexadecimal positional number systems. A *positional number system* is characterized by a radix (or base) r, which is an integer greater than or equal to 2, and by a set of r digits, which are numbered from 0 to $r - 1$. For example, for radix 8, the digits range from 0 to 7.

In a positional number system, a number is encoded as a vector of n digits in which each digit is weighted according to its position in the vector. An n-bit integer A is represented in a positional number system as follows:

$$A = (a_{n-1} a_{n-2} a_{n-3} \ldots a_1 a_0) \tag{1.1}$$

where $0 \leq a_i \leq r - 1$. The high-order and low-order digits are a_{n-1} and a_0, respectively.

The number in Equation 1.1 (also referred to as a vector or operand) can represent positive integer values in the range 0 to $r^n - 1$. Thus, a positive integer A is written as

$$A = a_{n-1} r^{n-1} + a_{n-2} r^{n-2} + a_{n-3} r^{n-3} + \ldots + a_1 r^1 + a_0 r^0 \tag{1.2}$$

The value for A can be represented more compactly as

$$A = \sum_{i=0}^{n-1} a_i r^i \tag{1.3}$$

The expression of Equation 1.2 can be extended to include fractions. For example,

$$A = a_{n-1} r^{n-1} + \ldots + a_1 r^1 + a_0 r^0 + a_{-1} r^{-1} + a_{-2} r^{-2} + a_{-m} r^{-m} \tag{1.4}$$

Equation 1.4 can be represented as

$$A = \sum_{i=-m}^{n-1} a_i r^i \tag{1.5}$$

The number system requires exactly r digits. For example, radix 10 consists of digits 0 through 9. The highest digit value is one less than the radix. Adding 1 to the highest digit value produces a sum of zero and a carry of 1 to the next higher-order column.

1.1.1 Binary Number System

The radix is 2 in the *binary number system*; therefore, only two digits are used: 0 and 1. The low-value digit is 0 and the high-value digit is $(r - 1) = 1$. The binary number system is the most conventional and easily implemented system for internal use in a digital computer; therefore, most digital computers use the binary number system. There is a disadvantage when converting to and from the externally used decimal system; however, this is compensated for by the ease of implementation and the speed of execution in binary of the four basic operations: addition, subtraction, multiplication, and division. The radix point is implied within the internal structure of the computer; that is, there is no specific storage element assigned to contain the radix point.

The weight assigned to each position of a binary number is as follows:

$$2^{n-1} 2^{n-2} \ldots\ 2^3\ 2^2\ 2^1\ 2^0 \bullet 2^{-1} 2^{-2} 2^{-3}\ \ldots\ 2^{-m}$$

where the integer and fraction are separated by the radix point (binary point). The decimal value of the binary number 1101.101_2 is obtained by using Equation 1.4, where $r = 2$ and $a_i \in \{0,1\}$ for $-m \leq i \leq n - 1$. Therefore,

$$\begin{array}{cccccccc} 2^3 & 2^2 & 2^1 & 2^0 & \bullet & 2^{-1} & 2^{-2} & 2^{-3} \\ 1 & 1 & 0 & 1 & . & 1 & 0 & 1_2 \end{array} = (1 \times 2^3) + (1 \times 2^2) + (0 \times 2^1) + (1 \times 2^0) + (1 \times 2^{-1}) + (0 \times 2^{-2}) + (1 \times 2^{-3})$$

$$= 13.625_{10}$$

1.1.2 Octal Number System

The radix is 8 in the *octal number system*; therefore, eight digits are used, 0 through 7. The low-value digit is 0 and the high-value digit is $(r - 1) = 7$. The weight assigned to each position of an octal number is as follows:

$$8^{n-1}\,8^{n-2} \ldots\ 8^3\ 8^2\ 8^1\ 8^0 \bullet 8^{-1}\,8^{-2}\,8^{-3} \ldots\ 8^{-m}$$

where the integer and fraction are separated by the radix point (octal point). The decimal value of the octal number 217.6_8 is obtained by using Equation 1.4, where $r = 8$ and $a_i \in \{0,1,2,3,4,5,6,7\}$ for $-m \leq i \leq n - 1$. Therefore,

$$\begin{array}{cccccl} 8^2 & 8^1 & 8^0 & \bullet & 8^{-1} & \\ 2 & 1 & 7 & . & 6_8 & = (2 \times 8^2) + (1 \times 8^1) + (7 \times 8^0) + (6 \times 8^{-1}) \\ & & & & & = 143.75_{10} \end{array}$$

When a count of 1 is added to 7_8, the sum is zero and a carry of 1 is added to the next higher-order column on the left.

Binary-coded octal Each octal digit can be encoded into a corresponding binary number. The highest-valued octal digit is 7; therefore, three binary digits are required to represent each octal digit. This is shown in Table 1.1, which lists the eight decimal digits (0 through 7) and indicates the corresponding octal and binary-coded octal (BCO) digits. Table 1.1 also shows octal numbers of more than one digit.

1.1.3 Decimal Number System

The radix is 10 in the *decimal number system*; therefore, ten digits are used, 0 through 9. The low-value digit is 0 and the high-value digit is $(r - 1) = 9$. The weight assigned to each position of a decimal number is as follows:

$$10^{n-1}\,10^{n-2} \ldots\ 10^3\ 10^2\ 10^1\ 10^0 \bullet 10^{-1}\,10^{-2}\,10^{-3} \ldots\ 10^{-m}$$

where the integer and fraction are separated by the radix point (decimal point). The value of 7537_{10} is immediately apparent; however, the value is also obtained by using Equation 1.4, where $r = 10$ and $a_i \in \{0,1,2,3,4,5,6,7,8,9\}$ for $-m \leq i \leq n - 1$. That is,

$$\begin{array}{ccccl} 10^3 & 10^2 & 10^1 & 10^0 & \\ 6 & 3 & 5 & 7_{10} & = (6 \times 10^3) + (3 \times 10^2) + (5 \times 10^1) + (7 \times 10^0) \end{array}$$

When a count of 1 is added to decimal 9, the sum is zero and a carry of 1 is added to the next higher-order column on the left.

Table 1.1 Binary-coded octal numbers

Decimal	Octal	Binary-coded octal
0	0	000
1	1	001
2	2	010
3	3	011
4	4	100
5	5	101
6	6	110
7	7	111
8	10	001 000
9	11	001 001
.	.	.
.	.	.
.	.	.
172	254	010 101 100
385	601	110 000 001

Binary-coded decimal Each decimal digit can be encoded into a corresponding binary number. The highest-valued decimal digit is 9, which requires four bits in the binary representation. Therefore, four binary digits are required to represent each decimal digit. This is shown in Table 1.2, which lists the ten decimal digits (0 through 9) and indicates the corresponding binary-coded decimal (BCD) digits. Table 1.2 also shows BCD numbers of more than one decimal digit.

1.1.4 Hexadecimal Number System

The radix is 16 in the *hexadecimal number system*; therefore, 16 digits are used, 0 through 9 and A through F, where A, B, C, D, E, and F correspond to decimal 10, 11, 12, 13, 14, and 15, respectively. The low-value digit is 0 and the high-value digit is $(r - 1) = 15$ (F). The weight assigned to each position of a hexadecimal number is as follows:

$$16^{n-1}\,16^{n-2}\,\ldots\;16^3\,16^2\,16^1\,16^0\bullet 16^{-1}\,16^{-2}\,16^{-3}\,\ldots\;16^{-m}$$

where the integer and fraction are separated by the radix point (hexadecimal point). The decimal value of the hexadecimal number $6A8C.D416_{16}$ is obtained by using Equation 1.4, where $r = 16$ and $a_i \in \{0,1,2,3,4,5,6,7,8,9,A,B,C,D,E,F\}$ for $-m \le i \le n - 1$. Therefore,

$$\begin{array}{l} 16^3\ 16^2\ 16^1\ 16^0 \bullet \quad 16^{-1}16^{-2}16^{-3}16^{-4} \\ 6\ \ A\ \ 8\ \ C\ .\ \ D\ \ 4\ \ 1\ \ 6 = (6 \times 16^3) + (10 \times 16^2) + (8 \times 16^1) \\ \qquad + (12 \times 16^0) + (13 \times 16^{-1}) + (4 \times 16^{-2}) \\ \qquad + (1 \times 16^{-3}) + (6 \times 16^{-4}) \\ \qquad = 27{,}276.82846_{10} \end{array}$$

When a count of 1 is added to hexadecimal F, the sum is zero and a carry of 1 is added to the next higher-order column on the left.

Table 1.2 Binary-coded decimal numbers

Decimal	Binary-coded decimal
0	0000
1	0001
2	0010
3	0011
4	0100
5	0101
6	0110
7	0111
8	1000
9	1001
10	0001 0000
11	0001 0001
12	0001 0010
.	.
.	.
.	.
124	0001 0010 0100
365	0011 0110 0101

Binary-coded hexadecimal Each hexadecimal digit corresponds to a 4-bit binary number as shown in Table 1.3. All 2^4 values of the four binary bits are used to represent the 16 hexadecimal digits. Table 1.3 also indicates hexadecimal numbers of more than one digit.

The emphasis of this book is the analysis and synthesis of sequential machines, and thus, it is assumed that a reader will have an adequate background not only in combinational logic but also in conversion between different radices. Therefore, conversion between radices will not be presented, other than stating the following general method:

To convert a number in radix 10 to any other radix r, repeatedly divide the integer by radix r, then repeatedly multiply the fraction by radix r. To convert any nondecimal number A_{ri} in radix ri to another nondecimal number A_{rj} in radix rj, first convert the number A_{ri} to decimal using Equation 1.4, then convert the decimal number to radix rj by using repeated division and/or repeated multiplication.

Conversion from any radix to radix 10 is easily accomplished by using Equation 1.2 or 1.4 for integers, or Equation 1.4 or 1.5 for numbers consisting of integers and fractions. Table 1.4 lists the digits used for the four number systems that were described in Section 1.1.

Table 1.3 Binary-coded hexadecimal numbers

Decimal	Hexadecimal	Binary-coded hexadecimal
0	0	0000
1	1	0001
2	2	0010
3	3	0011
4	4	0100
5	5	0101
6	6	0110
7	7	0111
8	8	1000
9	9	1001
10	A	1010
11	B	1011
12	C	1100
13	D	1101
14	E	1110
15	F	1111
.	.	.
.	.	.
.	.	.
124	7C	0111 1100
365	16D	0001 0110 1101

Table 1.4 Digits used for binary, octal, decimal, and hexadecimal number systems

0	1	2	3	4	5	6	7	8	9	A	B	C	D	E	F
Binary															
Octal															
Decimal															
Hexadecimal															

1.2 Number Representations

For convenience, only integers will be considered in this section, but the principles also apply to fractions. Since both positive and negative numbers are used in computers, an encoding scheme must be devised in which both types of numbers are distributed as evenly as possible. There must also be an easy method to distinguish between positive and negative numbers, such as a sign test, and to detect the number zero.

For a signed radix number, the high-order (leftmost) digit is usually reserved for the sign of the number. Consider the following number A with radix r:

$$A = (a_{n-1}a_{n-2}a_{n-3} \ldots a_1a_0)_r$$

where the sign digit a_{n-1} has the value

$$A = \begin{cases} 0 & \text{if } A \geq 0 \\ r-1 & \text{if } A < 0 \end{cases} \tag{1.6}$$

The remaining digits in A indicate either the true magnitude or the magnitude in a complemented form. There are three conventional ways to represent signed numbers in a positional number system: sign magnitude, diminished-radix complement, and radix complement.

1.2.1 Sign Magnitude

In this representation, an integer has the following decimal range:

$$-(r^{n-1} - 1) \text{ to } +(r^{n-1} - 1) \tag{1.7}$$

where the number zero is considered to be positive. Thus, a positive number A is represented as

$$A = (0\, a_{n-2} a_{n-3} \;\cdots\; a_1 a_0)_r \tag{1.8}$$

and a negative number with the same absolute value as

$$-A = [(r-1)\, a_{n-2} a_{n-3} \;\cdots\; a_1 a_0]_r \tag{1.9}$$

In sign-magnitude notation, the positive version A differs from the negative version $-A$ only in the sign digit position. The magnitude portion $a_{n-2} a_{n-3} \;\cdots\; a_1 a_0$ is identical for both positive and negative numbers of the same absolute value.

1.2.2 Diminished-Radix Complement

This is the *(r – 1) complement* in which an integer has the following decimal range:

$$-(r^{n-1} - 1) \text{ to } +(r^{n-1} - 1) \tag{1.10}$$

which is the same as the range for sign-magnitude integers and where the number zero is again considered to be positive. Thus, a positive number A is represented as

$$A = (0\, a_{n-2} a_{n-3} \;\cdots\; a_1 a_0)_r \tag{1.11}$$

and a negative number as

$$A' = [(r-1)\, a_{n-2}' a_{n-3}' \;\cdots\; a_1' a_0']_r \tag{1.12}$$

where

$$a_i' = (r-1) - a_i \tag{1.13}$$

Example 1.1 Obtain the diminished-radix complement (9s complement) of 08752.43_{10}, where 0 is the sign digit indicating a positive number. The 9s complement is obtained by using Equation 1.12 and Equation 1.13. When a number is complemented in any form, the number is negated. Therefore, the sign of the complemented radix 10 number is $(r - 1) = 9$. The remaining digits of the number are obtained by using Equation 1.13 such that, each digit in the complemented number is obtained by subtracting the given digit from 9. Therefore, the 9s complement of 08752.43_{10} is

$9 - 0 = 9, 9 - 8 = 1, 9 - 7 = 2, 9 - 5 = 4, 9 - 2 = 7, \bullet\, 9 - 4 = 5, 9 - 3 = 6$
$= 91247.56_{10}$,

where the sign digit is $(r - 1) = 9$. If the above answer is negated, then the original number will be obtained. Thus, the 9s complement of 91247.56_{10} $= 08752.43_{10}$; that is, the 9s complement of -1247.56_{10} is $+8752.43_{10}$, as written in conventional sign magnitude notation for radix 10.

1.2.3 Radix Complement

This is the *r complement*, where an integer has the following decimal range:

$$-(r^{n-1}) \text{ to } + (r^{n-1} - 1) \qquad (1.14)$$

where the number zero is positive. A positive number A is represented as

$$A = (0\, a_{n-2} a_{n-3} \ \ldots\ a_1 a_0)_r \qquad (1.15)$$

and a negative number as

$$(A')_{+1} = \{[(r - 1)\, a_{n-2}' a_{n-3}' \ \ldots\ a_1' a_0'] + 1\}_r \qquad (1.16)$$

Thus, the radix complement is obtained by adding 1 to the diminished-radix complement; that is, $(r - 1) + 1 = r$. Note that all three number representations have the same format for positive numbers and differ only in the way that negative numbers are represented.

Example 1.2 Obtain the radix complement (10s complement) of 08752.43_{10}. Determine the 9s complement as in Example 1.1, then add 1. The 10s complement of 08752.43_{10} is

$9 - 0 = 9, 9 - 8 = 1, 9 - 7 = 2, 9 - 5 = 4, 9 - 2 = 7, \bullet\, 9 - 4 = 5, 9 - 3 = 6 + 1$
$= 91247.57_{10}$.

To verify that the radix complement of 08752.43_{10} is 91247.57_{10}, the sum of the two numbers should equal zero for radix 10. This is indeed the case, as shown below. The previous three number representations are summarized in Table 1.5 for radix r.

$$\begin{array}{rr} & 08752.43 \\ +) & \underline{91247.57} \\ & 00000.00 \end{array}$$

Table 1.5 Number representations for positive and negative integers of the same absolute value for radix *r*

Number representation	Positive numbers	Negative numbers
Sign magnitude	$0\ a_{n-2}a_{n-3}\ \ldots\ a_1a_0$	$(r-1)\ 0\ a_{n-2}a_{n-3}\ \ldots\ a_1a_0$
Diminished-radix complement	$0\ a_{n-2}a_{n-3}\ \ldots\ a_1a_0$	$(r-1)\ a_{n-2}'a_{n-3}'\ \ldots\ a_1'a_0'$
Radix complement	$0\ a_{n-2}a_{n-3}\ \ldots\ a_1a_0$	$(r-1)\ a_{n-2}'a_{n-3}'\ \ldots\ a_1'a_0' + 1$

Overflow Overflow occurs when the result of an arithmetic operation (usually addition) exceeds the word size of the machine; that is, the sum is not within the representable range of numbers provided by the number representation. For numbers in 2s complement representation, the range is from -2^{n-1} to $+2^{n-1}-1$. For two n-bit numbers

$$A = a_{n-1}a_{n-2}a_{n-3}\ \ldots\ a_1a_0$$

$$B = b_{n-1}\ b_{n-2}\ b_{n-3}\ \ldots\ b_1b_0$$

a_{n-1} and b_{n-1} are the sign bits of operands A and B, respectively. Overflow can be detected by either of the following two equations:

$$\text{Overflow} = (a_{n-1} \bullet b_{n-1} \bullet s_{n-1}') + (a_{n-1}' \bullet b_{n-1}' \bullet s_{n-1})$$

$$\text{Overflow} = c_{n-1} \oplus c_{n-2} \tag{1.17}$$

where the symbol "•" is the logical AND operator, the symbol "+" is the logical OR operator, the symbol "⊕" is the exclusive-OR operator as defined in Section 1.3, and c_{n-1} and c_{n-2} are the carry bits out of positions $n-1$ and $n-2$, respectively.

Thus, overflow produces an erroneous sign reversal and is possible only when both operands have the same sign. An overflow cannot occur when adding two operands of different signs, since adding a positive number to a negative number produces a result that falls within the limit specified by the two numbers.

1.3 Boolean Algebra

In 1854, George Boole introduced a systematic treatment of the logic operations AND, OR, and NOT, which is now called boolean algebra. The symbols (or operators) used for the algebra and the corresponding function definitions are listed in Table 1.6. The table also includes the exclusive-OR function, which is characterized by the three operations of AND, OR, and NOT. Table 1.7 illustrates the truth tables for the boolean operations AND, OR, NOT, exclusive-OR, and exclusive-NOR, where z_1 is the result of the operation.

Table 1.6 Boolean operators for variables x_1 and x_2

Operator	Function	Definition
$\bullet$	AND	$x_1 \bullet x_2$ (Also $x_1 x_2$)
$+$	OR	$x_1 + x_2$
$'$	NOT (negation)	x_1'
$\oplus$	Exclusive-OR	$(x_1 x_2') + (x_1' x_2)$

Table 1.7 Truth table for AND, OR, NOT, exclusive-OR, and exclusive-NOR operations

AND		OR		NOT		Exclusive-OR		Exclusive-NOR	
$x_1 x_2$	z_1	$x_1 x_2$	z_1	x_1	z_1	$x_1 x_2$	z_1	$x_1 x_2$	z_1
0 0	0	0 0	0	0	1	0 0	0	0 0	1
0 1	0	0 1	1	1	0	0 1	1	0 1	0
1 0	0	1 0	1			1 0	1	1 0	0
1 1	1	1 1	1			1 1	0	1 1	1

The AND operator, which corresponds to the boolean product, is also indicated by the symbol "$\wedge$" ($x_1 \wedge x_2$) or by no symbol if the operation is unambiguous. Thus, $x_1 x_2$, $x_1 \bullet x_2$, and $x_1 \wedge x_2$ are all read as "x_1 AND x_2." The OR operator, which corresponds to the boolean sum, is also specified by the symbol "$\vee$." Thus, $x_1 + x_2$ and $x_1 \vee x_2$ are both read as "x_1 OR x_2." The symbol for the complement (or negation) operation is usually specified by the prime "$'$" symbol immediately following the variable (x_1'), by a bar over the variable ($\overline{x}_1$), or by the symbol "$\neg$" ($\neg x_1$). This book will use the symbols defined in Table 1.6.

Boolean algebra is a deductive mathematical system which can be defined by a set of variables, a set of operators, and a set of axioms (or postulates). An *axiom* is a statement that is universally accepted as true; that is, the statement needs no proof, because its truth is obvious. The axioms of boolean algebra form the basis from which the theorems and other properties can be derived.

Most axioms and theorems are characterized by two laws. Each law is the dual of the other. The principle of duality specifies that the *dual* of an algebraic expression can be obtained by interchanging the binary operators • and + and by interchanging the identity elements 0 and 1. Since the primary emphasis of this book is the analysis and synthesis of sequential machines, the axioms and theorems of combinational logic are presented without proof. Some proofs are left as exercises.

Boolean algebra is an algebraic structure consisting of a set of elements B, together with two binary operators • and + and a unary operator ', such that the following axioms are true, where the notation $x_1 \in X$ is read as "x_1 is an element of the set X":

Axiom 1: Boolean set definition The set B contains at least two elements x_1 and x_2, where $x_1 \neq x_2$.

Axiom 2: Closure laws For every $x_1, x_2 \in B$,

(a) $x_1 + x_2 \in B$
(b) $x_1 \bullet x_2 \subset B$

Axiom 3: Identity laws There exists two unique *identity elements* 0 and 1, where 0 is an identity element with respect to the boolean sum and 1 is an identity element with respect to the boolean product. Thus, for every $x_1 \in B$,

(a) $x_1 + 0 = 0 + x_1 = x_1$
(b) $x_1 \bullet 1 = 1 \bullet x_1 = x_1$

Axiom 4: Commutative laws The commutative laws specify that the order in which the variables appear in a boolean expression is irrelevant — the result is the same. Thus, for every $x_1, x_2 \in B$,

(a) $x_1 + x_2 = x_2 + x_1$
(b) $x_1 \bullet x_2 = x_2 \bullet x_1$

Axiom 5: Associative laws The associative laws state that three or more variables can be combined in an expression using boolean multiplication or addition and that the order of the variables can be altered without changing the result. Thus, for every $x_1, x_2, x_3 \in B$,

(a) $(x_1 + x_2) + x_3 = x_1 + (x_2 + x_3)$
(b) $(x_1 \bullet x_2) \bullet x_3 = x_1 \bullet (x_2 \bullet x_3)$

Axiom 6: Distributive laws The distributive laws for boolean algebra are similar, in many respects, to those for college algebra. The interpretation, however, is different and is a function of the boolean product and the boolean sum. This is a very useful axiom in minimizing boolean functions. For every $x_1, x_2, x_3 \in B$,

(a) The operator + is distributive over the operator • such that,
$x_1 + (x_2 \bullet x_3) = (x_1 + x_2) \bullet (x_1 + x_3)$

(b) The operator • is distributive over the operator + such that,
$x_1 \bullet (x_2 + x_3) = (x_1 \bullet x_2) + (x_1 \bullet x_3)$

Axiom 7: Complementation laws For every $x_1 \in B$, there exists an element x_1' (called the complement of x_1), where $x_1' \in B$, such that,

(a) $x_1 + x_1' = 1$
(b) $x_1 \bullet x_1' = 0$

The theorems presented below are derived from the axioms and are listed in pairs, where each relation in the pair is the dual of the other.

Theorem 1: 0 and 1 associated with a variable Every variable in boolean algebra can be characterized by the identity elements 0 and 1. Thus, for every $x_1 \in B$,

(a) $x_1 + 1 = 1$
(b) $x_1 \bullet 0 = 0$

Theorem 2: 0 and 1 complement The 2-valued boolean algebra has two distinct identity elements 0 and 1, where $0 \neq 1$. The operations using 0 and 1 are as follows:

$0 + 0 = 0$ $\quad$ $0 + 1 = 1$
$1 \bullet 1 = 1$ $\quad$ $1 \bullet 0 = 0$

A corollary to Theorem 2 specifies that element 1 satisfies the requirements of the complement of element 0, and vice versa. Thus, each identity element is the complement of the other.

(a) $0' = 1$
(b) $1' = 0$

Theorem 3: Idempotent laws Idempotency relates to a nonzero mathematical quantity which, when applied to itself for a binary operation, remains unchanged. Thus, if $x_1 = 0$, then $x_1 + x_1 = 0 + 0 = 0$ and if $x_1 = 1$, then $x_1 + x_1 = 1 + 1 = 1$. Therefore, one of the elements is redundant and can be discarded. The dual is true for the operator •. The idempotent laws eliminate redundant variables in a boolean expression and can be extended to any number of identical variables. This law is also referred to as the law of tautology, which precludes the needless repetition of the variable. For every $x_1 \in B$,

(a) $x_1 + x_1 = x_1$
(b) $x_1 \bullet x_1 = x_1$

Theorem 4: Involution law The involution law states that the complement of a complemented variable is equal to the variable. There is no dual for the involution law. The law is also called the law of double complementation. Thus, for every $x_1 \in B$,

$$x_1'' = x_1$$

Theorem 5: Absorption law 1 This version of the absorption law states that some 2-variable boolean expressions can be reduced to a single variable without altering the result. Thus, for every $x_1, x_2 \in B$,

(a) $x_1 + (x_1 \bullet x_2) = x_1$
(b) $x_1 \bullet (x_1 + x_2) = x_1$

Theorem 6: Absorption law 2 This version of the absorption law is used to eliminate redundant variables from certain boolean expressions. Absorption law 2 eliminates a variable or its complement and is a very useful law for minimizing boolean expressions.

(a) $x_1 + (x_1' \bullet x_2) = x_1 + x_2$
(b) $x_1 \bullet (x_1' + x_2) = x_1 \bullet x_2$

Theorem 7: DeMorgan's laws DeMorgan's laws are also useful in minimizing boolean functions. DeMorgan's laws convert the complement of a sum term or a product term into a corresponding product or sum term, respectively. For every $x_1, x_2 \in B$,

(a) $(x_1 + x_2)' = x_1' \bullet x_2'$
(b) $(x_1 \bullet x_2)' = x_1' + x_2'$

Parts (a) and (b) of DeMorgan's laws represent expressions for NOR and NAND gates, respectively. DeMorgan's laws can be generalized for any number of variables, such that,

(a) $(x_1 + x_2 + \ldots + x_n)' = x_1' \bullet x_2' \bullet \ldots \bullet x_n'$
(b) $(x_1 \bullet x_2 \bullet \ldots \bullet x_n)' = x_1' + x_2' + \ldots + x_n'$

When applying DeMorgan's laws to an expression, the operator • takes precedence over the operator +. For example, use DeMorgan's law to complement the boolean expression $x_1 + x_2x_3$.

$$(x_1 + x_2x_3)' = [x_1 + (x_2x_3)]'$$
$$= x_1'(x_2' + x_3')$$

Note that: $(x_1 + x_2x_3)' \neq x_1' \bullet x_2' + x_3'$

Minterm A minterm is the boolean product of n variables and contains all n variables of the function exactly once, either true or complemented. For example, for the function $z_1(x_1, x_2, x_3)$, $x_1x_2'x_3$ is a minterm.

Maxterm A maxterm is a boolean sum of n variables and contains all n variables of the function exactly once, either true or complemented. For example, for the function $z_1(x_1, x_2, x_3)$, $(x_1 + x_2' + x_3)$ is a maxterm.

Product term A product term is the boolean product of variables containing a subset of the possible variables or their complements. For example, for the function $z_1(x_1, x_2, x_3)$, $x_1'x_3$ is a product term, because it does not contain all the variables.

Sum term A sum term is the boolean sum of variables containing a subset of the possible variables or their complements. For example, for the function $z_1(x_1, x_2, x_3)$, $(x_1' + x_3)$ is a sum term, because it does not contain all the variables.

Sum of minterms A sum of minterms is an expression in which each term contains all the variables, either true or complemented. For example,

$$z_1(x_1, x_2, x_3) = x_1'x_2x_3 + x_1x_2'x_3' + x_1x_2x_3$$

is a boolean expression in a sum-of-minterms form. This particular form is also referred to as a *minterm expansion*, a *standard sum of products*, a *canonical sum of products*, or a *disjunctive normal form*. Since each term is a minterm, the expression for z_1 can be written in a more compact sum-of-minterms form as $z_1(x_1, x_2, x_3) = \Sigma_m(3,4,7)$, where each term is converted to its minterm value. For example, the first term in the expression is $x_1'x_2x_3$, which corresponds to binary 011, representing minterm 3.

Sum of products A sum of products is an expression in which at least one term does not contain all the variables; that is, at least one term is a proper subset of the possible variables or their complements. For example,

$$z_1(x_1, x_2, x_3) = x_1'x_2x_3 + x_2'x_3' + x_1x_2x_3$$

is a sum of products for the function z_1, because the second term does not contain the variable x_1.

Product of maxterms A product of maxterms is an expression in which each term contains all the variables, either true or complemented. For example,

$$z_1(x_1, x_2, x_3) = (x_1' + x_2 + x_3)(x_1 + x_2' + x_3')(x_1 + x_2 + x_3)$$

is a boolean expression in a product-of-maxterms form. This particular form is also referred to as a *maxterm expansion*, a *standard product of sums*, a *canonical product of sums*, or a *conjunctive normal form*. Since each term is a maxterm, the expression for z_1 can be written in a more compact product-of-maxterms form as $z_1(x_1, x_2, x_3) = \Pi_M(0,3,4)$, where each term is converted to its maxterm value.

Product of sums A product of sums is an expression in which at least one term does not contain all the variables; that is, at least one term is a proper subset of the possible variables or their complements. For example,

$$z_1(x_1, x_2, x_3) = (x_1' + x_2 + x_3)(x_2' + x_3')(x_1 + x_2 + x_3)$$

is a product of sums for the function z_1, because the second term does not contain the variable x_1.

Summary of boolean algebra axioms and theorems Table 1.8 provides a summary of the axioms and theorems of boolean algebra. Each of the laws listed in the table is presented in pairs, where applicable, in which each law in the pair is the dual of the other.

Table 1.8 Summary of boolean algebra axioms and theorems

Axiom or theorem	Definition
Axiom 1: Boolean set definition	$x_1, x_2 \in B$
Axiom 2: Closure laws	(a) $x_1 + x_2 \in B$
	(b) $x_1 \bullet x_2 \in B$

Continued on next page

Table 1.8 Summary of boolean algebra axioms and theorems

Axiom or theorem	Definition
Axiom 3: Identity laws	(a) $x_1 + 0 = 0 + x_1 = x_1$
	(b) $x_1 \bullet 1 = 1 \bullet x_1 = x_1$
Axiom 4: Commutative laws	(a) $x_1 + x_2 = x_2 + x_1$
	(b) $x_1 \bullet x_2 = x_2 \bullet x_1$
Axiom 5: Associative laws	(a) $(x_1 + x_2) + x_3 = x_1 + (x_2 + x_3)$
	(b) $(x_1 \bullet x_2) \bullet x_3 = x_1 \bullet (x_2 \bullet x_3)$
Axiom 6: Distributive laws	(a) $x_1 + (x_2 \bullet x_3) = (x_1 + x_2) \bullet (x_1 + x_3)$
	(b) $x_1 \bullet (x_2 + x_3) = (x_1 \bullet x_2) + (x_1 \bullet x_3)$
Axiom 7: Complementation laws	(a) $x_1 + x_1' = 1$
	(b) $x_1 \bullet x_1' = 0$
Theorem 1: 0 and 1 associated with a variable	(a) $x_1 + 1 = 1$
	(b) $x_1 \bullet 0 = 0$
Theorem 2: 0 and 1 complement	(a) $0' = 1$
	(b) $1' = 0$
Theorem 3: Idempotent laws	(a) $x_1 + x_1 = x_1$
	(b) $x_1 \bullet x_1 = x_1$
Theorem 4: Involution law	$x_1'' = x_1$
Theorem 5: Absorption law 1	(a) $x_1 + (x_1 \bullet x_2) = x_1$
	(b) $x_1 \bullet (x_1 + x_2) = x_1$
Theorem 6: Absorption law 2	(a) $x_1 + (x_1' \bullet x_2) = x_1 + x_2$
	(b) $x_1 \bullet (x_1' + x_2) = x_1 \bullet x_2$
Theorem 7: DeMorgan's laws	(a) $(x_1 + x_2)' = x_1' \bullet x_2'$
	(b) $(x_1 \bullet x_2)' = x_1' + x_2'$

1.4 Minimization Techniques

This section will present various techniques for minimizing a boolean function. A boolean function is an algebraic representation of digital logic. Each term in an expression represents a logic gate and each variable in a term represents an input to a logic gate. It is important, therefore, to have the fewest number of terms in a boolean equation and the fewest number of variables in each term. A boolean equation with a minimal number of terms and variables reduces not only the number of logic gates, but also the delay required to generate the function.

1.4.1 Algebraic Minimization

The number of terms and variables that are necessary to generate a boolean function can be minimized by algebraic manipulation. Since there are no specific rules or algorithms to use for minimizing a boolean function, the procedure is inherently heuristic in nature. The only method available is an empirical procedure utilizing the axioms and theorems which is based solely on experience and observation without reference to theoretical principles. The examples which follow illustrate the process for minimizing a boolean function using the axioms and theorems of boolean algebra.

Example 1.3 Minimize the following expression using boolean algebra: $(x_1 + x_2)(x_1 + x_3)$

$(x_1 + x_2)(x_1 + x_3) = (x_1 + x_2)x_1 + (x_1 + x_2)x_3$	Distributive law
$= x_1x_1 + x_1x_2 + x_1x_3 + x_2x_3$	Distributive law
$= x_1 + x_1x_2 + x_1x_3 + x_2x_3$	Idempotent law
$= x_1(1 + x_2 + x_3) + x_2x_3$	Distributive law
$= x_1 + x_2x_3$	Theorem 1

Example 1.4 Convert the following expression to a sum-of-products form: $[(x_1 + x_2)' + x_3]'$

$[(x_1 + x_2)' + x_3]' = [x_1'x_2' + x_3]'$	DeMorgan's law
$= (x_1 + x_2)x_3'$	DeMorgan's law
$= x_1x_3' + x_2x_3'$	Distributive Law

Example 1.5 Minimize the following expression using boolean algebra: $x_1'x_2x_3 + x_1x_2'x_3' + x_1'x_2'x_3' + x_1x_2'x_3 + x_1x_2x_3$

$x_1'x_2x_3 + x_1x_2'x_3' + x_1'x_2'x_3' + x_1x_2'x_3 + x_1x_2x_3$	
$= x_2x_3(x_1' + x_1) + x_1x_2'x_3' + x_1'x_2'x_3' + x_1x_2'x_3$	Distributive law
$= x_2x_3 + x_1x_2'x_3' + x_1'x_2'x_3' + x_1x_2'x_3$	Complementation law
$= x_2x_3 + x_2'x_3'(x_1 + x_1') + x_1x_2'x_3$	Distributive law
$= x_2x_3 + x_2'x_3' + x_1x_2'x_3$	Complementation law
$= x_2x_3 + x_2'(x_3' + x_1x_3)$	Distributive law
$= x_2x_3 + x_2'(x_3' + x_1)$	Absorption law 2
$= x_2x_3 + x_2'x_3' + x_1x_2'$	Distributive law

Example 1.6 Apply DeMorgan's law to the following expression: $(x_1x_2' + x_3'x_4 + x_5x_6)'$

$$(x_1x_2' + x_3'x_4 + x_5x_6)' = (x_1x_2')' (x_3'x_4)' (x_5x_6)'$$
$$= (x_1' + x_2)(x_3 + x_4')(x_5' + x_6')$$

Example 1.7 Minimize the following function to a sum-of-products form using boolean algebra:

$$z_1 = x_1'x_2x_3' + x_2(x_1' + x_3) + x_2x_3'(x_1 + x_2)'$$

$$z_1 = x_1'x_2x_3' + x_1'x_2 + x_2x_3 + x_2x_3'(x_1'x_2')$$ Distributive law and Commutative law and DeMorgan's law

$$= x_1'x_2x_3' + x_1'x_2 + x_2x_3$$ Complementation law

$$= x_1'x_2(x_3' + 1) + x_2x_3$$ Distributive law

$$= x_1'x_2 + x_2x_3$$ Theorem 1

1.4.2 Karnaugh Maps

A Karnaugh map provides a geometrical representation of a boolean function. The Karnaugh map is arranged as an array of squares (or cells) in which each square represents a binary value of the input variables. The map is a convenient method of obtaining a minimal number of terms with a minimal number of variables per term for a boolean function. A Karnaugh map presents a clear indication of function minimization without recourse to boolean algebra and will generate a minimized expression in either a sum-of-products form or a product-of-sums form.

Figure 1.1 shows Karnaugh maps for two, three, four, and five variables. Each square in the maps corresponds to a unique minterm. The maps for three or more variables contain column headings that are represented in the *Gray code* format; the maps for four or more variables contain column and row headings that are represented in Gray code. Using the Gray code to designate column and row headings permits physically adjacent squares to be also logically adjacent; that is, to differ by only one variable. Map entries that are adjacent can be combined into a single term. For example, the expression $z_1 = x_1x_2'x_3 + x_1x_2x_3$, which corresponds to minterms 5 and 7 in Figure 1.1 (b), reduces to $z_1 = x_1x_3(x_2' + x_2) = x_1x_3$ using the distributive and complementation laws. Thus, if 1s are entered in minterm locations 5 and 7, then the two minterms can be combined into the single term x_1x_3.

Similarly, in Figure 1.1 (c), if 1s are entered in minterm locations 4, 6, 12, and 14, then the four minterms combine as x_2x_4'. That is, only variables x_2 and x_4' are common to all four squares — variables x_1 and x_3 are discarded by the complementation law. The minimized expression obtained from the Karnaugh map can be verified algebraically by listing the four minterms as a sum-of-minterms expression, then applying the appropriate laws of boolean algebra as shown below.

$$x_1'x_2x_3'x_4' + x_1'x_2x_3x_4' + x_1x_2x_3'x_4' + x_1x_2x_3x_4'$$

$$= x_1'x_2x_4'(x_3' + x_3) + x_1x_2x_4'(x_3' + x_3)$$

$$= x_1'x_2x_4' + x_1x_2x_4'$$

$$= x_2x_4'(x_1' + x_1)$$

$$= x_2x_4'$$

Alternatively,

$$x_1'x_2x_3'x_4' + x_1'x_2x_3x_4' + x_1x_2x_3'x_4' + x_1x_2x_3x_4'$$

$$= x_2x_4'(x_1'x_3' + x_1'x_3 + x_1x_3' + x_1x_3)$$

$$= x_2x_4'$$

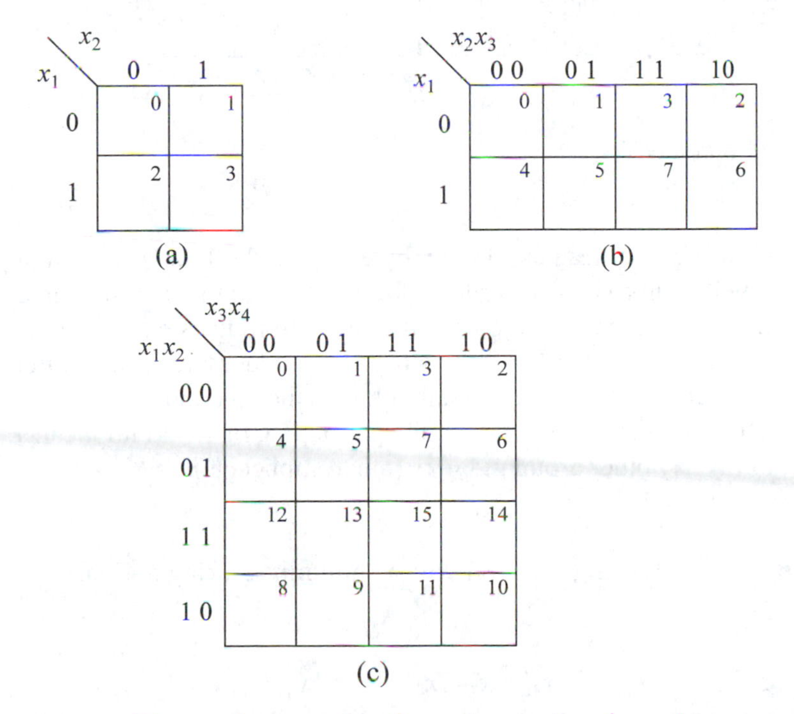

Figure 1.1 Karnaugh maps showing minterm locations: (a) two vartiables; (b) three variables; (c) four variables; (d) five variables; and (e) alternative map for five variables.

$x_5 = 0$

x_1x_2 \ x_3x_4	00	01	11	10
00	0	2	6	4
01	8	10	14	12
11	24	26	30	28
10	16	18	22	20

$x_5 = 1$

x_1x_2 \ x_3x_4	00	01	11	10
00	1	3	7	5
01	9	11	15	13
11	25	27	31	29
10	17	19	23	21

(d)

x_1x_2 \ $x_3x_4x_5$	000	001	011	010	110	111	101	100
00	0	1	3	2	6	7	5	4
01	8	9	11	10	14	15	13	12
11	24	25	27	26	30	31	29	28
10	16	17	19	18	22	23	21	20

(e)

Figure 1.1 (Continued)

When minimizing a boolean expression by grouping the 1s in a Karnaugh map, the result will be in a sum-of-products form; grouping the 0s results in a product-of-sums form. Each product term in a sum-of-products expression is specified as an *implicant* of the function, since the product term implies the function. That is, if the product term is equal to 1, then the function is also equal to 1. Specifically, a *prime implicant* is a unique grouping of 1s (an implicant) that does not imply any other grouping of 1s (other implicants).

Example 1.8 The following function will be minimized using a 4-variable Karnaugh map:

$$z_1(x_1, x_2, x_3, x_4) = x_2x_3' + x_2x_3x_4' + x_1x_2'x_3 + x_1x_3x_4'$$

The minimized result will be obtained in both a sum-of-products form and a product-of-sums form. To plot the function in the Karnaugh map, 1s are entered in the minterm locations that represent the product terms. For example, the term x_2x_3' is represented by the 1s in minterm locations 4, 5, 12, and 13. Only variables x_2 and x_3' are common to these four minterm locations. The term $x_2x_3x_4'$ is entered in minterm locations 6 and 14. When the function has been plotted, a minimal set of prime implicants can be obtained to represent the function. The largest grouping of 1s should always be combined, where the number of 1s in a group is a power of 2. The grouping of 1s is shown in Figure 1.2 and the resulting equation in Equation 1.18 in a sum-of-products notation.

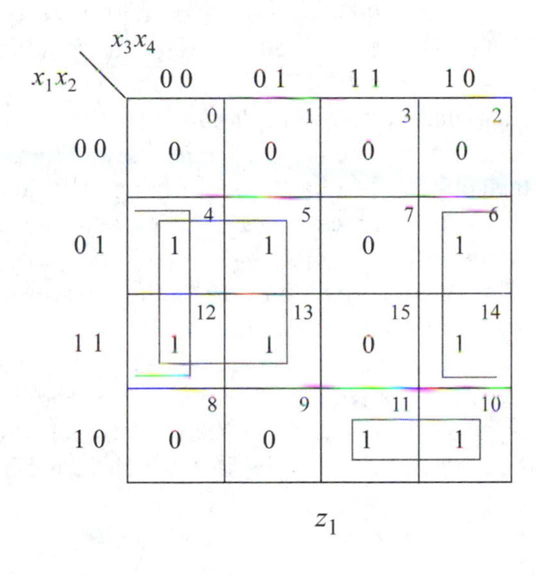

Figure 1.2 Karnaugh map representation of the function $z_1(x_1,x_2,x_3,x_4) = x_2x_3' + x_2x_3x_4' + x_1x_2'x_3 + x_1x_3x_4'$.

$$z_1(x_1,x_2,x_3,x_4) = x_2x_3' + x_2x_4' + x_1x_2'x_3 \tag{1.18}$$

The minimal product-of-sums expression can be obtained by combining the 0s in Figure 1.2 to form sum terms in the same manner as the 1s were combined to form product terms. However, since 0s are being combined, each sum term must equal 0. Thus, the four 0s in row $x_1x_2 = 00$ in Figure 1.2 combine to yield the sum term $(x_1 + x_2)$. In a similar manner, the remaining 0s are combined to yield the product-of-sums expression shown in Equation 1.19. When combining 0s to obtain sum terms, treat a variable value of 1 as false and a variable value of 0 as true. Thus, minterm locations 7 and 15 have variables $x_2x_3x_4 = 111$, providing a sum term of $(x_2' + x_3' + x_4')$.

$$z_1(x_1,x_2,x_3,x_4) = (x_1 + x_2)(x_2 + x_3)(x_2' + x_3' + x_4') \tag{1.19}$$

Equation 1.18 and Equation 1.19 both specify the conditions where z_1 is equal to 1. For example, consider the first term of Equation 1.18. If $x_2\, x_3 = 10$, then Equation 1.18 yields $z_1 = 1 + \cdots + 0$ which generates a value of 1 for z_1. Applying $x_2\, x_3 = 10$ to Equation 1.19 will cause every term to be equal to 1, such that, $z_1 = (1)\,(1)\,(1) = 1$.

Figure 1.1 (d) illustrates a 5-variable Karnaugh map. To determine adjacency, the left map is superimposed on the right map. Any cells that are then physically adjacent are also logically adjacent and can be combined. Since x_5 is the low-order variable, the left map contains only even-numbered minterms; the right map is characterized by odd-numbered minterms. If 1s are entered in minterm locations 28, 29, 30, and 31, the four cells combine to yield the term $x_1\, x_2\, x_3$.

Figure 1.1 (e) illustrates an alternative configuration for a Karnaugh map for five variables. The map hinges along the vertical centerline and folds like a book. Any squares that are then physically adjacent are also logically adjacent. For example, if 1s are entered in minterm locations 24, 25, 28, and 29, then the four squares combine to yield the term $x_1\, x_2\, x_4'$.

Some minterm locations in a Karnaugh map may contain unspecified entries which can be used as either 1s or 0s when minimizing the function. These "*don't care*" entries are indicated by a dash (–) in the map. A typical situation which includes "don't care" entries is a Karnaugh map used to represent the BCD numbers. This requires a 4-variable map in which minterm locations 10 through 15 contain unspecified entries, since digits 10 through 15 are invalid for BCD.

Example 1.9 A minimized equation will be derived which is asserted whenever a BCD digit is even. All even BCD digits are plotted on a Karnaugh map as shown in Figure 1.3 for function z_1. The unspecified entries in minterm locations 10, 12, and 14 are assigned a value of 1; all remaining unspecified entries are assigned a value of 0. The equation for z_1 is shown in Equation 1.20.

x_1x_2 \ x_3x_4	0 0	0 1	1 1	1 0
0 0	1 (0)	0 (1)	0 (3)	1 (2)
0 1	1 (4)	0 (5)	0 (7)	1 (6)
1 1	– (12)	– (13)	– (15)	– (14)
1 0	1 (8)	0 (9)	– (11)	– (10)

z_1

Figure 1.3 Karnaugh map to represent even-numbered BCD digits.

$$z_1 = x_4' \tag{1.20}$$

To obtain the equation which specifies BCD digits that are evenly divisible by 3, 1s are entered in minterm locations 0, 3, 6, and 9 to yield Equation 1.21.

$$z_1 = x_1x_4 + x_2'x_3x_4 + x_2x_3x_4' + x_1'x_2'x_3'x_4' \tag{1.21}$$

Map-entered variables Variables may also be entered in a Karnaugh map as map-entered variables, together with 1s and 0s. A map of this type is more compact than a standard Karnaugh map, but contains the same information. A map containing map-entered variables is particularly useful in analyzing and synthesizing synchronous sequential machines. When variables are entered in a Karnaugh map, two or more squares can be combined only if the squares are adjacent and contain the same variable(s).

Example 1.10 The following boolean equation will be minimized using a 3-variable Karnaugh map with x_4 as a map-entered variable:

$$z_1(x_1,x_2,x_3,x_4) = x_1x_2'x_3x_4' + x_1x_2 + x_1'x_2'x_3'x_4' + x_1'x_2'x_3'x_4$$

Note that instead of $2^4 = 16$ squares, the map of Figure 1.4 contains only $2^3 = 8$ squares, since only three variables are used in constructing the map. To facilitate plotting the equation in the map, the variable that is to be entered is shown in parenthesis as follows:

$$z_1(x_1,x_2,x_3,x_4) = x_1x_2'x_3(x_4') + x_1x_2 + x_1'x_2'x_3'(x_4') + x_1'x_2'x_3'(x_4)$$

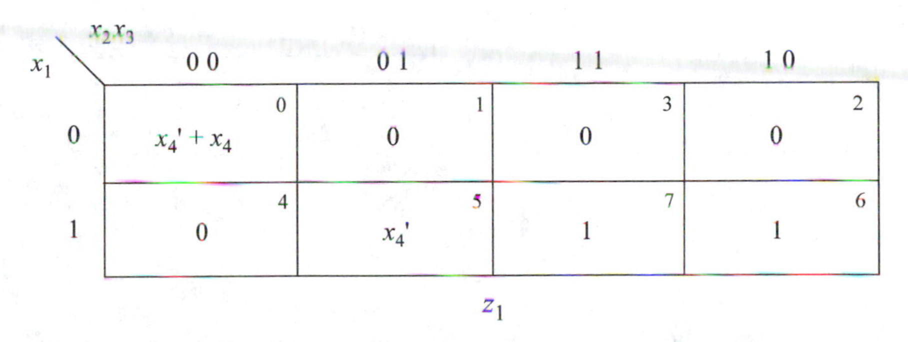

Figure 1.4 Karnaugh map using x_4 as a map-entered variable for Example 1.10.

The first term in the equation for z_1 is $x_1x_2'x_3$ (x_4') and indicates that the variable x_4' is entered in minterm location 5 ($x_1x_2'x_3$). The second term x_1x_2 is plotted in the usual manner: 1s are entered in minterm locations 6 and 7. The third term specifies that the variable x_4' is entered in minterm location 0 ($x_1'x_2'x_3'$). The fourth term also applies to minterm 0, where x_4 is entered. The expression in minterm location 0, therefore, is $x_4' + x_4$.

To obtain the minimized equation for z_1 in a sum-of-products form, 1s are combined in the usual manner; variables are combined only if the minterm locations containing the variables are adjacent and the variables are identical. Consider the expression $x_4' + x_4$ in minterm location 0. Since $x_4' + x_4 = 1$, minterm 0 equates to $x_1'x_2'x_3'$. The entry of 1 in minterm location 7 can be restated as $1 + x_4'$ without changing the value of the entry (Theorem 1). This allows minterm locations 5 and 7 to be combined as $x_1x_3x_4'$. Finally, minterms 6 and 7 combine to yield the term x_1x_2. The minimized equation for z_1 is shown in Equation 1.22.

$$z_1 = x_1'x_2'x_3' + x_1x_3x_4' + x_1x_2 \tag{1.22}$$

Example 1.11 Given the Karnaugh map shown in Figure 1.5, the minimized equation for z_1 will be obtained using x_4 as a map-entered variable. The variable x_4 in minterm locations 0 and 2 can be combined as $x_1'x_3'x_4$. The expression $x_4 + x_4'$ (= 1) in minterm location 2 combines as $x_1'x_2x_3'$. The variable x_4' in locations 4, 5, and 7 combine as the following two terms: $x_1x_2'x_4'$ (minterm locations 4 and 5) and $x_1x_3x_4'$ (minterm locations 5 and 7). The resulting equation for z_1 is shown in Equation 1.23.

x_1 \ x_2x_3	0 0	0 1	1 1	1 0
0	0: x_4	1: 0	3: 0	2: $x_4 + x_4'$
1	4: x_4'	5: $x_4 + x_4'$	7: x_4'	6: 0

z_1

Figure 1.5 Karnaugh map using x_4 as a map-entered variable for Example 1.11.

$$z_1 = x_1'x_3'x_4 + x_1'x_2x_3' + x_1x_3x_4' + x_1x_2'x_4' + x_1x_2'x_3 \tag{1.23}$$

Example 1.12 The following boolean equation will be minimized using x_4 and x_5 as map-entered variables:

$$z_1 = x_1'x_2'x_3'(x_4x_5') + x_1'x_2 + x_1'x_2'x_3'(x_4x_5) + x_1x_2'x_3'(x_4x_5) + x_1x_2'x_3 + x_1x_2'x_3'(x_4') + x_1x_2'x_3'(x_5')$$

Figure 1.6 shows the map entries for Example 1.12. The expression $x_4x_5' + x_4x_5$ in minterm location 0 reduces to x_4; the 1 entry in minterm location 2 can be expanded to $1 + x_4$ without changing the value in location 2. Therefore, locations 0 and 2 combine as $x_1'x_3'x_4$. The expression $x_4x_5 + x_4' + x_5'$ in minterm location 4 reduces to 1. Thus, the 1 entries in the map combine in the usual manner to yield Equation 1.24.

x_1 \ x_2x_3	0 0	0 1	1 1	1 0
0	(0) $x_4x_5' + x_4x_5$	(1) 0	(3) 1	(2) 1
1	(4) $x_4x_5 + x_4' + x_5'$	(5) 1	(7) 0	(6) 0

z_1

Figure 1.6 Karnaugh map using x_4 and x_5 as map-entered variables for Example 1.12.

$$z_1 = x_1'x_3'x_4 + x_1'x_2 + x_1x_2' \tag{1.24}$$

Example 1.13 As a final example of map-entered vartiables, consider the Karnaugh map shown in Figure 1.7. The x_3 entry and the "don't care" combine to yield $x_1'x_3$. Every square contains the variable x_4 if the "don't care" is given the value x_4 and the 1 entry is assigned the value of $x_4 + x_4'$. Therefore, the second term is simply x_4. Next, the "don't care" and the 1 entry in minterm location 3 are combined to yield x_2. The final equation is shown in Equation 1.25.

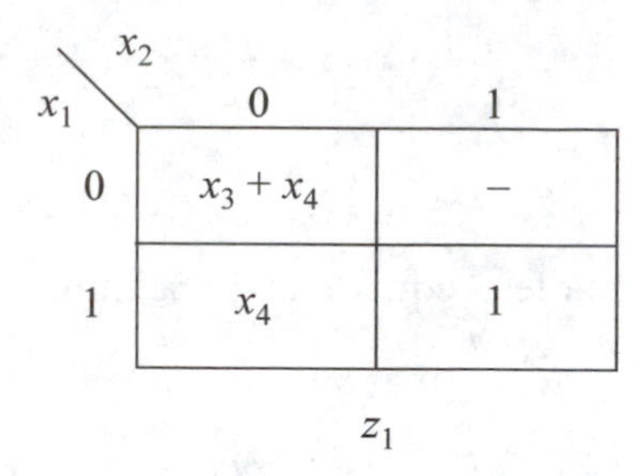

Figure 1.7 Karnaugh map for Example 1.13.

$$z_1 = x_1'x_3 + x_4 + x_2 \tag{1.25}$$

Karnaugh maps are ideally suited for 2-, 3-, 4-, or 5-variable boolean functions. The 6-variable Karnaugh map shown in Figure 1.8 tends to be unwieldy when plotting complex boolean functions where a large number of 1s are scattered throughout the map. The maps are superimposed, one upon the other, to determine adjacency requirements. A more systematic minimization technique is recommended for boolean functions with six or more variables. This is the Quine-McCluskey algorithm.

1.4.3 Quine-McCluskey Algorithm

The Quine-McCluskey algorithm is a tabular method of obtaining a minimal set of prime implicants that represents the boolean function. Because the process is inherently algorithmic, the technique is easily implemented with a computer program. The method consists of two steps: first obtain a set of prime implicants for the function; then obtain a minimal set of prime implicants that represents the function.

The rationale for the Quine-McCluskey method relies on the repeated application of the distributive and complementation laws. For example, for a 4-variable function, minterms $x_1x_2x_3'x_4$ and $x_1x_2x_3'x_4'$ are adjacent because they differ by only one variable. The two minterms can be combined, therefore, into a single product term as follows:

$$
\begin{aligned}
& x_1 x_2 x_3' x_4 + x_1 x_2 x_3' x_4' \\
&= x_1 x_2 x_3'(x_4 + x_4') \\
&= x_1 x_2 x_3'
\end{aligned}
$$

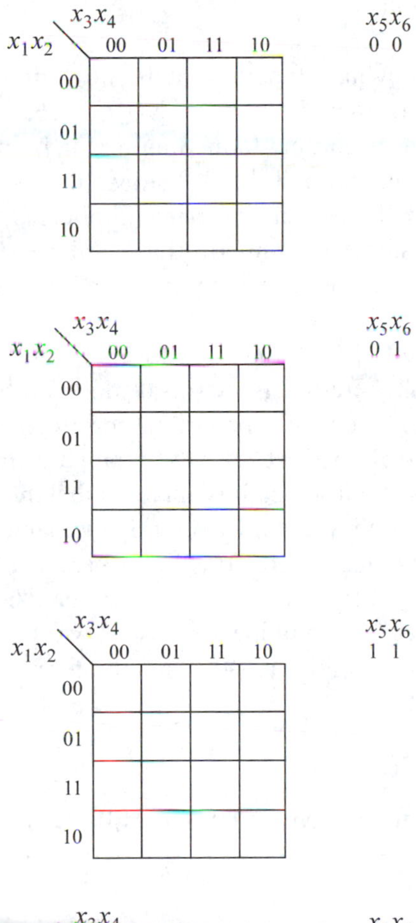

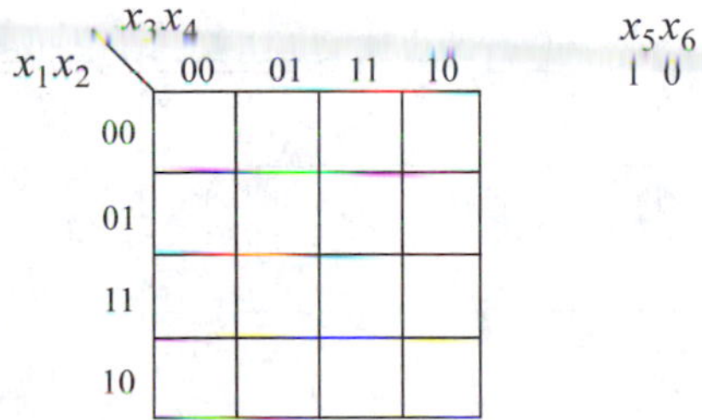

Figure 1.8 Six-variable Karnaugh map.

The resulting product term is specified as $x_1x_2x_3x_4 = 110–$, where the dash (–) represents the variable that has been removed. The process repeats for all minterms in the function. Two product terms with dashes in the same position can be further combined into a single term if they differ by only one variable. Thus, the terms $x_1x_2x_3x_4 = 110–$ and $x_1x_2x_3x_4 = 100–$ combine to yield the term $x_1x_2x_3x_4 = 1–0–$, which corresponds to x_1x_3'.

The minterms are initially grouped according to the number of 1s in the binary representation of the minterm number. Comparison of minterms then occurs only between adjacent groups of minterms in which the number of 1s in each group differs by one. Minterms in adjacent groups that differ by only one variable can then be combined.

Example 1.14 The following function will be minimized using the Quine-McCluskey method: $z_1(x_1, x_2, x_3, x_4) = \Sigma_m(0,1,3,6,7,8,9,14)$. The first step is to list the minterms according to the number of 1s in the binary representation of the minterm number. Table 1.9 shows the listing of the various groups. Minterms that combine cannot be prime implicants; therefore, a check (✓) symbol is placed beside each minterm that combines with another minterm. When all lists in the table have been processed, the terms that have no check marks are prime implicants.

Consider List 1 in Table 1.9. Minterm 0 differs by only one variable with each minterm in Group 1. Therefore, minterms 0 and 1 combine as 000–, as indicated in the first entry in List 2 and minterms 0 and 8 combine to yield –000, as shown in the second row of List 2. Next, compare minterms in List 1, Group 1 with those in List 1, Group 2. It is apparent that the following pairs of minterms combine because they differ by only one variable: (1,3), (1,9), and (8,9) as shown in List 2, Group 1. Minterms 1 and 3 are in adjacent groups and can combine because they differ by only one variable. The resulting term is 00–1. Minterms 1 and 6 cannot combine, because they differ by more than one variable. Minterms 1 and 9 combine as –001 and minterms 8 and 9 combine to yield 100–.

Table 1.9 Minterms listed in groups for Example 1.14

List 1			List 2			List 3		
Group	Minterms	$x_1x_2x_3x_4$	Group	Minterms	$x_1x_2x_3x_4$	Group	Minterms	$x_1x_2x_3x_4$
0	0	0 0 0 0 ✓	0	0,1	0 0 0 – ✓	0	0,1,8,9	– 0 0 –
				0,8	– 0 0 0 ✓			
1	1	0 0 0 1 ✓	1	1,3	0 0 – 1			
	8	1 0 0 0 ✓		1,9	– 0 0 1 ✓			
				8,9	1 0 0 – ✓			
2	3	0 0 1 1 ✓	2	3,7	0 – 1 1			
	6	0 1 1 0 ✓		6,7	0 1 1 –			
	9	1 0 0 1 ✓		6,14	– 1 1 0			
3	7	0 1 1 1 ✓						
	14	1 1 1 0 ✓						

In a similar manner, minterms in the remaining groups are compared for possible adjacency. Note that those minterms that combine differ by a power of 2 in the decimal value of their minterm number. For example, minterms 6 and 14 combine as –110, because they differ by a power of 2 ($2^3 = 8$). Note also that the variable x_1 which is removed is located in column 2^3, where the binary weights of the four variables are $x_1 x_2 x_3 x_4 = 2^3\ 2^2\ 2^1\ 2^0$.

List 3 is derived in a similar manner to that of List 2. However, only those terms that are in adjacent groups and have dashes in the same column can be compared. For example, the terms 0,1 (000–) and 8,9 (100–) both contain dashes in column x_4 and differ by only one variable. Thus, the two terms can combine into a single product term as $x_1 x_2 x_3 x_4 = -00-$ ($x_2' x_3'$). If the dashes are in different columns, then the two terms do not represent product terms of the same variables and thus, cannot combine into a single product term.

When all comparisons have been completed, some terms will not combine with any other term. These terms are indicated by the absence of a check symbol and are designated as prime implicants. For example, the term $x_1 x_2 x_3 x_4 = 00{-}1$ ($x_1' x_2' x_4$) in List 2 cannot combine with any term in either the previous group or the following group. Thus, $x_1' x_2' x_4$ is a prime implicant. The following terms represent prime implicants: $x_1' x_2' x_4$, $x_1' x_3 x_4$, $x_1' x_2 x_3$, $x_2 x_3 x_4'$ and $x_2' x_3'$.

Some of the prime implicants may be redundant, since the minterms covered by a prime implicant may also be covered by one or more other prime implicants. Therefore, the second step in the algorithm is to obtain a minimal set of prime implicants that covers the function. This is accomplished by means of a *prime implicant chart* as shown in Figure 1.9 (a). Each column of the chart represents a minterm; each row of the chart represents a prime implicant. The first row of Figure 1.9 (a) is specified by the minterm grouping of (1,3), which corresponds to prime implicant $x_1' x_2' x_4$ (00–1). Since prime implicant $x_1' x_2' x_4$ covers minterms 1 and 3, an × is placed in columns 1 and 3 in the corresponding prime implicant row. The remaining rows are completed in a similar manner. Consider the last row which corresponds to prime implicant $x_2' x_3'$. Since prime implicant $x_2' x_3'$ covers minterms 0, 1, 8, and 9, an × is placed in the minterm columns 0, 1, 8, and 9.

A single × appearing in a column indicates that only one prime implicant covers the minterm. The prime implicant, therefore, is an *essential prime implicant*. In Figure 1.9 (a), there are two essential prime implicants: $x_2 x_3 x_4'$ and $x_2' x_3'$. A horizontal line is drawn through all ×s in each essential prime implicant row. Since prime implicant $x_2 x_3 x_4'$ covers minterm 6, there is no need to have prime implicant $x_1' x_2 x_3$ also cover minterm 6. Therefore, a vertical line is drawn through all ×s in column 6, as shown in Figure 1.9 (a). For the same reason, a vertical line is drawn through all ×s in column 1 for the second essential prime implicant $x_2' x_3'$.

The only remaining minterms not covered by a prime implicant are minterms 3 and 7. Minterm 3 is covered by prime implicants $x_1' x_2' x_4$ and $x_1' x_3 x_4$; minterm 7 is covered by prime implicants $x_1' x_3 x_4$ and $x_1' x_2 x_3$, as shown in Figure 1.9 (b). Therefore, a minimal cover for minterms 3 and 7 consists of the *secondary essential prime implicant* $x_1' x_3 x_4$. The complete minimal set of prime implicants for the function z_1 is shown in Equation 1.26. The minimized expression for z_1 can be verified by plotting the function on a Karnaugh map, as shown in Figure 1.10.

$$z_1(x_1,x_2,x_3,x_4) = x_2x_3x_4' + x_2'x_3' + x_1'x_3x_4 \tag{1.26}$$

Prime implicants		Minterms 0	1	3	6	7	8	9	14
1,3	$(x_1'x_2'x_4)$		×	×					
3,7	$(x_1'x_3x_4)$			×		×			
6,7	$(x_1'x_2x_3)$				×	×			
* 6,14	$(x_2x_3x_4')$				×				⊗
* 0,1,8,9	$(x_2'x_3')$	⊗	×				×	×	

(a)

Prime implicants		Minterms 0	1	3	6	7	8	9	14
1,3	$(x_1'x_2'x_4)$			×					
3,7	$(x_1'x_3x_4)$			×		×			
6,7	$(x_1'x_2x_3)$					×			

(b)

Figure 1.9 Prime implicant chart for Example 1.14: (a) essential and nonessential prime implicants; and (b) secondary essential prime implicants with minimal cover for remaining prime implicants.

Functions which include unspecified entries ("don't cares") are handled in a similar manner. The tabular representation of step 1 lists all the minterms, including "don't cares." The "don't care" conditions are then utilized when comparing minterms in adjacent groups. In step 2 of the algorithm, only the minterms containing specified entries are listed — the "don't care" minterms are not used. Then the minimal set of prime implicants is found as described in Example 1.14.

x_1x_2 \ x_3x_4	0 0	0 1	1 1	1 0
0 0	1 (0)	1 (1)	1 (3)	0 (2)
0 1	0 (4)	0 (5)	1 (7)	1 (6)
1 1	0 (12)	0 (13)	0 (15)	1 (14)
1 0	1 (8)	1 (9)	0 (11)	0 (10)

z_1

Figure 1.10 Karnaugh map for Example 1.14.

Example 1.15 The function may not always contain an essential prime implicant, or the secondary essential prime implicants may not be intuitively obvious, as they were in Example 1.14. The technique for obtaining a minimal cover of secondary prime implicants is called the *Petrick algorithm* and can best be illustrated by an example. Given the prime implicant chart of Figure 1.11 for function z_1, it is obvious that there are no essential prime implicants, since no minterm column contains a single ×.

Prime implicants	Minterms m_i	m_j	m_k	m_l	m_m
pi_1	×	×			×
pi_2	×		×	×	
pi_3		×		×	
pi_4			×		×

Figure 1.11 Prime implicant chart for Example 1.15.

It is observed that minterm m_i is covered by prime implicants pi_1 or pi_2; m_j is covered by pi_1 or pi_3; m_k is covered by pi_2 or pi_4; m_l is covered by pi_2 or pi_3; and m_m is covered by pi_1 or pi_4. Since the function is covered only if all minterms are covered, Equation 1.27 represents this requirement.

$$\text{Function is covered} = (pi_1 + pi_2)(pi_1 + pi_3)(pi_2 + pi_4)(pi_2 + pi_3)(pi_1 + pi_4) \quad (1.27)$$

Equation 1.27 can be reduced by boolean algebra or by a Karnaugh map to obtain a minimal set of prime implicants that represents the function. Figure 1.12 illustrates the Karnaugh map in which the sum terms of Equation 1.27 are plotted. The map is then used to obtain a minimized expression that represents the different combinations of prime implicants in which all minterms are covered. Equation 1.28 lists the product terms specified as prime implicants in a sum-of-products notation.

$pi_1 pi_2$ \ $pi_3 pi_4$	0 0	0 1	1 1	1 0
0 0	0	0	0	0
0 1	0	0	1	0
1 1	1	1	1	1
1 0	0	0	1	0

Figure 1.12 Karnaugh map in which the sum terms of Equation 1.27 are entered as 0s.

$$\text{Function is covered} = pi_1 pi_2 + pi_2 pi_3 pi_4 + pi_1 pi_3 pi_4 \quad (1.28)$$

The first term of Equation 1.28 represents the fewest number of prime implicants to cover the function. Thus, function z_1 will be completely specified by the expression $z_1 = pi_1 pi_2$. From any covering equation, the term with the fewest number of variables is chosen to provide a minimal set of prime implicants. Assume, for example, that prime implicant $pi_1 = x_i x_j' x_k$ and that $pi_2 = x_l' x_m x_n$. Thus, the sum-of-products expression is $z_1 = x_i x_j' x_k + x_l' x_m x_n$.

1.5 Logic Symbols

The logic symbols used in digital design have undergone some radical changes in recent years. A new standard emerged in the 1980s to replace the distinctive-shape symbols of previous standards. The new standard is ANSI/IEEE Std. 91-1984 and is currently used for military and semiconductor documentation and by several major computer manufacturers.

The standard defines "an international language by which it is possible to determine the functional behavior of a logic circuit as described on a logic or circuit diagram with minimal reference to supporting documentation." Thus, all logic function symbols are uniquely defined and convey all necessary information with which to completely understand the functional operation of the device.

The standard provides uniform-shape symbols for all logic gates. Figure 1.13 shows the logic gate symbols used for both the ANSI/IEEE Std. 91-1984 uniform-shape symbols and the corresponding distinctive-shape symbols. The symbols for logic macros such as, multiplexers, decoders, and encoders will be covered in their respective sections of this chapter. The *polarity symbol* "◺" in Figure 1.13 indicates an active-low assertion on either an input or an output of a logic symbol. The polarity symbol points in the direction of signal flow.

By convention, input signals enter the logic symbol on the left and exit on the right. Since logic gates are identical in shape, a qualifying symbol is inserted within the rectangle to specify the function of the gate. The *qualifying function symbols* are defined as follows:

Symbol	Description
&	The AND function of two or more inputs. All inputs must be at their active voltage level to assert the output at its active voltage level.
≥1	The OR function of two or more inputs. One or more inputs must be at an active voltage level to assert the output at its active voltage level.
1	The NOT (invert) function. Only one input enters the logic symbol. The input voltage is inverted from high to low or from low to high.
=1	The exclusive-OR function. Only one of the two inputs is active to assert the output at its active voltage level. Thus, if x_1 and x_2 are the inputs and z_1 is the output, then $z_1 = x_1x_2' + x_1'x_2$. If the output has an active-low polarity symbol, then this represents the exclusive-NOR function. The exclusive-NOR function is also referred to as the equality function, where $z_1 = x_1x_2 + x_1'x_2'$.

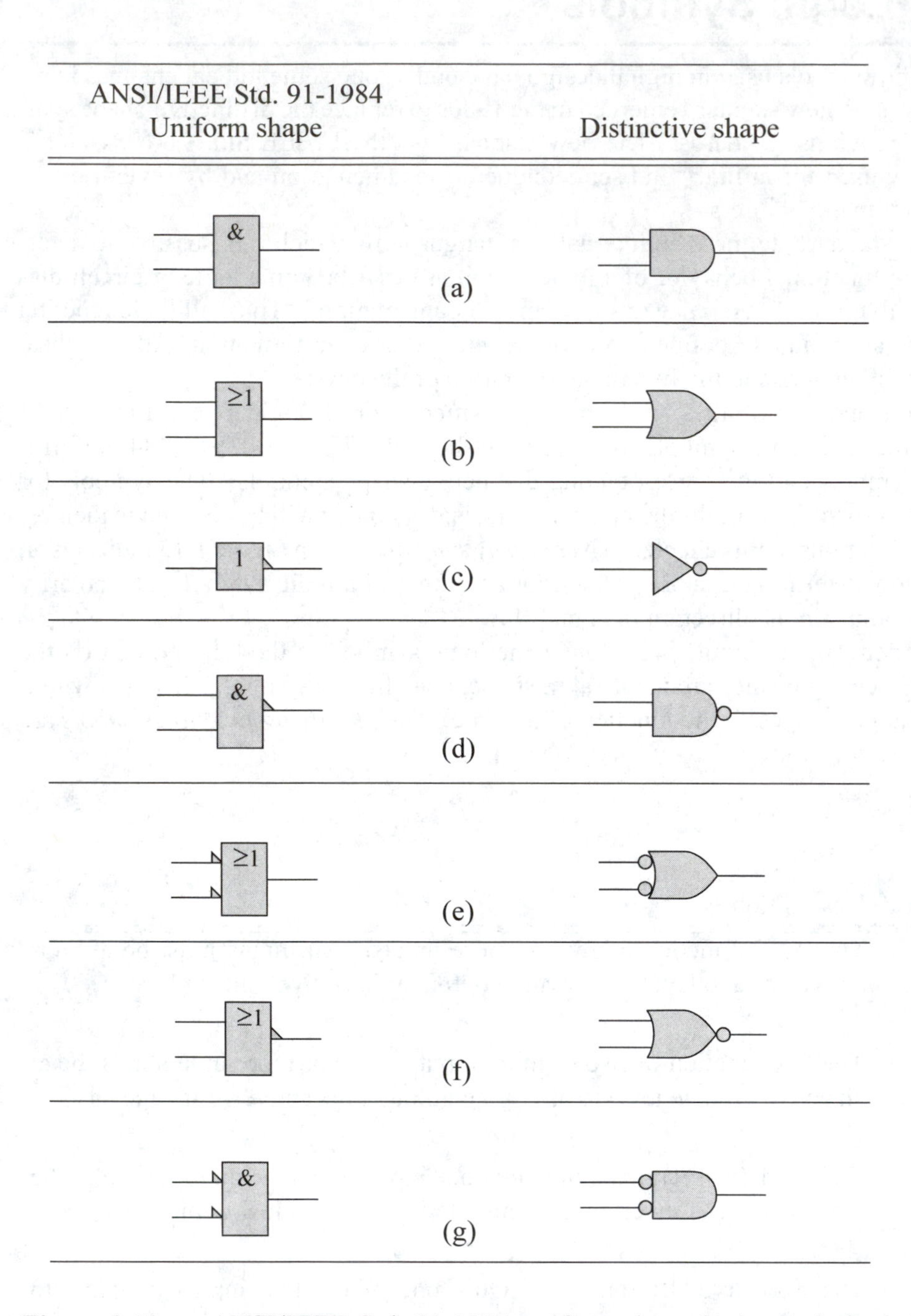

Figure 1.13 ANSI/IEEE Std. 91-1984 uniform-shape logic symbols and the corresponding distinctive-shape logic symbols: (a) AND gate; (b) OR gate; (c) NOT (inverter); (d) NAND gate for the AND function; (e) NAND gate for the OR function; (f) NOR gate for the OR function; (g) NOR gate for the AND function; (h) exclusive-OR function; and (i) exclusive-NOR function.

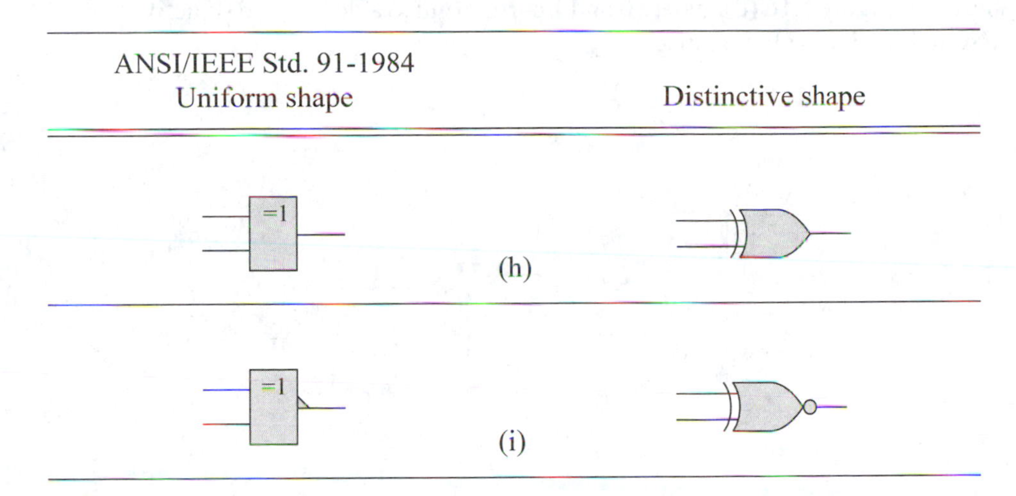

Figure 1.13 (Continued)

The structure of the logic symbols in ANSI/IEEE Std. 91-1984 provides a means to present logic functions in a more compact form. For example, the function $z_1 = x_1x_2'x_3 + x_1'x_2x_3'$ is depicted two ways in Figure 1.14: using the distinctive-shape symbols and the uniform-shape symbols in a compact structure. Figure 1.15 illustrates the logic to represent the function $z_1 = x_1'x_2 + x_3 + x_4'$.

The logic symbols of the ANSI/IEEE Std. 91-1984 will be used in this chapter to provide examples from which the reader can gain familiarity with the symbols and their functions. The remaining chapters will use the distinctive shape symbols. Inputs and outputs should always be placed on opposite sides of the logic symbol, with the inputs entering from the left. There should be no signals entering or leaving the top or bottom of the symbol. The AND gates in Figure 1.14 (b) represent an array. The array can extend downward to include additional gates, where each gate represents a product term in a sum-of-products notation. The array may also consist of OR gates to represent a product-of-sums form.

1.6 Analysis of Combinational Logic

A logic gate is a physical device that implements a logical operation. The symbol used to represent a gate shows the physical characteristics such as, input and output assertion levels and also the logical function such as, AND, OR, and NOT. Figure 1.16 (a) shows an AND gate with the corresponding physical truth table in Figure 1.16 (b), where L and H refer to a low and high voltage level, respectively. Thus, only when both inputs are at their respective high voltage levels will the output be at a high voltage level. The word *high* as used here, does not necessarily mean a positive voltage level, but merely the more positive of two voltage levels. The logical truth table is shown in Figure 1.16 (c), where 0 and 1 correspond to a deasserted (inactive) or asserted (active) value, respectively.

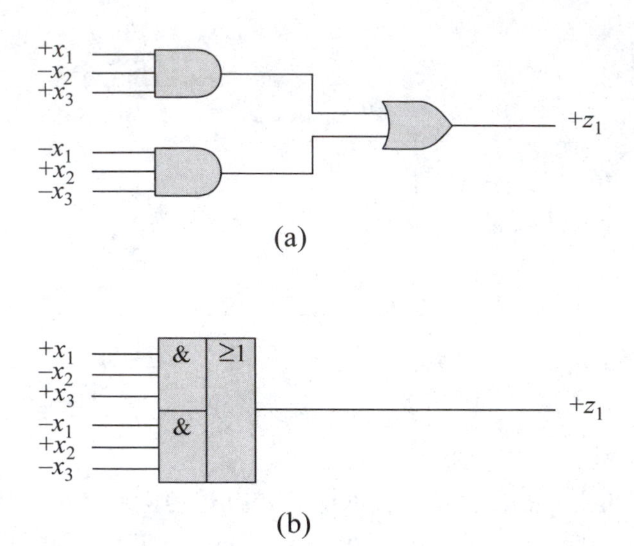

Figure 1.14 Logic diagram for function $z_1 = x_1x_2'x_3 + x_1'x_2x_3'$: (a) distinctive-shape symbols; and (b) uniform-shape symbols in a compact representation.

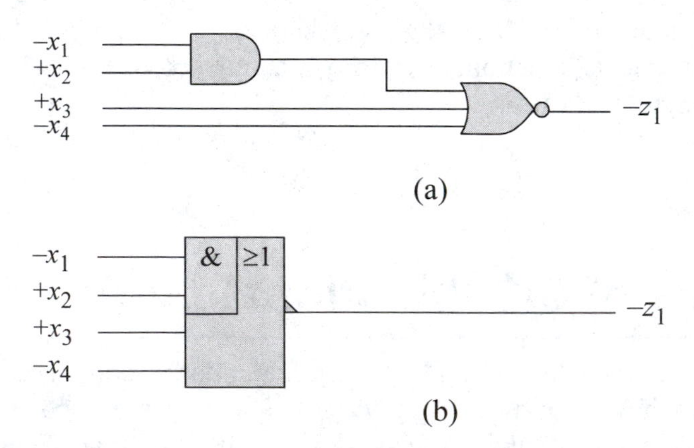

Figure 1.15 Logic diagram for function $z_1 = x_1'x_2 + x_3 + x_4'$: (a) distinctive-shape symbols; and (b) uniform-shape symbols in a compact representation.

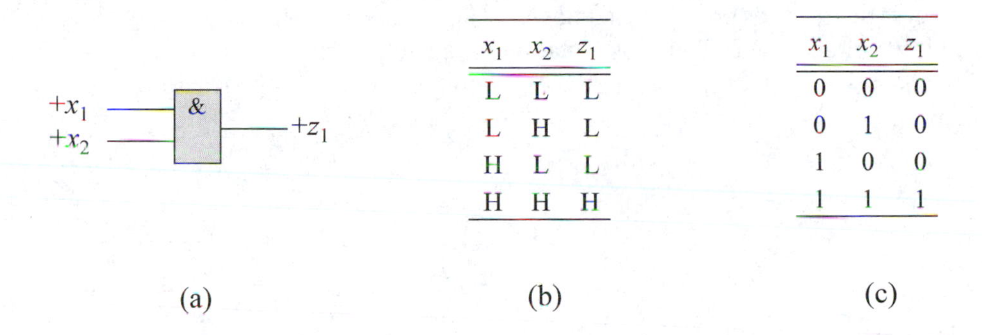

x_1	x_2	z_1
L	L	L
L	H	L
H	L	L
H	H	H

x_1	x_2	z_1
0	0	0
0	1	0
1	0	0
1	1	1

Figure 1.16 Logical AND function: (a) gate symbol; (b) physical truth table; and (c) logical truth table.

Figure 1.17 (a) illustrates an OR gate with the corresponding physical and logical truth tables in Figure 1.17 (b) and Figure 1.17 (c), respectively. If either input (or both) is at a high voltage level, then the output is at a high voltage level. Figure 1.18 depicts the gate symbol, physical truth table, and logical truth table for a NAND gate.

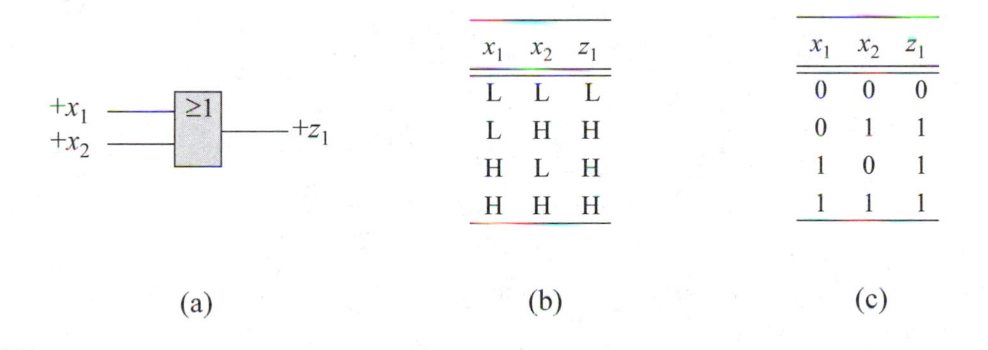

x_1	x_2	z_1
L	L	L
L	H	H
H	L	H
H	H	H

x_1	x_2	z_1
0	0	0
0	1	1
1	0	1
1	1	1

Figure 1.17 Logical OR function: (a) gate symbol; (b) physical truth table; and (c) logical truth table.

The +/– symbols that precede an input or output variable indicate the active voltage level of the variable. Thus, $+x_1$ or $-x_1$ indicates that variable x_1 is active at a high or low voltage level, respectively. The NAND gates in Figure 1.19 portray two different boolean equations. In Figure 1.19 (a), if both inputs are asserted at a high voltage level, then output z_1 is asserted at a low voltage level. Thus, the equation for an active-low output is $z_1 = x_1 x_2$. In Figure 1.19 (b), however, if x_1 is asserted (producing a high level at the gate input — which requires a high level input) and x_2 is deasserted (producing a high level at the gate input), then z_1 is asserted at a low voltage level. Thus, the equation for an active-low output is $z_1 = x_1 x_2'$. That is, z_1 is asserted (active) if x_1 is asserted (active) and x_2 is deasserted (inactive).

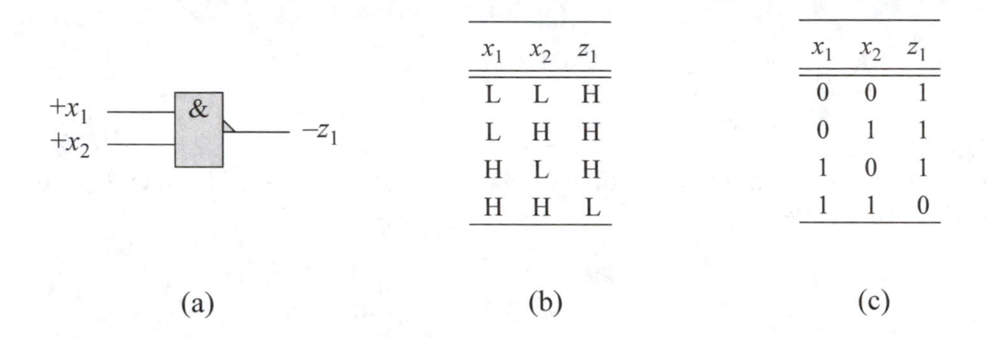

x_1	x_2	z_1
L	L	H
L	H	H
H	L	H
H	H	L

(b)

x_1	x_2	z_1
0	0	1
0	1	1
1	0	1
1	1	0

(c)

Figure 1.18 Logical AND function with an active-low output: (a) NAND gate symbol; (b) physical truth table; and (c) logical truth table.

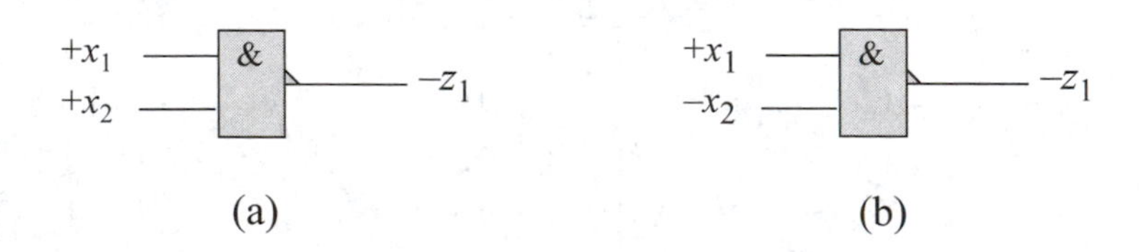

Figure 1.19 NAND gates representing two different equations: (a) $z_1 = x_1 x_2$; and (b) $z_1 = x_1 x_2'$.

Figure 1.20 illustrates a 2-level, multigate network using only NAND gates. The output of gate 1will be low if the expression $x_1' x_2'$ is true. Since gate 3 requires that at least one input be at a low voltage level to assert the output, the low output from gate 1 propagates through gate 3 to assert output z_1. The polarity symbol "◺" may be considered as complementation; therefore, the path through gates 1 and 3 is equivalent to the expression $(x_1' x_2')''$, which reduces to $x_1' x_2'$ by the involution law.

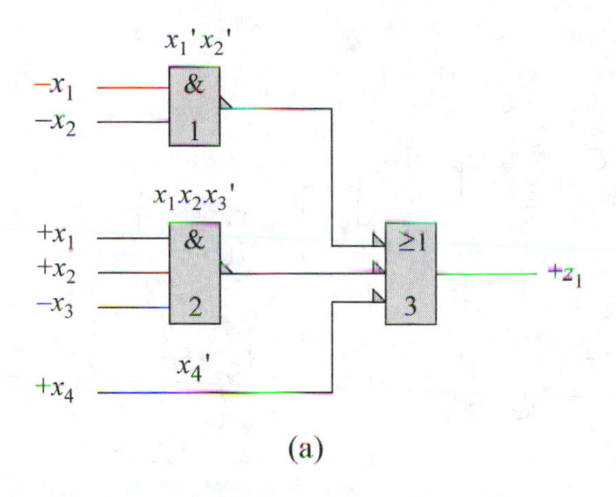

(a)

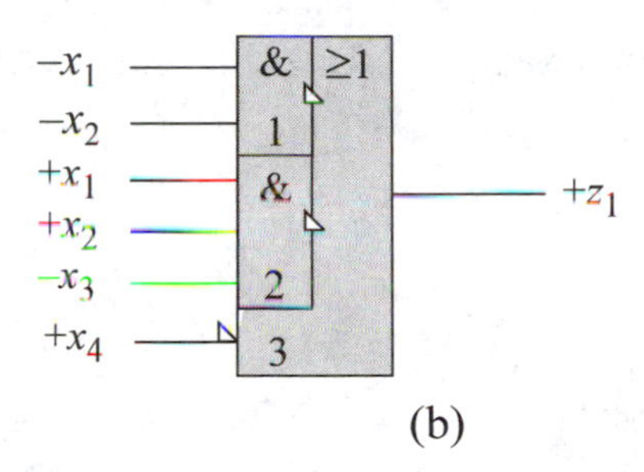

(b)

Figure 1.20 Multigate network using NAND gates only: (a) $z_1 = x_1'x_2' + x_1x_2x_3' + x_4'$; and (b) the same implementation using a compact format.

Similarly, the output of gate 2 is active low if the expression $x_1x_2x_3'$ is true. Thus, z_1 is asserted high for $(x_1x_2x_3')'' = x_1x_2x_3'$. The fourth input variable is $+x_4$, which specifies that x_4 is active at a high voltage level. In order for the low input requirement to be met for gate 3, input x_4 must be inactive. Thus, if x_4' is true, then z_1 is asserted. The complete equation for z_1, therefore, is $z_1 = x_1'x_2' + x_1x_2x_3' + x_4'$. Figure 1.20 (b) represents the same circuit, but in a more compact format.

Example 1.16 Given the logic diagram shown in Figure 1.21, the equation for output z_1 will be obtained. The circuit is implemented using NAND gates only. The output level is shown for each gate with the corresponding expression that generates the indicated voltage level. For example, the output of gate 3 will be at a high voltage level if the expression $(x_1 + x_2)x_3'$ is true. Similarly, the output of gate 5 will be at a high voltage level if the expression $x_1'x_2' + x_4$ is true. The complete equation for output z_1 is shown in Equation 1.29.

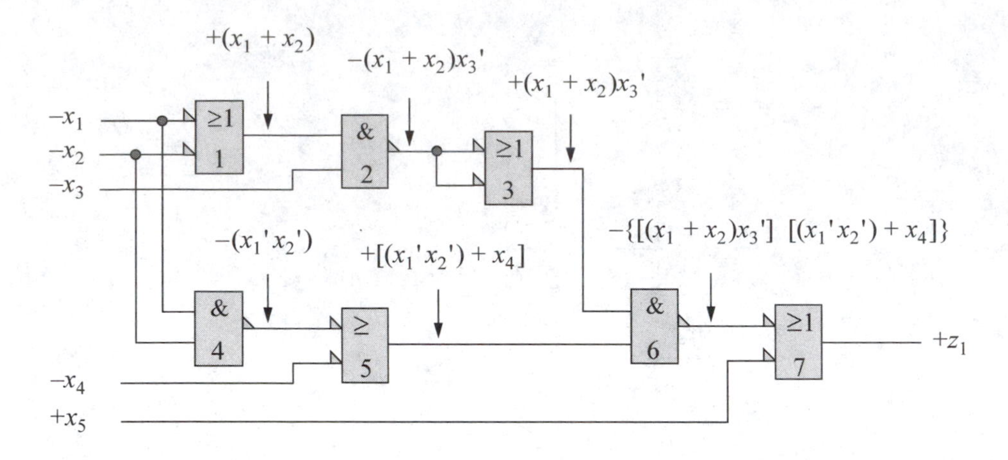

Figure 1.21 Logic diagram for Example 1.16 using only NAND gates. Output $z_1 = \{[(x_1 + x_2)x_3'] \; [(x_1'x_2') + x_4]\} + x_5'$.

$$z_1 = \{[(x_1 + x_2)x_3'] \; [(x_1'x_2') + x_4]\} + x_5' \tag{1.29}$$

Example 1.17 Given the logic diagram shown in Figure 1.22, the equation for output z_1 will be obtained. The circuit uses only NOR gates in its implementation. The output level is shown for each gate with the corresponding expression that represents the functionality of the circuit at that point. For example, the output of gate 3 is at a high voltage level if the expression $(x_1 + x_3')x_2'x_4$ is true. Likewise, the output of gate 5 is at a high level if the expression $x_2(x_3' + x_4')$ is true. The complete equation for z_1 is shown in Equation 1.30.

1.7 Synthesis of Combinational Logic

Synthesis of combinational logic consists of translating a set of network specifications into minimized boolean equations and then to generate a logic diagram from the equations using the logic primitives of AND, OR, and NOT. The equations are independent of any logic family and portray the functional operation of the network. The logic primitives can be realized by either AND gates, OR gates, and inverters, or by *functionally complete gates* such as, NAND or NOR gates.

The synthesis procedure is relatively straightforward. First, the equation is implemented with AND/OR/INVERT symbols without regard for logical complementation. Then, the negative assertion levels are assigned to the gates, where applicable, using the polarity symbol "⊾." This step establishes the logic family that is used for the implementation. The inputs are usually available in both high and low assertions.

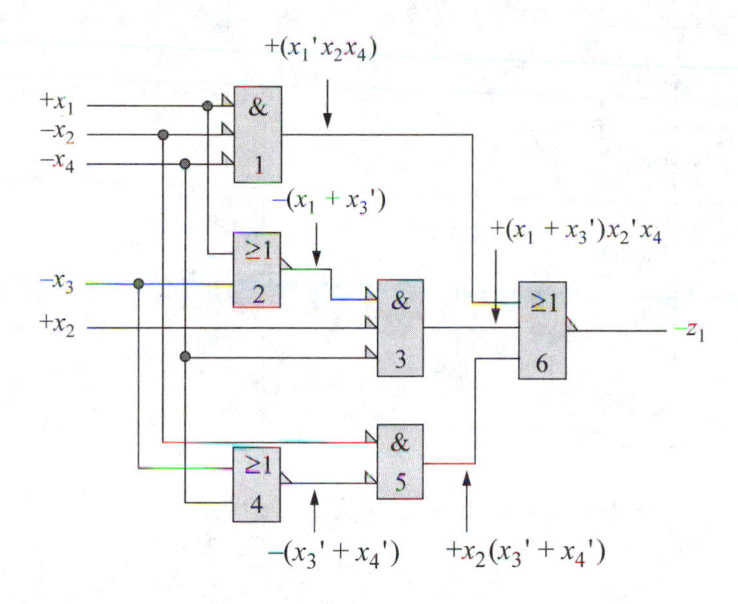

Figure 1.22 Logic diagram for Example 1.17 using only NOR gates. Output $z_1 = x_1'x_2x_4 + x_2'x_4(x_1 + x_3') + x_2(x_3' + x_4')$.

$$z_1 = x_1'x_2x_4 + x_2'x_4(x_1 + x_3') + x_2(x_3' + x_4') \tag{1.30}$$

Example 1.18 Equation 1.31 will be synthesized using NAND gates only. The first term $x_1x_2'x_3'x_4$ is represented by gate 1 in Figure 1.23. The inputs are shown at their active voltage levels such that, all gate inputs are high when the term $x_1x_2'x_3'x_4$ is true. That is, gate 1 will generate a low output when x_1 and x_4 are asserted and x_2 and x_3 are deasserted. The second expression $x_1'(x_2x_3x_4' + x_2'x_3')$ is implemented by gates 2, 3, 4, and 5.

$$z_1 = x_1x_2'x_3'x_4 + x_1'(x_2x_3x_4' + x_2'x_3') + x_1x_3(x_5 \oplus x_6)' \tag{1.31}$$

Since only NAND gates are used in the implementation, the exclusive-NOR function is decomposed into a sum-of-products form $x_5x_6 + x_5'x_6'$ and implemented by gates 6, 7, and 8. If only NOR gates were used in the implementation of Example 1.18, then the active level of all inputs would be complemented and output z_1 would be active at a low voltage level.

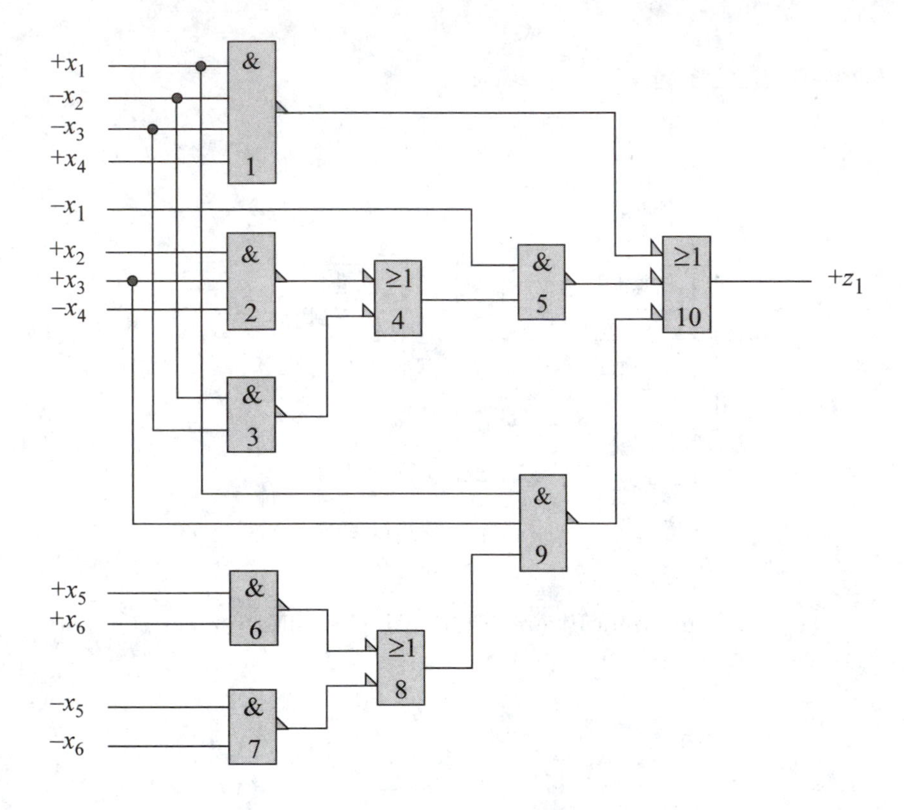

Figure 1.23 NAND gate implementation of Equation 1.31 for Example 1.18.

Example 1.19 Equation 1.32 will be synthesized using NOR gates only. The equation will first be represented as a sum of products, then converted to a minimal product-of-sums form for implementation.

$$z_1 = x_1x_3'x_4 + [(x_1 + x_2)' + x_3]' + (x_2 \oplus x_4')' \tag{1.32}$$

The equation can be represented as a sum of products as follows:

$$
\begin{aligned}
z_1 &= x_1x_3'x_4 + [(x_1 + x_2)' + x_3]' + (x_2 \oplus x_4')' \\
&= x_1x_3'x_4 + (x_1'x_2' + x_3)' + x_2x_4' + x_2'x_4 \\
&= x_1x_3'x_4 + (x_1 + x_2)x_3' + x_2x_4' + x_2'x_4 \\
&= x_1x_3'x_4 + x_1x_3' + x_2x_3' + x_2x_4' + x_2'x_4
\end{aligned}
$$

The equation can now be plotted on a Karnaugh map, as shown in Figure 1.24, then converted to a product-of-sums form, as shown in Equation 1.33. The logic diagram is shown in Figure 1.25 using only NOR gates. The inputs are available in both high and low assertion.

x_1x_2 \ x_3x_4	0 0	0 1	1 1	1 0
0 0	0	1	1	0
0 1	1	1	0	1
1 1	1	1	0	1
1 0	1	1	1	0

z_1

Figure 1.24 Karnaugh map for Equation 1.32 of Example 1.19.

$$z_1 = (x_1 + x_2 + x_4)(x_2 + x_3' + x_4)(x_2' + x_3' + x_4') \tag{1.33}$$

1.8 Multiplexers

A multiplexer is a logic macro device that allows digital information from two or more data inputs to be directed to a single output. Data input selection is controlled by a set of select inputs that determine which data input is gated to the output. The select inputs are labeled $s_0, s_1, s_2, \cdots, s_i, \cdots, s_{n-1}$, where s_0 is the low-order select input with a binary weight of 2^0 and s_{n-1} is the high-order select input with a binary weight of 2^{n-1}. The data inputs are labeled $d_0, d_1, d_2, \cdots, d_j, \cdots, d_{2^n-1}$. Thus, if a multiplexer has n select inputs, then the number of data inputs will be 2^n and will be labeled d_0 through d_{2^n-1}. For example, if $n = 2$, then the multiplexer has two select inputs s_0 and s_1 and four data inputs d_0, d_1, d_2, and d_3.

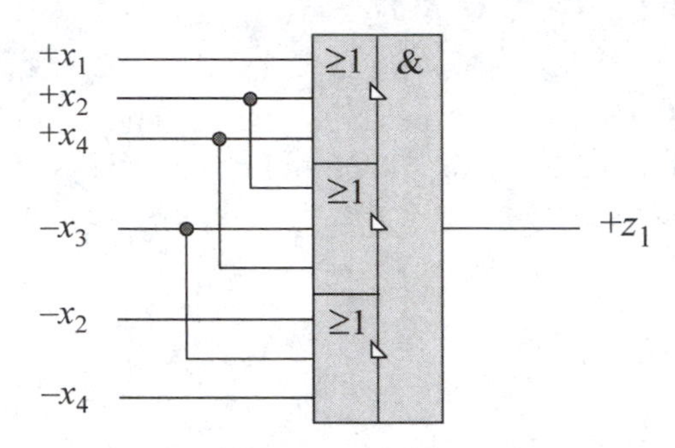

Figure 1.25 NOR gate implementation of Equation 1.33 for Example 1.19.

Figure 1.26 shows four typical multiplexers drawn in the ANSI/IEEE Std. 91-1984 format. The truth tables for the 4:1 (four-to-one) and 8:1 multiplexers are shown in Table 1.10 and Table 1.11, respectively. The truth tables for the 2:1 and 16:1 multiplexers are derived in a similar manner. Consider the 4:1 multiplexer in Table 1.10. If $s_1 s_0 = 00$, then data input d_0 is selected and its value is propagated to the multiplexer output z_1. Similarly, if $s_1 s_0 = 01$, then data input d_1 is selected and its value is directed to the multiplexer output.

The equation that represents output z_1 in the 4:1 multiplexer of Figure 1.26 (b) is shown in Equation 1.34. Output z_1 assumes the value of d_0 if $s_1 s_0 = 00$, as indicated by the term $s_1 ' s_0 ' d_0$. Likewise, z_1 assumes the value of d_1 when $s_1 s_0 = 01$, as indicated by the term $s_1 ' s_0 d_1$.

The symbols shown in Figure 1.26 represent single multiplexers controlled by dedicated select inputs. The symbology is different, however, when more than one multiplexer is controlled by the same set of select inputs. The multiplexer symbol shown in Figure 1.27 illustrates four 2:1 multiplexers with a common control block. Each multiplexer shares a common select input s_0 and a common enable. Only when the enable (EN) input is at a low voltage level will the selected data input of each multiplexer be propagated to the corresponding output.

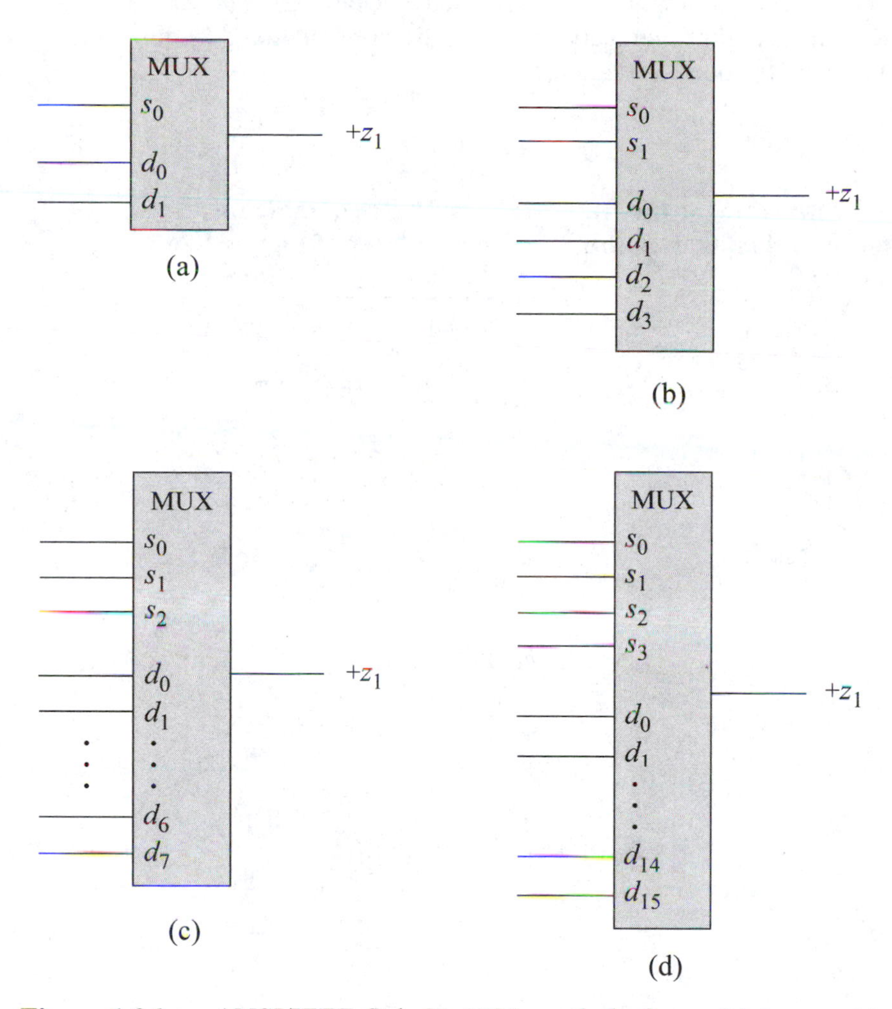

Figure 1.26 ANSI/IEEE Std. 91-1984 symbols for multiplexers: (a) 2:1 multiplexer; (b) 4:1 multiplexer; (c) 8:1 multiplexer; and (d) 16:1 multiplexer.

1.9 Decoders

A decoder is a combinational logic macro that is characterized by the following property: For every valid combination of inputs, a unique output is generated. In general, a decoder has n binary inputs and m mutually exclusive outputs, where $2^n \geq m$. An n:m (n-to-m) decoder is shown in Figure 1.28, where the label DX specifies a demultiplexer. Each output represents a minterm that corresponds to the binary representation of the input vector. Thus, $z_i = m_i$, where m_i is the ith minterm of the n input variables. For example, if $n = 3$ and $x_1x_2x_3 = 101$, then output z_5 is asserted. A decoder with n inputs, therefore, has a maximum of 2^n outputs. Because the outputs are mutually exclusive, only one output is active for each different combination of the inputs. The decoder outputs may be asserted high or low. Decoders have many applications in digital engineering, ranging from instruction decoding to memory addressing to code conversion.

Table 1.10 Truth table for the 4:1 multiplexer of Figure 1.26 (b)

Select inputs $s_1 s_0$	Data input selected
0 0	d_0
0 1	d_1
1 0	d_2
1 1	d_3

Table 1.11 Truth table for the 8:1 multiplexer of Figure 1.26 (c)

Select inputs $s_2 s_1 s_0$	Data input selected
0 0 0	d_0
0 0 1	d_1
0 1 0	d_2
0 1 1	d_3
1 0 0	d_4
1 0 1	d_5
1 1 0	d_6
1 1 1	d_7

$$z_1 = s_1{'}s_0{'}d_0 + s_1{'}s_0 d_1 + s_1 s_0{'}d_2 + s_1 s_0 d_3 \tag{1.34}$$

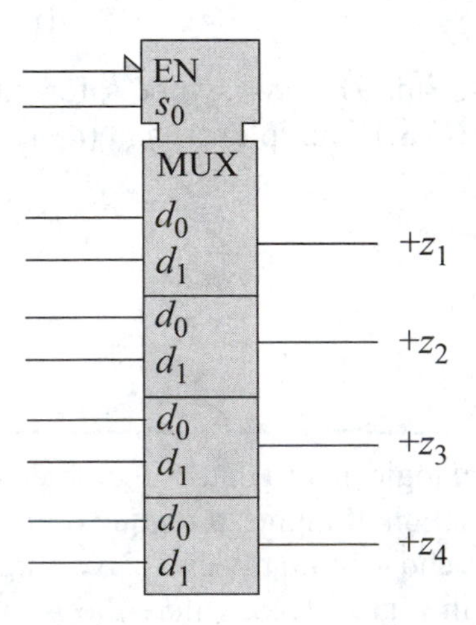

Figure 1.27 ANSI/IEEE Std. 91-1984 logic symbol for four 4:1 multiplexers with a common control block.

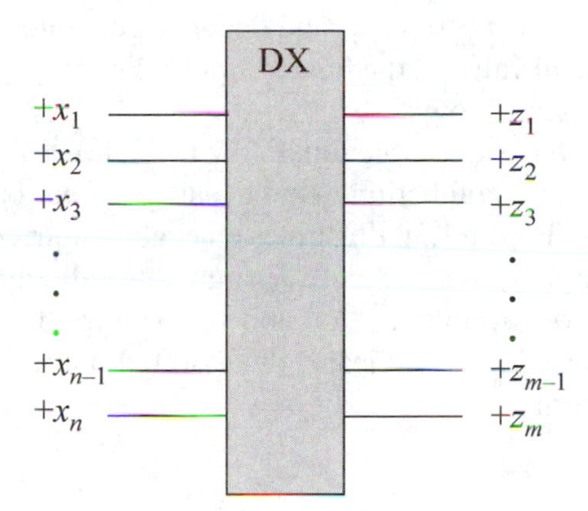

Figure 1.28 An $n{:}m$ decoder.

Figure 1.29 illustrates the logic symbol for a 2:4 decoder, where x_1 and x_2 are the binary input variables and z_1, z_2, z_3, and z_4 are the output variables. Input x_2 is the low-order variable. Since there are two inputs, each output corresponds to a different minterm of two variables as shown in the truth table of Table 1.12.

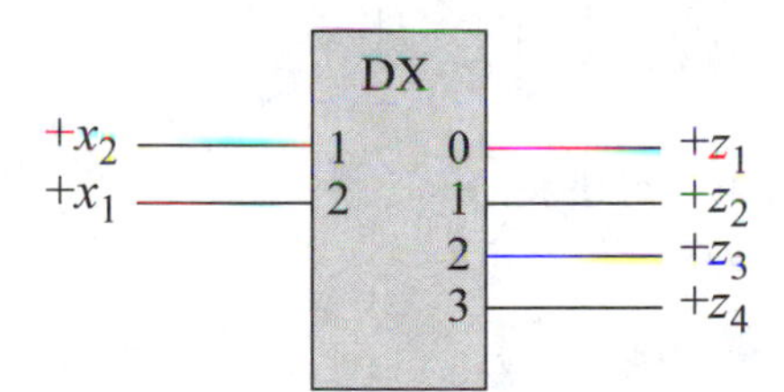

Figure 1.29 Logic symbol for a 2:4 decoder.

Table 1.12 Truth table for the 2:4 decoder of Figure 1.29

Inputs x_1x_2	Outputs $z_1z_2z_3z_4$	Minterm decoding function
0 0	1 0 0 0	$x_1'x_2'$
0 1	0 1 0 0	$x_1'x_2$
1 0	0 0 1 0	x_1x_2'
1 1	0 0 0 1	x_1x_2

A 3:8 decoder is shown in Figure 1.30 which decodes a binary number into the corresponding octal number. The three inputs are x_1, x_2, and x_3 with binary weights of 2^2, 2^1, and 2^0, respectively. The decoder generates an output that corresponds to the decimal value of the binary inputs. For example, if $x_1x_2x_3 = 110$, then output z_6 is asserted low.

A decoder may also have an enable function which allows the selected output to be asserted. The enable function may be a single input or an AND gate with two or more inputs. Figure 1.30 illustrates an enable input consisting of an AND gate with three inputs. The 3:8 decoder generates all eight minterms z_0 through z_7 of three binary variables x_1, x_2, and x_3. The truth table for the decoder is shown in Table 1.13 and indicates the asserted output that represents the corresponding minterm.

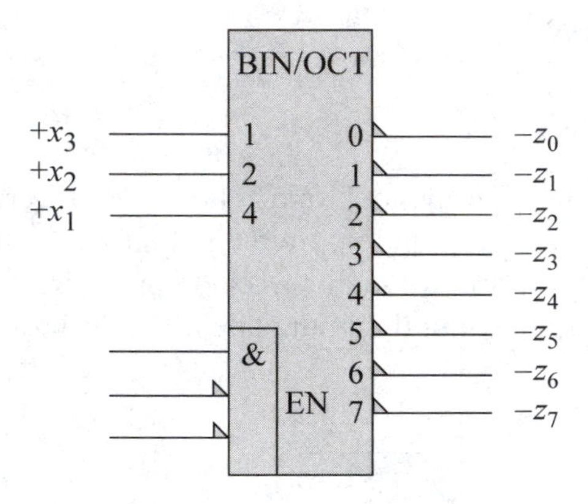

Figure 1.30 A binary-to-octal decoder.

Table 1.13 Truth table for the 3:8 decoder of Figure 1.30

$x_1 x_2 x_3$	z_0	z_1	z_2	z_3	z_4	z_5	z_6	z_7
0 0 0	**0**	1	1	1	1	1	1	1
0 0 1	1	**0**	1	1	1	1	1	1
0 1 0	1	1	**0**	1	1	1	1	1
0 1 1	1	1	1	**0**	1	1	1	1
1 0 0	1	1	1	1	**0**	1	1	1
1 0 1	1	1	1	1	1	**0**	1	1
1 1 0	1	1	1	1	1	1	**0**	1
1 1 1	1	1	1	1	1	1	1	**0**

Example 1.20 One use for a decoder is to implement a boolean function. The equation shown in Equation 1.35 can be synthesized with a 3:8 decoder and one OR gate as shown in Figure 1.31. The term x_1x_2 in Equation 1.35 corresponds to minterms 6 ($x_1x_2x_3'$) and 7 ($x_1x_2x_3$). The terms $x_1x_2'x_3'$ and $x_1'x_2x_3$ are represented by minterm outputs 4 and 3, respectively. Output z_1, therefore, is implemented by ORing the appropriate outputs of the decoder.

$$z_1 = x_1x_2 + x_1x_2'x_3' + x_1'x_2x_3 \tag{1.35}$$

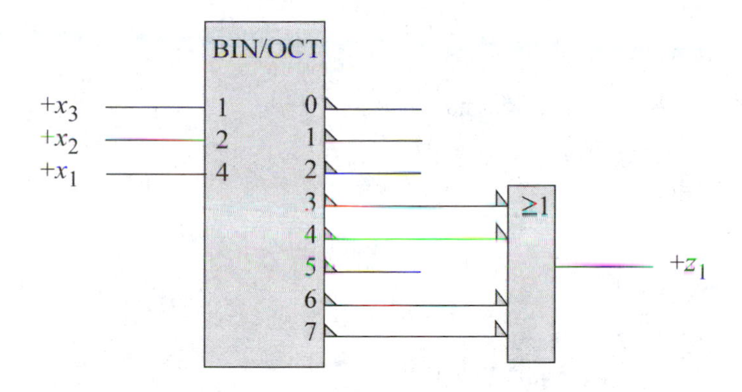

Figure 1.31 Implementation of $z_1 = x_1x_2 + x_1x_2'x_3' + x_1'x_2x_3$ using a decoder and additional logic for Example 1.20.

A 4:10 decoder is shown in Figure 1.32, which converts the BCD format to decimal. The four inputs represent the ten decimal digits in binary; the ten outputs represent the ten mutually exclusive decimal digits. Of the 2^4 possible combinations of four bits, only ten combinations are required: 0000 to 1001. The remaining combinations 1010 through 1111 are invalid for BCD. Decoders will be utilized in a later chapter for synthesizing synchronous sequential machines.

Example 1.21 The equations shown in Equation 1.36 can be synthesized by a 4:16 decoder and additional logic gates. The logic implementation is shown in Figure 1.33.

$$\begin{aligned} z_1(x_1,x_2,x_3,x_4) &= \Sigma_m(1,2,3,4,5,6) \\ z_1(x_1,x_2,x_3,x_4) &= \Sigma_m(7,8,9,10,11,12) \end{aligned} \tag{1.36}$$

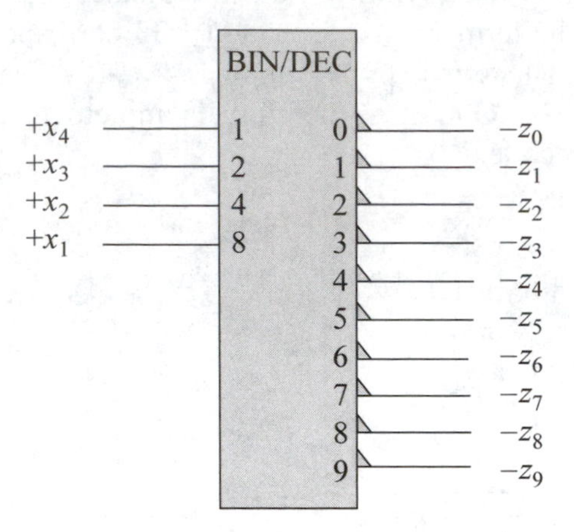

Figure 1.32 A BCD-to-decimal decoder.

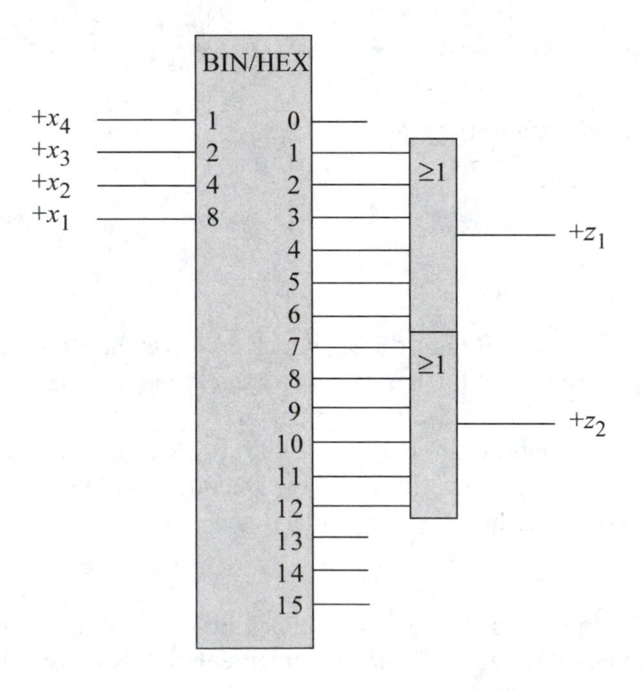

Figure 1.33 Implementation of $z_1(x_1,x_2,x_3,x_4) = \Sigma_m(1,2,3,4,5,6)$ and $z_1(x_1,x_2,x_3,x_4) = \Sigma_m(7,8,9,10,11,12)$ using a decoder and additional logic for Example 1.21.

1.10 Encoders

An encoder is a macro logic circuit with n mutually exclusive inputs and m binary outputs, where $n \leq 2^m$. The function of an encoder can be considered to be the inverse of a decoder; that is, the mutually exclusive inputs are encoded into a corresponding binary number. A general block diagram for an n:m encoder is shown in Figure 1.34. An encoder is also referred to as a code converter. In the label of Figure 1.34, X corresponds to the input code and Y corresponds to the output code. The general qualifying label X/Y is replaced by the input and output codes, respectively such as, OCT/BIN for an octal-to-binary code converter. Only one input x_i is asserted at a time. The decimal value of x_i is encoded as a binary number which is specified by the m outputs. An 8:3 octal-to-binary encoder and a 10:4 BCD-to-binary encoder are shown in Figure 1.35 (a) and (b), respectively. In Figure 1.35 (b), if input 7 is asserted, then the output values are $z_1z_2z_3z_4 = 0111$, where the binary weight of the outputs is $z_1z_2z_3z_4 = 2^3 2^2 2^1 2^0$.

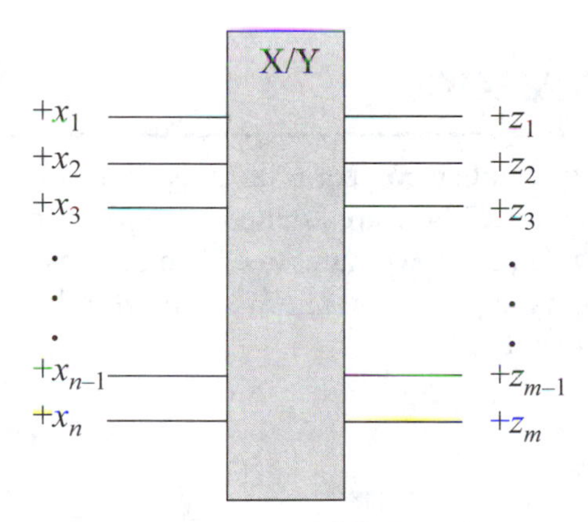

Figure 1.34 An n:m encoder or code converter.

It was stated previously that encoder inputs are mutually exclusive. There may be situations, however, where more than one input can be active at a time. Then a priority must be established to select and encode a particular input. This is referred to as a *priority encoder*. Usually the input with the highest valued subscript is selected as highest priority for encoding. Thus, if x_i and x_j are active simultaneously and $i < j$, then x_j has priority over x_i. For example, assume that the octal-to-binary encoder of Figure 1.35 (a) is a priority encoder. If inputs x_1, x_5, and x_7 are asserted simultaneously, then the outputs will indicate the binary equivalent of decimal 7 such that, $z_3z_2z_1 = 111$.

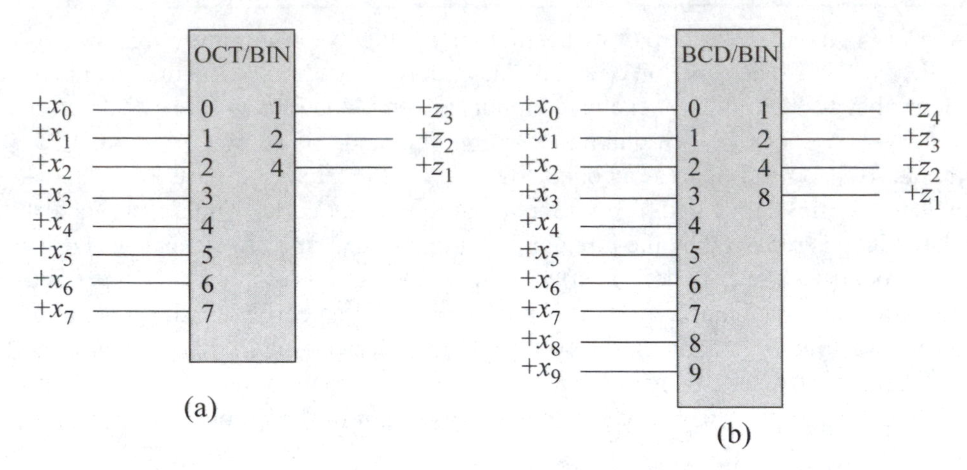

Figure 1.35 Encoders: (a) octal-to-binary; and (b) BCD-to-binary.

1.11 Comparators

A comparator is a logic circuit that compares the magnitude of two n-bit binary numbers X_1 and X_2. Therefore, there are $2n$ inputs and three outputs that indicate the relative magnitude of the two numbers. The outputs are mutually exclusive, specifying $X_1 < X_2$, $X_1 = X_2$, or $X_1 > X_2$. Figure 1.36 shows a general block diagram of a comparator.

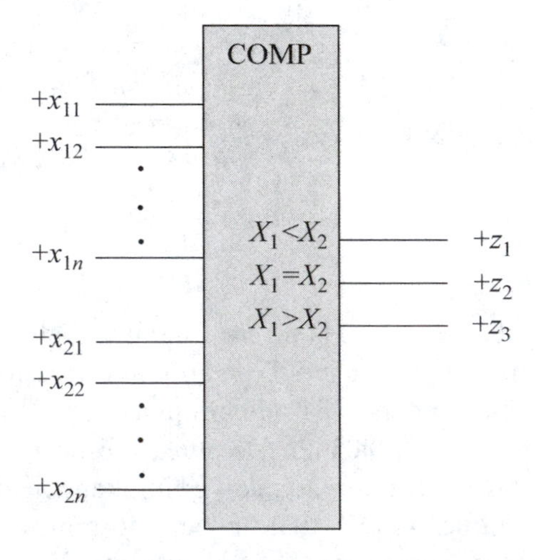

Figure 1.36 General block diagram of a comparator.

If two or more comparators are connected in cascade, then three additional inputs are required for each comparator. These additional inputs indicate the relative magnitude of the previous lower-order comparator inputs and specify $X_1 < X_2, X_1 = X_2$, or $X_1 > X_2$ for the previous stage. Cascading comparators usually apply only to commercially available comparator integrated circuits. For example, two 8-bit numbers can be compared by using two 4-bit comparators as shown in Figure 1.37. The magnitude inputs for the low-order comparator are assigned the following values: $(X_1 < X_2) = 0$, $(X_1 = X_2) = 1$, and $(X_1 > X_2) = 0$. This assignment initializes the compare operation by specifying that the two operands are equal at that point in the comparison. Any other assignment would have a negative effect on the comparison operation.

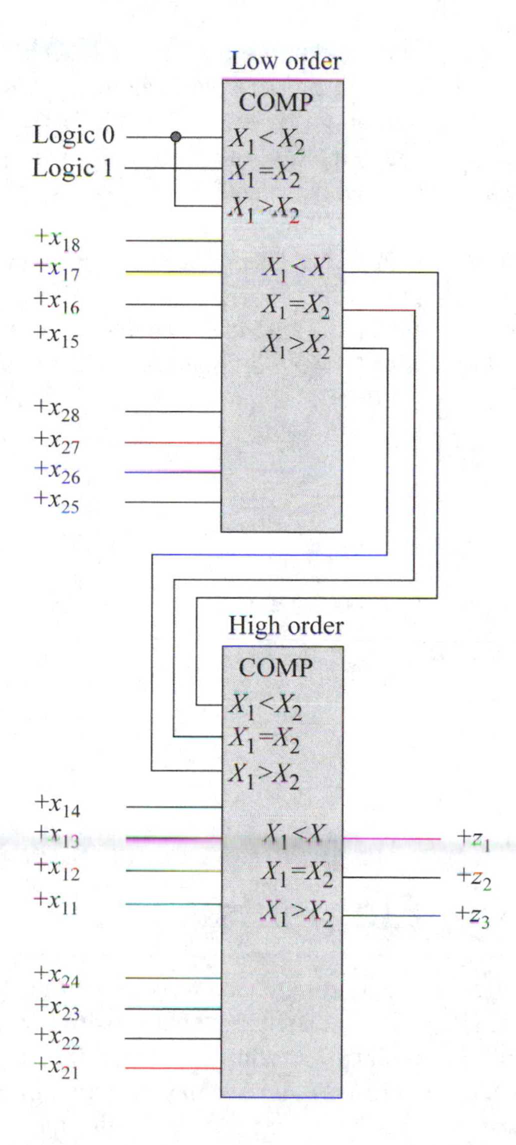

Figure 1.37 Two 4-bit comparators to compare two 8-bit binary numbers.

The design of a comparator is relatively straightforward. Consider two 3-bit unsigned operands $X_1 = x_{11}x_{12}x_{13}$ and $X_2 = x_{21}x_{22}x_{23}$, where x_{13} and x_{23} are the low-order bits of X_1 and X_2, respectively. Three equations will now be derived to represent the three outputs; one equation each for $X_1 < X_2, X_1 = X_2$, and $X_1 > X_2$. Comparison occurs in a left-to-right manner beginning at the high-order bits of the two operands.

Operand X_1 will be less than X_2 if $x_{11}\ x_{21} = 01$. Thus, X_1 cannot be more than 011 while X_2 cannot be less than 100, indicating that $X_1 < X_2$. Therefore, the first term of the equation for $X_1 < X_2$ is $x_{11}'x_{21}$. If, however, $x_{11} = x_{21}$, then the relative magnitude depends on the values of x_{12} and x_{22}. The equality of two bits is represented by the exclusive-NOR function, also called the *equality function*. Thus, the second term in the equation for $X_1 < X_2$ is $(x_{11} \oplus x_{21})'x_{12}'x_{22}$. The analysis continues in a similar manner for the remaining bits of the two operands.

The equation for $X_1 < X_2$ is shown in Equation 1.37. The equality of X_1 and X_2 is true if and only if each bit-pair is equal, where $x_{11} = x_{21}, x_{12} = x_{22}$, and $x_{13} = x_{23}$; that is, $X_1 = X_2$ if and only if $x_{1i} = x_{2i}$ for $i = 1, 2, 3$. This is indicated by the boolean product of three equality functions as shown in Equation 1.37 for $X_1 = X_2$. The final equation, which specifies $X_1 > X_2$, is obtained in a manner analogous to that for $X_1 < X_2$. If the high-order bits are $x_{11}x_{21} = 10$, then it is immediately apparent that $X_1 > X_2$. Using the equality function with the remaining bits yields the equation for $X_1 > X_2$ as shown in Equation 1.37. The design process is modular and can be extended to accommodate any size operands in a well-defined regularity. Two n-bit operands will contain column subscripts of 11 through $1n$ and 21 through $2n$, where n specifies the low-order bits.

$$(X_1 < X_2) = x_{11}'x_{21} + (x_{11} \oplus x_{21})'x_{12}'x_{22} + (x_{11} \oplus x_{21})'(x_{12} \oplus x_{22})'x_{13}'x_{23}$$

$$(X_1 = X_2) = (x_{11} \oplus x_{21})'(x_{12} \oplus x_{22})'(x_{13} \oplus x_{23})'$$

$$(X_1 > X_2) = x_{11}x_{21}' + (x_{11} \oplus x_{21})'x_{12}x_{22}' + (x_{11} \oplus x_{21})'(x_{12} \oplus x_{22})'x_{13}x_{23}' \quad (1.37)$$

1.12 Storage Elements

This section will review the operating characteristics of the *SR* latch, *D* flip-flop, *JK* flip-flop, and *T* flip-flop. A latch is a level-sensitive storage element in which a change to an input signal affects the output directly without recourse to a clock input. The set (s) and reset (r) inputs may be active high or active low. The *D*, *JK*, and *T* flip-flops, however, are triggered on the application of a clock signal and are positive- or negative-edge-triggered devices.

1.12.1 *SR* Latch

The *SR* latch is usually implemented using either NAND gates or NOR gates, as shown in Figure 1.38 (a) and Figure 1.38 (b), respectively. When a negative pulse (or level) is applied to the –Set input of the NAND gate latch, the output $+y_1$ becomes active at a high voltage level. This high level is also connected to the input of NAND gate 2. Since the set and reset inputs cannot both be active simultaneously, the reset input is at a high level, providing a low voltage level on the output of gate 2 which is fed back to the input of gate 1. The negative feedback, therefore, provides a second set input to the latch. The original set pulse can now be removed and the latch will remain set. Concurrent set and reset inputs represent an invalid condition, since both outputs will be at the same voltage level; that is, outputs $+y_1$ and $-y_1$ will both be at the more positive voltage level — an invalid state for a bistable device with complementary outputs.

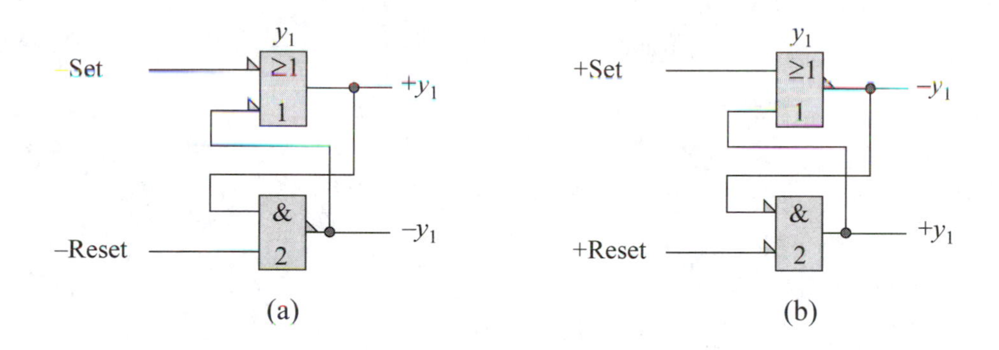

Figure 1.38 *SR* latches: (a) using NAND gates; and (b) using NOR gates.

If the NAND gate latch is set, then a low voltage level on the –Reset input will cause the output of gate 2 to change to a high level which is fed back to gate 1. Since both inputs to gate 1 are now at a high level, the $+y_1$ and $-y_1$ outputs will change to a low and high level, respectively, which is the reset state for the latch. The *characteristic table* for an *SR* latch is shown in Table 1.14, where $Y_{j(t)}$ and $Y_{k(t+1)}$ are the present state and next state of the latch, respectively. The excitation equation is shown in Equation 1.38.

1.12.2 *D* Flip-Flop

A *D* flip-flop is an edge-triggered device with one data input and one clock input. Figure 1.39 illustrates a positive-edge-triggered *D* flip-flop. The $+y_1$ output will assume the state of the *D* input at the next positive clock transition. After the occurrence of the clock's positive edge, any change to the D input will not affect the output until the next active clock transition. The characteristic table for a D flip-flop is shown in Table 1.15 and the corresponding excitation equation in Equation 1.39.

Table 1.14 ***SR*** **latch characteristic table**

Data inputs $S\ R$	Present state $Y_{j(t)}$	Next state $Y_{k(t+1)}$
0 0	0	0
0 0	1	1
0 1	0	0
0 1	1	0
1 0	0	1
1 0	1	1
1 1	0	Invalid
1 1	1	Invalid

$$Y_{k(t+1)} = S + R' \, Y_{j(t)} \tag{1.38}$$

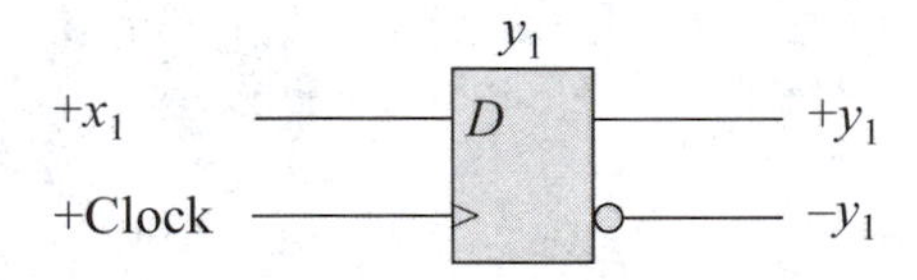

Figure 1.39 A positive-edge-triggered D flip-flop.

Table 1.15 ***D*** **flip-flop characteristic table**

Data input D	Present state $Y_{j(t)}$	Next state $Y_{k(t+1)}$
0	0	0
0	1	0
1	0	1
1	1	1

$$Y_{k(t+1)} = D \tag{1.39}$$

1.12.3 *JK* Flip-Flop

The *JK* flip-flop is also an edge-triggered storage device. The active clock transition can be either the positive or negative edge. Figure 1.40 illustrates a negative-edge-triggered *JK* flip-flop. The functional characteristics of the *JK* data inputs are defined in Table 1.16. The characteristic table of Table 1.17 lists the next state $Y_{k(t+1)}$ for each combination of *J*, *K*, and the present state $Y_{j(t)}$ based on the functional characteristics of *J* and *K*. Table 1.18 shows an excitation table in which a particular state transition predicates a set of values for *J* and *K*. This table is especially useful in the synthesis of synchronous sequential machines.

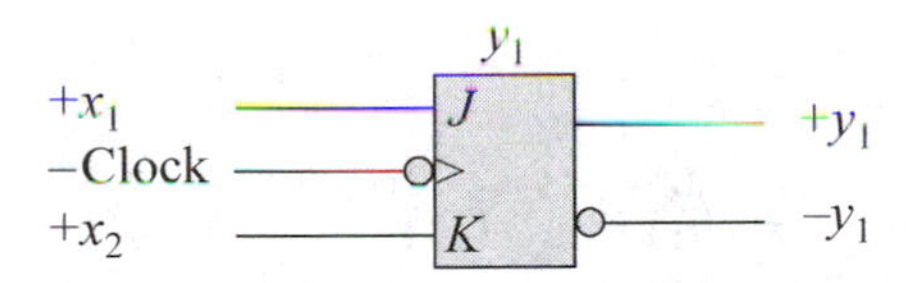

Figure 1.40 A negative-edge-triggered *JK* flip-flop.

Table 1.16 *JK* functional characteristic table

J K	Function
0 0	No change
0 1	Reset
1 0	Set
1 1	Toggle

Table 1.17 *JK* flip-flop characteristic table

Data inputs *J K*	*Present state* $Y_{j(t)}$	*Next state* $Y_{k(t+1)}$
0 0	0	0
0 0	1	1
0 1	0	0
0 1	1	0
1 0	0	1
1 0	1	1
1 1	0	1
1 1	1	0

Table 1.18 Excitation table for a *JK* flip-flop

Present state $Y_{j(t)}$	Next state $Y_{k(t+1)}$	Data inputs JK
0	0	0 –
0	1	1 –
1	0	– 1
1	1	– 0

The excitation equation for a *JK* flip-flop is derived from Table 1.18 and is shown in Equation 1.40.

$$Y_{k(t+1)} = Y_{j(t)}' J + Y_{j(t)} K' \tag{1.40}$$

1.12.4 *T* Flip-Flop

The toggle (*T*) flip-flop is shown in Figure 1.41 as a positive-edge-triggered device. When the *T* input is at a logic 1 level, the flip-flop will toggle (change state) at the next active clock transition. The characteristic table is shown in Table 1.19 and the excitation table in Table 1.20. The corresponding excitation equation is shown in Equation 1.41.

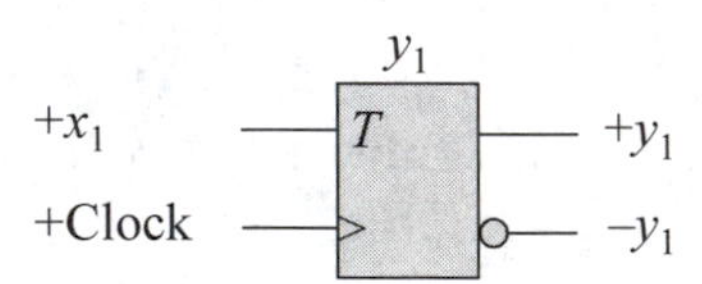

Figure 1.41 A *T* flip-flop.

Table 1.19 *T* flip-flop characteristic table

Data input T	Present state $Y_{j(t)}$	Next state $Y_{k(t+1)}$
0	0	0
0	1	1
1	0	1
1	1	0

Table 1.20 *T* flip-flop excitation table

Present state $Y_{j(t)}$	Next state $Y_{k(t+1)}$	Data input T
0	0	0
0	1	1
1	0	1
1	1	0

$$Y_{k(t+1)} = Y_{j(t)}{}'\,T + Y_{j(t)}\,T' \tag{1.41}$$

1.13 Programmable Logic Devices

Combinational logic can also be implemented using programmable logic devices (PLDs). PLDs are prefabricated ICs in which fused and hard-wired interconnections are used. PLDs implement 2-level switching functions by means of an AND array and an OR array. There are three main types of PLDs: *programmable read-only memories* (PROMs), *programmable array logic* (PAL) devices, and *programmable logic array* (PLA) devices. A related device is a *field-programmable gate array* (FPGA) which implements complex logic functions by means of a configuration program stored in an on-chip memory. The program establishes the interconnection of internal logic blocks and I/O blocks.

The basic organization of a PLD consists of an AND array driving an OR array as shown in Figure 1.42. There is a set of inputs X_i containing n input signals and a set of outputs Z_i containing m output signals. The amount of programming capability depends upon the type of PLD that is used. For example, a PROM contains a fixed AND array and a programmable OR array; a PAL contains a programmable AND array and a fixed OR array; a PLA contains both a programmable AND array and a programmable OR array. Both PAL and PLA architectures have versions which contain storage elements in conjunction with combinational logic. PLDs can be used in the synthesis of both combinational and sequential logic networks.

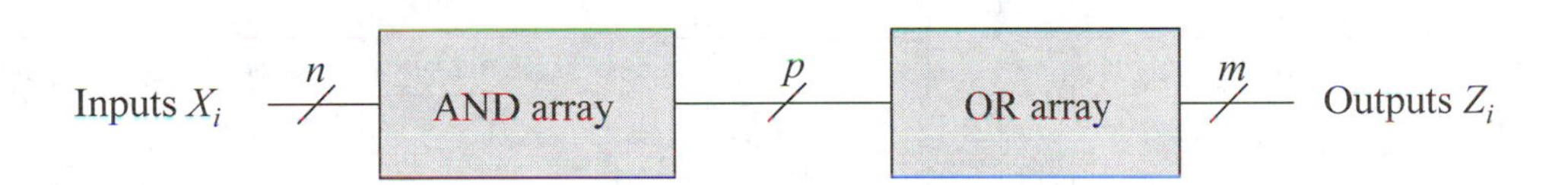

Figure 1.42 Basic organization of a programmable logic device.

1.13.1 Programmable Read-Only Memories

A PROM is a storage device in which the information is permanently stored; that is, the data remains valid even after power is turned off. PROMs are used for application programs, tables, code conversion, control store for microprogram sequencers, and other functions in which the stored data is not changed. The organization of a PROM is essentially the same as that for other PLDs: an input vector (an address) connects to an AND array which in turn connects to an OR array which generates the output vector (or word) for the PROM.

In general, a PROM contains n inputs and m outputs. Because the inputs function as an address, there are 2^n unique addresses to select one of 2^n words. The AND array decodes the address to select a specific word in memory. Thus, the interconnections in the AND array are fixed by the manufacturer and cannot be programmed by the user. The OR array, however, is programmable. The interconnections in the OR array are programmed by the user using special internal circuitry and a programming device to indicate the bit configuration of each word in memory. Each interconnection functions as a fuse; thus, the fuse can be left intact (indicating a logic 1) or opened (indicating a logic 0).

The basic organization of a PROM is shown in Figure 1.43. The AND array is fixed and the OR array is programmable. The dot at the intersection of two lines in the AND array indicates a *hard-wired* connection, and thus, cannot be programmed. The × at the intersection of two lines in the OR array represents a *fused* connection in which the fuse is intact. An intersection with no × indicates an open fuse. This convention is used for all programmable logic devices in this book.

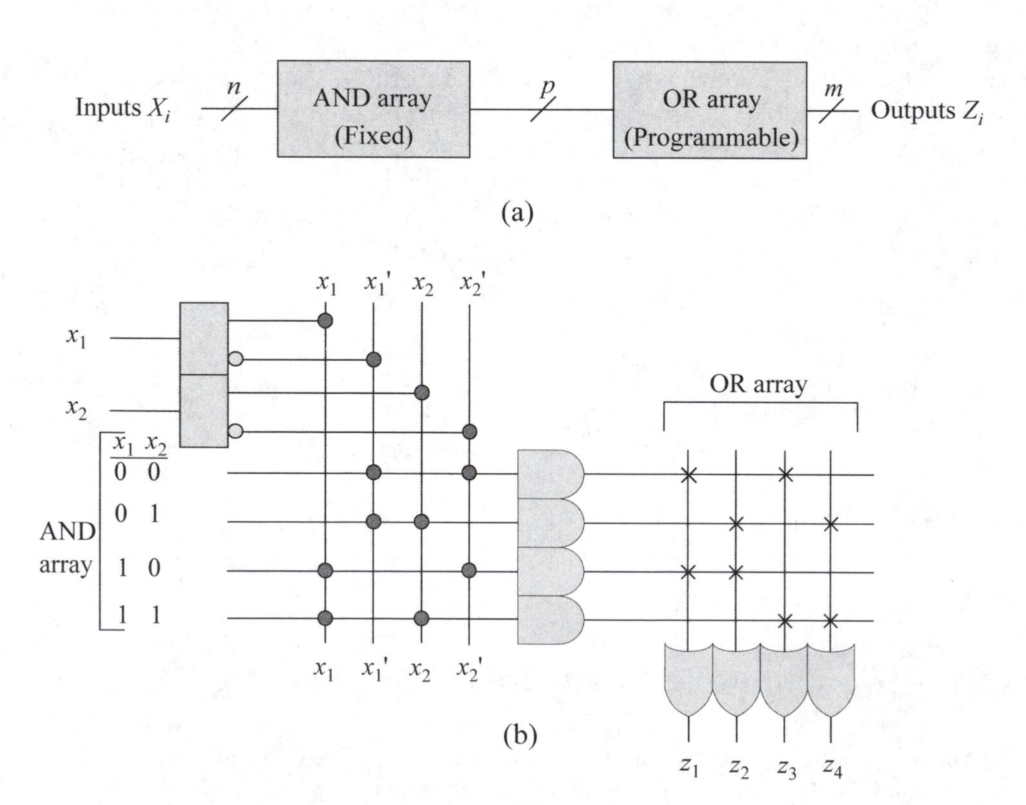

Figure 1.43 Basic organization of a PROM: (a) block diagram; and (b) implementation using two inputs for the address and four outputs.

All unused inputs in the AND array correspond to an open or *floating* input. Thus, all unused AND gate inputs must generate a logic 1 so that the output of the AND gate will be a function of the "hard-wired" connections only. The PROM illustrated in Figure 1.43 contains four words with four bits per word as shown in Table 1.21. When the PROM address is $x_1x_2 = 10$, the output word is $z_1z_2z_3z_4 = 1100$, which may be a data constant depending on the application of the PROM at that address.

Table 1.21 PROM program

Address x_1x_2	Outputs $z_1z_2z_3z_4$
0 0	1 0 1 0
0 1	0 1 0 1
1 0	1 1 0 0
1 1	0 0 1 1

Although a PROM is primarily used to store application programs, code conversions, tables, and data constants, it can also be used to implement logic functions. For example, output z_1 of Figure 1.43 is asserted when $z_1 = x_1'x_2' + x_1x_2'$. The number of outputs can be increased without increasing the number of inputs. The two input variables in Figure 1.43 are decoded into four lines by means of four AND gates that represent the AND array. By convention, only one line is drawn as the input to each AND gate. This single line, however, represents four lines, one line each for x_1, x_1', x_2, and x_2'. The input buffer drivers generate the true and complement of each input variable. Each AND gate output represents one of the four minterms of two variables. The four outputs of the AND array decoder are connected through fuses to each OR gate. A PROM generates outputs that are in a sum-of-minterms form.

Example 1.22 A PROM will be used to implement combinational logic in a sum-of-minterms form. Consider the truth table of Table 1.22. There are two inputs that select one of four words, where each word consists of three bits z_1, z_2, and z_3. Each output can be represented as a sum-of-minterms. Although the functions can be more easily implemented with discrete gates, the example serves to illustrate the technique for PROM programming to implement combinational logic functions.

Table 1.22 PROM programming for Example 1.22

Address x_1x_2	Outputs $z_1z_2z_3$
0 0	1 1 0
0 1	0 1 1
1 0	1 0 1
1 1	0 0 0

Figure 1.44 shows the PROM organization and programming for Example 1.22. The fuses are shown either intact or open to correspond to the appropriate entries in the truth table, where a 1 and 0 specify an intact or open fuse, respectively. For example, z_1 is connected by intact fuses to AND gate outputs corresponding to $x_1'x_2' + x_1x_2'$. Any unused OR gate input must not contribute to the generated function; therefore, all open-fused OR gate inputs must provide a logic 0 to the gate. This example demonstrates the general procedure for implementing combinational logic in a PROM. The number of inputs and outputs determines the size of the PROM. The truth table is then generated, which specifies the programming requirements of the device. No minimization is necessary, since the size of the PROM is determined by the number of inputs and outputs.

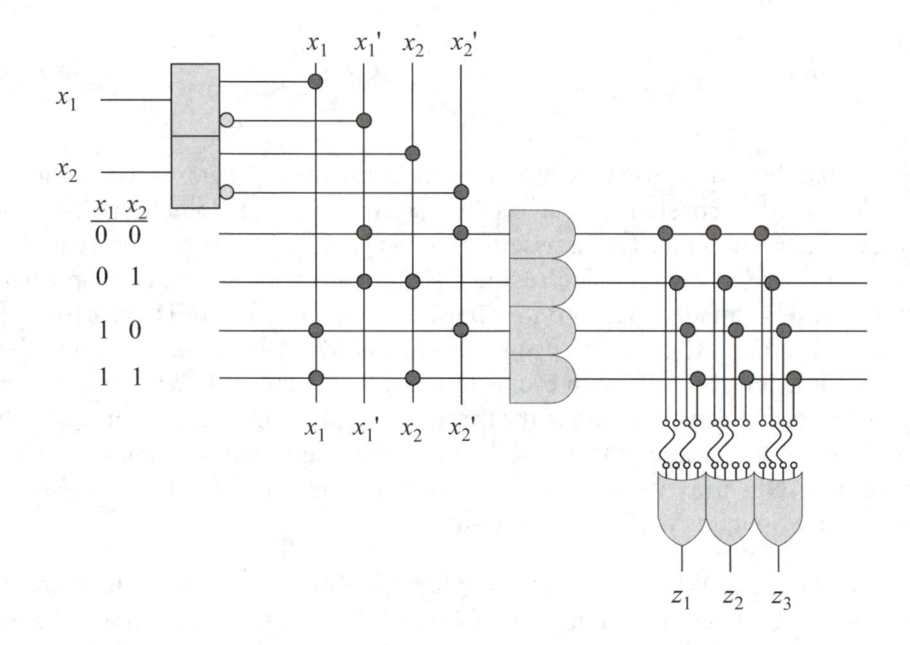

Figure 1.44 PROM organization to implement the functions $z_1 = x_1'x_2' + x_1x_2'$, $z_2 = x_1'x_2' + x_1'x_2$, and $z_3 = x_1'x_2 + x_1x_2'$.

1.13.2 Programmable Array Logic

A PAL device confirms to the general structure of a PLD. The number of AND gates and OR gates is variable, depending on the part number of the commercially-available PAL. Some PALs contain 10 dedicated inputs and eight outputs, some of which have dual I/O functions. Each output section consists of eight programmable AND gates connected to a fixed OR gate. There are eight output sections for a total of 64 AND gates. In many cases, the outputs are also fed back through separate buffers (drivers) to the programmable AND array. The basic organization of a PAL is shown in Figure 1.45. The AND array is programmable and the OR array is fixed. The requirement that a PROM must have 2^n AND gates in the AND array for n inputs is not a restriction for a PAL — the number of AND gates in the AND array is not a function of the number of inputs. By convention, the logic symbols in the AND array use a single horizontal line connected to the gate input, where the single line represents all of the inputs to the gate. The intersecting vertical lines in the AND array represent the device inputs, both true and complemented.

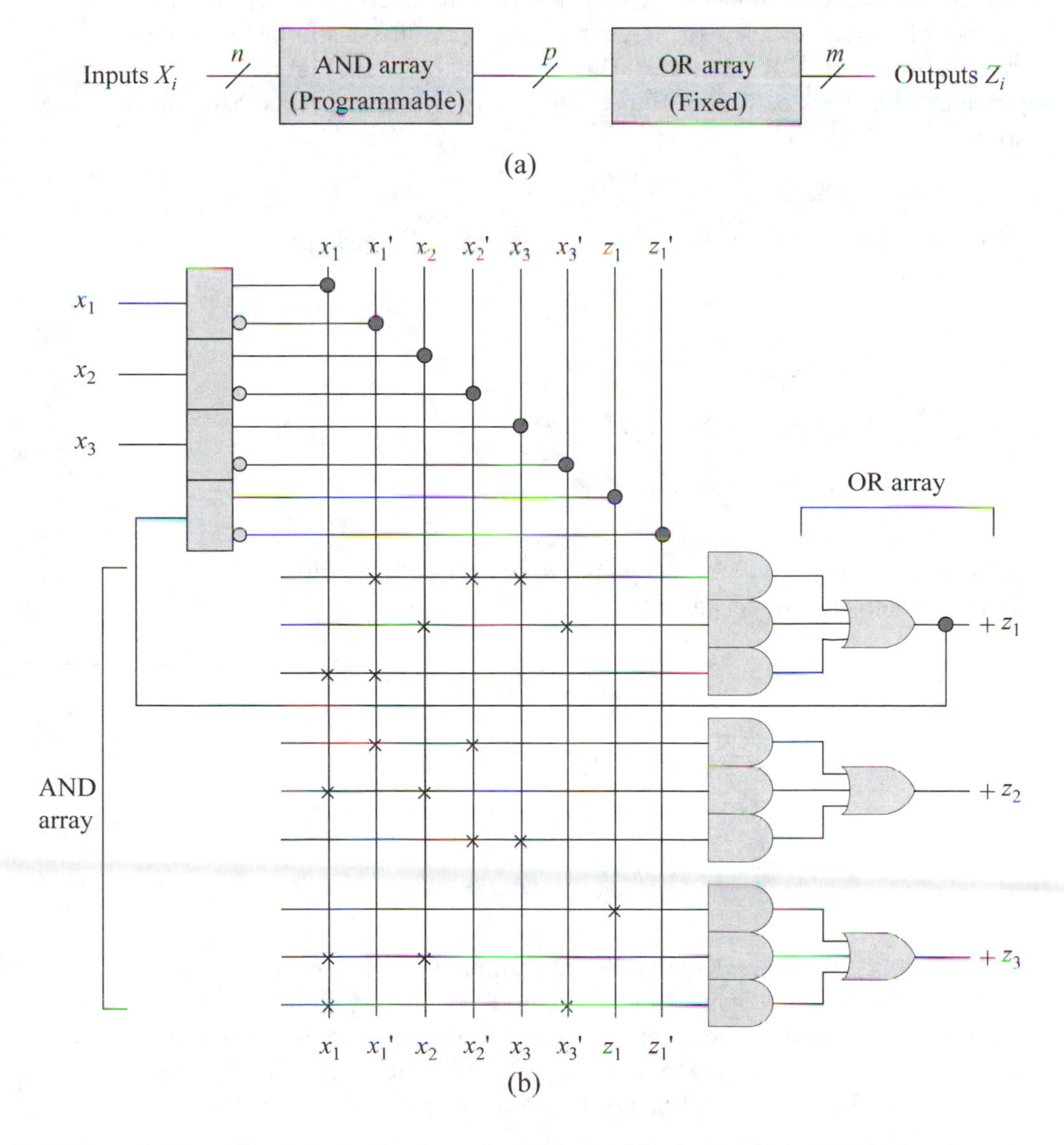

Figure 1.45 Basic organization of a PAL: (a) block diagram; and (b) implementation using three inputs and three outputs.

Figure 1.45 illustrates a PAL device consisting of three inputs and three outputs. Each input is connected to a buffer-driver which generates both true and complemented outputs of the corresponding input. The device consists of nine AND gates forming the programmable AND array and three OR gates which form the fixed OR array. Each AND gate has eight fused programmable inputs as shown by the eight vertical lines intersecting each horizontal line. The horizontal line is called the *product line* and symbolizes the multiple-input configuration of the AND gate. The output of each AND gate is the corresponding *product term*.

When designing with a PAL, the boolean expressions should be minimized, if necessary, so that the number of product terms in each expression does not exceed the number of AND gates in each AND-OR structure. If the number of product terms is too large, then the function can still be realized by utilizing two AND-OR structures. This concept is shown in Figure 1.45 which provides a feedback path from output z_1 to a driver which connects to the AND array.

Example 1.23 The following boolean expressions will be implemented using the PAL of Figure 1.45:

$$z_1(x_1, x_2, x_3) = \Sigma_{\mathrm{m}}(1, 2, 6)$$

$$z_2(x_1, x_2, x_3) = \Sigma_{\mathrm{m}}(0, 1, 5, 6, 7)$$

$$z_3(x_1, x_2, x_3) = \Sigma_{\mathrm{m}}(1, 2, 4, 6, 7)$$

Using boolean algebra or Karnaugh maps, the above functions convert to the following sum-of-products forms:

$$z_1 = x_1'x_2'x_3 + x_2x_3'$$

$$z_2 = x_1'x_2' + x_1x_2 + x_2'x_3$$

$$z_3 = x_1'x_2'x_3 + x_2x_3' + x_1x_2 + x_1x_3'$$

$$= z_1 + x_1x_2 + x_1x_3'$$

The three functions are realized by the programmable AND array shown in Figure 1.45. Function z_1 contains only two terms; thus, the bottom AND gate is not required. To insure that the output of the lower AND gate does not contribute to output z_1, the fuses for inputs x_1 and x_1' are left intact. Thus, the AND gate receives both the true and complemented values of input x_1, so that the output of the gate is $x_1x_1' = 0$.

Function z_2 is programmed in a similar manner. In this case, all three AND gates are utilized. Function z_3 requires four product terms; however, the first two terms are the same as the two terms that represent z_1. Therefore, using output z_1, function z_3 can be reduced to three terms, as shown in Figure 1.45.

1.13.3 Programmable Logic Array

The basic organization of a PLA is shown in Figure 1.46. Both the AND array and the OR array are programmable. Since both arrays are programmable, the PLA has more programming capability and thus, more flexibility than the PROM or PAL.

The output function in a PLA is limited only by the number of AND gates in the AND array, since all AND gates can be programmed to connect to all OR gates. The output function in a PAL, however, is restricted not only by the number of AND gates in the AND array, but also by the fixed connections from the AND array outputs to the OR array.

The PLA in Figure 1.46 is programmed to generate the following boolean functions for z_1, z_2, z_3, and z_4:

$$z_1 = x_1 x_2' + x_1' x_2$$

$$z_2 = x_1 x_3 + x_1' x_3'$$

$$z_3 = x_1 x_2' + x_1 x_3 + x_1' x_2' x_3'$$

$$z_4 = \text{Logic } 1$$

Output z_4 is a logic 1, because the inputs to the AND gate whose output connects to the OR gate have all fuses open. Thus, the inputs to the AND gate become high voltage levels due to internal circuitry, and generate a high level on the output.

This section has presented only a cursory introduction to PLDs. PLDs are also discussed in Chapter 4 which illustrates their use in the synthesis of synchronous sequential machines.

As can be seen from the above examples, the output function may not use all of the input variables x_i. Combinational logic networks can be characterized in terms of fundamental logic operations such as, AND, OR, and NOT. This also includes the functionally complete set of logic gates, NAND and NOR. The output symbols z_i may be asserted either high or low. Combinational logic is used extensively to represent the next-state function and the output function for both synchronous and asynchronous sequential machines.

This chapter on combinational logic has been presented as a review of basic analysis and synthesis techniques. Included was a cursory review of number systems and number representations. Boolean algebra was introduced which provides the basic mathematical tools to analyze and synthesize various switching networks. Karnaugh maps and map-entered-variables were presented as one method of minimizing boolean functions with up to five variables. For functions containing more than five variables, the Quine-McCluskey algorithm is a preferred method to obtain a minimal sum-of-products expression. Combinational logic macros such as, multiplexers, decoders, encoders, and comparators provide additional flexibility in synthesizing combinational networks. Storage elements such as, *SR* latches, *D*, *JK*, and *T* flip-flops were also reviewed, since they have a pivotal usage in both synchronous sequential machines and asynchronous sequential machines. Finally, a review of PLDs was presented illustrating a hardware programmable approach to combinational logic synthesis.

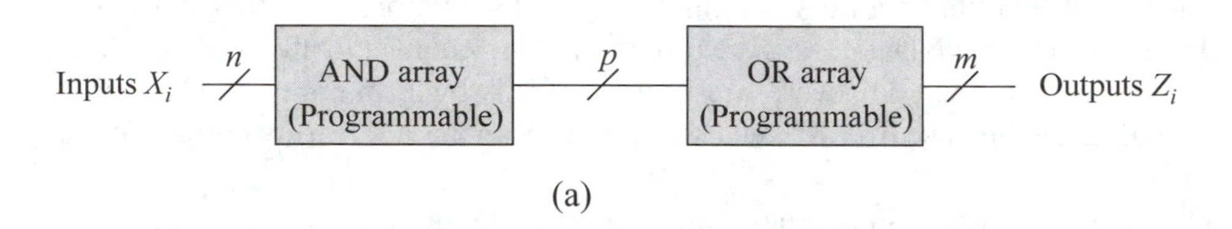

(a)

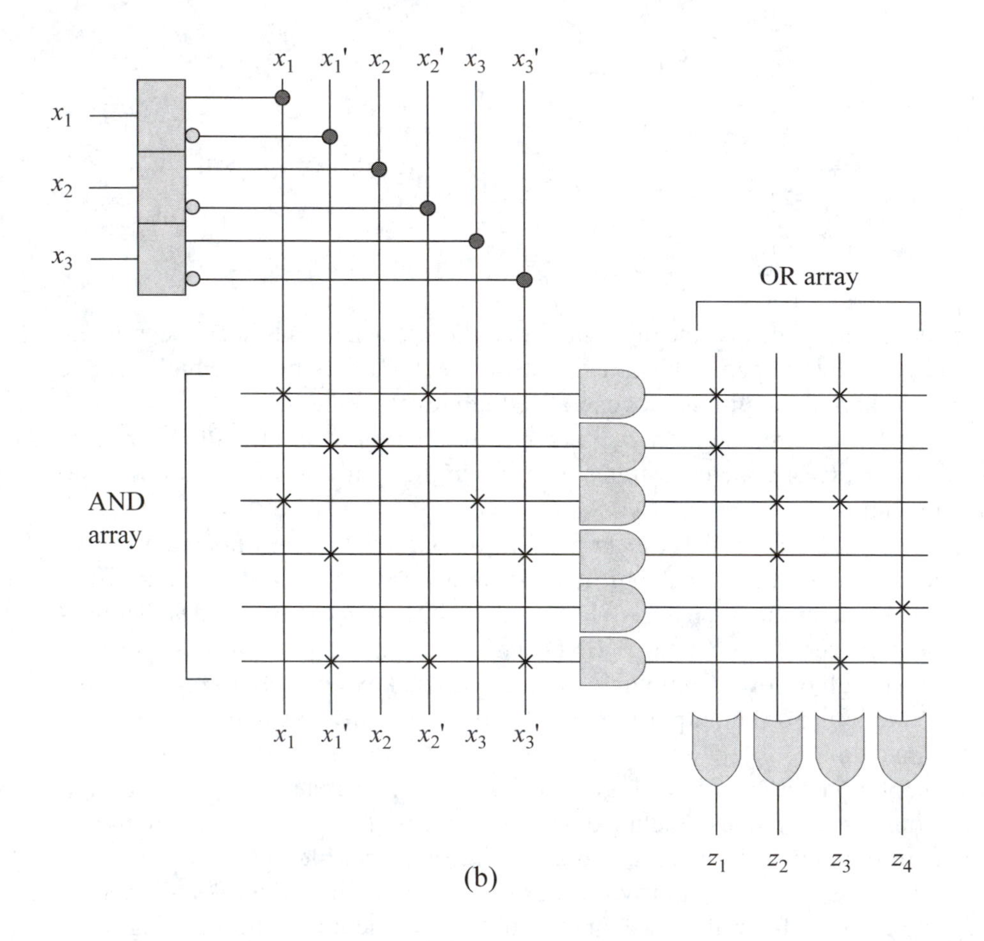

(b)

Figure 1.46 Basic organization of a PLA; (a) block diagram; and (b) implementation using three inputs and four outputs.

1.14 Problems

1.1 Given a positional number system for radix 12
(a) How many digits are there?
(b) What is the high-order digit?

1.2 Obtain a sum-of-products expression using Equation1.4 for the following numbers:
(a) 389_{10}
(b) 1231.03_{4}

1.3 Obtain the decimal value of the following numbers:
(a) 1110.1011_{2}
(b) 431.32_{5}

1.4 Generate the decimal integers 0 to 15 for the following positional number systems:
(a) Radix 4
(b) Radix 12

1.5 Convert Π to radix 2 using 13 binary digits.

1.6 Complete the tables shown below for the decimal numbers shown, using the indicated number representations for radix 2 using eight bits.
(a)

	$+127_{10}$	-127_{10}
Sign magnitude		
Diminished-radix complement		
Radix complement		

(b)

	$+54_{10}$	-54_{10}
Sign magnitude		
Diminished-radix complement		
Radix complement		

1.7 Convert the following integers to radix 3 and radix 7 with the same numerical value:
10_{10} and 111_{4}

1.8 Convert the following unsigned binary numbers to decimal:
(a) 11110000.101_2
(b) 10000001.111_2

1.9 The numbers shown below are in 2s complement representation. Convert the numbers to sign-magnitude representation for radix 2 with the same numerical value using eight bits.

2s complement	Sign magnitude
0111 1111	
1000 0001	
0000 1111	
1111 0001	
1111 0000	

1.10 The binary numbers shown below are in 2s complement representation for radix 2. Determine whether the indicated arithmetic operations produce an overflow.

(a)

```
    1001 1000
-)  0010 0010
```

(b)

```
    0011 0110
-)  1110 0011
```

1.11 Obtain the radix complement of $F8B6_{16}$.

1.12 Obtain the diminished-radix complement of 0778_9.

1.13 Use boolean algebra to prove that $x_1 + x_1 = x_1$, which is the idempotent law with respect to boolean addition.

1.14 Prove that $x_1 + x_1'x_2 = x_1 + x_2$, which is absorption law 2.

1.15 Minimize the following boolean function for three variables:

$$z_1(x_1, x_2, x_3) = x_1x_2 + x_1x_3 + x_1x_2'x_3 + x_1'x_2x_3 + x_1'x_2$$

1.16 Minimize the following boolean function for four variables:

$$z_1(x_1, x_2, x_3, x_4) = x_1x_2x_3x_4 + x_1x_2x_3'x_4 + x_1x_2'x_3x_4 + x_1x_2'x_3x_4'$$

1.17 Minimize the following boolean function for three variables:

$$z_1(x_1, x_2, x_3) = x_1'x_2'x_3' + x_1'x_2'x_3 + x_1'x_2x_3 + x_1x_2'x_3$$

1.18 Use boolean algebra to minimize the following function:

$$z_1(x_1, x_2, x_3, x_4) = x_1'x_2'x_3'x_4 + x_2x_3'x_4' + x_1'x_2'x_3x_4 + x_1'x_2x_3x_4' + x_1x_2'x_4 + x_2x_3x_4'$$

1.19 Indicate whether the following statement is true or false:

$$x_1'x_2'x_3 + x_1x_2x_3 = x_3$$

1.20 Using the laws of boolean algebra, change the expression shown below into the exclusive-NOR format.

$$z_1 = x_1'x_2'x_3'x_4' + x_1'x_2x_3'x_4 + x_1x_2x_3x_4 + x_1x_2'x_3x_4'$$

1.21 Plot the following function on a Karnaugh map and obtain the minimized expression in both a sum-of-products and a product-of-sums form:

$$z_1(x_1, x_2, x_3) = x_1'x_2 + x_1'x_2'x_3 + x_1'x_2x_3 + x_1x_2x_3$$

1.22 Obtain a minimized expression for a function z_1 which will indicate when a 4-bit binary number $x_1x_2x_3x_4$ satisfies the expression $12 \leq z_1 < 3$, where x_4 is the low-order variable.

1.23 Obtain the minimized expression for z_1 in both a sum-of-products and a product-of-sums format using the following Karnaugh map:

x_1x_2 \ x_3x_4	0 0	0 1	1 1	1 0
0 0	1	0	1	1
0 1	1	1	0	1
1 1	0	0	0	1
1 0	0	0	1	1

z_1

1.24 Obtain the minimized expression for z_1 in both a sum-of-products and a product-of-sums format using the following Karnaugh map:

$x_5 = 0$

x_1x_2 \ x_3x_4	0 0	0 1	1 1	1 0
0 0	1	0	0	0
0 1	0	0	1	1
1 1	0	1	1	1
1 0	1	0	0	0

$x_5 = 1$

x_1x_2 \ x_3x_4	0 0	0 1	1 1	1 0
0 0	0	0	0	1
0 1	0	0	1	1
1 1	0	1	1	0
1 0	0	0	0	1

z_1

1.25 Plot the function shown below on a Karnaugh map using x_3 as a map-entered variable. Then obtain the minimized expression for z_1 in a sum-of-products notation.

$$z_1(x_1, x_2, x_3) = \Sigma_m(0,4,6,7)$$

1.26 Plot the function shown below on a Karnaugh map using x_4 as a map-entered variable. Then obtain the minimized expression for z_1 in a sum-of-products notation.

$$z_1(x_1, x_2, x_3, x_4) = x_1'x_2'x_3'x_4 + x_1'x_2x_3'x_4 + x_1x_2x_3'x_4 + x_1x_2'x_3x_4 + x_1x_2'x_3x_4'$$

1.27 Given the Karnaugh map shown below, obtain the minimized expression for z_1.

$x_1 \backslash x_2x_3$	0 0	0 1	1 1	10
0	A	0	0	A+A'
1	A'	A+A'	A'	0

z_1

1.28 Use the Quine-McCluskey algorithm to obtain the minimized equation for the following function:

$$z_1(x_1, x_2, x_3, x_4) = \Sigma_m(0,2,6,7,8,9,10,13,15)$$

1.29 Use the Quine-McCluskey algorithm to obtain the minimized equation for the following function:

$$z_1(x_1, x_2, x_3, x_4) = \Sigma_m(2,6,7,8,10,13,14) + \Sigma_d(0)$$

1.30 Given the logic diagram shown below, obtain the representative equation for z_1 as derived directly from the logic. Then change the equation to a sum-of-products notation.

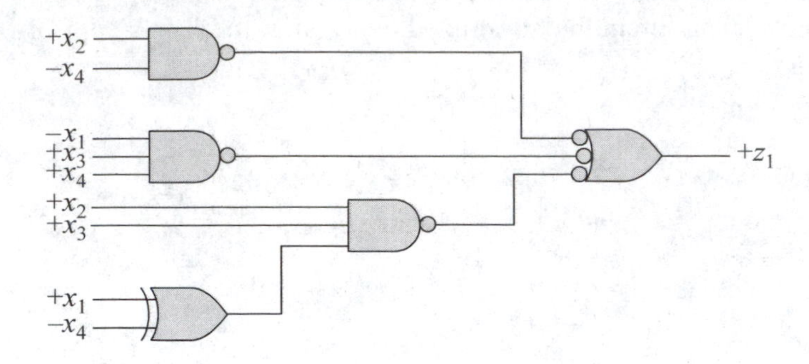

1.31 Given the logic diagram shown below, obtain the representative equation for z_1 as derived directly from the logic. Then change the equation to a sum-of-products notation.

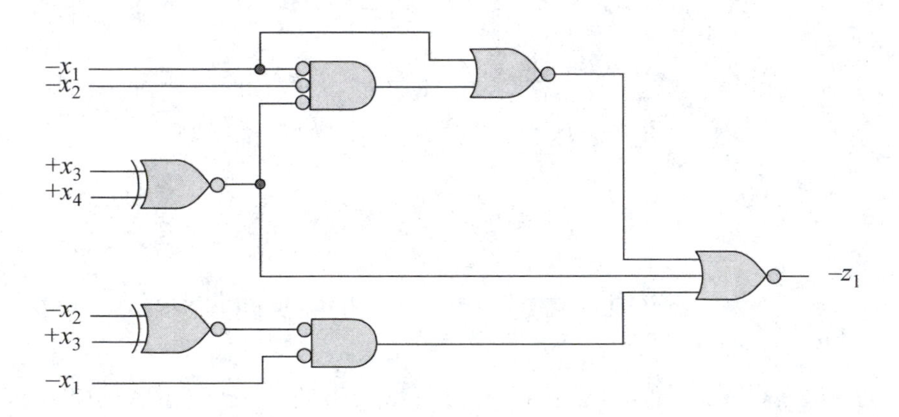

1.32 Synthesize the equation shown below using NAND gates only. Do not minimize the equation.

$$z_1 = x_1x_2'x_3 + x_3x_4'(x_2 \oplus x_4)' + (x_2x_3'x_4)'$$

1.33 Synthesize the equation shown below using exclusive-OR gates only.

$$\begin{aligned} z_1 = {} & x_1'x_2'x_3'x_4 + x_1'x_2'x_3x_4' + x_1'x_2x_3'x_4' \\ & + x_1'x_2x_3x_4 + x_1x_2x_3'x_4 + x_1x_2x_3x_4' \\ & + x_1x_2'x_3'x_4' + x_1x_2'x_3x_4 \end{aligned}$$

1.34 Using only a 4:1 multiplexer, synthesize the function $z_1(x_1,x_2,x_3) = \Sigma_m(2,4,5,6)$. The problem is simplified if x_3 is used as a map-entered variable.

1.35 Use two 4:1 multiplexers and any additional logic to implement an 8:1 multiplexer. Each 4:1 multiplexer has an active-low enable input.

1.36 Use two 2:4 decoders to decode three variables. Each decoder has an active-low enable.

1.37 Implement an octal-to-binary encoder using only NOR gates. The outputs are to be active low.

1.38 Obtain the truth table and the minimized output equations for an octal-to-binary priority encoder.

1.39 A majority circuit is a combinational logic circuit whose output is a logic 1 if the majority of the inputs are a logic 1; otherwise, the output is a logic 0. Obtain the boolean equation for $z_1(x_1,x_2,x_3)$ that is implemented by the majority (Maj) circuits shown below. The equation for z_1 is to be in a minimal sum-of-products form.

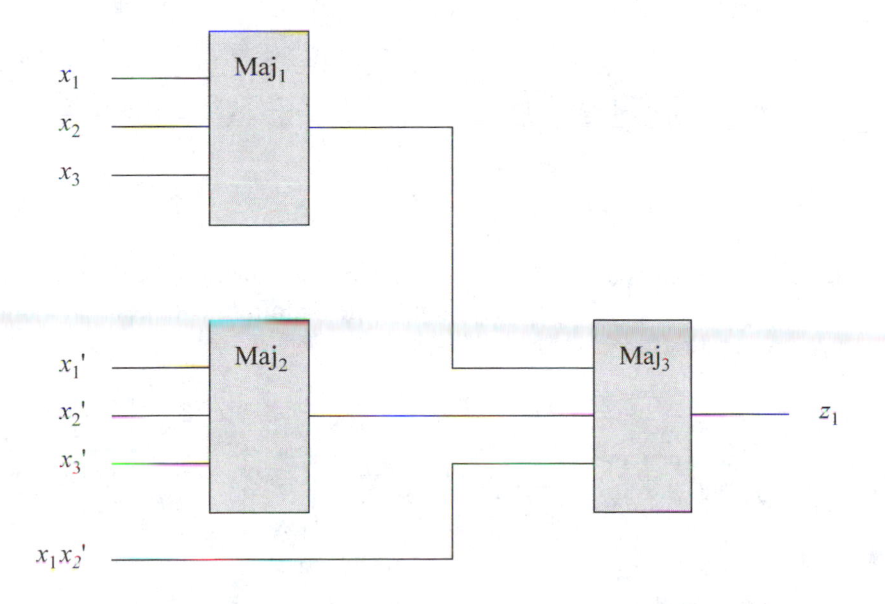

1.40 The waveforms shown below for $+s$ and $+r$ are applied to the *SR* latch. Obtain the waveform for output $+y_1$. Assume that the latch is reset initially.

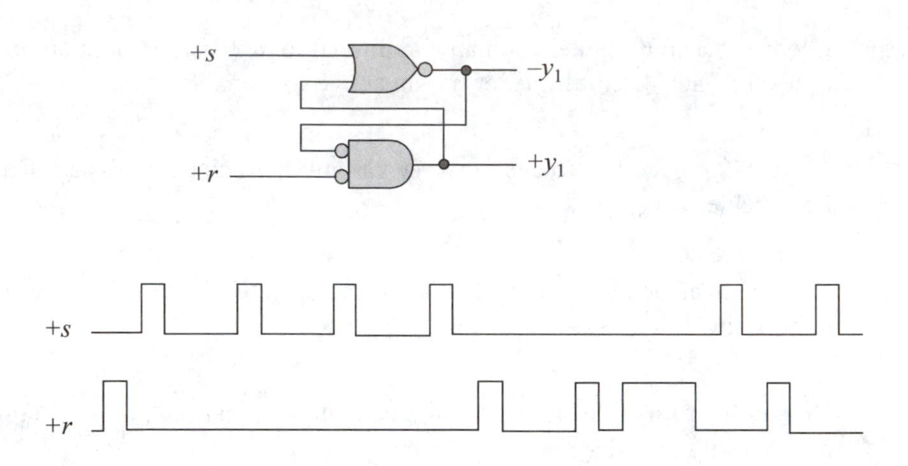

1.41 Draw the waveform for output z_1 of a negative-edge-triggered *JK* flip-flop using the waveforms shown below. Assume that the flip-flop is reset initially.

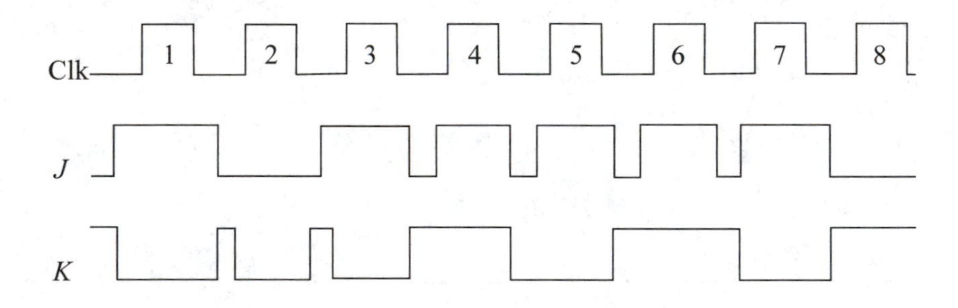

1.42 One stage of a shift register is shown below using a *D* flip-flop. Modify this stage using a *T* flip-flop, while maintaining the same functional operation.

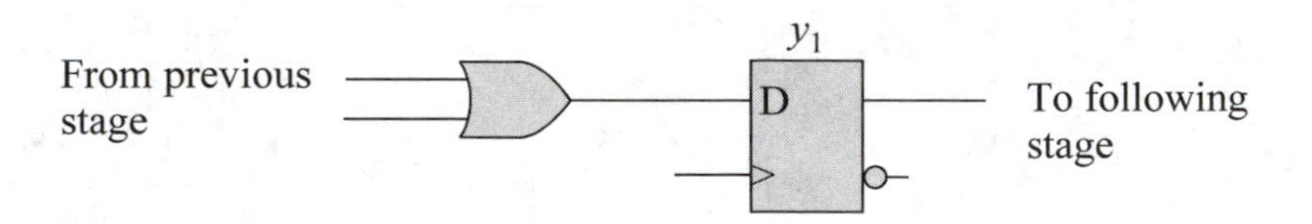

2.1 Sequential Circuits
2.2 Classes of Sequential Machines
2.3 Methods of Analysis
2.4 Complete and Incomplete Synchronous Sequential Machines
2.5 Problems

2

Analysis of Synchronous Sequential Machines

Logic circuits in digital systems may be either combinational or sequential. Thus far, the emphasis has been on combinational logic. A *combinational* logic circuit is one whose present outputs are determined entirely from the present inputs. The outputs of combinational logic circuits can be expressed in terms of fundamental logical operations such as, AND, OR, and NOT. This chapter will analyze synchronous sequential machines; that is, analyze existing machines by applying certain techniques to observe the behavior of the machines. Synchronization is achieved by *clock* pulses, where a clock is a device that produces pulses at regular intervals. Several different machines will be analyzed. The methods used for analyzing will consist of state diagrams, state tables, present-state maps, input maps, output maps, and timing diagrams.

Before defining sequential logic and synchronous sequential machines, the concept of a state will be defined. A *state* is a set of values that is measured at different locations within the machine. The values correspond to the set or reset condition of storage elements. The storage elements may be clocked flip-flops or SR latches. A storage element can store one bit of information, and therefore, has two stable states: 0 and 1. A machine is stable in a particular state when no input signals are changing; that is, the input variables and the clock, if applicable, are stable.

If there are p synchronous storage elements, then the state of the machine is the p-tuple of the storage element states and consists of 0s and 1s, where each bit is the value of a particular storage element cell. Thus, there are 2^p possible states, some of which may be unused. For example, if there are five storage elements, then possible states (also called state codes) are: 01011, 10010, 11100, etc.

Sequential circuits are encountered not only in computer design, but also in other design applications. Elevator control systems, for example, store information pertaining to the floors that require service and they also service the floors in the correct sequence. Traffic control systems are another example of sequential circuits in operation in our daily lives.

There are several applications of sequential circuits (or sequential machines) in computer design. For example, a pipelined processor is controlled by a sequence of six different states:

- In state 1, the instruction is fetched from storage.
- In state 2, the operand address is generated.
- In state 3, the operand address is translated into a real address, if necessary.
- In state 4, the operand is fetched.
- In state 5, the operation is executed, as defined by the instruction.
- In state 6, the result is stored.

Writing on a disk sector is controlled by a sequence of different states. The address field of the sector may consist of four states:

- In state 1, the preamble is written, which is used for clock synchronization.
- In state 2, the sector address is written, which consists of the cylinder number, sector number, and head number.
- In state 3, a check character is written for error detection and correction.
- In state 4, the postamble is written. The postamble is used to separate two fields.

The data field of the sector may also consist of four states:

- In state 1, the preamble is written.
- In state 2, the data that is to be stored in the sector is written.
- In state 3, a check character is written for error detection and correction.
- In state 4, the postamble is written.

Other applications for sequential machines include detection of a particular bit sequence in a stream of serial bits, or detection of a particular word in a stream of serial words.

2.1 Sequential Circuits

Digital computers and most digital systems use combinational logic. However, additional logic circuits are necessary to store data and to operate on the data. These circuits include counters, shift registers, and sequence generators, which control the operation of the system. These circuits are called sequential circuits and may be synchronous (clocked) or asynchronous (not clocked).

A sequential logic circuit consists of combinational logic and storage elements. The circuit is called sequential, because operations are performed in sequence. A sequential circuit can assume a finite number of internal states and can, therefore, be regarded as a *finite-state machine* (or simply a state machine) with a finite number of inputs and a finite number of outputs. In general, there are two types of sequential circuits: synchronous sequential machines and asynchronous sequential machines.

Synchronous sequential machines respond to signals only at discrete intervals of time; that is, clock pulses control the operation of the circuit. Clock pulses are synchronization signals provided by a clock, usually an astable multivibrator, which generates a periodic series of pulses. The storage elements are affected only at the positive or negative transition of the clock pulses.

A *synchronous sequential machine* is a machine whose present outputs are a function of the present state only or the present state and present inputs. The present states are determined by the sequence of previous inputs (the input history) and the previous states. The sequence of previous inputs is the order in which the inputs occurred. A requirement of a synchronous sequential machine is that state changes occur only when the machine is clocked, either on the positive or negative transition of the clock. Thus, input changes do not affect the present state of the machine until the occurrence of the next active clock transition.

A synchronous sequential machine is deterministic; that is, if the input sequence is repeated, then the output sequence is repeated. Synchronous sequential machines also include iterative networks such as, registers and counters, where the individual cells or stages are repeated.

Asynchronous sequential machines are not clocked, but respond to input signals and storage element outputs without a regular or predictable time relationship to specified events. The outputs of an asynchronous sequential machine depend upon the order, or sequence, of the input signals and can be affected at any time. The storage elements are comprised of latches. Thus, an asynchronous sequential machine may be considered to consist of combinational logic with feedback signals.

2.1.1 Machine Alphabets

A synchronous sequential machine is a mathematical model of a sequential logic circuit and consists of the following three alphabets:

Input alphabet There is a set of external inputs consisting of n binary variables

$$\{x_1, x_2, \cdots, x_n\}$$

where $x_i = 0$ or 1 for $1 \leq i \leq n$. The inputs generate 2^n input combinations of ordered n-tuples. Each combination of the input variables is referred to as a vector (or symbol) of the input alphabet. Thus, the input alphabet X is the set of 2^n input symbols such that,

$$X = \{X_0, X_1, X_2, \cdots, X_{2^n-2}, X_{2^n-1}\} \tag{2.1}$$

For example, if a machine has three input variables, x_1, x_2, and x_3, then the input alphabet consists of eight symbols X_0 through X_7:

$$X = \{000, 001, 010, \cdots, 110, 111\}$$

where

$$\begin{aligned}
X_0 &= x_1x_2x_3 = 000 \\
X_1 &= x_1x_2x_3 = 001 \\
X_2 &= x_1x_2x_3 = 010 \\
&\vdots \\
X_{2^n-2} = X_6 &= x_1x_2x_3 = 110 \\
X_{2^n-1} = X_7 &= x_1x_2x_3 = 111
\end{aligned}$$

Some vectors of the input alphabet may not be used.

State alphabet There is a set of present internal state variables consisting of p synchronous storage elements

$$\{y_1, y_2, \cdots, y_p\}$$

where $y_i = 0$ or 1 for $1 \le i \le p$. The storage elements generate 2^p possible states, although the machine may not use all states.

Each combination of the storage element values is referred to as a state (or *state code*) of the state alphabet. Thus, the state alphabet Y is the set of 2^p states such that,

$$Y = \{Y_0, Y_1, Y_2, \cdots, Y_{2^p-2}, Y_{2^p-1}\} \tag{2.2}$$

For example, if a machine has four storage elements y_1, y_2, y_3, and y_4, then the state alphabet consists of 16 states Y_0 through Y_{15}:

$$Y = \{0000, 0001, 0010, \cdots, 1110, 1111\}$$

where

$$Y_0 = y_1y_2y_3y_4 = 0000$$
$$Y_1 = y_1y_2y_3y_4 = 0001$$
$$Y_2 = y_1y_2y_3y_4 = 0010$$
$$\vdots$$
$$Y_{2^p-2} = Y_{14} = y_1y_2y_3y_4 = 1110$$
$$Y_{2^p-1} = Y_{15} = y_1y_2y_3y_4 = 1111$$

where 0 indicates a reset storage element and 1 indicates a set storage element. The state of the machine, therefore, is the p-tuple of the storage element states.

Output alphabet There is a set of outputs consisting of m binary variables

$$\{z_1, z_2, \cdots, z_m\}$$

where $z_i = 0$ or 1 for $1 \leq i \leq m$. The outputs generate 2^m output combinations of ordered m-tuples. Each combination of the output variables is referred to as a vector (or symbol) of the output alphabet. Thus, the output alphabet Z is the set of 2^m output vectors such that,

$$Z = \{Z_0, Z_1, Z_2, \cdots, Z_{2^m-2}, Z_{2^m-1}\} \tag{2.3}$$

For example, if a machine has two outputs z_1 and z_2, then the output alphabet consists of four symbols Z_0 through Z_3:

$$Z = \{00, 01, 10, 11\}$$

where

$$Z_0 = z_1z_2 = 00$$
$$Z_1 = z_1z_2 = 01$$
$$Z_2 = z_1z_2 = 10$$
$$Z_{2^m-1} = Z_3 = z_1z_2 = 11$$

Some symbols of the output alphabet may not be used.

The present external inputs $x_1, x_2, \cdots, x_n$ and the present values of the state variables $y_1, y_2, \cdots, y_p$, which were obtained at clock (t), combine to produce the present outputs $z_1, z_2, \cdots, z_m$ and also the next state of the machine at the occurrence of the next clock ($t + 1$).

The following notation will be used to represent the specified vectors and states:

$X_{i(t)}$ is the present input vector, where $X_{i(t)} \in X$.
$Y_{j(t)}$ is the present state, where $Y_{j(t)} \in Y$.
$Y_{k(t+1)}$ is the next state, where $Y_{k(t+1)} \in Y$.
$Z_{r(t)}$ is the present output vector, where $Z_{r(t)} \in Z$.

2.1.2 Formal Definition of a Synchronous Sequential Machine

Before formally defining a synchronous sequential machine, the *Cartesian product* of two sets will be defined. For any two sets S and T, the Cartesian product of S and T is written as $S \times T$ and is the set of all *ordered pairs* of S and T, where the first member of the ordered pair is an element of S and the second member is an element of T.

Ordered pairs are of the form $<s, t>$, where $s \in S$ and $t \in T$. Angle brackets (< >) are used to indicate an ordered collection of items. An ordered pair (2-tuple) consists of two items in a known fixed order, where the ordering of the items is important. An example of an ordered pair is the representation of a point in a 2-dimensional plane. Thus, $<1, 2>$, $<1, 3>$, $<2, 4>$ all represent different points in a plane.

An ordered triple (3-tuple) is an ordered pair whose first member is an ordered pair. For example, an ordered triple is of the form

$$<<x_1, x_2>, x_3>$$

An ordered quadruple (4-tuple) is an ordered pair whose first member is an ordered triple. Thus, an ordered quadruple can be written as

$$<<<x_1, x_2>, x_3>, x_4>$$

The process can be continued to define an ordered n-tuple. An ordered n-tuple is an ordered pair whose first member is an ordered ($n - 1$)-tuple. For example, an ordered n-tuple is written as

$$<<<x_1, x_2>, x_3>, x_4, \cdots, x_{n-1}>, x_n>$$

Example 2.1 Relations are operations on sets of ordered pairs. If R is a relation on S and $<s, t>$ is an ordered pair in the set R written as sRt, then the relation $\leq$ on the set $S = \{1, 2, 3\}$ is

$$R = \{<1, 1>, <2, 2>, <3, 3>, <1, 2>, <1, 3>, <2, 3>\}$$

Example 2.2 Let $S = \{s_1, s_2, s_3\}$ and $T = \{t_1, t_2, t_3, t_4, t_5\}$. If R is a relation on $S \times T$ as specified by the graph of Figure 2.1, then the relation R is the set

$$R = \{<s_1, t_1>, <s_1, t_3>, <s_1, t_4>, <s_2, t_2>, <s_3, t_2>\}$$

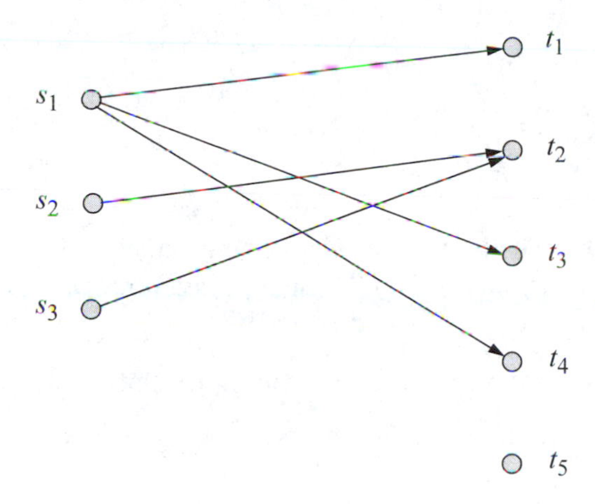

Figure 2.1 Graph of relation R on $S \times T$.

If R is a binary relation on the Cartesian product $S \times T$, then an example of a set of ordered pairs is

$$R = \{<s_1, t_1>, <s_1, t_2>, <s_2, t_2>, <s_3, t_2>, <s_1, t_3>\}$$

In predicate notation, the Cartesian product is written as

$$S \times T = \{<s, t> \mid s \in S \wedge t \in T\} \tag{2.4}$$

and is read as "The Cartesian product of S and T is the set of all ordered pairs of s and t such that $s \in S$ and $t \in T$."

Example 2.3 Let $S = \{s_1, s_2\}$ and $T = \{t_1, t_2, t_3\}$. Then the Cartesian product of S and T is

$$S \times T = \{<s_1, t_1>, <s_1, t_2>, <s_1, t_3>, <s_2, t_1>, <s_2, t_2>, <s_2, t_3>\}$$

Thus, the general classification of a synchronous sequential machine M can be formally defined as the 5-tuple

$$M = (X, Y, Z, \delta, \lambda) \tag{2.5}$$

where

1. X is a nonempty finite set of inputs.
2. Y is a nonempty finite set of states.
3. Z is a nonempty finite set of outputs.
4. δ is the next-state function which maps the Cartesian product of $X \times Y$ into Y.
5. λ is the output function which maps the Cartesian product of $X \times Y$ into Z.

There is sometimes confusion as to the precise meaning of "into," "onto," and "one-to-one" mappings.

1. The function $f(X, Y) : X \rightarrow Y$ is a one-to-one mapping if distinct elements in X are mapped to distinct elements in Y; that is, no two different elements in X have the same image in Y.

2. The function $f(X, Y) : X \rightarrow Y$ is an onto mapping if every element in Y is an image of at least one element in X.

3. If a mapping is not onto, then it is into.

Figure 2.2 illustrates some examples of different types of mappings for the function $f(X, Y) : X \rightarrow Y$. The next-state function δ in Equation 2.5 maps the Cartesian product of X and Y into Y, and is written as

$$\delta (X, Y) : X \times Y \rightarrow Y \tag{2.6}$$

The colon (:) in Equation 2.6 is read as "which"; the symbol ($\times$) is the Cartesian product. Thus, Equation 2.6 is read as "δ is a function of X and Y which maps the Cartesian product of X and Y into Y." The words "function" and "mapping" are considered to be synonymous. The next-state function δ can be further defined as

$$\delta (X, Y) : X_{i(t)} \times Y_{j(t)} \rightarrow Y_{k(t+1)} \tag{2.7}$$

where $Y_{j(t)}$ is the present state of the machine and $Y_{k(t+1)}$ is the next state of the machine. Both $Y_{j(t)}$ and $Y_{k(t+1)}$ are elements of the set Y. The subscripts (t) and $(t + 1)$ are used simply to differentiate between the present state and the next state.

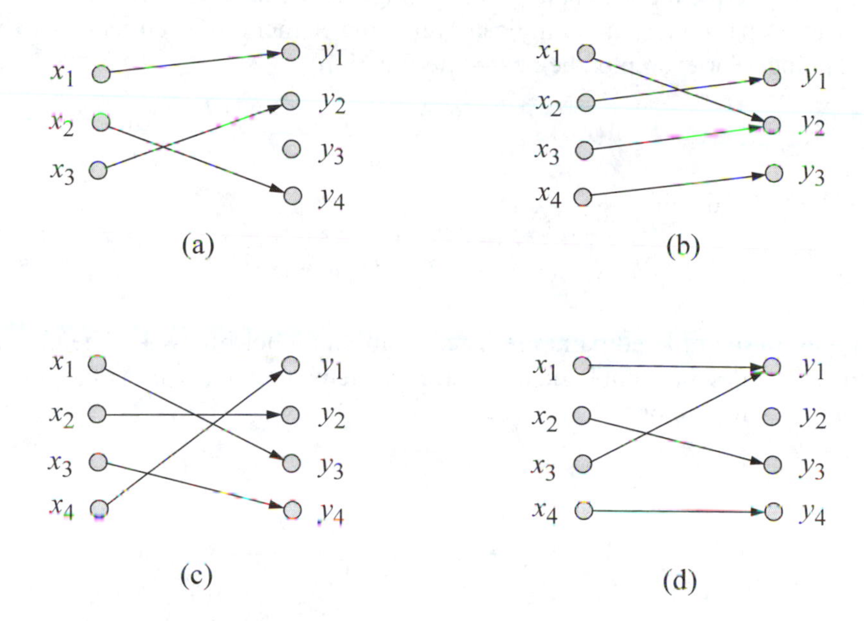

Figure 2.2 Example of different types of mappings or functions: (a) one-to-one into, but not onto; (b) onto, but not one-to-one; (c) one-to-one onto; (d) not one-to-one and not onto; therefore, into.

The present state of a machine is the present value of the storage elements. The next state of a machine is the value of the storage elements after the occurrence of the next clock pulse and is dependent upon the present inputs and the present state.

As defined in Section 2.1.1, a synchronous sequential machine is deterministic. The deterministic attribute can now be defined in terms of the machine alphabets and the next-state function δ. The next state $Y_{k(t+1)}$ is uniquely determined by the present inputs $X_{i(t)}$ and the present state $Y_{j(t)}$. Thus, the next state can be expressed as

$$Y_{k(t+1)} = \delta(X_{i(t)}, Y_{j(t)}) \tag{2.8}$$

The next-state function $\delta(X, Y)$ can be written in predicate notation as

$$\delta(X, Y) = \{<X_i, Y_j> \mid X_i \in X \wedge Y_j \in Y\} \tag{2.9}$$

and is the set containing all ordered pairs $<X_i, Y_j>$, where $X_i \in X$ and $Y_j \in Y$.

The function δ determines the next state of the machine as a function of the present input vector and the present state of the machine. The state of the storage elements and the values of the input variables at time (t) are transformed by the mapping function δ to generate the next state of the storage elements at time $(t + 1)$.

The cartesian product simply indicates all of the elements of each set that are used in the mapping. For example, the next-state function

$$\delta(X, Y) : X \times Y \rightarrow \mathrm{Y}$$

contains the ordered pairs

$$<X_1, Y_1>, <X_1, Y_2>, \cdots, <X_n, Y_p>$$

Although all of the ordered pairs are listed, some may not be used. The next-state function δ generates the input logic to the storage elements and can be as simple as one AND gate as shown below.

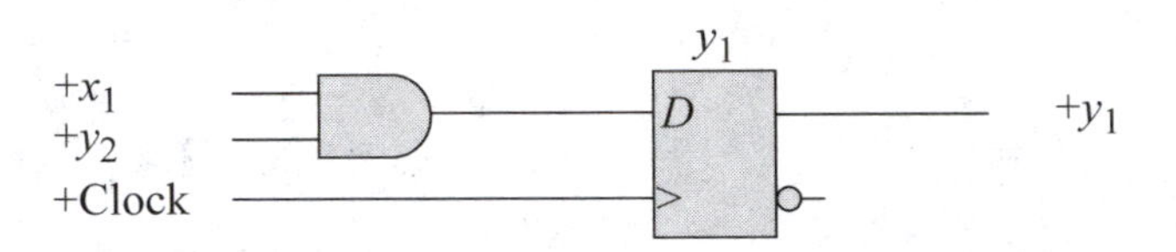

The output function λ maps the Cartesian product of X and Y into Z, and is written as

$$\lambda(X, Y) : X \times Y \rightarrow Z \tag{2.10}$$

In general, the value of the present output $Z_{r(t)}$ is a function of the present state $Y_{j(t)}$ and the present inputs $X_{i(t)}$ and can be expressed as

$$Z_{r(t)} = \lambda(X_{i(t)}, Y_{j(t)}) \tag{2.11}$$

where λ is the output function. The machine transforms an input vector into an output vector.

The next-state function δ represents the next state of the machine, whereas, the output function λ is a function of the present state of the machine. The outputs, therefore, cannot be presently indicated for the next state, because the machine is not yet in the next state.

General model The general model of a synchronous sequential machine is shown in Figure 2.3 and consists of one or more of the following units:

1. Input logic which consists combinational logic that transforms the input vector and the present state into the next state by means of the δ next-state function.

2. Synchronous storage elements which indicate the present state of the machine.

3. Output logic which consists primarily of combinational logic that transforms the present state and the input vector or the present state only into the present output by means of the λ output function. In some situations, the λ output logic may require a flip-flop. In general, this occurs if the output signals are required to be active for the duration of one clock cycle. A flip-flop may then be needed to latch the output for the requisite duration. The flip-flop is not considered part of the synchronous sequence activity of the machine, however. Further discussion on problems associated with output logic will deferred until Chapter 3.

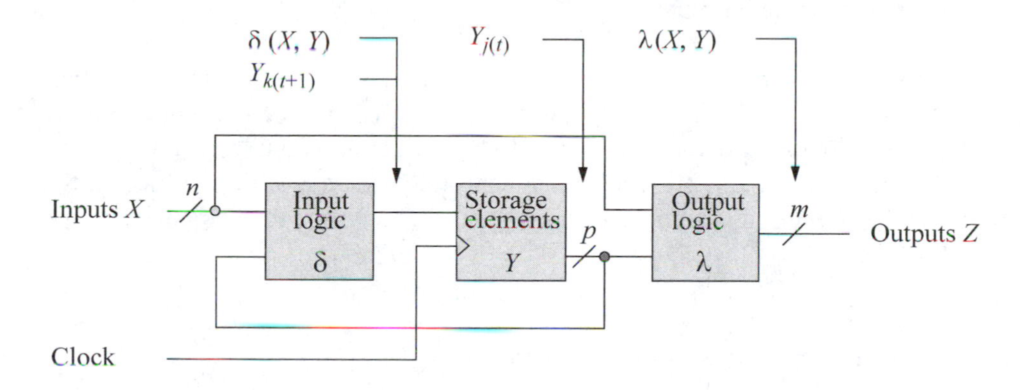

Figure 2.3 General model of a synchronous sequential machine.

Thus, the combinational logic generates a next state $Y_{k(t+1)} \in Y$ and a unique output vector $Z_{r(t)} \in Z$ for each input vector $X_{i(t)} \in X$ and present state $Y_{j(t)} \in Y$. The state of the storage elements and the value of the input vector immediately preceding the $(t + 1)$ clock pulse are combined by the next-state function δ to generate storage element inputs which will determine the next state of the storage elements.

The output function λ maps ordered pairs consisting of the input vector and the present state $<X_{i(t)}, Y_{j(t)}>$ into the output alphabet of the machine. The machine of Figure 2.3 indicates that the output function λ is an $(n + p)$-input, m-output switching function. If the input signals of Figure 2.3 do not connect to the output logic, then this indicates that the output function λ is a p-input, m-output switching function, in which the output vector is a function of the present state only.

When inputs are synchronized with the system clock, they are clocked on the alternate transition to that of the storage elements. That is, if the storage elements are positive-edge-triggered devices, then the input data are clocked on the negative transition of the clock. Likewise, if the storage elements are negative-edge-triggered devices, then the input data are clocked on the positive transition of the clock. This allows the input data to become stable before the active clock transition occurs at the clock input of the storage element.

Synchronizing input data with the system clock prevents the storage elements from becoming metastable. *Metastable* is a condition in which the storage element is not stable in either the set or reset state. This condition results when the input data changes state very close to the active clock transition of an edge-triggered storage element. The output of the storage element may approach the active voltage level, then either remain at that level or return to the inactive voltage level. In the latter case, the resulting unwanted narrow pulse (or glitch) can cause a machine malfunction.

The metastable condition results because the input data was not stable within the specified setup time for the storage element. The *setup time* is the time interval between the application of data at the data input terminal and a subsequent active transition at the clock input terminal.

Summary of machine alphabets Table 2.1 summarizes the machine alphabets for synchronous sequential machines. The table assumes that there are three binary input variables, four storage elements, and two binary output variables. The model of Figure 2.3 can be used to represent both synchronous and asynchronous sequential machines. However, for asynchronous machines, the storage elements are not clocked.

Table 2.1 Summary of machine alphabets

Input alphabet *X* $X = \{X_0, X_1, X_2, \cdots, X_{2^n-2}, X_{2^n-1}\}$

↑ $X_1 = x_1'x_2'x_3 = 001$ (for $n = 3$)

↑ Binary variable

State alphabet *Y* $Y = \{Y_0, Y_1, Y_2, \cdots, Y_{2^p-2}, Y_{2^p-1}\}$

↑ $Y_1 = y_1'y_2'y_3'y_4 = 0001$ (for $p = 4$)

↑ Storage element

Output alphabet *Z* $Z = \{Z_0, Z_1, Z_2, \cdots, Z_{2^m-2}, Z_{2^m-1}\}$

↑ $Z_1 = z_1'z_2 = 01$ (for $m = 2$)

↑ Binary variable

2.2 Classes of Sequential Machines

This section will present classes of sequential machines. With the exception of the last class of machines (asynchronous sequential machines), all other machines are synchronous, where changes to states and outputs occur only in synchronization with clock pulses. The storage elements in synchronous sequential machines are bistable multivibrators such as, *D* flip-flops and *JK* flip-flops. The storage elements in asynchronous sequential machines are *SR* latches. Some of the machines in the following sections contain no data inputs; however, all of the machines produce outputs. Each machine consists of at least one of the three logical units that were defined in Section 2.1.2:

1. Input combinational logic
2. Synchronous storage elements
3. Output combinational logic

If input combinational logic is used, then the δ next-state function maps the present state or the present state and the present inputs into the state alphabet. That is, the δ mapping generates the input equations for the storage elements. If output combinational logic is used, then the λ function maps the present state or the present state and the present inputs into the output alphabet. Thus, the λ mapping generates outputs from the machine.

2.2.1 Combinational Logic

Before proceeding to the first class of synchronous sequential machines, the λ output logic will be discussed briefly. The output logic is primarily combinational in nature and represents the rightmost block of Figure 2.3, which is redrawn in Figure 2.4 showing only the λ output logic function. Although combinational logic is not sequential in nature, it is presented here to add completeness for the discussion which follows. By including combinational logic as a separate classification, all logic design can be characterized by the classes of machines presented in this section.

These machines also have no storage elements. The machine consists only of output combinational logic as shown in Figure 2.4. The output function λ maps the present inputs into the output alphabet of the machine as shown in Equation 2.12. The present outputs are determined only by the present inputs as shown in Equation 2.13.

$$\lambda(X) : X_{i(t)} \rightarrow Z \tag{2.12}$$

$$Z_{r(t)} = \lambda(X_{i(t)}) \tag{2.13}$$

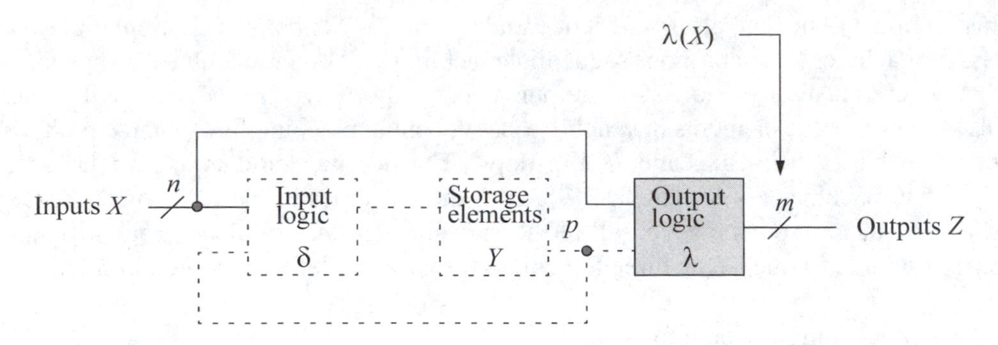

Figure 2.4 Combinational logic.

The binary input variables may be combined with any of the boolean operators, AND, OR, and NOT to generate boolean sums or boolean products in any combination. Thus, the λ output function combines the binary input variables of an input symbol X_i to generate an output symbol Z_r, which consists of boolean sums and/or products as shown below.

X_i [x_1, x_2, x_3] → Output logic λ → Z_r

An example of a λ output function is given by the following boolean sum-of-products equation, where λ(X) such that,

$$z_1 = x_1 x_2 + x_3' x_4 \quad (2.14)$$

The corresponding logic diagram is shown below.

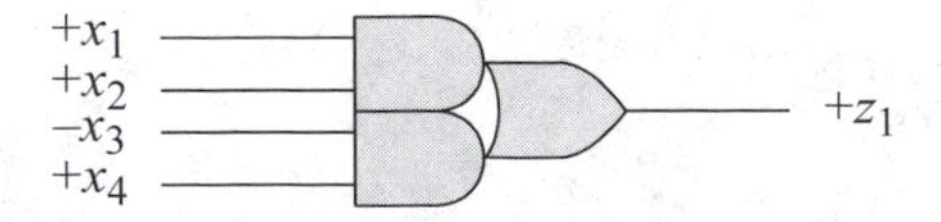

The output function can also be expressed as a product of sums. The previous sum-of-products expression (Equation 2.14) can be transformed into a product-of-sums expression by repeated application of the distributive law and the commutative law as follows:

$$\begin{aligned} z_1 &= x_1 x_2 + x_3' x_4 \\ &= (x_1 x_2 + x_3')(x_1 x_2 + x_4) \\ &= (x_1 + x_3')(x_2 + x_3')(x_1 + x_4)(x_2 + x_4) \end{aligned} \tag{2.15}$$

Like the previous example, this expression is also a 2-level network and can be implemented using five NOR gates, as shown in Figure 2.5.

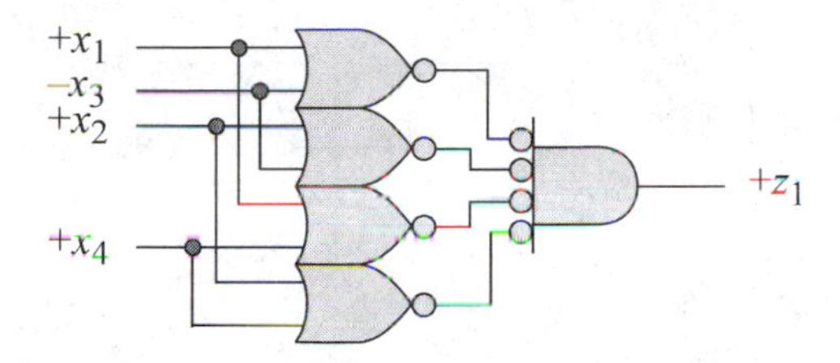

Figure 2.5 Two-level implementation of the output function λ for Equation 2.15.

Another example of a λ output function is the equivalence function (or exclusive-NOR), where λ(*X*) such that,

$$\begin{aligned} z_1 &= (x_1 \oplus x_2)' \\ &= x_1 x_2 + x_1' x_2' \end{aligned}$$

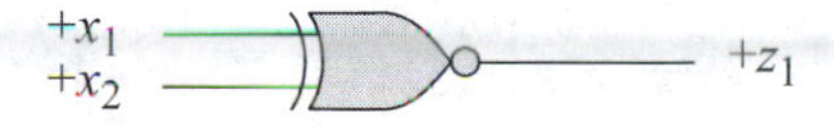

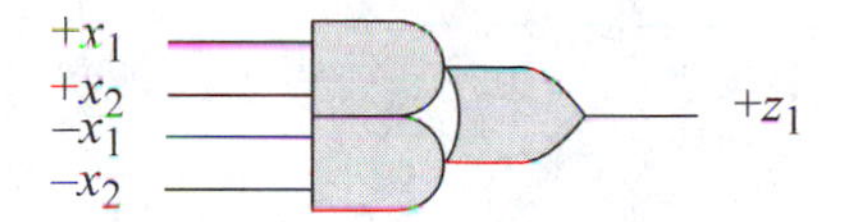

The output logic is not restricted to a sum-of-products or a product-of-sums expression, but can include multiple-level networks and multiple-output networks. The expression

$$z_1(x_1, x_2, x_3, x_4) = \Sigma_m(1,5,6,12,14) \tag{2.16}$$

can be implemented as either a 2-level or a 3-level network. Equation 2.16 is represented by the Karnaugh map and the 2-level logic diagram shown in Figure 2.6 (a) and (b), respectively.

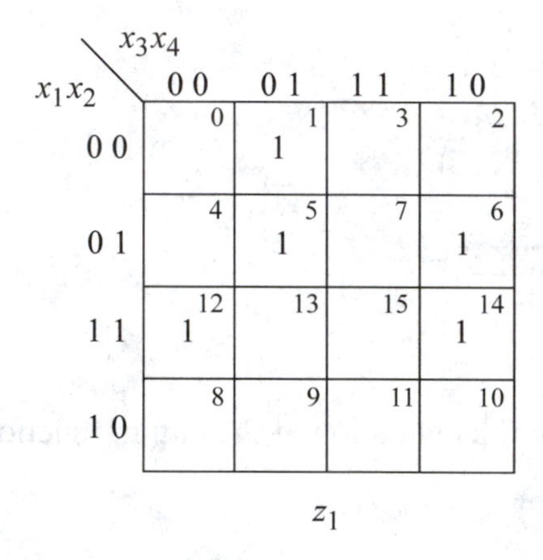

$$z_1 = x_1'x_3'x_4 + x_2x_3x_4' + x_1x_2x_4'$$

(a)

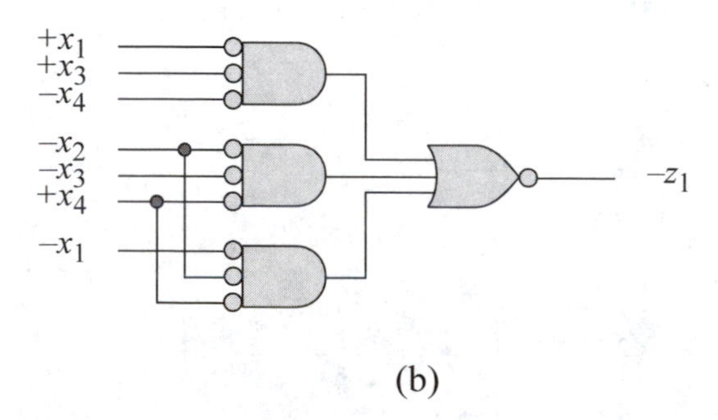

(b)

Figure 2.6 Two-level implementation of the λ output function for Equation 2.16: (a) Karnaugh map; (b) logic diagram.

The minimum expression for Equation 2.16 can be obtained from the Karnaugh map as

$$z_1 = x_1'x_3'x_4 + x_2x_3x_4' + x_1x_2x_4' \tag{2.17}$$

By application of the distributive law, Equation 2.17 can be expressed as

$$z_1 = x_1'x_3'x_4 + x_2x_4'(x_1 + x_3) \tag{2.18}$$

which uses slightly less logic than the equivalent circuit of Figure 2.6, but requires one additional gate delay.

2.2.2 Registers

In their simplest form, registers contain only storage elements. They may, however, consist of input logic and output logic as shown in Figure 2.7. If input logic is specified in the design, then the next-state function δ is a function of the input alphabet X, and maps X into Y, as shown in Equation 2.19. Also, δ is a function of the present input vector $X_{i(t)}$, and maps $X_{i(t)}$ into the next state $Y_{k(t+1)}$, as shown in Equation 2.20. Equation 2.21 defines the next state.

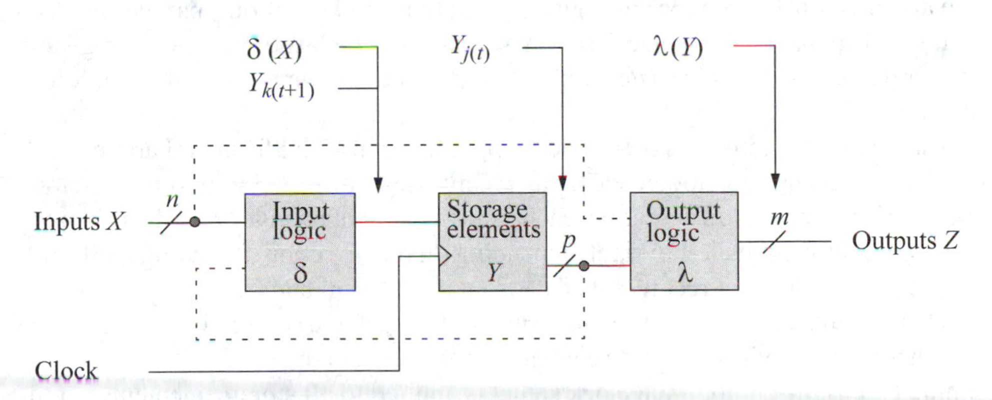

Figure 2.7 Register general block diagram.

$$\delta(X) : X \rightarrow Y \tag{2.19}$$
$$\delta(X_{i(t)}) : X_{i(t)} \rightarrow Y_{k(t+1)} \tag{2.20}$$
$$Y_{k(t+1)} = \delta(X_{i(t)}) \tag{2.21}$$

When the output function λ is implemented in a register, then λ is a function of the state alphabet Y and maps Y into the output alphabet Z, as shown in Equation 2.22. The present output vector $Z_{r(t)}$ is shown in Equation 2.23.

$$\lambda(Y) : Y \rightarrow Z \tag{2.22}$$

$$Z_{r(t)} = \lambda(Y_{j(t)}) \tag{2.23}$$

Registers can be Moore- or Mealy-type machines depending on the configuration of the output logic. Moore and Mealy machines are presented in Section 2.2.4 and Section 2.2.5, respectively. Registers are used to store binary information in a digital system. The information can consist of a single bit or of several bits which define instructions or operands. An n-bit register consists of n storage elements and can store n bits of binary information, one bit in each storage element.

The input alphabet X consists of 2^n input vectors of binary information. As defined in Section 2.1.2, these vectors are elements of the set X, where

$$X = \{X_0, X_1, X_2, \cdots, X_{2^n-2}, X_{2^n-1}\}$$

and X_i is a set of n bits $\in \{0,1\}$, which represents the ith vector of the input alphabet. The storage elements may be D flip-flops, JK flip-flops, or SR latches. There are many different types of registers: parallel-in, parallel-out; parallel-in, serial-out; serial-in, parallel-out; and serial-in, serial-out. The last three types of registers are also referred to as *shift registers*, in which the contents can be shifted either left or right.

Data are loaded into a register either in parallel or in serial format. Parallel loading is faster, because all storage elements receive new information during one clock pulse. During a serial load operation, each storage element receives new data from the element to its immediate left or right, depending on the direction of loading (shifting). The first storage element receives its data from an external source.

When the machine is connected to form a shift register, a right-shift register is one in which the output of each storage element is connected to the input of the element to its immediate right. A common clock signal is applied to all storage elements simultaneously. The only external data input for the machine is to the input of the leftmost storage element. There is always an output from the rightmost stage; however, the other stages may not always have their outputs made available to the rest of the system. In a synchronous right-shift register, all bits are shifted right one bit position during each clock pulse.

Shifting can also be implemented without storage elements. This is a faster approach to shifting, in which 0 through $n - 1$ bits are shifted in one clock period. The synthesis of this type of shifter, together with synchronous shift registers, will be described in detail in Chapter 3.

A typical register for eight bits is shown in Figure 2.8. This is a parallel-in, parallel-out register which consists of the next-state function δ, storage elements $y_0 - y_7$, and the output function λ. Both the δ and λ mappings use combinational logic consisting of eight AND gates for each of the two functions. The clock pulse sets the register to the contents of the input vector $x_0, x_1, x_2, \cdots, x_7$.

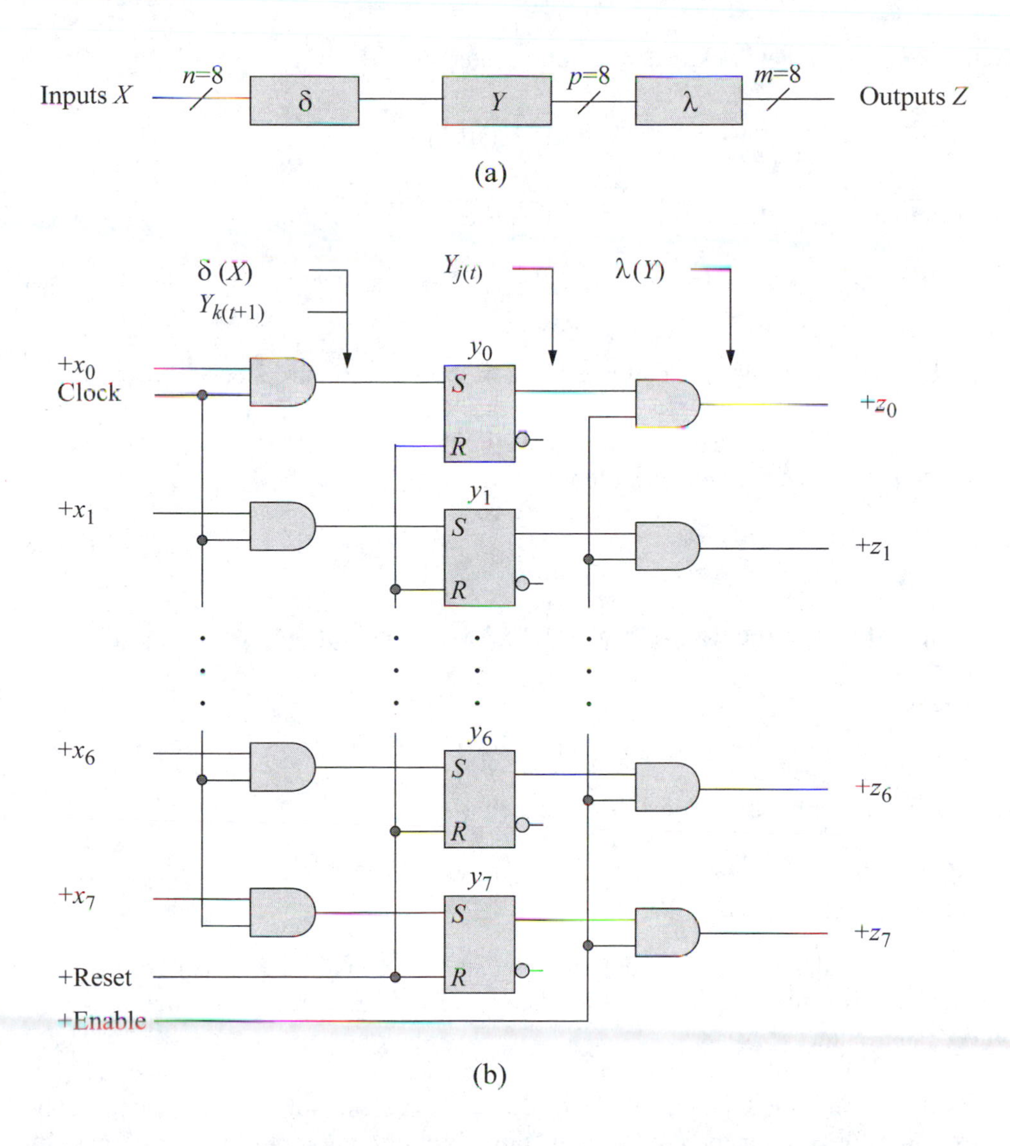

Figure 2.8 Register implemented as a parallel-in, parallel-out register: (a) block diagram; and (b) logic diagram using latches for the storage elements.

Because the storage elements are latches, the input data must not change while the clock is positive. An Enable signal is one of the inputs to the λ output function combinational logic and transforms the state of the register to the output vector $z_0, z_1, z_2, \cdots, z_7$.

Regarding Figure 2.8, the next-state function δ can be expressed generally as

$$\delta(X) : X \rightarrow Y \tag{2.24}$$

With reference to the clock, δ can be represented as

$$\delta(X_{i(t)}) : X_{i(t)} \rightarrow Y_{k(t+1)} \tag{2.25}$$

In particular, state Y_j has a next state which is a function of an input vector X_i.

$$Y_{k(t+1)} = \delta(X_{i(t)}) \tag{2.26}$$

Thus, for the parallel-in, parallel-out register shown in Figure 2.8, the next state is equal to the present input vector after the clock pulse is asserted such that, $Y_{k(t+1)} = X_{i(t)}$. Also, a particular storage element y_j has a next value y_k which is dependent upon a binary input variable x_i.

$$Y_{k(t+1)} = \delta(X_{i(t)}) \tag{2.27}$$

Still with reference to the register of Figure 2.8, the present output vector $Z_{r(t)}$ is expressed as

$$Z_{r(t)} = Y_{j(t)} \bullet \text{Enable} \tag{2.28}$$

A specific output binary value z_r is expressed as

$$z_r = y_j \bullet \text{Enable} \tag{2.29}$$

Since the state alphabet and the output alphabet both contain the same number of elements in their respective sets ($2^3 = 8$), then $j = r$, and λ is a one-to-one mapping of Y onto Z, such that, $z_r = y_r$.

Figure 2.9 also illustrates a synchronous register, in this case, a serial-in, serial-out shift register implemented with D flip-flops. There is no logic shown for either the δ next-state function or the λ output function, although the input logic is implied by the D input of the flip-flop. The shift register is reset initially, then information is shifted into stage$_0$ one bit per clock pulse. The positive transition of the clock signal causes the following shift sequence to occur:

$$x_0 \rightarrow y_0$$
$$y_j \rightarrow y_{j+1}$$
$$y_7 \text{ is lost}$$

Referring to Figure 2.9, the next-state function δ can be expressed generally as

$$\delta(X) : X \rightarrow Y \tag{2.30}$$

With reference to the clock, δ can be represented as

$$\delta(X_{i(t)}) : X_{i(t)} \rightarrow Y_{k(t+1)} \tag{2.31}$$

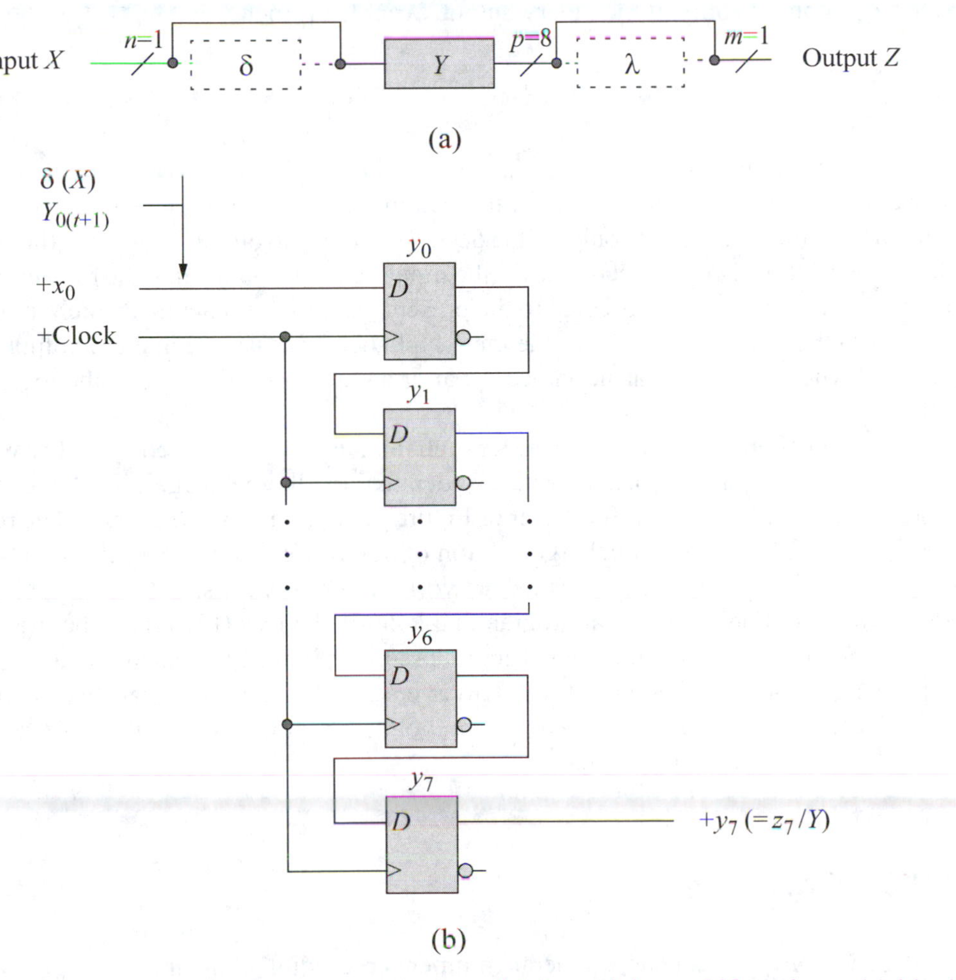

Figure 2.9 Register implemented as a serial-in, serial-out shift register: (a) block diagram; and (b) logic diagram.

A particular state Y_j has a next state Y_k which is a function of the previous state and the input vector X_i (=x_0), as follows:

$$Y_{k(t+1)} = \{x_0, Y_{j(t)}\} \tag{2.32}$$

where the braces indicate concatenation. That is, x_0 is concatenated with the previous state of the shift register after one shift cycle. A specific storage element y_j has a next value which is the same as the present value of storage element y_{j-1} and is expressed as

$$y_{j(t+1)} = y_{(j-1)t} \tag{2.33}$$

The present output vector $Z_{r(t)}$ is expressed as

$$Z_{r(t)} = Y_{j(t)} \tag{2.34}$$

Since $Z_{r(t)}$ consists only of the binary output variable z_1, then

$$Z_{r(t)} = Y_{j(t)} = y_7 = z_7 = x_{0(t+8)} \tag{2.35}$$

The configuration of a register, as shown in Figure 2.7, can also be classified as a special case of a Moore machine (refer to Section 2.2.4), because the output vector is a function of the present state only. If the design contains no output logic, then the output of the machine becomes active when the input vector is loaded into the register. In that case, the output vector is equal to the present state of the machine, which in turn is equal to the input vector. Thus, the input alphabet X is mapped into the output alphabet Z when the clock transition occurs, after an appropriate delay for the logic to stabilize.

Many registers are *iterative networks* such that, each stage (or cell) together with its storage element and associated gates, is identical to all other stages. Both the register of Figure 2.8 and the shift register of Figure 2.9 are iterative networks. The register is designed using a parallel organization of iterative cells, whereas, the serial-in, serial-out shift register uses a cascaded network of iterative cells.

It was stated previously that programmable logic devices (PLDs) can be used in the implementation of combinational logic circuits. Some PLDs also contain flip-flops in conjunction with logic gates and inverters and thus, a register can be designed entirely using only PLDs. As stated in a foregoing paragraph, the synthesis of registers will be deferred until Chapter 3.

2.2.3 Counters

This section will present some general comments regarding counters. The details of synthesis will be covered in Chapter 3. Counters are one of the simplest types of sequential machines, requiring only one input in most cases. The single input is a clock pulse. Although most counters can be categorized as a type of Moore machine, counters are of sufficient importance to warrant a separate classification. A *counter* is constructed from one or more flip-flops that change state in a prescribed sequence upon the application of a series of clock pulses. The sequence of states in a counter may generate a binary count, a binary-coded decimal (BCD) count, or any other counting sequence. The counting sequence does not have to be sequential.

Counters are used for counting the number of occurrences of an event and for general timing sequences. A block diagram of a synchronous counter is shown in Figure 2.10. The diagram depicts a typical counter consisting of combinational input logic for the δ next-state function, storage elements, and combinational output logic for the λ output function. Input logic is required when an initial count must be loaded into the counter. The input logic then differentiates between a clock pulse that is used for loading and a clock pulse that is used for counting. Not all counters are implemented with input and output logic, however. Some counters contain only storage elements that are connected in cascade.

There are two classifications of counters with respect to the clock inputs. The first type is referred to as a *synchronous counter*, in which the outputs are characterized by simultaneous state changes for all storage elements whose input conditions effect a change of state. The clock signal is applied to all storage elements concurrently. The second type is specified as an *asynchronous counter* (or ripple counter), because the clock input to the counter is connected to the first storage element only. The clock inputs for all other storage elements are connected to the outputs of the preceding stage; that is, the clock signal for $stage_j$ is generated by the output of $stage_{j-1}$.

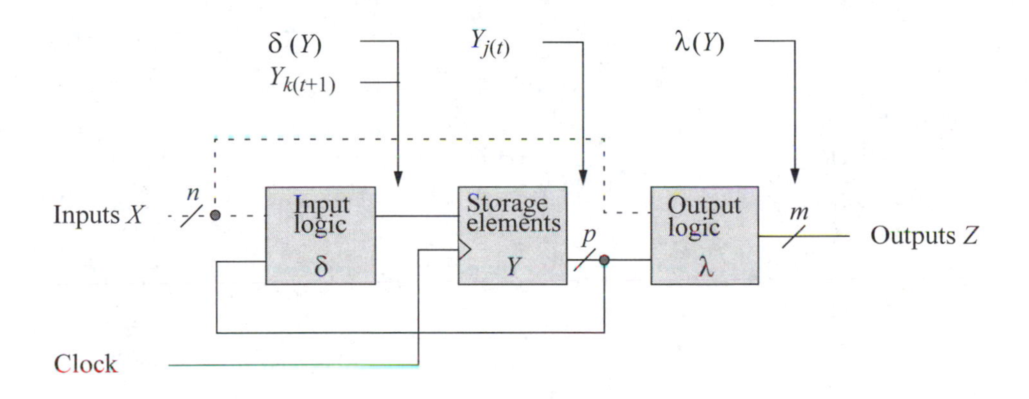

Figure 2.10 Counter block diagram.

Counters can be designed as count-up counters, in which the counting sequence increases numerically, or as count-down counters, in which the counting sequence decreases numerically. A counter may also be designed as both a count-up counter and a count-down counter, the mode of operation being controlled by a separate input.

Any counting sequence is a valid sequence for a counter, depending upon the application. Binary counters are the most common, and increment or decrement in a binary sequence, such as, 0000, 0001, 0010, 0011, . . . , 1110, 1111, 0000, . . . , which represents a modulo-16 binary counter, where the terminal count is followed by the initial count.

The counting sequence for a BCD counter is modulo 10: 0000, 0001, 0010, 0011, . . . , 1000, 1001, 0000, ... There are also counters that are classified as Gray code counters or as Johnson counters. A *Gray code* counter has the unique characteristic where only one stage of the counter changes state with each clock pulse, thus providing a glitch-free counting sequence, as shown in Table 2.2 for four bits, where y_4 is the low-order bit. A *Johnson counter* is designed by connecting the complement of the output of the final storage element to the input of the first storage element. This feedback connection generates the singular sequence of states shown in Table 2.3 for four bits, where y_4 is the low-order bit. A 4-bit Johnson counter produces a total of eight states; a 5-bit counter yields ten states. In general, an n-stage Johnson counter will generate $2n$ states, where n is the number of storage elements in the counter.

Table 2.2 Four-bit Gray code

y_1	y_2	y_3	y_4
0	0	0	0
0	0	0	1
0	0	1	1
0	0	1	0
0	1	1	0
0	1	1	1
0	1	0	1
0	1	0	0
1	1	0	0
1	1	0	1
1	1	1	1
1	1	1	0
1	0	1	0
1	0	1	1
1	0	0	1
1	0	0	0
0	0	0	0

Table 2.3 Four-bit Johnson counter

y_1	y_2	y_3	y_4
0	0	0	0
1	0	0	0
1	1	0	0
1	1	1	0
1	1	1	1
0	1	1	1
0	0	1	1
0	0	0	1
0	0	0	0

Regarding the diagram of the synchronous sequential machine of Figure 2.10, the output logic is used primarily to provide the state of the machine to external hardware under control of an enable signal. This can be in the form of AND gates or 3-state drivers. Otherwise, the output of the machine is obtained directly from the storage elements.

The next-state function δ of Figure 2.10 is determined by the state alphabet Y only, and maps Y into Y, as indicated in the general equation of Equation 2.36. A unique state $Y_{j(t)}$ is mapped into the next state as shown in Equation 2.37. Thus, the next state $Y_{k(t+1)}$ is determined by the present state only as shown in Equation 2.38.

$$\delta(Y) : Y \rightarrow Y \tag{2.36}$$

$$\delta(Y) : Y_{j(t)} \rightarrow Y_{k(t+1)} \tag{2.37}$$

$$Y_{k(t+1)} = \delta(Y_{j(t)}) \tag{2.38}$$

The output function λ is determined by Y only, and maps the state alphabet Y into the output alphabet Z, as specified by the general equation of Equation 2.39. The present output vector $Z_{r(t)}$ is expressed as shown in Equation 2.40.

$$\lambda(Y) : Y \rightarrow Z \tag{2.39}$$

$$Z_{r(t)} = \lambda(Y_{j(t)}) \tag{2.40}$$

If there is no logic associated with the output vector Z_r, then the λ output function is a one-to-one mapping of Y_j onto Z_r. A mapping that is one-to-one onto is an isomorphism. When a mapping is an *isomorphism*, the two structures (or sets) are identical except for the names that are assigned to the elements of the sets and to the operations being performed. Two machines are isomorphic if they differ only in the labels (or names) that are used to denote the elements of the sets and have the same δ and λ functions, which differ only in the labels that are used to denote these functions.

An n-bit (n-stage) binary counter requires n storage elements, and generates 2^n different combinations of n bits. The range of an n-bit counter is from 0 to $2^n - 1$. Each state has only one next state, and the states of the machine are in sequence; however, not necessarily in ascending or descending order. Like registers, counters may also be designed with iterative stages.

Figure 2.11 illustrates a synchronous modulo-8 count-up binary counter. Synchronous counters are faster than asynchronous counters, because the clock pulse is transmitted to all stages simultaneously. The counter of Figure 2.11 is implemented with *JK* flip-flops as the storage elements, where each flip-flop is wired in toggle mode. The clock inputs to the storage elements are negative edge-triggered inputs; thus, the counter is incremented on each negative transition of the clock pulse. The counter is designed as a modulo-8 counter; therefore, after being reset initially, the counting sequence is 000, 001, 010, . . . 110, 111, 000, ...

Referring to Figure 2.11, the next-state function δ is dependent only upon the present state of the counter. Thus, in general

$$\delta(Y) : Y \rightarrow Y \tag{2.41}$$

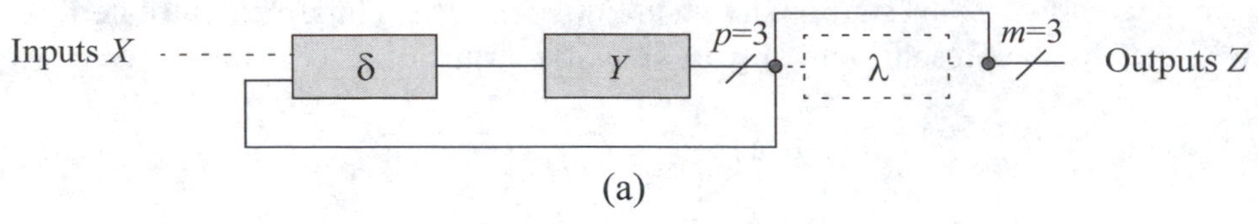

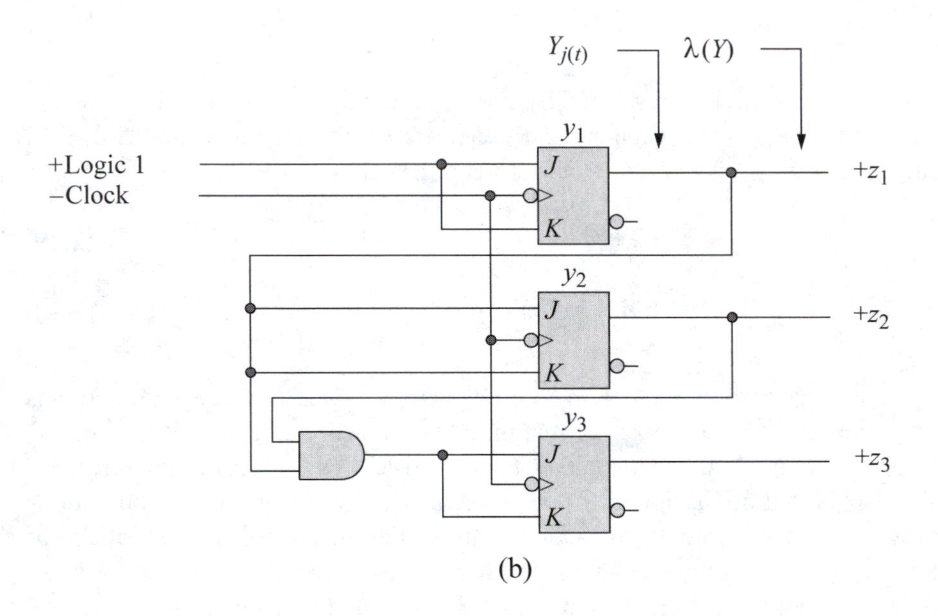

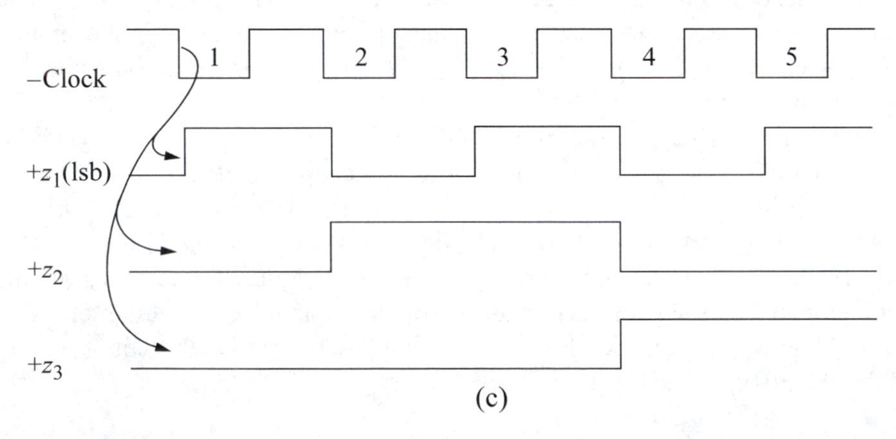

Figure 2.11 Counter implemented as a synchronous modulo-8 count-up binary counter using *JK* flip-flops: (a) block diagram; (b) logic diagram; and (c) timing diagram.

A specific state Y_j can be expressed as

$$\delta(Y_{j(t)}) : Y_{j(t)} \rightarrow Y_{k(t+1)} \tag{2.42}$$

such that

$$Y_{k(t+1)} = \delta(Y_{j(t)}) \tag{2.43}$$

The output function λ is dependent upon the present state only. Thus, in general

$$\lambda(Y) : Y \rightarrow Z \tag{2.44}$$

For a specific state Y_j of the state alphabet Y,

$$\lambda(Y_j) : Y_j \rightarrow Z_r \tag{2.45}$$

Thus,

$$Z_r = \lambda(Y_j) \tag{2.46}$$

Since there is no output logic for this particular counter, the present output vector $Z_{r(t)}$ is equal to the present state of the counter such that,

$$Z_{r(t)} = Y_{j(t)} \tag{2.47}$$

The logic for the δ mapping is represented by the following input equations:

$$\begin{aligned} Jy_1 &= Ky_1 = 1 \\ Jy_2 &= Ky_2 = y_1 \\ Jy_3 &= Ky_3 = y_1 y_2 \end{aligned}$$

The λ mapping is represented by the following output equations:

$$\begin{aligned} z_1 &= y_1 \\ z_2 &= y_2 \\ z_3 &= y_3 \end{aligned}$$

An asynchronous counter is shown in Figure 2.12. The storage elements within the counter do not change state simultaneously, because the clock input signal is not connected to the clock input of all flip-flops — only the first flip-flop receives the external clock signal. Therefore, the operating speed of the counter is considerably slower than the speed of the synchronous counter of Figure 2.11, although the counting range and the counting sequence remain the same.

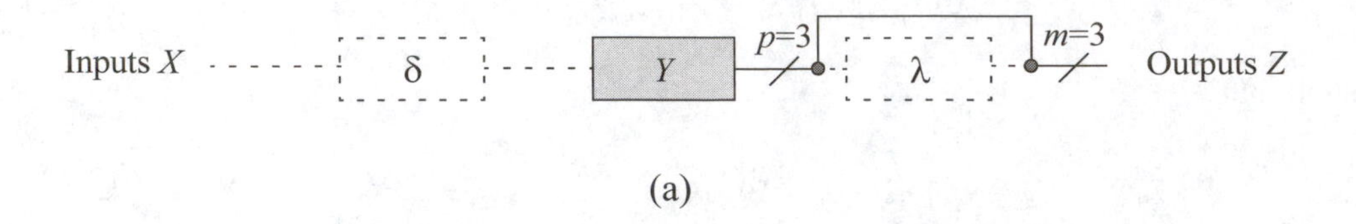

(a)

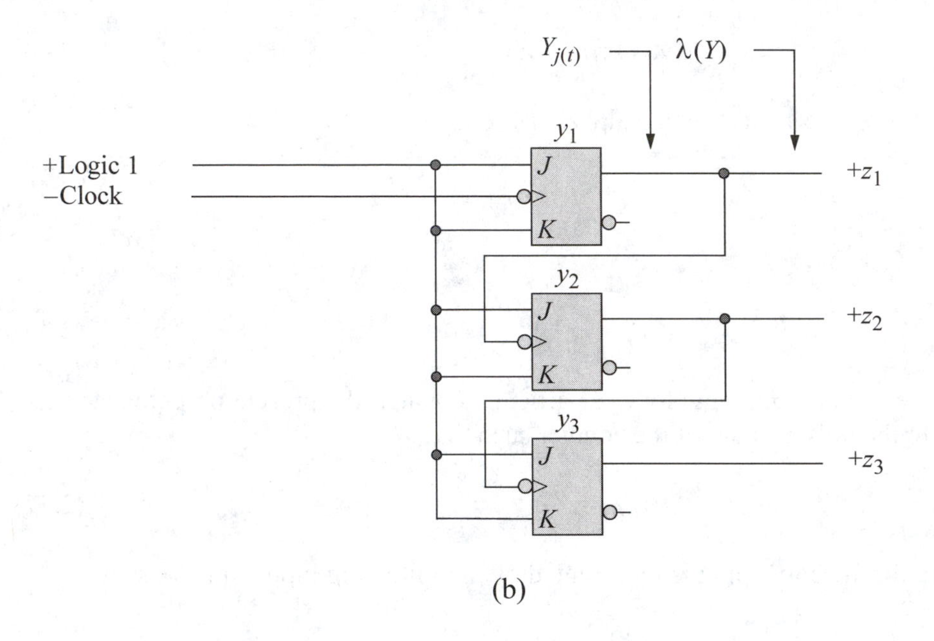

(b)

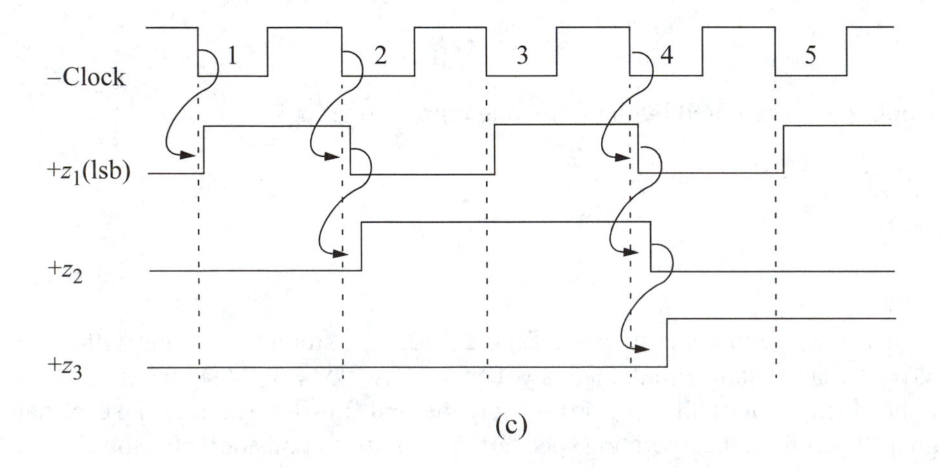

(c)

Figure 2.12 Counter implemented as an asynchronous modulo-8 count-up binary counter using *JK* flip-flops: (a) block diagram; (b) logic diagram; and (c) timing diagram showing accumulative delays.

An asynchronous counter is still regarded as a synchronous sequential machine, because the machine is clocked by an external signal. However, the internal wiring provides an asynchronous operation, since only the first stage receives the external clock. Programmable logic devices that contain storage elements can also be used in the implementation of a counter.

2.2.4 Moore Machines

Moore machines are synchronous sequential machines in which the output function λ produces an output vector Z_r which is determined by the present state only, and is not a function of the present inputs. The general configuration of a Moore machine is shown in Figure 2.13. The next-state function δ is an $(n + p)$-input, p-output switching function. The output function λ is a p-input, m-output switching function. If a Moore machine has no data input, then it is referred to as an *autonomous* machine. Autonomous circuits are independent of the inputs. The clock signal is not considered as a data input. An autonomous Moore machine is an important class of synchronous sequential machines, the most common application being a counter, as discussed in the previous section. A Moore machine may be synchronous or asynchronous, however, this section pertains to synchronous organizations only.

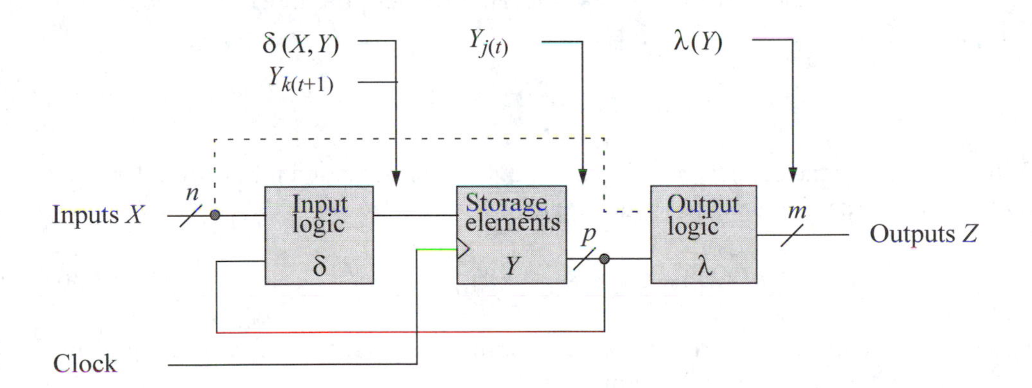

Figure 2.13 Moore synchronous sequential machine in which the outputs are a function of the present state only.

A Moore machine is a 5-tuple and can be formally defined as follows:

$$M = (X, Y, Z, \delta, \lambda) \tag{2.48}$$

where

1. X is a nonempty finite set of inputs such that,

$$X = \{X_0, X_1, X_2, \cdots, X_{2^n-2}, X_{2^n-1}\}$$

2. Y is a nonempty finite set of states such that,

$$Y = \{Y_0, Y_1, Y_2, \cdots, Y_{2^p-2}, Y_{2^p-1}\}$$

3. Z is a nonempty finite set of outputs such that,

$$Z = \{Z_0, Z_1, Z_2, \cdots, Z_{2^m-2}, Z_{2^m-1}\}$$

4. $\delta(X, Y) : X \times Y \rightarrow Y$

5. $\lambda(Y) : Y \rightarrow Z$

The next state of a Moore machine is determined by both the present inputs and the present state. Thus, the next-state function δ is a function of the input alphabet X and the state alphabet Y, and maps the Cartesian product of X and Y into Y, as shown in Equation 2.49.

$$\delta(X, Y) : X \times Y \rightarrow Y \tag{2.49}$$

With reference to the clock signal, Equation 2.49 can be restated for a particular input vector X_i and a particular state Y_j such that,

$$\delta(X, Y) : X_{i(t)} \times Y_{j(t)} \rightarrow Y_{k(t+1)} \tag{2.50}$$

where $X_{i(t)} \in X$ and $Y_{j(t)} \in Y$ are the present inputs and present state, respectively, and $Y_{k(t+1)} \in Y$ is the next state. Therefore,

$$Y_{k(t+1)} = \delta(X_{i(t)}, Y_{j(t)}) \tag{2.51}$$

The state of the storage elements and the values of the binary input variables are combined by the δ mapping to uniquely determine the next state of the machine.

The output alphabet Z is a function of the state alphabet Y only, therefore,

$$Z = \{\lambda(Y_j) \mid 0 \leq j \leq Y_{2^p-1}\} \tag{2.52}$$

The output vector Z_r is determined entirely by the present state of the machine, and is independent of the external inputs. Thus, when an input vector is applied to the inputs of a Moore machine, the resulting output vector does not occur until after the active clock transition.

The output function λ maps the state alphabet Y into the output alphabet Z. This can be expressed in general as

$$\lambda(Y) : Y \rightarrow Z \tag{2.53}$$

The corresponding present output $Z_{r(t)}$ becomes

$$Z_{r(t)} = \lambda(Y_{j(t)}) \tag{2.54}$$

where $Z_{r(t)} \in Z$.

A simple Moore machine is shown in Figure 2.14. There is one binary input variable x_1 for this machine, therefore, $n = 1$. Thus, the input alphabet is $X = \{X_0, X_1\}$, where

	x_1
$X_0 =$	0
$X_1 =$	1

There are two storage elements y_1 and y_2, therefore, $p = 2$. Thus, the state alphabet is $Y = \{Y_0, Y_1, Y_2, Y_3\}$, where

	$y_1 y_2$
$Y_0 =$	0 0
$Y_1 =$	0 1
$Y_2 =$	1 0
$Y_3 =$	1 1

There is one binary output variable z_1, therefore, $m = 1$. Thus, the output alphabet is $Z = \{Z_0, Z_1\}$, where

	z_1
$Z_0 =$	0
$Z_1 =$	1

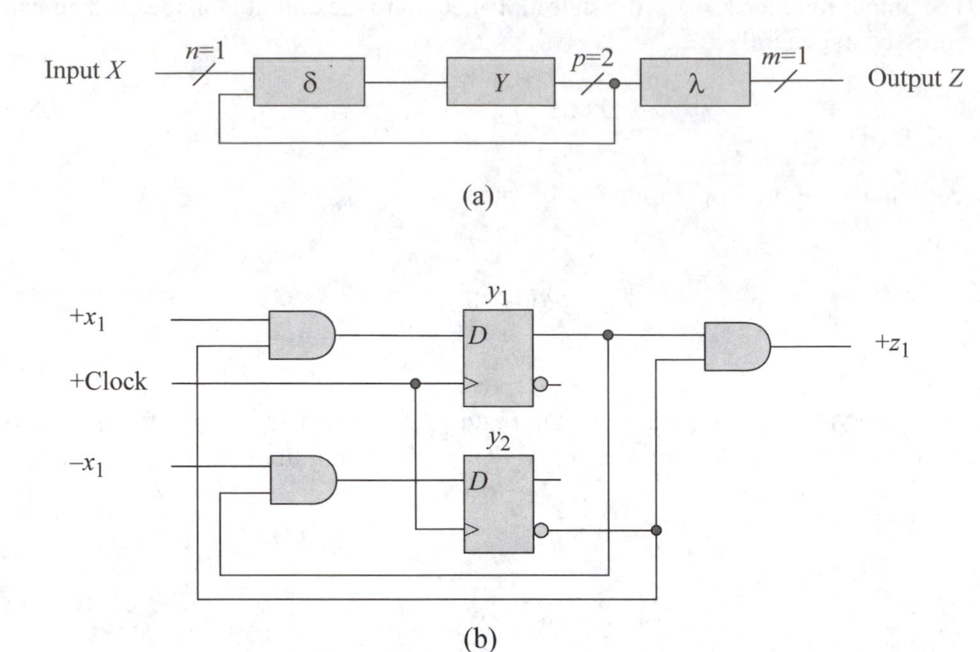

Figure 2.14 Moore machine: (a) block diagram; and (b) logic diagram.

The operation of the machine can be analyzed by the tabular representation shown in Table 2.4, which is a next-state table. The δ logic determines the next state (or path) of the machine depending upon the input vector. Table 2.4 is derived from Equation 2.50, which specifies that

$$\delta(X, Y) : X_{i(t)} \times Y_{j(t)} \rightarrow Y_{k(t+1)}$$

for $0 \leq i \leq 1$ and $0 \leq j \leq 3$. Using Equation 2.50, the first row of Table 2.4 is represented as

$$\delta(X_0, Y_0) : X_0 \times Y_0 \rightarrow Y_0$$

$$\delta(X_0, Y_0) = Y_0 \qquad 2.55)$$

and the ordered pairs of X and Y are <0, 00>, <0, 01>, <0, 10>, <0, 11>, <1, 00>, <1, 01>, <1, 10>, <1, 11>. However, the machine will never enter state $Y_3 = y_1 y_2 =$ 11 as a next state. Table 2.4 can also be obtained using Equation 2.51, which generates the first row as follows:

$$Y_{k(t+1)} = \delta(X_{i(t)}, Y_{j(t)})$$

$$Y_0 = \delta(X_0, Y_0)$$

Table 2.4 Next-state table for Figure 2.14

Present state y_1y_2	Input x_1	Next state y_1y_2	Output z_1
$0\ 0 = Y_0$	$0 = X_0$	$0\ 0 = Y_0$	$0 = Z_0$
$0\ 0 = Y_0$	$1 = X_1$	$1\ 0 = Y_2$	$0 = Z_0$
$0\ 1 = Y_1$	$0 = X_0$	$0\ 0 = Y_0$	$0 = Z_0$
$0\ 1 = Y_1$	$1 = X_1$	$0\ 0 = Y_0$	$0 = Z_0$
$1\ 0 = Y_2$	$0 = X_0$	$0\ 1 = Y_1$	$1 = Z_1$
$1\ 0 = Y_2$	$1 = X_1$	$1\ 0 = Y_2$	$1 = Z_1$
$1\ 1 = Y_3$	$0 = X_0$	$0\ 1 = Y_1$	$0 = Z_0$
$1\ 1 = Y_3$	$1 = X_1$	$0\ 0 = Y_0$	$0 = Z_0$

The machine that is represented by Table 2.4 is obviously a Moore machine, because the output does not depend upon the input, but only upon the present state. Table 2.5 lists the δ mappings for Figure 2.14. The next-state mapping $\delta(X_i, Y_j) = Y_3$ for $0 \leq i \leq 1$ and $0 \leq j \leq 3$ is not defined for this machine; that is, the machine is incompletely specified for a next state of Y_3 (refer to Section 2.4). Table 2.6 lists the λ mappings for Figure 2.14, which are derived from Equation 2.53, where $\lambda(Y) : Y_{j(t)} \rightarrow Z_{r(t)}$ for $0 \leq j \leq 3$ and $0 \leq r \leq 1$. Using the above derived equation with $j = 0$, the λ output mapping for the first row is obtained as follows:

$$\lambda(Y_0) : Y_0 \rightarrow Z_0$$

$$\lambda(Y_0) = Z_0$$

Table 2.6 can also be obtained using Equation 2.54, as follows:

$$Z_{r(t)} = \lambda(Y_{j(t)})$$

$$Z_0 = \lambda(Y_0)$$

Equation 2.51 specifies the next state for the machine of Figure 2.14 as $Y_{k(t+1)} = \delta(X_{i(t)}, Y_{j(t)})$. Therefore, the next states for storage elements y_1 and y_2 are obtained from the same δ mapping, as follows:

$$y_{1(t+1)} = \delta(x_{1(t)}, y_{2(t)}) = x_1 y_2' = Dy_1 \tag{2.56}$$

$$y_{2(t+1)} = \delta(x_{1(t)}, y_{1(t)}) = x_1' y_1 = Dy_2 \tag{2.57}$$

Table 2.5 δ mappings for Figure 2.14

$\delta(X_0, Y_0) = Y_0$
$\delta(X_1, Y_0) = Y_2$
$\delta(X_0, Y_1) = Y_0$
$\delta(X_1, Y_1) = Y_0$
$\delta(X_0, Y_2) = Y_1$
$\delta(X_1, Y_2) = Y_2$
$\delta(X_0, Y_3) = Y_1$
$\delta(X_1, Y_3) = Y_0$

Table 2.6 λ mappings for Figure 2.14

$\lambda(Y_0) = Z_0 = 0$
$\lambda(Y_1) = Z_0 = 0$
$\lambda(Y_2) = Z_1 = 1$
$\lambda(Y_3) = Z_0 = 0$

Equation 2.54 specifies the present output for the machine as $Z_{r(t)} = \lambda(Y_{j(t)})$. The present value of the binary output variable z_1 is obtained from the same λ mapping, which is represented by the AND gate that generates z1. Thus,

$$z_{1(t)} = \lambda(y_1, y_2) = y_1 y_2' \tag{2.58}$$

A Moore machine, however, does not have to be implemented with λ logic. In that case, the output transformation is a one-to-one mapping of Y onto Z such that, $Z = Y$. That is, the output alphabet is equal to the state alphabet, and requires no additional transformation to generate the outputs.

Machine alphabets can be expressed with tables. Boolean functions for the δ and λ mappings can also be expressed in tabular form, but are usually represented as equations or Karnaugh maps. Like other classes of machines, Moore machines may also be designed using only programmable logic devices.

2.2.5 Mealy Machines

Mealy machines are synchronous sequential machines in which the output function λ produces an output vector $Z_{r(t)}$ which is determined by both the present input vector $X_{i(t)}$ and the present state of the machine $Y_{j(t)}$. The general configuration of a Mealy machine is shown in Figure 2.15. The next-state function δ is an $(n + p)$-input, p-output switching function. The output function λ is an $(n + p)$-input, m-output switching function. A Mealy machine is not an autonomous machine, because the outputs are a function of the input signals. A Mealy machine may be synchronous or asynchronous, however, this section pertains to synchronous organizations only.

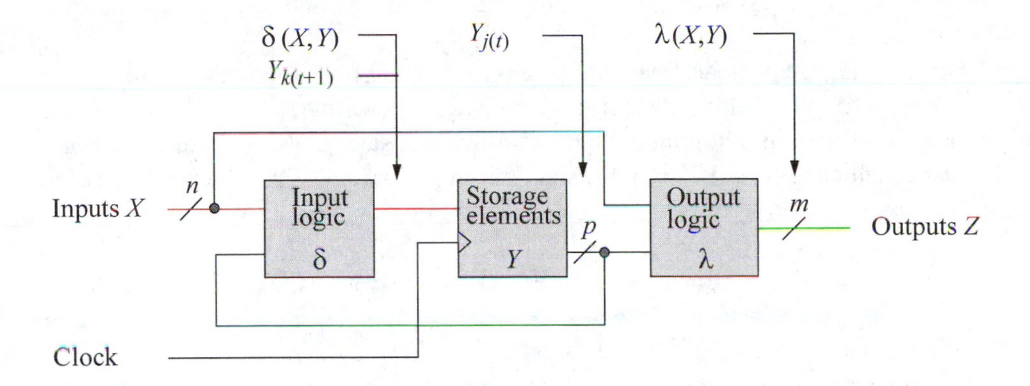

Figure 2.15 Mealy machine in which the outputs are a function of both the present state and the present inputs.

A Mealy machine is a 5-tuple and can be formally defined as follows:

$$M = (X, Y, Z, \delta, \lambda) \tag{2.59}$$

where

1. X is a nonempty finite set of inputs such that,

$$X = \{X_0, X_1, X_2, \cdots, X_{2^n-2}, X_{2^n-1}\}$$

2. Y is a nonempty finite set of states such that,

$$Y = \{Y_0, Y_1, Y_2, \cdots, Y_{2^p-2}, Y_{2^p-1}\}$$

3. Z is a nonempty finite set of outputs such that,

$$Z = \{Z_0, Z_1, Z_2, \cdots, Z_{2^m-2}, Z_{2^m-1}\}$$

4. $\delta(X, Y) : X \times Y \rightarrow Y$

5. $\lambda(X, Y) : Y \rightarrow Z$

The definitions for Mealy and Moore machines are the same, except for part 5, which shows that the outputs of a Mealy machine are a function of both the inputs and the present state, whereas, the outputs of a Moore machine are a function of the present state only. A Moore machine, therefore, is a special case of a Mealy machine.

The next-state function δ is used for both Mealy and Moore machines; however, to maintain completeness within this section, the derivation of the next-state equations will be reiterated.

The next state of a Mealy machine is determined by both the present inputs and the present state in the same manner as for a Moore machine. Thus, the next-state function δ is a function of the input alphabet X and the state alphabet Y, and maps the Cartesian product of X and Y into Y, as shown in Equation 2.49, which is rewritten here for convenience.

$$\delta(X, Y) : X \times Y \rightarrow Y \qquad (2.49)$$

With reference to the clock signal, Equation 2.49 can be restated for a particular input vector X_i and a particular state Y_j, thus producing the same equation as for a Moore machine (Equation 2.50), which is rewritten here.

$$\delta(X, Y) : X_{i(t)} \times Y_{j(t)} \rightarrow Y_{k(t+1)} \qquad (2.50)$$

where $X_{i(t)} \in X$ and $Y_{j(t)} \in Y$ are the present inputs and present state, respectively, and $Y_{k(t+1)} \in Y$ is the next state. Therefore, the next-state equation for a Mealy machine is the same as the next-state equation for a Moore machine (Equation 2.51), and is re-written below.

$$Y_{k(t+1)} = \delta(X_{i(t)}, Y_{j(t)}) \qquad (2.51)$$

The state of the storage elements and the values of the binary input variables are com-bined by the δ mapping to uniquely determine the next state of the machine.

The output alphabet Z is a function of the input alphabet X and of the state alpha-bet Y, therefore,

$$Z = \{\lambda(X_i, Y_j) \mid 0 \leq i \leq 2^n-1 \text{ and } 0 \leq j \leq 2^p-1\} \qquad (2.60)$$

The output vector Z_r is determined by both the present state of the machine and the ex-ternal inputs. Thus, when an input vector is applied to the inputs of a Mealy machine, the resulting output vector may be asserted immediately, after appropriate logic de-lays. Therefore, a Mealy machine operates slightly faster than a corresponding Moore machine.

The output function λ maps the Cartesian product of the input alphabet X and the state alphabet Y into the output alphabet Z. This can be expressed in general as

$$\lambda(X, Y) : X \times Y \rightarrow Z \tag{2.61}$$

The corresponding present output $Z_{r(t)}$ becomes

$$Z_{r(t)} = \lambda(X_{i(t)}, Y_{j(t)}) \tag{2.62}$$

where $Z_{r(t)} \in Z$.

A simple Mealy machine is shown in Figure 2.16. There is one binary input variable x_1 for this machine, therefore, $n = 1$. Thus, the input alphabet is $X = \{X_0, X_1\}$, where

	x_1
$X_0 =$	0
$X_1 =$	1

There are two storage elements y_1 and y_2, therefore, $p = 2$. Thus, the state alphabet is $Y = \{Y_0, Y_1, Y_2, Y_3\}$, where

	$y_1 y_2$
$Y_0 =$	0 0
$Y_1 =$	0 1
$Y_2 =$	1 0
$Y_3 =$	1 1

There is one binary output variable z_1, therefore, $m = 1$. Thus, the output alphabet is $Z = \{Z_0, Z_1\}$, where

	z_1
$Z_0 =$	0
$Z_1 =$	1

The operation of the machine can be analyzed by the tabular representation shown in Table 2.7, which is the next-state table. The δ logic determines the next state (or path) of the machine depending upon the input vector. Table 2.7 is derived from Equation 2.50 which specifies that

$$\delta(X, Y) : X_{i(t)} \times Y_{j(t)} \rightarrow Y_{k(t+1)}$$

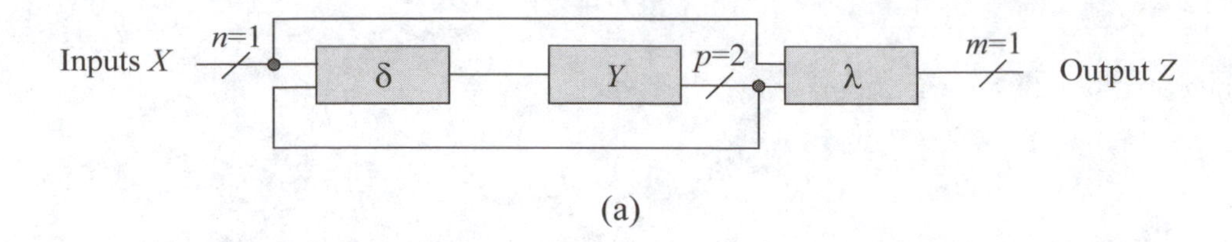

(a)

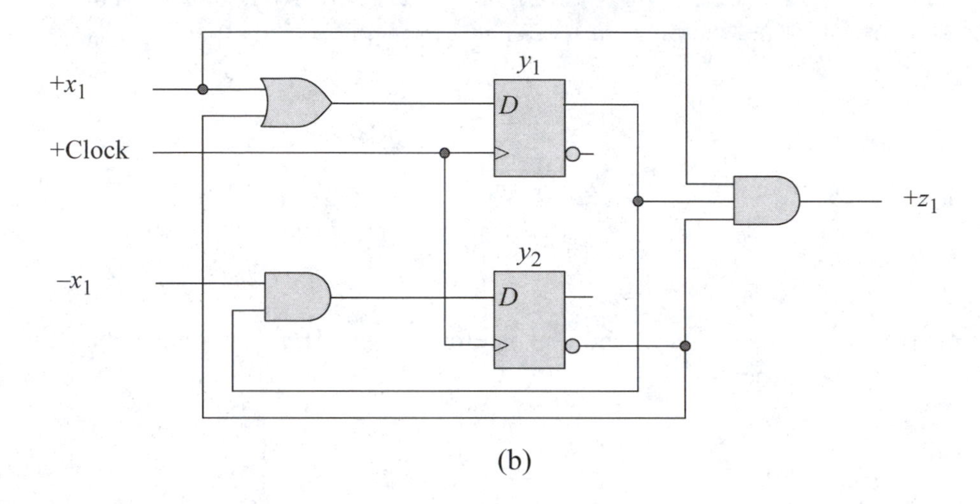

(b)

Figure 2.16 Mealy machine: (a) block diagram; and (b) logic diagram.

for $0 \leq i \leq 1$ and $0 \leq j \leq 3$. Using Equation 2.50, the first row of Table 2.7 is represented as

$$\delta(X_0, Y_0) : X_0 \times Y_0 \rightarrow Y_2$$

$$\delta(X_0, Y_0) = Y_2 \tag{2.63}$$

and the ordered pairs of X and Y are <0, 00>, <0, 01>, <0, 10>, <0, 11>, <1, 00>, <1, 01>, <1, 10>, <1, 11>. All states of the state alphabet are valid next states. Table 2.7 can also be obtained using Equation 2.51, which generates the first row as follows:

$$Y_{k(t+1)} = \delta(X_{i(t)}, Y_{j(t)})$$

$$Y_2 = \delta(X_0, Y_0) \tag{2.64}$$

The machine that is represented by Table 2.7 is obviously a Mealy machine, because the output depends upon both the present state and the present input.

Table 2.7 Next-state table for Figure 2.16

Present state y_1y_2	Input x_1	Next state y_1y_2	Output z_1
$0\ 0 = Y_0$	$0 = X_0$	$1\ 0 = Y_2$	$0 = Z_0$
$0\ 0 = Y_0$	$1 = X_1$	$1\ 0 = Y_2$	$0 = Z_0$
$0\ 1 = Y_1$	$0 = X_0$	$0\ 0 = Y_0$	$0 = Z_0$
$0\ 1 = Y_1$	$1 = X_1$	$1\ 0 = Y_2$	$0 = Z_0$
$1\ 0 = Y_2$	$0 = X_0$	$1\ 1 = Y_3$	$0 = Z_0$
$1\ 0 = Y_2$	$1 = X_1$	$1\ 0 = Y_2$	$1 = Z_1$
$1\ 1 = Y_3$	$0 = X_0$	$0\ 1 = Y_1$	$0 = Z_0$
$1\ 1 = Y_3$	$1 = X_1$	$1\ 0 = Y_2$	$0 = Z_0$

Table 2.8 lists the δ mappings for Figure 2.16. The next-state mapping $\delta(X_i, Y_j) = Y_k$ for $0 \le i \le 1$, $0 \le j \le 3$, and $0 \le k \le 3$ is defined for all states of this machine; that is, the machine is completely specified for all next states $Y_{k(t+1)}$.

Table 2.9 lists the λ mappings for Figure 2.16, which are derived from Equation 2.61, where $\lambda(X, Y) : X_{i(t)} \times Y_{j(t)} \rightarrow Z_{r(t)}$ for $0 \le i \le 1$, $0 \le j \le 3$, and $0 \le r \le 1$. Using the above derived equation with $i = 0$ and $j = 0$, the λ output mapping for the first and sixth rows are obtained as follows:

$$\lambda(X_0, Y_0) : X_0 \times Y_0 \rightarrow Z_0$$
$$\lambda(X_0, Y_0) = Z_0 \quad \text{(First row)}$$

$$\lambda(X_1, Y_2) : X_1 \times Y_2 \rightarrow Z_1$$
$$\lambda(X_1, Y_2) = Z_1 \quad \text{(Sixth row)}$$

Table 2.9 can also be obtained using Equation 2.62, as follows:

$$Z_{r(t)} = \lambda(X_{i(t)}, Y_{j(t)})$$
$$Z_0 = \lambda(X_0, Y_0) \quad \text{(First row)}$$
$$Z_1 = \lambda(X_1, Y_2) \quad \text{(Sixth row)}$$

Equation 2.51 specifies the next state for Moore machines and also for the Mealy machine of Figure 2.16 as

$$Y_{k(t+1)} = \delta(X_{i(t)}, Y_{j(t)})$$

Table 2.8 δ mappings for Figure 2.16	Table 2.9 λ mappings for Figure 2.16
$\delta(X_0, Y_0) = Y_2$	$\lambda(X_0, Y_0) = Z_0 = 0$
$\delta(X_1, Y_0) = Y_2$	$\lambda(X_1, Y_0) = Z_0 = 0$
$\delta(X_0, Y_1) = Y_0$	$\lambda(X_0, Y_1) = Z_0 = 0$
$\delta(X_1, Y_1) = Y_2$	$\lambda(X_1, Y_1) = Z_0 = 0$
$\delta(X_0, Y_2) = Y_3$	$\lambda(X_0, Y_2) = Z_0 = 0$
$\delta(X_1, Y_2) = Y_2$	$\lambda(X_1, Y_2) = Z_1 = 1$
$\delta(X_0, Y_3) = Y_1$	$\lambda(X_0, Y_3) = Z_0 = 0$
$\delta(X_1, Y_3) = Y_2$	$\lambda(X_1, Y_3) = Z_0 = 0$

Therefore, the next states for storage elements y_1 and y_2 for the Mealy machine of Figure 2.16 are obtained from the same δ mapping, as follows:

$$y_{1(t+1)} = \delta(x_{1(t)}, y_{2(t)}) = x_1 + y_2' = Dy_1$$

$$y_{2(t+1)} = \delta(x_{1(t)}, y_{1(t)}) = x_1' y_1 = Dy_2 \tag{2.65}$$

Equation 2.62 specifies the present output for the machine as $Z_{r(t)} = \lambda(X_{i(t)}, Y_{j(t)})$. The present value of the binary output variable z_1 is obtained from the same λ mapping, which is represented by the AND gate that generates z_1. Thus,

$$z_{1(t)} = \lambda(x_1, y_1, y_2) = x_1 y_1 y_2' \tag{2.66}$$

A Mealy machine must be implemented with the λ function, which is a boolean representation of combinational logic, in most cases. The output transformation, therefore, is a mapping of X and Y into Z. Three examples of Mealy machines will now be presented.

Example 2.4 Using the Mealy machine of Figure 2.16, obtain the next-state sequence and the output sequence for the following δ and λ functions, respectively:

(a) The next-state sequence is $\delta(X_0 X_1 X_1 X_0, Y_2) = Y_3 Y_2 Y_2 Y_3$

(b) The output sequence is $\lambda(X_1 X_0 X_1 X_1, Y_2) = Z_1 Z_0 Z_1 Z_1 = 1011$

Example 2.5 A serial binary adder adds two binary operands in serial mode. Let x_1 and x_2 be two n-bit binary operands as shown below.

$$x_1 = x_{1(n-1)} x_{1(n-2)} x_{1(n-3)} \cdots x_{1(1)} x_{1(0)}$$
$$x_2 = x_{2(n-1)} x_{2(n-2)} x_{2(n-3)} \cdots x_{2(1)} x_{2(0)}$$

where $x_{1(0)}$ and $x_{2(0)}$ are the low-order bits of x_1 and x_2, respectively.

The serial adder first performs the operation $[x_{1(0)} + x_{2(0)}]$, which generates a sum bit z_1 and a carry bit z_2. The sum bit is shifted into a serial-in, parallel-out shift register and the carry bit is added to the next two operand bits $x_{1(1)}$ and $x_{2(1)}$, which generates a new sum bit and a new carry bit. The process repeats for all n bits of both operands. The logic diagram for this machine is shown in Figure 2.17. This is a Mealy machine, because at least one output is a direct function of the inputs. The flip-flop is clocked on the positive transition of the clock and the operands change state on the negative transition of the clock.

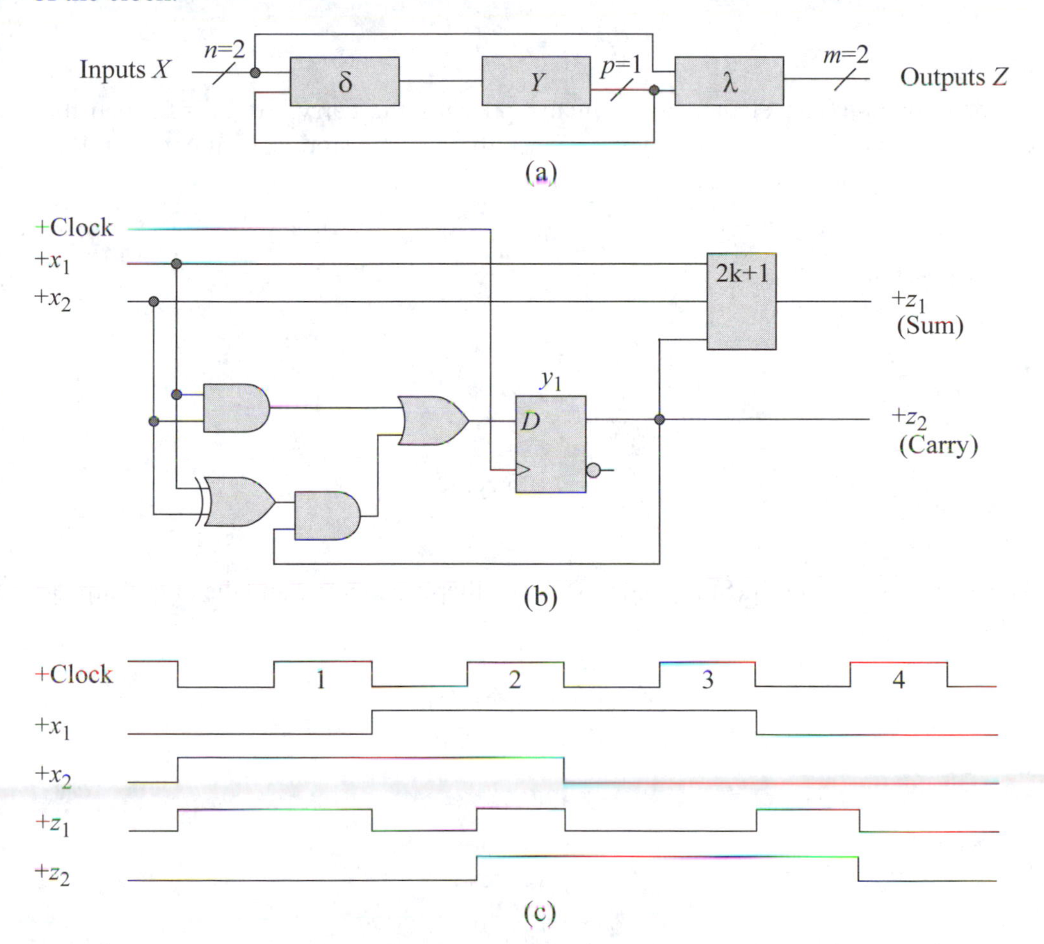

Figure 2.17 A Mealy machine implemented as a serial adder: (a) block diagram; (b) logic diagram; and (c) timing diagram.

There are two input operands $x_{1(i)}$ and $x_{2(i)}$ for $0 \le i \le n-1$, therefore, $n = 2$ and the input alphabet X is

$$X = \{X_0, X_1, X_2, X_3\}$$

where

	$x_1 x_2$
$X_0 =$	0 0
$X_1 =$	0 1
$X_2 =$	1 0
$X_3 =$	1 1

There is one storage element y_1, which represents the carry. If $y_1 = 0$, then the carry = 0; if $y_1 = 1$, then the carry = 1. Since there is one storage element, therefore $p = 1$ and the state alphabet Y is

$$Y = \{Y_0, Y_1\}$$

where

	y_1
$Y_0 =$	0
$Y_1 =$	1 (Carry)

There are two outputs z_1 (Sum) and z_2 (Carry), therefore, $m = 2$ and the output alphabet Z is

$$Z = \{Z_0, Z_1, Z_2, Z_3\}$$

where

	Sum z_1	Carry z_2
$Z_0 =$	0	0
$Z_1 =$	0	1
$Z_2 =$	1	0
$Z_3 =$	1	1

A *full adder* has three inputs and two outputs as shown in Table 2.10. The carry-out at time (t) becomes the carry-in at time $(t+1)$ and is added to the next two operand bits. The sum and carry equations are specified in Equation 2.67 and Equation 2.68, respectively. For high-speed parallel operation, the first lines of Equation 2.67 and Equation 2.68 are used. These equations require more logic; however, only two gate delays are needed, resulting in a faster operation.

Table 2.10 Truth table for a full adder

x_1	x_2	c_{in} (1)	Sum z_1	Carry z_2
0	0	0	0	0
0	0	1	1	0
0	1	0	1	0
0	1	1	0	1
1	0	0	1	0
1	0	1	0	1
1	1	0	0	1
1	1	1	1	1

(1) c_{in} is the carry z_2 from the previous add cycle.

The operation of the machine is described by the next-state table shown in Table 2.11, where $Y_0 = y_1 = 0$ represents a carry of 0, and $Y_1 = y_1 = 1$ represents a carry of 1.

$$\begin{aligned} \text{Sum } (z_1) &= x_1'x_2'c_{in} + x_1'x_2c_{in}' + x_1x_2'c_{in}' + x_1x_2c_{in} \\ &= c_{in}(x_1'x_2' + x_1x_2) + c_{in}'(x_1'x_2 + x_1x_2') \\ &= x_1 \oplus x_2 \oplus c_{in} \end{aligned} \tag{2.67}$$

$$\begin{aligned} \text{Carry}(z_2) &= x_1'x_2c_{in} + x_1x_2'c_{in} + x_1x_2c_{in}' + x_1x_2c_{in} \\ &= c_{in}(x_1'x_2 + x_1x_2') + x_1x_2 \\ &= x_1x_2 + (x_1 \oplus x_2)\, c_{in} \end{aligned} \tag{2.68}$$

Table 2.11 Next-state table for Figure 2.17

Present state y_1	Input x_1x_2	Next state y_1	Output z_1 (Sum)	z_2 (Carry)
$0 = Y_0$ (No carry)	$0\ 0 = X_0$	$0 = Y_0$	0	$0 = Z_0$
$0 = Y_0$	$0\ 1 = X_1$	$0 = Y_0$	1	$0 = Z_2$
$0 = Y_0$	$1\ 0 = X_2$	$0 = Y_0$	1	$0 = Z_2$
$0 = Y_0$	$1\ 1 = X_3$	$1 = Y_1$	0	$1 = Z_1$
$1 = Y_1$ (Carry)	$0\ 0 = X_0$	$0 = Y_0$	1	$0 = Z_2$
$1 = Y_1$	$0\ 1 = X_1$	$1 = Y_1$	0	$1 = Z_1$
$1 = Y_1$	$1\ 0 = X_2$	$1 = Y_1$	0	$1 = Z_1$
$1 = Y_1$	$1\ 1 = X_3$	$1 = Y_1$	1	$1 = Z_3$

Example 2.6 Figure 2.18 also illustrates a serial adder, but with no feedback from the storage element to the δ logic. The next-state table is the same as for Example 2.5; however, the timing diagram is slightly different.

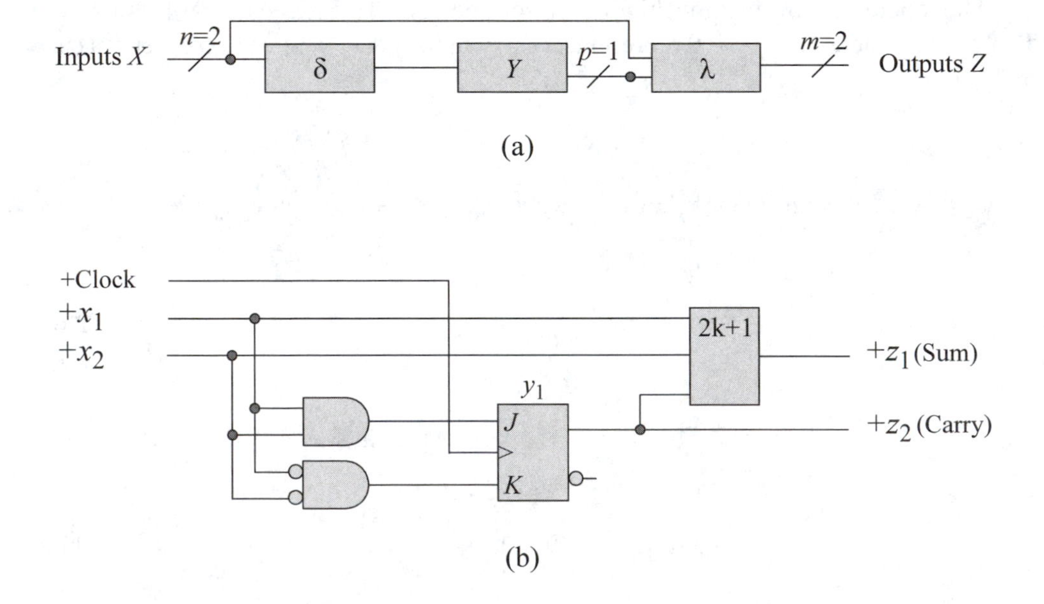

Figure 2.18 A Mealy machine implemented as a full adder: (a) block diagram; and (b) logic diagram.

Summary of equations Table 2.12 lists the next-state and output equations for Moore and Mealy machines. Like other classes of machines, Moore and Mealy machines can also be designed using only programmable logic devices.

Table 2.12 Summary of equations for Moore and Mealy machines

Moore and Mealy	$\delta(X, Y) : X \times Y \rightarrow Y$	(2.50)
	$\delta(X, Y) : X_{i(t)} \times Y_{j(t)} \rightarrow Y_{k(t+1)}$	(2.51)
	$Y_{k(t+1)} = \delta(X_{i(t)}, Y_{j(t)})$	(2.52)
Moore	$\lambda(Y) : Y \rightarrow Z$	(2.54)
	$Z_{r(t)} = \lambda(Y_{j(t)})$	(2.55)
Mealy	$\lambda(X, Y) : X \times Y \rightarrow Z$	(2.62)
	$Z_{r(t)} = \lambda(X_{i(t)}, Y_{j(t)})$	(2.63)

2.2.6 Asynchronous Sequential Machines

The general configuration of an asynchronous sequential machine is shown in Figure 2.19. The logic that is represented by the δ next-state function and the λ output function is not always necessary; that is, there may be no input logic that is exclusively a function of the inputs or there may be no output logic for the machine. An example of this type of machine is an *SR* latch. There is a large variation of configurations for asynchronous sequential machines: there may be no input or output logic, as previously stated; there may be input logic, but no output logic; there may be no input logic, but there may be output logic; there may be both input and output logic; the output logic may or may not be a direct function of the inputs.

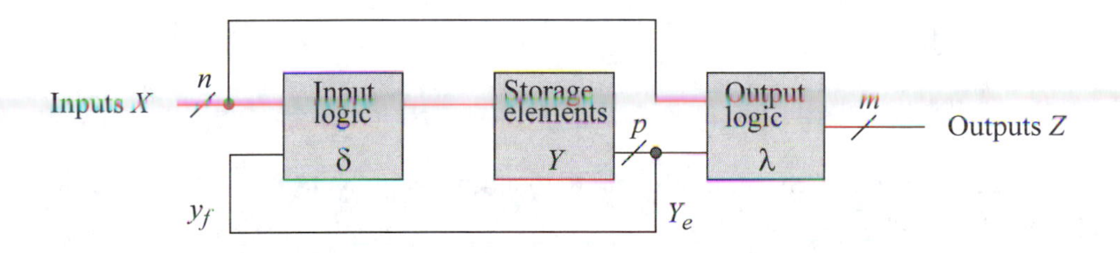

Figure 2.19 An asynchronous sequential machine.

The main distinction between synchronous sequential machines and asynchronous sequential machines is that there is no clock input signal for an asynchronous sequential machine; therefore, latches are used as the storage elements. The operation of this type of machine depends only on the order (or sequence) in which the inputs change and can be affected at any instant of time. The inputs are not synchronized with a system clock, and may change state at random times. Because time is an element in sequential circuits, the outputs are a function not only of the present inputs, but also of the previous machine states.

One other important distinction between synchronous sequential machines and asynchronous sequential machines is that the outputs of the storage elements are labeled as excitation variables Y_e, which are then fed back through logic gates to latch the storage elements in a particular state, either 0 or 1. After the machine has stabilized, these feedback signals are called feedback variables y_f.

A simple asynchronous sequential machine is illustrated in Figure 2.20 using NOR logic. The output of the y_1 storage element is active low, and is the excitation variable Y_e, which becomes the feedback variable y_f after the logic has stabilized.

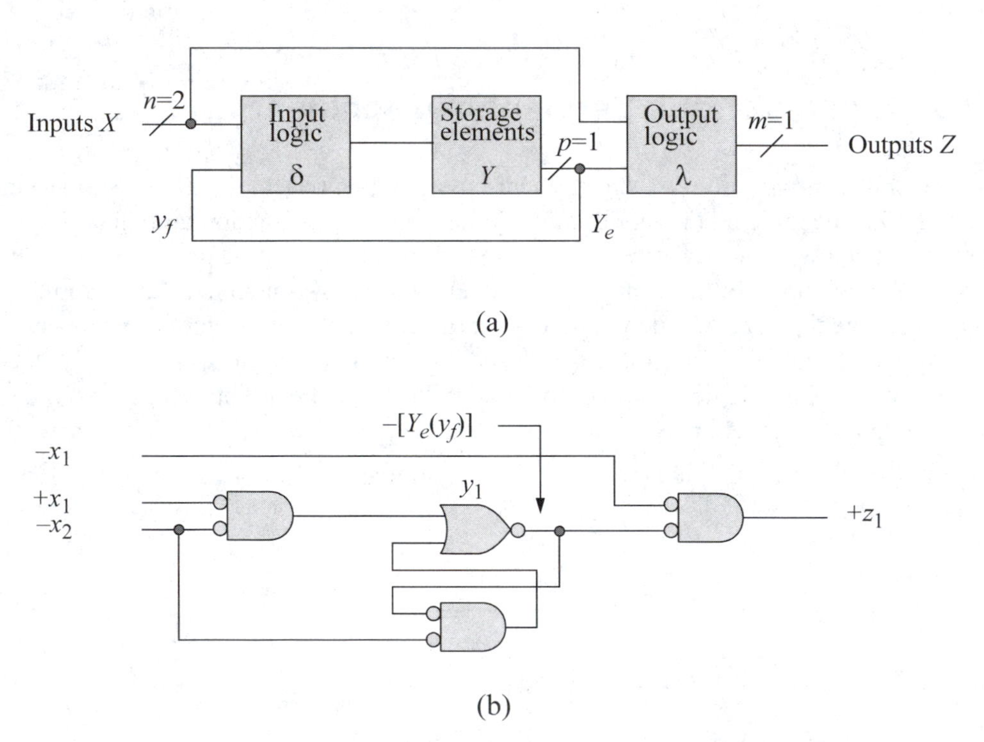

Figure 2.20 An asynchronous sequential machine using NOR logic: (a) block diagram; and (b) logic diagram.

No further analysis of asynchronous sequential machines will be considered in this section. The analysis and synthesis of asynchronous sequential machines is considerably different than that for synchronous sequential machines and will be deferred until later chapters. The analysis of asynchronous sequential machines will be presented in Chapter 5 and the synthesis will be presented in Chapter 6. Asynchronous sequential machines were included in this section simply to complete the list of machines that are used for logic design. All logic design can be represented by one of the six classes of machines described in this section.

2.2.7 Additional Definitions for Synchronous Sequential Machines

This section contains a list of definitions that are encountered in the analysis and synthesis of synchronous sequential machines.

State machine In the analysis of some synchronous sequential machines, the primary interest may be in the properties of the state transitions and not in the outputs. In this case, the machine has no output logic; therefore, the outputs are taken directly from the storage element outputs. When a synchronous sequential machine does not contain logic for the output function λ, then a distinction is made between this type of machine and other synchronous sequential machines. This is a special class of a Moore machine and is sometimes referred to as a state machine or a finite-state machine, although all synchronous sequential machines are regarded as state machines.

With the restrictive property of no λ logic, a state machine is characterized by the following 3-tuple:

$$M = (X, Y, \delta) \tag{2.69}$$

where

1. X is a nonempty finite set of inputs such that,
   $X = \{X_0, X_1, X_2, \cdots, X_{2^n-2}, X_{2^n-1}\}$

2. Y is a nonempty finite set of states such that,
   $Y = \{Y_0, Y_1, Y_2, \cdots, Y_{2^p-2}, Y_{2^p-1}\}$

3. $\delta(X, Y) : X \times Y \rightarrow Y$

The general model of a state machine is shown in Figure 2.21 and consists of one or more of the following units:

1. Input logic which consists of combinational logic that transforms the input vector and the present state into the next state by means of the next-state function δ.

2. Synchronous storage elements which indicate the present state of the machine.

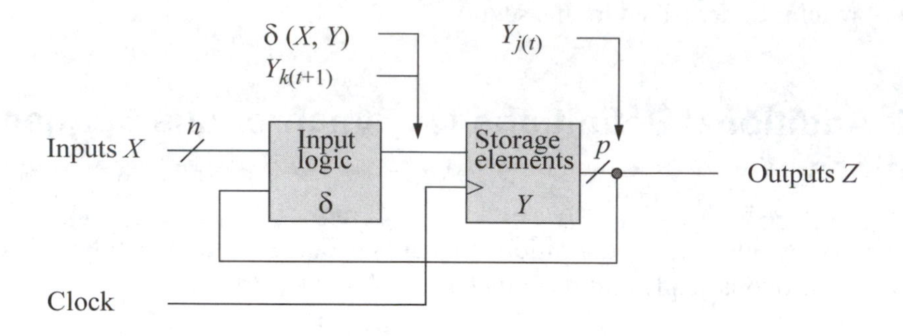

Figure 2.21 General model of a synchronous state machine.

A PROM may be considered as this type of state machine. The δ function consists of address generation logic and the p outputs are obtained from the storage elements in the PROM, which are the result of programming the device. The feedback signals from the output to the input logic are used for address modification, where the number of feedback lines is $\leq p$.

Submachine Given a synchronous sequential machine M as the following 5-tuple:

$$M = (X, Y, Z, \delta, \lambda) \tag{2.70}$$

then machine M', which is defined by the following 5-tuple:

$$M' = (X', Y', Z', \delta', \lambda') \tag{2.71}$$

is a submachine of M if and only if

$$\begin{aligned} &X' \subseteq X \\ &Y' \subseteq Y \\ &Z' \subseteq Z \\ &\delta'(X', Y') : X' \times Y' \rightarrow Y' \\ &\lambda'(X', Y') : X' \times Y' \rightarrow Z' \end{aligned} \tag{2.72}$$

where $X' \subseteq X$ indicates that the input alphabet set X' is a subset of the input alphabet set X where every element of X' is also an element of X. If $X' \subseteq X$ and $X' \neq X$, then X' is referred to as a proper subset of X, denoted as $X' \subset X$. The set X', therefore, is not restricted to be a proper subset of X. The same statements are true for the state and output alphabets. The submachine described above refers to a 5-tuple. However, the same rationale also applies to the 3-tuple state machine of Equation 2.69.

Terminal state For a synchronous sequential machine M, if no input sequence exists which will cause a state transition from a given state $Y_j \in Y$ to any other state $Y_k \in Y$, then state Y_j is called a terminal state. This condition for Y_j is expressed in Equation 2.73.

$$\delta(X_i, Y_j) : X_i \times Y_j \nrightarrow Y_k \tag{2.73}$$

Also, the design of some machines may not permit a transition from certain state subsets of the state alphabet Y to a particular state subset of Y, even though the machine does not contain any terminal states.

Strongly connected machines If, for every pair of states $Y_j \in Y$ and $Y_k \in Y$ of a machine M, there exists an input sequence which causes a transition from Y_j to Y_k, then M is a strongly connected machine. That is, a machine is strongly connected if and only if there is a transition from state Y_j to any other state Y_k by applying a particular sequence of inputs as indicated in Equation 2.74.

$$\delta(X_i, Y_j) = Y_k \qquad 2.74)$$

Thus, any nontrivial machine which has terminal states is not strongly connected.

Submachines can also be strongly connected. A synchronous sequential machine M and a submachine M' are defined as follows:

$$M = (X, Y, Z, \delta, \lambda)$$

$$M' = (X', Y', Z', \delta', \lambda') \tag{2.75}$$

A strongly connected submachine M' of M contains a subset $Y' \in Y$ for which every state in Y' (where $Y' \subseteq Y$) can be entered from every other state in Y' by state transitions that never leave Y'. That is, for every pair of states $Y_j' \in Y'$ and $Y_k' \in Y'$, there exists an input sequence which causes a transition from Y_j' to Y_k', where all transitions occur within the state alphabet Y'. This is indicated by Equation 2.76, which states that the next-state function δ' maps an input vector $X_i' \in X'$ and the present state $Y_j' \in Y'$ into the next state $Y_k' \in Y'$.

$$\delta' (X_i', Y_j') = Y_k' \tag{2.76}$$

It can also be stated that, if machine M is strongly connected, then $X' = X$ implies that $Y' = Y$.

Machine homomorphism The concept of homomorphism will be defined, then this mapping will be extended to include synchronous sequential machines. A simple algebraic structure is written as $< S, ^{o} >$, where ^{o} is an operation on the set S. Two different algebraic structures may share similar characteristics. Let one algebraic structure be defined on a set S and a similar structure be defined on a set S', where ^{o} and $*$ are operations on the sets S and S', respectively.

Thus, $< S, ^{o} >$ and $< S', * >$ are algebraic structures with operations ^{o} and $*$, respectively. Examples of operations are: $\leq$, $=$, $+$, $\oplus$, etc. A function h which maps S to S', and is written as $h: S \rightarrow S'$, is a homomorphism from algebraic structure $< S, ^{o} >$ to algebraic structure $< S', * >$ if, for every $s_1, s_2 \in S$, Equation 2.77 is true.

$$h(s_1 \, ^{o} \, s_2) = h(s_1) * h(s_2) \tag{2.77}$$

where ^{o} is an operation in S and $*$ is an operation in S'.

The definition of homomorphism can be restated as follows: a homomorphism of S to S' means a mapping

$$S \rightarrow S' : s \rightarrow s'$$

such that,

1. Every $s_i \in S$ has a unique image $s_i' \in S'$
2. If $s_1 \rightarrow s_1'$ and $s_2 \rightarrow s_2'$, then $s_1 \, ^{o} \, s_2 \rightarrow s_1' * s_2'$
3. Every $s' \in S'$ is an image

This defines a homomorphism of S *onto* S' where S' is a homomorphic image of S.

The homomorphism h is illustrated in Figure 2.22, which shows two algebraic structures $< S, ^{o} >$ and $< S', * >$ having similar characteristics. The algebraic structure $< S', * >$ is considered to simulate the structure $< S, ^{o} >$. The algebraic structure $< S, ^{o} >$ is called the domain of the function h and the structure $< S', * >$ is called the range of the function h.

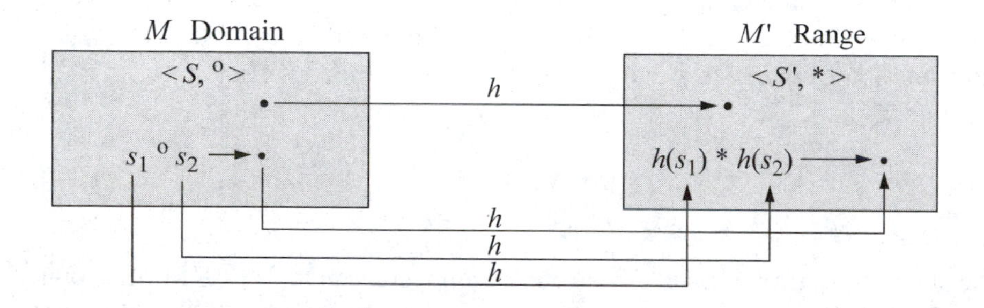

Figure 2.22 Two algebraic structures $< S, ^{o} >$ and $< S', * >$ where the function h is a homomorphism.

A homomorphism preserves some of the characteristics of the domain structure. The definition of a homomorphism states that performing the operation '$\circ$' on two elements of the domain and then applying the homomorphism yields the same result as obtained by applying the homomorphism on each of the elements first, and then performing the operation '*' in the range.

The concept of homomorphism can also be applied to synchronous sequential machines. The synchronous sequential machine homomorphism is a mapping which preserves the structure of the next-state function δ and the output function λ of the machine. Three mappings are required for a machine homomorphism: one mapping from an input set to an input set, one mapping from a state set to a state set, and one mapping from an output set to an output set.

Let a synchronous sequential machine M be defined as follows:

$$M = (X, Y, Z, \delta, \lambda)$$

Then, a machine M'

$$M' = (X', Y', Z', \delta', \lambda')$$

is a *homomorphic image* of machine M if and only if the following three *onto* mappings (functions) exist:

$$\begin{array}{ll} h_1 : X \rightarrow X' & \text{Input alphabets of } M \text{ and } M' \text{, respectively} \\ h_2 : Y \rightarrow Y' & \text{State alphabets of } M \text{ and } M' \text{, respectively} \\ h_3 : Z \rightarrow Z' & \text{Output alphabets of } M \text{ and } M' \text{, respectively} \end{array} \tag{2.78}$$

where h_1, h_2, and h_3 are homomorphic functions which map X onto X', Y onto Y', and Z onto Z', respectively such that,

$$h_2 [\delta(X_i, Y_j)] = \delta'[h_1(X_i), h_2(Y_j)] \tag{2.79}$$

$$h_3 [\lambda(X_i, Y_j)] = \lambda'[h_1(X_i), h_2(Y_j)] \tag{2.80}$$

where

$$\delta(X_i, Y_j) = Y_k$$

$$\delta'[h_1(X_i), h_2(Y_j)] = Y_k'$$

$$\lambda(X_i, Y_j) = Z_r$$

$$\lambda'[h_1(X_i), h_2(Y_j)] = Z_r'$$

and $X_i \in X$, $Y_j \in Y$, and δ and δ' are the next-state functions of machines M and M', respectively. The output functions of machines M and M' are λ and λ', respectively.

The input sequences for machines M and M' are identical. We are interested, therefore, in obtaining homomorphisms for the state transitions and the output sequences. Equation 2.79 illustrates that identical results can be obtained by applying the homomorphism and the next-state function for both machines. Since M' is a homomorphic image of M, then performing the function δ on X_i and Y_j, and then applying the homomorphism h_2, yields the same results as obtained by first applying the homomorphisms h_1 and h_2 on X_i and Y_j separately, and then performing the function δ'. The same rationale is valid for the λ and λ' functions. Because machine M' is a homomorphic image of machine M, the operations of the two machines are indistinguishable to an observer who can see only the inputs and outputs of M and M'.

Figure 2.23 illustrates machine homomorphism using Equation 2.79 and Equation 2.80. The outputs of the two machines are identical. In Figure 2.23 (a), machine M applies the next-state function δ to the two elements X_i and Y_j and then applies the homomorphism h_2 to the result as indicated by $h_2[\delta(X_i, Y_j)]$. In Figure 2.23 (b), machine M' first applies the homomorphisms h_1 and h_2 separately to the elements X_i and Y_j, respectively, and then uses the next-state function δ' as indicated by $\delta'[h_1(X_i), h_2(Y_j)]$, which yields the same result.

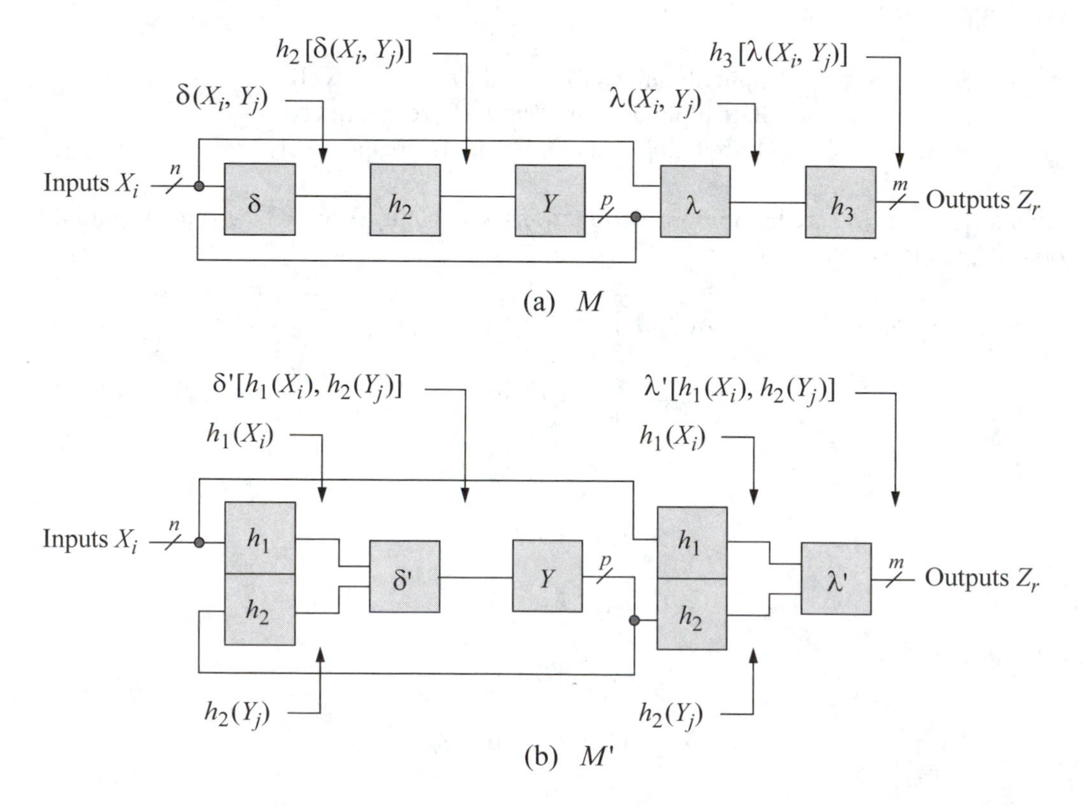

Figure 2.23 Block diagram of machine homomorphism: (a) machine M; and (b) machine M', which is a homomorphic image of machine M.

An identical method is used for the λ and λ' mappings and for the homomorphism h_3. In Figure 2.23 (a), machine M applies the output function λ to the two elements X_i and Y_j, and then applies the homomorphism h_3 to the result as indicated by $h_3[\lambda(X_i, Y_j)]$. In Figure 2.23 (b), machine M' first applies the homomorphisms h_1 and h_2 separately to the elements X_i and Y_j, respectively, and then uses the output function λ' as indicated by $\lambda'[h_1(X_i), h_2(Y_j)]$.

A similar procedure is used to derive an equation which represents a homomorphic image for two machines M and M', where

$$M = (X, Y, \delta)$$

$$M' = (X', Y', \delta')$$

Example 2.7 Let $S = \{s, s^2, s^3, s^4, \cdots, s^{11}, s^{12}\}$ and $S' = \{s^2, s^4, s^6, s^8, s^{10}, s^{12}\}$. Then $h : s^n \rightarrow s^{2n}$ is a homomorphism of S onto S'.

Example 2.8 Let machine M be a modulo-8 adder and machine M' be a modulo-4 adder. The modulo-4 adder is a homomorphic image of the modulo-8 adder. Figure 2.24 shows the homomorphism of M onto M'.

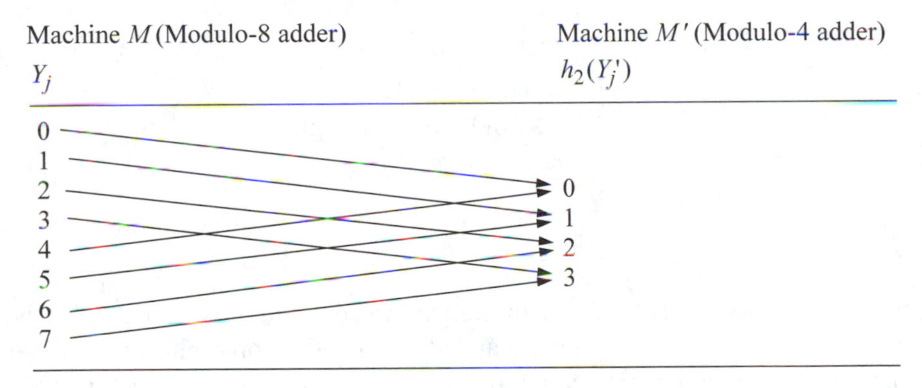

Figure 2.24 Onto mapping of $Y \rightarrow Y'$, where machine M' is a homomorphic image of machine M.

The output sequence for the modulo-8 adder is represented by Y_j and the output sequence for the modulo-4 adder is represented by Y_j'. The modulo-4 adder is a homomorphic image of the modulo-8 adder, because the mapping h_2 is a mapping of Y onto

Y'. The adders have no λ output function; therefore, only Equation 2.79 is applicable. Equation 2.79 states that

$$h_2[\delta(X_i, Y_j)] = \delta'[h_1(X_i), h_2(Y_j)]$$

where δ is the add function for the modulo-8 adder, δ' is the add function for the modulo-4 adder, and h_1 and h_2 represent the homomorphisms $h_1: X_i \rightarrow X_i'$ and $h_2: Y_j \rightarrow Y_j'$. The input vector X_i is added to the present state Y_j. Numerical examples are shown below.

(i) Let $X_i = 1$ and $Y_j = 1$. Then

$$\begin{aligned} h_2[\delta(1, 1)] &= \delta'[h_1(1), h_2(1)] \\ h_2[2] &= \delta'[1, 1] \\ 2 &= 2 \end{aligned}$$

(ii) Let $X_i = 2$ and $Y_j = 4$. Then

$$\begin{aligned} h_2[\delta(2, 4)] &= \delta'[h_1(2), h_2(4)] \\ h_2[6] &= \delta'[2, 0] \\ 2 &= 2 \end{aligned}$$

(iii) Let $X_i = 2$ and $Y_j = 7$. Then

$$\begin{aligned} h_2[\delta(2, 7)] &= \delta'[h_1(2), h_2(7)] \\ h_2[1] &= \delta'[2, 3] \\ 1 &= 1 \end{aligned}$$

Machine isomorphism An isomorphism was defined in Section 2.2.3. The definition will be restated here and expanded to include isomorphisms between machines. Let the function g be a homomorphism from algebraic structure $\langle S, {}^{\circ}\rangle$ to algebraic structure $\langle S', *\rangle$, such that,

$$g(s_1 \,{}^{\circ}\, s_2) = g(s_1) * g(s_2) \tag{2.81}$$

Then the function g is an isomorphism if the mapping for g is one-to-one onto S'.

When the homomorphism is also an isomorphism, then the two structures are identical except for the labels that are assigned to the elements of S and S', and to the names of the operations (or compositions) '${}^{\circ}$' and '*'. The two structures $\langle S, {}^{\circ}\rangle$ and $\langle S', *\rangle$ are said to be isomorphic and are indistinguishable from each other.

The definition of isomorphism can be restated as follows: An isomorphism of S to S' means a mapping

$$S \rightarrow S' : s \rightarrow s'$$

such that,

1. Every $s \in S$ has a unique image $s' \in S'$.
2. If $s_1 \rightarrow s_1'$ and $s_2 \rightarrow s_2'$, then $(s_1 \circ s_2) \rightarrow (s_1' * s_2')$.
3. Every $s' \in S'$ is an image and the mapping is one-to-one; that is, $s \rightarrow s' \rightarrow s$.

Then S and S' are isomorphic and the mapping is an isomorphism.

Let two synchronous sequential machines be defined as follows:

$$M = (X, Y, Z, \delta, \lambda)$$

$$M' = (X', Y', Z', \delta', \lambda')$$

The machines are isomorphic if and only if the following three one-to-one onto mappings exist:

$$\begin{aligned} g_1 &: X \rightarrow X' \\ g_2 &: Y \rightarrow Y' \\ g_3 &: Z \rightarrow Z' \end{aligned} \tag{2.82}$$

where g_1, g_2, and g_3 are isomorphic functions which map X one-to-one onto X', Y one-to-one onto Y', and Z one-to-one onto Z' such that,

$$g_2[\delta(X_i, Y_j)] = \delta'[g_1(X_i), g_2(Y_j)] \tag{2.83}$$

$$g_3[\lambda(X_i, Y_j)] = \lambda'[g_1(X_i), g_2(Y_j)] \tag{2.84}$$

where $X_i \in X$, $Y_j \in Y$, δ and δ' are the next-state functions of machines M and M', respectively, and λ and λ' are the output functions of machines M and M', respectively.

If the homomorphism is a one-to-one onto mapping, then machine M' is isomorphic to machine M. If one machine can be obtained from another machine by relabeling the states, then the machines are isomorphic to each other; that is, two machines are isomorphic if they differ only in the labels (or names) that are used to denote the elements of the state alphabet sets and have the same δ and λ functions, which also differ only in the labels that are used to denote these functions.

The isomorphic mappings g_1, g_2, and g_3 are combinational logic circuits, which may be considered as 1-state sequential machines. If machine M' is to be isomorphic to machine M, then M' is modified to simulate M as shown in Figure 2.25. This is accomplished by placing combinational logic that corresponds to g_1 at the input to machine M' that maps X into X', and by placing combinational logic that corresponds to g_3^{-1} (the inverse of g_3) that maps Z' into Z.

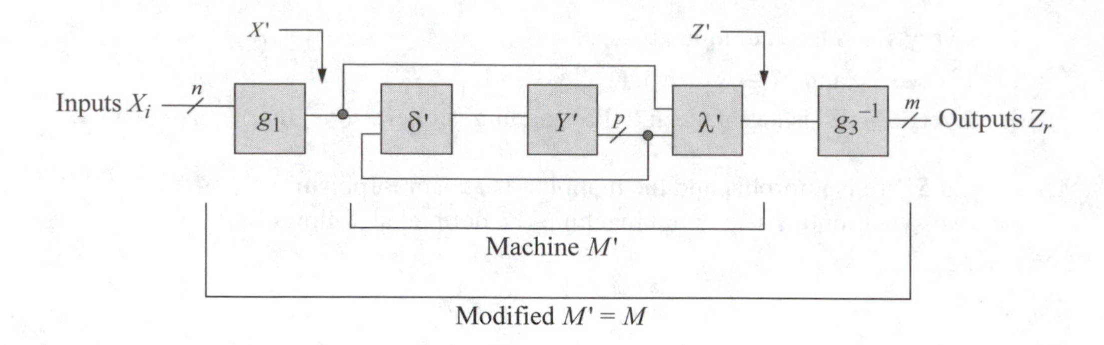

Figure 2.25 Machine M' is modified to simulate machine M.

Example 2.9 Let two machines M and M' be defined as shown in Figure 2.26, where M is a modulo-4 adder and M' is a modulo-5 multiplier with the zero element omitted. The zero element is not used, because the mapping would not be one-to-one. To be an isomorphism, the mapping must be both one-to-one and onto; that is, $s \rightarrow s'$. The example shown in Figure 2.26 (a) is not one-to-one, because Y_2 has an image of Y_1', Y_3', and Y_4'; therefore, the machine of Figure 2.26 (a) is not an isomorphism. Referring to Figure 2.26 (a), the following mappings should apply to all transformations for the mapping to be an isomorphism:

$$\begin{aligned} Y_0 &\rightarrow Y_1' \\ Y_1 &\rightarrow Y_2' \\ Y_2 &\rightarrow Y_3' \\ Y_3 &\rightarrow Y_4' \end{aligned}$$

as obtained from the first row. That is, Y_0 must always map to Y_1', Y_1 must always map to Y_2', etc. However, this does not always occur in Figure 2.26 (a). If the elements are rearranged, however, as shown in Figure 2.24 (b), then the two machines are isomorphic.

Equivalent machines Machines are equivalent if their functional operation is identical; that is, if the same input sequence generates the same output sequence. Let two synchronous sequential machines be defined as follows:

$$M = (X, Y, Z, \delta, \lambda)$$

$$M' = (X', Y', Z', \delta', \lambda')$$

Machine M (modulo-4 adder)

	$+_4$	X_0	X_1	X_2	X_3
	X_0	Y_0	Y_1	Y_2	Y_3
X	X_1	Y_1	Y_2	Y_3	Y_0
	X_2	Y_2	Y_3	Y_0	Y_1
	X_3	Y_3	Y_0	Y_1	Y_2

Machine M' (modulo-5 multiplier)

	$\times_5$	X_1'	X_2'	X_3'	X_4'
	X_1'	Y_1'	Y_2'	Y_3'	Y_4'
X'	X_2'	Y_2'	Y_4'	Y_1'	Y_3'
	X_3'	Y_3'	Y_1'	Y_4'	Y_2'
	X_4'	Y_4'	Y_3'	Y_2'	Y_1'

(a)

	$+_4$	X_0	X_1	X_2	X_3
	X_0	Y_0	Y_1	Y_2	Y_3
X	X_1	Y_1	Y_2	Y_3	Y_0
	X_2	Y_2	Y_3	Y_0	Y_1
	X_3	Y_3	Y_0	Y_1	Y_2

	$\times_5$	X_1'	X_3'	X_4'	X_2'
	X_1'	Y_1'	Y_3'	Y_4'	Y_2'
X'	X_2'	Y_3'	Y_4'	Y_2'	Y_1'
	X_3'	Y_4'	Y_2'	Y_1'	Y_3'
	X_4'	Y_2'	Y_1'	Y_3'	Y_4'

(b)

M

				4	5	6
0	1	2	3	0	1	2

M'

					5	6	7	8	9	10	11	12	13	14	15	16
0	1	2	3	4	0	1	2	3	4	0	1	2	3	4	0	1

(c)

Figure 2.26 Machine isomorphism between machine M and machine M': (a) not an isomorphism, because the mapping is not one-to-one; (b) isomorphism, because the mapping is one-to-one and onto; and (c) alignment of the elements in M and M' for their respective modulus.

The two machines, M and M', are equivalent ($M \equiv M'$) if and only if every state in M has a corresponding equivalent state in M', and every state in M' has a corresponding equivalent state in M. That is, for each state $Y_j \in Y$ there is a state $Y_j' \in Y'$ such that, $Y_j \equiv Y_j'$, and conversely, for each state $Y_j' \in Y'$ there is a state $Y_j \in Y$ such that, $Y_j' \equiv Y_j$. Two states Y_i and Y_j of a machine M are equivalent if, for every input sequence, the output sequence of M is the same whether the machine starts in state Y_i or in state Y_j.

The concept of equivalence applies to one, two, or more machines. Let three synchronous sequential machines be defined as follows:

$$M = (X, Y, Z, \delta, \lambda)$$

$$M' = (X', Y', Z', \delta', \lambda')$$

$$M'' = (X'', Y'', Z'', \delta'', \lambda'')$$

Thus, equivalence between machines has the following three properties:

1. $M \equiv M$ for each machine M.
2. If $M \equiv M'$, then $M' \equiv M$.
3. If $M \equiv M'$, and $M' \equiv M''$, then $M \equiv M''$.

Summary of additional definitions Table 2.13 lists the essential equations and characteristics for synchronous sequential machine terminology that were defined in this section.

Table 2.13 Summary of additional definitions for synchronous sequential machines

Machine	$M = (X, Y, Z, \delta, \lambda)$	(2.5)
State machine	$M = (X, Y, \delta)$	(2.69)
Submachine	$M = (X, Y, Z, \delta, \lambda)$	(2.70)
	$M' = (X', Y', Z', \delta', \lambda')$	(2.71)
	M' is a submachine of M if and only if $X' \subseteq X$ $Y' \subseteq Y$ $Z' \subseteq Z$ $\delta'(X', Y') : X' \times Y' \rightarrow Y'$ $\lambda'(X', Y') : X' \times Y' \rightarrow Z'$	(2.72)
Terminal state	$\delta(X_i, Y_j) : X_i \times Y_j \nrightarrow Y_k$ Y_j is a terminal state	(2.73)
Strongly connected machine	$\delta(X_i, Y_j) = Y_k$	(2.74)
	$\delta'(X_i', Y_j') = Y_k'$	(2.76)

Continued on next page

Table 2.13 Summary of additional definitions for synchronous sequential machines

Homomorphism	$h(s_1 \circ s_2) = h(s_1) * h(s_2)$	(2.77)
Machine homomorphism	$M = (X, Y, Z, \delta, \lambda)$ $M' = (X', Y', Z', \delta', \lambda')$ M' is a homomorphic image of machine M if and only if the following three onto mappings (functions) exist: $h_1 : X \rightarrow X'$ (Input alphabets of M and M') $h_2 : Y \rightarrow Y'$ (State alphabets of M and M') $h_3 : Z \rightarrow Z'$ (Output alphabets of M and M') such that, $h_2[\delta(X_i, Y_j)] = \delta'[h_1(X_i), h_2(Y_j)]$ $h_3[\lambda(X_i, Y_j)] = \lambda'[h_1(X_i), h_2(Y_j)]$ The homomorphism h may be an onto, a one-to-one into, or a one-to-one onto mapping.	 (2.78) (2.79) (2.80)
Isomorphism	$g(s_1 \circ s_2) = g(s_1) * g(s_2)$ The isomorphism is a one-to-one onto mapping.	(2.81)
Machine isomorphism	$M = (X, Y, Z, \delta, \lambda)$ $M' = (X', Y', Z', \delta', \lambda')$ Machine M' is isomorphic to machine M if and only if following three one-to-one onto mappings exist: $g_1 : X \rightarrow X'$ $g_2 : Y \rightarrow Y'$ $g_3 : Z \rightarrow Z'$ such that, $g_2[\delta(X_i, Y_j)] = \delta'[g_1(X_i), g_2(Y_j)]$ $g_3[\lambda(X_i, Y_j)] = \lambda'[g_1(X_i), g_2(Y_j)]$	 (2.82) (2.83) (2.84)
Equivalent machines	$M = (X, Y, Z, \delta, \lambda)$ $M' = (X', Y', Z', \delta', \lambda')$ $M \equiv M'$ if and only if each state $Y_j \in Y$ has an equivalent state $Y_j' \in Y'$ and each state $Y_j' \in Y'$ has an equivalent state $Y_j \in Y$.	

2.3 Methods of Analysis

Analysis is the methodical investigation of a problem and the decomposition of the problem into smaller related units for further detailed study. The problem, in this case, is a synchronous sequential machine which will be studied using various analytical techniques. These techniques include a next-state table, a present-state map, next-state maps, input maps and their associated input equations, output maps and equations, a timing diagram, and a state diagram.

Understanding a synchronous sequential machine through analysis is ideal preparation for later synthesizing (or designing) sequential machines. Different machines will be presented in this section and, as each machine is analyzed, the various units of the machine will be identified with the corresponding equations that were established in Section 2.2. First, the techniques, or methods, that are used in the analysis procedure will be defined.

2.3.1 Next-State Table

The next-state table was introduced in Section 2.2.4 in analyzing the operation of the Moore machine of Figure 2.14. The next-state table is a convenient method of describing the operation of a machine in tabular form. The table lists all possible present states and input values, together with the next state and present output.

Table 2.14 shows a typical next-state table for a machine with two *D* flip-flops. All combinations of two variables are listed under the present-state heading. The machine is assumed to be reset initially, which is represented by the first two rows, where $y_1y_2 = 00$. In this example, each pair of rows in the table corresponds to a state in the machine. For example, the first and second rows represent state $y_1y_2 = 00$. The state can also be expressed as a state name, such as "*a*."

Table 2.14 Typical next-state table for a Moore synchronous sequential machine using *D* flip-flops

State name	Present state y_1y_2	Input x_1	Flip-flop inputs Dy_1 Dy_2	Next state y_1y_2	Output z_1
a	0 0	0	0 0	0 0	0
	0 0	1	0 1	0 1	0
b	0 1	0	0 1	0 1	0
	0 1	1	1 0	1 0	0
c	1 0	0	1 0	1 0	1
	1 0	1	1 1	1 1	1
d	1 1	0	1 1	1 1	0
	1 1	1	0 0	0 0	0

The entries in Table 2.14 denote the state transitions and output that correspond to a given sequence of inputs. In the first row, the present state is *a* ($y_1y_2 = 00$). If $x_1 = 0$, the machine remains in state *a* and the present output $z_1 = 0$. If, however, $x_1 = 1$ in state *a*, then the machine moves to state *b* ($y_1y_2 = 01$) at the next assertion of the clock and state *b* becomes the new present state. No indication is given in the next-state table as to the active assertion of the clock; the machine may be clocked on either the positive or negative clock transition.

Since *D* flip-flops are used in the synchronous sequential machine of Table 2.14, the next state is identical to the values of Dy_1 and Dy_2 after the active clock transition. Output z_1 is active when the present state is $y_1y_2 = 10$. The output is a function of the state of the machine only; thus, the next-state table represents a Moore machine.

2.3.2 Present-State Map

The present-state map is also useful in analyzing synchronous sequential machines. Although not used as frequently as next-state maps, the present-state map adds completeness to the analysis procedure. For the 2-flip-flop machine of Table 2.14, there are four possible states as shown in Figure 2.27. The state names are chosen arbitrarily and are a function of the flip-flops y_1y_2. State *a* is established as $y_1y_2 = 00$; state *b* as $y_1y_2 = 01$, etc. Output z_1 is active in state *c* as shown in the upper right corner of minterm location $y_1y_2 = 10$.

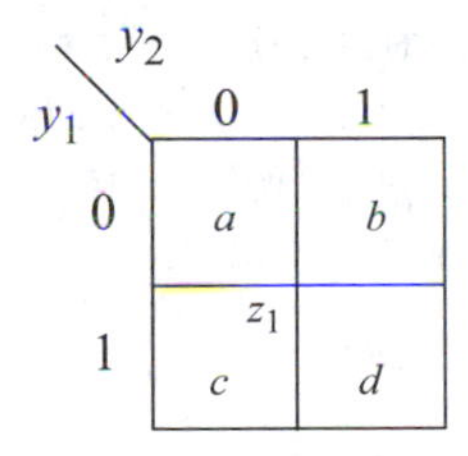

Figure 2.27 Present-state map for the machine of Table 2.14.

2.3.3 Next-State Map

The next-state map is simply the next-state table represented in Karnaugh map form as shown in Figure 2.28 for the machine of Table 2.14. The information that is specified in the next-state map can also be obtained from the logic diagram, if one is given. Since there are two flip-flops in the implementation of this machine, there are two next-state maps — one for each flip-flop. The map contains eight squares to accommodate the two flip-flops and input x_1.

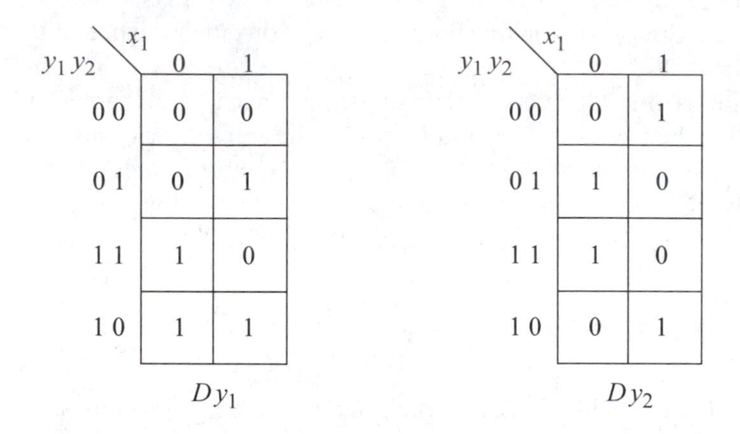

Figure 2.28 Next-state maps for the machine of Table 2.14. These are the same as input maps when *D* flip-flops are used.

For a *D* flip-flop, the next state corresponds to the present value of the *D* input. Referring to Table 2.14 and Figure 2.28, if the present state is $y_1y_2 = 00$ and $x_1 = 0$, then $Dy_1 = 0$ and the next state for y_1 will be 0. For a present state of $y_1y_2 = 00$ and $x_1 = 1$, then $Dy_1 = 0$ and the next state for $y_1 = 0$. Similarly, in state *c*, where $y_1y_2 = 10$, $Dy_1 = 1$ regardless of the value of x_1; therefore, the next state for y_1 is 1. Using this procedure, the information contained in the next-state table is transferred to the next-state maps.

2.3.4 Input Map

The input map represents the δ next-state function from which equations are generated for the data input logic of the flip-flop. Because *D* flip-flops are used in the implementation of the machine shown in Table 2.14, the next-state maps also specify the input maps; thus, the two types of maps are identical. In Figure 2.28, the input maps for y_1 and y_2 yield the following input equations:

$$Dy_1 = x_1'y_1 + y_1y_2' + x_1y_1'y_2$$

$$Dy_2 = x_1'y_2 + x_1y_2' \qquad (2.85)$$

The δ next-state logic can be implemented directly from the equations of Equation 2.85.

Using the same next-state table as Table 2.14, but replacing the D flip-flops with JK flip-flops yields the next-state table of Table 2.15. The next-state maps are identical to those shown in Figure 2.28. However, the input maps change due to the characteristics of a JK flip-flop, which are shown in Figure 2.29.

Table 2.15 Typical next-state table for a Moore machine using *JK* flip-flops

State name	Present state $y_1 y_2$	Input x_1	Flip-flop inputs Jy_1	Ky_1	Jy_2	Ky_2	Next state $y_1 y_2$	Output z_1
a	0 0	0	0	– (1)	0	–	0 0	0
	0 0	1	0	–	1	–	0 1	0
b	0 1	0	0	–	–	0	0 1	0
	0 1	1	1	–	–	1	1 0	0
c	1 0	0	–	0	0	–	1 0	1
	1 0	1	–	0	1	–	1 1	1
d	1 1	0	–	0	–	0	1 1	0
	1 1	1	–	1	–	1	0 0	0

(1) A dash (–) specifies a "don't care" condition.

State transition From	To	Values of J	K
0 →	0	0	–
0 →	1	1	–
1 →	0	–	1
1 →	1	–	0

Figure 2.29 State transition table for a JK flip-flop. A dash (–) specifies a "don't care" condition.

The input maps for the machine of Table 2.15 are shown in Figure 2.30. There are two maps for each flip-flop, one for the J input and one for the K input. The maps are constructed directly from Table 2.15. Figure 2.30 also specifies the JK input equations, which define the logic for the δ next-state function.

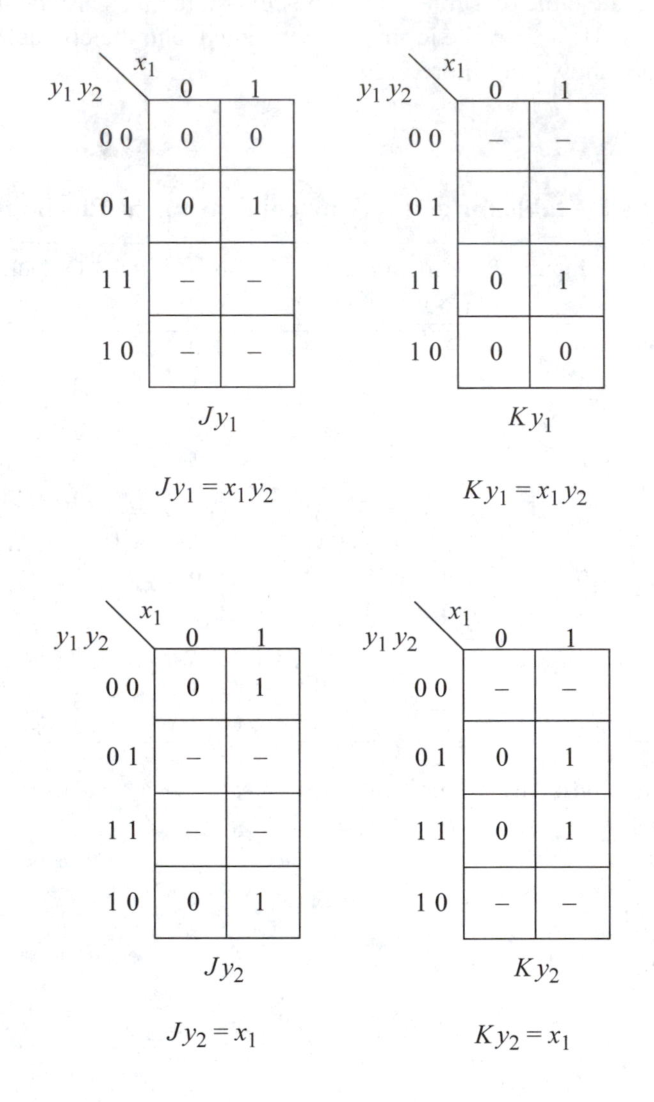

Figure 2.30 Input maps for the machine of Table 2.15.

2.3.5 Output Map

The output map represents the λ output function from which equations are generated for the output logic of the machine. The output map for the machines of Table 2.14 and Table 2.15 is shown in Figure 2.31. Output z_1 is asserted in state $y_1y_2 = 10$ regardless of the value of x_1, which corresponds to the definition of a Moore machine.

Figure 2.31 Output map for the machine of Table 2.14 and Table 2.15.

2.3.6 Timing Diagram

Another useful tool for analyzing synchronous sequential machines is a timing diagram (or waveforms) which illustrates the voltage levels of the inputs, storage elements, and outputs as the machine progresses through a sequence of states. Using the Moore machine of Table 2.15 and assuming an initial reset state of $y_1 y_2 = 00$, the timing diagram for this machine is shown in Figure 2.32 for an arbitrary input sequence of $x_1 = 1101$. The *JK* flip-flops are clocked on the negative clock transition. To assure that the flip-flops do not become *metastable*, any changes to input x_1 will occur on the positive clock transition. This guarantees that the *JK* inputs will be stable before the negative clock transition, thus meeting the setup requirements for the flip-flop. Metastability, a condition of instability on the output of a flip-flop caused by a change to the data input at or near the active clock transition, is discussed in detail in Chapter 3.

The machine begins in state *a*, where $y_1 y_2 = 00$. Since Figure 2.30 indicates that both Jy_1 and Ky_1 require y_2 to be set (or asserted), therefore, $Jy_1 Ky_1 = 00$ in state *a*. Also, the values of Jy_2 and Ky_2 are the same as the value for x_1. Thus, $Jy_2 Ky_2 = 11$ at the positive clock transition in state *a*, as illustrated in the timing diagram of Figure 2.32.

At the negative clock transition at the end of state *a*, both flip-flops are clocked and the following events occur:

$y_1 = 0$, because $Jy_1 Ky_1 = 00$ (no change)
$y_2 = 1$, because $Jy_2 Ky_2 = 11$ (toggle)

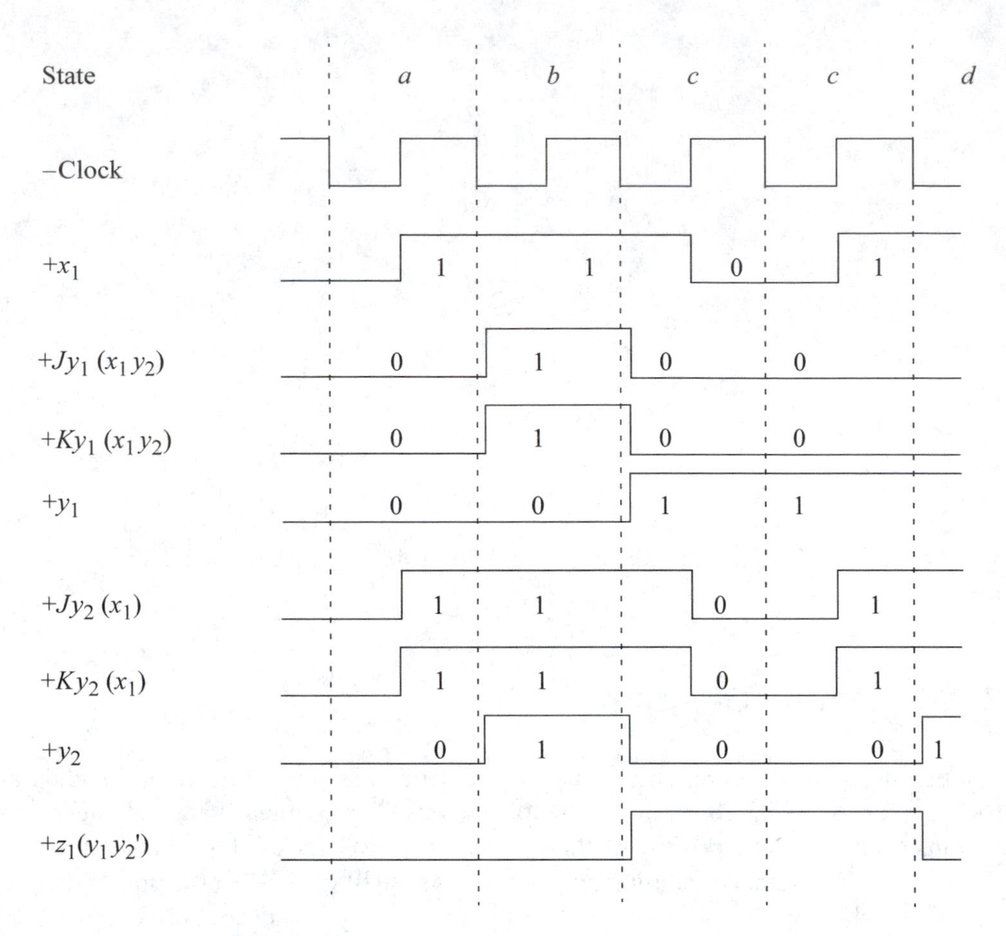

Figure 2.32 Timing diagram for the Moore machine of Table 2.15 for an arbitrary input sequence of $\boldsymbol{x}_1 = 1101$.

The machine then enters state b where $y_1y_2 = 01$. The delay between the negative clock transition and the time when y_2 is asserted in Figure 2.32 is due to the propagation delay of the flip-flop.

In state b, x_1 remains asserted and the following events occur at the negative clock transition at the end of state b:

$$y_1 = 1, \text{ because } Jy_1Ky_1 = 11 \text{ (toggle)}$$
$$y_2 = 0, \text{ because } Jy_2Ky_2 = 11 \text{ (toggle)}$$

The machine then enters state c where $y_1y_2 = 10$. Output z_1 is asserted in state c regardless of the value of x_1. As before, the flip-flops are clocked on the negative clock transition and the flip-flop outputs become stable after the appropriate propagation delay of the devices. Input x_1 becomes deasserted at the positive clock transition of state c; flip-flop y_2 is already deasserted. Therefore, when the negative clock transition occurs at the end of state c, the following conditions exist:

$$Jy_1 = Ky_1 = x_1 y_2 = 0$$
$$Jy_2 = Ky_2 = x_1 = 0$$

Therefore, the negative clock transition at the end of state c causes the following events to occur:

$$y_1 = 1, \text{ because } Jy_1 Ky_1 = 00 \text{ (no change)}$$
$$y_2 = 0, \text{ because } Jy_2 Ky_2 = 00 \text{ (no change)}$$

Thus, the machine remains in state c where $y_1 y_2 = 10$. At the next negative clock transition, $x_1 = 1$ and the following events occur:

$$y_1 = 1, \text{ because } Jy_1 Ky_1 = 00 \text{ (no change)}$$
$$y_2 = 1, \text{ because } Jy_2 Ky_2 = 11 \text{ (toggle)}$$

The machine enters state d where $y_1 y_2 = 11$. From Table 2.15 it can be seen that the next state will be either state d if $x_1 = 0$ or state a if $x_1 = 1$.

When analyzing a synchronous sequential machine, a timing diagram shows more detail than any other analytical method — the state times are precisely defined and the propagation delays are clearly illustrated. Since the clock is an astable multivibrator, the clock has no stable level. The clock signal in Figure 2.32 is specified as –Clock, where the minus sign indicates that the negative clock transition is used to clock the flip-flops which are negative-edge-triggered devices. All other signals are active high (+).

2.3.7 State Diagram

A *state diagram* is a directed graph which is used in conjunction with the state table. The state diagram portrays the same information as the state table, but presents a graphical representation in which the state transitions are more easily followed. The state diagrams that are used in this book are similar to flow chart diagrams in which the transition sequences and thus, the operational characteristics of the machine, are clearly delineated. Two symbols are used: a state symbol and an output symbol.

The *state symbol* is designated by a circle as shown in Figure 2.33. These nodes (or vertices) correspond to the state of the machine; the state name, such as state a, is placed inside the circle. The connecting directed lines between states correspond to the allowable state transitions. There are one or more entry paths and one or more exit paths as indicated by the arrows, unless the vertex is a *terminal state*, in which case there is no exit.

The flip-flop names are positioned alongside the state symbol. In Figure 2.33, the machine is designed using three flip-flops which are designated as $y_1 y_2 y_3$, where y_3 is the low-order flip-flop. Directly beneath the flip-flop names, the *state code* is specified. The state code represents the state of the individual flip-flops. In Figure 2.33, the state code is 101, which corresponds to $y_1y_2'y_3$. If an input causes a transition from state *a* to another state, this is indicated by placing the name of the input variable adjacent to the exit arrow as shown in Figure 2.33 for x_1 and x_1'.

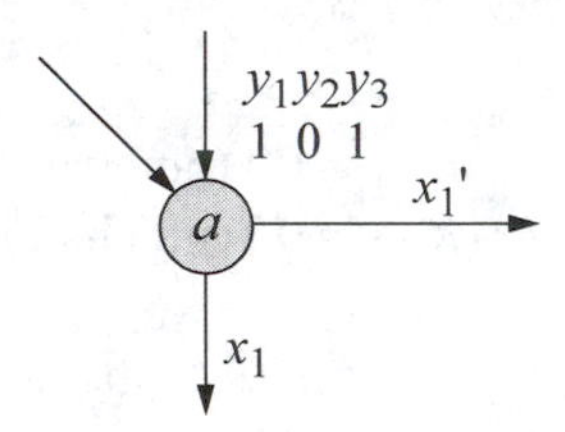

Figure 2.33 State diagram state symbol indicating state *a*.

The *output symbol* is represented by a rectangle and is placed immediately following the state symbol, as shown in Figure 2.34 (a) for a Moore machine, or placed immediately after an input variable that causes the output to become active, as shown in Figure 2.34 (b) for a Mealy machine. Figure 2.34 (a) specifies a Moore machine in which output z_1 is a function of the present state only; that is, state *b*, where $y_1y_2 = 01$. Figure 2.34 (b) indicates a Mealy machine in which output z_1 is a function of both the present state *b* ($y_1y_2 = 01$) and input x_1.

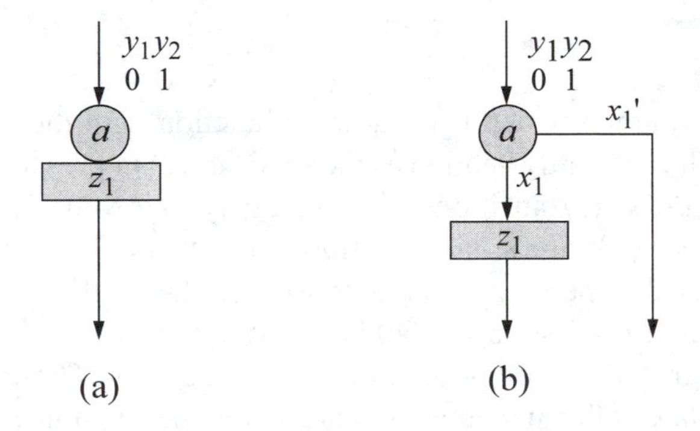

Figure 2.34 State diagram output symbol indicating output z_1: (a) Moore machine; and (b) Mealy machine.

For a Moore machine, the outputs can be asserted for segments of the clock period rather than for the entire clock period only. This is illustrated in Figure 2.35 where the positive clock transitions define the clock cycles, and hence, the state times. Two clock cycles are shown, one for the present state $Y_{j(t)}$ and one for the next state $Y_{k(t+1)}$.

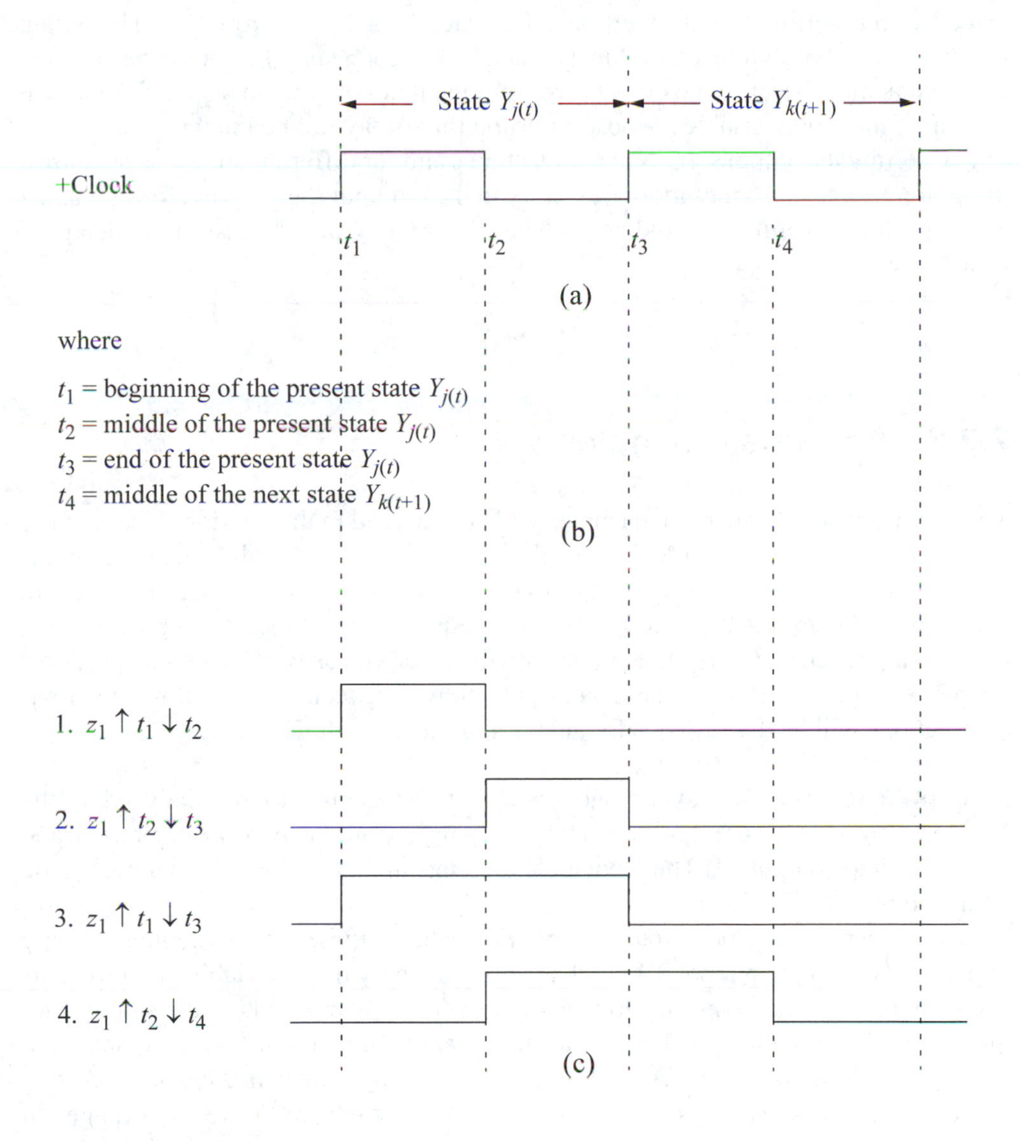

Figure 2.35 Output assertion/deassertion times for Moore machines: (a) clock pulses; (b) definition of assertion/deassertion times; and (c) assertion/deassertion statements with corresponding asserted outputs.

The leading edge of the clock pulse, which defines the beginning of the present state, is labeled t_1. The leading edge may be a positive or negative clock transition and is used for clocking positive- or negative-edge-triggered devices, respectively. All assertion/deassertion times are referenced to the present state $Y_{j(t)}$. Time t_2 occurs at the middle of the present state; time t_3 occurs at the end of the present state; and time t_4 occurs at the midpoint of the next state $Y_{k(t+1)}$. The assertion of an output is indicated by an up-arrow (↑); deassertion is indicated by a down-arrow (↓). The output assertion/deassertion times for a Mealy machine cannot be uniquely specified as for a Moore machine, because the outputs are contingent not only upon a specific state but also upon the input variables, whose assertion times may not be known.

Asserting the output signals at various times and for different durations, as shown in Figure 2.35 (c), provides more flexibility in the λ output logic. Waveforms 2 and 4 are especially useful in avoiding glitches. Glitches are discussed in detail in Chapter 3.

2.3.8 Analysis Examples

Three synchronous sequential machines will be analyzed in this section. The first is a Moore implementation with two *D* flip-flops; the second is a Mealy machine also implemented with two *D* flip-flops; the third uses *JK* flip-flops and contains both Moore and Mealy outputs. For all three synchronous sequential machines, the input alphabet, the state alphabet, and the output alphabet will be derived as well as the mappings for the δ next-state function and the λ output function. The techniques outlined in previous sections will be the vehicles for analysis of these machines.

Example 2.10 The first synchronous sequential machine to be analyzed is the Moore machine of Figure 2.36 consisting of a single input x_1, two *D* flip-flops y_1 and y_2, and a single output z_1. The input alphabet, state alphabet, and output alphabet are tabulated in Table 2.16.

Since there is only one input, the input alphabet is quite simple, containing only two input vectors (or symbols) X_0 and X_1 and one binary input variable x_1. Two storage elements y_1 and y_2 specify four states Y_0, Y_1, Y_2, and Y_3 where the storage elements assume four unique values. State names are assigned to the states as shown in Table 2.16 where state *a* corresponds to $y_1y_2 = 00$. The state names do not necessarily have to be in ascending sequence when tabulated; they are, however, arranged in an ascending systematic sequence when entered in the state diagram, as will be shown later in this example. The output alphabet consists of two output symbols Z_0 and Z_1 and one binary output variable z_1.

Equation 2.51 defines the next state of a Moore machine as

$$Y_{k(t+1)} = \delta(X_{i(t)}, Y_{j(t)})$$

The next-state functions for y_1 and y_2 are shown in Equation 2.86 as obtained from the logic diagram of Figure 2.36. Because *D* flip-flops are used, the next state of the machine is determined solely by the *D* inputs Dy_1 and Dy_2.

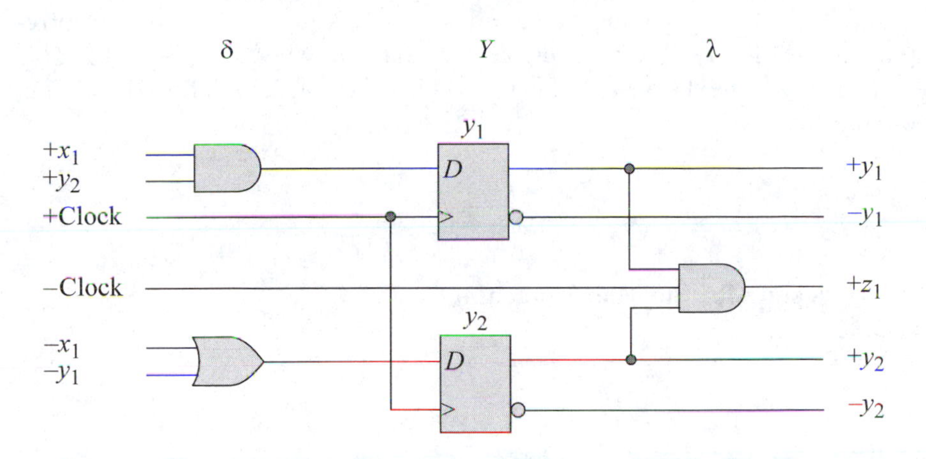

Figure 2.36 Moore machine of Example 2.10.

Table 2.16 Input alphabet, state alphabet, and output alphabet for the Moore machine of Figure 2.36

Input alphabet		State alphabet			Output alphabet	
Symbol	x_1	State name	Symbol	$y_1 y_2$	Symbol	z_1
X_0	0	*a*	Y_0	0 0	Z_0	0
X_1	1	*b*	Y_1	0 1	Z_1	1
		d	Y_2	1 0		
		c	Y_3	1 1		

$$Y_{k(t+1)} = \begin{cases} \delta y_1(x_1, y_2) = x_1 y_2 \\ \delta y_2(x_1, y_1) = x_1{}' + y_1{}' \end{cases} \tag{2.86}$$

The next states are listed in tabular form in Table 2.17, where Y_0, Y_1, Y_2, and Y_3 are the next states of the machine as a function of y_1 and y_2. The same information is shown in a slightly different format in Table 2.18, which lists all combinations of the three variables: input x_1 and storage elements y_1, y_2. By assigning the values listed in Table 2.18 to x_1, y_1, and y_2, the logic diagram of Figure 2.36 will generate the corresponding next states for Y_0, Y_1, Y_2, and Y_3. For example, when $x_1y_1y_2 = 101$, Dy_1 and $Dy_2 = 1$. Thus, the next state is Y_3 ($y_1y_2 = 11$). Equation 2.54 defined the output of a Moore machine as

$$Z_{r(t)} = \lambda(Y_{j(t)})$$

Table 2.17 Next state for the Moore machine of Figure 2.36

	$Y_{k(t+1)}$	
Next-state symbol	$y_1(x_1y_2)$	$y_2(x_1' + y_1')$
Y_0	0	0
Y_1	0	1
Y_2	1	0
Y_3	1	1

Table 2.18 Next-state for the Moore machine of Figure 2.36 with reference to x_1, y_1, and y_2

Next-state symbol	Input x_1	Storage element y_1y_2
Y_1	0	0 0
Y_1	0	0 1
Y_1	0	1 0
Y_1	0	1 1
Y_1	1	0 0
Y_3	1	0 1
Y_0	1	1 0
Y_2	1	1 1

The λ output function for the Moore machine of Figure 2.36 is shown in Equation 2.87.

$$Z_1 = \lambda(y_1y_2) = y_1y_2 = z_1 \tag{2.87}$$

A more formal presentation for the δ next-state function and the λ output function is expressed in Table 2.19. All combinations of mappings are illustrated for both δ and λ. To verify that $\delta(X_1, Y_3) = Y_2$ for example, where $X_1 = x_1 = 1$ and $Y_3 = y_1y_2 = 11$, apply the signals $x_1y_1y_2 = 111$ to the logic diagram of Figure 2.36. It is seen that $Dy_1 = x_1y_2 = 1$ and $Dy_2 = x_1' + y_1' = 0$. The next state, therefore, is $y_1y_2 = 10$ (Y_2). Output z_1 is asserted only when $\lambda(Y_3) = Z_1$; that is, $Y_3 = y_1y_2 = 11$.

Table 2.19 δ next-state function and λ output function for the Moore machine of Figure 2.36

δ next-state function	λ output function
$\delta(X_0, Y_0) = Y_1$	$\lambda(Y_0) = Z_0$
$\delta(X_0, Y_1) = Y_1$	$\lambda(Y_1) = Z_0$
$\delta(X_0, Y_2) = Y_1$	$\lambda(Y_2) = Z_0$
$\delta(X_0, Y_3) = Y_1$	$\lambda(Y_3) = Z_1$
$\delta(X_1, Y_0) = Y_1$	
$\delta(X_1, Y_1) = Y_3$	
$\delta(X_1, Y_2) = Y_0$	
$\delta(X_1, Y_3) = Y_2$	

A complete next-state table for the Moore machine of Figure 2.36 is shown in Table 2.20. Begin in the reset state where $y_1y_2 = 00$, then determine the next state when $x_1 = 0$ then 1, successively. Repeat this procedure for all remaining states. To illustrate the procedure, assume that the machine is in state d ($y_1y_2 = 10$). If $x_1 = 0$, then $Dy_1 = x_1y_2 = 00 = 0$ and $Dy_2 = x_1' + y_1' = 1 + 0 = 1$. Thus, the next state is $y_1y_2 = 01$ and the machine enters state b. However, if $x_1 = 1$, then $Dy_1 = x_1y_2 = 10 = 0$ and $Dy_2 = x_1' + y_1' = 0 + 0 = 0$ and the machine enters state a ($y_1y_2 = 00$).

The present-state map, next-state maps, input maps, and output map are presented in Figure 2.37. These maps are derived from either the next-state table or from the logic diagram. For example, in Table 2.20, states b and d specify that Dy_1 is asserted only when both x_1 and y_2 are asserted. Therefore, using x_1 as a map-entered variable, the input map for Dy_1 contains the entry x_1 in column y_2. In the same manner, the input map for y_2 and the output map for z_1 are derived.

The equation for Dy_2 can be derived by one of two methods:

1. Read directly from the map without changing the minterm entries. This yields $Dy_2 = y_1' + y_1x_1'$. Then using the absorption law, $Dy_2 = x_1' + y_1'$, or
2. Change the map entries as follows:

$y_1 \backslash y_2$	0	1
0	$x_1 + x_1'$	$x_1 + x_1'$
1	x_1'	x_1'

This does not change the minterm values. Since every square contains x_1', therefore, $Dy_2 = x_1'$. Now reassign $x_1 + x_1'$ as a value of 1, then combine the two 1s in row $y_1 = 0$ as $Dy_2 = y_1'$. Thus, $Dy_2 = x_1' + y_1'$.

Table 2.20 Next-state table for the Moore machine of Figure 2.36

State name	Present state $y_1 y_2$	Input x_1	Flip-flop inputs Dy_1 Dy_2	Next state $y_1 y_2$	Output z_1
a	0 0	0	0 1	0 1	0
	0 0	1	0 1	0 1	0
b	0 1	0	0 1	0 1	0
	0 1	1	1 1	1 1	0
d	1 0	0	0 1	0 1	0
	1 0	1	0 0	0 0	0
c	1 1	0	0 1	0 1	1
	1 1	1	1 0	1 0	1

The timing diagram is shown in Figure 2.38 using an arbitrary input sequence of $x_1 = 1011$. The machine is reset initially to state *a*. Using the timing diagram in conjunction with the logic diagram of Figure 2.36 or the next-state table of Table 2.20, the machine proceeds to state *b* at the next positive clock transition, regardless of the value of x_1. At the positive clock transition at the end of state *b*, the following conditions exist:

$$x_1 = 0 \qquad y_1 = 0 \qquad y_2 = 1$$

and the machine remains in state *b* ($y_1 y_2 = 01$), because

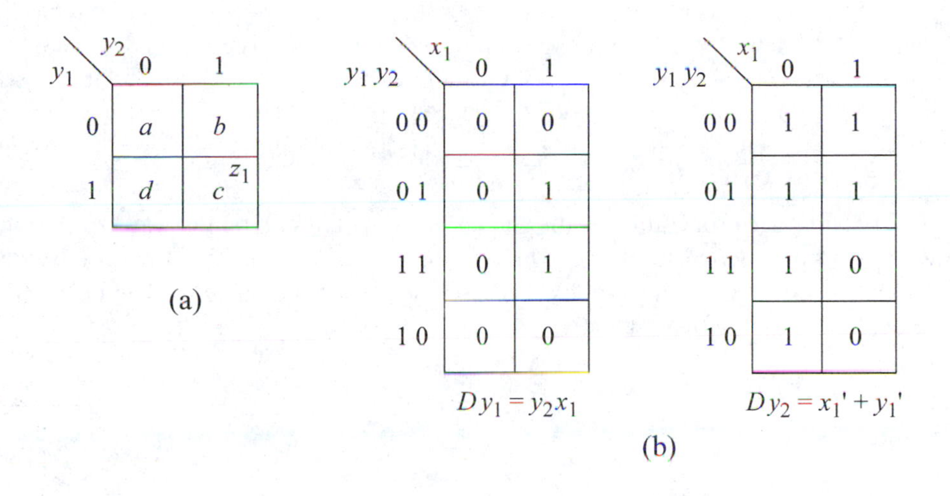

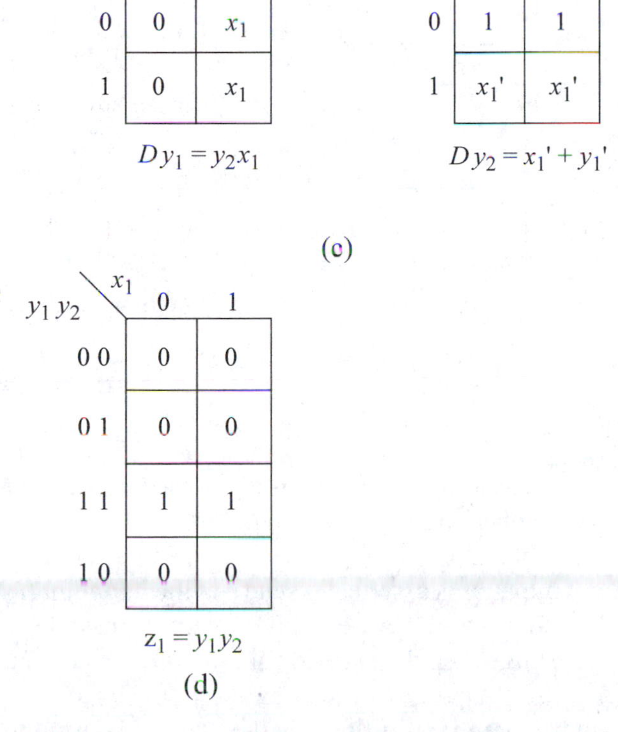

Figure 2.37 Karnaugh maps for the Moore machine of Figure 2.36: (a) present-state map; (b) input maps (or next-state maps for *D* flip-flops); (c) input maps using x_1 as a map-entered variable; and (d) output map.

$$Dy_1 = x_1 y_2 = 01 = 0$$

$$Dy_2 = x_1' + y_1' = 1 + 1 = 1$$

Using a similar procedure for the remaining states, it can be determined that the state transition sequence for $x_1 = 1011$ is *abbcd* · · · . Output z_1 is asserted when $y_1 y_2$ clock = 110; that is, $y_1 y_2$ clock' causes z_1 to be active during the last half of the clock cycle in which $y_1 y_2 = 11$.

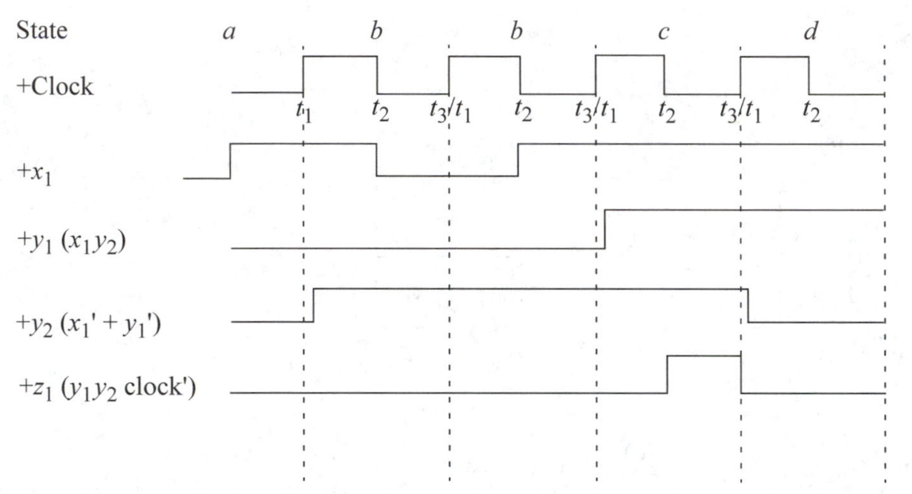

Figure 2.38 Timing diagram for the Moore machine of Figure 2.36 using an arbitrary input sequence of $\boldsymbol{x}_1 = 1011$.

The final analysis technique is the state diagram, which is a graphical representation of the functions δ and λ. The state diagram, illustrated in Figure 2.39, is derived directly from the next-state table of Table 2.20. Notice the assertion and deassertion statements adjacent to z_1 in state *c*. Output z_1 is asserted in state *c* at the midpoint (t_2) of the cycle and is deasserted at the end (t_3) of the cycle. This prevents a possible erroneous output (or glitch) from occurring when the state transition is from state *d* to state *b*.

Output glitches can occur when two or more flip-flops change state and the output assertion is at time t_1. When the active clock transition triggers the machine to initiate a state transition from state *d* to state *b*, both flip-flops change state. The machine then enters a period of instability until the machine stabilizes in state *b* ($y_1 y_2 = 01$). If y_2 is faster at setting than y_1 is at resetting, then the machine will momentarily enter state *c* ($y_1 y_2 = 11$). Because state *c* contains a Moore-type output, whenever the machine enters state c, output z_1 will be asserted. Thus, an erroneous output will be generated on z_1 as the machine passes through transient state c for a state transition sequence of $d \rightarrow b$. Glitches are covered in more detail in Chapter 3.

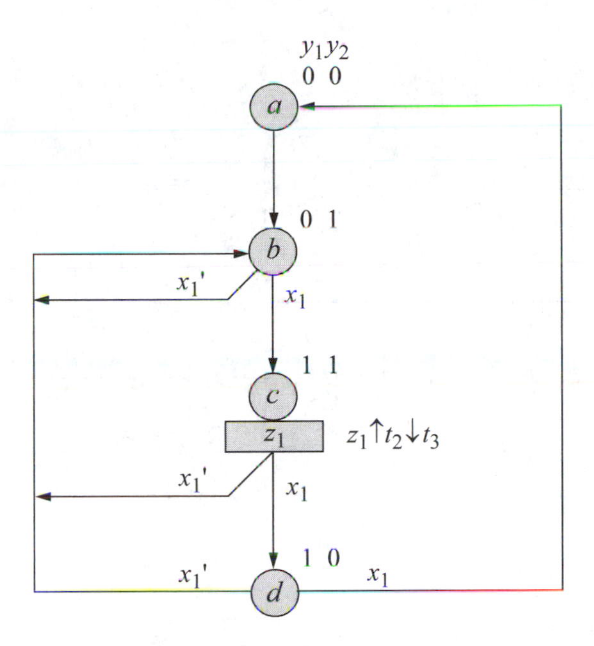

Figure 2.39 State diagram for the Moore machine of Figure 2.36.

The state diagram presents a much clearer state transition sequence than either the logic diagram or the timing diagram. The state diagram is hardware-independent; that is, it shows the complete operation of the machine at a glance for all possible input sequences, but does not indicate the type of storage elements or gates that are used in the implementation. The logic diagram shows the physical realization of the machine; however, the sequence of state transitions is obtained only after a tedious examination of inputs and present states. The timing diagram is an illustration of actual waveforms that would be observed on an oscilloscope for a given input sequence.

Example 2.11 This example analyzes the Mealy machine shown in Figure 2.40. There are two inputs x_1 and x_2, two D flip-flops y_1 and y_2, and one output z_1. Output z_1 is asserted when $y_1y_2x_2 = 101$. The input alphabet, state alphabet, and output alphabet are shown in Table 2.21.

The δ next-state functions for y_1 and y_2 are expressed in Equation 2.88 as obtained from the logic diagram of Figure 2.40. The next state of the machine is the same as the concatenated D inputs of y_1y_2; that is, the next state is Dy_1Dy_2.

The λ output function is presented in Equation 2.89 and is derived from the boolean equation for the output variable of Figure 2.40.

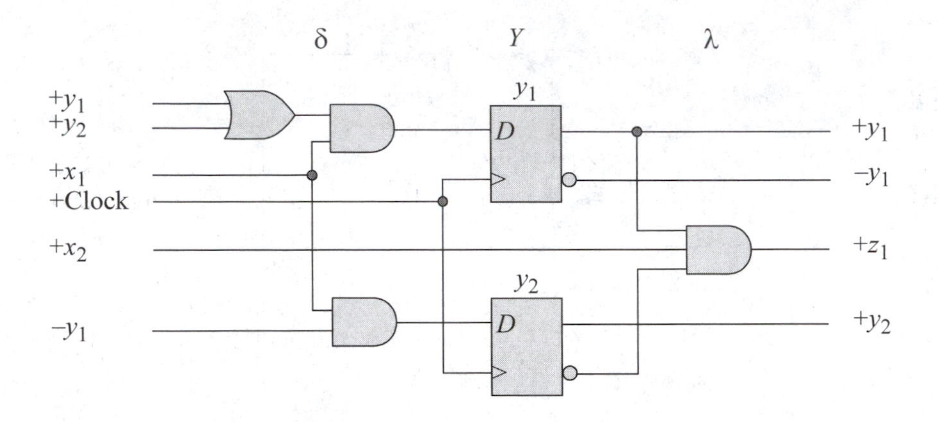

Figure 2.40 Mealy machine of Example 2.11.

Table 2.21 Input alphabet, state alphabet, and output alphabet for the Mealy machine of Figure 2.40

Input alphabet		State alphabet			Output alphabet	
Symbol	Input x_1x_2	State name	Symbol	Storage element y_1y_2	Symbol	Output z_1
X_0	0 0	*a*	Y_0	0 0	Z_0	0
X_1	0 1	*b*	Y_1	0 1	Z_1	1
X_2	1 0	*d*	Y_2	1 0		
X_3	1 1	*c*	Y_3	1 1		

$$Y_{k(t+1)} = \begin{cases} \delta y_1(x_1, y_1y_2) = x_1(y_1 + y_2) \\ \delta y_2(x_1, y_1) = x_1y_1' \end{cases} \tag{2.88}$$

$$Z_1 = \lambda(y_1y_2x_2) = y_1y_2'x_2 = z_1 \tag{2.89}$$

Table 2.22 lists the next states in tabular form and Table 2.23 depicts the same information with all combinations of the four variables x_1, x_2, y_1, and y_2. Many states have the same symbol, because the next state is not a function of x_2, although x_2 is listed in the table. Input x_2 is required only for the assertion of output z_1. By applying the values of x_1, y_1, and y_2 of Table 2.23 to the corresponding input signals of Figure 2.40, the state symbols Y_i of Table 2.23 are obtained. For example, when $x_1y_1y_2 = 100$, then $Dy_1Dy_2 = 01$, yielding state symbol Y_1. Also, when $x_1y_1y_2 = 101$, then $Dy_1Dy_2 = 11$, yielding state symbol Y_3. Thus, the state diagram can be derived directly from the information contained in Table 2.23.

Table 2.22 Next states for the Mealy machine of Figure 2.40

	$Y_{k(t+1)}$	
State	y_1	y_2
symbol	$x_1(y_1+y_2)$	x_1y_1'
Y_0	0	0
Y_1	0	1
Y_2	1	0
Y_3	1	1

Table 2.23 Next states for the Mealy machine of Figure 2.40 with reference to x_1, x_2, y_1, and y_2

Next-state symbol	Inputs x_1x_2	Storage elements y_1y_2
Y_0	0 0	0 0
Y_0	0 0	0 1
Y_0	0 0	1 0
Y_0	0 0	1 1
Y_0	0 1	0 0
Y_0	0 1	0 1
Y_0	0 1	1 0
Y_0	0 1	1 1

Continued on next page

Table 2.23 Next states for the Mealy machine of Figure 2.40 with reference to x_1, x_2, y_1, and y_2

Next-state symbol	Inputs x_1x_2	Storage elements y_1y_2
Y_1	1 0	0 0
Y_3	1 0	0 1
Y_2	1 0	1 0
Y_2	1 0	1 1
Y_1	1 1	0 0
Y_3	1 1	0 1
Y_2	1 1	1 0
Y_2	1 1	1 1

The next-state table for the machine of Figure 2.40 is shown in Table 2.24. The table is derived from the logic diagram of Figure 2.40 by assuming an initial reset condition ($y_1y_2 = 00$) and then applying all possible combinations of input signals to determine the flip-flop inputs and thus, the next state. For example, in state b ($y_1y_2 = 01$), if $x_1x_2 = 10$, then $Dy_1Dy_2 = 11$, because

$$Dy_1 = x_1(y_1 + y_2) = 1(0 + 1) = 1$$

$$Dy_2 = x_1y_1' = 11 = 1$$

and the next state is c ($y_1y_2 = 11$).

The input and output maps, state diagram, and timing diagram for the Mealy machine of Figure 2.40 will be left as exercises. See problems 2.33, 2.34, and 2.35, respectively.

Example 2.12 The third synchronous sequential machine to be analyzed is the Mealy machine of Figure 2.41. Output z_2 is a Moore-type output, since the assertion of z_2 is independent of the inputs. Output z_1 is a Mealy-type output, since the assertion of z_1 is a function of both the present state and the present input. That is, z_1 is asserted when the machine is in state $y_1y_2 = 11$ and x_1 is active. The entire machine, therefore, will be analyzed as a Mealy machine. The alphabets are listed in Table 2.25. Note that no two z_i are active concurrently. Thus, the output alphabet is reduced, consisting of only three symbols rather than $2^2 = 4$ symbols.

Table 2.24 Next-state table for the Mealy machine of Figure 2.40

State name	Present state $y_1 y_2$	Inputs $x_1 x_2$	Flip-flop inputs Dy_1	Dy_2	Next state $y_1 y_2$	Output z_1
a	0 0	0 0	0	0	0 0	0
		0 1	0	0	0 0	0
		1 0	0	1	0 1	0
		1 1	0	1	0 1	0
b	0 1	0 0	0	0	0 0	0
		0 1	0	0	0 0	0
		1 0	1	1	1 1	0
		1 1	1	1	1 1	0
d	1 0	0 0	0	0	0 0	0
		0 1	0	0	0 0	1
		1 0	1	0	1 0	0
		1 1	1	0	1 0	1
c	1 1	0 0	0	0	0 0	0
		0 1	0	0	0 0	0
		1 0	1	0	1 0	0
		1 1	1	0	1 0	0

Since the machine is implemented with *JK* flip-flops, the next state is not identical to the data inputs as in the machines of Example 2.10 and Example 2.14, in which *D* flip-flops were used as the storage elements. For a *JK* flip-flop, the next state may also depend upon the present state. Therefore, one additional step is required: Use the excitation characteristics of the *JK* flip-flop to determine the next state of the machine.

The next-state excitation equation for a *JK* flip-flop is defined in Equation 2.90.

$$Y_{k(t+1)} = JY_{j(t)}' + K'Y_{j(t)} \tag{2.90}$$

The *JK* equations for y_1 and y_2 are obtained from the *JK* input logic of Figure 2.41 and are listed below in Equation 2.91. The δ next-state functions for y_1 and y_2 — which are derived from the *JK* input equations by substituting Equation 2.91 into Equation 2.90 — are shown in Equation 2.92.

$$\begin{aligned} Jy_1 &= x_1 y_2 & Ky_1 &= x_1 y_2 \\ Jy_2 &= x_1 \oplus x_2 & Ky_2 &= 1 \end{aligned} \tag{2.91}$$

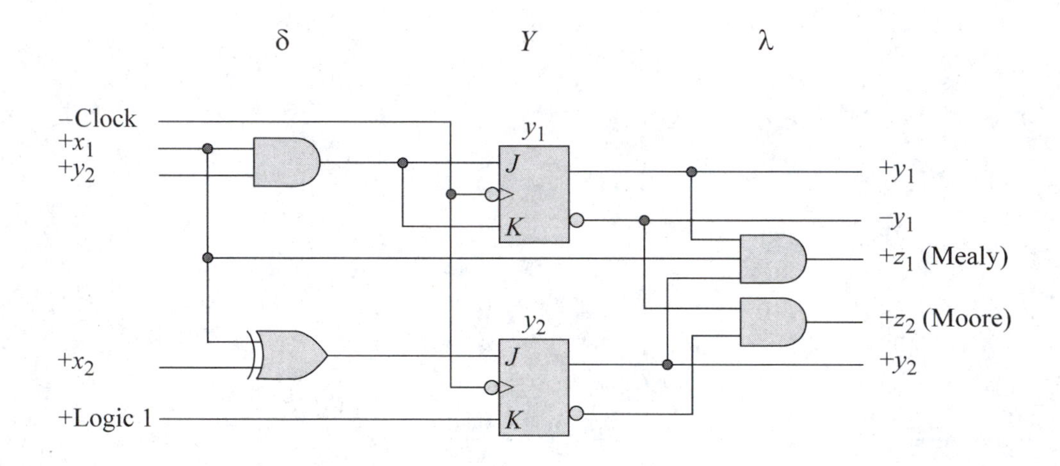

Figure 2.41 Mealy machine of Example 2.12.

Table 2.25 Input alphabet, state alphabet, and output alphabet for the Mealy machine of Figure 2.41

Input alphabet		State alphabet			Output alphabet	
Symbol	Input x_1x_2	State name	Symbol	Storage element y_1y_2	Symbol	Output z_1
X_0	0 0	*a*	Y_0	0 0	Z_1	0 1
X_1	0 1	*b*	Y_1	0 1	Z_0	0 0
X_2	1 0	*c*	Y_2	1 0	Z_0	0 0
X_3	1 1	*d*	Y_3	1 1	Z_2	1 0

$$\begin{aligned}\delta y_1 (X, Y) &= [(x_1 y_2) y_1' + (x_1 y_1)' y_1] \\ &= [x_1 y_2 y_1' + (x_1' + y_1') y_1] \\ &= (x_1 y_1' y_2 + x_1' y_1)\end{aligned}$$

$$\begin{aligned}\delta y_2 (X, Y) &= [(x_1 \oplus x_2) y_2' + (1)' y_2] \\ &= (x_1 x_2' y_2' + x_1' x_2 y_2') \qquad (2.92)\end{aligned}$$

The next-state table is illustrated in Table 2.26 and is derived from either the equations of Equation 2.92 or from the logic diagram of Figure 2.41. Using the logic diagram, assume an initial reset condition of $y_1 y_2 = 00$, then apply all possible combinations of the input signals to obtain the next state. For example, in state a ($y_1 y_2 = 00$), whenever $y_2 = 0$, Jy_1 and Ky_1 will both be 0; that is, $Jy_1 Ky_1 = x_1 y_2 = 0 = 00$. Thus, the next state for y_1 will be unchanged at a value of 0 for the four rows that represent state a.

Since $Jy_2 = x_1 \oplus x_2$, then $Jy_2 = 1$ when x_1 and x_2 are set to different values; Ky_2 is always a logic 1. Therefore, the first four rows of Table 2.26 show the next state for y_2 as established by the following criteria:

$y_2 = 0$, because $x_1 x_2 = 00$ which yields $Jy_2 Ky_2 = 01$ (Reset)
$y_2 = 1$, because $x_1 x_2 = 01$ which yields $Jy_2 Ky_2 = 11$ (Toggle)
$y_2 = 1$, because $x_1 x_2 = 10$ which yields $Jy_2 Ky_2 = 11$ (Toggle)
$y_2 = 0$, because $x_1 x_2 = 11$ which yields $Jy_2 Ky_2 = 01$ (Reset)

The remaining rows in the table are completed in the same manner.

Table 2.26 Next-state table for the Mealy machine of Figure 2.41

State name	Present state $y_1 y_2$	Inputs $x_1 x_2$	Flip-flop inputs Jy_1 Ky_1	Jy_2 Ky_2	Next state $y_1 y_2$	Outputs $z_1 z_2$
a	0 0	0 0	0 0	0 1	0 0	0 1
		0 1	0 0	1 1	0 1	0 1
		1 0	0 0	1 1	0 1	0 1
		1 1	0 0	0 1	0 0	0 1

Continued on next page

Table 2.26 Next-state table for the Mealy machine of Figure 2.41

State name	Present state y_1y_2	Inputs x_1x_2	Flip-flop inputs Jy_1	Ky_1	Jy_2	Ky_2	Next state y_1y_2	Outputs z_1z_2
b	0 1	0 0	0	0	0	1	0 0	0 0
		0 1	0	0	1	1	0 0	0 0
		1 0	1	1	1	1	1 0	0 0
		1 1	1	1	0	1	1 0	0 0
c	1 0	0 0	0	0	0	1	1 0	0 0
		0 1	0	0	1	1	1 1	0 0
		1 0	0	0	1	1	1 1	0 0
		1 1	0	0	0	1	1 0	0 0
d	1 1	0 0	0	0	0	1	1 0	0 0
		0 1	0	0	1	1	1 0	0 0
		1 0	1	1	1	1	0 0	1 0
		1 1	1	1	0	1	0 0	1 0

The input maps for y_1 are shown in Figure 2.42 and are derived from the logic diagram or the next-state table. Using state b ($y_1y_2 = 01$) as an example, the values for Jy_1 and Ky_1 are the same as the values for x_1. Therefore, with x_1 as a map-entered variable, minterm location $y_1y_2 = 01$ in the Karnaugh maps for Jy_1 and Ky_1 contains the variable x_1. The input maps for y_2 and the output maps for z_1 and z_2, the state diagram, and the timing diagram are left as exercises. See problems 2.36, 2.37, and 2.38, respectively.

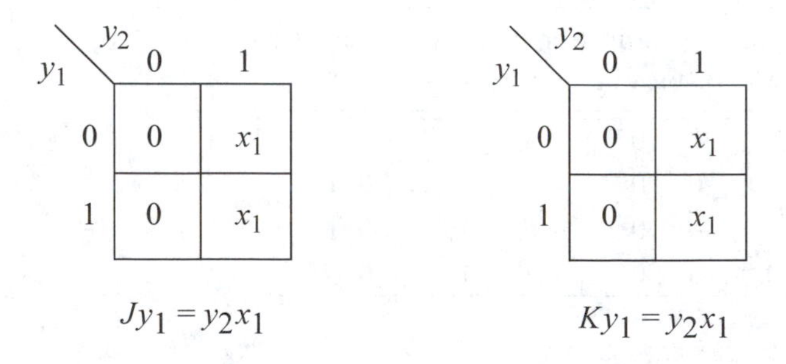

Figure 2.42 Input maps for y_1 for the Mealy machine of Figure 2.41.

2.4 Complete and Incomplete Synchronous Sequential Machines

This section discusses complete synchronous sequential machines where each present input and each present state generates a next state; that is, the state transitions are completely defined for all possible inputs and for all possible present states and produce specific outputs. Incomplete synchronous sequential machines are also presented, in which some combination of input variables and present states are not specified. Incompleteness also exists where the state transitions are completely defined, but certain combinations of states and inputs produce output values that are not critical. These output signals can be classified as "don't care" outputs and can be left unspecified.

2.4.1 Complete Synchronous Sequential Machines

A complete synchronous sequential machine was defined in Section 2.1.3 as the following 5-tuple:

$$M = (X, Y, Z, \delta, \lambda)$$

where

1. X is a nonempty finite set of inputs.
2. Y is a nonempty finite set of states.
3. Z is a nonempty finite set of outputs.
4. δ is the next-state function which maps the Cartesian product of $X \times Y$ into Y.
5. λ is the output function which maps the Cartesian product of $X \times Y$ into Z.

A complete synchronous sequential machine specifies that, for each input $X_{i(t)}$ and each present state $Y_{j(t)}$, the next state $Y_{k(t+1)}$ and the present output $Z_{r(t)}$ both exist, where

$$Y_{k(t+1)} = \delta(X_{i(t)}, Y_{j(t)})$$

$$Z_{r(t)} = \lambda(Y_{j(t)}) \qquad \text{(Moore machine)}$$

$$Z_{r(t)} = \lambda(X_{i(t)}, Y_{j(t)}) \qquad \text{(Mealy machine)}$$

Thus, the machine is completely specified.

2.4.2 Incomplete Synchronous Sequential Machines

Where a complete synchronous sequential machine is completely defined with reference to next states and outputs, an incomplete synchronous sequential machine may have some states Y_j and some inputs X_i for which $\delta(X_{i(t)}, Y_{j(t)})$, $\lambda(Y_{j(t)})$, or $\lambda(X_{i(t)}, Y_{j(t)})$ are not defined. Thus, some combinations of states and inputs are not possible, or certain input sequences will never occur. Also, some output symbols may never be asserted such that, the output alphabet of the machine is a subset of Z. This was evident in Example 2.12 for the Mealy machine of Figure 2.41, where $Z = \{Z_0, Z_1, Z_2\}$; Z_3 was not specified. These conditions then produce unspecified next states or outputs for the machine.

Incomplete machines occur in synthesis when it is determined that not all states of Y or not all outputs of Z are required to meet the machine specifications. This will result in fewer states, because some states are unused and may be considered as unspecified, or "don't care" states.

An incomplete synchronous sequential machine is defined as the following 5-tuple:

$$M = (X, Y, Z, \delta, \lambda)$$

where

1. X is a nonempty finite set of inputs.
2. Y is a nonempty finite set of states.
3. Z is a nonempty finite set of outputs.
4. δ is the next-state function which maps a *subset* of the Cartesian product of $X \times Y$ into Y.
5. λ is the output function which maps a *subset* of the Cartesian product of $X \times Y$ into Z.

Characteristics 4 and 5 indicate that for an incompletely specified synchronous sequential machine, there may exist states Y_j and inputs X_i for which $\delta(X_{i(t)}, Y_{j(t)})$ and $\lambda(Y_{j(t)})$ for a Moore machine, or $\delta(X_{i(t)}, Y_{j(t)})$ and $\lambda(X_{i(t)}, Y_{j(t)})$ for a Mealy machine, are undefined.

When the next state is unspecified, the complete behavior of the machine cannot be predicted for all possible state and input sequences. To prevent this occurrence, the unspecified next states can be replaced by a *terminal* state whose outputs are some predefined values. In most cases, however, these unused states are simply "don't care" states which may then be used to simplify the machine during synthesis.

Example 2.13 The state diagram of Figure 2.43 illustrates an incompletely specified Moore machine containing a single input x_1 and two outputs z_1 and z_2. Input x_1 consists of a sequence of 2-bit words with a bit space between words, as shown below,

$$x_1 = \left| b_1 b_2 \right| \quad \left| b_1 b_2 \right| \quad \left| b_1 b_2 \right| \cdots$$

where $b_i = 0$ or 1. The machine asserts z_1 whenever an input word is $x_1 = 11$; z_2 is asserted whenever $x_1 = 00$. For all other values of the 2-bit words, both outputs remain deasserted. However, three state levels must be maintained so that the machine will remain synchronized with all subsequent words. That is, state a will always check the first bit of each word.

The operation of the machine is unspecified for present states $y_1y_2y_3 = 010$ and 111. These are unused — or "don't care" states — and should normally never be the next state for any state transition. If, however, due to noise or other electrical disturbance or even a hardware failure, one of the unused states is entered, then the machine can proceed to a terminal state and indicate this fact by asserting the outputs of the terminal state to some unique values. This is illustrated in Table 2.27, where states t' and t are the unused states. The next state for states t' and t is state t, the terminal state. State t asserts outputs z_1 and z_2 to a value of 11.

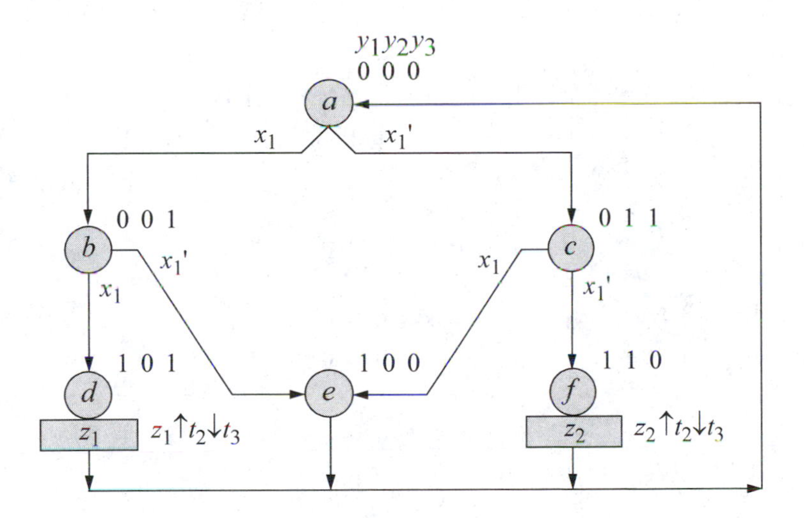

Figure 2.43 Incompletely specified Moore machine of Example 2.13. Unused states are $y_1y_2y_3 = 010$ and 111.

The input maps of Figure 2.44 (a) are derived from either the state diagram of Figure 2.43 or from the next-state table of Table 2.27. For example, using the state diagram and considering state a ($y_1y_2y_3 = 000$), the next state for y_1 is 0, regardless of the value of x_1. Thus, $Dy_1 = 0$ for minterm location $y_1y_2y_3 = 000$. The next state for y_2 is 1 if $x_1 = 0$. Since we are interested only where the storage elements have a value of 1, therefore, $Dy_2 = x_1'$ for minterm location $y_1y_2y_3 = 000$. Also, the next state for y_3 is 1, regardless of the value of x_1. Using the same rationale, the next state for flip-flops $y_1y_2y_3$ as the machine leaves state b, is as follows:

$$Dy_1 = 1, \qquad Dy_2 = 0, \qquad Dy_3 = x_1$$

The remaining map entries are obtained from the state diagram in the manner described above or from the next-state table to yield the input maps of Figure 2.44 (a).

Table 2.27 Next-state table for the Moore machine of Figure 2.43

State name	Present state $y_1y_2y_3$	Input x_1	Flip-flop inputs Dy_1	Dy_2	Dy_3	Next state $y_1y_2y_3$	Outputs z_1z_2
a	0 0 0	0	0	1	1	0 1 1	0 0
	0 0 0	1	0	0	1	0 0 1	0 0
b	0 0 1	0	1	0	0	1 0 0	0 0
	0 0 1	1	1	0	1	1 0 1	0 0
t′	0 1 0	0	1	1	1 *	1 1 1 *	1 1
	0 1 0	1	1	1	1 *	1 1 1 *	1 1
c	0 1 1	0	1	1	0	1 1 0	0 0
	0 1 1	1	1	0	0	1 0 0	0 0
e	1 0 0	0	0	0	0	0 0 0	0 0
	1 0 0	1	0	0	0	0 0 0	0 0
d	1 0 1	0	0	0	0	0 0 0	1 0
	1 0 1	1	0	0	0	0 0 0	1 0
f	1 1 0	0	0	0	0	0 0 0	0 1
	1 1 0	1	0	0	0	0 0 0	0 1
t	1 1 1	0	1	1	1 *	1 1 1 *	1 1
	1 1 1	1	1	1	1 *	1 1 1 *	1 1

* Unused states are terminal states.

The output maps of Figure 2.44 (b) are easily obtained from the state diagram, since z_1 is asserted only when the machine enters state *d*; the assertion is t_2, the deassertion is t_3. Output z_2 is active only in state *f* using the same assertion/deassertion statement as z_1.

In this example, the behavior of the incompletely specified machine was described by another machine whose state transitions were completely specified. This was accomplished by introducing a terminal state t as the next state for all unused states.

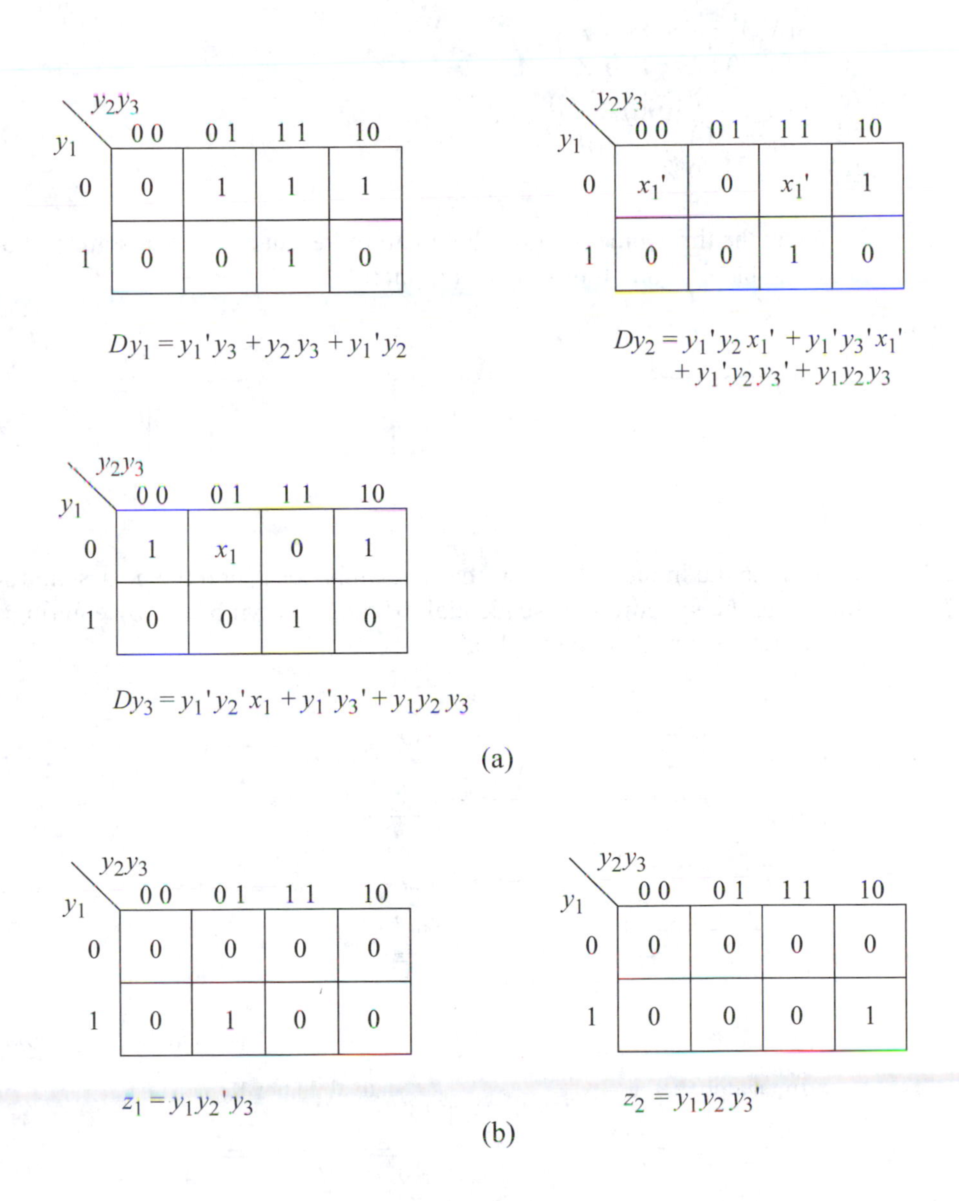

Figure 2.44 Karnaugh maps for the Moore machine of Figure 2.43: (a) input maps; and (b) output maps.

2.5 Problems

2.1 List the five characteristics that define a synchronous sequential machine.

2.2 Indicate which expressions below represent Moore and/or Mealy machines.

(a) $\delta(X, Y) : X \times Y \rightarrow Y$
(b) $\lambda(X, Y) : X \times Y \rightarrow Z$
(c) $Y_{k(t+1)} = \delta(X_{i(t)}, Y_{j(t)})$
(d) $Z_{r(t)} = \lambda(X_{i(t)}, Y_{j(t)})$
(e) $\lambda(Y) : Y \rightarrow Z$

2.3 Connect the three blocks shown below to represent a Mealy machine. Label all three blocks and all inputs and outputs.

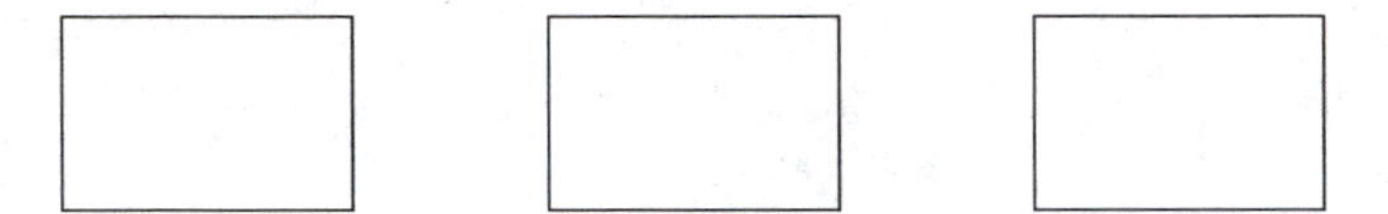

2.4 Determine the input alphabet X, the state alphabet Y, and the next-state function δ for the synchronous sequential machine shown below by entering the appropriate symbols in the tables.

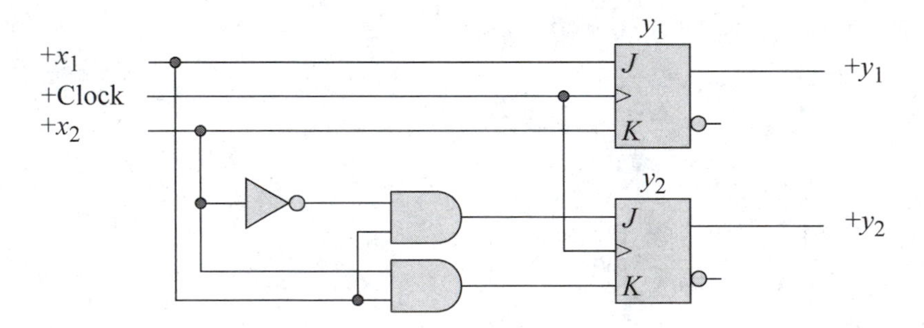

Input alphabet X	
X	x_1x_2
X_0	
X_1	
X_2	
X_3	

State alphabet Y	
Y	y_1y_2
Y_0	
Y_1	
Y_2	
Y_3	

Next-state function δ. The entries are to be a function of Y; for example, Y_0.

δ	X_0	X_1	X_2	X_3
Y_0	Y_0			
Y_1		Leave blank		
Y_2		Leave blank		
Y_3				

2.5 The logic circuit shown below is expressed as the 7-tuple ($x_1, x_2, a_1, a_2, a_3, a_4, z_1$), where a_1, a_2, a_3, a_4 are the outputs of their corresponding gates. The input alphabet X is defined as

$$X = \{X_0, X_1, X_2, X_3\}$$

Determine which input vectors produce a stable state for the binary output variable z_1, where the input vectors are

$$X_0 = x_1x_2 = 00$$
$$X_1 = x_1x_2 = 01$$
$$X_2 = x_1x_2 = 10$$
$$X_3 = x_1x_2 = 11$$

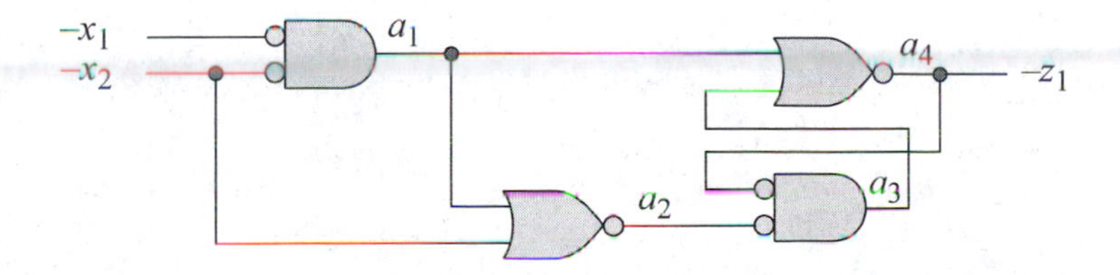

2.6 Indicate which of the state diagrams shown below are strongly connected.

(a)

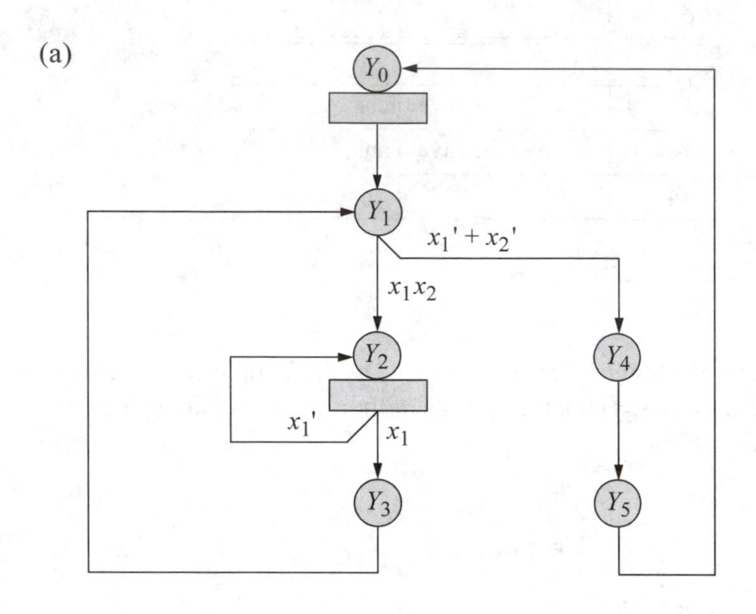

(b)

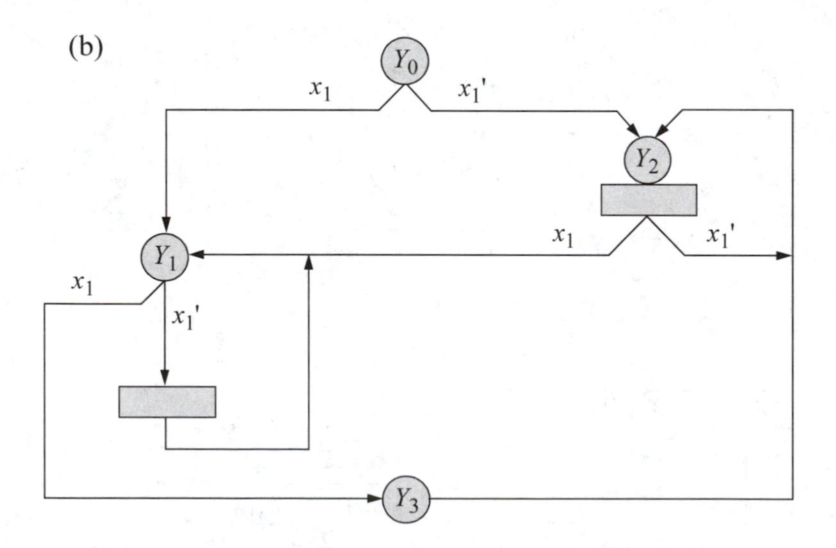

(c)

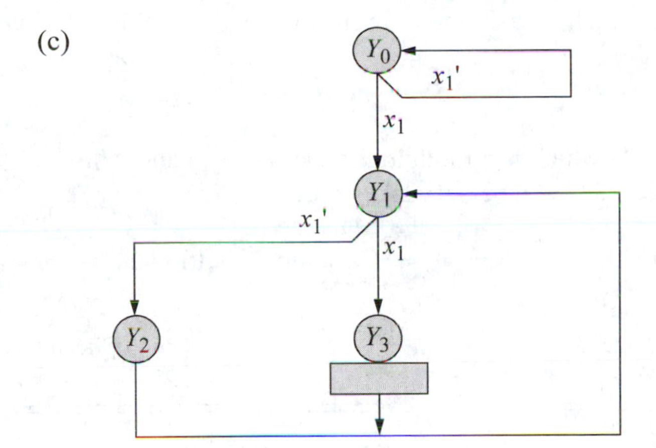

(d)

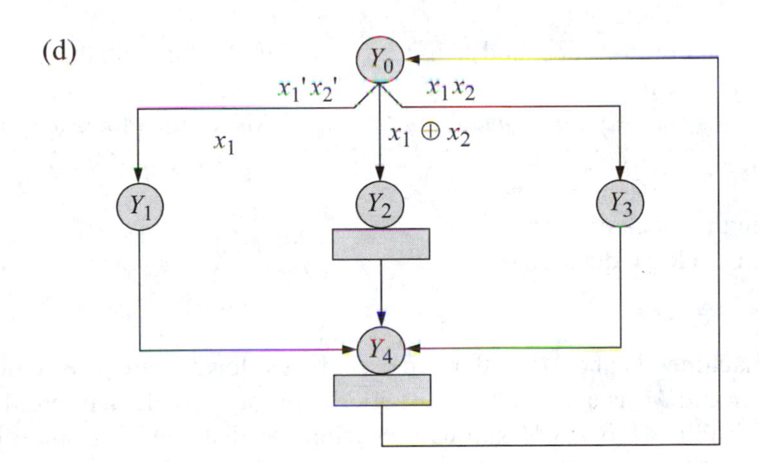

2.7 If machine M is strongly connected, show that the homomorphic image M' of M is also strongly connected.

2.8 If machine M' is a homomorphic image of machine M, and machine $(M')'$ is a homomorphic image of machine M', show that $(M')'$ is the homomorphic image of M. The three machines are defined as follows:

$M = (X, Y, Z, \delta, \lambda)$
$M' = (X', Y', Z', \delta', \lambda')$
$(M')' = [(X')', (Y')', (Z')', (\delta')', (\lambda')']$

2.9 Obtain the homomorphic mappings and equation for machine M', where $M' = (X', Y', \delta')$, which is a homomorphic image of machine M, where $M = (X, Y, \delta)$.

2.10 Let two machines, M which is a modulo-2 adder $(M, +2)$ and M' which is a modulo-3 multiplier $(M', \times 3)$, be defined as follows:

Machine M
(Modulo-2 adder)

	+2	Augend X_0	X_1
Addend	X_0	0	1
	X_1	1	0

Machine M'
(Modulo-3 multiplier)

	×3	Multiplicand X_1	X_2
Multiplier	X_1	1	2
	X_2	2	1

where M is a modulo-2 adder and M' is a modulo-3 multiplier with the zero element omitted. Modify machine M' so that M' simulates machine M; that is, machine M' is isomorphic to machine M. Use a PAL for the modulo-3 multiplier.

(a) Obtain the block diagram.
(b) Obtain the logic diagram.

2.11 Let two machines M and M' be defined as shown below, where M is a modulo-4 adder and M' is a modulo-5 multiplier with the zero element omitted. Modify machine M' so that M' simulates machine M; that is, M' is isomorphic to machine M. Use a ROM for the modulo-5 multiplier.

Machine M
(Modulo-4 adder)

	+4	X: X_0	X_1	X_2	X_3
X	X_0	Y_0	Y_1	Y_2	Y_3
	X_1	Y_1	Y_2	Y_3	Y_0
	X_2	Y_2	Y_3	Y_0	Y_1
	X_3	Y_3	Y_0	Y_1	Y_2

Machine M'
(Modulo-5 multiplier)

	×5	X': X_1'	X_2'	X_3'	X_4'
X'	X_1'	Y_1'	Y_2'	Y_3'	Y_4'
	X_2'	Y_2'	Y_4'	Y_1'	Y_3'
	X_3'	Y_3'	Y_1'	Y_4'	Y_2'
	X_4'	Y_4'	Y_3'	Y_2'	Y_1'

2.12 A mechanical combination lock is a sequential machine. Although the lock is not synchronous, it can still be considered as a sequential machine in which a unique sequence of inputs produces an output. Generate a state diagram for a combination lock which opens when a correct sequence of three integers is received.

2.13 Obtain the state diagram for the Moore synchronous sequential machine shown below. The flip-flops are reset initially; that is, $y_1 y_2 = 00$.

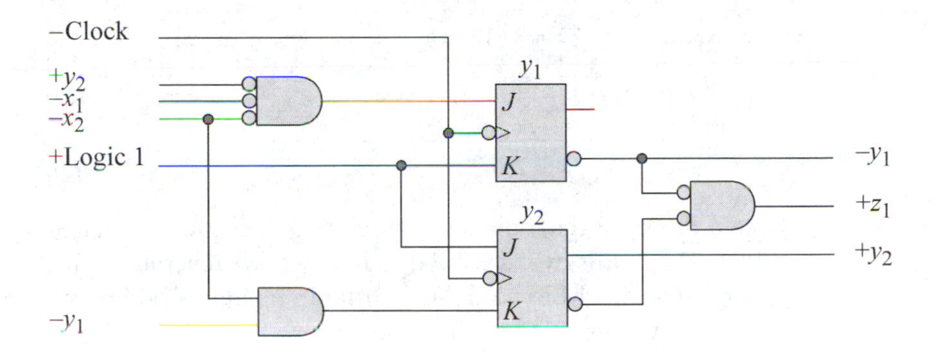

2.14 Obtain the state diagram for the Mealy synchronous sequential machine shown below. Assume that the flip-flop is reset to zero initially. The flip-flop is clocked on the positive clock transition; data changes on the negative clock transition.

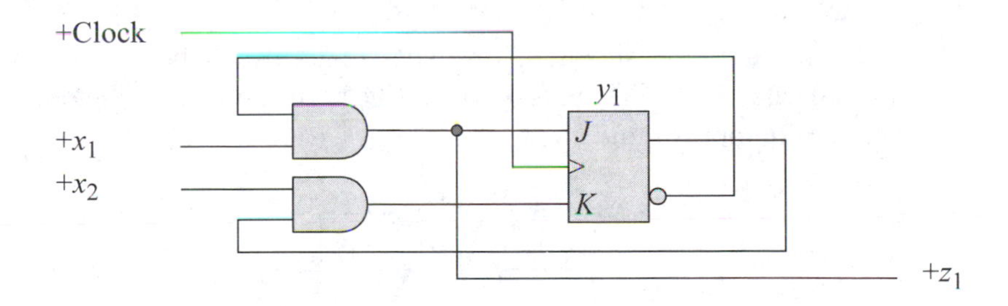

2.15 Draw the state diagram for the Mealy synchronous sequential machine shown below. The flip-flop is reset initially; that is, $y_1 = 0$.

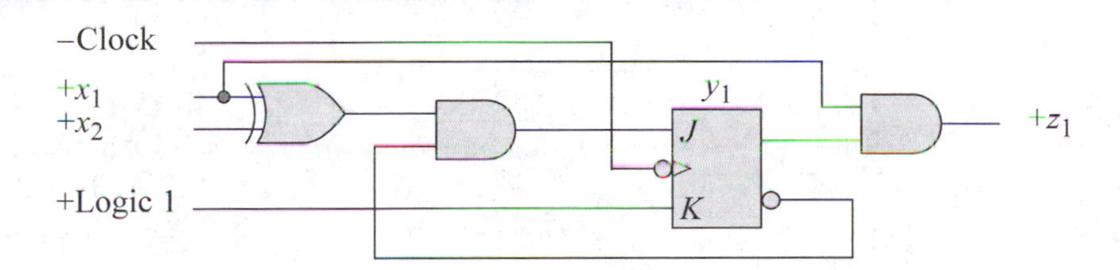

2.16 Obtain the state diagram for the Mealy synchronous sequential machine shown below. The flip-flops are reset initially; that is, $y_1y_2 = 00$.

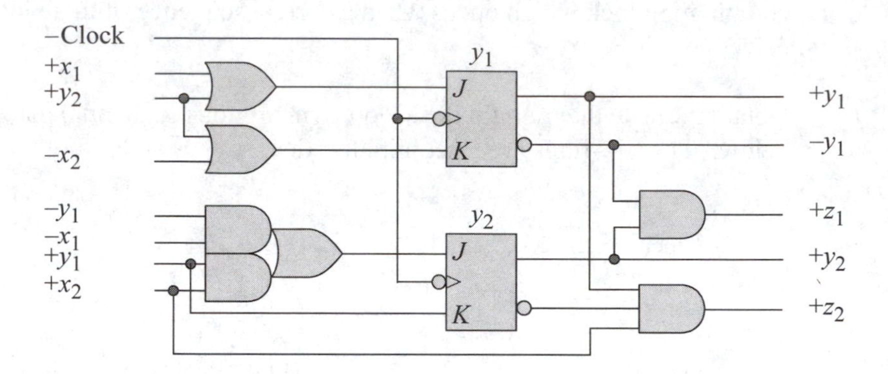

2.17 Obtain the state diagram for a synchronous sequential machine which receives 4-bit words over a serial data line x_1 then generates a fifth bit, which is appended to the 4-bit word. The fifth bit maintains odd parity over all five bits. The five bits are then transmitted over a 5-bit parallel bus. The format of the serial data line is shown below with one bit space between words, where b_i is 0 or 1.

$$x_1 = \left| \; b_1b_2b_3b_4 \; \right| \quad \left| \; b_1b_2b_3b_4 \; \right| \quad \left| \; b_1b_2b_3b_4 \; \right| \cdots$$

2.18 A state diagram is shown below which reads a sequence of 3-bit words over a serial data line x_1. There is one bit space between words. Determine the function of outputs z_1 and z_2.

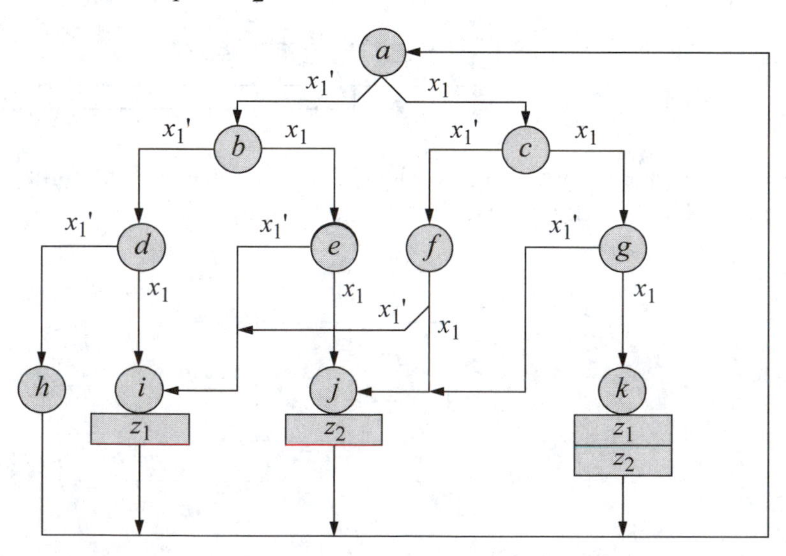

2.19 Two 3-bit unsigned binary numbers X_j and X_k are defined as follows:

$$X_j = x_{j(2)}\, x_{j(1)}\, x_{j(0)}$$
$$X_k = x_{k(2)}\, x_{k(1)}\, x_{k(0)}$$

where $x_{j(0)}$ and $x_{k(0)}$ are the two low-order bits of X_j and X_k, respectively. The state diagram shown below performs an operation on the two operands under control of a counter i. States b, d, and e are terminal states, each of which generates an output. The outputs indicate the relationship between X_j and X_k. Determine the result of the operation for each terminal state as a function of the operands X_j and X_k.

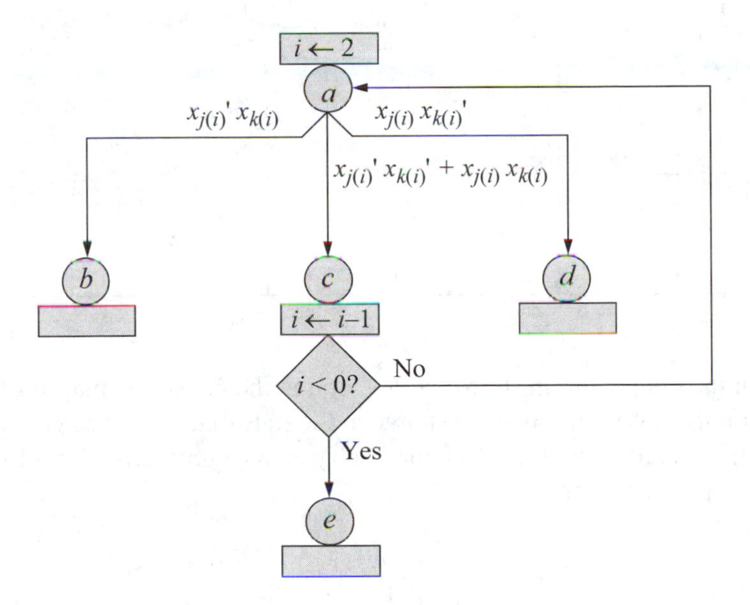

2.20 Obtain the next-state map for flip-flop y_1 for the state diagram shown below. The map entries are to be in a minimum sum-of-products form. Use x_1, x_2, and x_3 as map-entered variables.

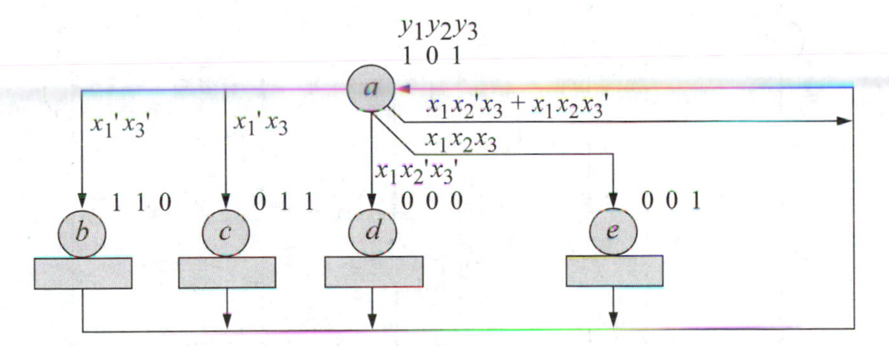

2.21 Obtain the next-state map for flip-flop y_3 for the state diagram shown below. The map entries are to be in a minimum sum-of-products form. Use x_1, x_2, and x_3 as map-entered variables.

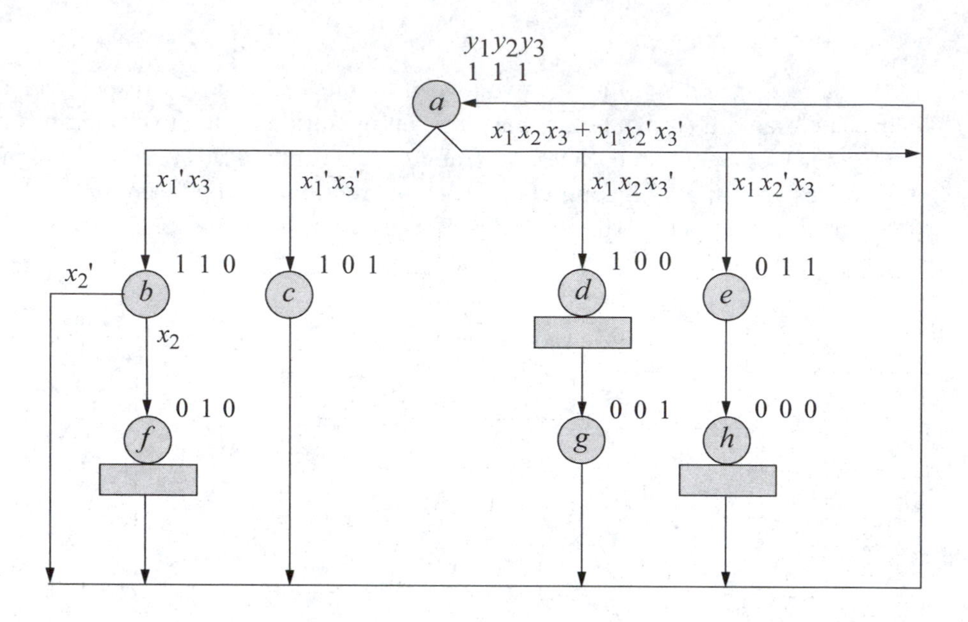

2.22 Given the state diagram shown below, obtain the next-state map for flip-flop y_4 for only those states that are shown in the state diagram. Use x_1, x_2, and x_3 as map-entered variables. Each map entry is to be represented in a minimum sum-of-products form.

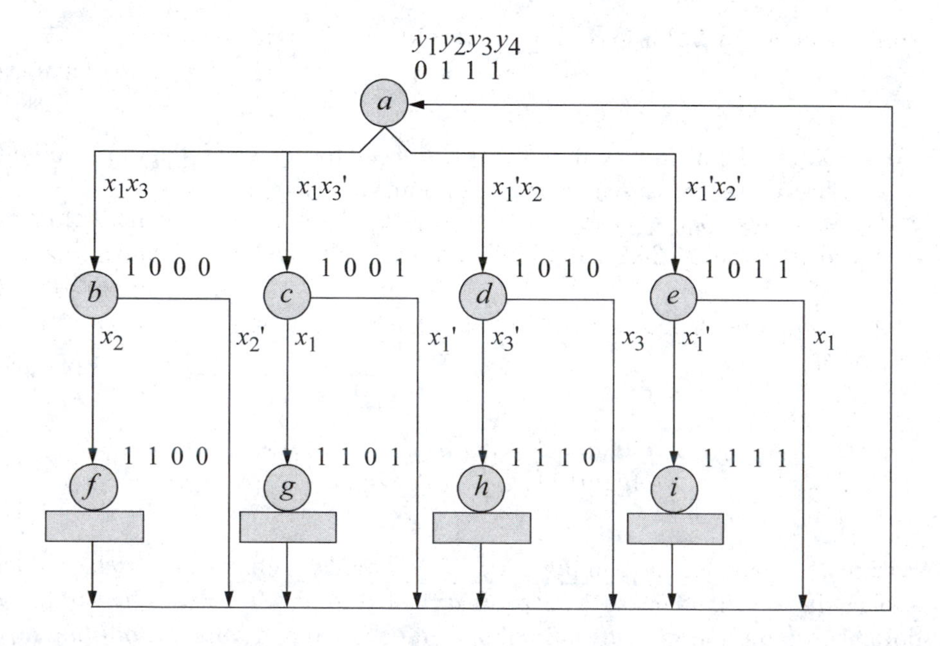

2.23 Given the next-state table shown below, obtain the *JK* input maps for flip-flop y_2 only. Use x_1 as a map-entered variable.

Present state y_1y_2	Input x_1	Next state y_1y_2	Present output z_1
0 0	0	0 0	0
0 1	0	0 0	0
1 0	0	1 0	0
1 1	0	1 0	1
0 0	1	1 1	0
0 1	1	1 1	0
1 0	1	0 1	0
1 1	1	0 1	1

2.24 Construct one set of input maps using *JK* flip-flops for the state diagram shown below for flip-flop y_3 only. Use x_1, x_2, and x_3 as map-entered variables.

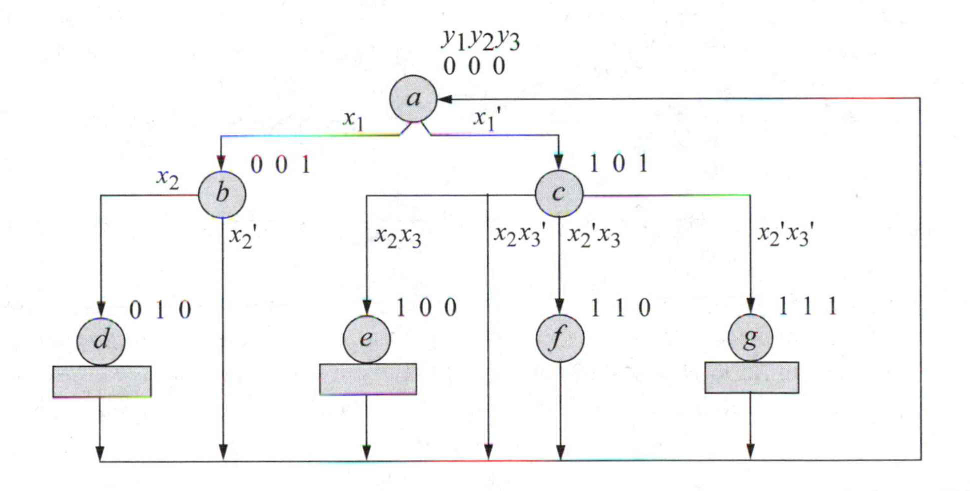

2.25 Obtain the *JK* input equations for flip-flop y_3 for state a ($y_1y_2y_3 = 101$) using the state diagram shown below. Use x_1, x_2, and x_3 as map-entered variables. The equations are to be in a minimum sum-of-products form.

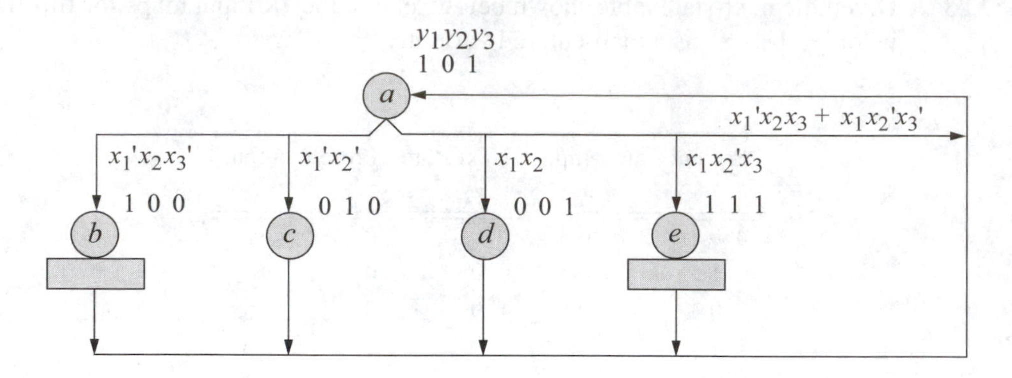

2.26 Obtain the *JK* input equations for flip-flop y_1 for state *a* ($y_1y_2y_3 = 101$) only, using the state diagram shown below. Use x_1, x_2, and x_3 as map-entered variables. The equations are to be in a minimum sum-of-products form.

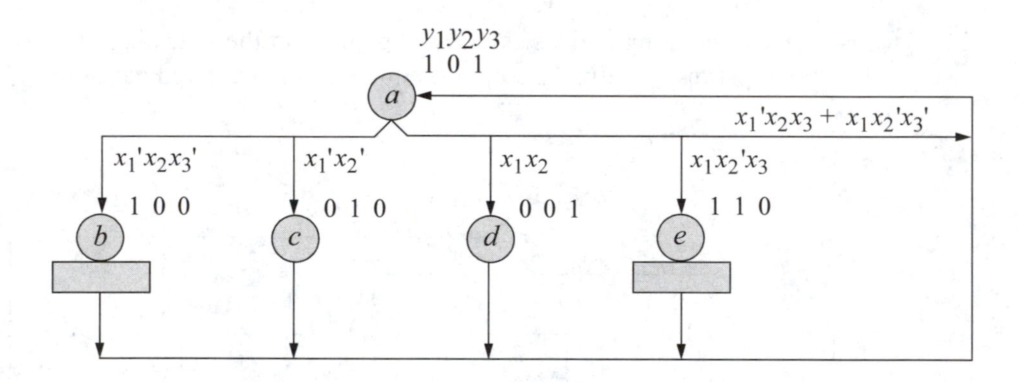

2.27 Given the synchronous sequential machine shown below, draw the timing diagram for y_1, y_2, and y_3 for four clock cycles. Assume that the flip-flops are reset initially; that is, $y_1y_2y_3 = 000$.

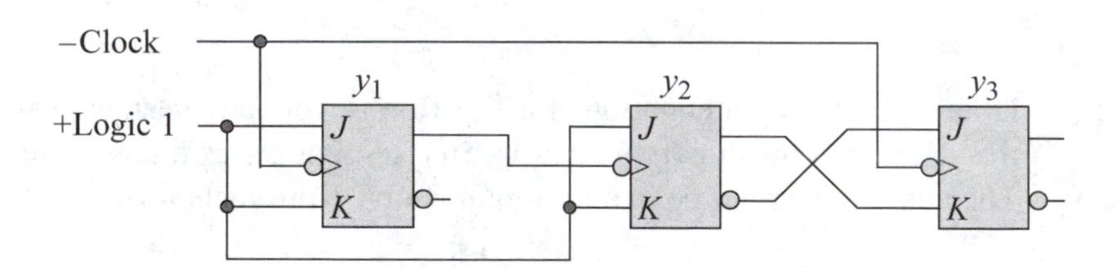

2.28 A Moore synchronous sequential machine uses a serial-in, parallel-out shift register in its implementation as shown below. Draw the waveforms for outputs z_1, z_2, and z_3 for five clock cycles. When the serial input data is at a high (+) voltage level, then a 1 is shifted into the shift register, otherwise, a 0 is shifted into the register. Assume that the shift register is reset initially.

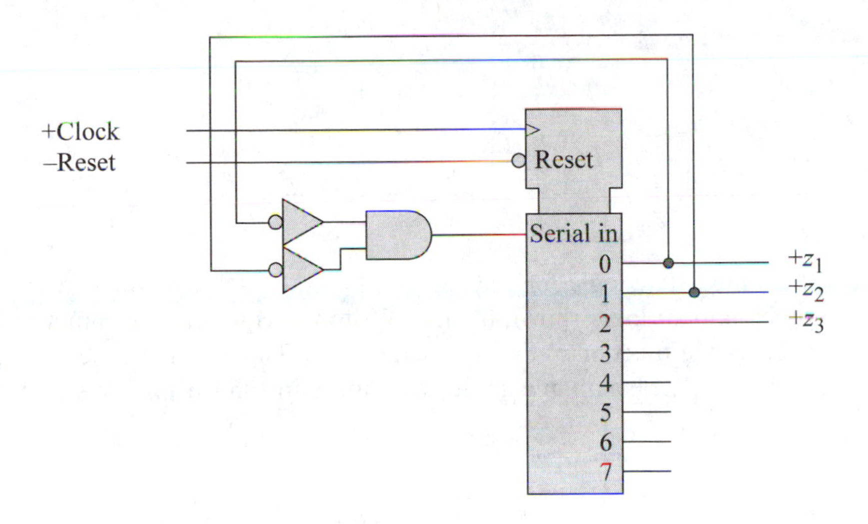

2.29 Complete the excitation table shown below for output z_1 of the of the logic diagram at time $t+1$. The JK flip-flop is clocked on the negative clock transition. Changes to inputs x_1 and x_2 occur on the positive clock transition. A 1 in the excitation table represents a high (+) voltage level; a 0 represents a low (–) voltage level. Variable $z_{1(t)}$ indicates the value of z_1 before the flip-flop is clocked; $z_{1(t+1)}$ indicates the value of z_1 after the flip-flop is clocked.

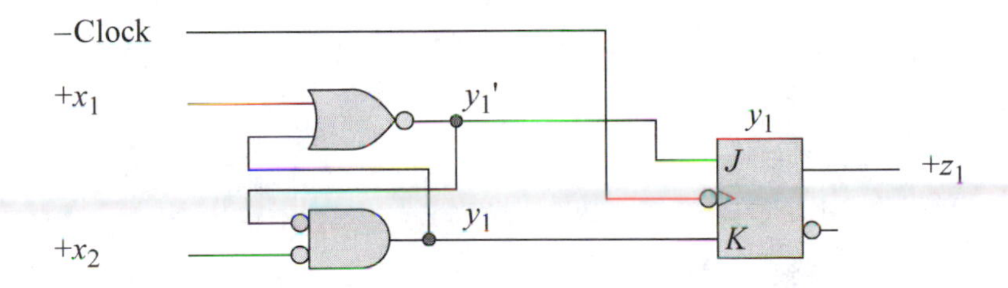

x_1x_2	$y_1'y_1$	$z_{1(t)}z_{1(t+1)}$
0 0	0 1	0
0 1		0
1 0		0
1 1		0
0 0	1 0	1
0 1		1
1 0		1
1 1		1

2.30 The δ next-state logic for a synchronous sequential machine consists of the combinational logic shown below. Write the equation for output z_1 which specifies the function of the next-state logic. The output is asserted high (+) as shown. The equation is to be in a minimum sum-of-products form.

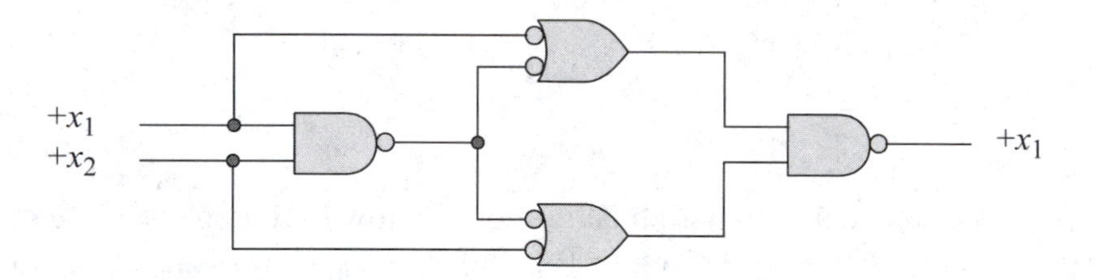

2.31 Three state flip-flops are shown below. The λ output logic consists of two majority circuits. A majority circuit is a combinational logic circuit whose output is a logic 1 (+) if the majority of the inputs are a logic 1; otherwise, the output is a logic 0 (–). Obtain the boolean equation for output z_1 in a minimized sum-of-products form.

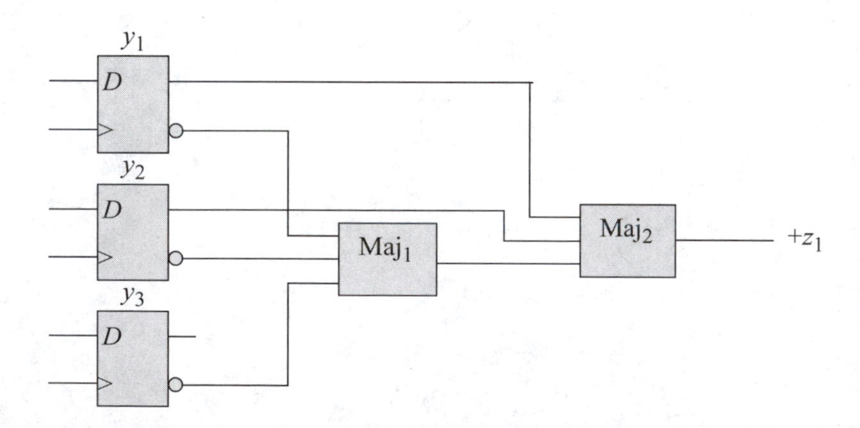

2.32 Given the excitation table shown below for an *MN* flip-flop, obtain the input equations for flip-flop y_3 of a counter which consists of three *MN* flip-flops. States $Y_{j(t)}$ and $Y_{k(t+1)}$ represent the states of the flip-flop before and after the clock transition, respectively. The counter counts in the sequence shown below. The equations are to be in a minimum sum-of-products form.

Excitation table

$Y_{j(t)}$	$Y_{k(t+1)}$	MN
0	0	– 1
0	1	– 0
1	0	1 –
1	1	0 –

Counting sequence

$y_1 y_2$	y_3
0 0	0
0 0	1
0 1	1
0 1	0
1 1	0
1 1	1
1 0	1
1 0	0
0 0	0

Problems 2.33, 2.34, and 2.35 refer to Example 2.11. The logic diagram is shown below.

2.33 Obtain the input and output maps for the Mealy machine shown below. Use x_1 and x_2 as map-entered variables.

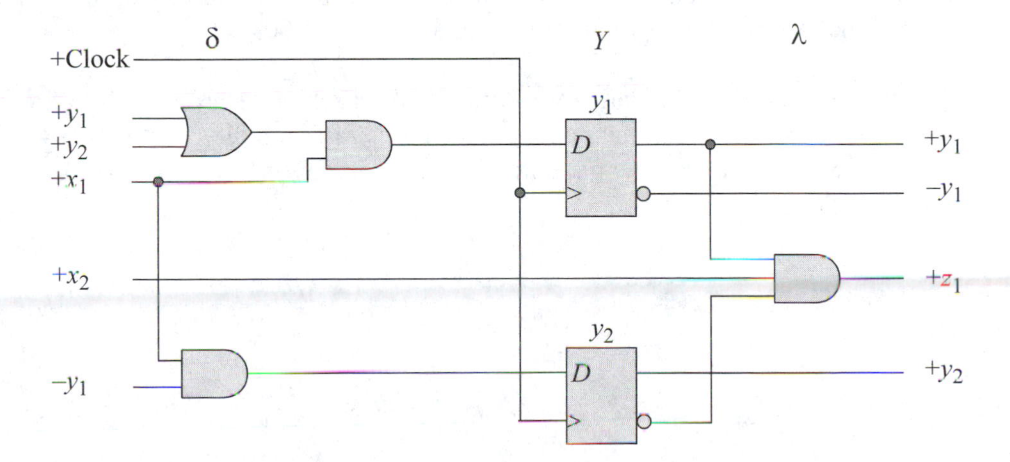

2.34 Generate the state diagram for the Mealy machine of problem 2.33. Assume that the machine is reset initially; that is, $y_1y_2 = 00$.

2.35 Generate the timing diagram for the Mealy machine of problem 2.33 for four clock cycles. Assume that the machine is reset initially; that is, $y_1y_2 = 00$. Let x_1 and x_2 assume the following input sequence: $x_1 = 1110$, $x_2 = 0001$.

Problems 2.36, 2.37, and 2.38 refer to Example 2.12. The logic diagram is shown below.

2.36 Obtain the input maps for y_1 and y_2 and the output maps for z_1 and z_2 for the Mealy machine shown below. Use x_1 and x_2 as map-entered variables.

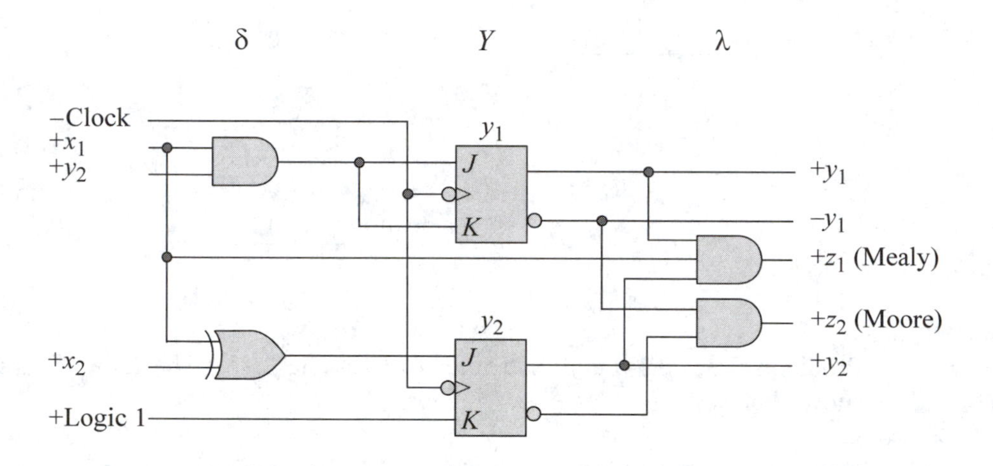

2.37 Generate the state diagram for the Mealy machine of problem 2.36. Assume that the machine is reset initially; that is, $y_1y_2 = 00$.

2.38 Generate a timing diagram for the Mealy machine of problem 2.36 for four clock cycles. Assume that the machine is reset initially; that is, $y_1y_2 = 00$. Let x_1 and x_2 assume the following sequence: $x_1 = 0011$, $x_2 = 0110$.

3.1 Synthesis Procedure
3.2 Synchronous Registers
3.3 Synchronous Counters
3.4 Moore Machines
3.5 Mealy Machines
3.6 Moore-Mealy Equivalence
3.7 Output Glitches
3.8 Problems

3

Synthesis of Synchronous Sequential Machines 1

Techniques for synthesizing (designing) synchronous sequential machines are introduced. A detailed procedure is presented to synthesize a synchronous sequential machine from a given set of machine specifications. The design process begins with a set of specifications and culminates with a logic diagram or a list of boolean functions from which the logic can be designed.

Unlike combinational logic, which can be completely specified by a truth table, a synchronous sequential machine requires a state diagram or state table for its precise description. The state diagram depicts the sequence of events that must occur in order for the machine to perform the functions which are defined in the machine specifications. A *synchronous sequential machine* consists of storage elements, usually flip-flops, and δ next-state combinational logic that connects to the flip-flop data inputs. The machine may also contain combinational logic for the λ output function. In some cases, the output logic may require one or more storage elements, depending on the assertion and deassertion of the output signals. This will be elaborated in a later section.

The number of flip-flops is determined by the number of states required by the machine. The combinational logic is derived directly from either the state diagram or from the state table. When the type and quantity of storage elements has been determined, the design process proceeds in a manner analogous to that of combinational logic design.

The design of any machine, whether combinational or sequential, must ultimately consider the manufacturing cost of the machine. Thus, the requirement for hardware minimization is of paramount importance. Hardware minimization is realized by reducing (or minimizing) the number of states in the machine, thus minimizing the number of storage elements and logic gates, while maintaining the input-output requirements. Reducing the number of states in a machine may not always reduce the number of flip-flops, since the number of eliminated states may not reduce the total state count by a power of two. Recall that the number of possible state is 2^p, where p is the number of storage elements in the state alphabet Y.

For example, if a 16-state machine (requiring $p = 4$ storage elements) is reduced to a 12-state machine, then four storage elements are still required. There will be no reduction in the number of flip-flops until the number of states has been reduced to at least eight ($2^p = 2^3 = 8$ flip-flops). However, the increased number of unused states may result in less combinational logic, because these "don't care" states can be combined with other machine states in a Karnaugh map, resulting in a reduction of combinational logic.

A proper choice of state code assignments may also reduce the number of gates in the δ next-state function logic. Since there are p storage elements, the binary values of these p-tuples can usually be chosen such that the combinational input logic is minimized. A judicious choice of state codes permits more entries in the Karnaugh map to be combined. Since the map entries represent the input logic, combining a greater number of minterm locations results in input equations with fewer terms and fewer variables per term. Thus, an overall reduction in logic gates is realized for the δ next-state logic.

The synthesis procedure utilizes a hierarchical method — also referred to as a *top-down* approach — for machine design. This is a systematic and orderly procedure that commences with the machine specifications and advances down through increasing levels of detail to arrive at a final logic diagram. Thus, the machine is decomposed into modules which are independent of previous and following modules, yet operate together as a cohesive system to satisfy the machine's requirements.

3.1 Synthesis Procedure

This section develops a detailed method for designing synchronous sequential machines using various types of storage elements. The hierarchical design algorithm is shown below.

1. Develop a state diagram from the problem definition, which may be either a word description and/or a timing diagram.

2. Check for equivalent states and then eliminate redundant states.

3. Assign state codes for the storage elements in the form of a binary p-tuple.

4. Generate a next-state table.

5. Select the type of storage element to be used, then generate the input maps for the δ next-state function and derive the input equations.

6. Generate the output maps for the λ output function and derive the output equations.

7. Design the logic diagram using the input equations, the storage elements, and the output equations.

A critical step in synthesizing synchronous sequential machines is the derivation of the state diagram. The *state diagram* specifies the machine performance and gives a clear indication of the state transitions and the output assertion, both of which are a function of the input sequence. If the state diagram is correct, then the remaining steps are relatively straightforward and will result in a logic circuit that performs according to the machine specifications. If, however, the state diagram does not reflect the exact performance of the machine, then the remaining steps — although correct in themselves — will not result in a machine that adheres to the prescribed specifications.

Methods will be described to identify equivalent states, after which the redundant states can be eliminated and the state diagram redrawn as a reduced state diagram. State codes are assigned according to rules which will be defined in a later section. If p storage elements are required to implement the machine, then a state code in the form of a binary p-tuple is assigned to each state in the state diagram. The next-state table is then derived from the state diagram. This step is not always necessary and can be eliminated in many cases, especially when D flip-flops are used. A next-state map may also prove useful for completeness.

When the type of storage elements has been determined, the input maps can then be obtained using the next-state table, and from these maps the corresponding input equations. An alternative approach is to derive the input maps from the state diagram directly. Both methods will be described. The output maps can be derived directly from the state diagram. The λ output logic is usually combinational, but may require storage elements, depending upon the assertion/deassertion specifications. Finally, the logic diagram is designed using the input equations, the storage elements, and the output equations.

3.1.1 Equivalent States

Before exemplifying the steps of the synthesis algorithm, two methods will be presented for determining equivalent states. When equivalent states have been found, all but one are redundant and should be eliminated before implementing the state diagram with hardware. At each node in the state diagram, two events occur: the outputs (if applicable) for the present state are generated as a function of the present state only (Moore) or the present state and inputs (Mealy); the next state is determined as a function of the present state only or the present state and inputs.

When deriving a state diagram from machine specifications, some states might be included which contain no new information that is pertinent to the machine's performance. These are classified as redundant states and should be eliminated since they may increase the amount of logic that is required and thus, the cost of the machine.

If the state diagram is sufficiently small, then redundant states can be easily recognized as the state diagram is being constructed. For larger state diagrams, redundant states may inadvertently be inserted due to the complexity of the machine specifications. During construction of the state diagram, it is best to obtain a diagram that accurately reflects the machine specifications regardless of the number of states that are included. When a correct state diagram has been established, a simple algorithm can then be utilized to find equivalent states. Redundant states can then be eliminated, yielding a reduced state diagram which still completely characterizes the behavior of the machine.

Two states Y_i and Y_j of a machine are equivalent if, for every input sequence, the output sequence when started in state Y_i is identical to the output sequence when started in state Y_j. Therefore, two states Y_i and Y_j are equivalent if and only if the following two conditions are true:

1. For every input sequence X_i, the output sequence Z_i is the same whether the machine begins in state Y_i or Y_j; that is, $\lambda(Y_i, X_i) = \lambda(Y_j, X_i)$, where $\lambda(Y_i, X_i)$ is the output from present state Y_i with input vector X_i and $\lambda(Y_j, X_i)$ is the output from present state Y_j with input vector X_i.

2. Both states Y_i and Y_j have the same or equivalent next state; that is, $\delta(Y_i, X_i) \equiv \delta(Y_j, X_i)$, where $\delta(Y_i, X_i)$ is the next state for a present state of Y_i with input vector X_i and $\delta(Y_j, X_i)$ is the next state for a present state of Y_j with input vector X_i and the symbol $\equiv$ specifies equivalence.

State equivalence satisfies the *equivalence relation* properties:

Reflexive For every state Y_i in the machine, $Y_i \equiv Y_i$; that is, Y_i is related to itself.

Symmetric For every pair of states Y_i and Y_j in the machine, if $Y_i \equiv Y_j$, then $Y_j \equiv Y_i$; that is, the order of the relation is not important.

Transitive For any three states Y_i, Y_j, and Y_k in the machine, if $Y_i \equiv Y_j$ and $Y_j \equiv Y_k$, then $Y_i \equiv Y_k$; that is, $Y_i \equiv Y_j \equiv Y_k$.

Row matching Finding equivalent states and then eliminating redundant states will be explained using the machine illustrated in Figure 3.1. The technique used in the first method is a *row-matching* procedure employing an approach that is more heuristic than algorithmic. The state diagram in Figure 3.1 depicts a Moore machine with the corresponding next-state table shown in Table 3.1. The machine examines a serial 3-bit word on an input line x_1 and generates an output z_1 whenever the 3-bit word is 111 or 000. There is one bit space (one clock period) between contiguous words during which time output z_1 is asserted, as shown below.

$$x_1 = \quad \cdots \left| b_1\, b_2\, b_3 \right| \underset{z_1}{\uparrow} \left| b_1\, b_2\, b_3 \quad \cdots \right|$$

where b_i is 0 or 1.

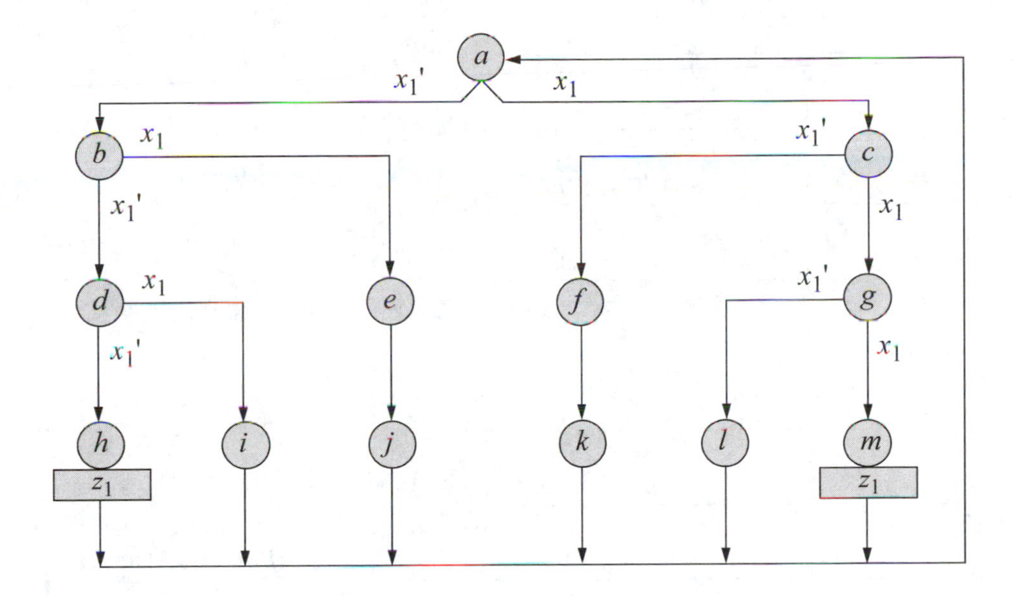

Figure 3.1 Moore machine to detect a sequence of $\boldsymbol{x_1}$ = 111 or 000.

By carefully considering the machine specifications as the state diagram is being generated, it is relatively easy — in this example — to obtain a state diagram that has no redundant states. However, in order to illustrate the techniques used to identify equivalent states and then to eliminate redundant states, superfluous states have been deliberately inserted.

Using the two rules for equivalence, states *i*, *j*, *k*, and *l* of Table 3.1 are seen to be equivalent, because they all have state *a* as the next state and all have output $z_1 = 0$. By convention, the lowest number or the lowest ranked letter is retained for the state name and all other equivalent states are given this state name. Therefore, change all *j*s, *k*s, and *l*s to *i* wherever they appear in the state diagram. This will eliminate rows *j*, *k*, and *l* from the next-state table. The renamed states are indicated by a slash followed by the new name. Although state *m* also has a next state of *a*, output $z_1 = 1$; thus, state *m* is not equivalent to states *i*, *j*, *k*, and *l*. It is also apparent that states *h* and *m* are equivalent: both have state *a* as the next state and both assert output z_1. Therefore, replace all *m*s with *h*s in Table 3.1.

Table 3.1 Next-state table for the Moore machine of Figure 3.1 using the row-matching technique to find equivalent states

Present state	Input x_1	Next state	Output z_1
a	0	*b*	0
	1	*c*	0
b	0	*d*	0
	1	*e*	0
c	0	*f* / *e*	0
	1	*g*	0
d	0	*h*	0
	1	*i*	0
e	0	*j* / *i*	0
	1	*j* / *i*	0
f / *e*	0	*k* / *i*	0
	1	*k* / *i*	0
g	0	*l* / *i*	0
	1	*m* / *h*	0
h	0	*a*	1
	1	*a*	1
i	0	*a*	0
	1	*a*	0
j / *i*	0	*a*	0
	1	*a*	0
k / *i*	0	*a*	0
	1	*a*	0
l / *i*	0	*a*	0
	1	*a*	0
m / *h*	0	*a*	1
	1	*a*	1

Equivalent states (Eliminate state *f*) — states *e* and *f*

Equivalent states (Eliminate state *m*) — states *h* and *m*

Equivalent states (Eliminate states *j*, *k*, and *l*) — states *i*, *j*, *k*, and *l*

After the equivalent states have been renamed, then again check for equivalent states using the modified state names to determine if any further equivalences exist. In states *d* and *g*, output $z_1 = 0$. Therefore, states *d* and *g* are equivalent if $h \equiv i$. The outputs for states *h* and *i*, however, are different; thus, states *h* and *i* are not equivalent and consequently states *d* and *g* are not equivalent.

Continuing with the examination of Table 3.1, it is observed that states *e* and *f* are now equivalent, because in both states the next state is *i* and output $z_1 = 0$. Therefore, change all *f*s to *e*s wherever *f*s appear in Table 3.1. Further inspection produces no additional equivalent states. The reduced next-state table is shown in Table 3.2; the reduced state diagram with only eight states is illustrated in Figure 3.2.

Table 3.2 Reduced next-state table for the Moore machine of Figure 3.1

Present state	Input x_1	Next state	Output z_1
a	0	b	0
	1	c	0
b	0	d	0
	1	e	0
c	0	e	0
	1	g	0
d	0	h	0
	1	i	0
e	0	i	0
	1	i	0
g	0	i	0
	1	h	0
h	0	a	1
	1	a	1
i	0	a	0
	1	a	0

Implication table The second method for determining equivalent states is by use of an *implication table*. This technique is more algorithmic in nature than the previous method, because it follows a set of well-defined rules to find equivalent states in a finite number of steps. The example of Table 3.1 will again be used, this time to illustrate the steps required to find equivalent states using an implication table. For convenience and clarity, Table 3.1 is reproduced in Table 3.3.

The implication table is a lower-left triangular matrix whose rows are labeled with the state names in ascending sequence with the exception of the first state. The columns are also labeled with the state names in ascending sequence with the exception of the last state. The reason for omitting the first state name in the upper-left corner of the matrix is because $a \equiv a$, negating the necessity of inserting a square to determine if a is equivalent to a. The same rationale applies to omitting the last state name in the lower-right corner of the matrix. The square at the intersection of a row-column pair is marked with an $\times$ if the corresponding states are not equivalent or marked with the symbol $\equiv$ if the states are equivalent.

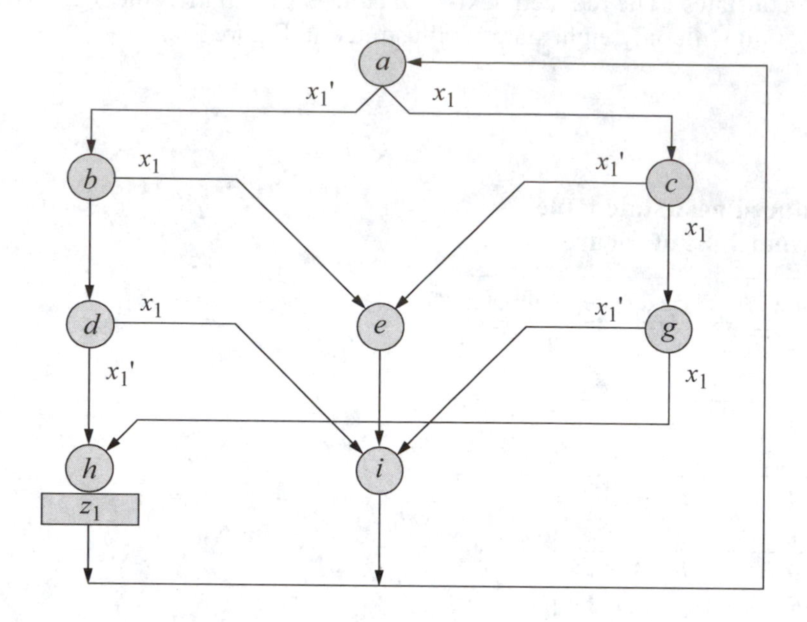

Figure 3.2 Reduced state diagram for the Moore machine of Figure 3.1.

Table 3.3 Next-state table for the Moore machine of Figure 3.1

Present state	Input x_1	Next state	Output z_1
a	0	*b*	0
	1	*c*	0
b	0	*d*	0
	1	*e*	0
c	0	*f*	0
	1	g	0
d	0	*h*	0
	1	*i*	0
e	0	*j*	0
	1	*j*	0
f	0	*k*	0
	1	*k*	0
g	0	*l*	0
	1	*m*	0

Continued on next page

Table 3.3 Next-state table for the Moore machine of Figure 3.1

Present state	Input x_1	Next state	Output z_1
h	0	a	1
	1	a	1
i	0	a	0
	1	a	0
j	0	a	0
	1	a	0
k	0	a	0
	1	a	0
l	0	a	0
	1	a	0
m	0	a	1
	1	a	1

The first step is to construct a chart of the form shown in Figure 3.3. The chart contains a square for every possible pair of states. The square in row f, column c, for example, corresponds to state pair (f,c). Thus, the squares in the first column correspond to state pairs (b,a), (c,a), (d,a), etc. Squares above the diagonal are not required, because they represent the symmetric property of equivalence; that is, if $c \equiv d$, then $d \equiv c$. Thus, only one of the state pairs is required. Also, squares such as (a,a), (b,b), etc. are omitted, since the state pair within the parentheses is obviously equivalent.

To fill in the first column of the chart, row a of the next-state table shown in Table 3.3 is compared with each of the remaining rows. Consider rows a and b. Since output $z_1 = 0$ for states a and b, then

$$a \equiv b \text{ if and only if } b \equiv d\ (x_1 = 0) \text{ and } c \equiv e\ (x_1 = 1)$$

This is shown in Figure 3.3 where the implied pairs (b,d) and (c,e) are placed in the square for state pair (a,b). An *implied pair* of states indicates that equivalence is implied for the pair but has not yet been verified. During later steps in the procedure, implied pairs will be shown to be either equivalent or nonequivalent. Since this is a Moore machine, the value of x_1 is not significant when comparing states for equivalence. Next, state pair (a,c) is considered. Output $z_1 = 0$ for states a and c; therefore,

$$a \equiv c \text{ if and only if } b \equiv f \text{ and } c \equiv g$$

This fact is indicated in the table for state pair (a,c). The process continues for each square in the implication table. When two states are not equivalent, an × is placed in the square that corresponds to the state pair under consideration, for example, (a,h), where the outputs differ. When two states are equivalent, this is indicated by placing the equivalence symbol ≡ within the matrix square that corresponds to the state pair under consideration. For example, states h and m are equivalent, because output $z_1 = 1$ for both states and both states have state a as their next state. Therefore, the equivalence symbol ≡ is placed in state pair (h,m).

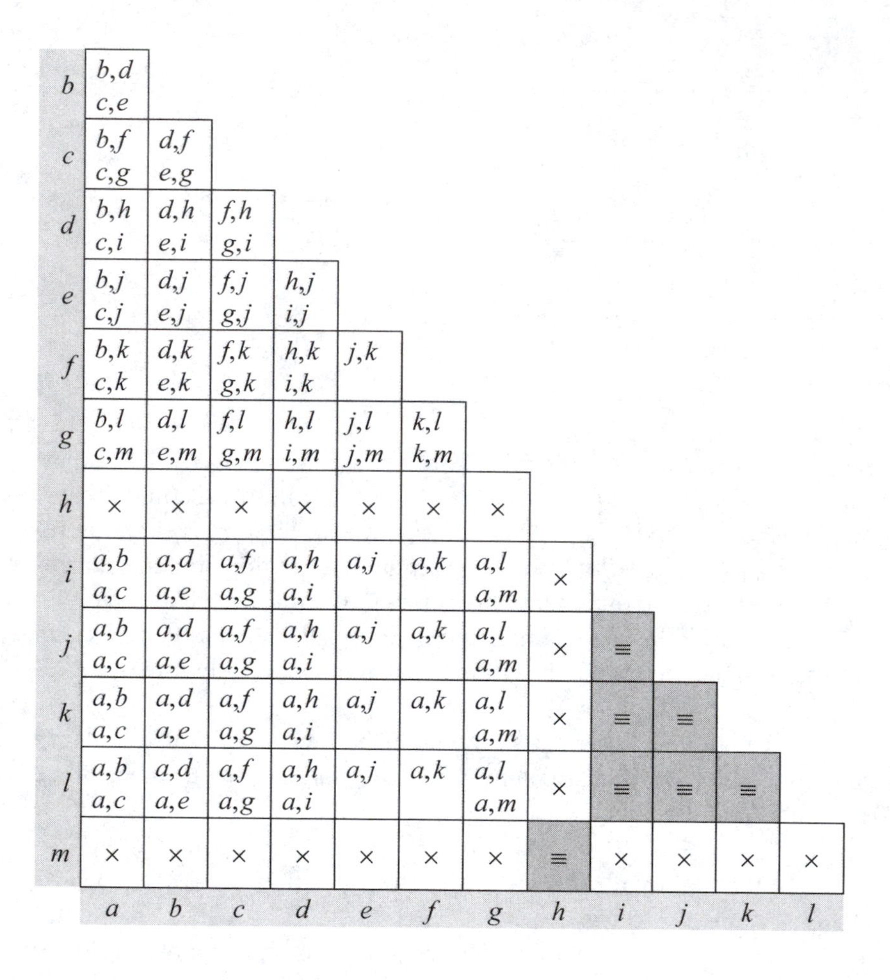

b	*b,d* *c,e*											
c	*b,f* *c,g*	*d,f* *e,g*										
d	*b,h* *c,i*	*d,h* *e,i*	*f,h* *g,i*									
e	*b,j* *c,j*	*d,j* *e,j*	*f,j* *g,j*	*h,j* *i,j*								
f	*b,k* *c,k*	*d,k* *e,k*	*f,k* *g,k*	*h,k* *i,k*	*j,k*							
g	*b,l* *c,m*	*d,l* *e,m*	*f,l* *g,m*	*h,l* *i,m*	*j,l* *j,m*	*k,l* *k,m*						
h	×	×	×	×	×	×	×					
i	*a,b* *a,c*	*a,d* *a,e*	*a,f* *a,g*	*a,h* *a,i*	*a,j*	*a,k*	*a,l* *a,m*	×				
j	*a,b* *a,c*	*a,d* *a,e*	*a,f* *a,g*	*a,h* *a,i*	*a,j*	*a,k*	*a,l* *a,m*	×	≡			
k	*a,b* *a,c*	*a,d* *a,e*	*a,f* *a,g*	*a,h* *a,i*	*a,j*	*a,k*	*a,l* *a,m*	×	≡	≡		
l	*a,b* *a,c*	*a,d* *a,e*	*a,f* *a,g*	*a,h* *a,i*	*a,j*	*a,k*	*a,l* *a,m*	×	≡	≡	≡	
m	×	×	×	×	×	×	×	≡	×	×	×	×
	a	*b*	*c*	*d*	*e*	*f*	*g*	*h*	*i*	*j*	*k*	*l*

Figure 3.3 Implication table for the Moore machine of Table 3.3 after the first pass. The symbol × indicates nonequivalent states; the symbol ≡ indicates equivalent states.

Self-implied pairs are redundant and need not be inserted. A self-implied pair is a state pair within a matrix square that is the same as the state pair under consideration. For example, in the entry shown below for state pair (s,x), the entry (s,x) in the matrix square can be eliminated, since (s,x) is a self-implied pair.

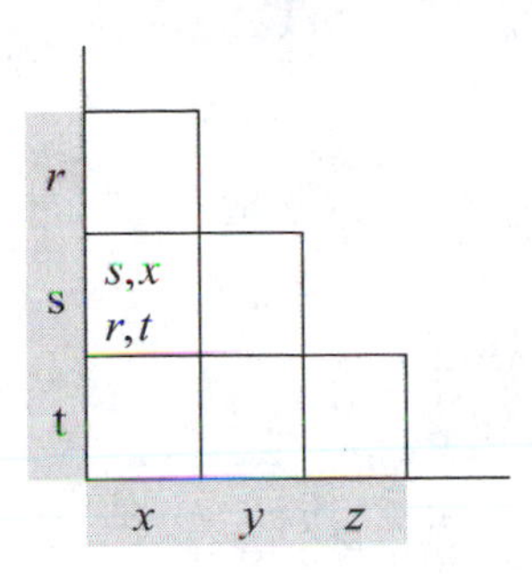

Thus,

$$s \equiv x \text{ if and only if } s \equiv x \text{ and } r \equiv t$$

can be reduced to

$$s \equiv x \text{ if and only if } r \equiv t$$

When all state pairs in the row-column coordinates have been compared, each square in the implication table contains implied pairs, the symbol × indicating that the state pair is not equivalent, or the symbol ≡ indicating that the state pair is equivalent. The implied pairs in each square are now checked for equivalence. If one of the implied pairs in a square is not equivalent, then the state pair under consideration is not equivalent. For example, using state pair (s,x) shown below, if implied pair (h,m) is not equivalent, then $s \neg\equiv x$, where the symbol $\neg\equiv$ is read as "not equivalent."

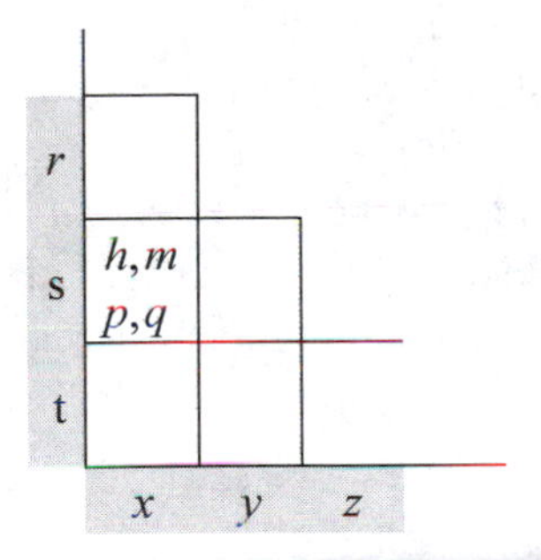

The second pass to find equivalent states is now performed. Beginning with state pair (a,b), continue down column a until each state pair (a,b) through (a,m) has been found either equivalent or not equivalent. It is not always possible, however, to immediately determine equivalency. For example,

$$a \equiv b \text{ if and only if } b \equiv d \textit{ and } c \equiv e$$

At this point, the equivalency of state pairs (b,d) and (c,e) is not immediately evident; thus, their equivalency will be resolved at a later step.

Next, check state pair (a,c) for equivalence. Since

$$a \equiv c \text{ if and only if } b \equiv f \text{ and } c \equiv g$$

and since it is unknown at this time whether $b \equiv f$ or $c \equiv g$, we proceed to state pair (a,d), which specifies that

$$a \equiv d \text{ if and only if } b \equiv h \text{ and } c \equiv i$$

Since $b \neg\equiv h$ (the outputs are different), therefore, $a \neg\equiv d$ and an × is placed in the square for state pair (a,d). Continue in this manner for all columns in Figure 3.3. Notice that state pair (d,e) specifies that

$$d \equiv e \text{ if and only if } h \equiv j \text{ and } i \equiv j$$

This statement is true only if both state pairs (h,j) and (i,j) are equivalent. If either state pair is not equivalent, then $d \neg\equiv e$. Although state pair (i,j) is equivalent, state pair (h,j) is not equivalent. Therefore, $d \neg\equiv e$. The process of finding equivalent states continues until no more ×s can be inserted; that is, until no more nonequivalent states can be found. The result of the second pass is shown in Figure 3.4.

Now begin a third pass, using the implication table of Figure 3.4, to determine equivalency between the remaining state pairs. The results are presented in Figure 3.5. Finally, a fourth pass yields Figure 3.6, which illustrates the following equivalent states:

$$e \equiv f$$

$$h \equiv m$$

$$i \equiv j \equiv k \equiv l$$

The results obtained by the implication table are identical to those obtained by the row-matching procedure. Although the implication table method is tedious, it guarantees a reduced state diagram.

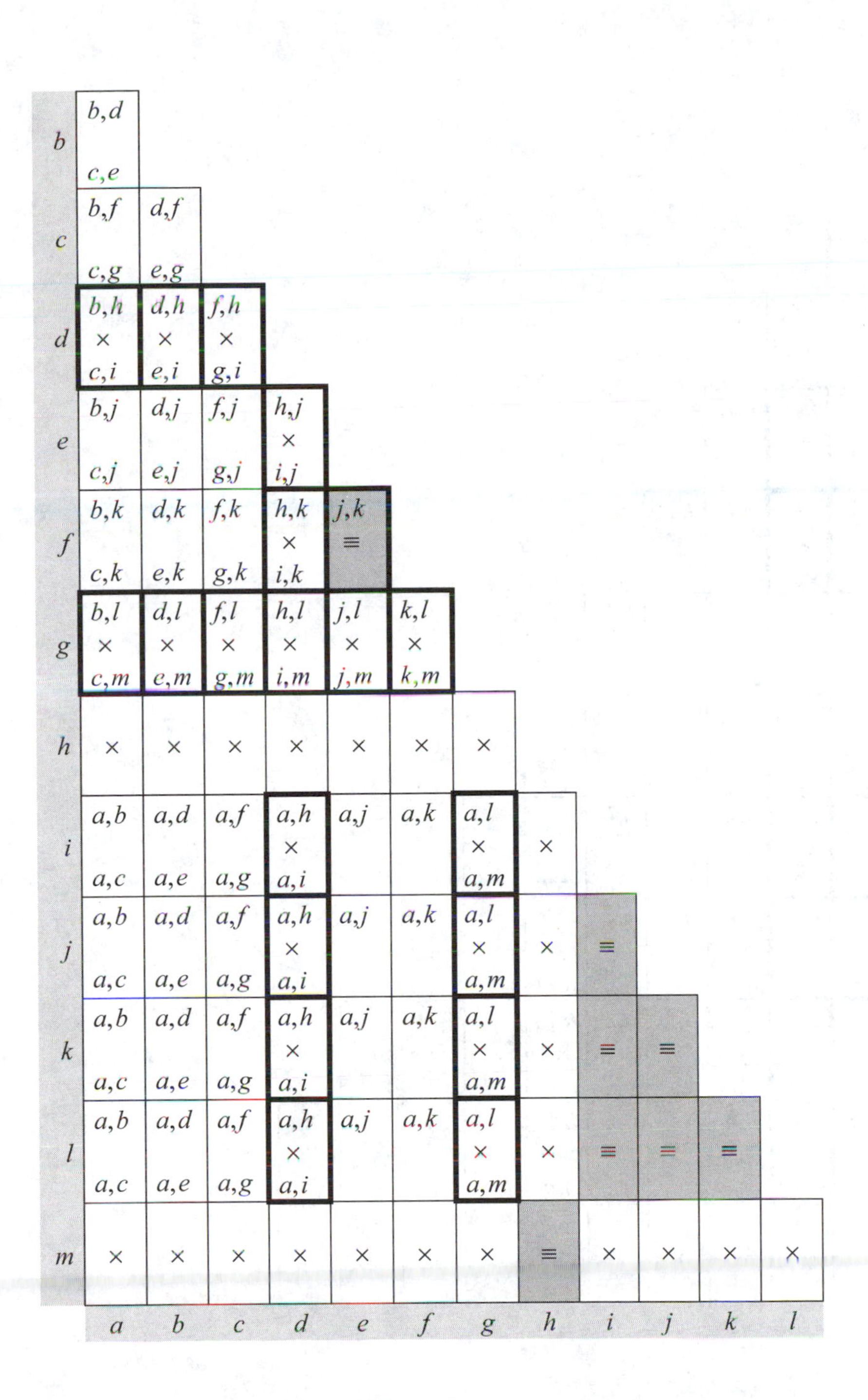

Figure 3.4 Implication table for the Moore machine of Table 3.3 after the second pass. The results of the second pass are shown in bold-lined squares. The symbol × indicates nonequivalent states; the symbol ≡ indicates equivalent states.

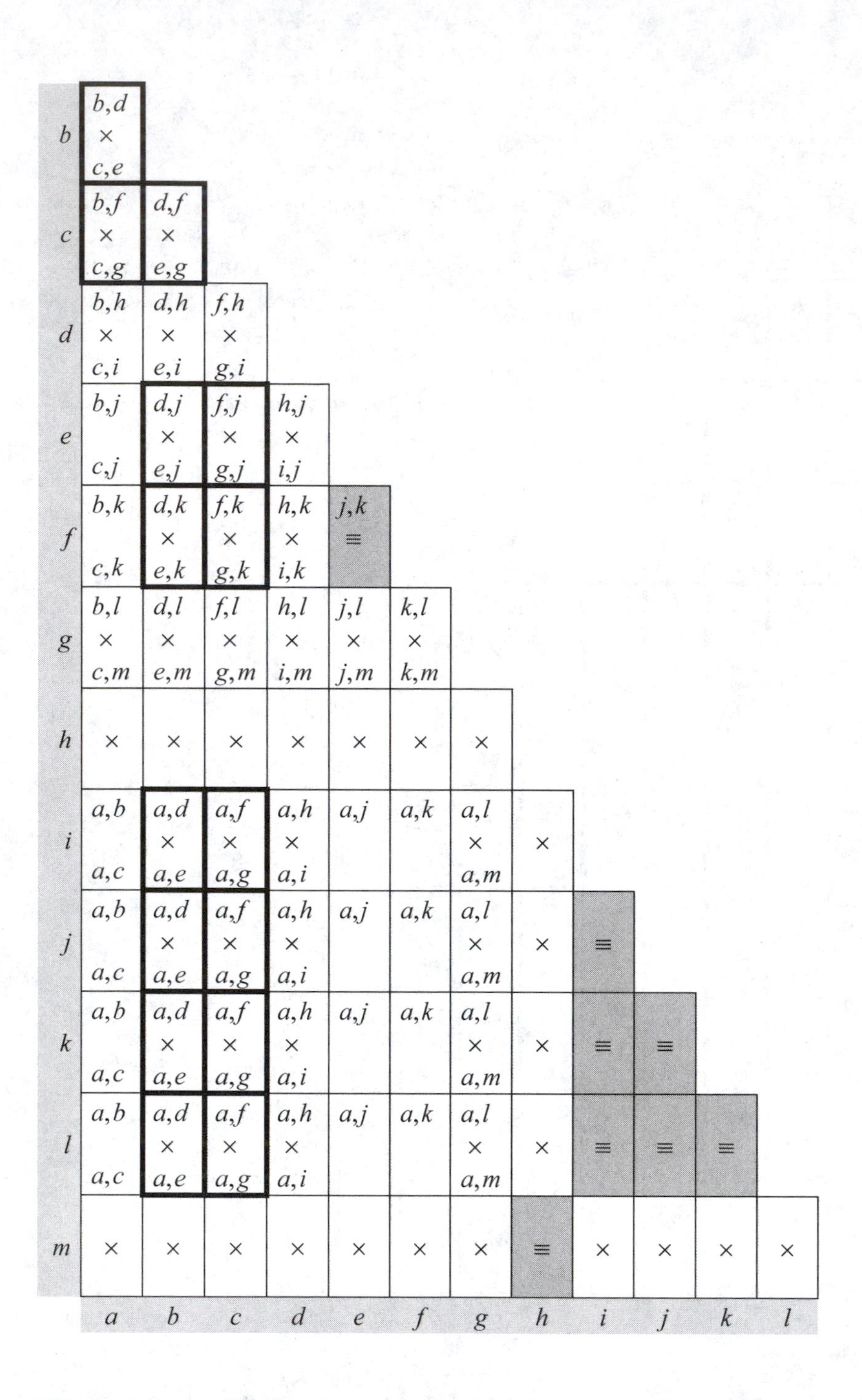

	a	b	c	d	e	f	g	h	i	j	k	l
b	b,d × c,e											
c	b,f × c,g	d,f × e,g										
d	b,h × c,i	d,h × e,i	f,h × g,i									
e	b,j c,j	d,j × e,j	f,j × g,j	h,j × i,j								
f	b,k c,k	d,k × e,k	f,k × g,k	h,k × i,k	j,k ≡							
g	b,l × c,m	d,l × e,m	f,l × g,m	h,l × i,m	j,l × j,m	k,l × k,m						
h	×	×	×	×	×	×	×					
i	a,b a,c	a,d × a,e	a,f × a,g	a,h × a,i	a,j	a,k	a,l × a,m	×				
j	a,b a,c	a,d × a,e	a,f × a,g	a,h × a,i	a,j	a,k	a,l × a,m	×	≡			
k	a,b a,c	a,d × a,e	a,f × a,g	a,h × a,i	a,j	a,k	a,l × a,m	×	≡	≡		
l	a,b a,c	a,d × a,e	a,f × a,g	a,h × a,i	a,j	a,k	a,l × a,m	×	≡	≡	≡	
m	×	×	×	×	×	×	×	≡	×	×	×	×

Figure 3.5 Implication table for the Moore machine of Table 3.3 after the third pass. The results of the third pass are shown in bold-lined squares. The symbol × indicates nonequivalent states; the symbol ≡ indicates equivalent states.

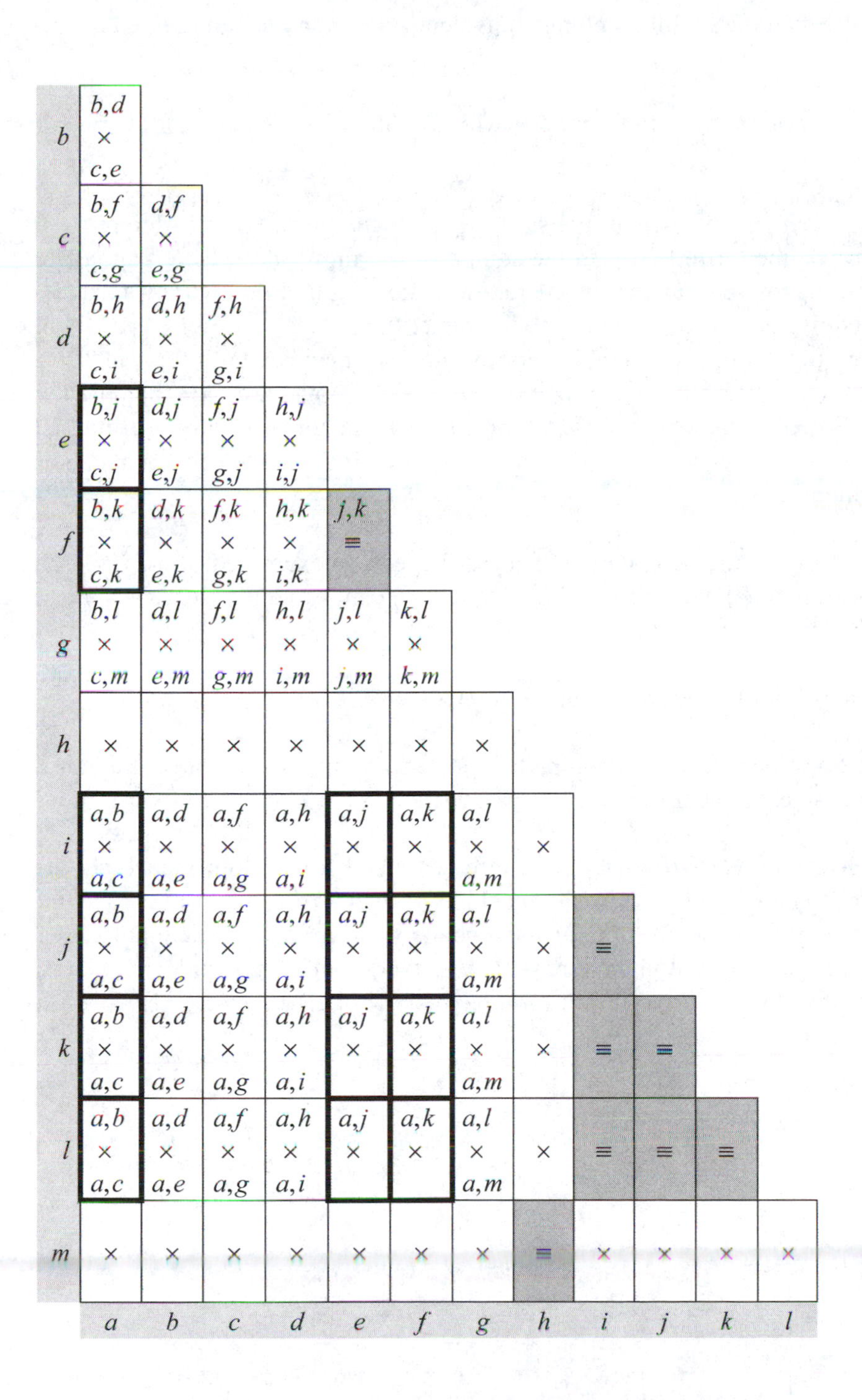

	a	b	c	d	e	f	g	h	i	j	k	l
b	b,d × c,e											
c	b,f × c,g	d,f × e,g										
d	b,h × c,i	d,h × e,i	f,h × g,i									
e	b,j × c,j	d,j × e,j	f,j × g,j	h,j × i,j								
f	b,k × c,k	d,k × e,k	f,k × g,k	h,k × i,k	j,k ≡							
g	b,l × c,m	d,l × e,m	f,l × g,m	h,l × i,m	j,l × j,m	k,l × k,m						
h	×	×	×	×	×	×	×					
i	a,b × a,c	a,d × a,e	a,f × a,g	a,h × a,i	a,j ×	a,k ×	a,l × a,m	×				
j	a,b × a,c	a,d × a,e	a,f × a,g	a,h × a,i	a,j ×	a,k ×	a,l × a,m	×	≡			
k	a,b × a,c	a,d × a,e	a,f × a,g	a,h × a,i	a,j ×	a,k ×	a,l × a,m	×	≡	≡		
l	a,b × a,c	a,d × a,e	a,f × a,g	a,h × a,i	a,j ×	a,k ×	a,l × a,m	×	≡	≡	≡	
m	×	×	×	×	×	×	×	≡	×	×	×	×

Figure 3.6 Implication table for the Moore machine of Table 3.3 after the fourth pass. The results of the fourth pass are shown in bold-lined squares. The symbol × indicates nonequivalent states; the symbol ≡ indicates equivalent states. Equivalent states are (e,f), (h,m), and (i,j,k,l).

The implication method for finding equivalent states can be summarized as follows:

1. Construct a chart similar to Figure 3.3 which contains one square for every pair of states.

2. Compare each pair of rows in the state table. If the outputs are different for a pair of rows, then insert an × in the square of the implication table that corresponds to the row-pair under consideration, indicating that the two rows (states) are not equivalent. If the outputs are the same, then place the implied pair(s) in the square of the implication table. For example, if the next states for s and t are x and y, then (x,y) is an implied pair. Also, if the outputs are the same and the next states are the same or equivalent, or if the square contains only self-implied pairs, then insert the symbol ≡ in the square, indicating that the row-column pair is equivalent.

3. Continue through the table until all squares have been examined. If square (s,t) contains an implied pair (x,y), and if square (x,y) contains an ×, then $s \neg\equiv t$ and an × is inserted in square (s,t).

4. Repeat step 3 until no more ×s can be entered.

5. When the last pass has been completed, any square that does not contain an × represents an equivalent row-column pair written as $s \equiv t$.

Although the above two procedures were described for a Moore machine, the methods work equally well for a Mealy machine. When a row-column pair is compared for a Mealy machine, the outputs must be the same for each same input value.

There is one additional comment regarding row-column equivalent pairs; that is, interdependence. In the partial implication table shown below, $a \equiv i$. Also, $j \equiv b$ if $c \equiv k$.

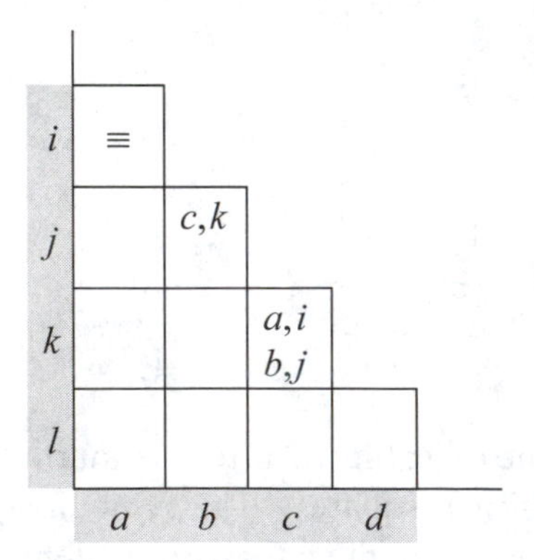

Examining state pair (c,k) for equivalence, we see that $c \equiv k$ if $a \equiv i$ and $b \equiv j$. Since it is already known that $a \equiv i$, the equivalence of c and k depends only on the equivalence of b and j. Conversely the equivalence of b and j depends only on the equivalence of c and k. It can be stated, therefore, that $c \equiv k$ if $b \equiv j$ if $c \equiv k$. This is an equivalence condition that is based upon a mutual dependence and is called *interdependence*. Thus, there is an interdependence between state-pairs (b,j) and (c,k). Therefore, $b \equiv j$ and $c \equiv k$. This can be verified by examining the next-state table of the machine and observing that the output sequence beginning in state b will be the same as the output sequence beginning in state j for the same input sequence. The same rationale is true for states c and k.

3.2 Synchronous Registers

An ordered set of storage elements and associated combinational logic, where each cell performs an identical or similar operation, is called a *register*. Each cell of a register stores one bit of binary information. There are many different types of synchronous registers, including parallel-in, parallel-out; parallel-in, serial-out; serial-in, parallel-out; and serial-in, serial-out registers. In the synthesis of synchronous registers, it is not always necessary to use the formalized design procedure described in Section 3.1. Usually, intuitive reasoning and experience are sufficient requisites for the synthesis of these elementary storage devices. The state diagram provides sufficient information to implement the registers directly.

The next state of a register is usually a direct correspondence to the input vector, whose binary variables connect to the flip-flop data inputs, either directly or through δ next-state logic. Most registers are used primarily for temporary storage of binary data, either signed or unsigned, and do not modify the data internally; that is, the state of the register is unchanged until the next active clock transition. Other registers may modify the data in some elementary manner such as, shifting left or shifting right, where a left shift of one bit corresponds to a multiply-by-two operation and a right shift of one bit corresponds to a divide-by-two operation.

An n-bit register requires n storage elements, either SR latches, D flip-flops, or JK flip-flops. There are 2^n different states in an n-bit register, where each n-tuple corresponds to a unique state of the register.

3.2.1 Parallel-In, Parallel-Out Registers

The simplest register, and the most prevalent, is the *parallel-in, parallel-out* (PIPO) register used for temporary storage of binary data. The synthesis procedure is not required for this type of register. Figure 3.7 illustrates a p-bit register designed with D flip-flops containing only storage elements. There is a one-to-one correspondence between the input alphabet X, the state alphabet Y, and the output alphabet Z. The values of the present inputs $X_{i(t)}$ become the next state $Y_{k(t+1)}$ of the register at the next active clock transition as shown in Equation 3.1.

$$Y_{k(t+1)} = \delta(X_{i(t)}) \tag{3.1}$$

Similarly, the value of the present state $Y_{j(t)}$ becomes the present output $Z_{r(t)}$ as shown in Equation 3.2.

$$Z_{r(t)} = Y_{j(t)} \tag{3.2}$$

The register of Figure 3.7 represents a simple Moore machine where the output $Z_{r(t)}$ is a function of the present state $Y_{j(t)}$ only. The clock signal is obtained from logic that is external to the register and which allows a single clock pulse to be generated only when the register is to be loaded from a new set of inputs. A pulse-generating circuit of this type is designed in the chapter describing the synthesis of asynchronous sequential machines.

A typical application for a PIPO register is shown in Figure 3.8. Assume that the contents of register 1 are to be added to the contents of register 2 and that the sum replaces the contents of register 1 as follows:

$$\text{Register 1} \leftarrow [\text{register 1}] + [\text{register 2}]$$

where the brackets indicate "the contents of." Three microoperations are required for the addition:

1. Place the contents of register 1 on the data bus and allow the next active clock transition to load register *Y*.

2. Place the contents of register 2 on the data bus, set $\text{carry}_{\text{in}} = 0$, specify an add operation to the adder, and allow the next active clock transition to load register *Z*.

3. Place the contents of register *Z* on the data bus and allow the next active clock transition to load register 1.

The contents of register 1 are placed on the data bus and then loaded in parallel into register *Y*. The output of register *Y* is connected directly to the *A* input of the adder. Then the contents of register 2 are placed on the data bus and appear at the inputs of register *Y* and the *B* input of the adder. Register *Y*, however, is not clocked at this time and retains its present state. Since the adder contains only combinational logic, the sum appears at the input of register Z after a specific propagation delay. At the next active clock transition, register Z is loaded from the adder outputs and assumes the sum of registers 1 and 2. The contents of register Z are then placed on the data bus and register 1 is loaded from the bus at the next active clock transition.

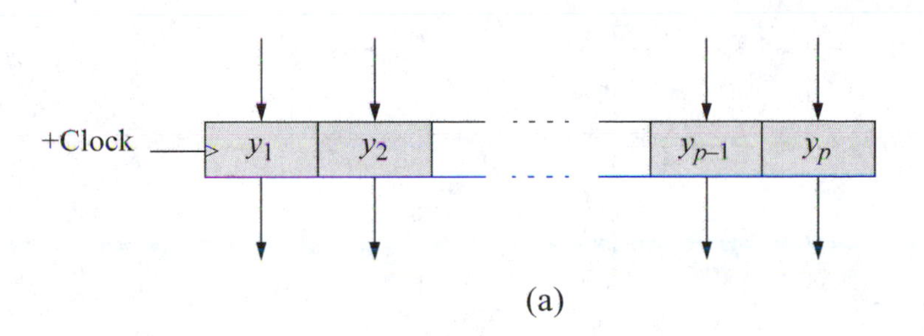

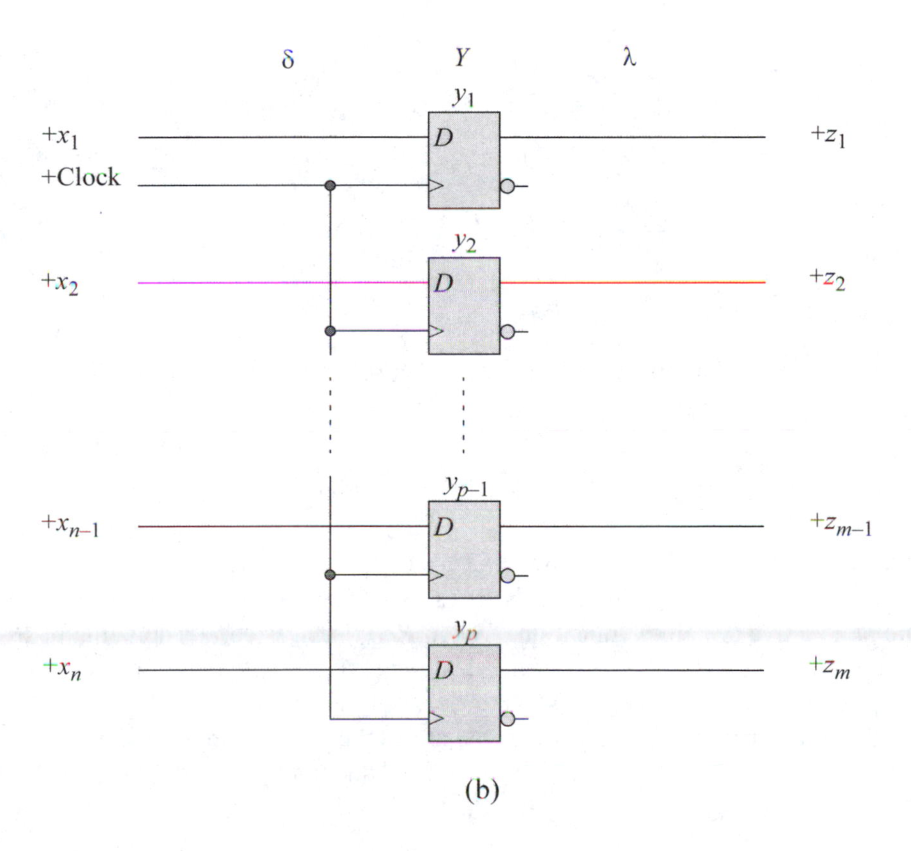

Figure 3.7 Parallel-in, parallel-out register: (a) block diagram; and (b) logic diagram. The clock signal is active only when the register is to be loaded.

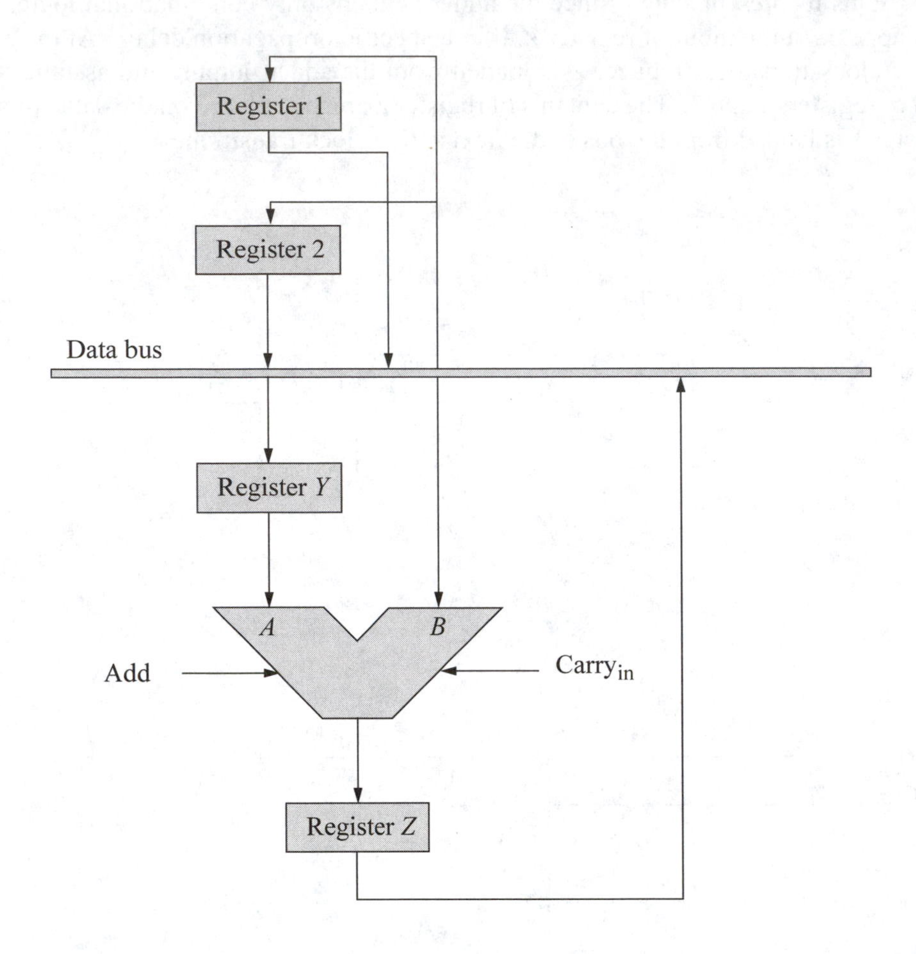

Figure 3.8 Typical application of a parallel-in, parallel-out register.

A typical register Ri is shown below. Ri_{in} is a control signal from a microinstruction register, and when active high (+), signifies that Ri will be loaded from the data bus at the next active (+) clock transition. When Ri_{out} is asserted, the contents of Ri are placed on the data bus. All registers are designed with 3-state output buffers that permit the register contents to be placed on the bus only when the corresponding Ri_{out} signal is active; otherwise, the register outputs are in the high impedance state.

The three microoperations can now be described succinctly by the following notation:

1. $R1_{out}$, Y_{in}
2. $R2_{out}$, $C_{in} = 0$, Add, Z_{in}
3. Z_{out}, $R1_{in}$

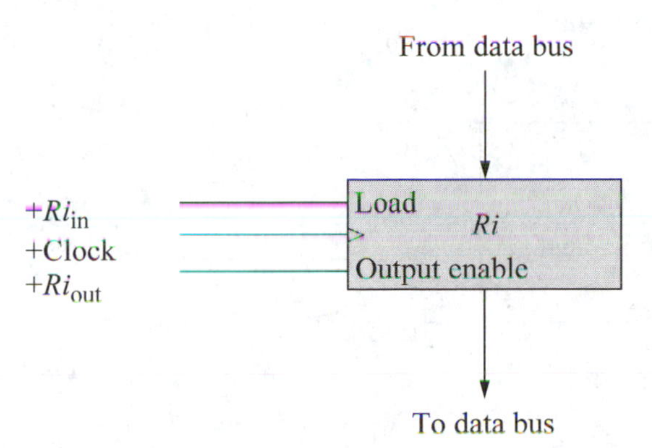

An alternative approach to a PIPO is shown in Figure 3.9. The δ next-state logic not only provides the next state for the register by gating the input vector X_i, but also controls the time at which the register changes state. In this design, the register is clocked continuously by the system clock, which is a free-running astable multivibrator. The register is loaded, however, only when the Load control signal is active high (+). When the Load input is inactive low (–), the data inputs of each flip-flop are $JK = 00$, which causes no change to the state of the machine. Thus, the register remains in its present state until the Load input changes to an active high voltage level. The new input vector X_i then replaces the previous state of the register.

3.2.2 Parallel-In, Serial-Out Registers

A *parallel-in, serial-out* (PISO) register accepts binary input data in parallel and generates binary output data in serial form. The binary data can be shifted either left or right under control of a shift direction signal and a clock pulse, which is applied to all flip-flops simultaneously. The register shifts left or right one bit position at each active clock transition. Bits shifted out one end of the register are lost unless the register is cyclic, in which case, the bits are shifted (or rotated) into the other end.

If the register is a PISO right-shift device, then two conditions determine the value of the bits shifted into the vacated positions on the left. If the binary data represents an unsigned number, then 0s are shifted into the vacated positions. If the binary data represents a signed number — with the high-order bit specified as the sign of the number, where a 0 bit represents a positive number and a 1 bit represents a negative number — then the sign bit extends right one bit position for each active clock transition.

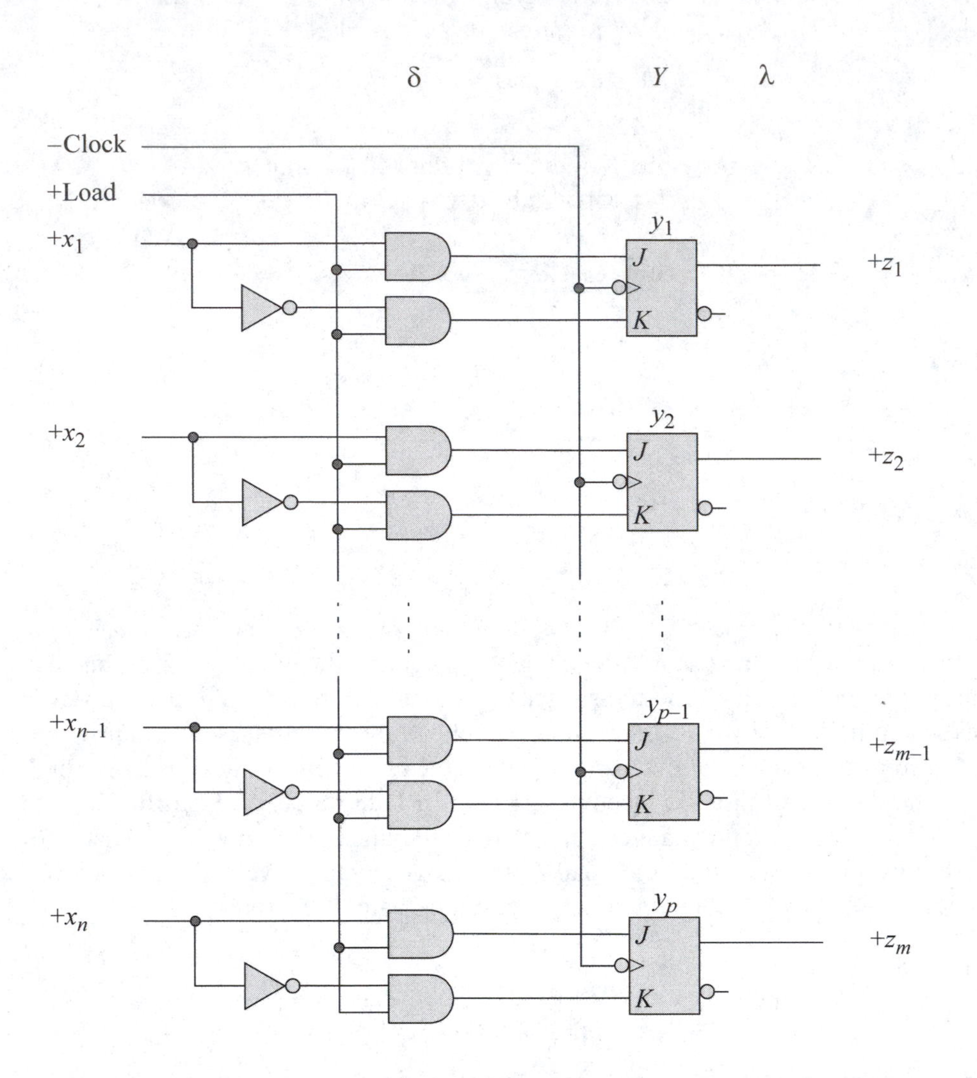

Figure 3.9 Parallel-in, parallel-out register. The clock signal is a continuous train of free-running pulses. Loading is controlled by the Load input.

The procedure for the synthesis of a PISO register is relatively straightforward and the formalized method can be circumvented. An example best illustrates the design process. A 4-bit register receives a parallel input vector in the form of binary bits x_1, x_2, x_3, and x_4. This operand is stored in four D flip-flops y_1, y_2, y_3, and y_4, as shown in Figure 3.10. Upon application of a clock signal, the operand shifts right one bit position. The serial output z_1 is generated from the output of flip-flop y_4. Zeros fill the vacated positions on the left. The state diagram is illustrated in Figure 3.11.

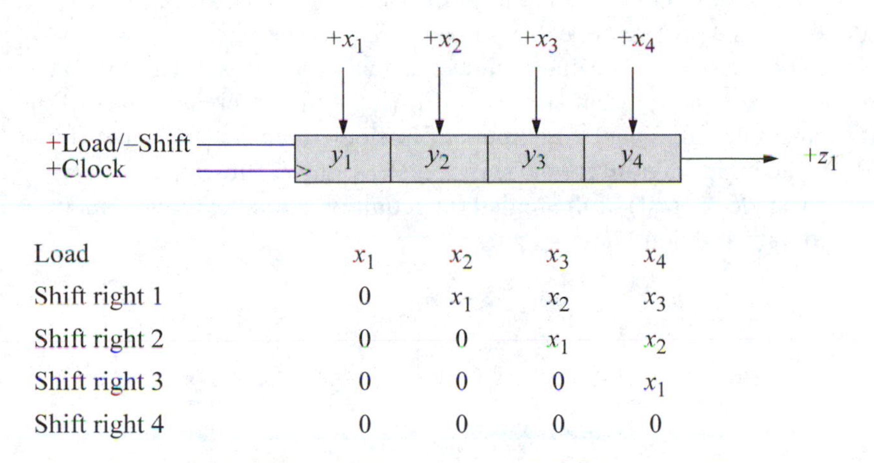

Load	x_1	x_2	x_3	x_4
Shift right 1	0	x_1	x_2	x_3
Shift right 2	0	0	x_1	x_2
Shift right 3	0	0	0	x_1
Shift right 4	0	0	0	0

Figure 3.10 A 4-bit parallel-in, serial-out shift register.

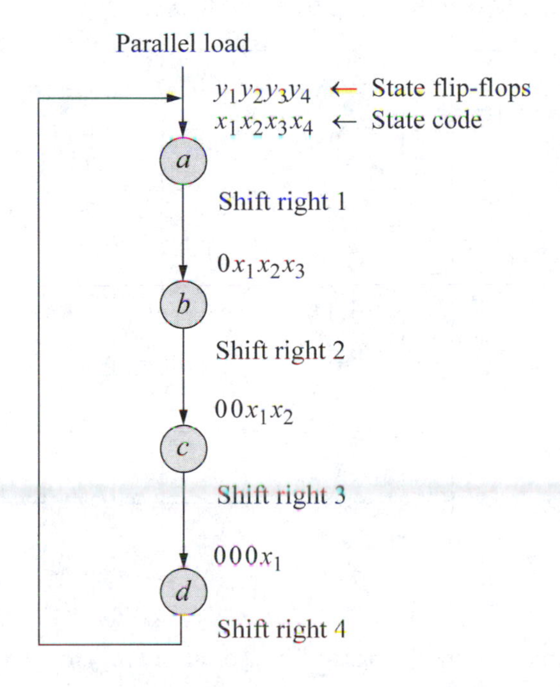

Figure 3.11 State diagram for the parallel-in, serial-out register of Figure 3.10.

Upon completion of the load cycle, $y_i = x_i$. During the shift sequence, $y_i = y_{i-1}$ or 0, depending on the shift count. After four shift cycles, the state of the register is $y_1y_2y_3y_4 = 0000$, and the process repeats with a new input vector X_i.

Careful examination of the state diagram reveals that each clock pulse shifts in 0s from the left and replaces the present state of a flip-flop with the present state of the flip-flop to its immediate left. Thus, the output of flip-flop y_i connects to the data input of flip-flop y_{i+1}. The logic diagram for the shift register using D flip-flops is shown in Figure 3.12. Each stage (or cell) of the register is required to load external data or to shift in the bit from the previous stage.

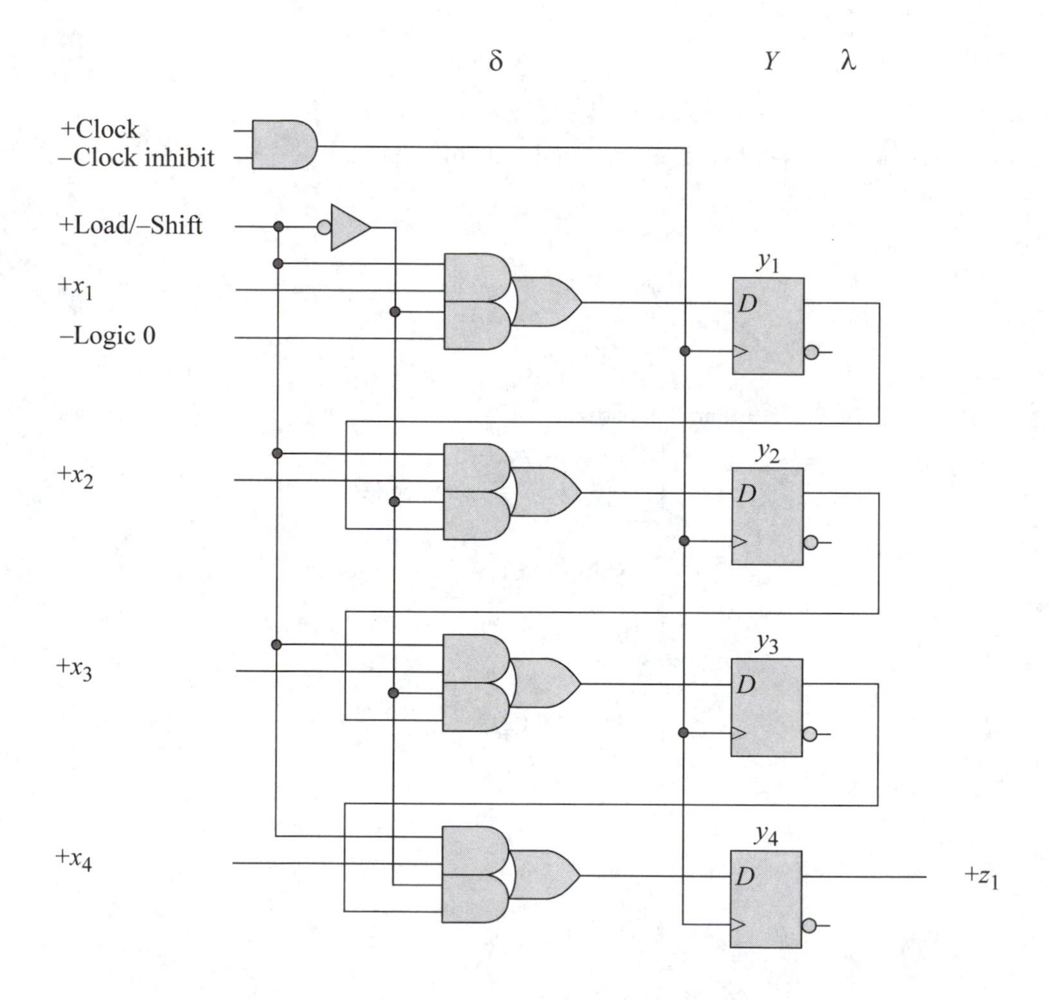

Figure 3.12 Implementation of a parallel-in, serial-out register using D flip-flops.

The parallel-in or serial-out modes are established by the +Load/–Shift input. When this input signal is at a high voltage level, the parallel inputs are enabled and synchronous loading occurs on the next low-to-high clock transition. When the +Load/–Shift input is low, the register is configured in a right-shift mode, with 0s shifted into the leftmost stage. A 1-bit serial shift occurs with each clock pulse. The function of the –Clock Inhibit input signal is to permit the system clock to be free-running, while preventing the register from loading or shifting. The –Clock Inhibit signal is changed to the low level while the +Clock input is low. The register can also be designed with *JK* flip-flops or *SR* latches. Since every stage of the register is identical, it is classified as a sequential iterative network.

One application of a PISO register is to convert data from a parallel bus into serial data for use by a single-track device, such as a disk drive, as shown in Figure 3.13. The serialization process occurs during a write operation. A PISO register can be designed using synthesis techniques described in Chapter 6.

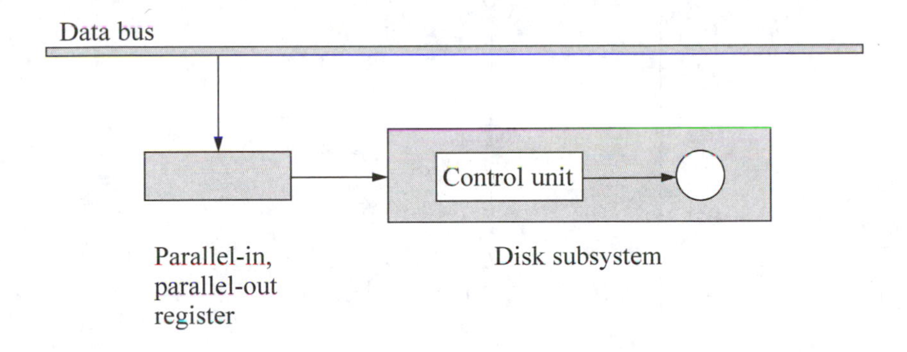

Figure 3.13 Serializing parallel data for a disk subsystem.

3.2.3 Serial-In, Parallel-Out Registers

The *serial-in, parallel-out* (SIPO) register is another typical synchronous iterative network containing p identical cells. Data enters the register from the left and shifts serially to the right through all p stages, one bit position per clock pulse. After p shifts, the register is fully loaded and the bits are transferred in parallel to the destination.

An example of a 4-bit SIPO register is shown in Figure 3.14, in which four bits of serial data, x_1, x_2, x_3, and x_4 are shifted into the register from the left, where x_4 is the first bit entered. The state diagram for the device is shown in Figure 3.15. The initial state of the register is either unknown or reset to $y_1y_2y_3y_4 = 0000$. During the shift sequence, $y_1 = x_i$ and $y_i = y_{i-1}$. After four shift cycles, the state of the register is $y_1y_2y_3y_4 = x_1x_2x_3x_4$ and the 4-bit word is transferred in parallel to a destination.

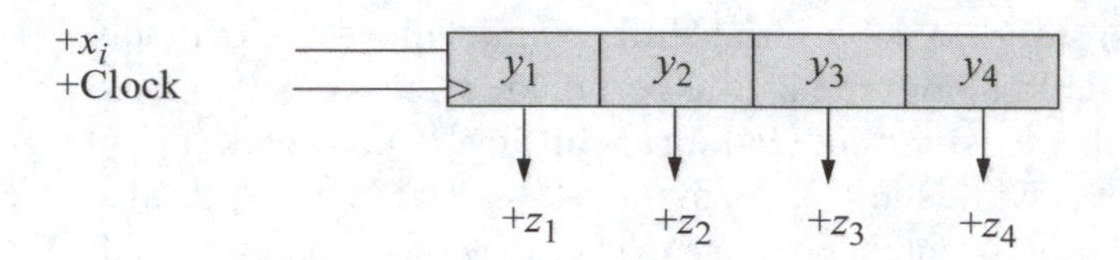

Figure 3.14 A 4-bit serial-in, parallel-out register.

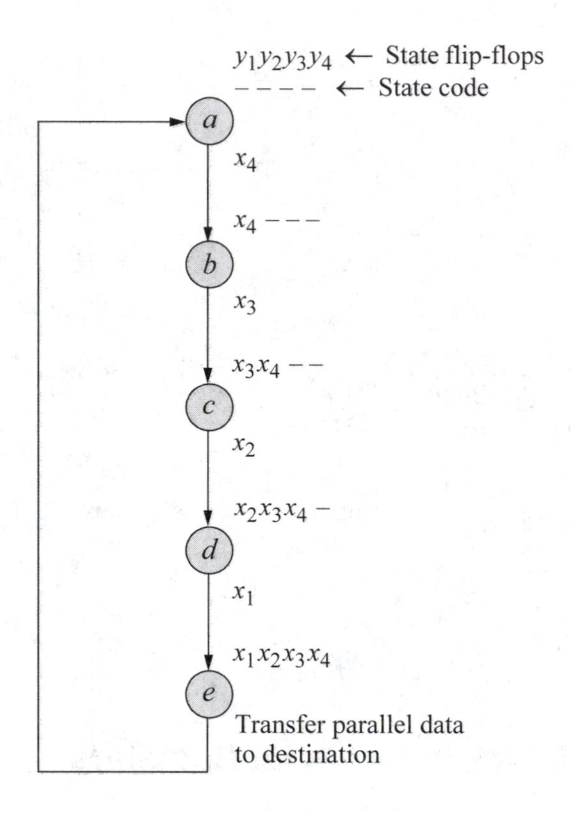

Figure 3.15 State diagram for the serial-in, parallel-out register of Figure 3.14.

Like the PISO register, the synthesis of a SIPO register is intuitively obvious and can be designed from the state diagram without any intermediate steps. The data input of each flip-flop is connected directly to the output of the preceding flip-flop with the exception of flip-flop y_1, which receives the external serial binary data. Figure 3.16

shows the implementation of the register, using *JK* flip-flops. *D* flip-flops or *SR* latches are equally acceptable storage elements. Each stage of the machine is required to perform only one function: Store the state of the storage element to its immediate left. Data bits at the serial input are changed at the positive clock transition to allow bit x_i to be stable at the *JK* inputs of flip-flop y_1 before the active negative clock transition.

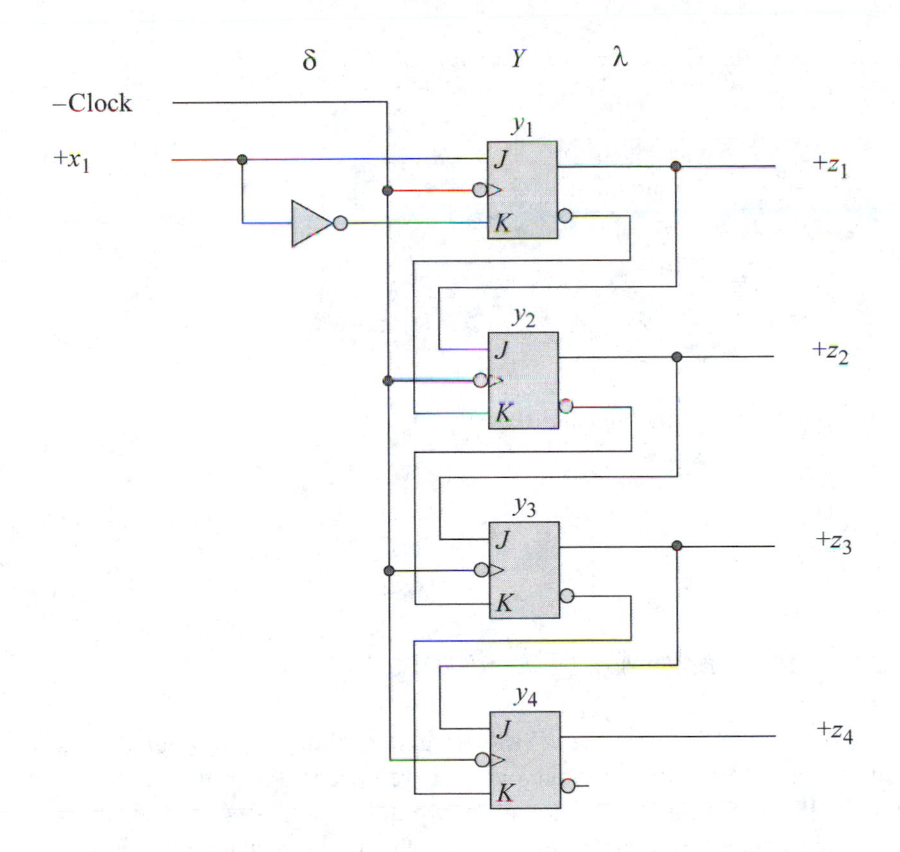

Figure 3.16 Implementation of a 4-bit serial-in, parallel-out register using *JK* flip-flops.

A typical application of a serial-in, parallel-out register is to deserialize binary data from a single-track peripheral subsystem as illustrated in Figure 3.17. The resulting word of parallel bits is placed on the system data bus.

A second useful application of a SIPO register is to generate a sequence of non-overlapping pulses for system timing. This provides a simple, yet effective state machine, where each pulse represents a different state. A small amount of additional logic is required as shown in Figure 3.18 (a). The machine outputs are presented in

Figure 3.18 (b). The machine is initially reset to $y_1y_2y_3y_4 = 0000$. Whenever $y_1y_2y_3 = 000$, a 1 bit will be shifted into flip-flop y_1 at the next positive clock transition. If either y_1, y_2, or y_3 = 1, then a 0 bit will be shifted into flip-flop y_1, and $y_i = y_{i-1}$ at the next positive clock transition. Thus, the required four nonoverlapping pulses are generated.

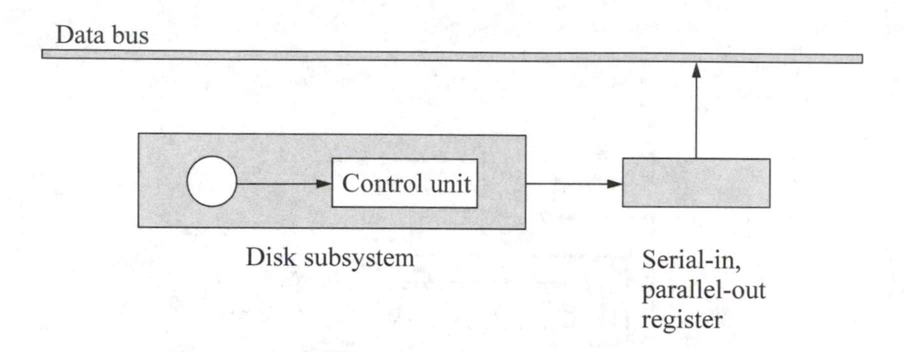

Figure 3.17 A serial-in, parallel-out register to deserialize data from a disk subsystem.

3.2.4 Serial-In, Serial-Out Registers

The synthesis of a *serial-in, serial-out* (SISO) register is identical to that of a SIPO register, with the exception that only one output is required. The rightmost flip-flop provides the single output for the register as shown in Figure 3.19 (a). The state diagram is the same as that of Figure 3.15, except that the output in state *e* consists of serial, not parallel data. Figure 3.19 (b) depicts the logic diagram for a 4-bit SISO register using *JK* flip-flops.

An important application of a SISO register is shown in Figure 3.20. A serial bit stream is read from a disk drive and converted into parallel bits by means of a SIPO register. When eight bits have been shifted into the register, the bytes are shifted in parallel into a matrix of SISO registers, where each bit is shifted into a particular column. The SISO register, in this application, performs the function of a first-in, first-out (FIFO) queue and acts as a buffer between the disk drive and the system I/O data bus. Information is read from a disk sector into the FIFO and then transferred to a destination by means of the data bus. The destination may be a CPU register or a storage location if direct-memory access is implemented. The mode of transfer is bit parallel, byte serial and is either synchronous, where the transfer rate is determined by a system clock, or asynchronous, where the transfer rate is determined by the disk control unit.

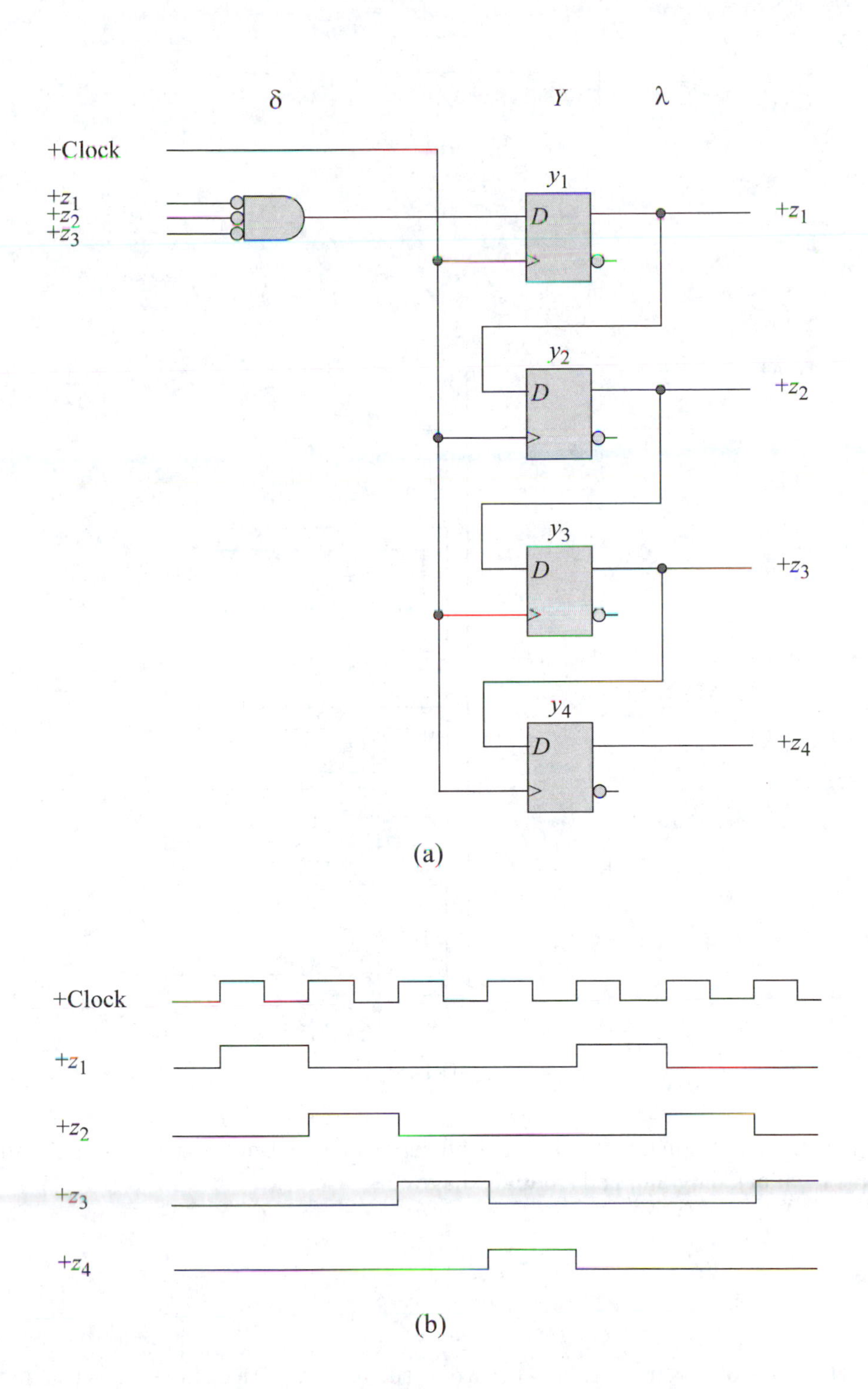

Figure 3.18 A serial-in, parallel-out register configured to generate a sequence of nonoverlapping pulses: (a) logic diagram; and (b) timing diagram.

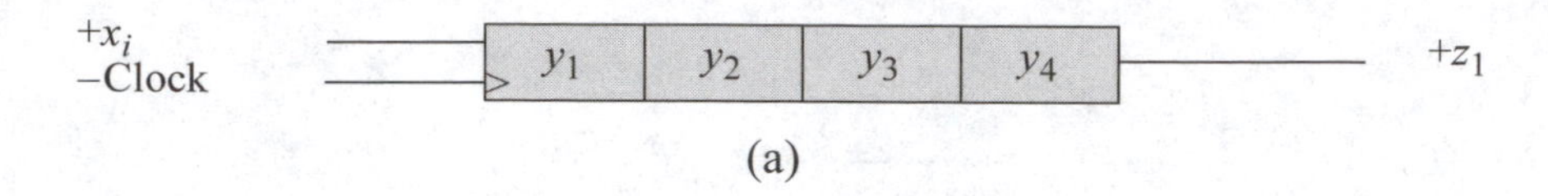

(a)

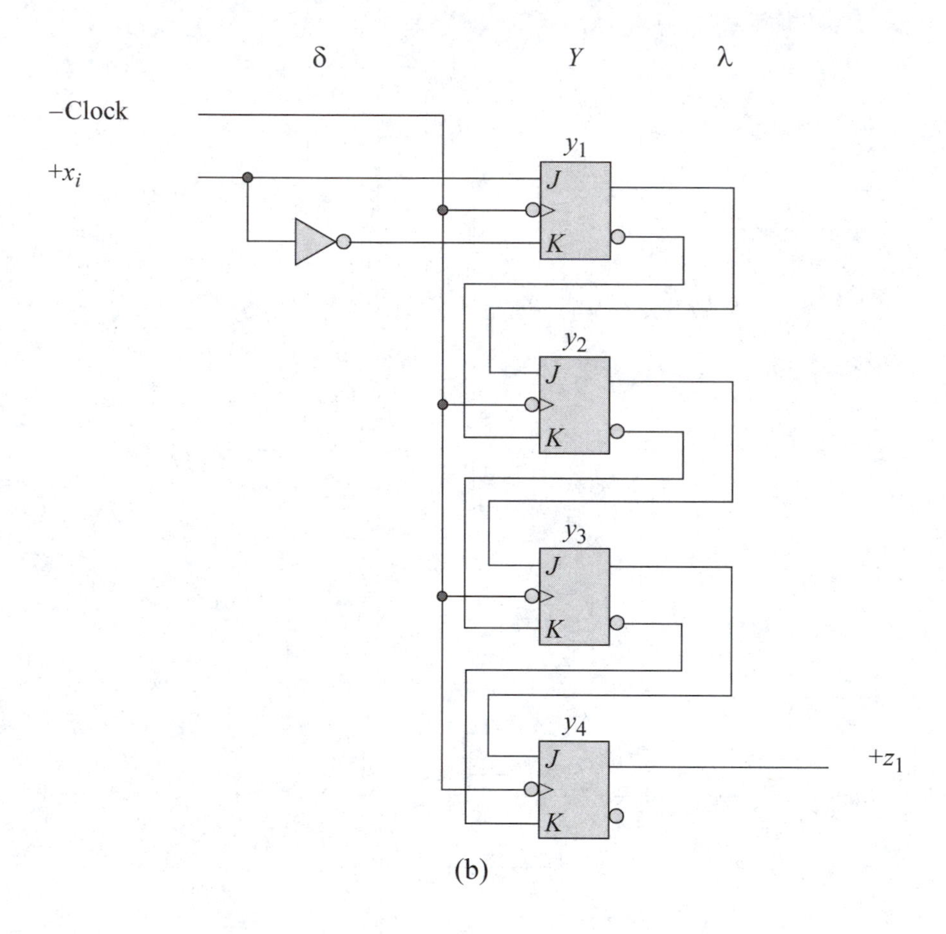

(b)

Figure 3.19 Implementation of a 4-bit serial-in, serial-out register using *JK* flip-flops: (a) block diagram; and (b) logic diagram.

The mode of data transfer between a disk subsystem and a destination is usually in *burst mod*e, in which the disk control unit remains logically connected to the system bus for the entire data transfer sequence; that is, until the complete sector has been transferred. Some systems, however, do not allow burst mode transfer, because this would prevent other peripheral devices from gaining control of the bus. This presents no problem when the disk control unit contains a FIFO. In this situation, data continues to be read from the disk and is transferred to the FIFO, where the bytes are retained until the disk control unit again gains control of the bus. The FIFO prevents data from being lost while the control unit is arbitrating for bus control.

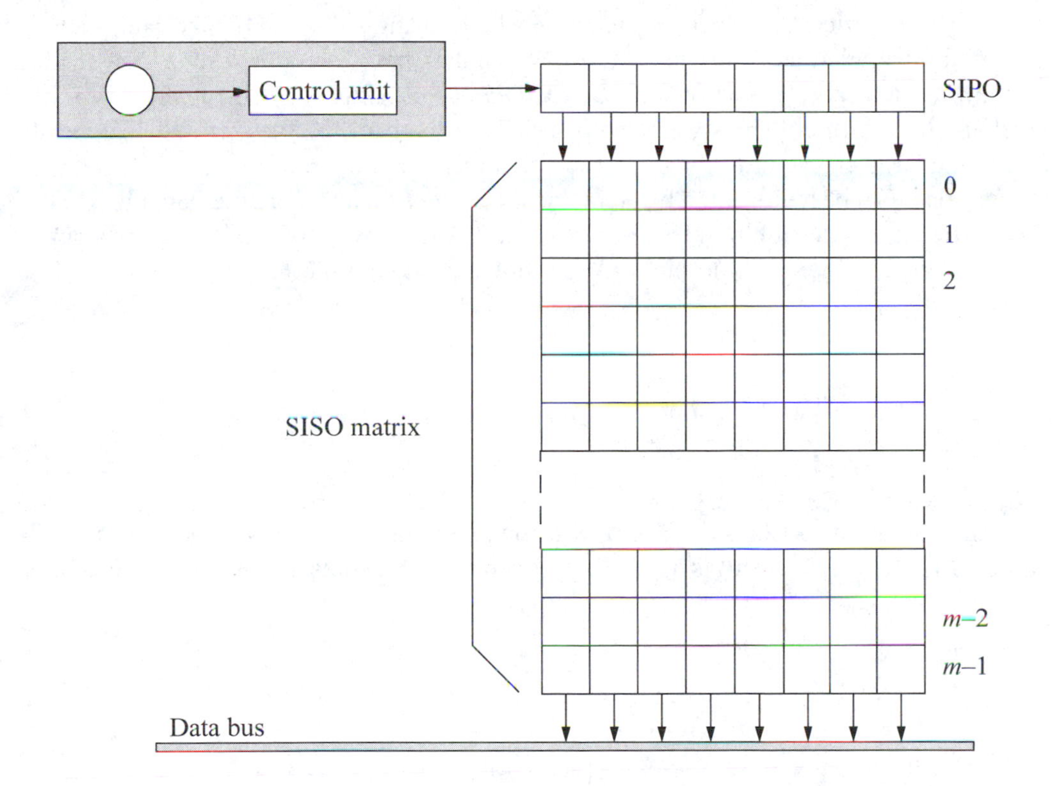

Figure 3.20 Application of a serial-in, serial-out register as an m-byte FIFO buffer.

The same implementation of a SISO register matrix can be used as an instruction queue in a CPU instruction pipeline. The CPU prefetches instructions from memory during unused memory cycles and stores the instructions in the FIFO queue. Thus, an instruction stream can be placed in the instruction queue to wait for decoding and execution by the processor. Instruction queueing provides an effective method to increase system throughput.

3.2.5 Linear Feedback Shift Registers

Another significant application of shift registers is in the field of error detection and correction. To treat the subject of *linear feedback shift registers* adequately would require more than one section in a chapter. Indeed, an entire chapter would not be unwarranted to present a thorough discourse on the theory and mathematical derivations of these unique devices. The intent of this section, therefore, is to provide only an overview of some fundamental attributes of linear feedback shift registers and to illustrate their use as a pragmatic solution to error detection and correction.

These devices are especially useful for large strings of serial binary data, such as found on single-track storage devices. When data bytes are stored on a disk sector, they are usually followed by a *cyclic redundancy check* (CRC) character. The CRC character is formed by passing the bits serially through a CRC generator, then appending the CRC character to the end of the bit stream, as shown in Figure 3.21. The binary data can be represented by a *binary polynomial* $P(X)$, where the data bits are the coefficients of the polynomial. For example, the data byte 1001 1011 can be written as a seventh degree polynomial $P(X)$ as shown in Equation 3.3.

$$P(X) = 1x^7 + 0x^6 + 0x^5 + 1x^4 + 1x^3 + 0x^2 + 1x^1 + 1x^0 \tag{3.3}$$

If the polynomial $P(X)$ is divided by a second polynomial, called a *generator polynomial* $G(X)$, then the result is a *quotient polynomial* $Q(X)$ plus a *remainder polynomial* $R(X)$; that is,

$$\frac{P(X)}{G(X)} = Q(X) + R(X)$$

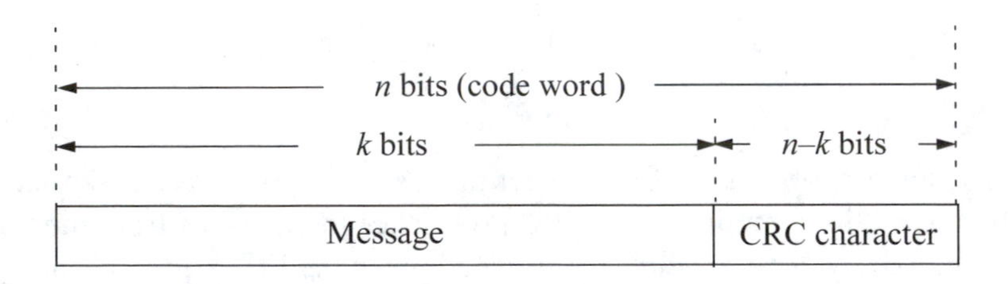

Figure 3.21 Format for an n-bit code word.

The CRC character that is added to the end of the binary bit stream is the remainder $R(X)$. The CRC generator performs a continuous division of the bit stream polynomial $P(X)$ by the generator polynomial $G(X)$. The degree (highest exponent value) of $G(X)$ is

$$0 < G(X) < P(X)$$

Also, $G(X)$ has a nonzero coefficient for the X^0 term.

Division of the data polynomial by the generator polynomial is performed by modulo-2 division. Since modulo-2 is a linear operation, the register used to generate and check CRC characters is called a *linear feedback shift register*.

Consider the previous data stream of

Bit position	7	6	5	4	3	2	1	0	
Data	1	0	0	1	1	0	1	1	(LSB)

which yields a data (or message) polynomial of $P(X) = 1x^7 + 0x^6 + 0x^5 + 1x^4 + 1x^3 + 0x^2 + 1x^1 + 1x^0$, or simply $x^7 + x^4 + x^3 + x + 1$ (LSB). Assume a generator polynomial of $x^4 + x + 1$. The CRC check character is formed by dividing the shifted message polynomial by the generator polynomial. The code word that is transmitted to the destination during a write operation consists of n bits: a message segment of k bits, and a CRC check bit segment of $n - k$ bits, which is the remainder polynomial. When the bit stream is received at the destination during a read operation, both the data bits and the CRC bits are shifted through an identical CRC generator/checker circuit. Because the received bit stream now includes the remainder polynomial, the division operation — produced by shifting — results in a remainder of zero if no errors occurred during transmission or reception; that is, $R(X) = 0$.

To encode a message polynomial $P(X)$, multiply $P(X)$ by $x(n - k)$, because the transmitted code word will have 12 bits: eight message bits and four check bits ($n - k = 12 - 8 = 4$). Thus, the exponent of each message bit will be increased by x^4. The code word is then divided by the generator polynomial $G(X)$ to yield a quotient $Q(X)$ and a remainder $R(X)$, which is appended to the end of the message bits. This encoding process is expressed in Equation 3.4.

$$\frac{x^{n-k}P(X)}{G(X)} = Q(X) + \frac{R(X)}{G(X)} \tag{3.4}$$

To elucidate the encoding method, assume a message of $P(X) = 1001\ 1011 = x^7 + x^4 + x^3 + x + 1$ and a generator polynomial of $G(X) = x^4 + x + 1$. The transmitted message is multiplied by x^4 — the high-order variable of the generator polynomial — so that the code word will contain n bits $[k + (n - k)]$. This corresponds to a left shift of $n - k$ bits. This new message of n bits is then divided by the generator polynomial to produce an n-bit code word containing a k-bit message and its corresponding $n - k$

check bits, which is the CRC character. The encoding technique is shown in Figure 3.22 (a).

The remainder polynomial $R(X) = x^2 + 1$ is appended to the end of the message and $P'(X) + R(X)$ is transmitted. Thus, the transmitted code word $C(X)$ is

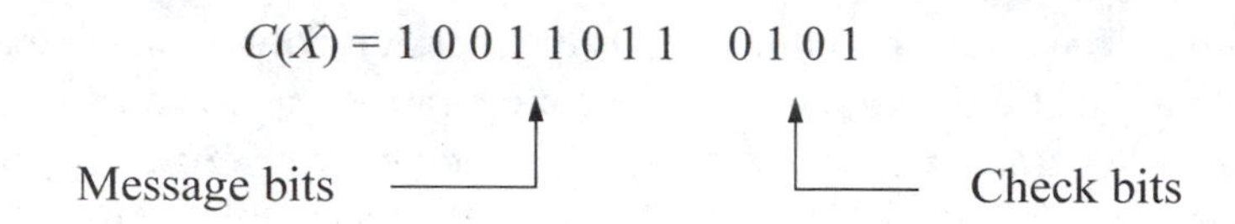

When the code word is read back from the destination, the received polynomial $C(X)$ — containing the appended remainder — is divided by the generator polynomial $G(X)$. If no error has occurred in transmission, then the remainder $R(X)$ will be zero. This is illustrated in Figure 3.22 (b).

The hardware for encoding a message polynomial is shown in Figure 3.23 (a), and consists of a linear feedback shift register for use with the generator polynomial $G(X)$. Since modulo-2 addition is performed, the shift register contains two exclusive-OR gates. The message bits enter the register from the left beginning with the high-order bit. The gating logic shown at the output of flip-flop y_4 circulates the shifted polynomial through the feedback paths until the code word polynomial is generated, then allows the CRC character to be transmitted.

Figure 3.23 (b) tabulates the sequence of shifts which will generate the remainder $R(X)$. The shift register is reset to zero initially. Recall that the message polynomial was shifted left $n - k = 12 - 8 = 4$ bit positions to allow four bits for the CRC character. This is represented by shift counts 9 through 12. After 12 shifts, the remainder polynomial $R(X) = x^2 + 1 = 0101$.

When the code word is read back from the destination, it is processed through the same shift register. Because the remainder is also shifted through the register, the result will be zero if no error has occurred during transmission. The sequence of shifts is tabulated in Figure 3.24 (a) and indicates that the received code word — and thus the message — was free from errors.

Figure 3.24 (b) represents a shift sequence in which the received code word contained an error in bit position 7. The remainder is nonzero as indicated in shift count 12 and is expressed by the error polynomial $E(X)$. Using $E(X)$, a table lookup procedure specifies the bit in error. To correct the error in this example, the x^7 term is added to the received code word using modulo-2 addition, resulting in the original code word containing the message and check bits.

The primitive generator polynomial $x^4 + x + 1$ is only one of many primitive polynomials used for error detection and correction. A popular generator polynomial for disk drives and binary synchronous communication is $x^{16} + x^{15} + x^2 + 1$.

Encode the message

Message polynomial $P(X)$ = 1001 1011 $= x^7 + x^4 + x^3 + x + 1$
Generator polynomial $G(X) = x^4 + x + 1$

Multiply the message polynomial by x^4
The shifted message polynomial $P'(X)$ $= x^4(x^7 + x^4 + x^3 + x + 1)$
$= x^{11} + x^8 + x^7 + x^5 + x^4$

Divide $P'(X)$ by $G(X)$.

$$
\begin{array}{r|l}
 & x^7 \quad + x \quad + 1 \\
\hline
x^4 + x + 1 & x^{11} + x^8 + x^7 + x^5 + x^4 \\
 & \underline{x^{11} + x^8 + x^7} \\
 & \qquad\qquad x^5 + x^4 \\
 & \qquad\qquad \underline{x^5 + x^2 + x} \\
 & \qquad\qquad\quad x^4 + x^2 + x \\
 & \qquad\qquad\quad \underline{x^4 \qquad + x + 1} \\
 & \qquad\qquad\qquad + x^2 \qquad + 1
\end{array}
$$

Therefore, the transmitted code word $= x^{11} + x^8 + x^7 + x^5 + x^4 + \underline{x^2 + 1}$

CRC check bits ⟶ ↑

(a)

Check the received code word

Received code word $= x^{11} + x^8 + x^7 + x^5 + x^4 + x^2 + 1$
Divide the received code word by the generator polynomial $G(X)$

$$
\begin{array}{r|l}
 & x^7 \quad + x \quad + 1 \\
\hline
x^4 + x + 1 & x^{11} + x^8 + x^7 + x^5 + x^4 + x^2 + 1 \\
 & \underline{x^{11} + x^8 + x^7} \\
 & \qquad\qquad + x^5 + x^4 + x^2 + 1 \\
 & \qquad\qquad \underline{+ x^5 \qquad + x^2 + 1} \\
 & \qquad\qquad\qquad + x^4 + x + 1 \\
 & \qquad\qquad\qquad \underline{+ x^4 + x + 1} \\
 & \qquad\qquad\qquad\qquad\qquad 0
\end{array}
$$

(b)

Figure 3.22 Encoding the message polynomial $P(X) = x^7 + x^4 + x^3 + x + 1$ for transmission using the generator polynomial $G(X) = x^4 + x + 1$: (a) divide $P'(X)$ by $G(X)$ to yield $R(X)$; and (b) checking the received polynomial $C(X)$ for errors.

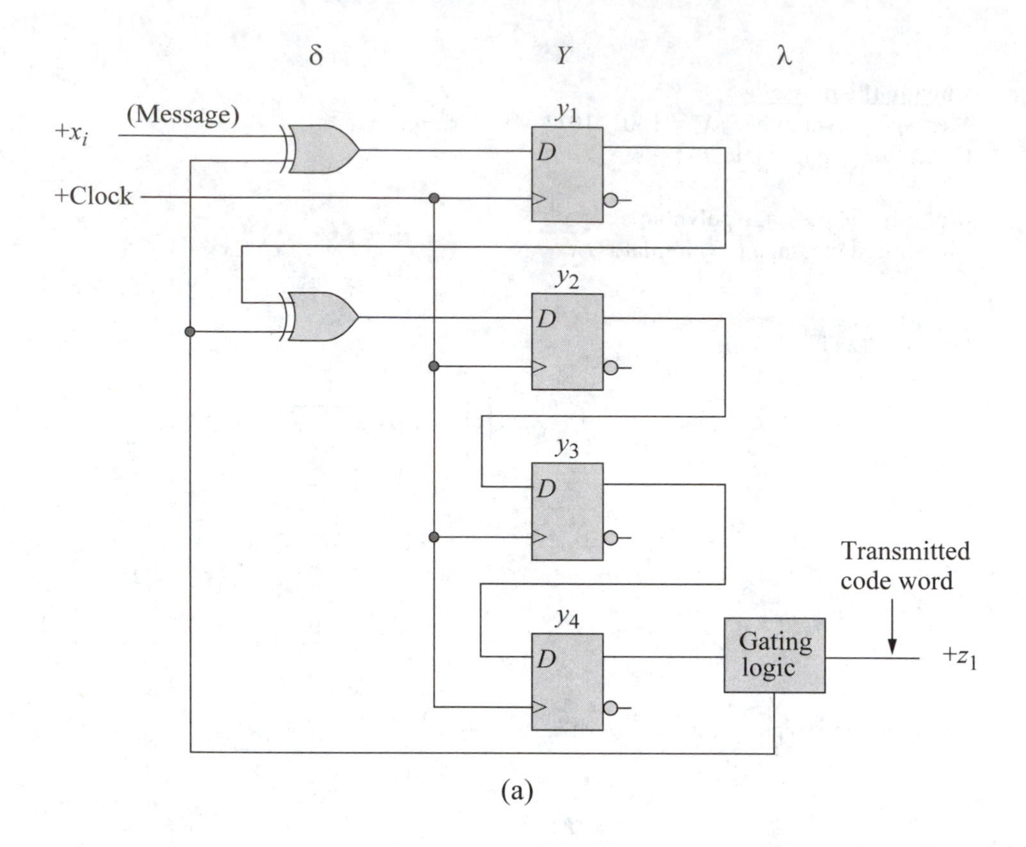

(a)

Shift	Message $P'(X)$	y_1	y_2	y_3	y_4	(High order)
		0	0	0	0	
1	1	1	0	0	0	
2	0	0	1	0	0	
3	0	0	0	1	0	
4	1	1	0	0	1	
5	1	0	0	0	0	
6	0	0	0	0	0	
7	1	1	0	0	0	
8	1	1	1	0	0	
9	0	0	1	1	0	
10	0	0	0	1	1	
11	0	1	1	0	1	
12	0	1	0	1	0	$= R(X) = x^2 + 1 = 0101$

(b)

Figure 3.23 Linear feedback shift register for generator polynomial $x^4 + x + 1$: (a) implementation; and (b) shift sequence to generate the remainder polynomial $R(X)$ for message 1001 1011.

Shift	Received code word	y_1	y_2	y_3	y_4 (High order)
		0	0	0	0
1	1	1	0	0	0
2	0	0	1	0	0
3	0	0	0	1	0
4	1	1	0	0	1
5	1	0	0	0	0
6	0	0	0	0	0
7	1	1	0	0	0
8	1	1	1	0	0
9	0	0	1	1	0
10	1	1	0	1	1
11	0	1	0	0	1
12	1	0	0	0	0 = No error

(a)

Assume that an error occurred in bit position 7 of the received code word.

Bit position =	11	10	9	8	7	6	5	4	3	2	1	0
Received code word =	1	0	0	1	**0**	0	1	1	0	1	0	1
	Message								Check bits			

Therefore, The received code word polynomial is $x^{11} + x^8 + x^5 + x^4 + x^2 + 1$.

Shift	Received code word	y_1	y_2	y_3	y_4 (High order)
		0	0	0	0
1	1	1	0	0	0
2	0	0	1	0	0
3	0	0	0	1	0
4	1	1	0	0	1
5	0	1	0	0	0
6	0	0	1	0	0
7	1	1	0	1	0
8	1	1	1	0	1
9	0	1	0	1	0
10	1	1	1	0	1
11	0	1	0	1	0
12	1	1	1	0	1 $= R(X) = E(X)\,x^7$

(b)

Figure 3.24 Tabulation of shift sequences: (a) for received code word containing no error; and (b) for received code word containing an error. By adding x^7 to the received code word, the original code word is obtained.

3.2.6 Combinational Shifter

Before leaving the topic of shift registers, another type of shifter will be presented. Although not sequential in structure, it is used extensively in high-speed processors, specifically for machines with long word sizes such as, 32- or 64-bit operands. The shifter accomplishes all shift operations, whether left or right, algebraic or logical, by shifting left only. This results in considerable hardware savings, especially for 64-bit operands.

There are four basic shift operations: *shift left logical* (SLL) and *shift left algebraic* (SLA) for unsigned and signed operands, respectively; *shift right logical* (SRL) and *shift right algebraic* (SRA) for unsigned and signed operands, respectively.

Left shift count Assume a 6-bit, left-shift count field for radix 2 as follows:

$$C = c_5 c_4 c_3 c_2 c_1 c_0 \tag{3.5}$$

where C is the shift count and c_i is a shift bit, either 0 or 1, with a binary weight of 2^i. For example, let the left shift count be

$$\begin{array}{rcccccc} & 2^5 & 2^4 & 2^3 & 2^2 & 2^1 & 2^0 \\ C = & 0 & 0 & 1 & 1 & 0 & 0_2 \end{array}$$

The shift count can be expressed as

$$\begin{aligned} C &= c_5 \times 2^5 + c_4 \times 2^4 + c_3 \times 2^3 + c_2 \times 2^2 + c_1 \times 2^1 + c_0 \times 2^0 \\ &= 0 \times 2^5 + 0 \times 2^4 + 1 \times 2^3 + 1 \times 2^2 + 0 \times 2^1 + 0 \times 2^0 \\ &= 8 + 4 \\ &= 12 \end{aligned}$$

The shift count expansion shown above can be represented in a more compact mathematical form as a summation of terms as shown in Equation 3.6.

$$C = \sum_{i=0}^{n-1} c_i 2^i \tag{3.6}$$

where $c_i = 0$ or 1, and $2^n - 1$ is the maximum shift count. If $n = 6$, then the maximum shift count is $2^6 - 1 = 63$.

Right shift count The right shift count can be considered as the negative of the left shift count. Thus, given a right shift count, the equivalent left shift count is the 2s complement (negation) of the right shift count. That is, a right shift can be realized by shifting left an amount that is the 2s complement of the right shift count. For example, let the right shift count be

$$\begin{array}{rcccc} & 2^3 & 2^2 & 2^1 & 2^0 \\ C = & 0 & 0 & 1 & 1_2 \end{array}$$

Therefore,

$$\begin{aligned} C &= c_3 \times 2^3 + c_2 \times 2^2 + c_1 \times 2^1 + c_0 \times 2^0 \\ &= 0 \times 2^3 + 0 \times 2^2 + 1 \times 2^1 + 1 \times 2^0 \\ &= 2 + 1 \\ &= 3 \end{aligned}$$

The same results can be obtained by shifting left an amount that is the 2s complement of 0011_2. Thus, the equivalent left shift is

$$\begin{array}{rcccc} & 2^3 & 2^2 & 2^1 & 2^0 \\ C = & 1 & 1 & 0 & 1_2 \end{array}$$

and is expressed as

$$\begin{aligned} C &= c_3' \times 2^3 + c_2' \times 2^2 + c_1' \times 2^1 + c_0 \times 2^0 \\ &= 1 \times 2^3 + 1 \times 2^2 + 0 \times 2^1 + 1 \times 2^0 \\ &= 8 + 4 + 1 \\ &= 13 \end{aligned}$$

A more compact notation for the *r*s complement of a number *A* is specified in Equation 3.7.

$$(A)'_{+1} = r^n - A \tag{3.7}$$

For radix 2, Equation 3.7 becomes

$$(A)'_{+1} = 2^n - A \tag{3.8}$$

Let A' be the $r - 1$ (1s) complement, then $(A)' + 1$ is the $[(r - 1) + 1] = rs$ (2s) complement. Thus, for a right shift count of $C = 0011_2$, the equivalent left shift count is

$$
\begin{aligned}
(A)'_{+1} &= 2^4 - C \\
&= 16 - 3 \\
&= 13 \\
&= 1101_2
\end{aligned}
$$

If Equation 3.7 is analyzed, it is seen that the 2s complement of a number can be obtained by leaving the low-order 0s unchanged, then subtracting the first nonzero low-order digit from r, then subtracting all remaining digits from $r - 1$. For radix 2, Equation 3.8 specifies that the 2s complement of a binary number can be obtained as follows:

> Scanning the number from right to left, keep the low-order 0s and the first 1 unchanged, then complement (invert) all remaining bits.

Eight-bit operands will be used for illustrative purposes for the shift operations described below, although the methods apply to any size operand. A left shift of one bit position is equivalent to multiplying the number by two; a right shift of one bit position is equivalent to dividing the number by two.

Shift left logical (SLL) The logical shift operations are much simpler to implement than the arithmetic shift operations. For SLL, the high-order bit of the unsigned operand is shifted out of the left end of the shifter for each shift cycle. Zeros are entered from the right and fill the vacated low-order bit positions. Figure 3.25 depicts an 8-bit register with a left shift count of 3 (011).

Shift left algebraic (SLA) SLA operates on signed operands in 2s complement representation for radix 2. The numeric part of the operand is shifted left the number of bit positions specified in the shift count field. The sign remains unchanged and does not participate in the shift operation. All remaining bits participate in the left shift. For the 8-bit register operand shown in Figure 3.26, the bits are shifted out of the high-order numeric bit position and 0s are shifted in to the vacated register positions on the right. The left shift count is 011. An overflow occurs if a bit shifted out of the high-order numeric position is different than the sign bit.

Shift right logical (SRL) As mentioned previously, any right shift operation can be implemented by shifting left an amount that is the 2s complement of the right shift count. This is expressed in Equation 3.9.

$$\text{Left shift count} = (\text{right shift count})' + 1 \tag{3.9}$$

The equivalent left shift operation is implemented in two levels of hardware. Figure 3.27 shows an 8-bit operand that is to be shifted right three bit positions. Since the right shift count is 011, the equivalent left shift count is 100 + 1, which is the 2s complement of the right shift count. In level A, the operand is offset to the right by a number of bit positions equal to the operand length, minus one bit. The "minus one bit" represents a left shift of one bit position when 2s complementing the right shift count; that is, it is the "+1" in Equation 3.9. This built-in left shift of one bit position reduces the amount hardware by one cell. The remaining high-order bit positions are set to zero, because the operation is a logical right shift of an unsigned number.

In level B, the operand is shifted left by an amount equal to the equivalent left shift count minus 1; that is, the 1s complement of the right shift count. The resultant operand in level B is identical to the shifted operand obtained by the original right shift operation.

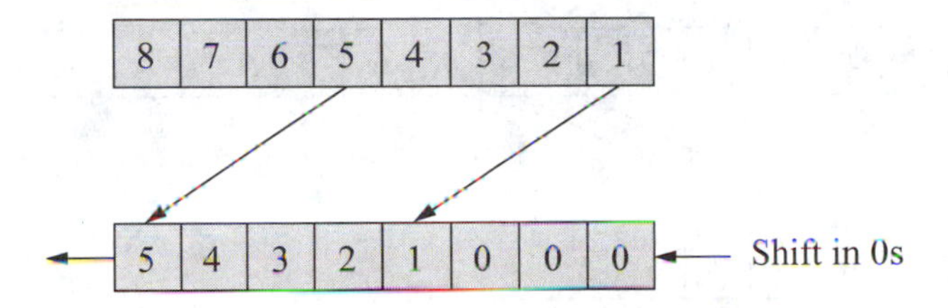

Figure 3.25 Shift left logical 3 (011) bit positions.

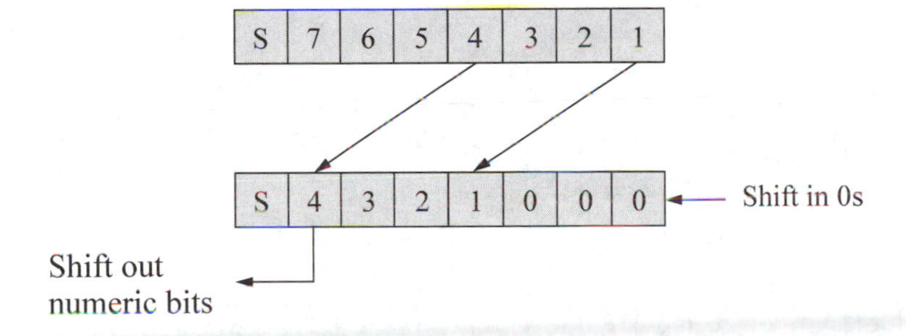

Figure 3.26 Shift left arithmetic 3 (011) bit positions. The sign bit does not shift.

Shift right algebraic (SRA) The numeric part of the signed operand is shifted right the number of bits specified by the shift count. The sign of the operand remains unchanged. All numeric bits participate in the right shift. The sign bit propagates right to fill in the vacated high-order numeric bit positions. When the operation is executed by shifting left, it is identical to SRL with the exception that the high-order bits in level A are set to the value of the sign bit. Figure 3.28 illustrates a SRA operation with a right shift count of 5 (101). The equivalent left shift count is 010 + 1, or simply 010 after the "+1" left shift has been implemented. The operand in level B is identical to the operand that would have been obtained from the original SRA operation.

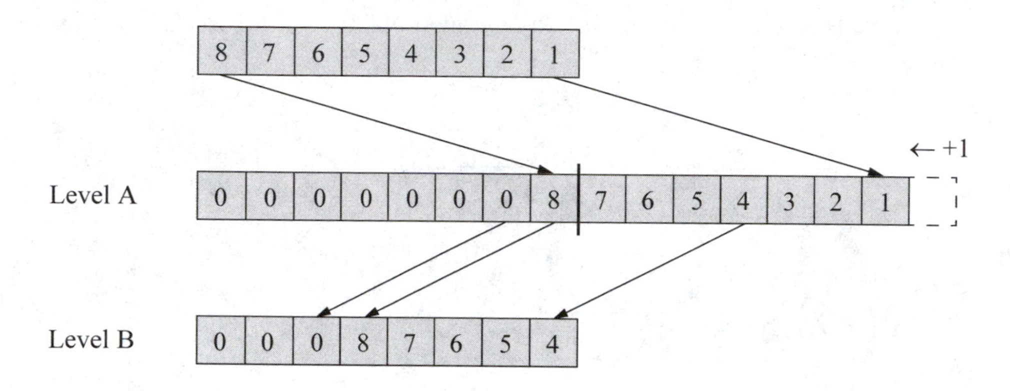

Figure 3.27 Shift right logical 3 (011) bit positions. The equivalent left shift is 100 + 1.

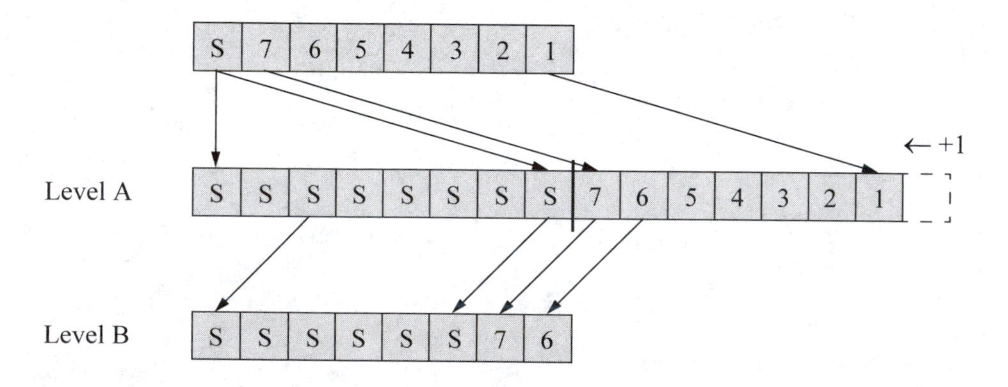

Figure 3.28 Shift right algebraic 5 (101) bit positions. The equivalent left shift is 010 + 1.

3.3 Synchronous Counters

Counters are essential devices used in the design of digital systems. Counters have a finite number of states and represent simple Moore machines in most cases. The λ output logic is usually a function of the present state only; that is, $\lambda(Y_{j(t)})$. The state of the counter is interpreted as an integer with respect to a modulus. A number A modulo n is defined as the remainder after dividing A by n. Some counters contain a set of binary input variables from which the counter achieves an initial state.

A clock input signal causes the counter flip-flops to react only at selected discrete intervals of time; in some cases, the clock input occurs randomly. Using the clock pulses to initiate state changes, the machine counts in either an ascending or descending sequence of states. Other counters have no inputs except a clock pulse and are usually reset to an initial state of $y_1 y_2 \ldots y_p = 00 \ldots 0$. In general, a p-stage counter counts modulo 2^p.

This section will discuss only synchronous counters; asynchronous counters are inherently slow, because of the ripple effect caused by the output of stage y_i functioning as the clock input for stage y_{i+1}. The maximum time for an asynchronous counter to change state occurs when all flip-flops are set and the count increments from $2^p - 1$ to zero.

The synchronous sequential machines in this section are associated with a set of transformations on a set of states and follow a prescribed sequence of states under control of a clock input signal. When the active clock transition occurs at the input, the state of the machine changes to some predetermined value as defined by the machine specifications. Counting sequentially is completely arbitrary, although the next state is usually an increment or decrement by one, or a state in which only one flip-flop changes state, as in a Gray code counter.

3.3.1 Modulo-8 Counter

The simplest counter is the *binary counter* which counts in an increasing binary sequence from $y_1 y_2 \ldots y_p = 00 \ldots 0$ to $y_1 y_2 \ldots y_p = 11 \ldots 1$, then returns to zero. Using the synthesis procedure described in Section 3.1, a modulo-8 counter will be designed using D flip-flops.

Step 1: State diagram The state diagram for a modulo-8 counter is shown in Figure 3.29. Unlike conventional Moore and Mealy state diagrams, which detect code words or bit sequences, the state diagram for a counter is relatively straightforward. The counter is initially reset to $y_1 y_2 y_3 = 000$, then increments by one at each positive clock transition until state h ($y_1 y_2 y_3 = 111$) is reached. At the next positive clock transition, the counter sequences to state a ($y_1 y_2 y_3 = 000$).

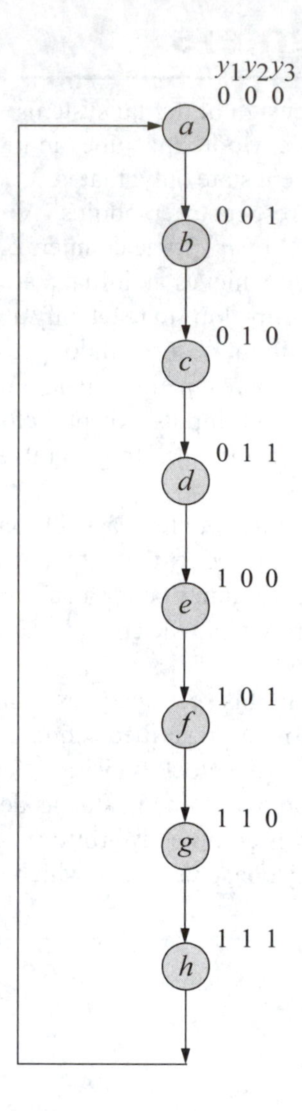

Figure 3.29 State diagram for a modulo-8 counter.

When designing a modulo-m counter, if the modulus is an integral power of 2, such as, 2,4,8,16, ... , then the number of flip-flops required to implement the counter is

$$n = \log_2 m \tag{3.10}$$

where n is the number of flip-flops and m is the modulus. The logarithm of a positive number with a positive base greater than 1 is the exponent of the power to which the base must be raised to equal the number. From this definition of a logarithm, the following equation is obtained:

$$m = 2^n \tag{3.11}$$

Equation 3.10 and Equation 3.11 state the same relation in different ways. Thus, for a modulo-8 counter, three flip-flops ($y_1 y_2 y_3$) are required, since $8 = 2^3$.

Step 2: Equivalent states Since every state of a modulo-8 counter is required, there are no equivalent states in the state diagram of Figure 3.29.

Step 3: State code assignment The state codes 000 through 111 are predefined for a modulo-8 counter as shown in the state diagram.

Step 4: Next-state table The next-state table lists all possible present states and next states, as shown in Table 3.4. The next state $y_{k(t+1)}$ for a D flip-flop is the same as the present value of the D input before the flip-flop is clocked. Thus, $y_{i(t+1)} = Dy_{i(t)}$. No binary input variables are required and the counter outputs are taken directly from the state flip-flop outputs.

Table 3.4 Next-state table for the modulo-8 counter of Figure 3.29

Present state	Next state	Flip-flop inputs		
$y_1 y_2 y_3$	$y_1 y_2 y_3$	Dy_1	Dy_2	Dy_3
0 0 0	0 0 1	0	0	1
0 0 1	0 1 0	0	1	0
0 1 0	0 1 1	0	1	1
0 1 1	1 0 0	1	0	0
1 0 0	1 0 1	1	0	1
1 0 1	1 1 0	1	1	0
1 1 0	1 1 1	1	1	1
1 1 1	0 0 0	0	0	0

Step 5: Input maps The input maps, or excitation maps, represent the δ next-state logic for the flip-flop inputs. The input maps can be obtained from either the state diagram or the next-state table. Since the counter contains three flip-flops, there will be three input maps, one for each flip-flop. The input maps are shown in Figure 3.30 and specify the input logic for flip-flops y_1, y_2, and y_3.

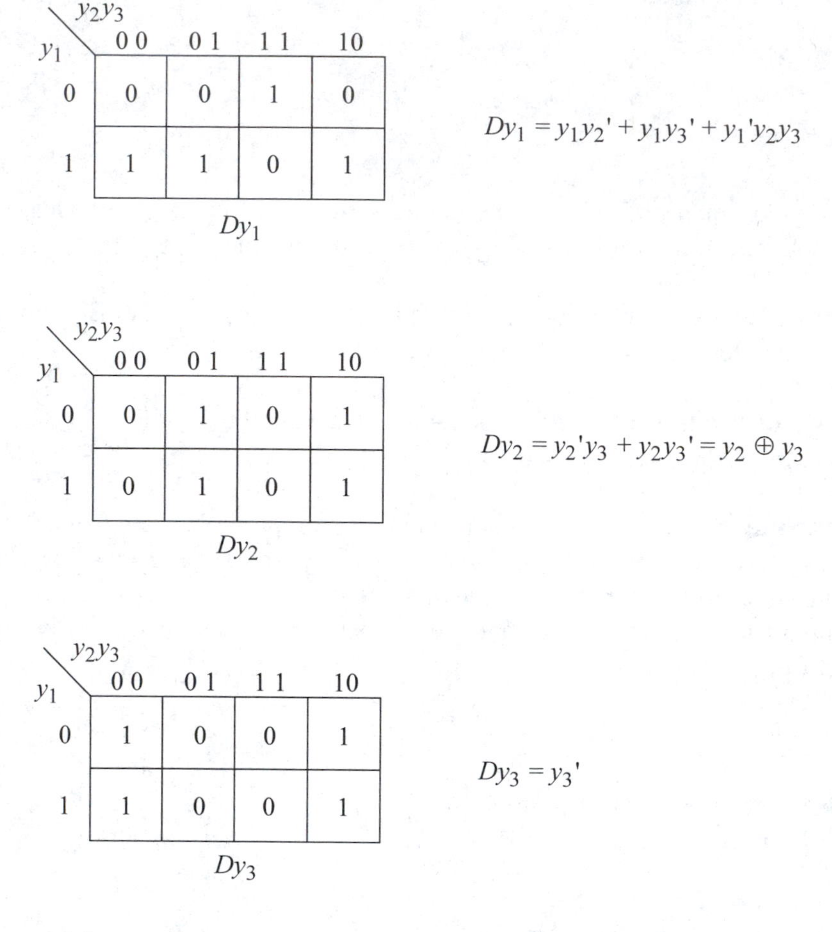

Figure 3.30 Input maps for the modulo-8 counter of Figure 3.29 using *D* flip-flops.

Referring to the state diagram of Figure 3.29 and beginning in state *a* ($y_1y_2y_3 = 000$), the next state for y_1 is 0. Therefore, a 0 is entered in minterm location 000 for flip-flop y_1. Zeros are also entered in locations 001 and 010 for y_1. In state *d* ($y_1y_2y_3 = 011$), the next state for y_1 is 1. Thus, a 1 is entered in minterm location 011 for y_1. In the same manner, next-state values are entered in the remaining minterm locations for y_1. Similarly, entries for the input maps representing flip-flops y_2 and y_3 are derived. Because the modulus for a modulo-8 counter is an integral power of 2, there are no unused ("don't care") states in the input maps. The input equations for flip-flops y_1, y_2, and y_3 are presented in Equation 3.12.

$$
\begin{aligned}
Dy_1 &= y_1y_2' + y_1y_3' + y_1'y_2y_3 \\
Dy_2 &= y_2'y_3 + y_2y_3' \\
&= y_2 \oplus y_3 \\
Dy_3 &= y_3'
\end{aligned} \tag{3.12}
$$

Step 6: Output maps Output maps are not required for this modulo-8 counter.

Step 7: Logic diagram The logic diagram is obtained from the input equations of Equation 3.12 and is shown in Figure 3.31. Note that flip-flop y_3 is connected in toggle mode and that flip-flop y_2 is set to 1 only when $y_2 \neq y_3$. The clock pulse is applied to the clock input of all flip-flops simultaneously. By careful analysis of the logic diagram, the timing diagram of Figure 3.32 is realized. As is evident, flip-flop y_3 changes state (toggles) on the rising edge of each clock pulse, because its complemented output is connected to the D input. Flip-flop y_2 is set to 1 only when the states of y_2 and y_3 are different. On the rising edge of clock pulse 2, $y_2y_3 = 01$, thus flip-flop y_2 is set to 1 after an appropriate propagation delay through internal logic. At the positive transition of clock pulse 3, the conditions for setting y_2 are again met ($y_2y_3 = 10$). Since y_2 is already set, the flip-flop remains set after the occurrence of clock pulse 3.

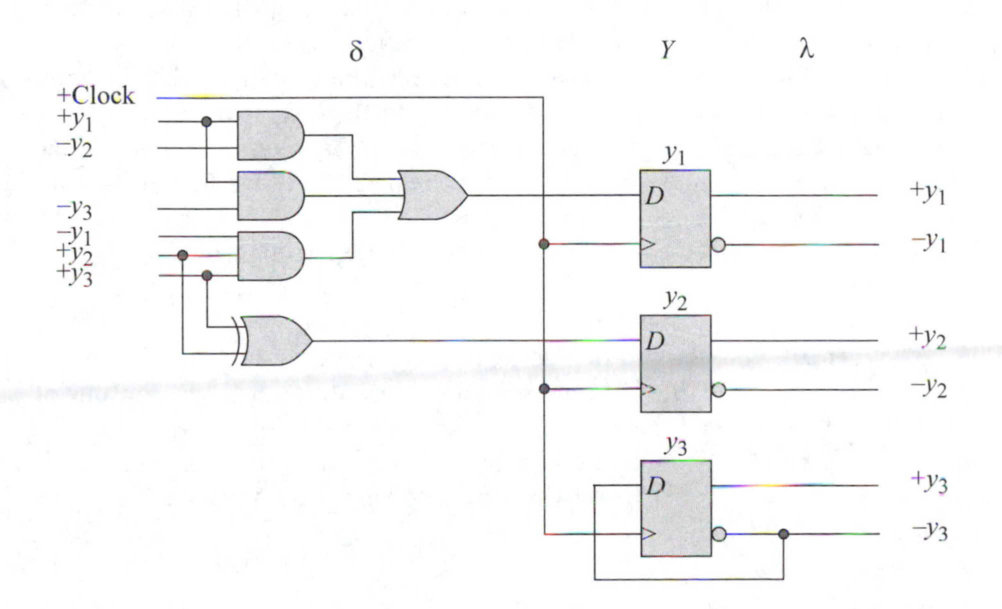

Figure 3.31 Logic diagram for the modulo-8 counter of Figure 3.29 using D flip-flops, where y_3 is the low-order stage.

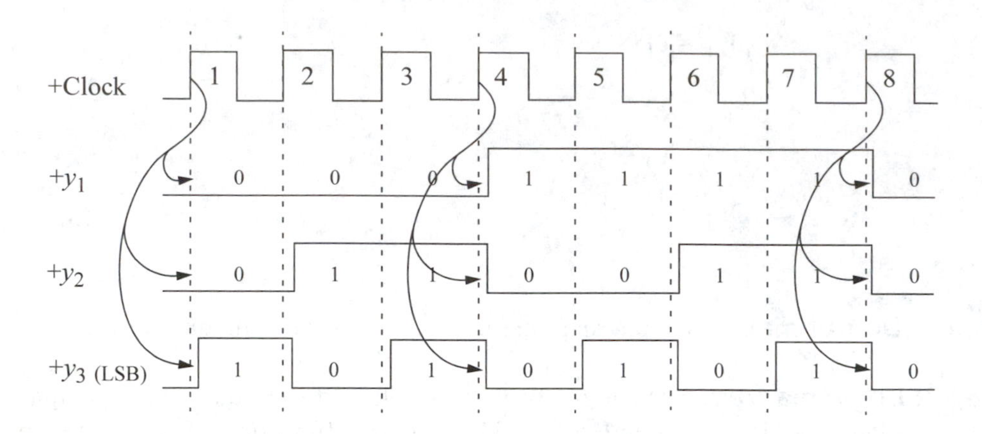

Figure 3.32 Timing diagram for the modulo-8 counter of Figure 3.31 using positive-edge-triggered *D* flip-flops.

Three conditions cause the *D* input of flip-flop y_1 to be set to a value of 1: $y_1y_2 = 10$, $y_1y_3 = 10$, or $y_1y_2y_3 = 011$. The positive transition of clock pulse 4 sets flip-flop y_1 to 1, because $y_1y_2y_3 = 011$ at that instant. Clock pulse 5 maintains y_1 in the set condition, because two conditions are met: $y_1y_2 = 10$ and $y_1y_3 = 10$, either of which generates a value of 1 for Dy_1. At the occurrence of clock pulses 6 and 7, $y_1y_2 = 10$ and $y_1y_3 = 10$, both of which maintain flip-flop y_1 in the set state. It is not until clock pulse 8 that all terms of Dy_1, as represented in Equation 3.12, are zero. Thus, the positive transition of clock pulse 8 resets flip-flop y_1.

The same counter will now be synthesized using *JK* flip-flops. An appreciable amount of savings for a counter of this size can be realized by using *JK* flip-flops instead of *D* flip-flops. This is due to the inherent added logic functions supplied by the *JK* data inputs. Before proceeding with the synthesis, the excitation table for a *JK* flip-flop will be derived.

The operation of a *JK* flip-flop is expressed by the *characteristic table* of Table 3.5. When $JK = 00$, the flip-flop does not change state at the following active clock transition. A value of $JK = 01$ represents a reset condition; whereas, a value of $JK = 10$ represents a set condition during the next clock period. A value of $JK = 11$ specifies a toggle condition where the flip-flop changes state (or is complemented) from 0 to 1 or from 1 to 0. Table 3.5 provides the following characteristic equation for a *JK* flip-flop:

$$Y_{k(t+1)} = K'Y_{j(t)} + JY_{j(t)}'$$

Table 3.5 Characteristic table for a *JK* flip-flop

Flip-flop inputs JK	Present state $Y_{j(t)}$	Next state $Y_{k(t+1)}$
0 0	0	0
0 0	1	1
0 1	0	0
0 1	1	0
1 0	0	1
1 0	1	1
1 1	0	1
1 1	1	0

Upon examination of Table 3.5, the characteristic table can be reduced as shown in Table 3.6, which represents the next state for given values of J and K. When $JK = 00$, the next state is the same as the present state. When $JK = 11$, the next state is the complement of the present state. A third useful attribute of a JK flip-flop is the excitation table shown in Table 3.7. The excitation table is obtained from Table 3.5 and indicates the values of J and K for a given state transition sequence. For example, the first and third rows of Table 3.5 specify that a state transition of $Y_{j(t)} = 0$ to $Y_{k(t+1)} = 0$ can be obtained by either $JK = 00$ or 01. This is tabulated in Table 3.7, which indicates that if $Y_{j(t)}Y_{k(t+1)} = 00$, then the value of J must be 0, but the value of K can be either 0 or 1 — a "don't care" condition. That is, if $JK = 00$, then the next state is the same as the present state of 0; if $JK = 01$, then the flip-flop is reset to a state of 0. In both cases, the next state is 0. Using the same rationale, a state transition from 0 to 1 is obtained by either a value of $JK = 10$ (set) or a value of $JK = 11$ (toggle), which specifies that J must be 1, but K can be a 0 or 1.

Table 3.6 Reduced characteristic table for a *JK* flip-flop

Flip-flop inputs JK	Next state $Y_{k(t+1)}$
0 0	$Y_{j(t)}$
0 1	0
1 0	1
1 1	$Y_{j(t)}'$

Table 3.7 Excitation table for a *JK* flip-flop

Present state $Y_{j(t)}$	Next state $Y_{k(t+1)}$	Flip-flop inputs $J\ K$
0	0	0 –
0	1	1 –
1	0	– 1
1	1	– 0

The synthesis of the modulo-8 counter of Figure 3.29, using *JK* flip-flops, will now be resumed.

Step 1: State diagram Steps 1, 2, and 3 of the synthesis procedure remain unchanged.

Step 2: Equivalent states

Step 3: State code assignment

Step 4: Next-state table Using the state diagram of Figure 3.29 and the excitation table of Table 3.7, the next-state table of Table 3.8 is generated.

Table 3.8 Next-state table for the modulo-8 counter of Figure 3.29 using *JK* flip-flops

Present state $y_1\ y_2\ y_3$	Next state $y_1\ y_2\ y_3$	Jy_1	Ky_1	Jy_2	Ky_2	Jy_3	Ky_3
0 0 0	0 0 1	0	–	0	–	1	–
0 0 1	0 1 0	0	–	1	–	–	1
0 1 0	0 1 1	0	–	–	0	1	–
0 1 1	1 0 0	1	–	–	1	–	1
1 0 0	1 0 1	–	0	0	–	1	–
1 0 1	1 1 0	–	0	1	–	–	1
1 1 0	1 1 1	–	0	–	0	1	–
1 1 1	0 0 0	–	1	–	1	–	1

(Flip-flop inputs: Jy_1 through Ky_3)

Step 5: Input maps The input maps of Figure 3.33 can be obtained from either the state diagram directly, using the *JK* excitation table, or from the next-state table. The *JK* input equations are shown in Equation 3.13, which clearly indicate that only one AND gate (y_2y_3) is required for the δ next-state logic.

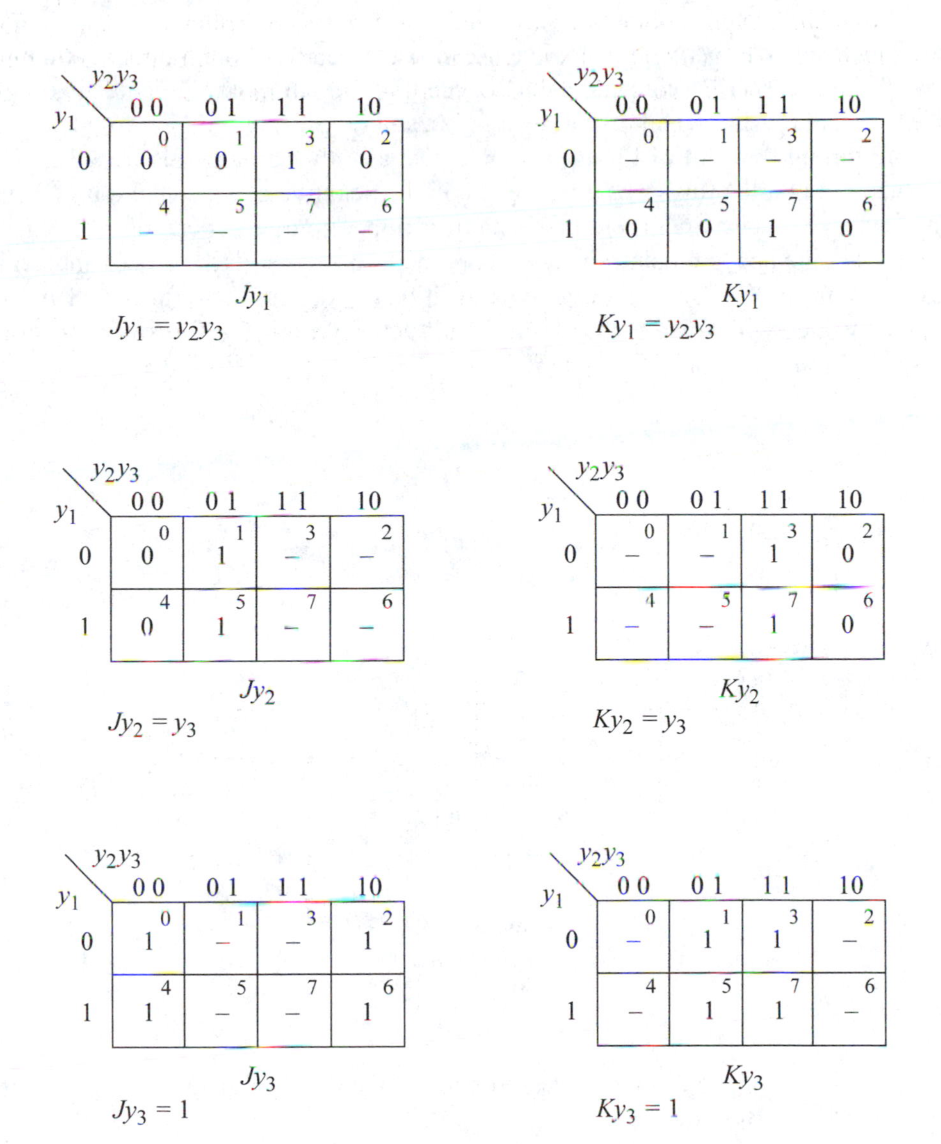

Figure 3.33 Input maps for the modulo-8 counter of Figure 3.29 using *JK* flip-flops.

$$
\begin{aligned}
Jy_1 &= y_2 y_3 \qquad & Ky_1 &= y_2 y_3 \\
Jy_2 &= y_3 \qquad & Ky_2 &= y_3 \\
Jy_3 &= 1 \qquad & Ky_3 &= 1
\end{aligned}
\tag{3.13}
$$

Step 6: Output maps The outputs are obtained directly from the flip-flop outputs, thus no output maps are required.

Step 7: Logic diagram The logic diagram, obtained from the input equations, is shown in Figure 3.34. Notice that the low-order flip-flop y_3 is implemented in toggle mode ($JK = 11$). Also, when $y_3 = 1, y_2$ toggles, and when $y_2y_3 = 11, y_1$ toggles. The resulting timing diagram of Figure 3.35 is essentially the same as Figure 3.32, with the exception that the flip-flops are clocked on the negative clock transition. Circuit delays have again been incorporated so that flip-flop output levels can be more easily visualized at the active clock transition. Circuit action takes place only on the negative edge of the clock. Once the negative transition has occurred, no further change of the input values will cause a change in circuit activity until the following negative clock transition.

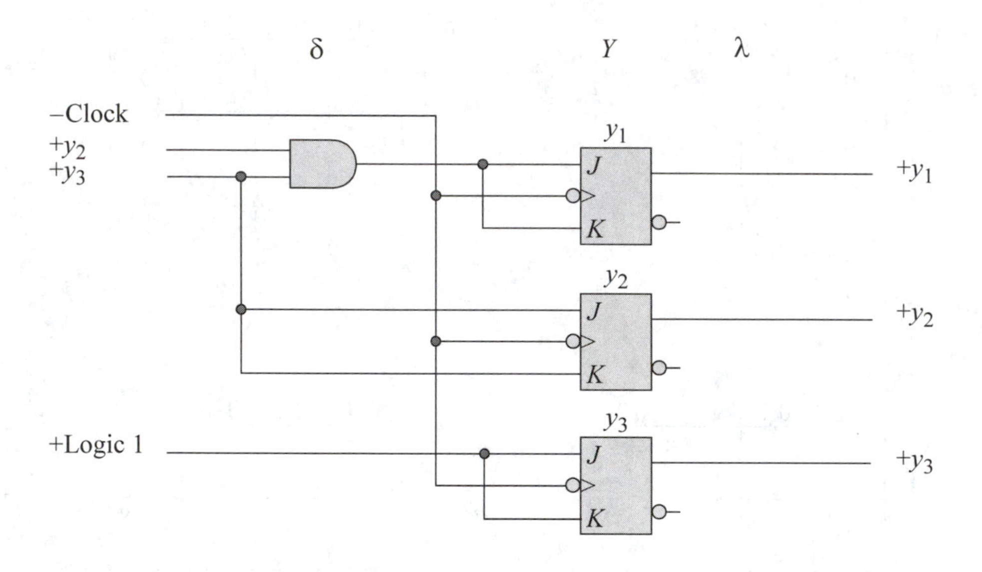

Figure 3.34 Logic diagram for the modulo-8 counter of Figure 3.29 using *JK* flip-flops, where y_3 is the low-order stage.

It may appear that flip-flop y_1 will toggle on the negative edge of clock pulse 3, because y_1 toggles if $y_2y_3 = 11$. At the instant clock pulse 3 changes from high to low, however, the conditions to toggle y_2 are not met, because $y_2y_3 = 10$. It is only after the negative edge has occurred that the conditions to toggle y_1 are met ($y_2y_3 = 11$), but by then the active negative transition has already taken place. Flip-flop y_1 must wait until clock pulse 4 in order to toggle.

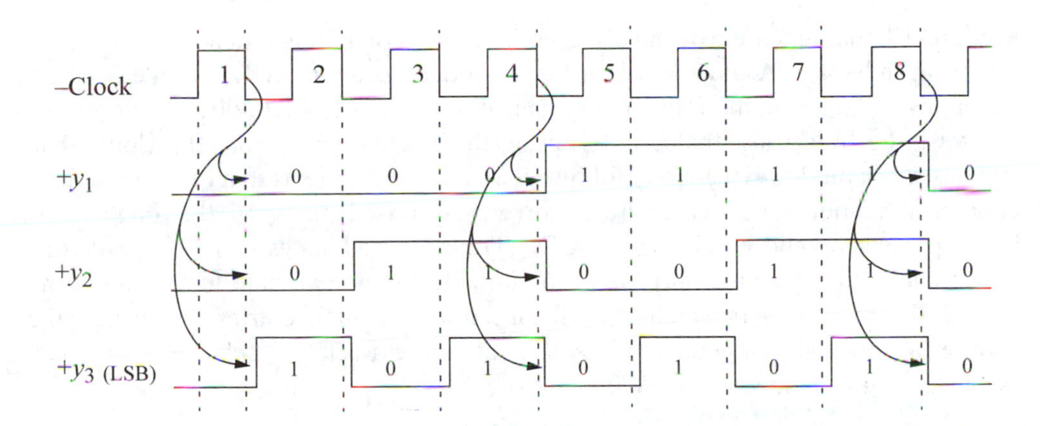

Figure 3.35 Timing diagram for the modulo-8 counter of Figure 3.34 using negative-edge-triggered *JK* flip-flops.

Flip-flop y_2 toggles whenever $y_3 = 1$. With the propagation delays shown, it is apparent that the negative edge of the clock pulse toggles y_2 only at clock pulses 2, 4, 6, and 8. Clock pulse 3 will not toggle y_2, because at the instant clock pulse 3 changes from high to low, y_3 is still reset.

As the counter progresses through the counting sequence, the machine specifications may require an indication when a particular state has been reached. This is easily implemented with an AND gate as shown in Figure 3.36 to detect, for example, state $f(y_1y_2y_3 = 101)$. The inclusion of the +Clock signal prevents an erroneous output caused by a momentary transition through state $y_1y_2y_3 = 101$ when two or more flip-flops change state for a state transition sequence that does not end in state *f*. This technique is discussed further in Section 3.7.

+y_1
–y_2
+y_3
+Clock
+State $f(y_1y_2y_3 = 101)$

Figure 3.36 Implementation to detect state $f(y_1y_2y_3 = 101)$ for the counter of Figure 3.29.

3.3.2 Modulo-10 Counter

Modulo-10 counters are extensively used in digital computers when counting is required in radix 10. A modulo-10, or binary-coded decimal (BCD) decade counter, generates ten states in the following sequence: 0000, 0001, 0010, 0011, 0100, 0101, 0110, 0111, 1000, 1001, 0000, ... Thus, each decade requires four flip-flops. The synthesis of a modulo-10 counter follows the same procedure as that of the modulo-8 counter in Section 3.3.1. There are, however, six unused states, 1010 through 1111, that represent invalid numbers for BCD. These unused states can be regarded as "don't care" states for the purpose of minimizing the δ next-state logic, unless the counter is self-starting, in which case, all unused states contain entries which cause the counter to proceed to a predetermined state at the next active clock transition. The synthesis begins below, using *JK* flip-flops.

Step 1: State diagram The state diagram for the modulo-10 counter is shown in Figure 3.37. The counter will be self-starting; that is, if the counter enters an unused state due to noise or any other transient condition, the next clock pulse will return the counter to a predetermined valid state. It is unlikely that the machine will enter an unused state; however, the possibility does exist. Digital systems enter unused states only under adverse environmental conditions such as, electrical noise, power supply voltage outside the specified operating range, or a hardware malfunction. If any of these situations occur, then the performance of the entire system is in jeopardy, not just the counter. All unused states, *k* through *p*, will sequence to $y_1y_2y_3y_4 = 0000$ as the self-starting state. The unused states correspond to minterms 10 through 15.

Step 2: Equivalent states Every state in Figure 3.37 is unique, including the unused states. Even though the unused states sequence to the same next state ($y_1y_2y_3y_4 = 0000$), their outputs are different. The outputs correspond to the state code of the individual states for this type of Moore machine. Therefore, no equivalent states exist.

Step 3: State code assignment The state codes are assigned in sequence, $y_1y_2y_3y_4 = 0000$ through 1001 for the valid digits, and $y_1y_2y_3y_4 = 1010$ through 1111 for the invalid BCD digits.

Step 4: Next-state table Using the excitation table for a *JK* flip-flop, reproduced below as Table 3.9, the next-state table is obtained as shown in Table 3.10. Note that states $y_1y_2y_3y_4 = 1010$ through $y_1y_2y_3y_4 = 1111$ proceed to state $y_1y_2y_3y_4 = 0000$ as the next state, since this is a self-starting counter and state $y_1y_2y_3y_4 = 0000$ was chosen as the self-starting state.

In the first row of Table 3.10, flip-flop y_1 has a next state of 0, providing a state transition of $y_1 = 0$ to $y_1 = 0$. Thus, using the excitation table for a *JK* flip-flop, $Jy_1Ky_1 = 0-$. The same is true for Jy_2Ky_2 and Jy_3Ky_3. Flip-flop y_4, however, changes from 0 to 1; thus, $Jy_4Ky_4 = 1-$. In a similar manner, the flip-flop input values are obtained for the remaining states $y_1y_2y_3y_4 = 0001$ through $y_1y_2y_3y_4 = 1111$.

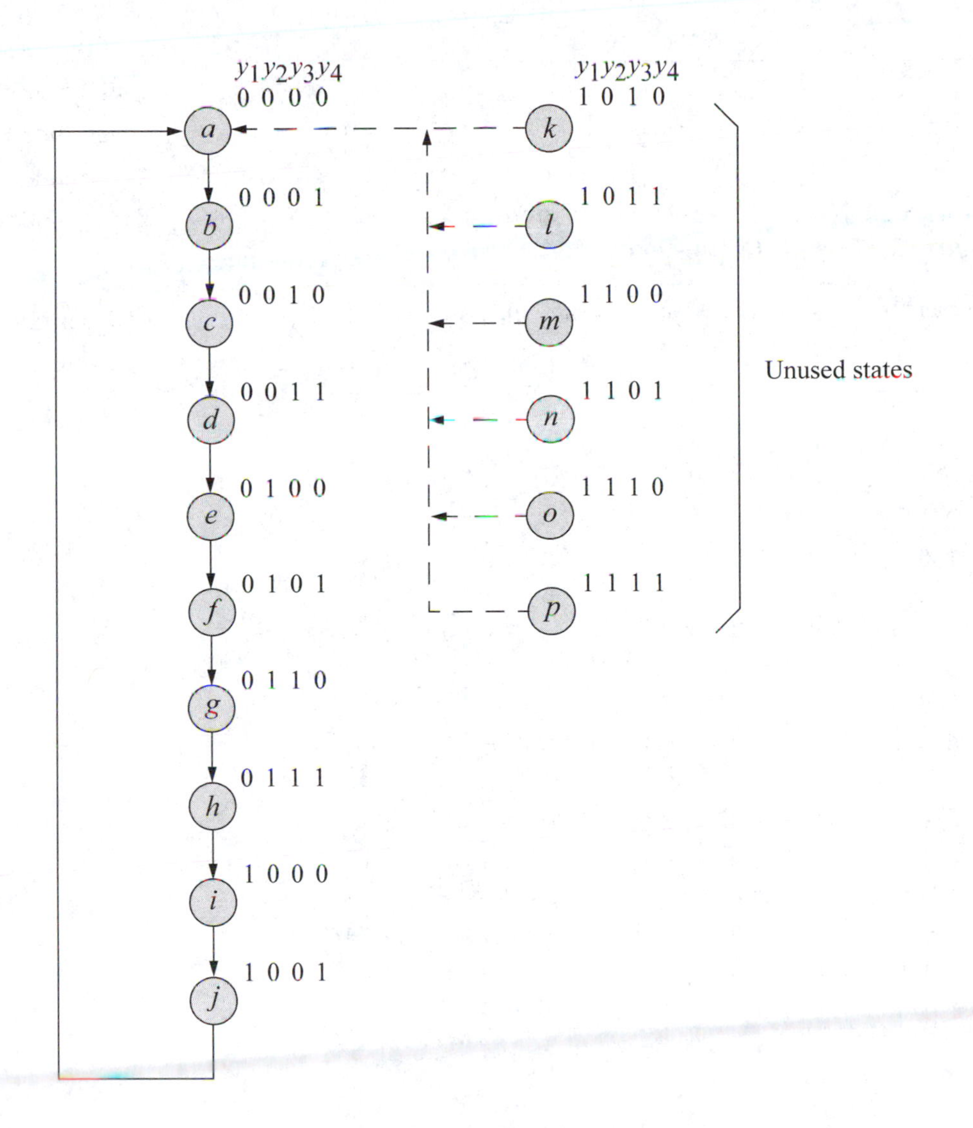

Figure 3.37 State diagram for a modulo-10 counter. Unused states *k* through *p* correspond to invalid BCD digits. The counter is self-starting to $y_1y_2y_3y_4 = 0000$ from any unused state.

Table 3.9 Excitation table for a *JK* flip-flop

Present state $Y_{j(t)}$	Next state $Y_{k(t+1)}$	Flip-flop inputs JK
0	0	0 –
0	1	1 –
1	0	– 1
1	1	– 0

Table 3.10 Next-state table for the modulo-10 counter of Figure 3.37 using *JK* flip-flops

Present state	Next state	Flip-flop inputs							
$y_1y_2y_3y_4$	$y_1y_2y_3y_4$	J_{y1}	K_{y1}	J_{y2}	K_{y2}	J_{y3}	K_{y3}	J_{y4}	K_{y4}
0 0 0 0	0 0 0 1	0	–	0	–	0	–	1	–
0 0 0 1	0 0 1 0	0	–	0	–	1	–	–	1
0 0 1 0	0 0 1 1	0	–	0	–	–	0	1	–
0 0 1 1	0 1 0 0	0	–	1	–	–	1	–	1
0 1 0 0	0 1 0 1	0	–	–	0	0	–	1	–
0 1 0 1	0 1 1 0	0	–	–	0	1	–	–	1
0 1 1 0	0 1 1 1	0	–	–	0	–	0	1	–
0 1 1 1	1 0 0 0	1	–	–	1	–	1	–	1
1 0 0 0	1 0 0 1	–	0	0	–	0	–	1	–
1 0 0 1	0 0 0 0	–	1	0	–	0	–	–	1
1 0 1 0	0 0 0 0	–	1	0	–	–	1	0	–
1 0 1 1	0 0 0 0	–	1	0	–	–	1	–	1
1 1 0 0	0 0 0 0	–	1	–	1	0	–	0	–
1 1 0 1	0 0 0 0	–	1	–	1	0	–	–	1
1 1 1 0	0 0 0 0	–	1	–	1	–	1	0	–
1 1 1 1	0 0 0 0	–	1	–	1	–	1	–	1

Step 5: Input maps The input maps of Figure 3.38 are derived either from the next-state table directly or from the state diagram in conjunction with the *JK* flip-flop excitation table. For example, using the state diagram, the entries for Jy_1 and Ky_1 for location $y_1y_2y_3y_4 = 0111$ are obtained as follows: Flip-flop y_1 changes from 0 to 1 as the machine progresses from state h to state i. From the excitation table, a transition from 0 to 1 results in $Jy_1Ky_1 = 1–$, as indicated in the input maps for Jy_1 and Ky_1. Four

sets of input maps are necessary: two maps, Jy_i and Ky_i, for each flip-flop y_i. The input equations are presented in Equation 3.14.

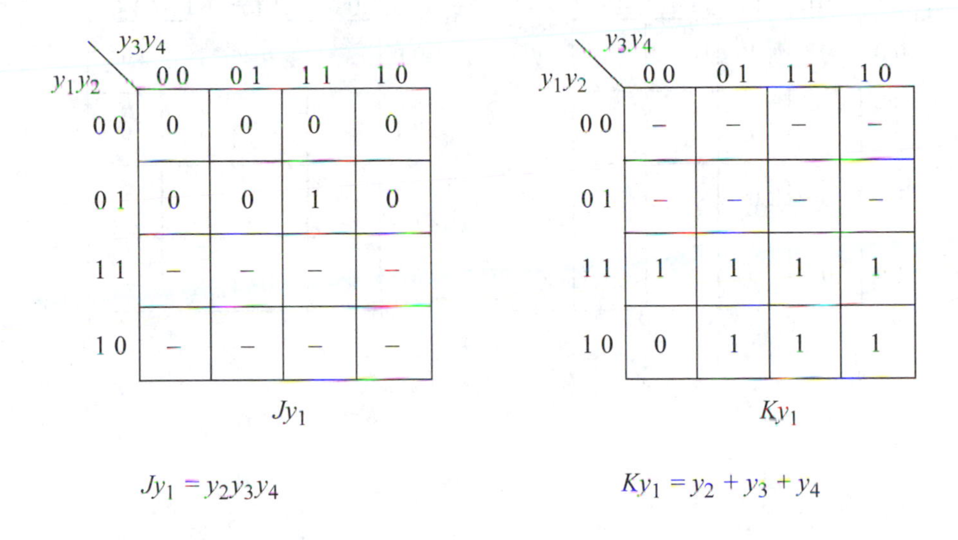

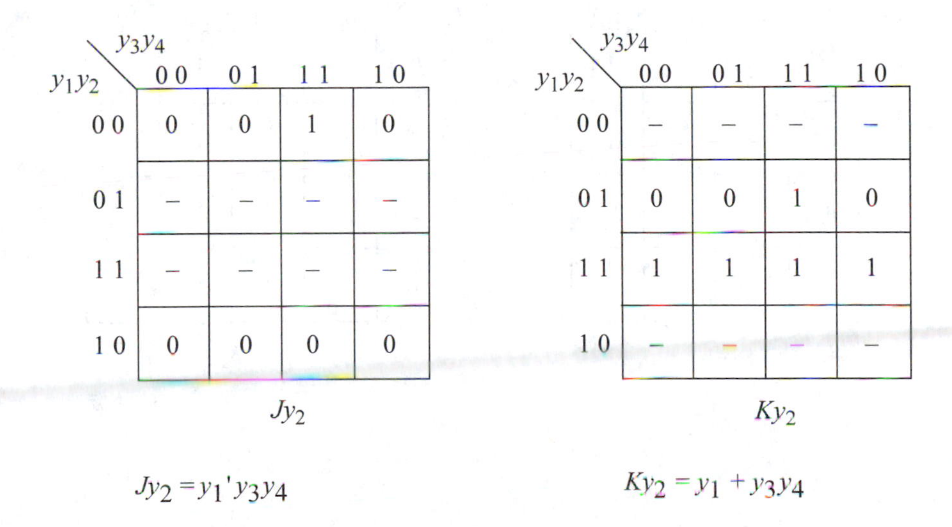

Figure 3.38 Input maps for the modulo-10 counter of Figure 3.37 using a self-starting state of $y_1y_2y_3y_4 = 0000$.

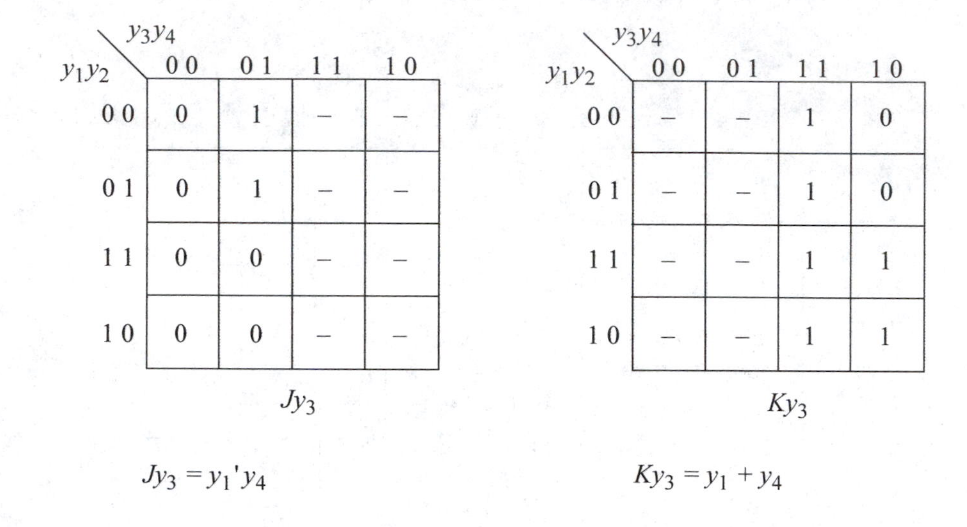

y_1y_2 \ y_3y_4	0 0	0 1	1 1	1 0
0 0	0	1	–	–
0 1	0	1	–	–
1 1	0	0	–	–
1 0	0	0	–	–

Jy_3

$Jy_3 = y_1'y_4$

y_1y_2 \ y_3y_4	0 0	0 1	1 1	1 0
0 0	–	–	1	0
0 1	–	–	1	0
1 1	–	–	1	1
1 0	–	–	1	1

Ky_3

$Ky_3 = y_1 + y_4$

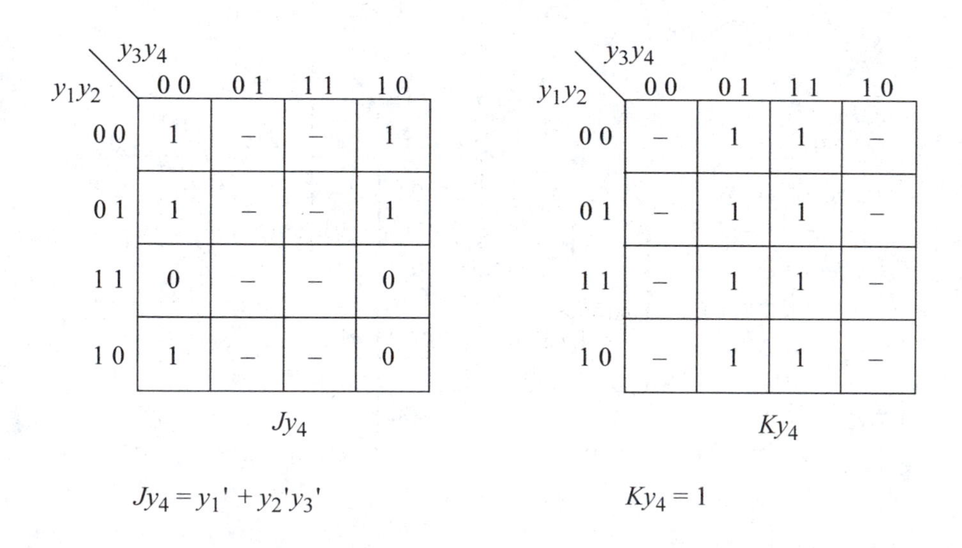

y_1y_2 \ y_3y_4	0 0	0 1	1 1	1 0
0 0	1	–	–	1
0 1	1	–	–	1
1 1	0	–	–	0
1 0	1	–	–	0

Jy_4

$Jy_4 = y_1' + y_2'y_3'$

y_1y_2 \ y_3y_4	0 0	0 1	1 1	1 0
0 0	–	1	1	–
0 1	–	1	1	–
1 1	–	1	1	–
1 0	–	1	1	–

Ky_4

$Ky_4 = 1$

Figure 3.38 (Continued)

$$
\begin{aligned}
Jy_1 &= y_2 y_3 y_4 & Ky_1 &= y_2 + y_3 + y_4 \\
Jy_2 &= y_1' y_3 y_4 & Ky_2 &= y_1 + y_3 y_4 \\
Jy_3 &= y_1' y_4 & Ky_3 &= y_1 + y_4 \\
Jy_4 &= y_1' + y_2' y_3' & Ky_4 &= 1
\end{aligned} \tag{3.14}
$$

Step 6: Output maps No output maps are required for a single 4-bit modulo-10 counter. If, however, the counter is one decade of a multi-decade counter, then λ output logic is necessary to indicate when decade$_i$ has attained a count of $y_1y_2y_3y_4 = 1001$. The next active clock transition will reset decade$_i$ and increment by one the next higher-order decade$_{i+1}$. The design of an *n*-digit BCD counter is constructed from *n* 4-bit modulo-10 counters.

Step 7: Logic diagram The logic diagram of Figure 3.39 is derived from the *JK* input equations of Equation 3.14. Although not shown in the logic diagram, it is assumed that the counter flip-flops have a set and reset function. The counter is reset initially to $y_1y_2y_3y_4 = 0000$. The timing diagram is illustrated in Figure 3.40 (a). The clock signal is supplied to all flip-flops simultaneously. State changes occur only on the negative clock transition. Using either Equation 3.14 or the logic diagram of Figure 3.39, the counter can be shown to increment through the modulo-10 counting sequence, then return to 0000 at the next negative clock transition.

Since the counter is self-starting, the next state should be 0000 from any invalid state 1010 through 1111. Assume that transient electrical noise causes the counter to change to state 1100; that is, $y_1y_2y_3'y_4'$. From these values of the state flip-flops, the *JK* input equations acquire the logical values shown in Equation 3.15. Thus, $Jy_iKy_i = 01$, which is a reset condition for all flip-flops and the counter will proceed to state $y_1y_2y_3y_4 = 0000$ at the next negative clock transition.

The counter logic can be minimized considerably by not allowing the self-starting attribute; that is, the unused states — corresponding to minterm locations 10 through 15, which represent invalid BCD digits — are treated as "don't care" states. The 1s in the input maps can now combine with more minterm locations to provide a minimal number of logic gates for the δ next-state function. The input maps with the additional "don't care" states are shown in Figure 3.41. The corresponding *JK* input equations are expressed in Equation 3.16. The logic diagram with no self-starting state is depicted in Figure 3.42 and is implemented from the *JK* input equations of Equation 3.16.

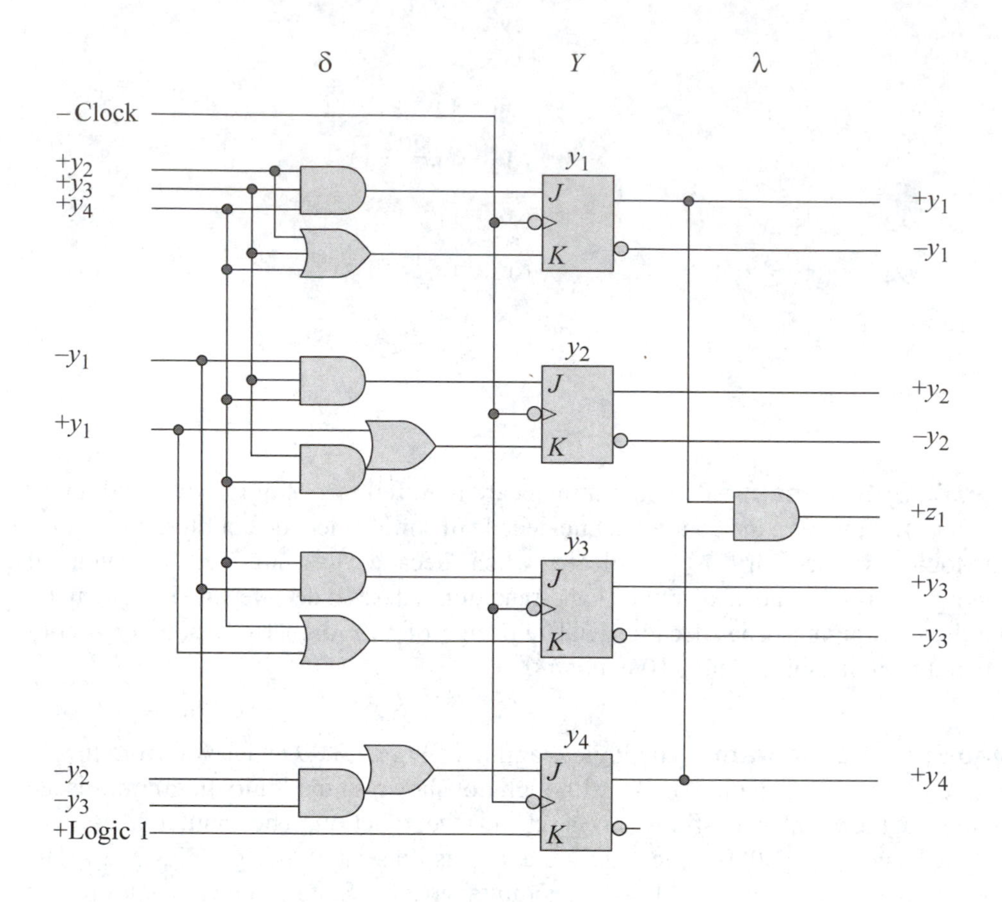

Figure 3.39 Logic diagram for the modulo-10 counter of Figure 3.37 using *JK* flip-flops, where flip-flop y_4 is the low-order stage. The counter has a self-starting state of $y_1y_2y_3y_4 = 0000$.

As stated previously, the λ output logic is not mandatory for a single-stage 4-bit modulo-10 counter. For a multi-stage counter, however, a signal must be made available from decade$_i$ to decade$_{i+1}$ to indicate when decade$_i$ has reached a terminal count of 1001. The logic diagram for a 3-digit, modulo-10 counter with a range of 000 to 999 is shown in Figure 3.43. Each stage of the counter contains the internal logic as shown in Figure 3.39.

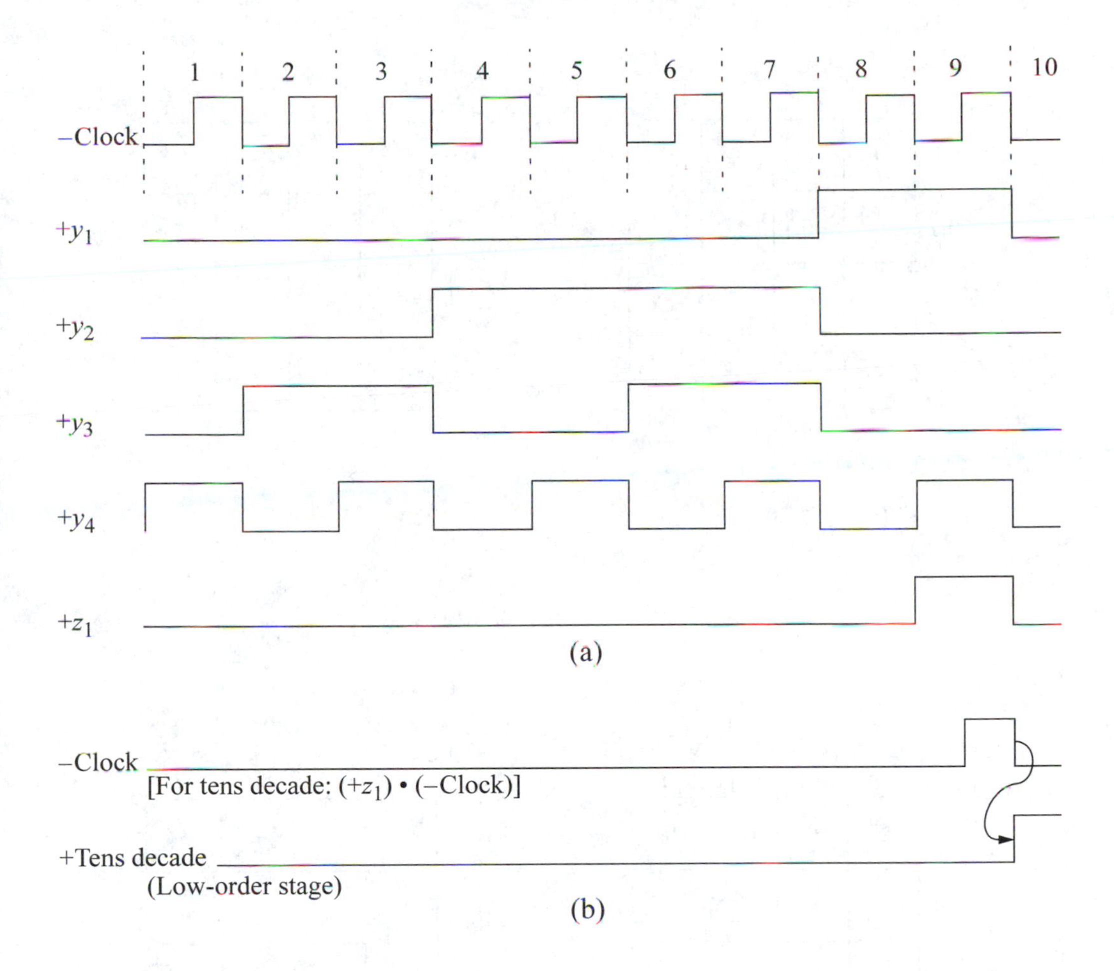

Figure 3.40 Timing diagram: (a) for the modulo-10 counter of Figure 3.39, where flip-flop y_4 is the low-order stage; and (b) clock generation for the tens decade of a multistage BCD counter.

$$\begin{aligned}
Jy_1 &= y_2y_3y_4 &= 0 \qquad Ky_1 &= y_2 + y_3 + y_4 = 1 \\
Jy_2 &= y_1'y_3y_4 &= 0 \qquad Ky_2 &= y_1 + y_3y_4 = 1 \\
Jy_3 &= y_1'y_4 &= 0 \qquad Ky_3 &= y_1 + y_4 = 1 \\
Jy_4 &= y_1' + y_2'y_3' &= 0 \qquad Ky_4 &= 1
\end{aligned} \tag{3.15}$$

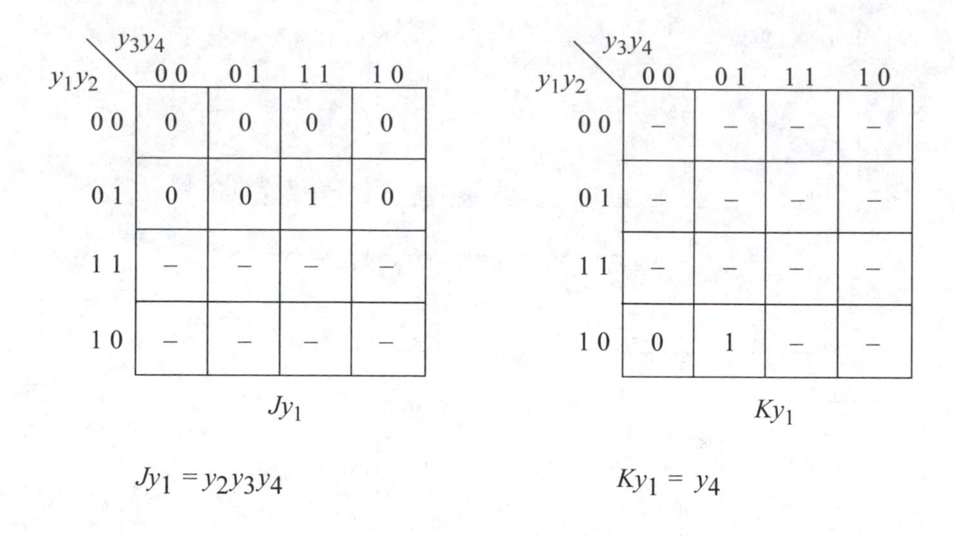

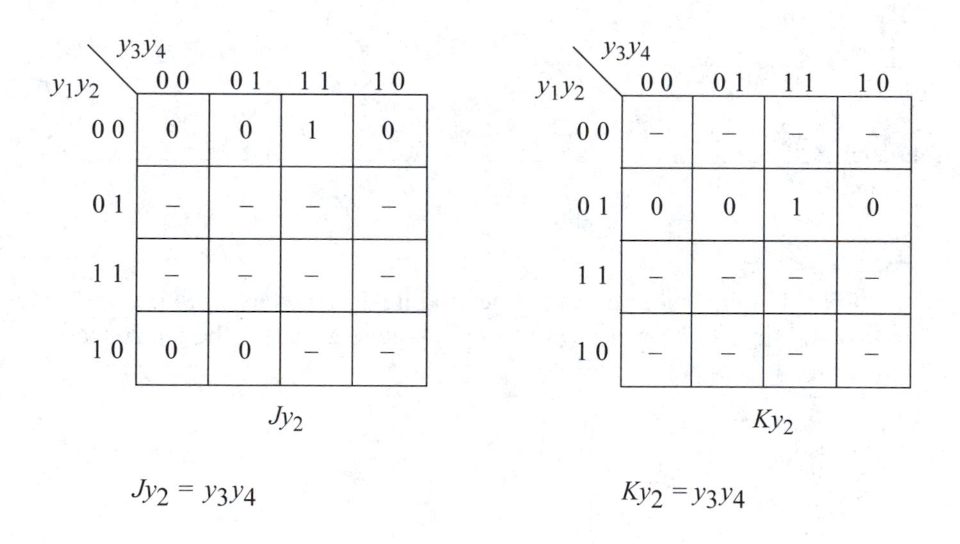

Figure 3.41 Input maps for the modulo-10 counter of Figure 3.37 with no self-starting state.

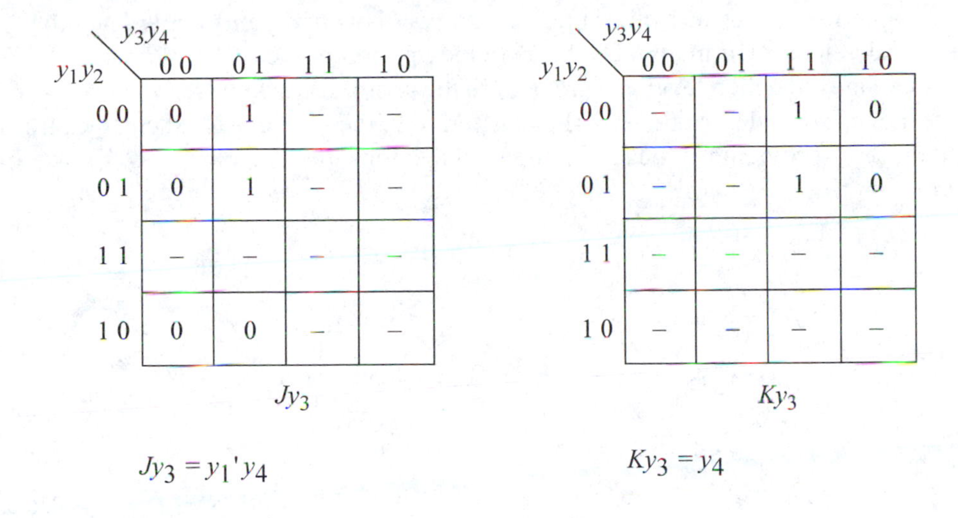

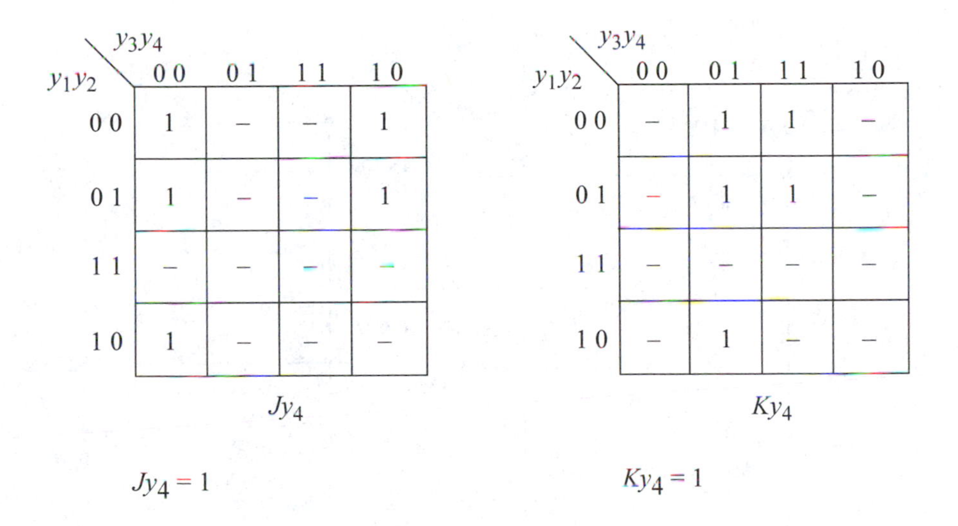

Figure 3.41 (Continued)

With the exception of the units decade, the clock pulse is transmitted to each successive decade only when the previous lower-order decade has reached a state of 1001. Because clocking occurs on the negative transition, each next higher-order decade receives a positive clock pulse of full duration. The clock pulse for the tens decade, for example, is generated when the units decade has reached a terminal count of

1001. During the first half of clock period 9 (Refer to Figure 3.40 for timing), the – Clock input to AND gate 1 of Figure 3.43 is low. This maintains a low level to the tens decade clock. During the last half of clock period 9, the –Clock input to AND gate 1 changes to a high level and propagates the positive clock pulse to the clock input of the tens decade. At the end of clock period 9, the –Clock signal changes from high to low, providing a negative edge to the tens decade clock, as shown in Figure 3.40 (b).

$$
\begin{aligned}
Jy_1 &= y_2 y_3 y_4 & Ky_1 &= y_4 \\
Jy_2 &= y_3 y_4 & Ky_2 &= y_3 y_4 \\
Jy_3 &= y_1' y_4 & Ky_3 &= y_4 \\
Jy_4 &= 1 & Ky_4 &= 1
\end{aligned}
\tag{3.16}
$$

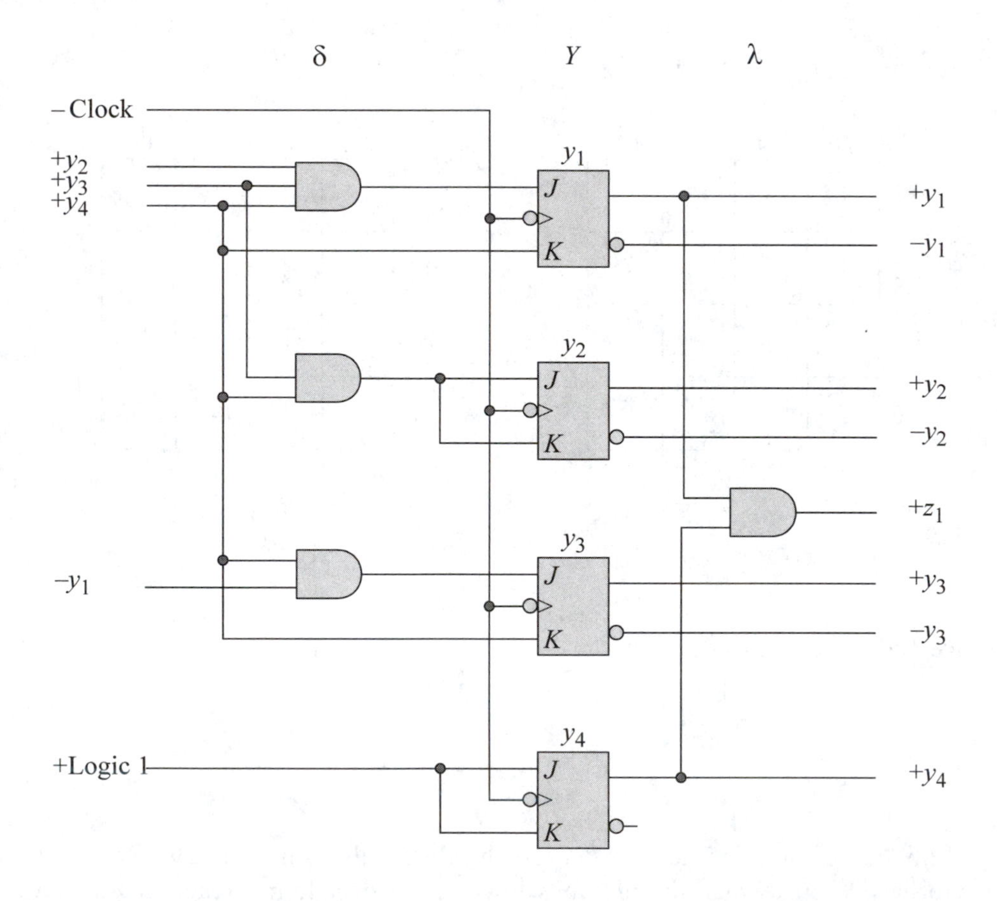

Figure 3.42 Logic diagram for the modulo-10 counter of Figure 3.37 using *JK* flip-flops, where flip-flop y_4 is the low-order stage. The counter is not self-starting.

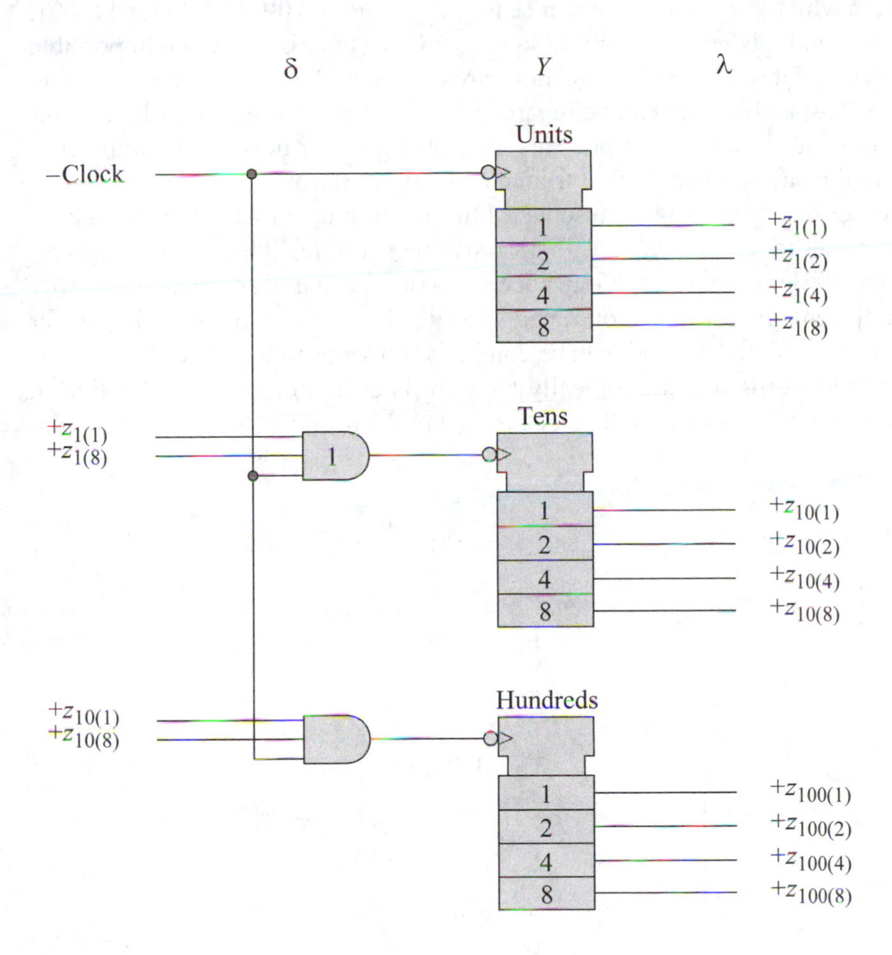

Figure 3.43 Three-digit modulo-1000 counter.

3.3.3 Johnson Counter

Thus far, the counters described in this chapter have had a counting sequence that increased in a binary manner from zero to some predefined upper limit. The modulo-8 counter used all $2^n = 2^3 = 8$ combinations of three variables. The modulo-10 counter, however, used only ten combinations of 16 possible states. Other counters are frequently utilized in digital computers. These are down-counters that count in a descending sequence from some preset state to zero. The synthesis of these counters is identical to the method presented in previous sections. The state diagram contains state codes that decrement by one for each succeeding state.

Still other counters can be designed for a unique application in which the counting sequence is neither entirely up nor entirely down. These have a nonsequential counting sequence that is prescribed by external requirements. Such a counter is shown in

Figure 3.44, in which the counting sequence is $y_1y_2y_3 = 000, 100, 110, 111, 011, 001,$ $000, \ldots$. The counter is reset initially to $y_1y_2y_3 = 000$. For six of the eight possible states for three variables, the state transitions are completely defined. The remaining two states are unspecified and can be regarded as "don't care" states in order to minimize the δ next-state logic. This presents no problem under normal operating conditions where the environment is free from electrical interference.

The counter of Figure 3.44 represents a Johnson counter in which any two contiguous state codes (or code words) differ by only one variable. It is similar, in this respect, to a Gray code counter. The Gray code concept is used in Karnaugh maps. Any physically adjacent minterm locations are also logically adjacent, because they differ in only one variable and therefore, can be combined into a term with fewer variables. Contiguous code words that are logically adjacent is an important consideration in eliminating output glitches and will be elaborated in detail in Section 3.7.

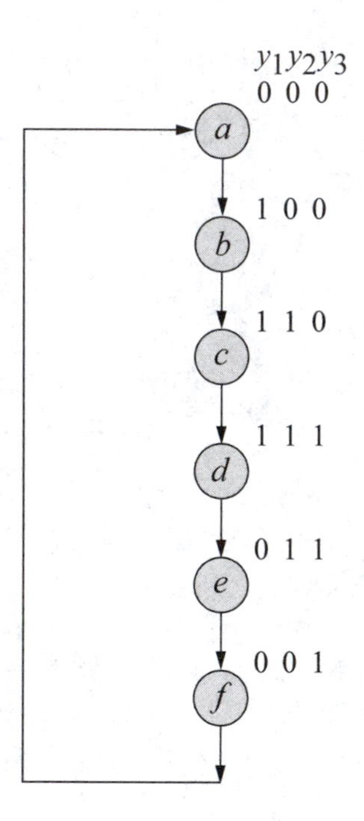

Figure 3.44 State diagram for a Johnson counter with a nonsequential counting sequence. There are two unused states: $y_1y_2y_3 = 010$ and 101. The Johnson counter is also referred to as a Möbius counter, because the output of the last stage is inverted and fed back to the first stage. August F. Möbius was a German mathematician who discovered a one-sided surface that is constructed from a rectangle by holding one end fixed, rotating the opposite end through 180 degrees, and applying it to the first end.

Using the synthesis procedure described in Section 3.2, the input maps are obtained using D flip-flops, as shown in Figure 3.45. The maps can be derived directly from the state diagram without the necessity of generating a next-state table. For example, from state b ($y_1y_2y_3 = 100$), the machine sequences to state c ($y_1y_2y_3 = 110$) where the next state for flip-flop y_1 is 1. Thus, a 1 is entered in minterm location $y_1y_2y_3 = 100$ for flip-flop y_1. Likewise, from state c the machine proceeds to state d where the next state for y_1 is 1; therefore, a 1 is entered in minterm location $y_1y_2y_3 = 110$ for flip-flop y_1. In a similar manner, the remaining entries are obtained for the input map for y_1, as well as for the input maps for y_2 and y_3. The input equations are listed in Equation 3.17.

The logic diagram is shown in Figure 3.46. The counting sequence is easily verified by asserting the appropriate input logic levels to the flip-flop D inputs for each state of the counter and then applying the active clock transition.

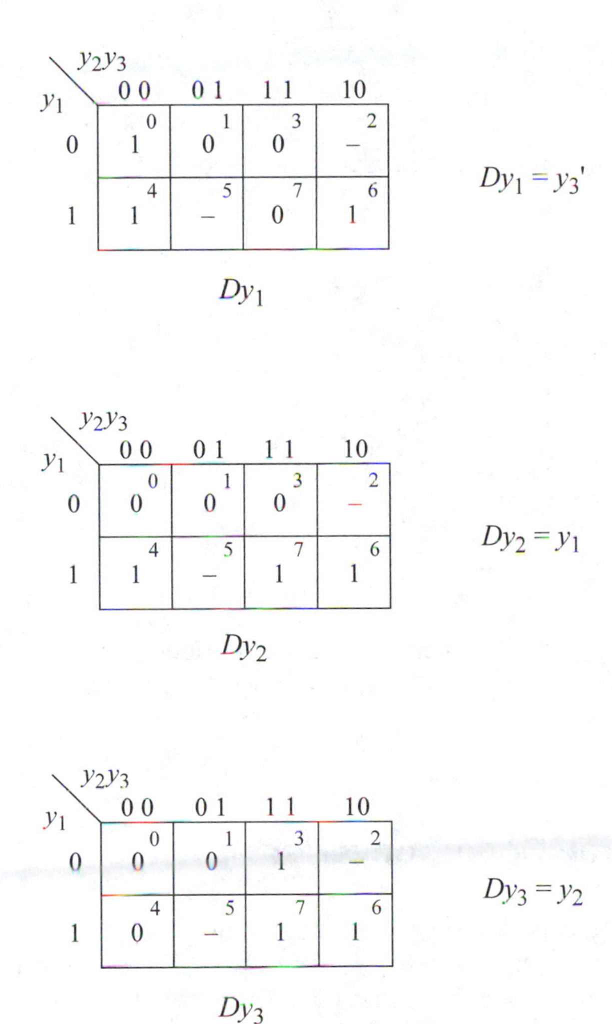

Figure 3.45 Input maps for the Johnson counter of Figure 3.44 using D flip-flops. The unused states are $y_1y_2y_3 = 010$ and 101.

$$Dy_1 = y_3'$$
$$Dy_2 = y_1$$
$$Dy_3 = y_2 \tag{3.17}$$

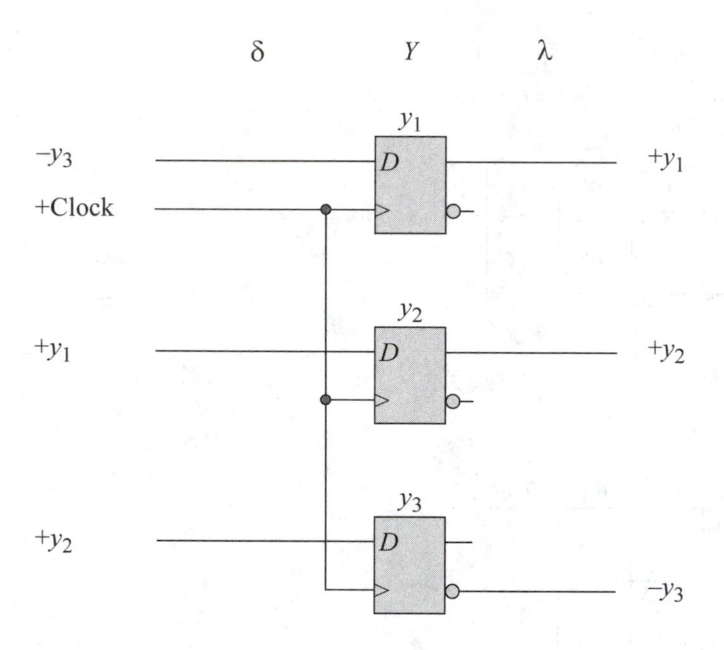

Figure 3.46 Logic diagram for the Johnson counter of Figure 3.44 using *D* flip-flops. The counting order is nonsequential. Flip-flop y_3 is the low-order stage.

3.3.4 Binary-to-Gray Code Converter

One final application of counters is presented for the design of code converters. Although a code converter is more easily implemented by means of a read-only memory (ROM), it is presented in this section to illustrate the versatility of counters. It may seem inappropriate to classify a code converter as a counter, but it falls into the general category of machines that are loaded in parallel and then sequence to a new state upon the application of an active clock transition. Unlike registers such as, parallel-in, parallel-out registers, where the next state is a function of the present inputs, a code converter of this type has a next state that is a function of the present state only.

The next-state table of Table 3.11 shows the relationship between the binary 8421 code and the Gray code. The Gray code belongs to a class of cyclic codes called reflective codes. Notice in the first four rows, that y_4 reflects across the reflecting axis; that is, y_4 in rows 2 and 3 is the mirror image of y_4 in rows 0 and 1. In the same manner, y_3 and y_4 reflect across the reflecting axis drawn under row 3. Thus, rows 4 through 7 reflect the state of rows 0 through 3 for y_3 and y_4. The same is true for y_2, y_3, and y_4 relative to rows 8 through 15 and rows 0 through 7.

Table 3.11 Next-state table for converting from the binary 8421 code to the Gray code

	Present state (Binary code $b_1\ b_2\ b_3\ b_4$)				Next state (Gray code $g_1\ g_2\ g_3\ g_4$)				
Row	y_1	y_2	y_3	y_4	y_1	y_2	y_3	y_4	
0	0	0	0	0	0	0	0	0	
1	0	0	0	1	0	0	0	1	
2	0	0	1	0	0	0	1	1	← y_4 is reflected
3	0	0	1	1	0	0	1	0	
4	0	1	0	0	0	1	1	0	← y3 and y_4
5	0	1	0	1	0	1	1	1	are reflected
6	0	1	1	0	0	1	0	1	
7	0	1	1	1	0	1	0	0	
8	1	0	0	0	1	1	0	0	← y_2, y_3, and y_4
9	1	0	0	1	1	1	0	1	are reflected
10	1	0	1	0	1	1	1	1	
11	1	0	1	1	1	1	1	0	
12	1	1	0	0	1	0	1	0	
13	1	1	0	1	1	0	1	1	
14	1	1	1	0	1	0	0	1	
15	1	1	1	1	1	0	0	0	

The Gray code is an unweighted code and a Gray code counter has significant applications in sequential machine testing. By applying the outputs of an n-bit Gray code counter to the inputs of a synchronous sequential machine under test, the machine's behavior can be more easily monitored, since only one input changes during each test cycle.

A state diagram is not relevant in this application, since the machine will not sequence through a series of states. Rather, a binary input vector X_i is loaded into the machine, and after a clock pulse is applied, the corresponding Gray code word becomes the next state. The input maps for the code converter are shown in Figure 3.47, using *JK* flip-flops. For convenience, the excitation table for a *JK* flip-flop is reproduced in Table 3.12. The method of generating two sample map entries will now be described. The method is representative of all entries. Consider the present state in row 3 ($y_1y_2y_3y_4 = 0011$) of Table 3.11. Flip-flop y_3 proceeds from a present state of 1 to a next state of 1. From Table 3.12, a 1-to-1 transition specifies the *JK* input values to be $JK = -\,0$. These values are entered in the map for Jy_3 and Ky_3. Now examine row 12 ($y_1y_2y_3y_4 = 1100$) of Table 3.11. Flip-flop y_3 moves from a present state of 0 to a next state of 1. This transition yields *JK* values of $JK = 1-$, which are entered in minterm location 1100 for Jy_3 and Ky_3, respectively. The input equations are listed in Equation 3.18.

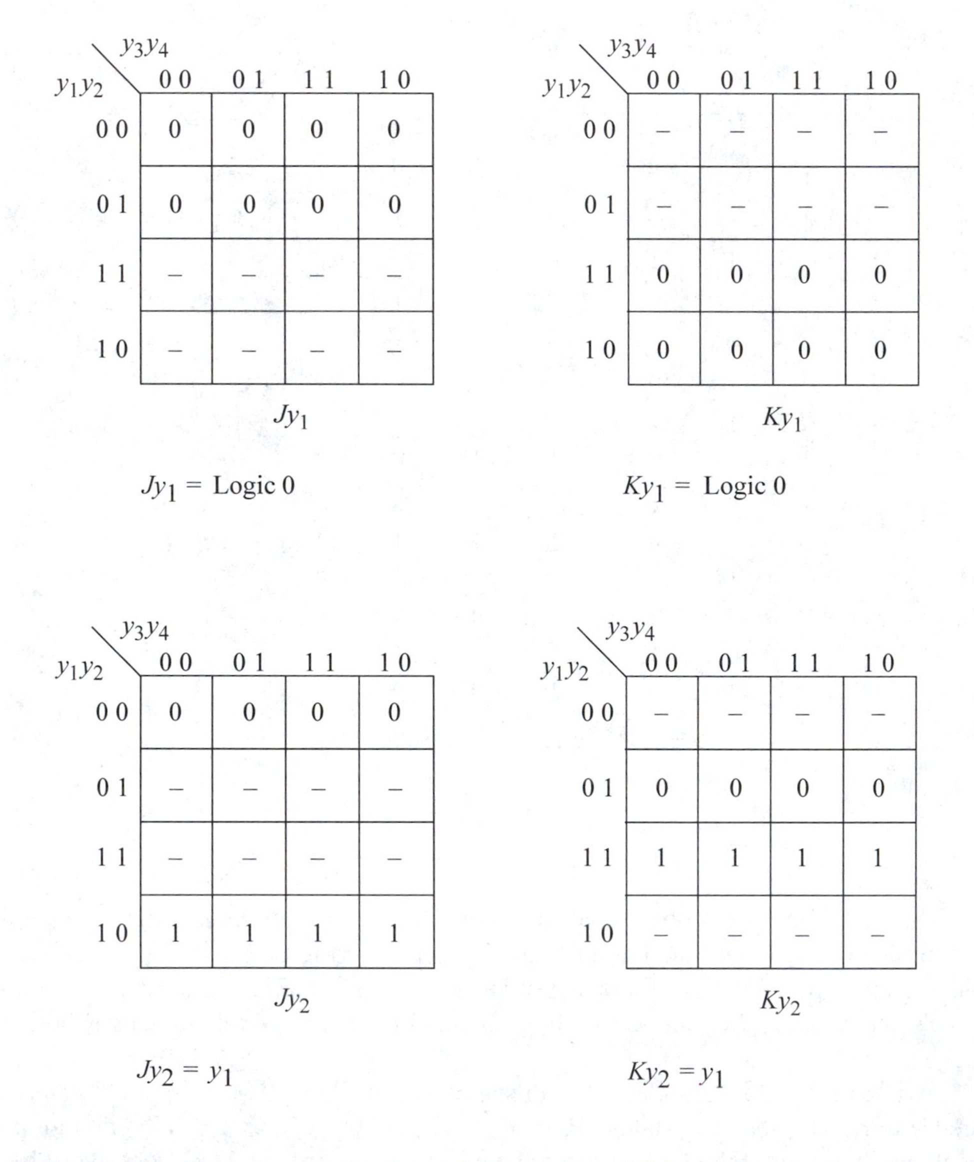

Figure 3.47 Input maps for the code converter of Table 3.11 which translates the binary 8421 code into the Gray code.

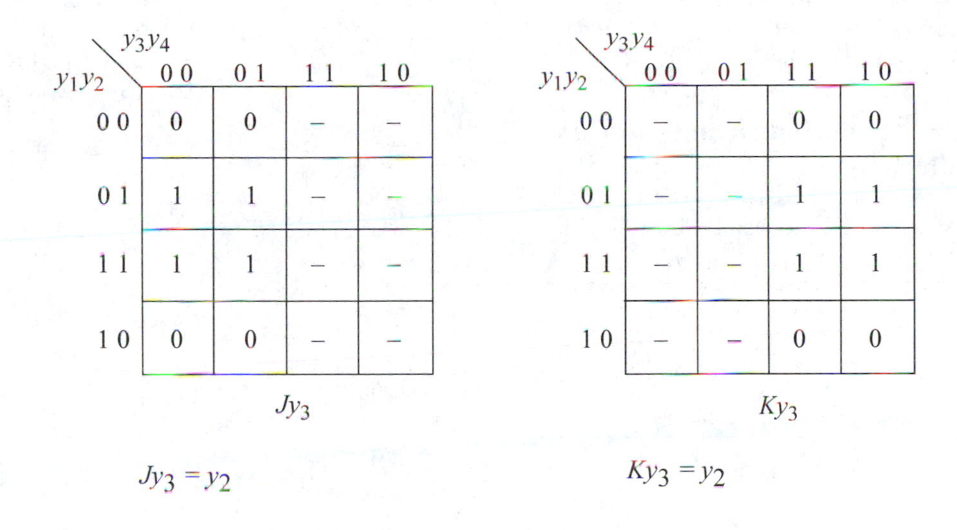

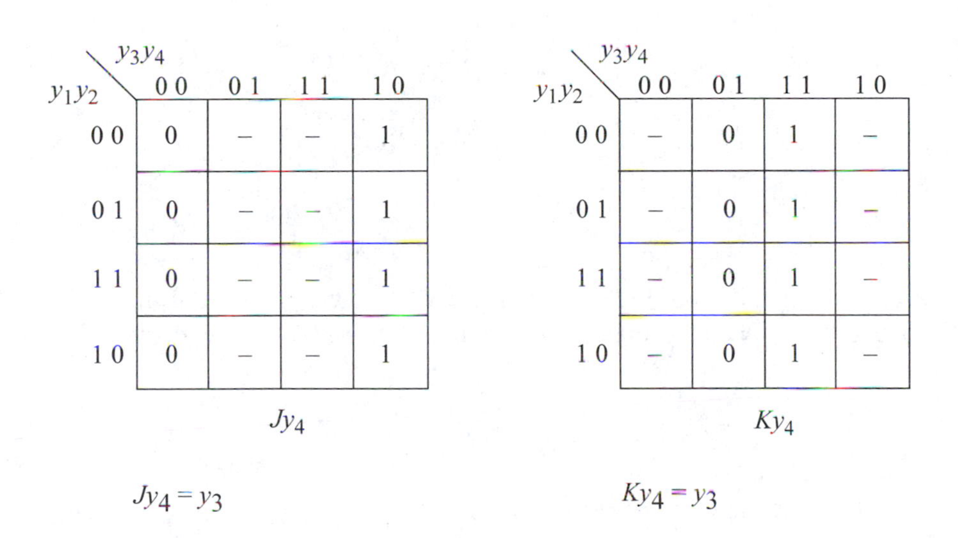

Figure 3.47 (Continued)

The logic diagram is shown in Figure 3.48. The binary code word $b_1\ b_2\ b_3\ b_4$ is loaded into the code converter by generating a positive pulse on the +Load signal. The next negative clock transition performs the requisite binary-to-Gray code translation. From Table 3.11, it is observed that the high-order bit of the Gray code word is the same as the high-order bit of the corresponding binary code word. Therefore, all the logic associated with flip-flop y_1 can be eliminated, and $g_1 = b_1$. If all four flip-flops are an integral part of a macro logic function, then flip-flop y_1 is retained and acts as a 1-bit parallel-in, parallel-out register.

Table 3.12 Excitation table for a *JK* flip-flop

Present-state $Y_{j(t)}$	Next state $Y_{k(t+1)}$	Flip-flop inputs JK
0	0	0 –
0	1	1 –
1	0	– 1
1	1	– 0

$$
\begin{aligned}
Jy_1 &= \text{Logic } 0 & Ky_1 &= \text{Logic } 0 \\
Jy_2 &= y_1 & Ky_2 &= y_1 \\
Jy_3 &= y_2 & Ky_3 &= y_2 \\
Jy_4 &= y_3 & Ky_4 &= y_3
\end{aligned} \tag{3.18}
$$

Upon further analysis of Figure 3.48, a procedure for converting from the binary 8421 code to the Gray code can be formulated. Let an n-bit binary code word be represented as

$$b_{n-1}\, b_{n-2}\, \cdots\, b_1\, b_0$$

and an n-bit Gray code word be represented as

$$g_{n-1}\, g_{n-2}\, \cdots\, g_1\, g_0$$

where b_0 and g_0 are the low-order bits of the binary and Gray codes, respectively. The ith Gray code bit g_i can be obtained from the corresponding binary code word by the following algorithm:

$$
\begin{aligned}
g_{n-1} &= b_{n-1} \\
g_i &= b_i \oplus b_{i+1}
\end{aligned}
$$

for $0 \leq i \leq n - 2$, where the symbol $\oplus$ denotes modulo-2 addition defined as:

$$0 \oplus 0 = 0$$

$$0 \oplus 1 = 1$$

$$1 \oplus 0 = 1$$

$$1 \oplus 1 = 0$$

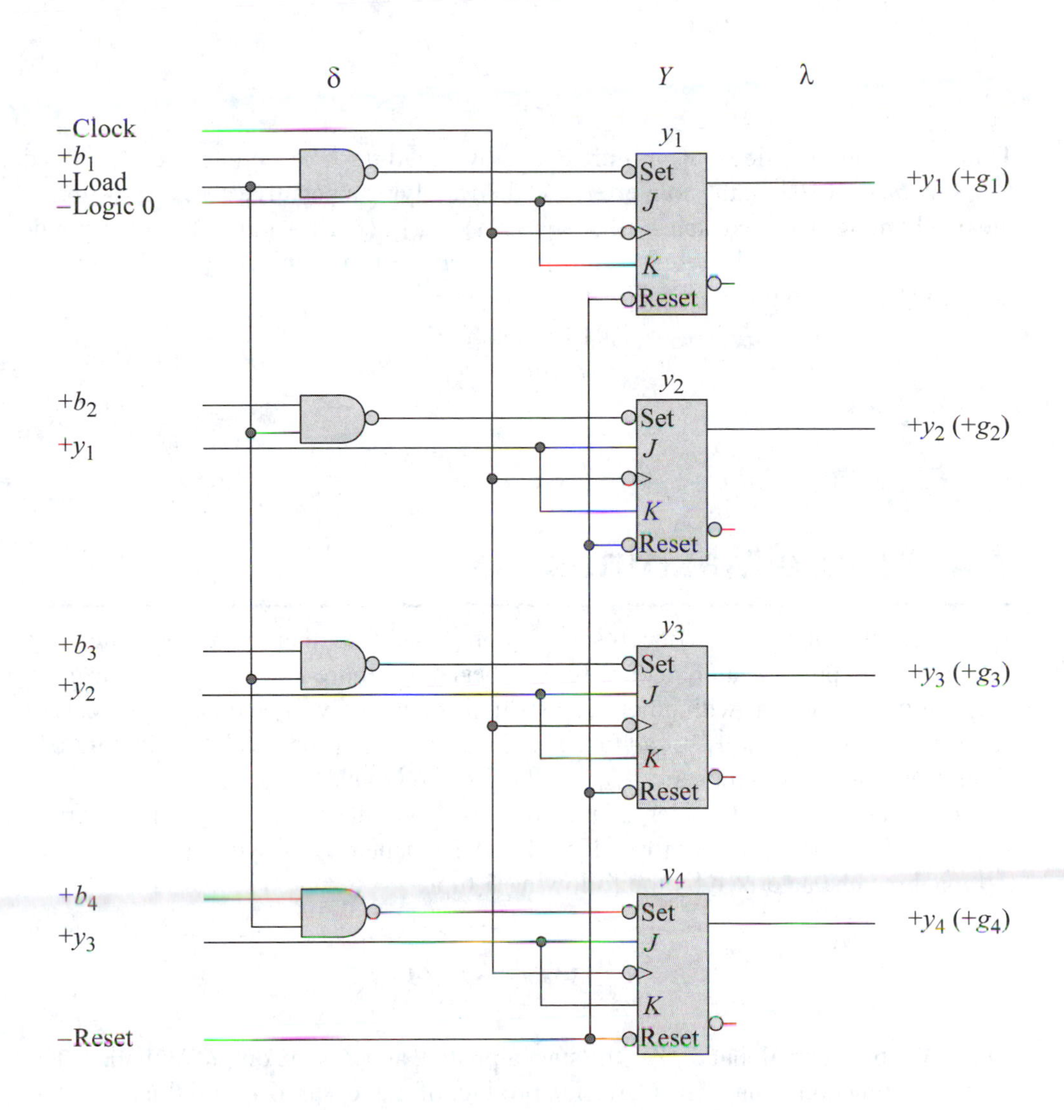

Figure 3.48 Logic diagram of the binary-to-Gray code converter of Table 3.11. Flip-flop y_4 is the low-order stage.

For example, using the algorithm, the 4-bit binary code word $b_3\, b_2\, b_1\, b_0 = 1010$ translates to the 4-bit Gray code word $g_3\, g_2\, g_1\, g_0 = 1111$ as follows:

$$g_3 = b_3 \qquad\qquad = 1$$
$$g_2 = b_2 \oplus b_3 = 0 \oplus 1 = 1$$
$$g_1 = b_1 \oplus b_2 = 1 \oplus 0 = 1$$
$$g_0 = b_0 \oplus b_1 = 0 \oplus 1 = 1$$

This can be verified in Figure 3.48 by loading the binary code word $b_3\, b_2\, b_1\, b_0 = 1010$ into the converter, and then applying a negative clock transition to the clock input. The next state will be $y_1 y_2 y_3 y_4$ = g1g2g3g4 = 1111, the corresponding Gray code word. The reverse algorithm to convert from the Gray code to the binary 8421 code is defined as follows:

$$b_{n-1} = g_{n-1}$$
$$b_i = b_{i+1} \oplus g_i$$

3.4 Moore Machines

This section extends the concepts of Moore machines that were introduced in Chapter 2 and presents a procedure for synthesizing Moore machines. The primary focus of this section, as with other sections in this chapter, will be on the synthesis of *deterministic synchronous sequential machines*, in which the next state is uniquely determined by the present state $Y_{j(t)}$ and the present inputs $X_{i(t)}$.

The Moore model of sequential machines is the result of a paper by E.F. Moore in 1956. A Moore machine was formally defined in Chapter 2 as a synchronous sequential machine characterized by the following 5-tuple:

$$M = (X, Y, Z, \delta, \lambda)$$

where X is the input alphabet, Y is the state alphabet, and Z is the output alphabet. The next-state function δ maps the Cartesian product of X and Y into Y, and thus, is determined by both the present inputs and the present state. The output function λ maps Y into Z such that, the output vector is a function of the present state only and is independent of the external inputs.

Moore machines are used in a wide variety of digital applications. These range from reception and transmission of both parallel input vectors and serial bit streams, which consist of fixed-word-length messages or a single block of text comprised of n bits. The operation of these machines is either synchronous or asynchronous.

A set of machine specifications for a Moore machine will be given, and from this detailed functional description two examples will be presented which illustrate the machine implementation, first with D flip-flops and then, for comparison, with JK flip-flops. Each step of the synthesis procedure will be delineated with emphasis on pragmatic design principles.

The steps in the synthesis of synchronous sequential machines, which were listed in Section 3.1, are illustrated in Example 3.1. In all examples in this chapter, the clock signal is a periodic waveform with a duty cycle of 50%, as shown below, where the duration from t_1 to t_3 represents one state time.

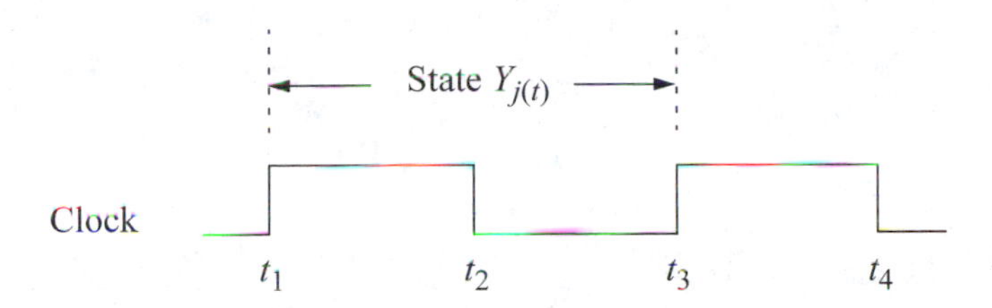

Example 3.1 This implementation will use D flip-flops as the storage elements. The machine specifications are described as follows:

Synthesize a Moore machine which accepts serial data in the form of 3-bit words on an input line x_1. There is one bit space between contiguous words, as shown below,

$$x_1 = \cdots \;\Big|\, b_1 b_2 b_3 \,\Big|\quad \Big|\, b_1 b_2 b_3 \,\Big|\quad \Big|\, b_1 b_2 b_3 \,\Big| \cdots$$

where $b_i = 0$ or 1. Whenever a word contains the bit pattern $b_1 b_2 b_3 = 111$, the machine will assert output z_1 during the bit time between words according to the following assertion/deassertion statement:

$$z_1 \uparrow t_2 \downarrow t_3$$

An example of a valid word in a series of words is shown below. Notice that the output signal is displaced in time with respect to the input sequence and occurs one state time later.

$x_1 = \cdots \; |0\,0\,1| \;\; |1\,0\,1| \;\; |0\,1\,1| \;\; |1\,1\,1| \;\; |0\,1\,0| \cdots$

Output $z_1 \uparrow t_2 \downarrow t_3$

Step 1: State diagram This is an extremely important step since all remaining steps depend upon a state diagram which correctly represents the machine specifications. Generating an accurate state diagram is thus a pivotal step in the synthesis of synchronous sequential machines. The state diagram for this example is illustrated in Figure 3.49, which graphically describes the machine's behavior. Seven states are required, providing four state levels — one level for each bit in the 3-bit words and one level for the bit space between words.

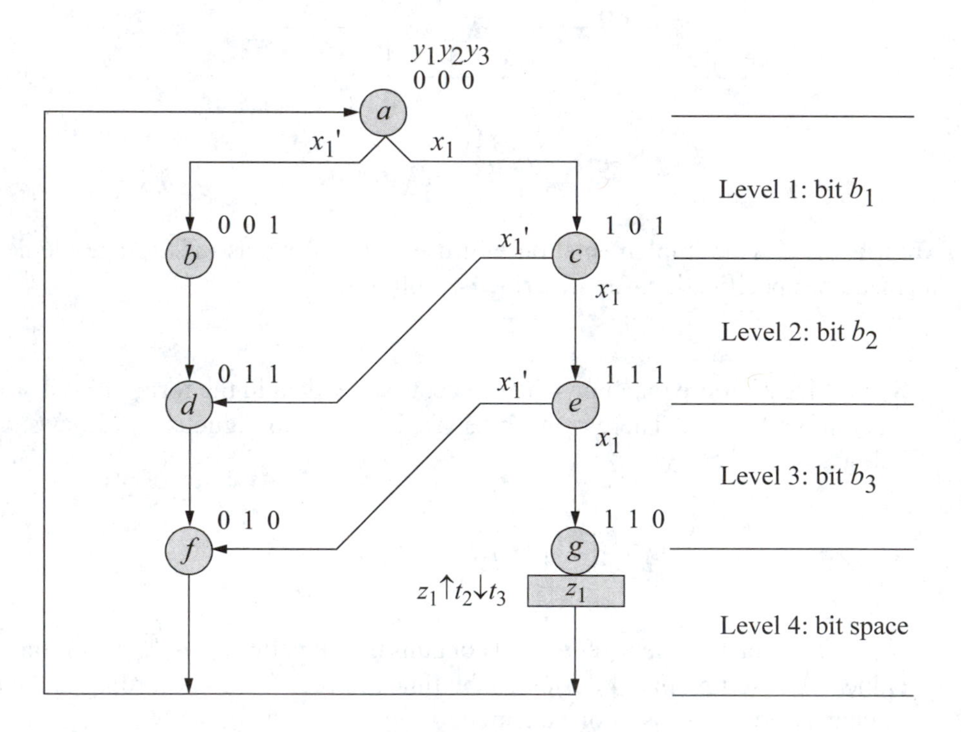

Figure 3.49 State diagram for the Moore machine of Example 3.1, which generates an output z_1 whenever a 3-bit word $x_1 = 111$. There is one unused state.

The machine is reset to an initial state which is labeled a in Figure 3.49. Since x_1 is a serial input line, the state transition can proceed in only one of two directions from state a, depending upon the value of x_1 : to state b if $x_1 = 0$ or to state c if $x_1 = 1$. Both paths represent a test of the first bit (b_1) of a word and both state transitions occur on the rising edge of the clock signal. Since the state transition from a to b occurs only if $x_1 = 0$, therefore, any bit sequence consisting of $b_1b_2b_3 = 000$ through 011 will proceed from state a to state b. Since this path will never generate an output (the first bit is invalid), there is no need to test the value of x_1 in states b or d. States b and d, together with state f, are required only to maintain four state levels in the machine, where the first three levels represent one bit each of the 3-bit word as follows: states a, b, and d correspond to bits b_1, b_2, and b_3, respectively. State f, which is the fourth level, corresponds to the bit space between words. This assures that the clocking will remain synchronized for the following word. From state d, the machine proceeds to state f and then returns to state a where the first bit of the next word is checked.

The second path from state a takes the machine to state c if $x_1 = 1$. Because this is the first bit of a possible valid 3-bit sequence, state c must also test the value of x_1, in this case bit b_2. If x_1 $(b_2) = 0$ in state c, then this represents an invalid word and output z_1 will not be asserted. Thus, any bit sequence consisting of $b_1b_2b_3 = 100$ or 101 will not assert z_1, but must still maintain four state levels. For either of the previous sequences, the state transition will be a, c, d, f, a.

If x_1 $(b_2) = 1$ in state c, then the machine proceeds to state e. This corresponds to a bit sequence of $x_1 = b_1b_2b_3 = 11-$, where b_2 is the second bit of a possible valid 3-bit sequence. Finally, in state e, x_1 $(b_3) = 0$ or 1. If $x_1 = 0$, then this represents an invalid sequence. Output z_1 is not asserted and the transition is to state f and then to state a where the first bit of the next word is checked. If, however, $x_1 = 1$ in state e, then this completes a valid sequence for a 3-bit word and the machine proceeds to state g where z_1 is asserted. The machine then returns to state a. In order to preserve synchronization between state a and the first bit of each word, it is important that any path taken in the state diagram always maintains four state levels.

The path a, c, e, g depicts a valid word which culminates in the assertion of output z_1 in state g. All other sequences of three bits will not generate an output, but each path must maintain four state levels. This guarantees that state a will always test the first bit of each 3-bit word. Output z_1 is a function of the state alphabet only, and thus, the state diagram represents a Moore machine.

Step 2: Equivalent states By carefully considering each state transition and by efficiently utilizing existing states while maintaining four state levels, the state diagram of Figure 3.49 is obtained in which there are no equivalent states. States b and c are not equivalent, because the next state for c is different than the next state for b if $x_1 = 1$. Thus, the second requirement for equivalent states is not met (refer to Section 3.1.1). The same is true for states d and e. Also, states f and g cannot be equivalent, because the outputs are different (the first requirement for equivalency). The absence of equivalent states can be verified by using either the row-matching technique or an implication table. In this example, however, it is intuitively obvious that there are no redundant states.

Step 3: State code assignment Whenever possible, state codes should be assigned such that there are a maximal number of adjacent 1s in the flip-flop input maps. This allows more minterm locations to be combined, resulting in minimized input equations in a sum-of-products form. State codes are adjacent if they differ in only one variable. For example, state codes $y_1y_2y_3 = 101$ and 100 are adjacent, because only y_3 changes. Thus, minterm locations 101 and 100 can be combined into one term. However, state codes $y_1y_2y_3 = 101$ and 110 are not adjacent, because two variables change: flip-flops y_2 and y_3.

The rules shown below are useful in assigning state codes such that there will be a maximal number of 1s in adjacent squares of the input maps, thus minimizing the δ next-state logic. It should be noted, however, that these rules do not guarantee a minimum solution with the fewest number of terms and the fewest number of variables per term. There may be several sets of state code assignments that meet the adjacency requirements, but not all will result in a minimally reduced set of input equations.

1. When a state has two possible next states, then the two next states should be adjacent; that is, if an input causes a state transition from state Y_i to either Y_j or Y_k, then Y_j and Y_k should be assigned adjacent state codes.

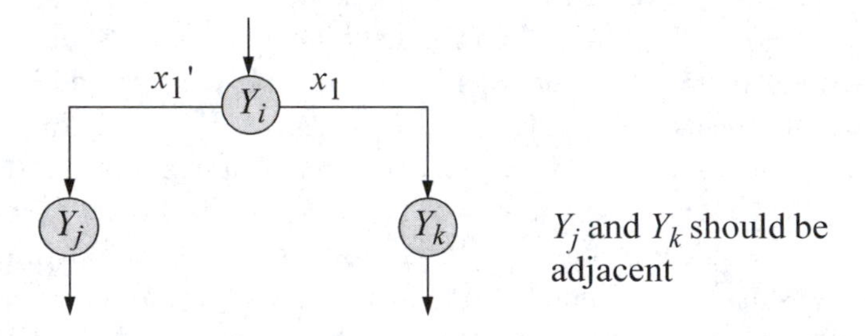

2. When two states have the same next state, the two states should be adjacent; that is, if Y_i and Y_j both have Y_k as a next state, then Y_i and Y_j should be assigned adjacent state codes.

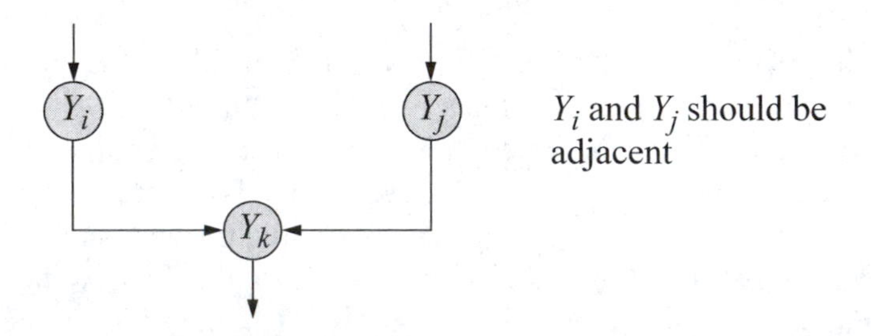

3. A third rule is useful in minimizing the λ output logic. States which have the same output should have adjacent state code assignments; that is, if states Y_i and Y_j both have z_1 as an output, then Y_i and Y_j should be adjacent. This allows for a larger grouping of 1s in the output map.

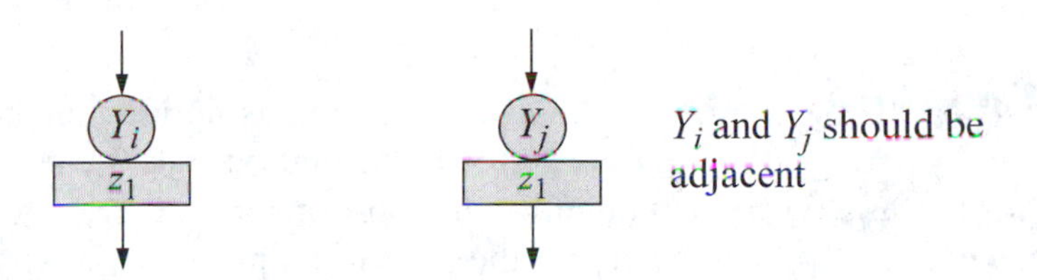

When applying the above rules for state code assignment, begin by using a Karnaugh map to locate adjacent states. Although the three rules simplify the procedure, trial and error techniques may still be required.

The Karnaugh map of Figure 3.50 is a present-state map which shows one possible state code assignment for the Moore machine of Figure 3.49. Assign state *a* as an initial reset state with a state code of $y_1y_2y_3 = 000$. Rule 1 for state code assignment specifies that the following state pairs should be adjacent to minimize the δ next-state logic: (*b*,*c*), (*d*,*e*), and (*f*,*g*). For example, in state *a*, input x_1 causes a state transition to state *b* if $x_1 = 0$ or to state *c* if $x_1 = 1$. Thus, states *b* and *c* should be adjacent. Similarly, state *c* has a next state of *d* or *e*, depending on the value of x_1, while state *e* has a next state of *f* or *g*. The adjacency of state pair (*f*,*g*) is determined by both Rule 1 and Rule 2. There is one unused state, which can be classified as a "don't care" state.

y_1 \ y_2y_3	00	01	11	10
0	*a*	*b*	*d*	*f*
1	–	*c*	*e*	*g*

Figure 3.50 Present-state map showing state code assignments for the Moore machine state diagram of Figure 3.49.

If there is a contention between minimizing the δ next-state logic or the λ output logic — where it is possible to make either two nonoutput states adjacent or two output states adjacent, but not both — then a greater savings in logic is usually realized by minimizing the δ next-state input logic rather than the λ output logic.

The assignment of a particular state code for any given state is arbitrary, provided that the principles of Rules 1, 2, and 3 are observed. Thus, the set of state codes assigned to states *a* through *g* in Figure 3.49 is only one of several different sets of codes. This representational state code assignment contains one unused state: $y_1y_2y_3$ = 100.

Step 4: Next-state table The next-state table is derived directly from the state diagram by listing all present states (including unused states), the next states for each present state as a function of input x_1, the flip-flop inputs that are required to advance the machine to the next state, and the present output. The next-state table for the Moore machine represented by the state diagram of Figure 3.49 is shown in Table 3.13. Unlike a Mealy machine, the next-state table asserts output z_1 as a function of the present state only. This is evident in state g ($y_1y_2y_3$ = 110) in which z_1 is active regardless of the value of input x_1. Since the storage elements for this design consist of D flip-flops, the flip-flop input values for Dy_1, Dy_2, and Dy_3 are the same as the next state for y_1, y_2, and y_3 for any given present state.

Table 3.13 Next-state table for the Moore machine of Figure 3.49 using *D* flip-flops

Present state			Input	Next state			Flip-flop inputs			Present output
y_1	y_2	y_3	x_1	y_1	y_2	y_3	Dy_1	Dy_2	Dy_3	z_1
0	0	0	0	0	0	1	0	0	1	0
0	0	0	1	1	0	1	1	0	1	0
0	0	1	0	0	1	1	0	1	1	0
0	0	1	1	0	1	1	0	1	1	0
0	1	0	0	0	0	0	0	0	0	0
0	1	0	1	0	0	0	0	0	0	0
0	1	1	0	0	1	0	0	1	0	0
0	1	1	1	0	1	0	0	1	0	0
1	0	0	0	–	–	–	–	–	–	–
1	0	0	1	–	–	–	–	–	–	–
1	0	1	0	0	1	1	0	1	1	0
1	0	1	1	1	1	1	1	1	1	0
1	1	0	0	0	0	0	0	0	0	1
1	1	0	1	0	0	0	0	0	0	1
1	1	1	0	0	1	0	0	1	0	0
1	1	1	1	1	1	0	1	1	0	0

Step 5: Input maps The input maps, also called excitation maps, represent the δ next-state logic for the flip-flop inputs. The input maps are constructed from either the state diagram or the next-state table and are shown in Figure 3.51 using input x_1 as a map-entered-variable. Refer to the input map for flip-flop y_1. Since the purpose of an input map is to obtain the flip-flop input equations by combining 1s in the minterm locations, the variable x_1 is entered as the value in minterm location $y_1y_2y_3 = 000$. That is, y_1 has a next value of 1 if and only if x_1 has a value of 1.

In state b, the next state for $y_1 = 0$, regardless of the value of x_1; thus, a 0 is inserted in minterm location $y_1y_2y_3 = 001$. In a similar manner, the entries for the remaining squares of y_1 are obtained and also for the input maps for y_2 and y_3. Consider state c ($y_1y_2y_3 = 101$) in the map for y_2. The next state for y_2 is 1 whether x_1 is 0 or 1. Thus, $x_1 + x_1'$ could be entered in minterm location $y_1y_2y_3 = 101$. However, since the expression $x_1 + x_1'$ equals 1, a value of 1 is inserted. That is, y_2 becomes 1 regardless of the path that is taken. Depending on whether the input equations are to be in a sum-of-products or a product-of-sums form, 1s or 0s are combined, accordingly. The input equations are listed in Equation 3.19.

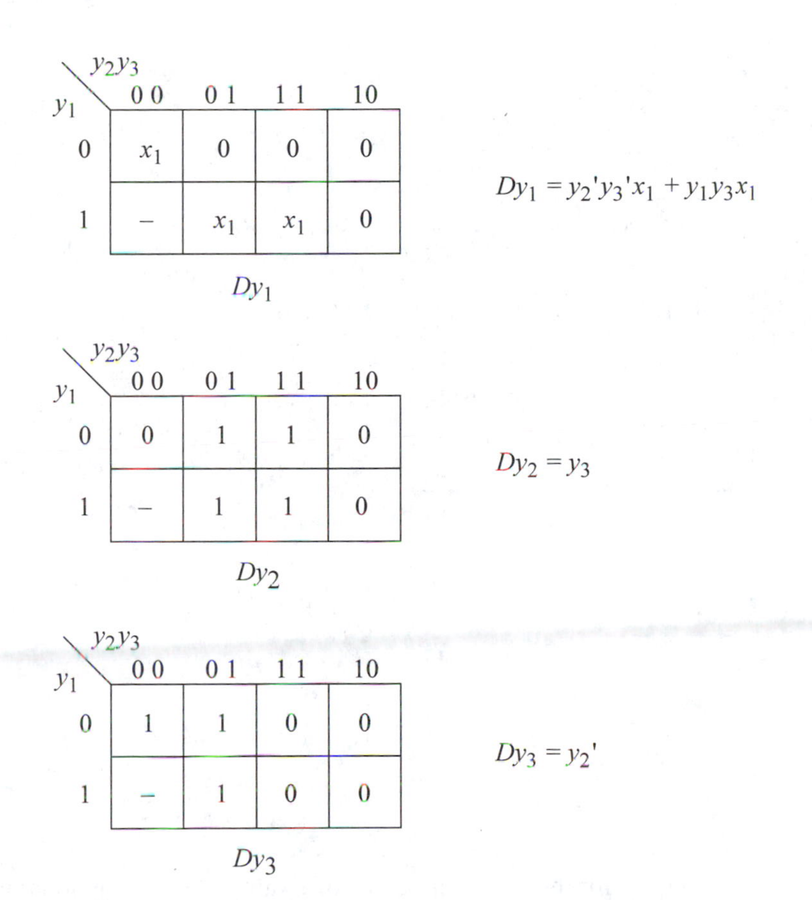

Figure 3.51 Input maps for the Moore machine of Figure 3.49.

$$Dy_1 = y_2'y_3'x_1 + y_1y_3x_1$$
$$Dy_2 = y_3$$
$$Dy_3 = y_2' \tag{3.19}$$

If state code adjacency was not maintained in Figure 3.49, then the input equations may contain more terms and/or more variables per term. To illustrate this, reassign the state codes in Figure 3.49 as follows: $a = 000$, $b = 001$, $c = 010$, $d = 011$, $e = 100$, $f = 101$, $g = 110$. The unused state is 111. Figure 3.49 is redrawn below as Figure 3.52 using the reassigned, nonadjacent state codes. The input maps are illustrated in Figure 3.53 using x_1 as a map-entered-variable. As is evident from Figure 3.53, the number of logic gates increases considerably if the synthesis process does not adhere to the adjacency rules.

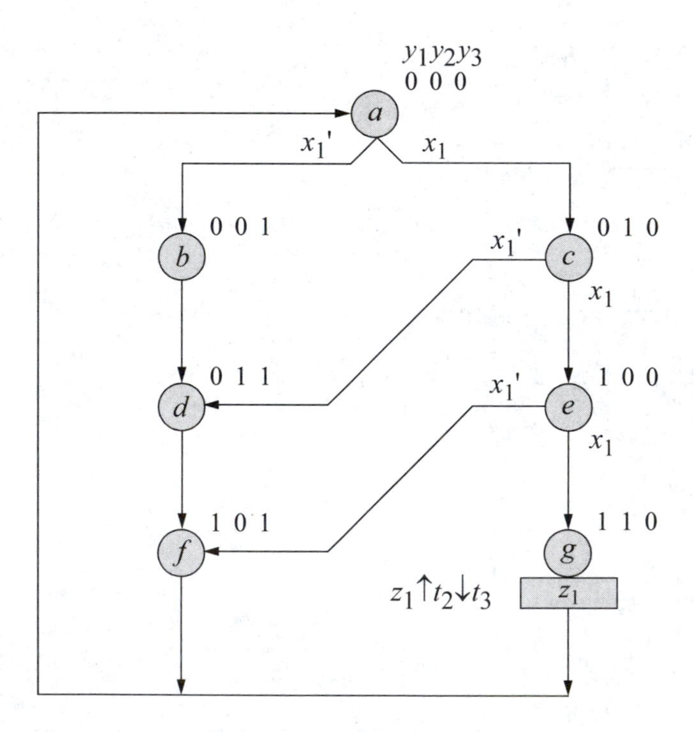

Figure 3.52 State diagram for the Moore machine of Example 3.1 using nonadjacent state codes for state pairs (*b*, *c*), (*d*, *e*), and (*f*, *g*). There is one unused state: 111.

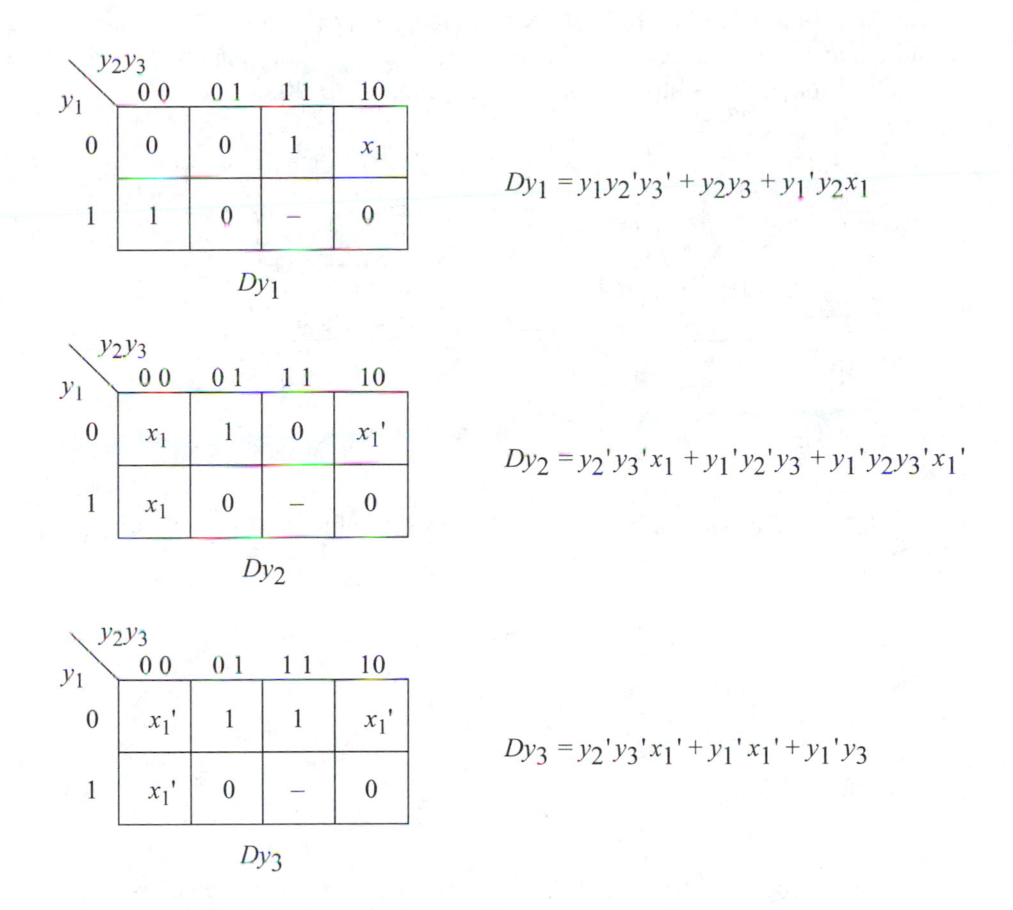

Figure 3.53 Input maps for the Moore machine of Figure 3.49 which is redrawn as Figure 3.52 using nonadjacent state codes.

Step 6: Output maps Outputs from a synchronous sequential machine can be asserted at a variety of different times depending on the machine specifications. In some cases, the output assertion and deassertion may not be specified, giving substantial flexibility in the design of the λ output logic. A contributing factor in considering the output design is the possibility of glitches. Glitches are spurious electronic signals caused by varying gate delays or improper design techniques, in which the design was not examined in sufficient detail using "worst case" circuit conditions. Glitches are more predominant in Moore machines where the outputs are a function of the state alphabet only. These narrow, unwanted pulses will be considered in more detail later in this chapter.

The output map for the Moore machine of Figure 3.49, using state code adjacency requirements, is obtained from either the state diagram or the next-state table. The output map is shown in Figure 3.54. Note that input x_1 is not used as a map-entered-variable, because the outputs for a Moore machine are a function of the present state only. The equation for output z_1 is shown in Equation 3.20.

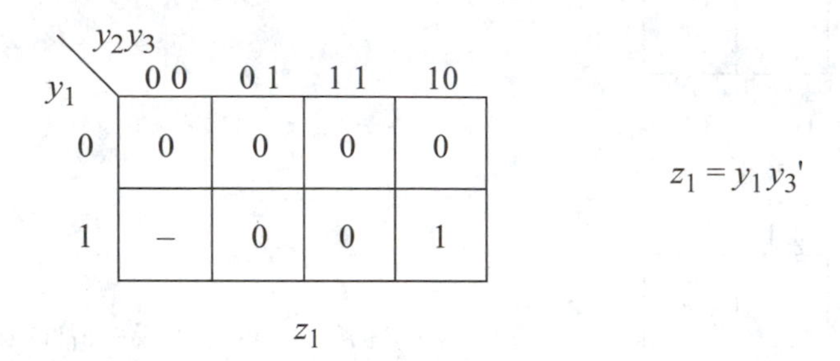

Figure 3.54 Output map for z_1 for the Moore machine of Figure 3.49.

$$z_1 = y_1 y_3' \tag{3.20}$$

The unused state $y_1y_2y_3 = 100$ can be used to minimize the equation for z_1, since the assertion of z_1 occurs at time t_2, long after the machine has stabilized. The state flip-flops are clocked on the positive clock transition at time t_1. A state change causes the machine to move from one stable state to another stable state. If more than one flip-flop changes state during this transition, then it is possible for the machine to momentarily pass through a transient state before reaching the destination stable state. This period of instability occurs immediately after the rising edge of the clock and has a duration of only a small percentage of the clock cycle. The machine has certainly stabilized by time t_2.

Since the assertion/deassertion statement for z_1 is $\uparrow t_2 \downarrow t_3$, the machine has stabilized in its destination state before the specified assertion of z_1. Thus, the "don't care" state $y_1y_2y_3 = 100$ can be used to minimize the equation for z_1 without regard for any momentary transition through state 100, which would otherwise produce a glitch on z_1. This problem of output glitches associated with state transitions will be covered in more detail in Section 3.7.

Step 7: Logic diagram The logic diagram is implemented from the input equations of Equation 3.19 using positive-edge-triggered D flip-flops, and the output equation of Equation 3.20, as shown in Figure 3.55. In Figure 3.55 (a), output z_1 is asserted at time t_2 and deasserted at time t_3 by the application of the –Clock signal to the active-high AND gate inputs. In Figure 3.55 (b), since NOR logic stipulates active low inputs for the AND operation, the +Clock signal is used to provide the requisite assertion and deassertion for z_1. The timing diagram of Figure 3.56 further illustrates the operation of the machine for the state transition sequence a, c, e, g with an input sequence of $x_1 = 111$. Since state g is coincident with the bit space between words, x_1 is not checked at that time, as indicated by the cross-hatched area.

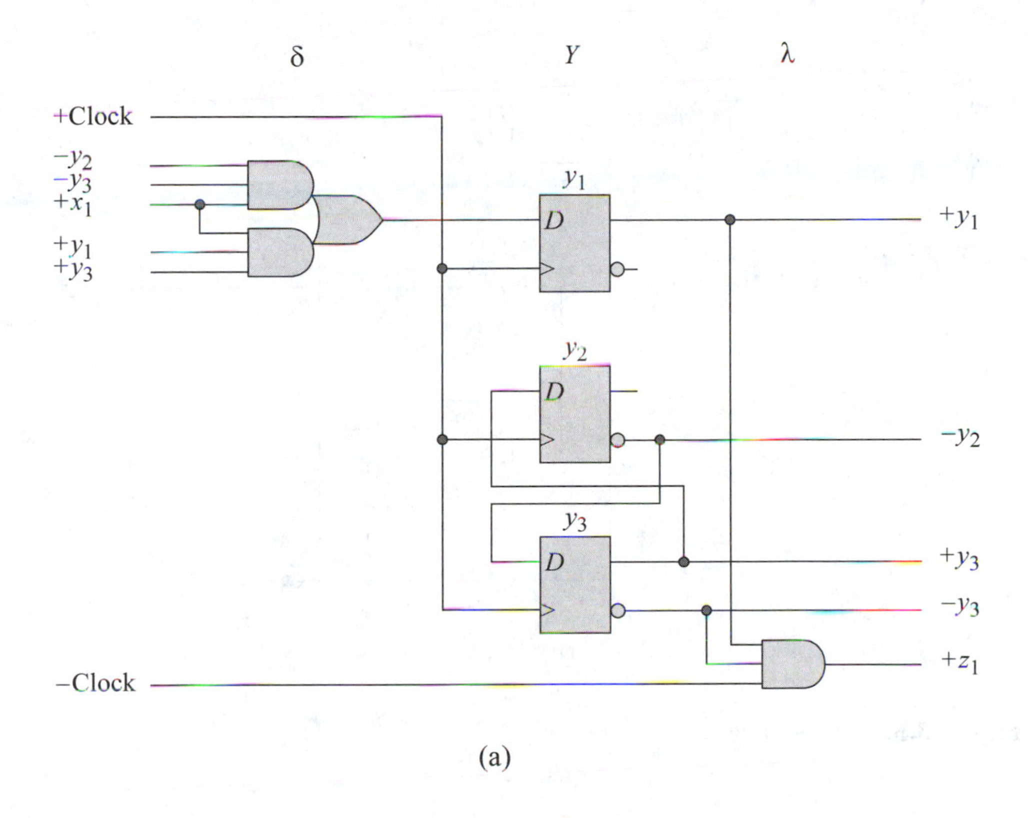

(a)

Figure 3.55 Logic diagram for the Moore machine of Figure 3.49 using D flip-flops. Output z_1 is asserted when an input sequence of 111 has been detected in a 3-bit word on a serial data line x_1. Output z_1 is asserted at time t_2 and deasserted at time t_3: (a) using AND and OR gates; and (b) using NOR gates.

For synchronous sequential machines, it will be assumed that any changes to the binary input variables will occur on the alternate clock transition to that which triggers the state flip-flops; that is, inputs will change at time t_2 for the Moore machine shown in Figure 3.55. This allows the input signals to stabilize at the flip-flop data inputs before the active clock transition occurs, thus allowing the flip-flop input setup conditions to be met. The *setup time* for a flip-flop is the time interval between the application of a signal at the data input and an active transition at the clock input.

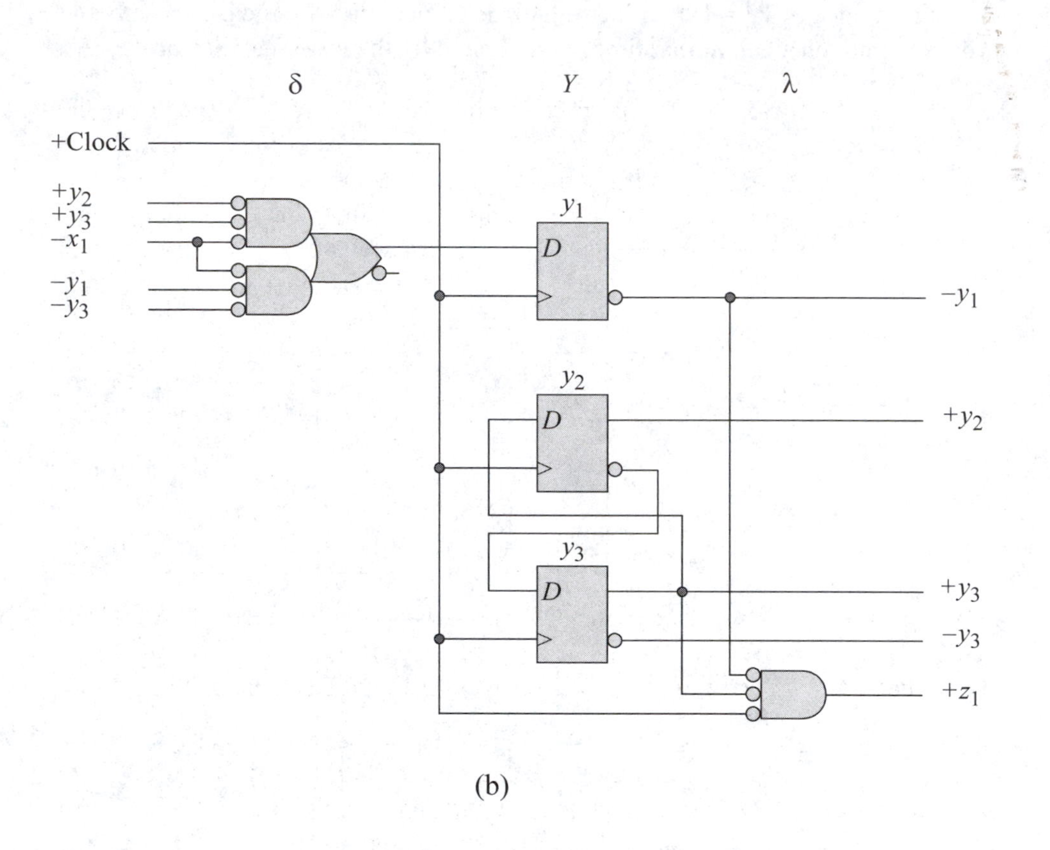

(b)

Figure 3.55 (Continued)

If the setup time is not observed, then a condition of metastability may result. A *metastable* condition in a flip-flop is characterized by a slight margin of instability during the active clock transition. In a metastable state, the flip-flop output may generate a nondigital signal for an unspecified time duration, then change to a logic 0 or 1 voltage level. In other situations, the output may switch to a high voltage level, remain at that level momentarily, then switch back to a low voltage level; or it may remain at the high voltage level. In a metastable condition, therefore, the operation of the flip-flop is unpredictable. If metastability occurs and a flip-flop fails to latch the assertion of an asynchronous input, then the output of the flip-flop will return to zero. The machine will then proceed along an incorrect path that is based upon the input being inactive. By allowing the binary input variables to change only on the alternate clock transition, this guarantees that the inputs will not change at or near the active clock transition, thus eliminating any possibility of metastability.

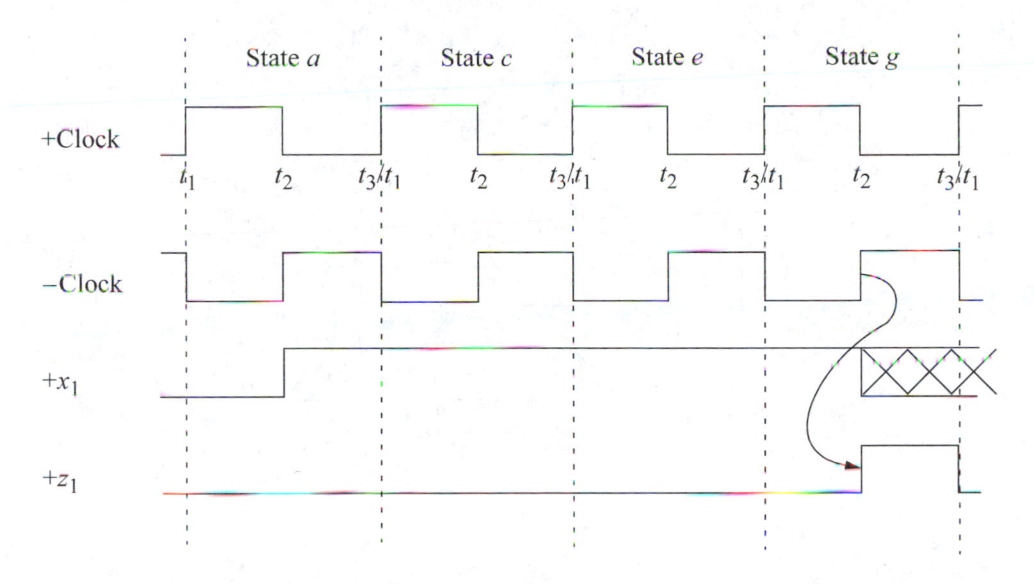

Figure 3.56 Timing diagram for the Moore machine of Figure 3.49 showing states a, c, e, and g for an input sequence of $x_1 = 111$. Output z_1 is asserted during the state time between words at time t_2 and deasserted at time t_3.

Figure 3.57 illustrates two assertion times for input x_1 using a D flip-flop with a positive-edge-triggered clock input. In Figure 3.57 (a), x_1 is synchronized with the system clock; therefore, changes to x_1 take place on the negative clock transition. The time interval $t_2 - t_3$ allows the D input to stabilize before the active clock transition; thus, the flip-flop output remains stable. Figure 3.57 (b) shows x_1 asserted near the active clock transition. Output y_1 may change to the high voltage level and remain at that level, or y_1 may fail to latch at the high voltage level and return to the low voltage level, resulting in a metastable condition. If y_1 does not remain set, then x_1 must remain asserted until the next positive clock transition so that the machine can detect the occurrence of x_1 and proceed to the next state.

Three techniques are used to handle asynchronous inputs:

1. Synchronize the input to the system clock.
2. Assign state codes so that a metastable condition will not result in an erroneous state transition.
3. Insert a redundant state in one path of an input-dependent branch.

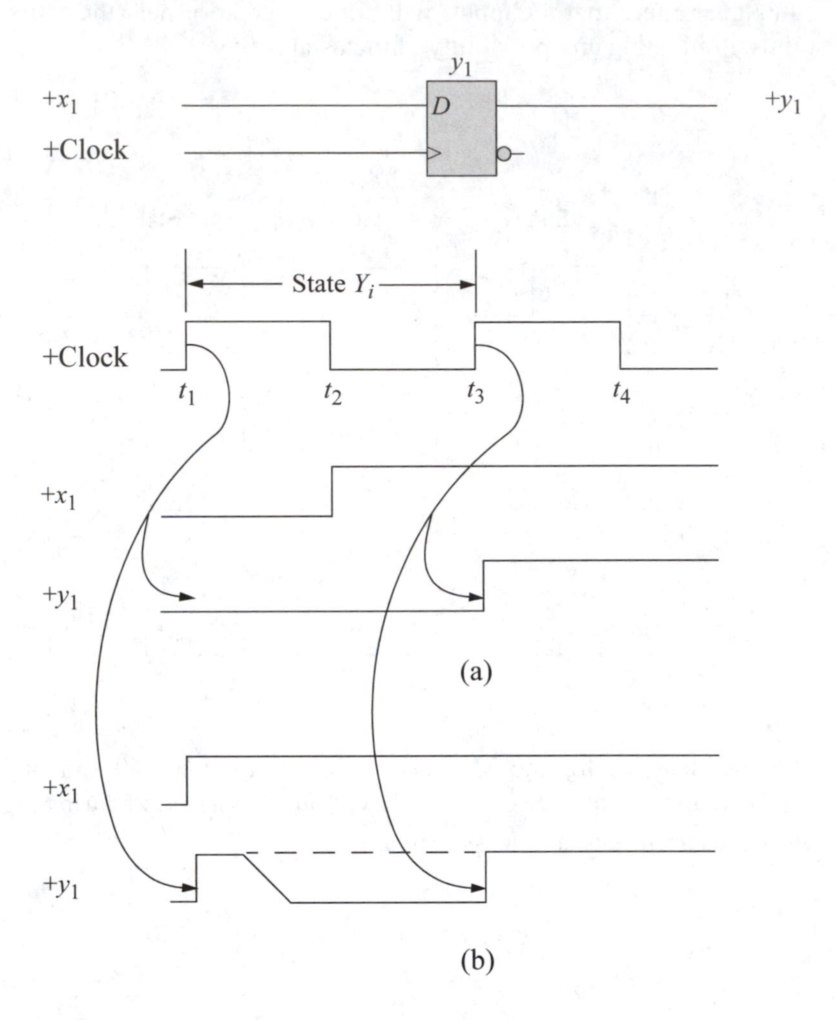

Figure 3.57 Assertion of binary input variable $\boldsymbol{x}_1$: (a) synchronous; and (b) asynchronous, where the setup time was not observed, resulting in metastability of the flip-flop output.

Figure 3.58 demonstrates a method to synchronize an asynchronous input x_1 to the system clock. Two flip-flops are necessary as shown in Figure 3.58 (a). Asynchronous input x_1 is connected to the D input of flip-flop y_A, whose clock input receives a clock signal that is four times the frequency of the system clock. A quadruple frequency provides an adequate number of sampling points for the asynchronous input. The active-high output of y_A is connected to the D input of flip-flop y_B, whose clock input receives the complemented system clock. The output of y_B represents a synchronized version of input x_1. The purpose of flip-flop y_A is to insure that Dy_B will always be stable before the active clock transition for flip-flop y_B, thus assuring that the setup requirements are met for Dy_B.

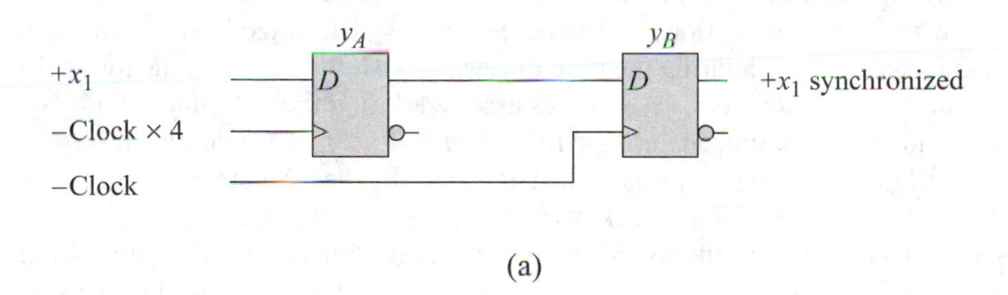

(a)

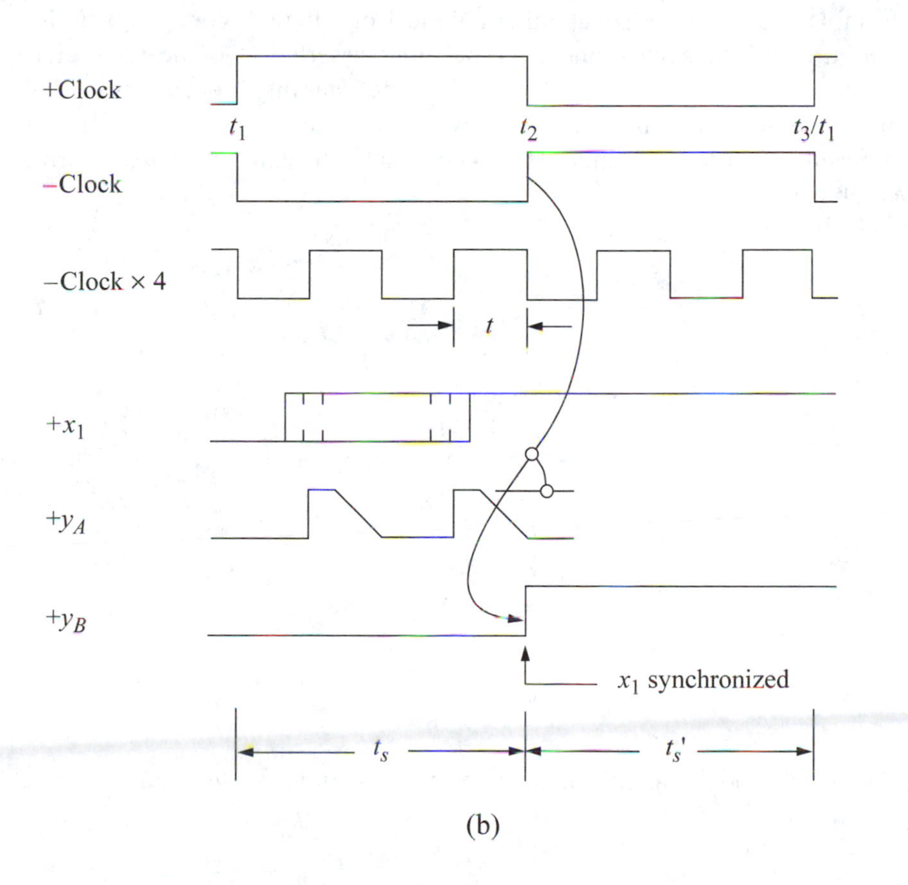

(b)

Figure 3.58 Synchronizing an asynchronous input x_1 to the system clock; (a) logic diagram; and (b) timing diagram. If x_1 is asserted during time t_s, and the setup requirements for time t are met, then a state transition for the machine will occur at time t_3/t_1 — the next positive transition of the system clock. If x_1 is asserted during time t_s', then any state transition must wait until the next active clock transition.

Time duration t in Figure 3.58 (b) is characterized by the following statement:

t is ≥ the metastable time of y_A plus the setup time of Dy_B.

If the conditions for time t are met, then any metastable condition generated by y_A will not adversely affect the output stability of flip-flop y_B. Thus, the output of y_B will never produce a metastable condition and can be used as a synchronized version of input x_1. If x_1 becomes asserted during the time interval $t_1 - t_2$, then a state transition will occur at time t_3/t_1. If, however, x_1 becomes asserted during the time interval $t_2 - t_3$, then x_1 will not be synchronized until the following t_2 time, and a state transition will not occur until the following t_3/t_1 time. The latter situation necessitates that x_1 remain active until the following system clock period.

A second technique for reliably using asynchronous inputs is shown in the partial state diagram of Figure 3.59. Asynchronous input x_1 is not synchronized to the system clock, but is used in its asynchronous mode. As shown in Figure 3.57, the output of a flip-flop whose data input changes at or near the active clock transition may become metastable and will stabilize at either a logic 1 or a logic 0 voltage level. Referring to Figure 3.59, if asynchronous x_1 becomes asserted at or near the clock transition, then y_1 may become metastable. If, after metastability, y_1 remains at a logic 0, then the machine remains in state *a*. If, however, y_1 enters a logic 1 state, then the machine proceeds to state *b*. It is important to note that both states *a* and *b* are correct next states for state a.

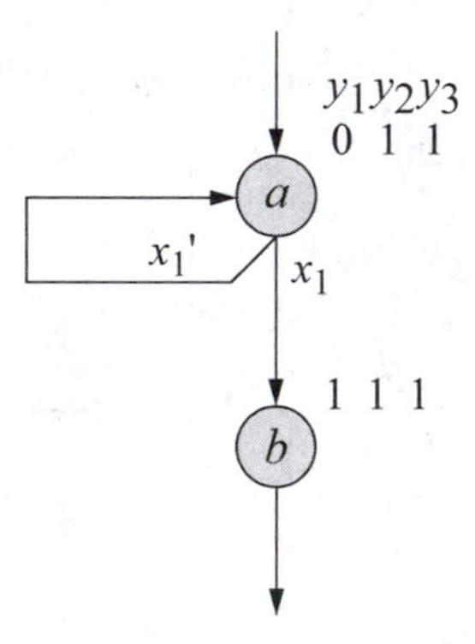

Figure 3.59 Method for managing an asynchronous input x_1 to assure a correct state transition.

This correct state transition was achieved by assigning adjacent state codes to states *a* ($y_1y_2y_3 = 011$) and *b* ($y_1y_2y_3 = 111$) where only the affected flip-flop y_1 changes state. If the assertion of x_1 results in flip-flop y_1 achieving stability in the logic 0 level after being metastable, then x_1 must remain active until the following clock period in order for its occurrence to be registered. If system performance cannot guarantee the assertion of x_1 until the following clock period, then the duration of the active level of the input can be increased by using a technique described in Chapter 6 on the synthesis of asynchronous sequential machines.

The third situation is where an input-dependent branch occurs. The method shown in the partial state diagram of Figure 3.60 can be used. Figure 3.60 (a) portrays a branch from state *a* to state *b* if an asynchronous input $x_1 = 1$, or to state *c* if $x_1 = 0$. Figure 3.60 (b) presents a method to handle the asynchronous input by inserting an intermediate state *a*' in the path from state *a* to state *c*.

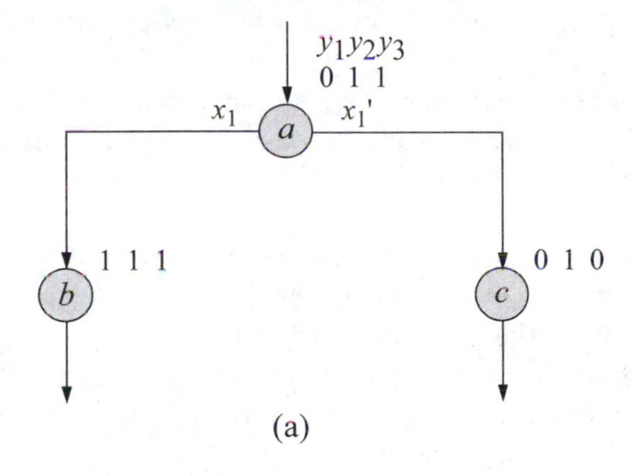

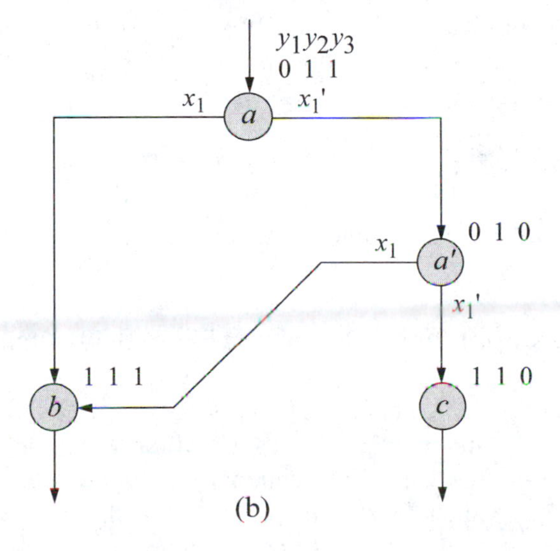

Figure 3.60 Partial state diagram depicting a two-way branch dependent upon an asynchronous input x_1: (a) required branch; and (b) branch modified for a possible metastable condition of flip-flop y_1.

If x_1 becomes active sufficiently early so that the setup conditions for flip-flop y_1 are met, then the machine proceeds from state *a* to state *b*, which is the correct sequence. However, if x_1 becomes asserted near the clock transition, resulting in a metastable condition for y_1, then the machine may move to state *c*, which is an incorrect sequence if $x_1 = 1$. The machine may not detect the assertion of x_1 and may never enter state *b*.

To compensate for a possible metastable condition, an intermediate state *a'* is inserted in the state diagram as shown in Figure 3.60 (b). There are three cases to consider. If x_1 is asserted so that the setup conditions are met, then the machine proceeds from state *a* to state *b* in a normal manner. If x_1 becomes asserted near the active clock transition, then flip-flop y_1 may become metastable and remain reset. The machine then moves to state *a'*, but not to state *c*. Input x_1 will remain active for at least one more clock cycle, thus allowing a correct sequence from state *a* to state *b* through state *a'*. Finally, if x_1 does not become active during state *a* or near the next clock transition, the machine proceeds from state *a'* to state *c*, which is a correct state transition for an inactive input x_1.

The extra state *a'* must be judiciously inserted such that the machine specifications are still maintained. Since the state diagram now contains one additional state time, the input characteristics must be carefully considered to assure proper machine performance.

A necessary constraint for a machine with asynchronous inputs is that the specific state where a metastable state is possible cannot be directly associated with an output. Before returning to a logic 0, the metastable output may have reached a sufficiently high voltage level to trigger the receiving logic circuit. This could result in an output glitch. Managing asynchronous inputs usually requires additional logic; however, the increased system reliability justifies the added cost.

Example 3.2 To elucidate the hardware differences between a Moore machine using *D* flip-flops and one implemented with *JK* flip-flops for functionally equivalent machines, Example 3.2 synthesizes the Moore machine of Figure 3.49 using *JK* flip-flops.

Step 1: State diagram The same as Figure 3.49 of Example 3.1.

Step 2: Equivalent states The same as Example 3.1 in which there are no equivalent states.

Step 3: State code assignment The same as Figure 3.49 of Example 3.1.

Step 4: Next-state table The next-state table is shown in Table 3.14 and is obtained directly from the state diagram. Since this is a Moore machine, output z_1 is asserted in state *g* ($y_1y_2y_3 = 110$) and is independent of the value of x_1. Table 3.15 is used to obtain the *JK* input values for a particular state transition.

In row $y_1y_2y_3 = 000$ of Table 3.14, y_1 has a present state of 0 and a next state of 0 when $x_1 = 0$. Thus, the state transition for y_1 is from 0 to 0, yielding *JK* input values of $Jy_1Ky_1 = 0-$. When $x_1 = 1$, y_1 proceeds from 0 to 1, providing *JK* input values of

$Jy_1Ky_1 = 1-$. In the same manner, the flip-flop input values for Jy_1Ky_1, Jy_2Ky_2, and Jy_3Ky_3 are obtained as shown in Table 3.14.

Table 3.14 Next-state table for the Moore machine of Figure 3.49 using *JK* flip-flops

Present state			Input	Next state			Flip-flop inputs						Present output
y_1	y_2	y_3	x_1	y_1	y_2	y_3	Jy_1	Ky_1	Jy_2	Ky_2	Jy_3	Ky_3	z_1
0	0	0	0	0	0	1	0	–	0	–	1	–	0
0	0	0	1	1	0	1	1	–	0	–	1	–	0
0	0	1	0	0	1	1	0	–	1	–	–	0	0
0	0	1	1	0	1	1	0	–	1	–	–	0	0
0	1	0	0	0	0	0	0	–	–	1	0	–	0
0	1	0	1	0	0	0	0	–	–	1	0	–	0
0	1	1	0	0	1	0	0	–	–	0	–	1	0
0	1	1	1	0	1	0	0	–	–	0	–	1	0
1	0	0	0	–	–	–	–	–	–	–	–	–	–
1	0	0	1	–	–	–	–	–	–	–	–	–	–
1	0	1	0	0	1	1	–	1	1	–	–	0	0
1	0	1	1	1	1	1	–	0	1	–	–	0	0
1	1	0	0	0	0	0	–	1	–	1	0	–	1
1	1	0	1	0	0	0	–	1	–	1	0	–	1
1	1	1	0	0	1	0	–	1	–	0	–	1	0
1	1	1	1	1	1	0	–	0	–	0	–	1	0

Table 3.15 Excitation table for a *JK* flip-flop

Present-state $Y_{j(t)}$	Next state $Y_{k(t+1)}$	Flip-flop inputs $J\,K$
0	0	0 –
0	1	1 –
1	0	– 1
1	1	– 0

Step 5: Input maps Three pairs of input maps are required for the state flip-flops as shown in Figure 3.61 and are obtained from the next-state table of Table 3.14 or from the state diagram of Figure 3.49 in conjunction with the *JK* flip-flop excitation table. The unused state is $y_1y_2y_3 = 100$ and is indicated by the "don't care" symbol in all input maps.

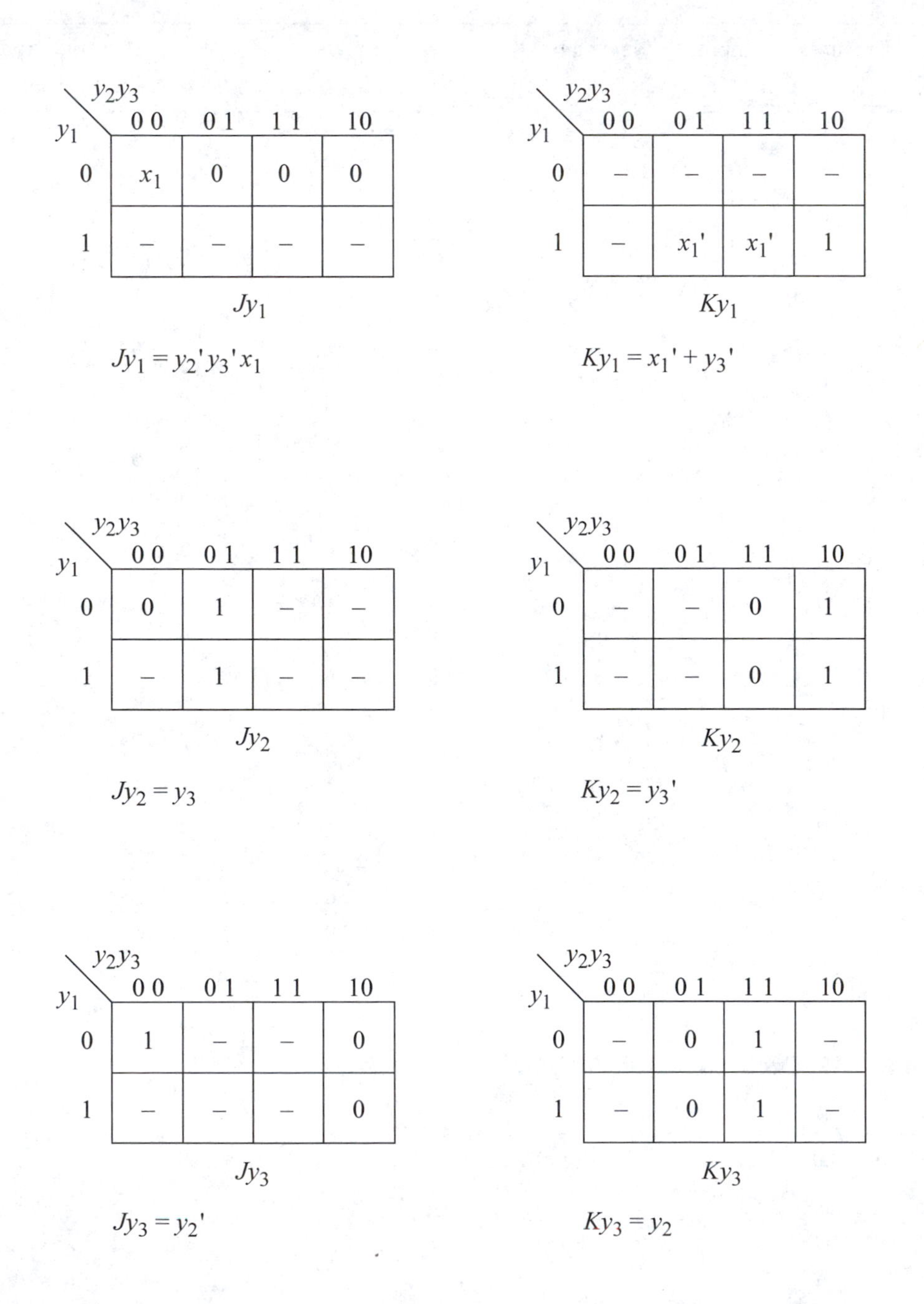

Figure 3.61 Input maps for the Moore machine of Figure 3.49 using *JK* flip-flops. Input x_1 is a map-entered variable.

Referring to Table 3.14 for state $y_1y_2y_3 = 000$, the values for Jy_1 are the same as the values for x_1; therefore, the entries for $y_1y_2y_3 = 000$ in the input map for y_1 are $Jy_1Ky_1 = x_1 -$. In state $y_1y_2y_3 = 101$, the values for Ky_1 are the complement of the values for x_1; therefore, the entries for Jy_1Ky_1 are $-x_1'$. The remaining entries for all maps are obtained in a similar manner, yielding the input equations of Equation 3.21.

$$
\begin{aligned}
Jy_1 &= y_2'y_3'x_1 & \qquad Ky_1 &= x_1' + y_3' \\
Jy_2 &= y_3 & \qquad Ky_2 &= y_3' \\
Jy_3 &= y_2' & \qquad Ky_3 &= y_2
\end{aligned}
\tag{3.21}
$$

Note the equation for Ky_1. The x_1' term was obtained by generating a slightly different, but equivalent, version of the input map for Ky_1 as shown in Figure 3.62. The entry for state 110 is 1, but can be expanded to $x_1 + x_1'$ by the complementation law of Axiom 7. Alternatively, the 1 can be replaced by $1 + x_1'$ as specified by Theorem 1. Therefore, every minterm location contains the entry x_1', including the "don't care" states. This yields the x_1' term for Ky_1. Now combine x_1' with the x_1 term in location 110 to yield a 1, resulting in the y_3' term of Ky_1. Alternatively, simply use the 1 in minterm location 110 to obtain the y_3' term.

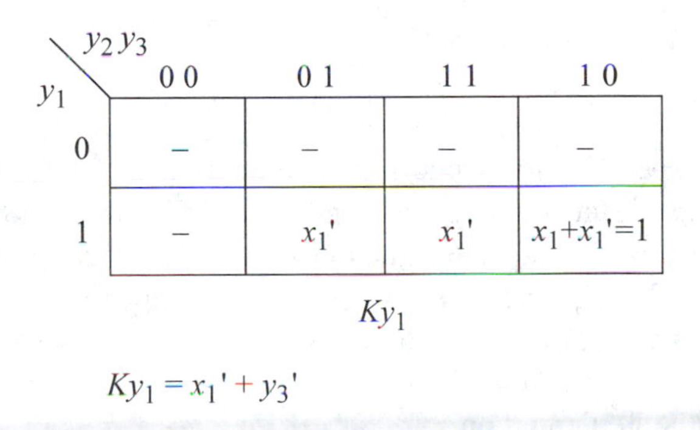

Figure 3.62 Expanded version of the input map for Ky_1 from Figure 3.61.

Step 6: Output map The output map is generated from either the state diagram or the next-state table and is shown in Figure 3.63. Like the Moore design using D flip-flops, the unused state combines with location 110, resulting in Equation 3.22. As stated previously, the unused state can be utilized in the minimization process only because the assertion for z_1 occurs after the machine has stabilized.

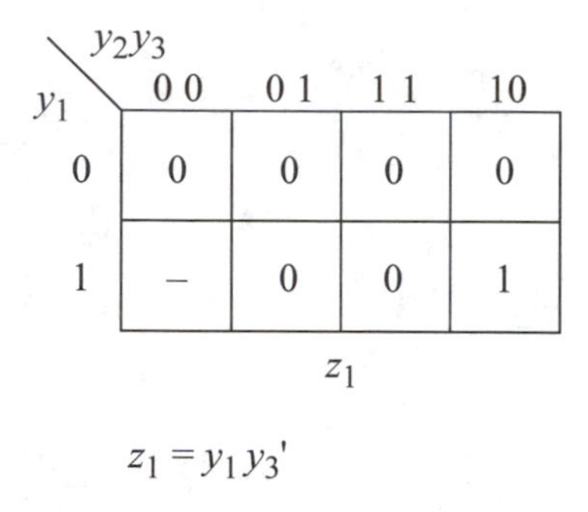

Figure 3.63 Output map for z_1 for the Moore machine of Figure 3.49.

$$z_1 = y_1y_3' \tag{3.22}$$

If the machine specifications required z_1 to be asserted at time t_1 rather than at time t_2, then the unused state could not be used to minimize the λ output logic. A transition from state *a* to state *c* may cause the machine to pass through the transient state $y_1y_2y_3 = 100$ if flip-flop y_1 sets before y_3 sets. Assigning a value of 1 to the "don't care" state assigns a voltage level of logic 1 to z_1 if a state sequence passes through $y_1y_2y_3 = 100$. Since the transition from state *a* to state *c* is not associated with an output, a glitch would be generated on z_1.

Step 7: Logic diagram The logic diagram, implemented with *JK* flip-flops, is presented in Figure 3.64 and is derived from the input and output equations of Equation 3.21 and Equation 3.22, respectively. The storage elements are negative-edge-triggered *JK* flip-flops. The system clock changes to the positive level at time t_2, thus asserting output z_1 if the machine is in state *g* ($y_1y_2y_3 = 110$). It is assumed, of course, that the machine will never stabilize in the unused state $y_1y_2y_3 = 100$; thus y_2 is not a factor in asserting output z_1.

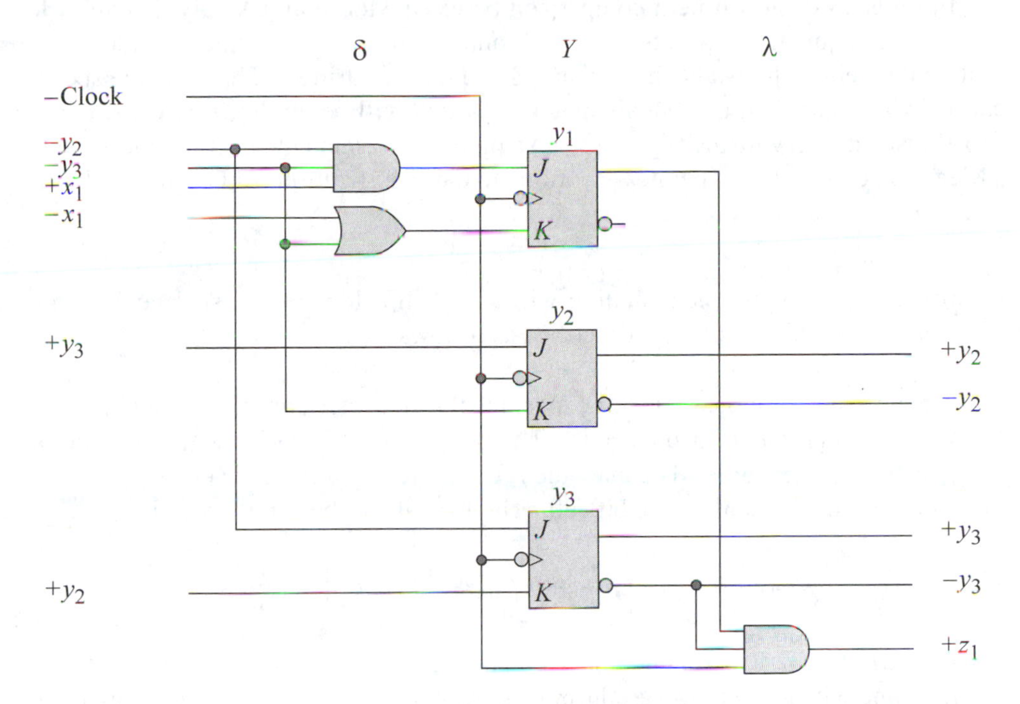

Figure 3.64 Logic diagram for the Moore machine ofFigure 3.49 using *JK* flip-flops. Output z_1 is asserted when an input sequence of 111 has been detected in a 3-bit word on a serial data line x_1. Output z_1 is asserted at time t_2 and deasserted at time t_3.

3.5 Mealy Machines

A Mealy machine is also an important class of finite-state machine and represents an alternative model that is more widely used than a Moore machine. The Mealy class of synchronous sequential machines is the result of a paper by G. H. Mealy in 1955 on the synthesis of sequential circuits. A Mealy machine was formally defined in Chapter 2 as a synchronous sequential machine characterized by the following 5-tuple:

$$M = (X, Y, Z, \delta, \lambda)$$

where X is the input alphabet, Y is the state alphabet, and Z is the output alphabet. The next-state function δ maps the Cartesian product of X and Y into Y, and thus, is determined by both the present inputs and the present state. The output function λ maps the Cartesian product of X and Y into Z such that, the output vector is a function of both the present inputs and the present state. This is the underlying difference between Moore and Mealy machines — *the outputs of a Moore machine are directly related to the present state only, whereas, the outputs of a Mealy machine are a function of both the present state and the present inputs.*

In order to demonstrate a comparison between Moore and Mealy machines for functionally equivalent circuits, the synchronous sequential machine of Example 3.1 will be the vehicle to establish similarities and dissimilarities. The machine specifications will be applied to a Mealy machine, which will be implemented first with *D* flip-flops and then with *JK* flip-flops. Example 3.3 presents the design procedure for a Mealy machine which operates according to the specifications of Example 3.1, with some minor modifications.

Example 3.3 This implementation will use *D* flip-flops as the storage elements. The machine specifications are described as follows:

> Design a synchronous sequential machine that accepts serial data on an input line x_1 which consists of 3-bit words. The words are contiguous with no space between adjacent words. The machine is controlled by a periodic clock, where one clock period is equal to one bit cell. The format for the 3-bit words is
>
> $$x_1 = \cdots \left| b_1b_2b_3 \right| b_1b_2b_3 \left| b_1b_2b_3 \right| \cdots$$
>
> where $b_i = 0$ or 1. Whenever a word contains the bit pattern $b_1b_2b_3 = 111$, the machine will assert output z_1 during the b_3 bit cell according to the following assertion/deassertion statement: $z_1 \uparrow t_2 \downarrow t_3$. Thus, z_1 is active for the last half of bit cell b_3.

An example of a valid word in a series of words is as follows:

$$x_1 = \cdots \left| 001 \right| 101 \left| 011 \right| 101 \left| 111 \right| 010 \right| \cdots$$

$$z_1 \uparrow t_2 \downarrow t_3$$

Notice that the output signal — compared with the Moore machine — occurs during the third bit period of a valid word. The Moore output was displaced in time by one clock period.

Step 1: State diagram The state diagram for this Mealy machine is graphically depicted in Figure 3.65, where output z_1 is asserted during the third state time of a word as a function of input x_1. Because the machine may be required to test all three bits of any word, it is important that any path taken from state *a* back to state *a* always maintains three state levels as shown. The machine is reset to an initial state of $y_1y_2y_3 = 000$. The state diagram should be drawn in such a manner that the paths — or state transitions — are easily discernible, thus giving a clear illustration of machine activity.

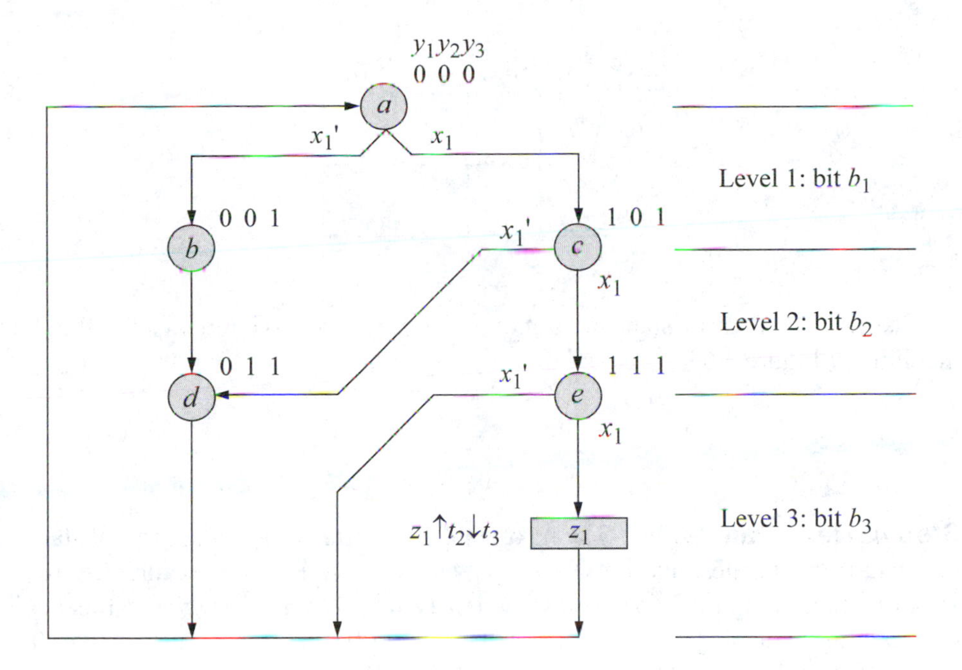

Figure 3.65 State diagram for the Mealy machine of Example 3.3, which generates an output z_1 whenever a 3-bit word $x_1 = 111$. Unused states are: $y_1y_2y_3 = 010$, 100, and 110.

Step 2: Equivalent states When generating the simple state diagram of Figure 3.65, it was easy to determine that state *c* must proceed to state *d* if $x_1 = 0$, rather than to a new state (*f*) at the same level as states *d* and *e*. State *f* would subsequently be found to be equivalent to state *d* and thereby eliminated as a redundant state. Therefore, the state diagram of Figure 3.65 has no equivalent states and is a reduced state diagram as shown.

Step 3: State code assignment Rule 1 for state code adjacency prescribes that the state codes for *b* and *c* be adjacent, because both are destination states resulting from a branch in state *a*. Likewise, the state codes for states *d* and *e* should be adjacent. Rule 2 also applies to states *d* and *e*, since both are source states for state *a*.

The state diagram of Figure 3.65 contains five states. Therefore, the number of storage elements is $p = 3$, which yields a total of $2^p = 2^3 = 8$ states, of which three are unused and will be treated as "don't care" states. State *a* is assigned a state code of $y_1y_2y_3 = 000$, which is an initialization, or reset, state. The Karnaugh map of Figure 3.66 is a present-state map showing one possible state code assignment for states *a*, *b*, *c*, *d*, and *e*, in which the adjacency requirements are met for state pairs (*b*,*c*) and (*d*,*e*).

$y_1 \backslash y_2y_3$	0 0	0 1	1 1	10
0	*a*	*b*	*d*	–
1	–	*c*	*e*	–

Figure 3.66 Present-state map showing state code assignments for the Mealy machine of Figure 3.65.

Step 4: Next-state table The next-state table lists all possible present states, input values, next states, and present outputs. The table is obtained directly from the state diagram of Figure 3.65 and is shown in Table 3.16. To derive the table, begin in state *a* in the state diagram. In state *a* ($y_1y_2y_3 = 000$), x_1 has a value of either 0 or 1. If $x_1 = 0$, then the next state is *b* ($y_1y_2y_3 = 001$). The next state $Y_{k(t+1)}$ for a *D* flip-flop is the same as the present input $Dy_{k(t)}$ before the flip-flop is clocked. Thus,

$$y_{1(t+1)} = Dy_{1(t)}$$

$$y_{2(t+1)} = Dy_{2(t)}$$

$$y_{3(t+1)} = Dy_{3(t)}$$

Therefore, the values of the flip-flops *D* inputs are the same as the next-state values, which yields $Dy_1Dy_2Dy_3 = 001$ in the first row of Table 3.16. If $x_1 = 1$ in state *a*, then the machine proceeds to state *c* ($y_1y_2y_3 = 101$). In state *a*, output $z_1 = 0$, regardless of the value of x_1.

In a similar manner, the remaining entries in Table 3.16 are obtained. Notice that the unused states have "don't care" conditions for both the next states and the outputs. This may help to minimize the δ next-state logic (the flip-flop input logic) and the λ output logic. If the machine is functioning properly, then the unused states will never be entered. Thus, there is no inherent risk in assigning "don't care" values for both the next states and the outputs of all unused states.

In states *b*, *d*, and *e*, input x_1 is not a factor in determining the next state; the machine moves to the next state regardless of the value of x_1. In state *e*, however, x_1 determines whether output z_1 is asserted. The assertion/deassertion statement $z_1 \uparrow t_2 \downarrow t_3$ indicates that z_1 is active for the last half of the clock period. Although the next-state table is a precursor for the input maps which follow, the table can be circumvented when *D* flip-flops are utilized, because the values of the *D* inputs are the same as those specified in the next-state column. The next state for each flip-flop is shown in the state diagram.

Table 3.16 Next-state table for the Mealy machine of Figure 3.65 using *D* flip-flops

Present state			Input	Next state			Flip-flop inputs			Present output
y_1	y_2	y_3	x_1	y_1	y_2	y_3	Dy_1	Dy_2	Dy_3	z_1
0	0	0	0	0	0	1	0	0	1	0
0	0	0	1	1	0	1	1	0	1	0
0	0	1	0	0	1	1	0	1	1	0
0	0	1	1	0	1	1	0	1	1	0
0	1	0	0	–	–	–	–	–	–	–
0	1	0	1	–	–	–	–	–	–	–
0	1	1	0	0	0	0	0	0	0	0
0	1	1	1	0	0	0	0	0	0	0
1	0	0	0	–	–	–	–	–	–	–
1	0	0	1	–	–	–	–	–	–	–
1	0	1	0	0	1	1	0	1	1	0
1	0	1	1	1	1	1	1	1	1	0
1	1	0	0	–	–	–	–	–	–	–
1	1	0	1	–	–	–	–	–	–	–
1	1	1	0	0	0	0	0	0	0	0
1	1	1	1	0	0	0	0	0	0	1

Step 5: Input maps The input maps represent the δ next-state logic for the flip-flop inputs. The input maps can be derived from either the next-state table or the state diagram. Since there are three flip-flops, there will be three input maps, one for each flip-flop. The maps denote the input equations for the D input of each flip-flop: Dy_1, Dy_2, and Dy_3, as shown in Figure 3.67. The input variable x_1 is used as a map-entered variable. Dashes are entered in the three unused states of each map.

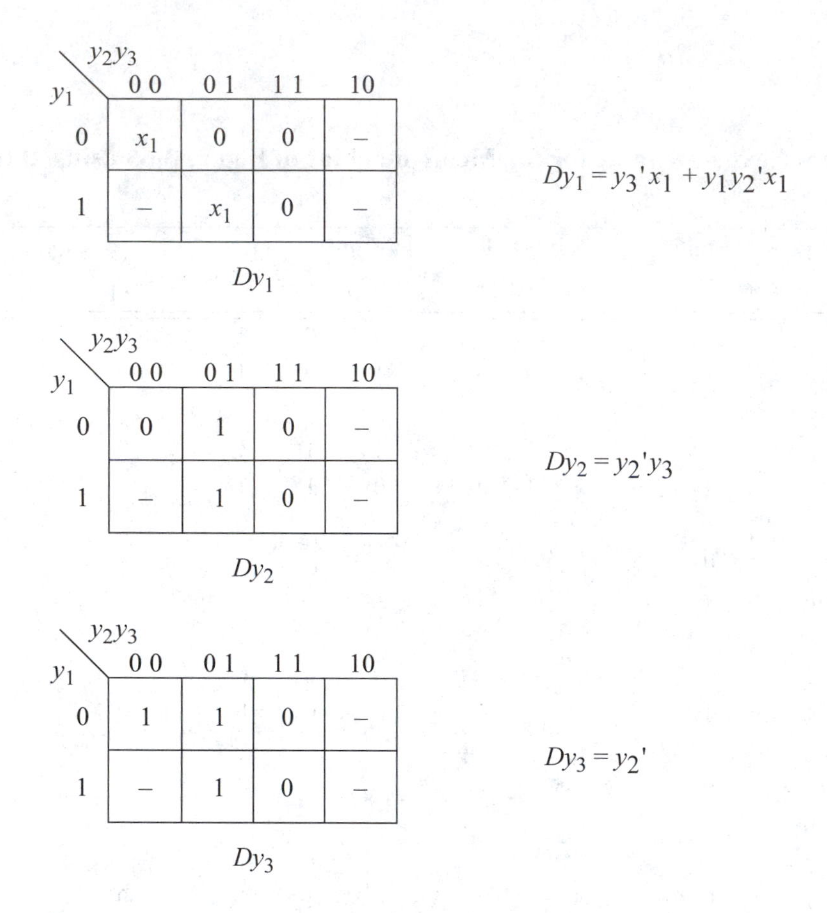

Figure 3.67 Input maps for the Mealy machine of Figure 3.65 using *D* flip-flops and adjacent state codes for state pairs (*b*, *c*) and (*d*, *e*).

Consider each flip-flop in turn. In state $y_1y_2y_3 = 000$, the next state for y_1 is 0 if $x_1 = 0$, or 1 if $x_1 = 1$. Thus, the value for y_1 becomes 1 only if $x_1 = 1$. Therefore, x_1 is inserted in location $y_1y_2y_3 = 000$, indicating that x_1 specifies the value of y_1 during the next clock period. In state $y_1y_2y_3 = 001$, the next state for y_1 is 0 regardless of the value of x_1; therefore, a 0 is inserted in minterm location $y_1y_2y_3 = 001$. This procedure is continued, examining each state in turn — using either the state diagram or the next-state table — until each square in the map for Dy_1 contains an entry. The minterm locations combine in Figure 3.67, yielding the equation for Dy_1 in a minimum sum-of-products form as shown in Equation 3.23. In the same manner, the maps for Dy_2 and Dy_3 are constructed, producing input equations as shown in Equation 3.23. A total of four logic gates are required for the δ next-state logic: three gates for Dy_1, one gate for Dy_2, and no gates for Dy_3.

$$Dy_1 = y_3'x_1 + y_1y_2'x_1$$
$$Dy_2 = y_2'y_3$$
$$Dy_3 = y_2' \tag{3.23}$$

The input equations represented by Equation 3.23 were obtained using adjacent state codes in Figure 3.65, where the following state pairs were adjacent according to Rules 1 and 2: (*b*,*c*) and (*d*,*e*). If, however, the state pairs were not adjacent, then the input equations may contain more terms and/or more variables per term. To illustrate the importance of adjacency, arbitrarily reassign the state codes in Figure 3.65 as follows: $a = 000$, $b = 001$, $c = 111$, $d = 011$, and $e = 100$. The unused states are now 010, 101, and 110.

The state diagram of Figure 3.65 is redrawn below as Figure 3.68 using the reassigned, nonadjacent state codes. The input maps for this new state diagram are shown in Figure 3.69 using x_1 as a map-entered variable. A total of eight logic gates are now required to implement the δ next-state logic: three gates for Dy_1, three gates for Dy_2 (sharing the $y_1'y_3'x_1$ term with Dy_1), and two gates for Dy_3 (sharing the $y_1y_2x_1'$ term with Dy_2). It is evident, therefore, that applying Rules 1 and 2 when assigning state codes results in a significant saving of gates, even for the small machine of Figure 3.65. The choice of nonadjacent state codes was arbitrary; however, other choices increase the δ next-state logic accordingly. Adjacent state code assignments will be used for the remainder of Example 3.3.

Step 6: Output maps Outputs from a Moore machine may be considered as unconditional outputs, because they depend only on the present state of the machine, whereas, outputs from a Mealy machine are conditional — they are a function not only of the present state, but also of the present inputs. A synchronous sequential machine may contain both Moore and Mealy output signals.

The output map for the Mealy machine of Figure 3.65 is derived from the state diagram or the next-state table. Output z_1 occurs only in state *e* ($y_1y_2y_3 = 111$) and only if x_1 is asserted. This is illustrated in the output map of Figure 3.70, using x_1 as a map-entered variable. In all other states, z_1 is inactive. Each square in the map corresponds to a state in the state diagram. The entries in the squares, or minterm locations, depict the values of z_1 during that state time, where a state time is specified as one clock period. For example, z_1 is inactive during state *a*; therefore, a 0 is inserted in minterm location $y_1y_2y_3 = 000$.

For the same reason, a 0 is entered in states *b*, *c*, and *d*, having state codes of 001, 101, and 011, respectively. Since z_1 is active in state *e* only if x_1 is also active, the input variable x_1 is entered in minterm location 111, which corresponds to state *e*. Thus, z_1 assumes the value of x_1 during state *e*. The assertion and deassertion of z_1 is not shown in the output map, because the map represents only the value of z_1 for each state, or clock period. The assertion or deassertion in the middle of a clock cycle cannot be indicated.

Since the unused states can never be entered if the machine is functioning properly, these states can be used as "don't care" entries in the output map. The equation for the λ output logic for z_1 is obtained by combining logically adjacent map entries to yield

$$z_1 = y_1 y_2 x_1 \tag{3.24}$$

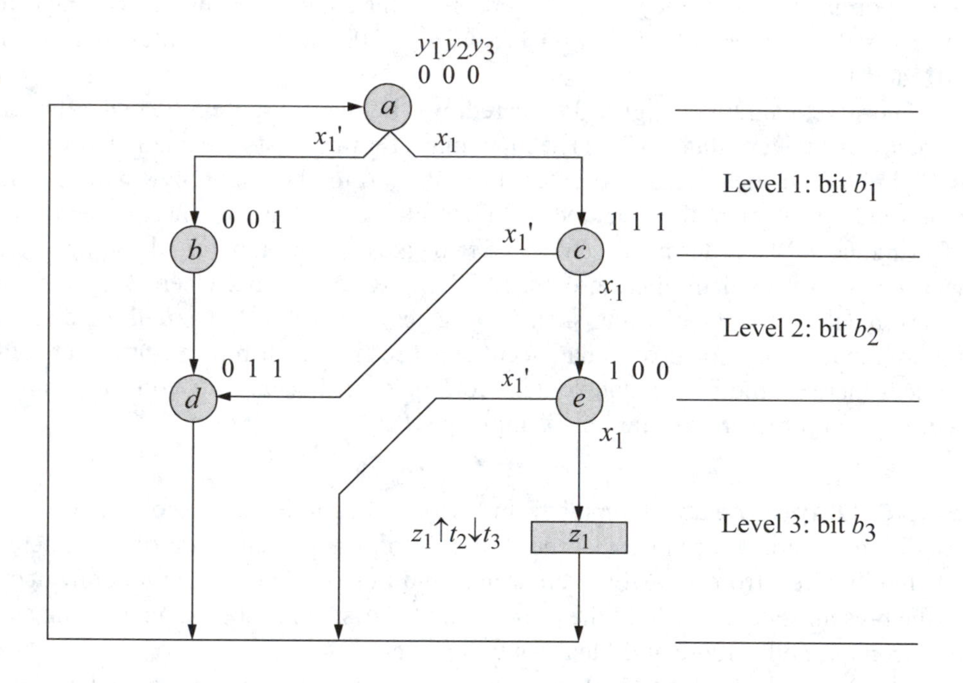

Figure 3.68 State diagram for the Mealy machine of Example 3.3, using nonadjacent state codes for state pairs (b, c) and (d, e). Unused states are: $y_1y_2y_3 = 010$, 101, and 110.

Step 7: Logic diagram The logic diagram is implemented using the input equations of Equation 3.23, three D flip-flops, and the output equation of Equation 3.24. The application of these boolean equations produces the logic diagram of Figure 3.71. The AND gate which generates output z_1 at an active high (+) voltage level has a fanin of four. The fourth input is the complemented version of the clock signal. During state *e* with x_1 active, when –Clock changes from a low to a high voltage level at time t_2, all inputs of the AND gate are at a high level, thus asserting z_1. Using –Clock as a gating input provides the correct assertion/deassertion for z_1; that is, $z_1 \uparrow t_2 \downarrow t_3$.

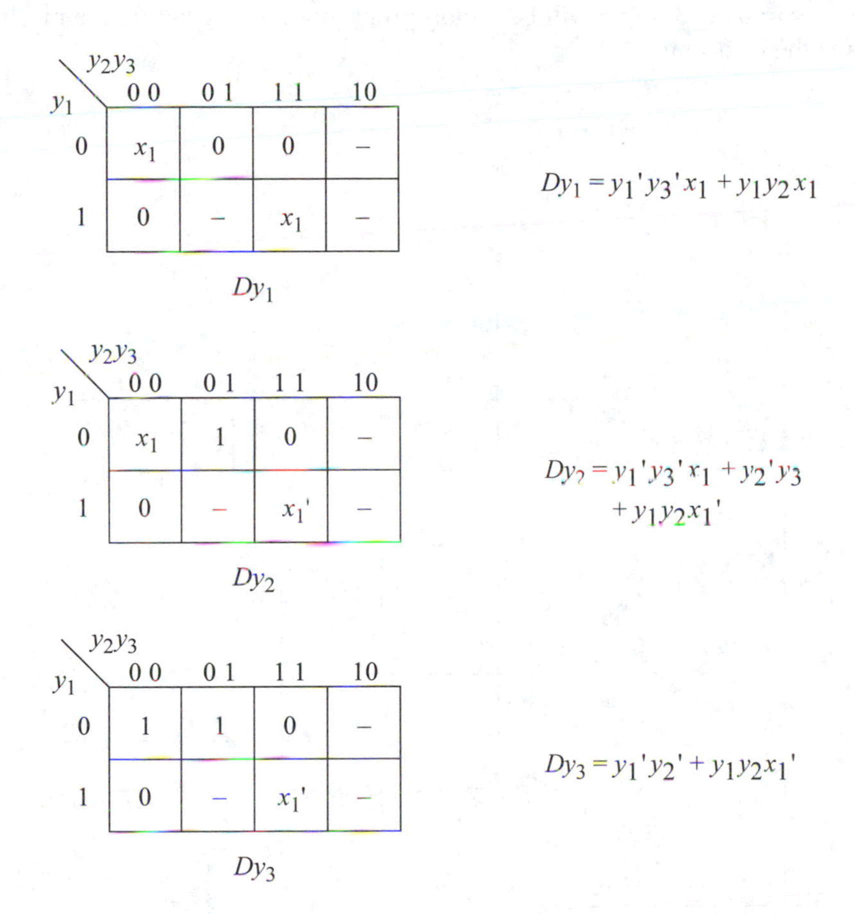

Figure 3.69 Input maps for the Mealy machine of Figure 3.65 using *D* flip-flops and nonadjacent state codes for state pairs (*b*, *c*) and (*d*, *e*).

y_1 \ y_2y_3	00	01	11	10
0	0	0	0	–
1	–	0	x_1	–

z_1

Figure 3.70 Output map for the Mealy machine of Figure 3.65, using input x_1 as a map-entered variable.

By systematically applying different sequences for input x_1, the machine can be shown to operate in the manner prescribed by the machine specifications. As an example, waveforms are shown in the timing diagram of Figure 3.72, illustrating the state transitions for an input sequence of $x_1 = 100111$. Circuit delays are assumed but are not shown, allowing the state transitions to be more easily discerned. After the positive clock transition, there will be a short propagation delay before the machine stabilizes in the next state.

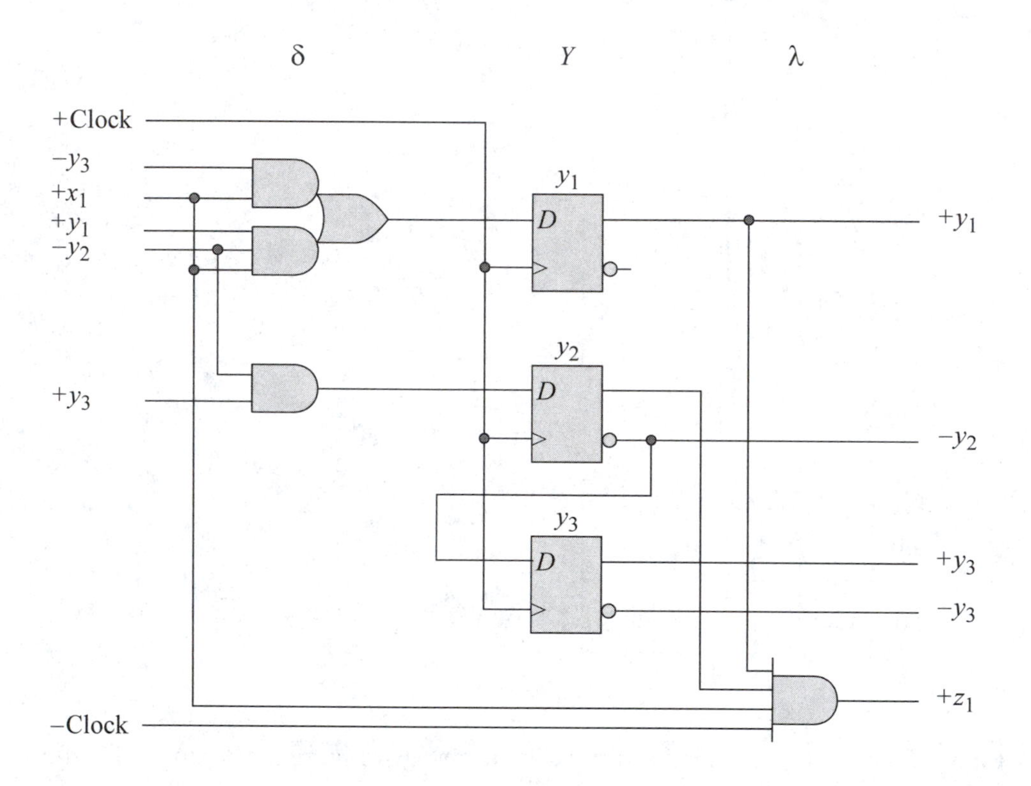

Figure 3.71 Logic diagram for the Mealy machine of Figure 3.65 using *D* flip-flops. Output z_1 is asserted when an input sequence of 111 has been detected in a 3-bit word on a serial data line x_1. Output z_1 is asserted at time t_2 and deasserted at time t_3.

The state diagram of Figure 3.65 and the logic diagram of Figure 3.71 will be used in conjunction with the timing diagram for the description which follows. The machine is initially reset to state *a* ($y_1y_2y_3 = 000$). Input x_1 changes to an active high voltage level during state *a* at time t_2 which causes a transition from state *a* to state *c* at the next positive clock edge. During state *c*, x_1 becomes inactive, resulting in a transition to state *d*. Entering state *d* signifies that an invalid word has been detected and that z_1 will not be asserted. There is no need, therefore, to test the value of x_1 in state *d*, consequently eliminating unnecessary logic circuits. The machine then moves to state *a* to test the first bit of the next word.

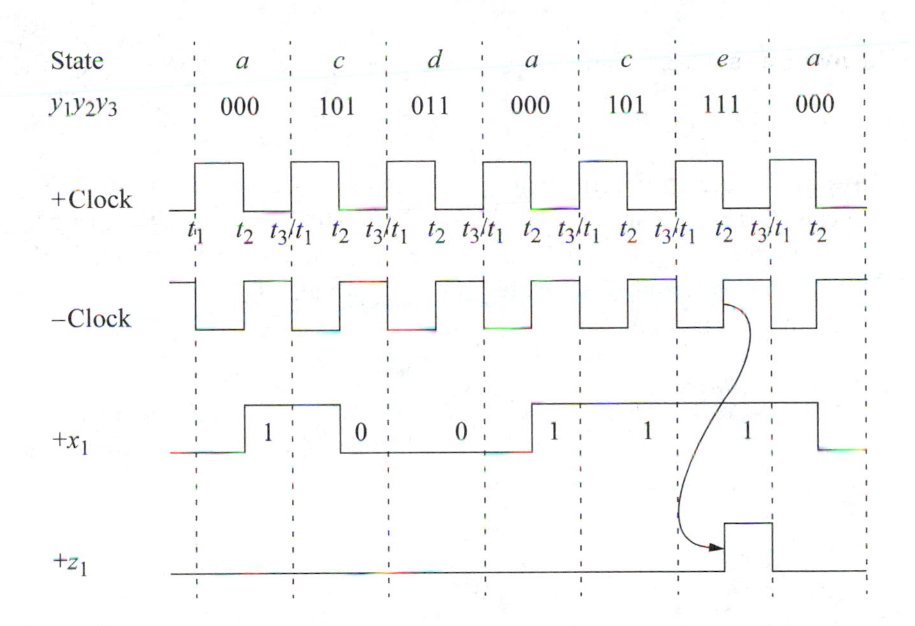

Figure 3.72 Timing diagram for the Mealy machine of Figure 3.65 for an input sequence of $x_1 = 100111$ with a resulting assertion of output z_1 during state *e*. Time t_3/t_1 indicates the end of a state (t_3) or the beginning of a state (t_1).

The state transition sequence can also be observed in the logic diagram. Consider the transition from state *d* to state *a*. When the present state is *d* ($y_1y_2y_3 = 011$), the flip-flop data inputs are $Dy_1Dy_2Dy_3 = 000$; therefore, the next state will be *a* ($y_1y_2y_3 = 000$). With a present state of $y_1y_2y_3 = 000$ and with x_1 asserted, the flip-flop data inputs are $Dy_1Dy_2Dy_3 = 101$, causing a transition to state *c* at the next positive clock edge. Likewise, in state *c*, with x_1 active, $Dy_1Dy_2Dy_3 = 111$, and the machine proceeds to state *e* ($y_1y_2y_3 = 111$). Finally, in state *e*, with x_1 still active — indicating a valid word — output z_1 is asserted. Whether x_1 is 0 or 1, the next state is *a* and the process repeats. Using the logic diagram, the correct logic levels for Dy_1, Dy_2, and Dy_3, can be easily determined for any input sequence.

Example 3.4 In this example, the Mealy machine of Example 3.3 is synthesized using *JK* flip-flops. Thus, the state diagram of Figure 3.65 applies also to this implementation, because the diagram depicts the flow — or state transitions — for a machine and is independent of the type of storage elements or the logic family that is used.

Step 1: State diagram Same as Figure 3.65 of Example 3.3.

Step 2: Equivalent states Same as Example 3.3 in which there are no equivalent states.

Step 3: State code assignment The results obtained from Step 1 through Step 3 of the synthesis procedure using *D* flip-flops remain unchanged for *JK* flip-flops. Therefore, the same state diagram will be used (Step 1), which is also a reduced state diagram (Step 2). The state codes that were previously assigned (Step 3) are also applicable for this implementation.

Step 4: Next-state table Using the state diagram of Example 3.3 together with the *JK* flip-flop excitation table — both reproduced below for convenience in Figure 3.73 and Table 3.17, respectively — the next-state table of Table 3.18 is obtained.

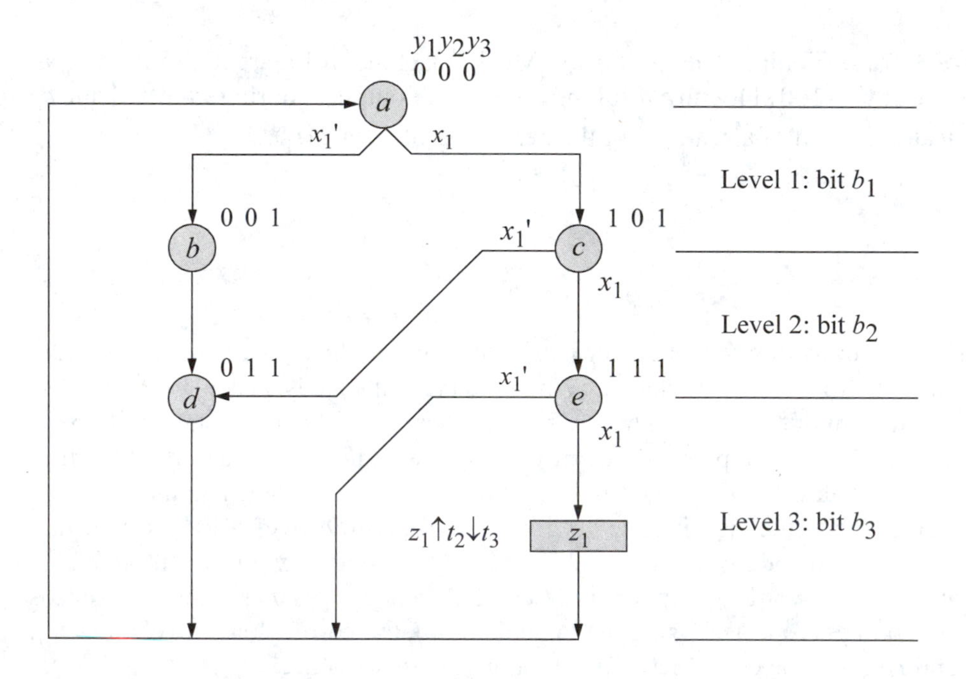

Figure 3.73 State diagram for the Mealy machine of Example 3.4, which generates an output z_1 whenever a 3-bit word $x_1 = 111$. Unused states are: $y_1y_2y_3 = 010$, 100, and 110.

Table 3.17 Excitation table for a *JK* flip-flop

Present-state $Y_{j(t)}$	Next state $Y_{k(t+1)}$	Flip-flop inputs $J\ K$
0	0	0 –
0	1	1 –
1	0	– 1
1	1	– 0

Table 3.18 Next-state table for the Mealy machine of Figure 3.73 using *JK* flip-flops

Present state y_1	y_2	y_3	Input x_1	Next state y_1	y_2	y_3	Flip-flop inputs Jy_1	Ky_1	Jy_2	Ky_2	Jy_3	Ky_3	Present output z_1
0	0	0	0	0	0	1	0	–	0	–	1	–	0
0	0	0	1	1	0	1	1	–	0	–	1	–	0
0	0	1	0	0	1	1	0	–	1	–	–	0	0
0	0	1	1	0	1	1	0	–	1	–	–	0	0
0	1	0	0	–	–	–	–	–	–	–	–	–	–
0	1	0	1	–	–	–	–	–	–	–	–	–	–
0	1	1	0	0	0	0	0	–	–	1	–	1	0
0	1	1	1	0	0	0	0	–	–	1	–	1	0
1	0	0	0	–	–	–	–	–	–	–	–	–	–
1	0	0	1	–	–	–	–	–	–	–	–	–	–
1	0	1	0	0	1	1	–	1	1	–	–	0	0
1	0	1	1	1	1	1	–	0	1	–	–	0	0
1	1	0	0	–	–	–	–	–	–	–	–	–	–
1	1	0	1	–	–	–	–	–	–	–	–	–	–
1	1	1	0	0	0	0	–	1	–	1	–	1	0
1	1	1	1	0	0	0	–	1	–	1	–	1	1

The generation of a representative row ($y_1y_2y_3 = 101$ for $x_1 = 0$) is presented. In state c ($y_1y_2y_3 = 101$), the state transition for flip-flop y_1 is from 1 to 0 if $x_1 = 0$. A transition from 1 to 0 stipulates that $Jy_1Ky_1 = -1$. When $x_1 = 0$, flip-flop y_2 changes from 0 to 1, which specifies JK input values for y_2 as $Jy_2Ky_2 = 1-$. Finally, y_3 remains set as the machine proceeds from state c to state d. Thus, a 1 to 1 transition for y_3 determines the values of J and K to be $Jy_3Ky_3 = -0$. The above sets of input values for Jy_1Ky_1, Jy_2Ky_2, and Jy_3Ky_3 are entered in the appropriate columns of Table 3.18 for state $y_1y_2y_3 = 101$ when $x_1 = 0$.

Since the unused states 010, 100, and 110 can never be entered in a normal operating environment, the JK inputs as well as the output can be treated as "don't care" conditions for these states.

Step 5: Input maps The input maps presented in Figure 3.74 are constructed from the JK input values contained in the next-state table or from the state diagram in conjunction with the JK flip-flop excitation table. If the next-state table is used, then the values for J and K are transferred directly from a present-state row in the table to the corresponding minterm locations in the input maps. Since the next state is a function of input x_1, the input maps include x_1 as a coordinate variable. Referring to Table 3.18, in state $y_1y_2y_3 = 000$ for $x_1 = 0$, the values for Jy_1 and Ky_1 are $0-$, respectively. These values are inserted in the input maps for Jy_1 and Ky_1 in minterm location $y_1y_2y_3x_1 = 0000$.

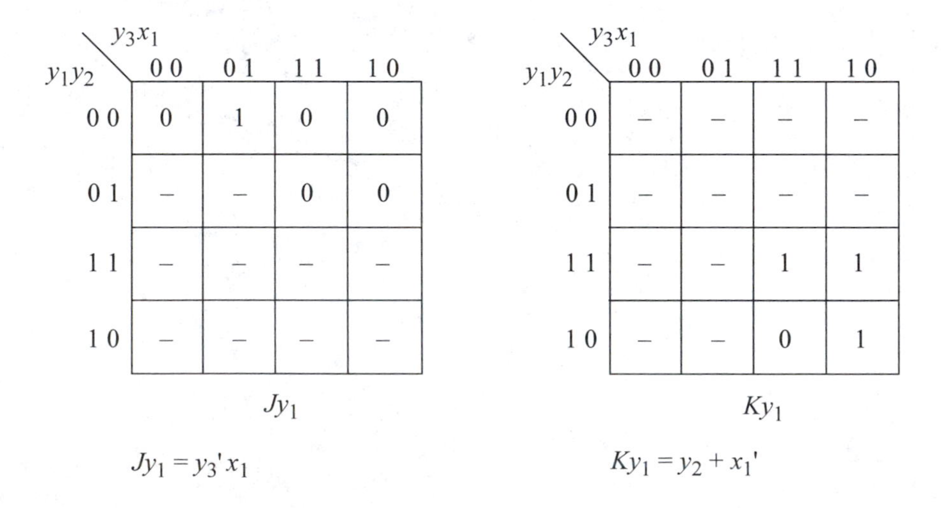

Figure 3.74 Input maps for the Mealy machine of Figure 3.73 using JK flip-flops and adjacent state codes for state pairs (b, c) and (d, e).

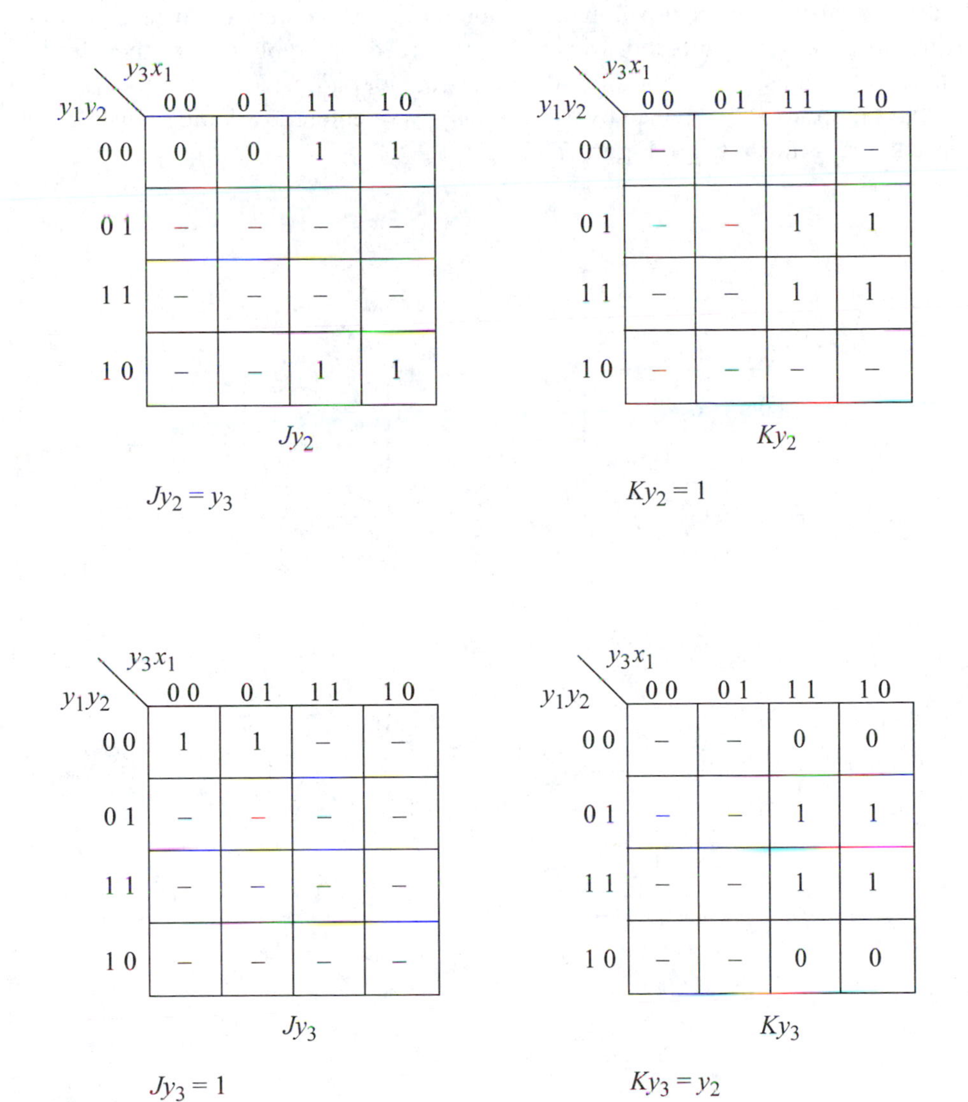

Figure 3.74 (Continued)

Likewise, state $y_1y_2y_3 = 000$ with $x_1 = 1$ specifies that $Jy_1Ky_1 = 1-$. These values are entered in the input maps in location 0001 for Jy_1 and Ky_1. The remaining minterm locations for Jy_1 and Ky_1 are completed in the same manner. The "don't care" entry is inserted for the unused states of 010, 100, and 110. Similarly, the input maps for y_2 and y_3 are constructed.

In many cases — especially where design time is to be kept to a minimum — the next-state table can be bypassed. The input maps are then derived directly from the state diagram using the *JK* flip-flop excitation table. There may be instances, however, in this abbreviated method, where a partial next-state table is required. In that event, the table consists only of the state under consideration. Also, a more compact set of input maps can be obtained by using x_1 as a map-entered variable. This new set of input maps is illustrated in Figure 3.75.

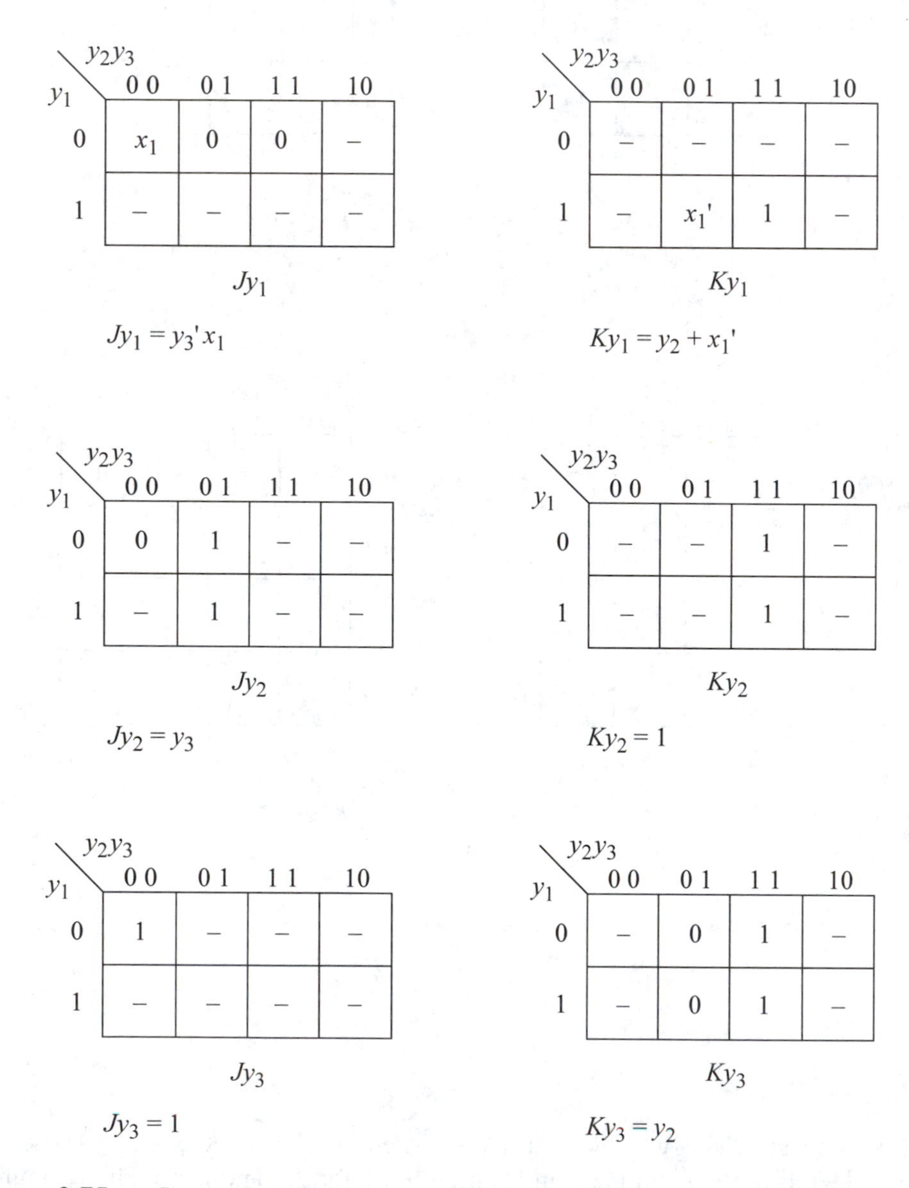

Figure 3.75 Input maps for the Mealy machine of Figure 3.73 using *JK* flip-flops and adjacent state codes for state pairs (b, c) and (d, e). Input x_1 is a map-entered variable.

To construct the input maps from the state diagram, first enter all unused states with the "don't care" symbol by placing dashes in locations 010, 100, and 110. Refer to Figure 3.73 for the procedure which follows. In state *a* ($y_1y_2y_3 = 000$) with $x_1 = 0$, the state transition for y_1 is from 0 to 0 (state *b*) and from 0 to 1 (state *c*) if $x_1 = 1$. Therefore, y_1 assumes the value of x_1 for any transition from state *a*. Since the values for Jy_1 and Ky_1 are not readily apparent, a partial next-state table is required for state *a* only, as shown below in Table 3.19. Also, the only next state under consideration is for y_1.

Table 3.19 Partial next-state table to determine the values for Jy_1Ky_1 in state *a* ($y_1y_2y_3 = 000$)

Present state $y_1y_2y_3$	Input x_1	Next state y_1	Flip-flop inputs Jy_1	Ky_1
$\underline{0}$ 0 0	0	$\underline{0}$	0	–
$\underline{0}$ 0 0	1	$\underline{1}$	1	–
			x_1	–

Then, using the excitation table for *JK* flip-flops, the values for Jy_1 and Ky_1 can be determined. When $x_1 = 0$, y_1 has a transition from 0 to 0; therefore, $Jy_1Ky_1 = 0–$. When $x_1 = 1$, y_1 changes from 0 to 1 at the next active clock transition, producing input values of $Jy_1Ky_1 = 1–$. The final values for Jy_1 and Ky_1 can now be determined. The values for Jy_1 are identical to those of input x_1; therefore, $Jy_1Ky_1 = x_1$ –. These values for Jy_1 and Ky_1 are then inserted in the input maps in minterm location $y_1y_2y_3 = 000$.

In state *b* ($y_1y_2y_3 = 001$), the machine proceeds to state *d* — regardless of the value of x_1 — where y_1 remains reset. Thus, the transition for y_1 is from 0 to 0, yielding flip-flop input values of $Jy_1Ky_1 = 0–$.

In state *c*, however, the next value of y_1 is again established by x_1. The transition for y_1 is from 1 to 0 if $x_1 = 0$, or from 1 to 1 if $x_1 = 1$. The partial next-state table of Table 3.20 tabulates the next-state values for y_1 as a function of x_1 and provides the requisite entries for Jy_1 and Ky_1. When $x_1 = 0$, the transition for y_1 is from 1 to 0, yielding input values of $Jy_1Ky_1 = –1$. When $x_1 = 1$, a 1-to-1 transition occurs for y_1, yielding input values of $Jy_1Ky_1 = –0$. Referring to Table 3.20, the entry for Jy_1 is a "don't care" condition, whereas, Ky_1 has values that are complementary to those of x_1; therefore, $Ky_1 = x_1'$.

With continued reference to the state diagram, it is observed that in state *d* ($y_1y_2y_3 = 011$) y_1 has a next-state value of 0, regardless of the value of x_1. Thus, the transition for y_1 is from 0 to 0, producing input values of $Jy_1Ky_1 = 0–$.

Finally, state *e* ($y_1y_2y_3 = 111$) provides a state transition from 1 to 0 for y_1. Even though z_1 is asserted if $x_1 = 1$, this is not a determining factor for the next state of y_1. Flip-flop y_1 acquires a state of 0, independent of the value of x_1. Thus, $Jy_1Ky_1 = –1$.

Table 3.20 Partial next-state table to determine the values for Jy_1Ky_1 in state c ($y_1y_2y_3$ 101)

Present state $y_1y_2y_3$	Input x_1	Next state y_1	Flip-flop inputs Jy_1	Ky_1
<u>1</u> 0 1	0	<u>0</u>	–	1
<u>1</u> 0 1	1	<u>1</u>	–	0
			–	x_1'

Referring to the input map of Figure 3.75 for Ky_1, the 1 in minterm location $y_1y_2y_3 = 111$ can be regarded as $x_1 + x_1'$ when determining the input equation. Thus, every square contains x_1', yielding the x_1' term in the equation for Ky_1. The x_1 term must also be covered. This is easily accomplished by considering location 111 as simply a 1 (using x_1' a second time), yielding the term y_2 in the equation for Ky_1.

Using the same procedure, the remaining input maps are derived. The equations for Jy_1Ky_1, Jy_2Ky_2, and Jy_3Ky_3 are identical to those obtained by using the more tedious approach of a complete next-state table. The set of input equations is listed in Equation 3.25. The use of map-entered variables in the determination of input or output equations from a Karnaugh map usually shortens the design process, because the corresponding maps are more compact and are constructed from a fewer number of steps.

$$
\begin{aligned}
Jy_1 &= y_3'x_1 & \qquad Ky_1 &= y_2 + x_1' \\
Jy_2 &= y_3 & \qquad Ky_2 &= 1 \\
Jy_3 &= 1 & \qquad Ky_3 &= y_2
\end{aligned}
\tag{3.25}
$$

Step 6: Output map The output map is unchanged from the map shown in Figure 3.70, but is duplicated here for convenience in Figure 3.76, using x_1 as a map-entered variable.

Step 7: Logic diagram The logic diagram, shown in Figure 3.77, is implemented from the input equations of Equation 3.25, three negative-edge-triggered *JK* flip-flops, and the output equation shown in Figure 3.76. Output z_1 is asserted at time t_2 and deasserted at time t_3 by the application of the –Clock signal to the positive-input AND gate which generates z_1. When the –Clock signal attains a high voltage level at time t_2 in state d ($y_1y_2y_3 = 111$), z_1 is asserted, assuming that x_1 is already asserted.

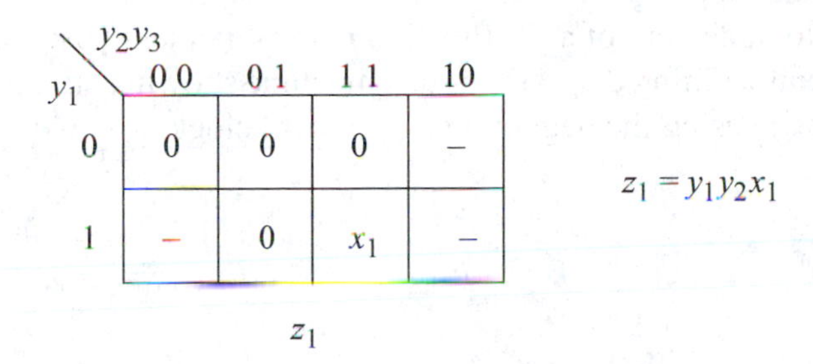

Figure 3.76 Output map for the Mealy machine of Figure 3.65, using input x_1 as a map-entered variable.

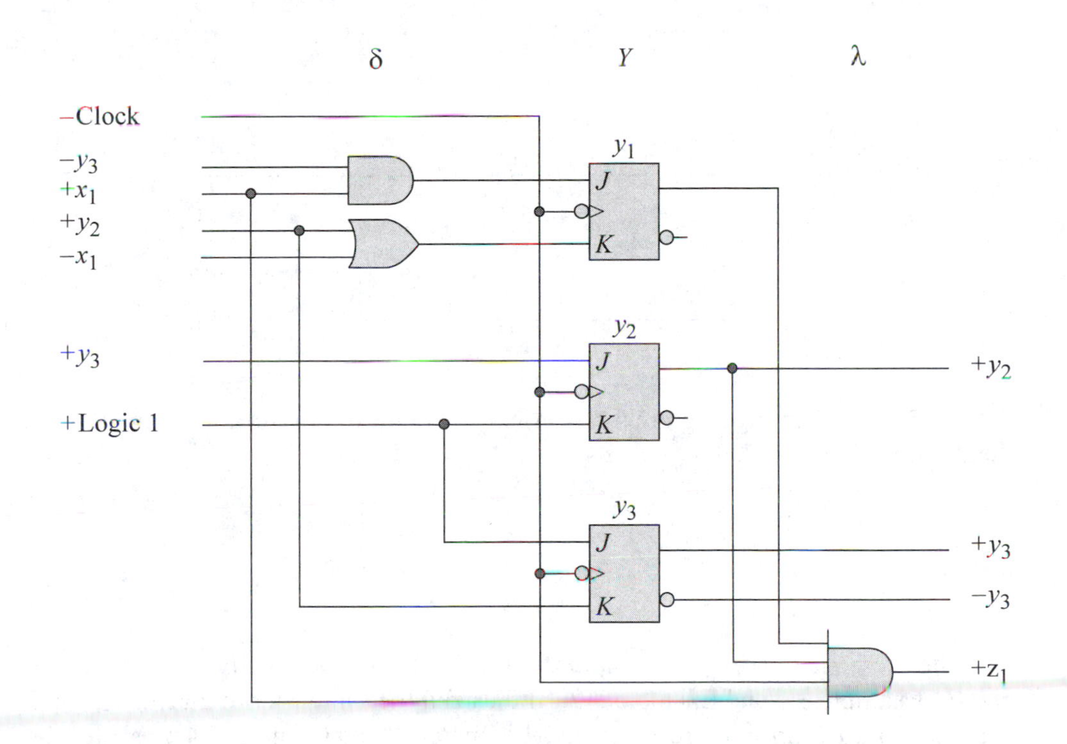

Figure 3.77 Logic diagram for the Mealy machine of Figure 3.73 using *JK* flip-flops. Output is asserted when an input sequence of 111 has been detected in a 3-bit word on a serial data line x_1. Output z_1 is asserted at time t_2 and deasserted at time t_3.

Comparing the logic diagram of Figure 3.71 using *D* flip-flops with the diagram of Figure 3.77 using *JK* flip-flops, it is evident that fewer logic gates are required for the *JK* flip-flop implementation. This is due to the additional logic capabilities achieved by the *JK* data inputs. Although both designs perform to the same specifications, the inherent logic feature of a *JK* flip-flop results in a less costly circuit.

Figure 3.78 presents a timing diagram for this machine showing states *c* and *e* and the assertion of output z_1 using the negative-edge system clock.

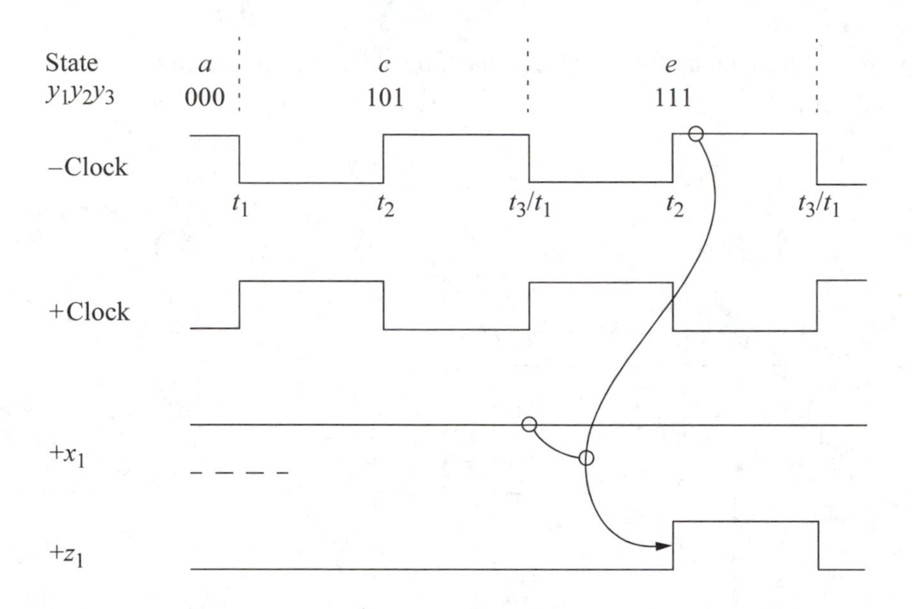

Figure 3.78 Timing diagram for the Mealy machine of Figure 3.73 using the logic diagram of Figure 3.77.

The previous four examples represent the same, or functionally equivalent, machine. Example 3.1 and Example 3.2 synthesized the same Moore machine with *D* flip-flops and *JK* flip-flops, respectively. Example 3.3 and Example 3.4 presented design techniques for a functionally equivalent Mealy machine using *D* flip-flops and *JK* flip-flops, respectively. The Mealy machine was modified slightly to accommodate the λ output logic as a function of both the present state and the present input. The equations for all four machines are summarized in Table 3.21.

To provide a hardware comparison for the amount of logic required for the δ next-state logic for the four examples, assume the following: four 2-input gates per integrated circuit and three 3-input gates per integrated circuit. The number of integrated circuits (ICs) required for the δ next-state logic for the different versions are:

Example 3.1	Moore	(*D* flip-flops)	11/12 ICs
Example 3.2	Moore	(*JK* flip-flops)	10/12 ICs
Example 3.3	Mealy	(*D* flip-flops)	14/12 ICs
Example 3.4	Mealy	(*JK* flip-flops)	10/12 ICs

Table 3.21 Summary of the δ next-state equations and the λ output equations for the Moore and Mealy machines of Example 3.1 through Example 3.4

δ next-state equations		λ output equations
Example 3.1 Moore machine using *D* flip-flops		
$Dy_1 = y_2'y_3'x_1 + y_1y_3x_1$		$z_1 = y_1y_3'$
$Dy_2 = y_3$		
$Dy_3 = y_2'$		
Example 3.2 Moore machine using *JK* flip-flops		
$Jy_1 = y_2'y_3'x_1$	$Ky_1 = x_1' + y_3'$	$z_1 = y_1y_3'$
$Jy_2 = y_3$	$Ky_2 = y_3'$	
$Jy_3 = y_2'$	$Ky_3 = y_2$	
Example 3.3 Mealy machine using *D* flip-flops		
$Dy_1 = y_3'x_1 + y_1y_2'x_1$		$z_1 = y_1y_2x_1$
$Dy_2 = y_2'y_3$		
$Dy_3 = y_2'$		
Example 3.4 Mealy machine using *JK* flip-flops		
$Jy_1 = y_3'x_1$	$Ky_1 = y_2 + x_1'$	$z_1 = y_1y_2x_1$
$Jy_2 = y_3$	$Ky_2 = 1$	
$Jy_3 = 1$	$Ky_3 = y_2$	

In this simple example of a synchronous sequential machine, both Moore and Mealy machines use the same number of flip-flops. The Mealy machine, however, may need fewer storage elements in general than an equivalent Moore machine. This is because of the additional number of states in a Moore machine that are necessary to generate the output as a function of the present state only.

3.6 Moore-Mealy Equivalence

Moore and Mealy machines have been described in detail Chapter 2 and synthesized in Section. 3.4 and Section 3.5. The only difference between the two machines occurs in the λ output logic: A Moore-type output is a function of the present state only, whereas, a Mealy-type output is a function of both the present state and the present input. There must, therefore, be an equivalence between the two machines. Both should recognize the same set of input sequences and generate the same outputs.

3.6.1 Mealy-to-Moore Transformation

A Mealy machine can be transformed into a corresponding Moore machine, where both machines accept the same set of input vectors. A general procedure using a next-state table will be presented followed by an example. To obtain a Moore machine from the Mealy machine of Table 3.22, it is necessary to generate two new states for each state of the Mealy model in which different output values occur for different inputs.

Table 3.22 Next-state table for a Mealy machine

	Present state	Input	Next state	Output
State name	y_1 y_2	x_1	y_1 y_2	z_1
a	0 0	0	0 0 (*a*)	0
	0 0	1	1 1 (*d*)	1
b	0 1	0	1 0 (*c*)	1
	0 1	1	0 1 (*b*)	1
c	1 0	0	0 1 (*b*)	0
	1 0	1	0 0 (*a*)	0
d	1 1	0	1 1 (*d*)	1
	1 1	1	1 0 (*c*)	0

For example, state *a* can be entered from state *a* when $x_1 = 0$, or from state *c* when $x_1 = 1$. However, since different outputs are associated with state *a*, state *a* is replaced with two equivalent states: a_0 where $z_1 = 0$, and a_1 where $z_1 = 1$. Every transition to state *a* where $z_1 = 0$ is now directed to state a_0 and every transition to state *a* where $z_1 = 1$ is now directed to state a_1. The same procedure is applied to state *d*.

The following method is used to derive an equivalent Moore next-state table:

State *a*: In state *a*, the machine is operating as a Mealy machine, therefore, state *a* is changed to an equivalent Moore operation. When $x_1 = 0$ and $z_1 = 0$, the machine is directed to a new state where $z_1 = 0$, regardless of the value of x_1. When $x_1 = 1$ and $z_1 = 1$, the machine is directed to a new state where $z_1 = 1$, regardless of the value of x_1.

State *b*: There is no change required to state *b*, because the output is not a function of the input.

State *c*: There is no change required to state *c*, because the output is not a function of the input.

State *d*: In state *d*, the machine is again operating as a Mealy machine; therefore, state *d* is changed to an equivalent Moore operation such that, when $x_1 = 0$ and $z_1 = 1$, the machine is directed to a new state where $z_1 = 1$, irrespective of the value of x_1. Also, when $x_1 = 1$ and $z_1 = 0$, the machine is directed to a new state where $z_1 = 0$, irrespective of the value of x_1.

Thus, a different next-state table is obtained as shown in Table 3.23. States a_0 and a_1 are both directed to the same next state for the same values of x_1. The same is true for states d_0 and d_1. Note that a particular state in Table 3.23 has the same output value for z_1 regardless of the input value of x_1. This is the desired characteristic for a Moore machine.

It is now simply a matter of redefining the state codes to accommodate six states. This is shown in Table 3.24. The assignment of state codes for the three storage elements $y_1y_2y_3$ is arbitrary at this time.

Thus, the next-state table of Table 3.24 can be used to design a Moore machine from the corresponding Mealy machine of Table 3.22. A similar procedure is used to transform a Moore machine into a Mealy machine such that, both accept the same input sequences.

Every state in a Mealy machine which generates an output must be transformed into two new states. The state in which the output occurred no longer produces an output as a function of the input, instead the input causes a state transition to another state whose sole function is to produce an output. Thus, the required Moore equivalence is obtained. When transforming from a Mealy to a Moore model, the resulting Moore machine usually requires more states, and thus more storage elements, than a comparable Mealy machine. Although the λ output function for a Moore machine is simpler than that for a Mealy machine, the latter is more widely used.

Table 3.23 Expanded next-state table for the Mealy machine of Table 3.22

State name	Present state y_1 y_2	Input x_1	Next state y_1 y_2	Output z_1
a_0	0 0	0	0 0 (a_0)	0
	0 0	1	1 1 (d_1)	0
a_1	0 0	0	0 0 (a_0)	1
	0 0	1	1 1 (d_1)	1
b	0 1	0	1 0 (c)	1
	0 1	1	0 1 (b)	1
c	1 0	0	0 1 (b)	0
	1 0	1	0 0 (a_0)	0
d_0	1 1	0	1 1 (d_1)	0
	1 1	1	1 0 (c)	0
d_1	1 1	0	1 1 (d_1)	1
	1 1	1	1 0 (c)	1

Example 3.5 A state diagram for a Mealy machine which detects an input sequence on a serial data line x_1 is shown in Figure 3.79. Whenever $x_1 = 1101$ anywhere in the bit stream, z_1 is asserted at time t_2 and deasserted at time t_3. The characteristics of state d are preserved by two new states, d and e, in the equivalent Moore state diagram of Figure 3.80. State d no longer produces an output in the Moore model. The output is generated in state e if $x_1 = 1$ in state d. The addition of the extra state (e) permits conformation to the definition of a Moore machine.

Table 3.25 presents the next-state table for the Mealy machine of Figure 3.79. Since different outputs are associated with state d depending upon the input, state d is replaced by two states which perform an equivalent operation: state d with an output of 0 and state e with an output of 1, as shown in Table 3.26.

Table 3.24 Next-state table for an equivalent Moore machine that represents the Mealy machine of Table 3.22

	Present state	Input	Next state	Output
State name	$y_1\ y_2\ y_3$	x_1	$y_1\ y_2\ y_3$	z_1
$a\ (a_0)$	0 0 0	0	0 0 0 (a)	0
	0 0 0	1	1 0 1 (f)	0
$b\ (a_1)$	0 0 1	0	0 0 0 (a)	1
	0 0 1	1	1 0 1 (f)	1
$c\ (b)$	0 1 0	0	0 1 1 (d)	1
	0 1 0	1	0 1 0 (c)	1
$d\ (c)$	0 1 1	0	0 1 0 (c)	0
	0 1 1	1	0 0 0 (a)	0
$e\ (d_0)$	1 0 0	0	1 0 1 (f)	0
	1 0 0	1	0 1 1 (d)	0
$f\ (d_1)$	1 0 1	0	1 0 1 (f)	1
	1 0 1	1	0 1 1 (d)	1

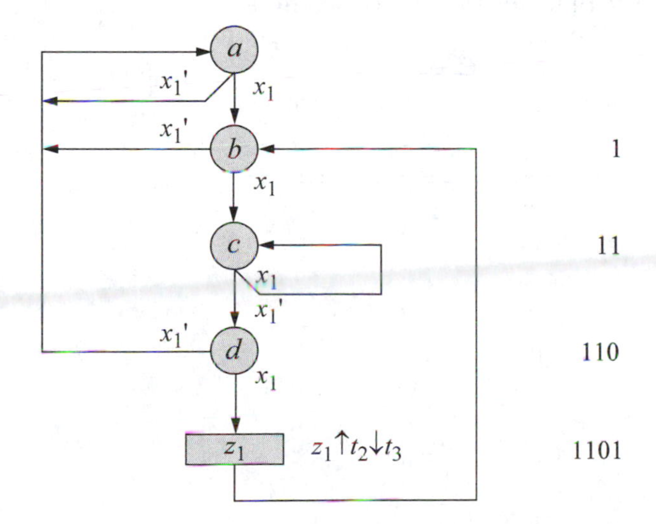

Figure 3.79 State diagram for a Mealy machine to detect an input sequence of x = 1101.

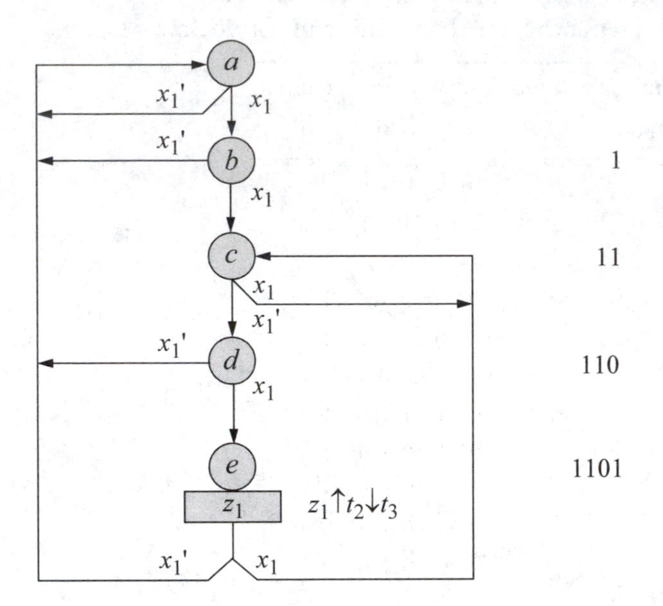

Figure 3.80 Equivalent Moore state diagram for the Mealy machine of Figure 3.79. Output z_1 is asserted whenever the input sequence is $\boldsymbol{x}_1 = 1101$.

Table 3.25 Next-state table for the Mealy machine of Figure 3.79

Present state	Present input x_1	Next state	Present output z_1
a	0	*a*	0
	1	*b*	0
b	0	*a*	0
	1	*c*	0
c	0	*d*	0
	1	*c*	0
d	0	*a*	0
	1	*b*	1

Table 3.26 Next-state table for the equivalent Moore machine of Figure 3.80

Present state	Present input x_1	Next state	Present output z_1
a	0	a	0
	1	b	0
b	0	a	0
	1	c	0
c	0	d	0
	1	c	0
d	0	a	0
	1	e	0
e	0	a	1
	1	c	1

The same technique applies to multiple outputs, as shown for the Mealy machine in Figure 3.81. This machine detects two different word configurations on a serial input line x_1, as shown below.

$$x_1 = \cdots \left| b_1 b_2 b_3 \right| b_1 b_2 b_3 \left| b_1 b_2 b_3 \right. \cdots$$

If $x_1 = 110$, then assert z_1
If $x_1 = 101$, then assert z_2 $\quad z_1/z_2$

The information on input x_1 consists of 3-bit words. There is no space between adjacent words. Outputs z_1 and z_2 are asserted during the third bit time of a valid word. The corresponding Moore model is shown in Figure 3.82 for the word format shown below.

$$x_1 = \cdots \left| b_1 b_2 b_3 \right| \; \left| b_1 b_2 b_3 \right| \; \left| b_1 b_2 b_3 \right| \cdots$$

If $x_1 = 110$, then assert z_1
If $x_1 = 101$, then assert z_2 $\quad z_1/z_2$

As in the case of a single output, every state in the Mealy model which generates an output must be changed to an equivalent two-state sequence. Thus, in Figure 3.81, states *c* and *d* are each replaced by state pairs (*c*, *e*) and (*d*, *f*), respectively, as shown in the equivalent Moore state diagram of Figure 3.82. Outputs z_1 and z_2 are now asserted in the bit space between words, since separate states were added to perform the required function of output generation. Table 3.27 and Table 3.28 contain the next-state tables for the Mealy and Moore machines of Figure 3.81 and Figure 3.82, respectively.

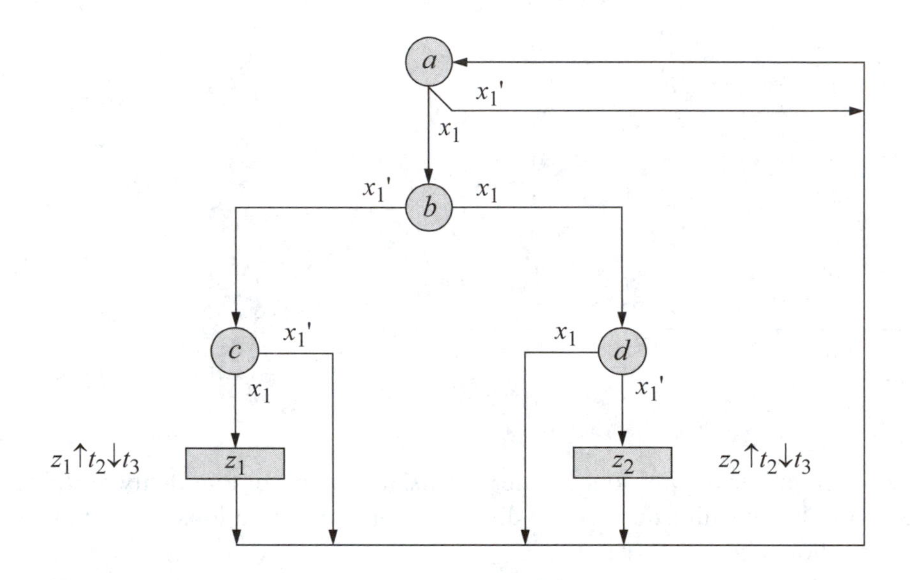

Figure 3.81 State diagram for a Mealy machine. Output z_1 is asserted whenever $x_1 = 101$; output z_2 is asserted whenever $x_1 = 110$.

It is possible for the output of the Moore machine to be valid for the entire state time; however, the output was asserted at time t_2 and deasserted at time t_3 to maintain compatible output assertion with the Mealy machine. The timing diagram of Figure 3.83 shows the assertion and deassertion of output z_1 for a valid input sequence of $x_1 = 1101$ for the Mealy model of Figure 3.79 and the Moore model of Figure 3.80. Although both machines detect the same valid input sequence and generate the output at the same assertion/deassertion times, the output for the Moore machine is delayed by one clock cycle. The delayed output should present no problem if it is interpreted at the correct time displacement.

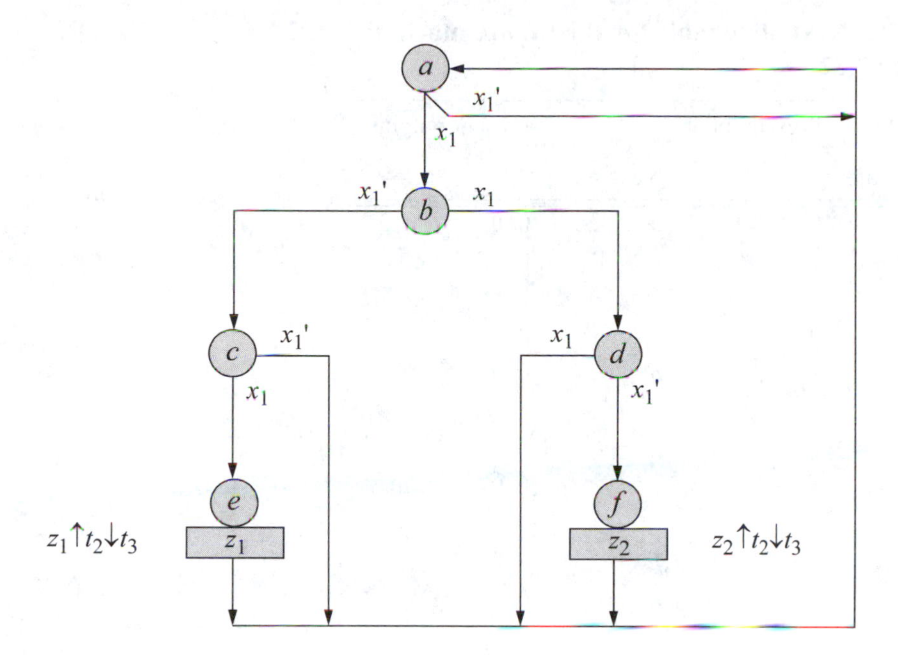

Figure 3.82 Equivalent Moore state diagram for the machine of Figure 3.81. Output z_1 is asserted whenever $x_1 = 101$; output z_2 is asserted whenever $x_1 = 110$.

Table 3.27 Next-state table for the Mealy machine of Figure 3.81

Present state	Present input x_1	Next state	Present outputs z_1 z_2
a	0	*a*	0 0
	1	*b*	0 0
b	0	*c*	0 0
	1	*d*	0 0
c	0	*a*	0 0
	1	*a*	1 0
d	0	*a*	0 1
	1	*a*	0 0

Table 3.28 Next-state table for the Moore machine of Figure 3.82

Present state	Present input x_1	Next state	Present outputs z_1 z_2
a	0	a	0 0
	1	b	0 0
b	0	c	0 0
	1	d	0 0
c	0	a	0 0
	1	e	0 0
d	0	f	0 0
	1	a	0 0
e	0	a	1 0
	1	a	1 0
f	0	a	0 1
	1	a	0 1

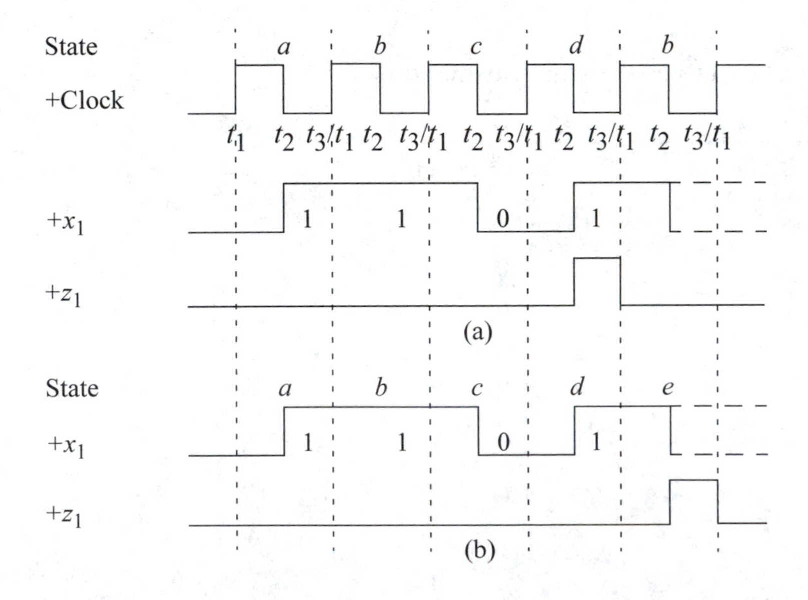

Figure 3.83 Timing diagram for a valid input sequence of $x_1 = 1101$ showing the relationship between the assertion of output z_1 for: (a) the Mealy machine of Figure 3.79; and (b) the Moore machine of Figure 3.80.

3.6.2 Moore-to-Mealy Transformation

The output that is associated with a state in a Moore machine can be related to the input for an equivalent state in a Mealy machine. This is illustrated in the transformation diagram of Figure 3.84.

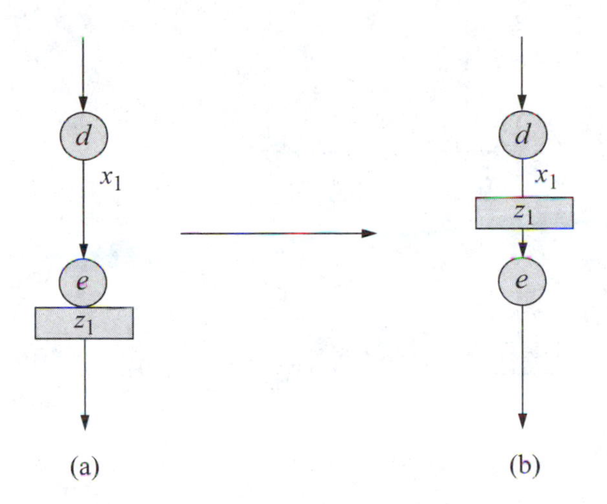

Figure 3.84 Moore-to-Mealy transformation of states: (a) Moore output; and (b) Mealy output.

Figure 3.85 shows a Mealy machine which was transformed from the Moore machine of Figure 3.80. The state diagram, however, yields an equivalent Mealy machine with five states rather than four as in the original machine of Figure 3.79. Using the procedure for finding equivalent states, it can be verified that states b and e are equivalent. Using state e as a redundant state, the original four states of Figure 3.79 are retained.

A Mealy machine usually requires less states than a corresponding Moore machine. Also, *JK* flip-flops usually require less δ next-state logic than *D* flip-flops, due to the built-in logic functions of the *JK* inputs. Therefore, a Mealy machine implemented with *JK* flip-flops will ordinarily produce a machine with a minimal amount of logic.

3.7 Output Glitches

A *glitch* in synchronous sequential machines is any false or spurious electronic signal. These narrow, unwanted pulses wreak havoc in digital systems if the glitch occurs on an output signal. For example, the output signal of a logic circuit may connect to the clock input of a counter and produce erroneous results. Therefore, eliminating output glitches is extremely important, even at the expense of additional logic.

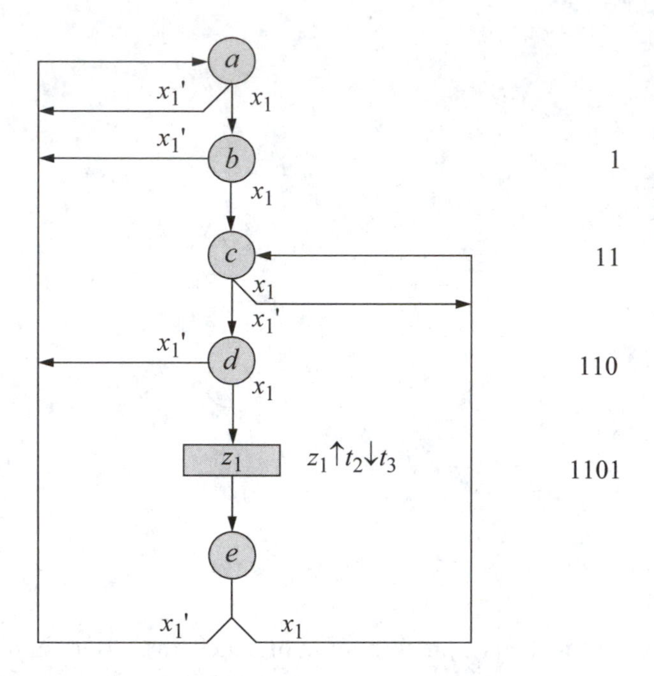

Figure 3.85 State diagram for Mealy machine which was transformed from the Moore machine of Figure 3.80. Output z_1 is asserted whenever the input sequence is $x_1 = 1101$.

In synchronous systems, glitches can occur in the time period between the active clock transition and circuit stabilization. This is shown in Figure 3.86 and is indicated by the time duration Δt. It is during this time, when the machine is changing states, that the outputs are susceptible to glitches. Due to varying propagation delays of the storage elements' internal logic, the machine may enter an unstable, or transient, state. Although momentary in duration, this transient state can cause an output glitch in both Moore and Mealy machines if the output is decoded directly from the p-tuple state codes. If the outputs are enabled at time t_2, then glitches that occur during the period of instability are of no consequence — the machine has long since stabilized. Four methods will be presented to eliminate output glitches:

1. State code assignment
2. Storage elements
3. Complemented clock
4. Delayed clock

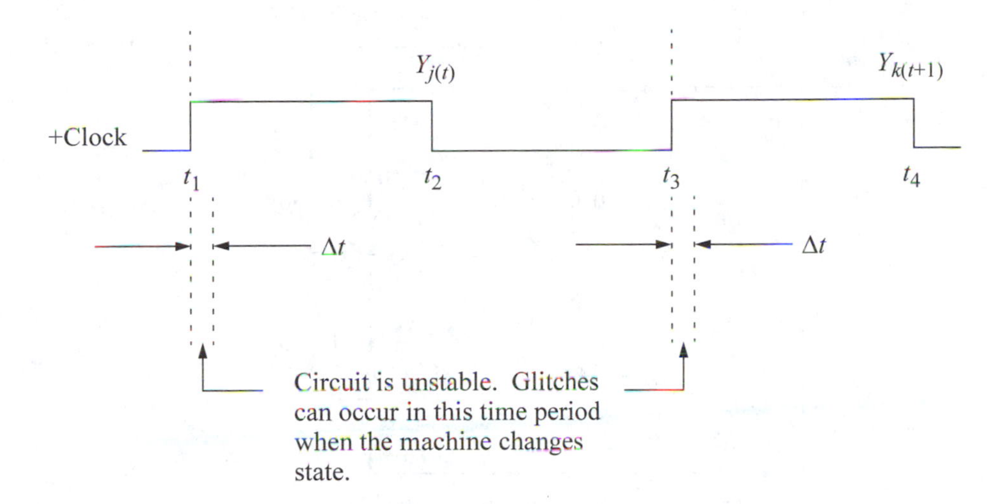

Figure 3.86 Illustration of time interval during which glitches are possible for synchronous sequential machines.

When two or more flip-flops change state, they may not all change to their next state at precisely the same time. This is due to different propagation delays in the internal structure of the flip-flops. The Moore machine of Figure 3.87 illustrates the effect of these diverse circuit delays when a state change is initiated. State adjacency was not maintained in order to demonstrate different conditions during a state transition sequence. Also, the δ next-state logic is irrelevant to this discussion and is indicated simply as a block with the corresponding boolean equations.

Because output z_1 is asserted at time t_1, the system is susceptible to glitches for certain state transitions. If the machine is presently in state *a*, an active level on input x_1 will cause a transition to state *c* at the next positive clock transition. In proceeding from state *a* to state *c*, however, two flip-flops change state: y_1 changes from 1 to 0 and y_2 changes from 0 to 1. Thus, it is possible for the machine to pass through either of two transient states before stabilizing in state *c*, as shown in Figure 3.88. Flip-flop y_3 remains reset during the entire state transition. If y_1 resets before y_2 sets, then the machine will sequence through state *b* ($y_1y_2y_3 = 000$). The machine does not generate an output for state *b*; therefore, no output glitch occurs while the machine is passing through this transient state. If, however, flip-flop y_2 sets before y_1 resets, then the machine will pass through state *e* ($y_1y_2y_3 = 110$) momentarily before stabilizing in state *c* ($y_1y_2y_3 = 010$). Because output z_1 has a direct correspondence with state *e*, whenever state *e* is entered — either as a transient state or as a stable state — z_1 is asserted. In this case, a glitch will appear on output z_1 for a duration equal to the length of time that the machine remains in state *e*, as shown in Figure 3.87 (c).

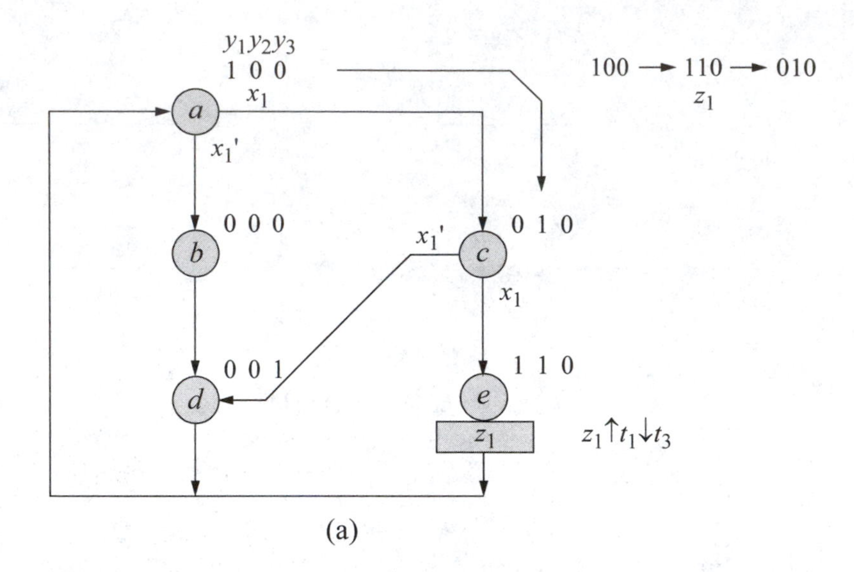

(a)

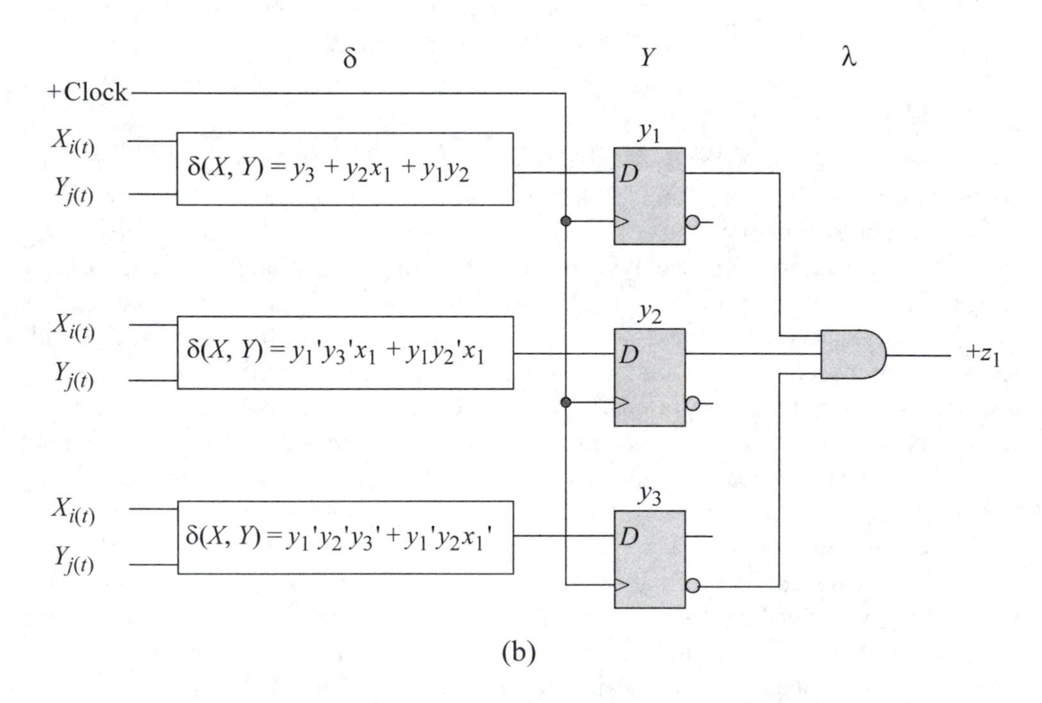

(b)

Figure 3.87 Circuit to illustrate the effect of different propagation delays in a Moore machine: (a) state diagram with unused states $y_1y_2y_3 = 011$, 101, and 111; (b) logic diagram; and (c) timing diagram.

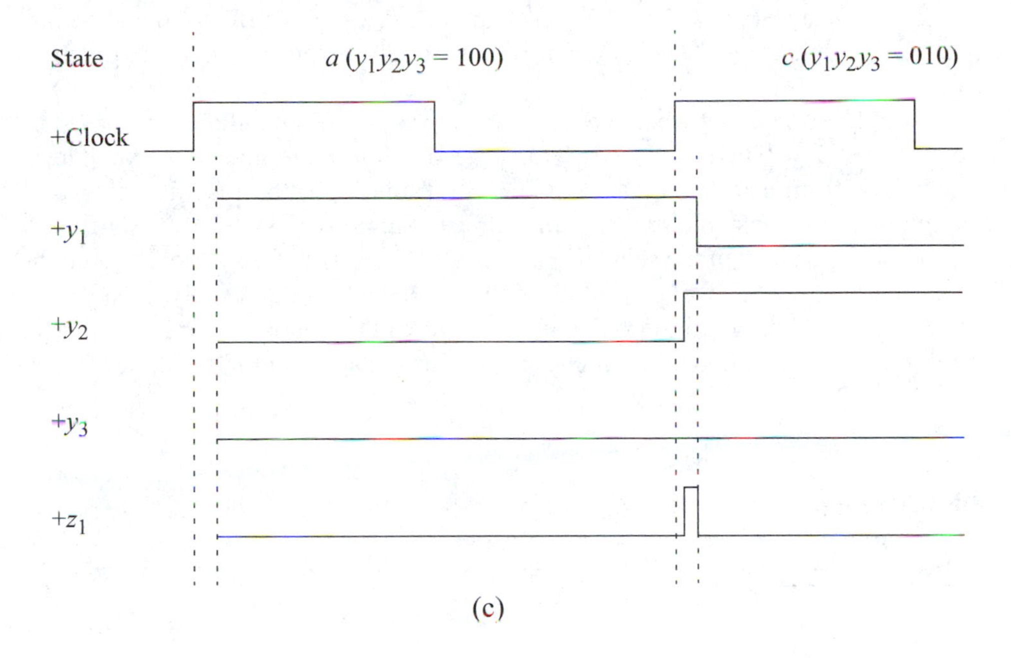

(c)

Figure 3.87 (Continued)

	y_1	y_2	y_3
State *a*	1	0	0
Transient state *b*	0	0	0
Transient state *e*	1	1	0
State *c*	0	1	0

Figure 3.88 A state transition sequence that may cause a glitch on output z_1 for the Moore machine of Figure 3.87.

The only other sequences which cause two or more flip-flops to change state, resulting a possible path through a transient state, are as follows:

State *c* ($y_1y_2y_3 = 010$) to state *d* ($y_1y_2y_3 = 001$)

State *d* ($y_1y_2y_3 = 001$) to state *a* ($y_1y_2y_3 = 100$)

In the first sequence, flip-flop y_1 remains reset for the entire transition. Since y_1 must be set to enter state *e* (y1y2y3 = 110), this will not cause a glitch on z_1. In the second sequence, y_2 remains reset, again denying access to state *e*, thus preventing a glitch on z_1.

It should be noted that whenever two flip-flops change state, there are three possible paths which will take the machine to the destination stable state. For example, in the sequence from state *a* to state *c* shown in Table 3.29, the machine may pass through one of two unstable states, or may proceed directly to state *c*. If flip-flop y_1 resets before y_2 sets, then the sequence is *a*, *b*, *c*. If y_2 sets before y_1 resets, then the sequence is *a*, *e*, *c*. If flip-flops y_1 and y_2 change states simultaneously, then the sequence is *a* to *c* directly. In the following sections of this chapter, four methods will be presented to eliminate output glitches in synchronous sequential machines.

Table 3.29 Sequence from state *a* to state *c*

State *a*	Transient state	State *c*	
$y_1y_2y_3$	$y_1y_2y_3$	$y_1y_2y_3$	Transition sequence
1 0 0 →	0 0 0 (*b*) →	0 1 0	$a \rightarrow b \rightarrow c$
1 0 0 →	1 1 0 (*e*) →	0 1 0	$a \rightarrow e \rightarrow c$
1 0 0 →	→	0 1 0	$a \rightarrow c$

3.7.1 Glitch Elimination Using State Code Assignment

It is often possible to reassign state codes to avoid output glitches. This may result in additional δ next-state logic, however, if state code adjacency cannot be maintained. But if the primary concern is to eliminate output glitches, then additional logic is inconsequential in order to produce a more reliable machine. Output glitches can occur when two or more flip-flops change state and only when the machine specifications require that the outputs be asserted at time t_1.

The state diagram of Figure 3.87 (a) is reproduced in Figure 3.89, in which state codes have been reassigned to yield a machine that is inherently free of glitches. To verify that output z_1 will not produce a glitch, it is necessary to check all state transition sequences to determine if any sequence passes through state *e* for a transition that does not include output z_1. If only one flip-flop changes state, then the *p*-tuple decoder for the λ output logic for z_1 will not generate a spurious signal. Glitches are possible only when two or more flip-flops change state during a state transition sequence.

The path from state *a* to state *b* presents no problem, because only flip-flop y_1 changes state. The same is true for the path from state *a* to state *c* (only y_3 changes), from state *b* to state *d* (only y_3 changes), and from state *c* to state *d* (only y_1 changes). Two flip-flops change state when the machine proceeds from state *d* to state *a*;

however, y_2 remains reset — it must be set in order to enter state e and assert z_1. The transition from state c to state e involves a change of two flip-flops. The machine may sequence through the transient states listed in Table 3.30. None of the transient states, however, will produce an output; state d ($y_1y_2y_3 = 001$) is not associated with an output and $y_1y_2y_3 = 111$ is an unused state. Therefore, the path from state c to state e presents no problem in this single-output Moore machine. The final path from state e to state a causes all three flip-flops to change state. All possible transient states in this path either have no outputs or are unused states. Thus, the machine presented in Figure 3.89 will operate reliably relative to the unconditional output z_1. Output z_1 is classified as unconditional, because its assertion depends only on the present state and not on the present inputs. Outputs in a Mealy machine are designated as conditional, because they are a function of both the present state and the present inputs.

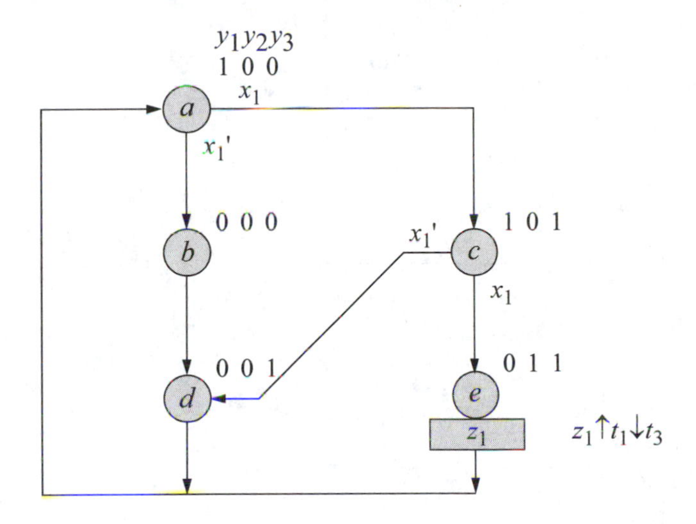

Figure 3.89 The Moore machine of Figure 3.87 (a) with reassigned state codes. No output glitches are possible. Unused states are: $y_1y_2y_3 = 010$, 110, and 111.

Table 3.30 State transition sequences for Figure 3.89 from state c to state e

Start state $y_1y_2y_3$	Transient state $y_1y_2y_3$		End state $y_1y_2y_3$	Comments
1 0 1 (c) →	0 0 1 (d)	→	0 1 1 (e)	State d has no output
1 0 1 (c) →	1 1 1 (Unused)	→	0 1 1 (e)	
1 0 1 (c) →		→	0 1 1 (e)	

The state diagram of Figure 3.90 presents a machine with two Moore-type outputs. In analyzing Figure 3.90 (a), three paths are found which may generate glitches on outputs z_1 or z_2, depending on the propagation delays of flip-flops y_1, y_2, and y_3. The paths under consideration are shown in Table 3.31.

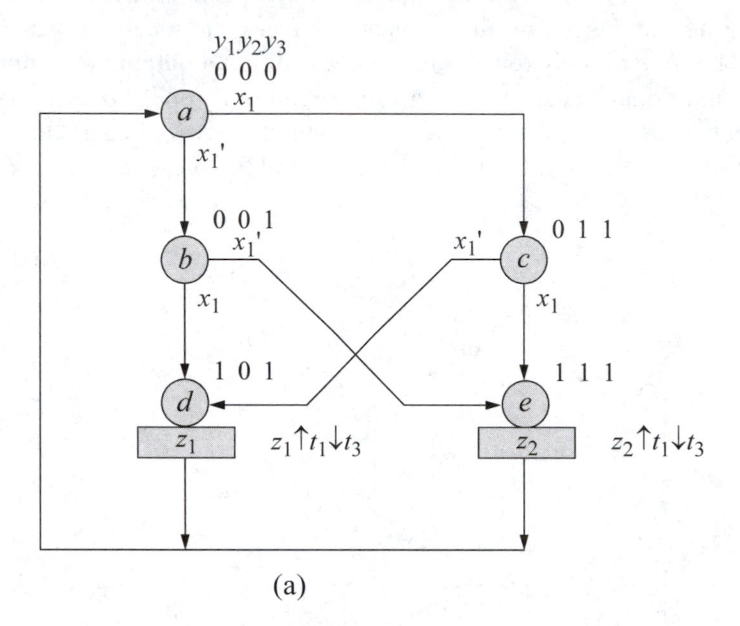

(a)

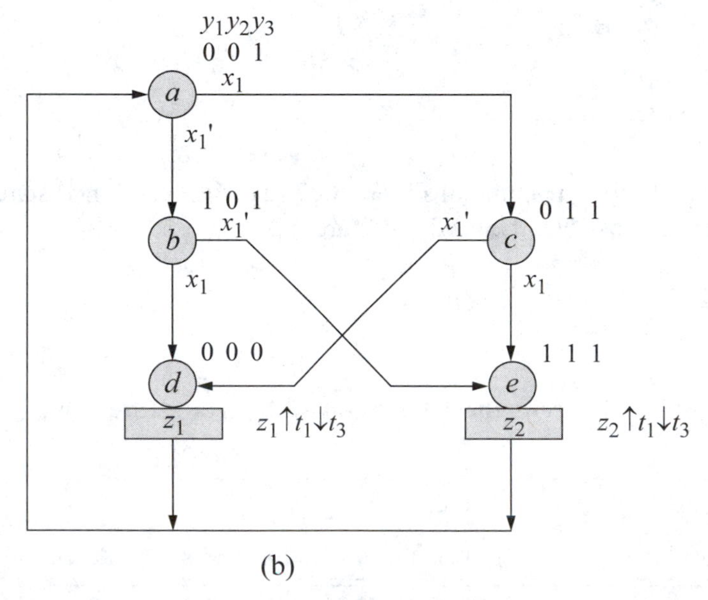

(b)

Figure 3.90 State diagram for a Moore machine with multiple outputs: (a) incorrect state code assignment that may produce glitches on outputs z_1 and z_2. Unused states are $y_1y_2y_3$ = 010, 100, and 110; and (b) correct state code assignment for glitch-free operation. Unused states are $y_1y_2y_3$ = 010, 100, and 110.

Table 3.31 State transition sequences for Figure 3.90 (a) that may produce glitches on outputs z_1 and z_2

Start state $y_1y_2y_3$	Transient state $y_1y_2y_3$	End state $y_1y_2y_3$	Comments
0 0 1 (b) →	0 1 1 (c) →	1 1 1 (e)	State c has no output
0 0 1 (b) →	1 0 1 (d) →	1 1 1 (e)	Glitch on z_1
0 0 1 (b) →	→	1 1 1 (e)	
0 1 1 (c) →	0 0 1 (b) →	1 0 1 (d)	State b has no output
0 1 1 (c) →	1 1 1 (e) →	1 0 1 (d)	Glitch on z_2
0 1 1 (c) →	→	1 0 1 (d)	
1 1 1 (e) →	0 0 1 (b) →	0 0 0 (a)	State b has no output
1 1 1 (e) →	0 1 0 (Unused) →	0 0 0 (a)	
1 1 1 (e) →	0 1 1 (c) →	0 0 0 (a)	State c has no output
1 1 1 (e) →	1 0 0 (Unused) →	0 0 0 (a)	
1 1 1 (e) →	1 0 1 (d) →	0 0 0 (a)	Glitch on z_1
1 1 1 (e) →	1 1 0 (Unused) →	0 0 0 (a)	

The path from state b ($y_1y_2y_3 = 001$) to state e ($y_1y_2y_3 = 111$) will produce a glitch on output z_1 if flip-flop y_1 sets before y_2 sets. The duration of the glitch depends on the propagation delays of the two flip-flops. If the delays differ only slightly, then the duration of the glitch will be extremely small. The duration — and thus the voltage level — may be of insufficient magnitude to propagate through the λ output logic for z_1. This cannot be guaranteed, however, and the path from state b to state e should be considered as a hazardous state transition.

The transition from state c to state d will cause output z_2 to glitch if flip-flop y_1 sets before y_2 resets. Another possible path for this sequence is through state b. This occurs if flip-flop y_2 resets before y_1 sets. This presents no spurious output on z_1 or z_2, however, because state b is not associated with an output. Finally, the transition from state e to state a results in all flip-flops changing state. Thus, the machine may pass through state d and assert output z_1.

Figure 3.90 (b) depicts the state diagram of Figure 3.90 (a) with a reassignment of state codes. Since the rules for state code adjacency could not be adhered to, trial-and-error techniques were used to obtain the resulting state codes. This is only one of several viable state code assignments that will produce glitch-free outputs for z_1 and z_2.

The present-state map can also be effectively used in analyzing and assigning state codes. The present-state map for Figure 3.90 (a) is shown in Figure 3.91 (a) illustrating the three paths that could cause a glitch on either output z_1 or z_2. Referring to Figure 3.91 (a), map 1, there are two possible paths from state *b* to state *e*, as designated by the arrows. One path takes the machine through state *c*, which is not associated with an output; therefore, no glitch is possible. The other path passes through transient state *d*, as previously shown. Since state *d* asserts output z_1 unconditionally, a reassignment of state codes is necessary.

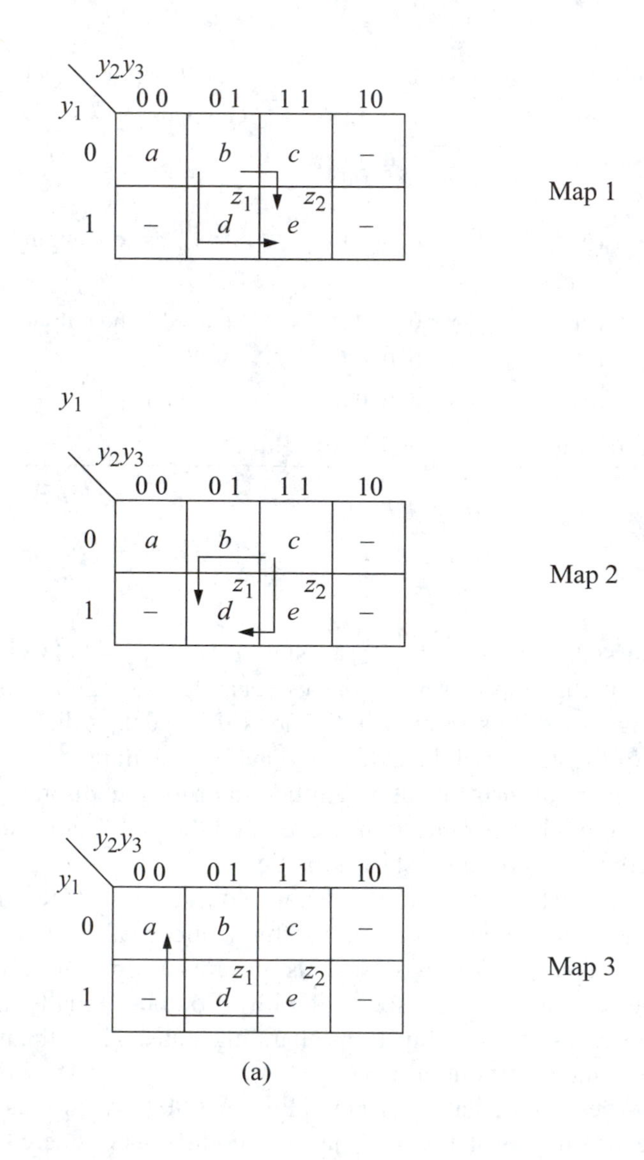

Figure 3.91 Present-state maps: (a) for Figure 3.90 (a) showing three paths where glitches are possible; and (b) for Figure 3.90 (b) which yields a glitch-free operation.

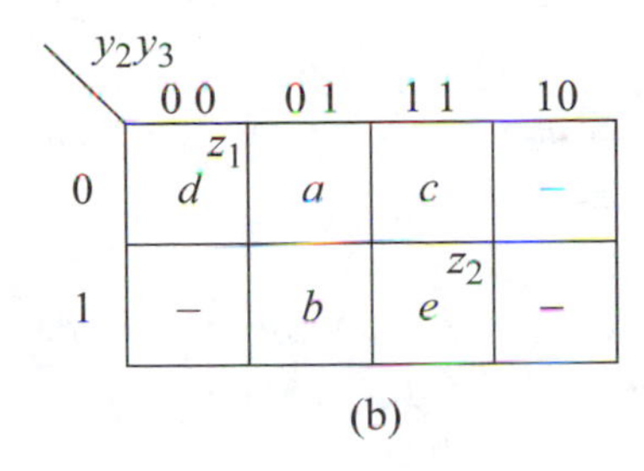

(b)

Figure 3.91 (Continued)

The remaining two hazardous transitions are shown in Figure 3.91 (a), maps 2 and 3. In map 2, the arrows indicate a transition from state c to state d through transient state b or e. If the sequence is through state b, then no glitch occurs on the outputs. If, however, the machine sequences through state e on the way to state d, then a glitch may appear on output z_2. In map 3, the sequence is from state e to state a. Because all three flip-flops change state, all other states — whether assigned or unused — are potential transient states. The only exposure to spurious outputs is through state d, where z_1 is active. Figure 3.91 (b) shows the present-state map for Figure 3.90 (b) with reassigned state codes, yielding glitch-free operation.

Before leaving this section on glitch elimination by state code assignment, one additional topic is presented. The previous examples illustrated glitches that were generated on Moore machines, where the only requirement for a glitch was that the transient state contain an output. Glitches are also possible on Mealy machines; however, two requirements must now be met:

1. The transient state must contain an output.
2. The input must be in the correct asserted level to generate an output for the logic family that is used.

Figure 3.92 depicts a state diagram for a Mealy machine with a single output z_1. Due to an improper choice of state codes, z_1 may glitch as the machine sequences from state c ($y_1y_2y_3 = 101$) through transient state e ($y_1y_2y_3 = 111$) to state d ($y_1y_2y_3 = 011$). The transient state may be entered because both y_1 and y_2 change state. The justification for proceeding from state c to state d was the active level of x_1. Since x_1 remains active throughout the transition, this meets the two requirements for output assertion in a Mealy machine; thus, z_1 will produce an erroneous output as the machine passes through state e. This is graphically depicted in the timing diagram of Figure 3.93.

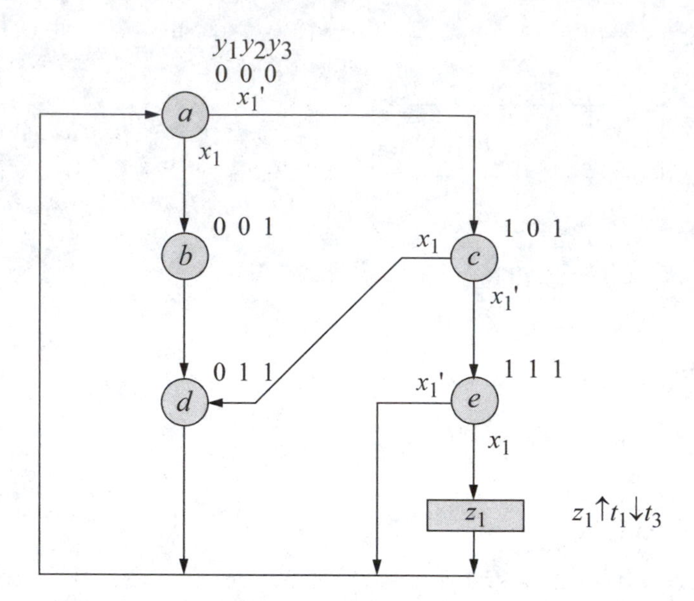

Figure 3.92 Mealy machine with an improper choice of state codes which may produce a glitch on output z_1 during a transition from state c to state d.

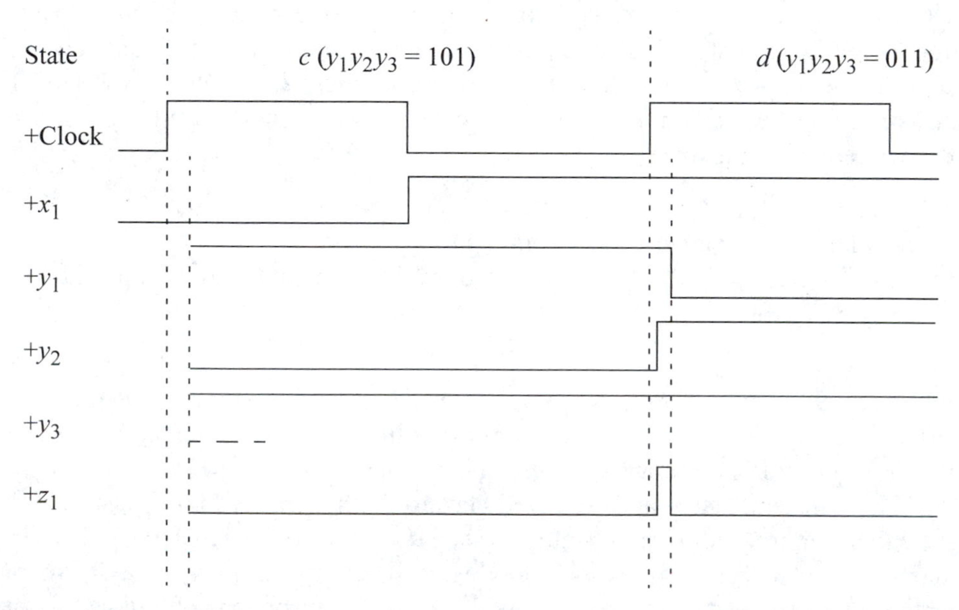

Figure 3.93 Timing diagram showing a glitch on output z_1 for the Mealy machine of Figure 3.92 during a transition from state c to state d.

3.7.2 Glitch Elimination Using Storage Elements

It may not be possible to reassign state codes to eliminate output glitches. For example, Figure 3.94 presents a state diagram for a Moore machine in which any combination of state code assignments may still result in an output glitch. This can be verified by permuting the state codes so that all state transitions have been examined for all possible sets of codes. In this machine, reassignment of state codes to eliminate output glitches is not possible, because all $2^p = 2^2 = 4$ state codes have been assigned and parallel paths exist for state transitions *b*, *c*, *a* and *b*, *d*, *a*. Thus, unused states cannot be utilized to avoid a state transition in which both flip-flops change state. A glitch may occur as the machine moves from state *b* ($y_1y_2 = 01$) to state *d* ($y_1y_2 = 10$) if flip-flop y_1 sets before y_2 resets. The input and output maps are shown in Figure 3.95.

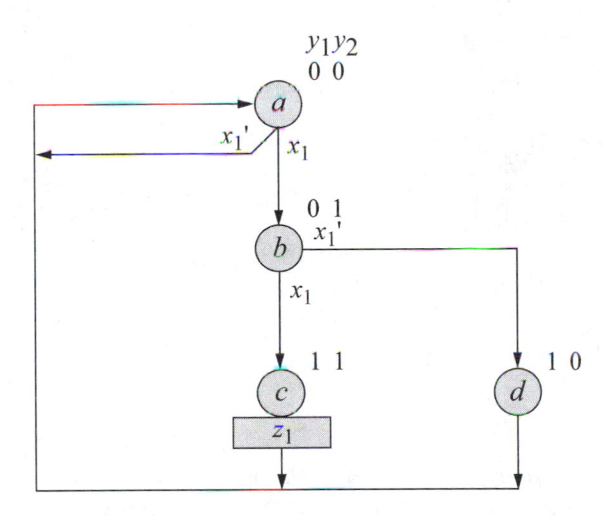

Figure 3.94 Moore machine with an inherent output glitch for a state transition from state *b* to state *d*.

If the machine specifications require that output z_1 be asserted for a duration of one clock period, but the precise time of assertion is not specified, then a *D* flip-flop in the λ output logic will satisfy this requirement. The following two assertion/deassertion statements stipulate the assertion of z_1 for one clock cycle:

1. $z_1 \uparrow t_1 \downarrow t_3$
2. $z_1 \uparrow t_2 \downarrow t_4$

Both produce an output pulse of the requisite duration; however, statement 2 delays the output by one-half clock cycle.

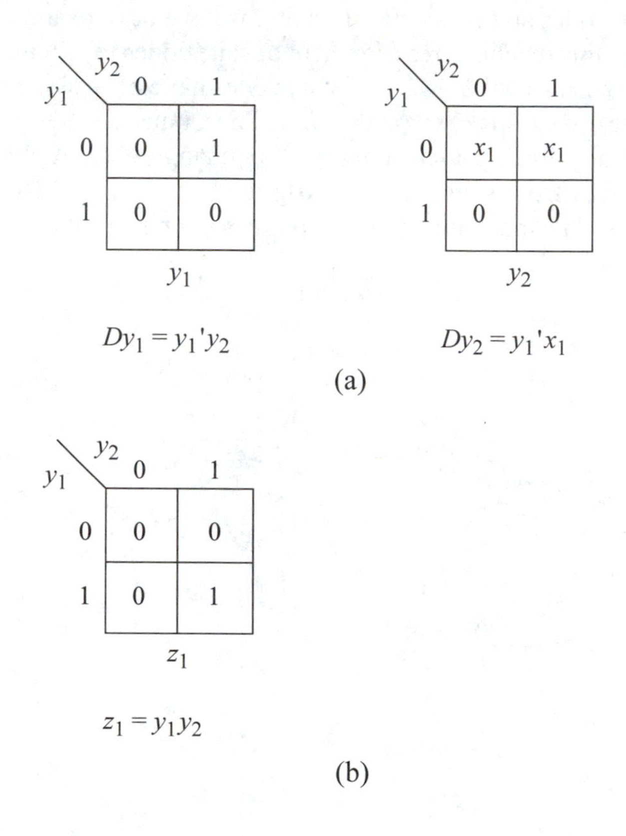

Figure 3.95 Karnaugh maps for the Moore machine of Figure 3.94 using x_1 as a map-entered variable: (a) input maps; and (b) output map.

Figure 3.96 (a) shows the logic diagram for statement 1, where $z_1 \uparrow t_1 \downarrow t_3$. Note that a delayed clock signal is used for the clock input of flip-flop z_1. Referring to the timing diagram of Figure 3.96 (b), the clock for flip-flop z_1 is delayed past the time (Δt) when glitches could occur. Thus, although the output of the AND gate that decodes z_1 may glitch, the active clock transition for flip-flop z_1 will not occur until after the glitch has returned to a logic 0. Delay circuits usually require a separate driver and receiver. The total delay of the clock signal for flip-flop z_1, including the inverter driver, the delay circuit, and the inverter receiver, must satisfy the following timing requirement:

Clock delay for flip-flop $z_1 \geq$ the propagation delay of flip-flops y_1 and y_2 plus the AND gate propagation delay for Dz_1 plus the setup time for Dz_1.

The time duration of Δt is insignificant compared with the duration of the clock cycle; therefore, output z_1 is still considered to be asserted at time t_1 and deasserted at time t_3.

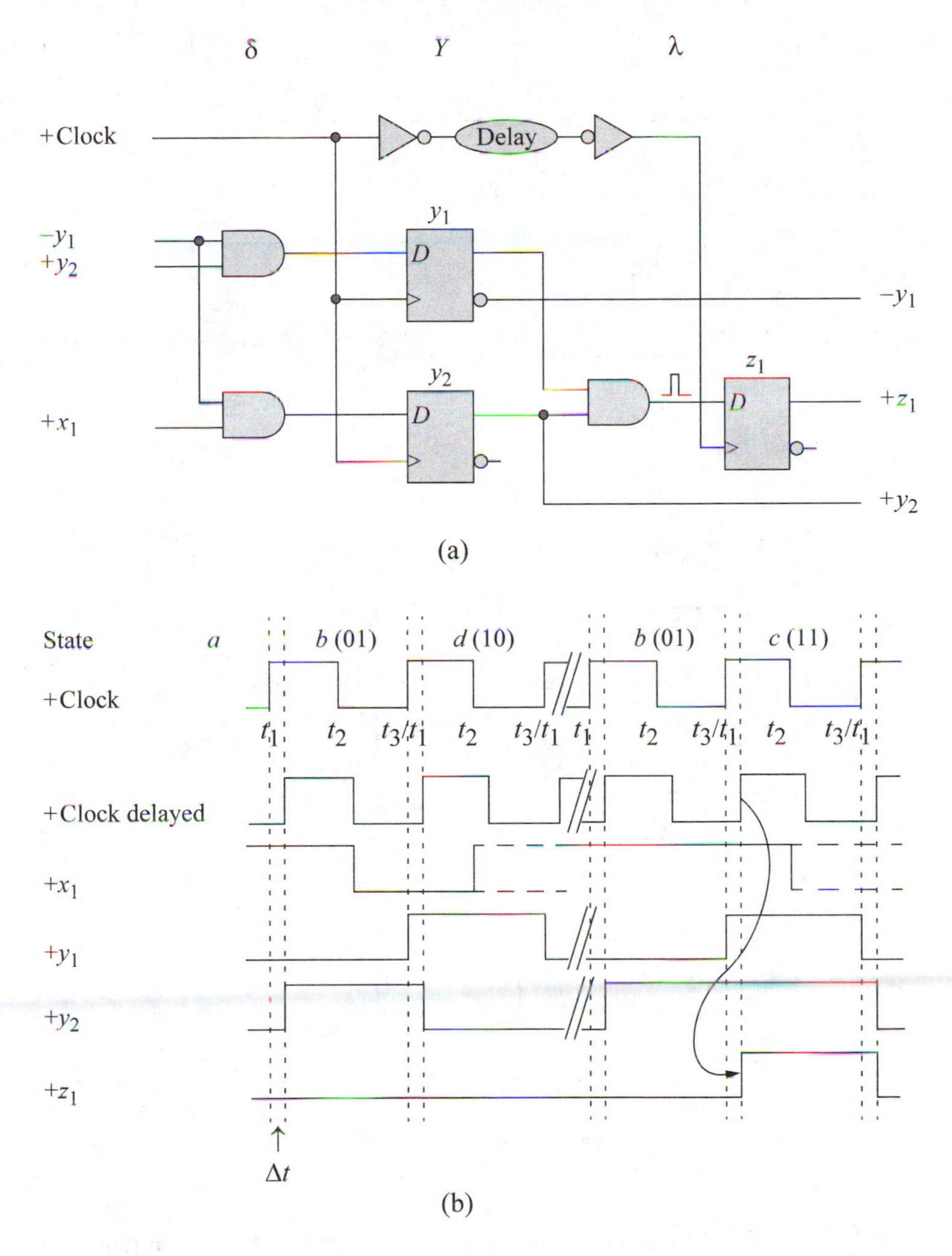

Figure 3.96 The Moore machine of Figure 3.94: (a) logic diagram where $z_1 \uparrow t_1 \downarrow t_3$; and (b) timing diagram. Glitches may occur during time Δt.

An alternative approach to asserting z_1 for a duration of one clock cycle, irrespective of the assertion time, is shown in Figure 3.97. This technique represents statement 2 above, and asserts z_1 at time t_2, long past the time when glitches could occur. The complement of the clock provides a positive active clock transition for flip-flop z_1 at time t_2 and resets z_1 at t_4. Glitches may still be present at input Dz_1 during Δt as shown in Figure 3.97 (a); however, the active clock transition does not take place until t_2, as shown in the timing diagram of Figure 3.97 (b).

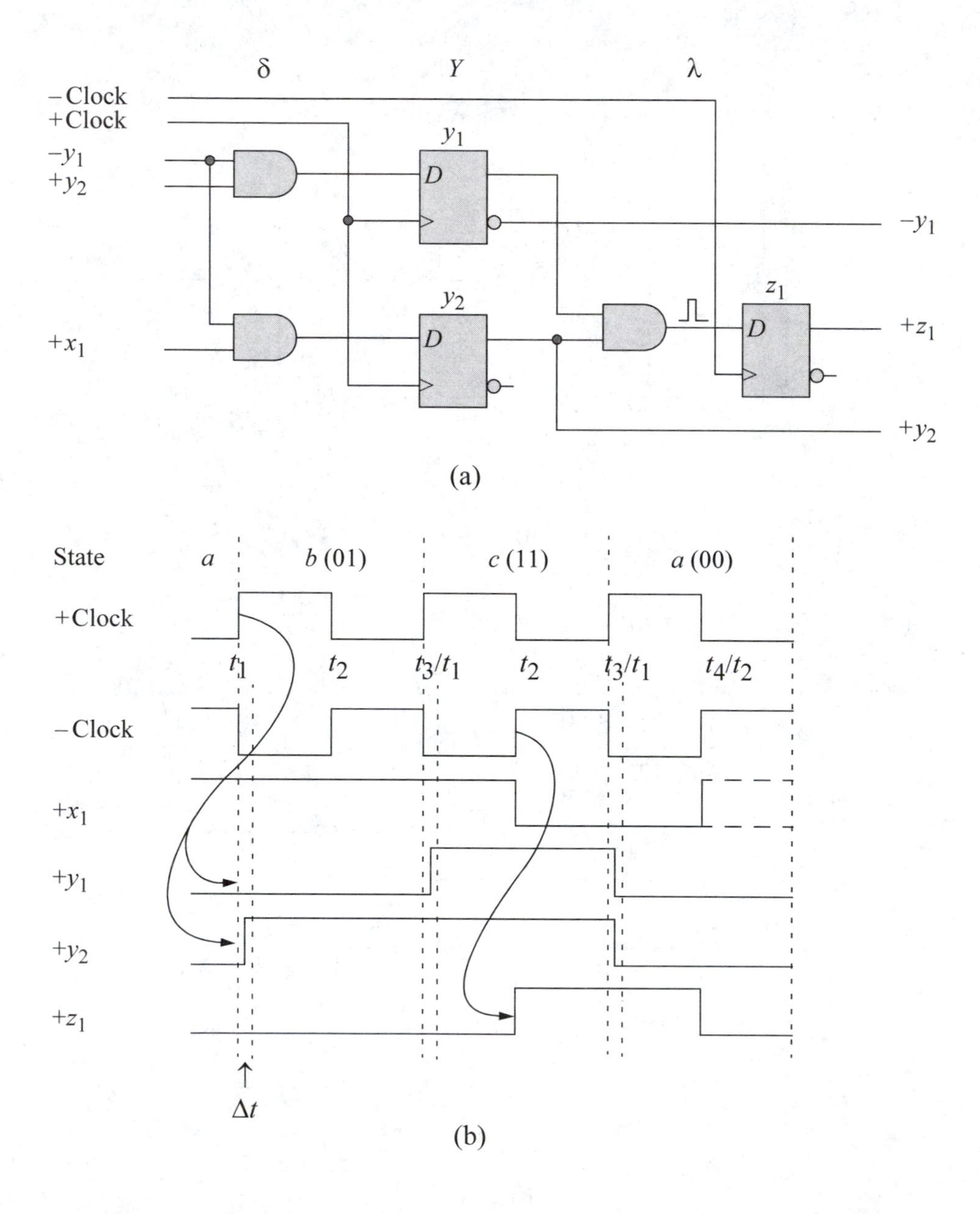

Figure 3.97 The Moore machine of Figure 3.94: (a) logic diagram where $z_1 \uparrow t_2 \downarrow t_4$; and (b) timing diagram. Glitches may occur during time Δt.

Therefore, even though a glitch occurs at the Dz_1 input during Δt, it is not latched, because the spurious output will have returned to a logic 0 before time t_2, which is the next active clock transition for flip-flop z_1.

Figure 3.98 illustrates a Mealy machine in which a glitch is possible on output z_1 if z_1 is asserted at time t_1 and deasserted at time t_3. The machine moves from state c ($y_1y_2 = 11$) to state a ($y_1y_2 = 00$) irrespective of the value of input x_1. If, however, x_1 is asserted during this state transition, then a glitch may appear on z_1 if flip-flop y_1 resets before y_2 resets. The problem can be resolved in the same manner as for the Moore machine by using a delayed clock signal to set a D flip-flop for z_1. The timing diagram showing the occurrence of a potential glitch if no flip-flop were used and the λ output logic necessary to resolve the problem are shown in Figure 3.99 (a) and Figure 3.99 (b), respectively.

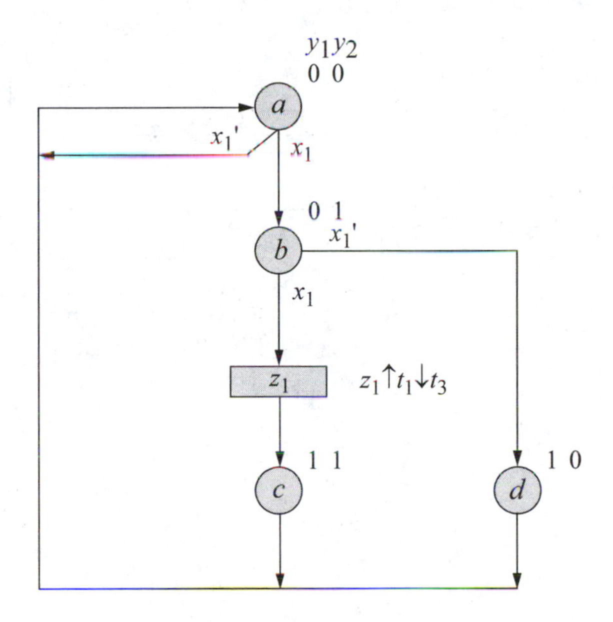

Figure 3.98 Mealy machine with an inherent output glitch for a state transition from state c to state a if input $x_1 = 1$.

3.7.3 Glitch Elimination Using Complemented Clock

The simplest and most inexpensive method of eliminating output glitches is to include the complement of the machine clock in the implementation of the λ output logic. The output logic will consist of an AND gate which decodes the p-tuple state codes. One input of the AND gate is connected to the complement of the machine clock; that is, the negation of the clock signal which drives the state flip-flops. This will generate an output signal that is only one-half the duration of the clock cycle, but guarantees that the output is free from any erroneous assertions. The output is asserted at time t_2 and deasserted at time t_3.

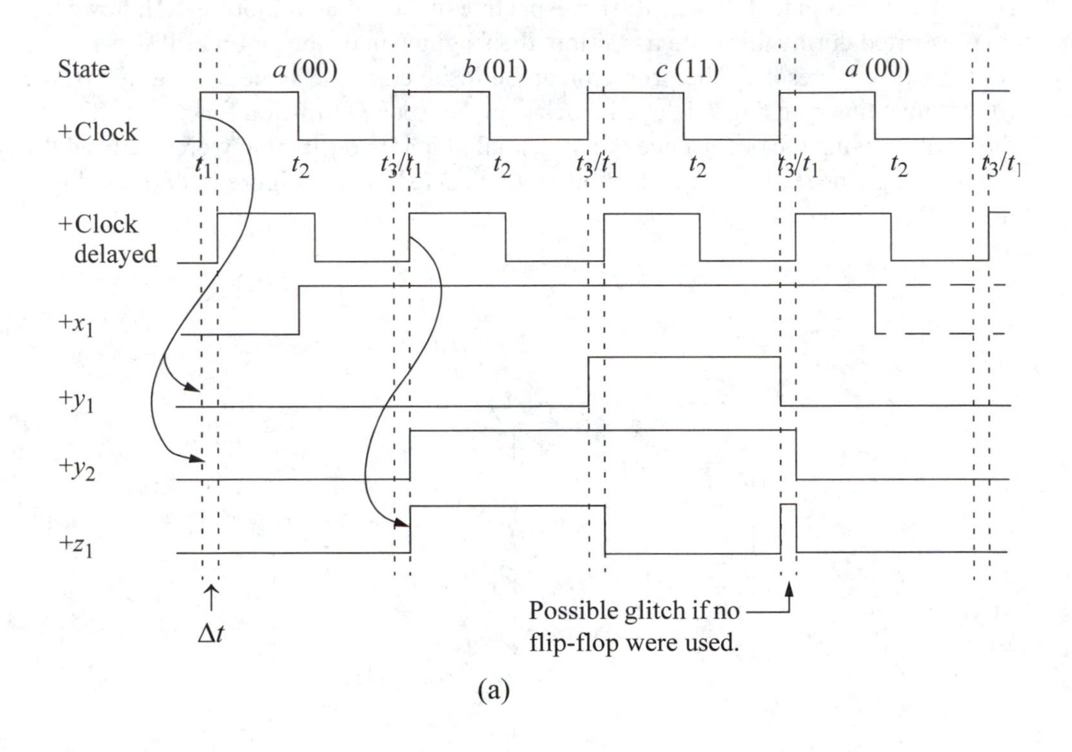

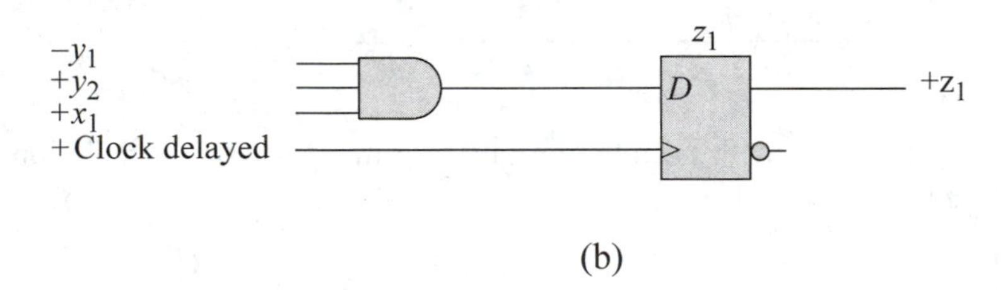

Figure 3.99 Mealy machine of Figure 3.98 where $z_1 \uparrow t_1 \downarrow t_3$: (a) timing diagram showing possible glitch on output z_1; and (b) λ output logic to resolve the glitch problem.

Figure 3.100 (a) shows a general block diagram for a Moore machine using the complement of the machine clock as an output gating function. A glitch that is caused by a state transition in which two or more flip-flops change state has no effect on the output — the glitch has returned to a logic 0 before the active level of the complemented clock occurs. Since the output assertion is for only one-half a clock period, storage elements are not required for the λ output logic. The timing diagram of Figure 3.100 (b) shows output vector Z_r asserted at time t_2 as a result of the high level of the machine clock complement. This method also applies to Mealy machines; the inputs, however, must remain active during the last half of the clock cycle.

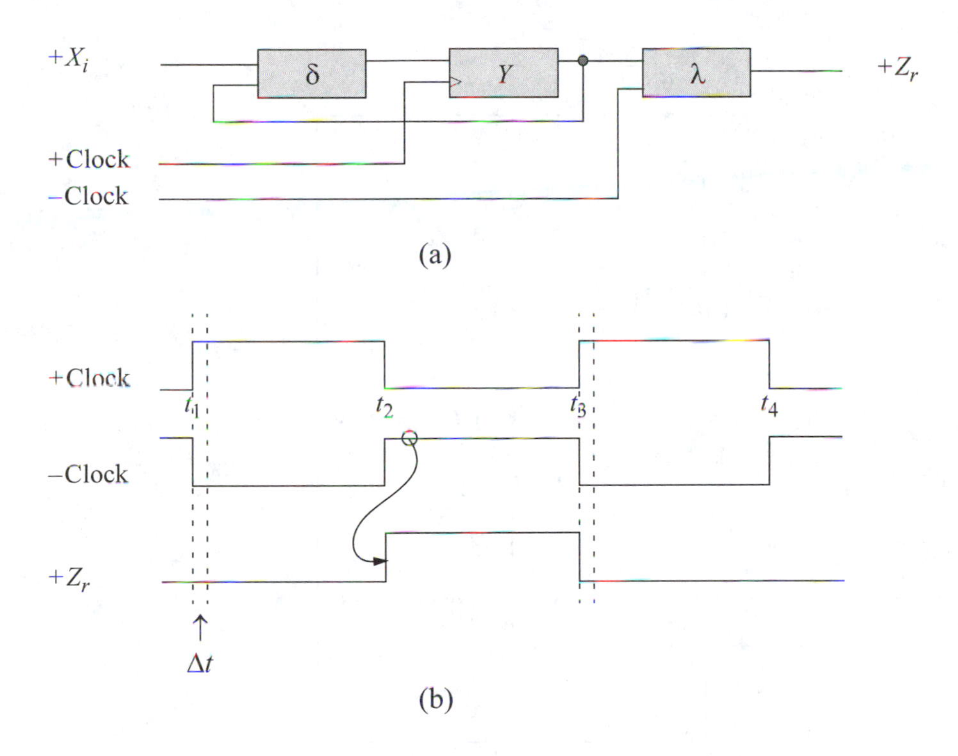

Figure 3.100 A generalized Moore model which uses the complement of the machine clock as a gating function to eliminate output glitches. The output vector Z_r is asserted at time t_2 and deasserted at t_3: (a) block diagram; and (b) timing diagram.

The multiple-output Moore machine, depicted by the state diagram of Figure 3.101 (a), contains a plethora of possible output glitches if the assertion/deassertion statement for the two outputs z_1 and z_2 is $\uparrow t_1 \downarrow t_3$. Table 3.32 lists the state transition sequences that may cause glitches on the outputs. The four possible output glitches are rendered ineffective, however, by including the machine clock complement in the output decoder (AND function using NOR logic) for z_1 and z_2. Outputs z_1 and z_2 are asserted at time t_2 and deasserted at t_3. Figure 3.101 (b) and Figure 3.101 (c) show the input maps using *JK* flip-flops and the output maps, respectively. The logic diagram is presented in Figure 3.101 (d) in which the δ next-state logic is represented by boolean input equations. The λ output logic for z_1 and z_2 is implemented with NOR logic, necessitating active low inputs for the AND operation. Therefore, the complemented machine clock signal, labeled +Clock, is connected to both output AND gates. The timing diagram for states *c* and *d* is depicted in Figure 3.101 (e).

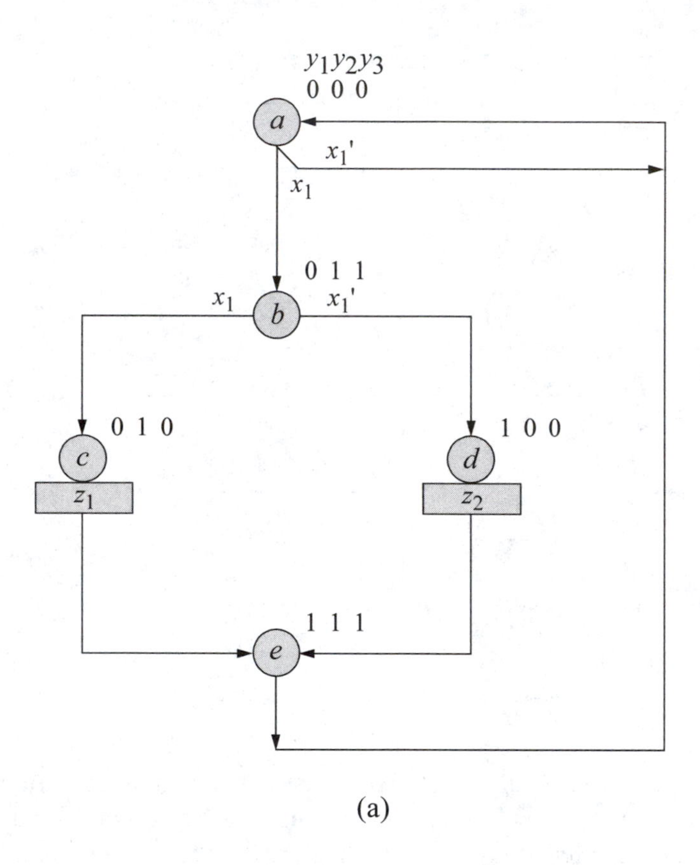

(a)

Figure 3.101 Multiple-output Moore machine in which four possible output glitches are rendered ineffective by including the machine clock complement in the output decoder for z_1 and z_2. Outputs z_1 and z_2 are assserted at time t_2 and deasserted at t_3: (a) state diagram; (b) input maps using *JK* flip-flops; (c) output maps; (d) logic diagram; and (e) timing diagram.

3.7.4 Glitch Elimination Using Delayed Clock

One final technique is presented to circumvent the negative effects of glitches. If the machine specifications require that outputs be asserted at time t_1 and deasserted at t_2, then glitches are again possible, because output assertion occurs at the active clock transition. This technique applies to both Moore and Mealy machines.

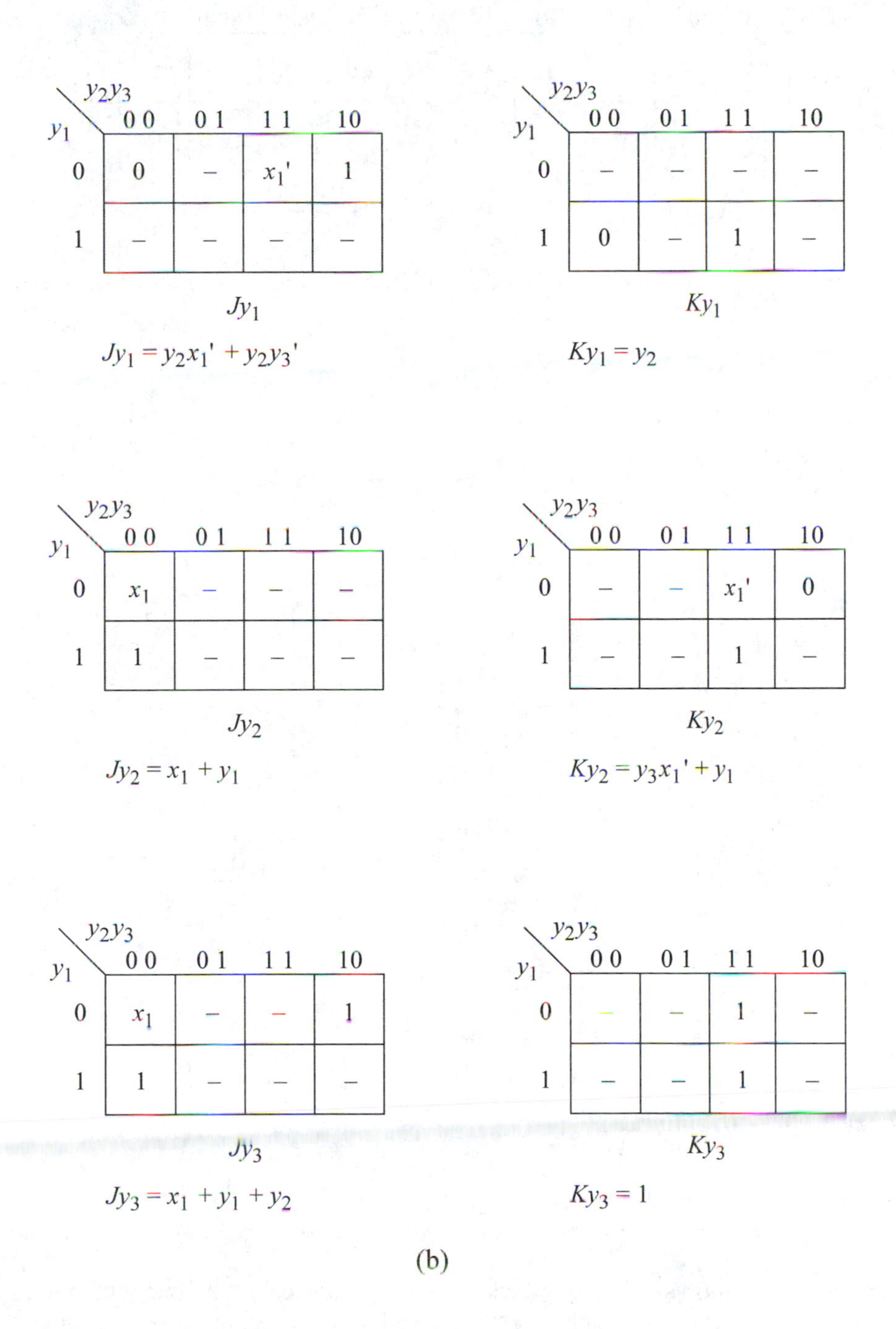

Figure 3.101 (Continued)

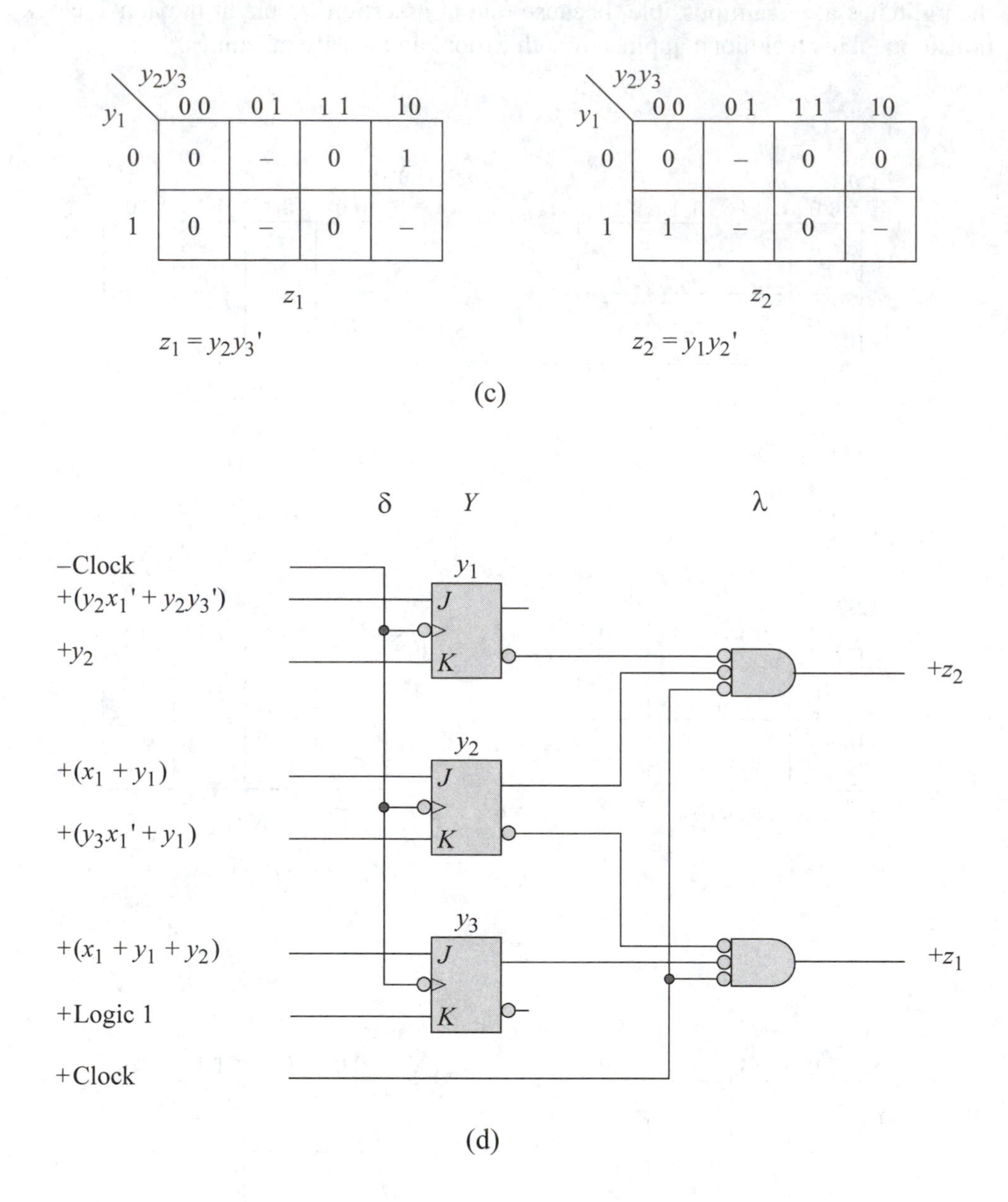

Figure 3.101 (Continued)

Figure 3.102 (a) shows a general block diagram for a Mealy machine which uses the active level of the delayed machine clock to enable the λ output logic. The state flip-flops are clocked by the +Clock signal, whereas the λ output logic is enabled by the +Clock delayed signal. The duration of the delay circuit must be equal to or greater than the time Δt — the time when glitches can occur — as shown in Figure 3.102 (b). The machine has stabilized at the termination of the Δt period. The delay circuit can be either a delay element with a dedicated driver and receiver or simply an even number of inverters.

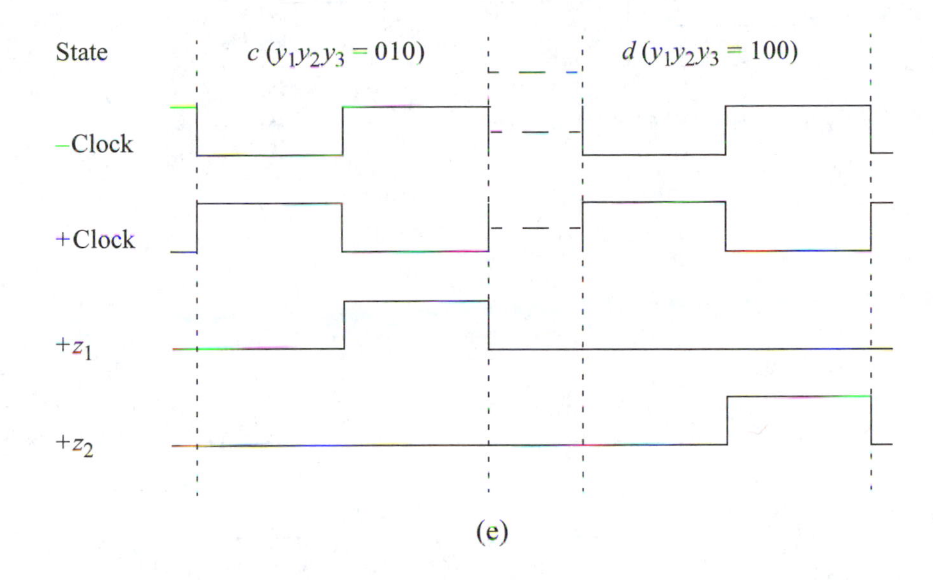

(e)

Figure 3.101 (Continued)

Table 3.32 State transition sequences for Figure 3.101 (a) that may produce glitches on outputs z_1 and z_2

Start state $y_1y_2y_3$	Transient state $y_1y_2y_3$	End state $y_1y_2y_3$	Comments
0 0 0 (*a*) →	0 1 0 (*c*)	→ 0 1 1 (*b*)	Glitch on z_1
0 1 1 (*b*) →	0 1 0 (*c*)	→ 1 0 0 (*d*)	Glitch on z_1
1 1 1 (*e*) →	0 1 0 (*c*)	→ 0 0 0 (*a*)	Glitch on z_1
1 1 1 (*e*) →	1 0 0 (*d*)	→ 0 0 0 (*a*)	Glitch on z_2

Since the outputs are asserted for one-half a clock period, storage elements are not required. Storage elements are necessary only when the outputs are active for a duration of one clock period, either $\uparrow t_1 \downarrow t_3$ or $\uparrow t_2 \downarrow t_4$. Since both clock and complemented clock are active for only one-half a clock cycle, they cannot be used to enable combinational λ output logic for one clock period.

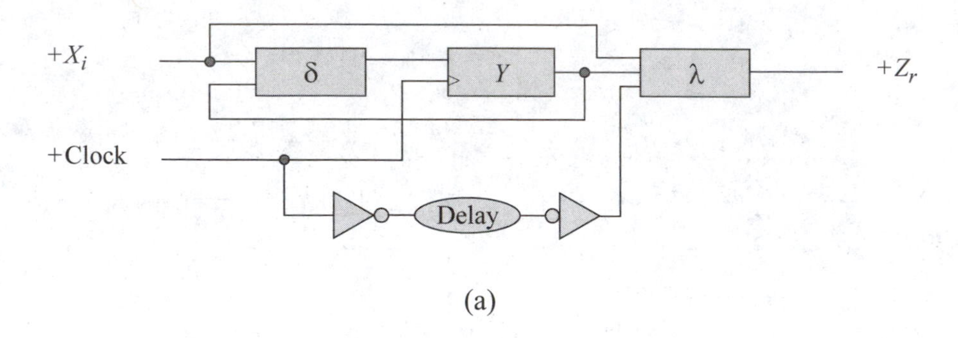

(a)

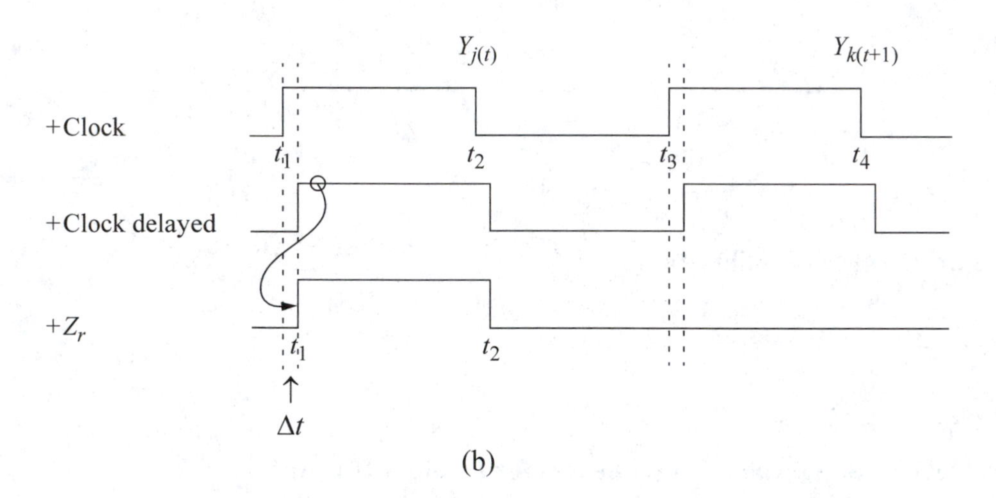

(b)

Figure 3.102 A general Mealy machine using clock delayed as an output gating function, where the output vector $Z_r \uparrow t_1 \downarrow t_2$: (a) block diagram; and (b) timing diagram.

The simple Moore machine of Figure 3.103 will be used to illustrate the delayed clock technique for glitch-free output assertion. The specified assertion and deassertion times for output z_1 are t_1 and t_2, respectively. Since the output logic for z_1 is simply an AND gate to decode $y_1 y_2$, a glitch is possible on z_1 for a state transition from state *b* to state *d* if flip-flop y_1 sets before y_2 resets.

The glitch can be avoided if the AND gate is enabled by the machine clock, delayed by a time increment of Δt. Thus, any momentary transition through state *c* will not affect z_1, because the active high level of the +Clock delayed signal will not be at the more positive voltage level until after time Δt, the time during which glitches could occur. The duration of Δt is quite small in relation to the clock cycle, so that the assertion and deassertion of z_1 is still considered to be $\uparrow t_1 \downarrow t_2$.

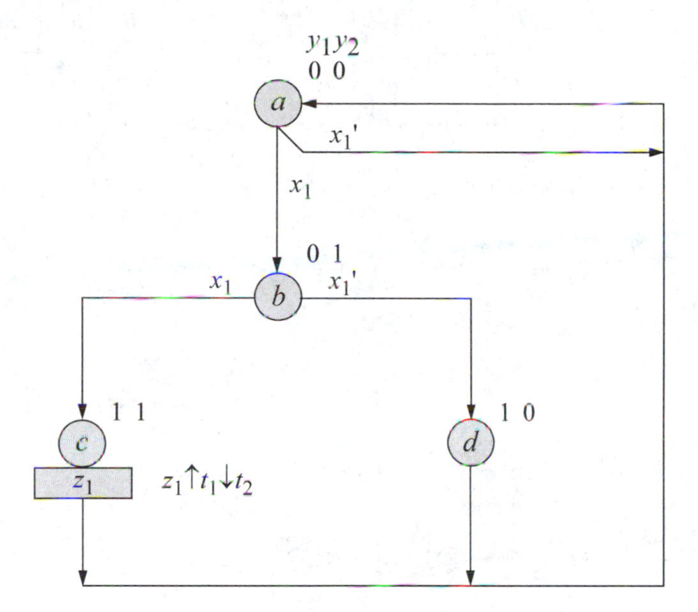

Figure 3.103 Moore machine where a glitch is possible for a transition from state *b* to state *d*.

This is illustrated in the logic diagram of Figure 3.104 (c) and depicted in the timing diagram of Figure 3.104 (d). Referring to Figure 3.104 (d), during a transition from state *b* to state *d*, flip-flop y_1 sets before y_2 resets. This places a high voltage level on the second and third inputs of the AND gate that generates z_1, as shown in Figure 3.104 (c). If the +Clock delayed signal — which is at a low voltage level at this time — was not connected to the first input of the AND gate, then a glitch would occur on output z_1. Disabling the λ output logic during the time period of Δt by delaying the active clock transition as a gating function, provides an output that is free from spurious signals caused by varying circuit propagation delays.

The machine is designed using the synthesis procedure described earlier. The input and output maps are shown in Figure 3.104 (a) and Figure 3.104 (b), respectively, which are constructed from the state diagram using x_1 as a map-entered variable.

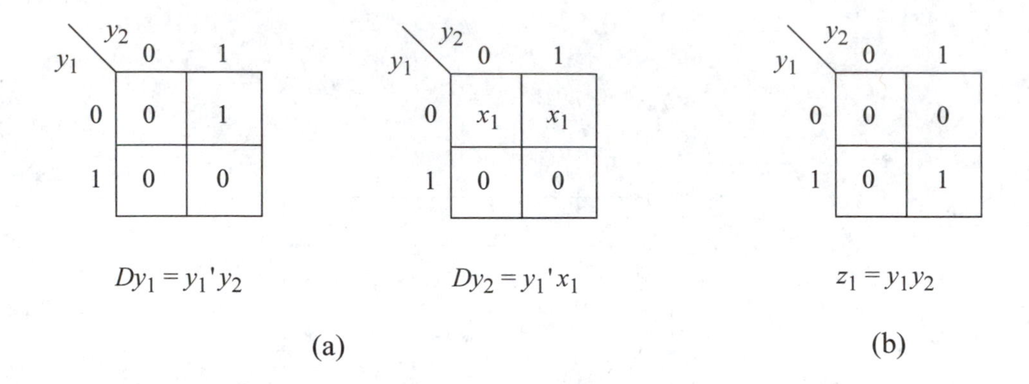

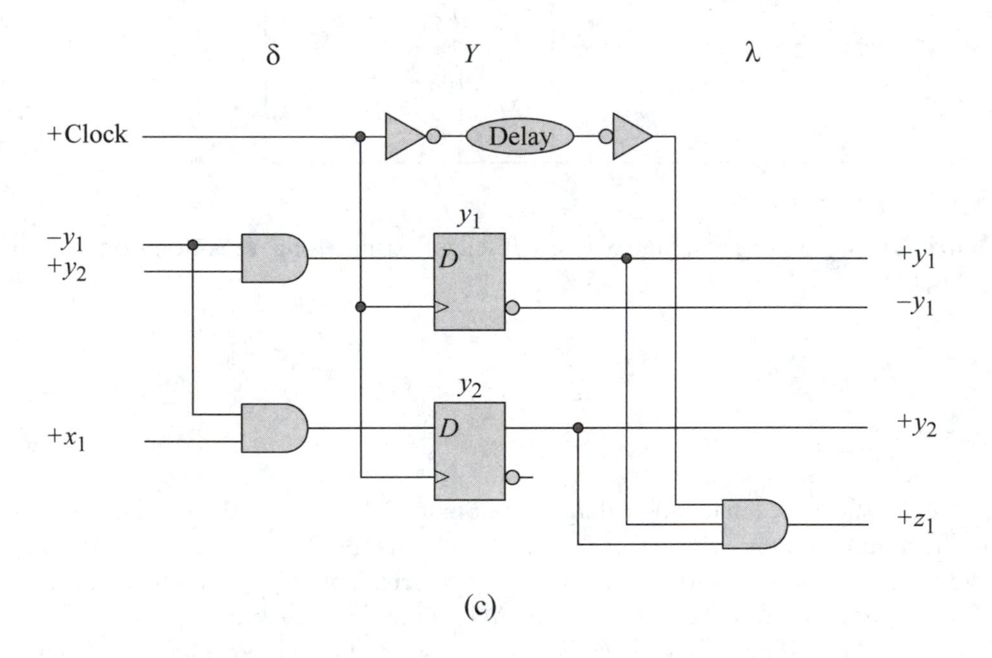

Figure 3.104 Moore machine of Figure 3.103 showing glitch elimination using a delayed clock signal for output enabling: (a) input maps; (b) output map; (c) logic diagram in which the assertion/deassertion of z_1 is $\uparrow t_1 \downarrow t_2$; and (d) timing diagram.

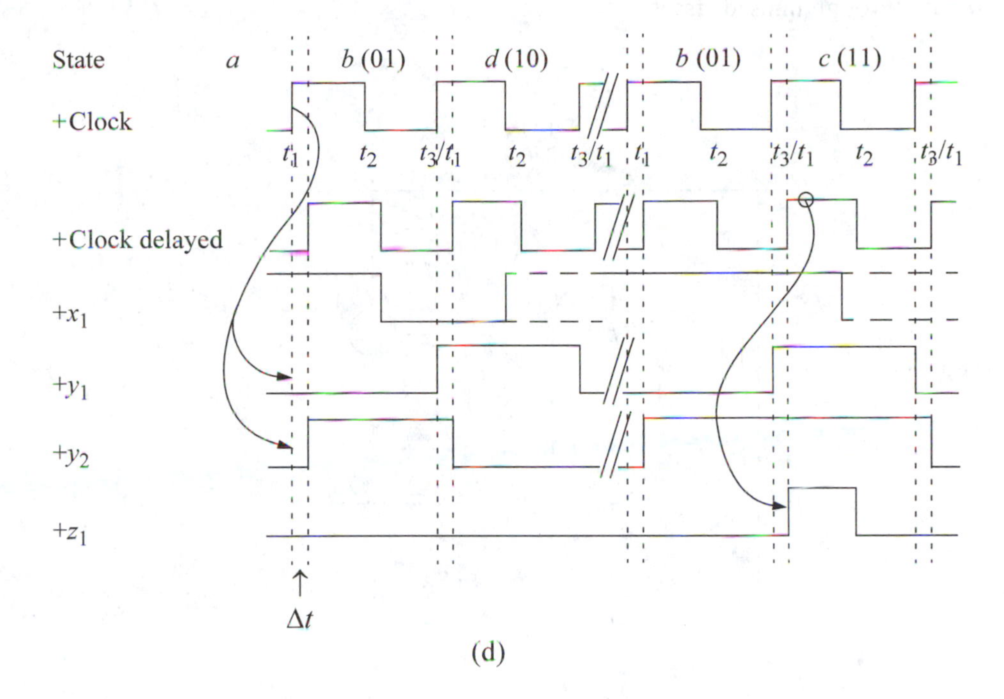

Figure 3.104 (Continued)

3.7.5 Glitches and Output Maps

Care must be taken when using "don't care" states in an output map in order to minimize the λ output logic. Any state transition which causes the machine to pass through an unused state in the output map for a transition that does not include the output under consideration, necessitates that a 0 be placed in that minterm location. Since the unused state is a transient state, the output will glitch as the machine passes through that state en route to the destination state.

Consider the state diagram and output map for the Moore machine shown in Figure 3.105. The input variable(s) have been omitted since they are not pertinent to the present discussion. When a reduced state diagram has been derived, the next step is to assign state codes. The state codes should be assigned so that no outputs will glitch as the machine progresses through the required state transition sequences. The state code assignment of Figure 3.105 results in an output that is free from transient assertions. This can be verified by checking all possible state transitions.

There is a potential problem only when a state transition causes two or more flip-flops to change state. This occurs only for the state changes shown in Table 3.33, none of which will produce a glitch on output z_1. The two paths, c, e and e, a, both include z_1, therefore, no glitch will be generated. Additionally, the transitions *c*, *e* and *e*, *a* pass through unused states.

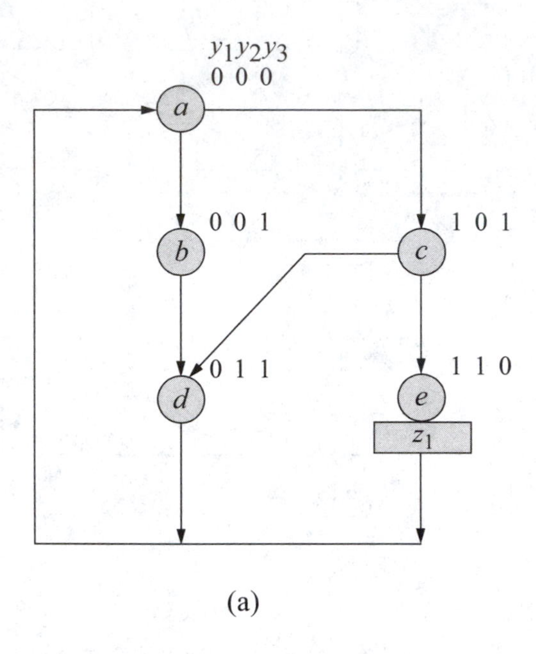

(a)

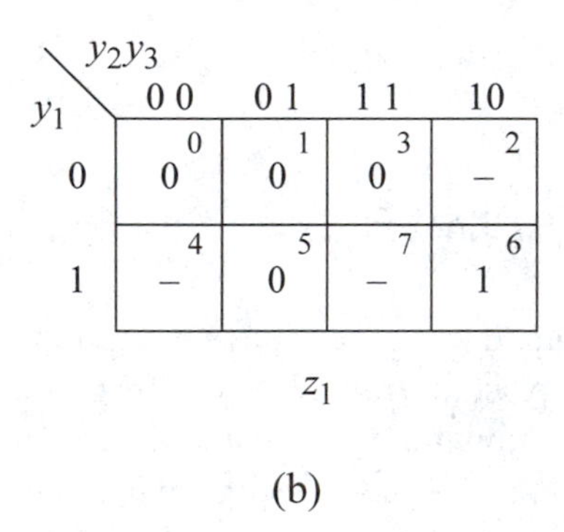

(b)

Figure 3.105 Moore machine: (a) state diagram. Unused states are $y_1y_2y_3$ = 010, 100, and 111; and (b) output map.

Referring to Figure 3.105 (b), the equation for output z_1 can be minimized by combining the 1 in minterm 6 with one of the unused states in minterm 2, 4, or 7. Before using a "don't care" state, however, the state diagram must be checked to be certain that no state transition will occur that causes the machine to pass through the unused state en route to a destination state for a path that does not include z_1. Every path in the state diagram must be checked for this possible occurrence. If one path is found that may cause the machine to pass through an unused state that does not involve z_1, then a 0 must be inserted in the corresponding unused state.

Table 3.33 State transitions for Figure 3.105 in which two or more flip-flops change state

Start state $y_1y_2y_3$	Transient state $y_1y_2y_3$	End state $y_1y_2y_3$	Comments
0 0 0 (*a*) →	0 0 1 (*b*) →	1 0 1 (*c*)	No output for *b*
0 0 0 (*a*) →	1 0 0 (Unused) →	1 0 1 (*c*)	
0 1 1 (*d*) →	0 0 1 (*b*) →	0 0 0 (*a*)	No output for *b*
0 1 1 (*d*) →	0 1 0 (Unused) →	0 0 0 (*a*)	
1 0 1 (*c*) →	0 0 1 (*b*) →	0 1 1 (*d*)	No output for *b*
1 0 1 (*c*) →	1 1 1 (Unused) →	0 1 1 (*d*)	
1 0 1 (*c*) →	1 0 0 (Unused) →	1 1 0 (*e*)	
1 0 1 (*c*) →	1 1 1 (Unused) →	1 1 0 (*e*)	
1 1 0 (*e*) →	0 1 0 (Unused) →	0 0 0 (*a*)	
1 1 0 (*e*) →	1 0 0 (Unused) →	0 0 0 (*a*)	

In Figure 3.105, three paths exist where two flip-flops change state, creating a condition for a possible glitch on z_1. The three paths, taken from Table 3.33, are shown in Table 3.34.

Table 3.34 Potential transient states for Figure 3.105

$y_1y_2y_3$	$y_1y_2y_3$	$y_1y_2y_3$
0 1 1 (*d*) →	0 1 0 (Unused) →	0 0 0 (*a*)
0 0 0 (*a*) →	1 0 0 (Unused) →	1 0 1 (*c*)
1 0 1 (*c*) →	1 1 1 (Unused) →	0 1 1 (*d*)

As indicated above, all three unused states are potential transient states. It is possible, therefore, that the machine may sequence through one or more unused states at some time during its operation. Since a minterm value in the output map represents the output value for that state, a 0 must be entered in all unused states for this design. Thus, the equation for output z_1 must contain all three flip-flop variables as shown in Equation 3.26.

$$z_1 = y_1 y_2 y_3' \tag{3.26}$$

To illustrate the necessity for all unused states in this example to contain a logic 0, consider combining minterms 6 and 7 to yield the following equation for z_1:

$$z_1 = y_1 y_2 \tag{3.27}$$

Thus, whenever $y_1 y_2 = 11$, z_1 is asserted. This condition could occur for a transition from state *c* to state *d*. But the path from state *c* to state *d* does not include z_1. Therefore, flip-flop y_3 must be included in the equation for z_1, as shown in Equation 3.26.

The timing diagram of Figure 3.106 graphically illustrates the erroneous output that may occur by omitting flip-flop y_3 from Equation 3.27 and reinforces the necessity to include all three flip-flops in the equation for z_1. If flip-flop y_2 sets before y_1 resets, then there will be a momentary transition through the unused state $y_1 y_2 y_3 = 111$ before the machine stabilizes in state *d*. This unstable state will generate a spurious output on z_1. Similarly, combining minterms 2 and 6 to yield Equation 3.28 may result in an output glitch, because flip-flop y_1 was omitted from the equation. The same rationale applies to the combination of minterms 4 and 6.

$$z_1 = y_2 y_3' \tag{3.28}$$

The example presented in this section was chosen to stress the importance of not only optimizing state code assignments, but also of combining unused states discriminately.

The problem of spurious output assertions in this example can be easily solved by reassigning state codes, as shown in Figure 3.107 (a). The following two conditions must be verified before implementing the λ output logic for this design, and for any synchronous sequential machine:

1. No state transition will produce a glitch on the outputs caused by differing flip-flop delays.
2. No unused state that is used to minimize the output logic will produce a glitch on the outputs caused by differing flip-flop delays.

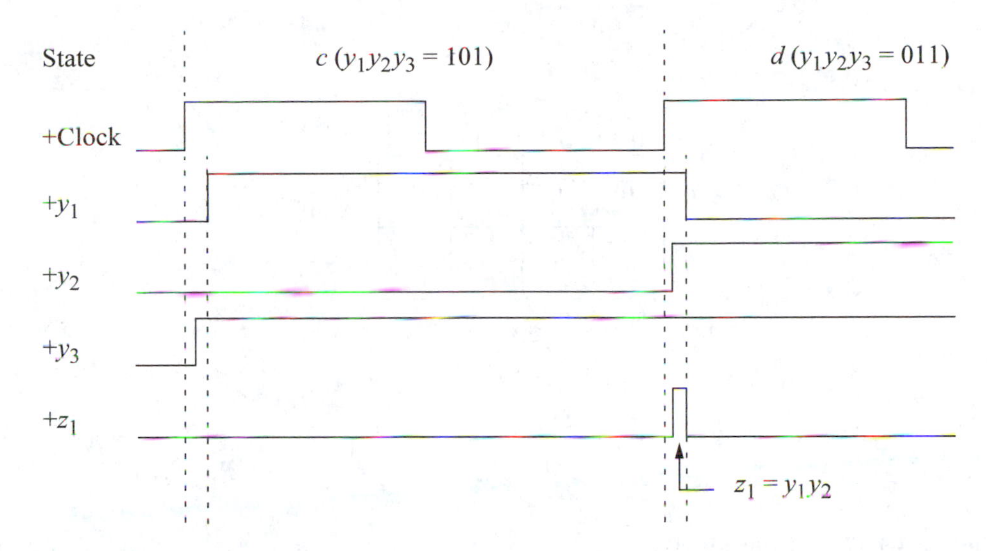

Figure 3.106 Timing diagram illustrating a spurious assertion for output z_1 when minterm 7 was assigned a value of 1 in the output map of Figure 3.105 (b) in order to minimize the equation for z_1 by combining minterms 6 and 7.

By checking all possible state transitions in Figure 3.107 (a), it can be verified that no glitch will appear on output z_1. Also, although it is possible for the machine to pass through all three unused states, no unused state will cause a glitch on z_1 when combined with the 1 in minterm location $y_1y_2y_3 = 010$.

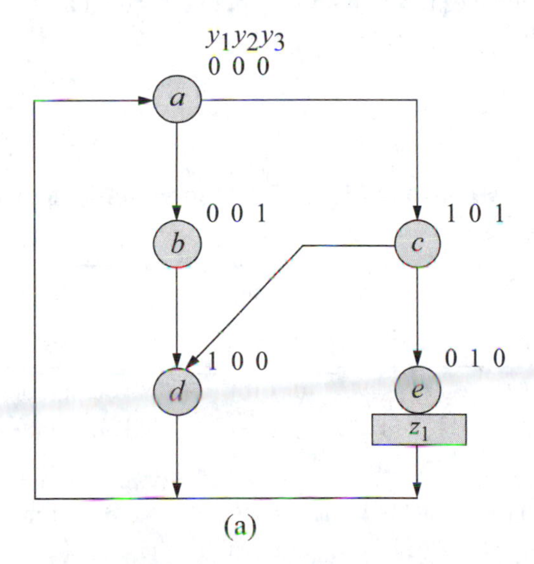

Figure 3.107 State diagram for the Moore machine of Figure 3.105 with reassigned state codes that will not produce a glitch on output z_1: (a) state diagram. Unused states are $y_1y_2y_3$= 011, 110, and 111; and (b) output map. All unused states can be used for z_1.

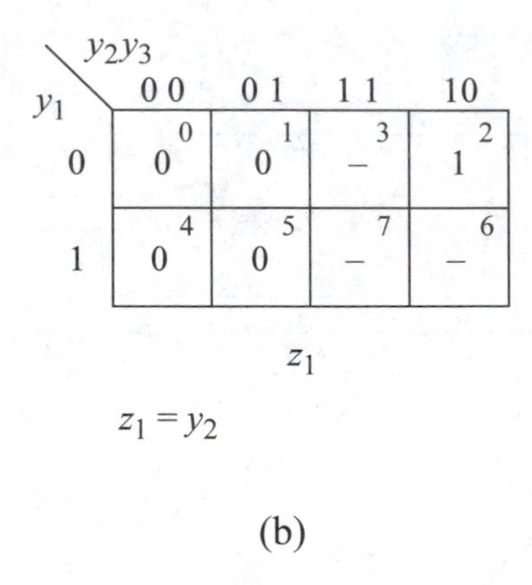

(b)

Figure 3.107 (Continued)

Table 3.35 lists all transitions where two or more flip-flops change state. The transition from state *c* to state *e* may cause the machine to sequence through one of the unused states. This will not result in a glitch on z_1, however. The transition will simply assert z_1 earlier, as shown in Figure 3.108 (a) for the sequence $y_1y_2y_3$ = 101 (*c*) → $y_1y_2y_3$ = 011 (Unused) → $y_1y_2y_3$ = 010 (*e*). This situation will occur if flip-flops y_1 and y_2 change states before y_3 changes state. Figure 3.108 (b) shows the assertion for z_1 if the machine sequences through non-output states or if all three flip-flops change states simultaneously.

Table 3.35 State transitions for Figure 3.107 in which two or more flip-flops change states

Start state $y_1y_2y_3$		Transient state $y_1y_2y_3$		End state $y_1y_2y_3$	Comments
0 0 0 (*a*)	→	0 0 1 (*b*)	→	1 0 1 (*c*)	No output for *b*
0 0 0 (*a*)	→	1 0 0 (*d*)	→	1 0 1 (*c*)	No output for *d*
0 0 1 (*b*)	→	0 0 0 (*a*)	→	1 0 0 (*d*)	No output for *a*
0 0 1 (*b*)	→	1 0 1 (*c*)	→	1 0 0 (*d*)	No output for *c*
1 0 1 (*c*)	→	0 0 0 (*a*)	→	0 1 0 (*e*)	No output for *a*
1 0 1 (*c*)	→	0 0 1 (*b*)	→	0 1 0 (*e*)	No output for *b*

Continued on next page

Table 3.35 State transitions for Figure 3.107 in which two or more flip-flops change states

Start state $y_1y_2y_3$		Transient state $y_1y_2y_3$		End state $y_1y_2y_3$	Comments
1 0 1 (c)	→	0 1 1 (Unused)	→	0 1 0 (e)	
1 0 1 (c)	→	1 0 0 (d)	→	0 1 0 (e)	No output for d.
1 0 1 (c)	→	1 1 0 (Unused)	→	0 1 0 (e)	
1 0 1 (c)	→	1 1 1 (Unused)	→	0 1 0 (e)	

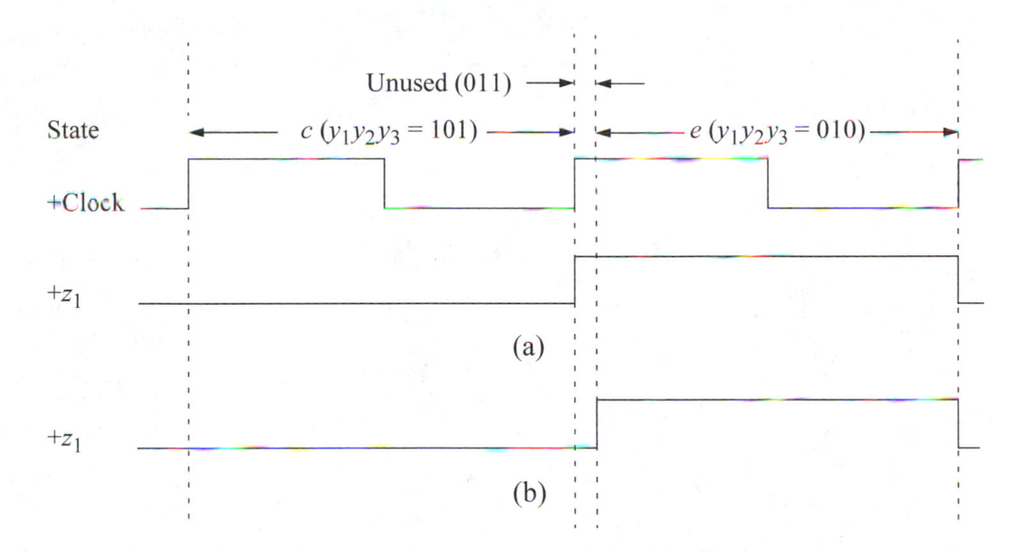

Figure 3.108 Assertion times for output z_1 for the Moore machine of Figure 3.107 for a state transition sequence c, e: (a) c (101) → unused (011) → e (010); and (b) c (101) → e (010).

3.7.6 Compendium of Output Glitches

Figure 3.109 is a pictorial summary illustrating different techniques to eliminate possible glitches in the λ output logic. All four assertion/deassertion statements are presented, together with the hardware implementation that is required to nullify the effects of glitches. Although Figure 3.109 depicts Moore outputs, Mealy outputs are easily accommodated by the inclusion of X_i, the input alphabet. Assume that the present state $Y_{j(t)} = y_1y_2 = 10$ and that output z_1 is to be asserted in this state. The state flip-flops are clocked on the positive clock transition. The λ output logic is shown with both AND and NOR logic implementation.

Figure 3.109 (a) presents $z_1 \uparrow t_1 \downarrow t_2$. Because glitches can occur in the Δt time period after the active (positive) clock transition, the clock signal must be delayed before enabling the AND or NOR gates. This design results in an output pulse that is one-half a clock period and occurs in the first half of state $Y_{j(t)}$.

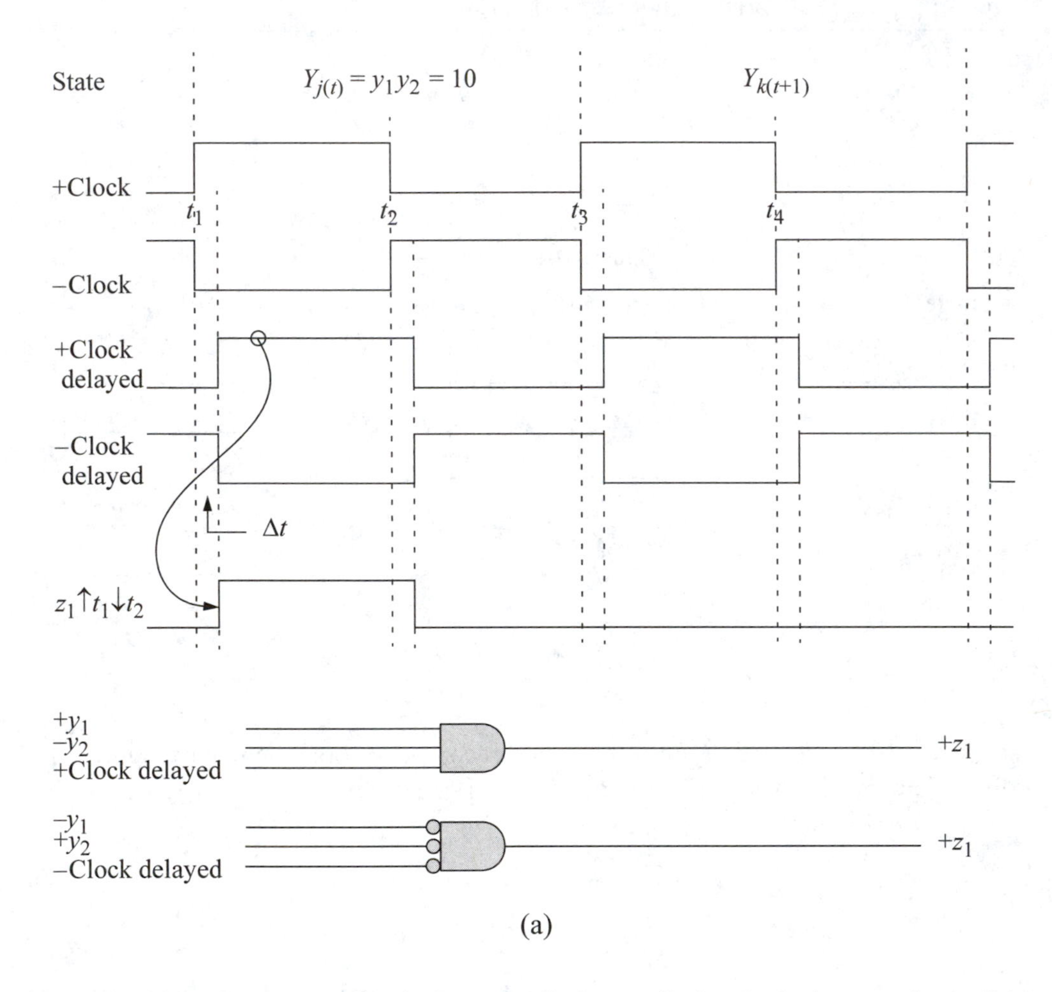

Figure 3.109 Summary of techniques to eliminate glitches in the λ output logic: (a) $z_1 \uparrow t_1 \downarrow t_2$; (b) $z_1 \uparrow t_1 \downarrow t_3$; (c) $z_1 \uparrow t_2 \downarrow t_3$; and (d) $z_1 \uparrow t_2 \downarrow t_4$.

Figure 3.109 (b) shows the waveform and logic for $z_1 \uparrow t_1 \downarrow t_3$. This method is used when the machine specifications require an output pulse that is to be active for the entire clock period (state time). A storage element is mandatory — the clock signal level by itself is not of sufficient duration to assert z_1 for the complete state time.

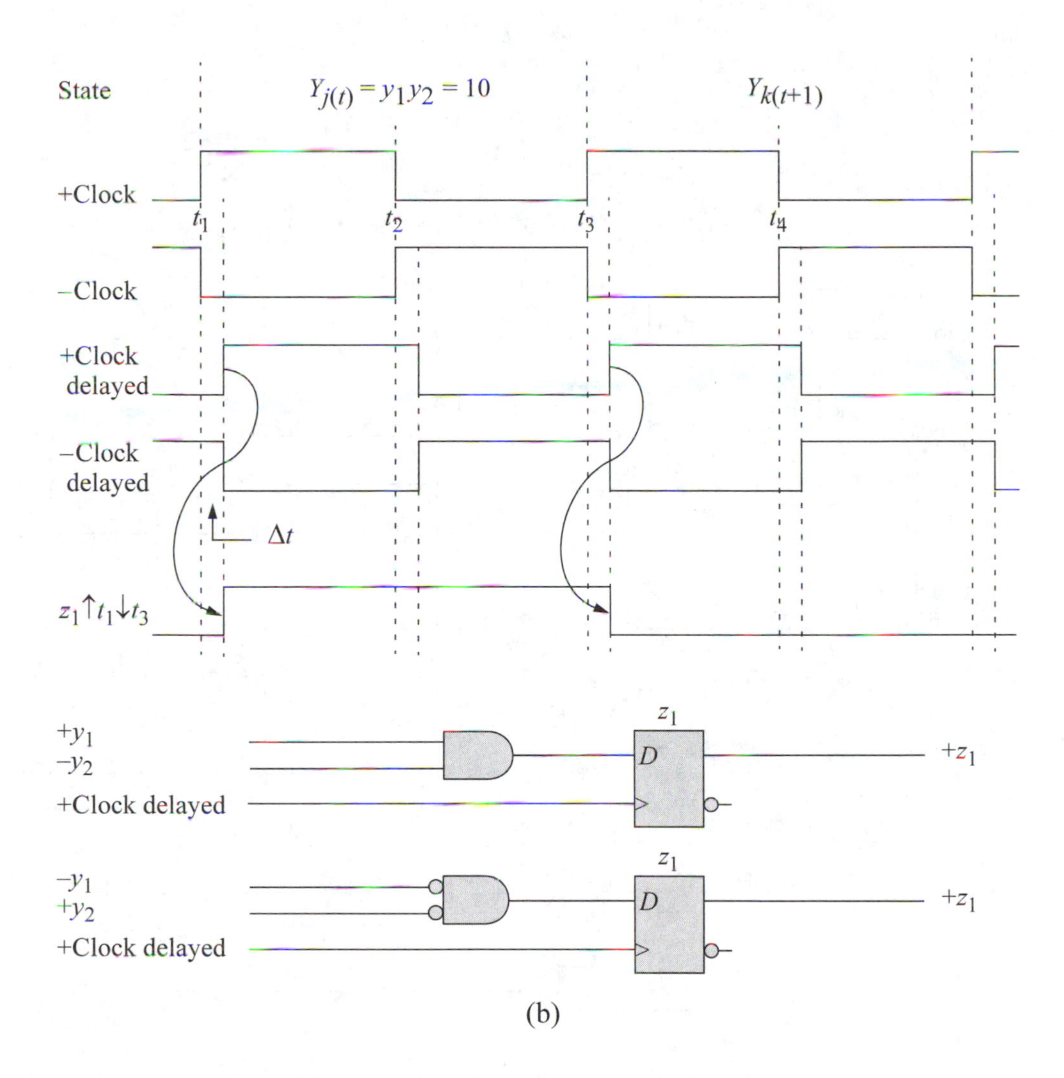

(b)

Figure 3.109 (Continued)

Figure 3.109 (c) depicts $z_1 \uparrow t_2 \downarrow t_3$. As stated previously, this technique is the simplest hardware implementation to eliminate output glitches — a single AND gate is sufficient. A delayed clock signal is not necessary, because time t_2 occurs well past time Δt, the time interval during which glitches may occur. The AND gate implementation requires active high (+) inputs; therefore, the –Clock signal is used as an enabling input, since –Clock is at the more positive voltage level during time t_2 to t_3. When a NOR gate is used to realize an AND function, the inputs are active low (–).

Thus, if NOR logic is used to activate z_1, then the +Clock signal becomes the enabling input, since it is at the more negative voltage level during time t_2 to t_3.

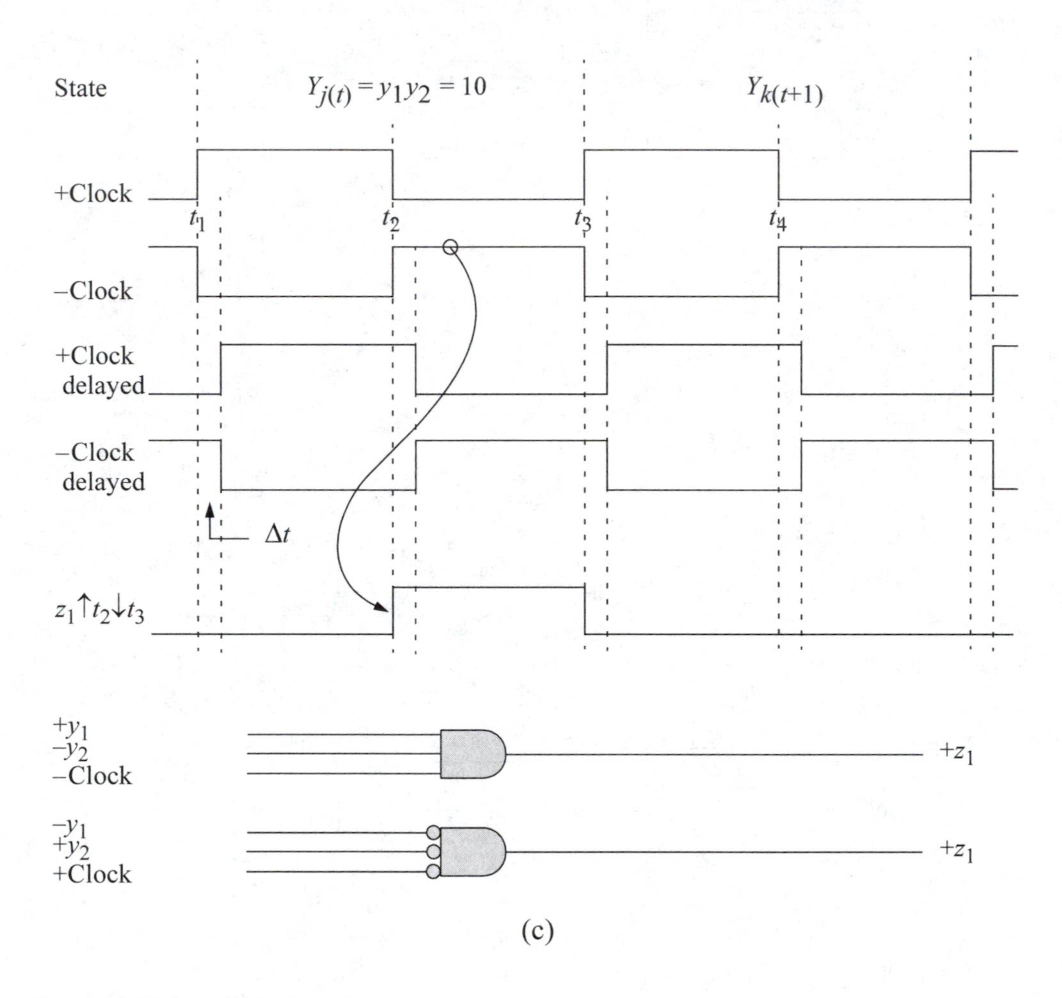

(c)

Figure 3.109 (Continued)

Finally, Figure 3.109 (d) shows $z_1 \uparrow t_2 \downarrow t_4$. The output is asserted for one clock period, but delayed by one-half a clock period. The *D* flip-flop is clocked on the positive clock transition. This occurs at time t_2 when the –Clock signal generates a low-to-high transition. The next positive transition of –Clock will not take place until time t_4, at which time the voltage level at Dz_1 will be low if $Y_{j(t)}$ is no longer the present state.

Flip-flop z_1 will then be reset and output z_1 will be deasserted. Whenever the assertion of z_1 is specified at time t_2, a delayed clock signal is not required, because time t_2 occurs well past the time when glitches could occur.

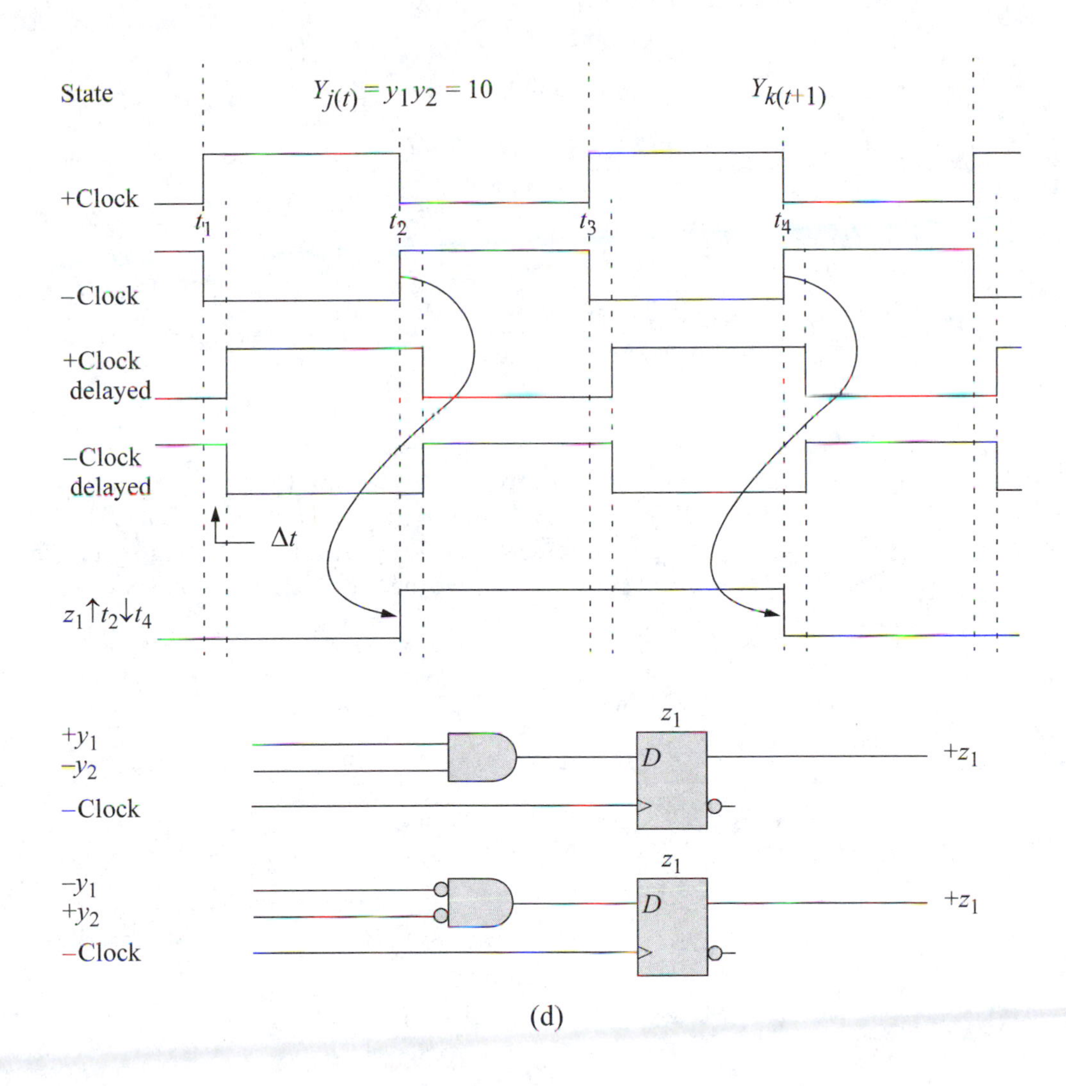

(d)

Figure 3.109 (Continued)

3.8 Problems

3.1 The output sequence for flip-flop y_1, which is shown below, is to be generated using one *JK* flip-flop. Determine sequence of values for *J* and *K* which will generate the output for y_1. The *JK* inputs are asserted in clock period $t_{(i)}$ in order to generate the output for y_1 in clock period $t_{(i+1)}$. Use the "don't care" symbol where applicable.

	Clock period					
	t_1	t_2	t_3	t_4	t_5	t_6
y_1	?	0	1	0	0	1
J						
K						

where ? is an unknown state of 0 or 1.

3.2 A synchronous sequential machine has inputs x_1 and x_2. The signal sequence for the inputs is shown below. Derive the input maps (with only x_1 and x_2 as coordinates) for a *JK* flip-flop that will generate the output sequence shown for flip-flop y_1. The inputs are asserted in clock period $t_{(i)}$ in order to generate the output for y_1 in clock period $t_{(i+1)}$. Use the "don't care" symbol where applicable.

	Clock period				
	t_1	t_2	t_3	t_4	t_5
x_1	1	1	0	0	–
x_2	1	0	1	0	–
y_1	?	1	0	0	1
J					
K					

where ? is an unknown state of 0 or 1.

3.3 A synchronous sequential machine uses a shift register as part of its implementation. Stage$_i$ of the shift register is shown below using a *D* flip-flop as the storage element. Modify stage$_i$ by using a *T* flip-flop as the storage element, while maintaining the same functional operation. Use the least amount of logic.

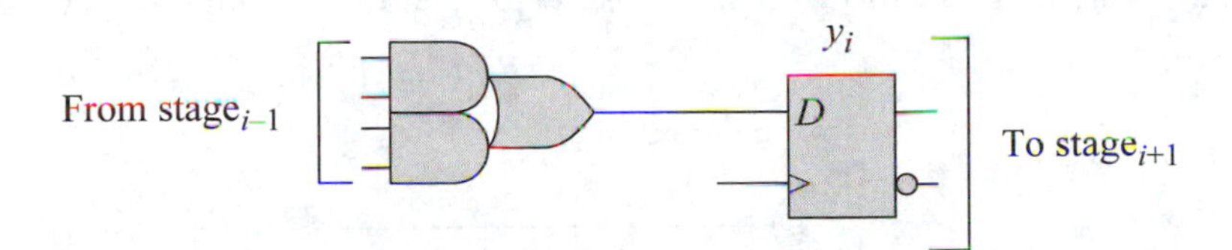

3.4 Find all equivalent states for the next-state table shown below.

Present state	Input x_1	Next state	Output z_1
a	0	b	0
	1	g	0
b	0	g	0
	1	a	1
c	0	h	0
	1	g	1
d	0	c	0
	1	a	1
e	0	h	0
	1	c	0
f	0	c	0
	1	e	1
g	0	d	0
	1	g	1
h	0	c	0
	1	a	1

3.5 Design a Moore machine using *JK* flip-flops whose operation is defined by the state diagram shown below. Use x_1, x_2, and x_3 as map-entered variables.

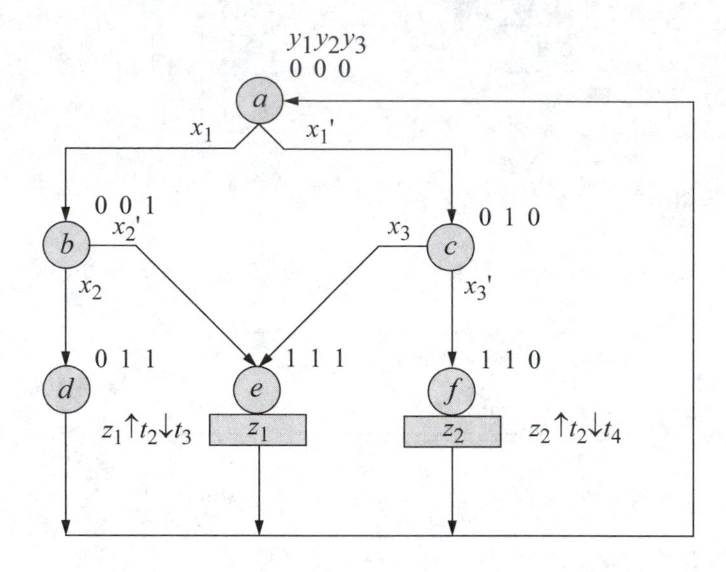

3.6 Obtain the input equations, in a minimum sum-of-products form, for flip-flop y_1 only, using a *JK* flip-flop for the state diagram shown below.

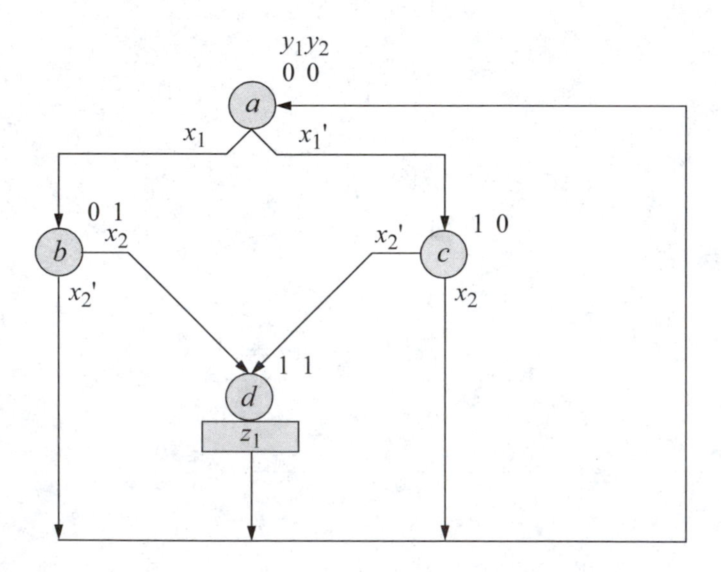

3.7 Obtain the input equations for Jy_1 and Ky_1 for state a only, for the partial state diagram shown below.

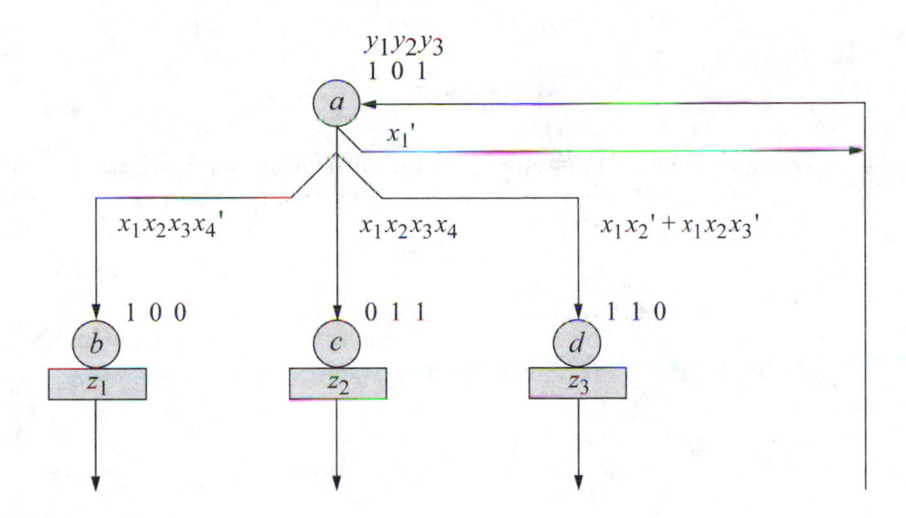

3.8 Determine the counting sequence for the counters shown below in parts (a) and (b). The counters are reset initially; that is, $y_1y_2 = 00$, where y_2 is the low-order flip-flop.

(a)

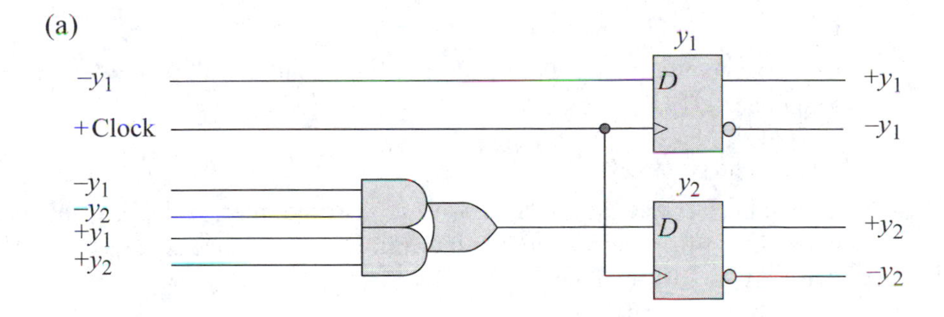

(b)

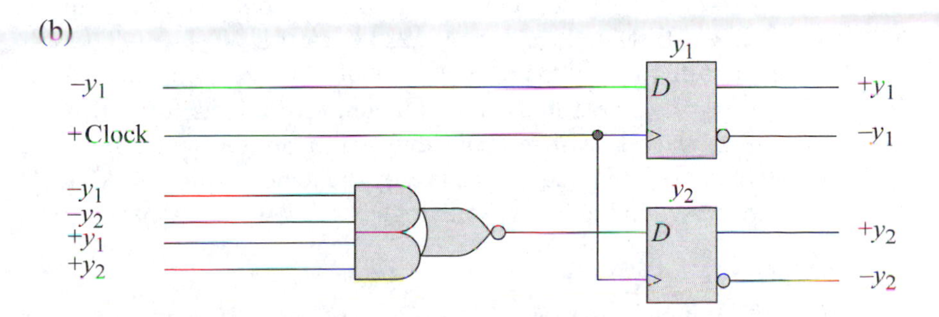

3.9 Obtain the input equations for a counter which counts in the sequence shown below. The counter uses *D* flip-flops and has a self-starting state of 0000.

$y_1y_2y_3y_4$ = 0000, 1001, 0001, 1011, 0010, 1100, 0011, 1101, 1000, 1110, 1010, 1111, 0000, $\cdots$.

3.10 Design a modulo-11 counter with no self-starting state: (a) using *D* flip-flops; and (b) using *JK* flip-flops. Observe the difference in the δ next-state logic of the two designs.

3.11 Design a synchronous counter using *JK* flip-flops that counts in the sequence shown below. The counter is not self-starting.

$y_1y_2y_3$ = 000, 100, 010, 001, 100, $\cdots$.

3.12 Obtain the input equations for flip-flops y_1 and y_4 only, for a BCD counter which counts in the sequence shown below. The equations are to be in minimum form. Use *JK* flip-flops. There is no self-starting state.

$y_1y_2y_3y_4$ = 0000, 0001, 0010, 0011, 0100, 0101, 0110, 0111, 1000, 1001, 0000, $\cdots$.

3.13 Design a counter using *JK* flip-flops which counts in the following decimal sequence: 0, 1, 3, 7, 6, 4, 0, $\cdots$.

3.14 Derive the input equations for a 3-bit Johnson counter, which is implemented with *JK* flip-flops and counts in the following sequence: $y_1y_2y_3$ = 000, 100, 110, 111, 011, 001, 000, $\cdots$.

3.15 Design a 4-bit Gray code counter using *JK* flip-flops. The counting sequence is as follows: 0, 1, 3, 2, 6, 7, 5, 4, 12, 13, 15, 14, 10, 11, 9, 8, 0, $\cdots$. The counter is initially reset to $y_1y_2y_3y_4$ = 0000. Show the equations for Jy_1, Ky_1 through Jy_4, Ky_4, first in a minimum sum-of-products form, then in exclusive-OR/NOR form, where applicable. Compare the speed of operation for the two forms of input equations if only the logic primitives AND, OR, and NOT were used in the δ next-state logic.

3.16 Obtain the input equation for flip-flop y_2 only, for a modulo-8 up/down counter which counts in the sequence shown below. The *Direction* input determines the direction of counting:

If *Direction* = 0, then count up
If *Direction* = 1, then count down

The counter is implemented with *D* flip-flops. Obtain the input equation first in a minimum sum-of-products form, then in exclusive-OR/NOR form.

$y_1y_2y_3$	*Direction*
0 0 0	0
0 0 1	0
0 1 0	0
0 1 1	0
1 0 0	0
1 0 1	0
1 1 0	0
1 1 1	1
1 1 0	1
1 0 1	1
1 0 0	1
0 1 1	1
0 1 0	1
0 0 1	1
0 0 0	0
.	
.	
.	

3.17 Generate a reduced state diagram for a Moore machine which generates an output z_1 whenever a serial, 4-bit binary word on an input line x_1 is greater than or equal to six. The first bit received in each word is the high-order bit. There is no space between words. Output z_1 is asserted during the fourth bit of a word. Do not assign state codes.

3.18 Generate a reduced state diagram for a Moore machine to detect an input pattern of exactly one group of one or more consecutive 1s on a serial input line x_1. An unconditional output z_1 is asserted when the first 1 occurs and remains asserted for all additional 1s in the group and for all 0s following the group unless another group of 1s occurs. If another group of 1s occurs, then output z_1 is deasserted and the machine enters and remains in a terminal state which generates an error output.

For example, a valid input sequence is 000011110000 · · · 0, because there is a single group of 1s. An invalid sequence is 000010011100 · · · 0, because there is more than one group of 1s. Do not assign state codes.

3.19 Generate a reduced state diagram for a Moore machine which will generate odd parity for each 5-bit word that is transmitted on a serial data line x_1. The format for the words is as follows:

$$\left| b_1 b_2 b_3 b_4 z_1 \right| b_1 b_2 b_3 b_4 z_1 \left| b_1 b_2 b_3 b_4 z_1 \right| \cdots \;\; |$$

where $b_i = 0$ or 1, and z_1 is the odd parity bit. There is no space between words.

3.20 Generate a reduced state diagram for a Moore machine which will load three bits of serial data from a synchronous input line x_1 into a shift register. The timing diagram for input x_1 and output z_1 is shown below. State codes are to be assigned so that no glitches will appear on the output. The state diagram must contain all the information that is required for correct operation of the machine. The state flip-flops are clocked on the positive transition of the clock.

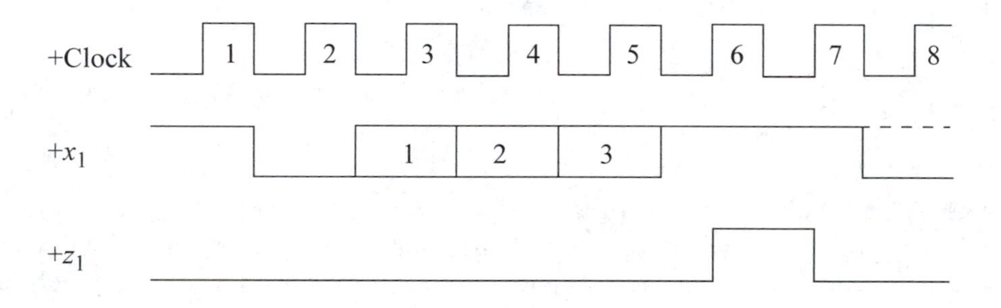

3.21 Generate a reduced state diagram for a Moore machine that has three parallel inputs, x_1, x_2, and x_3 and three outputs, z_1, z_2, and z_3 as shown below. The inputs constitute a 3-bit word. There is one bit space between words as shown below. The outputs are asserted in the bit space between words. Do not assign state codes. The machine operates as follows:

(a) Output $z_1 = 1$ if the 3-bit word contains an odd number of 1s.
(b) Output $z_2 = 1$ if the 3-bit word contains two or more 1s.
(c) Output $z_3 = 1$ if the 3-bit word contains no 1s.

x_1	x_1	x_1	...
x_2	x_2	x_2	...
x_3	x_3	x_3	...

x_1, x_2, x_3 → Synchronous sequential machine → z_1, z_2, z_3

3.22 Design a simple Moore machine with two inputs, one storage element, and one output, where $\delta(x_1, x_2, y_1)$ and $\lambda(y_1)$. Use any type of logic gates and storage elements.

3.23 Generate a reduced state diagram for a Mealy machine which produces a conditional output z_1 whenever a serial data line x_1 contains a sequence of three consecutive 1s. Overlapping sequences are allowed. Do not assign state codes.

3.24 Generate a reduced state diagram for a Mealy machine which detects a 4-bit word of 1001 on a serial input line x_1. If a correct sequence is detected, then a conditional output z_1 is generated. There is no spacing between words.

3.25 Generate a reduced state diagram for a Mealy machine which generates an output z_1 whenever a serial data line x_1 contains a 4-bit word with an odd number of 1s. The format for the serial data line is shown below, with no space between words. Do not assign state codes.

$$x_1 = \left| b_1b_2b_3b_4 \right| b_1b_2b_3b_4 \left| b_1b_2b_3b_4 \right| \cdots \qquad \text{where } b_i = 0 \text{ or } 1$$

3.26 Design a simple Mealy machine using one negative-edge-triggered *JK* flip-flop and any additional logic, where

$$X = \{X_0, X_1, X_2, X_3\} \quad \text{for } n = 2$$
$$Y = \{Y_0, Y_1\} \quad \text{for } p = 1$$
$$Z = \{Z_0, Z_1\} \quad \text{for } m = 1$$

Use only elements X_1 and X_2 of the set X.

3.27 Transform the Mealy machine, shown in the next-state table below, to a corresponding Moore machine.

Next-state table for a Mealy machine

State name	Present state y_1y_2	Input x_1	Next state y_1y_2	Output z_1
a	0 0	0	1 0 (c)	1
	0 0	1	0 1 (b)	1
b	0 1	0	0 0 (a)	1
	0 1	1	1 1 (d)	0
c	1 0	0	0 1 (b)	0
	1 0	1	0 0 (a)	0
d	1 1	0	1 1 (d)	1
	1 1	1	1 0 (c)	0

3.28 Select state codes for states d and e for the Moore machine shown below so that there will be no output glitches. Consider all state transitions.

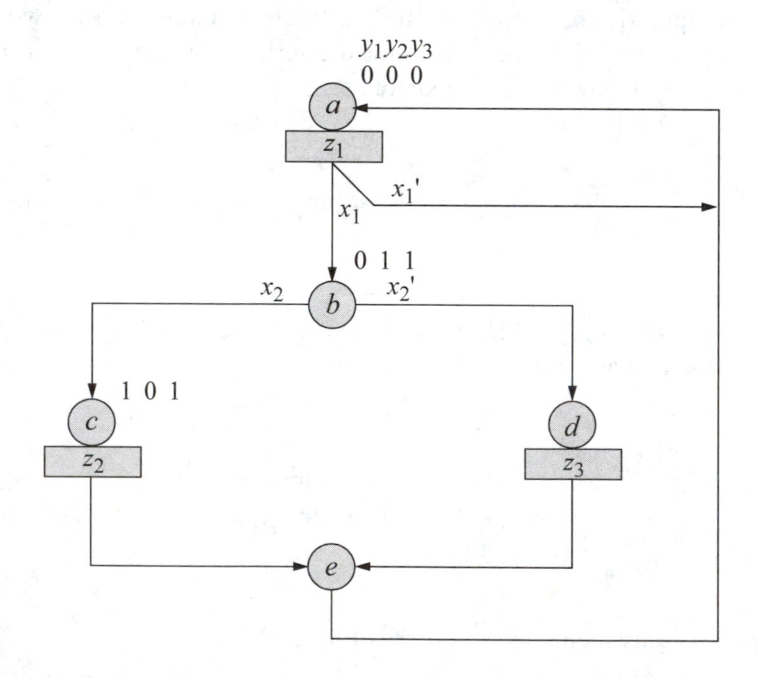

3.29 Select state codes for states c and d for the Moore machine shown below so that there will be no output glitches. Consider all state transitions.

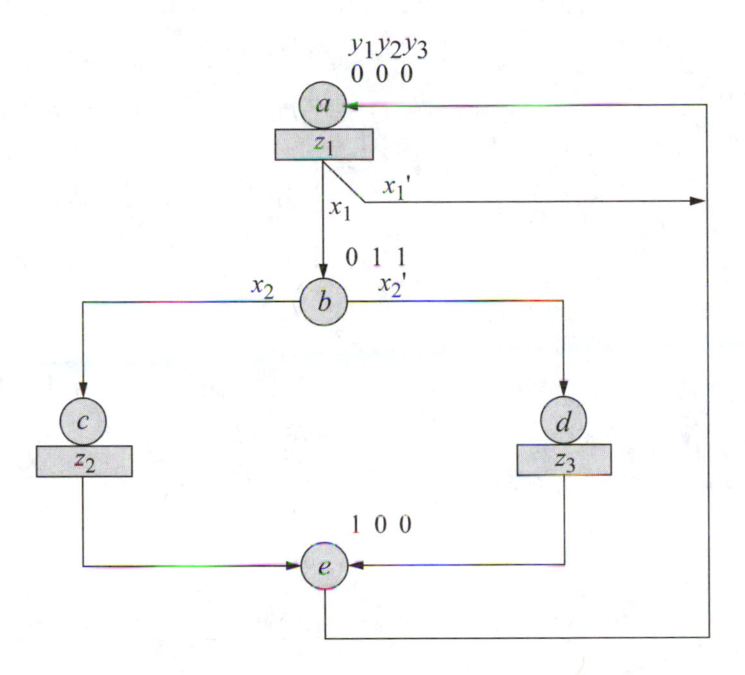

3.30 Given the state diagram shown below, indicate the number of glitches that may occur in the three outputs z_1, z_2, and z_3.

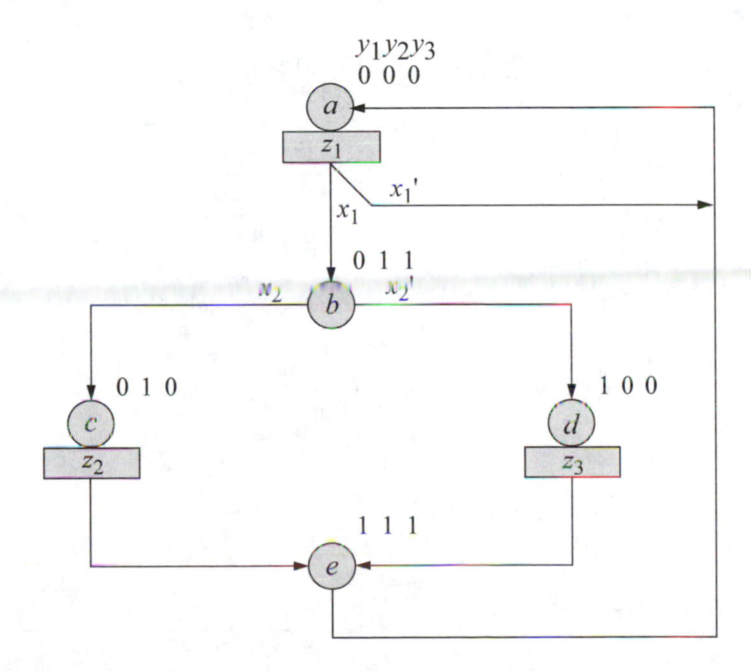

3.31 Select state codes for states c and e for the Moore machine shown below so that there will be no output glitches. Consider all state transitions.

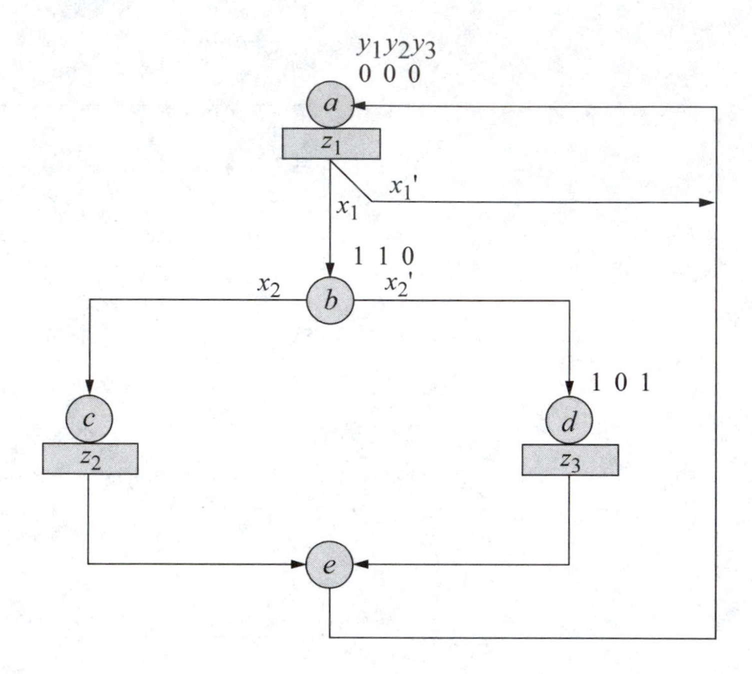

3.32 Select state codes for states c and d for the Moore machine shown below so that there will be no output glitches. Consider all state transitions.

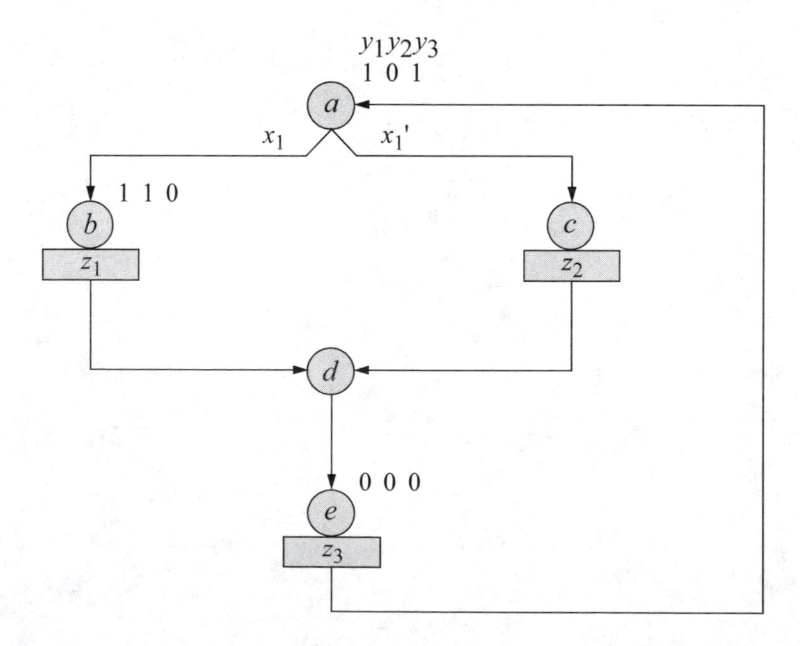

3.33 Select state codes for states b and e for the Moore machine shown below so that there will be no output glitches. Consider all state transitions.

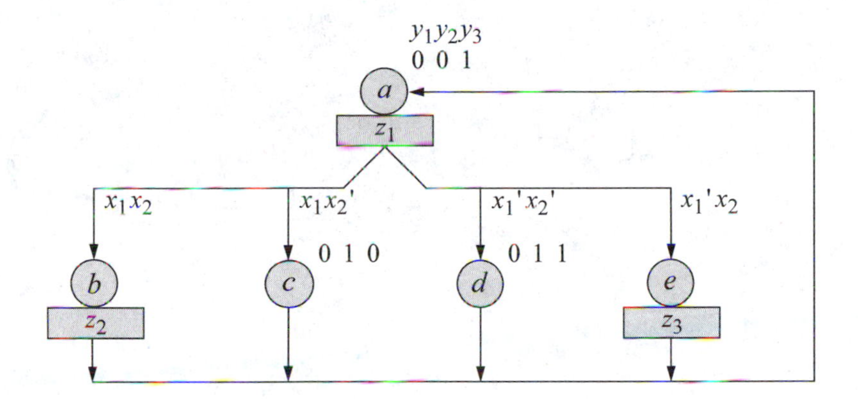

3.34 Assign state codes to the state diagram shown below so that there will be no output glitches. Consider all state transitions.

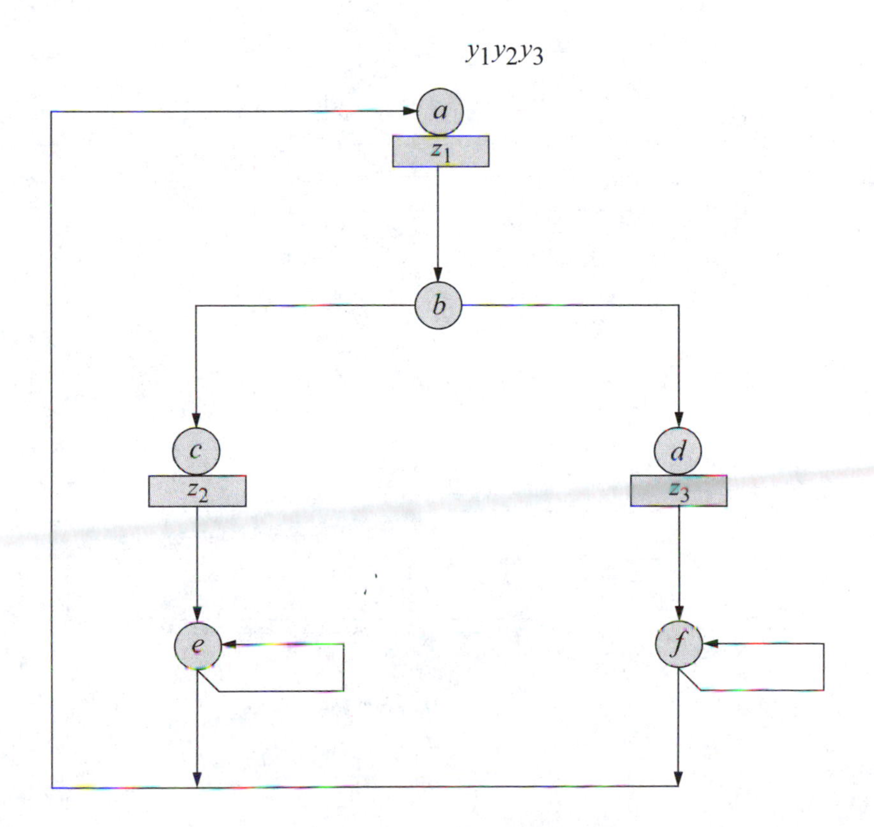

3.35 The λ output logic for z_1 of a synchronous sequential machine is shown below. Indicate the assertion and deassertion for z_1. The state flip-flops are clocked on the positive clock transition. Assume that no glitches will occur. The –Reset line provides an initial reset pulse when the power is turned on.

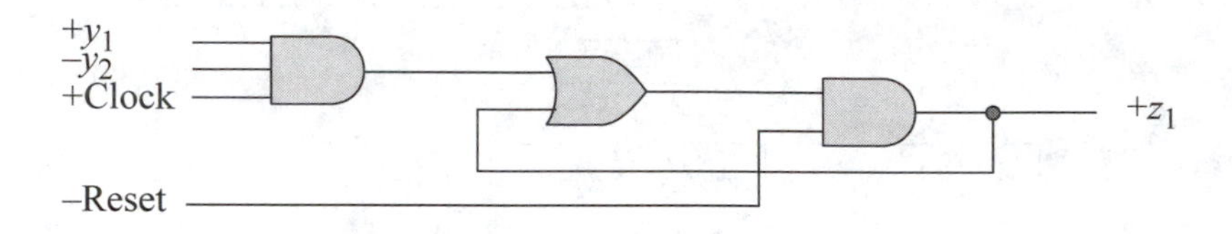

3.36 The λ output logic for output z_1 is shown below for two separate cases. Indicate the correct assertion/deassertion statement for each of the two cases. The state flip-flops are clocked on the positive transition of the clock. The delay circuit is used to delay the clock until the logic has stabilized. The –Reset line resets the logic when the power is turned on.

(a)

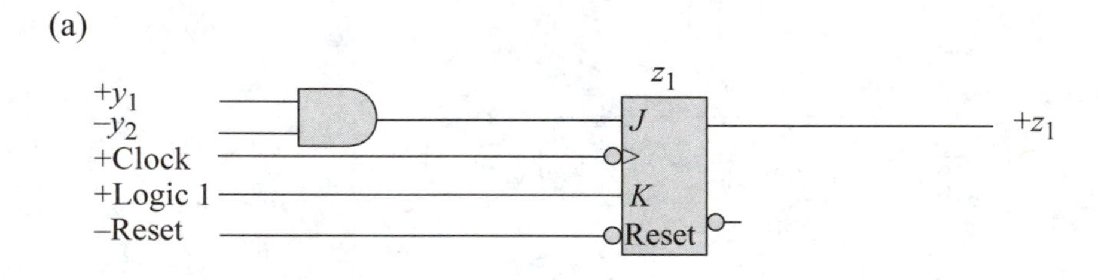

(b)

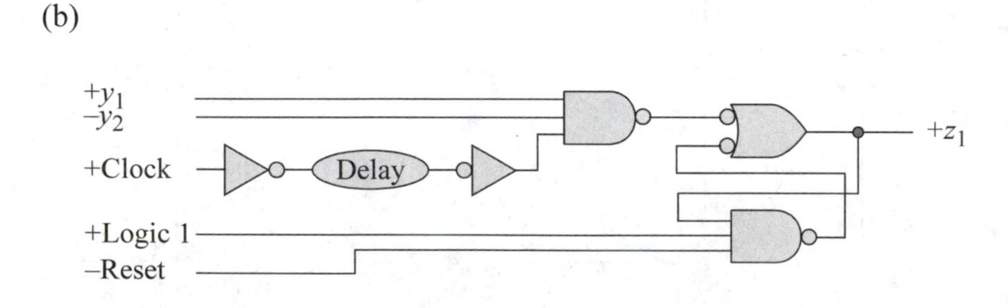

3.37 The λ output logic for z_1 is shown below for four separate cases. Determine the assertion/deassertion statement for z_1 for each of the four cases. The delay circuit is used to delay the clock for the output logic until the machine has stabilized. The state flip-flops y_1, y_2, and y_3 are clocked on the positive clock transition.

(a)

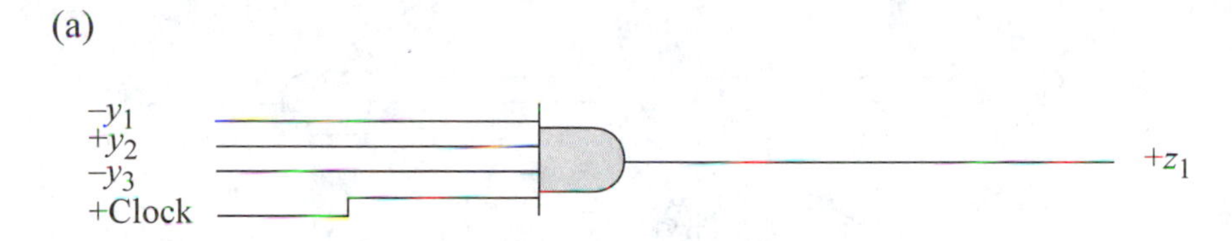

(b)

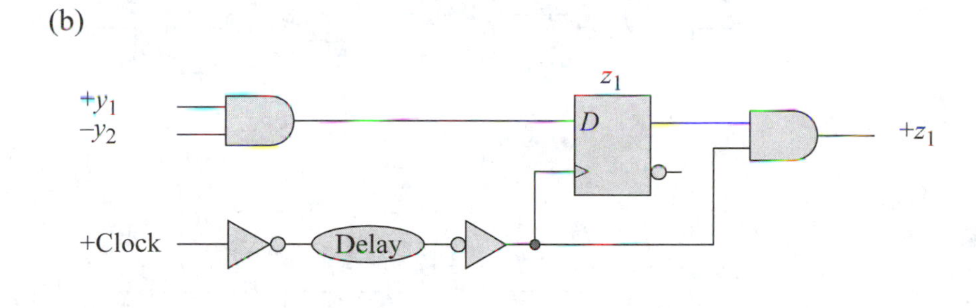

(c)

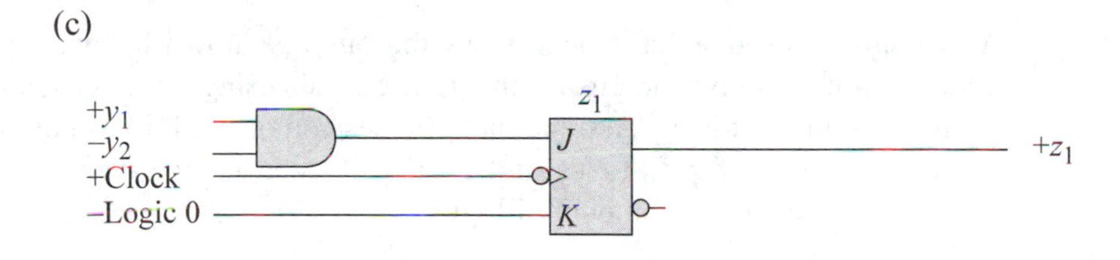

(d)

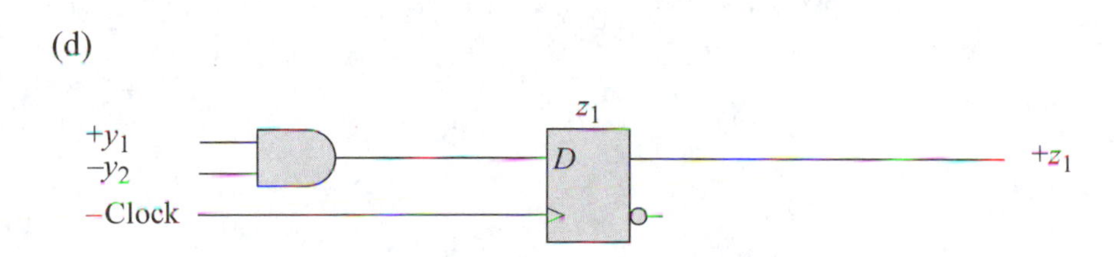

3.38 The λ output logic for z_1 is shown below for three separate cases. Determine the assertion/deassertion statement for z_1 for each of the three cases. The delay circuit is used to delay the clock for the output logic until the machine has stabilized. The state flip-flops y_1 and y_2 are clocked on the positive clock transition.

(a)

$+y_1$
$+y_2$
+Clock
Delay
$+z_1$

(b)

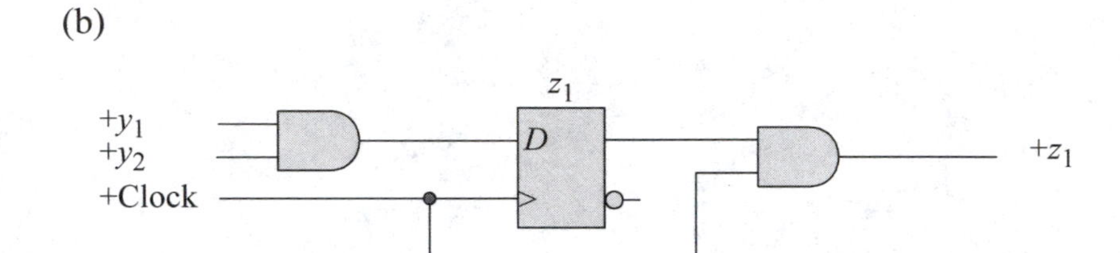

(c)

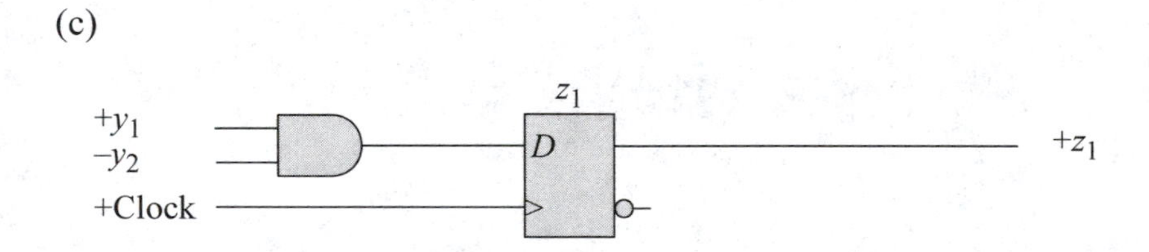

3.39 A synchronous sequential machine has the outputs shown below. Design three separate λ output circuits for the three outputs using the assertion/deassertion statements given. Assume that glitches can occur, but cannot be tolerated. The state flip-flops y_1 and y_2 are clocked on the positive clock transition. Use the least amount of logic.

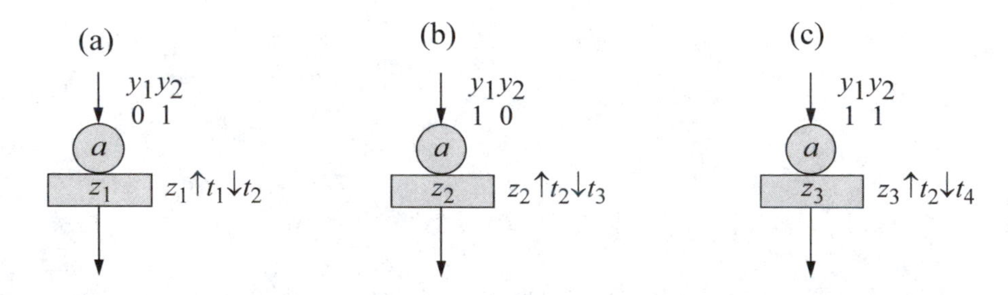

3.40 A partial state diagram for two different synchronous sequential machines with four states each is shown in parts (a) and (b) below. Design the λ output logic for parts (a) and (b) using the assertion/deassertion statements shown. Assume that no glitches will occur. The state flip-flops are clocked on the positive transition of the clock. Use the least amount of logic.

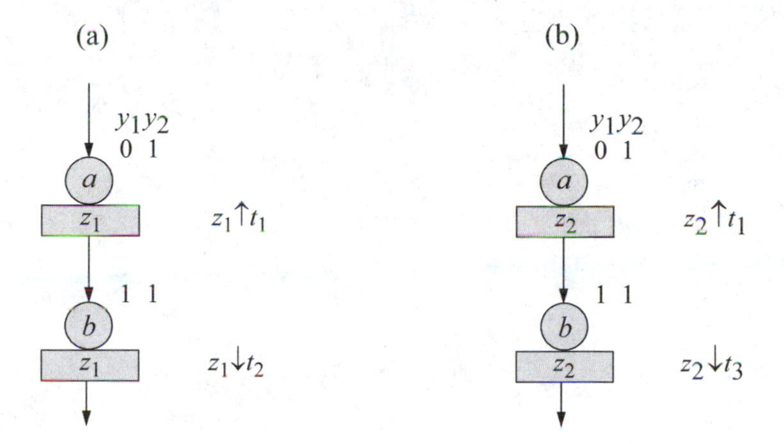

3.41 A partial state diagram for a synchronous sequential machine with four states is shown below. Design the λ output logic using the assertion/deassertion statements shown. Assume that glitches are possible, but cannot be tolerated. The state flip-flops are clocked on the positive clock transition. Use the least amount of logic.

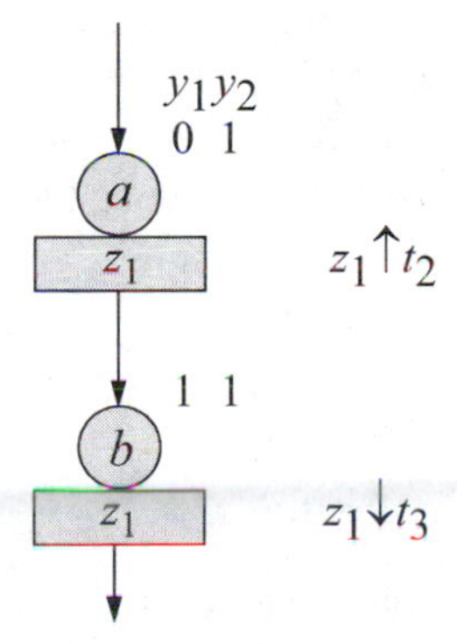

4

4.1 *Multiplexers for δ Next-State Logic*
4.2 *Decoders for λ Output Logic*
4.3 *Programmable Logic Devices*
4.4 *Microprocessor-Controlled Sequential Machines*
4.5 *Sequential Iterative Machines*
4.6 *Error Detection in Synchronous Sequential Machines*
4.7 *Problems*

Synthesis of Synchronous Sequential Machines 2

In the previous chapter, machines were synthesized with logic primitives such as, AND, OR, and NOT for both the δ next-state logic and the λ output logic. In this chapter, input and output functions consisting primarily of logic macros such as, multiplexers and decoders will be introduced. Programmable logic devices (PLDs) will be presented as an alternative implementation to further illustrate the synthesis procedure. Microprocessors, which provide considerable flexibility in the design and operation of synchronous sequential machines, will be discussed and their relative merits expounded by means of a design example. Sequential iterative networks such as, shift registers and modulo-*n* binary counters, will be further investigated and a method established to convert from a combinational iterative network to a functionally equivalent sequential network. The chapter will conclude with a presentation on error detection in synchronous sequential machines.

4.1 Multiplexers for δ Next-State Logic

A data multiplexer is a functional logic device that permits two or more data sources to share a common transmission medium, where each data source retains its own independent channel. A multiplexer is essentially a data selector that operates as an electronic switch by connecting one of n inputs to a single output. A multiplexer has a set of p select input lines defined as $S = s_{p-1}, s_{p-2}, \ldots, s_1, s_0$, where p is the number of storage elements of the machine and s_0 is the low-order select input. There is also a set of 2^p data input lines which are labeled from 0 to $2^p - 1$ and defined as $D = d_0, d_1, d_2, \ldots, d_{2^p-1}$. There is one output z_1, although some multiplexers have two complementary outputs: z_1 and z_1'.

The p select inputs have a range from 0 to $2^p - 1$. Their purpose is to select an input signal from the set of data inputs d_0 through d_{2^p-1}. The bit configuration of the select inputs represents an p-bit binary number which corresponds to the subscript of the selected input signal. Thus, if the p-bit select number has a value of i, then input d_i is transferred to output z_1.

A four-to-one (4:1) multiplexer is shown in Figure 4.1. The four combinations of s_1 and s_0 select, in turn, data inputs d_0, d_1, d_2, and d_3 as shown in Equation 4.1 and transfers the corresponding data input to output z_1. In general, output z_1 of a 2^p:1 multiplexer

$$z_1 = s_1's_0'd_0 + s_1's_0d_1 + s_1s_0'd_2 + s_1s_0d_3 \tag{4.1}$$

can be expressed as follows:

$$z_1 = \sum_{i=0}^{2^p-1} S_i d_i \tag{4.2}$$

where S_i is the ith minterm of the selection variables $s_{p-1}, s_{p-2}, \ldots, s_1, s_0$. Thus, data input d_0 is selected if the select inputs correspond to minterm $s_1s_0 = 00$. Inputs d_1, d_2, and d_3 are selected for minterms $s_1s_0 = 01$, 10, and 11, respectively. There is a one-to-one correspondence, therefore, between a minterm and a multiplexer data input signal. A p-variable Karnaugh map graphically depicts a set of 2^p minterms, ranging in value from 0 to $2^p - 1$. It is this rationale that allows an input map for a synchronous sequential machine to be implemented with multiplexers.

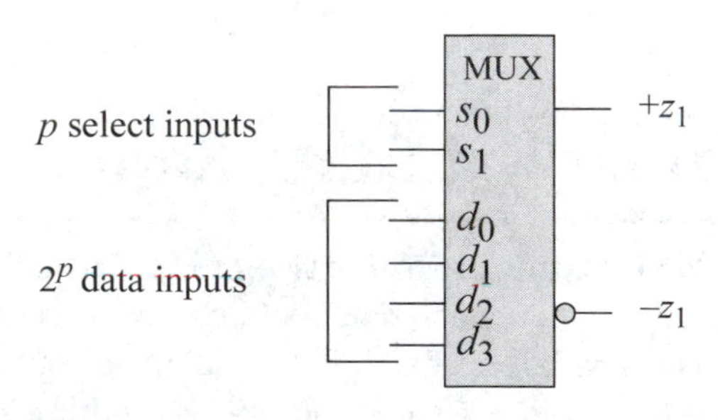

Figure 4.1 Four-to-one multiplexer.

When the δ next-state logic is implemented with multiplexers rather than with logic gates, the number of integrated circuits is minimized — if noncustom technology is used — and the design time is reduced. Also, since the multiplexer design is preestablished and cannot be altered, there is no need to maintain state code adjacency in order to minimize the δ next-state logic. The assignment of state codes is then influenced solely by the λ output logic, adding considerable flexibility in state code assignment. State codes must be chosen, however, so that there are no output glitches as the machine proceeds through all possible state transitions.

4.1.1 Linear-Select Multiplexers

A general block diagram for a synchronous sequential machine using multiplexers for the δ next-state logic is shown in Figure 4.2. The combinational logic which connects to the input of the multiplexer array is either very elementary or nonexistent. This will become evident in a synthesis example.

In this method, one multiplexer is needed for each state flip-flop and each multiplexer has p select inputs, where p represents the number of storage elements. Therefore, this technique requires $p(2^p{:}1)$ multiplexers. A machine with 12 states, requiring four storage elements, would have four 16:1 multiplexers. Since most multiplexers have a single output, a design of this type is most easily implemented with D flip-flops.

The active-high output of each state flip-flop is connected to a corresponding select input line; that is, $y_1, y_2, \ldots, y_p$ connect to $s_{p-1}, \ldots, s_1, s_0$, respectively, where y_p is the low-order flip-flop and s_0 is the low-order select input. Thus, if $p = 3$, then the following state flip-flop-to-select-input connections are necessary: y_3 connects to s_0, y_2 connects to s_1, and y_1 connects to s_2. Since the flip-flop outputs connect to the multiplexer select inputs in a one-to-one mapping, this type of connection can be referred to as *linear selection*. The examples which follow illustrate the principles of synthesis using multiplexers for the δ next-state logic.

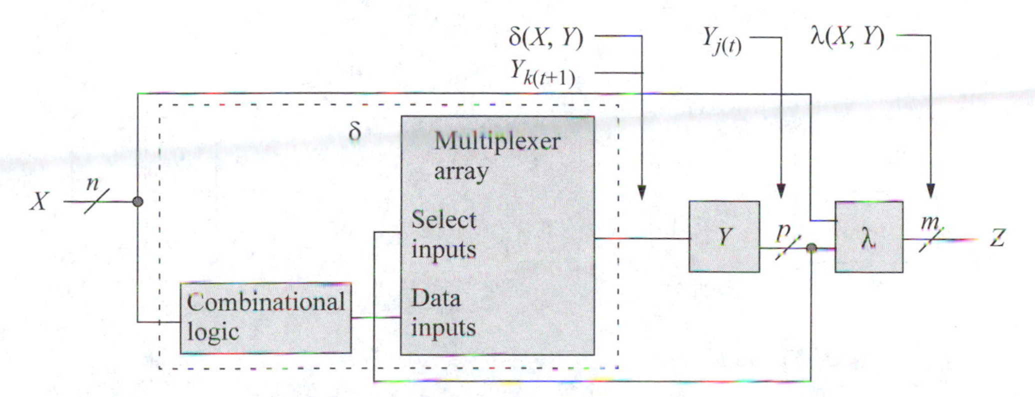

Figure 4.2 General block diagram of a synchronous sequential machine using multiplexers for the δ next-state logic.

Example 4.1 Using multiplexers for the δ next-state logic, a Moore machine will be synthesized that operates according to the state diagram of Figure 4.3. State codes are assigned to eliminate output glitches. Since the input logic is implemented with multiplexers, *D* flip-flops will be used as the storage elements. The input and output maps are shown in Figure 4.4 (a) and (b), respectively, using x_1 as a map-entered variable. Input equations are not necessary, because all possible minterms containing two variables are available at the output of the multiplexers.

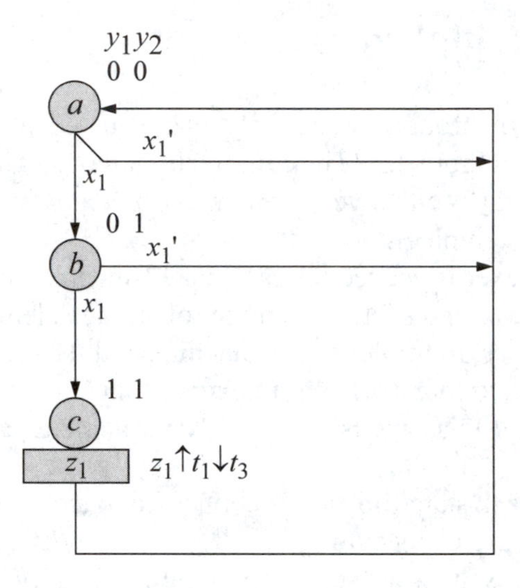

Figure 4.3 Moore machine to assert output z_1 if an input sequence of $x_1 = 11$ is detected. There is one unused state: $y_1y_2 = 10$.

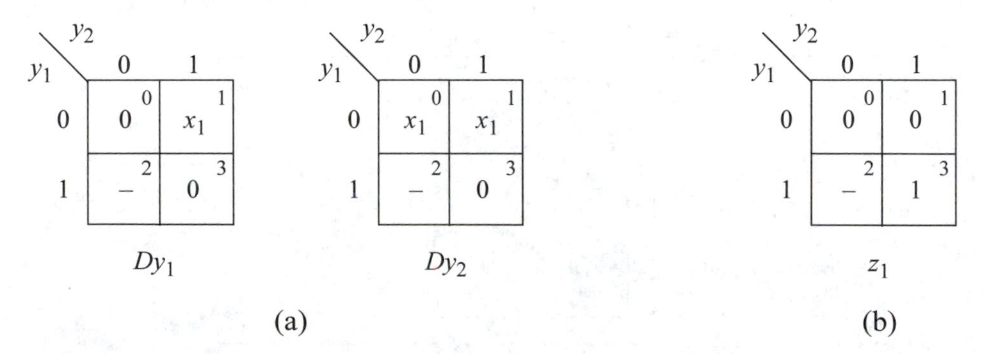

Figure 4.4 Input and output maps for the Moore machine of Figure 4.3.

Since there is a one-to-one correspondence between the minterms of a Karnaugh map and the data inputs of a multiplexer, designing the input logic is relatively straightforward: Simply assign the values of the minterms in the input map to the corresponding multiplexer data inputs with the same subscript. The logic diagram is shown in Figure 4.5. Flip-flop y_2 is the low-order flip-flop; therefore, it connects to the low-order select input s_0. Assigning a value of logic 0 to the unused state $y_1y_2 = 10$ in the input map for Dy_1, allows MUX y_1 inputs d_0, d_2, and d_3 to be connected to a logic 0. Since minterm 1 assumes the value of x_1, d_1 is connected to input x_1. Similarly, MUX y_2 inputs d_0 and d_1 connect to input x_1; d_2 and d_3 both connect to a logic 0.

Since there are only two distinct entries in the map for Dy_1 and Dy_2 — assuming a 0 value for minterm $y_1y_2 = 10$ — the utilization of a 4:1 multiplexer is inefficient for this example. Both multiplexers could be replaced by two 2:1 multiplexers. A technique to accomplish this will be presented in a later section.

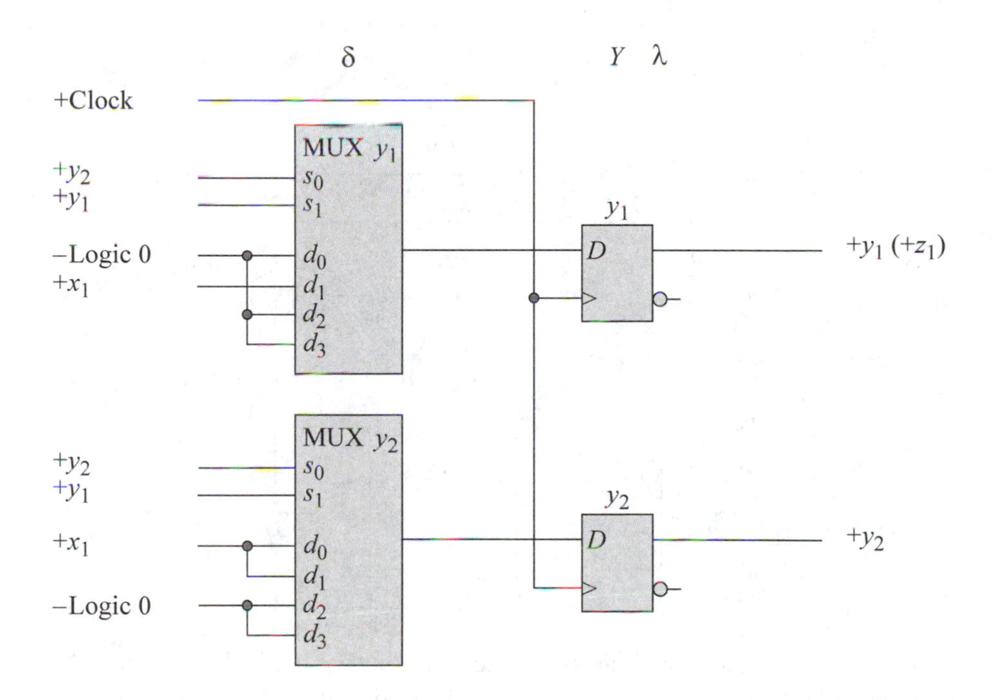

Figure 4.5 Implementation of the Moore machine of Figure 4.3 using multiplexers for the δ next-state logic.

Example 4.2 A Mealy machine will be designed using multiplexers for the δ next-state logic that operates according to the state diagram of Figure 4.6. The machine receives 3-bit words on a serial input line x_1, in the format shown below, and generates two outputs z_1 and z_2 during the third bit cell for specific bit sequences.

$$x_1 = \cdots \mid b_1 b_2 b_3 \mid b_1 b_2 b_3 \mid b_1 b_2 b_3 \mid \cdots$$

$z_1/z_2 \uparrow t_2 \downarrow t_4$

Both outputs are asserted at time t_2 and deasserted at t_4 when their respective input sequences have been detected as follows:

If $x_1 = 000$, then $z_1 \uparrow t_2 \downarrow t_4$
If $x_1 = 111$, then $z_2 \uparrow t_2 \downarrow t_4$

There is no bit space between words.

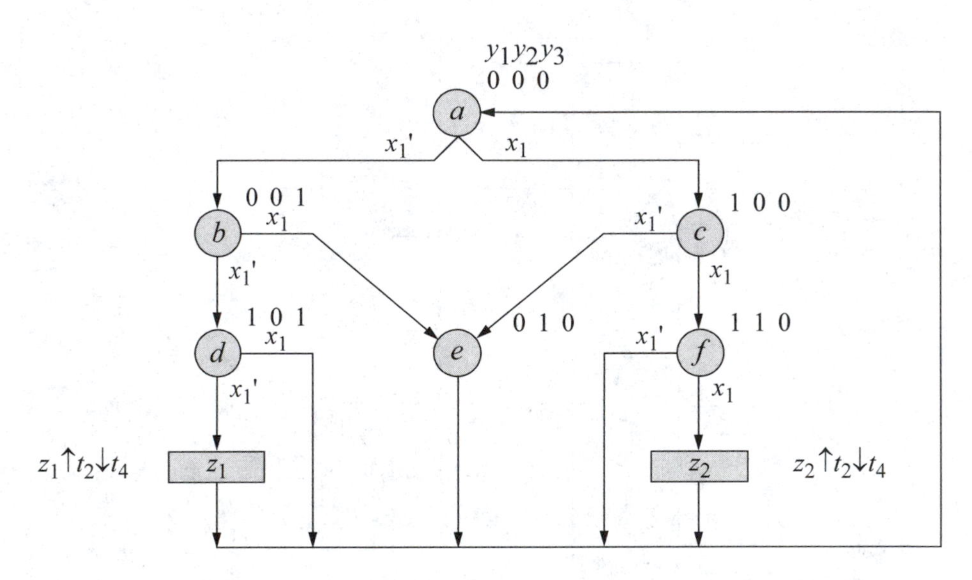

Figure 4.6 Mealy machine that asserts output z_1 if a 3-bit word $x_1 = 000$ is detected and asserts output z_2 if $x_1 = 111$. Unused states are $y_1y_2y_3 = 011$ and 111.

State code adjacency is not a prerequisite for input logic minimization, because the logic is implemented with multiplexers, which are unalterable logic macros. If discrete logic gates were utilized, then the following state pairs should be adjacent according to Rules 1 and 2 of Section 3.4: (*b*,*c*), (*d*,*e*), and (*e*,*f*). State codes were selected so that there would be no glitches on output z_1 or z_2 for all possible state transitions, considering the value of x_1. Using the synthesis procedure of Section 3.1, the next-state table and input maps are obtained, as shown in Table 4.1 and Figure 4.7, respectively.

There are three state flip-flops; therefore, $p = 3$ and $p(2^p{:}1) = 3(8{:}1)$ multiplexers are required for the input logic. Flip-flop y_3 connects to select input s_0, because both are low-order variables. Flip-flops y_2 and y_1 connect to select inputs s_1 and s_2, respectively, as shown in the logic diagram of Figure 4.8. The unused states $y_1y_2y_3 =$ 011 and 111 in each input map can be assigned a value of logic 0, although any other value would suffice, since the machine will never enter an unused state under normal operating conditions.

Table 4.1 Next-state table for the Mealy machine of Figure 4.6 using *D* flip-flops

Present state $y_1\ y_2\ y_3$	Input x_1	Next state $y_1\ y_2\ y_3$	Flip-flop inputs $Dy_1\ Dy_2\ Dy_3$	Present outputs $z_1\ z_2$
0 0 0	0	0 0 1	0 0 1	0 0
0 0 0	1	1 0 0	1 0 0	0 0
0 0 1	0	1 0 1	1 0 1	0 0
0 0 1	1	0 1 0	0 1 0	0 0
0 1 0	0	0 0 0	0 0 0	0 0
0 1 0	1	0 0 0	0 0 0	0 0
0 1 1	0	– – –	– – –	– –
0 1 1	1	– – –	– – –	– –
1 0 0	0	0 1 0	0 1 0	0 0
1 0 0	1	1 1 0	1 1 0	0 0
1 0 1	0	0 0 0	0 0 0	1 0
1 0 1	1	0 0 0	0 0 0	0 0
1 1 0	0	0 0 0	0 0 0	0 0
1 1 0	1	0 0 0	0 0 0	0 1
1 1 1	0	– – –	– – –	– –
1 1 1	1	– – –	– – –	– –

Since the input logic is implemented with multiplexers, there is essentially no logic synthesis — simply connect the data inputs to the appropriate signal, or value, as specified by the input maps for each flip-flop. Thus, for MUX y_1, data inputs d_0 and d_4 connect to input x_1 as indicated by minterms 0 and 4 of the input map for Dy_1.

Likewise, data input d_1 connects to the complemented input x_1' as specified by minterm location 1. All remaining data inputs for MUX y_1 connect to a logic 0 voltage level. Similarly, the data inputs for MUX y_2 and MUX y_3 connect to the appropriate signal or voltage level as specified by the minterm locations in the input maps for Dy_2 and Dy_3, respectively.

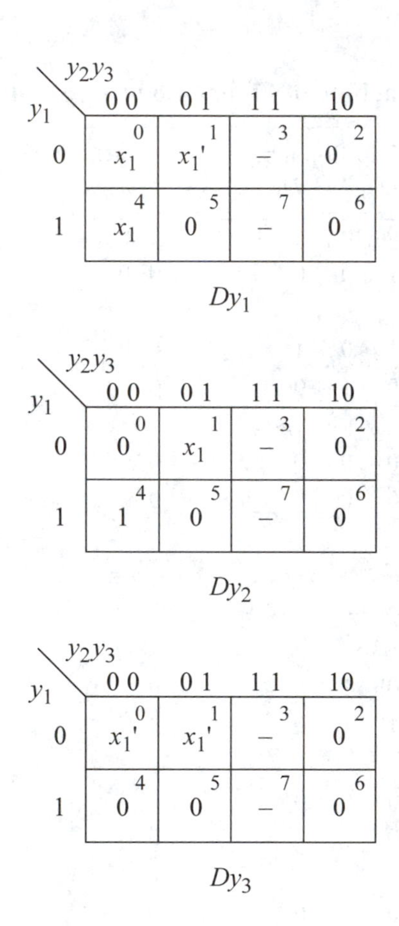

Figure 4.7 Input maps for the Mealy machine of Figure 4.6 using input x_1 as a map-entered variable.

Outputs z_1 and z_2 require assertion for one clock period, but delayed by one-half a clock period; that is, $z_1 \uparrow t_2 \downarrow t_4$ and $z_2 \uparrow t_2 \downarrow t_4$. The output maps, obtained from either the state diagram or the next-state table, are shown in Figure 4.9. By meticulous analysis of the state diagram, it can be determined that no state transition will cause the machine to pass through unused state $y_1y_2y_3 = 111$. Therefore, $y_1y_2y_3 = 111$ can be used to minimize the output equation for both z_1 and z_2, resulting in the equations of Equation 4.3.

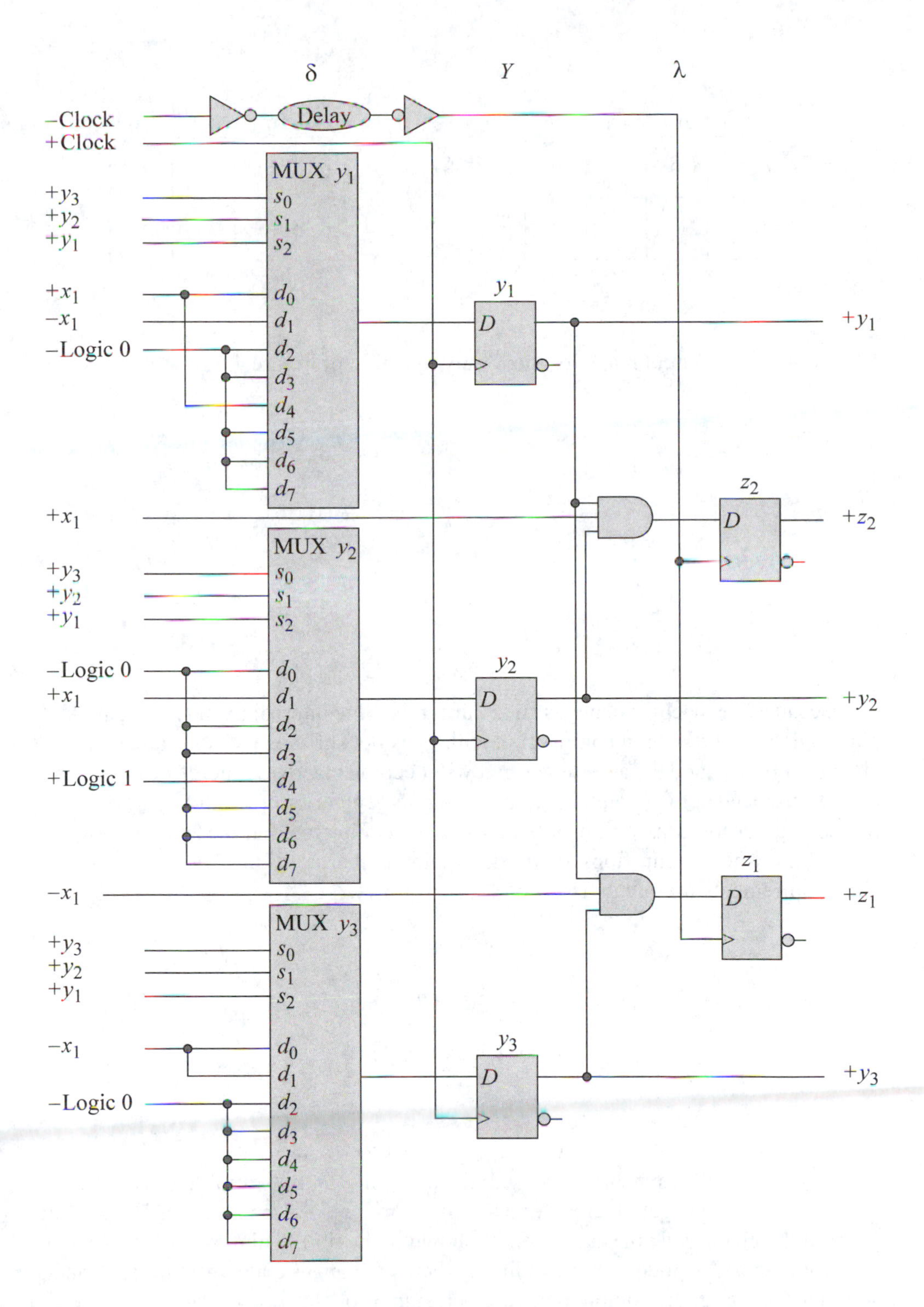

Figure 4.8 Logic diagram for the Mealy machine of Figure 4.6 using multiplexers for the δ next-state logic.

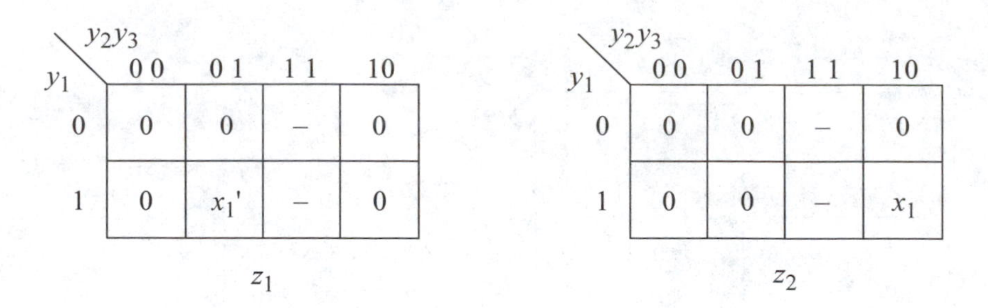

Figure 4.9 Output maps for the Mealy machine of Figure 4.6.

$$z_1 = y_1y_3x_1'$$

$$z_2 = y_1y_2x_1 \tag{4.3}$$

Because the machine changes state during the time interval from t_2 to t_4, an AND gate to decode the state in conjunction with x_1 is not sufficient — the state code will change at time t_3 and the asserted output will become inactive. Therefore, storage elements are needed to guarantee that both outputs remain asserted and conform to the requisite assertion/deassertion statements. This is illustrated in the λ output logic of Figure 4.8, where *D* flip-flops are used to maintain the assertion of outputs z_1 and z_2. The requirements for output assertion are as follows:

$$z_1 = y_1y_3x_1' \bullet (\text{Clock delayed})' \uparrow$$

$$z_2 = y_1y_2x_1 \bullet (\text{Clock delayed})' \uparrow$$

Since input x_1 is synchronized with the machine clock — implying that changes to x_1 occur on the negative clock transition — input x_1 may change state at the mid-point of the clock cycle in state *d* or *f*, as shown in the timing diagram of Figure 4.10. To avoid a possible metastable condition when x_1 changes state, a small delay is inserted at time t_2 in the complemented clock signal for flip-flops z_1 and z_2, since the complemented clock triggers the output flip-flops. The delay allows the next value of x_1 to be stable at inputs Dz_1 and Dz_2 before the output flip-flops are clocked by the positive transition of the –Clock delayed signal. This assures that only a valid input sequence will be detected. The rising edge of the –Clock delayed signal still occurs at time t_2, meeting the assertion requirements for outputs z_1 and z_2.

In this example, as in the previous example, a 2^p:1 multiplexer is extremely inefficient for the synthesis of the δ next-state logic. The logic for flip-flop inputs Dy_1 and Dy_2 could be generated by two 4:1 multiplexers, while input Dy_3 could be obtained from a 2:1 multiplexer. This deviation from multiplexer linear selection will be discussed in detail in a later section.

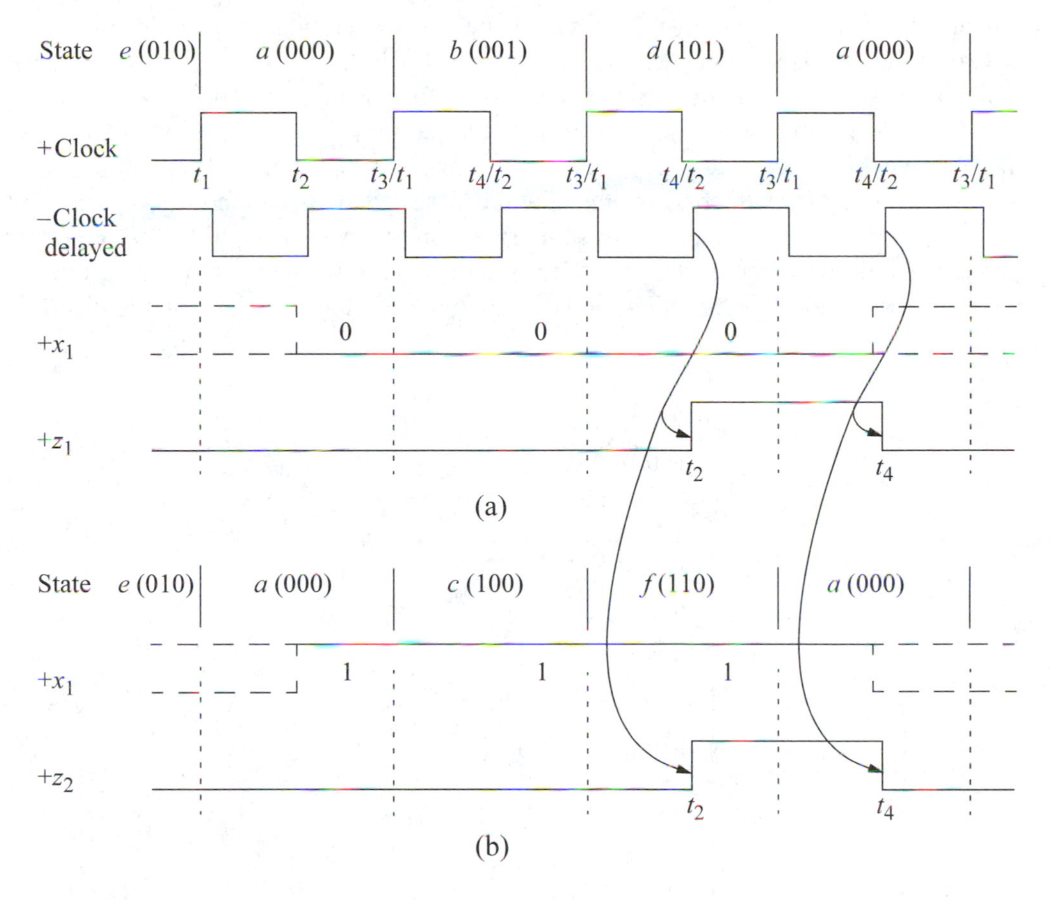

Figure 4.10 Timing diagram for the Mealy machine of Figure 4.6: (a) output $z_1 \uparrow t_2 \downarrow t_4$ for an input sequence of $x_1 = 000$; and (b) output $z_2 \uparrow t_2 \downarrow t_4$ for an input sequence of $x_1 = 111$.

Example 4.3 The examples are intended to reinforce existing design methodologies and to introduce new concepts. In the previous two examples, the δ next-state logic consisted only of multiplexers. In this example, discrete logic primitives together with multiplexers constitute the next-state function. A Moore synchronous sequential machine is to be designed that operates according to the following sequence diagram:

1. $a\,(x_1 \oplus x_2)' \rightarrow b\,(x_1'x_2) \rightarrow c\,(z_1{\uparrow}t_1{\downarrow}t_3)$
2. $a\,(x_1 \oplus x_2)' \rightarrow b\,(x_1 x_2') \rightarrow e\,(z_2{\uparrow}t_1{\downarrow}t_3)$

There are two sequences in the machine specifications: one to assert output z_1, and one to assert output z_2. Both sequences begin in state *a* and proceed to state *b* if inputs x_1 and x_2 are the same value — either both 0s or both 1s. In sequence 1, the machine proceeds to state *b* if the conditions are met in state *a*. In state *b*, if $x_1x_2 = 01$, then the machine moves to state *c* and asserts output z_1 for one clock period. In sequence 2, if the conditions are met in state *a*, the transition is again to state *b*; however, a valid input sequence in state *b* is now $x_1x_2 = 10$. Under these conditions, the machine proceeds to state *e* and generates output z_2 for one clock period.

Figure 4.11 graphically depicts the state diagram for a Moore machine that realizes the machine specifications. Because the machine represents a Moore model and outputs z_1 and z_2 are asserted at time t_1 and deasserted at t_3, state code assignments must be carefully chosen so that no state transition will produce a glitch on the outputs. If the outputs were asserted at time t_2, then any arbitrary state code assignment would suffice, since the assertion time occurs long after the machine has stabilized.

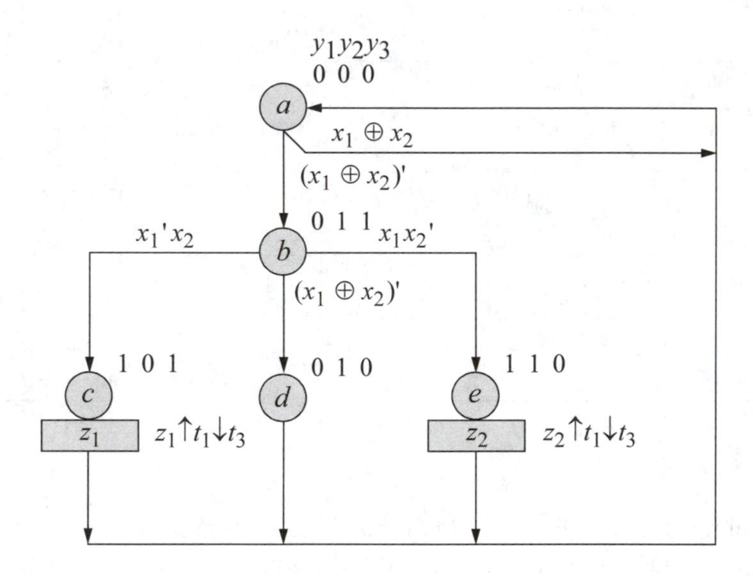

Figure 4.11 State diagram for a Moore machine that confirms to the machine specifications of Example 4.3. Unused states are: $y_1y_2y_3 = 001$, 100, and 111.

The present-state map provides a useful mechanism to determine state codes in which certain transient codes must be avoided. Using the state code adjacency rules and a trial-and-error technique, a present-state map can be generated which presents a clear visualization of state transition sequences. For example, the present-state map of Figure 4.12 depicts one possible state code assignment. It can be verified — directly from the map — that no state transition will pass through states c ($y_1y_2y_3$ = 101) or e ($y_1y_2y_3$ = 110) as transient states.

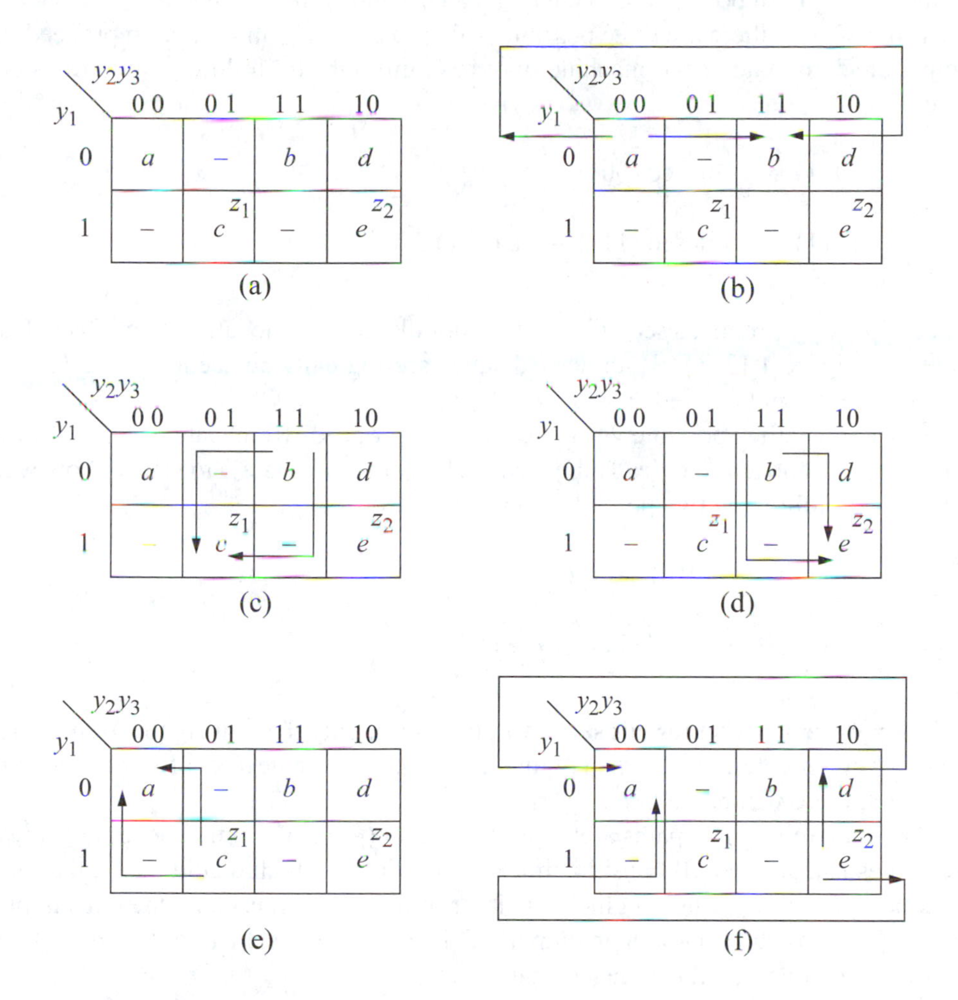

Figure 4.12 Present-state map for the Moore machine of Figure 4.11: (a) completed map showing state code assignments; (b) possible transitions for sequence (a,b); (c) possible transitions for sequence (b,c); (d) possible transitions for sequence (b,e); (e) possible transitions for sequence (c,a); and (f) possible transitions for sequence (e,a).

Refer to the state diagram of Figure 4.11 and the present-state maps of Figure 4.12 (b) through Figure 4.12 (f) for the discussion which follows. From state a, there are two possible transitions: (a,a) or (a,b). In proceeding from state a to state b, the machine may pass through two transient states as follows:

a (000) → Unused (001) → b (011)

a (000) → d (010) → b (011)

Since flip-flop y_1 remains reset, transitions occur only in the top row ($y_1 = 0$) of the present-state map. The machine may pass through either of two logically adjacent states, $y_1y_2y_3 = 001$ or 010, as it moves from state a to state b, as shown in Figure 4.12 (b). In both cases, the output states c and e are not entered.

From state b, there are three possible paths: (b,c), (b,d), and (b,e). In proceeding from state b to state c, the machine may pass through the following two transient states:

$$b\,(011) \rightarrow \text{Unused}\,(001) \rightarrow c\,(101)$$

$$b\,(011) \rightarrow \text{Unused}\,(111) \rightarrow c\,(101)$$

Since flip-flop y_3 remains set, only the four middle minterm locations are affected, as shown in Figure 4.12 (c). Both unused states are logically adjacent to state b. The output state e is not entered.

Only one path is possible when the machine proceeds from state b to state d, because states b and d are logically adjacent. The path from state b to state e, however, contains the following two possible transient states:

$$b\,(011) \rightarrow d\,(010) \rightarrow e\,(110)$$

$$b\,(011) \rightarrow \text{Unused}\,(111) \rightarrow e\,(110)$$

Flip-flop y_2 remains set for this sequence; therefore, only the four rightmost minterm locations are affected, as shown in Figure 4.12 (d). Since state d is not associated with an output, there will be no glitch on z_1.

The remaining three paths are: (c,a), (d,a), and (e,a). The only concern for these sequences is that the (c,a) transition must not produce a glitch on output z_2, the (d,a) transition must not produce a glitch on either of the two outputs, and the (e,a) transition must not produce a glitch on output z_1. The path from state c to state a contains the following two possible transient states:

$$c\,(101) \rightarrow \text{Unused}\,(001) \rightarrow a\,(000)$$

$$c\,(101) \rightarrow \text{Unused}\,(100) \rightarrow a\,(000)$$

Flip-flop y_2 remains reset for this transition; therefore, the leftmost four minterm locations are the only squares affected, as shown in Figure 4.12 (e). Since states d and a are logically adjacent, output glitches are not possible.

The final state transition sequence from state e to state a may pass through the following two transient states:

$$e\,(110) \rightarrow d\,(010) \rightarrow a\,(000)$$

$$e\,(110) \rightarrow \text{Unused}\,(100) \rightarrow a\,(000)$$

Flip-flop y_3 remains reset; therefore, only the two leftmost and the two rightmost minterm locations are affected, as shown in Figure 4.12 (f). In all of the above sequences where two flip-flops change state, a third transition is possible when the flip-flops change state simultaneously. This obviously presents no hazard for the two outputs.

The input maps, obtained from the state diagram, are presented in Figure 4.13. The minterm entries in the input map for Dy_i specify either a 0 value if every next state for flip-flop y_i is 0, a 1 value if every next state for y_i is 1, or an expression which indicates the path(s) that must be taken in order for y_i to have a next-state value of 1. For example, refer to the state diagram of Figure 4.11 and the input map for Dy_1. In state a ($y_1y_2y_3 = 000$), the next value for y_1 is 0, regardless of the path taken. Therefore, a 0 is entered in minterm location 0. In state b ($y_1y_2y_3 = 011$), there are three possible paths, only two of which take y_1 to a next value of 1, as indicated by the following state transitions:

1. $b\,(011) \rightarrow c\,(101)$, if $x_1x_2 = 01$

2. $b\,(011) \rightarrow d\,(010)$, if $x_1x_2 = 00$ or 11

3. $b\,(011) \rightarrow e\,(110)$, if $x_1x_2 = 10$

In transition 1, the next value for flip-flop y_1 is 1 if the expression $x_1'x_2$ is true; therefore, $x_1'x_2$ is entered in minterm location 3. In transition 2, the next value for y_1 is 0; therefore, the expression $(x_1 \oplus x_2)'$ is ignored, since the map entries indicate only those requirements that generate a next value of 1. In transition 3, y_1 has a next value of 1 if the expression x_1x_2' is true; therefore, x_1x_2' is entered in minterm location 3. Since flip-flop y_1 assumes a next value of 1 if either $x_1'x_2$ or x_1x_2' is true, the exclusive-OR of the two expressions is entered in minterm location 3. In the same manner, the remaining entries for Dy_1 are obtained. Similarly, input maps Dy_2 and Dy_3 are completed.

When boolean expressions are entered in the input maps, the expressions should be minimized, if possible. For example, in state b ($y_1y_2y_3 = 011$) of the map for Dy_2, flip-flop y_2 has a next value of 1 for the two transitions (b, d) and (b, e), which require the expression $(x_1 \oplus x_2)' + x_1x_2'$ to be entered in minterm location 3 of the map for Dy_2. The expression can be minimized, however, as follows:

$$\begin{aligned} & (x_1 \oplus x_2)' + x_1x_2' & \\ &= x_1x_2 + x_1'x_2' + x_1x_2' & \text{DeMorgan's theorem} \\ &= x_1 + x_1'x_2' & \text{Distributive axiom} \\ &= x_1 + x_2' & \text{Absorption theorem} \end{aligned}$$

Therefore, the expression $x_1 + x_2'$ is entered in minterm location 3 of the map for Dy_2. Input equations are not necessary, because multiplexers are the logic macros for the δ next-state logic.

$y_1 \backslash y_2y_3$	0 0	0 1	1 1	1 0
0	0 (0)	– (1)	$x_1'x_2 + x_1x_2' = x_1 \oplus x_2$ (3)	0 (2)
1	– (4)	0 (5)	– (7)	0 (6)

Dy_1

$y_1 \backslash y_2y_3$	0 0	0 1	1 1	1 0
0	$(x_1 \oplus x_2)'$ (0)	– (1)	$x_1 + x_2'$ (3)	0 (2)
1	– (4)	0 (5)	– (7)	0 (6)

Dy_2

$y_1 \backslash y_2y_3$	0 0	0 1	1 1	1 0
0	$(x_1 \oplus x_2)'$ (0)	– (1)	$x_1'x_2$ (3)	0 (2)
1	– (4)	0 (5)	– (7)	0 (6)

Dy_3

Figure 4.13 Input maps for the Moore machine of Figure 4.11

The output maps are shown in Figure 4.14. None of the unused states can be used to minimize the equation for output z1, because they may produce a glitch on z_1 for certain transitions. Minterm location $y_1y_2y_3 = 001$ is a possible transient state for the sequence (a,b) which does not include z_1; minterm location $y_1y_2y_3 = 100$ is a possible transient state for the sequence (e,a) which does not include z_1; and minterm location

$y_1y_2y_3 = 111$ may produce a glitch on output z_1 for the sequence (b,e) which also does not include z_1. None of the applicable unused states can be used to minimize the equation for z_2, because of possible transient states for sequences (c,a) and (b,c).

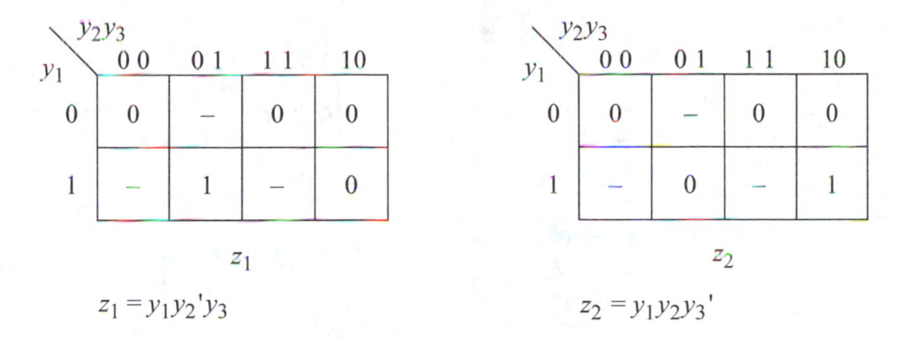

Figure 4.14 Output maps for the Moore machine of Figure 4.11.

The logic diagram of Figure 4.15 is synthesized directly from the input maps, in which minterm locations 0 through 7 correspond to the multiplexer data inputs d_0 through d_7, respectively. The logic that is necessary for data input d_3 of MUX y_1 is obtained from the expression in minterm 3 of the input map for Dy_1. This is the sum-of-products form for the exclusive-OR function. The logic for data input d_3 of MUX y_2 is obtained from the expression $(x_1 \oplus x_2)' + x_1x_2'$, which reduces to $x_1 + x_2'$. The logic primitives connected to the multiplexer data inputs are represented by the combinational logic block in Figure 4.2.

4.1.2 Nonlinear-Select Multiplexers

In the previous section, the δ next-state logic was implemented with linear-select multiplexers. Although the machines functioned correctly according to the specifications, the designs illustrated an inefficient use of the 2^p:1 multiplexers. Smaller multiplexers with fewer data inputs could be effectively utilized with a corresponding reduction in machine cost.

If the number of unique entries in an input map for flip-flop y_i satisfies the expression of Equation 4.4, where u is the number of unique entries and p is the number of storage elements, then at most a $(2^p \div 2)$:1 multiplexer will satisfy the requirements of Dy_i.

$$1 < u \geq (2^p \div 2) \tag{4.4}$$

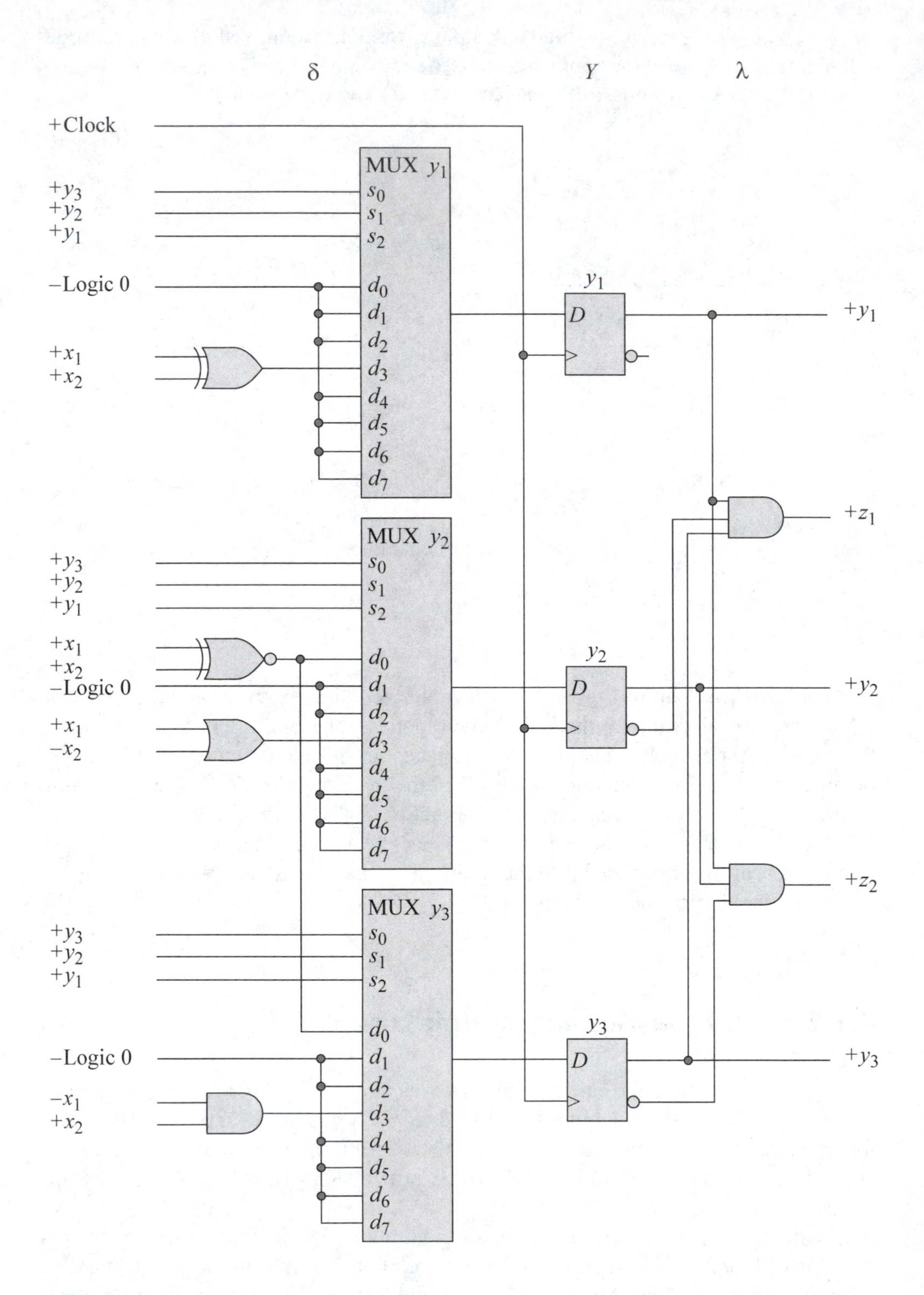

Figure 4.15 Logic diagram for the Moore machine of Figure 4.11 using combinational logic and a multiplexer array for the δ next-state logic.

If, however, $u > 2^p \div 2$, then a 2^p:1 multiplexer is necessary. The largest multiplexer with which to economically implement the input logic is a 16:1 multiplexer, and then only if the number of distinct entries in the input map warrants a multiplexer of this size. Other techniques, such as a PLD implementation, would make more efficient use of current technology.

A general block diagram for a synchronous sequential machine using nonlinear-select multiplexers for the δ next-state logic is shown in Figure 4.16. In this method, one multiplexer is still required for each state flip-flop, but the number of multiplexer select inputs p' is less than p, such that, $1 \leq p' < p$.

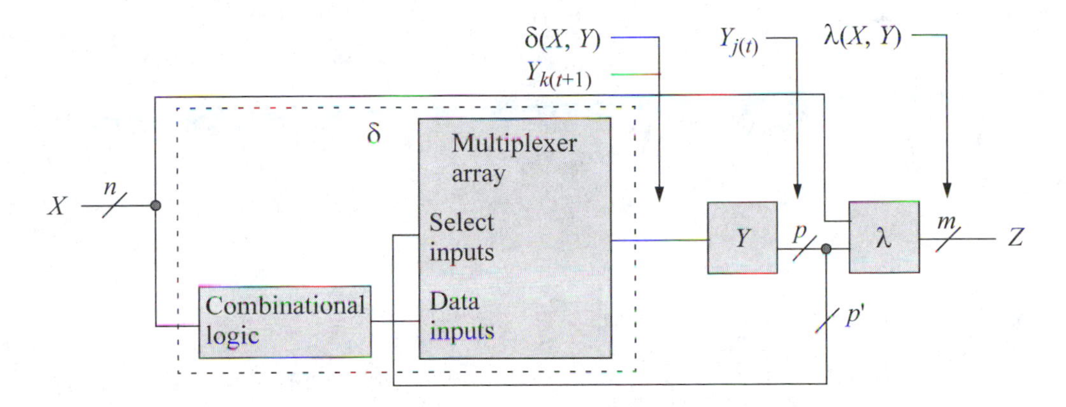

Figure 4.16 General block diagram for a synchronous sequential machine using nonlinear-select multiplexers for the δ next-state logic. The number of multiplexer select inputs p' is less than the number of state flip-flops p.

If a multiplexer has unused data inputs — corresponding to unused states in the input map — then these unused inputs can be connected to logically adjacent multiplexer inputs. The resulting linked set of inputs can be addressed by a common select variable. Thus, in a 4:1 multiplexer, if data input $d_2 = 1$ and d_3 = "don't care," then d_2 and d_3 can both be connected to a logic 1. The two inputs can now be selected by $s_1 s_0 = 10$ or 11; that is, $s_1 s_0 = 1–$. Also, multiplexers containing the same number of data inputs should be addressed by the same select input variables, if possible. This permits the utilization of noncustom technology, where multiplexers in the same integrated circuit share common select inputs.

Example 4.4 Given the input map for Dy_1 in Figure 4.17 (a), the δ next-state logic for flip-flop y_1 will be obtained using a nonlinear-select multiplexer. If linear selection is used, then an 8:1 multiplexer is required. The input map contains only three distinct entries, however: 1, 0, and x_1 — the "don't care" entry can be set to any value. Therefore, a 4:1 multiplexer is sufficient for this implementation. The minterm values and their corresponding entries are listed in Table 4.2. The table is derived from either the next-state table, where applicable, or the input map. The present-state columns correspond to minterm numbers 0 through 7. The flip-flop input column Dy_1 represents the multiplexer output. Since a 4:1 multiplexer is required, two select inputs are necessary, either y_1y_2, y_1y_3, or y_2y_3, where the low-order flip-flop y_2 or y_3 connects to the low-order multiplexer select input s_0.

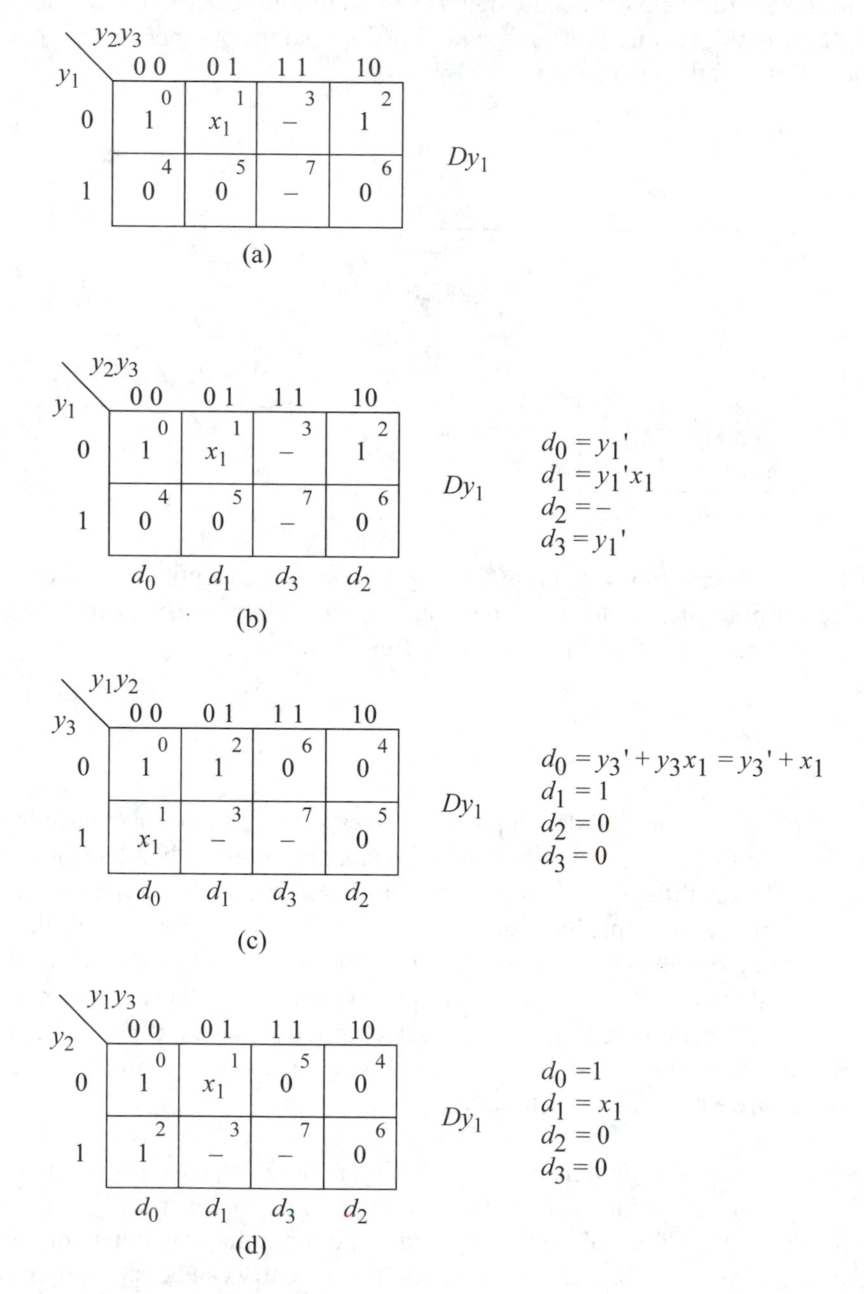

Figure 4.17 Input maps for Example 4.4: (a) for use with Table 4.2; (b) using y_2y_3 as select inputs; (c) using y_1y_2 as select inputs; and (d) using y_1y_3 as select inputs.

Table 4.2 Tabular representation of the input map for flip-flop y_1 of Example 4.4

Minterm	Present state $y_1y_2y_3$	Flip-flop y_1 input (Multiplexer y_1 output) Dy_1
0	0 0 0	1
1	0 0 1	x_1
2	0 1 0	1
3	0 1 1	– (x_1)
4	1 0 0	0
5	1 0 1	0
6	1 1 0	0
7	1 1 1	– (0)

Examining Table 4.2, it is apparent that the multiplexer output for Dy_1 will be a logic 1 when $y_1y_3 = 00$ for minterms 0 and 2. An attempt should be made, therefore, to use only y_1 and y_3 as select inputs, wherever possible. If the unused state $y_1y_2y_3 = 011$ is assigned a value of x_1, then the multiplexer output will assume the value of x_1 when $y_1y_3 = 01$. Thus far, only flip-flops y_1 and y_3 have been used to select the multiplexer data inputs. If the unused state $y_1y_2y_3 = 111$ is assigned a value of logic 0, then whenever $y_1y_3 = 1–$, the multiplexer output will be a logic 0. These steps are summarized in Table 4.3. The same results can be obtained by examining the input map and combining minterm pairs (0,2) and (1,3).

Table 4.3 Assignment of multiplexer select inputs for Dy_1 of Example 4.4

Multiplexer select inputs s_1 (y_1)	s_0 (y_3)	Flip-flop y_1 input (Multiplexer y_1 output) Dy_1
0	0	1
0	1	x_1
1	0	0
1	1	0

For comparison, Figure 4.18 (a) and (b) show two multiplexer configurations: an 8:1 linear-select multiplexer and a 4:1 nonlinear-select multiplexer. The 4:1 multiplexer reduces the input logic for Dy_1 by one-half. If additional logic is necessary in order to achieve the desired smaller multiplexer, then there may be no advantage over the linear-select multiplexer approach.

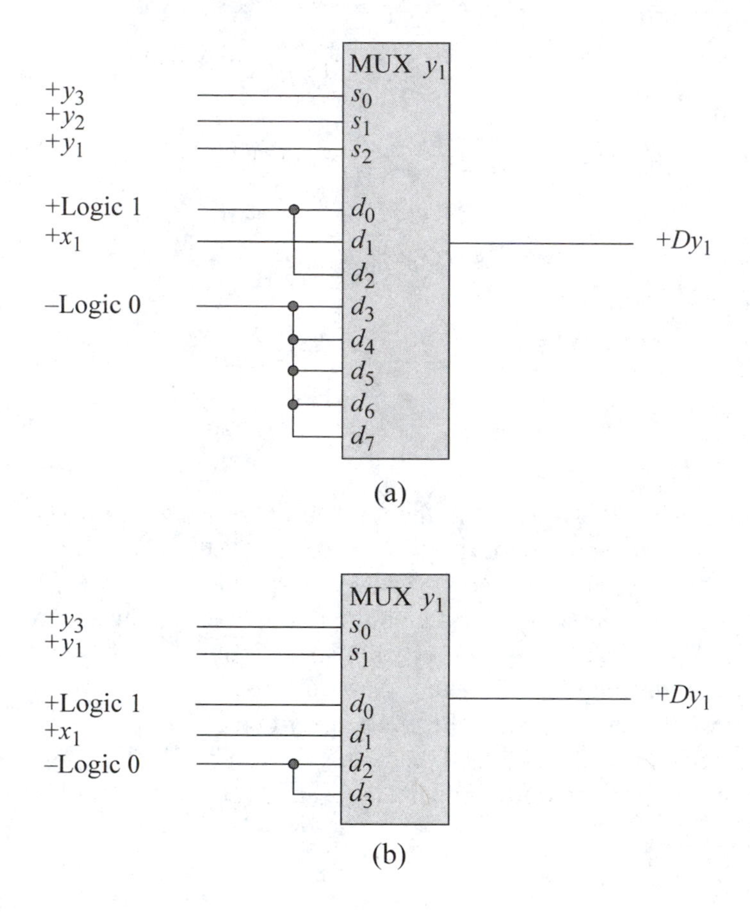

Figure 4.18 The δ next-state logic for the input map of Figure 4.17 using multiplexers: (a) linear-select multiplexer; and (b) nonlinear-select multiplexer.

An alternative approach which achieves the same results is shown in Figure 4.17 (b), (c), and (d). In this approach, all three permutations of the maps are used. The permutation which yields the fewest number of logic gates is then selected. In Figure 4.17 (b), let flip-flops y_2 and y_3 connect to multiplexer select inputs s_1 and s_0, respectively. Thus, when $y_2y_3 = 00$, input d_0 is selected and corresponds to $d_0 = y_1'(1) + y_1(0) = y_1'$; that is, minterms 0 and 4 are equivalent to y_1'. In the same manner, when

$y_2y_3 = 01$, input d_1 is selected and corresponds to $d_1 = y_1'x_1$. Similarly, $d_3 = -$, and $d_2 = y_1'$. In this permutation, one AND gate is required for multiplexer input d_1.

The input map is now permuted to connect y_1y_2 to select inputs s_1s_0, respectively, as shown in Figure 4.17 (c). The minterm locations are changed accordingly. When $y_1y_2 = 00$, input d_0 is selected such that, $d_0 = y_3' + y_3x_1 = y_3' + x_1$ by the absorption law. In the same manner, $d_1 = 1, d_2 = 0$, and $d_3 = 0$. For this permutation, one OR gate is required for multiplexer input d_0.

The input map is now permuted so that y_1y_3 connect to select inputs s_1s_0, respectively. The minterm locations are changed accordingly. In this permutation, $d_0 = 1$, $d_1 = x_1$, $d_2 = 0$, and $d_3 = 0$, requiring no gates for the multiplexer data inputs. This permutation would be then chosen for the design.

Example 4.5 Given the input map for flip-flop y_1 as shown in Figure 4.19 (a), the δ next-state logic for Dy_1 will be implemented using a nonlinear-select multiplexer. There are three solutions to this example: one implementation using only a 4:1 multiplexer and two using logic primitives and a 4:1 multiplexer. The input map is represented in tabular form in Table 4.4. One nonminimized approach uses y_1 and y_3 as multiplexer select inputs, combining minterms 0 and 2 as follows:

$$\begin{aligned} & y_1'y_2'y_3'x_1 + y_1'y_2y_3'x_1' \\ &= y_1'y_3'(y_2'x_1 + y_2x_1') \qquad \text{Distributive law} \\ &= y_1'y_3'(y_2 \oplus x_1) \end{aligned}$$

Thus, if $y_1y_3 = 00$, then data input d_0 is realized by the exclusive-OR function. Next, a value of 0 is assigned to minterms 1 and 3, and a value of 1 to minterm 4. The resulting assignment table for the select inputs is shown in Table 4.5. The third row of Table 4.5 was obtained by combining minterms 4 and 6 of Table 4.4, using y_1 and y_3 as select inputs. The last row of Table 4.5 was realized by noting that the values for minterms 5 and 7 correspond to y_2'. The resulting multiplexer configuration is illustrated in Figure 4.20 (a).

The minimized approach for the input logic uses flip-flops y_1 and y_2 as select inputs for s_1 and s_0, respectively. Unused minterm 1 is assigned the value of x_1; then $y_1y_2 = 00 = x_1$ for Dy_1. Next, a value of x_1' is assigned to unused minterm 3; thus, $y_1y_2 = 01 = x_1'$ for Dy_1. Assigning a logic 1 value to unused minterm 4, produces $y_1y_2 = 10 = 1$ for Dy_1. Finally, minterms 6 and 7 combine with y_1 and y_2 to yield $y_1y_2 = 11 = y_3'$. This procedure is summarized in Table 4.6 and the resulting multiplexer configuration is shown in Figure 4.20 (b). The operation of the multiplexers can be verified by assigning flip-flops $y_1y_2y_3$ the values 000 through 111, in turn, and confirming the value of Dy_1 in Figure 4.19. Both implementation methods were presented to illustrate different techniques for synthesizing with nonlinear-select multiplexers.

Alternatively, the input map can be represented by the original and two permutations, as shown in Figure 4.19 (b), (c), and (d) with the resulting multiplexer input equations. Figure 4.19 (c) yields a solution that requires no gates, the same as Table 4.6.

$y_1 \backslash y_2y_3$	00	01	11	10
0	(0) x_1	(1) –	(3) –	(2) x_1'
1	(4) –	(5) 1	(7) 0	(6) 1

Dy_1

(a)

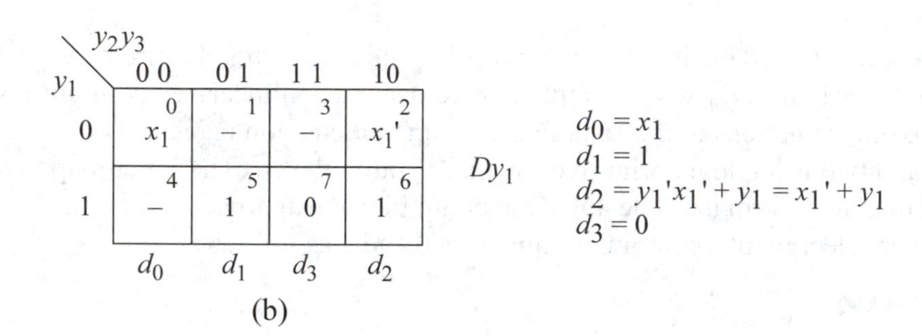

(b)

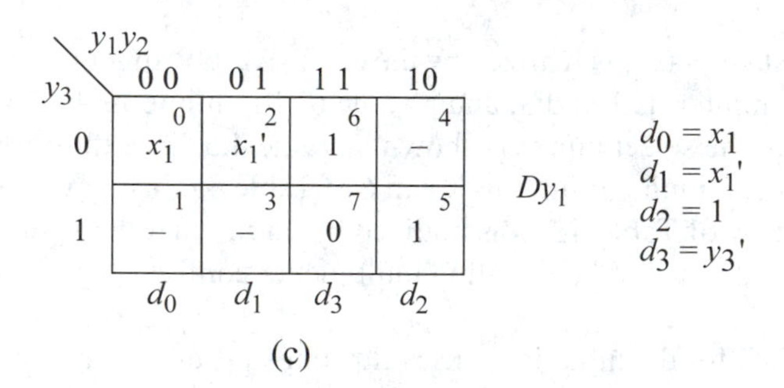

Dy_1

$d_0 = x_1$
$d_1 = x_1'$
$d_2 = 1$
$d_3 = y_3'$

(c)

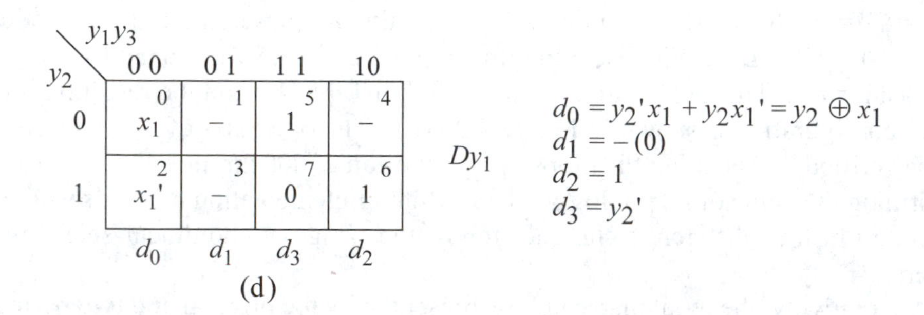

(d)

Figure 4.19 Input maps for flip-flop y_1 in Example 4.5: (a) for use with Table 4.4; (b) using y_2y_3 as select inputs; (c) using y_1y_2 as select inputs; and (d) using y_1y_3 as select inputs.

Table 4.4 Tabular representation of the input map for flip-flop y_1 of Example 4.5

Minterm	Present state $y_1 y_2 y_3$	Flip-flop y_1 input (Multiplexer y_1 output) Dy_1
0	0 0 0	x_1
1	0 0 1	–
2	0 1 0	x_1'
3	0 1 1	–
4	1 0 0	–
5	1 0 1	1
6	1 1 0	1
7	1 1 1	0

Table 4.5 Assignment of multiplexer select inputs for Dy_1 of Example 4.5 using a non-minimized approach

Multiplexer select inputs s_1 (y_1)	s_0 (y_3)	Flip-flop y_1 input (Multiplexer y_1 output) Dy_1
0	0	$y_2 \oplus x_1$
0	1	0
1	0	1
1	1	y_2'

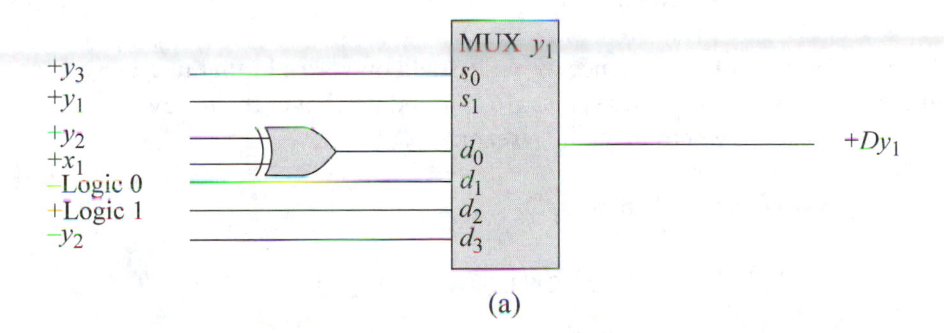

Figure 4.20 The δ next-state logic for the input map of Figure 4.19: (a) nonminimized configuration; and (b) minimized configuration.

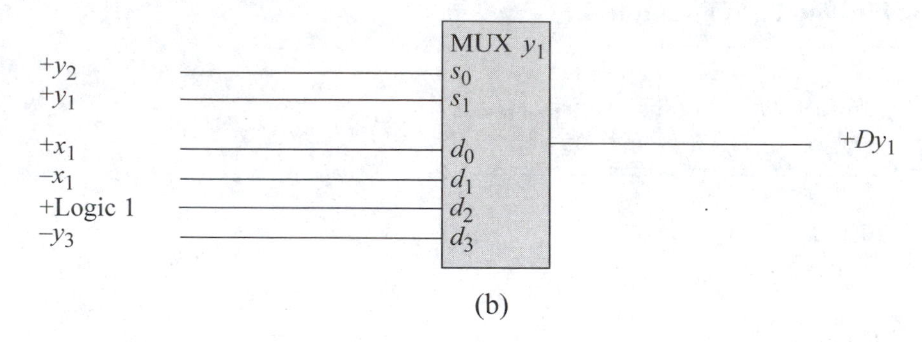

Figure 4.20 (Continued)

Table 4.6 Assignment of multiplexer select inputs for Example 4.5 using a minimized approach

Multiplexer select inputs s_1 (y_1)	s_0 (y_2)	Flip-flop y_1 input (Multiplexer y_1 output) Dy_1
0	0	x_1
0	1	x_1'
1	0	1
1	1	y_3'

Example 4.6 As a final example in this section, Example 4.6 presents a complete synthesis procedure for a Moore machine using multiplexers that operates in accordance with the machine specifications described below.

> The machine receives a sequence of binary data on a serial input line x_1 and generates two outputs z_1 and z_2, that are asserted whenever the following sequences appear anywhere in the bit stream:
>
> If $x_1 = 110$, then $z_1 \uparrow t_1 \downarrow t_3$
>
> If $x_1 = 101$, then $z_2 \uparrow t_2 \downarrow t_3$
>
> The input data is not in the form of n-bit words, but consists of a continuous bit stream of binary data. Valid overlapping sequences are possible such that, the input sequence that asserts output z_1 may be part of a valid sequence that

asserts output z_2, as shown below. The converse is true for a z_2, z_1 sequence. The outputs should not generate spurious signals during any state transition sequence.

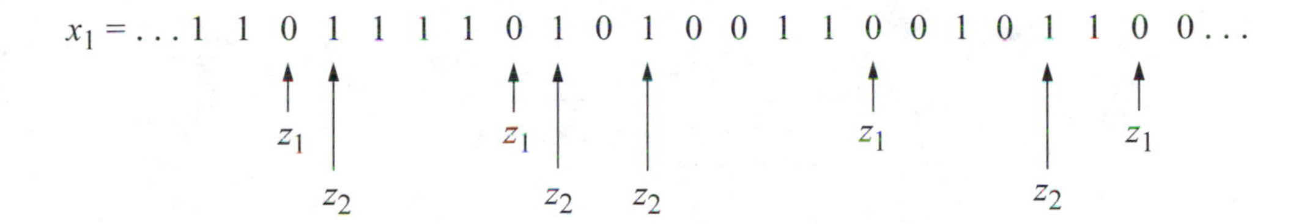

The δ next-state logic is to be implemented with multiplexers using the least amount of logic.

Step 1: State diagram The state diagram of Figure 4.21 graphically portrays the machine's behavior in accordance with the machine specifications. The machine remains in state *a* until the first 1 appears on input x_1. Note that state *c* is the state that is entered whenever there are two consecutive 1s anywhere in the bit stream, signifying the beginning bit configuration of a valid input sequence to assert output z_1. Also, state *f* is the state that is entered whenever the sequence x_1 = ... 10 occurs, signifying the beginning bit configuration of a valid input sequence to assert output z_2.

The state transition sequence (*a*,*b*,*c*,*d*,*e*), which corresponds to x_1 = 1101, illustrates an example of overlapping sequences: 110 and 101. Since no valid input sequence contains consecutive 0s, the machine returns to state *a* to begin checking for a new valid input sequence whenever x1 = ... 00.

Step 2: Equivalent states The rules for identifying equivalent states from Section 3.1.1 are restated as follows:

Two states are equivalent if and only if the following two statements apply:

1. Both states have the same output for the same input value.
2. Both states have the same or equivalent next state.

In this example, there are no situations where both rules are true concurrently; therefore, there are no equivalent states in the state diagram of Figure 4.21. The state transition sequence (*a*,*b*,*c*,*d*,*e*) is required in order to detect the two valid overlapping sequences 110 and 101, which appear as ... 1101. Likewise, the sequence (*a*,*b*,*f*,*e*,*c*,*d*) is required in order to detect the two valid overlapping sequences 101 and 110, which appear as 10110.

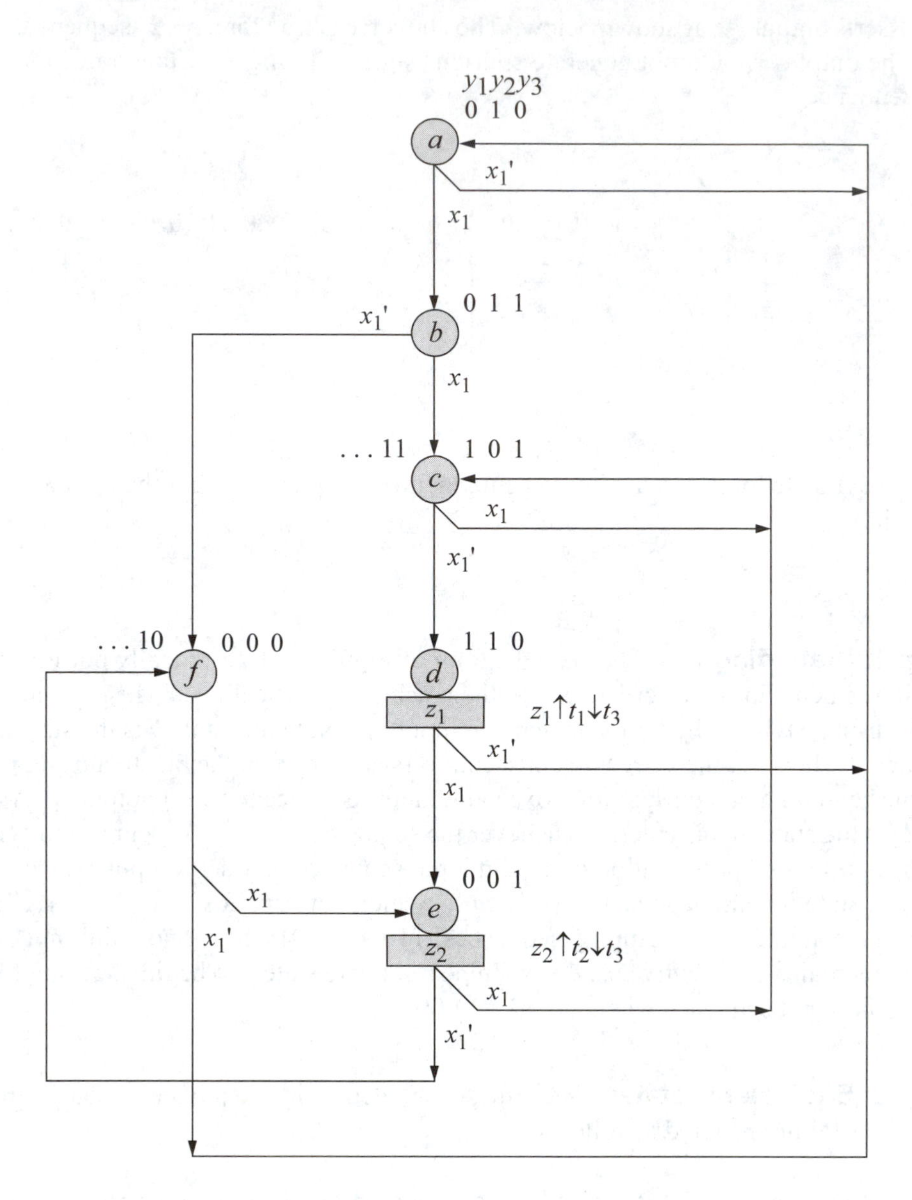

Figure 4.21 State diagram for the Moore machine of Example 4.6 to detect input sequences of $x_1 = 110$ and 101. Unused states are: $y_1y_2y_3 = 100$ and 111.

Step 3: State code assignment Using the rules for state code adjacency (see Section 3.4) and trial-and-error techniques, the state codes are assigned as shown in the state diagram of Figure 4.21 and in the present-state map of Figure 4.22. An important consideration in this design is the prevention of glitches on outputs z_1 and z_2. Because z_2 is asserted at time t_2, glitches are not possible on this output, since the machine has stabilized by the midpoint of the clock period. Care must be taken, however, to insure that a glitch will not be generated on output z_1, since assertion for this output occurs at time t_1 — the time at which the machine changes state. Judiciously choosing state codes, while verifying that no state transition resulting from these codes causes the machine to pass through transient state d (and assert z_1), yields the state codes of Figure 4.21.

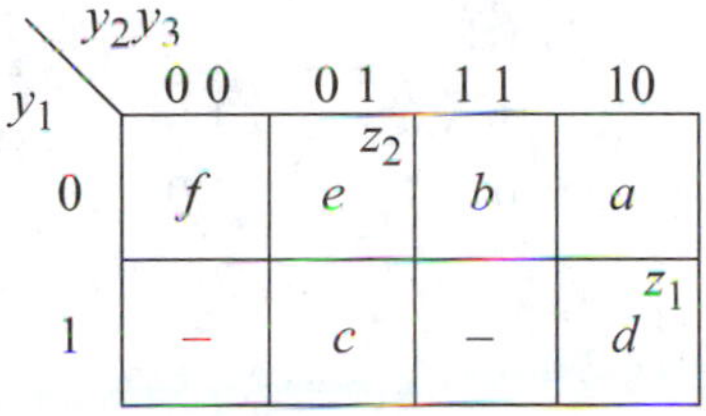

y_1 \ y_2y_3	00	01	11	10
0	f	e (z_2)	b	a
1	–	c	–	d (z_1)

Figure 4.22 Present-state map for the Moore machine of Figure 4.21.

Step 4: Next-state table The next-state table, derived directly from the state diagram, is shown in Table 4.7. All present states are listed, including unused states, the next state for each present state as a function of input x_1, inputs for the D flip-flops, and the present outputs.

Table 4.7 Next-state table for the Moore machine of Figure 4.21 using *D* flip-flops

Present state $y_1y_2y_3$	Input x_1	Next state $y_1y_2y_3$	Flip-flop inputs $Dy_1\ Dy_2\ Dy_3$	Present outputs z_1z_2
0 0 0	0	0 1 0	0 1 0	0 0
0 0 0	1	0 0 1	0 0 1	0 0
0 0 1	0	0 0 0	0 0 0	0 1
0 0 1	1	1 0 1	1 0 1	0 1
0 1 0	0	0 1 0	0 1 0	0 0
0 1 0	1	0 1 1	0 1 1	0 0
0 1 1	0	0 0 0	0 0 0	0 0
0 1 1	1	1 0 1	1 0 1	0 0
1 0 0	0	– – –	– – –	– –
1 0 0	1	– – –	– – –	– –

Table 4.7 Next-state table for the Moore machine of Figure 4.21 using *D* flip-flops

Present state $y_1y_2y_3$	Input x_1	Next state $y_1y_2y_3$	Flip-flop inputs $Dy_1\,Dy_2\,Dy_3$	Present outputs z_1z_2
Continued on next page				
1 0 1	0	1 1 0	1 1 0	0 0
1 0 1	1	1 0 1	1 0 1	0 0
1 1 0	0	0 1 0	0 1 0	1 0
1 1 0	1	0 0 1	0 0 1	1 0
1 1 1	0	– – –	– – –	– –
1 1 1	1	– – –	– – –	– –

Step 5: Input maps The input maps of Figure 4.23 (a) are constructed from either the state diagram or the next-state table. No equations are listed for the *D* flip-flop inputs, because multiplexers will be used for the δ next-state logic. Consider the input map for Dy_1. In state f ($y_1y_2y_3 = 000$) of the state diagram, flip-flop y_1 has a next value of 0, regardless of the value of x_1; therefore, a 0 is entered in minterm location 0. In state e ($y_1y_2y_3 = 001$), flip-flop y_1 has a next value of 1 only if $x_1 = 1$, necessitating an entry of x_1 in minterm location 1. In state c ($y_1y_2y_3 = 101$), flip-flop y_1 has a next value of 1, regardless of the path taken; therefore, a 1 is entered in minterm location 5. The input maps for Dy_2 and Dy_3 are completed in a similar manner. The input maps of Figure 4.23 (b) and (c) show the permuted maps for Dy_1 and Dy_2, respectively, together with the multiplexer data input equations.

Step 6: Output maps The output maps are obtained from either the state diagram or the next-state table. Figure 4.24 illustrates the output maps for this example. Since this is a Moore machine, input x_1 is not used as a map-entered variable, because the outputs are a function of the present state only. Therefore, a 1 is entered in the appropriate minterm locations for outputs z_1 and z_2.

The logic for any machine should always be minimized. Consider the output map for z_1. If minterm locations 6 and 7 can be combined, then the equation for z_1 will contain one term with only two variables such that, $z_1 = y_1y_2$. Before finalizing the equation for z_1, however, all state transitions must be checked to insure that no state change will cause the machine to pass through transient state $y_1y_2y_3 = 111$ for a path that does not include z_1. If one such path can be found, then a 0 must be entered in minterm location 7 in the output map for z_1, thus negating the use of minterm 7 to minimize z_1.

At least one path exists that may pass through the unused state $y_1y_2y_3 = 111$. In state b ($y_1y_2y_3 = 011$), the machine proceeds to state c ($y_1y_2y_3 = 101$) if $x_1 = 1$. Due to varying flip-flop propagation delays, if flip-flop y_1 sets before y_2 resets, then the machine will momentarily pass through the unused state $y_1y_2y_3 = 111$, producing a glitch on output z_1. This occurs because flip-flop y_3 was omitted as a variable in the equation for z_1. Therefore, z_1 cannot be minimized using minterm 7.

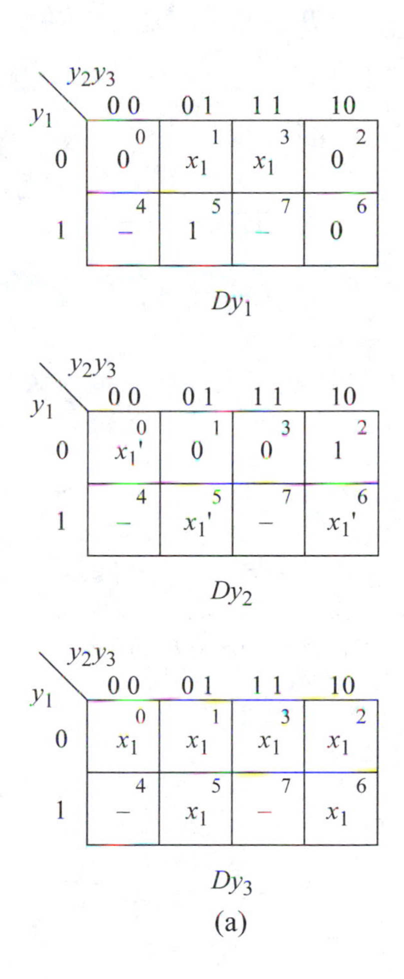

Figure 4.23 Input maps for the Moore machine of Figure 4.21 using D flip-flops: (a) using linear-select multiplexers; (b) using nonlinear-select multiplexers for Dy_1; and (c) using nonlinear-select multiplexers for Dy_2.

Consider, now, using minterm 4 to minimize z_1 such that, $z_1 = y_1y_3'$. The only state transition sequences that may pass through the unused state $y_1y_2y_3 = 100$ are (c,d) and (d,e), neither of which will produce a glitch on z_1. The path from state c to state d may pass through state $y_1y_2y_3 = 100$; however, state d is the end state for this sequence and z_1 is asserted in state d. Thus, output z_1 may be asserted early, as shown in Figure 4.25 (a), but this presents no output hazard.

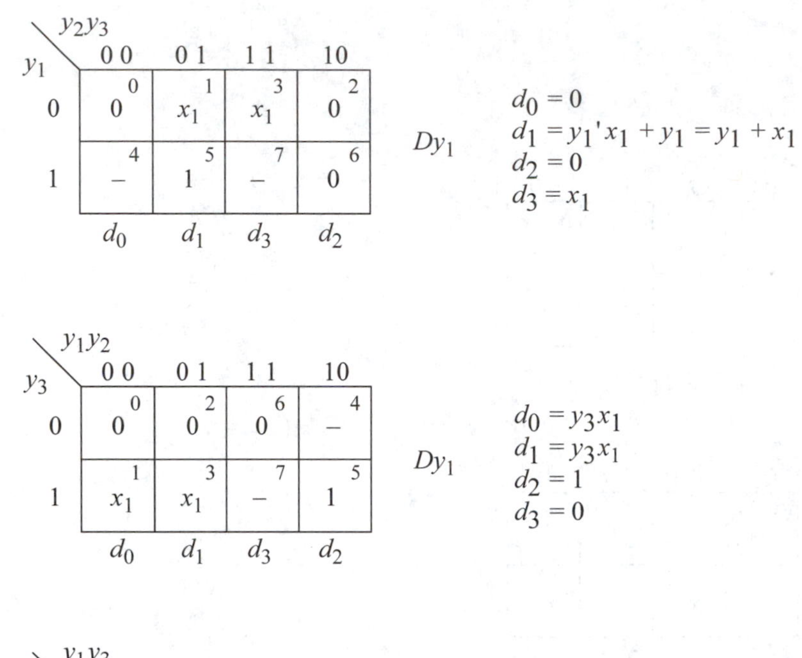

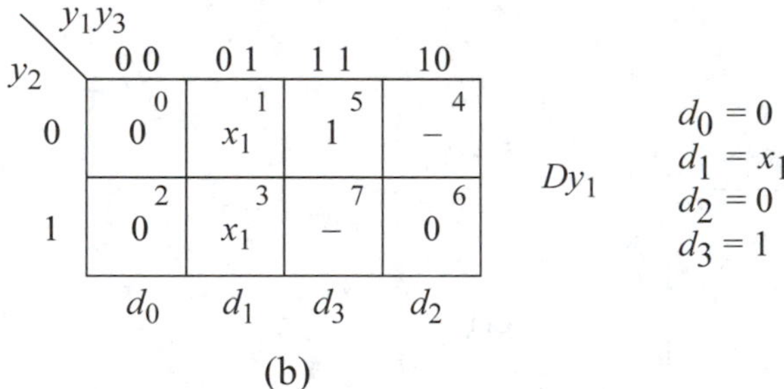

$d_0 = 0$
$d_1 = x_1$
$d_2 = 0$
$d_3 = 1$

(b)

Figure 4.23 (Continued)

Similarly, the path from state d to state e may pass through state $y_1y_2y_3 = 100$; however, state d is the starting state for this sequence and z_1 is asserted in state d. Thus, z_1 may be deasserted late, as shown in Figure 4.25 (b), but no glitch will appear on z_1. Therefore, a value of 1 may be entered in minterm location 4 to minimize the equation for z_1, as shown in Equation 4.5. No minimization is possible for output z_2.

Step 7: Logic diagram Linear selection implies that three 8:1 multiplexers be used in the implementation. However, the input maps for Dy_1 and Dy_2 of Figure 4.23 have only three unique entries. Therefore, the expression of Equation 4.4 is true; that is, $1 < u \leq (2^p \div 2)$, where u is the number of unique entries and p is the number of flip-flops. The input maps for Dy_1 and Dy_2 can be implemented with $(2^p \div 2)$:1 = 4:1 multiplexers. Equation 4.4 is not true, however, for the input map for Dy_3 — which has only one unique entry, because the conditions for the left inequality are not met: 1 is not less than u.

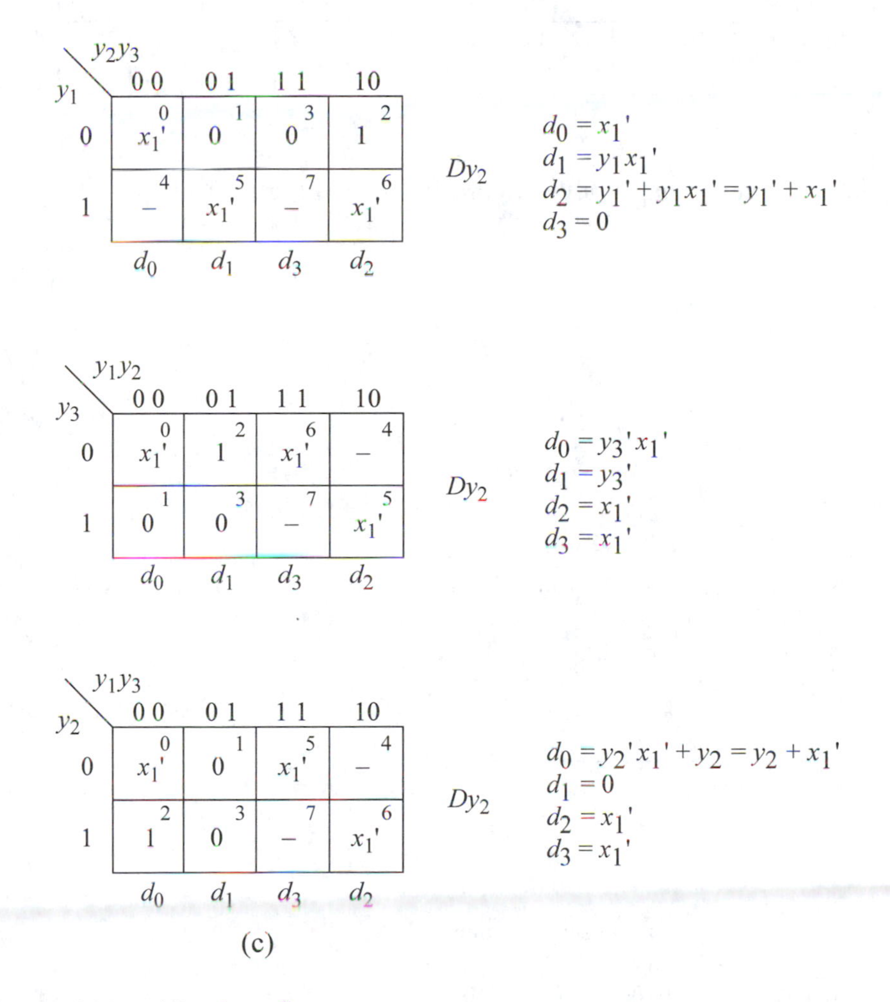

(c)

Figure 4.23 (Continued)

Table 4.8 is a representative tabulation of the input map for Dy_1. Assigning a value of 0 to minterm 4 and a value of 1 to minterm 7, the multiplexer select input assignment table of Table 4.9 is produced, using only flip-flops y_1 and y_3 as select inputs. A similar procedure yields the corresponding tables for input map Dy_2 as shown in Table 4.10 and Table 4.11, also using y_1y_3 as select inputs. Referring again to Table 4.8 and Table 4.9, it is now apparent that if values of 1 and 0 had been chosen for minterms 4 and 7, respectively, then this would be inconsistent with the precept to use the same flip-flop variables as select inputs — this choice of values uses y_1y_2 as select inputs.

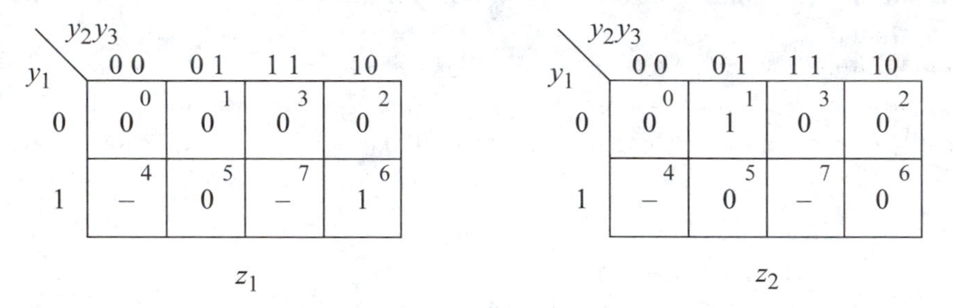

Figure 4.24 Output maps for the Moore machine of Figure 4.21.

$$z_1 = y_1y_3'$$

$$z_2 = y_1'y_2'y_3 \qquad (4.5)$$

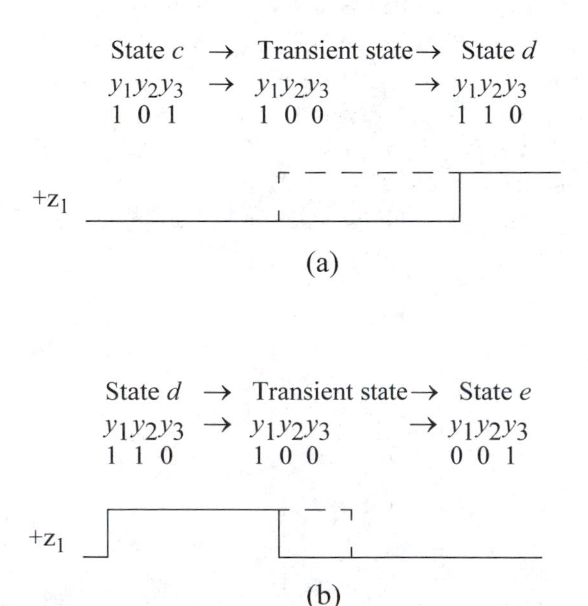

Figure 4.25 Assertion/deassertion for output z_1 (not drawn to scale) during transient state activity: (a) possible early assertion of z_1 for the (c,d); and (b) possible late deassertion of z_1 for the transition (d,e).

Table 4.8 Tabular representation of the input map for flip-flop y_1 of Figure 4.23

Minterm	Present state $y_1y_2y_3$	Flip-flop y_1 input (Multiplexer y_1 output) Dy_1
0	0 0 0	0
1	0 0 1	x_1
2	0 1 0	0
3	0 1 1	x_1
4	1 0 0	– (0)
5	1 0 1	1
6	1 1 0	0
7	1 1 1	– (1)

Table 4.9 Assignment of multiplexer select inputs for Dy_1 of Figure 4.23

Multiplexer select inputs s_1 (y_1)	s_0 (y_3)	Flip-flop y_1 input (Multiplexer y_1 output) Dy_1
0	0	0
0	1	x_1
1	0	0
1	1	1

Not all input maps partition as uniformly as the input map for Dy_1, as is evident by Table 4.10 for Dy_2. Using flip-flops y_1 and y_3 as select inputs for multiplexer y_2 allows rows 0 through 3 of Table 4.10 to be partitioned as shown in the first row of Table 4.11. (The absorption law reduces the expression $y_2'x_1' + y_2$ to $x_1' + y_2$.) Thus, when $y_1y_3 = 00$, the value for Dy_2 will be either x_1' if $y_2 = 0$, or 1 if $y_2 = 1$. This represents a sum-of-products expression for input d_0 of multiplexer y_2. When $y_1y_3 = 01$, Dy_2 receives a value of logic 0. If the unused states that correspond to minterms 4 and 7 are assigned a value of x_1', then whenever $y_1y_3 = 1–$, the input to Dy_2 is x_1'. The input map for Dy_3 indicates that the D input for flip-flop y_3 is simply the value of x_1 — assuming that minterms 4 and 7 are assigned the value of x_1. The logic diagram for this Moore machine is shown in Figure 4.26 using multiplexers and one OR gate for the input logic.

Table 4.10 Tabular representation of the input map for flip-flop y_2 of Figure 4.23

Minterm	Present state $y_1 y_2 y_3$	Flip-flop y_2 input (Multiplexer y_2 output) Dy_2
0	0 0 0	x_1'
1	0 0 1	0
2	0 1 0	1
3	0 1 1	0
4	1 0 0	$-(x_1')$
5	1 0 1	x_1'
6	1 1 0	x_1'
7	1 1 1	$-(x_1')$

Table 4.11 Assignment of multiplexer select inputs for Dy_2 of Figure 4.23

Multiplexer select inputs s_1 (y_1)	s_0 (y_3)	Flip-flop y_2 input (Multiplexer y_2 output) Dy_2
0	0	$y_2'x_1' + y_2 = x_1' + y_2$
0	1	0
1	0	x_1'
1	1	x_1'

Since output z_1 is active for the entire clock period in state *d*, the decoding for z_1 is simply a 2-input AND gate. Output z_2, however, must include the –Clock signal, because z_2 is asserted at time t_2 and is implemented with a positive-input AND gate. Figure 4.27 illustrates a timing diagram for two valid overlapping sequences, where x_1 = ... 11011. Correct operation of the machine can be verified by beginning in state *a* and applying all possible sequences to input x_1 and observing the state of the flip-flops and the two outputs.

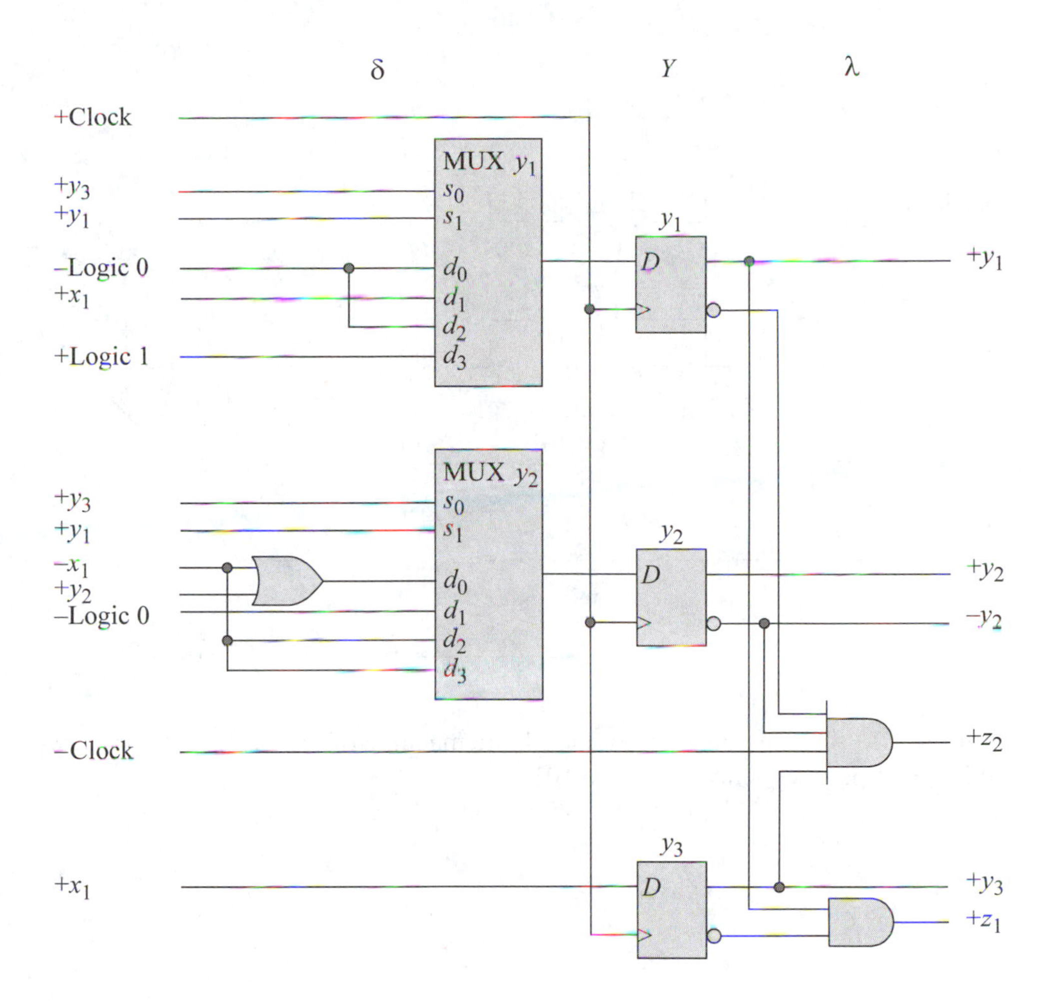

Figure 4.26 Logic diagram for the Moore machine of Figure 4.21 using multiplexers and logic gates for the δ next-state logic.

The alternative approach using permutations of the input maps will now be presented in more detail. This is a more straightforward technique to obtain the least amount of logic for the data inputs of a nonlinear-select multiplexer. This approach is more direct, because the columns of the permuted input maps correspond directly to the multiplexer data inputs. This method assigns the values of the variables associated with each column heading to specific multiplexer select inputs.

For example, if a 3-variable Karnaugh map for state flip-flops y_1, y_2, and y_3 has only four unique entries in the eight possible minterm locations, then a 4:1 multiplexer would be sufficient for the δ next-state logic. Thus, instead of three select inputs of an 8:1 multiplexer connecting to the outputs of the three state flip-flops in a linear-select mode, two select inputs of a 4:1 multiplexer connect to two of the three state flip-flops in a nonlinear-select mode.

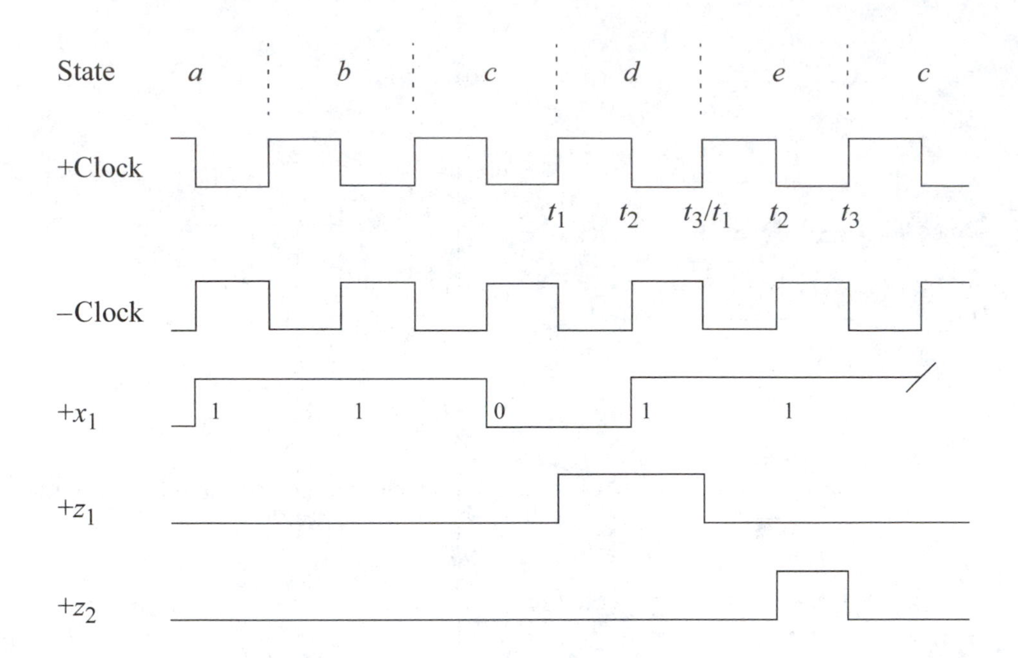

Figure 4.27 Timing diagram for the Moore machine of Figure 4.21 showing two valid overlapping sequences: 110 and 101.

The three flip-flops are permuted to provide three sets of two flip-flops per set. Thus, the two select inputs s_1 and s_0 of the 4:1 multiplexer connect to the outputs of two state flip-flops in one of the following patterns: $s_1s_0 = y_1y_2$, $s_1s_0 = y_1y_3$, or $s_1s_0 = y_2y_3$. Input s_0 is the low-order select input; therefore, the low-order flip-flop of each pair of flip-flops connects to s_0.

Consider the input map for Dy_2 of Figure 4.7 redrawn in Figure 4.28. There are only three unique values: 0, 1, and x_1. Therefore, a 4:1 multiplexer could be used to represent the δ next-state logic. Let $s_1s_0 = y_2y_3$. The equations for the multiplexer data inputs are: $d_0 = y_1$, $d_1 = y_1'x_1$, $d_2 = 0$, and $d_3 = –$. One AND gate is required for the d_1 input.

Now permute the flip-flop variables so that the column headings refer to flip-flops y_1 and y_2, while maintaining the correct minterm locations. Let $s_1s_0 = y_1y_2$. The equations for the multiplexer data inputs are: $d_0 = y_3x_1$, $d_1 = 0$, $d_2 = y_3'$, and $d_3 = 0$. This permutation requires one AND gate for d_0.

The final permutation consists of column headings labeled y_1y_3 such that, $s_1s_0 = y_1y_3$. The equations for the multiplexer data inputs are: $d_0 = 0$, $d_1 = x_1$, $d_2 = y_2'$, and $d_3 = 0$. This permutation requires no additional logic and, therefore, represents the desired select input connections for a nonlinear-select multiplexer.

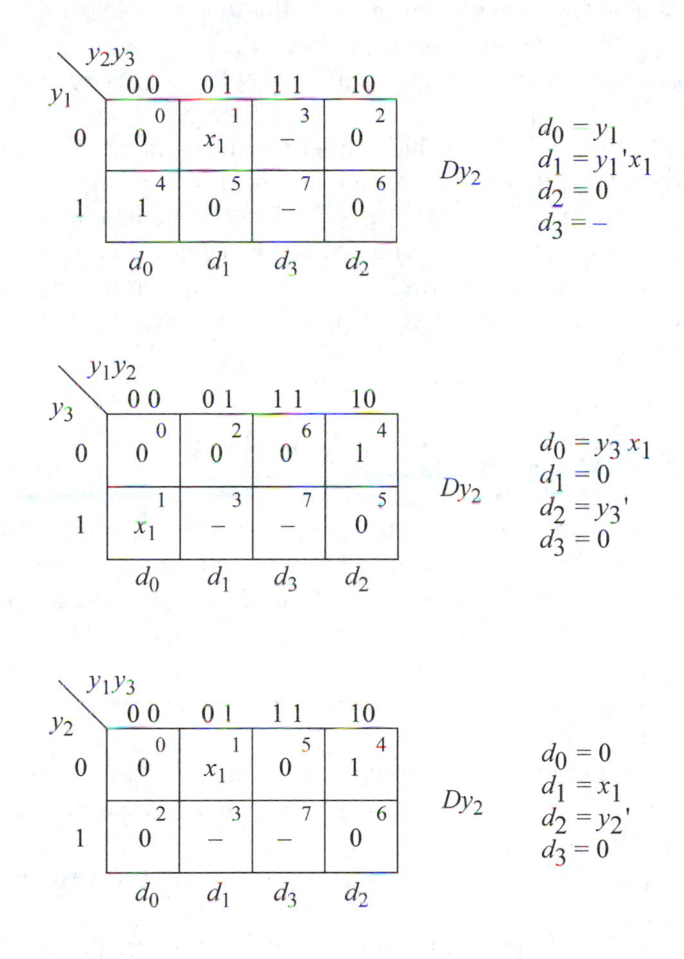

Figure 4.28 Input map for Dy_2 redrawn from Figure 4.7 showing two permutations.

As a final example of this technique, consider the input map for Dy_1 of Figure 4.23(b). Let $s_1s_0 = y_2y_3$. The multiplexer data input equations are: $d_0 = 0$, $d_1 = y_1'x_1 + y_1 = y_1 + x_1$ (absorption law), $d_2 = 0$, and $d_3 = x_1$. One OR gate is required for input d_1. Now permute the flip-flop variables so that the columns correspond to y_1y_2 and let $s_1s_0 = y_1y_2$. The multiplexer data input equations are: $d_0 = y_3x_1$, $d_1 = y_3x_1$, $d_2 = 1$, and $d_3 = 0$. One shared AND gate is required for inputs d_0 and d_1.

The final permutation specifies the column headings as y_1y_3. Therefore, let $s_1s_0 = y_1y_3$. The multiplexer data input equations are: $d_0 = 0$, $d_1 = x_1$, $d_2 = 0$, and $d_3 = 1$. This permutation requires no additional logic gates and is, therefore, the desired permutation for a nonlinear-select multiplexer with the minimal amount of additional logic.

This method is more exhaustive than the previous method, but provides all possibilities of multiplexer data input logic for a nonlinear-select multiplexer. It is then simply a matter of selecting the Karnaugh map permutation that provides the least amount of input logic. Then the state flip-flops specified in the column headings are connected to the multiplexer select inputs, where the low-order flip-flop is assigned to the low-order select input.

The same concept can be extended to include larger multiplexers. A 4-variable input map would normally require a 16:1 linear-select multiplexer. However, if the number of unique entries u is $5 \leq u \leq 8$, then an 8:1 multiplexer could be utilized in a nonlinear-select mode. The map can then be arranged so that each column contains three flip-flop variables which correlate to the eight data inputs d_0 through d_7. The column entries then represent the logic for the corresponding multiplexer data inputs.

4.2 Decoders for λ Output Logic

A *decoder* is a combinational macro logic circuit that translates a binary input number to an equivalent output number for a specific radix. There is a one-to-one correspondence between the outputs and the combinations of the input signals. In general, there are n input lines and m output lines, where $m = 2^n$. For each combination of the 2^n input values, only one unique output signal is active — all other outputs are inactive. Thus, a fundamental characteristic of a decoder is the mutual exclusiveness of the outputs.

In the previous section, the multiplexer data inputs were shown to correspond to the minterm locations in an input map. There is also a one-to-one correspondence between the outputs of a decoder and the minterms in an output map. The m output signals, labeled $f_0, f_1, \ldots, f_{m-1}$ correspond to the 2^n minterm values represented by the binary number on inputs $x_{n-1}, \ldots, x_1, x_0$ as shown in Figure 4.29, where x_0 is the low-order input variable. A decoder consisting of n inputs and 2^n outputs can be used to implement any boolean switching function, because each output represents a unique minterm. By connecting the appropriate decoder outputs to an OR gate, the desired function can be realized.

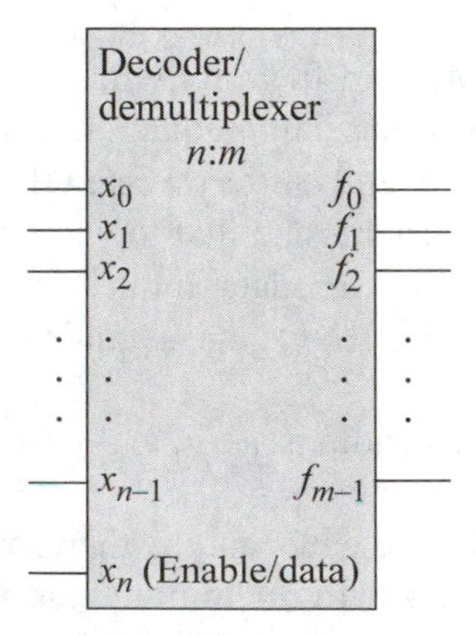

Figure 4.29 General block diagram of an $n \times m$ decoder matrix with an enable/data input.

A decoder is also classified as a *demultiplexer* (DX). Whether the device acts as a decoder or a demultiplexer depends on how the input lines are interpreted. If input x_n functions as an enable for the outputs, then the device is a decoder. However, if x_n is used as a data input, then the device functions as a demultiplexer, where the data input is directed to one of the m outputs as selected by the n inputs.

Different switching matrices can be designed for various values of n and m. Not all decoders have exactly 2^n outputs. For example, a binary coded decimal (BCD)-to-decimal decoder translates a 4-bit BCD number $x_3x_2x_1x_0$, where x_0 is the low-order input variable, to ten mutually exclusive outputs $f_0, f_1, \ldots, f_9$ that correspond to the binary input numbers. Each output function is defined as $f_i = X_i$, where X_i is the ith minterm of the n input variables. Thus, $f_i = 1$ if and only if the binary value of $x_3x_2x_1x_0$ is equal to X_i. Decoder outputs can be either active high or active low.

In the 3:8 decoder of Figure 4.30, each output corresponds to a different minterm of three variables. Additional logic capability can be achieved by including an AND gate as part of the decoder circuity. The purpose of the AND gate is to provide an enabling function for the outputs; that is, if the input conditions for the AND gate are met, then the appropriate output will be asserted, which is the octal equivalent of the binary input $x_2x_1x_0$. The truth table for the decoder of Figure 4.30 is shown in Table 4.12.

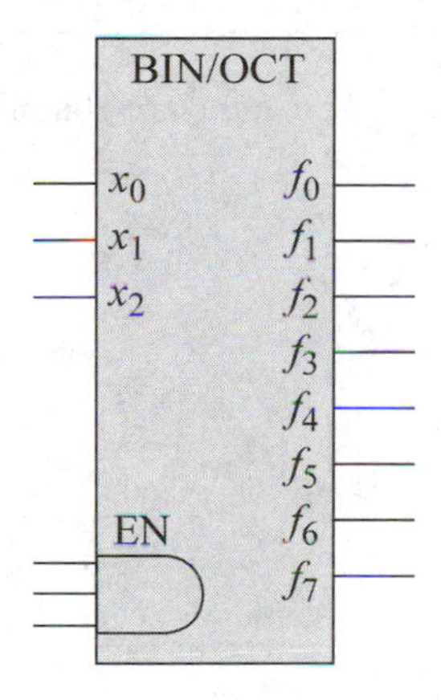

Figure 4.30 A binary-to-octal decoder with active-high outputs and an enable (EN) AND gate.

Example 4.7 A decoder for the λ output logic is economical only if the machine has sufficient multiple outputs to warrant the use of an n:2^n decoder. A Moore machine will be designed which will generate a sequence of six contiguous, nonoverlapping pulses. The six pulses are mutually exclusive and each pulse is active for one clock period, as shown in Figure 4.31. There must be no spurious signals on any of the outputs. Nonlinear-select multiplexers will be used, where applicable, for the δ next-state logic and decoders for the λ output logic. The assertion/deassertion statement for all outputs is $z_i{\uparrow}t_1{\downarrow}t_3$. A typical application of this type of Moore machine is as a ring counter, which functions as a finite-state machine to control the operation of an external digital system. Each output of the machine represents a different state.

Table 4.12 Truth table for the binary-to-octal decoder of Figure 4.30

	Decoder inputs			Decoder outputs[1]							
Minterm value	x_2	x_1	x_0	f_0	f_1	f_2	f_3	f_4	f_5	f_6	f_7
0	0	0	0	1	0	0	0	0	0	0	0
1	0	0	1	0	1	0	0	0	0	0	0
2	0	1	0	0	0	1	0	0	0	0	0
3	0	1	1	0	0	0	1	0	0	0	0
4	1	0	0	0	0	0	0	1	0	0	0
5	1	0	1	0	0	0	0	0	1	0	0
6	1	1	0	0	0	0	0	0	0	1	0
7	1	1	1	0	0	0	0	0	0	0	1

1. Assuming that all inputs to the enable AND gate are at a high voltage level.

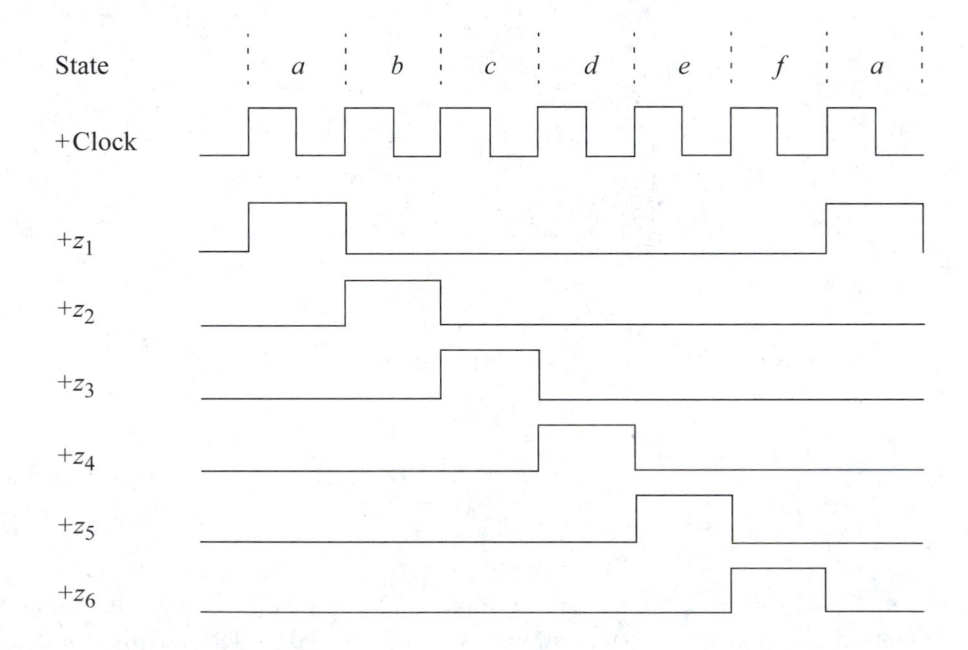

Figure 4.31 Timing diagram for the Moore machine of Example 4.7.

The state diagram is illustrated in Figure 4.32. In order to generate glitch-free outputs, a Gray code arrangement is used for the state code assignment. The state codes adhere to the definition of a Gray code, in which each code word differs from the preceding word in only one bit position. However, since only six out of the eight possible combinations of three variables are used, the code represented in Figure 4.32 differs slightly from the traditional Gray code in order to maintain logical adjacency between states *f* and *a*. Each state generates an output and each output is asserted for the entire state time.

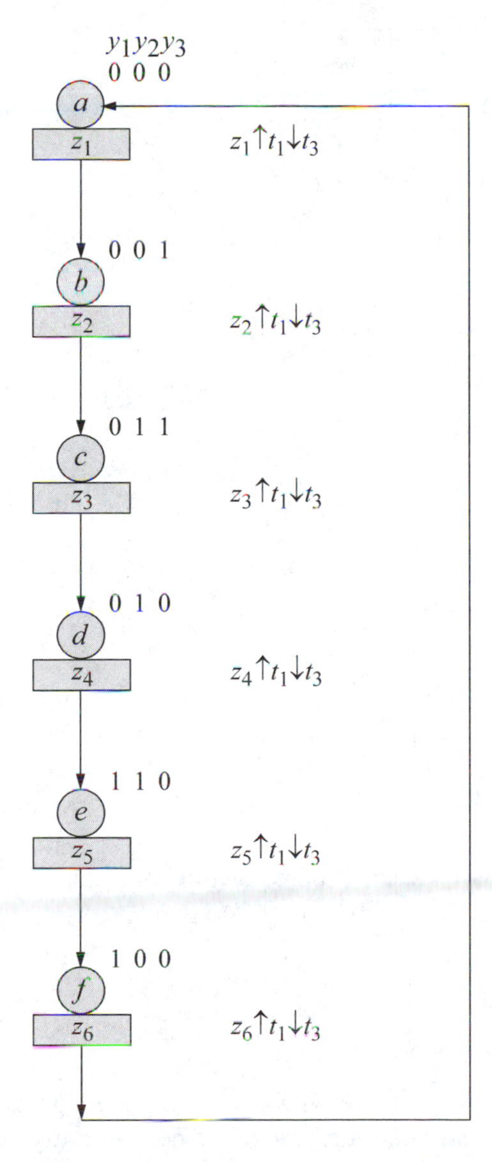

Figure 4.32 State diagram for the Moore machine of Example 4.7. Unused states are: $y_1y_2y_3$ = 101 and 111.

The input maps are derived from the state diagram in the usual manner and are shown in Figure 4.33. Since there are only two values (0 and 1) in any input map, multiplexers are not appropriate — logic gates will suffice for each flip-flop input. For brevity, the output maps are not shown. Each Moore output corresponds to a distinct decoder minterm and can be implemented directly from the state diagram. A 3:8 decoder is ideal for the λ output logic for this multiple output machine.

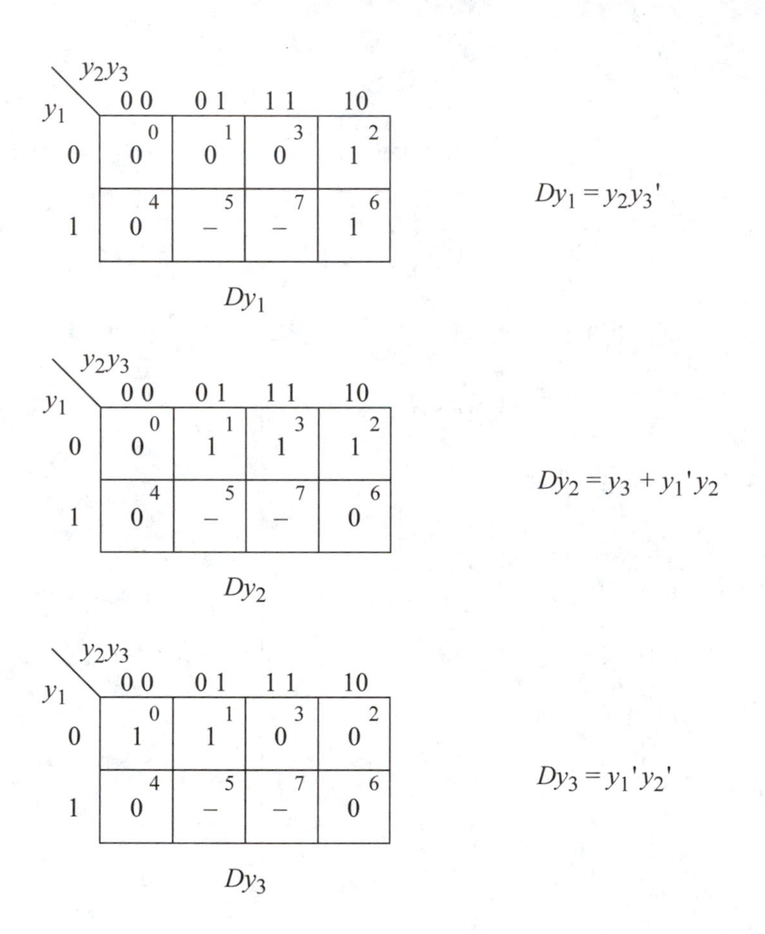

Figure 4.33 Input maps for the Moore machine of Figure 4.32.

The logic diagram is shown in Figure 4.34, using logic gates for the δ next-state logic and a 3:8 decoder for the λ output logic. The active-high output of each state flip-flop is connected to the appropriate decoder select/address input. Thus, y_1, y_2, and y_3 connect to x_2, x_1, and x_0, respectively, where y_3 and x_0 are the low-order variables. The uncommitted decoder outputs, f_5 and f_7, represent the unused states in the state diagram and input maps. Due to varying propagation delays in the internal logic matrix of the decoder, it is possible to have a slight overlap of two output pulses. If this overlap exists, it will be of extremely short duration and should not adversely affect the operation of the machine. There will, however, be no output glitches. An enabling function is not necessary for these outputs; therefore, if the decoder includes an output enable AND gate, then all inputs of the gate must be connected to a logic 1 for their respective active voltage levels.

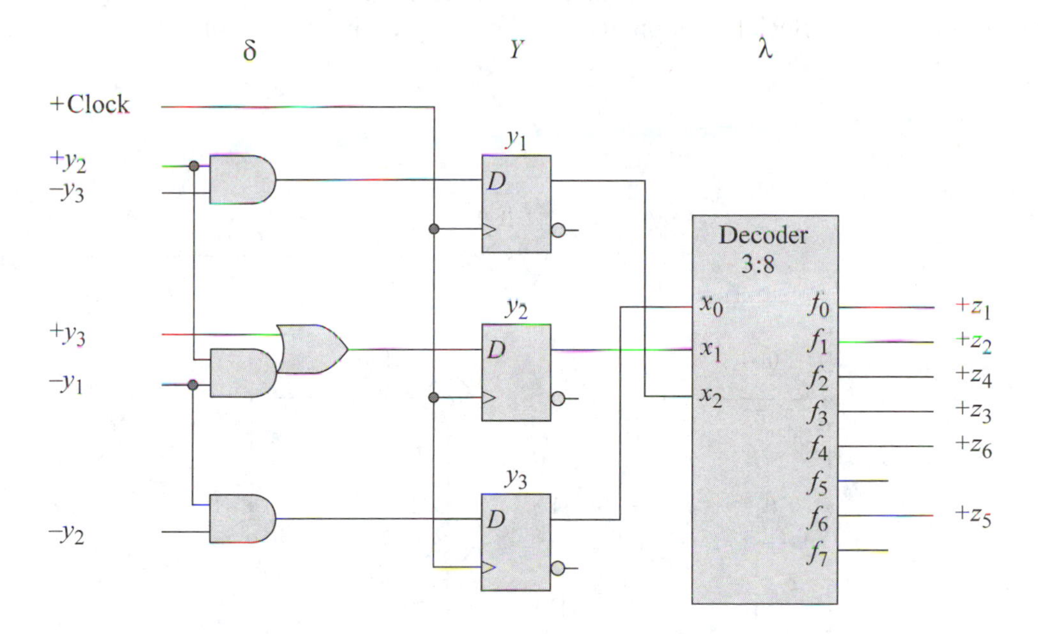

Figure 4.34 Logic diagram for the Moore machine of Figure 4.32.

Example 4.8 This example presents a Mealy machine that examines 3-bit words on a serial input line x_1. The format for the serial data is as follows:

$$x_1 = \cdots \left| b_1 b_2 b_3 \right| b_1 b_2 b_3 \left| b_1 b_2 b_3 \right| \cdots$$

where $b_i = 0$ or 1. The machine generates four different outputs, depending upon the bit configuration of the received data:

If $x_1 = 001$, then assert output z_1
If $x_1 = 011$, then assert output z_2
If $x_1 = 101$, then assert output z_3
If $x_1 = 111$, then assert output z_4

For all other bit patterns, the four outputs are inactive. The assertion/deassertion statement for all outputs is: $z_i \uparrow t_2 \downarrow t_3$. There are several noteworthy details in this example which will become evident as the synthesis process develops. Multiplexers will be used for the δ next-state logic, where applicable, and a decoder will be used to implement the λ output logic. As always, the design procedure will stress minimization of logic functions and reliability of system performance.

The serial words are contiguous with no space between words. The state diagram is developed by generating a path for each of the eight bit sequences (000 through 111) while maintaining three state levels for each word, as shown in Figure 4.35. The input data is synchronized with the machine clock; therefore, any changes to the data occur on the alternate clock transition to that which triggers the state flip-flops.

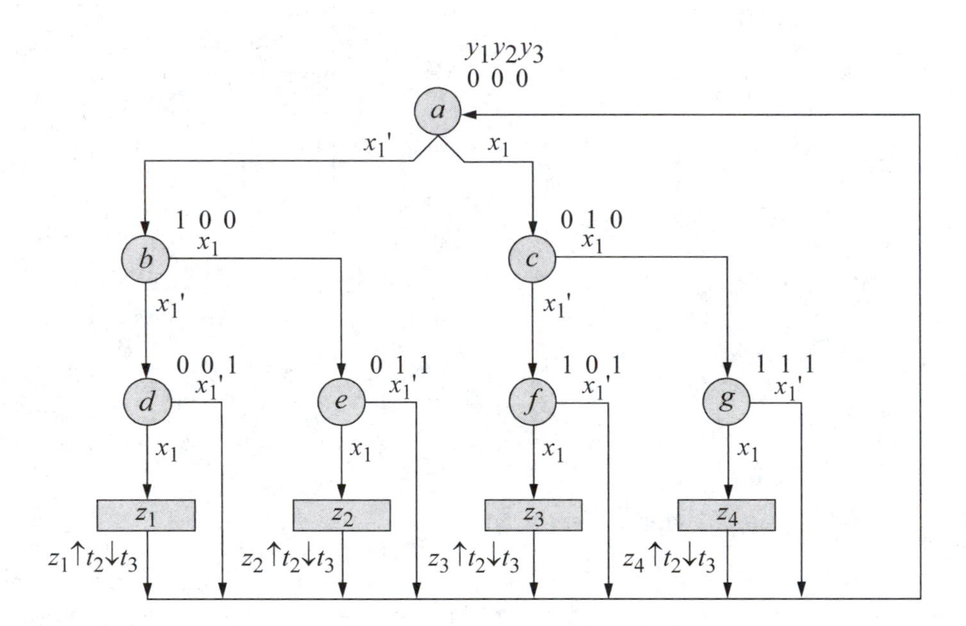

Figure 4.35 State diagram for the Mealy machine of Example 4.8. There is one unused state: $y_1y_2y_3 = 110$.

Since the assertion for all outputs is at time t_2, there is no need to be concerned with output glitches caused by state changes. This provides considerable freedom in assigning state codes. It was stated previously that a decoder can also be used as a demultiplexer, in which the inputs $x_{n-1}, \ldots, x_1, x_0$ function as select lines. Referring to these inputs as decoder select lines, therefore, is not inappropriate.

Three state flip-flops would normally require a 3:8 decoder, where flip-flops y_1, y_2, and y_3 connect to the decoder select inputs x_2, x_1, and x_0, respectively. There are, however, only four outputs from this machine. This suggests that a 2:4 decoder would suffice for the output logic, denoting a method of selection (or addressing) that is analogous to that used for nonlinear-select multiplexers. An attempt should be made, therefore, to use either y_1y_2, or y_1y_3, or y_2y_3 as decoder select inputs. Thus, the four output states should have state codes assigned as follows:

$$y_1y_2y_3 = 00–,\ 01–,\ 10–,\ 11–,\ \text{or}$$
$$y_1y_2y_3 = 0–0,\ 0–1,\ 1–0,\ 1–1,\ \text{or}$$
$$y_1y_2y_3 = –00,\ –01,\ –10,\ –11$$

Any of the above state code assignments permits the 2-variable minterms y_iy_j to generate the appropriate four outputs. Coincidental with the output state code assignment, consideration must be given to using input multiplexers with the fewest number of data inputs. It is also desirable to have a minimal number of logic gates for the multiplexer data inputs. Considering all of the above requirements for minimization, the state code assignment is obtained as shown in the state diagram of Figure 4.35. This is only one of several possible assignments.

The four output states *d*, *e*, *f*, and *g* correspond to $y_1y_2 = 00$, 01, 10, and 11, respectively. Therefore, y_1 connects to the decoder select input x_1 and y_2 connects to x_0. In each of the four output states, flip-flop y_3 is set to 1 so that y_3 will be a common input variable for all four outputs, thus maintaining a minimal number of λ logic functions. There is one unused state: $y_1y_2y_3 = 110$.

The input maps, obtained from the state diagram, are shown in Figure 4.36. The map for Dy_1 contains only three unique entries, implying that a 4:1 multiplexer would satisfy the input requirements. Three approaches are feasible for the multiplexer-designed input logic for flip-flop y_1. The first method uses flip-flops y_2 and y_3 as select inputs. This results in the assignment shown in Table 4.13, which is derived directly from the input map for Dy_1. The second approach uses flip-flops y_1 and y_3 as select inputs. This assignment is shown in Table 4.14. The third approach uses flip-flops y_1 and y_2 as select inputs, resulting in one AND gate ($y_3'x_1'$) for data input d_0. As is evident, all three approaches require the same amount of logic: one 4:1 multiplexer and one 2-input logic gate. Consider the input map for Dy_2. If the unused state $y_1y_2y_3 = 110$ is assigned a value of x_1, then the input logic is best obtained by a single AND gate, as shown in Equation 4.6.

The input map for Dy_3 contains only two unique entries: 0 and 1. This suggests that perhaps a 2:1 multiplexer with one select input s_0 could be used. Let $s_0 = y_1$. The multiplexer data input equations are: $d_0 = y_2y_3'$ and $d_1 = y_3'$. Permute the flip-flop variables so that $s_0 = y_2$. The data input equations are: $d_0 = y_1y_3'$ and $d_1 = y_3'$. Permute the flip-flop variables again so that $s_0 = y_3$. The data input equations are: $d_0 = y_1 + y_2$ and $d_1 = 0$. All three permutations require one logic gate plus a 2:1 multiplexer, which is equivalent to one-half of a noncustom integrated circuit.

$y_1 \backslash y_2y_3$	0 0	0 1	1 1	10
0	[0] x_1'	[1] 0	[3] 0	[2] 1
1	[4] 0	[5] 0	[7] 0	[6] –

Dy_1

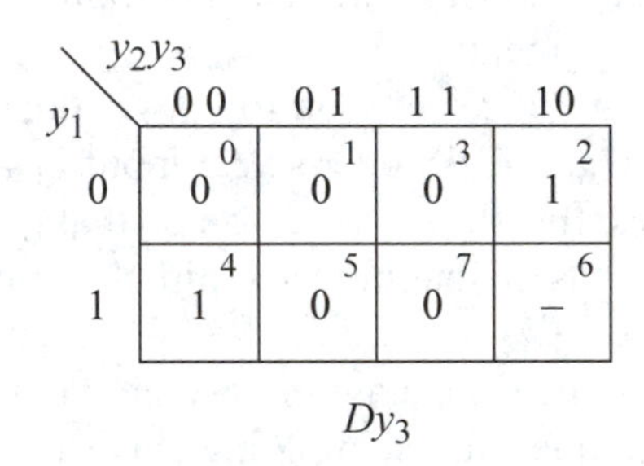

Dy_2

$y_1 \backslash y_2y_3$	0 0	0 1	1 1	10
0	[0] 0	[1] 0	[3] 0	[2] 1
1	[4] 1	[5] 0	[7] 0	[6] –

Dy_3

Figure 4.36 Input maps for the Mealy machine of Figure 4.35.

Table 4.13 Assignment of multiplexer select inputs for Dy_1 of Figure 4.36

Multiplexer select inputs s_1 (y_2)	s_0 (y_3)	Flip-flop y_1 input (Multiplexer y_1 output) Dy_1
0	0	$x_1'y_1'$
0	1	0
1	0	1[1]
1	1	0

1. Assuming that unused minterm 6 = 1.

Table 4.14 Assignment of multiplexer select inputs for Dy_1 of Figure 4.36

Multiplexer select inputs s_1 (y_1)	s_0 (y_3)	Flip-flop y_1 input (Multiplexer y_1 output) Dy_1
0	0	$y_2'x_1' + y_2 = y_2 + x_1'$
0	1	0
1	0	0[1]
1	1	0

1. Assuming that unused minterm 6 = 0.

$$Dy_2 = y_3'x_1 \tag{4.6}$$

If it is assumed that there are two 4:1 multiplexers per integrated circuit, then this permits two multiplexers — one each for y_1 and y_3 — to be allocated to the same integrated circuit, which provides common select inputs. Similar to the input map for Dy_1, three approaches are practicable for a multiplexer-designed input for flip-flop y_3. The first method uses flip-flops y_2 and y_3 as select inputs. This results in the assignment shown in Table 4.15, which is derived directly from the input map. The second approach uses flip-flops y_1 and y_3 as select inputs. This assignment is shown in Table 4.16. The third approach uses flip-flops y_1 and y_2 as select inputs, which yields the following multiplexer data input equations for y_3: $d_0 = 0$, $d_1 = y_3'$, $d_2 = y_3'$, and $d_3 = 0$. All three permutations result in no additional logic gates for the multiplexer data inputs. Therefore, nonlinear-select multiplexers will be used with common select inputs, where $s_1 s_0 = y_2 y_3$.

Table 4.15 Assignment of multiplexer select inputs for Dy_3 of Figure 4.36

Multiplexer select inputs s_1 (y_2)	s_0 (y_3)	Flip-flop y_3 input (Multiplexer y_3 output) Dy_3
0	0	y_1
0	1	0
1	0	1[1]
1	1	0

1. Assuming that unused minterm 6 = 1.

Table 4.16 Assignment of multiplexer select inputs for Dy_3 of Figure 4.36

Multiplexer select inputs		Flip-flop y_3 input (Multiplexer y_3 output)
s_1 (y_1)	s_0 (y_3)	Dy_3
0	0	y_2
0	1	0
1	0	1[1]
1	1	0

1. Assuming that unused minterm 6 = 1.

The output maps are shown in Figure 4.37. Because this example portrays a Mealy machine, input x_1 is used as a map-entered variable. Table 4.17 illustrates the conditions that are necessary to generate the four outputs. The term y_3x_1 is common to all outputs and will be a component of the output enabling function for the decoder.

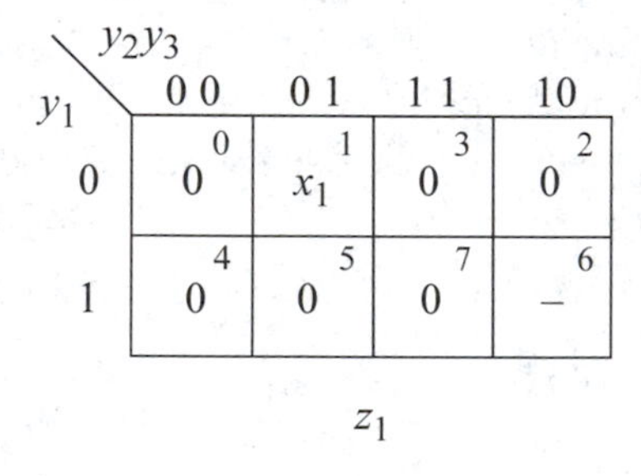

y_1 \ y_2y_3	0 0	0 1	1 1	10
0	0 (0)	0 (1)	x_1 (3)	0 (2)
1	0 (4)	0 (5)	0 (7)	– (6)

z_2

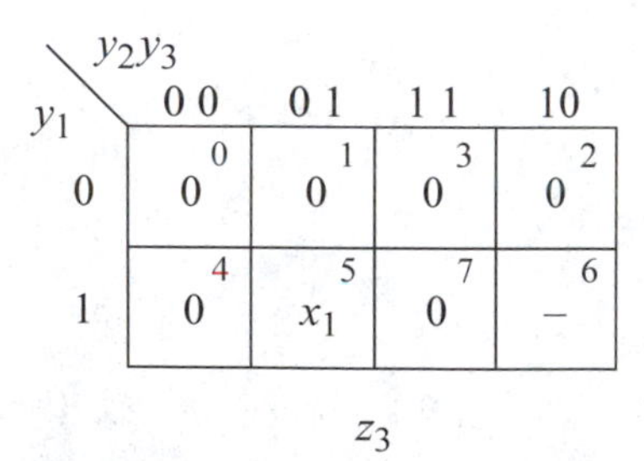

y_1 \ y_2y_3	0 0	0 1	1 1	10
0	0 (0)	0 (1)	0 (3)	0 (2)
1	0 (4)	0 (5)	x_1 (7)	– (6)

z_4

Figure 4.37 Output maps for the Mealy machine of Figure 4.35.

Table 4.17 Conditions for asserting z_1, z_2, z_3, and z_4 for the Mealy machine of Figure 4.35

Flip-flops		Outputs			
y_1	y_2	z_1	z_2	z_3	z_4
0	0	y_3x_1	0	0	0
0	1	0	y_3x_1	0	0
1	0	0	0	y_3x_1	0
1	1	0	0	0	y_3x_1

The logic diagram is illustrated in Figure 4.38 and is generated from the input maps of Figure 4.36 using the selection assignment of Table 4.13 and Table 4.15 for Dy_1 and Dy_3, respectively. The select inputs for MUX y_1 and MUX y_3 are identical and incorporate the nonlinear-select technique to utilize smaller multiplexers. The D input for flip-flop y_2 is simply the AND function as specified in Equation 4.6.

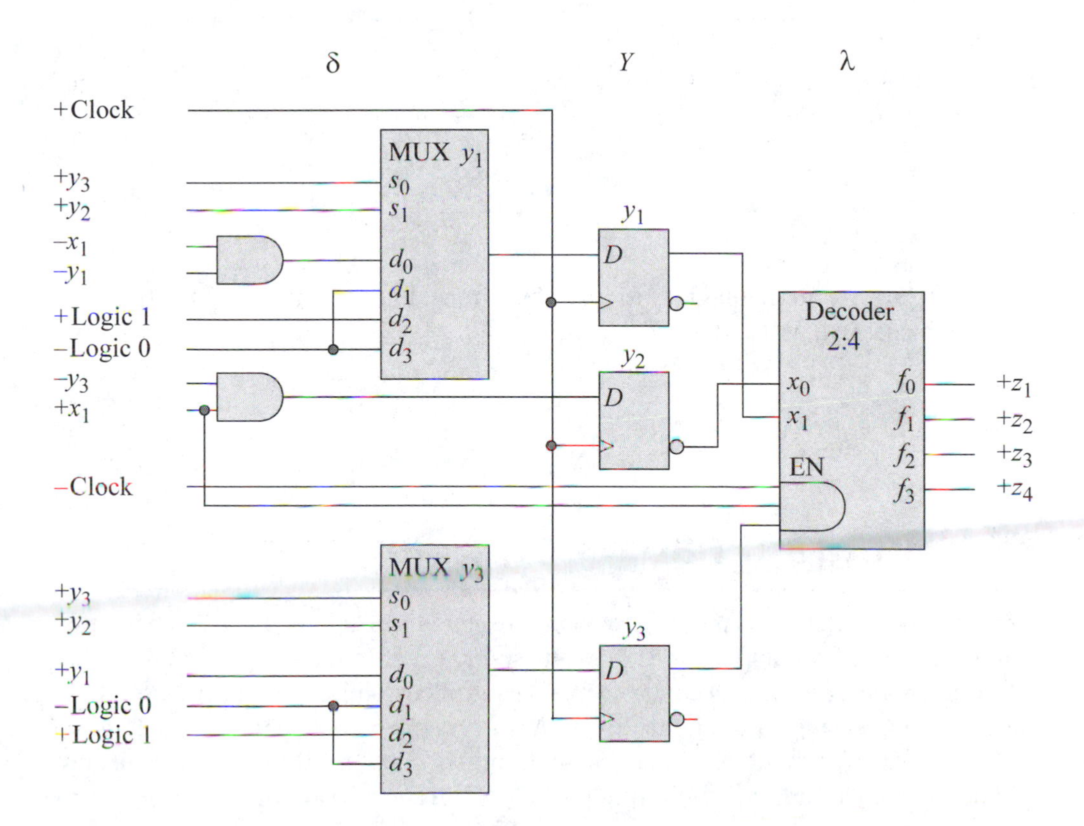

Figure 4.38 Logic diagram for the Mealy machine of Figure 4.35. All outputs are asserted at time t_2 and deasserted at t_3.

Flip-flops y_1 and y_2 connect to the decoder address inputs x_1 and x_0, respectively, to generate z_1, z_2, z_3, and z_4 contingent upon the input conditions of the enable AND gate. The enable gate contains three inputs: the –Clock signal to provide assertion at time t_2 and deassertion at t_3; the $+x_1$ signal to generate the Mealy-type outputs; and the $+y_3$ signal which is a necessary condition to assert the outputs in their respective states. Figure 4.39 depicts a timing diagram for state transition sequences *a,b,d* and *a,c,f*, which assert outputs z_1 and z_3, respectively.

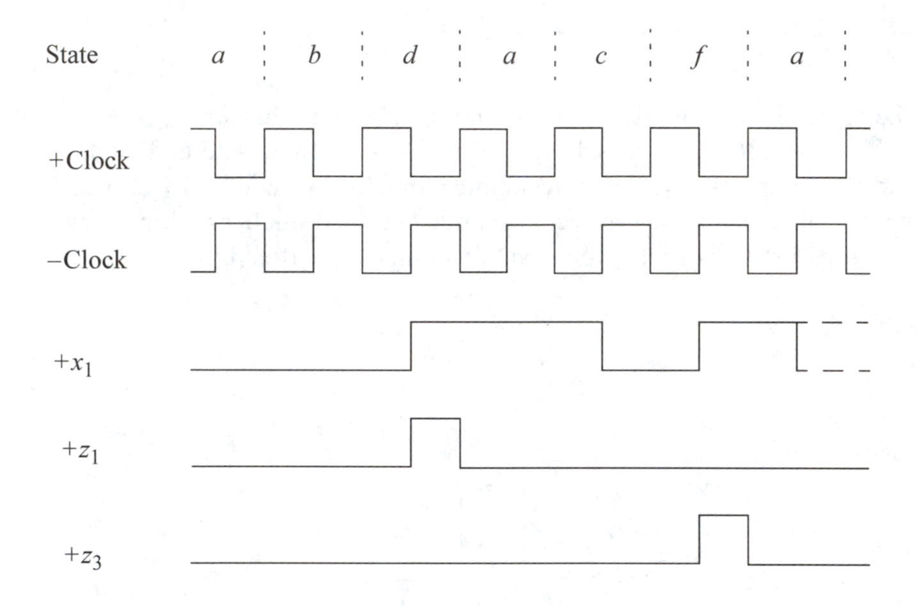

Figure 4.39 Timing diagram for the Mealy machine of Figure 4.35 for state transition sequences *a,b,d* and *a,c,f*.

4.3 Programmable Logic Devices

Programmable logic devices (PLDs) were presented in Chapter 1. This chapter extends the concepts to applications in synthesizing synchronous sequential machines. PLDs have achieved wide acceptance as a means of customizing logic circuits and are becoming increasingly popular in the synthesis of sequential machines. The large variety of PLD circuits yields higher integration than standard integrated circuits providing, in many cases, thousands of logic gates with accompanying storage elements, in a single package.

The propagation delay through a PLD decreases with each iteration of technological development, adding to its viability as an attractive alternative to standard logic circuits. PLDs offer a compromise between fixed-function devices and custom gate arrays. Field-programmable logic devices are presently available not only in transistor-transistor logic (TTL), but also in low-power complementary metal-oxide semiconductor (CMOS) technology and high-speed emitter-coupled logic (ECL).

Current software systems simplify the task of logic design using PLDs and also perform logic minimization and test vector generation for system simulation. The ultimate goal is to specify the input/output characteristics of a machine in a high-level language. The hardware-software system will then synthesize the machine to yield a minimized, functionally tested unit.

PLDs contain arrays of AND and OR logic gates in which the inputs are either prewired by the manufacturer or are fused so that the designer can establish the interconnection between circuits. Many PLDs also contain storage elements, making these devices ideally suited for both combinational and sequential logic design. A major advantage of synthesizing with PLDs is their field programmability attribute, which permits a fast response not only to initial design considerations but also to machine modifications.

The inherent versatility and ease of implementation of PLDs is complemented by a reduction in physical mounting space in contrast to an equivalent circuit using nonprogrammable logic devices. In some applications, state transition sequences may have to be altered in order to accommodate a change in machine specifications. These changes can be easily incorporated in PLD implementations by means of a hardware description language (HDL) such as, Verilog HDL or VHDL, which also update documentation automatically. A change in sequence, therefore, can be readily accomplished by simply reprogramming the PLD to incorporate the new design. This is in contrast to rewiring when using fixed-function devices.

The following sections illustrate the use of PLDs in the synthesis of synchronous sequential machines. The PLDs that will be presented are programmable read-only memories (PROMs), programmable array logic (PAL) devices, programmable logic arrays (PLAs), and field-programmable gate arrays (FPGAs).

4.3.1 Programmable Read-Only Memory

The concept of read-only memories (ROMs) for sequential machine synthesis and processor control is quite common. ROMs are also used extensively in developing new microprogram-controlled systems. Various versions of programmable ROM (PROM) devices are available which significantly reduce the turnaround time required for modifications to existing machines. Erasable PROMs (EPROMs) can be reprogrammed at the user's location, thereby enabling the designer to modify the stored program on site. There are two main types of EPROMs, distinguished by the method used for erasure: the electrically erasable PROM (EEPROM) in which the program is erased upon application of specific electrical signals and the ultraviolet erasable PROM (UVEPROM) in which the program is erased when exposed to ultraviolet radiation.

Figure 4.40 illustrates the internal organization of a PROM with two address inputs x_1 and x_2 and four outputs f_1, f_2, f_3, and f_4. Inputs x_1 and x_2 select one of four words using the AND array decoder: word 0, 1, 2, or 3 that corresponds to $x_1 x_2 = 00$, 01, 10, or 11, respectively. Thus, the AND array cannot be programmed, as indicated by the "hardwired" connection symbol " ● ." The OR array, however, is programmable. The symbol "×" indicates an intact fuse at the intersection of the AND gate product term and the OR gate input and provides a logic 1 to the specified OR gate input. The absence of an × indicates an open fuse, which provides a logic 0 to the OR gate input.

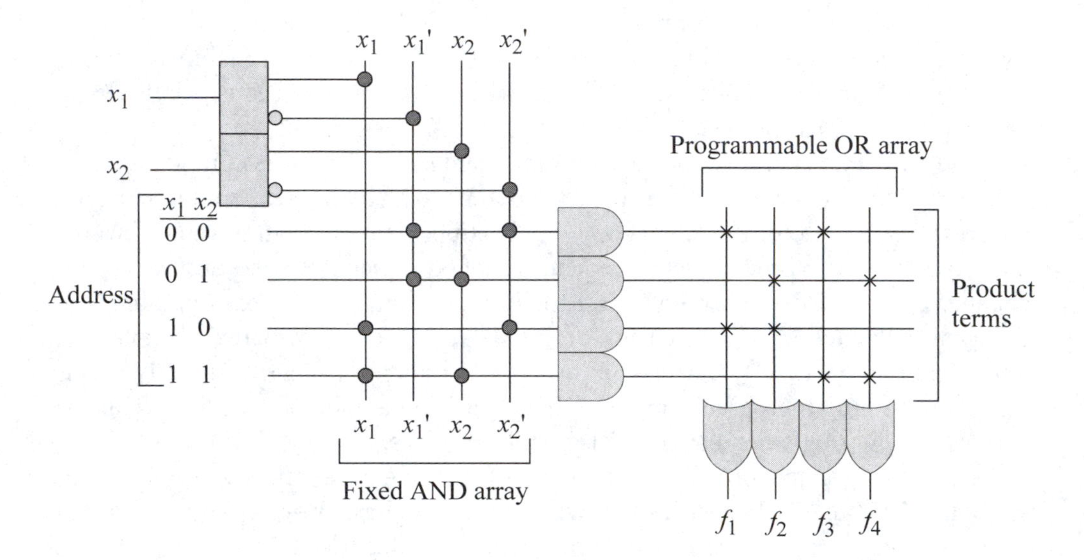

Figure 4.40 PROM organization with two address inputs x_1 and x_2 and four outputs f_1, f_2, f_3, and f_4.

The truth table for the PROM is shown in Table 4.18. The outputs are shown in a sum-of-products form in Equation 4.7. The 4-bit output word can be easily expanded to a larger word by simply concatenating additional OR gates in the OR array. The size (or capacity) of a PROM is characterized by the expression $2^n \times m$, where n represents the number of inputs and m specifies the number of outputs. Therefore, n inputs requires 2^n AND gates and m outputs requires m OR gates.

PROMs provide a simple, yet elegant method of implementing synchronous sequential machines. A general block diagram of a PROM-controlled Moore sequential machine is shown in Figure 4.41. A Mealy machine can be portrayed by inserting λ output logic as a function of the input alphabet X. The PROM address inputs are comprised of the machine inputs $x_1, x_2, \ldots, x_n$ concatenated with the outputs of the state flip-flops $y_1, y_2, \ldots, y_p$. The outputs of the PROM connect to an $(m + p)$-stage register, which is clocked on the positive transition of the machine clock. The outputs, therefore, are asserted at time t_1 and deasserted at t_3. Other versions of output assertion and deassertion are possible by employing the complement of the machine clock or by including λ output logic which is enabled by a particular clock phase; for example, $z_i \uparrow t_2 \downarrow t_3$.

Table 4.18 Truth table for the PROM of Figure 4.40

Address inputs x_1 x_2	Outputs f_1 f_2 f_3 f_4
0 0	1 0 1 0
0 1	0 1 0 1
1 0	1 1 0 0
1 1	0 0 1 1

$$f_1(x_1, x_2) = \Sigma_m(0, 2)$$
$$f_2(x_1, x_2) = \Sigma_m(1, 2)$$
$$f_3(x_1, x_2) = \Sigma_m(0, 3)$$
$$f_4(x_1, x_2) = \Sigma_m(1, 3) \tag{4.7}$$

The next state $Y_{k(t+1)}$ is generated by a unique word in the PROM as a function of the present inputs $X_{i(t)}$ and the present state $Y_{j(t)}$. The word that is addressed by $X_{i(t)} \cdot Y_{j(t)}$ appears at the outputs of the OR array as $f_1, f_2, \ldots, f_m, f_{m+1}, \ldots, f_{m+p}$. These outputs connect to the D inputs of the register such that, f_i connects to Dz_i and f_{m+i} connects to Dy_i. The register is loaded from the PROM on the positive clock transition. The present state is fed back to the PROM address inputs and, after an appropriate propagation delay through the PROM arrays, the next state and next outputs are available at the D inputs of the register. At the next positive clock transition, the next state is loaded into the register and becomes the present state.

Example 4.9 Given the state diagram of Figure 4.42, a Moore machine will be synthesized using a PROM for the δ next-state logic and a register for the state flip-flops and outputs, plus any additional logic that is necessary for the λ output logic. Step 1 (state diagram) of the synthesis procedure, therefore, is complete. As is evident from the state diagram, the machine has two serial inputs x_1 and x_2 and three outputs z_1, z_2, and z_3, which are asserted as shown. When the sequence $x_1 x_2 = 00$ occurs, z_1

is asserted; when the sequence $x_1x_2 = 11$ occurs, z_2 is asserted. Output z_3 is asserted during the fourth clock period, regardless of the path taken, indicating the end of any sequence. Figure 4.43 depicts valid sequences to assert z_1 and z_2 and a sequence in which only z_3 is asserted.

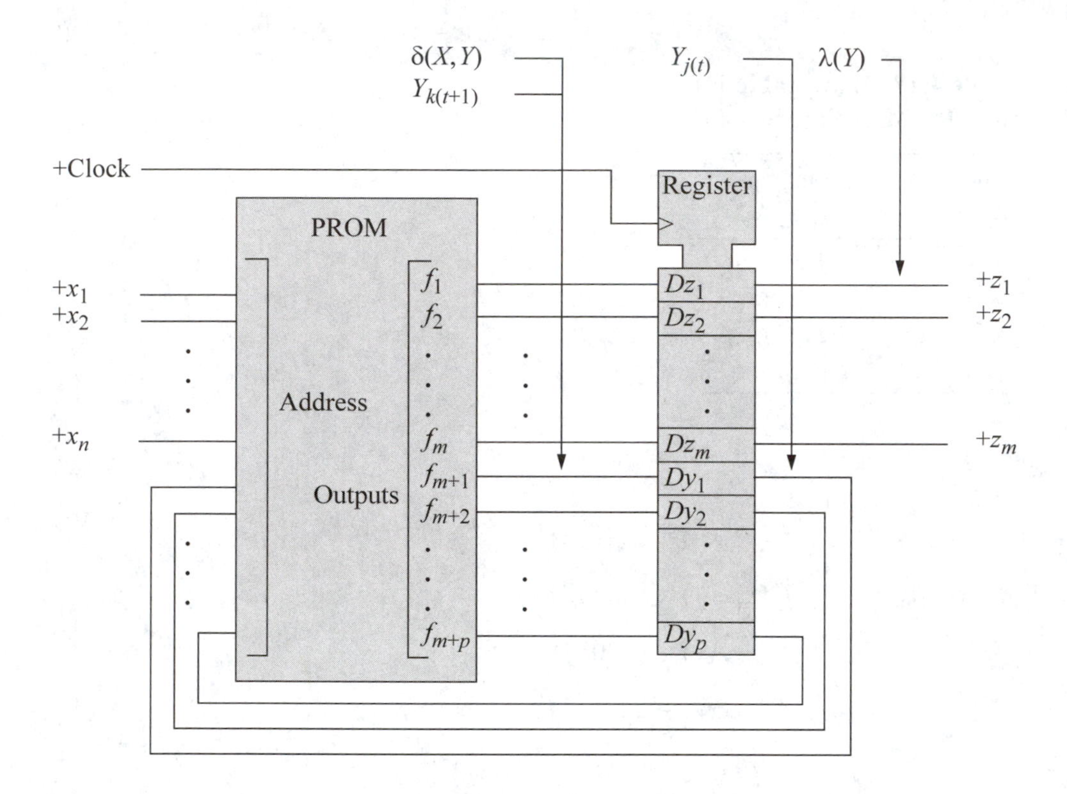

Figure 4.41 General block diagram of a $2^n \times (m + p)$ PROM-controlled Moore synchronous sequential machine.

Step 2 of the synthesis procedure checks for equivalent states. Section 3.1.1 specifies the necessary conditions for equivalence which are summarized as follows:

1. The two states under consideration must have the same outputs.
2. Both states must have the same or equivalent next state.

Both conditions must be true for equivalence to exist. The only potential equivalence is between states b and c. However, both states have different next states, depending upon the value of input x_2. Therefore, the state diagram contains no equivalent states.

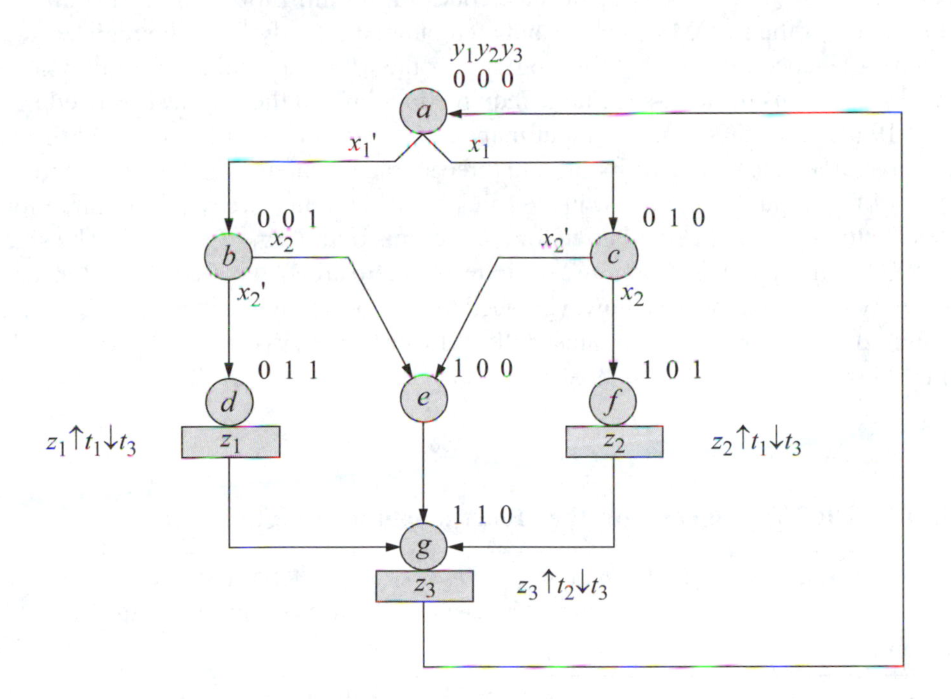

Figure 4.42 State diagram for a Moore machine to detect an input sequence of $x_1x_2 = 00$ or 11 to assert output z_1 or z_2, respectively. There is one unused state: $y_1y_2y_3 = 111$.

State	*a*	*b*	*d*	*g*	*a*	*c*	*f*	*g*	*a*	*b*	*e*	*g*	*a*
Input x_1	0	–	–	–	1	–	–	–	0	–	–	–	
Input x_2	–	0	–	–	–	1	–	–	–	1	–	–	
Output asserted			↑ z_1	↑ z_3			↑ z_2	↑ z_3				↑ z_3	

Figure 4.43 Illustration of valid sequences to assert outputs z_1 and z_2 and a sequence where only z_3 is asserted.

State codes are assigned in Step 3. The assignment of state codes is arbitrary in this example, since there is no need to minimize the δ next-state logic using state code adjacency — the input logic is implemented by the PROM program. Also, the λ

output logic does not require specific state codes for minimization, since the outputs are a function of the PROM program, and are obtained directly from the register.

Table 4.19 specifies the PROM program for the Moore machine of Figure 4.42. The table represents the next-state table required in Step 4 of the synthesis procedure. Table 4.19 also substitutes as the input maps and output maps of Steps 5 and 6, respectively. Since there are two inputs and three storage elements, the PROM address consists of five input lines: x_1, x_2, y_1, y_2, and y_3. Six outputs are required: three for the next state y_1, y_2, and y_3 and three for the presents outputs z_1, z_2 and z_3. The size of the PROM, therefore, is $2^n \times m = 2^5 \times 6$; that is, there are 32 words with six bits per word. Only the first 28 words, however, need to be programmed, since $y_1 y_2 y_3 = 111$ is an unused state, providing four unused PROM words: $y_1 y_2 y_3 x_1 x_2 = 11100$ through 11111.

Table 4.19 PROM program for the Moore machine of Figure 4.42

	PROM address		PROM outputs	
	Present state	Present inputs	Next state	Present outputs
State name	$y_1\ y_2\ y_3$	$x_1\ x_2$	$y_1\ y_2\ y_3$	$z_1\ z_2\ z_3$
a	0 0 0	0 0	0 0 1	0 0 0
	0 0 0	0 1	0 0 1	0 0 0
	0 0 0	1 0	0 1 0	0 0 0
	0 0 0	1 1	0 1 0	0 0 0
b	0 0 1	0 0	0 1 1	0 0 0
	0 0 1	0 1	1 0 0	0 0 0
	0 0 1	1 0	0 1 1	0 0 0
	0 0 1	1 1	1 0 0	0 0 0
c	0 1 0	0 0	1 0 0	0 0 0
	0 1 0	0 1	1 0 1	0 0 0
	0 1 0	1 0	1 0 0	0 0 0
	0 1 0	1 1	1 0 1	0 0 0
d	0 1 1	0 0	1 1 0	1 0 0
	0 1 1	0 1	1 1 0	1 0 0
	0 1 1	1 0	1 1 0	1 0 0
	0 1 1	1 1	1 1 0	1 0 0
e	1 0 0	0 0	1 1 0	0 0 0
	1 0 0	0 1	1 1 0	0 0 0
	1 0 0	1 0	1 1 0	0 0 0
	1 0 0	1 1	1 1 0	0 0 0

Continued on next page

Table 4.19 PROM program for the Moore machine of Figure 4.42

	PROM address		PROM outputs	
	Present state	Present inputs	Next state	Present outputs
State name	$y_1\ y_2\ y_3$	$x_1\ x_2$	$y_1\ y_2\ y_3$	$z_1\ z_2\ z_3$
f	1 0 1	0 0	1 1 0	0 1 0
	1 0 1	0 1	1 1 0	0 1 0
	1 0 1	1 0	1 1 0	0 1 0
	1 0 1	1 1	1 1 0	0 1 0
g	1 1 0	0 0	0 0 0	0 0 1
	1 1 0	0 1	0 0 0	0 0 1
	1 1 0	1 0	0 0 0	0 0 1
	1 1 0	1 1	0 0 0	0 0 1

Output glitches are not possible in this type of PROM implementation, because the PROM outputs are stable at the *D* inputs of the register long before the positive clock transition occurs. Thus, the output flip-flops will never enter a metastable condition.

The PROM is programmed directly from the state diagram. For example, refer to Table 4.19 in conjunction with the state diagram. In state *a* ($y_1y_2y_3 = 000$), the next state is *b* ($y_1y_2y_3 = 001$) if $x_1 = 0$, and *c* ($y_1y_2y_3 = 010$) if $x_1 = 1$. State *a* provides no outputs. In state *c* ($y_1y_2y_3 = 010$), the next state is *e* ($y_1y_2y_3 = 100$) if $x_2 = 0$, and *f* ($y_1y_2y_3 = 101$) if $x_2 = 1$. State *c* does not generate an output. The next state for state *d* ($y_1y_2y_3 = 011$) is state *g* ($y_1y_2y_3 = 110$), regardless of the values of x_1 and x_2. In state *d*, however, output z_1 is asserted while z_2 and z_3 are both inactive.

A standard 32 × 8 PROM is sufficient for this application. There will be four unused words and two unused PROM outputs. When the PROM program has been established, the device is then programmed by means of a PROM programmer — a unit that addresses each fuse location in turn and either opens the fuse or leaves it intact, depending upon the PROM output values specified in Table 4.19.

Step 7 of the synthesis procedure derives the logic diagram, which is shown in Figure 4.44. Two registers are depicted to illustrate the separate functions of the state register and the output register, although only a single 6-bit register is sufficient. The state flip-flop outputs are fed back to the PROM inputs to combine with the machine inputs to form the address function. Outputs z_1 and z_2 are produced directly from the output register, providing the requisite assertion and deassertion of t_1 and t_3, respectively, whereas, z_3 is generated by an AND gate using the –Clock signal as an enabling factor to assert z_3 at time t_2 and deassert z_3 at t_3. The timing diagram for a representative sequence to assert z_2 and z_3 is shown in Figure 4.45.

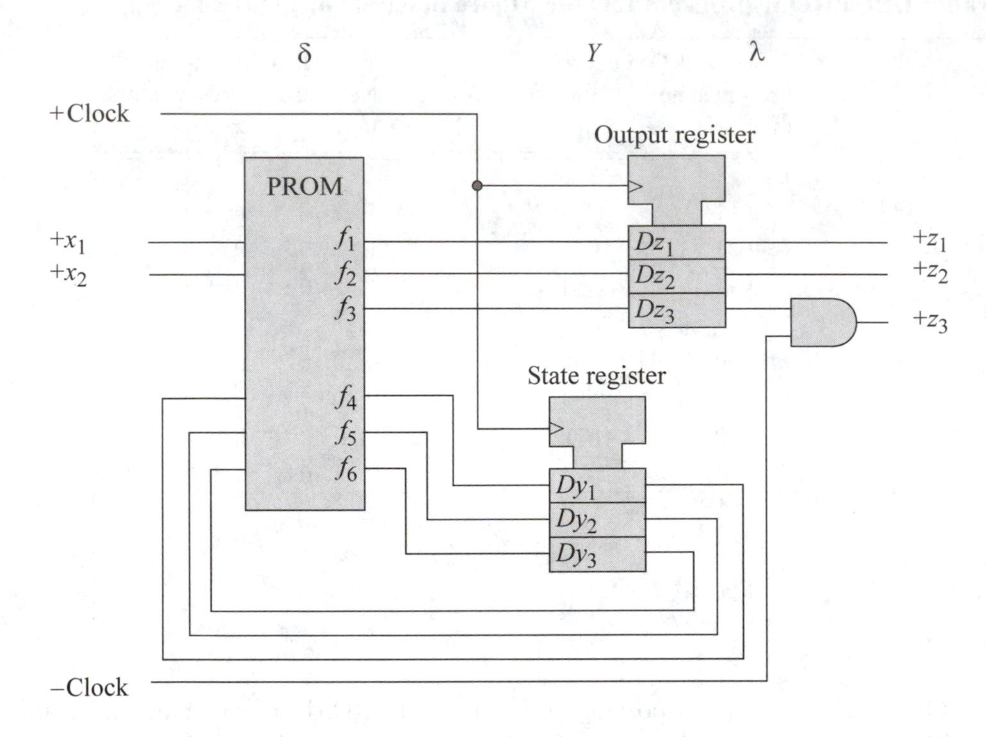

Figure 4.44 Logic diagram for the Moore machine of Figure 4.42 using a PROM for the δ next-state logic which is programmed according to the contents of Table 4.19.

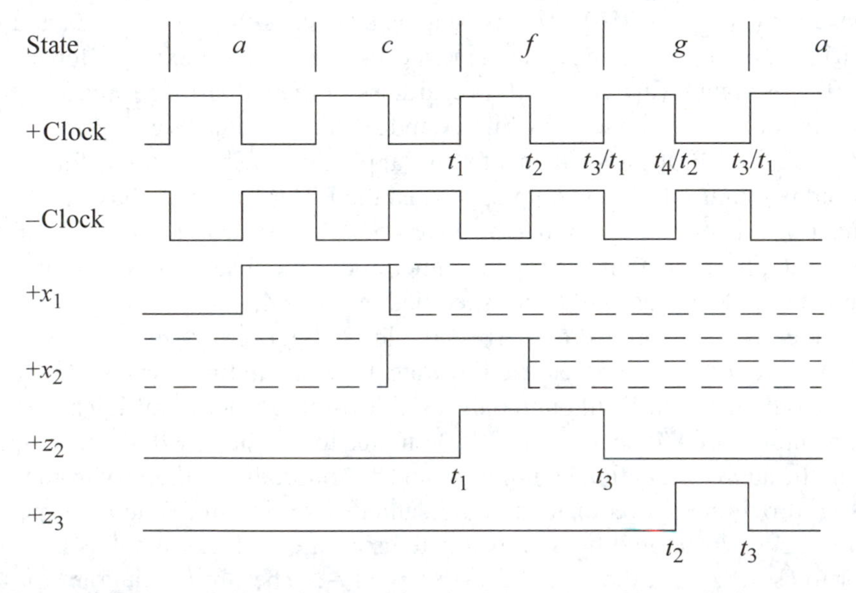

Figure 4.45 Timing diagram for the Moore machine of Figure 4.44 showing a valid sequence of $x_1x_2 = 11$ to assert output z_2. Output z_3 is asserted in state g.

4.3.2 Programmable Array Logic

A programmable array logic (PAL) device is structured with a programmable AND array and a nonprogrammable (fixed) OR array. The number of gates in the AND array is not a function of the number of inputs, as for PROMs. The AND array allows product terms to be programmed which then connect to a predefined OR array. The restriction of the prewired OR array is compensated by the wide variety of available PAL configurations. Specifying the configuration of the OR array, therefore, is simply a matter of device selection.

The organization of a basic PAL is shown in Figure 4.46. Each AND gate has $2n$ inputs, where n is the number of device inputs. The symbol "×" indicates an intact fuse, which connects a unique variable — either true or complemented — to one of the six AND gate inputs. The absence of an × indicates an open fuse, which supplies a logic 1 to the AND gate. Thus, the product terms consist only of the input variables specified by an ×.

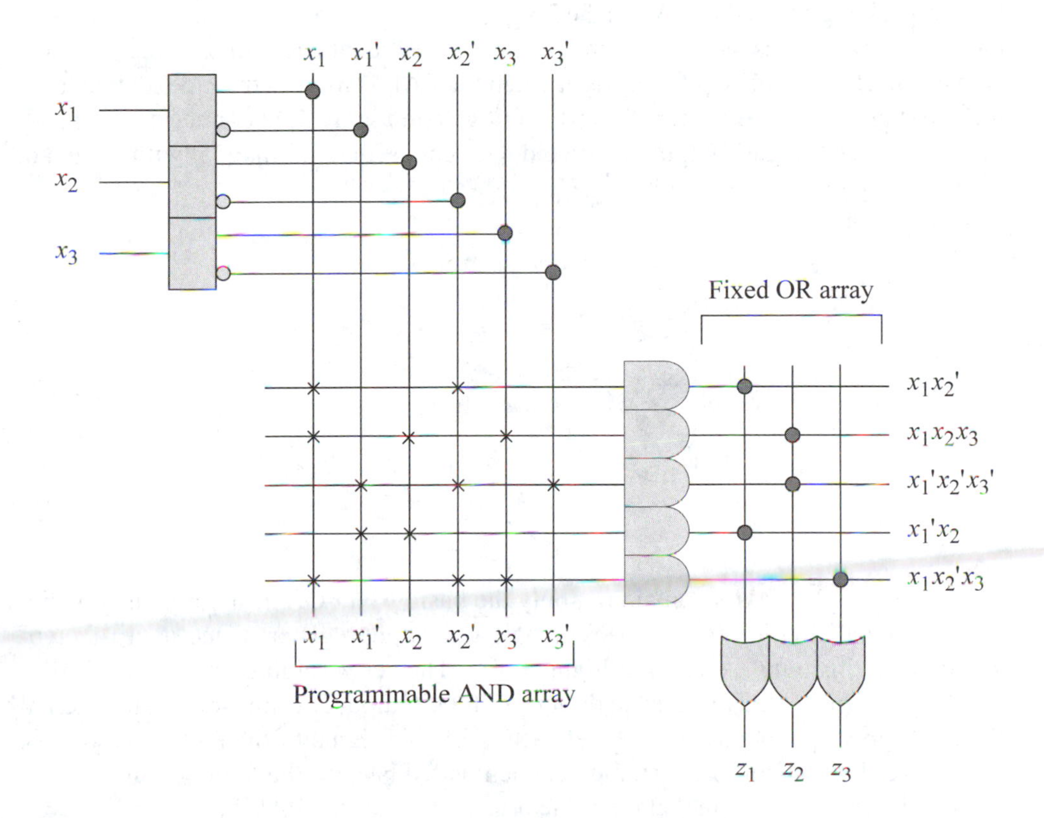

Figure 4.46 Organization of a PAL to implement the following functions: $z_1 = x_1x_2' + x_1'x_2$; $z_2 = x_1x_2x_3 + x_1'x_2'x_3'$; $z_3 = x_1x_2'x_3$.

All programmed terms of the three input variables x_1, x_2, and x_3 are available at the three outputs z_1, z_2, and z_3 in a sum-of-minterms or sum-of-products form. Each hardwired connection in the OR array indicates that the corresponding product term is a component of the appropriate output function. The output functions are indicated in Equation 4.8.

$$z_1 = x_1x_2' + x_1'x_2$$

$$z_2 = x_1x_2x_3 + x_1'x_2'x_3'$$

$$z_3 = x_1x_2'x_3 \tag{4.8}$$

Some PALs contain pins that provide both input and output functions. The PAL shown in Figure 4.47 contains a 3-state output driver that is enabled by a product term. When the product term is active, the sum-of-products function is transferred to output z_1. This function is also fed back to the input logic, providing a necessary requirement for synthesizing *SR* latches. When the 3-state driver is disabled, the output network of the driver operates as an input source. This bidirectional capability of the $-z_1(+x_4)$ signal is useful for shifting left or right when the PAL is used as the δ next-state logic for one stage of a shift register. Since the OR gate in Figure 4.47 connects to all product terms, each OR gate output of a PAL device can be diagrammatically illustrated in a more compact form as follows:

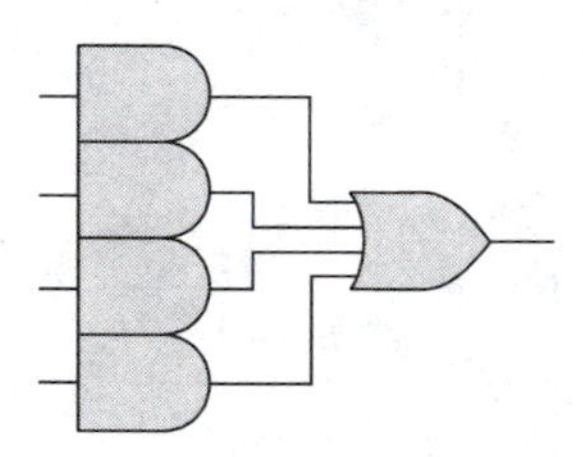

More complex PALs contain not only the basic AND-OR array organization but also additional circuitry for feedback signals and output registers. One stage of a typical PAL of this type is shown in Figure 4.48. The active-high output of the *D* flip-flop for y_1 is connected to a 3-state driver (buffer). Output z_1 is transmitted to external logic by applying a high voltage level to the +Enable signal of the PAL input buffer. The active- low output of the storage element is fed back to the δ input logic.

It is interesting to note that the basic organization of the PAL in Figure 4.48 conforms to the general structure of a synchronous sequential machine. The input drivers, together with the AND-OR arrays, constitute the δ next-state logic; flip-flop y_1 is the storage element; and the 3-state driver represents the λ output logic. The input equations obtained from input maps are in a sum-of-products form, which is the requisite format for programming the AND array. Thus, PALs and other PLDs are ideally suited for synthesizing synchronous sequential machines. To illustrate the synthesis procedure for machines that are implemented with PAL devices, two examples will be presented. The first is a machine depicting a Gray code counter; the second is a sequence detector with Mealy outputs.

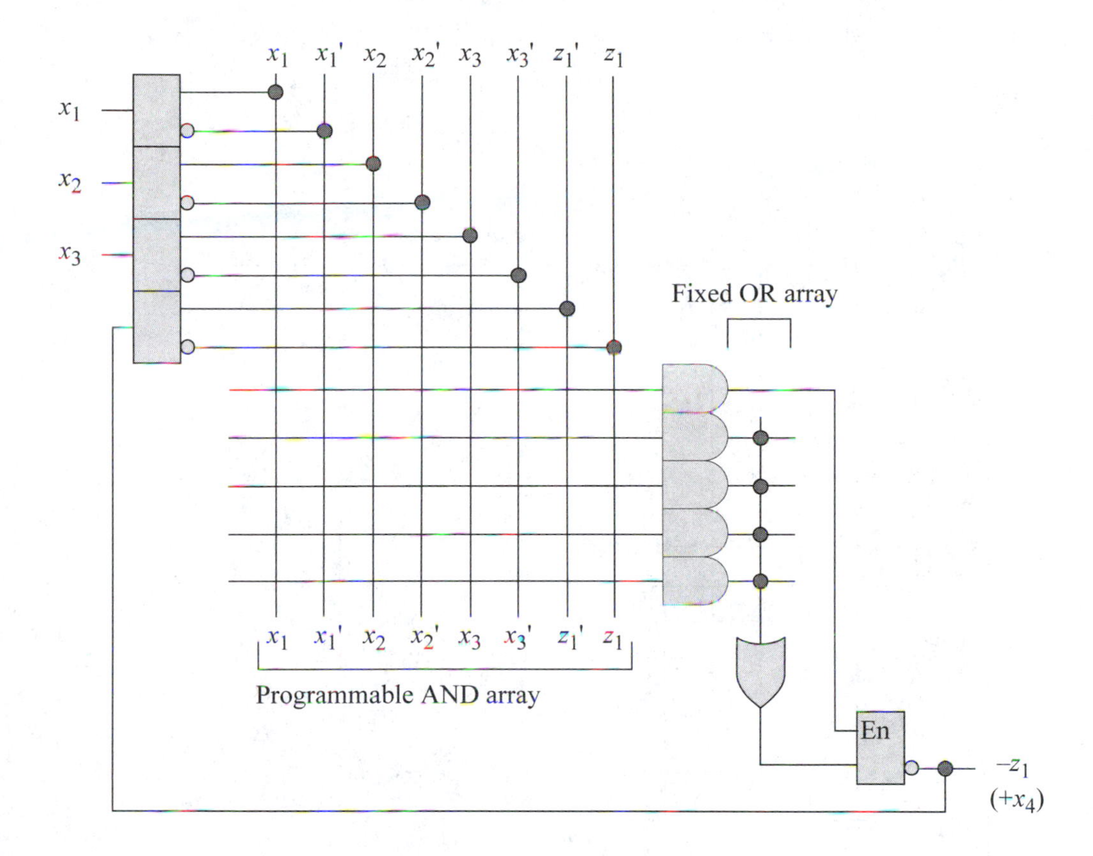

Figure 4.47 PAL circuit in which the 3-state output driver is enabled by a product term in the AND array. The output of the 3-state driver is bidirectional.

Example 4.10 A 3-bit Gray code counter will be synthesized that counts in the following sequence: 000, 001, 011, 010, 110, 111, 101, 100, 000, ... A single PAL device will be used for both the δ next-state logic and the storage elements, which consist of positive-edge-triggered D flip-flops. The state diagram is shown in Figure 4.49 and the input maps in Figure 4.50. The equations for flip-flops y_1, y_2, and y_3 are listed in Equation 4.9 in a sum-of-products form. If the state flip-flops do not have a reset input, then each term of Dy_1, Dy_2, and Dy_3 must contain a third variable — a Reset input x_1 — as shown in Equation 4.10. Thus, the counter operates only when the reset input is inactive.

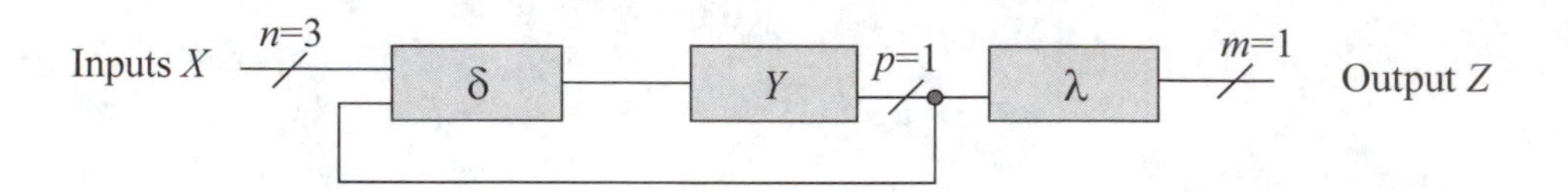

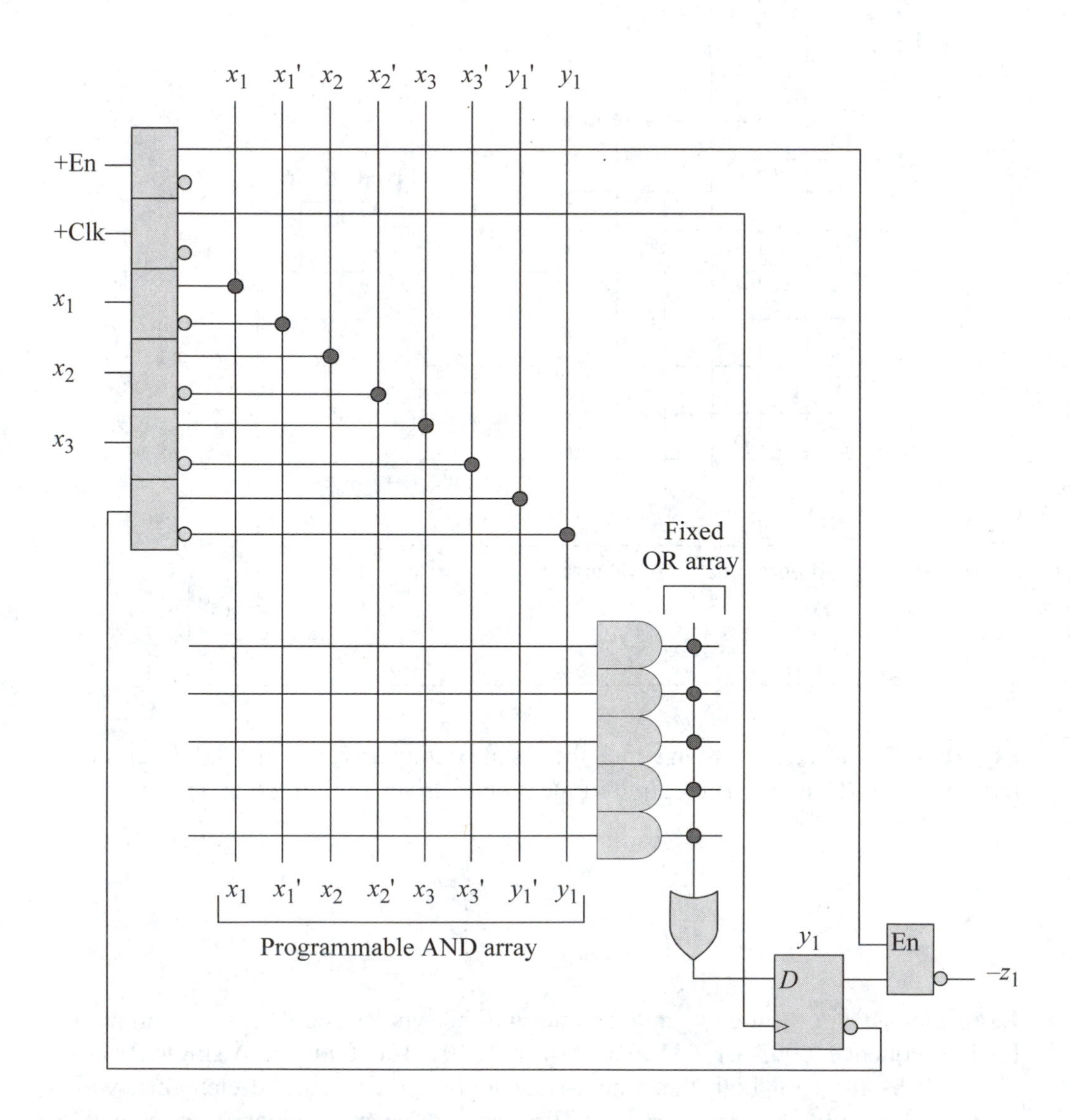

Figure 4.48 Typical PAL stage showing the AND-OR arrays with output flip-flop and feedback signal.

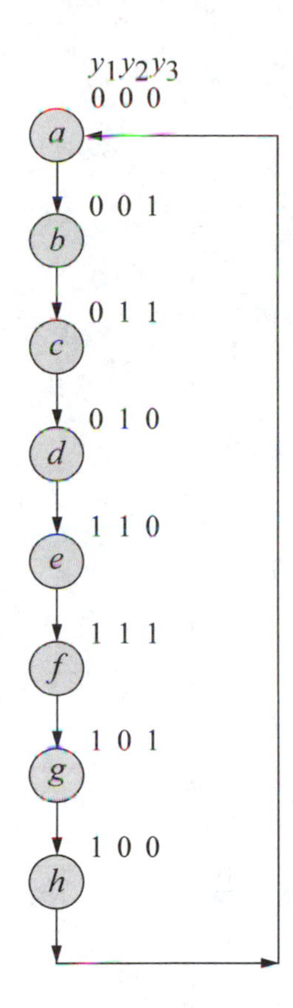

Figure 4.49 State diagram for a 3-bit Gray code counter.

The logic diagram using PAL technology is shown in Figure 4.51 and is programmed directly from the sum-of-products input equations of Equation 4.9. The only inputs to the counter are the +Clock signal and the +Enable signal. All remaining AND array inputs are feedback signals from the state flip-flops through drivers with inverting and noninverting outputs. The complemented output of flip-flop y_i is fed back to an input driver. Thus, when flip-flop y_i is set (=1), the complemented output of the driver generates a positive voltage level which is available to all AND gate inputs. If the input equation for y_i contains a flip-flop variable in its true form, then the complemented output of the driver is programmed to connect to the appropriate positive-input AND gate in the AND array. The δ next-state logic consists of the feedback drivers, the programmable AND array, and the fixed OR array. Outputs f_1, f_2, and f_3 of the OR gates connect to Dy_1, Dy_2, and Dy_3, respectively. The present state $Y_{j(t)}$ is fed back through drivers to the AND array and, after an appropriate propagation delay, appears at the flip-flop D inputs as the next state $Y_{k(t+1)}$. When the

+Enable signal is active, the state of the counter is available to external devices as active-low voltage levels.

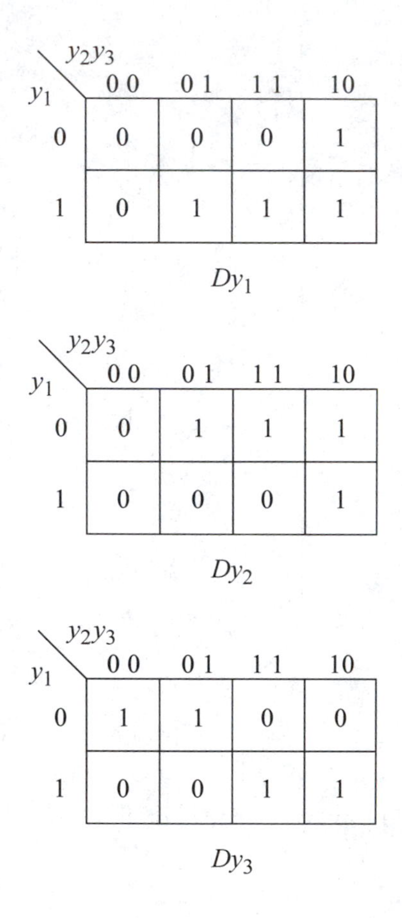

Figure 4.50 Input maps for the Gray code counter of Figure 4.49.

$$Dy_1 = y_1 y_3 + y_2 y_3'$$
$$Dy_2 = y_1' y_3 + y_2 y_3'$$
$$Dy_3 = y_1' y_2' + y_1 y_2 \tag{4.9}$$

$$Dy_1 = y_1 y_3 x_1' + y_2 y_3' x_1'$$
$$Dy_2 = y_1' y_3 x_1' + y_2 y_3' x_1'$$
$$Dy_3 = y_1' y_2' x_1' + y_1 y_2 x_1' \tag{4.10}$$

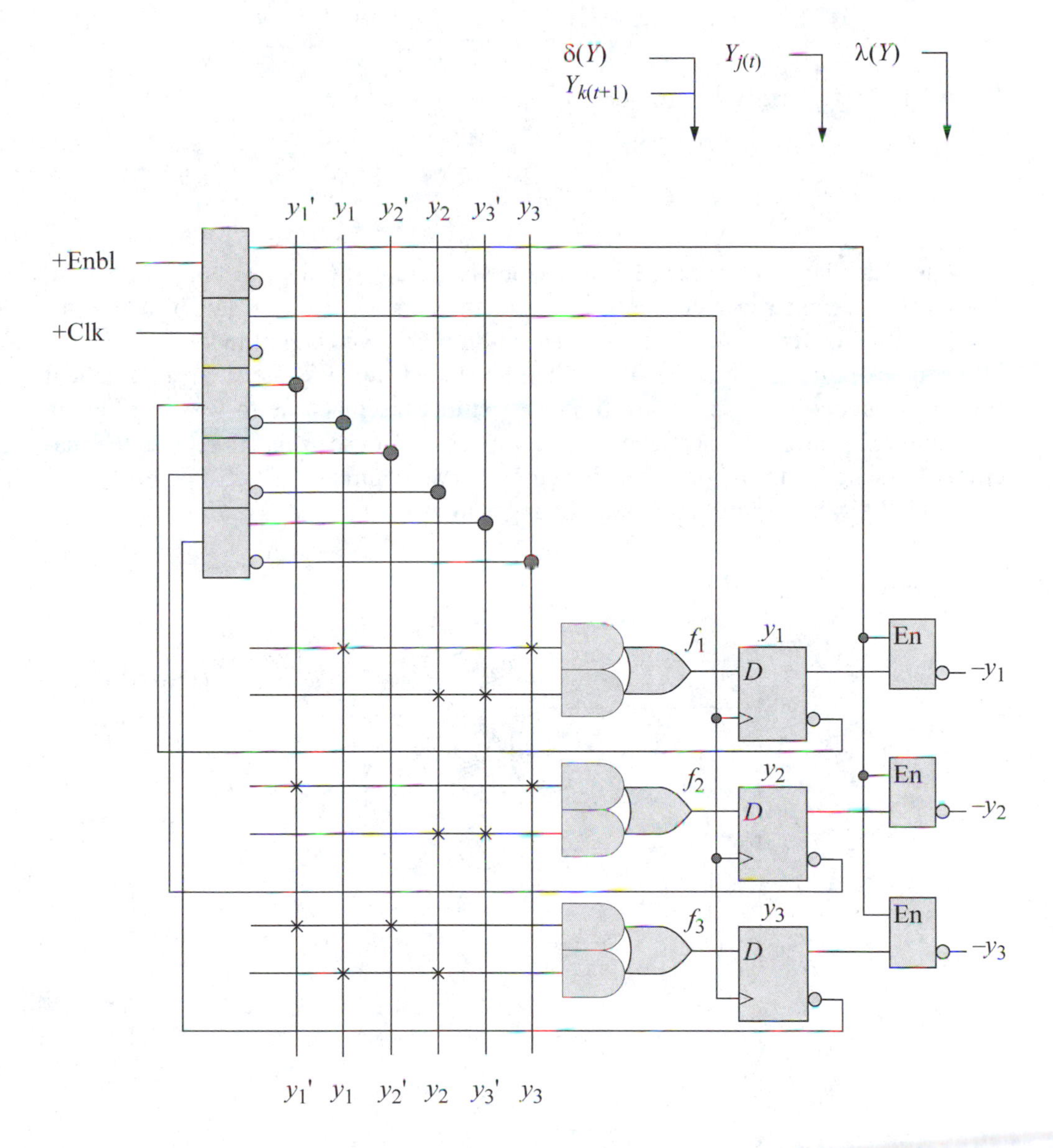

Figure 4.51 Logic diagram for the Gray code counter using a PAL device.

Example 4.11 PAL devices can be used to implement any synchronous sequential machine, including a sequence detector. A Mealy machine will be synthesized which checks for the sequence 01111110 on a serial input line x_1. The format for the input data is shown in Figure 4.52, where bit 0 is the low-order bit and is the first bit received. Input x_1 remains at a high voltage level until transmission is to begin, at which time x_1 assumes a low voltage level for one bit period, providing a negative transition.

$+x_1$	0	1	2	3	4	5	6	7

Figure 4.52 Format for an 8-bit sequence on a serial input line x_1.

The state diagram for this Mealy machine is graphically depicted in Figure 4.53. The machine remains in state a until the start of transmission is indicated by a high-to-low transition on input x_1. The bit sequence is then received beginning with bit 0, one bit per clock period. When the first 1 bit has been detected in state b, any subsequent 0 bit that occurs before six consecutive 1s returns the machine to state b to begin checking for a new valid sequence. Similarly, seven consecutive 1s returns the machine to state a to begin checking for a new valid sequence. Only when $x_1 = 01111110$ is output z_1 asserted. Changes to x_1 occur at time t_2.

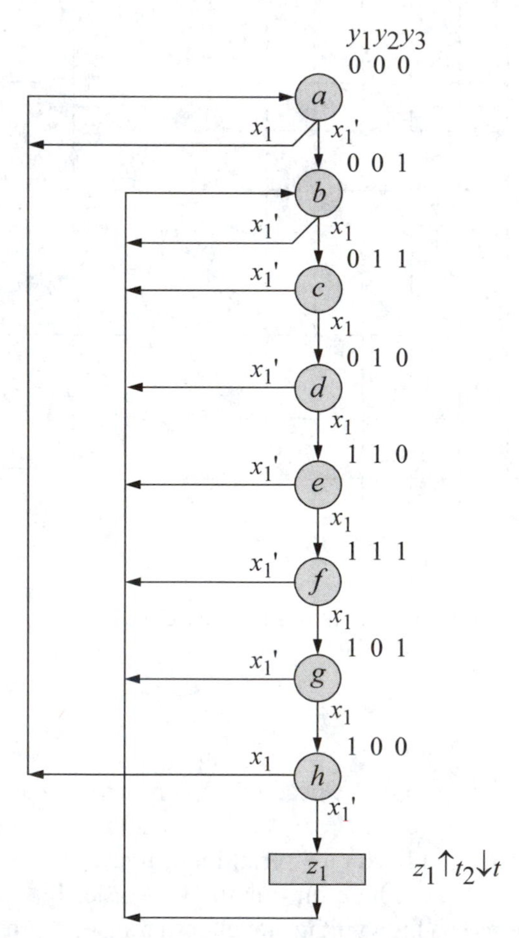

Figure 4.53 State diagram for the Mealy machine of Example 4.11.

The input maps, obtained from the state diagram, are shown in Figure 4.54. In state *a* ($y_1y_2y_3 = 000$), the next states for flip-flops y_1 and y_2 are 0 and 0, respectively, regardless of the value of x_1. The next state for y_3, however, is determined by the value of x_1; if $x_1 = 0$, then the next state for y_3 is 1, otherwise the next state is 0. In states *d*, *e*, *f*, and *g*, flip-flop y_1 has a next state of 1 only if $x_1 = 1$. Therefore, x_1 is entered in the map as a map-entered variable. Likewise, in states *b*, *c*, *d*, and *e*, flip-flop y_2 has a next value of 1 only if $x_1 = 1$. In the input map for flip-flop y_3, the next state is never an unconditional 0, irrespective of the path taken. The next state will be either an unconditional 1 or a value dependent upon x_1; that is, if $x_1 = 0$, then $y_3 = 1$. Since a logic $1 = x_1 + x_1'$, therefore, every minterm location in the map can be given a value of x_1'. This accounts for the x_1' term in the equation for Dy_3 of Equation 4.11. The x_1 term of the logic 1 expression must now be taken into account. This is very easily accomplished by reverting to the minterm value of 1, which generates the two remaining terms for Dy_3.

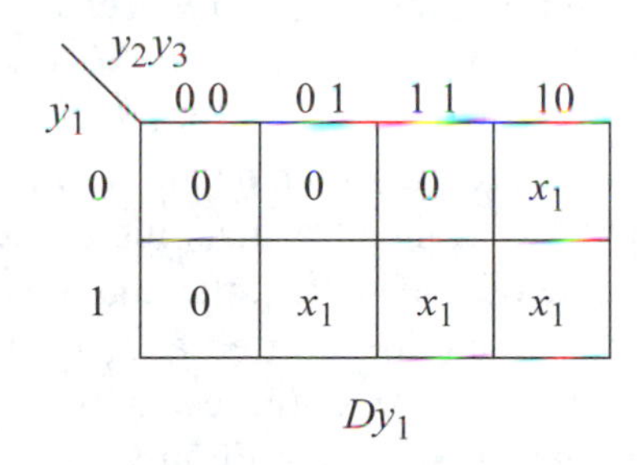

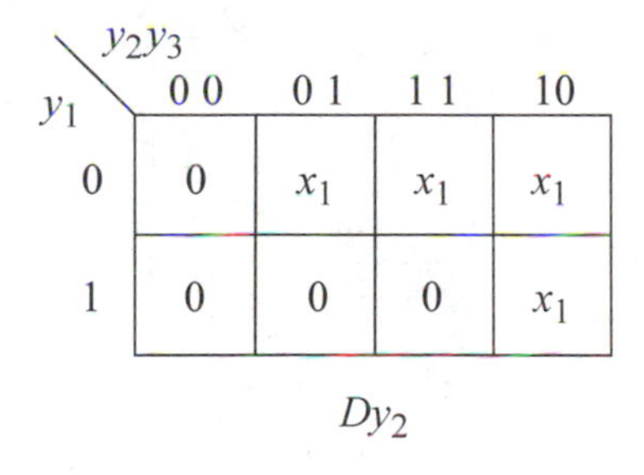

y_1

y_1 \ y_2y_3	00	01	11	10
0	x_1'	1	x_1'	x_1'
1	x_1'	x_1'	1	1

Dy_3

Figure 4.54 Input maps for the Mealy machine of Figure 4.53.

The input equations are listed in Equation 4.11 in a sum-of-products form for implementation using a PAL device. The term $y_2y_3'x_1$ in the equations for Dy_1 and Dy_2 cannot be shared in order to minimize logic as is possible when using discrete logic gates. There is no internal connection from the product term to the AND array.

$$Dy_1 = y_1y_3x_1 + y_2y_3'x_1$$

$$Dy_2 = y_1'y_3x_1 + y_2y_3'x_1$$

$$Dy_3 = x_1' + y_1'y_2'y_3 + y_1y_2 \qquad (4.11)$$

The output map is shown in Figure 4.55, as specified by the state diagram. The equation for output z_1 is shown in Equation 4.12. The PAL programming and logic diagram is illustrated in Figure 4.56. The PAL is programmed directly from the input and output equations of Equation 4.11 and Equation 4.12, respectively. There is no need to enable output z_1 with the –Clock signal to conform to the assertion/deassertion statement of $z_1\uparrow t_2\downarrow t_3$. The output assertion and deassertion times are determined inherently by the changes to the input variable x_1, which occur at time t_2. For example, in state *h*, if x_1 remains asserted at time t_2, then the machine proceeds to state *a*. If x_1 changes from a value of 1 to a value of 0 in state *h*, the change occurs at t_2, in which case, output z_1 will be asserted at t_2 and deasserted at t_3 when the machine leaves state *h*. The machine then proceeds to state *b*. A representative timing diagram for a valid input sequence of x_1 = 01111110 is shown in Figure 4.57.

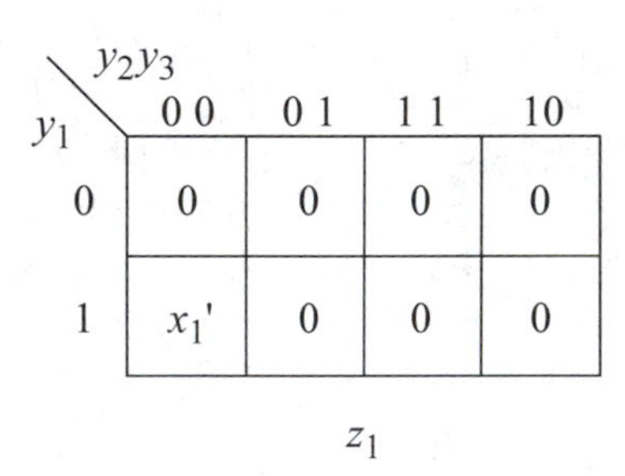

Figure 4.55 Output map for the Mealy machine of Figure 4.53.

$$z_1 = y_1y_2'y_3'x_1' \qquad (4.12)$$

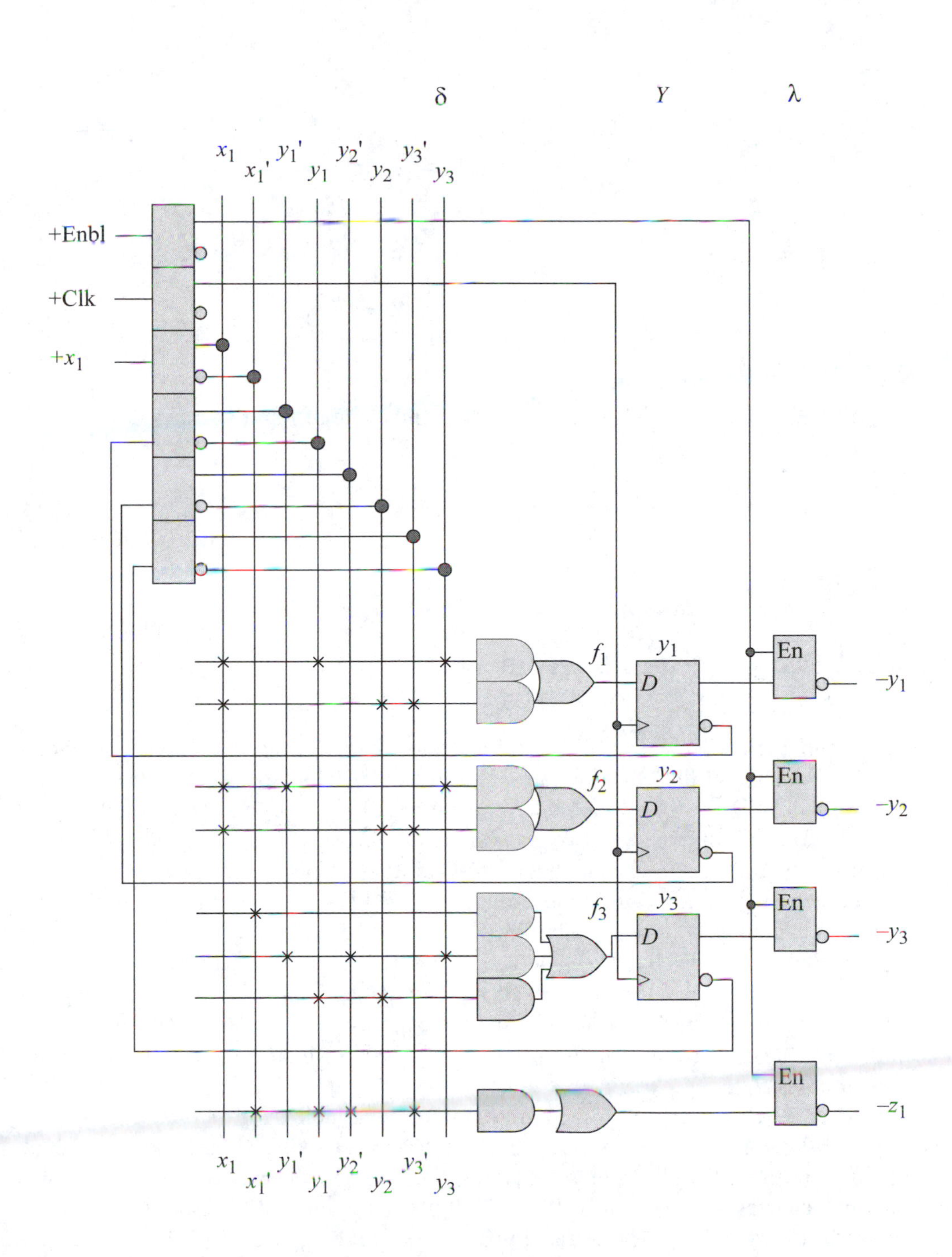

Figure 4.56 Logic diagram for the Mealy machine of Figure 4.53 using a PAL device.

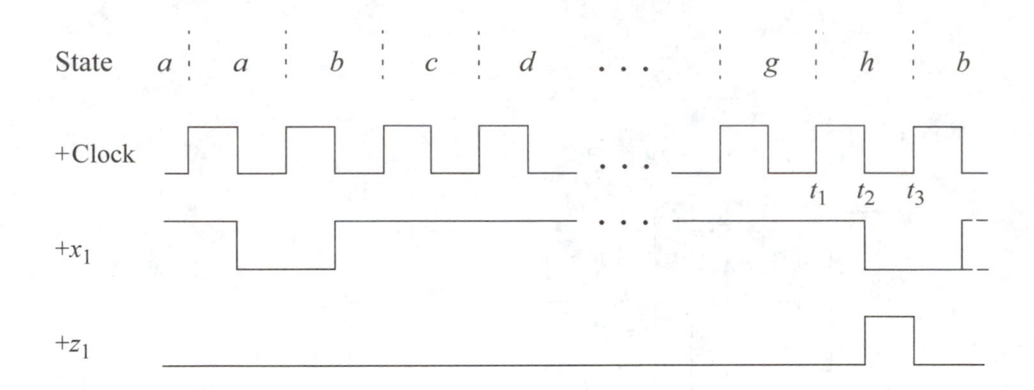

Figure 4.57 Timing diagram for the Mealy machine of Figure 4.53.

4.3.3 Programmable Logic Array

Programmable Logic Arrays (PLAs) offer a high degree of flexibility, because both the AND array and the OR array are programmable. Unlike the AND array for a PROM, the AND array for a PLA does not require 2^n AND gates to accommodate all combinations of n inputs. A PLA has n input variables, $x_1, x_2, \ldots, x_n$ and m output functions, $z_1, z_2, \ldots, z_m$. The number of AND gates — which determines the number of product terms — is variable, depending upon the PLA version. Like the previous PLDs, each input x_i connects to a buffer which increases the driving capability of the input and also provides both true and complemented values of the input signal. The OR array permits each OR gate to access any product term. Thus, the programmable OR array allows all OR gates to access the same product terms simultaneously. Each output z_i is generated from a sum-of-product expression which is a function of the input variables.

PLAs are characterized by the number of input variables, the number of product terms, and the number of output functions. All boolean expressions can be decomposed into a sum-of-products representation. For example, the exclusive-OR function $x_1 \oplus x_2$ equates to $x_1x_2' + x_1'x_2$. The multiple-input, multiple-output logic circuit shown in Figure 4.58 (a) is illustrated in an equivalent programmable representation in Figure 4.58 (b), using a PLA device to implement the equations shown in Equation 4.13. The symbol × indicates an intact fuse; the absence of an × indicates an open fuse, where the unconnected input assumes a logic 1 voltage level for an AND gate input and a logic 0 voltage level for an OR gate input.

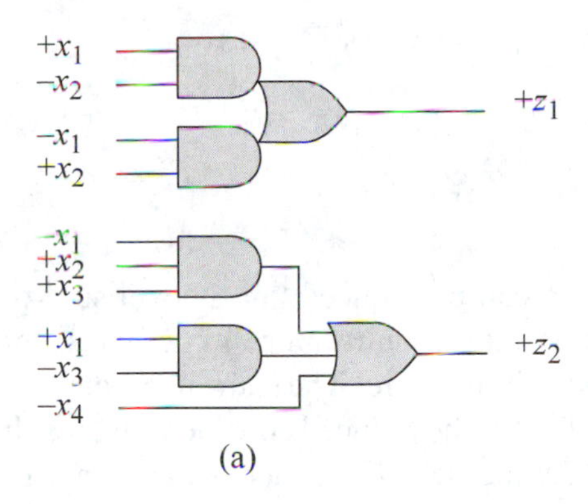

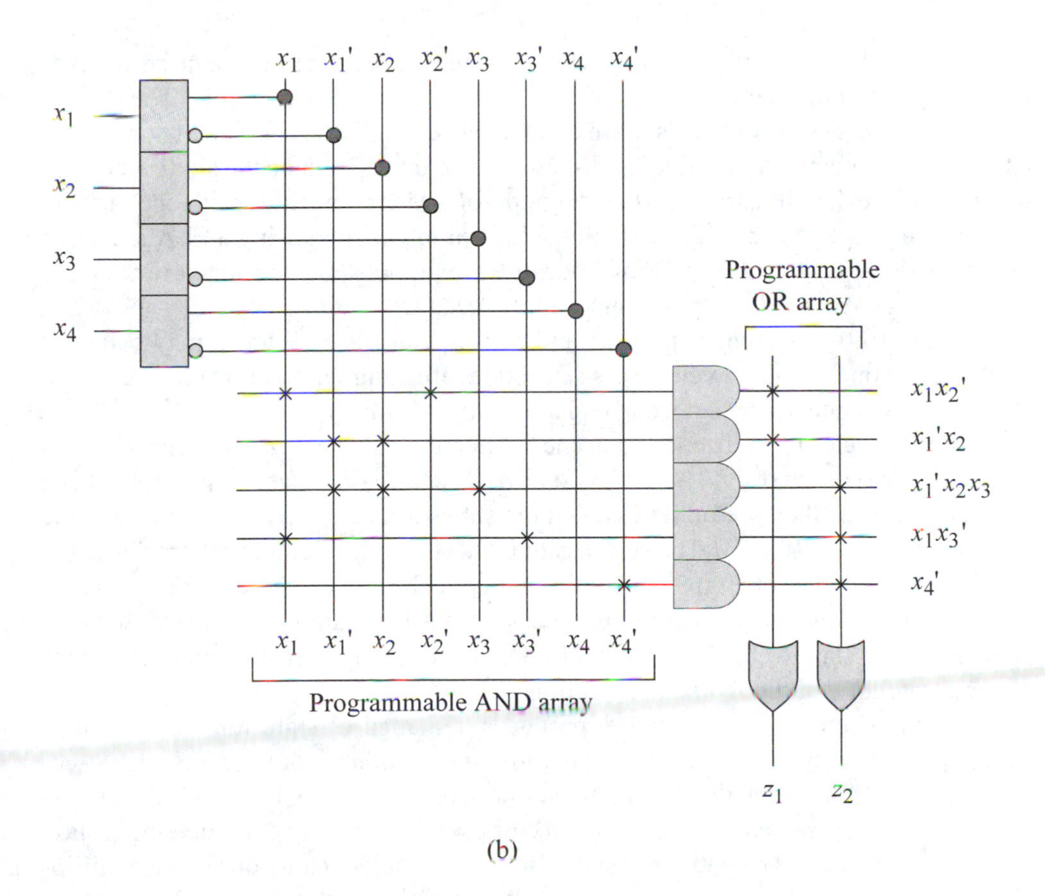

Figure 4.58 Implementation of a multiple-input, multiple-output, 2-level logic circuit: (a) conventional sum-of-products representation; and (b) equivalent circuit using a PLA device. Output $z_1 = x_1 x_2' + x_1' x_2$. Output $z_2 = x_1' x_2 x_3 + x_1 x_3' + x_4'$.

$$z_1 = x_1 x_2' + x_1' x_2$$

$$z_2 = x_1' x_2 x_3 + x_1 x_3' + x_4' \tag{4.13}$$

Multiple levels of logic can be reduced to a 2-level sum-of-products form. Although this implementation may require more logic gates, the high integration of PLAs easily absorbs the additional logic. The sum-of-products form permits the propagation delay through the PLA to be reduced considerably, resulting in a higher-speed function. Programmable devices usually reduce the amount of board space required, as well as the number of interconnections and the power supply size, thereby, reducing the cost of the machine.

Example 4.12 The synthesis of synchronous sequential machines can be realized using PLAs in a manner analogous to that used for PROMs and PALs. The assignment of state codes, however, is more crucial, since a judicious choice of state codes can reduce the number of product terms required, and thus, reduce the size of the PLA device. Figure 4.59 illustrates a state diagram for a Moore machine with one input x_1 and two outputs z_1 and z_2. The machine will be implemented using a PLA and three positive-edge-triggered *D* flip-flops. Because the outputs are asserted at time t_1 and deasserted at t_3, the λ output logic simply decodes states *c* ($y_1y_2y_3 = 111$) and *f* ($y_1y_2y_3 = 100$), asserting outputs z_1 and z_2, respectively. Glitches may occur on these Moore outputs, however, unless state codes are assigned such that no state transition will generate a transient state equal to state *c* or state *f*.

Using the rules for state code adjacency in Section 3.4, states *a* and *f* should be adjacent, because the next state for both is state *a* (Rule 2). Also, states *c* and *d* should be adjacent, because they are both possible next states for state *b* (Rule 1) and have the same next state, *e* (Rule 2). The state assignment shown in Figure 4.59 precludes the possibility of glitches on outputs z_1 and z_2. This can be verified by checking all paths to determine if any state transition produces a transient state that is identical to the state codes for states *c* or *f*. This condition can occur only if two or more flip-flops change state for a particular state transition.

The path from state *a* to state *b* produces a change of state for only one flip-flop (y_3). Similarly, the path from state *b* to state *d* produces only one change — y_2 changes from 0 to 1. Although the transition from state *b* to state *c* produces two changes, flip-flop y_3 remains set — y_3 must be reset for the machine to enter state *f* and assert output z_2. The path from state *d* to state *e* results in only one change of flip-flop variable (y_3). Both y_1 and y_3 change state when the machine proceeds from state *c* to state *e*; however, y_2 remains set, thus negating a glitch on z_2 in state *f*. The path from state *e* to state *f* produces a change to both y_1 and y_2, but flip-flop y_3 remains reset — y_3 must be set for the machine to enter state *c* and assert output z_1. Finally, the path from state f to state a occurs when only y_1 changes state. Thus, no state transition will produce a glitch on either output z_1 or z_2.

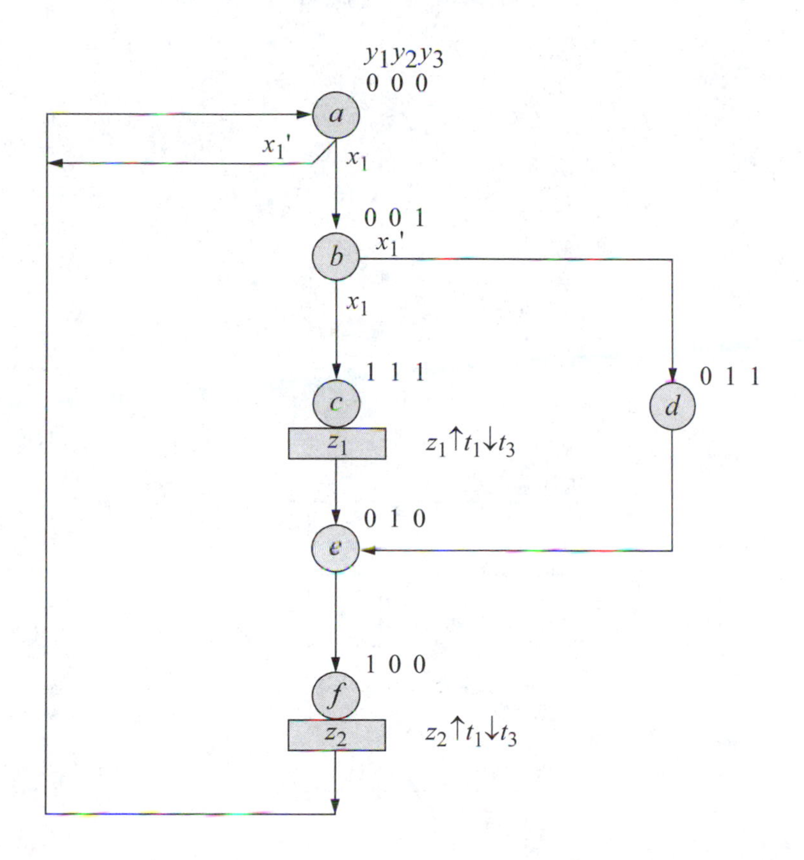

Figure 4.59 State diagram for the Moore machine of Example 4.12. Unused states are: $y_1y_2y_3 = 101$ and 110.

The input maps are shown in Figure 4.60 as obtained from the state diagram using input x_1 as a map-entered variable. A Karnaugh map yields a minimum sum-of-products expression, which is a requirement for generating output functions for a PLA device. Five product terms are required as shown in Equation 4.14.

The output maps are shown in Figure 4.61. The output equations can be minimized if the minterms for z_1 and z_2 are combined with unused minterm $y_1y_2y_3 = 101$ or 110. If these unused states are to be used for minimization, however, they must not function as transient states for any state sequence that does not include the corresponding output. The only transition that may pass through unused state $y_1y_2y_3 = 101$ is the path from state b to state c. This presents no hazard for output z_1, however, because this sequence includes z_1. Output z_1 may be asserted slightly early, but no glitch will be generated. Therefore, a 1 can be inserted in $y_1y_2y_3 = 101$ in order to minimize the equation for z_1. This is not true for z_2, however. The path from state b to state c does not include output z_2 in either the initial state or the destination state; therefore, a 0 must be inserted in state $y_1y_2y_3 = 101$ in the output map for z_2.

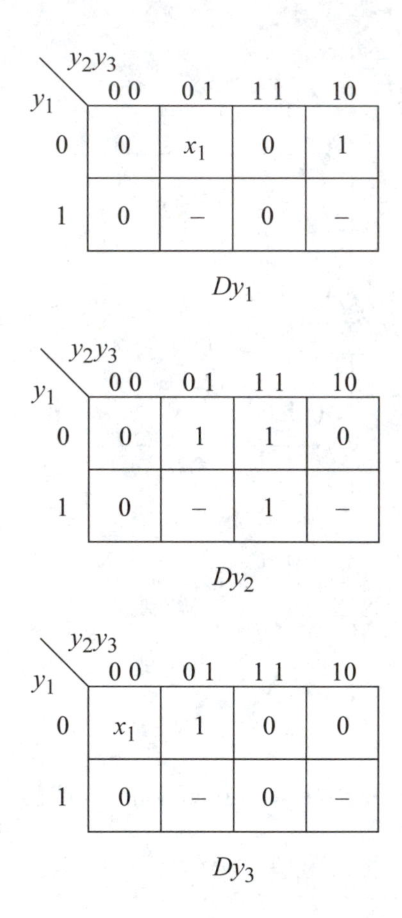

Figure 4.60 Input maps for the Moore machine of Figure 4.59.

$$
\begin{aligned}
Dy_1 &= y_2'y_3x_1 + y_2y_3' \\
Dy_2 &= y_3 \\
Dy_3 &= y_1'y_2'x_1 + y_2'y_3
\end{aligned}
\tag{4.14}
$$

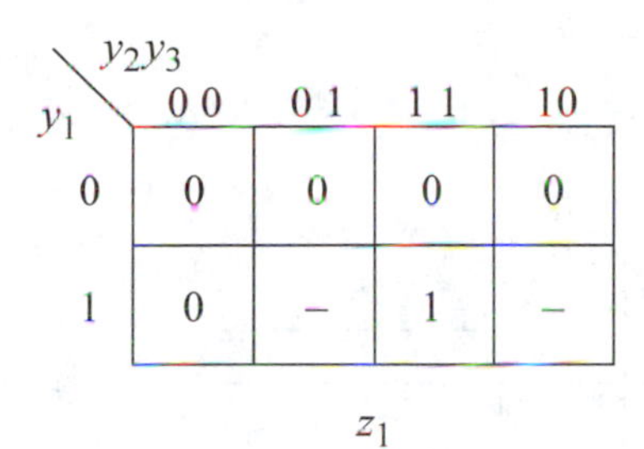

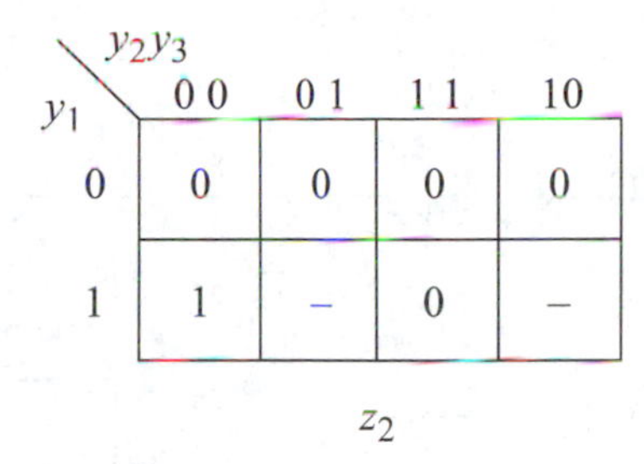

Figure 4.61 Output maps for the Moore machine of Figure 4.59. For output z_1, $y_1y_2y_3 = 101$ can be assigned a value of 1; $y_1y_2y_3 = 110$ must be 0, because of state sequence (e, f). For output z_2, $y_1y_2y_3 = 101$ must be 0, because of state sequence (b, c); $y_1y_2y_3 = 110$ must be 0, because of state sequence (c, e).

Since z_1 is already minimized, it is immaterial whether or not $y_1y_2y_3 = 110$ could be used to minimize z_1. This unused state could not be used in any case, however, because the path from state e to state f may pass through transient state $y_1y_2y_3 = 110$, and this sequence does not involve z_1. Next, consider $y_1y_2y_3 = 110$ to minimize z_2. The path from state c to state e may pass through this unused state for a sequence that does not include z_2. Therefore, a 0 must be inserted in location $y_1y_2y_3 = 110$ in the output map for z_2. The equations for outputs z_1 and z_2 are shown in Equation 4.15.

$$\begin{aligned} z_1 &= y_1y_3 \\ z_2 &= y_1y_2'y_3' \end{aligned} \tag{4.15}$$

The logic is shown in general block diagram format in Figure 4.62 (a) and in detail using a PLA with additional flip-flops in Figure 4.62 (b). To obtain the logic function for Dy_1, Dy_2, and Dy_3, the AND array is programmed according to the product terms of Equation 4.14 and the OR array is programmed to obtain the appropriate sum-of-products for the respective Dy_i input. In the same manner, the AND and OR arrays are programmed to generate outputs z_1 and z_2 according to Equation 4.15.

4.3.4 Field-Programmable Gate Array

Several versions of field-programmable gate arrays (FPGAs) are available, depending on the manufacturer. This section describes the organization and operation of a generalized FPGA and concludes with a synthesis example for a Moore machine.

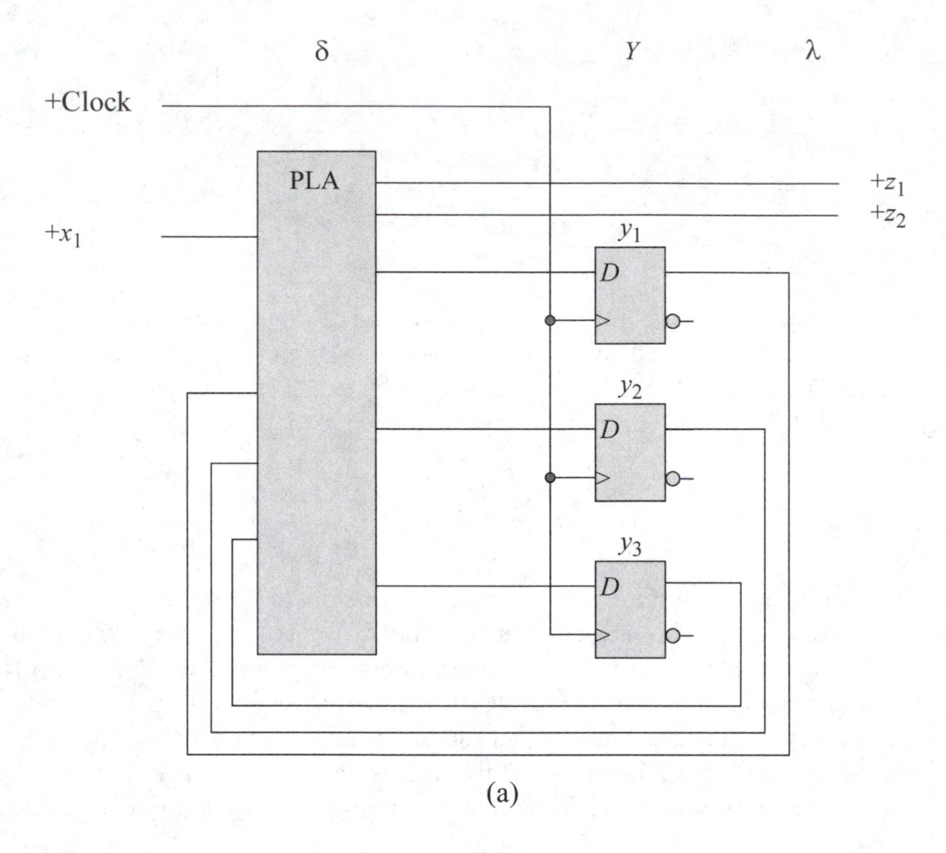

Figure 4.62 Implementation of the Moore machine of Figure 4.59: (a) block diagram; and (b) logic diagram using a PLA and three *D* flip-flops.

FPGAs are high-density gate arrays that can be configured by the user for specific applications. The organization of a typical FPGA, as shown in Figure 4.63, consists of a matrix of identical logic block elements encompassed by a perimeter of input/output (I/O) blocks. Each logic block element contains a programmable combinational logic function, storage elements, and output logic, as shown in Figure 4.64. The storage elements feed back to the combinational logic. Thus, the three major elements that comprise a synchronous sequential machine are available in this type of FPGA: δ next-state logic, storage elements, and λ output logic.

The combinational function utilizes a memory table-lookup technique which can be programmed to implement any boolean function of the input variables. The output of the combinational function connects to the storage elements. The output signals can be programmed to be a function of either the storage elements only or the storage elements and the present inputs, providing characteristics that are consistent with both Moore and Mealy machines.

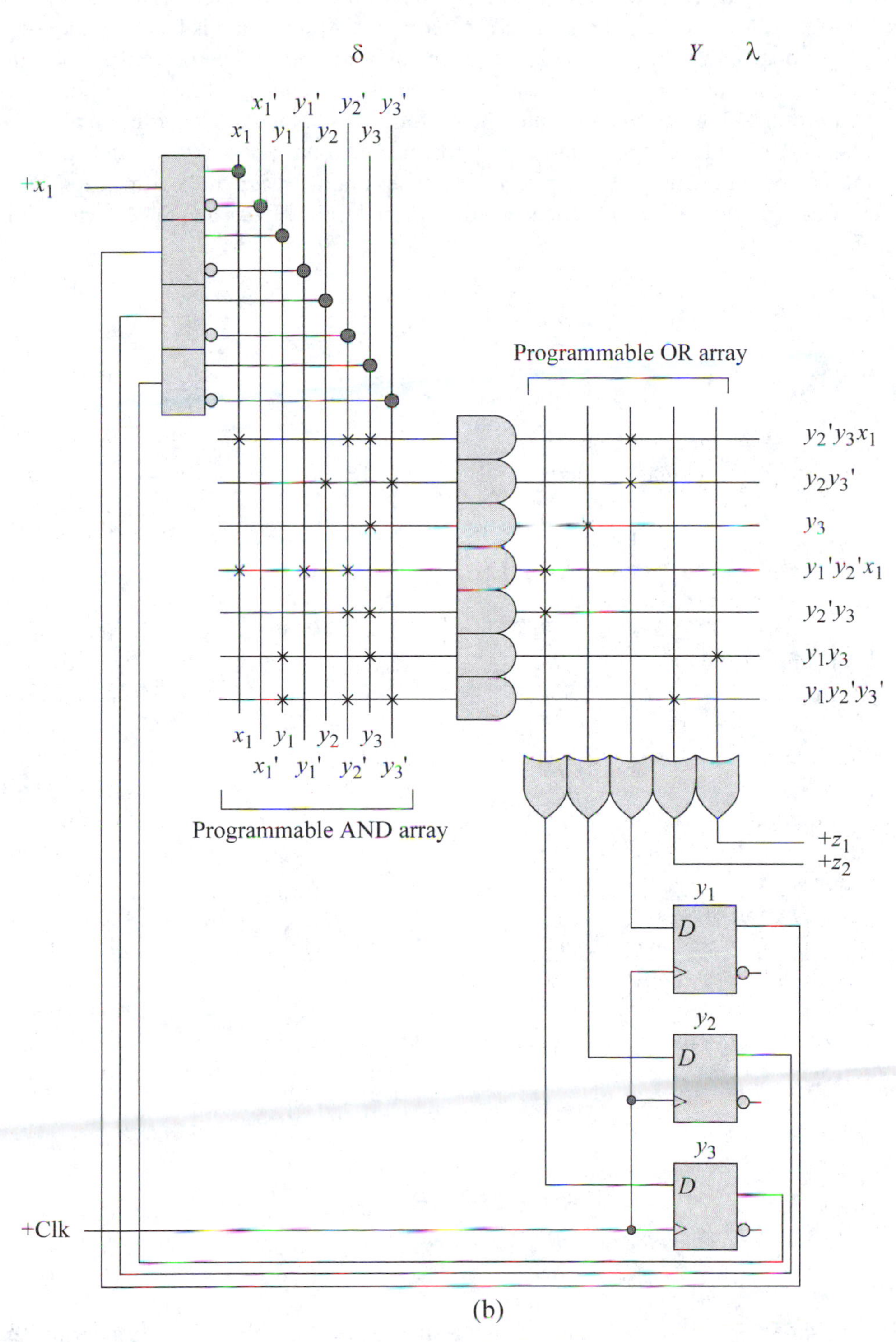

(b)

Figure 4.62 (Continued)

The fused PLDs discussed in previous sections provide a fixed organization, whereas, this particular version of FPGA provides an uncommitted organization that is configured by a resident program. The configuration program is loaded into a resident memory on the FPGA during the power-up sequence. The program conditions the combinational function in each array element to provide the required input logic for the storage elements and the output logic for the I/O blocks. The program also establishes the I/O blocks as input, output, or bidirectional circuits. The storage elements can be programmed to operate as *SR* latches or edge-triggered *D* flip-flops. The I/O blocks provide an interface between the internal logic blocks and the external pins of the array.

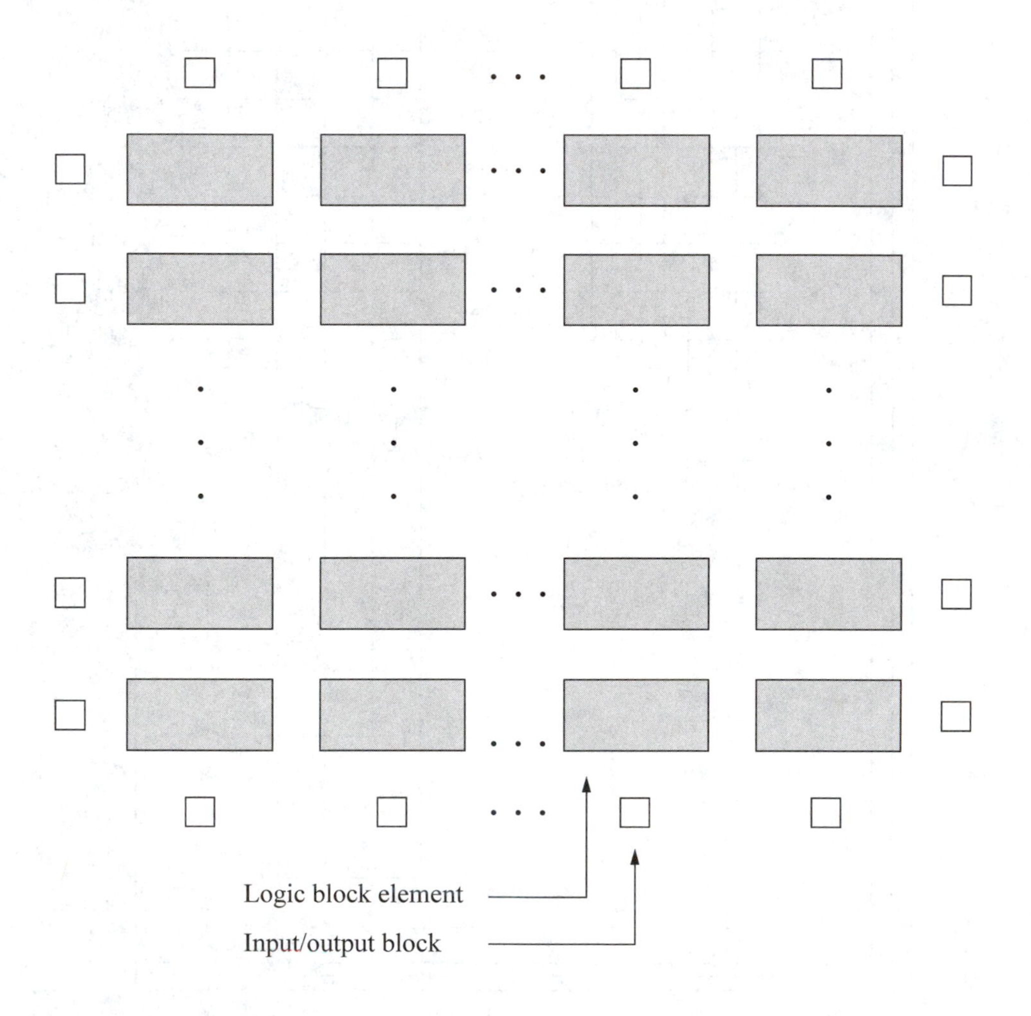

Figure 4.63 FPGA configuration using an array of logic blocks inside a perimeter of input/output blocks.

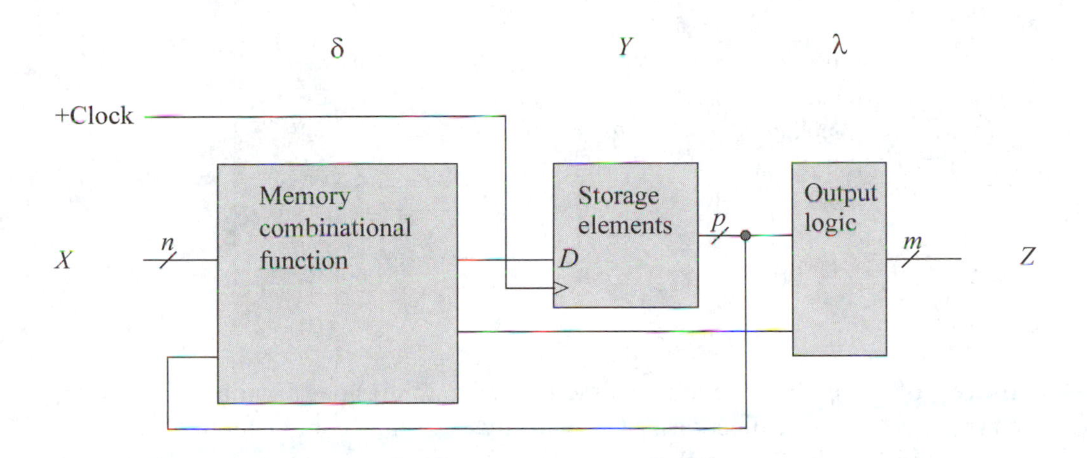

Figure 4.64 Typical logic block element of a programmable configurable FPGA.

Interconnection between logic blocks and between logic blocks and I/O blocks is accomplished by vertical and horizontal signal paths between the rows and columns of the logic blocks. Programmable switching matrices are provided at the intersection of the vertical and horizontal interconnect lines to provide a network for interblock communication.

This type of FPGA relies heavily on the software configuration program and, therefore, provides a general organization that can be adapted to meet the individual requirements of the user. The configuration program in the array's resident memory completely defines the operation of each logic block element. Assume that the memory combinational function of a logic block has five inputs, providing 5-variable minterms for any combinational function. The memory consists of a 32-by-1 array of storage cells. Thus, the five input variables form an address that accesses a unique 1-bit word in memory to produce a single output f_1. All address minterms can be programmed to assert f_1. A second output f_2 is available when the memory is partitioned into two segments; that is, two outputs are generated from two separate 4-variable functions. In this case, the addressing restriction requires that one input address variable x_i be common to both functions.

Figure 4.65 shows the two configurations of function generation. In Figure 4.65 (a) all five inputs form an address minterm to generate output $f_1(x_1, x_2, x_3, x_4, x_5)$. Figure 4.65 (b) illustrates a 2-function implementation. If input x_1 is the common input variable, then output f_1 is a function of x_1 and any 3-variable combination of x_2, x_3, x_4, and x_5. Similarly, f_2 is a function of x_1 and any different 3-variable combination of x_2, x_3, x_4, and x_5.

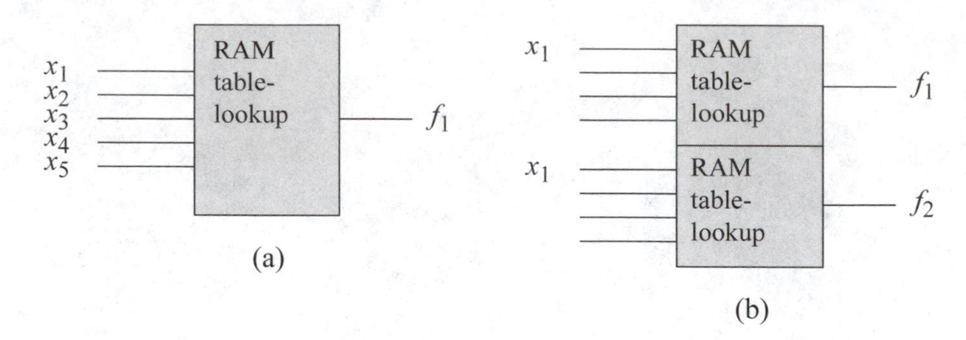

Figure 4.65 RAM table-lookup function generator: (a) one 5-variable function f_1; and (b) two 4-variable functions, $f_1(x_1$, and any 3-variable combination of x_2, x_3, x_4, and $x_5)$ and $f_2(x_1$, and any different 3-variable combination of x_2, x_3, x_4, and $x_5)$.

Example 4.13 Figure 4.66 illustrates a state diagram for a Moore machine containing two inputs x_1 and x_2, and two outputs z_1 and z_2. The machine will have a self-starting state of $y_1y_2y_3 = 000$. Thus, if either of the unused states ($y_1y_2y_3 = 100$ or 111) is entered due to noise or machine malfunction, then the outputs are made inactive and the machine returns to state a. Three flip-flops are required for this 6-state synchronous sequential machine.

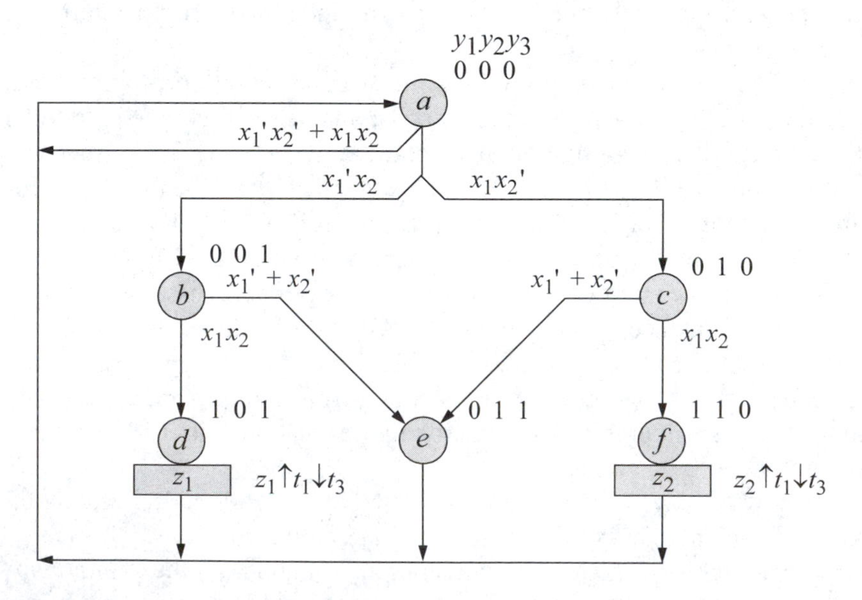

Figure 4.66 State diagram for the Moore machine of Example 4.13. Unused states are $y_1y_2y_3 = 100$ and 111.

Recall that outputs may produce erroneous values when the machine changes state at the active clock transition. Since both outputs are asserted at time t_1 and deasserted at t_3, state codes must be assigned such that no state transition will cause a glitch on z_1 or z_2. It is unlikely that a glitch will filter through the memory address decoder to the output; however, the exercise is worthwhile. The state codes shown in Figure 4.66 satisfy this requirement. The absence of glitches can be verified by checking all ten paths in Figure 4.66 for multiple changes of flip-flop states and observing whether state *d* ($y_1y_2y_3 = 101$) or state *f* ($y_1y_2y_3 = 110$) is entered as a transient state.

The input maps, obtained from the state diagram, are shown in Figure 4.67. Refer to the state diagram in conjunction with the input map for flip-flop y_1. In state *b*, the machine proceeds to state *d* only if $x_1x_2 = 11$. Thus, x_1x_2 is entered in minterm location $y_1y_2y_3 = 001$ in the map for flip-flop y_1. The same is true for the transition from state *c* to state *f*, necessitating an entry of x_1x_2 in minterm location $y_1y_2y_3 = 010$. With the exception of the unused states, all other map entries are assigned a value of 0, since the next state for y_1 in states *a*, *d*, *e*, and *f* is 0.

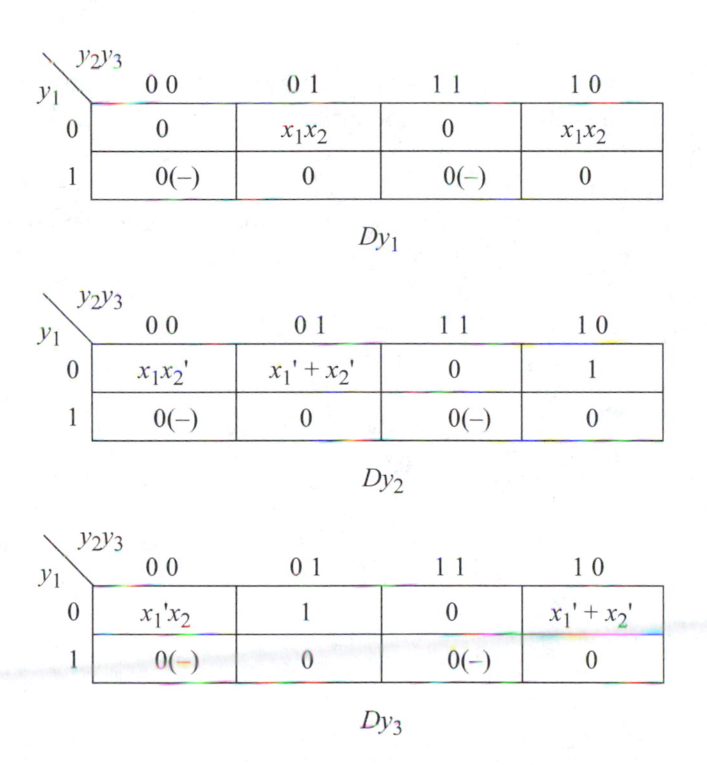

$y_1 \backslash y_2y_3$	0 0	0 1	1 1	1 0
0	0	x_1x_2	0	x_1x_2
1	0(–)	0	0(–)	0

Dy_1

$y_1 \backslash y_2y_3$	0 0	0 1	1 1	1 0
0	x_1x_2'	$x_1' + x_2'$	0	1
1	0(–)	0	0(–)	0

Dy_2

$y_1 \backslash y_2y_3$	0 0	0 1	1 1	1 0
0	$x_1'x_2$	1	0	$x_1' + x_2'$
1	0(–)	0	0(–)	0

Dy_3

Figure 4.67 Input maps for the Moore machine of Figure 4.66.

In constructing the input map for flip-flop y_2, three paths are possible from state *a*; however, only one path — from state *a* to state *c* — results in a next state of 1 for flip-flop y_2. This transition occurs if $x_1x_2 = 10$. Thus, the expression x_1x_2' is entered in minterm location $y_1y_2y_3 = 000$. The path from state *b* to state *e* results in a next state of 1 for y_2 if either $x_1 = 0$ or $x_2 = 0$. Thus, the expression $x_1' + x_2'$ is entered in minterm location 001 in the input map for y_2. In state *c*, the next state for y_2 is 1, regardless of the path taken. In the remaining three states, *d*, *e*, and *f*, the machine proceeds to state *a*, where $y_2 = 0$. In the same manner, the input map for flip-flop y_3 is constructed. The equations for Dy_1, Dy_2, and Dy_3 are shown in Equation 4.16. The unused states for y_1, y_2, and y_3 are set to 0, because minimization of the input equations is not necessary. The equations do not have to be minimized for this type of memory-configurable FPGA, since a minimal set of terms would not reduce the size of the memory unit.

$$Dy_1 = y_1'y_2'y_3x_1x_2 + y_1'y_2y_3'x_1x_2$$

$$Dy_2 = y_1'y_2'y_3'x_1x_2' + y_1'y_2'y_3x_1' + y_1'y_2'y_3x_2' + y_1'y_2y_3'$$

$$Dy_3 = y_1'y_2'y_3'x_1'x_2 + y_1'y_2'y_3 + y_1'y_2y_3'x_1' + y_1'y_2y_3'x_2' \quad (4.16)$$

The output maps for z_1 and z_2 are shown in Figure 4.68. Since the Moore outputs are a function of the present state only, there can be no more than three variables in the output equations. Thus, a single logic block element can accommodate the two outputs. The combinational memory unit is configured to generate functions f_1 and f_2, which correspond to z_1 and z_2, respectively. The output equations do not require minimization, since no reduction in internal logic is possible. The output equations are listed in Equation 4.17.

z_1:

$y_1 \backslash y_2y_3$	00	01	11	10
0	0	0	0	0
1	–	1	–	0

z_2:

$y_1 \backslash y_2y_3$	00	01	11	10
0	0	0	0	0
1	–	0	–	1

Figure 4.68 Output maps for the Moore machine of Figure 4.66.

$$z_1 = y_1 y_2' y_3$$

$$z_2 = y_1 y_2 y_3' \tag{4.17}$$

This type of FPGA operates in a manner analogous to that of a PROM-controlled synchronous sequential machine. That is, a resident program completely determines the machine's behavior as a function of the present state and the present inputs. The memory table-lookup program for this Moore machine is shown in Table 4.20, which is a tabular representation of the three input maps and the two output maps. Four logic block elements are required, one each for Dy_1, Dy_2, Dy_3, and z_1z_2. Each logic block contains one memory unit configured as 32 one-bit words. The table is established from the input and output equations of Equation 4.16 and Equation 4.17, respectively. For example, the equation for Dy_1 contains two 5-variable terms, $y_1'y_2'y_3x_1x_2$ and $y_1'y_2y_3'x_1x_2$. Thus, these two minterms address the memory unit associated with y_1 and each generates a logic 1 for Dy_1 when the appropriate inputs are active, as is evident in rows 8 and 12 of Table 4.20.

The equation for Dy_2 contains one 5-variable term, two 4-variable terms, and one 3-variable term, each of which generates a logic 1. The first term generates a logic 1 when $y_1y_2y_3x_1x_2 = 00010$. The second term produces two logic 1 entries, one each for minterms $y_1'y_2'y_3x_1'x_2$ and $y_1'y_2'y_3x_1'x_2'$. Although the variable x_2 is eliminated from both minterms by the distributive law, the product term must still generate a logic 1 for the two address minterms.

The same rationale is true for the third term of Dy_2. The fourth term, $y_1'y_2y_3'$, generates four logic 1 entries that correspond to the four minterms, $y_1'y_2y_3'x_1'x_2'$, $y_1'y_2y_3'x_1'x_2$, $y_1'y_2y_3'x_1x_2'$, and $y_1'y_2y_3'x_1x_2$. Some of the product terms for Dy_2 share common minterms, resulting in eight logic 1 entries for Dy_2. Eight logic 1 entries are also required for the sum-of-product expression for Dy_3. Outputs z_1 and z_2 are both 3-variable functions. This results in a logic 1 entry for the four minterm combinations where $y_1y_2y_3 = 101$ for z_1 and where $y_1y_2y_3 = 110$ for z_2. During the power-up sequence, the configuration program is loaded into the memory unit of each logic block element according to the format of Table 4.20.

The logic diagram is illustrated in Figure 4.69. Each state flip-flop is connected to the f_1 output of its corresponding memory element, which characterizes the operation of the flip-flop. Outputs z_1 and z_2 are asserted from the combinational logic functions generated by the configuration program in two segments of a single logic block element.

Refer to the state diagram of Figure 4.66, the table-lookup program of Table 4.20, and the logic diagram of Figure 4.69 for the discussion of machine operation which follows. The machine is reset to state *a*, then the values of inputs x_1 and x_2 are examined. If x_1 and x_2 are equal in value, either 00 or 11, then the machine remains in state *a*. This is represented by the expression $(x_1 \oplus x_2)' = x_1'x_2' + x_1x_2$. However, when the expression $x_1 \oplus x_2$ is true, the machine proceeds to state *b* if $x_1x_2 = 01$ or to state *c* if $x_1x_2 = 10$. If $y_1y_2y_3x_1x_2 = 00010$ in state *a*, then this minterm addresses a unique 1-bit word in the memory table-lookup unit for flip-flop y_2. The $f_1(Dy_2)$

output assumes a logic 1 value. At the next positive clock transition, flip-flop y_2 is set to 1 and the state code is $y_1y_2y_3 = 010$. In a similar manner, the machine progresses through the remaining states.

Table 4.20 Memory table-lookup program for the Moore machine of Figure 4.66

Memory inputs	Memory outputs				
$y_1y_2y_3x_1x_2$	$f_1(Dy_1)$	$f_1(Dy_2)$	$f_1(Dy_3)$	$f_1(z_1)$	$f_2(z_2)$
0 0 0 0 0	0	0	0	0	0
0 0 0 0 1	0	0	1	0	0
0 0 0 1 0	0	1	0	0	0
0 0 0 1 1	0	0	0	0	0
0 0 1 0 0	0	1	1	0	0
0 0 1 0 1	0	1	1	0	0
0 0 1 1 0	0	1	1	0	0
0 0 1 1 1	1	0	1	0	0
0 1 0 0 0	0	1	1	0	0
0 1 0 0 1	0	1	1	0	0
0 1 0 1 0	0	1	1	0	0
0 1 0 1 1	1	1	0	0	0
0 1 1 0 0	0	0	0	0	0
0 1 1 0 1	0	0	0	0	0
0 1 1 1 0	0	0	0	0	0
0 1 1 1 1	0	0	0	0	0
1 0 0 0 0	0	0	0	0	0
1 0 0 0 1	0	0	0	0	0
1 0 0 1 0	0	0	0	0	0
1 0 0 1 1	0	0	0	0	0
1 0 1 0 0	0	0	0	1	0
1 0 1 0 1	0	0	0	1	0
1 0 1 1 0	0	0	0	1	0
1 0 1 1 1	0	0	0	1	0
1 1 0 0 0	0	0	0	0	1
1 1 0 0 1	0	0	0	0	1
1 1 0 1 0	0	0	0	0	1
1 1 0 1 1	0	0	0	0	1
1 1 1 0 0	0	0	0	0	0
1 1 1 0 1	0	0	0	0	0
1 1 1 1 0	0	0	0	0	0
1 1 1 1 1	0	0	0	0	0

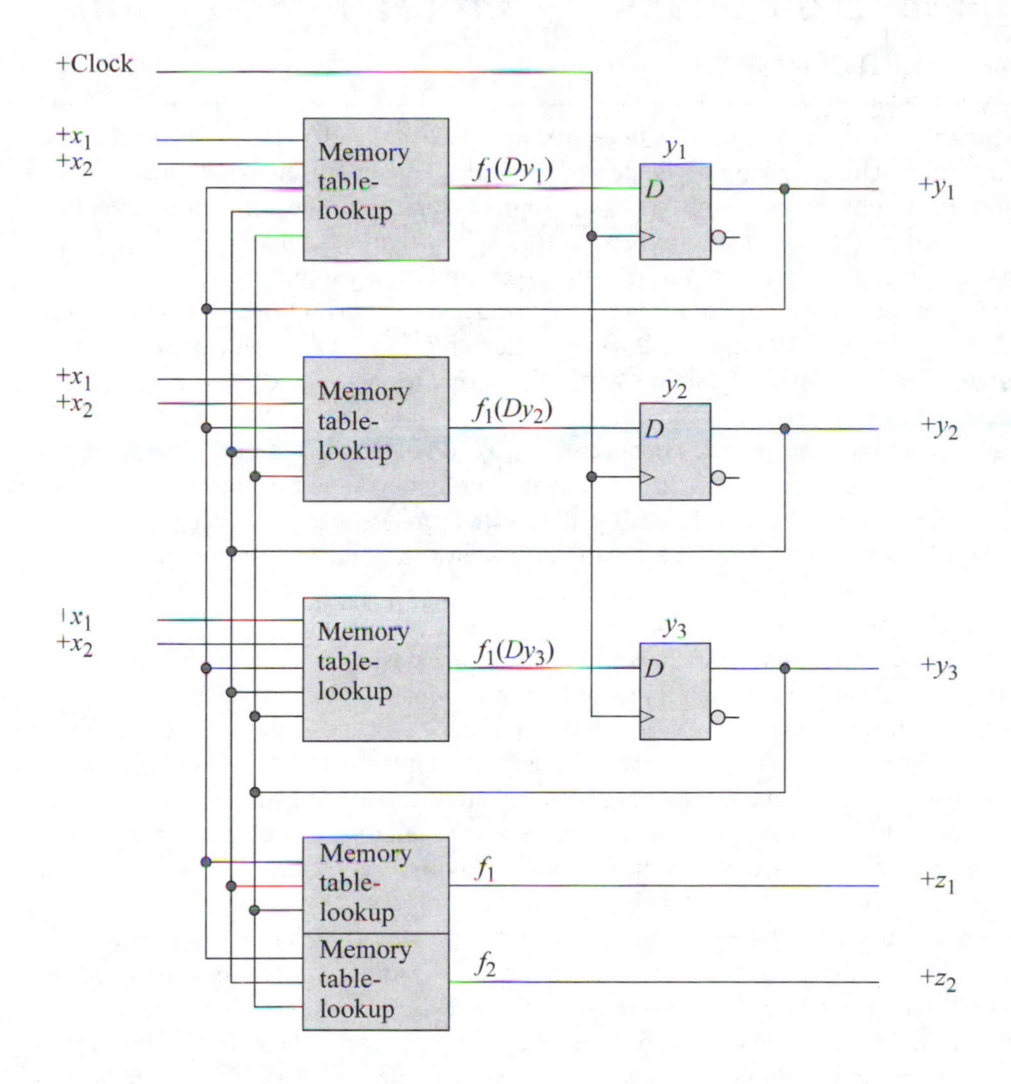

Figure 4.69 Logic diagram for the Moore machine of Figure 4.66 utilizing four logic block elements in a FPGA.

In state d, each flip-flop memory unit is addressed by $y_1y_2y_3$ = 101 concatenated with the present input values. One of four memory words is selected, depending on the values of x_1 and x_2. Each 1-bit word, however, provides a logic 0 to Dy_1, Dy_2, and Dy_3 such that, the next state is a (y1y2y3 = 000). Also, in state d, the combinational memory unit for output z_1 is programmed to generate a logic 1 for $f_1(z_1)$, asserting z_1 at time t_1. Output z_1 is deasserted when the machine changes to state a at t_3. Similarly, output z_2 is asserted at time t_1 in state f and deasserted when the machine sequences to state a.

4.4 Microprocessor-Controlled Sequential Machines

Microprocessors represent an alternative approach in synthesizing synchronous sequential machines. Many of the design concepts that were discussed in previous sections also apply to the synthesis of microprocessor-controlled sequential machines. Several high-performance microprocessors are currently available which include a stand-alone or integrated numeric coprocessor and input/output processor.

Microprocessors provide several advantages over conventional synchronous sequential machines. First, there is an execution unit. The execution unit adds considerable flexibility in synthesizing sequential machines because of the arithmetic and logic instructions that are available. For example, it is possible for a next state to be determined by an arithmetic operation, such as add or subtract, and then compare the result with a predefined quantity to activate a particular state transition.

A second advantage is the ability of the microprocessor to introduce variable time delays between states by means of software delays. The state times are no longer restricted to only one clock cycle. An increase in time between states can allow an external device to complete a task in one state before proceeding to the next state. This can also be accomplished in conventional machines using discrete logic circuits, but programmable state duration times adds an additional level of versatility.

The third advantage is of significant importance — a large number of states can be generated by a comparatively small amount of supplemental hardware. By simply adding output ports, the number of states that are available to external devices can be increased dramatically. If the additional states are not required by external circuitry, then no extra hardware is necessary — the additional states represent an increase in software only.

A fourth advantage is the ease of installing changes. If the machine specifications necessitate a change to the state diagram, then the modifications can usually be implemented in software. This is in contrast to hardwired machines where changes may require deleting and adding wires and integrated circuits. Thus, a microprocessor-controlled sequential machine is inherently more flexible than a corresponding hardwired machine.

A fifth advantage permits multiple machines to be configured with a single microprocessor. Each separate machine is under control of a different program in the microprocessor's memory. Also, each machine connects to external circuitry through unique I/O ports. There is little restriction, therefore, to the number of separate machines that can be implemented by one microprocessor.

A sixth advantage is that the operation of a machine can be uniquely defined by means of software. Once the hardware has been established, the functionality of the machine is simply a matter of programming the hardware to correspond to the machine specifications. Both Moore and Mealy machines are applicable to microprocessor implementation.

There is also a seventh advantage that applies more to large machines to which peripheral devices are attached. The interrupt structure of a microprocessor permits an external device to initiate an I/O operation by means of an interrupt. Each device has a unique interrupt vector which directs the processor to a program in memory containing an interrupt service routine to handle the interrupting device.

It may appear that the previous advantages would strongly favor a microprocessor-controlled sequential machine rather than a machine implemented with discrete components. There are, however, some significant disadvantages in using a microprocessor to synthesize a sequential machine. First, a microprocessor system is extremely expensive for small machines. The microprocessor requires support circuits such as, a clock generator, a read/write memory, a read-only memory, and input/output ports with associated address decoding. The cost for such a system is usually prohibitive for a machine with only a few states.

It was stated previously that the ability to program a sequential machine was an advantage. There is, however, a cost associated with programming — a programmer is required to write the application program (this can be done in parallel with the hardware design) and a programming device is necessary to program the ROM which will contain the application program. This represents a second disadvantage.

The third disadvantage is of major consideration. In contrast to a sequential machine synthesized with discrete logic elements, a microprocessor-controlled sequential machine is extremely slow. Several instruction cycles may be required for each state transition, during which time input signals are read from an input port in order to determine the next state. Outputs are available only when the corresponding bit is set in an output port. Input/output operations are accomplished by the relatively slow application program, either in I/O-mapped mode or memory-mapped mode. Input/output-mapped mode transfers data between a device register and an I/O port using instructions such as IN and OUT, whereas, memory-mapped mode transfers data using any memory-reference processor instruction.

4.4.1 General Considerations

A microprocessor is ideal for large machines if the machine can tolerate the relatively long duration between states. A microprocessor must read input signals from an input port, perform an operation on the inputs, then write to an output port to generate machine states and outputs. Figure 4.70 shows a general block diagram of a microprocessor-controlled synchronous sequential machine depicting the three characteristics of δ next-state logic, storage elements, and λ output logic. There is no need for a hardware feedback path from the present-state register (output port) to the data input register (input port) to generate the next state — the software generates the next state as a function of the present state only or the present state and the present inputs.

If the next state depends upon the assertion or deassertion of specific input variables, then the program will cause the microprocessor to read the contents of the input port to determine the value of the input variables. Information is transferred on the system bus, which consists of the address bus, the data bus, and the control bus. The address of the input port is placed on the address bus and decoded by the input port during a specific clock cycle. Control bus signals then enable the contents of the input port onto the data bus and then load the data into the microprocessor's accumulator. If the input conditions are met, then the software will generate the requisite state transition sequence and assert an output, if applicable.

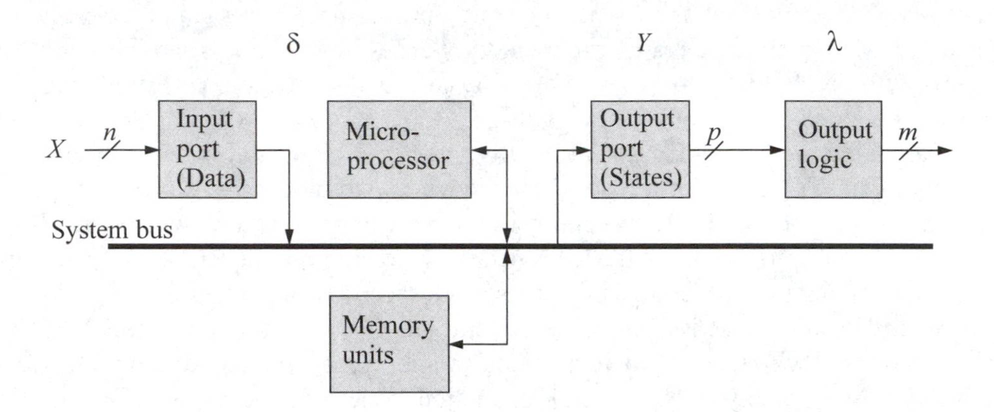

Figure 4.70 General block diagram of a microprocessor-controlled synchronous sequential machine.

The 8088 microprocessor will be the vehicle which will be used to demonstrate the concepts of synthesizing a microprocessor-controlled synchronous sequential machine. The venerable 8088 has been in widespread use in personal computers and is a core course in both hardware and software engineering in many universities. Although not a current high-performance processor, its architecture will suffice for an introductory level presentation on microprocessor-controlled sequential machines. The 8088 is also well-documented for both hardware and software systems. The synthesis procedures developed using the 8088 also apply to the more advanced 32- and 64-bit microprocessors.

Figure 4.71 gives more detail on the output port (state register) and the input port (data register), both of which are connected to the 8-bit data bus. It will be assumed that the machine states are required by external circuits. The present state is loaded from the microprocessor's accumulator into the state register by the OUT instruction. The state register bits are arbitrarily defined to specify a particular state. For example, state a is selected as bit 0 and state h as bit 7. The input variables are transferred from the input register to the microprocessor's accumulator on the bidirectional data bus by the IN instruction. Specific inputs are arbitrarily assigned to bits in the input register.

Since a particular state is defined by a unique bit in the state register, there is no need to assign state codes, thus, state code adjacency is not necessary for minimization. Also, output glitches are not relevant, because flip-flops are no longer decoded to generate a state. For example, Table 4.21 lists four states of a Moore machine as a function of two flip-flops y_1 and y_2, where output z_1 is asserted in state $y_1y_2 = 01$ and z_2 is asserted in state $y_1y_2 = 10$. If the machine proceeds from state *a* ($y_1y_2 = 00$) to state *b* ($y_1y_2 = 11$), then transient state *c* ($y_1y_2 = 01$) or *d* ($y_1y_2 = 10$) may be entered. An erroneous assertion may be produced on output z_1 or z_2 if discrete logic elements were used.

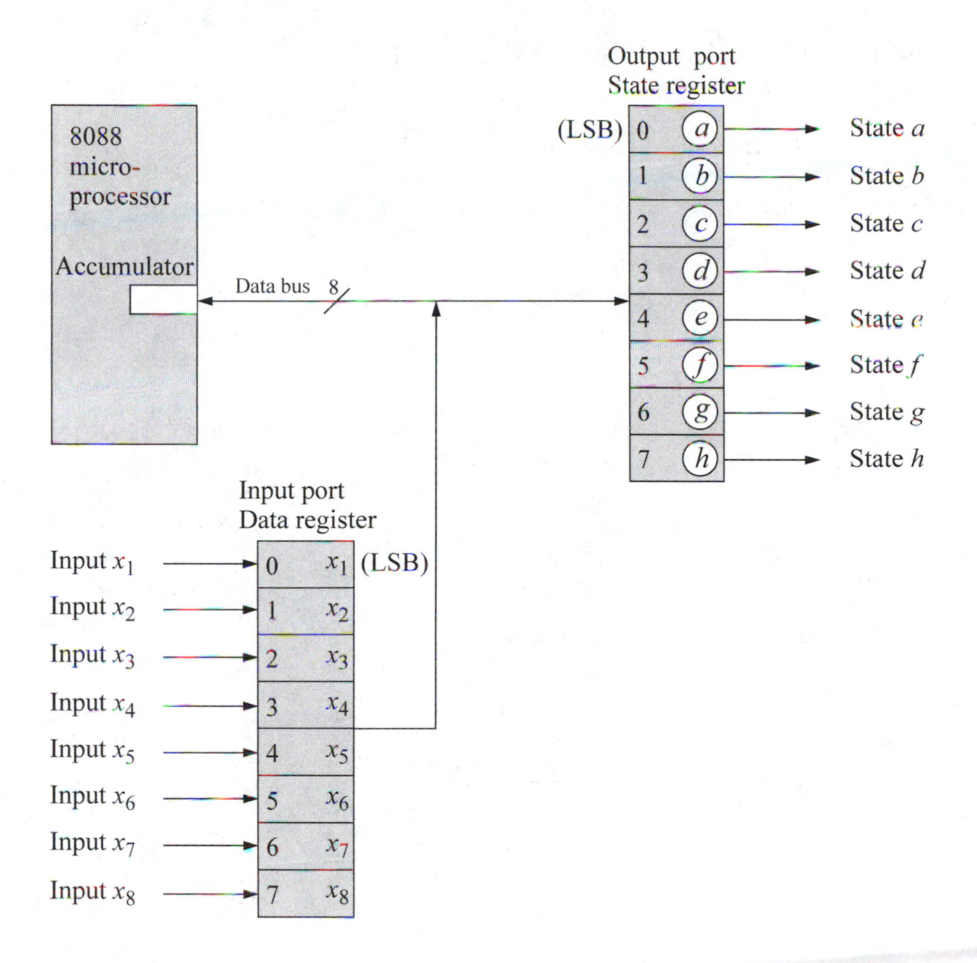

Figure 4.71 State register and data register with state name and input variable assignments.

The problem of glitches does not exist, however, when using a microprocessor to sequence from state *a* to state *b*, where the states reside in a state register. Figure 4.72 illustrates a state transition from state *a* to state *b* by means of two separate OUT instructions — assuming I/O-mapped I/O. To enter state *a*, the first OUT instruction loads the state register with 01H, where the letter H indicates a hexadecimal operand. This is shown in Figure 4.72 (a). To generate a transition from state *a* to state *b*, a second OUT instruction is executed. The microprocessor loads 02H into the state register. This operation accomplishes two objectives: first, the machine exits state *a* (bit 0 is reset); second, the machine enters state *b* (bit 1 is set).

Table 4.21 State code assignment for a 4-state Moore machine

State name	Present state $y_1\ y_2$	Present output $z_1\ z_2$
a	0 0	0 0
c	0 1	1 0
d	1 0	0 1
b	1 1	0 0

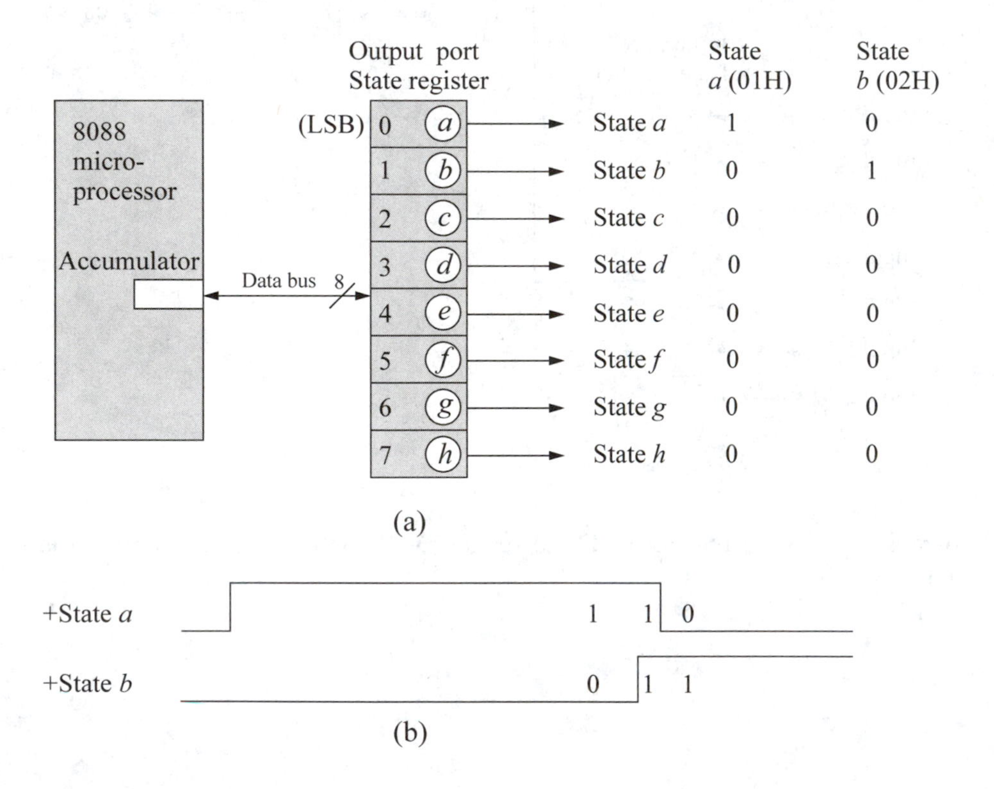

Figure 4.72 State transition from state *a* to state *b* in a microprocessor-controlled sequential machine: (a) block diagram showing output port contents for state *a* and state *b*; and (b) possible overlap of states *a* and *b* for a transition from state *a* to state *b*.

Since the state register is implemented with flip-flops, there may be a time variation in the set and reset characteristics of the flip-flops that represent states *a* and *b*. For example, state *b* may set before state *a* resets, as shown in Figure 4.72 (b). For a short duration, both states may be asserted; however, there will be no glitch — the machine will not pass through another state, because state flip-flops are not decoded to determine a particular state. Using a microprocessor implementation for a sequential machine, a state is represented by a single bit in a state register.

4.4.2 Mealy Machine Synthesis

Before presenting the architecture for a microprocessor-controlled sequential machine, an example will delineate the necessary steps for synthesizing a program from a set of machine specifications, in particular, a state diagram.

Example 4.14 A Mealy machine will be synthesized, which receives three parallel inputs x_1, x_2, and x_3, and generates two outputs z_1 and z_2. The following input sequences assert z_1 and z_2, accordingly:

If $x_1x_2x_3$ = 000, 111, 000, then assert z_1
If $x_1x_2x_3$ = 111, 000, 111, then assert z_2

The two valid sequences are shown in a different representation as follows:

$x_1 =$ ··· 010 ··· 101 ···
$x_2 =$ ··· 010 ··· 101 ···
$x_3 =$ ··· 010 ··· 101 ···

Assert z_1 (arrow to first 010 group)
Assert z_2 (arrow to first 101 group)

For all other sequences, z_1 and z_2 remain inactive. Assertion/deassertion statements are not necessary for the outputs, since their active state is controlled by the microprocessor — they are set under program control and remain in that state until reset by the program.

The bit assignments for the input and output ports are shown in Figure 4.73. The three inputs, x_1, x_2, and x_3, are assigned bit positions 0, 1, and 2, respectively, in the input port. The output port provides a register for all machine outputs. States *a* through *f* are assigned bit positions 0 through 5, respectively; outputs z_1 and z_2 reside in bits 6 and 7, respectively.

INPORT

7	6	5	4	3	2	1	0
–	–	–	–	–	x_3	x_2	x_1

OUTPORT

7	6	5	4	3	2	1	0
z_2	z_1	f	e	d	c	b	a

Figure 4.73 Bit assignments for the input and output port registers of Example 4.14.

The state diagram is shown in Figure 4.74. State codes are not necessary, because flip-flops are not decoded in order to define a particular state — a state is represented by a single bit in the output port (state register). Also, input and output maps are not applicable to a microprocessor-controlled sequential machine. The input maps are replaced by a program sequence, which reads input variables, analyzes their values, and sets a corresponding state in the state register. Moore outputs are set when the related state is entered; Mealy outputs are set after the state has been entered and the inputs evaluated.

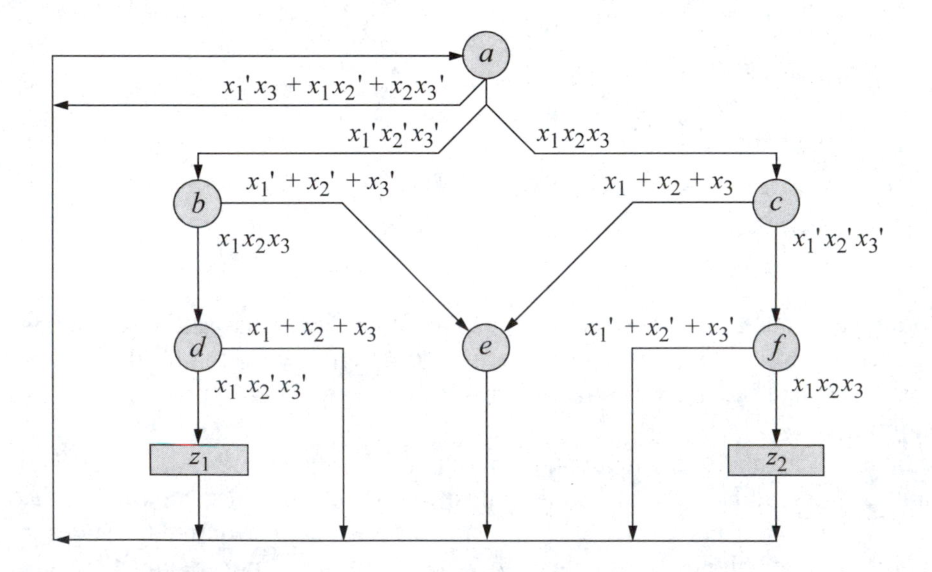

Figure 4.74 State diagram for the Mealy machine of Example 4.14.

The machine commences operation by entering state *a*, where the values of the parallel inputs are assessed. The machine proceeds to state *b* or *c* only if the input variables are $x_1x_2x_3$ = 000 or 111, respectively. For all other input combinations, the machine remains in state *a*. The requirements to remain in state *a* are tabulated in Figure 4.75 (a) and minimized using a Karnaugh map in Figure 4.75 (b). Alternatively, the machine remains in state *a* whenever the input variables satisfy the expression $(x_1'x_2'x_3')' \bullet (x_1x_2x_3)'$, which can be reduced by boolean algebra to the minimized sum-of-products expression shown in Figure 4.75 (b).

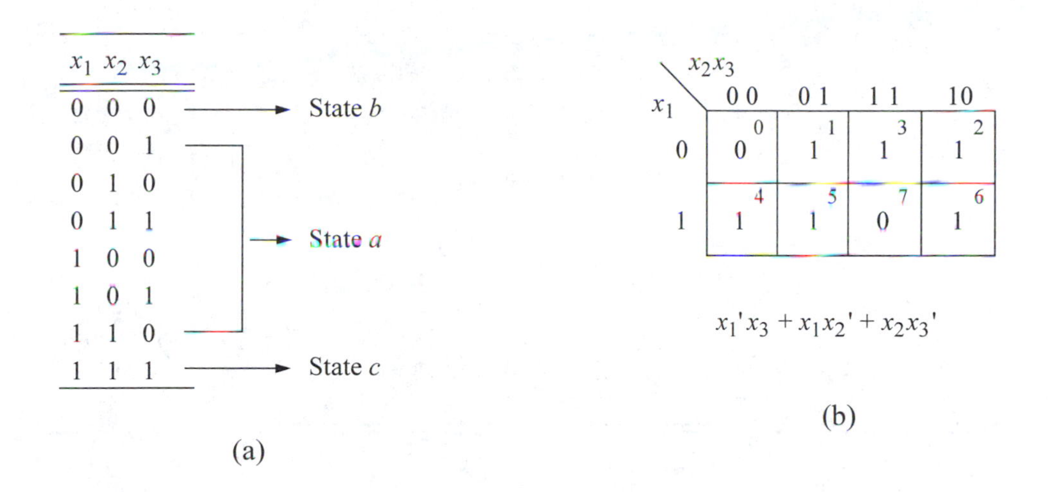

Figure 4.75 Input values for x_1, x_2, and x_3 which cause the Mealy machine of Figure 4.74 to remain in state *a*: (a) truth table; and (b) Karnaugh map.

To determine whether a transition to another state is relevant, the machine reads the state of the input port under program control. The contents are examined and the appropriate state transition is executed by setting the corresponding state to a value of 1 in the output port. Figure 4.76 lists the code segment that directs the machine in accordance with the state diagram. It is unnecessary to list the entire program, including the data and stack segments, since the program logic is the only consideration.

The addresses of the output and input ports have been predefined and are labelled OUTPORT and INPORT, respectively. To enter state *a*, the program must first load the hexadecimal value 01 into the accumulator byte register AL. Hexadecimal 01 corresponds to state *a* (OUTPORT bit 0). Then the contents of AL are transferred to the state register by the OUT instruction. At that time, the machine enters state *a*. The instructions for this sequence, as shown in Figure 4.76, are as follows:

```
STTA:   MOV   AL, 01H          ;Prepare to enter state a
        OUT   OUTPORT, AL      ;Exit previous state.
                               ;Enter state a
```

where the instruction format is *operation code destination, source*. In state *a*, the program examines the values of the three input variables x_1, x_2, and x_3 to determine the next state. If $x_1x_2x_3$ = 000 or 111, the machine proceeds to state *b* or *c*, respectively; otherwise, the machine remains in state *a* and again tests the values of the input variables. This is accomplished by the following instruction sequence:

```
STTA1:  IN    AL, INPORT       ;AL ← x1x2x3
        AND   AL, 07H          ;Mask off all bits except
                               ;... x1x2x3
        JZ    STTB             ;Proceed to state b if
                               ;... x1x2x3 = 000
        CMP   AL, 07H          Test for x1x2x3 = 111
        JE    STTC             ;Proceed to state c if
                               ;... x1x2x3 = 111
        JMP   STTA1            ;Otherwise, remain in state a
```

```
                . . .
STTA:   MOV   AL, 01H          ;Prepare to enter state a
        OUT   OUTPORT, AL      ;Exit previous state
                               ;... Enter state a
STTA1:  IN    AL, INPORT       ;AL ← x1x2x3
        AND   AL, 07H          ;Mask off all bits except
                               ;... x1x2x3
        JZ    STTB             ;Proceed to state b if
                               ;... x1x2x3 = 000
        CMP   AL, 07H          ;Test for x1x2x3 = 111
        JE    STTC             ;Proceed to state c if
                               ;... x1x2x3 = 111
        JMP   STTA1            ;Otherwise, remain in state a
```

Figure 4.76 8088 assembly language program for the Mealy machine of Figure 4.74.

```
STTB:   MOV   AL, 02H          ;Prepare for state transition
                               ;... from state a to state b
        OUT   OUTPORT, AL      ;Exit state a.  Enter state b
        IN    AL, INPORT       ;AL ← x1x2x3
        AND   AL, 07H          ;Mask off all bits except
                               ;... x1x2x3
        CMP   AL, 07H          ;Test for x1x2x3 = 111
        JE    STTD             ;Proceed to state d if
                               ;... x1x2x3 = 111
        JMP   STTE             ;Otherwise, proceed to state e
-----------------------------------------------------------------
STTC:   MOV   AL, 04H          ;Prepare for state transition
                               ;... from state a to state c
        OUT   OUTPORT, AL      ;Exit state a.  Enter state c
        IN    AL, INPORT       ;AL ← x1x2x3
        AND   AL, 07H          ;Mask off all bits except
                               ;... x1x2x3
        JZ    STTF             ;Proceed to state f if
                               ;... x1x2x3 = 000
        JMP   STTE             ;Otherwise, proceed to state e
-----------------------------------------------------------------
STTD:   MOV   AL, 08H          ;Prepare for state transition
                               ;... from state b to state d
        OUT   OUTPORT, AL      ;Exit state b.  Enter state d
        IN    AL, INPORT       ;AL ← x1x2x3
        AND   AL, 07H          ;Mask off all bits except
                               ;... x1x2x3
        JNZ   STTA             ;Proceed to state a if
                               ;... x1x2x3 ≠ 000
        MOV   AL, 48H          ;Prepare to assert z1 while
                               ;... remaining in state d
        OUT   OUTPORT, AL      ;Assert z1 in state d
        MOV   AL, 08H          ;Prepare to deassert z1 while
                               ;... remaining in state d
        OUT   OUTPORT, AL      ;Deassert z1
        JMP   STTA             ;Proceed to state a
```

Figure 4.76 (Continued)

```
STTE:   MOV   AL, 10H          ;Prepare to exit state b or c
                               ;... and enter state e
        OUT   OUTPORT, AL      ;Enter state a
        JMP   STTA             ;Proceed to state a
STTF:   MOV   AL, 20H          ;Prepare for state transition
                               ;... from state c to state f
        OUT   OUTPORT, AL      ;Exit state c.  Enter state f
        IN    AL,INPORT        ;AL ← x1x2x3
        AND   AL, 07H          ;Mask off all bits except
                               ;... x1x2x3
        CMP   AL, 07H          ;Test for x1x2x3 = 111
        JNZ   STTA             ;Proceed to state a if
                               ;... x1x2x3 ≠ 111
        MOV   AL, 0A0H         ;Prepare to assert z2 while
                               ;... remaining in state f
        OUT   OUTPORT, AL      ;Assert z2 in state f
        MOV   AL, 20H          ;Prepare to deassert z2 while
                               ;... remaining in state f
        OUT   OUTPORT, AL      ;Deassert z2
        JMP   STTA             ;Proceed to state a
```

Figure 4.76 (Continued)

If the state transition sequence is from state *a* to state *b*, then the machine exits state *a* (OUTPORT bit 0 = 0) and enters state *b* (OUTPORT bit 1 = 1) by the following instruction sequence:

```
STTB:   MOV   AL, 02H          ;Prepare for state transition
                               ;... from state a to state b
        OUT   OUTPORT, AL      ;Exit state a.  Enter state b
```

In state *b*, the machine proceeds to state *d* only if $x_1x_2x_3 = 111$, otherwise, the next state is *e*. That is, if $x_1x_2x_3 = 0{-}{-}, {-}0{-}$, or ${-}{-}0$, the program generates a transition to state *e*. This is accomplished by the following instruction sequence:

```
        IN    AL, INPORT       ;AL ← x1x2x3
        AND   AL, 07H          ;Mask off all bits except
                               ;... x1x2x3
        CMP   AL, 07H          ;Test for x1x2x3 = 111
        JE    STTD             ;Proceed to state d if
                               ;... x1x2x3 = 111
        JMP   STTE             ;Otherwise, proceed to state e
```

If the state transition is from state b to state d, then the machine exits state b and enters state d by the following two instructions:

```
STTD:   MOV   AL, 08H          ;Prepare for state transition
                               ;... from state b to state d
        OUT   OUTPORT, AL      ;Exit state b.  Enter state d
```

In state d, if $x_1x_2x_3 = 000$, then the Mealy-type output z_1 is asserted for two instruction cycles, then deasserted. The machine then proceeds to state a, where the process repeats. The duration of any assertion is a function of the clock frequency. Output z_1 is active only in state d and must be inactive before the machine proceeds to state a. This guarantees that state a will have no lingering output. If, however, $x_1x_2x_3 = 1--$, $-1-$, or $--1$, then the machine sequences to state a without asserting z_1. This series of events is executed by the following instructions:

```
        IN    AL, INPORT       ;AL ← x1x2x3
        AND   AL, 07H          ;Mask off all bits except
                               ;... x1x2x3
        JNZ   STTA             ;Proceed to state a if
                               ;... x1x2x3 ≠ 000
        MOV   AL, 48H          ;Prepare to assert z1 while
                               ;... remaining in state d
        OUT   OUTPORT, AL      ;Assert z1 in state d
        MOV   AL, 08H          ;Prepare to deassert z1 while
                               ;... remaining in state d
        OUT   OUTPORT, AL      ;Deassert z1
        JMP   STTA             ;Proceed to state a
```

If the state transition sequence is from state a to state c, then the program must reset state a (OUTPORT bit 0) and set state c (OUTPORT bit 2). The following instructions cause the machine to exit state a and enter state c:

```
STTC:   MOV   AL, 04H          ;Prepare for state transition
                               ;... from state a to state c
        OUT   OUTPORT, AL      ;Exit state a.  Enter state c
```

In state *c*, the input variables are examined to determine the next state. If $x_1x_2x_3 = 000$, then the next state is *f*. Any other combination of input variables will sequence the machine to state *e*. The following instructions determine the next state from state *c*:

```
        IN    AL, INPORT       ;AL ← x1x2x3
        AND   AL, 07H          ;Mask off all bits except
                               ;... x1x2x3
        JZ    STTF             ;Proceed to state f if
                               ;... x1x2x3 = 000
        JMP   STTE             ;Otherwise, proceed to state e
```

If the machine is presently in state *e*, then state *a* is the next state, regardless of the values of the input variables. The following instructions cause the machine to enter state *e* and then jump to the program sequence which causes a transition to state *a*:

```
STTE:   MOV   AL, 10H          ;Prepare to exit state b or c
                               ;... and enter state e
        OUT   OUTPORT, AL      ;Enter state e
        JMP   STTA             ;Proceed to state a
```

The final instruction sequence causes the machine to exit state *c* and enter state *f*. Similar to the code sequence for state *d*, instructions in state *f* assert the Mealy-type output z_2 for two instruction cycles, then deassert z_2 while remaining in state *f*. The program then generates a transition to state *a*, where the process repeats. The instructions to enter state *f* are as follows:

```
STTF:   MOV   AL, 20H          ;Prepare for state transition
                               ;... from state c to state f
        OUT   OUTPORT, AL      ;Exit state c.  Enter state f
```

The instruction sequence for input evaluation and output assertion in state *f* is as follows:

```
IN    AL, INPORT      ;AL ← x1x2x3
AND   AL, 07H         ;Mask off all bits except
                      ;... x1x2x3
CMP   AL, 07H         ;Test for x1x2x3 = 111
JNE   STTA            ;Proceed to state a if
                      ;... x1x2x3 ≠ 111
MOV   AL, 0A0H        ;Prepare to assert z2 while
                      ;... remaining in state f
OUT   OUTPORT, AL     ;Assert z2 in state f
MOV   AL, 20H         ;Prepare to deassert z2 while
                      ;... remaining in state f
OUT   OUTPORT, AL     ;Deassert z2
JMP   STTA            ;Proceed to state a
```

The time duration of each state is variable, depending on the number of instructions in the state. The state diagram of Figure 4.74 and the assembly language program of Figure 4.76 are reproduced concurrently in Figure 4.77. This parallel presentation facilitates an easier understanding of the machine's operation by placing the program segments adjacent to their respective states.

4.4.3 Machine State Augmentation

The state alphabet Y can be easily augmented by increasing the number of output ports. Figure 4.78 depicts a machine with n output ports yielding a total of $n \times$ eight states. Each output port is accessed by a unique address. A similar arrangement is possible for the input ports, although there are usually more states than input variables. Thus, only one input port is presented in Figure 4.78. An adverse condition is possible when two or more output ports constitute the state alphabet Y. The problem occurs when a state change is required between two states in different state registers. The present state $Y_{j(t)}$ bit is reset in state register$_i$ by an OUT instruction containing a 0 in the appropriate bit position of the output data. The program then executes an instruction sequence to enter the next state $Y_{k(t+1)}$. Until the next state is set, however, the machine is in an indeterminate state in which no present state is defined. This condition may or may not be significant depending on the application of state times to external circuitry. Additional logic can be added within the machine to resolve this circumstance. The design of the logic is left as an exercise.

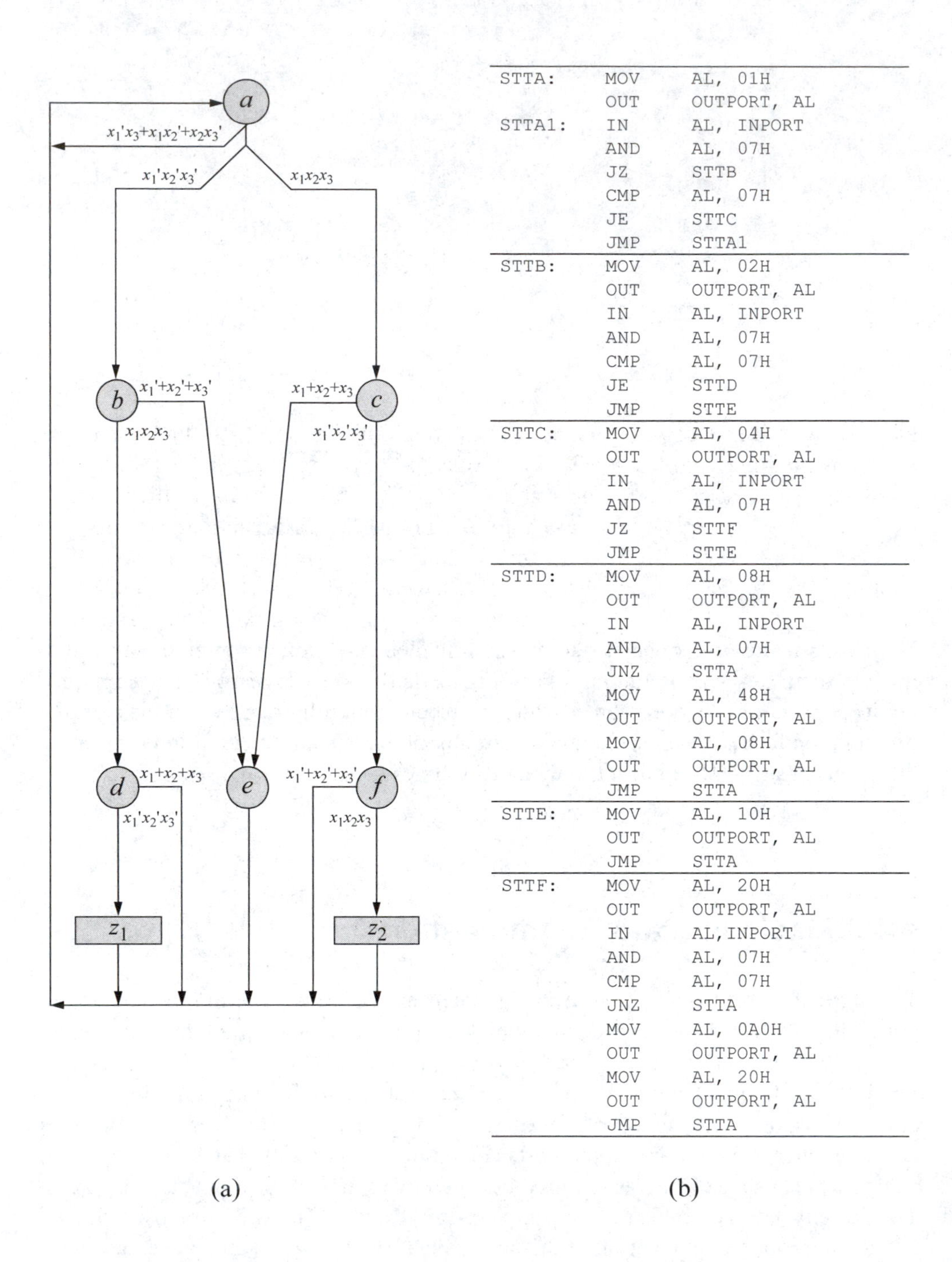


```
STTA:    MOV   AL, 01H
         OUT   OUTPORT, AL
STTA1:   IN    AL, INPORT
         AND   AL, 07H
         JZ    STTB
         CMP   AL, 07H
         JE    STTC
         JMP   STTA1
STTB:    MOV   AL, 02H
         OUT   OUTPORT, AL
         IN    AL, INPORT
         AND   AL, 07H
         CMP   AL, 07H
         JE    STTD
         JMP   STTE
STTC:    MOV   AL, 04H
         OUT   OUTPORT, AL
         IN    AL, INPORT
         AND   AL, 07H
         JZ    STTF
         JMP   STTE
STTD:    MOV   AL, 08H
         OUT   OUTPORT, AL
         IN    AL, INPORT
         AND   AL, 07H
         JNZ   STTA
         MOV   AL, 48H
         OUT   OUTPORT, AL
         MOV   AL, 08H
         OUT   OUTPORT, AL
         JMP   STTA
STTE:    MOV   AL, 10H
         OUT   OUTPORT, AL
         JMP   STTA
STTF:    MOV   AL, 20H
         OUT   OUTPORT, AL
         IN    AL,INPORT
         AND   AL, 07H
         CMP   AL, 07H
         JNZ   STTA
         MOV   AL, 0A0H
         OUT   OUTPORT, AL
         MOV   AL, 20H
         OUT   OUTPORT, AL
         JMP   STTA
```

Figure 4.77 Concurrent representation of the Mealy machine of Example 4.14: (a) state diagram; and (b) 8088 assembly language program.

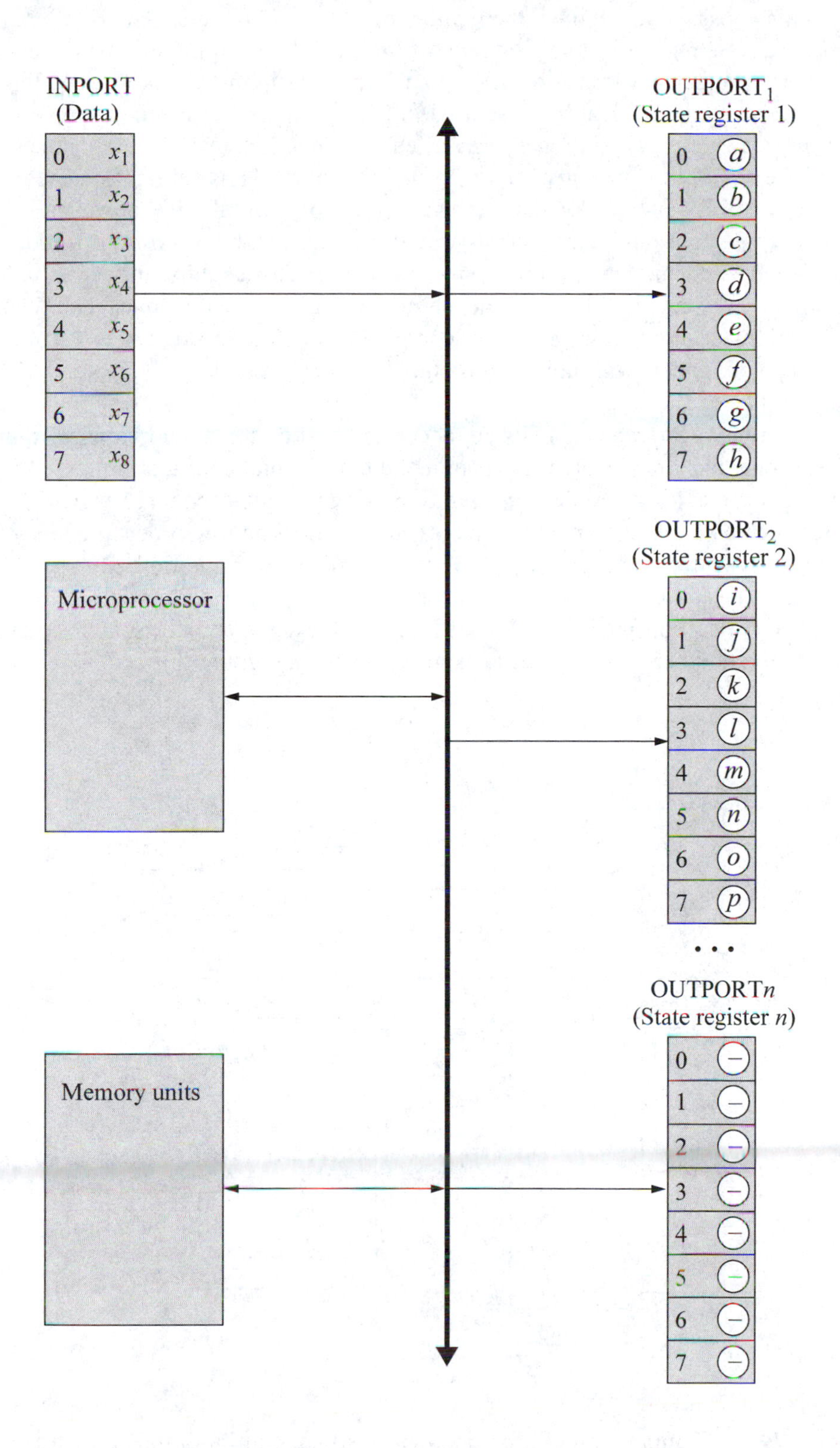

Figure 4.78 A microprocessor-controlled sequential machine with additional states.

It is also possible to augment the number of states without additional state registers. In the state register implementation of Figure 4.78, each bit signifies a unique state. This arrangement is called a horizontal configuration in which no decoding is used. A horizontal organization is desirable when the operating speed of state transitions is critical. A second approach is a vertical organization. A fully vertical method decodes the n bits of a state register with one n:2^n decoder. Thus, an 8-bit state register could conceivably generate 256 distinct states. A decoder of this size, however, is impractical for two reasons. First, a tree-structure of decoders is necessary in order to achieve the desired number of outputs, since no single decoder chip can accommodate 256 outputs. This greatly increases the amount of logic in the machine. The second reason is of lesser significance due to the comparatively slow speed of state transitions — the propagation delay through a multi-chip decoder is considerably greater than the propagation delay through a flip-flop.

An alternative approach is realized by combining horizontal and vertical organizations. Some bits of the state register utilize the horizontal configuration, while the remaining bits employ a vertical configuration using a single decoder. Figure 4.79 illustrates such an organization with one output register. The low-order five bits, representing states *a* through *e*, are depicted in a horizontal configuration. Bits 5, 6, and 7 connect to the address inputs of a 3:8 decoder such that, states *f* through *m* are generated by a vertical configuration. Thus, by adding one decoder to the state generation hardware, the number of available states increases from eight to thirteen.

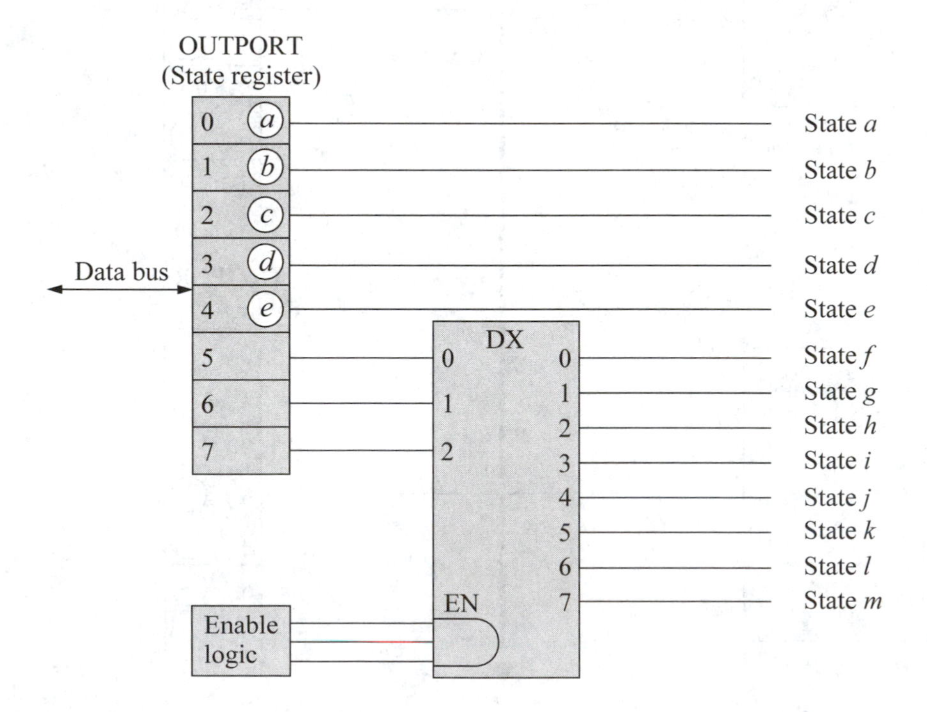

Figure 4.79 Combination of horizontal and vertical configurations for state generation in a microprocessor-controlled sequential machine.

State *f* (decoder output 0) cannot be used, however. If any one state *a* through *e* is set, then all remaining states will be reset, including bits 5, 6, and 7 of the state register, which will decode as zero and assert decoder output 0. This indicates that two states are active — one state *a* through *e* and state *f*. Two states should not be active simultaneously. Since decoder output 0 cannot be used to represent an active state, the total number of available states is 12. However, if the decoder incorporates an output-enable AND gate, then the decoder can be disabled whenever any state *a* through *e* is active, and enabled only when all horizontal states *a* through *e* are inactive. Decoder output 0, then, can be used as a valid state. The logic design which controls the output-enable AND gate is left as an exercise.

Assuming that the decoder can be correctly disabled, programming the machine to activate one state *a* through *e* is accomplished by setting the appropriate bit in the state register, while maintaining bits 5, 6, and 7 in a reset condition. To activate states *f* through *m*, states *a* through *e* must remain reset, while bits 5, 6, and 7 of the state register are set to values that correspond to states *f* through *m*. The state register is programmed to assert the 13 states according to the format of Table 4.22. Whenever the state register is programmed to assert states *f* through *m*, the decoder outputs must be considered invalid until the propagation delay through the decoder has been completed; otherwise, erroneous state assertions may occur until the outputs have stabilized.

Table 4.22 Format for programming the state register of Figure 4.79 to activate states *a* through *m*

OUTPORT bits		
7 6 5	4 3 2 1 0	Active state
0 0 0	0 0 0 0 1	*a*
0 0 0	0 0 0 1 0	*b*
0 0 0	0 0 1 0 0	*c*
0 0 0	0 1 0 0 0	*d*
0 0 0	1 0 0 0 0	*e*
0 0 0	0 0 0 0 0	*f*
0 0 1	0 0 0 0 0	*g*
0 1 0	0 0 0 0 0	*h*
0 1 1	0 0 0 0 0	*i*
1 0 0	0 0 0 0 0	*j*
1 0 1	0 0 0 0 0	*k*
1 1 0	0 0 0 0 0	*l*
1 1 1	0 0 0 0 0	*m*

4.4.4 Moore and Mealy Outputs

Machine outputs are asserted under program control. In particular, Moore outputs are asserted coincident with state activation. In its simplest form, a Moore output is identical in value to the state in which the output is asserted. For example, Figure 4.80 (a) depicts a state diagram segment for an 8-state machine in which output z_1 is asserted as a function of present state *d*. Figure 4.80 (b) shows the state register for the machine in which states *a* through *h* are assigned bit positions 0 through 7, respectively. No additional hardware is necessary to implement z_1. The same line that connects state *d* to external circuitry also asserts z_1; that is, state *d* and output z_1 are synonymous. When the machine enters state *d* under program control by setting the state register to a value of 08H, output z_1 is asserted simultaneously.

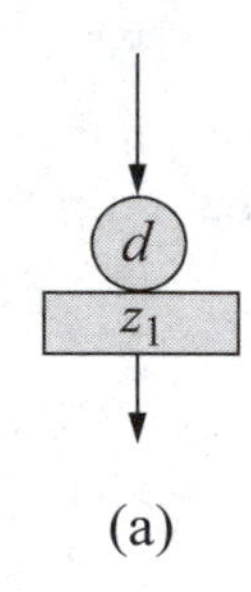

(a)

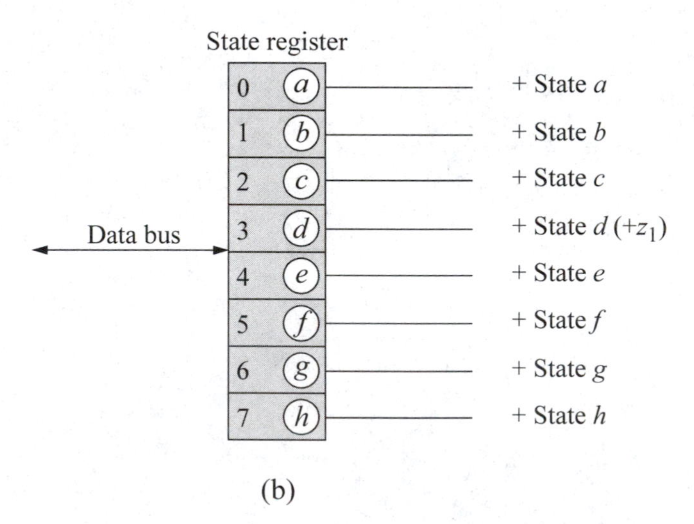

(b)

Figure 4.80 Moore output assertion in a microprocessor-controlled sequential machine: (a) state diagram segment; and (b) state register in which output z_1 is asserted when the machine enters state *d*.

To establish the assertion of Mealy outputs is a more involved process. When a state is entered that contains a Mealy output, the input variables must be examined to determine whether the output is to be asserted. This procedure was detailed in the assembly language program of Figure 4.76 for the Mealy machine of Figure 4.74.

Figure 4.81 (a) depicts a state diagram segment in which a Mealy output is asserted in state h if $x_1x_2 = 11$. Figure 4.81 (b) shows additional detail on the input and output ports for a generalized microprocessor-controlled sequential machine with Mealy outputs. The application program effects an entry into state h by addressing $OUTPORT_0$ coincident with a gated write pulse and by placing the value 80H on the data bus. The machine enters state h when the active-low Gated write pulse 1 signal produces a low-to-high transition at the positive-edge-triggered clock input of the $OUTPORT_0$ register. The program then reads the contents of the INPORT. A unique address selects the INPORT concurrent with a read pulse. It is assumed that supplementary circuitry stabilizes the input variables prior to the application of the read pulse. The INPORT is implemented with 3-state buffer/drivers that connect the input signals to the data bus. Thus, when the active-low Gated read pulse is at the more negative of two voltage levels, the Output Enable (OE) input gates the system inputs to the bidirectional data bus. When the Gated read pulse signal becomes inactive, the low-to-high transition loads the data bus into the accumulator of the microprocessor. The outputs of the INPORT then present a high impedance to the data bus.

If inputs $x_1x_2 = 11$, then the program addresses $OUTPORT_1$ and places the value 01H (z_1) on the data bus. On the rising edge of the active-low Gated write pulse 2 signal, the contents of the data bus are loaded into $OUTPORT_1$, thereby asserting z_1. In this application, the $OUTPORT_1$ register is dedicated to machine outputs only and can be used to implement both Moore and Mealy outputs. However, if the state register is not fully utilized, then the unused bit positions can accommodate machine outputs, negating the need for a separate output port.

4.4.5 System Architecture

Due to pin limitations on the 8088 microprocessor integrated circuit, the address/data bus is time multiplexed, as shown in Figure 4.82. During the first part of a bus cycle, the address/data pins $AD7$ through $AD0$ contain an address. The address is loaded into a transparent register and latched by the Address latch enable (ALE) pulse to provide address lines $A7$ through $A0$. Pins $AD7$ through $AD0$ are then used for data. Address lines $A15$ through $A8$ are not multiplexed and provide a constant address during the bus cycle. Address/status pins $A19/S6$ through $A16/S3$ are also time multiplexed.

The architecture of a typical microprocessor-controlled synchronous sequential machine is shown in Figure 4.83. This is a minimum mode configuration used for small systems in which the microprocessor generates all bus and control signals. The δ next-state logic is represented by the INPORT select decoder and the 3-state $INPORT_7$ buffer/driver. The state alphabet Y is depicted by the $OUTPORT_0$ register and the OUTPORT select decoder, which is shared with the λ output logic designated by $OUTPORT_1$.

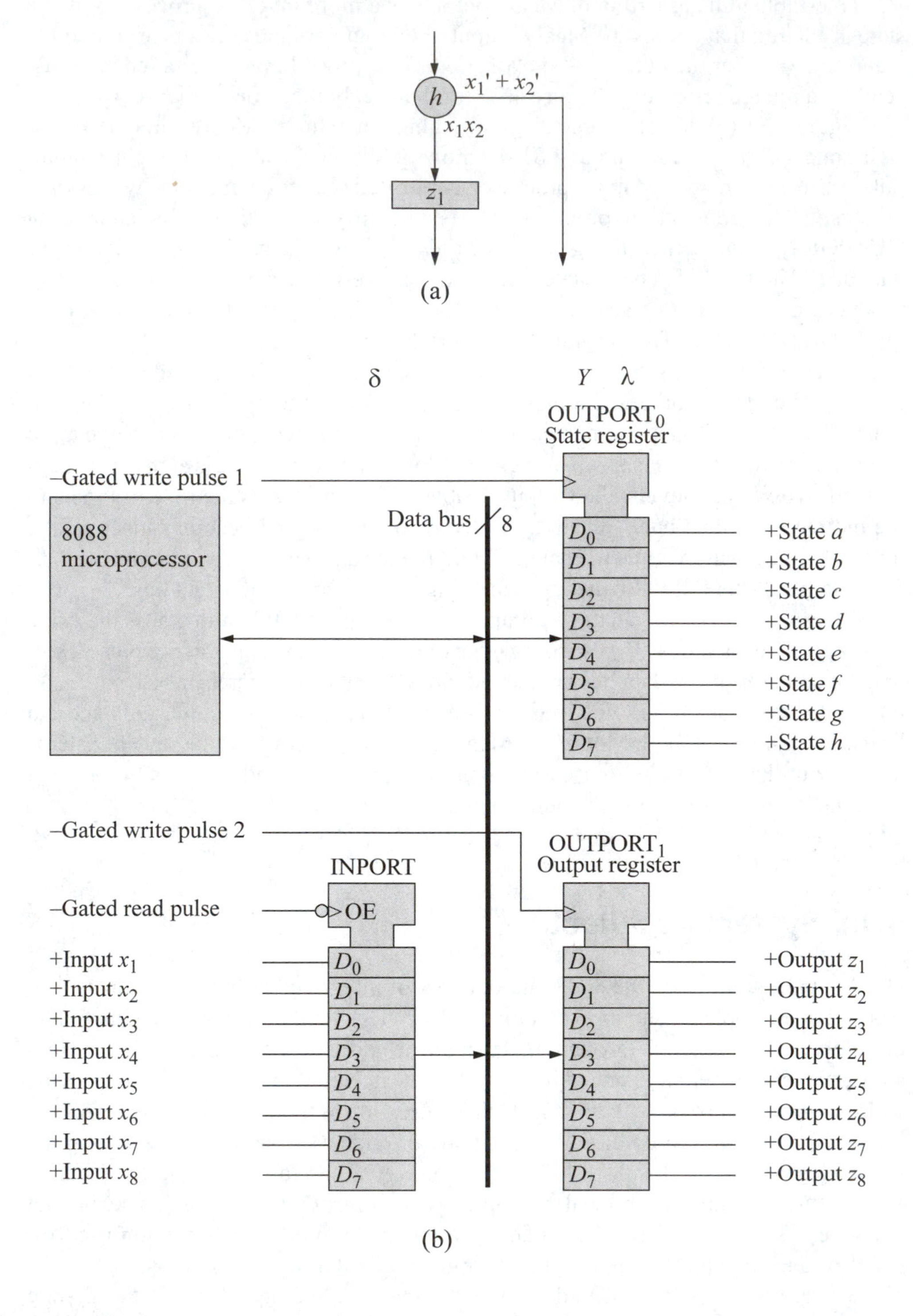

Figure 4.81 General block diagram showing the major components of a microprocessor-controlled sequential machine with Mealy outputs: (a) state diagram segment; and (b) input, state, and output ports.

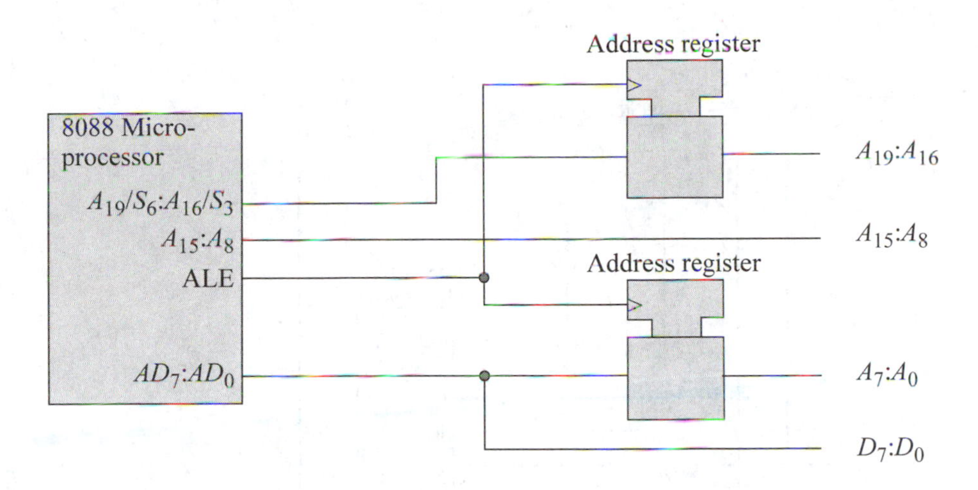

Figure 4.82 The 8088 microprocessor showing the multiplexed address/data bus.

The clock generator circuit provides basic system timing and synchronization. Accompanying the clock generator circuitry is an external crystal which operates at three times the desired processor frequency. A typical 8088 frequency is 5 MHz, requiring a crystal of 15 MHz.

The INPORT select decoder addresses the input port when the program executes an IN instruction. INPORT_7 can accommodate eight input variables $x_1, x_2, \ldots, x_8$. The input port address was arbitrarily chosen to be $A_7A_6A_5A_4 = 0111$. Thus, INPORT_7 is selected when address lines $A_7A_6A_5A_4 = 0111$, the +IO(–M) line is at the more positive voltage level (indicating an I/O operation), and the active-low Read pulse is asserted. When these conditions prevail, the output of the 3-state INPORT_7 buffer/driver is enabled and its contents are placed on the data bus coincident with the –Read pulse.

The waveforms for a read cycle are illustrated in Figure 4.84 (a). The contents of the data bus are loaded into the microprocessor's accumulator register AL on the rising edge of the –Read pulse signal. If the number of input variables does not exceed eight, then the decoder can be replaced by a single AND function to select INPORT_7. Implementing the INPORT selection logic with a decoder, however, allows for expansion of seven additional input ports, INPORT_0, INPORT_1, ... , INPORT_6, with no extra selection hardware. The input port selection logic is enabled only when the –Read signal is in its active state. Thus, if necessary, the same address can be used to select output ports, which are enabled by the active-low –Write pulse. Table 4.23 lists the addresses for eight possible input ports.

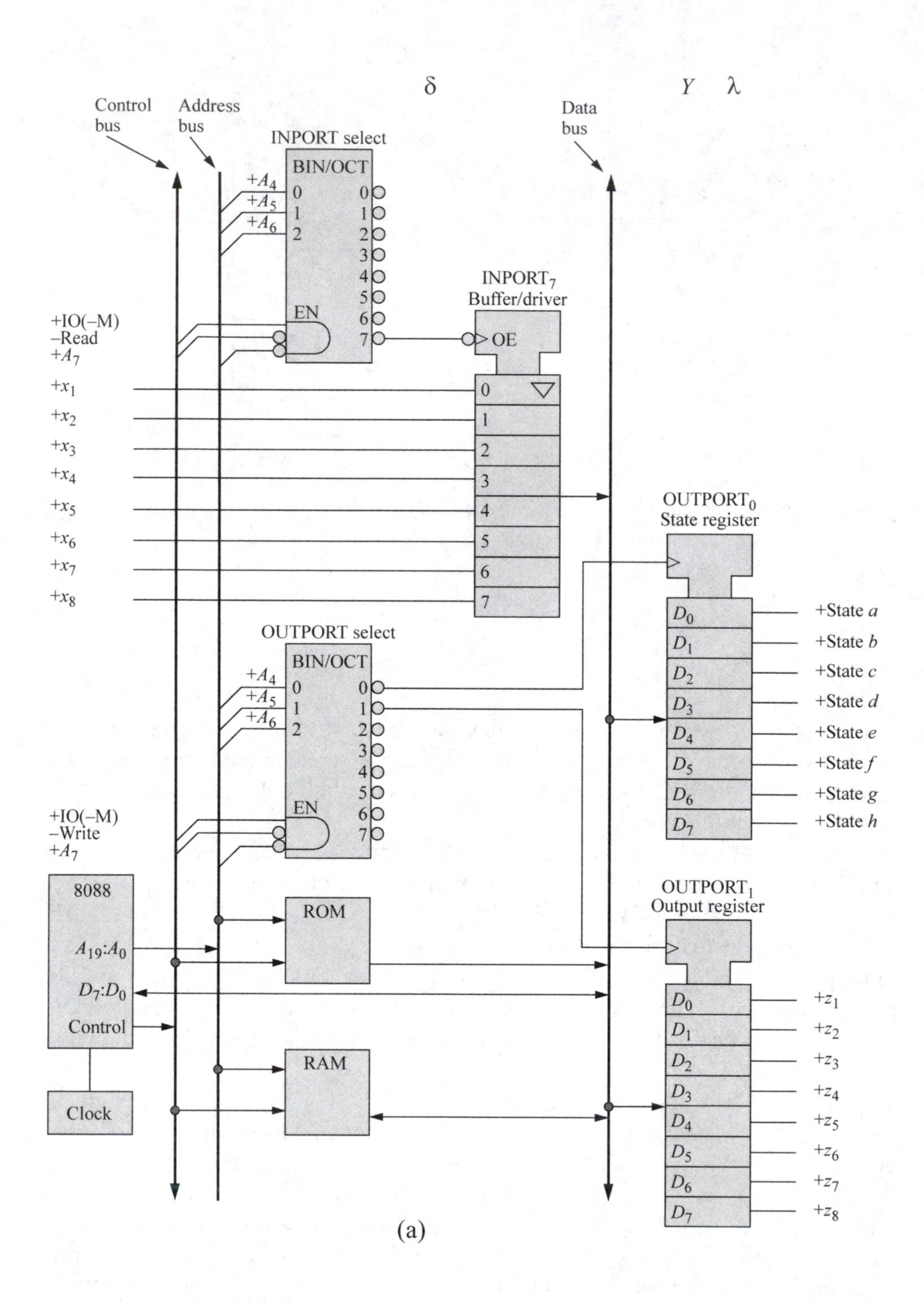

Figure 4.83 Architecture for an 8088-microprocessor-controlled synchronous sequential machine: (a) input/output port selection; and (b) memory selection.

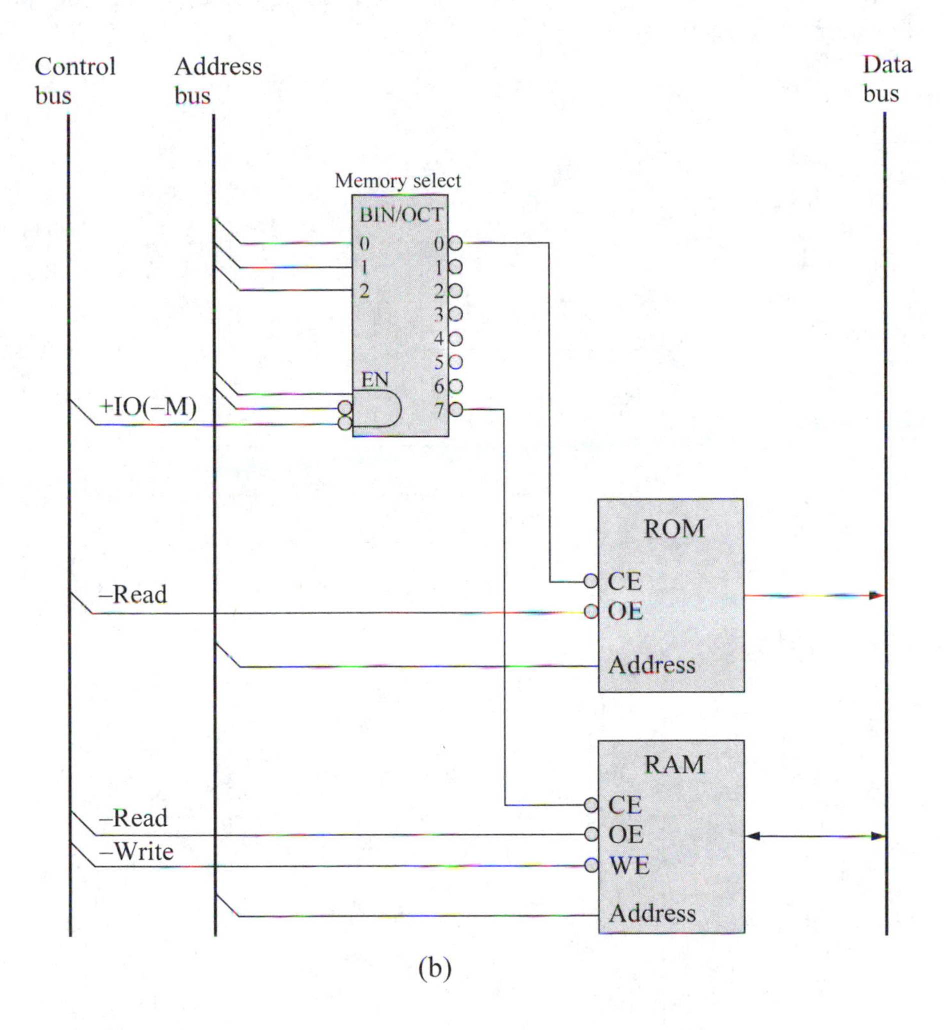

Figure 4.83 (Continued)

Output port selection is obtained in a similar manner, as defined in Table 4.24. When an OUT instruction is executed, the OUTPORT address is placed on the address bus. This is followed by the data from register AL, which is placed on the data bus. The data bus connects to all attached ports and memories, but only the addressed device is enabled. Then the –Write pulse is asserted as shown in Figure 4.84 (b), and, on the low-to-high transition, the contents of the data bus are loaded into the addressed output port. If OUTPORT_0 is selected, then the data specifies a present state $Y_{j(t)}$; if OUTPORT_1 is selected, then the data specifies a machine output z_i. Output ports are selected only when the +IO(–M) signal is at its more positive voltage level, indicating an I/O operation rather than a memory operation. Six additional output ports,

OUTPORT_2, OUTPORT_3, ... , OUTPORT_7, can be accommodated by decoder outputs 2 through 7, respectively.

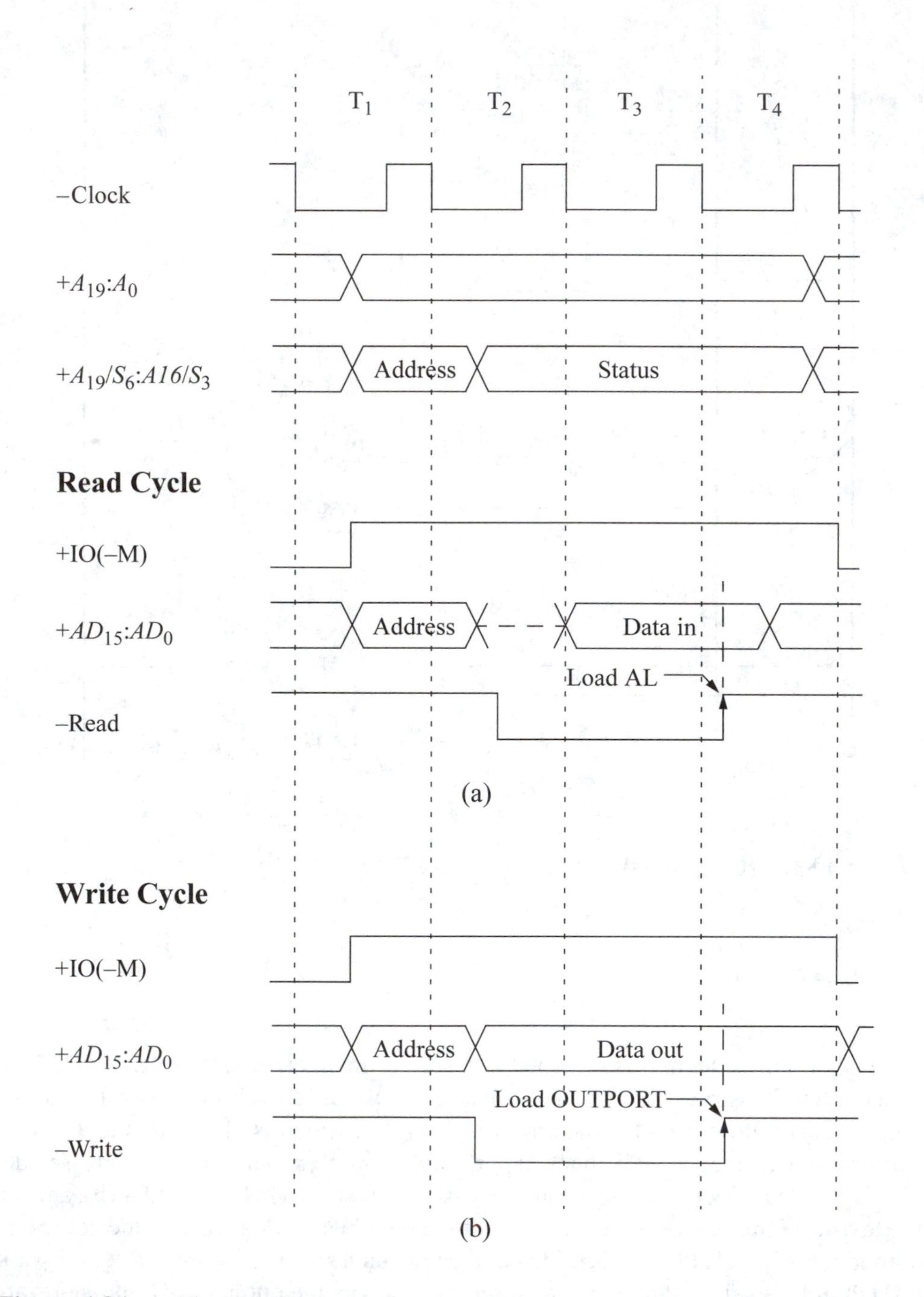

Figure 4.84 Timing diagram for the microprocessor-controlled synchronous sequential machine of Figure 4.83: (a) read cycle; and (b) write cycle.

Table 4.23 Input port addresses

Input port selected	Address bits A_7	$A_6A_5A_4$	$A_3A_2A_1A_0$
$INPORT_0$	0	0 0 0	0 0 0 0
$INPORT_1$	0	0 0 1	0 0 0 0
$INPORT_2$	0	0 1 0	0 0 0 0
$INPORT_3$	0	0 1 1	0 0 0 0
$INPORT_4$	0	1 0 0	0 0 0 0
$INPORT_5$	0	1 0 1	0 0 0 0
$INPORT_6$	0	0 1 0	0 0 0 0
$INPORT_7$	0	1 1 1	0 0 0 0

Table 4.24 Output port addresses

Output port selected	Address bits A_7	$A_6A_5A_4$	$A_3A_2A_1A_0$
$OUTPUT_0$	0	0 0 0	0 0 0 0
$OUTPUT_1$	0	0 0 1	0 0 0 0
$OUTPUT_2$	0	0 1 0	0 0 0 0
$OUTPUT_3$	0	0 1 1	0 0 0 0
$OUTPUT_4$	0	1 0 0	0 0 0 0
$OUTPUT_5$	0	1 0 1	0 0 0 0
$OUTPUT_6$	0	0 1 0	0 0 0 0
$OUTPUT_7$	0	1 1 1	0 0 0 0

The ROM and RAM memory devices are selected as shown in Figure 4.83 (b). The Memory select BIN/OCT decoder outputs are asserted low by a subset of the address bits and the +IO(–M) signal at its more negative potential. If the ROM is selected, then the chip is enabled (CE) by the appropriate subset of the address bits. The remaining address bits select a byte in the ROM. The –Read pulse then enables the ROM output (OE). The waveforms are similar to those shown in Figure 4.84 (a). The ROM contains the application program, such as the program shown in Figure 4.76.

Similarly, the RAM is addressed by the same Memory select decoder. For read operations, the RAM chip enable input receives an active-low signal from the address decoder output. The RAM outputs are then enabled by the –Read signal. Write operations generate an active-low write pulse which is connected to the Write Enable (WE) input of the RAM device. The contents of the data bus are loaded into an output port or the accumulator on the positive transition of the –Write pulse, as shown in Figure 4.84 (b). The RAM is used primarily for stack operations and temporary data storage.

The application program is derived directly from the state diagram or other machine specifications. In a typical operation, the machine enters a state by executing an OUT instruction, which accomplishes the following:

1. Set the +IO(–M) signal to its more positive potential.
2. Place the OUTPORT_0 address on bits $A7{:}A0$ of the address bus.
3. Place a byte of data that corresponds to the next state on the 8-bit data bus.
4. Assert the active-low –Write pulse.
5. OUTPORT_0 is loaded from the data bus on the positive transition of the –Write pulse. The machine then enters the next state.

To read the values of the input variables, the program executes an IN instruction, which accomplishes the following:

1. Set the +IO(–M) signal to its more positive potential.
2. Place the INPORT_7 address on bits $A7{:}A0$ of the address bus.
3. Assert the active-low –Read pulse.
4. The contents of INPORT_7 are placed on the data bus when the –Read pulse becomes active.
5. The accumulator register AL is loaded from the data bus on the positive transition of the –Read pulse.

Example 4.15 Figure 4.85 illustrates a simple state diagram which will be implemented using the architecture of Figure 4.83. The machine has both Moore- and Mealy-type outputs. The input and output port addresses are established by the following equate statements:

```
INPORT7     EQU     70H
OUTPORT0    EQU     00H
OUTPORT1    EQU     01H
```

The machine enters state *a* by the following instruction sequence:

```
STTA:   MOV     AL, 01H      ;Enter
        OUT     OUTPORT0     ;. . .state a
```

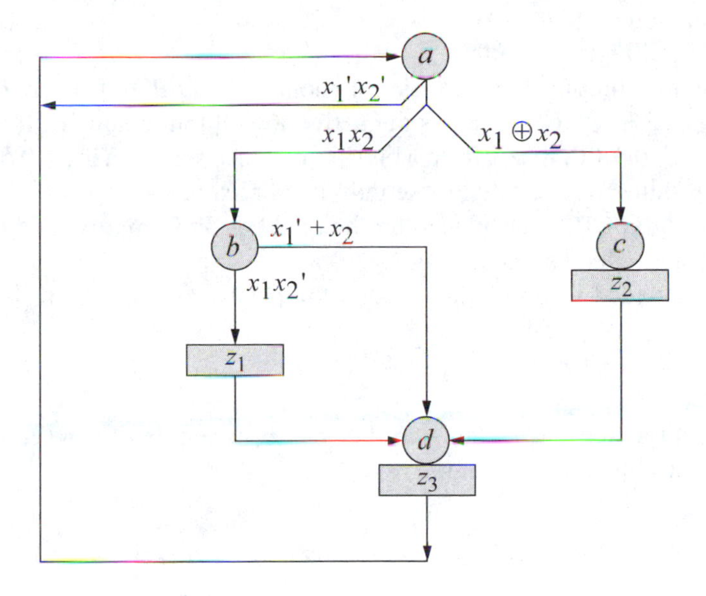

Figure 4.85 State diagram which is implemented using the architecture of Figure 4.83.

The following description is representative of all OUT instructions—only the port address changes:

1. The OUT instruction causes the +IO(–M) signal to be asserted high, disabling both memory devices.
2. Address bus $A7{:}A0$ = $\underline{0}$ $\underline{000}$ 0000.
3. Data bus $D7{:}D0$ = 0000 0001.
4. The –Write signal completes the enable function of the OUTPORT select decoder, asserting output 0 low.
5. When the –Write signal is deasserted, decoder output 0 generates a positive transition which loads the contents of the data bus into OUTPORT_0. The machine is now in state *a*.

The machine then reads the values of the input variables using the following instruction:

```
STTA1:   IN    AL, INPORT7    ;AL ← x1x2
```

The following description is representative of all IN instructions:

1. The IN instruction causes the +IO(–M) signal to be asserted high for a nonmemory input operation.
2. Address bus $A7{:}A0 = \underline{0}\ \underline{111}\ 0000$.
3. The –Read signal completes the enable function of the INPORT select decoder, asserting output 7 low. This enables the active-low output enable of $INPORT_7$. The outputs of $INPORT_7$ return from a high-impedance state to a logic 0 or 1 voltage level, depending on the state of the individual bits.
4. The contents of $INPORT_7$ are placed on the data bus and remain valid until the deassertion of the –Read signal.
5. Register AL is loaded from the data bus on the positive transition of the –Read signal.

The program then sequences the machine to state *b* or *c*, depending on the conditional jump instructions in state *a*, as shown below.

```
        AND   AL, 03H          ;Mask off all bits except x1x2
        JZ    STTA1            ;Jump to state a if x1x2 = 00
        CMP   AL, 03H          ;If x1x2 = 11, then
        JE    STTB             ;... proceed to state b
        JMP   STTC             ;Otherwise, proceed to state
                               ;... c, because x1x2 = 01 or 10
```

In state *b*, Mealy output z_1 is asserted if $x_1x_2 = 10$. This is demonstrated in the following instruction sequence:

```
STTB:   MOV   AL, 02H          ;Exit state a
        OUT   OUTPORT0         ;... Enter state b
        IN    AL, INPORT7      ;AL ← x1x2
        AND   AL, 03H          ;Mask off all bits except x1x2
        CMP   AL, 01H          ;Test for x1x2 = 10
        JZ    SETZ1            ;If true, go and assert z1
        JMP   STTD             ;If x1x2 ≠ 10, then
                               ;... proceed to state d

SETZ1:  MOV   AL, 01H          ;Prepare to assert z1
        OUT   OUTPORT1, AL     ;Assert z1 in state b
        SUB   AL, AL           ;Clear AL to deassert z1
        OUT   OUTPORT1, AL     ;Deassert z1
        JMP   STTD             ;Proceed to state d
```

The machine enters state *c* from state *a* and asserts Moore output z_2.

```
STTC:   MOV   AL, 04H           ;Exit state a
        OUT   OUTPORT0, AL      ;... Enter state c
        MOV   AL, 02H           ;Prepare to assert z2
        OUT   OUTPORT1, AL      ;Assert z2
        SUB   AL, AL            ;Clear AL to deassert z2
        OUT   OUTPORT1, AL      ;Deassert z2
        JMP   STTD              ;Proceed to state d
```

The machine enters state *d* from state *b* or *c* and asserts Moore output z_3.

```
STTD:   MOV   AL, 08H           ;Exit state b or c
        OUT   OUTPORT0, AL      ;... Enter state d
        MOV   AL, 04H           ;Prepare to assert z3
        OUT   OUTPORT1, AL      ;Assert z3
        SUB   AL, AL            ;Clear AL to deassert z3
        OUT   OUTPORT1, AL      ;Deassert z3
        JMP   STTA              ;Proceed to state a
```

The preceding program was presented to illustrate more graphically the operation of the machine at the component level. A machine with only eight states, however, as shown in Figure 4.83, can be synthesized more quickly and easily using discrete logic gates. Although a microprocessor-controlled machine possesses important attributes such as, programming flexibility and ease of expansion, the architecture is more appropriate for a machine with a large number of states. The ease of expansion is exemplified in the following section where a single microprocessor controls multiple machines.

4.4.6 Multiple Machines

The power and versatility of a microprocessor can be effectively utilized in the synthesis of multiple synchronous sequential machines. Although the microprocessor controls only one machine at a time, the machines appear to be operating concurrently. Figure 4.86 illustrates the input and output registers for two machines, A and B, each controlled by the same microprocessor. The organization assumes separate ports for machine outputs. Each of the six ports is accessed by a different address. The transition between the program segments that control each machine can be accomplished by intersegment CALL instructions or by interrupts. Intersegment CALL instructions transfer control between the code segment of machine A and the code segment of machine B. Figure 4.87 shows a simplified flowchart for the programs that control machines A and B. The application program for machine A (Program A) begins by entering the initial state. Program A then executes an intersegment CALL instruction to far procedure MACHB to invoke the program for machine B (Program B). The location of the present state in Program B was previously saved in a Jump table. Program B then jumps to the instruction sequence that applies to the present state.

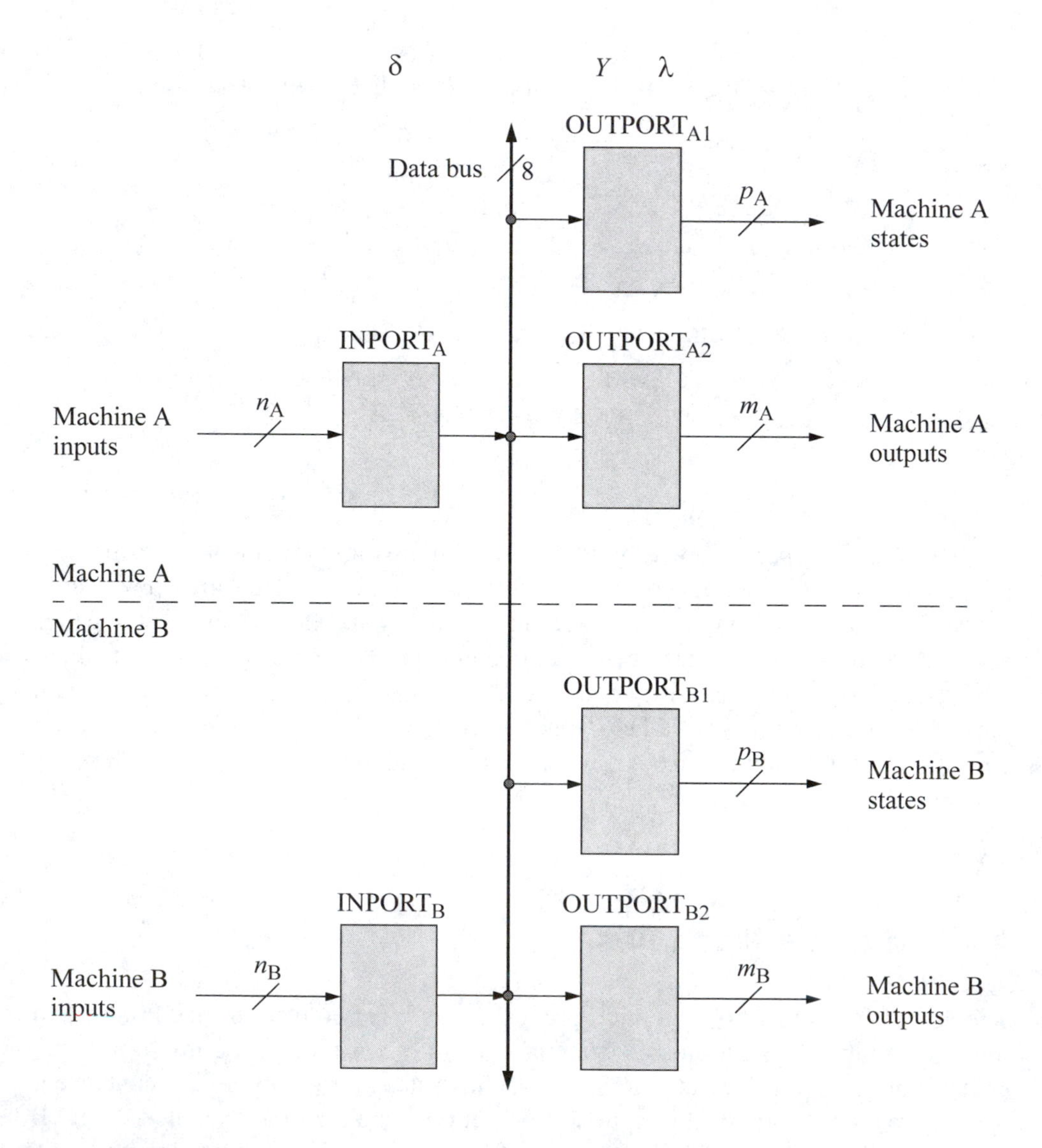

Figure 4.86 Multiple machines controlled by a single microprocessor.

In the present state code sequence, Program B examines the input variables, if applicable, to determine the next state. When the next state is entered, any Moore output that is associated with that state is asserted by setting the corresponding bit in the machine output register $OUTPORT_{B2}$. If a Mealy output is associated with the present state and the input conditions are met, then the appropriate bit is set in the $OUTPORT_{B2}$ register. The present state identity is saved so that Program B can locate the present state when the program is next invoked.

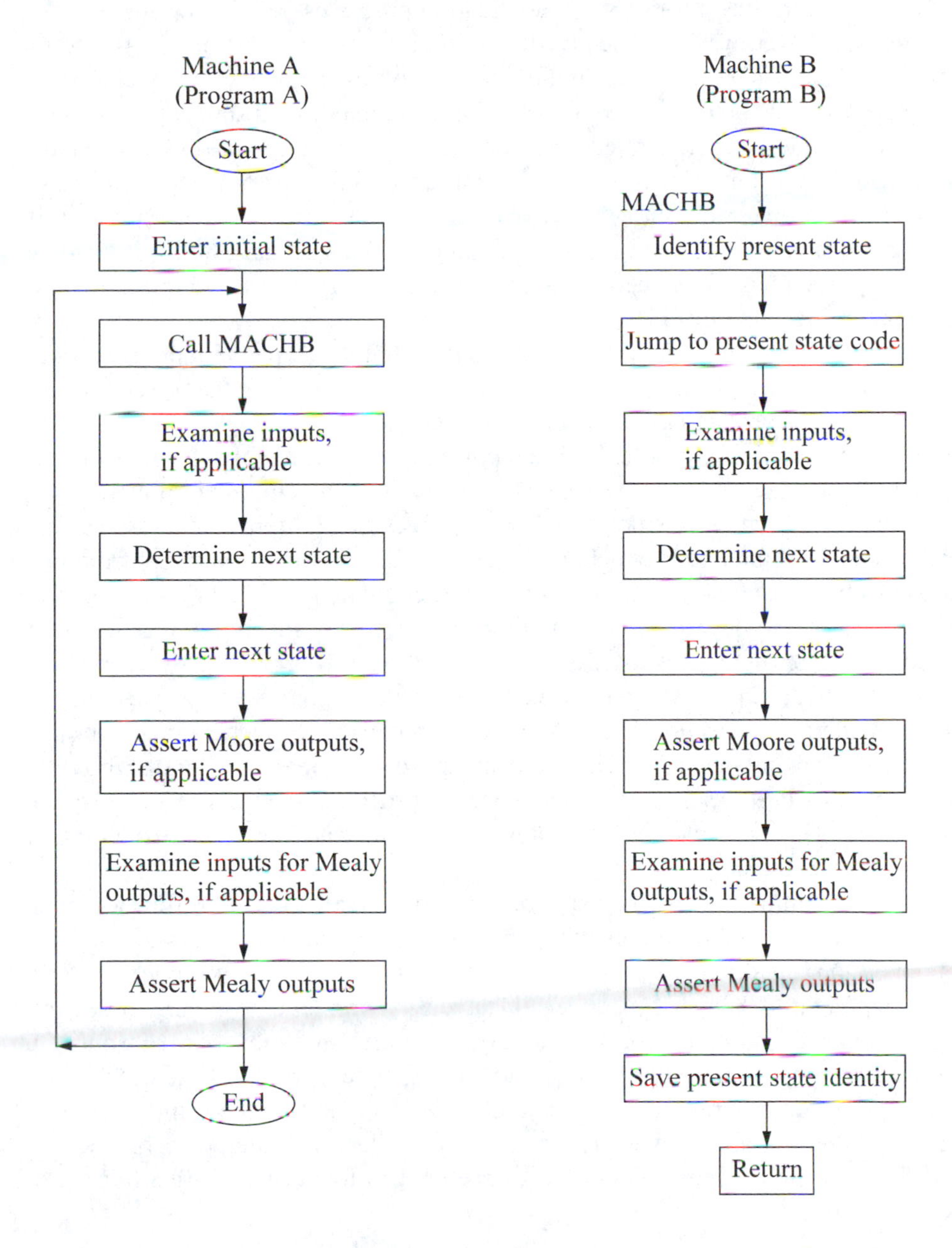

Figure 4.87 Simplified flowchart for the programs of two sequential machines controlled by the same microprocessor, using an intersegment CALL instruction.

When the return instruction (RET) is executed in Program B, the system returns control to Program A at the address following the CALL instruction. Program A then performs a similar instruction sequence to examine the input variables, if applicable, determine the next state, and set the appropriate outputs, if necessary. Program A then invokes Program B by means of another intersegment CALL instruction and the process repeats.

The interrupt structure of a microprocessor system represents an alternative approach to transferring control between application programs. An important criterion of an interrupt-driven system is the prioritization of machines. Machine A, for example, can be assigned a priority level higher than that of machine B. Both machines are initialized during the power-on sequence to some predetermined state, at which time two interrupt requests are generated. A priority arbitration unit in the system then determines which machine to acknowledge. Arbitration is necessary when there are simultaneous interrupt requests from two or more machines. When an interrupt acknowledge signal is received by the interrupting machine, an interrupt vector is placed on the data bus. The vector is a pointer to an interrupt service routine which handles the interrupting machine.

A second method of identifying interrupts is by polling. The main program polls each machine in turn and checks the state of an interrupt flag. If the flag is set, then the program transfers control to a procedure which services the interrupt. If the flag is reset, then the next machine is polled. When polling is used to identify the interrupting machine, priority is established by the order in which the machines are polled.

A third technique that is widely used is to connect the multiple machines in a configuration in which the interrupt acknowledge signal is propagated serially from machine to machine. The machine that receives the acknowledge signal first has highest priority. The priority level decreases with each succeeding machine. If machine A has a pending interrupt and is logically closest to the microprocessor — thereby establishing the highest priority — then the acknowledge signal is not propagated. If, however, machine A does not require service, then the acknowledge signal is propagated to the following machine. The interrupt approach to machine control provides considerable flexibility, especially when the microprocessor controls several sequential machines. The interrupt technique, however, requires additional control circuitry in the machine interface.

Any of the multibyte microprocessors currently available, or in development, would far exceed the requirements of a simple sequential machine. Neither memory management nor the diversity of processor instructions is necessary. Indeed, the architecture of Figure 4.83 could just as easily have been implemented with an 8085 microprocessor with no apparent degradation of performance. Most sequential machines would not require the 1 megabyte addressing capability of the 8088 or the memory segmentation properties for disjoint code, data, and stack segments. These mechanisms are available, however, for expansion to larger, more powerful sequential machines, which require the unique characteristics found in the 8088 microprocessor.

4.5 Sequential Iterative Machines

An *iterative machine* is an organization of identical cells (or elements) which are interconnected in an ordered manner. An iterative machine (or network) may consist of combinational logic arranged in a linear array in which signals between cells propagate in one direction only. A parity checker and comparator are examples of combinational iterative networks. Or, the iterative network may consist of sequential cells, such as found in shift registers and simple binary counters.

A general block diagram of a combinational iterative network is shown in Figure 4.88. There are k cells labeled $\text{cell}_1, \text{cell}_2, \ldots, \text{cell}_i, \ldots, \text{cell}_k$ and n external input variables for each cell, designated $x_{11}, x_{12}, \ldots, x_{1n}$ for cell_1, $x_{i1}, x_{i2}, \ldots, x_{in}$ for cell_i and $x_{k1}, x_{k2}, \ldots, x_{kn}$ for cell_k. There are p internal cell inputs that are generated as outputs from the previous cell. For a typical cell_i, these internal interconnections are labeled $y_{(i-1)1}, y_{(i-1)2}, \ldots, y_{(i-1)p}$. The information received from the previous cell can be considered as intercell carry signals. The network also contains m output signals, designated $z_{11}, z_{12}, \ldots, z_{1m}$ for cell_1, $z_{i1}, z_{i2}, \ldots, z_{im}$ for cell_i, and $z_{k1}, z_{k2}, \ldots, z_{km}$ for cell_k. The network outputs can be generated from each cell or from only the rightmost cell_k. The number of cells in a combinational iterative network is equal to the number of input sets that are applied to the network, where a typical set of inputs is $x_{i1}, x_{i2}, \ldots, x_{in}$.

In general, a combinational iterative network can be converted to a functionally equivalent sequential machine by using the logic of cell_i as the δ next-state logic of the sequential machine. That is, cell_i connects to the storage elements which in turn connect to the λ output logic. The storage elements also feed back to the δ next-state logic of cell_i.

In a combinational network, the operation is complete after an appropriate propagation delay, at which time the outputs are stable. In a functionally equivalent sequential machine, the operation is complete only when all inputs have been sequenced through the machine and the outputs have stabilized. Thus, k clock cycles are required to establish the final machine outputs, where k is the number of input sets. That is, one clock is necessary to process each set of inputs x_{ij}, where $1 \leq i \leq k$ and $1 \leq j \leq n$.

Example 4.16 The synthesis of a functionally equivalent sequential iterative network is best illustrated by an example. A Moore machine will be synthesized which generates an output z_1 if and only if there is exactly one 1 bit in an input vector X_i, which represents a k-bit parallel string of input data. The machine will be synthesized first as a combinational network, then as a functionally equivalent sequential machine.

Table 4.25 depicts the next-state table for a typical cell_i. Two outputs are required from cell_i which connect to the following cell_{i+1} and indicate the state of the network up to and including cell_i. The network states are defined as follows:

$y_{i1}y_{i2} = 00$ indicates no 1s detected
$y_{i1}y_{i2} = 01$ indicates a single 1 detected
$y_{i1}y_{i2} = 11$ indicates two or more 1s detected
$y_{i1}y_{i2} = 10$ is unused

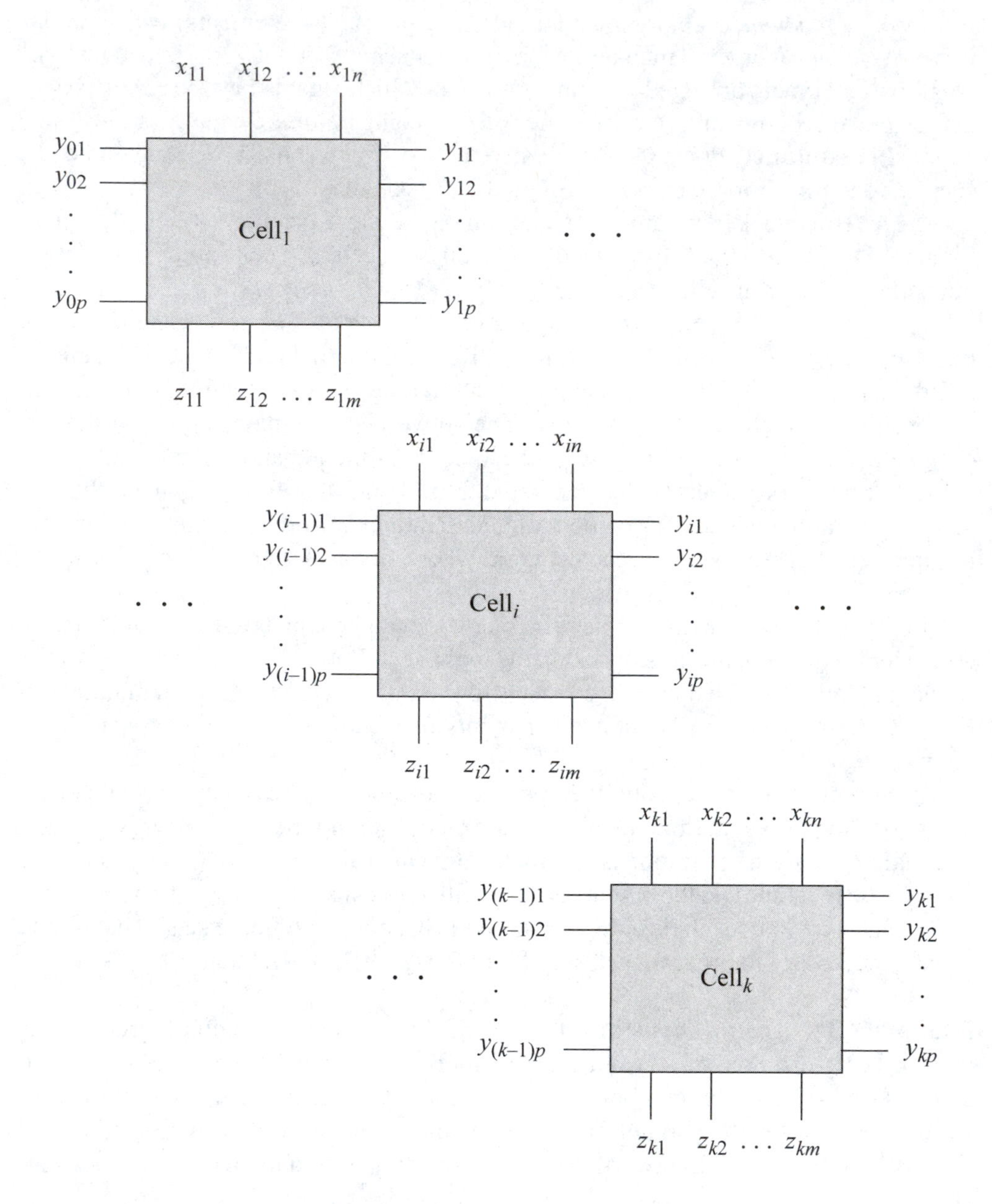

Figure 4.88 General block diagram of an iterative network.

A typical cell$_i$ is shown in Figure 4.89 (a). If $y_{(i-1)1}y_{(i-1)2} = 00$ and $x_i = 0$, then $y_{i1}y_{i2} = 00$, indicating that no 1 bits were detected up to and including cell$_i$. If, however, $x_i = 1$, then $y_{i1}y_{i2} = 01$, indicating that a single 1 bit was detected either by a previous cell or by cell$_i$. If $y_{(i-1)1}y_{(i-1)2} = 01$ and $x_i = 0$, then $y_{i1}y_{i2} = 01$, signifying that, up to and including cell$_i$, a single 1 bit was detected. If, however, $x_i = 1$, then $y_{i1}y_{i2} = 11$, signifying that two or more 1s have been detected. Output z_1 is asserted if $i = k$; that is, the rightmost cell$_k$ asserts z_1 if and only if there was a single 1 bit detected in the input vector $x_1, x_2, \ldots, x_i, \ldots, x_k$. Finally, if $y_{(i-1)1}y_{(i-1)2} = 11$ and $x_i = 0$ or 1, then $y_{i1}y_{i2} = 11$, denoting that two or more 1s were detected prior to or in association with cell$_i$.

Table 4.25 Next-state table for cell$_i$ for the Moore machine of Example 4.16

	Present state		Input	Next state		Output[1]
State description	$y_{(i-1)1}$	$y_{(i-1)2}$	x_i	y_{i1}	y_{i2}	z_1
No 1s	0	0	0	0	0	0
			1	0	1	0
A single 1	0	1	0	0	1	1
			1	1	1	0
Two or more 1s	1	1	0	1	1	0
			1	1	1	0
Unused	1	0	0	–	–	–
			1	–	–	–

(1) Output from cell$_k$

Figure 4.89 (b) shows the block diagram illustrated as a linear array of identical cells. In some cases, the two end cells may be slightly different. The leftmost cell of Figure 4.89 (b) contains intercell inputs y_{01} and y_{02} connected to a logic 0, indicating that no 1s were detected up to that point. The rightmost cell generates output z_1 as a function of y_{k1} and y_{k2}.

The equations which represent the logic of cell$_i$ are shown in Equation 4.18. The equations are obtained by plotting the entries of Table 4.25 on Karnaugh maps for y_{i1} and y_{i2} as shown below and then deriving the minimized equations. The logic diagram for cell$_i$ is shown in Figure 4.90 (a) and is synthesized from the input equations for y_{i1} and y_{i2} of Equation 4.18. By applying all valid combinations of $y_{(i-1)1}$, $y_{(i-1)2}$, and x_i to the cell inputs, the next states of Table 4.25 will be realized. Output z_1 is asserted in cell$_k$ only, when $y_{k1}y_{k2} = 01$.

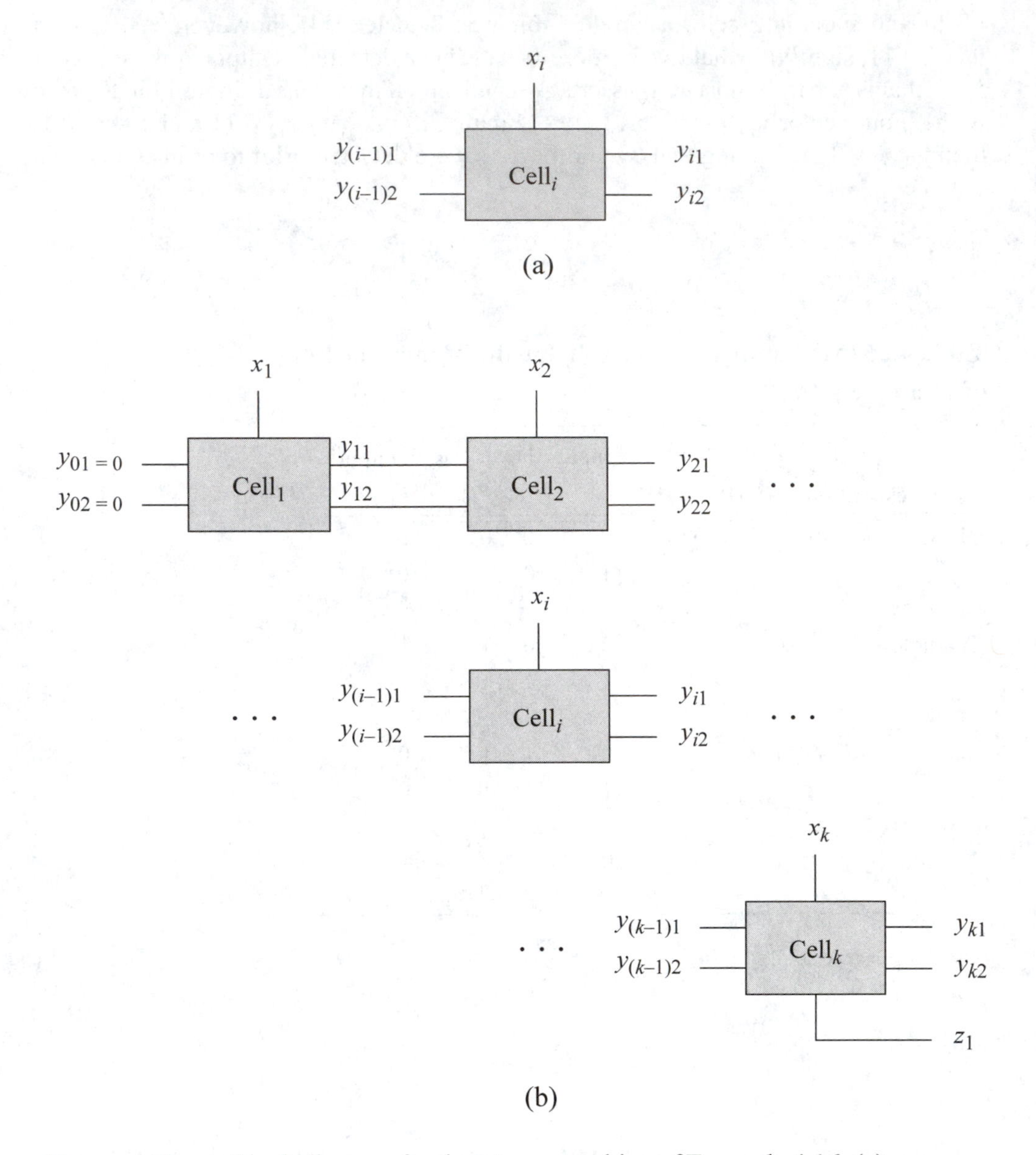

Figure 4.89 Block diagram for the Moore machine of Example 4.16: (a) representative cell$_i$; and (b) cell$_1$ through cell$_k$ corresponding to a k-bit input vector.

$$
\begin{aligned}
y_{i1} &= y_{(i-1)1} + y_{(i-1)2}\, x_i \\
y_{i2} &= y_{(i-1)2} + x_i \\
z_1 &= y_{k1}' y_{k2}
\end{aligned}
\tag{4.18}
$$

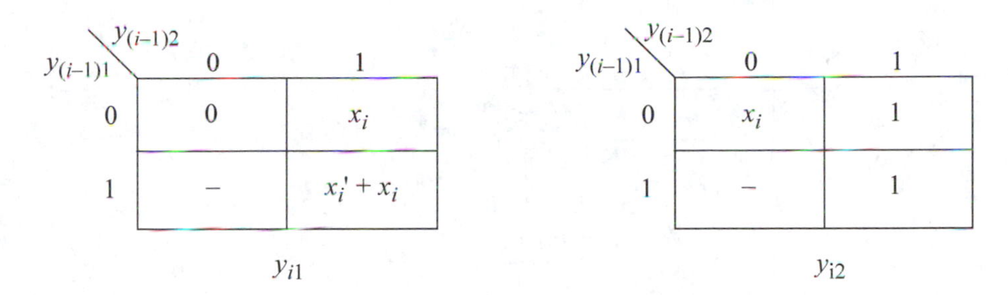

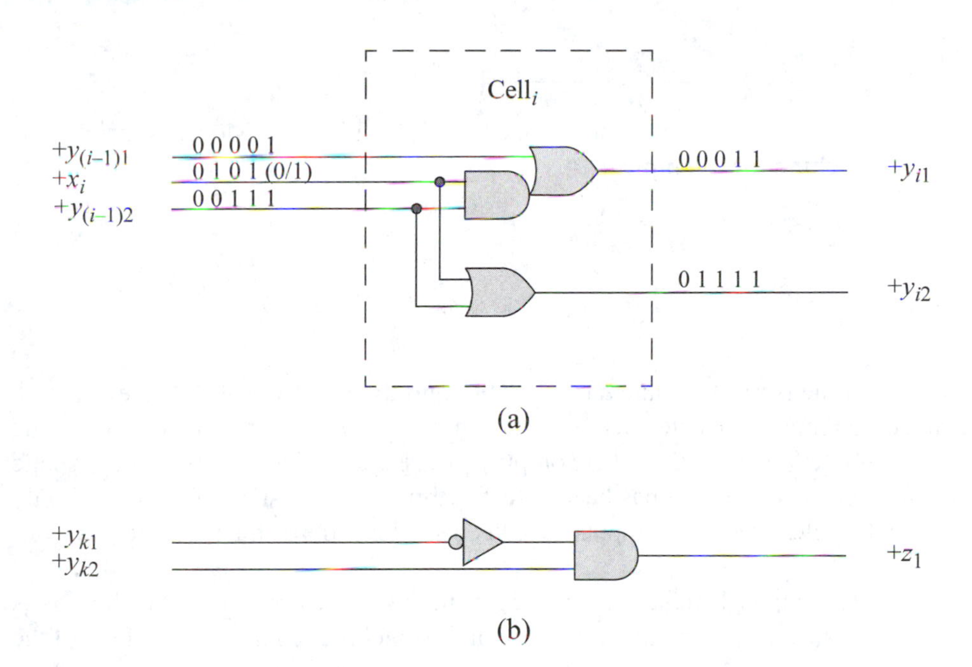

Figure 4.90 Illustrative logic for the Moore machine of Example 4.16: (a) logic diagram for a typical cell$_i$; and (b) logic diagram for output z_1 in cell$_k$.

The combinational iterative network of Example 4.16 will now be synthesized as a functionally equivalent sequential machine. The input vector $x_1, x_2, \ldots, x_i, \ldots, x_k$ is applied to the machine serially such that, one bit x_i is processed for each machine clock cycle. The usual design procedure for synchronous sequential machines will be followed. The next-state table of Table 4.25 also applies to the synthesis of an equivalent sequential machine. The state diagram is illustrated in Figure 4.91.

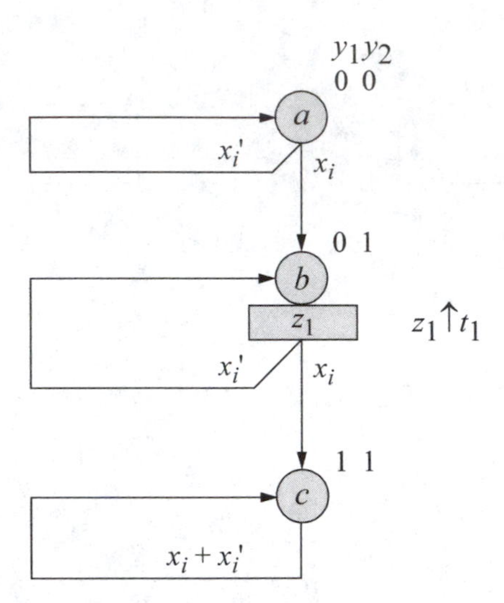

Figure 4.91 State diagram for a functionally equivalent Moore synchronous sequential machine for Example 4.16.

The machine is reset to state a ($y_1y_2 = 00$) and the input variable x_i is evaluated. The machine remains in state a as long as each bit $x_i = 0$. If $x_i = 1$, then the machine proceeds to state b ($y_1y_2 = 01$) where output z_1 is asserted. The assertion of z_1 in state b indicates that a single 1 bit has been detected thus far. The state of z_1 is valid only after k clock cycles when the k inputs have been sequenced into the machine and evaluated.

In state b, the next 1 bit sequences the machine to state c ($y_1y_2 = 11$) where z_1 is deasserted. State c is a terminal state in which the machine remains regardless of the subsequent bit pattern. Since the transition from state c ($y_1y_2 = 11$) to state a ($y_1y_2 = 00$) cannot occur, the machine will never generate state $y_1y_2 = 01$ as a transient state, which would produce a spurious output on z_1. Therefore, the λ output logic for z_1 is simply an AND gate to decode $y_1y_2 = 01$, negating the necessity of a complemented clock input to eliminate a glitch. Output z_1 is asserted at time t_1 in state b and remains asserted for all subsequent 0 bits.

The input maps of Figure 4.92 (a) are derived from the state diagram as in previous examples. In state a ($y_1y_2 = 00$), flip-flop y_1 has a next value of 0 regardless of the path taken. Flip-flop y_2 has a next value of 1 only if $x_i = 1$; therefore, input variable x_i is inserted in minterm location 0. Likewise, in state b ($y_1y_2 = 01$), y_1 has a next value of 1 only if $x_i = 1$. Flip-flop y_2, however, has a next value of 1 regardless of the value of x_i. Therefore, $x_i + x_i' = 1$ is placed in minterm location 1 of the input map for y_2. In state c ($y_1y_2 = 11$), both y_1 and y_2 have a next value of 1 for $x_i = 0$ or 1.

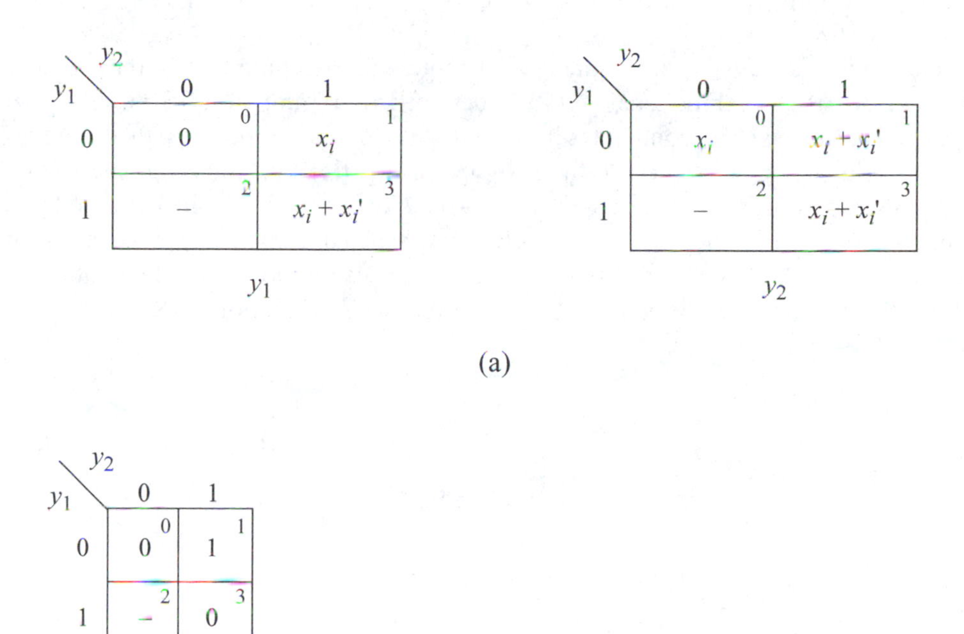

Figure 4.92 Karnaugh maps for the Moore machine of Figure 4.91: (a) input maps; and (b) output map.

Consider the input map for flip-flop y_1. Input x_i is common to minterm locations 1 and 3, both of which are in column $y_2 = 1$, yielding the term $y_2 x_i$. The expression $x_i + x_i' = 1$ in minterm 3 combines with the unused state to yield the term y_1. Regarding the input map for flip-flop y_2, every square contains the term x_i. Therefore, x_i is a component of the equation for y_2. Since the expression $x_i + x_i' = 1$ is common to minterm locations 1 and 3, the term y_2 is also a constituent part of the equation for y_2. The input equations for flip-flops y_1 and y_2 are listed in Equation 4.19.

$$
\begin{aligned}
y_1 &= y_2 x_i + y_1 \\
y_2 &= x_i + y_2 \\
z_1 &= y_1' y_2
\end{aligned}
\tag{4.19}
$$

The output map for z_1 is shown in Figure 4.92 (b). Output z_1 is asserted unconditionally in state b ($y_1y_2 = 01$). Thus, this Moore-type output is a function of y_1 and y_2 only, and is independent of the value of x_i. The equation for z_1 is specified in Equation 4.19.

The logic diagram for this synchronous sequential machine that is functionally equivalent to the combinational iterative network of Figure 4.89 (b) is shown in Figure 4.93. Typical cell$_i$ constitutes the δ next-state logic. The outputs of cell$_i$ are labeled y_{i1} and y_{i2} which connect to the D inputs of flip-flops y_1 and y_2, respectively. The flip-flop outputs are fed back to the δ next-state logic and also to the λ output logic to generate output z_1. The δ next-state logic is synthesized from the input equations of Equation 4.19, which are equivalent to the equations used in the implementation of cell$_i$ for the combinational iterative network, as shown in Equation 4.18.

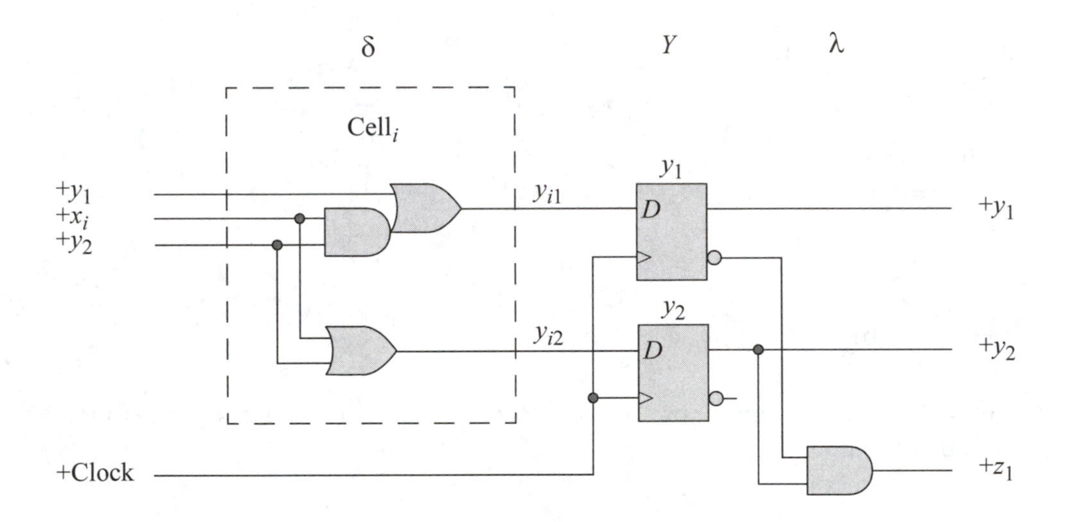

Figure 4.93 Logic diagram for a synchronous sequential machine that is functionally equivalent to the combinational iterative network of Figure 4.89 (b).

Example 4.16 has illustrated the design technique for a combinational iterative network using a unilateral linear array of identical cells. The same example was then converted to a functionally equivalent sequential machine where the logic of cell$_i$ from the iterative network was utilized as the δ next-state logic for the sequential machine. The synthesis procedure is similar for both implementations. The principal difference is that the input vector $x_1, x_2, \ldots, x_i, \ldots, x_k$ is received in parallel for the combinational network, whereas the input vector is received serially in an equivalent sequential machine, one bit per clock cycle.

The next-state table specifies the output state of a typical cell for an iterative network, while the same table specifies the next state for a corresponding sequential machine. In most cases, the logic represented by cell$_i$ can be used directly as the δ next-state logic of a corresponding sequential machine if D flip-flops are used as the storage elements. Other types of storage elements will require slightly different input equations, but are derived in a similar manner.

If speed is of paramount importance in the operation of a machine, then a combinational iterative network should be considered, even though the amount of additional logic may be significant, especially when the input vector is large. However, if a minimal amount of hardware is of primary consideration — with the speed of operation relegated to a lower priority — then a functionally equivalent synchronous sequential machine should be implemented.

4.6 Error Detection in Synchronous Sequential Machines

This section describes techniques of error detection for synchronous sequential machines. It is assumed that the design procedure resulted in a correctly synthesized machine which operates reliability according to the machine specifications and has no output glitches.

A fault in a synchronous sequential machine may alter the δ next-state function, the λ output function, or both. In order to detect errors in a sequential machine, the output alphabet Z can be encoded in an error detecting code C, such that, Z is a proper subset of C; that is, $Z \subset \text{C}$. The notation $Z \subset \text{C}$ implies that Z is a subset of C and that Z is not equal to C, as denoted by the expression

$$Z \subset \text{C} \Rightarrow (Z \subseteq \text{C}) \bullet (Z \neq \text{C})$$

where the notation "$Z \subset \text{C}$" indicates that Z is a proper subset of C; "$\Rightarrow$" is the implication symbol; "$(Z \subseteq \text{C})$" indicates that Z is a subset of C in which every element of Z is also an element of C; "•" indicates the logical AND operation; and "$(Z \neq \text{C})$" specifies that Z is not equal to C. Errors will be indicated by an output code that is not an element of the output alphabet. Thus, there will be no ambiguity between the error code and the output alphabet Z.

In many instances, error checking in synchronous sequential machines is analogous to error checking in combinational logic. Consider a synchronous sequential machine with n inputs and m outputs. The error detection circuitry may utilize the same n inputs and m outputs plus additional internal logic functions from the same machine, all of which are used as inputs to the error detection logic, as shown in Figure 4.94. The error signal may be latched to preserve the error indication. The signal can then be fed back as a level to disable the machine clock, maintaining the error condition for further analysis.

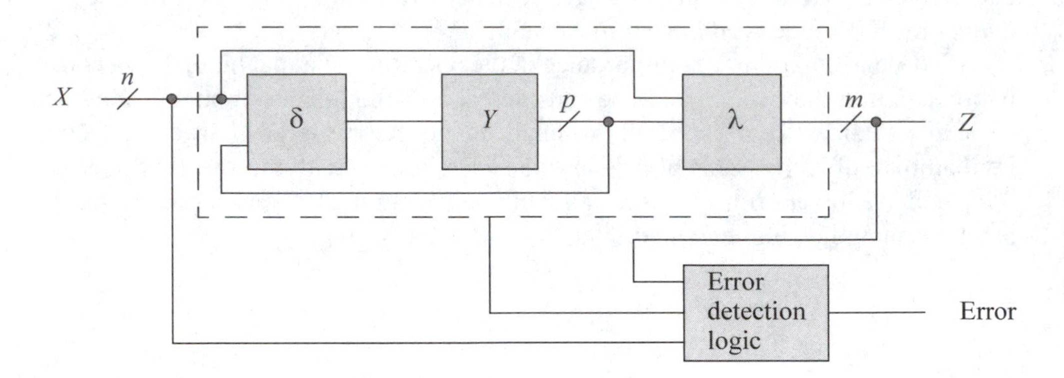

Figure 4.94 Generalized block diagram of error detection for a synchronous sequential machine.

An obvious error detection technique is to duplicate the existing machine and compare the outputs of both machines. This is similar in concept to machine homomorphism as discussed in Section 2.2.7, where

$$M = (X, Y, Z, \delta, \lambda)$$

is the original machine and

$$M' = (X', Y', Z', \delta', \lambda')$$

is the duplicate machine whose output is compared with machine M. Figure 4.95 illustrates the technique of error detection through duplication. Only the outputs from machine M are used as system outputs and connect to external circuitry. Although the method is simple in concept, the cost, however, may be prohibitive, especially for large machines.

Another error detection technique involves the use of an inverse machine M^{-1} which generates an output alphabet Z' that is identical to the original set of inputs X for machine M. The outputs of machine M connect to the inverse logic of machine M^{-1}. The output alphabet Z' of machine M^{-1} is compared to the input alphabet X of machine M. If $X \neq Z'$, then an error has occurred. Figure 4.96 illustrates the block diagram of error detection by means of *input regeneration*.

Another technique for detecting errors is to apply a check generating algorithm to the inputs. The outputs of the check generator, in conjunction with the machine outputs, are connected to a second checking circuit to detect errors. Figure 4.97 depicts a block diagram for this type of error detection.

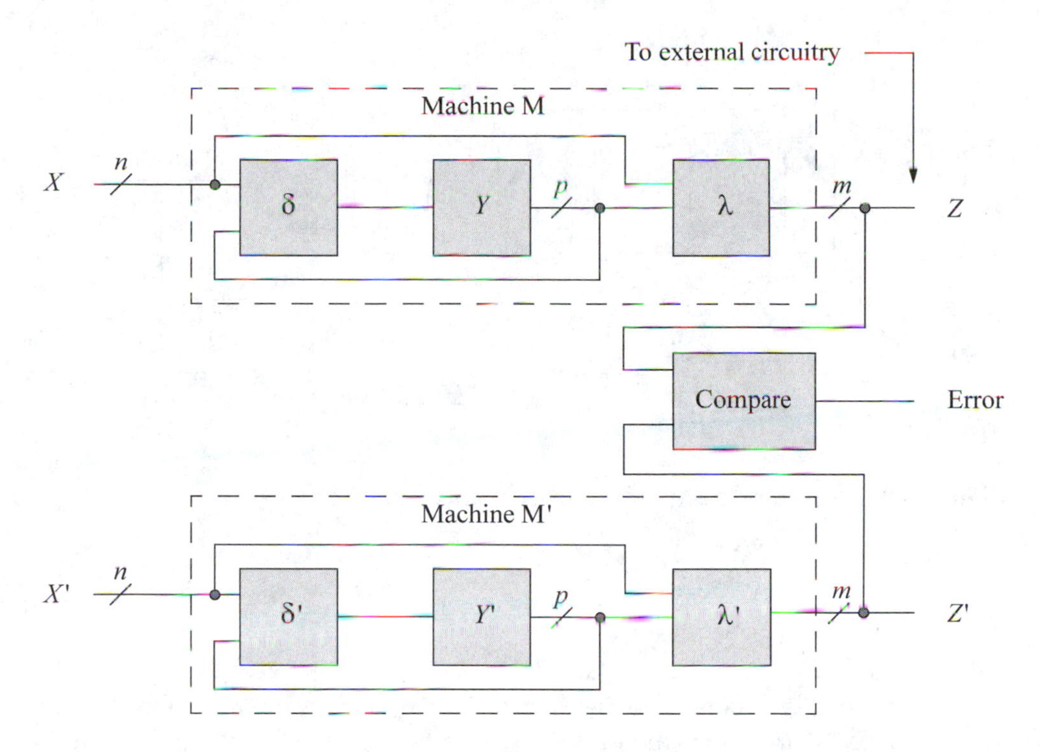

Figure 4.95 Block diagram of error detection for a synchronous sequential machine using a duplicate machine M'.

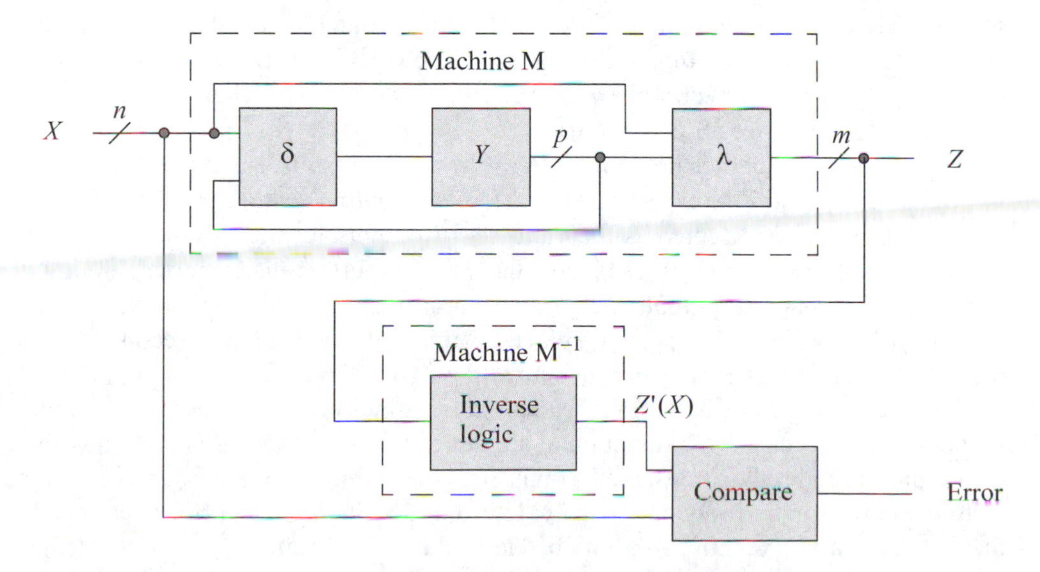

Figure 4.96 Error detection by means of input regeneration.

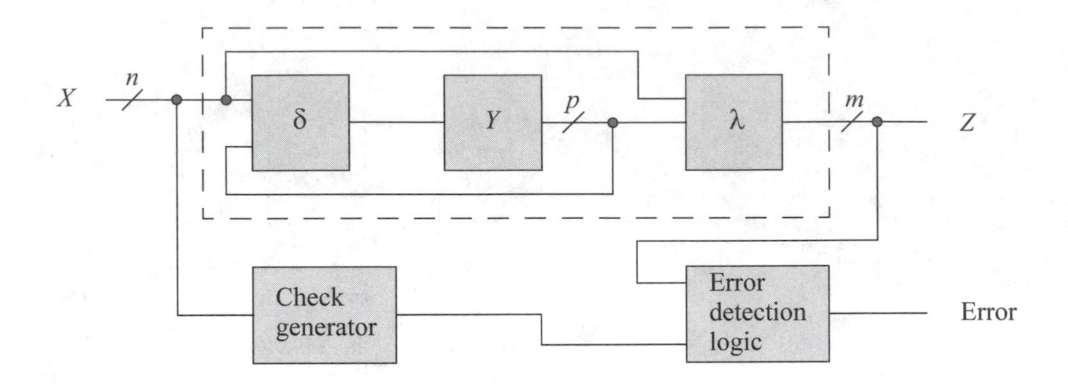

Figure 4.97 Error detection by means of check generation.

This method is analogous to residue checking or *parity prediction* for carry-lookahead adders. Parity prediction for carry lookahead adders is provided by the following equation: $P_S = P_A \oplus P_B \oplus P_C$. P_S is the actual parity of the sum. $P_A \oplus P_B \oplus P_C$ defines the predicted parity of the sum, where P_A and P_B represent the parity of operands *A* and *B*, respectively, and P_C is the parity of the carries. If the actual parity is equal to the predicted parity, then no error has occurred in the addition operation. However, if the actual parity is not equal to the predicted parity, then an error has been detected and an error flag is set.

Triple redundancy is a common technique for error detection, especially in wafer-scale integration where hundreds of thousands of logic gates are available for machine design. The logic of each major component of the machine is triplicated and the associated outputs are connected to a voting circuit which compares the three sets of outputs and selects one set for external use.

Let the three machines be defined as M, M', and M" with corresponding output vectors of Z_r, Z_r', and Z_r'', respectively. If at least two output vectors are identical, then one of the vectors is selected as the valid set of outputs for the machine. Thus, if $(Z_r = Z_r')$ or $(Z_r = Z_r'')$ or $(Z_r' = Z_r'')$, then the voting circuitry selects one set of outputs to represent the machine outputs. When an error occurs, a flag is set to indicate the defective machine. The wafer is then replaced when convenient, since a second fault in the same set of machines may produce invalid outputs.

Error checking for synchronous sequential machines involves checking an output sequence for correctness as a function of a unique input sequence. It is often possible to assign state codes that are elements of an error-detecting code. Thus, any deviation from the correct output sequence can be detected as a single- or multiple- bit error in the state code assignment. Errors can be classified as either an incorrect state change or as an incorrect output. Since an error in the δ next-state logic can result in an incorrect next state, the input combinational logic should also be checked. To be effective, errors should be detected in the same clock period in which the error occurred.

Parity checking is a convenient method of detecting incorrect state changes. If the machine is implemented with p storage elements, then a supplementary storage element can be added such that, the parity of all storage elements $1, 2, \ldots, p, p + 1$ will be odd. This is similar in concept to odd parity generation and checking for data buses, where each data byte contains a parity bit which maintains odd parity over all nine bits. Thus, the parity of all $p + 1$ storage elements will be odd except when an odd number of errors has occurred. In order to minimize the possibility of multiple errors, each storage element must have its own δ next-state logic; that is, combinational input logic cannot be shared with other storage elements that have common terms. Thus, only a single storage element will be in error due to a fault, resulting in an error detected due to even parity. Figure 4.98 shows a general block diagram of a synchronous sequential machine using parity checking for error detection.

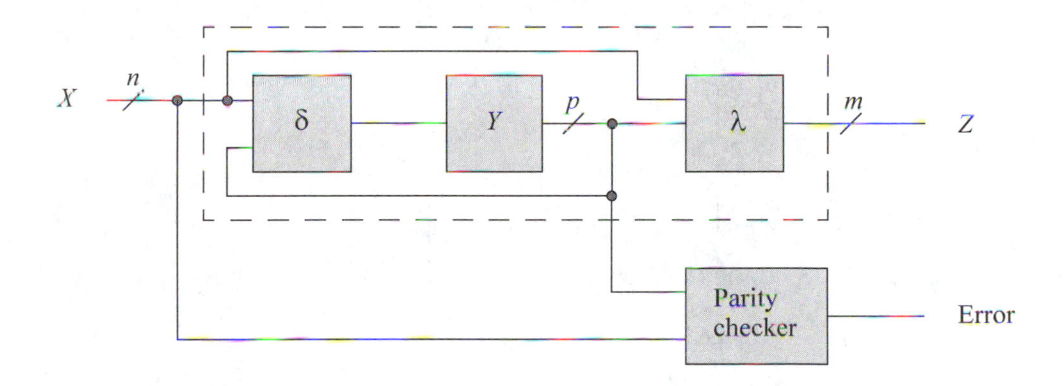

Figure 4.98 General block diagram of a synchronous sequential machine using parity checking for error detection.

Example 4.17 The synthesis of a parity-checked counter will now be presented. If each state code differs by only one bit between contiguous state codes, then parity checking is easily implemented. The Gray code meets this requirement. The state diagram for a 3-bit Gray code counter is shown in Figure 4.99 together with the state code assignment and the corresponding state of the parity flip-flop. The state codes are shown in the next-state table of Table 4.26, where y_1, y_2, and y_3 are the state flip-flops and y_p is the parity flip-flop which maintains odd parity over the four flip-flops. The parity flip-flop is set to a value of 1 initially and will change state with each clock pulse. If an incorrect state transition occurs, then the parity of $y_1 y_2 y_3 y_p$ will be even and an error will be indicated. Parity is checked by means of a parity checker.

The input maps are derived from either the state diagram or the next-state table and are shown in Figure 4.100 using D flip-flops. The input equations are listed in

Equation 4.20. The logic diagram is shown in Figure 4.101. In order to minimize the possibility of double errors, the term y_2y_3' in the equations for Dy_1 and Dy_2 will be duplicated. Thus, a fault in the logic represented by the term y_2y_3' in flip-flop y_1 or y_2, but not both, will result in an incorrect state transition for only one flip-flop.

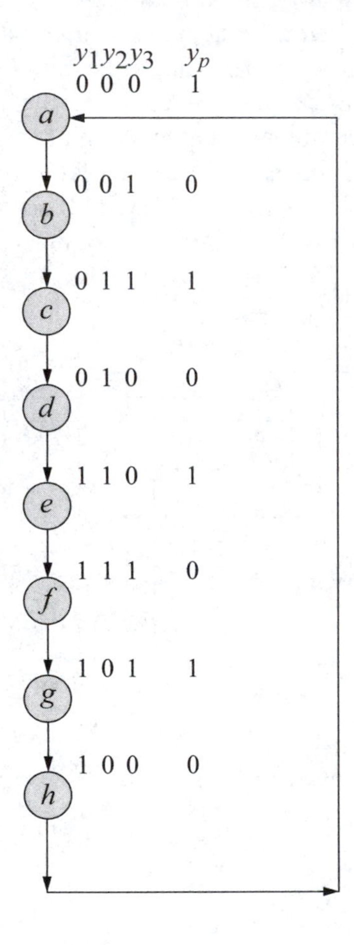

Figure 4.99 State diagram for a parity-checked, 3-bit Gray code counter.

The +Error signal in Figure 4.101 is latched in a D flip-flop to preserve the error indication. The Error flip-flop is clocked on the positive transition of the complemented clock, which occurs at time t_2. Clocking the Error flip-flop at time t_2 allows the output to be fed back to disable the machine clock before the next active positive clock transition. Thus, the machine will not change state and the error state will be maintained for further analysis.

Table 4.26 Next-state table for a parity-checked 3-bit Gray code counter

State name	Present state $y_1\ y_2\ y_3$	Parity flip-flop y_p	Next state $y_1\ y_2\ y_3$
a	0 0 0	1	0 0 1
b	0 0 1	0	0 1 1
c	0 1 1	1	0 1 0
d	0 1 0	0	1 1 0
e	1 1 0	1	1 1 1
f	1 1 1	0	1 0 1
g	1 0 1	1	1 0 0
h	1 0 0	0	0 0 0

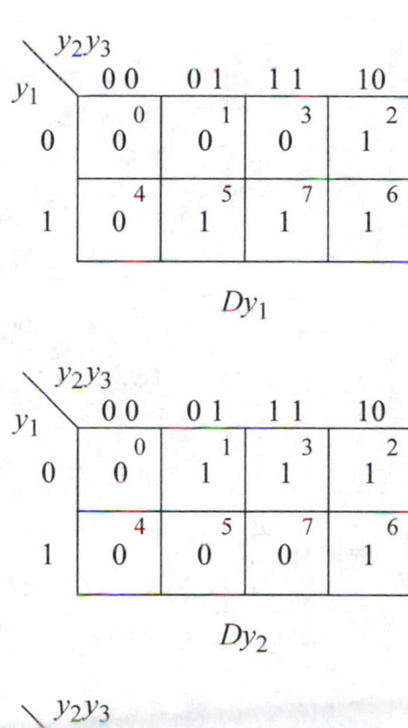

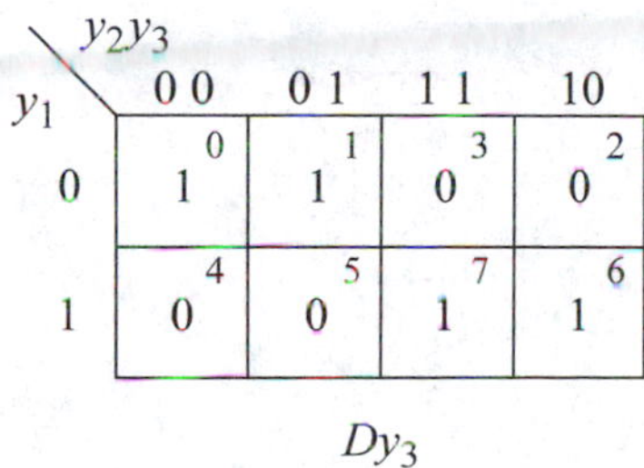

Figure 4.100 Input maps for the parity-checked Gray code counter of Figure 4.99.

$$Dy_1 = y_1y_3 + y_2y_3'$$
$$Dy_2 = y_1'y_3 + y_2y_3'$$
$$Dy_3 = y_1'y_2' + y_1y_2 \tag{4.20}$$

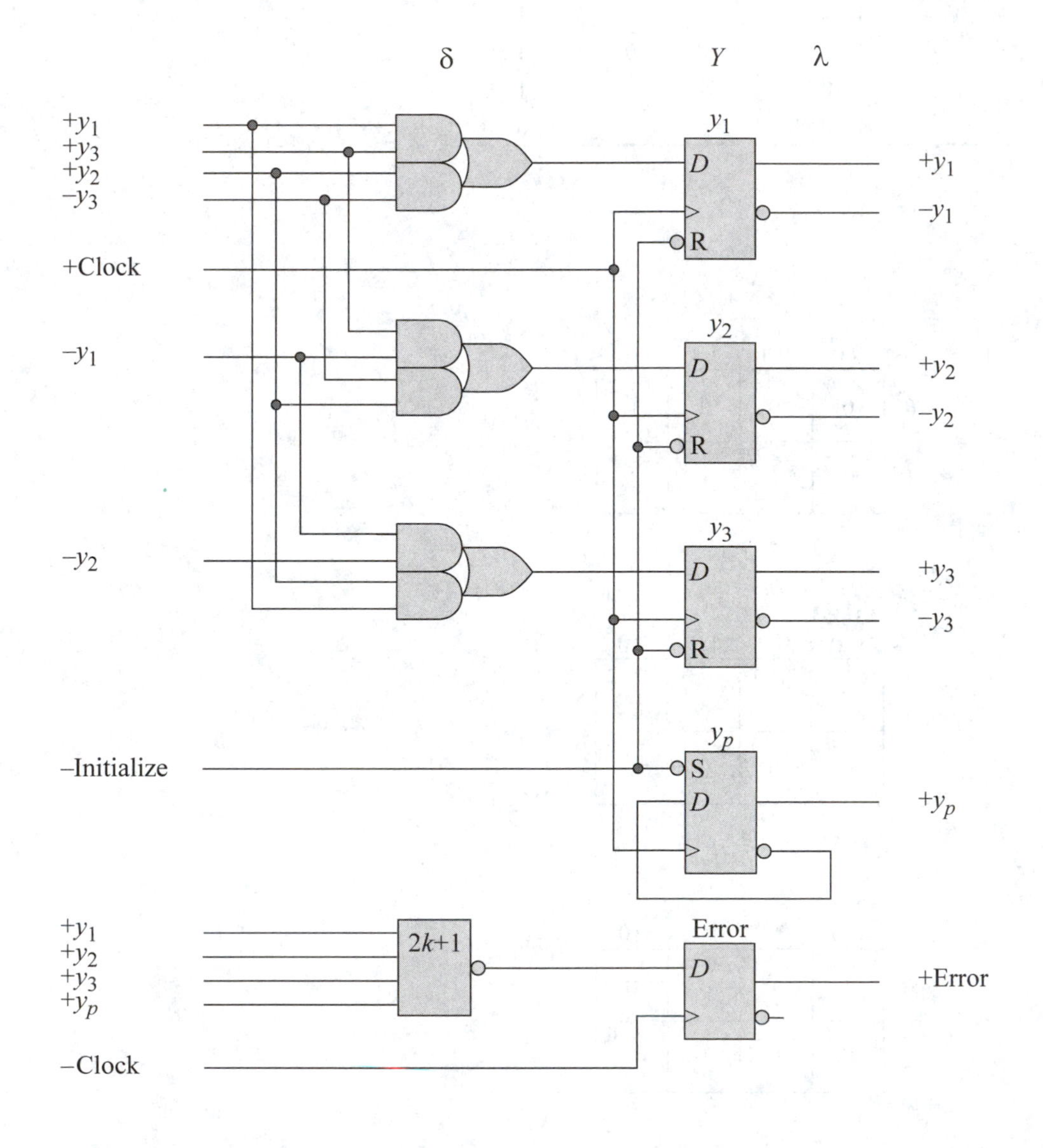

Figure 4.101 Logic diagram for the parity-checked 3-bit Gray code counter of Figure 4.99.

The counter is reset initially and the parity flip-flop is set; that is, the state of the four flip-flops is $y_1y_2y_3y_p = 0001$. The parity flip-flop y_p is configured in a toggle mode. Thus, each active clock transition will toggle y_p. Since only one counter flip-flop changes state between successive state codes, the parity of $y_1y_2y_3y_p$ will always be odd for correct machine operation. Parity is checked for correctness by the odd parity circuit $2k + 1$. If the inputs to the $2k + 1$ circuit consist of an odd number of high (+) levels (or 1s), then the output will be a low (–) level, indicating no error. If, however, an incorrect state transition occurs, such that the parity of $y_1y_2y_3y_p$ is even, then the output of the $2k + 1$ circuit will be a high (+) level, indicating an error. The $2k + 1$ circuit is equivalent to two exclusive-OR circuits whose outputs are connected to an exclusive-NOR circuit. Error detection assumes, of course, that the error detection logic is functioning correctly.

Error detection using parity checking applies not only to synchronous counters, but also to any type of synchronous sequential machine. The state codes do not necessarily have to maintain unit distance between successive state codes. This is not always possible in sequence-detecting Moore and Mealy machines. In any synchronous sequential machine, if a state transition causes an odd number of flip-flops to change state, then the parity flip-flop will be toggled. Conversely, if an even number of flip-flops change state, then the parity flip-flop is not toggled. The toggling function of the parity flip-flop can be determined from the state diagram or from the next-state table.

Example 4.18 The Moore machine of Example 3.1 in Section 3.4 will be redesigned to include error detecting circuitry using parity checking. The machine accepts serial data in the form of 3-bit words on an input line x_1. There is one bit space between contiguous words. Whenever x_1 contains the word 111, output z_1 will be asserted in the bit period between words. Output z_1 is asserted at time t_2 and deasserted at t_3.

The state diagram is reproduced in Figure 4.102 and includes the parity flip-flop y_p which maintains odd parity over the four flip-flops y_1, y_2, y_3, and y_p. The δ next-state input logic for the parity flip-flop provides the necessary conditions to set y_p to the appropriate value in order to provide odd parity for the next state $Y_{k(t+1)}$. The input logic for y_p is a function of the present inputs and the present state. The requisite states in Figure 4.102 are adjacent according to the state code adjacency rules of Section 3.4. It is not always possible, however, to assign adjacent state codes for all state transition sequences, especially for a sequential machine containing many diverse paths.

Table 4.27 presents the next-state table and shows not only the next state, but also the next value of the parity flip-flop y_p which provides odd parity over $y_1y_2y_3y_p$. The derivation of the input maps for flip-flops y_1, y_2, and y_3 and the output map for z_1 will be omitted, since these maps were discussed in detail in Section 3.4, Example 3.1. The equations, however, are shown in Equation 4.21.

The input map for flip-flop y_p is shown in Figure 4.103 and is derived from either the state diagram or the next-state table. The map entries specify the next state for y_p and use input x_1 as a map-entered variable. The entry of 1 in minterm locations 1, 2, and 6 is equivalent to the expression $x_1 + x_1'$. This permits combining minterm locations 0 and 1 to yield the term $y_1'y_2'x_1$, combining minterm locations 1 and 5 to yield $y_2'y_3x_1'$, and combining locations 6 and 7 to yield $y_1y_2x_1$. The entries of 1 in minterm locations 2 and 6 are then combined in the usual manner. The minimized input equation for y_p is shown in Equation 4.22 using a D flip-flop.

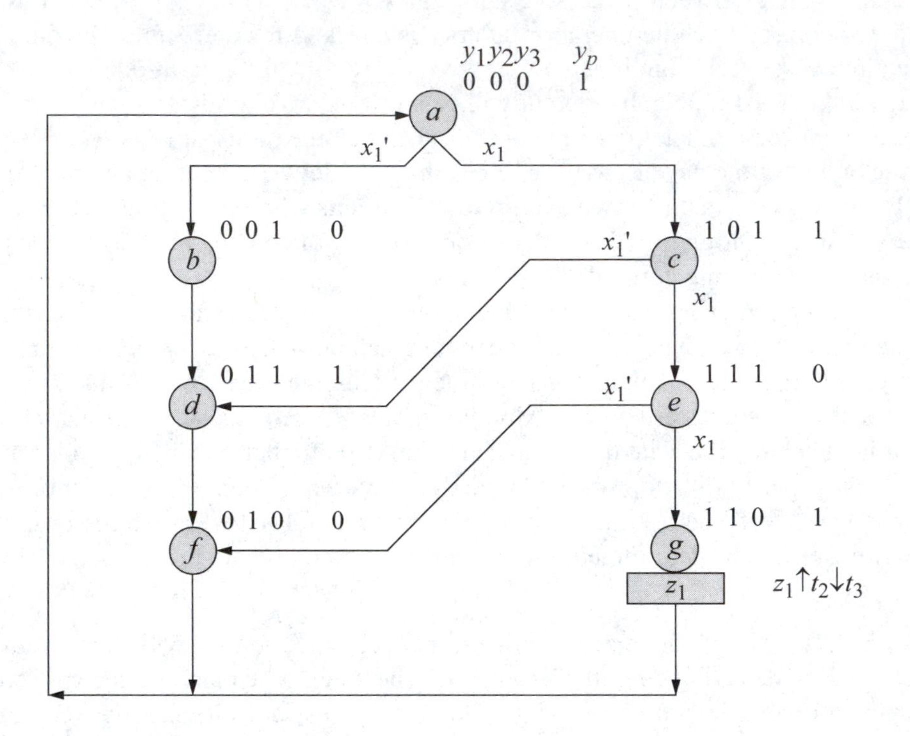

Figure 4.102 State diagram for the Moore machine of Example 4.18, which generates an output z_1 whenever a 3-bit word $x_1 = 111$. The parity flip-flop y_p is included for error detection. There is one unused state: 100.

Table 4.27 Next-state table for the parity-checked Moore machine of Figure 4.102

Present state $y_1\ y_2\ y_3$	Input x_1	Next state $y_1\ y_2\ y_3$	Parity flip-flop y_p	Flip-flop inputs $Dy_1\ Dy_2\ Dy_3$	Present output z_1
0 0 0	0	0 0 1	0	0 0 1	0
0 0 0	1	1 0 1	1	1 0 1	0
0 0 1	0	0 1 1	1	0 1 1	0
0 0 1	1	0 1 1	1	0 1 1	0
0 1 0	0	0 0 0	1	0 0 0	0
0 1 0	1	0 0 0	1	0 0 0	0

Continued on next page

Table 4.27 Next-state table for the parity-checked Moore machine of Figure 4.102

Present state $y_1\, y_2\, y_3$	Input x_1	Next state $y_1\, y_2\, y_3$	Parity flip-flop y_p	Flip-flop inputs $Dy_1\, Dy_2\, Dy_3$	Present output z_1
0 1 1	0	0 1 0	0	0 1 0	0
0 1 1	1	0 1 0	0	0 1 0	0
1 0 0	0	– – –	–	– – –	–
1 0 0	1	– – –	–	– – –	–
1 0 1	0	0 1 1	1	0 1 1	0
1 0 1	1	1 1 1	0	1 1 1	0
1 1 0	0	0 0 0	1	0 0 0	1
1 1 0	1	0 0 0	1	0 0 0	1
1 1 1	0	0 1 0	0	0 1 0	0
1 1 1	1	1 1 0	1	1 1 0	0

$$
\begin{aligned}
Dy_1 &= y_2'y_3'x_1 + y_1y_3x_1 \\
Dy_2 &= y_3 \\
Dy_3 &= y_2' \\
z_1 &= y_1y_3'
\end{aligned} \tag{4.21}
$$

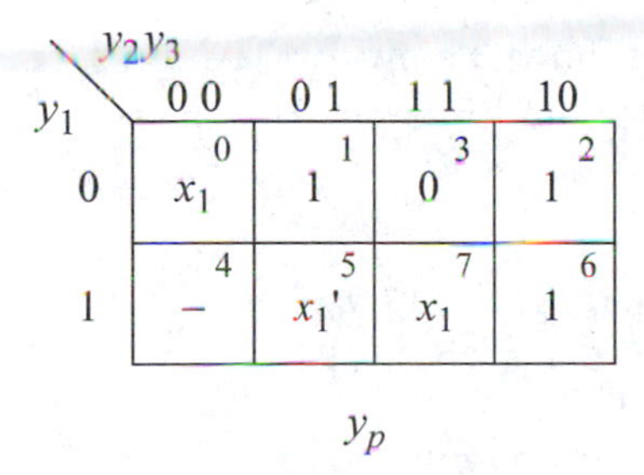

Figure 4.103 Input map for the parity flip-flop y_p for the Moore machine of Figure 4.102.

$$Dy_p = y_1{}'y_2{}'x_1 + y_2{}'y_3x_1{}' + y_1y_2x_1 + y_2y_3{}' \tag{4.22}$$

The logic diagram is shown in Figure 4.104. The logic is identical to that shown in Section 3.4, Example 3.1, except for the inclusion of an initialization signal and the error detection logic. The machine is initially reset to $y_1y_2y_3y_p = 0001$. The output of the $2k + 1$ logic element will be at a high voltage level if an error has occurred in a state transition sequence. By applying an appropriate bit sequence to input x_1, the logic of Figure 4.104 will operate according to the specifications presented in the state diagram. If the machine sequences through all states correctly, then no error will be indicated. Assume, however, that the sequence from state c ($y_1y_2y_3 = 101$) to state d ($y_1y_2y_3 = 011$), caused by $x_1 = 0$, is changed so that the machine sequences to state e ($y_1y_2y_3 = 111$) instead, due to a hardware malfunction. The input logic for flip-flop y_p will cause y_p to be set, because the term $y_2{}'y_3x_1{}'$ is true for a correct sequence to state d. The resulting values of the state flip-flops and the parity flip-flop will be $y_1y_2y_3y_p = 1111$, which will produce an error. The state of the +Error signal can then be stored in a D flip-flop which in turn can be used to disable the system clock.

In this section, a parity flip-flop and associated logic provided a means of detecting single-bit errors in state transition sequences. It must be emphasized, however, that the error checking logic itself may be at fault and may generate an error signal when none exists. This is true for any error detection scheme, whether it is Hamming code, cyclic redundancy check code, or parity prediction for carry lookahead adders. There is a practical and philosophical limit to the amount of checking that must be constructed — a point is reached where it becomes impractical to synthesize additional logic to check the checker.

4.7 Problems

4.1 Given the input map shown below for flip-flop y_1, design the δ next-state logic using a linear-select multiplexer. Use the least amount of logic.

y_1 \ y_2y_3	00	01	11	10
0	0 (cell 0)	0 (cell 1)	1 (cell 3)	1 (cell 2)
1	x_1x_2 (cell 4)	x_3 (cell 5)	$x_1 + x_2$ (cell 7)	x_1 (cell 6)

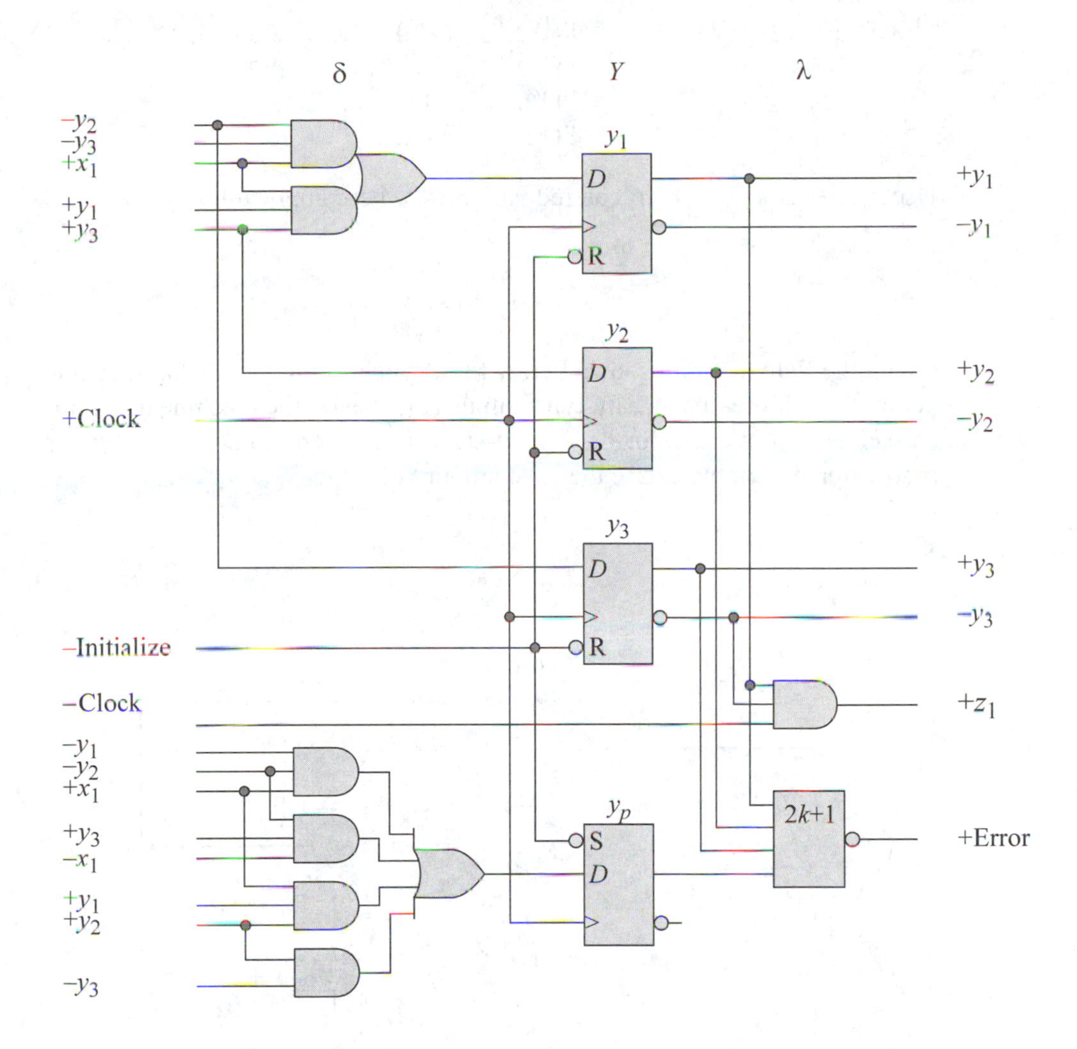

Figure 4.104 Logic diagram for the Moore machine of Figure 4.102, including error checking logic using a parity flip-flop y_p.

4.2 Given the following equations for Dy_1, Dy_2, and Dy_3, obtain the input maps, the input equations, and the logic diagram using linear-select multiplexers plus any logic primitives for the δ next-state logic.

$$Dy_1(y_1, y_2, y_3, x_1, x_2, x_3) = y_1'y_2'y_3 + y_1y_2'y_3'x_2'x_3' + y_1y_2'y_3'x_2x_3 + y_1y_2'y_3x_2 + y_1y_2'y_3x_3 + y_1y_2y_3'x_2x_3$$

$$Dy_2(y_1, y_2, y_3, x_1, x_2, x_3) = y_1'y_2'y_3x_2'x_3 + y_1y_2'y_3'x_2x_3 + y_1y_2'y_3'x_2x_3' + y_1y_2'y_3'x_2'x_3 + y_1y_2'y_3x_2'x_3' + y_1y_2y_3'$$

$$\begin{aligned} Dy_3(y_1, y_2, y_3, x_1, x_2, x_3) &= y_1'y_2'y_3'x_1x_2'x_3 + y_1'y_2'y_3'x_1x_2x_3' + \\ &= y_1'y_2'y_3x_2'x_3' + y_1'y_2'y_3x_2x_3 + \\ &= y_1y_2'y_3'x_2x_3 + y_1y_2'y_3x_2 + y_1y_2'y_3x_3 + \\ &= y_1y_2y_3'x_2'x_3' \end{aligned}$$

Use x_1, x_2, and x_3 as map-entered variables, where applicable.

4.3 Given the state diagram shown below for a synchronous sequential machine containing Moore- and Mealy-type outputs, synthesize the machine using linear-select multiplexers for the δ next-state logic. Use inputs x_1 and x_2 as map-entered variables. Use the least amount of logic.

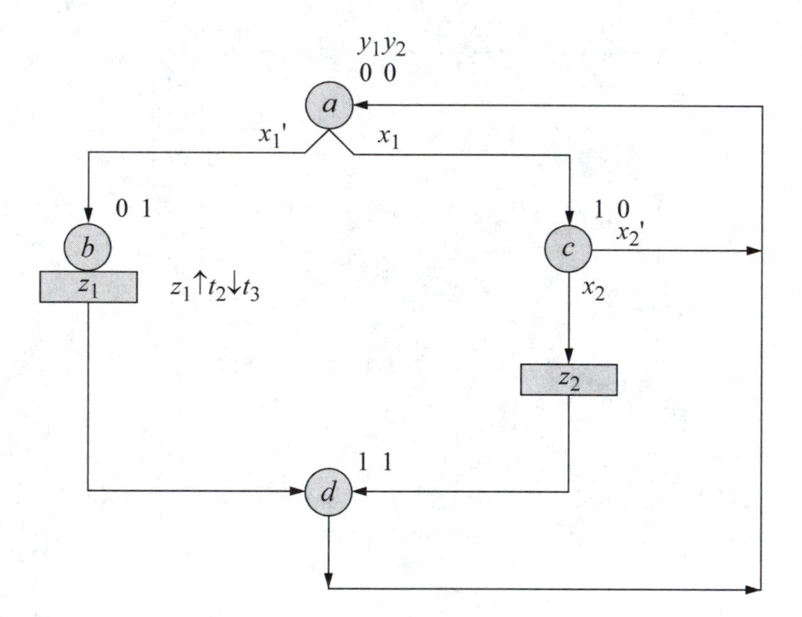

4.4 The state diagram for a Moore synchronous sequential machine is shown below. Implement the machine using linear-select multiplexers for the δ next-state logic, D flip-flops for the storage elements, and any additional logic for the λ output logic. Use x_1 and x_2 as map-entered variables.

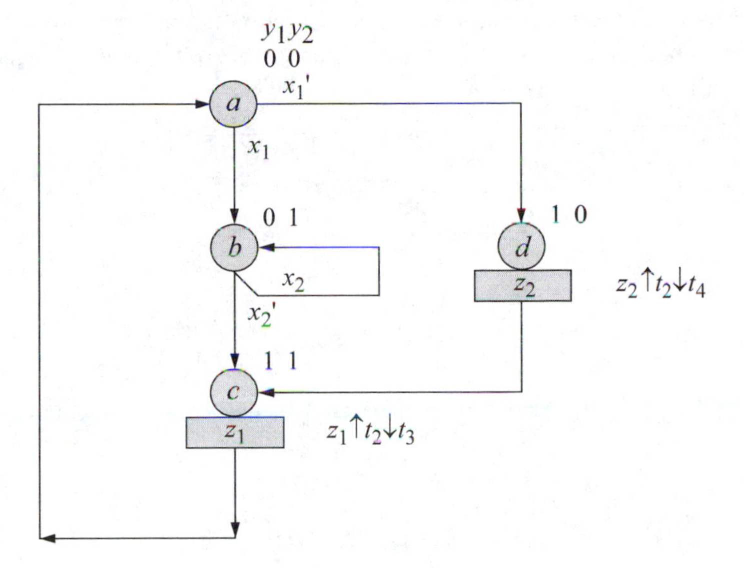

4.5 Given the state diagram shown below for a Moore synchronous sequential machine, derive the input maps for flip-flops y_1, y_2, and y_3 and the corresponding input equations. Synthesize the machine using linear-select multiplexers for the δ next-state logic, D flip-flops for the storage elements, and logic primitives for the λ output logic. Show a representative timing diagram to assert outputs z_1 and z_2.

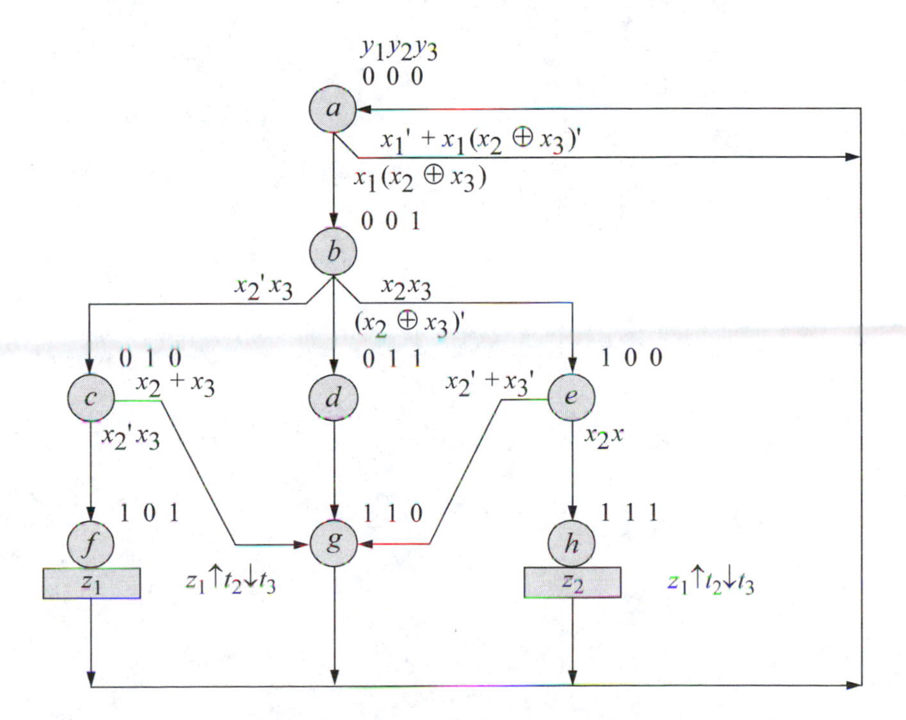

4.6 The truth table shown below represents the δ next-state logic for flip-flop y_1 of a synchronous sequential machine. Implement the logic using a nonlinear-select multiplexer and additional logic gates, if necessary. Use the least amount of logic.

y_1 y_2 y_3	Dy_1
0 0 0	x_1
0 0 1	1
0 1 0	1
0 1 1	1
1 0 0	–
1 0 1	0
1 1 0	–
1 1 1	–

4.7 The truth table shown below represents the δ next-state logic for flip-flop y_1 of a synchronous sequential machine. Implement the logic using a nonlinear-select multiplexer and additional logic gates, if necessary. Use the least amount of logic.

y_1	y_2	y_3	y_4	Dy_1
0	0	0	0	x_1'
0	0	0	1	1
0	0	1	0	x_1'
0	0	1	1	1
0	1	0	0	–
0	1	0	1	–
0	1	1	0	x_1
0	1	1	1	0
1	0	0	0	0
1	0	0	1	0
1	0	1	0	–
1	0	1	1	0
1	1	0	0	0
1	1	0	1	0
1	1	1	0	–
1	1	1	1	–

4.8 Given the input map shown below for flip-flop y_1 of a synchronous sequential machine, implement the δ next-state logic using a nonlinear-select multiplexer and additional logic gates, if necessary. Use the least amount of logic.

y_1 \ y_2y_3	0 0	0 1	1 1	1 0
0	1 (0)	1 (1)	x_1x_2 (3)	– (2)
1	– (4)	1 (5)	$x_1' + x_2'$ (7)	$(x_1x_2)'$ (6)

Dy_1

4.9 Given the input map shown below for flip-flop y_1 of a synchronous sequential machine, implement the δ next-state logic using a nonlinear-select multiplexer and additional logic gates, if necessary. Use the least amount of logic.

y_1 \ y_2y_3	0 0	0 1	1 1	1 0
0	1 (0)	1 (1)	x_1x_2 (3)	– (2)
1	– (4)	$(x_1x_2)' + x_1x_2$ (5)	$x_1' + x_2'$ (7)	$(x_1x_2)'$ (6)

Dy_1

4.10 Given the input map shown below for flip-flop y_1 of a synchronous sequential machine, implement the δ next-state logic using a nonlinear-select multiplexer and additional logic gates, if necessary. Obtain the least amount of logic using ECL.

y_1 \ y_2y_3	0 0	0 1	1 1	1 0
0	$(x_1x_2)'$ (0)	$x_1' + x_2'$ (1)	x_1x_2 (3)	– (2)
1	– (4)	1 (5)	1 (7)	1 (6)

Dy_1

4.11 Given the input maps shown below for flip-flops y_1 and y_2 of a synchronous sequential machine, implement the δ next-state logic using nonlinear-select multiplexers and additional logic gates, if necessary. Use the least amount of logic.

Dy_1 (minterm number in parentheses)

y_1 \ y_2y_3	0 0	0 1	1 1	10
0	(0) 0	(1) –	(3) 0	(2) 0
1	(4) 1	(5) –	(7) 0	(6) 1

Dy_2 (minterm number in parentheses)

y_1 \ y_2y_3	0 0	0 1	1 1	10
0	(0) 1	(1) 0	(3) x_1	(2) 0
1	(4) 1	(5) –	(7) –	(6) 0

4.12 Given the state diagram shown below for a Moore machine, synthesize the logic using nonlinear-select multiplexers for the δ next-state logic. Use the least amount of logic. Also synthesize the machine using discrete logic for the δ next-state logic and compare the two designs. Since no state transition sequence will pass through state c ($y_1y_2y_3 = 011$) as a transient state, output z_1 can be asserted at time t_1 and deasserted at t_2 with no possibility of glitches.

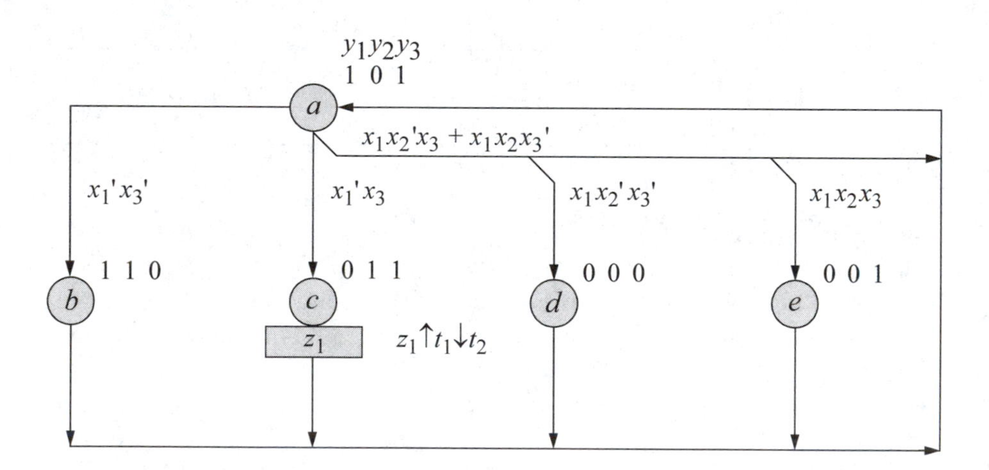

4.13 Given the state diagram shown below for a Moore machine, implement the design using nonlinear-select multiplexers for the δ next-state logic, D flip-flops for the storage elements, and a decoder for the λ output logic. Outputs z_1, z_2, and z_3 are asserted at time t_2 and deasserted at t_3.

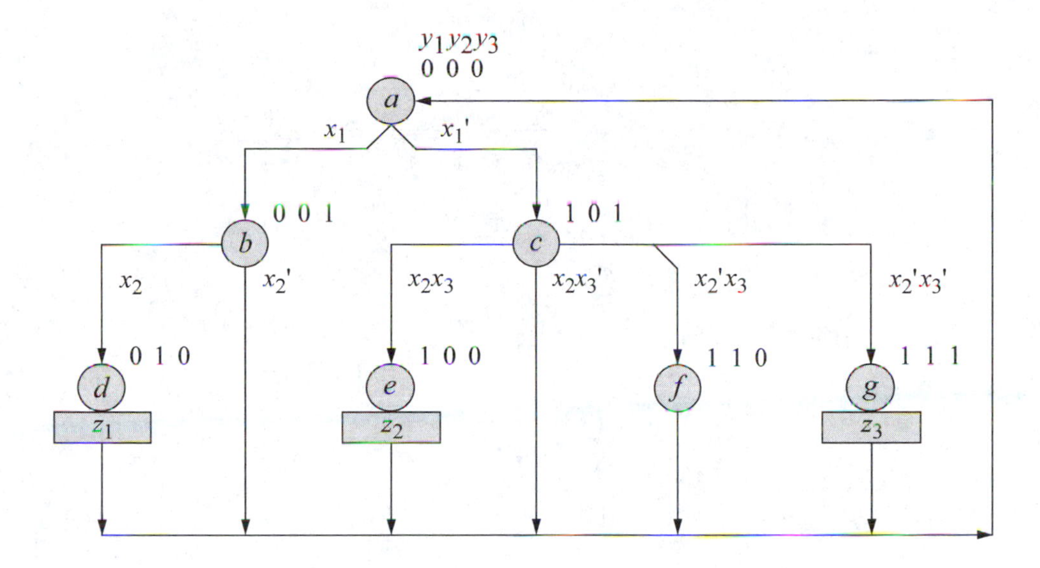

4.14 A Moore sequential machine receives 3-bit words on a parallel input bus as shown below, where x_3 is the low-order bit. There are five outputs z_1, z_2, z_3, z_4, and z_5, where the subscripts indicate the decimal value of the corresponding unsigned binary input word. Synthesize the machine using discrete NOR logic gates for the δ next-state logic, *JK* flip-flops for the storage elements, and a decoder for the λ output logic. The outputs are asserted at time t_2 and deasserted at t_3.

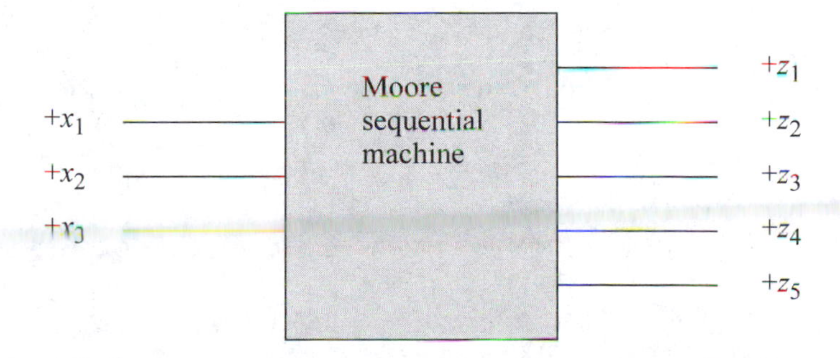

4.15 Given the state diagram shown below for a Moore machine, implement the machine using a PROM for the δ next-state logic and *D* flip-flops for the storage elements.

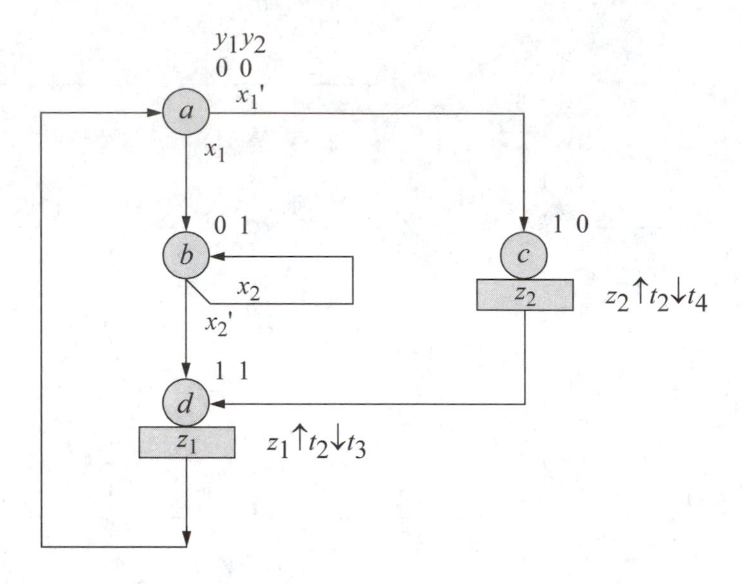

4.16 Given the state diagram shown below for a Moore machine, implement the machine using a PROM for the δ next-state logic and a parallel-in, parallel-out register for the storage elements. The machine has a self-starting state of $y_1y_2y_3 = 111$, which is also a terminal state.

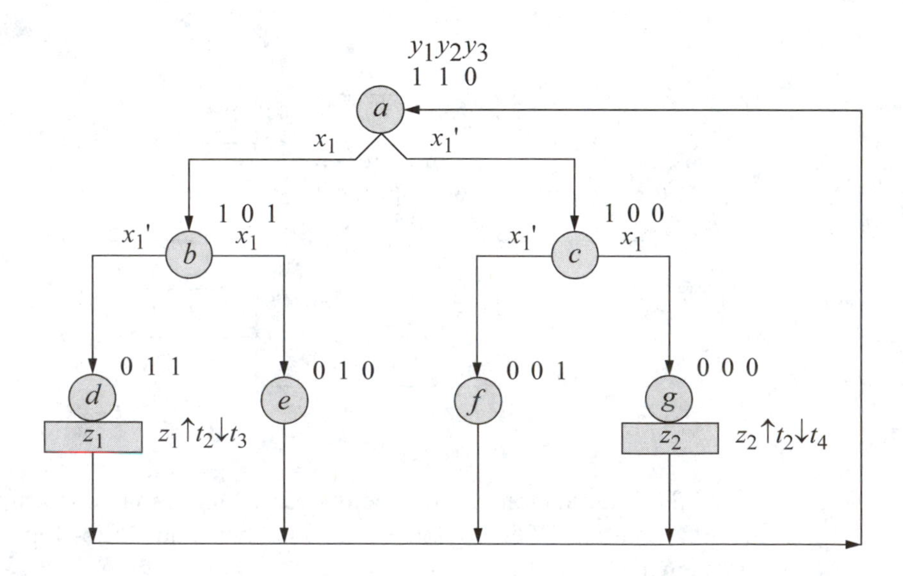

4.17 Given the state diagram shown below with both Moore- and Mealy-type outputs, implement the machine using a PROM for the δ next-state logic and a parallel-in, parallel-out register for the storage elements. Outputs z_1 and z_2 can be generated directly from the PROM and will be asserted at time t_2 and deasserted at t_3 in their respective states, depending on the value of x_1. Output z_3 is asserted at time t_2 and deasserted at t_3. The machine has a self-starting state of $y_1y_2y_3 = 000$.

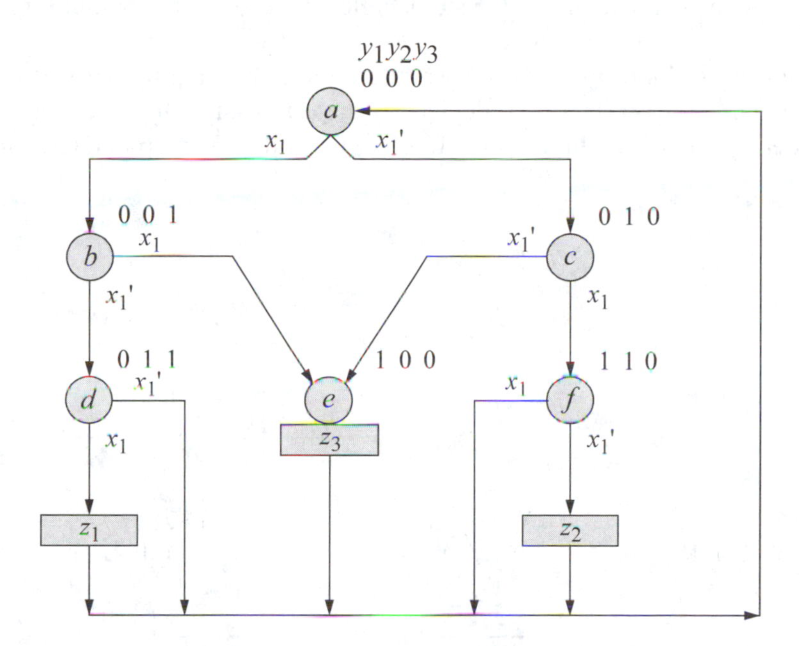

4.18 Synthesize a Moore sequential machine that receives a 4-bit word on a parallel input bus. The input bus is designated $x_1x_2x_3x_4$ as shown below. The parallel bus specifies two 2-bit operands, where x_1x_2 is the multiplicand and x_3x_4 is the multiplier. Inputs x_2 and x_4 are the low-order bits of the multiplicand and multiplier, respectively.

There are four outputs $x_1x_2x_3z_4$ which indicate the product when the multiplicand x_1x_2 is multiplied by the multiplier x_3x_4. Output z_4 is the low-order product bit. All outputs are asserted at time t_1 and deasserted at t_3. Design the Moore machine using a PROM for the δ next-state logic and a parallel-in, parallel-out register for the state flip-flops and the λ output logic.

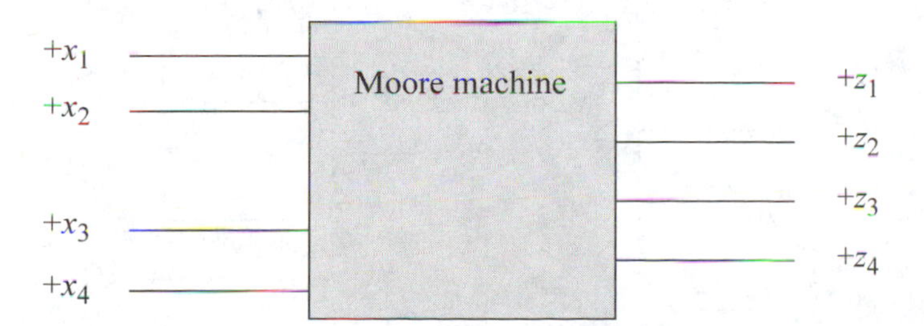

4.19 Design a 3-bit Johnson counter using a PAL device for the δ next-state logic, the storage elements, and the λ output logic. The counter counts in the following sequence: 000, 100, 110, 111, 011, 001, 000, ...

4.20 Design a 4-bit Moore sequential machine that operates according to the following sequence: $y_1y_2y_3y_4$ = 0000, 1000, 1100, 1110, 1111, 0000, ... Output z_1 is asserted unconditionally in state $y_1y_2y_3y_4$ = 1111. Use a PAL device for the δ next-state logic, the storage elements, and the λ output logic.

4.21 Design a Moore machine whose operation is defined by the state diagram shown below. Use a PAL device for the δ next-state logic, the storage elements, and the λ output logic. Use x_1, x_2, and x_3 as map-entered variables.

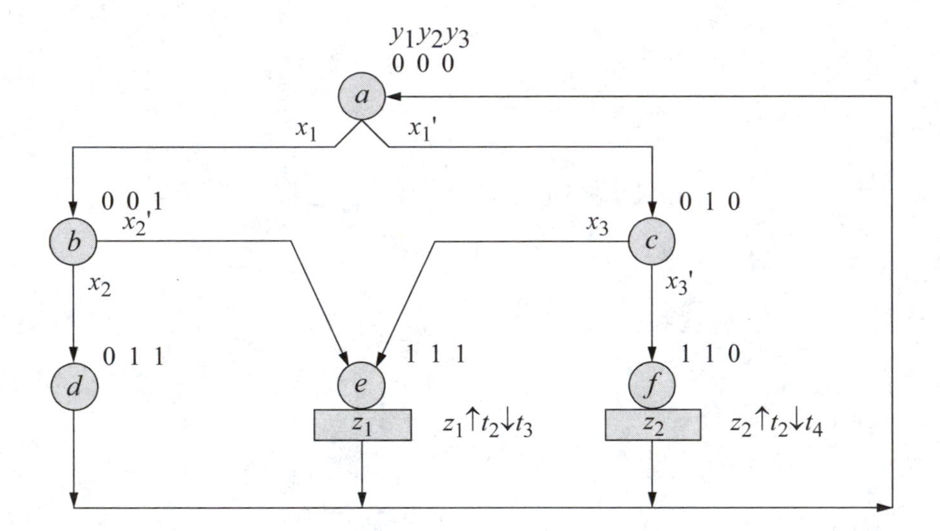

4.22 Synthesize a Moore sequential machine that receives a 4-bit word on a parallel input bus. The input bus is designated $x_1x_2x_3x_4$ as shown below. The parallel bus specifies two 2-bit operands, where x_1x_2 is the multiplicand and x_3x_4 is the multiplier. Inputs x_2 and x_4 are the low-order bits of the multiplicand and multiplier, respectively. There are four outputs $z_1z_2z_3z_4$ which indicate the product when the multiplicand x_1x_2 is multiplied by the multiplier x_3x_4. Output z_4 is the low-order product bit. All outputs are asserted at time t_2 and deasserted at t_3. Design the Moore machine using a PAL device for the δ next-state, the state flip-flops, and the λ output logic.

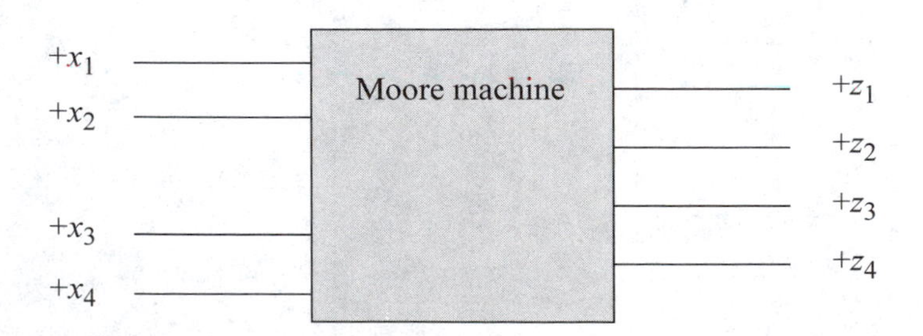

4.23 Implement the state diagram shown below for a Moore machine using a PLA device for the δ next-state logic, the state flip-flops, and the λ output logic. Use x_1 and x_2 as map-entered variables. The outputs are asserted at time t_1 and deasserted at t_3. Before implementing the design, be certain that no state transition sequence passes through states *a*, *c*, or *d* for a path that does not include the corresponding output.

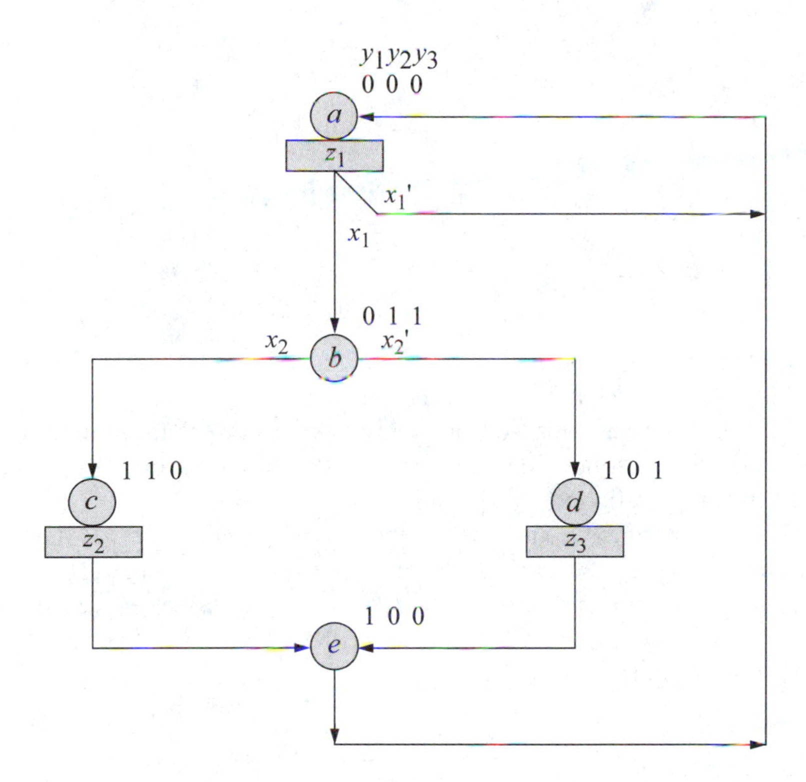

4.24 Design a synchronous counter that counts according to the specifications shown below. Input x_1 is a mode control line such that,

If $x_1 = 0$, then count up
If $x_1 = 1$, then count down

The counter has an initial state of $y_1y_2y_3 = 000$. Use a PLA device for the δ next-state logic, the state flip-flops, and the λ output logic. Use input x_1 as a map-entered variable. The counting sequence is shown on the next page.

y_1	y_2	y_3	x_1
0	0	0	0
0	0	1	0
0	1	0	0
0	1	1	0
1	0	0	0
1	0	1	0
1	1	0	0
1	1	1	0
. . .			

y_1	y_2	y_3	x_1
1	1	1	1
1	1	0	1
1	0	1	1
1	0	0	1
0	1	1	1
0	1	0	1
0	0	1	1
0	0	0	1
. . .			

4.25 Synthesize a Moore sequential machine that receives a 4-bit word on a parallel input bus. The input bus is designated $x_1x_2x_3x_4$ as shown below. The parallel bus specifies two 2-bit operands, where x_1x_2 is the multiplicand and x_3x_4 is the multiplier. Inputs x_2 and x_4 are the low-order bits of the multiplicand and multiplier, respectively.

There are four outputs $z_1z_2z_3z_4$ which indicate the product when the multiplicand x_1x_2 is multiplied by the multiplier x_3x_4. Output z_4 is the low-order product bit. All outputs are asserted at time t_2 and deasserted at t_3. Design the Moore machine using a PLA device for the δ next-state, the state flip-flops, and the λ output logic.

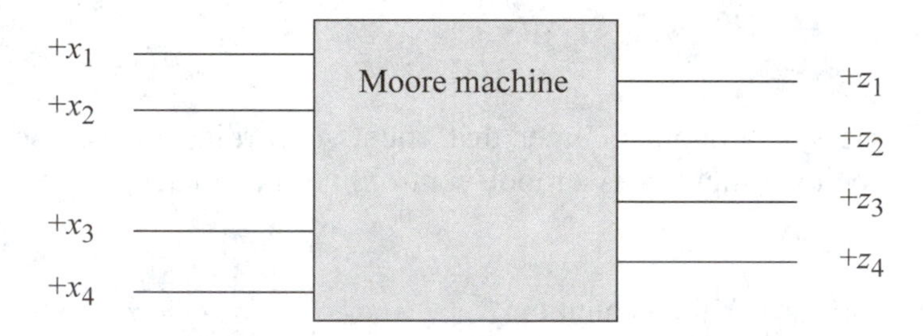

4.26 A microprocessor-controlled sequential machine has a 16-bit address bus, labeled $A15A14$ ... $A1A0$. Show the binary values of address bits $A15$, $A14$, $A13$, and $A12$ for the operations indicated in parts (a) and (b).

(a) Load state register 4 from the data bus.

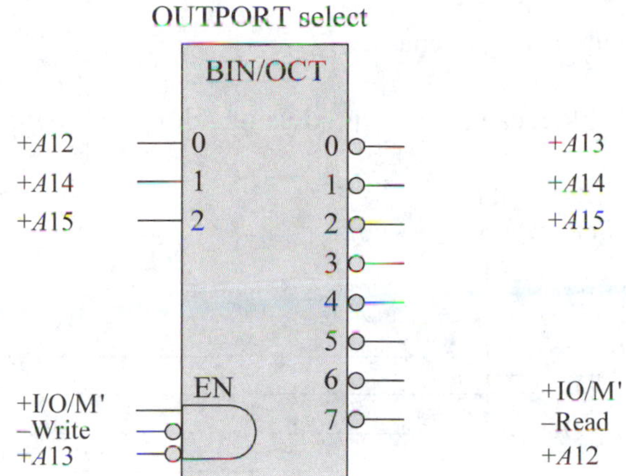

(b) Gate input register 6 to the data bus.

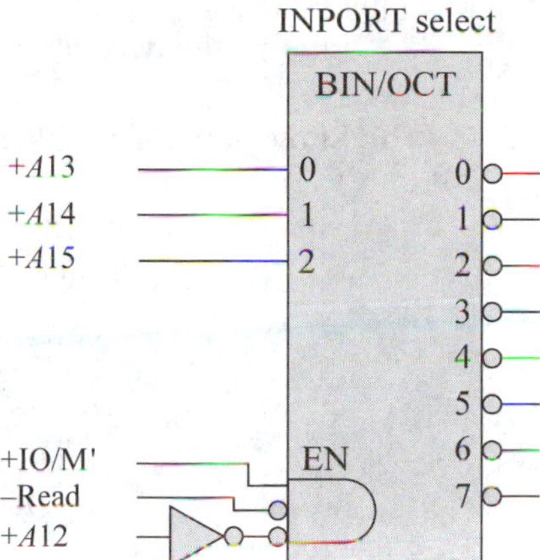

4.27 A microprocessor-controlled sequential machine has a 16-bit address bus, labeled $A15A14$ ... $A1A0$. Show the binary values of address bits $A15$, $A14$, $A13$, and $A12$ for the operations indicated in parts (a) and (b).

(a) Load state register 6 from the data bus.

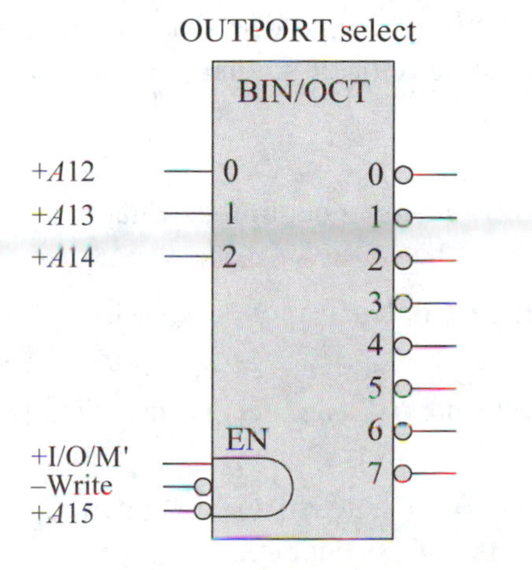

(b) Gate input register 3 to the data bus.

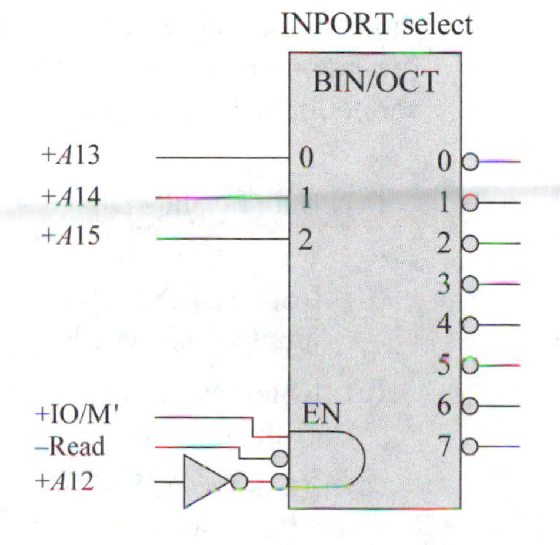

4.28 A microprocessor-controlled sequential machine contains an 8-bit state register as shown below. The machine must have only one state asserted at a time.

(a) What is the restriction when using the register and the 3:8 decoder, assuming that the decoder is always enabled?

(b) Show how the problem can be prevented using additional logic.

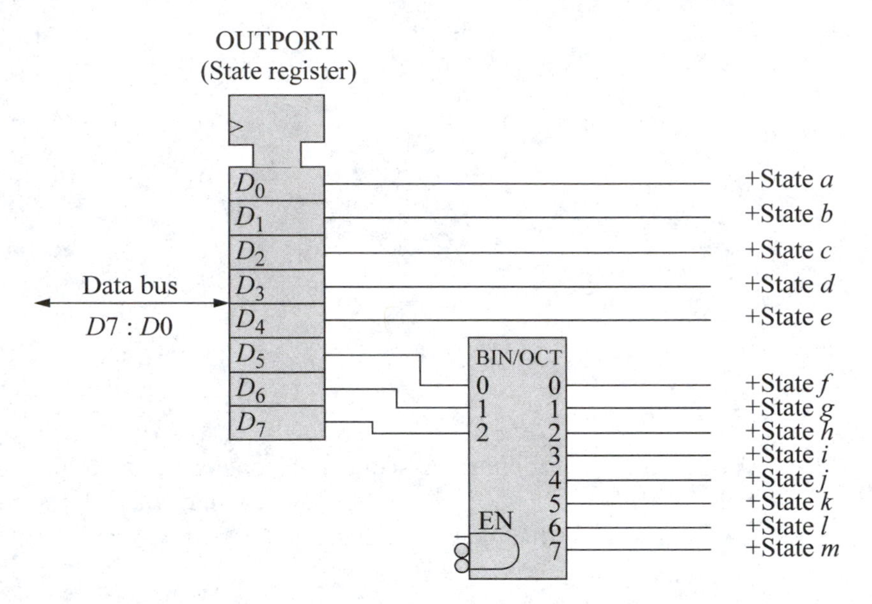

4.29 A microprocessor-controlled sequential machine has an 8-bit state register and a 4:16 decoder as shown below, providing a total of 20 states. All 20 states must be used; however, the machine must have only one state asserted at a time. Indicate which statements below will accomplish this by using additional logic in conjunction with the register and decoder. Assume that the programmer will correctly load the state register so that only one state is asserted for states *a*, *b*, *c*, and *d*.

(a) Data bus bits $D0 + D1 + D2 + D3$ connect to the –Enable input of the decoder.

(b) [(State *a*)' (State *b*)' + (State *c*)' (State *d*)']' connects to the –Enable input of the decoder.

(c) [(State *a*) (State *b*) (State *c*) (State *d*)]' connects to the –Enable input of the decoder.

(d) [(State *a*) + (State *b*) + (State *c*) + (State *d*)]' (State *e*) generates a new state *e* if the decoder is always enabled.

(e) [(State *a*) + (State *b*) + (State *c*) + (State *d*)] connects to the –Enable input of the decoder.

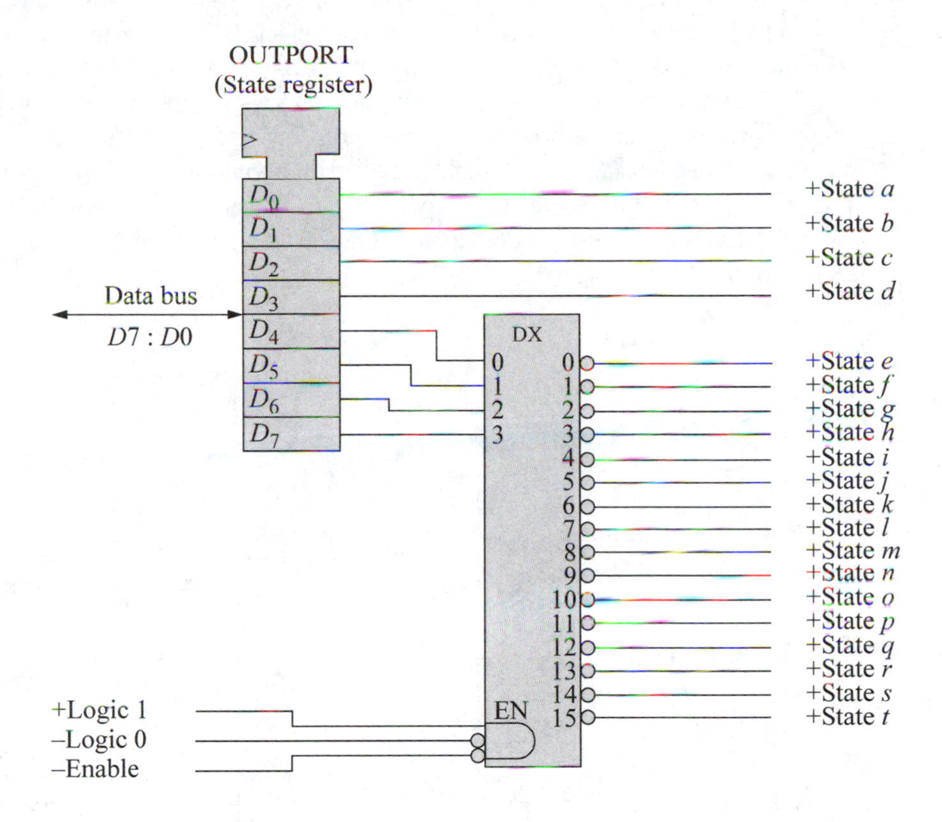

4.30 A microprocessor-controlled sequential machine contains one output port and one input port which are addressed as follows:

	$A15$	$A14$	$A13$ (LSB)
OUTPORT	0	0	1
INPORT	1	1	0

Design a logic circuit that will write to the output port and read from the input port using one 3:8 decoder and any additional logic gates. Use the least amount of logic.

4.31 Synthesize an n-bit odd parity checker which is to be implemented as a combinational iterative network. The output of the rightmost cell will be 1 if the number of 1s in the parallel input vector $x_1, x_2, \ldots, x_i, \ldots, x_n$ is odd, otherwise the output will be 0. Then convert the parity checker into a functionally equivalent synchronous sequential machine.

4.32 A microprocessor-controlled sequential machine has an 8-bit state register and a 3:8 decoder as shown below, providing a total of 13 states. All 13 states must be used; however, the machine must have only one state asserted at a time. Show how this can be accomplished by using additional logic gates.

Assume that the programmer will correctly load the state register so that only one state is asserted for states *a*, *b*, *c*, *d*, and *e*. Any additional logic that is required to guarantee that only one state is asserted, is transparent to the programmer. The decoder does not have an enable input. Use the least amount of logic.

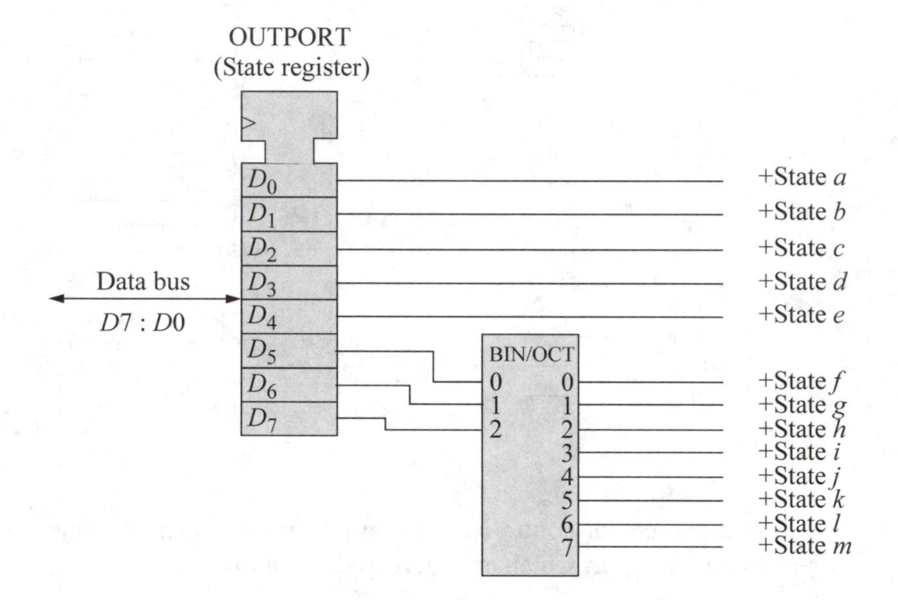

4.33 A state transition sequence for a microprocessor-controlled sequential machine requires a transition from state *h* to state *i* in the state registers shown below. The machine must always have one and only one state asserted. Design the necessary logic that will reset state *h* and set state *i* simultaneously. Use the least amount of logic. The following control signals are available:

+State register 1 address valid: The address for state register 1 is on the address bus and has been decoded.

+State register 2 address valid: The address for state register 2 is on the address bus and has been decoded.

–Read: This is an active-low read pulse from the microprocessor.

–Write: This is an active-low write pulse from the microprocessor. The positive transition of the pulse loads the state register from the data bus.

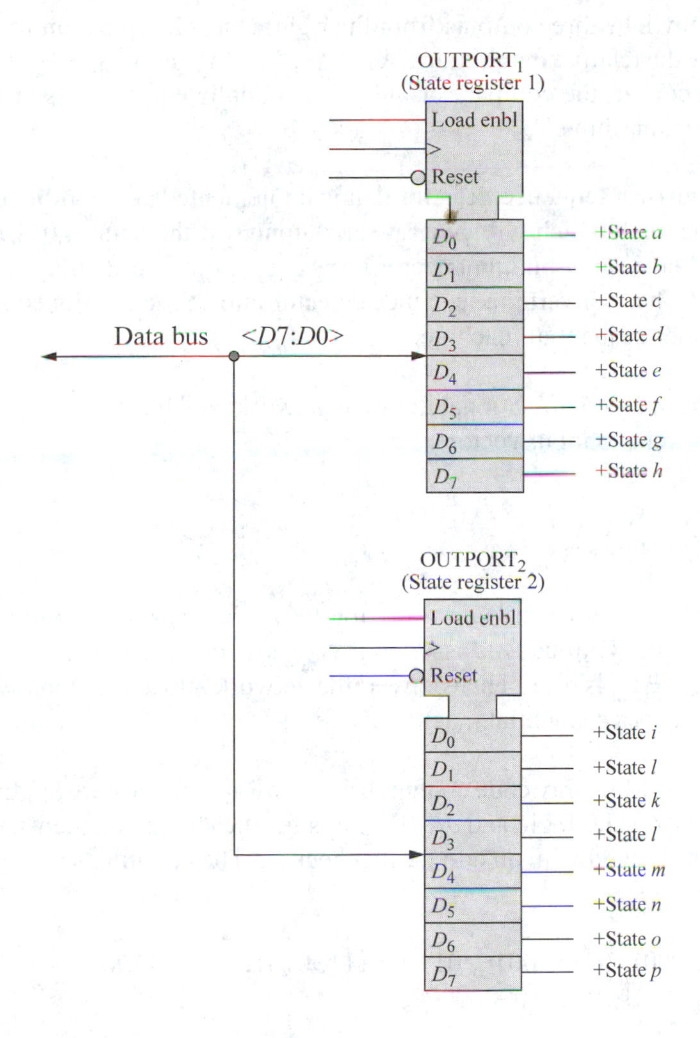

4.34 Synthesize a combinational iterative network which compares the following two unsigned n-bit binary numbers:

$$X_1 = x_{11}, x_{12}, \cdots, x_{1i}, \cdots, x_{1n} \text{ and}$$
$$X_2 = x_{21}, x_{22}, \cdots, x_{2i}, \cdots, x_{2n}$$

The circuit will indicate their relative magnitude. Use two intercell connections $y_{(i-1)1}$ and $y_{(i-1)2}$ that are encoded to indicate $X_1 < X_2$, $X_1 = X_2$, and $X_1 > X_2$ as follows:

$y_{(i-1)1}$	$y_{(i-1)2}$	Definition
0	0	$X_1 = X_2$
0	1	$X_1 > X_2$
1	0	$X_1 < X_2$

There will be three outputs from the rightmost cell: z_1, z_2, and z_3, which indicate the relative magnitudes: $X_1 < X_2$, $X_1 = X_2$, and $X_1 > X_2$, respectively. Then convert the comparator into a functionally equivalent synchronous sequential machine.

4.35 Synthesize a sequence detector that is implemented as a combinational iterative network which will generate an output z_1 if the pattern 0110 is detected anywhere in an n-bit input vector $x_1 \; x_2 \; ... \; x_i \; ... \; x_n$. Overlapping sequences are valid. Then convert the sequence detector into a functionally equivalent synchronous sequential machine.

4.36 Synthesize a combinational iterative network which compares each bit of the following two input vectors:

$$X_1 = x_{11}, x_{12}, \cdots, x_{1i}, \cdots, x_{1n} \text{ and}$$
$$X_2 = x_{21}, x_{22}, \cdots, x_{2i}, \cdots, x_{2n}$$

Output z_1 is asserted if $x_{1i} = x_{2i}$, for $1 \le i \le n$; that is, the expression $(x_{1i} \oplus x_{2i})'$ is true. Output z_2 is asserted if $x_{1i} \ne x_{2i}$, for $1 \le i \le n$; that is, the expression $x_{1i} \oplus x_{2i}$ is true. Then convert the network into a functionally equivalent synchronous sequential machine.

4.37 Design a 4-bit Gray code counter using NOR logic — where applicable — for the δ next-state logic and *JK* flip-flops for the storage elements. The operation is checked with an odd parity checker. The counting sequence is as follows:

0000, 0001, 0011, 0010, 0110, 0111, 0101, 0100, 1100, 1101, 1111, 1110, 1010, 1011, 1001, 1000, 0000, ...

4.38 Design a 4-bit Johnson counter using *D* flip-flops whose operation is checked with an odd parity checker. The counter counts in the following sequence: 0000, 1000, 1100, 1110, 1111, 0111, 0011, 0001, 0000, ...

4.39 Design a parity-checked Mealy synchronous sequential machine that generates an output z_1 whenever the sequence 1001 is detected on a serial data input line x_1. Overlapping sequences are valid. Output z_1 is asserted at time t_2 and deasserted at t_3. The parity flip-flop maintains odd parity over the state flip-flops and the parity flip-flop itself.

(a) Use *D* flip-flops as the storage elements.
(b) Use *JK* flip-flops as the storage elements.

5.1 Introduction
5.2 Fundamental-Mode Model
5.3 Methods of Analysis
5.4 Hazards
5.5 Oscillations
5.6 Races
5.7 Problems

5

Analysis of Asynchronous Sequential Machines

Another type of finite-state machine is an *asynchronous sequential machine*, where state changes occur on the application of input signals only — there is no machine clock. Like synchronous machines, the outputs of asynchronous machines are a function of either the present state only, or of the present state and the present inputs, corresponding to Moore and Mealy machines, respectively. The present state is directly related to the preceding sequence of inputs and states.

When an input variable changes in an asynchronous machine, the machine begins sequencing to a new state immediately. The input change may cause more than one storage element to alter its state, in which case, the machine may sequence through various transient states before entering a stable state. This is caused by different propagation delays in the logic gates and storage elements.

The operational speed of a synchronous machine is regulated by, and is coincident with, a machine clock. Thus, the machine can change states only on the active transition of the clock signal. Because the speed of an asynchronous machine is not limited by a clock frequency, the operating characteristics are usually faster than those of a corresponding synchronous machine. The operational speed is limited only by the propagation delay of the longest path. The procedure for synthesizing a reliable asynchronous sequential machine is much more difficult and challenging than for a comparable synchronous sequential machine.

In the operation of synchronous sequential machines, the effect of transient output signals caused by varying circuit delays was negated by selecting an appropriate clock signal — either true or complemented — for the λ output logic. This is not possible, however, for asynchronous machines, because there is no system clock. The values placed in the output maps, therefore, must be carefully chosen.

In some cases, a synchronous machine may require more hardware than a comparable asynchronous machine, thereby, increasing the cost. Also, the synchronization of data input variables may be impractical. Under these conditions, it may be more desirable to implement the machine specifications using an asynchronous approach.

Synchronous sequential machines occur more frequently in digital systems than their asynchronous counterparts, because of the ease with which the machines can be synthesized and the reliability imposed by a regularly occurring clock signal. In some cases, however, a machine clock may not be available, thus necessitating an asynchronous design. For extremely large synchronous machines using high-speed logic, the time required for the clock signal to propagate along the network of wires may be inordinately long. Thus, it is difficult to ensure the simultaneous arrival of the clock signal at all flip-flop clock inputs, which may hinder reliable synchronous operation. In this situation also, an asynchronous design may be more appropriate.

5.1 Introduction

The elements of a sequential machine, whether synchronous or asynchronous, introduce delays of various magnitudes. These elements include logic gates, flip-flops, and the interconnecting wires. Although the delay incurred by a network of wires is usually small in comparison to the propagation delay of gates and flip-flops (approximately one nanosecond per foot), the wire delay is significant when using high-speed logic circuits, such as emitter-coupled logic.

Asynchronous machines are used extensively in the interface control logic of peripheral devices which attach to asynchronous interfaces between a channel (or I/O Processor) and the device control unit. Since the interface contains no clock signal, the control unit utilizes randomly occurring interface control signals to transfer the data. During a write operation, the data bytes are transferred from the channel to the control unit interface logic which then synchronizes the data transfer rate to the speed of the peripheral device. During a read operation, the procedure is reversed.

Asynchronous sequential machines are implemented with Set/Reset (*SR*) latches as the storage elements. Thus, at least one *feedback* path is required in the synthesis of asynchronous machines. Figure 5.1 (a) illustrates an asynchronous machine which contains one feedback path from output z_1 to the input logic. Asynchronous machines are implemented in a sum-of-products form. Figure 5.1 (b) shows the same circuit redrawn in the conventional latch configuration.

A feedback path is a necessary but not a sufficient condition for an asynchronous device, because the feedback does not necessarily indicate storage capability. For example, the asynchronous machine of Figure 5.2 contains an *SR* latch with one feedback path. When inputs x_1 and x_2 are asserted, output z_1 becomes active and the latch will be set. However, when input x_2 is deasserted, the latch will be reset and output z_1 will be deasserted. Thus, the machine does not incorporate the storage characteristics exhibited by an asynchronous machine. Although the machine is implemented with a storage element, the output is a function of the present inputs only and operates, therefore, as a combinational circuit.

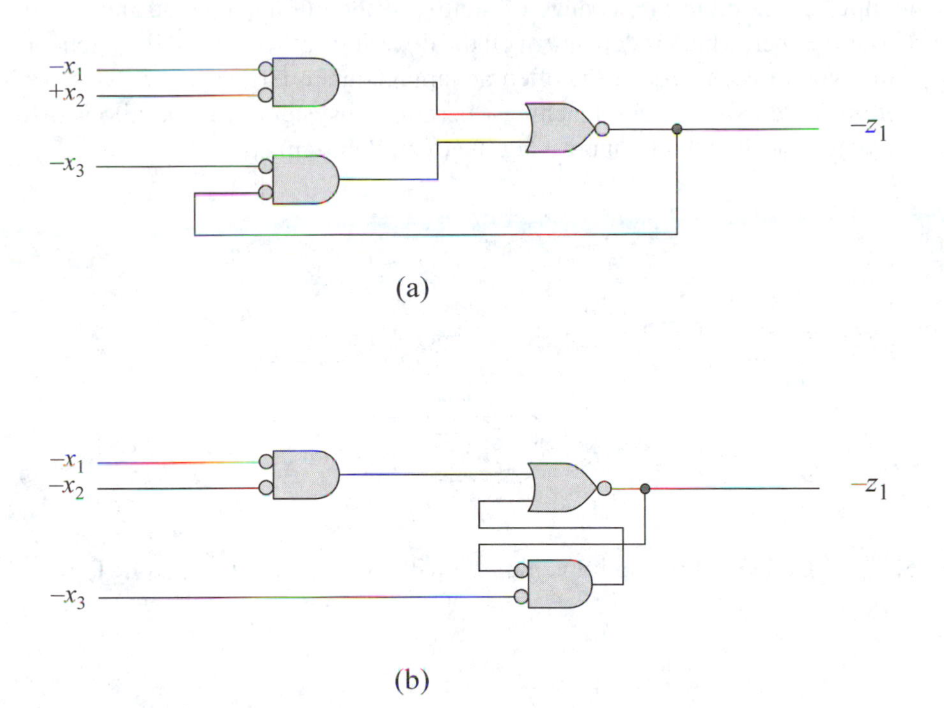

Figure 5.1 Asynchronous sequential machine with one feedback: (a) sum-of-products; and (b) conventional latch configuration.

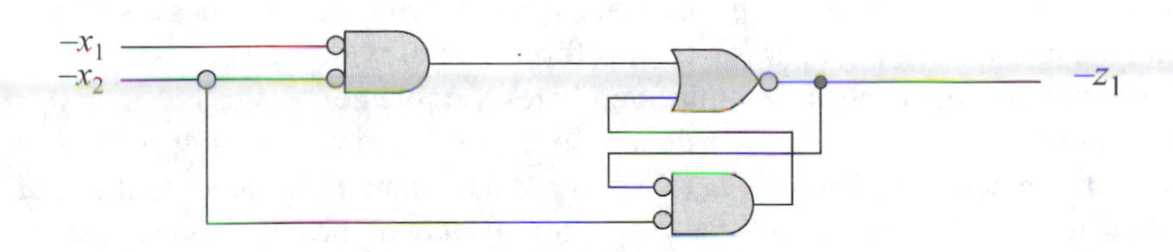

Figure 5.2 Asynchronous machine that operates as a combinational circuit.

The internal operation of a flip-flop is asynchronous in nature, even though the flip-flop is clocked. Consider, for example, the clocked *SR* flip-flop shown in

Figure 5.3. Assume that the flip-flop is set; that is, the $-y_1$ and $+y_2$ flip-flop outputs are active at a low voltage level and a high voltage level, respectively. If the set and reset inputs $x_s x_r = 01$, then at the next positive level of the +Clock signal, output $+y_1$ will be deasserted, providing a low input to the OR gate of the latch. After an appropriate propagation delay through the OR gate, the $-y_1$ signal will be deasserted. The change of state for the $-y_1$ and $+y_2$ outputs represents an asynchronous operation. Thus, the clock pulse initiates a change of state, but the outputs change state in an asynchronous manner, which is dependent on the differing gate delays of the latch circuitry. This type of asynchronism is often advantageous in large synchronous systems. Within large synchronous machines, certain subsystems can be allowed to operate asynchronously, thereby increasing the overall system speed.

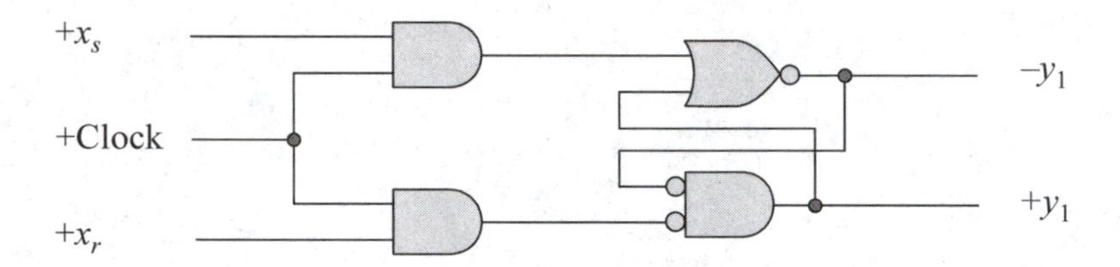

Figure 5.3 Clocked flip-flop implemented with an *SR* latch using NOR logic.

5.2 Fundamental-Mode Model

Both synchronous and asynchronous sequential machines use feedback paths in their implementation. The feedback in synchronous machines combines with the input variables to form the δ next-state logic, which provides the next state for the machine at the next active clock transition. The feedback in asynchronous machines also combines with the δ next-state logic; however, this formation provides the necessary conditions to implement *SR* latches, and thus, generates a stable next state.

There are two approaches to be considered for modeling and analyzing asynchronous sequential machines. The first is to insert a delay element in the output net of each logic gate. The delay represents the propagation delay of the gate. Thus, the corresponding gate is considered to have zero propagation delay. The second approach is to use a single delay element that represents the total delay of the entire machine and is placed in series with the feedback path. The analysis is simplified if the second approach is chosen, thereby specifying zero delay for all logic gates and interconnecting wires.

A general block diagram for an asynchronous sequential machine is shown in Figure 5.4. The input alphabet X consists of binary input variables $x_1, x_2, \cdots, x_n$ that can change value at any time and are represented as voltage levels rather than pulses. The state alphabet Y is characterized by p storage elements, where $Y_{1e}, Y_{2e}, \cdots, Y_{pe}$ are the *excitation* variables and $y_{1f}, y_{2f}, \cdots, y_{pf}$ are the *feedback* or *secondary* variables. The output alphabet Z is represented by $z_1, z_2, \cdots, z_m$. Both the δ next-state logic and the λ output logic are composed of combinational logic circuits. The delay element in Figure 5.4 represents the total delay of the machine from the time an input changes until the machine has stabilized in the next state, and is represented as a time delay of Δt. The time correlation between the excitation variables Y_{ie} and the feedback variables y_{if} is specified by Equation 5.1.

$$y_{if\,(t + \Delta t)} = Y_{ie\,(t)} \tag{5.1}$$

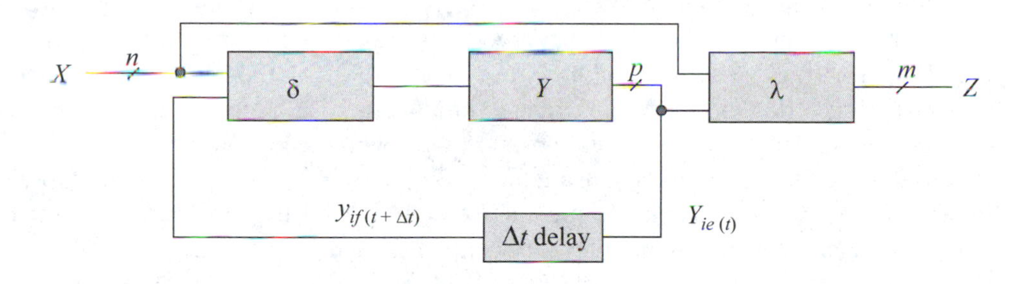

Figure 5.4 General block diagram of an asynchronous sequential machine.

For a given set of inputs, the machine will be in a stable state if and only if $y_{if} = Y_{ie}$, for $i = 1, 2, \ldots, p$. That is, after an appropriate delay, the feedback variables will be equal to the excitation variables, indicating that the machine has entered a *stable state*. When an input changes value, the δ next-state combinational logic produces a new set of values for the excitation variables $Y_{1e}, Y_{2e}, \cdots, Y_{pe}$. The machine then sequences through one or more unstable states. When the values of the feedback variables y_{1f}, $y_{2f}, \cdots, y_{pf}$ are equal to the values of the excitation variables $Y_{1e}, Y_{2e}, \cdots, Y_{pe}$, then the machine is stable in that state.

The transition from one stable state to another stable state is the result of a single change to an input variable. To ensure a deterministic operation, it is assumed that only one input changes at a time. If two inputs change simultaneously, then this will result in a *race* condition in which the machine may sequence to an incorrect next state. Race conditions are discussed in a later section.

Postulating that only one input changes value at a time should present no undo rigor in the synthesis procedure. It is always possible to affect a modification to the external circuitry so that only a single input changes value. This can be accomplished without altering the machine specifications. Also, when a binary input variable changes, the new value must be of sufficient duration to cause the machine to change state and proceed to a stable state. It is also assumed that no other input will change until the machine has entered a stable state. The previous two conditions specify a *fundamental-mode* operation. Thus, a fundamental-mode model has the following characteristics:

1. Only one input changes at a time.

2. No other input will change until the machine has sequenced to a stable state.

Figure 5.5 depicts an asynchronous sequential machine with which to begin the analysis procedure. Figure 5.5 (a) illustrates the network using AND/OR gates where the circuit delays are distributed throughout the machine. The voltage levels of the inputs will be such that their asserted or deasserted level will correspond to the required level of the logic gate inputs for the specified input equation: $z_1 = x_1x_2 + x_2y_{1f}$. Thus, if AND or NAND gates are used for the δ next-state logic, then the input variables will produce a high voltage level for the requisite input equation. However, if NOR gates are used, then the input variables will produce a low voltage level to satisfy the input equation, since the AND function using NOR logic requires low inputs, as shown in Figure 5.5 (b). If complemented inputs are required, then they must be generated within the machine so that their delay will contribute to the overall delay.

The fundamental-mode model of Figure 5.5 (c) offers the advantage of simplicity during analysis in which the logic gates have zero propagation delay. The combined delays of all the circuit elements are consolidated in the delay element Δt. Although an asynchronous sequential machine using distributed delays provides a more accurate model, the fundamental-mode model with a single consolidated delay is sufficiently accurate for analysis purposes.

The storage element output of a fundamental-mode model is the excitation variable Y_{1e}. After a delay of Δt, the same signal is specified as a feedback variable y_{1f}. The criterion for machine stability is established by the relationship between the excitation variable Y_{1e} and the feedback variable y_{1f} as a function of Δt. This characteristic is expressed in Equation 5.2, which states that the value of the feedback variable y_{1f} is equal to the value of the excitation variable Y_{1e} for $i = 1, 2, \ldots, p$ after a delay of Δt.

$$\begin{aligned} y_{if(t+\Delta t)} &= Y_{ie\,(t)} \quad \text{(Stable state)} \\ y_{if(t+\Delta t)} &\neq Y_{ie\,(t)} \quad \text{(Unstable state)} \end{aligned} \tag{5.2}$$

According to Equation 5.2, an asynchronous machine is stable in a particular state only when the feedback variable $y_{if(t+\Delta t)}$ is equal to the excitation variable $Y_{ie\,(t)}$. Conversely, the machine is unstable until all circuit delays have transpired; that is, while $y_{if(t+\Delta t)} \neq Y_{ie(t)}$, which denotes that the result of an input change is still in progress and the machine has not yet settled into a stable state. In certain situations, the machine will never sequence to a stable state. This occurrence will be discussed in the section on oscillations.

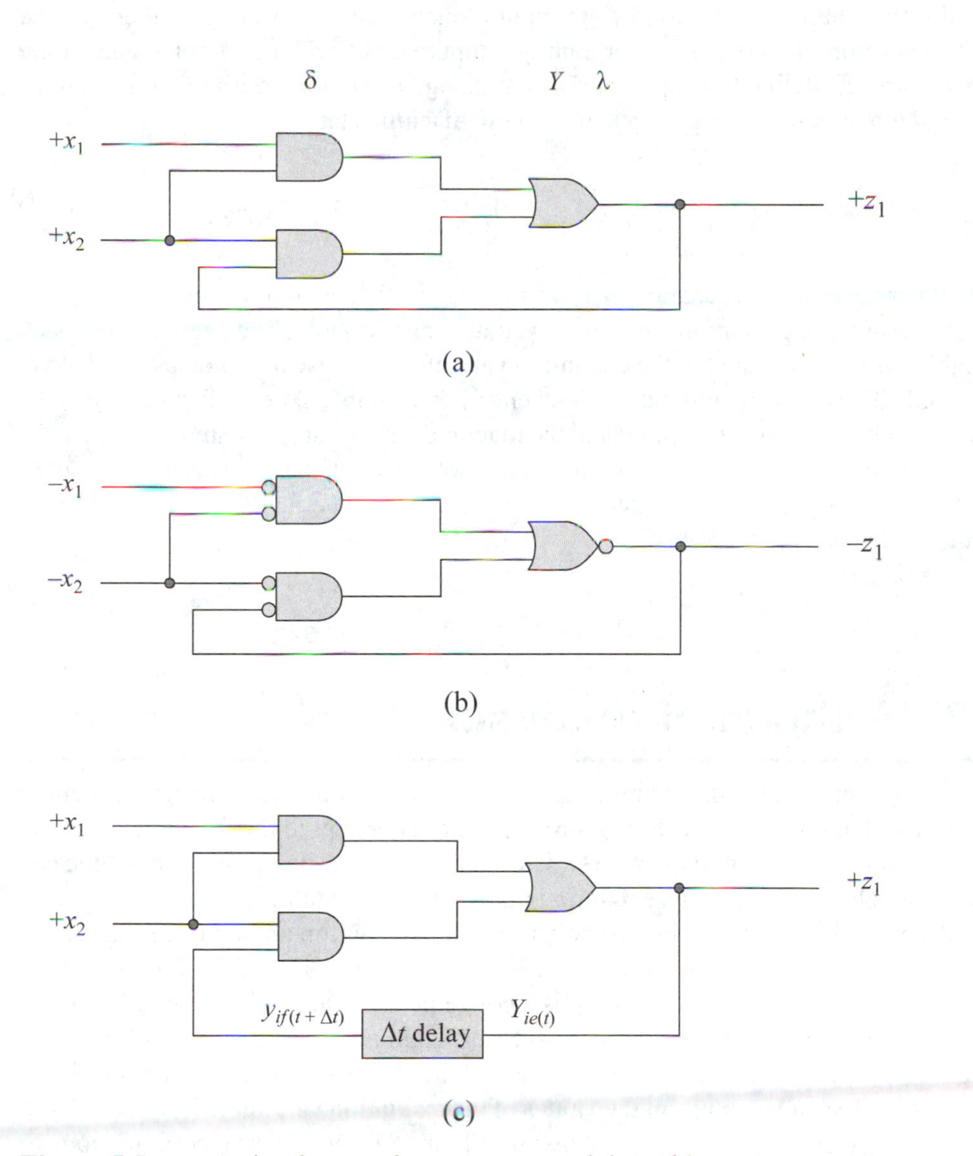

Figure 5.5 A simple asynchronous sequential machine: (a) conventional circuit with distributive delays using AND-OR gates; (b) conventional circuit with distributive delays using NOR logic; and (c) fundamental-mode model.

The functional notation $Y_{ie(t)}$ and $y_{if(t+\Delta t)}$ will subsequently be replaced by the more compact notation Y_{ie} and y_{if}, which corresponds to the excitation variable and the feedback variable, respectively. Their differences as functions of time, however, remain the same.

In general, the input alphabet of the machine is represented by the n-tuple set $X = \{X_0, X_1, X_2, \cdots, X_{2^n-1}\}$, where each X_i represents a unique input vector. The state alphabet is represented by the p-tuple set $Y = \{Y_0, Y_1, Y_2, \cdots, Y_{2^p-1}\}$, where each Y_j represents a unique state. The total state of an asynchronous sequential machine is specified by the concatenation of the input alphabet and the state alphabet; that is, $X_i Y_j$. Therefore, the total state for a unique input vector and a particular state is the $(n + p)$-tuple as defined in Equation 5.3. A change to an input variable will produce not only a new total state $X_i Y_j$, but also a new machine state Y_j.

$$(x_1, x_2, \cdots, x_n)(y_1, y_2, \cdots, y_p) \tag{5.3}$$

Whenever a single input changes, a new input vector is generated, which in turn produces a new set of excitation variables. At that instant in time, however, the feedback variables are not yet equal to the excitation variables, because the Δt delay has not yet occurred. Thus, $y_{if} \neq Y_{ie}$ and the machine enters an unstable state. After a delay of Δt, all circuit changes have transpired and the machine enters a stable state where $y_{if} = Y_{ie}$. The machine remains in this stable state until another input change causes a new set of excitation variables to be generated. The machine will exit a stable state only when an input variable changes value.

5.3 Methods of Analysis

This section presents additional mechanisms for analyzing asynchronous sequential machines. These analysis techniques are similar in concept to those used for analyzing synchronous sequential machines. The behavior of an asynchronous machine can be completely specified by an excitation map, a next-state table, a state diagram, and a flow table. Also included in the analysis are the excitation and output equations.

In general, the excitation variables are a boolean function of the inputs and the feedback variables. A Karnaugh map is a convenient method of representing the excitation variables. An excitation map is a Karnaugh map in which the columns are specified by the input variables and the rows by the feedback variables. The entries in the minterm locations represent the values of the excitation variables. The formats for representative excitation maps are shown in Figure 5.6 for different combinations of input and feedback variables.

Figure 5.6 (d) and Figure 5.6 (e) illustrate two methods for constructing a 5-variable Karnaugh map. The first method utilizes two identical, separate 4-variable maps. The low-order input variable x_3 is assigned values of 0 and 1 for the left and right maps, respectively. To determine adjacency between minterm locations, the left map is superimposed on the right map. Any minterm locations that are then physically adjacent are also logically adjacent and can be combined. The second method combines two 4-variable maps into a single 5-variable map. To determine adjacency between minterm locations in the single map, the map is hinged along the vertical centerline in a manner analogous to folding a book. Any minterm locations that are then physically adjacent are also logically adjacent and can be combined.

Figure 5.6 (f) and Figure 5.6 (g) illustrate two methods for constructing a 6-variable Karnaugh map. The first method utilizes identical, separate 5-variable maps in which the low-order input variable x_3 is assigned values of 0 and 1 for the left and right maps, respectively. Adjacency determination is identical to that described for the two separate maps of Figure 5.6 (d). The second method is similar in concept to the single map of Figure 5.6 (e). It is rare that an asynchronous sequential machine will extend beyond the 6-tuple $x_1x_2x_3y_{1f}y_{2f}y_{3f}$ comprised of three input and three feedback variables.

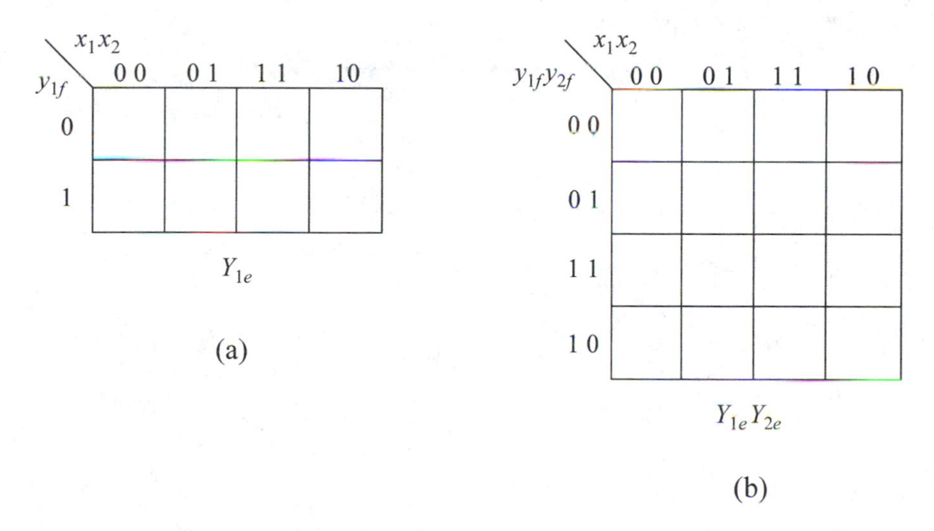

Figure 5.6 Formats for representative excitation maps: (a) two inputs and one excitation variable; (b) two inputs and two excitation variables; (c) two inputs and three excitation variables; (d) three inputs and two excitation variables; (e) alternative format for three inputs and two excitation variables; (f) three inputs and three excitation variables; and (g) alternative format for three inputs and three excitation variables.

Example 5.1 Figure 5.7 illustrates a simple asynchronous sequential machine which will be used as an introductory example for analysis. The output of the circuit is the excitation variable Y_{1e}, which also corresponds to the machine output z_1. After a delay of Δt, the feedback variable y_{1f} becomes equal to the excitation variable Y_{1e} and the machine is stable. The equation for Y_{1e} is obtained directly from the logic diagram as a function of the input variables x_1 and x_2 and of the feedback variable y_{1f}, as shown in Equation 5.4. The excitation variable is usually portrayed in the sum-of-products format. In this simple asynchronous machine, output z_1 has the same characteristics as the excitation variable Y_{1e}. Thus, the machine operates as a Moore model, because the output is a function of the state alphabet only and is independent of the input alphabet.

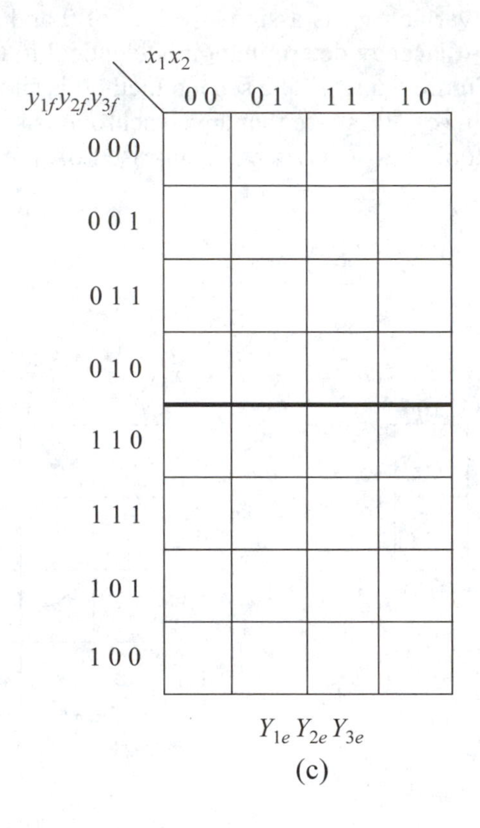

(c)

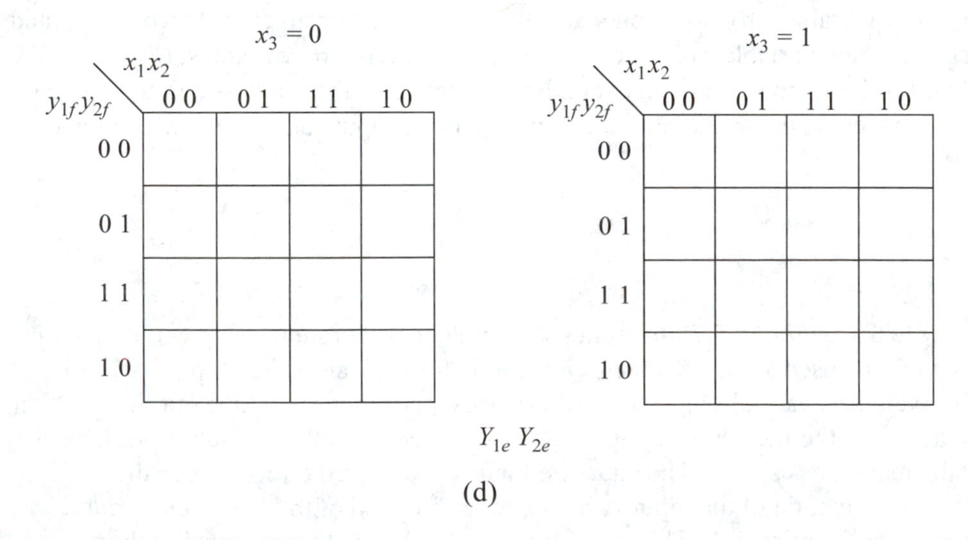

(d)

Figure 5.6 (Continued)

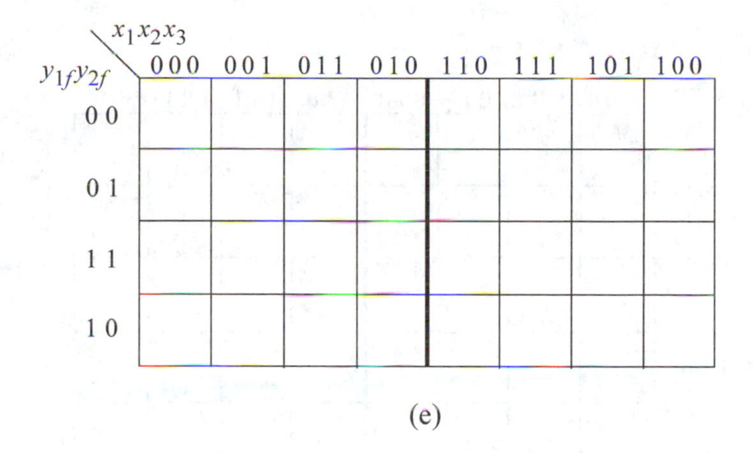

(e)

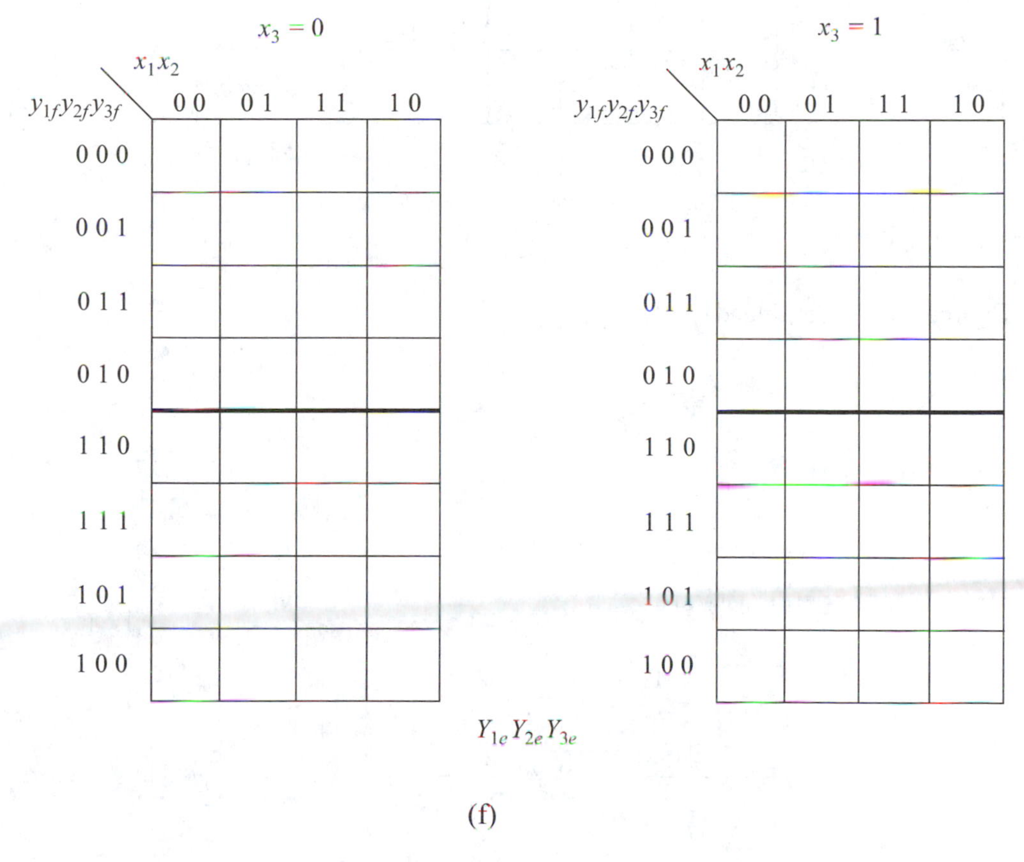

(f)

Figure 5.6 (Continued)

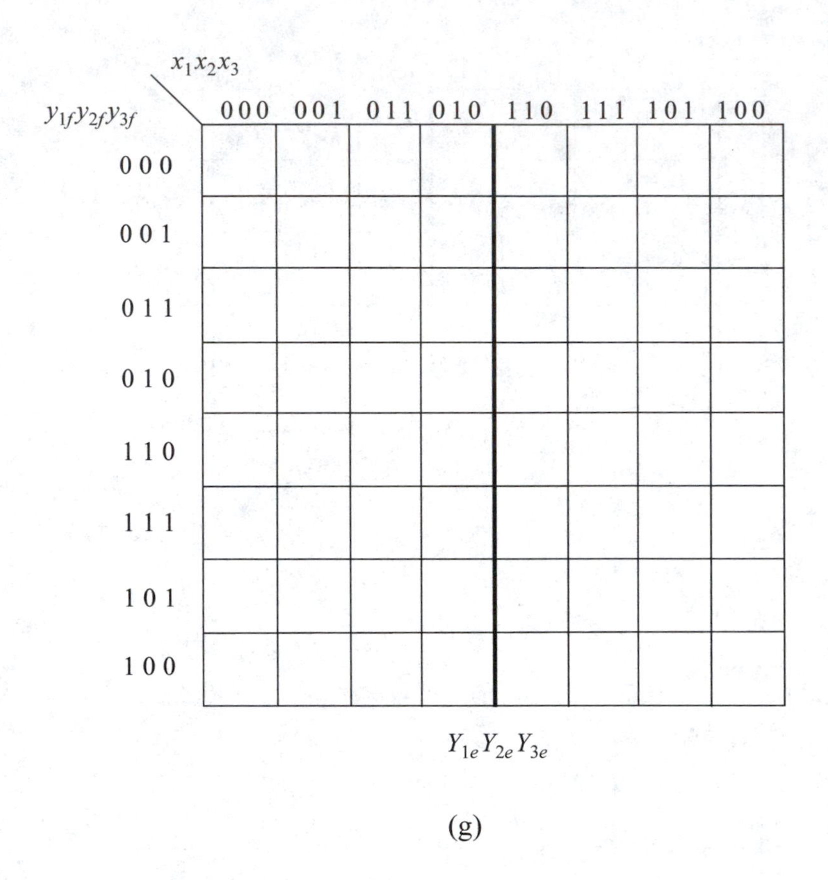

(g)

Figure 5.6 (Continued)

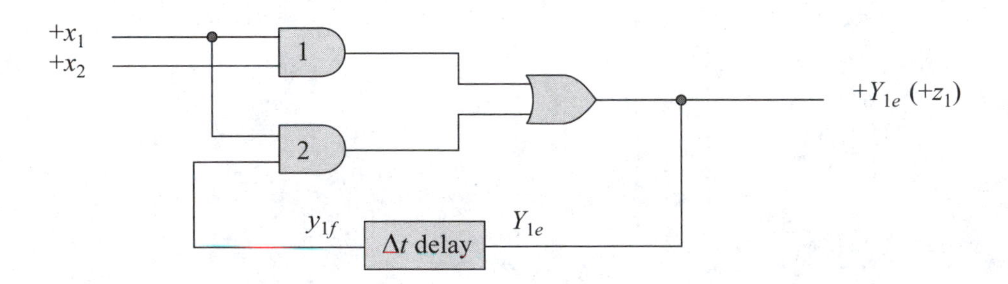

Figure 5.7 Asynchronous sequential machine with one feedback path.

$$Y_{1e} = x_1 x_2 + x_1 y_{1f}$$
$$z_1 = Y_{1e} \tag{5.4}$$

The excitation map is obtained by plotting the equation for Y_{1e} in a Karnaugh map, as shown in Figure 5.8 (a). The input variables x_1 and x_2 specify the columns of the map and are enumerated in the Gray code format in the same manner as for all Karnaugh maps of three variables. The feedback variable y_{1f} defines the values of the rows.

In order to characterize the operation of the machine, the stable states must be identified. Recall that an asynchronous sequential machine is stable in a particular state only if the feedback variable is equal to the excitation variable; that is, $y_{1f} = Y_{1e}$. By convention, a *stable state* is indicated by circling the corresponding map entry, as shown in Figure 5.8 (b). Since the values inserted in the minterm locations correspond to the values of Y_{1e}, any map entry that is equal to the row value of y_{1f} will indicate a stable state.

To facilitate the analysis of the machine, state names are inserted in the minterm locations together with the value of the excitation variable. When specifying a stable state, a circle is placed around the state variable or state name. Thus, in the row corresponding to $y_{1f} = 0$, states ⓐ, ⓑ, and ⓓ are stable, because in all three states $y_{1f} = Y_{1e} = 0$. Similarly, in the row corresponding to $y_{1f} = 1$, states ⓖ and ⓗ are stable, because in both states $y_{1f} = Y_{1e} = 1$.

The state transition sequences for the machine can now be determined. An asynchronous machine will remain in a stable state until an input changes. A change of a single input causes a horizontal movement in the excitation map. Assume that the machine is presently in stable state ⓐ ($y_{1f}x_1x_2 = 000$) and that input x_2 changes from 0 to 1. The machine will proceed horizontally to state ⓑ ($y_{1f}x_1x_2 = 001$) which is also stable, because $y_{1f} = Y_{1e} = 0$. The machine will remain in state ⓑ until an input again changes.

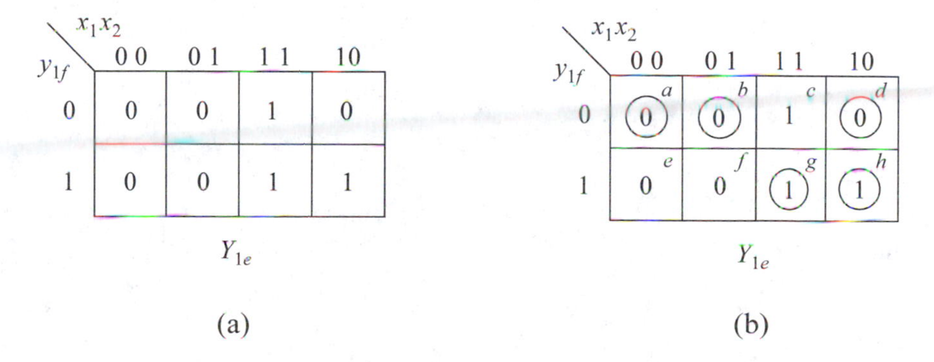

Figure 5.8 Karnaugh maps for the Moore machine of Figure 5.7: (a) excitation map; and (b) excitation map showing stable states, which are indicated by circled entries.

Similarly, if the machine is in state ⓐ and x_1 changes from 0 to 1, then the machine sequences horizontally in row $y_{1f} = 0$ and enters state ⓓ. Since the excitation value does not change, the feedback value will not change. Thus, the entire operation takes place only in row $y_{1f} = 0$. Note that it is not possible to sequence from state ⓓ ($y_{1f}x_1x_2 = 010$) to state ⓗ ($y_{1f}x_1x_2 = 110$) directly, because inputs x_1 and x_2 remain unchanged. In order to leave a stable state, an input must change value. State ⓗ can be entered only from state ⓖ. A state transition will occur from state ⓖ to state ⓗ if x_2 changes from 1 to 0. The machine will remain in state ⓗ until another change of input occurs, either to x_1 or x_2.

It is also not possible for the machine to sequence from state ⓑ to state ⓓ directly, because this transition requires a change of two input variables, a situation that is not allowed in asynchronous sequential machine operation. That is, $x_1x_2 = 01$ cannot change to $x_1x_2 = 10$ directly, or conversely. For the same reason, a simultaneous change from $x_1x_2 = 00$ to $x_1x_2 = 11$, or conversely, is not allowed.

If the machine is in state ⓑ and input x_1 changes from 0 to 1, then the machine enters state *c* which is unstable — or transient — because $y_{1f} \neq Y_{1e}$. This sequence is illustrated by the arrow from state ⓑ to state *c* in Figure 5.9. In state *c*, the excitation variable $Y_{1e} = 1$. Thus, after a delay of Δt, the feedback variable y_{1f} becomes equal to the excitation variable Y_{1e} and the machine moves to the row which corresponds to $y_{1f} = 1$. That is, the excitation value in state *c* is the value that the feedback will become after a delay of Δt. The machine then enters stable state ⓖ. The state transition sequence for this input change is ⓑ $\rightarrow c \rightarrow$ ⓖ. The sequence can be verified by examining the logic diagram of Figure 5.7.

When a single input changes, the machine can sequence to only those states that are contained in the column specified by the new input vector. Therefore, in state ⓑ, when input x_1 changes from 0 to 1, the machine moves horizontally to unstable state *c*, then vertically in column $x_1x_2 = 11$, and settles in state ⓖ. In unstable states, the inputs do not change — only the feedback variables change. To cause a feedback variable to change, the machine must pass through an unstable state.

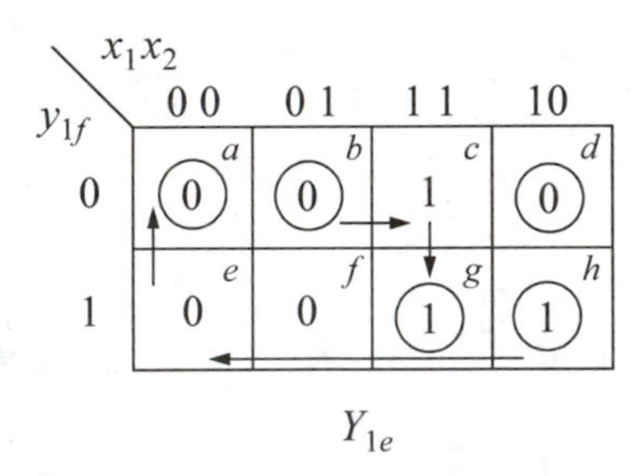

Figure 5.9 Representative state transition sequences for the asynchronous machine of Figure 5.7.

Assume that the machine is presently in state ⓗ ($y_{1f}x_1x_2 = 110$) where $y_{1f} = Y_{1e} = 1$. If input x_1 changes from 1 to 0, then the machine proceeds horizontally to state *e* where $x_1x_2 = 00$ and $Y_{1e} = 0$. The feedback variable has not yet changed;

therefore, the machine remains in row $y_{1f} = 1$ until a delay of Δt has transpired, at which time, the machine sequences from state e to state ⓐ. The machine remains in state ⓐ until another input change occurs. The state transition sequence for this change is ⓗ $\rightarrow e \rightarrow$ ⓐ.

In a similar manner, all possible state transition sequences can be traversed. This is accomplished by starting in any stable state and allowing a single input to change value. The next state will be either a stable state or an unstable state in a column represented by the new input vector. If the next state is stable, then the machine remains in the same feedback row, because the feedback variable will not change. Thus, the Δt delay does not actively participate in the determination of the next state.

If the next state is unstable, then the machine is in a transient state in the same feedback row and in the column specified by the new input values. There will be no vertical movement until the feedback variable begins to change in order to assume the value of the excitation variable. After a delay of Δt, the machine will move vertically in the same column and enter a row in which the value of the feedback variable is equal to the value of the previous excitation variable.

For large machines in which there is more than one feedback variable and, therefore, more than two feedback rows, the new state may also be either stable or unstable. The new state will be stable if $y_{if} = Y_{ie}$, for i = 1, 2, ... , p and the machine will remain in that state. If, however, $y_{if} \neq Y_{ie}$, then another vertical transition will occur and the machine will sequence to a new row in which the feedback variables are equal to the former excitation variables. This new state may again be either stable or unstable. The process of proceeding vertically in a column — either up or down — will continue until the machine sequences to a stable state. If no stable state exists in a particular column, then an oscillation will occur. Oscillations are discussed in a subsequent section.

The next-state table for the asynchronous Moore machine of Figure 5.7 is shown in Table 5.1 and is generated from the logic diagram or the excitation map. The present state is represented by the feedback variable y_{1f} and the next state by the excitation variable Y_{1e}. Whenever $y_{1f} = Y_{1e}$, the machine is in a stable state, as indicated by the circled states. States c, e, and f are unstable states; therefore, the entry for Y_{1e} is not circled.

Although a state diagram is usually not necessary in the synthesis of asynchronous sequential machines, it will be included in this analysis section for completeness. The state diagram is derived from either the excitation map or the next-state table and is shown in Figure 5.10. The conditions which caused a state change are indicated by the input variables, which are separated from the feedback variable by a slash.

Both stable and unstable states are represented in the state diagram. The unstable states are identified by a circle with dashed lines and sequence to the next state unconditionally. For large state diagrams, the unstable states can be omitted with no loss of clarity. Each stable state has two possible next states depending on which input variable changes. Thus, each stable state has two arrows exiting the state, whereas, an unstable state has only a single arrow exiting the state.

An alternative method of illustrating the flow of the machine is to tabulate the stable and unstable states as shown in the flow table of Figure 5.11. The flow table can be generated directly from the excitation map of Figure 5.8. The state names are chosen arbitrarily and only the stable states are assigned unique names. The flow table directly corresponds to the excitation map of Figure 5.8 (b). The excitation variables, however, are replaced with state names, allowing for easier interpretation of machine operation.

Table 5.1 Next-state table for the Moore machine of Figure 5.7

State name	Present state y_{1f}	Inputs x_1 x_2	Next state Y_{1e}	Output z_1
a	0	0 0	(0)	0
b	0	0 1	(0)	0
d	0	1 0	(0)	0
c	0	1 1	1	1
e	1	0 0	0	0
f	1	0 1	0	0
h	1	1 0	(1)	1
g	1	1 1	(1)	1

The flow table provides a convenient notational technique for analyzing machine operation. The table specifies the state transition sequences resulting from an input change. Thus, the behavior of the machine is completely specified by the flow table. The flow table is an extremely useful instrument for analysis and will be used extensively in the synthesis of asynchronous sequential machines.

The circled states in Figure 5.11 represent stable states; the uncircled states represent unstable states and indicate the name of the destination stable state to which the machine will sequence after a delay of Δt. Thus, if the machine is presently in state ⓑ and input x_1 changes from 0 to 1, then the operation proceeds horizontally in row $y_{1f} = 0$ and the machine will sequence through transient state d and proceed to stable state ⓓ after a delay of Δt.

Similarly, the transition from state ⓔ to state ⓐ is the result of input x_1 changing from 1 to 0, which causes the operation to move horizontally in row $y_{1f} = 1$ to unstable state a. After a delay of Δt, the machine will pass through transient state a and enter state ⓐ. All state transition sequences can be observed in the flow table, in which any transition is the result of a single input change. The state diagram of Figure 5.10 is redrawn as Figure 5.12 to correlate with the flow table of Figure 5.11.

In order to verify that the flow table summarizes the operation of a fundamental-mode machine, refer to the logic diagram of Figure 5.7 in conjunction with the flow table of Figure 5.12.

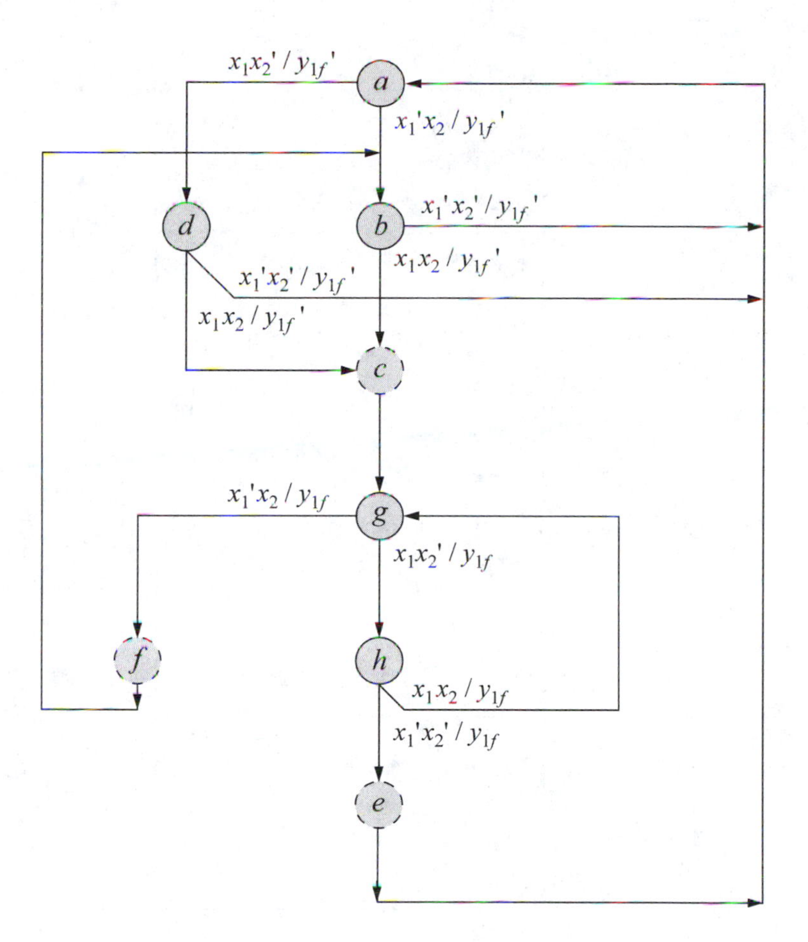

Figure 5.10 State diagram for the Moore machine of Figure 5.7.

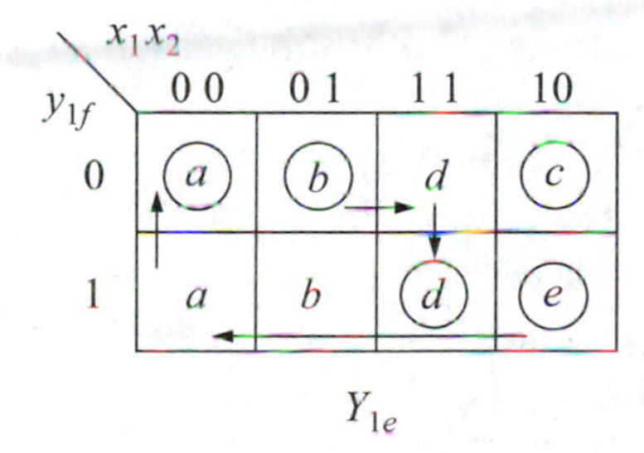

Figure 5.11 Flow table for the Moore machine of Figure 5.7.

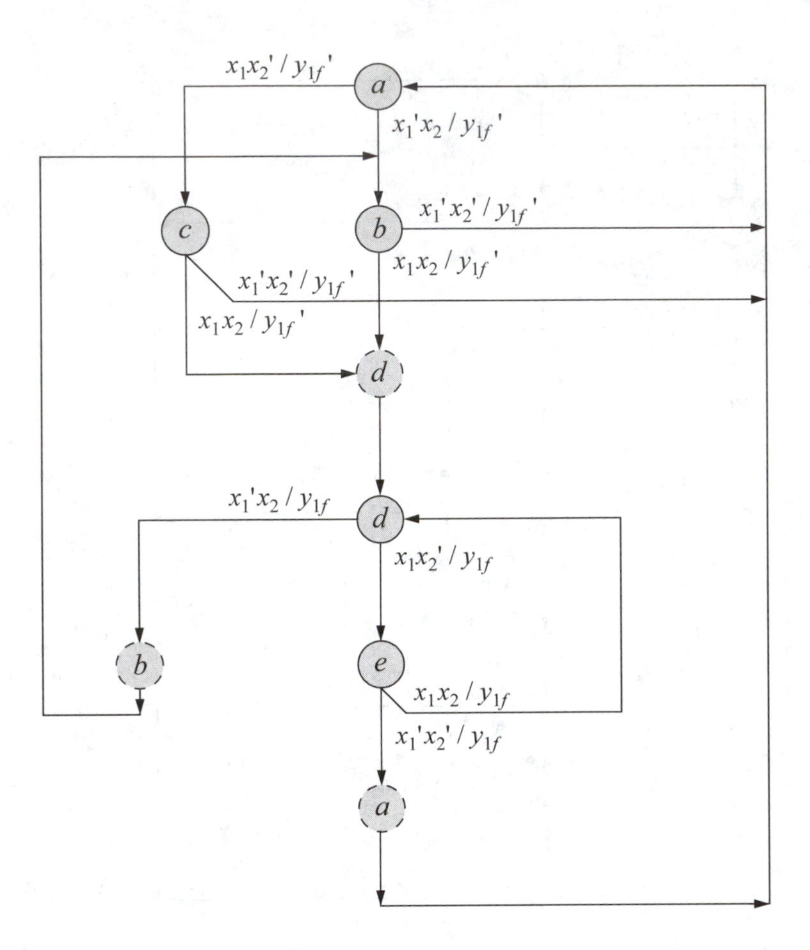

Figure 5.12 State diagram of Figure 5.10 redrawn to correlate with the state name assignment in the flow table of Figure 5.11.

Assume that the machine is reset to state ⓐ initially; that is, $y_{1f}x_1x_2 = 000$. If x_2 changes from 0 to 1, then the machine proceeds to a new stable state, ⓑ. The feedback variable will not change, because the outputs of both AND gates in Figure 5.7 are at a logic 0 potential. If input x_1 now changes from 0 to 1 ($y_{1f}x_1x_2 = 011$), then the outputs of AND gate 1 and the OR gate will be a logic 1 causing the excitation variable Y_{1e} to change from 0 to 1. The machine is now in an unstable condition, since $y_{1f} = 0$ and $Y_{1e} = 1$. After a delay of Δt, the feedback variable assumes a value of 1 and the machine enters state ⓓ. Provided that inputs x_1 and x_2 do not change from $x_1x_2 = 11$, the machine remains in state ⓓ with the latch set and $Y_{1e} = 1$.

While in stable state ⓓ, let x_2 change from 1 to 0. The transition proceeds horizontally to state ⓔ. The latch remains set with $Y_{1e} = 1$. If x_1 now changes from 1 to 0, then the excitation variable becomes 0 and the machine enters a transient state until the feedback value equals the excitation value. The state transition sequence for this input change is ⓔ $\rightarrow a \rightarrow$ ⓐ. In a similar manner, all other sequences can be observed by an appropriate change of inputs.

Example 5.2 Figure 5.13 shows another example of a fundamental-mode Moore asynchronous sequential machine with two input variables x_1 and x_2, one output $z_1 = Y_{1e}$, and one feedback path. The feedback connects to AND gates 2 and 3, both of which combine with the OR gate to form a latch. The machine can be completely described by listing all stable states, unstable states, and state sequences.

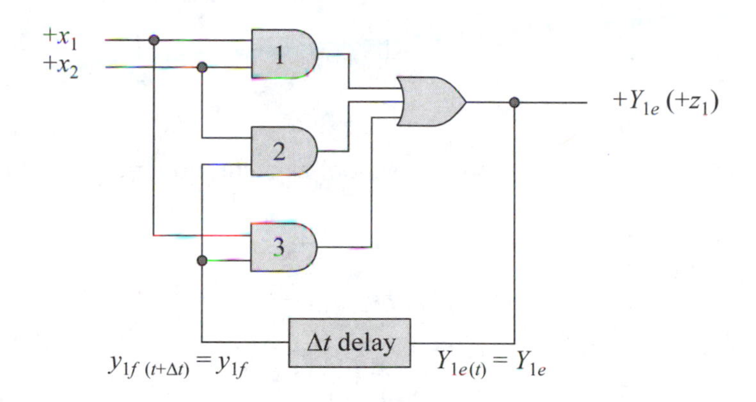

Figure 5.13 Fundamental-mode model of a Moore asynchronous sequential machine with two input variables and one feedback path.

Because Y_{1e} represents a combinational output for a particular value of t, the equation for Y_{1e} can be expressed by the boolean function shown in Equation 5.5. The terms containing $y_{1f}(t + \Delta t)$ specify that for a particular value of t, the feedback variable y_{1f} will be equal to the excitation variable Y_{1e} after a time delay of Δt.

$$Y_{1e(t)} = x_1 x_2 + x_2 y_{1f}(t + \Delta t) + x_1 y_{1f}(t + \Delta t)$$

$$Y_{1e} = x_1 x_2 + x_2 y_{1f} + x_1 y_{1f} \tag{5.5}$$

The excitation map is derived from Equation 5.5 or from the logic diagram and is shown in Figure 5.14. The circled entries specify a stable state, where $y_{1f} = Y_{1e}$. The flow table of Figure 5.15 is derived from the excitation map by replacing the 0 and 1 entries with state names. The stable state names in the flow table are circled to represent the corresponding stable states in the excitation map. The feedback variables are not usually specified in a flow table; however, they are included here for convenience for the discussion which follows. Using the flow table, the operation of the machine can be followed for each input change. Thus, the behavior of the machine is completely specified by the flow table.

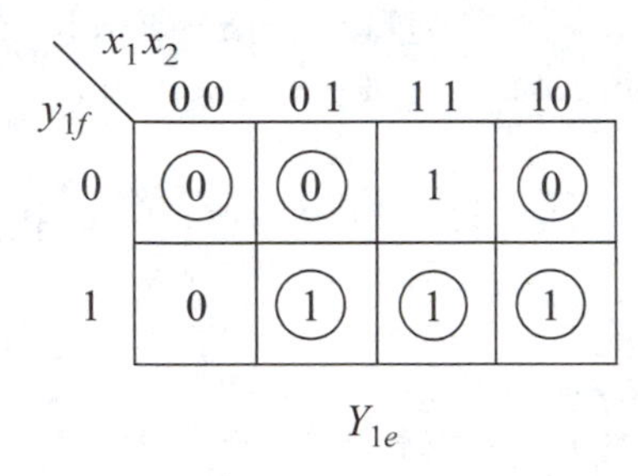

Figure 5.14 Excitation map for the Moore asynchronous sequential machine of Figure 5.13.

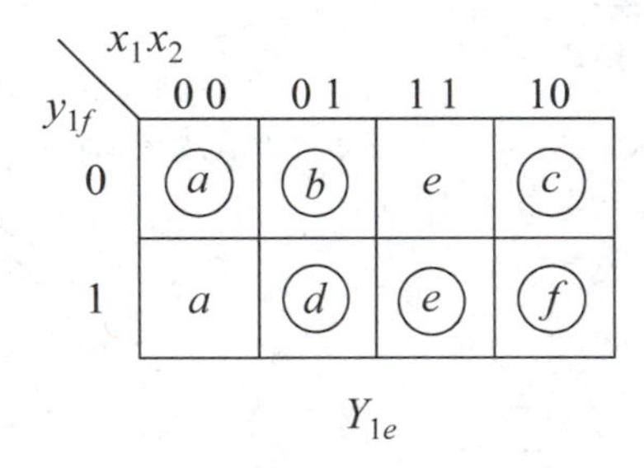

Figure 5.15 Flow table for the Moore machine of Figure 5.13.

By applying various input sequences to x_1 and x_2, the resulting sequence of excitation variables can be observed in the logic diagram and correlated in the flow table. For example, if the machine is presently in state ⓒ ($y_{1f}x_1x_2 = 010$), then $y_{1f} = Y_{1e} = 0$ and the latch is reset. If input x_2 changes from 0 to 1, then y_{1f} remains at a value of 0 — represented by unstable state e — until a time delay of Δt has occurred. During the time delay, the machine is changing from unstable state e to stable state ⓔ. After a delay of Δt has transpired, the machine is stable in state ⓔ with $y_{1f} = Y_{1e} = 1$. The latch is set through AND gate 1 and remains in the set state because of the feedback connection to AND gates 2 and 3.

In state ⓔ, a single change of an input variable will not alter the state of the machine. The machine will remain in the set state by means of AND gate 2 or 3, depending on which input changed value. Thus, the machine will proceed to either state ⓓ if x_1 changes value or to state ⓕ if x_2 changes value. The machine can be reset from states ⓓ or ⓕ through transient state a, but not from state ⓔ, since simultaneous input changes are not permitted. The machine can be reset by either of the following two sequences:

ⓓ → a → ⓐ
ⓕ → a → ⓐ

The state diagram of Figure 5.16 completes the analysis of this machine. The state diagram is most easily generated from the flow table. Stable states are encircled by solid lines; unstable states by dashed lines. The conditions for leaving a stable state are specified by the 3-tuple $x_1 x_2 / y_{1f}$, where x_1 and x_2 represent an input change and y_{1f} represents the feedback variable for the state in which the machine currently resides. For example, the machine will sequence from state ⓔ to state ⓓ if $x_1 x_2 = 01$ and $y_{1f} = 1$.

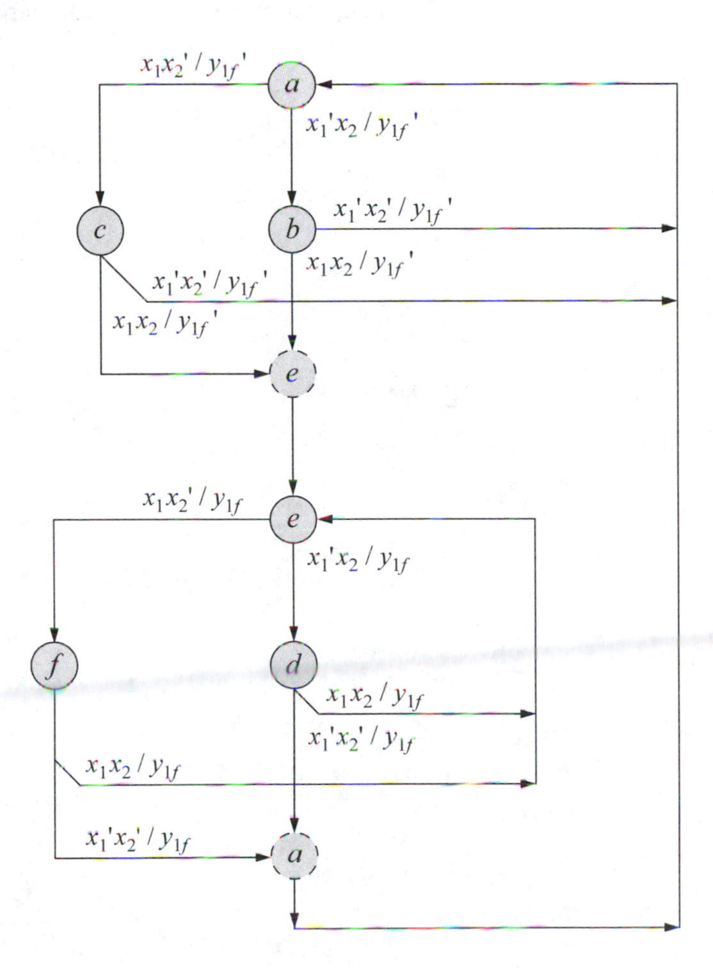

Figure 5.16 State diagram for the Moore asynchronous machine of a Figure 5.13.

Unlike synchronous sequential machines, edge-triggered, clocked flip-flops are not used as storage elements in asynchronous sequential machines; therefore, state codes are not necessary for establishing a particular state. That is, state flip-flops are not decoded to determine the present state of the machine. It is not the state code that determines the present state, but rather a state name in the flow table that corresponds to a unique set of input, excitation, and feedback variables. Thus, state names establish the present state even though the excitation and feedback variables remain unchanged for certain state transition sequences. There is, however, a commonality between synchronous and asynchronous machines — a specific sequence of states will produce a specific output.

Example 5.3 The logic diagram of Figure 5.17 is considerably more complex than either of the two previous examples. Figure 5.17 represents a Mealy asynchronous sequential machine where output z_1 is a function not only of the present state $y_{1f}y_{2f} = 11$, but also of the input variable x_2. The Δt time delay is not shown explicitly, but is implicit in the feedback paths of both excitation variables. The excitation map for the machine is derived through a two-step process: First, obtain the individual excitation maps for the variables Y_{1e} and Y_{2e}; then, combine the individual maps into a single excitation map that represents both excitation variables.

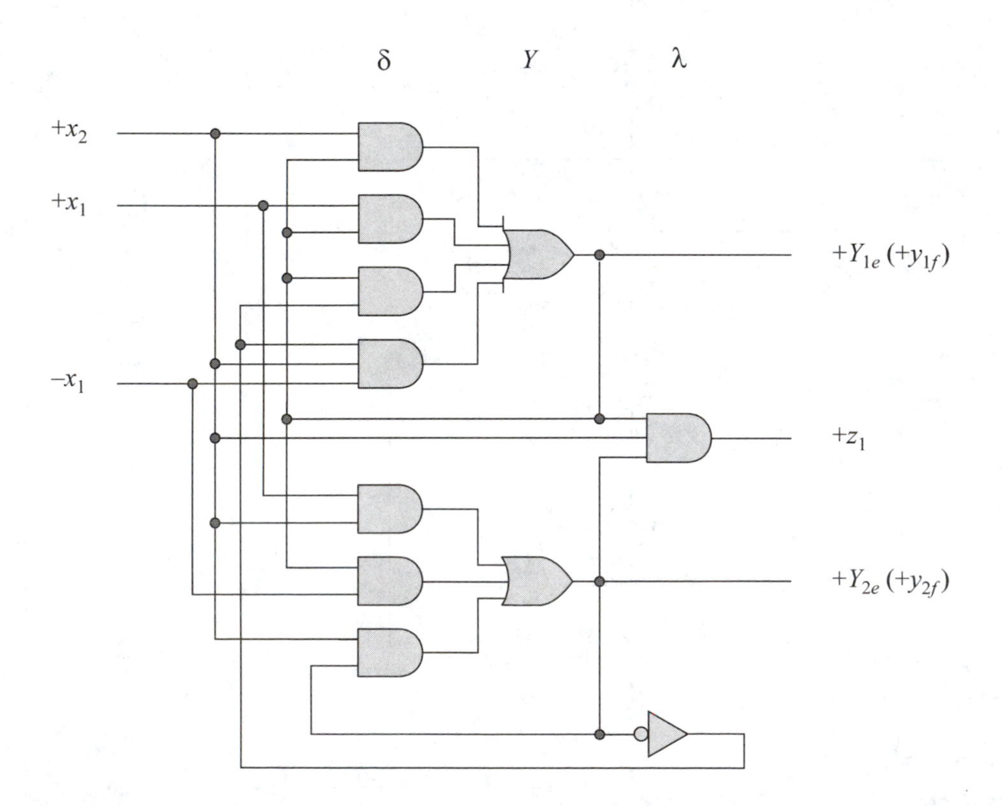

Figure 5.17 Logic diagram for an asynchronous Mealy machine with two feedback paths y_{1f} and y_{2f} and one output z_1.

The equations for the excitation variables are derived from the logic diagram and are shown in Equation 5.6. From Equation 5.6, the excitation maps for Y_{1e} and Y_{2e} can be constructed as depicted in Figure 5.18. The two maps are then combined into the single excitation map of Figure 5.19. The 2-tuple entries in the excitation map represent the values of the excitation variables Y_{1e} and Y_{2e}. Each minterm location is given a unique state name to facilitate the discussion which follows.

$$
\begin{aligned}
Y_{1e} &= y_{1f}x_2 + y_{1f}x_1 + y_{1f}y_{2f}' + y_{2f}'x_1'x_2 \\
Y_{2e} &= x_1x_2 + y_{1f}x_1' + y_{2f}x_2
\end{aligned}
\tag{5.6}
$$

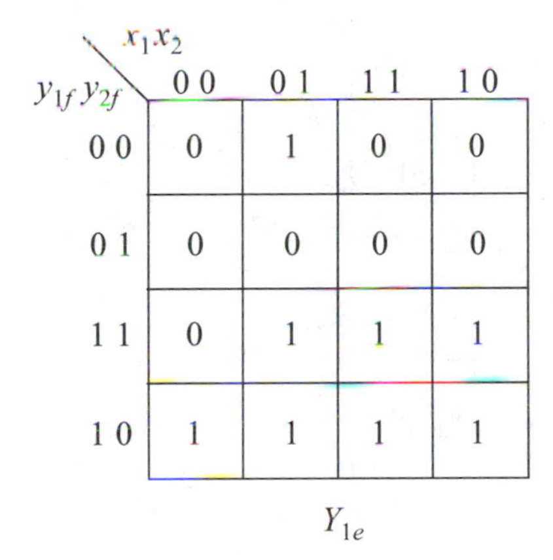

$y_{1f}y_{2f}$ \ x_1x_2	00	01	11	10
00	0	1	0	0
01	0	0	0	0
11	0	1	1	1
10	1	1	1	1

Y_{1e}

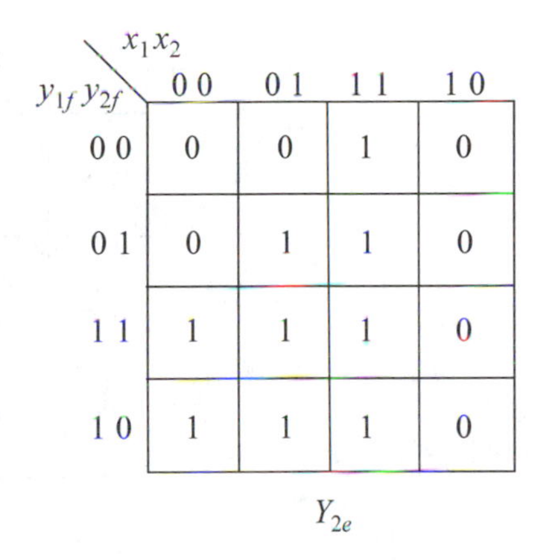

$y_{1f}y_{2f}$ \ x_1x_2	00	01	11	10
00	0	0	1	0
01	0	1	1	0
11	1	1	1	0
10	1	1	1	0

Y_{2e}

Figure 5.18 Individual excitation maps for the excitation variables Y_{1e} and Y_{2e} for the Mealy machine of Figure 5.17.

Different state transition sequences may result in dissimilar circuit delays due to different path lengths. Also, the delays through various circuit elements may change due to varying temperature, power supply output, and component aging. To distinguish between the various time delays resulting from an input change, each delay in a path is given a unique name, such as, Δt_1, Δt_2, ... , Δt_k for k total time delays.

In row $y_{1f}y_{2f} = 00$, there are two stable states: ⓐ ($y_{1f}y_{2f}x_1x_2 = 0000$) and ⓓ ($y_{1f}y_{2f}x_1x_2 = 0010$). If the machine is presently in state ⓐ and input x_2 changes from 0 to 1, then the operation proceeds to the right horizontally to unstable state b, where $Y_{1e}Y_{2e} = 10$. All remaining transitions resulting from this input change take place in column $x_1x_2 = 01$. Since the values of the excitation variables in state b are $Y_{1e}Y_{2e} = 10$, the feedback variables assume the same values after a delay of Δt_1 and the operation proceeds vertically down to row $y_{1f}y_{2f} = 10$. This new state n is also unstable, because $y_{1f}y_{2f} = 10$ and $Y_{1e}Y_{2e} = 11$. After a delay of Δt_2, the feedback variables become equal to the excitation variables and the machine sequences to row $y_{1f}y_{2f} = 11$ and enters stable state ⓙ where $y_{1f}y_{2f} = Y_{1e}Y_{2e} = 11$. The state transition sequence resulting from this input change is

ⓐ $\rightarrow b \rightarrow n \rightarrow$ ⓙ

The machine will remain in state ⓙ until an input again changes.

$y_{1f}y_{2f}$ \ x_1x_2	0 0	0 1	1 1	1 0
0 0	*a* (00)	*b* 10	*c* 01	*d* (00)
0 1	*e* 00	*f* (01)	*g* (01)	*h* 00
1 1	*i* 01	*j* (11)	*k* (11)	*l* 10
1 0	*m* 11	*n* 11	*o* 11	*p* (10)

$Y_{1e}Y_{2e}$

Figure 5.19 Excitation map for Y_{1e} and Y_{2e} for the Mealy asynchronous machine of Figure 5.17, obtained by combining the separate excitation maps of Figure 5.18.

In state ⓐ, if input x_1 changes from 0 to 1, then the operation proceeds horizontally to column $x_1x_2 = 10$ and enters state ⓓ. The machine does not pass through any transient states as a result of this input change. The state transition sequence is ⓐ $\rightarrow$ ⓓ. In state ⓓ, if x_2 changes from 0 to 1, then the machine moves to unstable state c where $Y_{1e}Y_{2e} = 01$. After a requisite delay of Δt, the feedback variables become equal to the excitation variables and the machine proceeds to the row where the feedback variables are $y_{1f}y_{2f} = 01$, while remaining in column $x_1x_2 = 11$. This new state is labeled ⓖ and is stable, because $y_{1f}y_{2f} = Y_{1e}Y_{2e} = 01$. The sequence for this operation is

ⓓ $\rightarrow c \rightarrow$ ⓖ

As a final sequence, assume that the machine is in stable state ⓟ and that input x_1 changes from 1 to 0. The operation proceeds to the left horizontally in row $y_{1f}y_{2f} = 10$ to unstable state m. In state m, the excitation values are $Y_{1e}Y_{2e} = 11$; therefore, after a delay of Δt_1, the feedback values become $y_{1f}y_{2f} = 11$ and the machine moves vertically up in column $x_1x_2 = 00$ to row $y_{1f}y_{2f} = 11$, which is unstable state i. In state i, the excitation values are $Y_{1e}Y_{2e} = 01$; therefore, after a delay of Δt_2 through the appropriate logic gates, the feedback values become $y_{1f}y_{2f} = 01$ and the machine enters unstable state e.

In state e, $Y_{1e}Y_{2e} = 00$. After a final delay of Δt_3, the feedback variables become equal to the excitation variables and the machine enters and remains in stable state ⓐ, where $y_{1f}y_{2f} = Y_{1e}Y_{2e} = 00$. This is the longest sequence that the machine will experience, because the feedback variables proceed through all 2^2 possible combinations of $y_{1f}y_{2f}$. The state transition sequence for this input change is

$$ⓟ \rightarrow m \rightarrow i \rightarrow e \rightarrow ⓐ$$

By applying the same technique, all remaining state transition sequences can be observed.

Chapter 6 will present a method of minimizing the number of logic gates for a machine of this type by inserting appropriate values for the excitation variables in the unstable states. For example, since states ⓕ, ⓙ, and ⓟ all sequence to state ⓐ, the excitation values in unstable states i and m could be replaced by $Y_{1e}Y_{2e} = 00$. This change would not only reduce the number of logic gates, but would also increase the operational speed of the machine. Assigning a value of $Y_{1e}Y_{2e} = 00$ to unstable state i will generate a race condition for the transition ⓙ $\rightarrow i \rightarrow$ ⓐ. The race, however, is not critical. Race conditions are discussed in Section 5.5.

Figure 5.19 can be converted to a flow table which permits the state transitions to be analyzed without regard for the feedback variables. The flow table is shown in Figure 5.20. The excitation variables are replaced with state names. The circled entries represent stable states and are assigned unique names. The uncircled entries represent unstable, or transient, states and indicate the destination stable state.

When a single change produces two or more successive excitation variable changes before entering a destination stable state, this is considered to be a ripple state transition sequence, because the effect of the input change ripples through the machine producing a sequence of transient state changes. The ripple effect is apparent in both the excitation map of Figure 5.19 and the flow table of Figure 5.20. In Figure 5.19, the ripple effect is caused by a transition from stable state ⓟ ($y_{1f}y_{2f}x_1x_2 = 1010$) to stable state ⓐ ($y_{1f}y_{2f}x_1x_2 = 0000$). The machine proceeds through a sequence of unstable excitation variables, as follows:

$$ⓟ \rightarrow m \rightarrow i \rightarrow e \rightarrow ⓐ$$

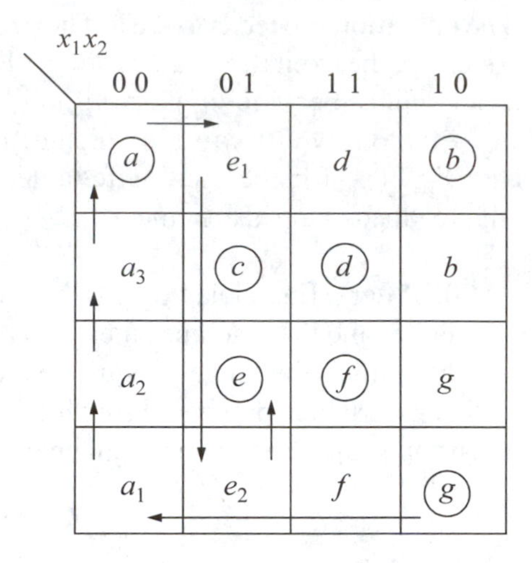

Figure 5.20 Flow table for the Mealy asynchronous machine of Figure 5.17.

The same sequence can be seen in the flow table of Figure 5.20, where the machine begins in stable state ⓖ and sequences to stable state ⓐ through intermediate states, as follows:

$$ⓖ \rightarrow a_1 \rightarrow a_2 \rightarrow a_3 \rightarrow ⓐ$$

A shorter ripple sequence occurs from state ⓐ to state ⓔ such that,

$$ⓐ \rightarrow e_1 \rightarrow e_2 \; ⓔ$$

All other sequences in the flow table are a direct transition through only one transient state.

The output map for z_1 is established from the logic diagram of Figure 5.17. The equation for z_1 is shown in Equation 5.7. Output z_1 is asserted in states ⓔ and ⓕ only, where $y_{1f}y_{2f}x_2 = 111$, as shown in Figure 5.21. It is sometimes possible to minimize the equation for an output by assigning values of 1 for particular transient states. In this example, however, no minimization is possible.

$$z_1 = y_{1f}y_{2f}x_2 \tag{5.7}$$

If a value of 1 is inserted in unstable state $y_{1f}y_{2f}x_1x_2 = 1110$, which is the transient state for a sequence from ⓕ to ⓖ, and a value of 1 inserted in unstable state $y_{1f}y_{2f}x_1x_2 = 1100$, which is a transient state for the sequence from ⓔ to ⓐ, then the equation for z_1 could be reduced to $z_1 = y_{1f}y_{2f}$. However, state $y_{1f}y_{2f}x_1x_2 = 1100$ is a common transient state for sequences ⓔ → ⓐ and ⓖ → ⓐ.

$y_{1f}y_{2f}$ \ x_1x_2	0 0	0 1	1 1	1 0
0 0	0 *a*	0	0	0 *b*
0 1	0	0 *c*	0 *d*	0
1 1	0	1 *e*	1 *f*	0
1 0	0	0	0	0 *g*

$Y_{1e}\,Y_{2e}$

Figure 5.21 Output map for the Mealy asynchronous machine of Figure 5.17.

Placing a value of 1 in the transient state for the sequence ⓕ → ⓖ poses no problem — z_1 will remain active for a slightly longer duration, but will not glitch. For the sequence ⓖ → ⓐ, however, z_1 is inactive in both the beginning state ⓖ and the ending state ⓐ. Therefore, z_1 should not be asserted in any transient state through which the machine passes in sequencing from stable state ⓖ to stable state ⓐ. Thus, a value of 0 for z_1 must be placed in unstable state $y_{1f}y_{2f}x_1x_2 = 1100$. Since no minimization is gained by adding a third 1 to the map, a value of 0 is also inserted in the unstable state $y_{1f}y_{2f}x_1x_2 = 1110$.

The state diagram for the asynchronous Mealy machine of Figure 5.17 is shown in Figure 5.22. The state diagram contains the same information as the flow table of Figure 5.20, from which it was derived, but in a graphical format. The state transitions are easily determined for all possible input sequences in which only one input variable changes value. The values of the input variables shown on the lines exiting the stable states indicate the input change that will initiate a corresponding state change. For example, if the machine is presently in state ⓐ ($x_1x_2 = 00$) and input x_2 changes from 0 to 1, then the machine proceeds through unstable states e_1 and e_2 to stable state ⓔ.

The state diagram, like its counterpart the flow table, completely defines the operation of the machine, but without regard for the values of the excitation or feedback variables. A more detailed analysis of the machine's operation requires the use of the excitation map. The excitation map provides the same information as the flow table and state diagram, and also specifies the values of the excitation and feedback variables for both stable and unstable states. The excitation map provides a more minute analysis of the machine, including the Δt delay parameter.

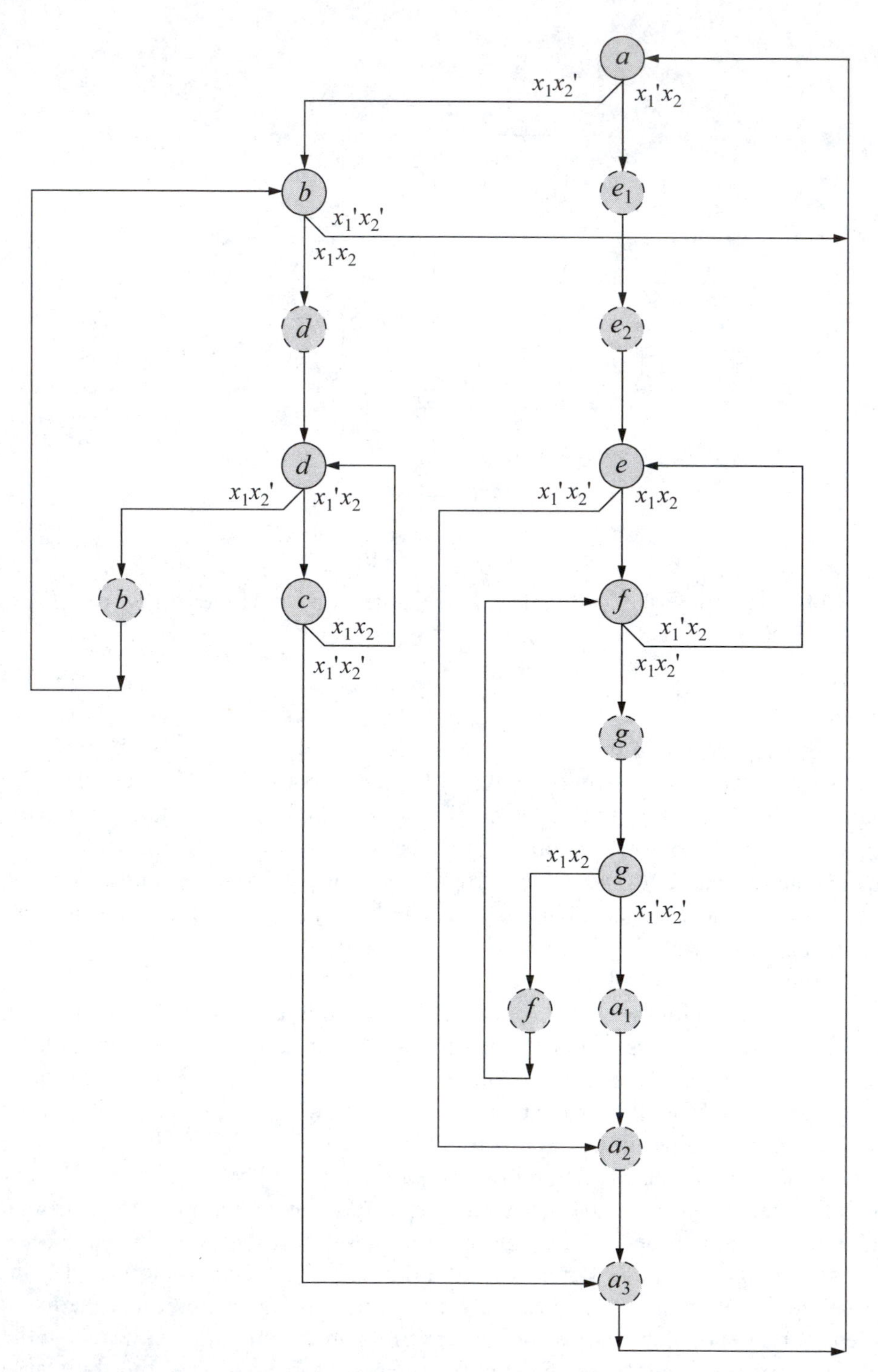

Figure 5.22 State diagram for the asynchronous Mealy machine of Figure 5.17.

A timing diagram completes the analysis of this Mealy machine. For convenience, the excitation map is reproduced below as Figure 5.23 so that it will be in close proximity to the timing diagram for ease of correlation. Assume that the machine is presently in state ⓐ ($y_{1f}y_{2f}x_1x_2 = 0000$). The first sequence to be examined is ⓐ $\rightarrow b \rightarrow n \rightarrow$ ⓙ, which results from input x_2 changing from 0 to 1. The arrows in the excitation map indicate the state transition sequences, which can also be followed in the waveforms of Figure 5.24.

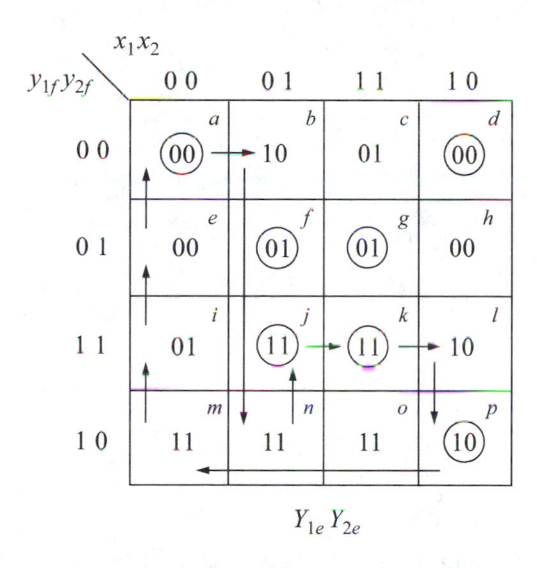

Figure 5.23 Excitation map for Y_{1e} and Y_{2e} for the Mealy asynchronous machine of Figure 5.17.

From state ⓐ, the machine proceeds to unstable state b, where $Y_{1e}Y_{2e} = 10$. After a delay of Δt_1, the feedback variables become equal to $y_{1f}y_{2f} = 10$. The time delay Δt_1 is shown in the timing diagram as the duration immediately preceding the assertion of y_{1f}. In state n, $Y_{1e}Y_{2e} = 11$. Thus, after a delay of Δt_2, the machine proceeds to stable state ⓙ. The time delay Δt_2 is the duration immediately preceding the assertion of y_{2f} in the timing diagram for this state transition sequence. The machine remains in state ⓙ until an input again changes.

The sequence ⓙ $\rightarrow$ ⓚ is comparatively less abstruse, since a change from 0 to 1 for input x_1 results in no change of excitation variables and, thus, no change of feedback variables. The machine remains in row $y_{1f}y_{2f} = 11$ in stable state ⓚ.

The sequence ⓚ $\rightarrow l \rightarrow$ ⓟ occurs when x_2 changes from 1 to 0 when the machine is in stable state ⓚ. The operation proceeds to state l where $Y_{1e}Y_{2e} = 10$. After a delay of Δt, the feedback variables change to $y_{1f}y_{2f} = 10$ and the machine moves to the corresponding row and enters stable state ⓟ. The time delay Δt is indicated in the timing diagram as the duration following the deassertion of x_2 and immediately preceding the deassertion of y_{2f}.

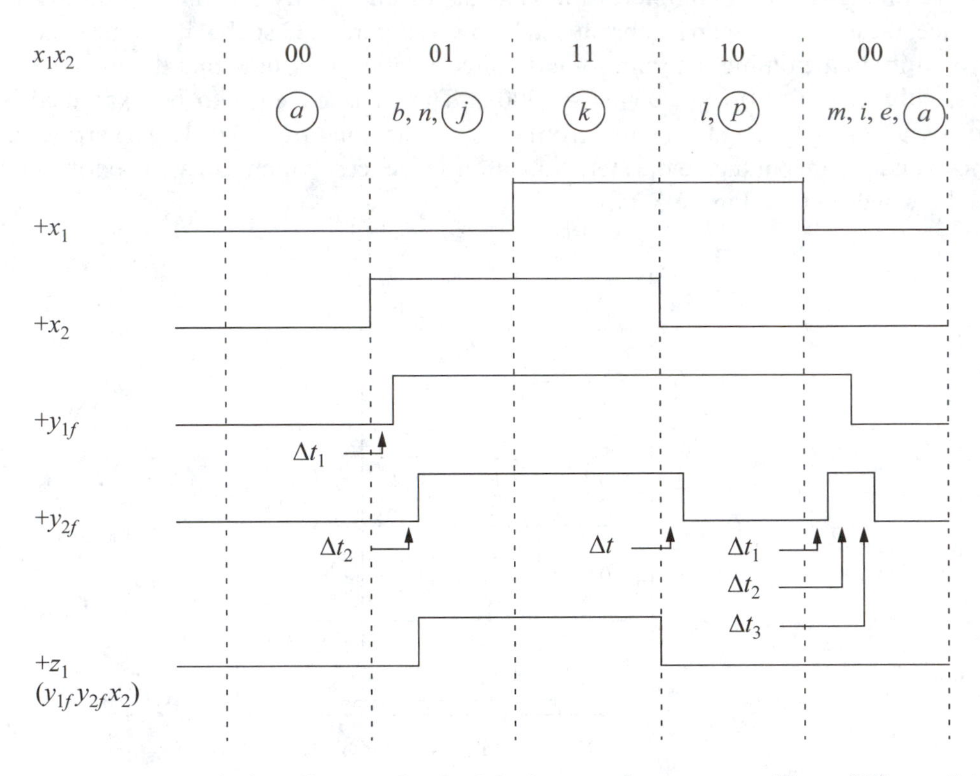

Figure 5.24 Timing diagram for the Mealy asynchronous machine of Figure 5.17 for the sequences Ⓐ $\rightarrow b \rightarrow n \rightarrow$ Ⓙ, Ⓙ $\rightarrow$ Ⓚ, Ⓚ $\rightarrow l \rightarrow$ Ⓟ, and Ⓟ $\rightarrow m \rightarrow i \rightarrow e \rightarrow$ Ⓐ.

The final sequence to be analyzed is Ⓟ $\rightarrow m \rightarrow i \rightarrow e \rightarrow$ Ⓐ. In state Ⓟ, if x_1 changes from 1 to 0, then a transition occurs horizontally to unstable state *m*, where $Y_{1e}Y_{2e} = 11$. After a delay of Δt_1, the feedback values become $y_{1f}y_{2f} = 11$ and the machine proceeds to unstable state *i*. The delay Δt_1 is indicated in the timing diagram as the time interval between the deassertion of x_1 and the assertion of y_{2f}. The assertion of y_{2f} corresponds to the value of 1 in the expression $y_{1f}y_{2f} = 11$. In state *i*, the excitation variables are $Y_{1e}Y_{2e} = 01$. Thus, after a delay of Δt_2, the feedback variables become $y_{1f}y_{2f} = 01$ and the machine moves to row $y_{1f}y_{2f} = 01$ and enters unstable state *e*. The delay Δt_2 is shown in the timing diagram as the time interval between the assertion of y_{2f} and the deassertion of y_{1f}.

The final transition occurs from unstable state *e* to stable state Ⓐ. In state *e*, the values of the excitation variables are $Y_{1e}Y_{2e} = 00$ which become the values of the feedback variables after a delay of Δt_3. At the completion of delay Δt_3, the machine resides in state Ⓐ, where $y_{1f}y_{2f} = Y_{1e}Y_{2e} = 00$. The delay Δt_3 is shown in the timing diagram as the time interval between the deassertion of feedback variable y_{1f} and the deassertion of feedback variable y_{2f}. After a delay of Δt_3, both feedback variables are zero, as shown in the timing diagram, and the machine is again in stable state ⓐ, as defined by the six-tuple $y_{1f}y_{2f}Y_{1e}Y_{2e}x_1x_2 = 000000$.

The assertion of feedback variable y_{2f} for two time delays in the timing diagram corresponds to the two unstable states, m and i, in which $Y_{2e} = 1$. That is, $Y_{1e}Y_{2e} = 11$ and 01 in states m and i, respectively. Therefore, after each of the appropriate delays has transpired, the feedback variable y_{2f} will have been active for two delay periods such that, $y_{1f}y_{2f} = 11$ and 01 for states i and e, respectively.

Example 5.4 As a final analysis example in this section, consider the equations of Equation 5.8, which represent the excitation variables and output for a Mealy asynchronous machine. There are three input variables and two feedback variables, necessitating a 5-variable map. The individual excitation maps for Y_{1e} and Y_{2e} are shown in Figure 5.25 and are obtained by plotting the excitation equations in their respective maps. The individual maps are then combined into a single excitation map for $Y_{1e}Y_{2e}$, as shown in Figure 5.26.

$$\begin{aligned} Y_{1e} &= x_1x_2 + y_{1f}x_2 + y_{1f}x_3 + y_{1f}y_{2f} \\ Y_{2e} &= x_1'x_2x_3 + y_{1f}x_1 + y_{2f}x_2 \\ z_1 &= y_{1f}y_{2f}x_2x_3' + y_{1f}y_{2f}x_1x_3 \end{aligned} \tag{5.8}$$

Each minterm location specifies the values of Y_{1e} and Y_{2e}, thus, the criterion for stability is easily determined. Any minterm location in which the values of the excitation variables are equal to the values of the feedback variables represents a stable state. In the top row of Figure 5.26, the following 5-tuples indicate stable states: $y_{1f}y_{2f}x_1x_2x_3 = 00000$, 00010, 00100, 00001, and 00101, as indicated by the circled entries for Y_{1e} and Y_{2e}. The remaining stable states are located in a similar manner.

A five-variable excitation map offers more possible state transition sequences than in any previous example. It is now feasible to begin a sequence in one map and end the sequence in the other map. For example, if the machine is presently in state $y_{1f}y_{2f}x_1x_2x_3 = 00010$ and input x_3 changes from 0 to 1, then the machine sequences through transient state $y_{1f}y_{2f}x_1x_2x_3 = 00011$ and enters stable state $y_{1f}y_{2f}x_1x_2x_3 = 01011$.

The operation of the machine is easier to follow if the excitation map is converted to a flow table, as shown in Figure 5.27. Each stable state is assigned a unique name which is circled. The unstable states are not circled and indicate either the destination stable state directly or the next transient state in a sequence of transient states that culminates in a stable state.

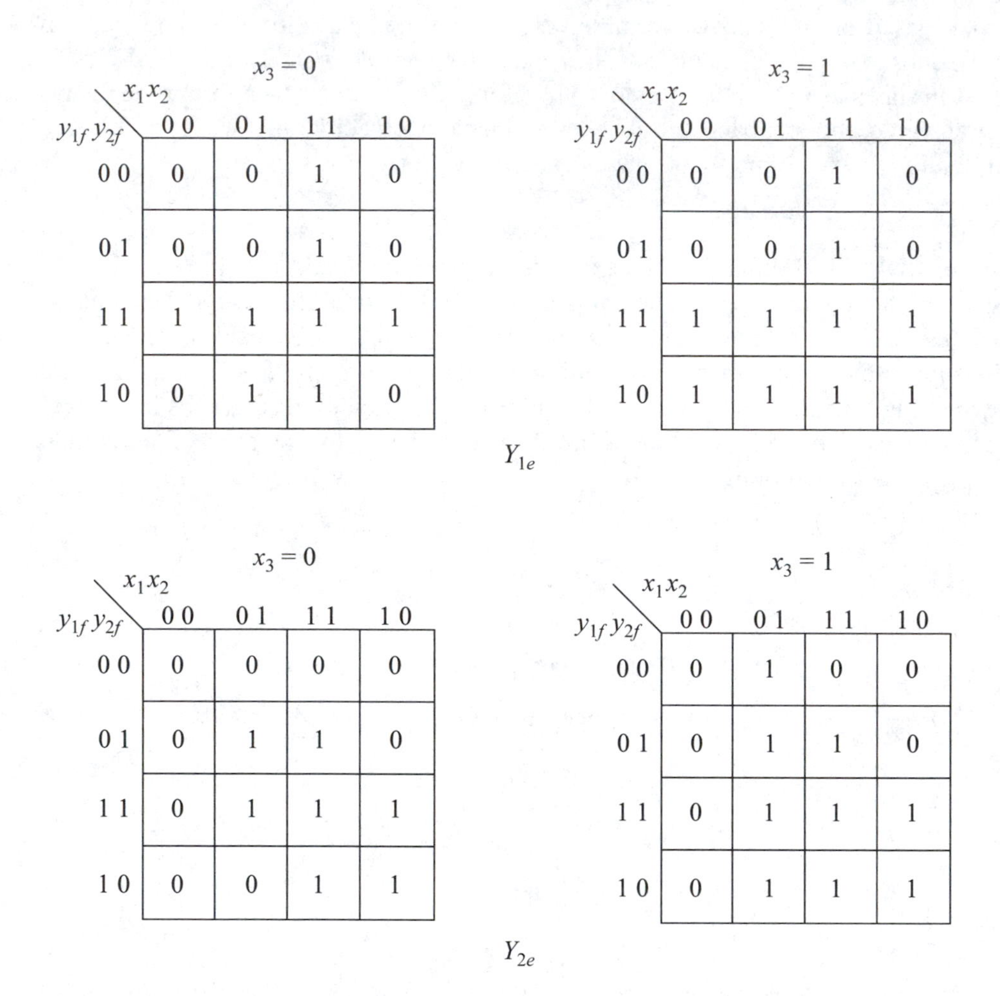

Y_{1e}, $x_3 = 0$

$y_{1f}y_{2f}$ \ x_1x_2	0 0	0 1	1 1	1 0
0 0	0	0	1	0
0 1	0	0	1	0
1 1	1	1	1	1
1 0	0	1	1	0

Y_{1e}, $x_3 = 1$

$y_{1f}y_{2f}$ \ x_1x_2	0 0	0 1	1 1	1 0
0 0	0	0	1	0
0 1	0	0	1	0
1 1	1	1	1	1
1 0	1	1	1	1

Y_{2e}, $x_3 = 0$

$y_{1f}y_{2f}$ \ x_1x_2	0 0	0 1	1 1	1 0
0 0	0	0	0	0
0 1	0	1	1	0
1 1	0	1	1	1
1 0	0	0	1	1

Y_{2e}, $x_3 = 1$

$y_{1f}y_{2f}$ \ x_1x_2	0 0	0 1	1 1	1 0
0 0	0	1	0	0
0 1	0	1	1	0
1 1	0	1	1	1
1 0	0	1	1	1

Figure 5.25 Excitation maps for Y_{1e} and Y_{2e} for Example 5.4.

For example, if the machine is currently in stable state ⓓ and input x_1 changes from 0 to 1, then the machine proceeds directly to state ⓕ by the following sequence: ⓓ $\rightarrow f \rightarrow$ ⓕ. Likewise, if the machine is in stable state ⓓ and input x_2 changes from 1 to 0, then the machine proceeds directly to state ⓐ by the following sequence: ⓓ $\rightarrow a \rightarrow$ ⓐ.

A sequence of transient states occurs if the machine begins in stable state ⓔ and input x_2 changes from 1 to 0. The machine progresses to state ⓐ by the following sequence: ⓔ $\rightarrow a_1 \rightarrow a_2 \rightarrow$ ⓐ. The flow table can also be used to indicate in which states the output is asserted, as shown in stable states ⓔ, ⓕ, ⓜ, and ⓝ.

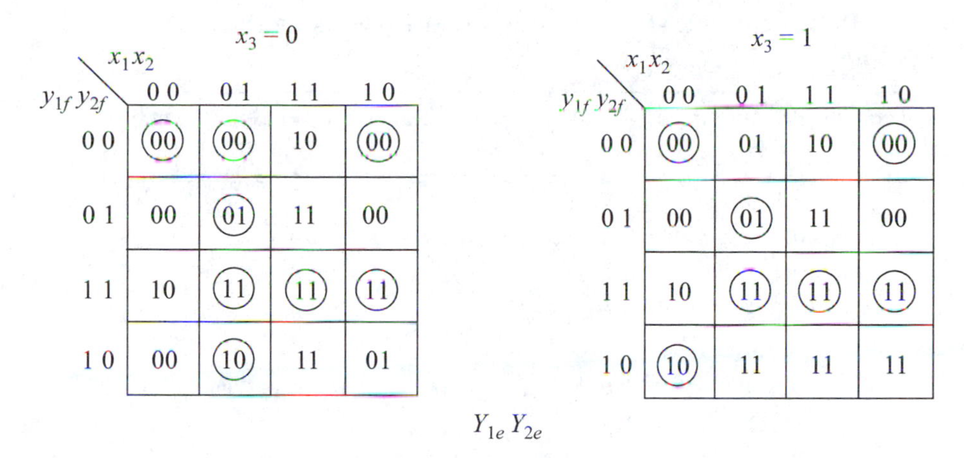

$x_3 = 0$

$y_{1f}y_{2f}$ \ x_1x_2	0 0	0 1	1 1	1 0
0 0	00	00	10	00
0 1	00	01	11	00
1 1	10	11	11	11
1 0	00	10	11	01

$x_3 = 1$

$y_{1f}y_{2f}$ \ x_1x_2	0 0	0 1	1 1	1 0
0 0	00	01	10	00
0 1	00	01	11	00
1 1	10	11	11	11
1 0	10	11	11	11

$Y_{1e}Y_{2e}$

Figure 5.26 Excitation map for Example 5.4 for $Y_{1e}Y_{2e}$ obtained by combining the individual excitation maps of Figure 5.25.

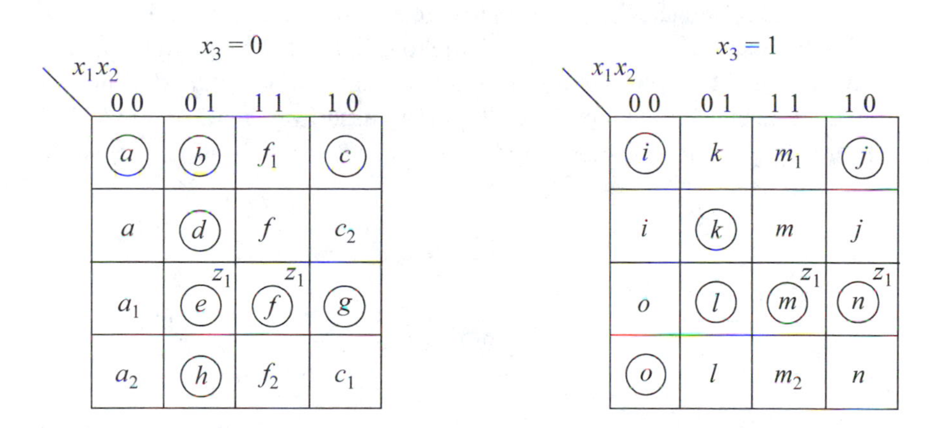

$x_3 = 0$

x_1x_2	0 0	0 1	1 1	1 0
	a	b	f_1	c
	a	d	f	c_2
	a_1	e z_1	f z_1	g
	a_2	h	f_2	c_1

$x_3 = 1$

x_1x_2	0 0	0 1	1 1	1 0
	i	k	m_1	j
	i	k	m	j
	o	l	m z_1	n z_1
	o	l	m_2	n

Figure 5.27 Flow table for the Mealy asynchronous machine of Figure 5.26.

Using the flow table, a representative sequence of state transitions is tabulated in Table 5.2. Table 5.2 lists some typical present stable states and the present inputs associated with the stable state. A single change of inputs causes the corresponding state transition sequence to be executed. Table 5.2 also lists the present output as a function of the present inputs and the present state.

Table 5.2 Representative state transition sequences using the flow table of Figure 5.27

Present stable state	Present inputs $x_1x_2x_3$	Next inputs $x_1x_2x_3$	State transition sequence	Present output z_1
ⓐ	0 0 0	0 0 1	ⓐ, ⓘ	0
ⓘ	0 0 1	0 1 1	ⓘ, k, ⓚ	0
ⓚ	0 1 1	0 0 1	ⓚ, i, ⓘ	0
ⓘ	0 0 1	0 0 0	ⓘ, ⓐ	0
ⓐ	0 0 0	1 0 0	ⓐ, ⓒ	0
ⓒ	1 0 0	1 0 1	ⓒ, ⓙ	0
ⓙ	1 0 1	1 1 1	ⓙ, m_1, m_2, ⓜ	0
ⓜ	1 1 1	1 0 1	ⓜ, ⓝ	1
ⓝ	1 0 1	0 0 1	ⓝ, o, ⓞ	1
ⓞ	0 0 1	0 0 0	ⓞ, a_2, ⓐ	0

The output map for z_1 is shown in Figure 5.28 and is plotted directly from Equation 5.8. For reference, the stable states are also indicated on the output map. As was stated previously, it is sometimes possible to minimize the equation for an output by inserting a value of 1 in certain minterm locations of an output map. Care must be taken, however, to ensure that there is no transitory assertion on the output. This technique is more appropriately discussed in the synthesis of asynchronous sequential machines, and will be deferred until Chapter 6.

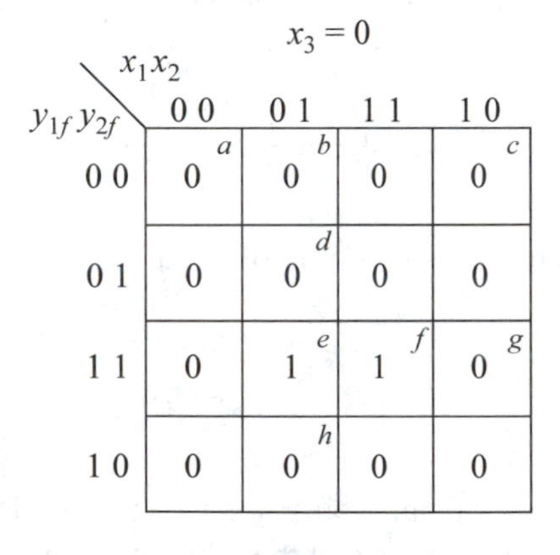

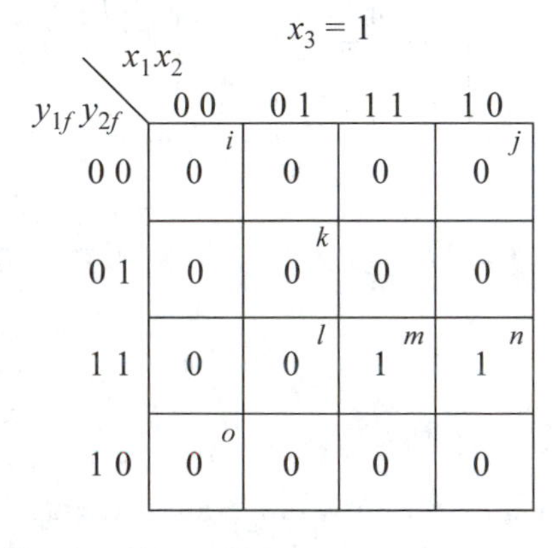

z_1

Figure 5.28 Output map for z_1 for the Mealy asynchronous machine of Figure 5.26.

The state diagram for Example 5.4 will be omitted from this analysis due to space limitations, since it requires 32 state symbols. The state diagram is derived from the flow table and contains the same information as the flow table, but in a graphical representation. The generation of a representative timing diagram is assigned as an exercise in the problem set of this chapter.

5.4 Hazards

The signal propagation delay must be considered in the analysis and synthesis of asynchronous sequential machines. When an input variable changes value, varying propagation delays caused by logic gates, wires, and different path lengths can produce erroneous transient signals on the outputs. These spurious signals are referred to as *hazards*. If the hazard occurs in the feedback path, then an incorrect state transition sequence may result.

Hazards are not usually a concern in synchronous sequential machines, because the clock negates the effects of a hazard. A hazard may appear at the input of a flip-flop, but is deasserted before the machine changes state at the next active clock transition. In asynchronous sequential machines, however, where the next state is not synchronized with a clock signal, a hazard can cause an incorrect next state or an erroneous output, and must be eliminated.

When a hazard occurs in the δ next-state logic, the machine may sequence to an invalid next state. If the hazard occurs in the λ output logic, then a glitch may appear on the output signal. An output glitch can cause significant problems, especially if the output is associated with input variables for an asynchronous sequential machine or with the clock input for a synchronous sequential machine.

These transitory signals generate a condition which is not specified in the expression for the machine, because boolean algebra does not take into account the propagation delay of switching circuits. In this section, hazards will be examined and methods presented for detecting and correcting these transient phenomena so that correct operation of an asynchronous sequential machine can be assured.

5.4.1 Static Hazards

Figure 5.29 illustrates a classic example of a combinational circuit with an inherent hazard. Although the network is combinational in structure, it may be an integral part of the δ next-state logic for an asynchronous sequential machine. The Karnaugh map which represents the circuit is shown in Figure 5.30 and the equation for output z_1 is shown in Equation 5.9. The map entries indicate that z_1 is asserted in minterm locations 1, 5, 6, and 7; thus, the function $z_1(x_1, x_2, x_3) = \Sigma m(1, 5, 6, 7)$. In particular, z_1 is active if $x_1 x_2 x_3 = 111$ or 101; that is, output z_1 is set to a value of 1 regardless of the value of input x_2.

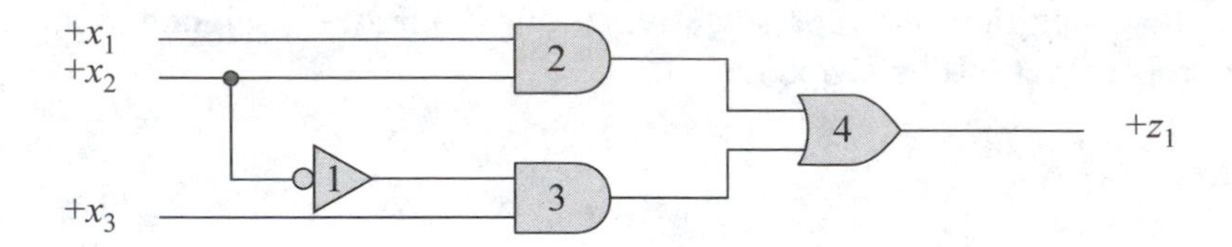

Figure 5.29 Combinational circuit which contains a potential static hazard.

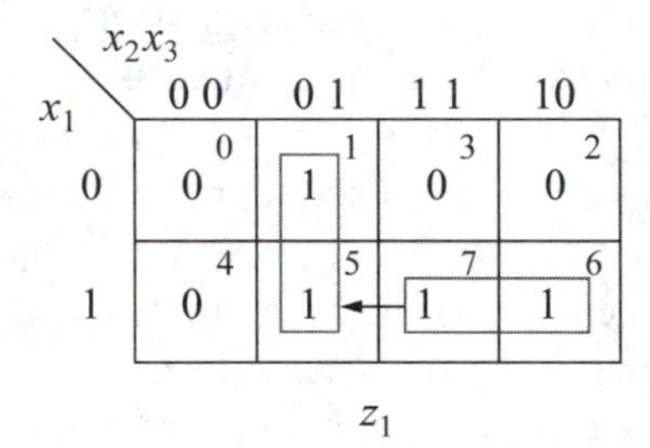

Figure 5.30 Karnaugh map corresponding to the circuit of Figure 5.29.

$$z_1 = x_1x_2 + x_2'x_3 \qquad (5.9)$$

Thus, if x_2 changes from 0 to 1 or from 1 to 0, the state of z_1 should remain constant (or static) during the transition. Any deviation from this static condition is referred to as a static hazard. If an input variable changes value causing an output to be deasserted momentarily when the output should remain asserted, then this is classified as a *static-1 hazard.* Conversely, when an input change causes an output to become asserted momentarily when it should remain deasserted, a *static-0 hazard* results. Figure 5.31 illustrates the two types of static hazards. In general, a static hazard is a singularity in which a single change to an input vector causes a momentary output transition to an incorrect state.

If two adjacent minterm locations both contain a value of 1 and are covered by the same prime implicant, then a single change within that grouping of 1s will not produce a hazard. For example, changing the input vector from $x_1x_2x_3 = 111$ to 110 will not cause a static hazard, because the corresponding 1s are covered by the term x_1x_2.

Thus, when x_3 changes from 1 to 0, the x_1x_2 term maintains the output at a logic 1. Therefore, all groups of 1s in a Karnaugh map must be connected by redundant prime implicants (or consensus terms) to avoid possible hazards.

Figure 5.31 Static hazards: (a) static-1 hazard; and (b) static-0 hazard.

Referring to Figure 5.29, when input x_2 changes from 1 to 0, the change is transmitted to the output along two paths: logic blocks 2 and 4; and logic blocks 1, 3, and 4. Let the propagation delay of blocks 1, 2, 3, and 4 be designated by the symbols Δt_1, Δt_2, Δt_3, and Δt_4, respectively. If the delay of block 2 is less than the combined delay of blocks 1 and 3 ($\Delta t_2 < \Delta t_1 + \Delta t_3$), then all inputs to block 4 will be at a low level for a duration of $[(\Delta t_1 + \Delta t_3) - \Delta t_2]$. If the block 4 inputs remain at a low level for a sufficient duration, such that, $\Delta t_4 < [(\Delta t_1 + \Delta t_3) - \Delta t_2]$, then output z_1 will generate a static-1 hazard.

Assume that the circuit of Figure 5.29 is stable in state $x_1x_2x_3 = 111$ and input x_2 changes from 1 to 0, as indicated by the arrow in Figure 5.30. In the analysis of asynchronous sequential machines, a variable and its complement must be treated as independent variables. The timing diagram of Figure 5.32 depicts the sequence of events that occur when x_2 changes from 1 to 0. The waveforms labeled INV1, AND 2, and AND 3 refer to the outputs of the inverter, AND gate 2, and AND gate 3, respectively.

The deassertion of x_2 is immediate. The new value of x_2 propagates to the output along two paths. After a coincident delay of Δt_1 and Δt_2 through the inverter and AND gate 2, the inverter output changes from a low to a high voltage level while the output of AND gate 2 changes to a low level. The output of AND gate 3 and the state of z_1 are unaffected at this time and remain at a low and high level, respectively.

AND gate 3 and the OR gate are in the same time relationship, thus, propagation delays Δt_3 and Δt_4 occur concurrently through both gates, causing the outputs of the gates to change to a high and low voltage level, respectively. Since the output of the circuit is obtained directly from the OR gate, z_1 changes from a high to a low level. After a final propagation delay of Δt_4 through the OR gate, the output of the OR gate reflects the previous values of the inputs that were in effect before the final propagation delay. Thus, output z_1 returns to its stable level of an active-high assertion, where $x_1x_2x_3 = 101$.

The glitch on output z_1, caused by input x_2 changing from a high to a low level, is a static hazard. The hazard results from a signal propagation delay through two different path lengths — one path through AND gate 2 and the OR gate; and a longer path through the inverter, AND gate 3, and the OR gate. There is no hazard when x_2 changes from a low to a high level, because the shorter path maintains z_1 at a high assertion. Static hazards are characterized by a single transient pulse. The pulse width represents the propagation delay through one logic gate.

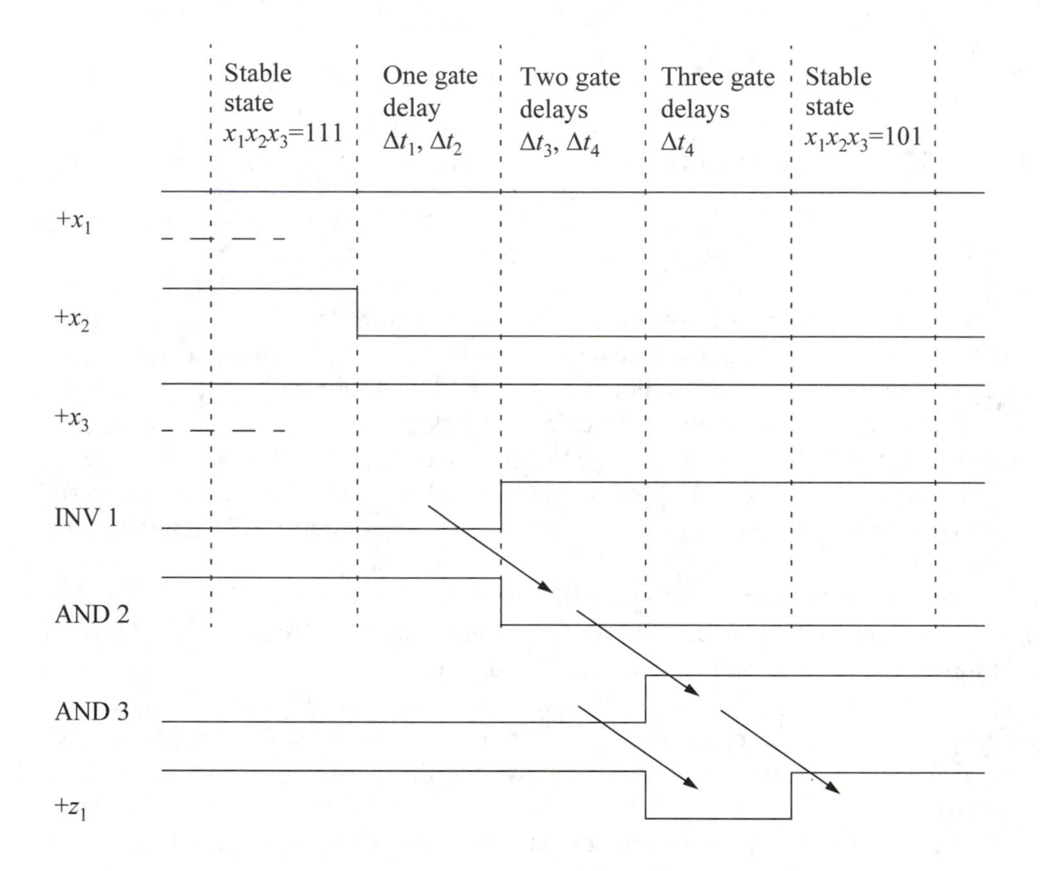

Figure 5.32 Timing diagram for the circuit of Figure 5.29, illustrating a static-1 hazard on output z_1.

Figure 5.33 shows an analogous approach to analyzing the circuit of Figure 5.29 for hazards. The top row of +/– symbols refers to the logic levels produced as a result of input x_2 changing from a high to a low level. The bottom row of +/– symbols specify the logic levels which result when x_2 changes from a low to a high level.

Refer to the top row of +/– symbols for the discussion which follows. Assume that x_2 changes from a high to a low level. The first column in each of the four sets of columns specifies a stable state, where $x_1x_2x_3 = 111$. The second column represents the state of the circuit after one gate delay, in which the inverter and AND gate 2 are in the same time relationship. The third column indicates the state of the circuit after two gate delays, in which AND gate 3 and the OR gate (from the top input) produce concurrent delays. At this time, output z_1 changes to a low voltage level. Finally, after three gate delays — the inverter, AND gate 3, and the OR gate (from the bottom input) — the circuit resumes a stable state, where z_1 returns to a high level.

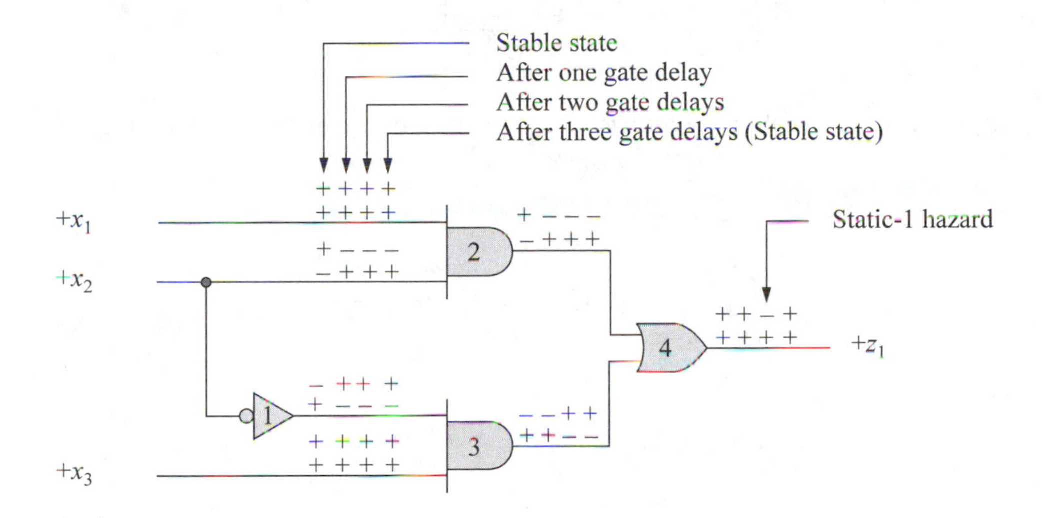

Figure 5.33 Alternative method for analyzing a circuit for static hazards.

Now consider the case where input x_2 changes from a low to a high logic level. If the initial state of the circuit is $x_1x_2x_3 = 101$ and x_2 changes from a low to a high level, then no hazard will occur on output z_1, as indicated by the bottom row of +/– symbols. There will never be time period when both inputs of the OR gate are at a low logic level. Thus, the circuit will not generate a static-1 hazard on z_1.

Because the map of Figure 5.30 specifies that output z_1 should remain at a constant high level when $x_1x_3 = 11$, the effects of the hazard can be eliminated by adding a third term to the equation for z_1. The output can be made independent of the value of x_2 by including the redundant prime implicant x_1x_3, which covers both the initial and terminal state of the transition. The redundant prime implicant will maintain the output at a constant high level during the transition.

The term x_1x_3 is called a *hazard cover*, since it covers the detrimental effects of the hazard. The effects of a static hazard can be negated by combining adjacent groups of 1s in a Karnaugh map as shown in Figure 5.34 for the network of Figure 5.29. The equation for z_1 is shown in Equation 5.10. The hazard will still occur at the OR gate inputs, but its effect will be negated due to the addition of AND gate 5, as shown in Figure 5.35. Although the circuit no longer contains a minimal number of gates, the operation is reliable.

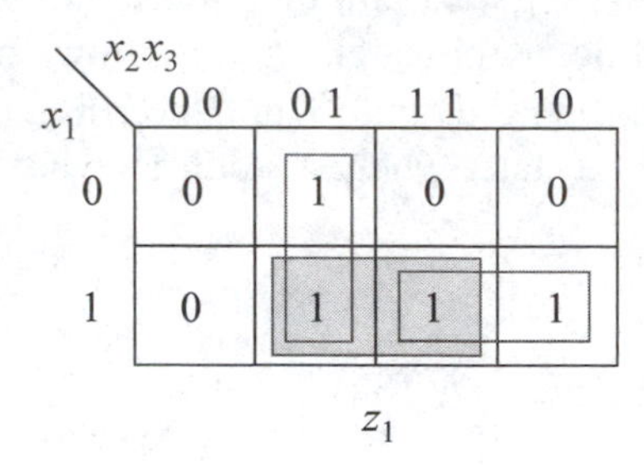

Figure 5.34 Negating the effects of a static hazard by combining adjacent groups of 1s.

$$z_1 = x_2'x_3 + x_1x_2 + x_1x_3 \tag{5.10}$$

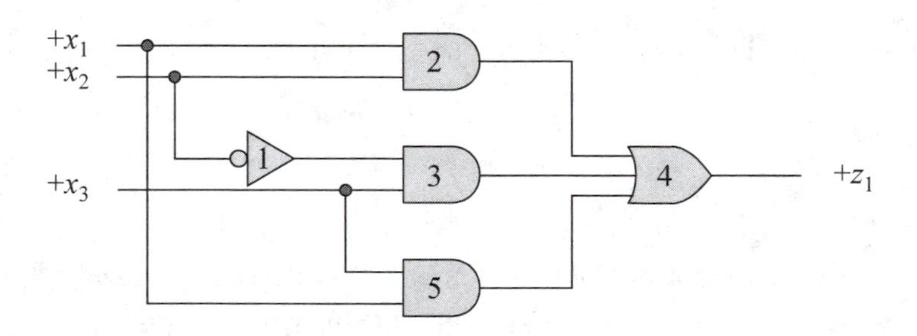

Figure 5.35 Eliminating the effects of a static hazard in the combinational network of Figure 5.29 by adding the term x_1x_3 to the equation for z_1.

Two input vectors are adjacent if they differ by only one variable. For example, $x_1x_2'x_3$ and $x_1x_2'x_3'$ are adjacent, because only x_3 changes value. A transition between adjacent input vectors which generate the same output value may cause a momentary erroneous transition in the output signal. These hazards are possible when two adjacent input combinations produce the same output, but are not combined by a common prime implicant.

A network is free of static hazards when all changes to the input vectors that specify a constant output are contained within a common prime implicant. That is, every grouping of 1s in a Karnaugh map must be combined with every adjacent group of 1s.

The same rationale also applies to a machine that is implemented as a product-of-sums. In this case, adjacent groupings of 0s must be connected by a common term.

The Karnaugh map of Figure 5.36 (a) specifies a circuit implementation in a sum-of-products form. The circuit is free of static hazards, because both groups of 1s are connected by the redundant prime implicant $x_1{}'x_3$. The equation for z_1 is shown in Equation 5.11. The logic diagram is illustrated in Figure 5.37 (a). Hazard-free operation of the circuit can be verified by using the procedures outlined previously in this section.

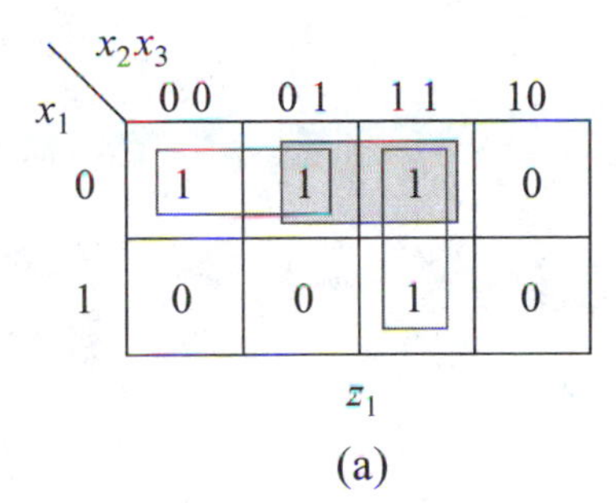

(a)

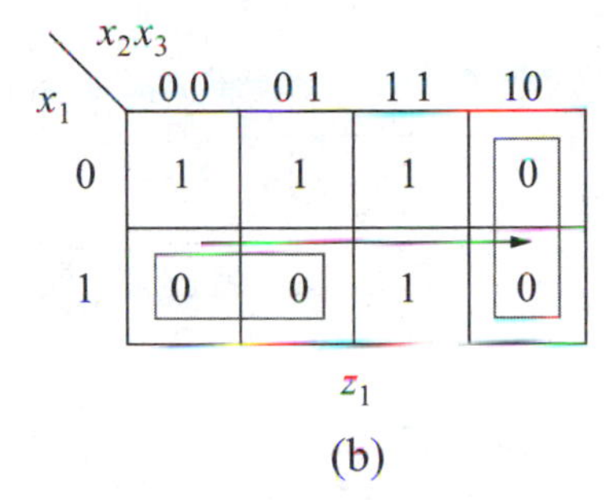

(b)

Figure 5.36 Karnaugh maps for hazard analysis: (a) sum-of-products form with no static-1 hazards; and (b) product-of-sums form with a potential static-0 hazard.

$$z_1 = x_1{}'x_2{}' + x_2x_3 + x_1{}'x_3 \tag{5.11}$$

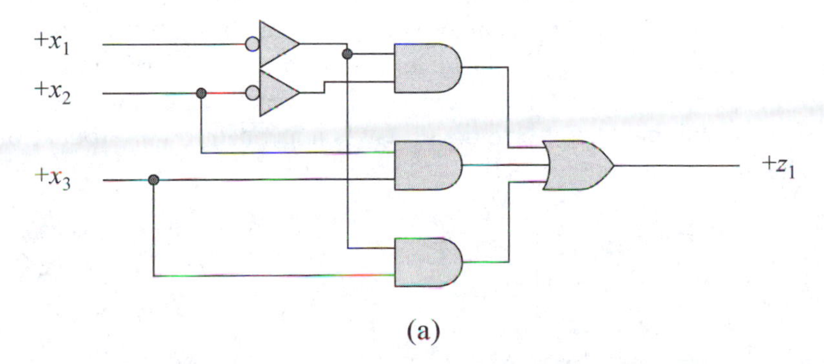

(a)

Figure 5.37 Equivalent circuits as specified by the Karnaugh maps of Figure 5.36: (a) sum-of-products implementation with no hazards; (b) product-of-sums implementation with a static hazard; and (c) product-of-sums implementation with no hazards.

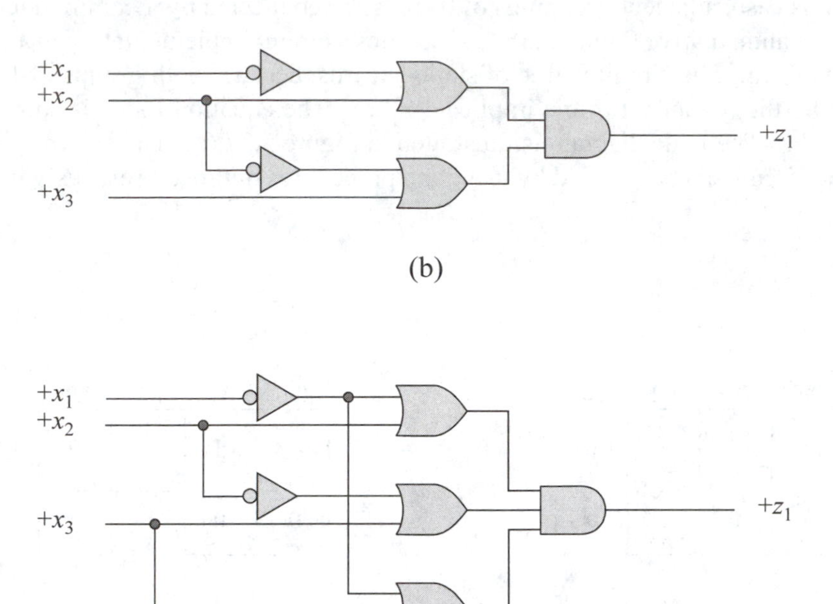

(c)

Figure 5.37 (Continued)

If z_1 is implemented as a product of sums by combining the 0s in the map, as shown in Figure 5.36 (b), then z_1 is specified by Equation 5.12. Equation 5.11 and Equation 5.12 both define the same function, which yields Equation 5.13. This does not necessarily mean, however, that the product-of-sums implementation is free of static hazards. For example, if input x_2 changes from 0 to 1, as shown by the arrow in Figure 5.36 (b), then a static-0 hazard will be generated on z_1 as shown in Figure 5.38. As in the sum-of-products implementation, the hazard can be eliminated by adding a redundant term. The additional sum term $(x_1' + x_3)$ is appended to the product-of-sums expression as shown in Equation 5.14.

$$z_1 = (x_1' + x_2)(x_2' + x_3) \tag{5.12}$$

$$\begin{aligned} z_1 &= x_1'x_2' + x_2x_3 + x_1'x_3 \\ &= (x_1' + x_2)(x_2' + x_3) \end{aligned} \tag{5.13}$$

$$z_1 = (x_1' + x_2)(x_2' + x_3)(x_1' + x_3) \tag{5.14}$$

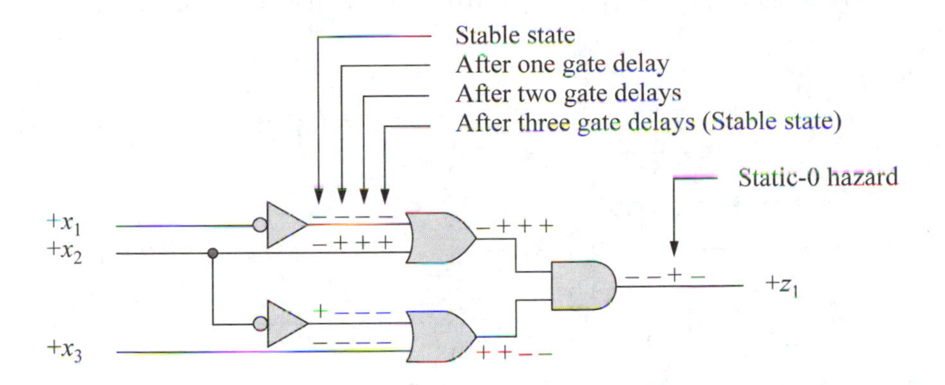

Figure 5.38 A static-0 hazard exhibited on output z_1 for the product-of-sums implementation of Figure 5.36 (b).

The resulting circuit, shown in Figure 5.37 (c), is functionally equivalent to the circuit of Figure 5.37 (a) and is free of static hazards. Output z_1 can be completely specified, therefore, by either the sum-of-products or the product-of-sums form as shown in Equation 5.15. By applying the product terms $x_1'x_2'$ or x_2x_3 to the equations of Equation 5.15, both equations will yield a value of 1 for z_1 as specified by the Karnaugh maps of Figure 5.36.

$$z_1 = x_1'x_2' + x_2x_3 + x_1'x_3$$
$$z_1 = (x_1' + x_2)(x_2' + x_3)(x_1' + x_3) \tag{5.15}$$

If a sum-of-products expression is made free of static hazards by including a redundant term, then the equivalent product-of-sums expression may also have to be formulated using a similar non-minimized form. The converse is also true.

As a second example, consider the circuit shown in Figure 5.39, which contains four input variables x_1, x_2, x_3, and x_4 and one output z_1. The procedure for analyzing a circuit for static hazards is the same regardless of the number of inputs. Since different path lengths exist from the inputs to the output, a hazard is possible on output z_1 when a single input changes value. Whether or not a hazard occurs depends on the relative propagation delays of the circuit elements. The equation for z_1 is shown in Equation 5.16 and the Karnaugh map in Figure 5.40.

The circuit will be analyzed from several different stable states to determine if static hazards exist when a single input changes value. Also, the type of hazard will be identified. Assume that the circuit is stable in state $x_1x_2x_3x_4 = 1101$ and that x_4 changes from 1 to 0. A hazard will not be generated on z_1, because the single input change occurs within the same prime implicant x_2x_3'. The same rationale applies if the network is stable in state $x_1x_2x_3x_4 = 1010$ and x_2 changes from 0 to 1.

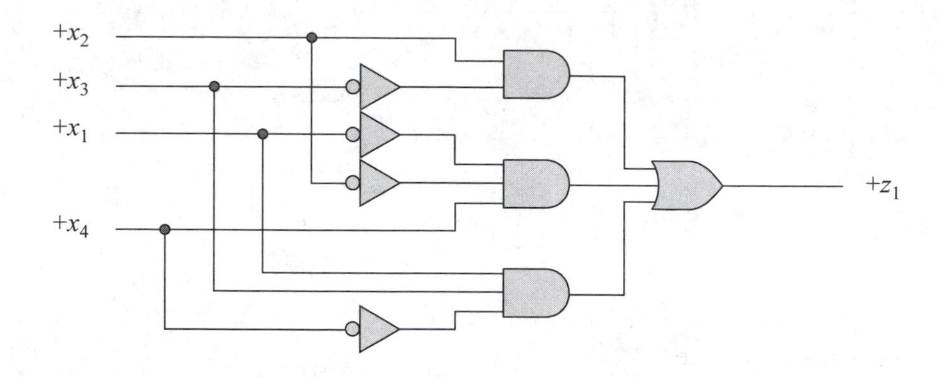

Figure 5.39 Circuit containing potential static hazards.

$$z_1 = x_2x_3' + x_1'x_2'x_4 + x_1x_3x_4' \tag{5.16}$$

x_1x_2 \ x_3x_4	0 0	0 1	1 1	1 0
0 0	0	1	1	0
0 1	1	1	0	0
1 1	1	1	0	1
1 0	0	0	0	1

z_1

Figure 5.40 Karnaugh map for the circuit of Figure 5.39.

If, however, the circuit is stable in state $x_1x_2x_3x_4 = 1110$ and x_3 changes from 1 to 0, then a static-1 hazard can appear on z_1. The circuit is analyzed in Figure 5.41

showing the output levels of each logic block at various propagation delay times. When the input vector changes from $x_1x_2x_3x_4 = 1110$ to 1100, the $+x_3$ input to AND gate 7 changes to a low logic level immediately. After one gate delay — through inverter 1 and AND gate 7 in parallel — the output of AND gate 7 is at a low level.

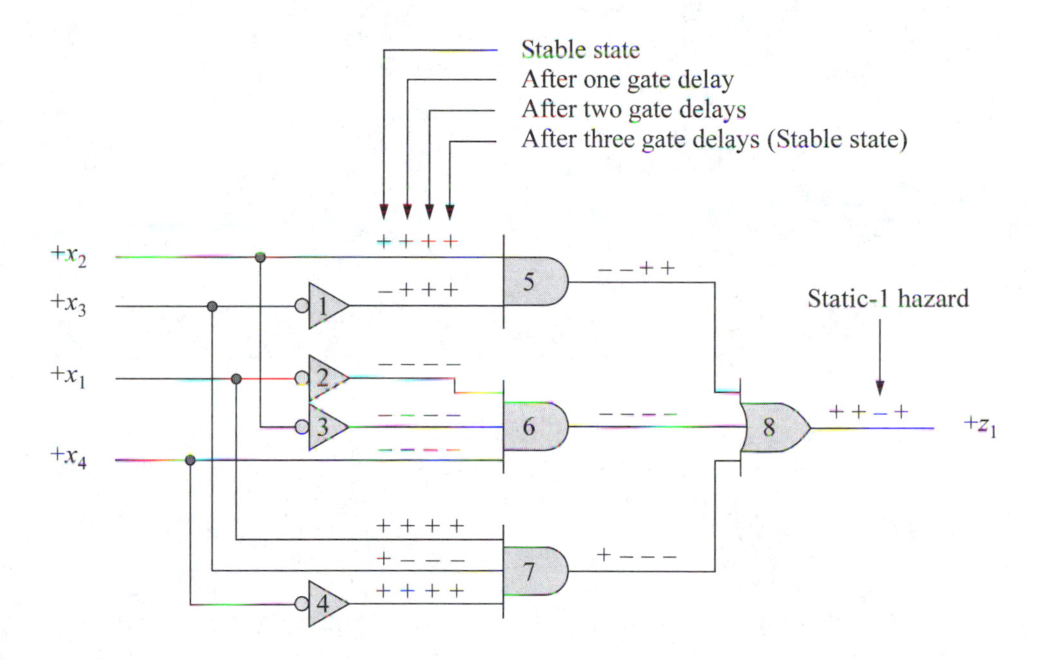

Figure 5.41 Figure 5.39 reproduced showing a static-1 hazard when the input vector changes from $x_1x_2x_3x_4 = 1110$ to 1100.

The output of AND gate 5 has not yet changed from a low to a high level as a result of x_3 changing value. The output of AND gate 6 remains at a constant low level. Thus, the outputs of all three AND gates, and hence the OR gate inputs, are at a low level. After a second delay through the OR gate, output z_1 changes from a high to a low level. After a third delay, the high level on the top input of the OR gate is propagated to the output and z_1 returns to its static condition.

This output transition from a high to a low level for one gate propagation delay results in a static-1 hazard on z_1. Therefore, the product terms x_2x_3' and $x_1x_3x_4'$ must be connected (or covered) by the redundant prime implicant $x_1x_2x_4'$. Because a hazard cover is necessary for the transition from $x_1x_2x_3x_4 = 1110$ to 1100, there is no need to check for a static hazard for the transition from $x_1x_2x_3x_4 = 1100$ to 1110.

The same circuit is reproduced in Figure 5.42 and analyzed from a different starting stable state. Assume that the circuit is stable in state $x_1x_2x_3x_4 = 0001$ and that x_2 changes from 0 to 1. It is evident from the +/– symbols that there will never be a time period when all three inputs to the OR gate are at a low logic level. Therefore, no static hazard exists on output z_1 resulting from this single input change.

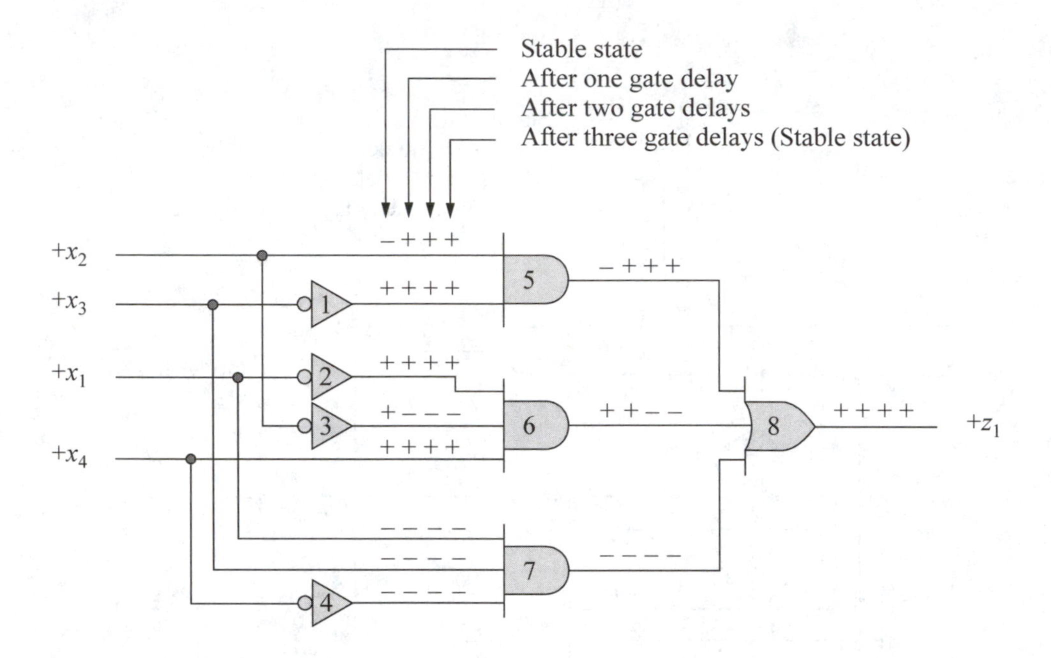

Figure 5.42 Figure 5.39 reproduced showing no static hazard when the input vector changes from $x_1x_2x_3x_4 = 0001$ to 0101.

Now consider the final transition during which output z_1 may generate a static hazard resulting from a single input change. Assume that the circuit is stable in state $x_1x_2x_3x_4 = 0101$ and x_2 changes from 1 to 0. Figure 5.39 is reproduced in Figure 5.43 for analysis. The effect of x_2 changing value results in a low logic level applied to AND gate 5 immediately. After one gate delay, the output of AND gate 5 changes from a high to a low level. During this same time duration, the output of inverter 3 changes to a high level. The output of AND gate 6, however, does not reflect this change until an additional delay through AND gate 6. After one gate delay, therefore, the outputs of all three AND gates are deasserted and remain at that level until the high level from inverter 3 propagates through AND gate 6, providing a high level on the OR gate input.

The effect of three low levels on the OR gate inputs is not realized until after a second delay through the OR gate, at which time output z_1 changes from a high to a low logic level. After this delay has transpired, the high level on the output of inverter 3 has propagated through AND gate 6 and the OR gate. At this time, output z_1 returns to a static high level. Thus, a change to input x_2 from a high to a low level results in a static-1 hazard on output z_1. The five separate input changes and the corresponding results on output z1 are as follows:

$x_1x_2x_3x_4 = 1101 \rightarrow 1100$ produces no static hazard on z_1
$x_1x_2x_3x_4 = 1010 \rightarrow 1110$ produces no static hazard on z_1
$x_1x_2x_3x_4 = 1110 \rightarrow 1100$ produces a static-1 hazard on z_1
$x_1x_2x_3x_4 = 0001 \rightarrow 0101$ produces no static hazard on z_1
$x_1x_2x_3x_4 = 0101 \rightarrow 0001$ produces a static-1 hazard on z_1

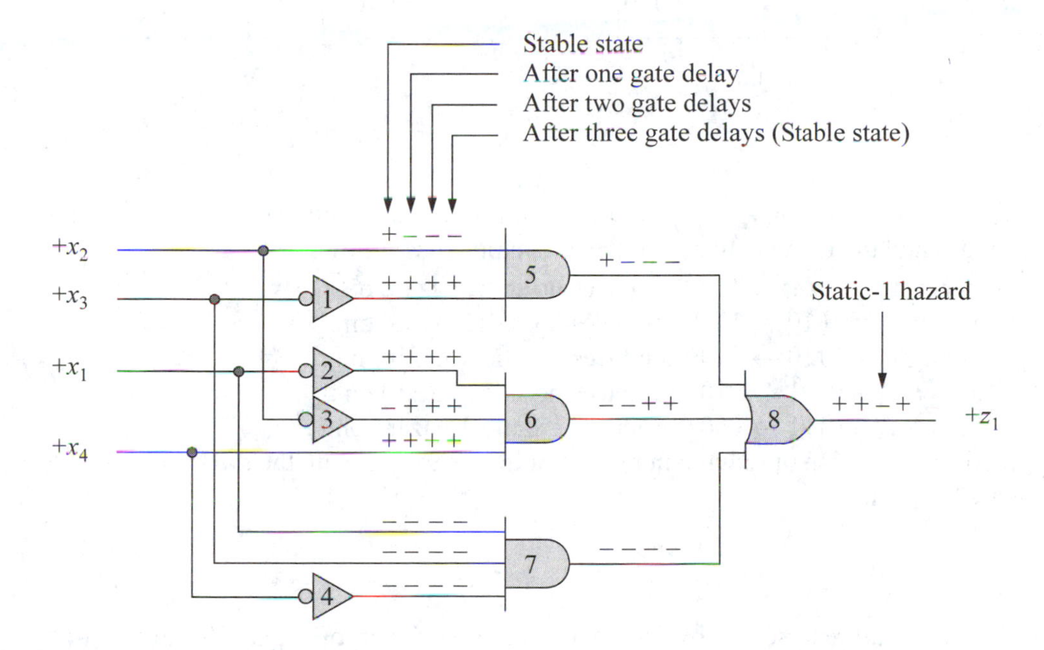

Figure 5.43 Figure 5.39 reproduced showing a static-1 hazard when the input vector changes from $x_1x_2x_3x_4 = 0101$ to 0001.

The Karnaugh map of Figure 5.40 is reproduced in Figure 5.44 (a) showing the previous five cases in which a single input change produces either no hazard or a static-1 hazard on output z_1. In order to eliminate all possible hazards in the logic circuit of Figure 5.39, all groups of 1s (essential prime implicants) must be connected as shown in the Karnaugh map of Figure 5.44 (b). Therefore, two redundant prime implicants must be added to the equation for z_1, as shown in Equation 5.17. The resulting change to the logic circuit is shown in Figure 5.45 with the addition of two AND gates.

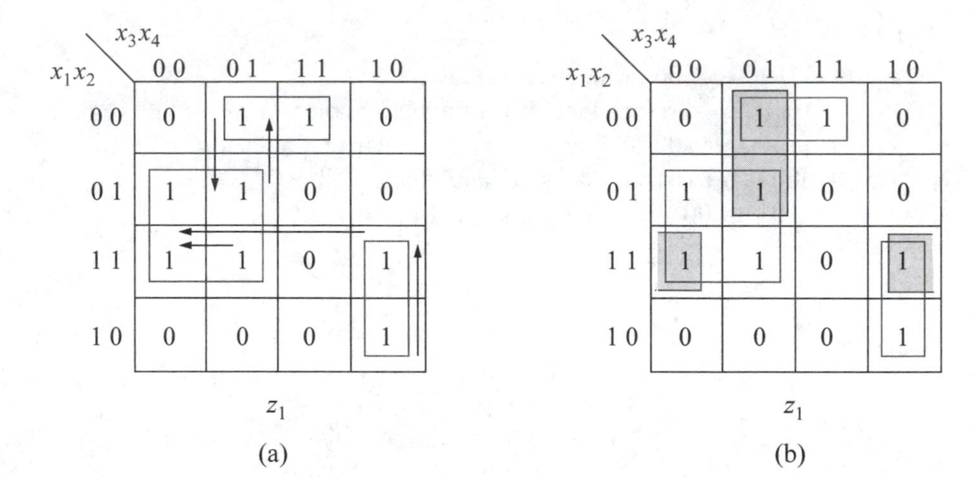

Figure 5.44 Karnaugh map for the circuit of Figure 5.39: (a) five separate input changes and the corresponding effects on output z_1 as follows:

$x_1x_2x_3x_4 = 1101 \rightarrow 1100$ produces no static hazard on z_1
$x_1x_2x_3x_4 = 1010 \rightarrow 1110$ produces no static hazard on z_1
$x_1x_2x_3x_4 = 1110 \rightarrow 1100$ produces a static-1 hazard on z_1
$x_1x_2x_3x_4 = 0001 \rightarrow 0101$ produces no static hazard on z_1
$x_1x_2x_3x_4 = 0101 \rightarrow 0001$ produces a static-1 hazard on z_1

and (b) the addition of redundant prime implicants to eliminate the effects of possible static hazards.

The circuit represented by Figure 5.45 requires fewer logic gates for the sum-of-products implementation than for an equivalent product-of-sums implementation. If the 0s were grouped in the Karnaugh map as shown in Figure 5.46 and all groups of 0s connected by redundant sum terms to eliminate possible hazards, then the product-of-sums equation for z_1 would be as shown in Equation 5.18. The equation clearly requires more logic gates than the equivalent sum-of-products expression of Equation 5.17, not only for the minimized function, but also for the hazard-free implementation.

$$z_1 = x_2x_3' + x_1'x_2'x_4 + x_1x_3x_4' + x_1'x_3'x_4 + x_1x_2x_4' \qquad (5.17)$$

Hazard cover (arrows pointing to $x_1'x_3'x_4$ and $x_1x_2x_4'$)

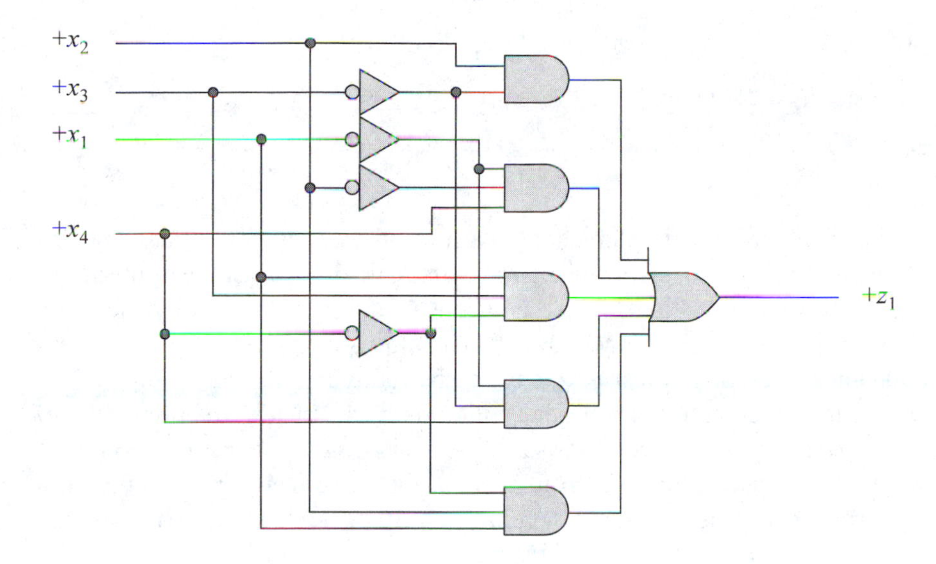

Figure 5.45 Logic circuit of Figure 5.39 reconfigured to eliminate the effects of potential static hazards by adding two AND gates to incorporate redundant prime implicants as hazard covers.

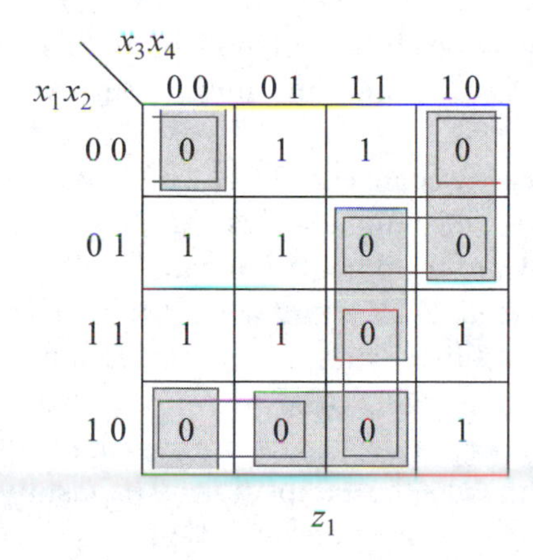

Figure 5.46 Karnaugh map for the circuit of Figure 5.39 in which the 0s are grouped to form a product-of-sums representation with no static-0 hazards on output z_1.

In Figure 5.46, the grouping of two 0s in locations $x_1x_2x_3x_4 = 0000$ and 1000 is required to avoid a possible static-0 hazard for the transition from $x_1x_2x_3x_4 = 0000$ to 1000. There is no hazard for the sequence $x_1x_2x_3x_4 = 1000$ to 0000.

$$z_1 = (x_1 + x_2 + x_4)(x_1 + x_2' + x_3')(x_1' + x_3' + x_4')(x_1' + x_2 + x_3)$$
$$(x_1 + x_3' + x_4)(x_2' + x_3' + x_4')(x_1' + x_2 + x_4')(x_2 + x_3 + x_4) \quad (5.18)$$

Hazard cover (arrows pointing to $(x_1 + x_3' + x_4)$, $(x_2' + x_3' + x_4')$, $(x_1' + x_2 + x_4')$, $(x_2 + x_3 + x_4)$)

In summary, if a circuit is implemented in a sum-of-products form, then a potential static-1 hazard exists. To analyze the circuit for static-1 hazards, the minterm locations that contain 1s are utilized. Two adjacent 1s that are covered by the same product term cannot produce a static-1 hazard. To eliminate all possible static-1 hazards, all adjacent groups of 1s must be connected by redundant product terms.

If a circuit is implemented in a product-of-sums form, then a potential static-0 hazard exists. To analyze the circuit for static-0 hazards, the minterm locations that contain 0s (maxterms) are utilized. Two adjacent 0s that are covered by the same sum term cannot produce a static-0 hazard. To eliminate all possible static-0 hazards, all adjacent groups of 0s must be connected by redundant sum terms.

5.4.2 Dynamic Hazards

Dynamic hazards, like static hazards, may also cause erroneous outputs in combinational circuits. If the combinational network is incorporated in the δ next-state logic of an asynchronous sequential machine, then an incorrect next state may result. The same rationale applies to dynamic hazards in the λ output combinational logic. A *dynamic hazard* is characterized by multiple output pulses resulting from a single change to the input vector.

When a single input variable changes value, an odd number of transitions may occur on the output signal, where the number of transitions is greater than one. Thus, the input change propagates toward the output signal along 3, 5, 7, ... paths. In order for a dynamic hazard to be realized, at least three different path lengths must be encountered. The first path will constitute a minimum propagation delay and cause the output to change value; the second path will cause an intermediate delay in which the output will return to its previous value; and the third path results in a maximum delay which causes the output to make a third transition. The last two transitions also generate a static hazard. Figure 5.47 illustrates two typical types of dynamic hazards.

(a) (b)

Figure 5.47 Dynamic hazards: (a) a triple transition from a low to a high voltage level; and (b) a triple transition from a high to a low voltage level.

Dynamic hazards can be eliminated in a manner analogous to that used for static hazards; that is, by adding redundant terms to the output equation. It may also be possible to change the form of the equation to eliminate a dynamic hazard. Figure 5.48 illustrates a nonminimized combinational circuit with an inherent dynamic hazard. The equation for output z_1 is shown in Equation 5.19. Let x_1, x_3, and x_4 be active low inputs and x_2 be active high. With all inputs active at their respective levels, the circuit is stable and z_1 is at a high logic level.

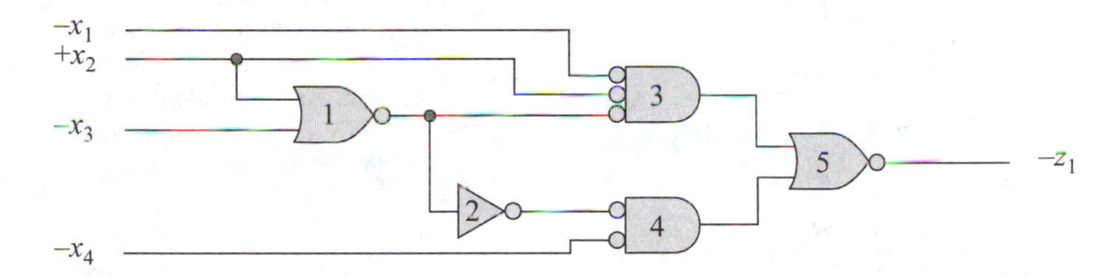

Figure 5.48 Combinational circuit with an inherent dynamic hazard.

$$z_1 = x_1 x_2' (x_2 + x_3') + x_2' x_3 x_4 \tag{5.19}$$

The timing diagram of Figure 5.49 depicts the sequence of events that occur when input x_2 changes from a high to a low logic level. The waveforms labeled OR1, INV2, AND3, AND4, and OR5 refer to the outputs of their respective logic blocks. The deassertion of x_2 is immediate. The new value of x_2 propagates to the output terminal along three paths.

The first path involves two gate delays: gates 3 and 5. After one gate delay, the output of gate 3 changes from a low to a high level. This change is reflected on output z_1 after a second gate delay through gate 5, at which time z_1 changes from a high to a low level.

The second path involves three gate delays: gates 1, 3, and 5. After one gate delay, the output of gate 1 changes from a low to a high level. This change is reflected on the output of gate 3 after a second delay (through gate 3) which now changes from a high to a low level. Both inputs to gate 5 are now at a low logic level. Therefore, after a third delay (through gate 5), output z_1 changes from a low to a high level.

The third path entails four gate delays: gate 1, the inverter, gate 4, and gate 5. After one gate delay, the output of gate 1 changes from a low to a high level. This change is manifested on the output of the inverter, which changes from a high to a low level. Both inputs to gate 4 are now at a low logic level. Therefore, after a third delay (through gate 4), the output of gate 4 changes from a low to a high level. This change encounters the fourth delay for this path as the signal propagates through gate 5 causing output z_1 to change from a high to a low logic level. The output eventually changes state, as it should, but only after the occurrence of two extra transitions. The resulting effect on z_1 from this single input change is a dynamic hazard with a triple change of state.

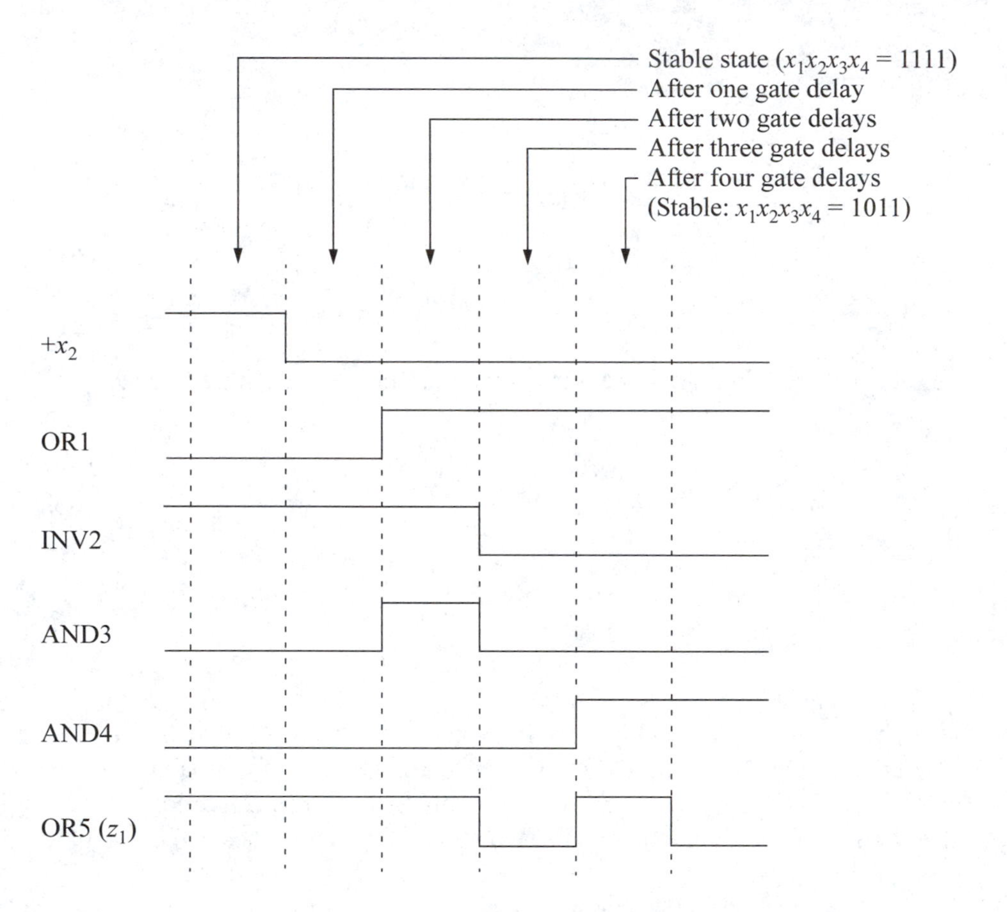

Figure 5.49 Timing diagram for the circuit of Figure 5.48 illustrating a dynamic hazard on output z_1.

Figure 5.50 shows an alternative approach to analyzing the circuit of Figure 5.48. The +/– symbols specify the logic levels which result when input x_2 changes from a high to a low level. The first column in each of the four sets of columns specifies a stable state in which $x_1x_2x_3x_4 = 1111$, where a 1 indicates an active voltage level, either high or low. The second column represents the state of the circuit after one gate delay, in which gates 1 and 3 are in the same time relationship.

The third column indicates the state of the circuit after two gate delays, in which gate 1 and the bottom input to gate 3 contribute to the delay. The second path which produces two gate delays when x_2 changes from a high to a low level consists of gates

3 and 5. This produces the first transition on output z_1. Logic gate 1 and the inverter also represent a similar two-gate propagation delay.

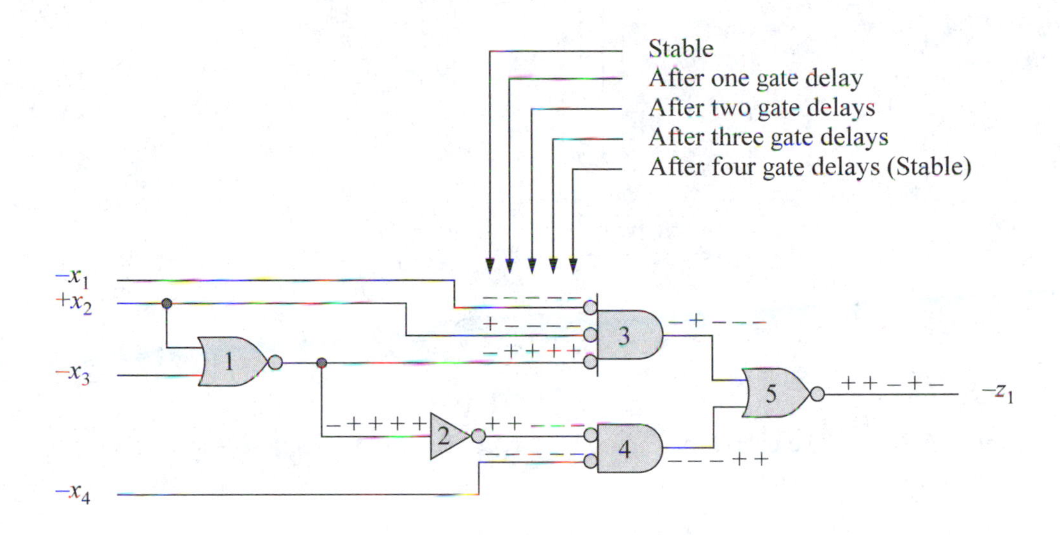

Figure 5.50 Alternative representation for analyzing a circuit for dynamic hazards.

The fourth column specifies the logic levels attained after three gate delays. Two different paths constitute separate durations of three gate delays: gates 1, 3, and 5; and gate 1, the inverter, and gate 4. Note that at the termination of the third propagation delay the outputs of gates 4 and 5 change from a low to a high level. This produces the second transition on output z_1.

The fifth column represents the fourth, and final, propagation delay, consisting of gate 1, the inverter, gate 4, and gate 5. At the end of the fourth propagation delay, the third transition occurs on the output, completing the sequence for a dynamic hazard on z_1.

The dynamic hazard can be eliminated by changing the form of Equation 5.19 to a sum-of-products representation as shown in Equation 5.20. The resulting circuit is shown in Figure 5.51. When input x_2 now changes from a high to a low logic level, output z_1 changes from a high to a low logic level after two gate delays and remains stable in that state.

$$\begin{aligned} z_1 &= x_1 x_2' x_2 + x_1 x_2' x_3' + x_2' x_3 x_4 \\ &= x_1 x_2' x_3' + x_2' x_3 x_4 \end{aligned} \quad (5.20)$$

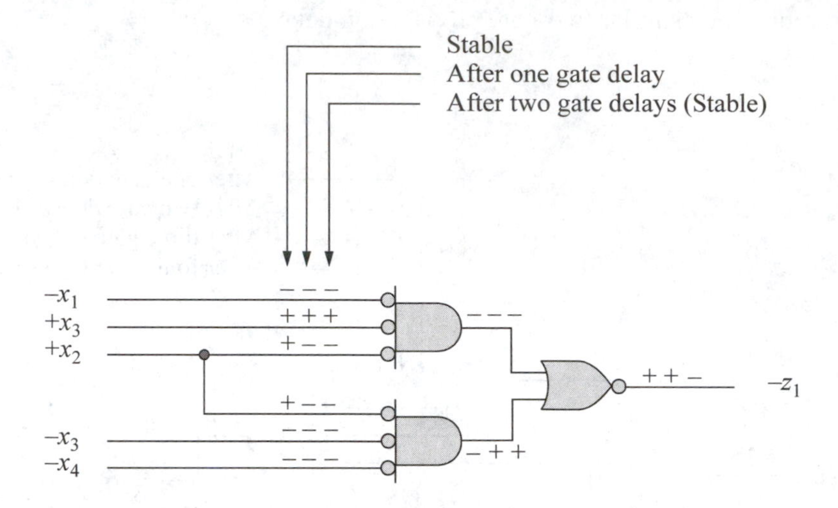

Figure 5.51 The circuit of Figure 5.48 reconfigured as a sum-of-products representation with no dynamic hazard.

Dynamic hazards, which are less common than static hazards, result from a single change to the input vector in which the change propagates to the output terminal along three or more different path lengths. Reconfiguring the logic equation, without changing the functionality of the circuit, will eliminate dynamic hazards. Dynamic hazards are also removed as a direct result of eliminating static hazards, in which redundant terms are added to the output equation.

5.4.3 Essential Hazards

An asynchronous sequential machine may be free of static and dynamic hazards in the δ next-state logic and the λ output logic, but may still sequence to an incorrect next state due to inordinately long propagation delays in certain circuit elements. The propagation delay through a logic gate may increase due to temperature change, power supply output variation, or component aging.

An *essential hazard* is caused not by an incorrectly synthesized network, but by the operation of the circuit itself; that is, it is inherent in the machine's design. Thus, the essential hazard cannot be eliminated by adding redundant terms in the network equation or by changing the form of the equation. Since an essential hazard is the result of excessive propagation delay in certain network elements, the effect of the hazard can be nullified by introducing appropriate delays in other components, such that the cause of the hazard will be negated.

Essential hazards occur specifically in fundamental-mode machines. The distribution of delays causes the machine to produce a steady-state hazard. An essential hazard is characterized by two propagation paths: one path affecting storage element y_j, the other y_k. An essential hazard is similar, in some respects, to critical races, which are described in a following section.

Essential hazards can be detected by analyzing the excitation map or flow table of an asynchronous sequential machine. The machine has a possible essential hazard if, beginning in a stable state, input x_i changes value and sequences the machine to a stable state that is different than the stable state reached after three successive changes to x_i. That is, an asynchronous sequential machine contains a possible essential hazard if a single change to an input variable x_i results in a transition from state S_j to state S_k, whereas, three consecutive changes to x_i results in a state transition sequence which terminates in state S_l, where $S_k \neq S_l$.

Thus, when the machine is implemented, there may be a series of propagation delays that will cause the machine to sequence to an incorrect next stable state resulting from a single change to input x_i. This incorrect state change occurs because the change to x_i propagates to different circuit elements at different times.

Consider the asynchronous sequential machine specified by the excitation maps and flow table of Figure 5.52 and Equation 5.21. The logic diagram is shown in Figure 5.53. If the machine is stable in state ⓐ ($y_{1f}y_{2f}x_1 = 000$) and input x_1 changes from a low to a high logic level, then an essential hazard is possible if the delay through the inverter is excessive; that is,

$$\text{Inverter delay} > \text{delay of gate 1} + \text{delay of gate 2} + \text{delay of gate 4}$$

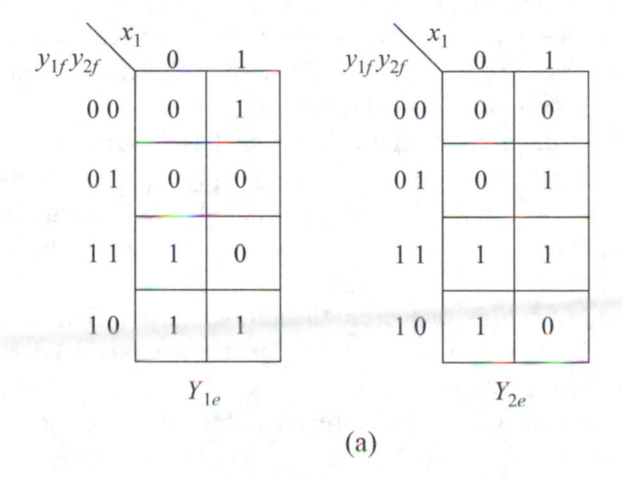

Figure 5.52 Karnaugh maps for an asynchronous sequential machine containing a possible essential hazard: (a) separate excitation maps for Y_{1e} and Y_{2e}; (b) combined excitation map for Y_{1e} and Y_{2e}; and (c) flow table.

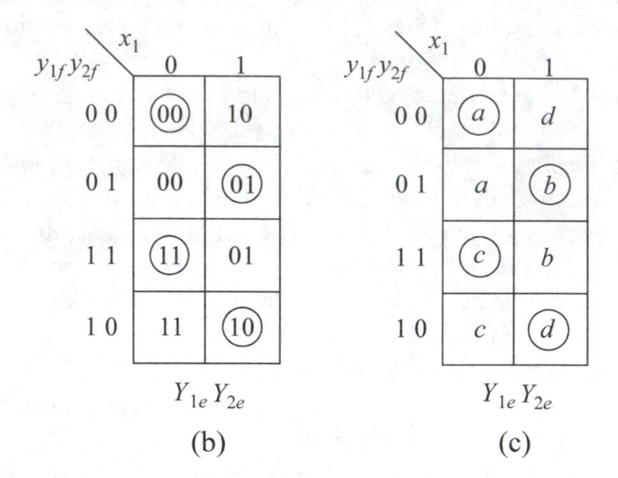

Figure 5.52 (Continued)

$$Y_{1e} = y_{2f}'x_1 + y_{1f}x_1'$$
$$Y_{2e} = y_{1f}x_1' + y_{2f}x_1 \quad (5.21)$$

The machine should sequence from state ⓐ ($y_{1f}y_{2f}x_1 = 000$) to state ⓓ ($y_{1f}y_{2f}x_1 = 101$) if the gate delays are within their respective specified range. However, the sequence described below is possible if the delay through the inverter greatly exceeds its maximum propagation delay.

Refer to the logic diagram of Figure 5.53, which is designed from the equations of Equation 5.21. If the machine is stable in state ⓐ ($y_{1f}y_{2f}x_1 = 000$) and input x_1 changes from a low to a high level, then the output of gate 1 changes to a low level after an appropriate delay. Latch y_1 will then set after a delay through gate 2. The circuit is now in a transient state of $y_{1f}y_{2f}x_1 = 101$.

The $+y_{1f}$ output of gate 2 connects to the input of gate 4. The inverter would normally have disabled gate 4 by this time and the machine would have stabilized in state $y_{1f}y_{2f}x_1 = 101$. The output of the inverter, however, is still at a high logic level due to excessive propagation delay. Therefore, the output of gate 4 changes to a low level and sets latch y_2 after a delay through gate 5. During this time period, the inverter propagates the change incurred by input x_1.

Even though the inverter has disabled the output of gate 4, latch y_2 remains set as a result of the feedback path ($+y_{2f}$) and the connection between the output of gate 6 and the input to gate 5. The inverter output also applies a low level to reset latch y_1. Since the output of gate 1 is disabled because of the $-y_{2f}$ signal from latch y_2, therefore latch y_1 is reset. After all propagation delays have elapsed, the machine is stable in state ⓑ ($y_{1f}y_{2f}x_1 = 011$), which is an incorrect terminal state for the state transition sequence ⓐ → ⓓ.

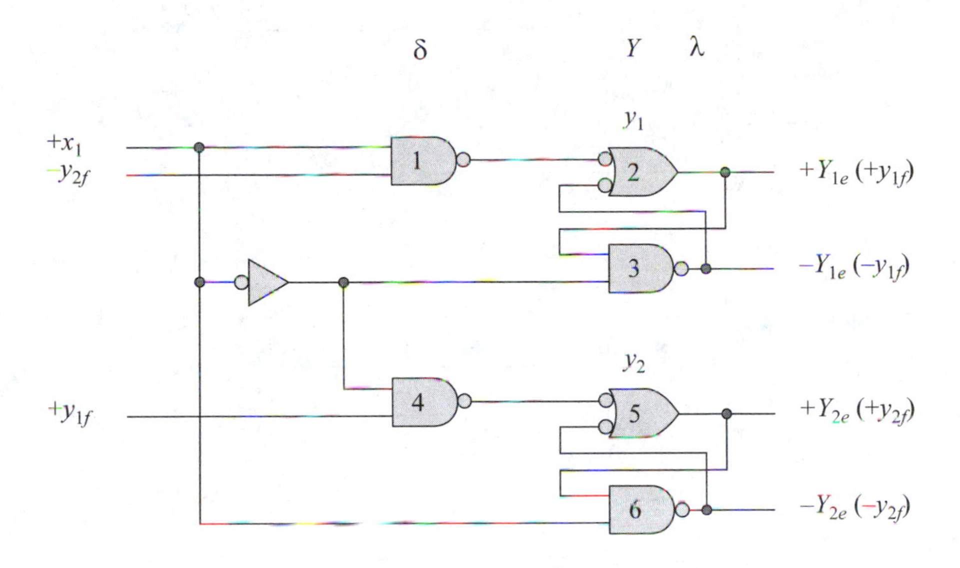

Figure 5.53 Asynchronous sequential machine with a possible essential hazard.

The hazard thus realized is essential; that is, the design of the machine cannot be changed to eliminate the hazard and still conform to the machine specifications. The essential hazard occurs because the change to input x_1 is transmitted to different parts of the circuit at different times. The change of state for y_1 is received at gate 4 before the inverter has propagated the change from x_1. The inverter output would normally have applied a low level to the input of gate 4, effectively preventing latch y_2 from being set.

The operation of the asynchronous sequential machine of Figure 5.53 can be further analyzed by the timing diagram of Figure 5.54. The signals refer to the output levels of their respective gates at various times as indicated by t_1 through t_{12}. Note that the inverter output does not reflect the change to input x_1 until time t_8, at which time a low logic level is applied to gate 3 of latch y_1. Meanwhile, latch y_2 is set, disabling the output of gate 1. Thus, after two delays, through gates 1 and 2, latch y_1 becomes reset. The machine enters stable state $y_1y_2 = 01$ instead of stable state $y_1y_2 = 10$.

Essential hazards can be eliminated in an asynchronous sequential machine — without affecting the logical operation of the machine — by inserting appropriate delay circuits in strategic locations within the machine. For example, if sufficient delay was added to the output of latch y_1, then the change to input x_1 would propagate to all the appropriate gates before the change to y_1 was received at those gates. Thus, the essential hazard is eliminated and proper operation of the machine is assured.

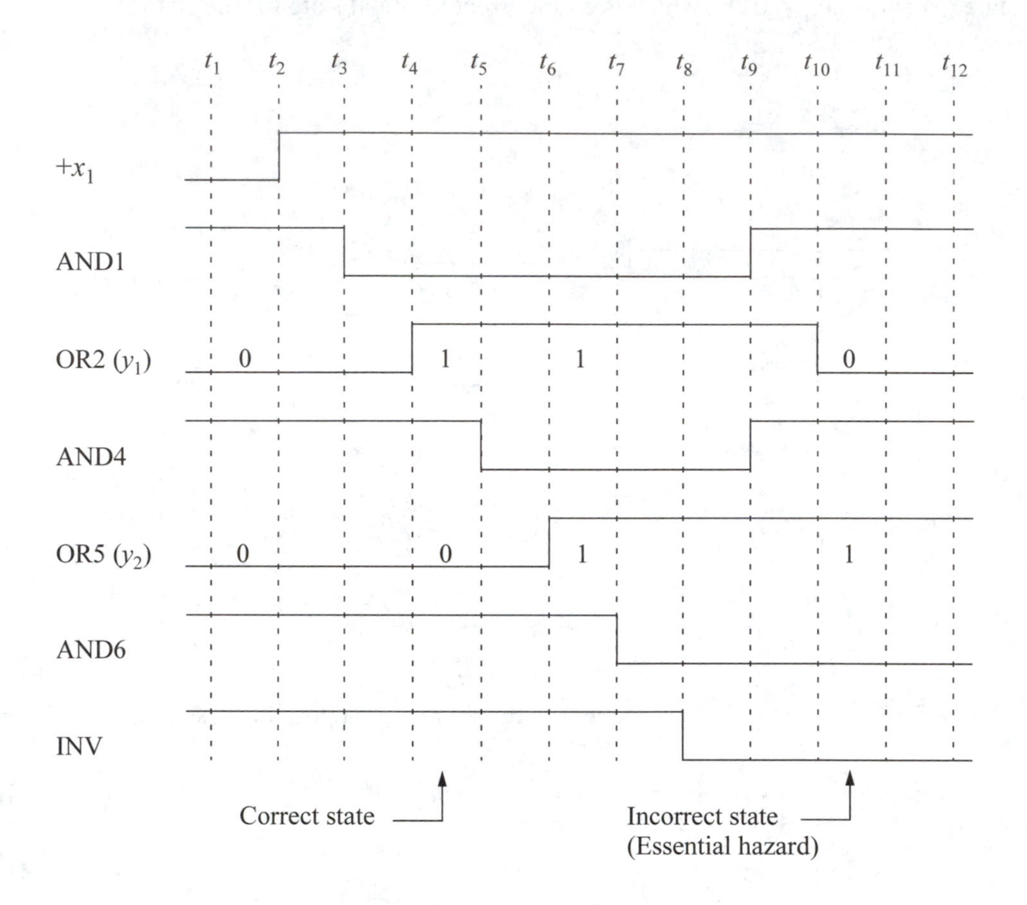

Figure 5.54 Timing diagram depicting an essential hazard for the asynchronous sequential machine of Figure 5.53 in which the machine sequences to an incorrect next stable state. The machine is initially stable in state $y_1y_2x_1 = 000$.

In Figure 5.52 (c), if the machine is presently in stable state ⓐ and x_1 changes from 0 to 1, then the next state should be state ⓓ. However, due to the essential hazard, the machine may sequence to state ⓑ instead, as indicated in the previous description. Beginning in state ⓐ, a single change to x_1 sequences the machine to state ⓓ. Again beginning in state ⓐ, three successive changes to x_1 sequences the machine to state ⓓ, then to state ⓒ,and finally to state ⓑ. Since a single change to x_1 results in a different final stable state than a triple change to x_1, the machine possesses an inherent essential hazard.

The essential hazard can be eliminated by inserting a delay circuit in the output of latch y_1 as shown in Figure 5.55. Delay components usually require a dedicated receiver and driver, which are shown as inverters in Figure 5.55. The combined delay of the driving inverter, the delay element, and the receiving inverter must be greater than the delay of the inverter in Figure 5.53. Adding delays, of course, decreases the operational speed of the machine.

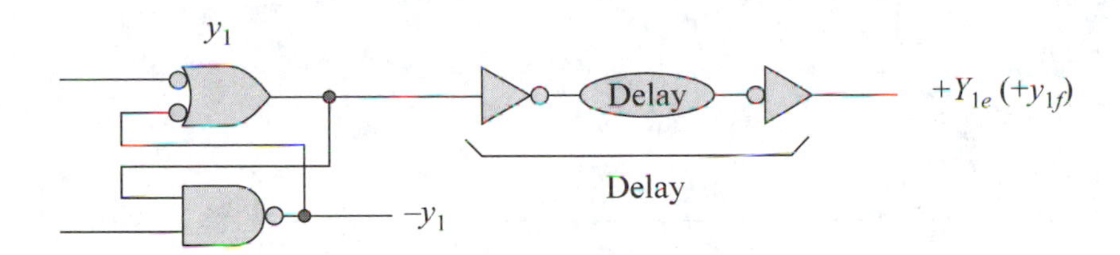

Figure 5.55 Delay circuit added to the output of latch y_1 of Figure 5.53 to eliminate an essential hazard.

Static, dynamic, and essential hazards do not present problems in synchronous sequential machines, because the active clock transition occurs after the hazard has become inactive, thus negating its effect on machine operation. A delay circuit in the output of a storage element is analogous to providing a clock signal to a synchronous sequential machine. A periodic clock signal introduces the requisite delay necessary to assure that the machine operates in a predictable manner. The essential hazard is not frequently encountered in asynchronous sequential machines, because the propagation delay of circuit components rarely exceeds the maximum delay specification.

5.4.4 Multiple-Order Hazards

Static and dynamic hazards may be considered as first-order hazards, since only one output is implicated. The essential hazard, however, may involve two or more state storage elements and generate a multiple-order hazard. An essential hazard can be classified as a *second-order hazard*. This occurs when a change to input x_i results in a change of state for state variable y_j, which in turn affects a change to state variable y_k before the change to x_i propagates to the y_k input logic. A second-order hazard of this type is graphically depicted in Figure 5.56 (a).

An essential hazard can also be considered as a *third-order hazard*. This occurs when a change to input x_i results in a change to state variable y_j, which in turn affects a change to y_k, which in turn causes a change to y_l before the change to x_i has propagated to the y_l input logic. A third-order hazard of this type is shown in Figure 5.56 (b).

Second- and third-order hazards suggest the possibility of still higher-order hazards. Higher-order hazards require that at least four state variables change state before the input change is propagated to the final state variable y_m. The excitation map or flow table can be examined directly to determine if a multiple-order hazard exists. This is a lengthy process, especially for two or more input variables, and assumes that certain delay criteria are met. Since essential hazards rarely occur in asynchronous sequential machines, the probability of higher-order hazards occurring is extremely unlikely.

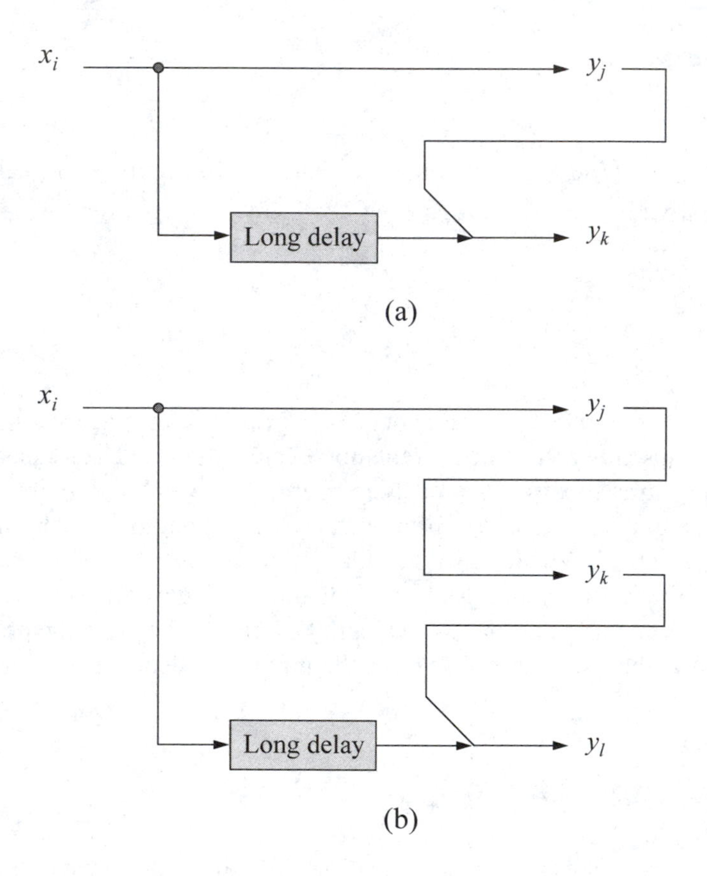

Figure 5.56 Graphs depicting multiple-order essential hazards: (a) a second-order essential hazard; and (b) a third-order essential hazard.

5.5 Oscillations

An *oscillation* occurs in an asynchronous sequential machine when a single input change results in an input vector in which there is no stable state. Consider the excitation map of Figure 5.57 for Y_{1e}. There are two input variables x_1 and x_2 and one feedback variable y_{1f}. If the machine is in stable state ⓑ ($y_{1f}x_1x_2 = 001$) where $y_{1f} = Y_{1e} = 0$ and x_1 changes from 0 to 1, then the machine sequences to transient state

c ($y_{1f}x_1x_2 = 011$). In state c, the excitation variable $Y_{1e} = 1$ and the feedback variable $y_{1f} = 0$. Thus, after a delay of Δt, the feedback variable becomes equal to the excitation variable and the machine proceeds to state g, where the feedback variable $y_{1f} = 1$. In state g, however, the excitation variable $Y_{1e} = 0$, designating state g as an unstable (or transient) state, because $y_{1f} \neq Y_{1e}$. After a further delay of Δt, the feedback variable becomes equal to the excitation variable and the machine sequences to state c, where the feedback variable $y_{1f} = 0$. Since the input vector $x_1x_2 = 11$ provides no stable state, the machine will oscillate between transient states c and g.

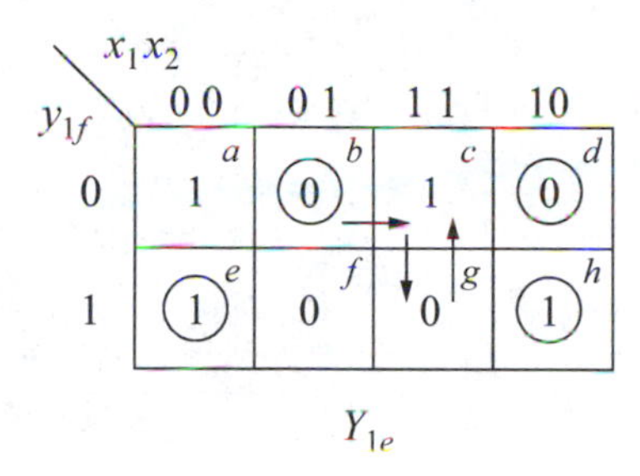

Figure 5.57 Excitation map for an asynchronous sequential machine containing an oscillation.

The equation for Y_{1e} is shown in Equation 5.22 and the logic diagram in Figure 5.58. The oscillation phenomenon can be illustrated in the logic diagram. Let the machine be stable in state ⓑ ($y_{1f}x_1x_2 = 001$) as indicated by the leftmost column of the +/– symbols. The outputs of gates 1, 2, and 3 are at a low logic level, generating a low output for gate 4. Thus, Y_{1e} is at a low logic level, which corresponds to the entry of 0 for Y_{1e} in state ⓑ of the excitation map.

$$Y_{1e} = x_1'x_2' + y_{1f}'x_1x_2 + y_{1f}x_2' \tag{5.22}$$

If input x_1 changes from 0 to 1, then the machine sequences to state c. Only gate 2 will affect the output of the machine for an input vector of $x_1x_2 = 11$ — the output of gates 1 and 3 remain at a constant low level because input x_2 is active. Thus, when x_1 becomes asserted, the output of gate 2 produces a high level, which in turn generates a high level for the output of OR gate 4, changing the value of Y_{1e} from 0 to 1, as represented in state c of the excitation map.

This change of state for Y_{1e} is fed back through the inverter as a low level to gate 2. The output of gate 2 then changes from a high to a low level. After a further delay, the output of gate 4, and thus Y_{1e}, changes from a high to a low level. This is represented by a value of 0 for Y_{1e} in state g of the excitation map.

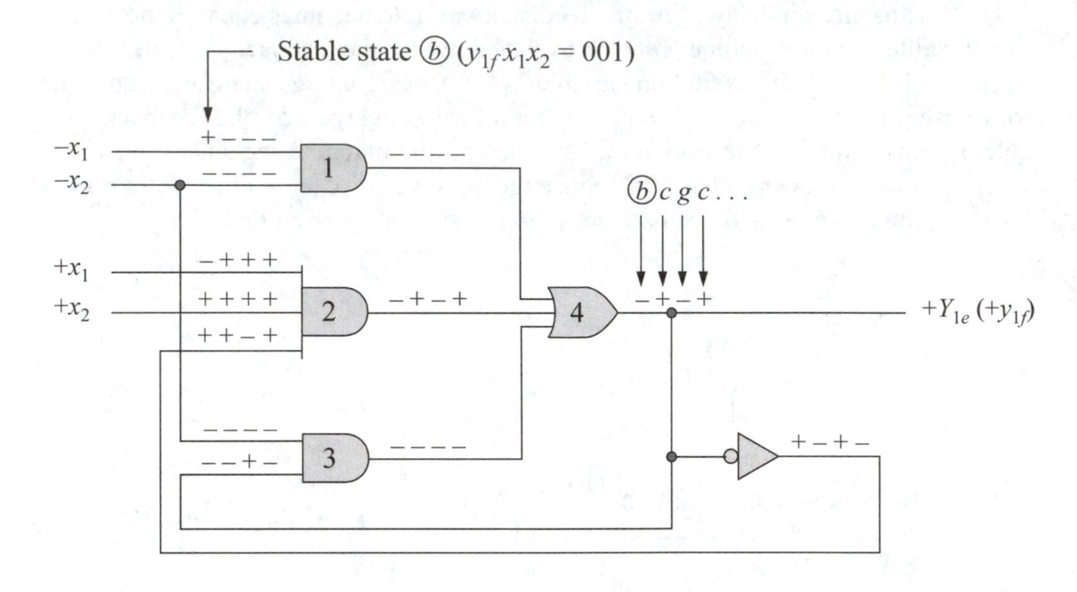

Figure 5.58 Logic diagram for the asynchronous sequential machine of Figure 5.57 which contains an oscillation.

This new change of state for Y_{1e} is fed back through the inverter as a high level to gate 2. The output of gate 2 then changes from a low to a high level. After a delay through gate 4, output Y_{1e} changes from a low to a high level and the machine enters unstable state c, where $Y_{1e} = 1$. Thus, an input vector of $x_1x_2 = 11$ causes the machine to oscillate between states c and g, causing Y_{1e} to alternate between high and low logic levels. This occurrence can be expressed as ⓑ $\rightarrow c \leftrightarrow g$.

As a further example, consider the excitation map of Figure 5.59 for excitation variables Y_{1e} and Y_{2e}. There are two inputs x_1 and x_2 and two feedback variables y_{1f} and y_{2f}. There will be at least one oscillation in the map, since column $x_1x_2 = 01$ contains no stable states. In fact, the map contains *multiple oscillations*. For example, if the machine is in state ⓔ ($y_{1f}y_{2f}x_1x_2 = 0100$) and input x_2 changes from 0 to 1, then the machine enters transient state f. In state f, the excitation variables are $Y_{1e}Y_{2e} = 00$. Thus, after a delay of Δt, the feedback variables become equal to the excitation variables, and the operation proceeds to row $y_{1f}y_{2f} = 00$ and enters transient state b ($y_{1f}y_{2f}x_1x_2 = 0001$). In state b, the excitation variables are $Y_{1e}Y_{2e} = 01$. Thus, after a delay of Δt, the feedback variables $y_{1f}y_{2f} = 01$ and the machine oscillates between transient states b and f. The oscillation can be expressed as ⓔ $\rightarrow f \leftrightarrow b$.

A similar situation occurs in stable state ⓔ when input x_1 changes from 0 to 1. Even though a column in an excitation map contains a stable state, oscillations are still possible. In this case, the machine sequences to transient state h, where $Y_{1e}Y_{2e} = 11$. After an appropriate delay, the feedback variables become $y_{1f}y_{2f} = 11$ and the machine enters transient state l. In state l, $Y_{1e}Y_{2e} = 01$, which sequences the machine back to state h. This oscillation is expressed as ⓔ $\rightarrow h \leftrightarrow l$.

$y_{1f}y_{2f}$ \ x_1x_2	0 0	0 1	1 1	1 0
0 0	a: 10	b: 01	c: 10	d: (00)
0 1	e: (01)	f: 00	g: 11	h: 11
1 1	i: 10	j: 01	k: (11)	l: 01
1 0	m: (10)	n: 11	o: 11	p: 11

$Y_{1e}Y_{2e}$

Figure 5.59 Excitation map for an asynchronous sequential machine which produces multiple oscillations.

A third oscillation occurs from stable state ⓚ when x_2 changes from 1 to 0. The machine sequences from state ⓚ to transient state l, then to transient state h, then back to state l, thus, ⓚ $\rightarrow l \leftrightarrow h$ represents this oscillation.

A fourth oscillation occurs from stable state ⓚ when x_1 changes from 1 to 0. The operation proceeds horizontally left to transient state j in row $y_{1f}y_{2f} = 11$. In state j, $Y_{1e}Y_{2e} = 01$. Thus, after a delay, $y_{1f}y_{2f} = 01$ and the machine sequences to transient state f. In state f, $Y_{1e}Y_{2e} = 00$. Thus, after a delay of Δt, the machine proceeds to state b, where $y_{1f}y_{2f} = 00$. In state b, however, $Y_{1e}Y_{2e} = 01$, which causes the machine to sequence to state f, where the oscillation repeats. This oscillation is specified by ⓚ$\rightarrow$ $j \rightarrow f \leftrightarrow b$.

A fifth oscillation develops from stable state ⓜ when input x_2 changes from 0 to 1. The operation sequences horizontally right to transient state n, where $Y_{1e}Y_{2e} = 11$. After a delay of Δt, the feedback variables become equal to the excitation variables and the machine proceeds to transient state j where $y_{1f}y_{2f} = 11$ and $Y_{1e}Y_{2e} = 01$. From state j, the operation is identical to the fourth oscillation. Thus, the oscillation produced by x_2 changing from 0 to 1 in stable state ⓜ can be expressed as ⓜ $\rightarrow n$ $\rightarrow j \rightarrow f \leftrightarrow b$.

The sixth and final oscillation also begins from stable state Ⓜ when x_1 changes from 0 to 1. The sequence for this oscillation is $\textcircled{m} \rightarrow p \rightarrow l \leftrightarrow h$. The complete set of oscillations is summarized by the expressions shown in Figure 5.60.

$$\textcircled{e} \rightarrow f \leftrightarrow b$$
$$\textcircled{e} \rightarrow h \leftrightarrow l$$
$$\textcircled{k} \rightarrow l \leftrightarrow h$$
$$\textcircled{k} \rightarrow j \rightarrow f \leftrightarrow b$$
$$\textcircled{m} \rightarrow n \rightarrow j \rightarrow f \leftrightarrow b$$
$$\textcircled{m} \rightarrow p \rightarrow l \leftrightarrow h$$

Figure 5.60 The complete set of oscillations exhibited by the asynchronous sequential machine represented by the excitation map of Figure 5.59.

An asynchronous sequential machine which has an oscillating characteristic can be used as an astable multivibrator to provide a clock signal to a synchronous sequential machine. An appropriate delay of Δt must be inserted into the network to provide the correct clock frequency. In the synthesis of most asynchronous sequential machines, however, the oscillation phenomenon must be avoided. The machine specifications can be modified slightly such that every input vector will provide at least one stable state. This modification should not drastically alter the general functional operation of the machine.

5.6 Races

In the analysis of synchronous sequential machines, if a change of state occurs between two states with nonadjacent state codes, then the machine may sequence through a transient state before entering the destination stable state. If the transient state contains a Moore-type output, then a transitory erroneous signal may be generated on the output. This glitch results from two or more variables changing state in a single state transition sequence in which the variables change values at different times.

A similar situation occurs in asynchronous sequential machines. If a single change to the input vector causes two or more excitation variables to change state, then multiple paths exist from the source stable state to the destination stable state. This is called a *race* condition.

5.6.1 Noncritical Races

Consider the excitation map of Figure 5.61. Assume that the machine is stable in state ⓕ ($y_{1f}y_{2f}x_1x_2 = 0101$). If input x_1 changes from 0 to 1, then the operation proceeds horizontally to state g ($y_{1f}y_{2f}x_1x_2 = 0111$). Because both excitation variables change value (from $Y_{1e}Y_{2e} = 01$ to 10), a race condition occurs. Three paths exist for this state transition sequence depending on the time at which the variables change value. Figure 5.62 illustrates the three possible paths for the sequence ⓕ → ⓞ. The path depicted by Figure 5.62 (a) occurs if Y_{2e} resets before Y_{1e} sets, yielding a transitory state of $Y_{1e}Y_{2e} = 00$. After a delay of Δt, the feedback variables assume the values of the excitation variables, such that, $y_{1f}y_{2f} = 00$. The machine then proceeds to state c in row $y_{1f}y_{2f} = 00$ where $Y_{1e}Y_{2e} = 10$. State c is a transient state; thus, after a further delay, $y_{1f}y_{2f} = 10$ and the machine sequences to stable state ⓞ.

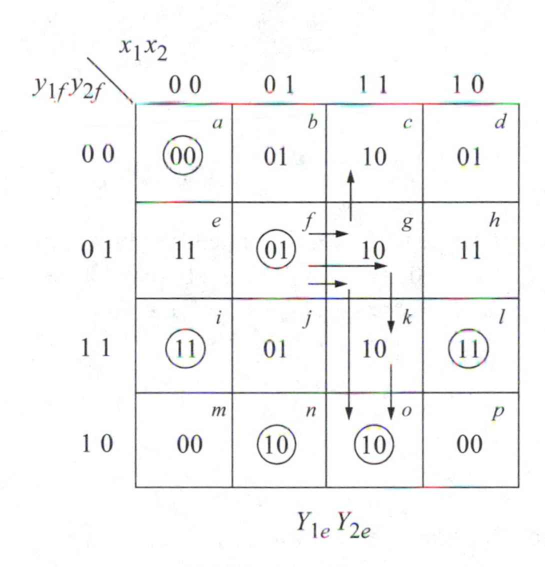

Figure 5.61 Excitation map for an asynchronous sequential machine illustrating a noncritical race condition when input x_1 changes from 0 to 1 in stable state ⓕ ($y_{1f}y_{2f}x_1x_2 = 0101$).

Figure 5.62 (b) illustrates the sequence that would occur if Y_{1e} sets before Y_{2e} resets. The machine momentarily enters state $Y_{1e}Y_{2e} = 11$. After a delay of Δt, state k is entered, because $y_{1f}y_{2f} = 11$. State k, however, is a transient state, where $Y_{1e}Y_{2e} = 10$. Thus, after an additional delay, the machine sequences to stable state ⓞ. Figure 5.62 (c) depicts the sequence that would occur if both excitation variables change state simultaneously. When x_1 changes from 0 to 1, the excitation variables change from Y1eY2e = 01 to 10. After a delay, the feedback variables become $y_{1f}y_{2f} = 10$ and the machine proceeds directly from stable state ⓕ to stable state ⓞ.

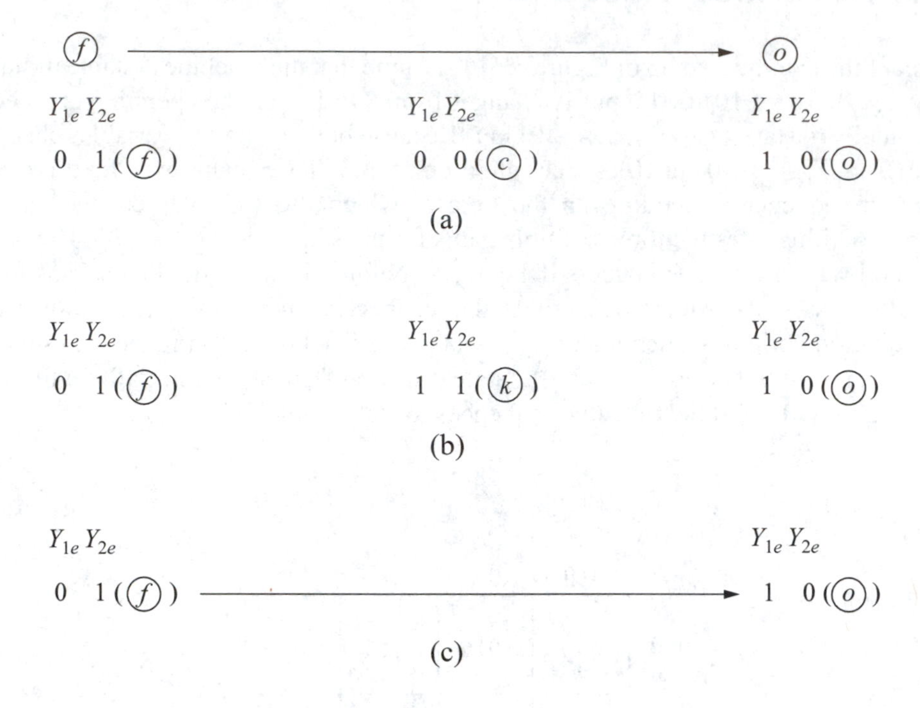

Figure 5.62 Three possible paths for the state transition sequence ⓕ ($y_{1f}y_{2f}x_1x_2$ = 0101) to ⓞ ($y_{1f}y_{2f}x_1x_2$ = 1011): (a) Y_{2e} resets before Y_{1e} sets; (b) Y_{1e} sets before Y_{2e} resets; and (c) Y_{1e} and Y_{2e} change value simultaneously.

Figure 5.63 depicts the same state transition sequence from ⓕ → ⓞ, but illustrates more detail for the intermediate steps. Figure 5.63 (a), (b), and (c) correspond to the paths shown in Figure 5.62 (a), (b), and (c), respectively. The three paths are summarized as follows:

ⓕ → g → c → ⓞ
ⓕ → g → k → ⓞ
ⓕ → g → ⓞ

Thus, three state transition sequences are possible, where each sequence results from a different propagation delay time through the excitation variable storage elements. Although a race condition exists, stable state ⓞ is the destination state for all three sequences. This is referred to as a *noncritical race* condition, since the destination stable state is the same regardless of the path taken.

$(f) \rightarrow g \rightarrow c \rightarrow (o)$

$(f)\ Y_{1e}Y_{2e} = 01 \rightarrow g\ (Y_{1e}Y_{2e} = 10)$

$\downarrow$

$Y_{1e}Y_{2e} = 00$ (Y_{2e} resets before Y_{1e} sets)

$\downarrow \Delta t_1$

$y_{1f}y_{2f} = 00$ (c: $Y_{1e}Y_{2e} = 10$)

$\downarrow \Delta t_2$

$y_{1f}y_{2f} = 10$ ((o): $Y_{1e}Y_{2e} = 10$)

(a)

$(f) \rightarrow g \rightarrow k \rightarrow (o)$

$(f)\ Y_{1e}Y_{2e} = 01 \rightarrow g\ (Y_{1e}Y_{2e} = 10)$

$\downarrow$

$Y_{1e}Y_{2e} = 11$ (Y_{1e} sets before Y_{2e} resets)

$\downarrow \Delta t_1$

$y_{1f}y_{2f} = 11$ (k: $Y_{1e}Y_{2e} = 10$)

$\downarrow \Delta t_2$

$y_{1f}y_{2f} = 10$ ((o): $Y_{1e}Y_{2e} = 10$)

(b)

$(f) \rightarrow g \rightarrow (o)$

$(f)\ Y_{1e}Y_{2e} = 01 \rightarrow g$ ($Y_{1e}Y_{2e} = 10$; Y_{1e} and Y_{2e} change state simultaneously)

$\downarrow \Delta t$

$y_{1f}y_{2f} = 10$ ((o): $Y_{1e}Y_{2e} = 10$)

(c)

Figure 5.63 Three possible paths for the state transition sequence (f) ($y_{1f}y_{2f}x_1x_2 = 0101$) to (o) ($y_{1f}y_{2f}x_1x_2 = 1011$): (a) Y_{2e} resets before Y_{1e} sets; (b) Y_{1e} sets before Y_{2e} resets; and (c) Y_{1e} and Y_{2e} change state simultaneously.

With respect to an input vector of $x_1x_2 = 11$ while in state ⓕ, the machine operates in a nondeterministic manner; that is, the precise path of circuit operation cannot be predicted. Different circuit delay magnitudes cause different paths to be realized providing a state transition sequence that is speed independent from stable state ⓕ to stable state ⓞ. The final state which the circuit attains does not depend on the order in which the excitation variables change state.

5.6.2 Cycles

If the excitation map of Figure 5.61 is changed slightly so that the excitation variables of state g are altered from $Y_{1e}Y_{2e} = 10$ to 11, then there will be only one path from state ⓕ to state ⓞ. Thus, when input x_1 changes from 0 to 1 in state ⓕ, the machine will execute the following sequence only:

$$ⓕ \rightarrow g \rightarrow k \rightarrow ⓞ$$

A sequence of this type, in which an asynchronous sequential machine proceeds through a unique sequence of unstable states and terminates in a stable state, is called a *cycle*. A noncritical race and a cycle, therefore, possess the same terminal characteristic — every sequence ends in the same stable state. Three paths are possible for a noncritical race, but only one path exists for a cycle.

5.6.3 Critical Races

Consider the excitation map of Figure 5.64. If the machine is presently in state ⓙ and input x_1 changes from 0 to 1, then three possible paths exist depending on the relative propagation delays of the storage elements and associated circuitry. The intended path is from state ⓙ to state ⓒ. Due to differing delay characteristics, however, the machine may terminate the sequence in either state ⓒ or state ⓞ. Figure 5.65 illustrates the three possible state transition sequences.

In Figure 5.65 (a), the storage element corresponding to excitation variable Y_{2e} resets before Y_{1e} resets; that is, $Y_{1e}Y_{2e} = 11$ changes to transitory state $Y_{1e}Y_{2e} = 10$. After a delay of Δt, the values of the feedback variables become equal to the values of the excitation variables. The machine then sequences to row $y_{1f}y_{2f} = 10$ while remaining in column $x_1x_2 = 11$ and enters stable state ⓞ. This is an incorrect destination state for the input vector.

Figure 5.65 (b) depicts a sequence in which Y_{1e} resets before Y_{2e} resets, taking the machine to a transitory state of $Y_{1e}Y_{2e} = 01$. After a delay, the feedback variables become equal to the excitation variables and the operation proceeds to row $y_{1f}y_{2f} = 01$ and enters transient state g. In state g, $Y_{1e}Y_{2e} = 00$. Therefore, after an additional delay, $y_{1f}y_{2f} = 00$ and the machine enters stable state ⓒ. Figure 5.65 (c) shows the sequence when both excitation variables change state simultaneously. The machine proceeds to transient state k and from there directly to state ⓒ after an appropriate delay. Depending on the relative propagation delays of the storage elements and associated circuitry, the destination stable state will be either state ⓒ or state ⓞ.

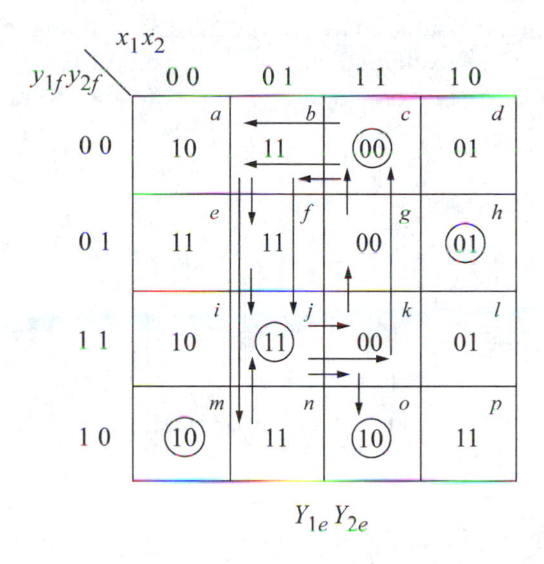

Figure 5.64 Excitation map for an asynchronous sequential machine illustrating a critical race condition when input x_1 changes from 0 to 1 in stable state ⓙ.

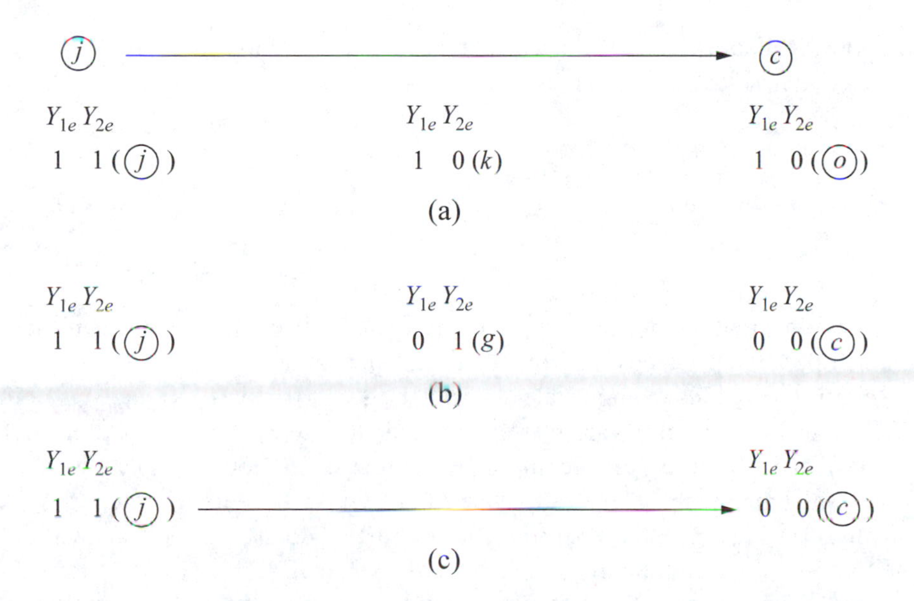

Figure 5.65 Three possible paths for the intended state transition sequence ⓙ ($y_{1f}y_{2f}x_1x_2 = 1101$) to ⓒ ($y_{1f}y_{2f}x_1x_2 = 0011$): (a) Y_{2e} resets before Y_{1e} resets; (b) Y_{1e} resets before Y_{2e} resets; and (c) Y_{1e} and Y_{2e} change value simultaneously.

The single change to the input vector results in a race condition, because both excitation variables change value ($Y_{1e}Y_{2e}$ = 11 to 00). However, since the destination stable state cannot be predicted, the machine will proceed to either stable state ⓒ or stable state ⓞ. This is termed a *critical race* condition and must be avoided.

As a final example, consider an asynchronous sequential machine whose operation is characterized by the excitation map of Figure 5.66. This map illustrates more subtle oscillations, noncritical races, and cycles. There are several sequences to analyze.

$y_{1f}y_{2f}$ \ x_1x_2	0 0	0 1	1 1	1 0
0 0	*a* 01	*b* (00)	*c* 10	*d* (00)
0 1	*e* 11	*f* 11	*g* (01)	*h* 10
1 1	*i* (11)	*j* 10	*k* 01	*l* 01
1 0	*m* 11	*n* 00	*o* 11	*p* 11

$Y_{1e}Y_{2e}$

Figure 5.66 Excitation map for an asynchronous sequential machine containing oscillations, noncritical races, and cycles.

First, assume that the machine is stable in state ⓖ. If input x_2 changes from 1 to 0, then the operation proceeds horizontally to transient state *h*. During this sequence, both excitation variables change state — from $Y_{1e}Y_{2e}$ = 01 to 10. If Y_{1e} sets before Y_{2e} resets, then the excitation variables enter transient state $Y_{1e}Y_{2e}$ = 11. After a delay of Δt, the feedback variables become equal to the excitation variables such that, $y_{1f}y_{2f}$ = 11 and the machine moves to state *l* in row $y_{1f}y_{2f}$ = 11. In state *l*, the excitation values are $Y_{1e}Y_{2e}$ = 01. Therefore, after a further delay, the feedback variables change to $y_{1f}y_{2f}$ = 01 and the machine proceeds to state *h*.

In state *h*, however, both excitation variables again change. If a simultaneous change does not occur, then the excitation variables may again change to $Y_{1e}Y_{2e}$ = 11, sequencing the machine to state *l* after an appropriate delay. Thus, an oscillation is possible between transient states *h* and *l* depending on the set and reset characteristics of the excitation variables. The oscillation results from a race condition caused by two changes of excitation variables ($Y_{1e}Y_{2e}$ = 01 to 10) from stable state ⓖ to transient state h. The oscillation can be described as ⓖ $\rightarrow h \leftrightarrow l$.

The same oscillation can occur from a different initial stable state. If the machine is in stable state ⓘ and input x_1 changes from 0 to 1, then the operation sequences horizontally to transient state l. This does not constitute a race condition, because only one excitation variable changes state ($Y_{1e}Y_{2e}$ = 11 to 01). The oscillation may occur as previously described, however, when the excitation variables in state l both change value in moving from state l to state h.

A noncritical race is also possible beginning in stable state ⓖ if x_2 changes from 1 to 0. If Y_{2e} resets before Y_{1e} sets, then the excitation variables enter transient state $Y_{1e}Y_{2e} = 00$. After a delay of Δt, the feedback variables become equal to the excitation variables such that, $y_{1f}y_{2f} = 00$. The machine then enters stable state ⓓ. This path is described as ⓖ $\rightarrow h \rightarrow$ ⓓ. Likewise, if Y_{1e} and Y_{2e} change from 01 to 11, then the machine moves to state l and then to state h where a double change of excitation values occurs. Assuming that an oscillation does not develop between transient states l and h, the excitation variables may change from $Y_{1e}Y_{2e} = 01$ to 00. The machine then proceeds to state ⓓ after an appropriate delay. This sequence is described as ⓖ $\rightarrow h \rightarrow l \rightarrow h \rightarrow$ ⓓ.

Similarly, if both excitation variables change state simultaneously, then the machine will sequence from state ⓖ to state p, and from there to state l, then to state h. Again, assuming that no oscillation will occur, the sequence terminates in stable state ⓓ. This sequence is described as ⓖ $\rightarrow h \rightarrow p \rightarrow l \rightarrow h \rightarrow$ ⓓ. All three state transition sequences terminate in the same stable state. Therefore, when input x_2 changes from 1 to 0 in state ⓖ, a noncritical race condition results. A similar noncritical race occurs beginning in state ⓘ when input x_1 changes from 0 to 1.

Figure 5.66 also contains several cycles, in which a unique path exists from the beginning stable state to the destination stable state. Three of the longer cycles are as follows:

ⓑ $\rightarrow a \rightarrow e \rightarrow$ ⓘ

ⓑ $\rightarrow c \rightarrow o \rightarrow k \rightarrow$ ⓖ

ⓖ $\rightarrow f \rightarrow j \rightarrow n \rightarrow$ ⓑ

An obvious restriction is imposed on the n-tuple input vector: After a single input changes value, the input vector must not change again until the machine has sequenced through all possible transient states from the beginning stable state to the destination stable state. For example, if the longest path results in propagation delays of Δt_1, Δt_2, Δt_3, and Δt_4, then the input must not change twice in the interval of time represented by the aggregate delay of the four circuit propagation delays. This restriction is necessary in order to reliably predict machine performance. Thus, a machine must be allowed to sequence to a stable state with respect to an input change before another change is applied to the input alphabet. The constraint thus imposed permits reliable behavior in a fundamental-mode operation.

Races can be avoided when it is possible to direct the machine through intermediate unstable states before reaching the destination stable state. This can be achieved by utilizing some of the unspecified entries in the excitation map. Also, it may be possible to add rows to the excitation map without increasing the number of excitation and feedback variables. The elimination of races will be discussed in Chapter 6.

5.7 Problems

5.1 Given the logic diagram shown below for a Moore asynchronous sequential machine, obtain the excitation map, the flow table, and the state diagram.

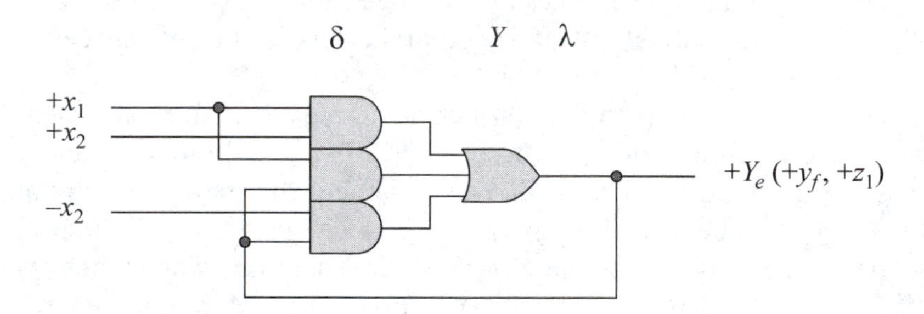

5.2 Given the logic diagram shown below for a Moore asynchronous sequential machine, obtain the excitation map, the flow table, and the state diagram. Note any unusual characteristics in the operation of the machine.

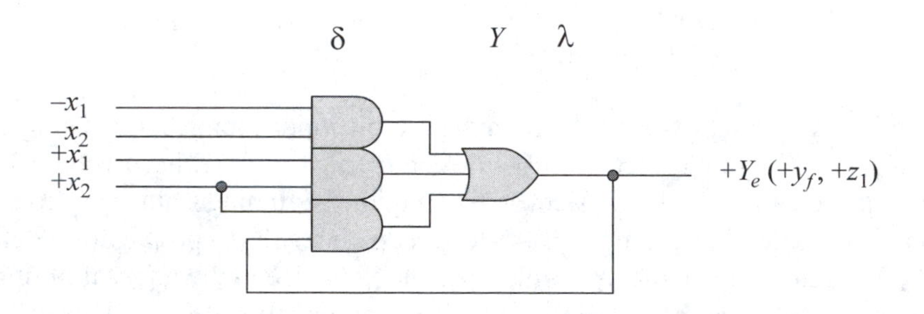

5.3 Given the logic diagram shown below for a Mealy asynchronous sequential machine, obtain the excitation map and the flow table.

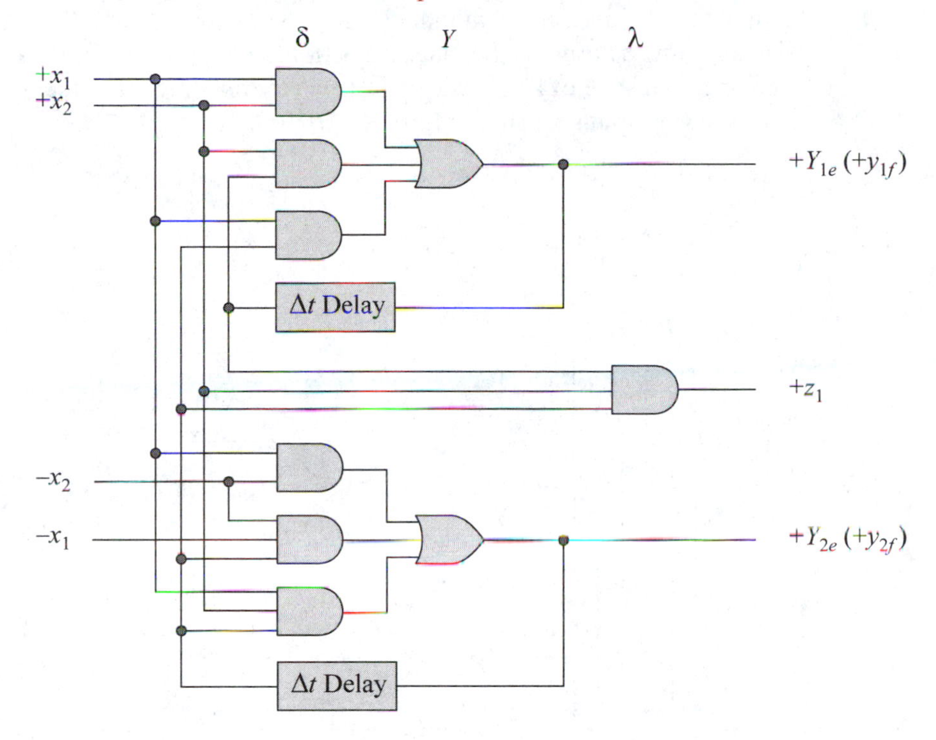

5.4 Analyze the Moore asynchronous sequential machine shown below by generating an excitation map and a flow table. Let the machine be initially reset to $y_{1f}y_{2f}x_1x_2 = 0000$. Determine the shortest input sequence which will assert output z_1.

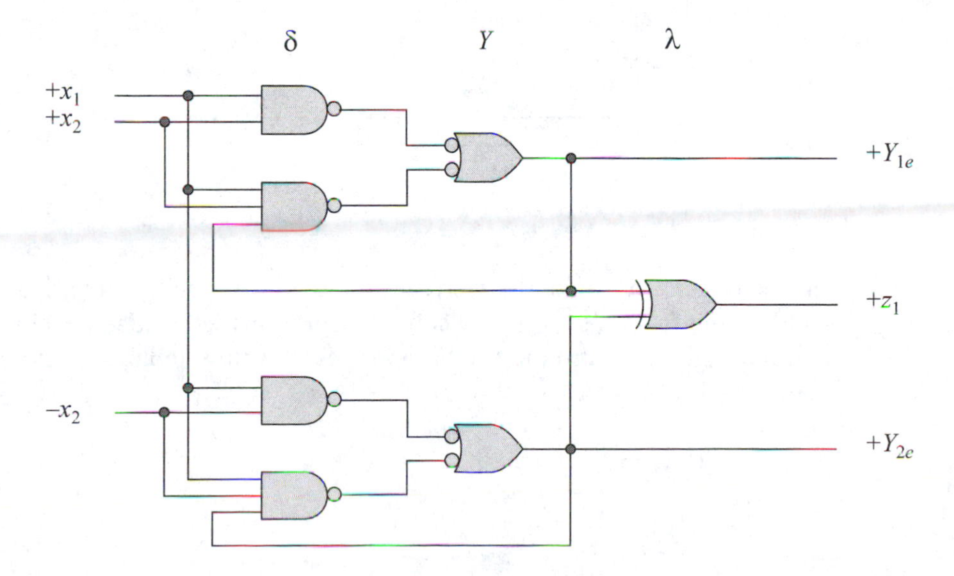

5.5 A Mealy asynchronous sequential machine is shown below with two feedback paths and one output z_1. Obtain the excitation maps and the flow table. Assume that the machine is initially set to $y_{1f}y_{2f}x_1x_2 = 1001$. From this initial condition, determine the state transition sequences that will assert z_1. From an initial state of $y_{1f}y_{2f}x_1x_2 = 0011$, determine the output sequence for the following input sequence: 11, 01, 11, 10, 00, 10.

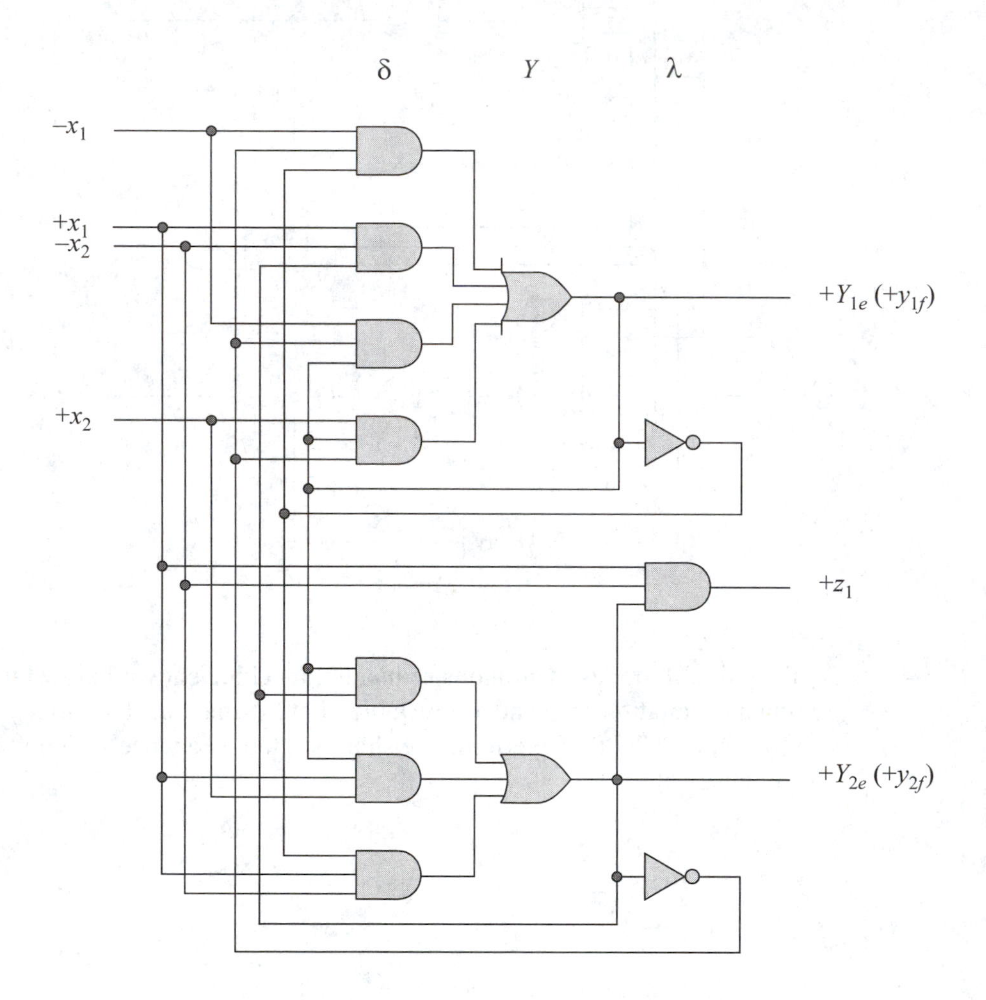

5.6 The equations shown below represent the logic for excitation variables Y_{1e} and Y_{2e}, and for output z_1 for a Mealy asynchronous sequential machine. Obtain the logic diagram, the excitation maps, the flow table, and the output map.

$$Y_{1e} = y_{1f}x_1 + y_{1f}x_2 + y_{1f}x_3 + y_{2f}x_3' + y_{1f}y_{2f}$$
$$Y_{2e} = x_2 + x_1x_3' + y_{2f}x_3'$$
$$z_1 = y_{1f}y_{2f}x_2 + y_{1f}y_{2f}'x_2'x_3$$

5.7 Using the excitation map shown below for a Mealy asynchronous sequential machine, draw a timing diagram for the following three sequences:

(j) → i → e → (a)
(a) → (d)
(d) → c → (g)

The equation for output z_1 is: $y_{1f}y_{2f}x_2$

$y_{1f}y_{2f}$ \ x_1x_2	0 0	0 1	1 1	1 0
0 0	*a* (00)	*b* 10	*c* 01	*d* (00)
0 1	*e* 00	*f* (01)	*g* (01)	*h* 00
1 1	*i* 01	*j* (11)	*k* (11)	*l* 10
1 0	*m* 11	*n* 11	*o* 11	*p* (10)

$Y_{1e}Y_{2e}$

5.8 The equations for excitation variables Y_{1e}, Y_{2e} and output z_1 are shown below for a Mealy asynchronous sequential machine.

$$Y_{1e} = x_1x_2' + y_{1f}y_{2f}'x_2 + y_{2f}x_1'x_2$$
$$Y_{2e} = x_1'x_2 + y_{2f}x_2$$
$$z_1 = y_{1f}y_{2f}x_2 + y_{1f}'y_{2f}'x_2$$

(a) Draw the logic diagram using only NOR logic with complementary outputs.

(b) Obtain the excitation maps for the individual excitation variables.

(c) Obtain the combined excitation map and indicate the stable states.

(d) Obtain the output map for z_1.

5.9 The logic diagram shown below represents a Moore asynchronous sequential machine using NOR logic with complementary outputs.

(a) Obtain the individual excitation maps for Y_{1e} and Y_{2e}.

(b) Obtain the combined excitation map for Y_{1e} and Y_{2e}, indicate the stable states, and show in which states output z_1 is asserted.

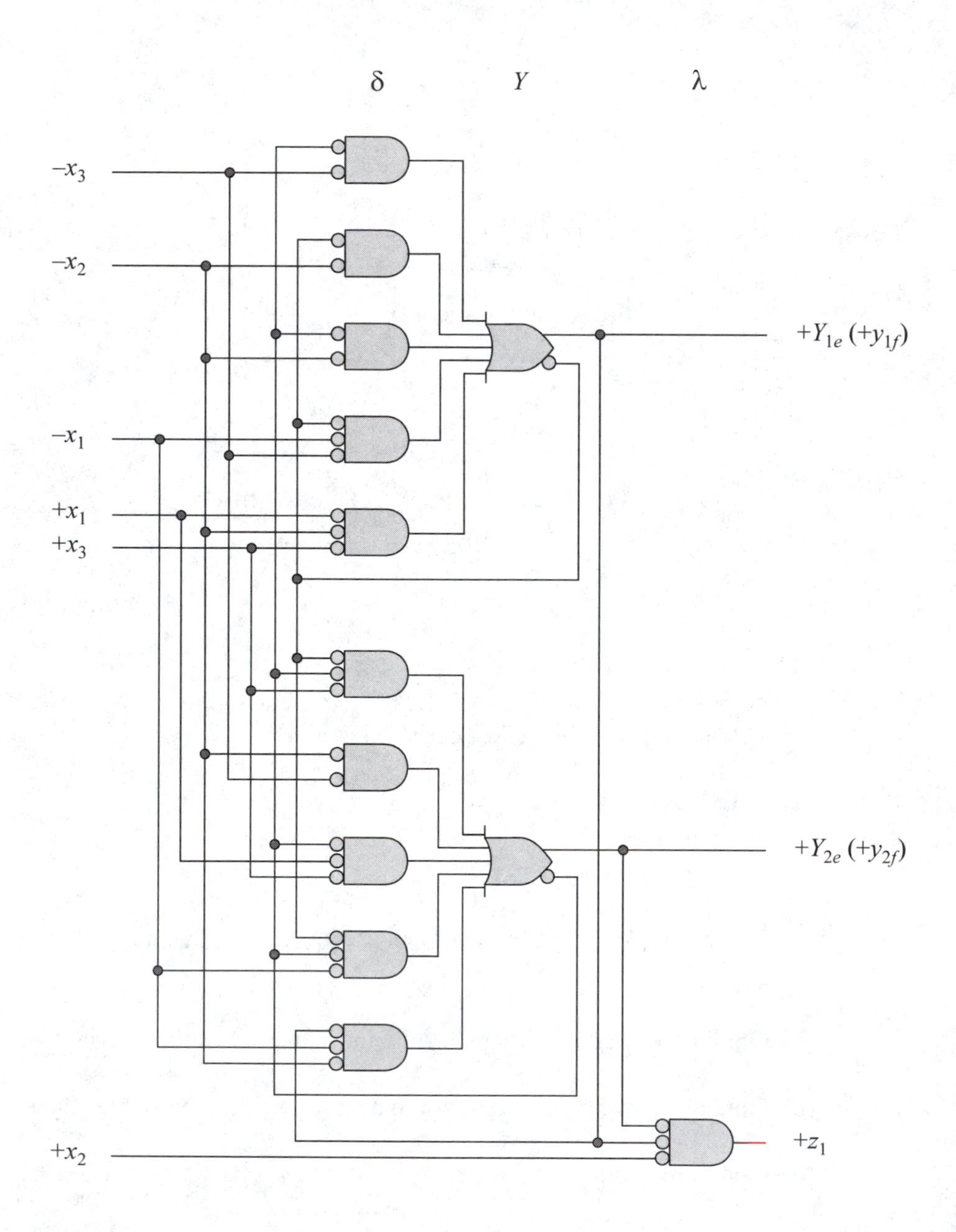

5.10 The excitation map for a Mealy asynchronous sequential machine is shown below for excitation variables Y_{1e} and Y_{2e}. Generate a timing diagram for the following state transition sequences, where each 5-tuple represents a stable state:

$y_{1f}y_{2f}x_1x_2x_3$ = 00000, 00010, 11110, 11111, 11101, 10001
The equation for the output is: $z_1 = y_{1f}y_{2f}x_2x_3' + y_{1f}y_{2f}x_1x_3$

$x_3 = 0$

$y_{1f}y_{2f}$ \ x_1x_2	0 0	0 1	1 1	1 0
0 0	*a* (00)	*b* (00)	10	*c* (00)
0 1	00	*d* (01)	11	00
1 1	10	*e* (11) z_1	*f* (11) z_1	*g* (11)
1 0	00	*h* (10)	11	01

Y_{1e}

$x_3 = 1$

$y_{1f}y_{2f}$ \ x_1x_2	0 0	0 1	1 1	1 0
0 0	*i* (00)	01	10	*j* (00)
0 1	00	*k* (01)	11	00
1 1	10	*l* (11)	*m* (11) z_1	*n* (11) z_1
1 0	*o* (10)	11	11	11

Y_{2e}

5.11 Given the equation shown below for z_1, identify all static-1 and static-0 hazards and determine the hazard covers.

$$z_1 = x_2'x_3' + x_1x_2$$

5.12 Given the equation shown below for z_1, identify all static-1 and static-0 hazards and determine the hazard covers.

$$z_1 = x_1'x_2x_3 + x_1x_3' + x_1'x_2'x_3$$

5.13 Given the Karnaugh shown below for function z_1, analyze all transitions resulting from a single change to the input vector and identify all static-1 and static-0 hazards. The inputs are active high. Draw the logic diagram in a sum-of-products form and a product-of-sums form, both containing no hazards.

x_1x_2 \ x_3x_4	0 0	0 1	1 1	1 0
0 0	1	1	0	0
0 1	1	1	0	1
1 1	0	0	0	1
1 0	0	0	0	0

z_1

5.14 Given the equation for z_1 shown below, identify all static-1 and static-0 hazards, determine the hazard cover terms, if applicable, and draw the hazard-free logic diagram as a sum-of-products and product-of-sums implementation. All inputs are active high.

$$z_1 = x_1'x_3' + [x_3 (x_2 + x_4') (x_1 + x_4)]$$

5.15 Given the equation for z_1 shown below, identify all static-1 and static-0 hazards and determine the hazard covers.

$$z_1 = x_2x_3x_5' + x_2x_3x_4' + x_1x_2'x_4$$

5.16 Given the equation for z_1 shown below, identify all static-1 hazards and determine the hazard covers.

$$z_1 = x_1'x_3'x_5' + x_2x_3'x_4x_5' + x_2x_3x_4x_5 + x_1x_3'x_4x_5$$

5.17 Given the expression for z_1 shown below, identify all static-1 and static-0 hazards.

$$z_1(x_1, x_2, x_3, x_4) = \Sigma_m(1, 5, 6, 7, 13, 14, 15)$$

5.18 Given the expression for z_1 shown below, identify all static-1 and static-0 hazards.

$$z_1(x_1, x_2, x_3, x_4) = \Sigma_m(0, 2, 5, 7, 8, 10, 13, 15)$$

5.19 Given the Karnaugh map shown below for function z_1, indicate which equations listed in parts (a) through (d) contain no static-1 hazards.

x_1x_2 \ x_3x_4	0 0	0 1	1 1	1 0
0 0	0	1	0	0
0 1	0	1	1	0
1 1	0	0	1	1
1 0	0	0	0	0

z_1

(a) $z_1 = \{[x_1'x_4(x_2 + x_3')]' \, [x_2x_3(x_1 + x_4)]'\}'$
(b) $z_1 = x_1'x_4(x_2 + x_3') + x_2x_3(x_1 + x_4)$
(c) $x_1 = x_1'x_4 + x_2x_3x_4 + x_1x_2x_3$
(d) $z_1 = \{[x_1'x_4(x_2'x_3)']' \, [x_2x_3(x_1'x_4')']'\}$

5.20 Given the Karnaugh map shown below, obtain the equation for function z_1 in a minimum sum-of-products form in which all possible static-1 hazards are covered. The equation is to contain the fewest number of terms.

x_1x_2 \ x_3x_4	0 0	0 1	1 1	1 0
0 0	0	1	0	0
0 1	0	1	1	0
1 1	0	0	1	0
1 0	0	0	–	1

z_1

5.21 Given the Karnaugh map shown below, obtain the equation for function z_1 in a minimum sum-of-products form in which all possible static-1 hazards are covered.

x_1x_2 \ x_3x_4	0 0	0 1	1 1	1 0
0 0	1	1	0	–
0 1	1	1	0	–
1 1	0	0	1	0
1 0	0	1	1	0

z_1

5.22 Determine the logic levels of output z_1 for the following equation when input x_2 changes from a low to a high logic level:

$$z_1 = x_1x_2'(x_2 + x_3') + (x_2 + x_3')'x_4$$

The circuit is implemented using only NAND logic and inverters. The circuit is stable in state $x_1x_2x_3x_4 = 1111$. Indicate whether the resulting output levels represent a static or dynamic hazard.

5.23 Given the logic circuit shown below, obtain the waveform for output z_1 when input x_2 changes from a high to a low logic level. The circuit is stable in state $x_1x_2x_3x_4x_5 = 11001$. Indicate whether the resulting output levels represent a static or dynamic hazard.

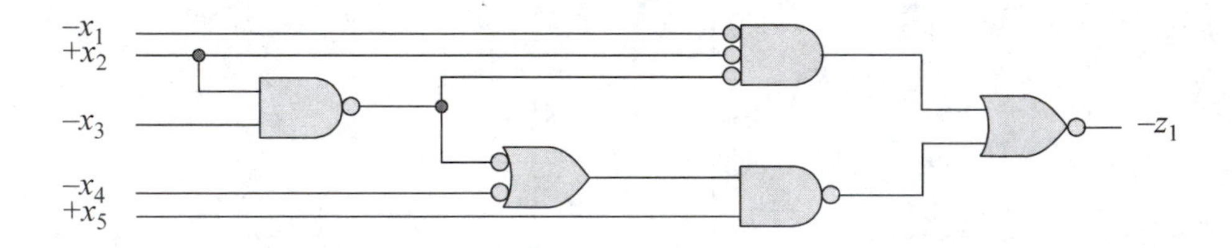

5.24 Given the following equations for Y_{1e} and Y_{2e}, determine if an essential hazard exists:

$$Y_{1e} = y_{2f}'x_1 + y_{1f}(y_{2f}' + x_1')$$
$$Y_{2e} = y_{1f}x_1' + y_{2f}(y_{1f} + x_1)$$

5.25 Given the following equations for Y_{1e} and Y_{2e}, determine if an essential hazard exists:

$$Y_{1e} = y_{2f}x_1' + y_{1f}x_1 + y_{1f}y_{2f}$$
$$Y_{2e} = y_{2f}x_1' + y_{1f}'x_1 + y_{1f}'y_{2f}$$

5.26 Given the following equations for Y_{1e} and Y_{2e}, determine if an essential hazard exists:

$$Y_{1e} = y_{2f}'x_1' + y_{1f}x_1 + y_{1f}y_{2f}'$$
$$Y_{2e} = y_{2f}x_1' + y_{1f}x_1 + y_{1f}y_{2f}$$

5.27 Given the following equations for Y_{1e} and Y_{2e}, determine if an essential hazard exists:

$$Y_{1e} = x_1$$
$$Y_{2e} = x_2$$

5.28 Given the excitation map shown below for excitation variables Y_{1e} and Y_{2e}, obtain all possible state transition sequences that result in one or more oscillations. Indicate the starting stable state and the path that is taken when a single input changes value; for example, ⓠ $\rightarrow r \leftrightarrow s$.

$y_{1f}y_{2f}$ \ x_1x_2	0 0	0 1	1 1	1 0
0 0	a: 01	b: 01	c: 01	d: 01
0 1	e: 01	f: 01	g: 00	h: 11
1 1	i: 01	j: 11	k: 01	l: 10
1 0	m: 11	n: 11	o: 10	p: 10

$Y_{1e}Y_{2e}$

5.29 Given the excitation map shown below for excitation variables Y_{1e} and Y_{2e}, obtain all possible state transition sequences that result in one or more oscillations. Indicate the starting stable state and the path that is taken when a single input changes value; for example, ⓠ $\rightarrow r \leftrightarrow s$.

$y_{1f}y_{2f}$ \ x_1x_2	0 0	0 1	1 1	1 0
0 0	*a* 01	*b* 01	*c* 01	*d* 01
0 1	*e* 01	*f* 11	*g* 01	*h* 10
1 1	*i* 11	*j* 11	*k* 10	*l* 10
1 0	*m* 01	*n* 01	*o* 00	*p* 11

$Y_{1e}Y_{2e}$

5.30 Given the excitation map shown below for excitation variables Y_{1e} and Y_{2e}, obtain all possible state transition sequences that result in one or more oscillations. Indicate the starting stable state and the path that is taken when a single input changes value; for example, ⓠ $\rightarrow r \leftrightarrow s$.

$y_{1f}y_{2f}$ \ x_1x_2	0 0	0 1	1 1	1 0
0 0	*a* 10	*b* 01	*c* 10	*d* 00
0 1	*e* 01	*f* 00	*g* 11	*h* 11
1 1	*i* 10	*j* 01	*k* 11	*l* 01
1 0	*m* 10	*n* 11	*o* 11	*p* 11

$Y_{1e}Y_{2e}$

5.31 Given the excitation map shown below for excitation variables Y_{1e} and Y_{2e}, specify all oscillations and indicate the path for each oscillation.

$x_3 = 0$

$y_{1f}y_{2f}$ \ x_1x_2	0 0	0 1	1 1	1 0
0 0	*a* 10	*b* 00	*c* 01	*d* 10
0 1	*e* 11	*f* 11	*g* 11	*h* 01
1 1	*i* 01	*j* 11	*k* 11	*l* 01
1 0	*m* 11	*n* 11	*o* 11	*p* 10

Y_{1e}

$x_3 = 1$

$y_{1f}y_{2f}$ \ x_1x_2	0 0	0 1	1 1	1 0
0 0	*q* 00	*r* 01	*s* 01	*t* 00
0 1	*u* 01	*v* 11	*w* 01	*x* 11
1 1	*y* 01	*z* 11	*aa* 10	*bb* 01
1 0	*cc* 00	*dd* 11	*ee* 10	*ff* 11

Y_{2e}

5.32 Identify all critical race conditions in the excitation map shown below. Although the elimination of race conditions is presented in Chapter 6, attempt to eliminate the critical race conditions in this problem by changing the excitation variables in certain transient states. Then determine if the resulting excitation map contains any possible hazard conditions.

$y_{1f}y_{2f}$ \ x_1x_2	0 0	0 1	1 1	1 0
0 0	*a* 00	*b* 10	*c* 00	*d* 11
0 1	*e* 11	*f* 01	*g* 10	*h* 11
1 1	*i* 11	*j* 01	*k* 10	*l* 11
1 0	*m* 00	*n* 01	*o* 10	*p* 10

$Y_{1e}Y_{2e}$

5.33 Given the excitation map shown below for excitation variables Y_{1e} and Y_{2e}, specify all noncritical and critical race conditions and indicate the paths for each condition.

$x_3 = 0$

$y_{1f}y_{2f}$ \ x_1x_2	0 0	0 1	1 1	1 0
0 0	*a* 00	*b* 01	*c* 11	*d* 00
0 1	*e* 00	*f* 11	*g* 10	*h* 01
1 1	*i* 01	*j* 11	*k* 10	*l* 00
1 0	*m* 10	*n* 11	*o* 10	*p* 11

y_{1e}

$x_3 = 1$

$y_{1f}y_{2f}$ \ x_1x_2	0 0	0 1	1 1	1 0
0 0	*q* 00	*r* 10	*s* 01	*t* 00
0 1	*u* 01	*v* 11	*w* 01	*x* 11
1 1	*y* 11	*z* 11	*aa* 01	*bb* 11
1 0	*cc* 01	*dd* 11	*ee* 10	*ff* 11

Y_{2e}

5.34 Given the excitation map shown below for an asynchronous sequential machine, list all possible transitions (paths) that could occur for noncritical races, critical races, and oscillations. Indicate the starting stable state and the path that is taken when a single input changes.

$y_{1f}y_{2f}$ \ x_1x_2	0 0	0 1	1 1	1 0
0 0	*a* 01	*b* 01	*c* 01	*d* 01
0 1	*e* 01	*f* 11	*g* 01	*h* 10
1 1	*i* 11	*j* 11	*k* 10	*l* 10
1 0	*m* 01	*n* 01	*o* 00	*p* 11

$Y_{1e}Y_{2e}$

5.35 Given the excitation map shown below for an asynchronous sequential machine, list all possible transitions (paths) that could occur for noncritical races, critical races, and oscillations. Indicate the starting stable state and the path that is taken when a single input changes.

$y_{1f}y_{2f}$ \ x_1x_2	0 0	0 1	1 1	1 0
0 0	*a* 10	*b* 01	*c* 10	*d* 00
0 1	*e* 01	*f* 00	*g* 11	*h* 11
1 1	*i* 10	*j* 01	*k* 11	*l* 01
1 0	*m* 10	*n* 11	*o* 11	*p* 11

$Y_{1e}Y_{2e}$

5.36 Given the excitation map shown below for an asynchronous sequential machine, list all possible transitions (paths) that could occur for noncritical races, critical races, and oscillations. Indicate the starting stable state and the path that is taken when a single input changes.

$y_{1f}y_{2f}$ \ x_1x_2	0 0	0 1	1 1	1 0
0 0	*a* 01	*b* 01	*c* 01	*d* 01
0 1	*e* 01	*f* 01	*g* 00	*h* 11
1 1	*i* 01	*j* 11	*k* 01	*l* 10
1 0	*m* 11	*n* 11	*o* 10	*p* 10

$Y_{1e}Y_{2e}$

5.37 Given the excitation map shown below for an asynchronous sequential machine, list all possible transitions (paths) that could occur for noncritical races, critical races, and oscillations. Indicate the starting stable state and the path that is taken when a single input changes.

$y_{1f}y_{2f}$ \ x_1x_2	00	01	11	10
00	*a* 00	*b* 00	*c* 10	*d* 00
01	*e* 10	*f* 00	*g* 01	*h* 01
11	*i* 11	*j* 10	*k* 01	*l* 11
10	*m* 11	*n* 11	*o* 10	*p* 10

$Y_{1e}Y_{2e}$

6.1 Introduction
6.2 Synthesis Procedure
6.3 Synthesis Examples
6.4 Problems

6

Synthesis of Asynchronous Sequential Machines

In previous chapters, the analysis and synthesis of synchronous sequential machines was presented in which a periodic clock signal controlled the operation of the machine. The effect of erroneous transient signals was nullified by the utilization of a system clock. In some situations, however, a synchronous sequential machine may be too expensive or too slow when compared to a functionally equivalent asynchronous sequential machine. Also, the synchronization of the binary input variables may not be possible. These diverse characteristics provide the rationale for the utilization of asynchronous sequential machines. Transient signals are handled differently in the synthesis procedure for asynchronous machines. Techniques will be presented in this chapter to synthesize fundamental-mode asynchronous sequential machines irrespective of the varying delays of circuit components.

The synthesis of asynchronous sequential machines is one of the most interesting and certainly the most challenging concepts of sequential machine design. In many situations, a synchronous clock is not available. The interface between an input/output processor (IOP) — or channel — and an input/output (I/O) subsystem control unit is an example of an asynchronous condition. Many large computer channels communicate with an I/O subsystem by means of a signal interlocking protocol on the interface. The control unit requests a word of data during a write operation by asserting an identifying epithet called a tag-in signal. The channel then places the word on the data bus and asserts an acknowledging tag-out signal. The device control unit accepts the data then deasserts the in tag, allowing the channel to deassert the corresponding out tag, completing the data transfer sequence for one word. An analogous situation occurs for a read operation in which the tag-in signal now indicates that a word is available on the data bus for the channel. The channel accepts the word and responds with the tag-out signal.

The data transfer sequence for the write and read operations was initiated, executed, and completed without utilizing a synchronizing clock signal. This technique permits not only a higher data transfer rate between the channel and an I/O device, but also allows the channel to communicate with I/O devices having a wide range of data transfer rates. The interface control logic in the device control unit is usually implemented as an asynchronous sequential machine. Even in large synchronous systems, it is often advantageous to allow certain subsystems to operate in an asynchronous manner, thereby increasing the overall speed of the system.

6.1 Introduction

Moore and Mealy attributes are also found in asynchronous sequential machines and possess similar characteristics to those encountered in synchronous sequential machines. The input alphabet X is a nonempty finite set of inputs such that,

$$X = \{X_0, X_1, X_2, \cdots, X_{2^n-1}\}$$

where n is the number of binary input variables. The state alphabet Y is a nonempty finite set of states such that,

$$Y = \{Y_0, Y_1, Y_2, \cdots, Y_{2^p-1}\}$$

where p is the number of storage elements. The machine can be in any one of 2^p exclusive states. Alternatively, if the machine has s distinct states, then the number of state variables must satisfy the expression $2^p \geq s$. The output alphabet Z is a nonempty finite set of outputs such that,

$$Z = \{Z_0, Z_1, Z_2, \cdots, Z_{2^m-1}\}$$

where m is the number of binary output variables. The output vector can be a function of the present state only, for Moore machines, or a function of both the present state and the present inputs for Mealy machines.

The δ next-state logic is a function of the input alphabet only, or a function of both the inputs and the present state. That is, δ is a function of X which maps X into Y such that, $\delta(X){:}X \rightarrow Y$ or δ is a function of X and Y which maps the Cartesian product of X and Y into Y such that, $\delta(X, Y){:}X \times Y \rightarrow Y$. The output can be a function of either the present state only, for Moore machines, where $\lambda(Y){:}Y \rightarrow Z$ or a function of both the present inputs and the present state for Mealy machines, which maps the Cartesian product of X and Y into Z such that, $\lambda(X, Y){:}X \times Y \rightarrow Z$. The equations for an asynchronous sequential machine are specified in Equation 6.1.

$$
\begin{array}{ll}
X = \{X_0, X_1, X_2, \cdots, X_{2^n-1}\} & \\
Y = \{Y_0, Y_1, Y_2, \cdots, Y_{2^p-1}\} & \\
Z = \{Z_0, Z_1, Z_2, \cdots, Z_{2^m-1}\} & \\
\delta(X){:}X \rightarrow Y & \\
\delta(X, Y){:}X \times Y \rightarrow Y & \\
\lambda(Y){:}Y \rightarrow Z & \text{(Moore)} \\
\lambda(X, Y){:}X \times Y \rightarrow Z & \text{(Mealy)}
\end{array}
\tag{6.1}
$$

The fundamental-mode model is shown in Figure 6.1. The input alphabet X consists of n binary variables $x_1, x_2, \ldots, x_n$ that change value in an asynchronous manner. Since the machine is characterized by a fundamental-mode operation, only one input variable x_i is allowed to change state at a time and no other input variable x_j will change state until the machine has sequenced to a stable state. This operating characteristic precludes the possibility of an inherent race condition.

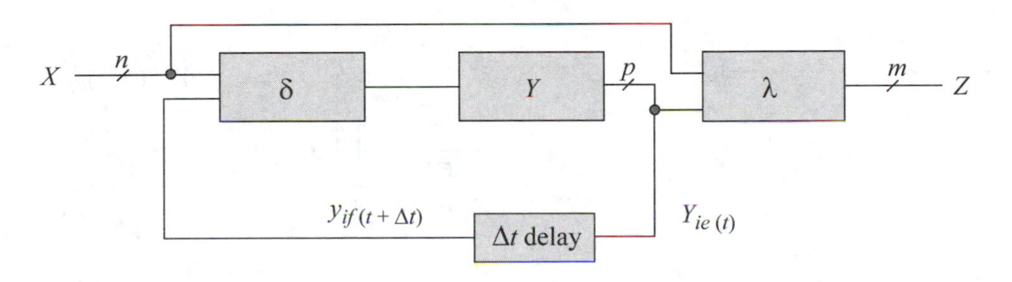

Figure 6.1 General block diagram of a fundamental-mode asynchronous sequential machine.

The state alphabet Y contains p storage elements in the form of *Set Reset* (*SR*) latches. The outputs of the latches represent the excitation variables $Y_{1e}, Y_{2e}, \ldots, Y_{pe}$. After a delay of Δt, these same outputs represent the feedback variables $y_{1f}, y_{2f}, \ldots, y_{pf}$ and connect to the AND gates of the *SR* latches. Thus, when the feedback variables become equal to the excitation variables, the machine enters a stable state where $y_{if} = Y_{ie}$. The output alphabet Z is represented by $z_1, z_2, \ldots, z_m$.

In Figure 6.1, the excitation variable Y_{ie} is fed back to the δ next-state logic, part of which constitutes the *SR* storage element. This feedback variable y_{if} connects to the product term in the sum-of-products expression that represents Y_{ie}, as shown in Figure 6.2 (c) and Equation 6.2, where $i = 1$. The feedback to the AND gate, together with the connection from the AND gate to the OR gate, forms the *SR* latch network.

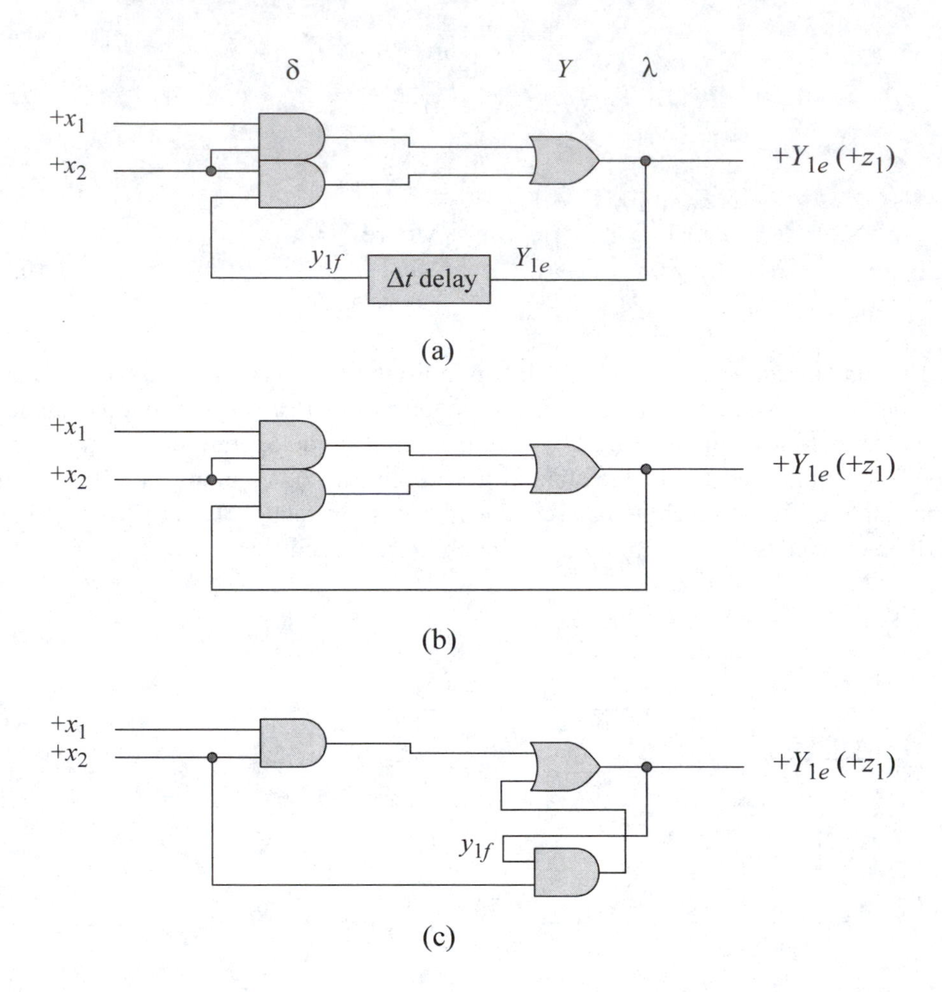

Figure 6.2 An asynchronous sequential machine: (a) fundamental-mode model; (b) conventional circuit with distributive delays; and (c) standard *SR* latch configuration.

$$Y_{1e} = x_1 x_2 + x_2 y_{1f} \tag{6.2}$$

6.2 Synthesis Procedure

The synthesis procedure for asynchronous sequential machines is similar in many respects to that described in Chapter 3 for synchronous sequential machines. This section develops a systematic method for the synthesis of fundamental-mode asynchronous sequential machines using *SR* latches as the storage elements. The machine operation is specified as a timing diagram and/or verbal statements. The synthesis procedure is summarized below. Each step of the procedure will be expounded in detail during the synthesis of several examples. The key step in the synthesis of synchronous sequential machines is the derivation of the state diagram, whereas, the key step in the synthesis of asynchronous sequential machines is the derivation of the primitive flow table. The synthesis procedure will result in a machine that operates according to the prescribed specifications. The solution, however, may not be unique.

1. **State diagram** The machine specifications are converted into a state diagram. A timing diagram and/or a verbal statement of the machine specifications is converted into a precise delineation which specifies the machine's operation for all applicable input sequences. This step is not a necessary requirement and is usually omitted; however, the state diagram characterizes the machine's operation in a graphical representation and adds completeness to the synthesis procedure.

2. **Primitive flow table** The machine specifications are converted to a state transition table called a primitive flow table. This is the least methodical step in the synthesis procedure and the most important. The primitive flow table depicts the state transition sequences and output assertions for all valid input vectors. The flow table must correctly represent the machine's operation for all applicable input sequences, even those that are not initially apparent from the machine specifications.

3. **Equivalent states** The primitive flow table may have an inordinate number of rows. The number of rows can be reduced by finding equivalent states and then eliminating redundant states. If the machine's operation is indistinguishable whether commencing in state Y_i or state Y_j, then one of the states is redundant and can be eliminated. The flow table thus obtained, is a *reduced primitive flow table.*

4. **Merger diagram** The merger diagram graphically portrays the result of the merging process in which an attempt is made to combine two or more rows of the reduced primitive flow table into a single row. The result of the merging technique is analogous to that of finding equivalent states; that is, the merging process can also reduce the number of rows in the table and hence, reduce the number of feedback variables that are required. Fewer feedback variables will result in a machine with less logic and therefore, less cost.

5. **Merged flow table** The merged flow table is constructed from the merger diagram. The table represents the culmination of the merging process in which two or more rows of a primitive flow table are replaced by a single equivalent row which contains one stable state for each merged row.

6. **Excitation maps and equations** An excitation map is generated for each excitation variable. Then the transient states are encoded, where applicable, to avoid critical race conditions. Appropriate assignment of the excitation variables for the transient states can minimize the δ next-state logic for the excitation variables. The operational speed of the machine can also be established at this step by reducing the number of transient states through which the machine must sequence during a cycle. Then the excitation equations are derived from the excitation maps. All static-1 and static-0 hazards are eliminated from the network for a sum-of-products or product-of-sums implementation, respectively.

7. **Output maps and equations** An output map is generated for each machine output. Output values are assigned for all nonstable states so that no transient signals will appear on the outputs. In this step, the speed of circuit operation can also be established. Then the output equations are derived from the output maps assuring that all outputs will be free of static-1 and static-0 hazards.

8. **Logic diagram** The logic diagram is implemented from the excitation and output equations using an appropriate logic family.

6.2.1 State Diagram

The state diagram is derived from a given set of machine specifications. The state diagram graphically depicts the functional operation of the machine and illustrates all possible state transition sequences. The stable states are represented by circles; the state names are inserted within the circles. Transient states are not shown in the state diagram initially, since their placement is not readily apparent in this step of the synthesis procedure. Directed arcs designate the state transition sequences. The state diagram for asynchronous sequential machines is similar to that used in synchronous sequential machines. There is one major difference, however: Since the state flip-flops are not encoded to determine the present state, state code assignment is not necessary.

To illustrate the construction of the state diagram, consider an asynchronous sequential machine which has two inputs x_1 and x_2 and one output z_1. The machine is reset initially such that, all excitation variables have a value of 0, output z_1 is deasserted, and inputs $x_1x_2 = 00$. The machine operates according to the input sequence criteria specified in Table 6.1. The circled entries in the input column represent the input vector in which the corresponding state is stable.

Figure 6.3 illustrates the state diagram that depicts the operating characteristics as specified in Table 6.1. The input vectors which cause the state transitions are labeled with bit combinations of x_1x_2. The output is a function of stable state ⓖ only. Each stable state in Figure 6.3 contains one or more entry arcs and three exit paths. The input vector that affected a sequence to state_i will cause the machine to remain in state_i until a new input vector occurs. Thus, in order to leave a stable state, an input must change value. All transient states have the same input values both entering and leaving the transient state. In order to induce a change to a feedback variable, the machine must pass through a transient state.

Table 6.1 Operating characteristics for a Moore asynchronous sequential machine

Present state	Inputs x_1x_2	Next state	Present output z_1
a	00	a	0
	01	b	0
	10	d	0
	11	Invalid	0
b	00	a	0
	01	b	0
	10	Invalid	0
	11	g	0
d	00	a	0
	01	Invalid	0
	10	d	0
	11	g	0
g	00	Invalid	1
	01	b	1
	10	h	1
	11	g	1
h	00	e	0
	01	Invalid	0
	10	h	0
	11	g	0
e	00	e	0
	01	b	0
	10	h	0
	11	Invalid	0

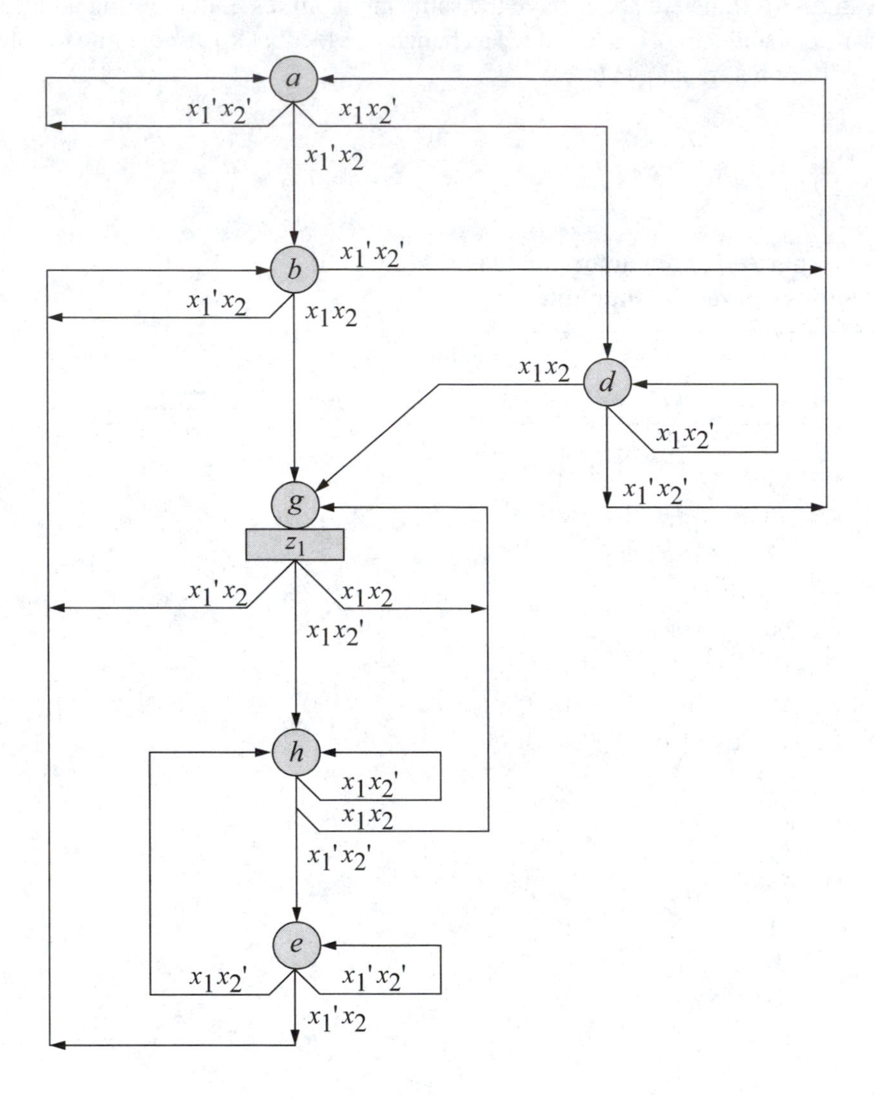

Figure 6.3 State diagram for a Moore asynchronous sequential machine that is characterized by Table 6.1.

Since the machine has two inputs, there are 2^2 resulting paths from each stable state. However, since simultaneous input changes are not allowed in the synthesis of asynchronous sequential machines, only three transitions are possible from each stable state. In Figure 6.3 for example, the machine sequences from state ⓐ to state ⓑ with an input vector of $x_1x_2 = 01$. The machine remains in state ⓑ as long as $x_1x_2 = 01$. In state ⓑ, an input vector of $x_1x_2 = 10$ is invalid; therefore, the remaining two vectors of $x_1x_2 = 00$ or 11 cause the machine to exit state ⓑ and proceed to state ⓐ or ⓖ, respectively.

The state diagram of Figure 6.3 contains two input variables and a total of six stable states. The six stable states can be arranged in an excitation map as shown in Figure 6.4, in which the four columns represent the four different input vectors and the two rows correspond to the two values of the feedback variable y_{1f}. The machine proceeds from state ⓐ ($x_1x_2 = 00$) to states ⓑ or ⓓ for an input vector of $x_1x_2 = 01$ or 10, respectively. Therefore, stable states ⓐ, ⓑ, and ⓓ can be placed in row $y_{1f} = 0$. In order for the machine to move from state ⓐ to states ⓔ, ⓖ, or ⓗ, the machine must sequence through a transient state that has an input vector of $x_1x_2 = 11$. Similarly, states ⓔ, ⓖ, and ⓗ are stable for input vectors of $x_1x_2 = 00$, 11, and 10, respectively, and can be placed in the second row of the excitation map. States ⓐ, ⓑ, and ⓓ can be accessed from the second row by a transition through a transient state which has an input vector of $x_1x_2 = 01$.

y_{1f} \ x_1x_2	0 0	0 1	1 1	10
0	*a* (0)	*b* (0)	*c* 1	*d* (0)
1	*e* (1)	*f* 0	*g* (1) z_1	*h* (1)

Y_{1e}

Figure 6.4 Excitation map for the Moore asynchronous sequential machine of Figure 6.3.

This is a heuristic approach to the derivation of an excitation map and is used here only as a means to further analyze the state diagram of Figure 6.3. The complete state diagram, including all transient states, is shown in Figure 6.5. It is not always possible to proceed directly from one stable state to any other stable state. For example, to move from state ⓐ to state ⓗ requires a transition through unstable state *c*, as indicated by the following state transition sequence: ⓐ → ⓑ → *c* → ⓖ → ⓗ for a requisite input sequence of $x_1x_2 = 00, 01, 11, 10$. This is illustrated in the excitation map of Figure 6.4 and the state diagram of Figure 6.5.

Correspondingly, a state transition sequence from state ⓔ, ⓖ, or ⓗ to any stable state ⓐ, ⓑ, or ⓓ requires a transition through transient state *f*. Output z_1 is active in state ⓖ only. From a reset condition, an input sequence of $x_1x_2 = 00, 01, 11$ will cause the machine to proceed from stable state ⓐ to stable state ⓑ, then through transient state *c* to stable state ⓖ, where output z_1 is asserted unconditionally.

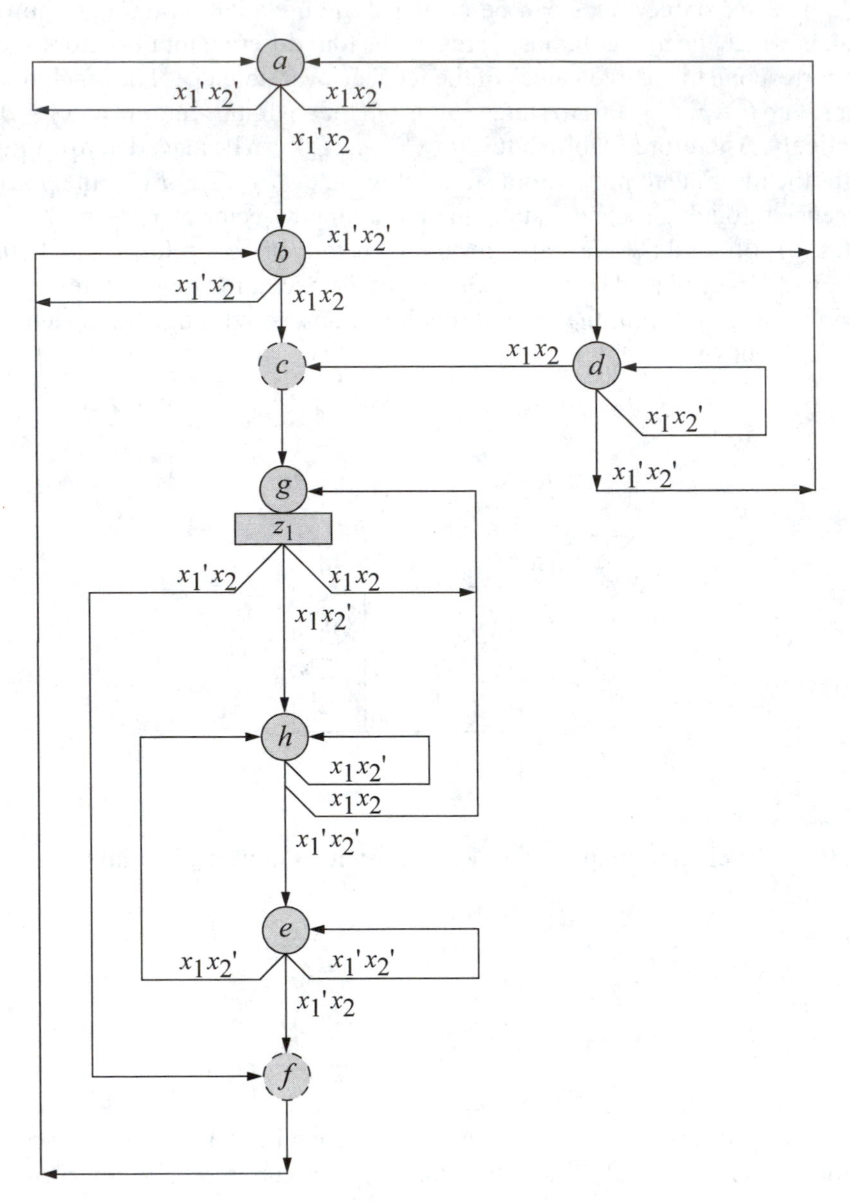

Figure 6.5 State diagram for a Moore asynchronous sequential machine that is characterized by Table 6.1 and indicates the transient states as shown in the excitation map of Figure 6.4.

6.2.2 Primitive Flow Table

The primitive flow table transforms the machine specifications into a tabular representation which specifies the state transition sequences for all valid input combinations. The derivation of the primitive flow table is the least systematic and the most crucial step in the synthesis procedure. The primitive flow table is characterized by having only one stable state in each row. This allows the output for that row to be uniquely specified as a function of a particular stable state. The stable states are indicated by circled entries, whereas, the unstable, or transient states, are uncircled. The transient states specify the next stable state for a particular state transition sequence. The unspecified entries are indicated by a dash (–) and are the result of the fundamental-mode operation or an input vector that is invalid from a particular stable state.

The primitive flow table can usually be reduced to a table with a fewer number of rows. However, it is important to derive the primitive flow table using as many rows as are necessary to represent the complete operating characteristics of the machine for all possible input sequences, rather than to attempt a row-reduction process at the same time as the table is being generated. Combining two steps into one step in constructing the primitive flow table may cause an obscure sequence to be overlooked. Thus, when deriving the primitive flow table, if it is not immediately apparent that the next state is equivalent to an existing state, then a new row is added and a new stable state is inserted. During the state reduction process, this added state will be found to be equivalent to a previously established state. All but one stable state in each set of equivalent states can be eliminated, yielding a reduced primitive flow table.

The examples which follow illustrate the construction of the primitive flow table from a timing diagram and/or a verbal specification. In order to prevent possible race conditions and associated timing problems when two or more inputs change value simultaneously, it will be assumed that only one input variable will change state at a time. Also, no additional changes to the input vector will occur until the machine has reached a stable state. This fundamental-mode operation provides the necessary attributes for the synthesis of a deterministic asynchronous sequential machine.

A verbal description of the machine's operation may be less precise than that of a corresponding primitive flow table. The very nature of the flow table causes the operation of the machine to be specified for every possible input sequence prescribed by the machine specifications. A timing diagram can be used to exemplify the verbal statement and graphically portray the output for each valid input sequence. The derivation of a representative primitive flow table is described below.

Example 6.1 Consider an asynchronous sequential machine with two inputs x_1 and x_2 and one output z_1. The initial conditions are: $x_1x_2z_1 = 000$. Output z_1 is asserted whenever $x_1x_2 = 11$ if and only if the input sequence was $x_1x_2 = 00, 01, 11$. The waveforms of Figure 6.6 depict some typical input sequences. Many other input sequences are possible and must be considered in order to thoroughly construct a primitive flow table. The number of different input sequences presented to a machine is finite, although the number is quite large in some situations. Usually only subsequences of longer sequences are shown in the waveforms. The primitive flow table is constructed by plotting the sequence of state changes as specified by the timing diagram, while allowing only one stable state per row.

Constructing the primitive flow table is a two-pass procedure. The first pass lists all stable states and their associated next states as obtained from the machine specifications such as, a timing diagram and/or a verbal description. These are the more obvious entries. The second pass establishes the entries for any unspecified transient states and may necessitate creating additional rows. The second pass also establishes subsequences that are not expressly specified by the timing diagram. Careful consideration must be given to these subsequences to assure that the machine operates according to the functional specifications.

Beginning at the leftmost section of the timing diagram, where the initial conditions are specified, proceed left to right assigning a stable state to each different combination of the input vector. Figure 6.7 shows the first step in the derivation of the primitive flow table. The column headings represent the input vector. The table entries specify the stable states, transient states, invalid state transitions which are represented as dashes, and outputs.

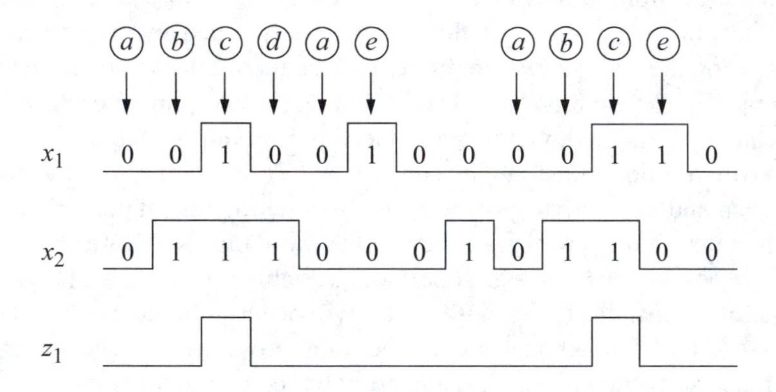

Figure 6.6 Timing diagram for the asynchronous sequential machine of Example 6.1.

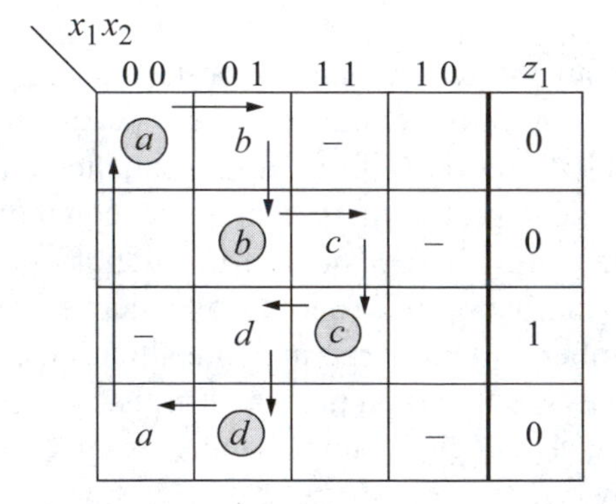

Figure 6.7 Initial steps in developing the primitive flow table for the asynchronous sequential machine of Example 6.1.

Assign stable state ⓐ to the initial condition of $x_1x_2z_1 = 000$ as shown in the timing diagram of Figure 6.6 and the first row (Row ⓐ) of the partial primitive flow table of Figure 6.7. Rows will be designated according to their stable state. Thus, the first row is specified as row ⓐ. The dash in column $x_1x_2 = 11$ of row ⓐ is the result of the fundamental-mode operation in which only one input variable is allowed to change state at a time. Since x_1 and x_2 will not change value simultaneously, the dash represents a "don't care" condition, which can be used in later steps to minimize the excitation and output maps. Output z_1 is inactive in stable state ⓐ. Thus, a 0 is entered in column z_1 of the primitive flow table.

The next new state occurs when input x_2 changes from 0 to 1. This new state is recorded as stable state ⓑ in both the timing diagram and the primitive flow table. The sequence is specified by placing an uncircled *b* in column $x_1x_2 = 01$ of row ⓐ. The uncircled entry is an unstable, or transient, state and indicates the path that is taken when the input vector changes from $x_1x_2 = 00$ to 01; that is, the uncircled entry specifies the destination stable state for a particular input change.

Once the inputs change, they will not change again until the machine has sequenced to a stable state. Therefore, from transient state *b*, the operation must proceed vertically in the same column to stable state ⓑ. This sequence is indicated by the arrows from state ⓐ through unstable state *b* to stable state ⓑ. A dash is inserted in column $x_1x_2 = 10$ of row ⓑ as required for fundamental-mode operation. Since output z_1 is inactive in state ⓑ, a 0 is inserted in column z_1 of row ⓑ.

From state ⓑ, the next new state occurs when x_1 changes from 0 to 1. This new state is designated as stable state ⓒ and represents the final state in a valid sequence of states to assert output z_1. This culmination of a valid sequence is shown in the timing diagram of Figure 6.6 in which z_1 is asserted for the duration of $x_1x_2 = 11$. The partial primitive flow table of Figure 6.7 depicts the same sequence and resulting output assertion in tabular form. The arrow exiting state ⓑ in Figure 6.7 shows the path taken for this state transition sequence. Beginning in state ⓑ, the operation proceeds to the right horizontally to column $x_1x_2 = 11$ where the machine enters transient state *c*. Transient state *c* designates the resulting stable state to which the machine will proceed after a delay of Δt. The machine will remain in column $x_1x_2 = 11$ until a stable state is entered and the input vector changes. In row ⓒ, a 1 is entered in column z_1 indicating the successful completion of a valid input sequence.

A new state is entered from state ⓒ when x_1 changes from 1 to 0. This state is designated stable state ⓓ, where $x_1x_2 = 01$. This does not represent the second state of a valid input sequence, because the beginning input vector $x_1x_2 = 00$ does not immediately precede the input combination $x_1x_2 = 01$. Therefore, states ⓑ and ⓓ are not equivalent. Stable state ⓑ represents the second input vector of a valid sequence, whereas, stable state ⓓ does not. Since $x_1x_2 \neq 11$, a 0 is inserted in column z_1 in row ⓓ.

In stable state ⓓ, if x_2 changes from 1 to 0, then the operation proceeds to the left horizontally and the machine enters transient state *a*. Transient state *a* designates stable state ⓐ as the destination state, which is the initial input combination for a valid input sequence to assert output z_1.

The completed primitive flow table for Example 6.1 is shown in Figure 6.8. In stable state ⓐ, if input x_1 changes from 0 to 1, then the machine sequences to a new stable state in column $x_1x_2 = 10$. This input vector does not represent a valid combination of inputs to activate output z_1. Therefore, a new state must be created. Since only one stable state is allowed in each row of a primitive flow table, a new row is added to accommodate stable state ⓔ, as shown in Figure 6.8. An entry of 0 is inserted in column z_1 for row ⓔ. If x_1 now returns to 0, then the machine proceeds to state ⓐ, which is the initial state for a valid input sequence.

x_1x_2 00	01	11	10	z_1
ⓐ	b	–	e	0
a	ⓑ	c	–	0
–	d	ⓒ	e	1
a	ⓓ	f	–	0
a	–	f	ⓔ	0
–	d	ⓕ	e	0

Figure 6.8 Complete primitive flow table for the asynchronous sequential machine of Example 6.1.

In state ⓔ, if the input vector changes from $x_1x_2 = 10$ to 11, then a new stable state is defined, labeled ⓕ. Although stable states ⓒ and ⓕ are in the same input column, stable state ⓒ represents the culmination of a valid input sequence in which output z_1 is asserted. In state ⓕ, however, the previous stable state (ⓔ) did not correspond to an input vector that was part of a valid sequence. Thus, the machine cannot proceed from state ⓔ to state ⓒ, necessitating that a new row (ⓕ) be created in the column which corresponds to the values of the input variables $x_1x_2 = 11$.

At this point, all apparent input combinations have been exhausted. Not all squares in the primitive flow table, however, contain an entry. Each square must contain a stable state, a transient state, or an invalid condition designated by a dash. When the primitive flow table has been partially developed with all necessary rows, then the subsequences can be identified. To complete a partially developed primitive flow table, locate an empty square and determine to which state the stable state of that row will sequence. Then insert the uncircled state name in that square.

Row ⓐ is complete. In row ⓑ, the machine can leave stable state ⓑ by either a change of inputs from $x_1x_2 = 01$ to 00 or to 11. If x_2 changes from 1 to 0, then the machine proceeds to column $x_1x_2 = 00$ in row ⓑ, which is the same criteria for stable state ⓐ. Thus, the machine will return to stable state ⓐ to begin checking for a valid sequence. Therefore, an uncircled *a* is inserted in column $x_1x_2 = 00$ of row ⓑ. An input change from $x_1x_2 = 01$ to 11 has already been described.

In stable state ⓒ, if x_2 changes from 1 to 0, then the machine proceeds to stable state ⓔ where z_1 is deasserted. Since column $x_1x_2 = 10$ does not represent a combination of inputs that is encountered in a valid input sequence, all entries in that column will contain either a transient state *e* or a dash.

In stable state ⓓ, if x_1 changes from 0 to 1, then the machine must enter a stable state in column $x_1x_2 = 11$ in which output z_1 is deasserted. This is because the state prior to state ⓓ was state ⓒ, which is a termination state for a valid input sequence. Therefore, stable state ⓓ, although corresponding to a valid input combination, was not reached by means of a valid sequence of inputs. Thus, the machine cannot proceed to state ⓒ, but must move from state ⓓ to state ⓕ in column $x_1x_2 = 11$. In stable state ⓓ, if the input vector changes from $x_1x_2 = 01$ to 00, then the machine enters stable state ⓐ to begin checking for a new valid input sequence.

In stable state ⓔ, if the input vector changes from $x_1x_2 = 10$ to 11, then the machine must enter a stable state in which output z_1 is deasserted. This is because state ⓔ did not correspond to a valid input vector. Thus, the next state cannot be state ⓒ, where z_1 is asserted, but must be state ⓕ, where z_1 is deasserted. The path from stable state ⓔ to stable state ⓐ occurs for an input sequence of $x_1x_2 = 10$ to 00. In state ⓐ, the inputs are again examined for a possible valid sequence.

The final row ⓕ cannot return to row ⓐ directly due to the fundamental-mode characteristic. The path from ⓕ to ⓐ must sequence through state ⓓ or state ⓔ. Although stable state ⓕ results from an input vector of $x_1x_2 = 11$, output z_1 is not asserted, because state ⓕ was not achieved from a valid input sequence.

A change to the input vector causes a horizontal move in the primitive flow table to the column whose heading corresponds to the new input values. When the new column is reached, an internal state change of feedback variables causes a vertical move in the new column. The machine will not exit from the new column until the input vector again changes. In order to move from a stable state, an input variable must change. A horizontal move must originate from a stable state.

The number of rows in a primitive flow table corresponds directly to the number of feedback variables. Some primitive flow tables have a large number of rows, necessitating several feedback variables. The number of rows, however, can usually be reduced significantly to yield a tractable number of feedback variables.

Generation of the primitive flow table is the most critical step in the synthesis of asynchronous sequential machines. Therefore, a sufficient number of examples will be presented so that expertise can be gained in this crucial step, each example providing a variation in the construction of the primitive flow table. If the primitive flow table is correctly constructed, then the remaining steps in the synthesis procedure, if properly executed, will result in a machine that meets the performance criteria of the machine specifications. If, however, the primitive flow table does not delineate all transitions for every possible input sequence, then the remaining steps, although correct in themselves, will result in a machine that does not operate according to the machine specifications.

Example 6.2 An asynchronous sequential machine has two inputs x_1 and x_2 and two outputs z_1 and z_2. Output z_1 will be asserted coincident with every second x_2 pulse, but only if x_1 is asserted for the duration of the pair of x_2 pulses. Output z_2 is asserted for every second x_2 pulse, but only if x_1 is deasserted for the duration of the pair of x_2 pulses. Input x_1 will not change state while x_2 is asserted. A representative timing diagram depicting the principal sequences is shown in Figure 6.9.

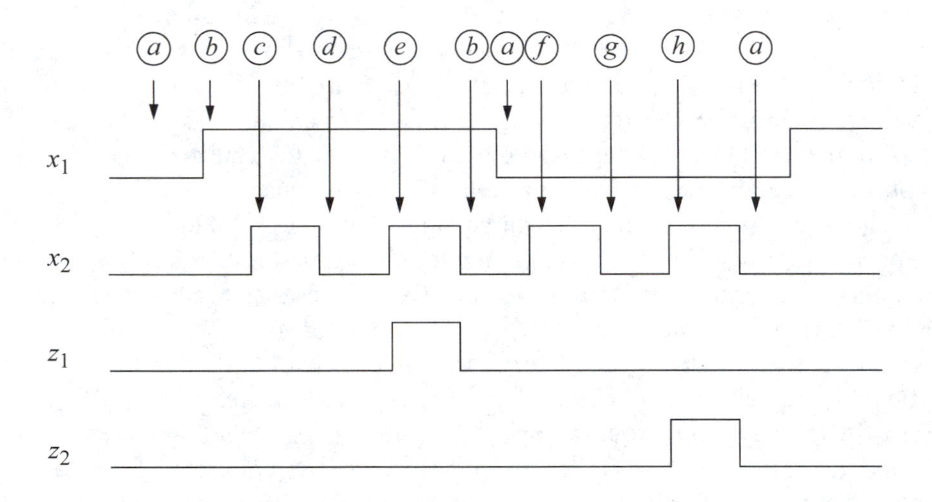

Figure 6.9 Representative timing diagram for the asynchronous sequential machine of Example 6.2.

Although the construction of the primitive flow table can begin with any input vector, it is preferable to start at the extreme left of the timing diagram where the machine is in a reset state. This is a more methodical approach and tends to substantially reduce the number of omitted subsequences. Any sequences that are overlooked when transforming the timing diagram into a tabular representation will be detected during the second pass. Filing in the empty transient states during the second pass requires consideration of every possible input sequence, even those that were not immediately apparent during the first pass.

Let stable state ⓐ be defined as $x_1x_2z_1z_2$ = 0000, as indicated in the timing diagram and by row ⓐ in the partially-completed primitive flow table of Figure 6.10. Input x_1 then changes from 0 to 1. This new input vector is denoted as state ⓑ and the operation proceeds horizontally from state ⓐ to transient state b, then vertically down to stable state ⓑ, where z_1z_2 = 00.

x_1x_2

00	01	11	10	z_1	z_2
ⓐ	f	–	b	0	0
a	–	c	ⓑ	0	0
–		ⓒ	d	0	0
	–	e	ⓓ	0	0
–		ⓔ	b	1	0
g	ⓕ		–	0	0
ⓖ	h	–		0	0
a	ⓗ		–	0	1

Figure 6.10 Partially completed primitive flow table for the asynchronous sequential machine of Example 6.2.

Proceeding to the right in the timing diagram, the input vector then changes from $x_1x_2 = 10$ to 11 and the machine sequences through transient state c then enters stable state ⓒ, where $z_1z_2 = 00$. State ⓒ represents the occurrence of the first x_2 pulse of a pair of x_2 pulses that will cause output z_1 to be asserted. Input x_2 then becomes inactive and the machine enters stable state ⓓ. This state transition sequence is from state ⓒ through transient state d to state ⓓ, where $z_1z_2 = 00$. Stable states ⓑ and ⓓ cannot be equivalent even though both states possess the same input/output characteristics: $x_1x_2z_1z_2 = 1000$. Stable state ⓑ is the precursor of the first x_2 pulse, whereas, stable state ⓓ is the preceding condition for the second x_2 pulse. If state ⓓ was replaced by state ⓑ, then output z_1 would never be asserted.

The next input change occurs when x_2 again becomes active. This new state is represented as stable state ⓔ and occurs when the input vector changes from $x_1x_2 = 10$ to 11. In state ⓔ, output z_1 is asserted for the duration of input x_2. Similar reasoning analogous to that used to establish the nonequivalence of states ⓑ and ⓓ also precludes the equivalence of stable states ⓒ and ⓔ. Stable state ⓒ indicates the assertion of the first x_2 pulse, whereas, stable state ⓔ denotes the assertion of the second x_2 pulse, which in turn activates output z_1.

In stable state ⓔ, the next change occurs when x_2 becomes inactive. This state is equivalent to state ⓑ, in which the machine is in a condition to detect the first x_2 pulse of a pair of x_2 pulses. In state ⓑ, the deassertion of x_1 returns the machine to state ⓐ. This completes one of the two main sequences.

The second main sequence begins in state ⓐ when input x_1 is inactive and the input vector changes from $x_1x_2 = 00$ to 01. This new state is designated as stable state ⓕ and indicates the occurrence of the first x_2 pulse of a pair of x_2 pulses when x_1 is deasserted. The machine then sequences through stable states ⓖ and ⓗ before returning to state ⓐ. For reasons similar to those previously presented, stable state pairs ⓕ, ⓗ and ⓐ, ⓖ are not equivalent. Since two or more inputs will not change state simultaneously in a fundamental-mode asynchronous sequential machine, dashes are inserted in the appropriate squares to denote this constraint. This completes the more obvious sequences.

For the discussion which follows, refer to the timing diagram of Figure 6.9, the partially completed primitive flow table of Figure 6.10, and the completed primitive flow table of Figure 6.11. The remaining six empty squares in Figure 6.10 represent transient states and must contain either an uncircled state that specifies the destination stable state or a dash that represents an invalid input sequence from a particular stable state. Rows ⓐ and ⓑ are complete.

x_1x_2 00	01	11	10	z_1	z_2
ⓐ	f	–	b	0	0
a	–	c	ⓑ	0	0
–	–	ⓒ	d	0	0
a	–	e	ⓓ	0	0
–	–	ⓔ	b	1	0
g	ⓕ	–	–	0	0
ⓖ	h	–	b	0	0
a	ⓗ	–	–	0	1

Figure 6.11 Complete primitive flow table for the asynchronous sequential machine of Example 6.2.

The blank square in column x1x2 = 01 of row (c) represents a transition that cannot occur from stable state (c). The verbal description specifies that input x_1 will not be deasserted while x_2 is active. Therefore, a dash is inserted in column $x_1x_2 = 01$ of row (c) indicating a "don't care" condition, as shown in Figure 6.11, which illustrates the complete primitive flow table.

In row (d), if x_1 changes from 1 to 0, then the machine must be reinitialized to stable state (a), since both inputs are now inactive. This is the reset condition from which the machine will sequence to examine the binary input variables in preparation for the subsequent assertion of z_1 or z_2. Therefore, an uncircled a is placed in column $x_1x_2 = 00$ of row (d), as shown in Figure 6.11. In row (e), column $x_1x_2 = 01$ must contain a dash for the same reason that justified the dash in column $x_1x_2 = 01$ of row (c) — an invalid sequence from a stable state in column $x_1x_2 = 11$.

The empty square in column $x_1x_2 = 11$ of row (f) represents an invalid sequence from stable state (f) — input x_1 will not change state while x_2 is asserted. Thus, a dash is inserted in column $x_1x_2 = 11$ of row (f). The same is true for the empty square in column $x_1x_2 = 11$ of row (h), since a change from $x_1x_2 = 01$ to 11 is invalid from any stable state in column $x_1x_2 = 01$, as specified in the verbal description of the machine specifications.

Finally, in stable state (g), if x_1 is asserted, then this represents the beginning condition in which the state of input x_2 is examined with respect to an active x_1 input. Thus, a change of input vector from $x_1x_2 = 00$ to 10 results in a state transition sequence from stable state (g) through transient state b to stable state (b). State (b) represents the state in which the machine anticipates the occurrence of the first x_2 pulse of a pair of x_2 pulses. The complete primitive flow table is shown in Figure 6.11.

The timing diagram of Figure 6.9 shows only the primary sequences as specified by the verbal description. Since inputs x_1 and x_2 are not synchronized by a periodic clock signal, their activation can occur randomly under the constraints of the machine specifications and the fundamental-mode of operation. Therefore, while traversing the timing diagram in a left-to-right direction, every valid state of the input vector should be considered, even those that are not explicitly shown. It is the inconspicuous sequences that tend to be overlooked.

Careful consideration of all primary sequences and subsequences will result in a primitive flow table that conforms precisely to the machine specifications. Also, by examining each state as potentially equivalent to a previously defined state, the completed primitive flow table may already be reduced to the fewest number of rows. Some state equivalences may not be readily apparent, however. This indeterminately small number of states will be eliminated in a later step.

Example 6.3 An asynchronous sequential machine has three inputs x_1, x_2, and x_3 and three outputs z_1, z_2, and z_3. When any input x_i is asserted, the corresponding output z_i is asserted. Output z_i will remain asserted — even after the deassertion of x_i — until another input z_j is asserted, at which time z_i is deasserted and z_j is asserted. Thus, the outputs are mutually exclusive; that is, the assertion of any input will assert its corresponding output and cause the deassertion of all other active outputs. When an input is asserted, there is a one-to-one correspondence between the input vector and the output vector. The inputs will not overlap, and two or more inputs will not change state simultaneously. Although the primitive flow table for this example can be constructed without the use of a timing diagram, that procedure will be left for the following example. The timing diagram of Figure 6.12 illustrates some typical sequences for Example 6.3.

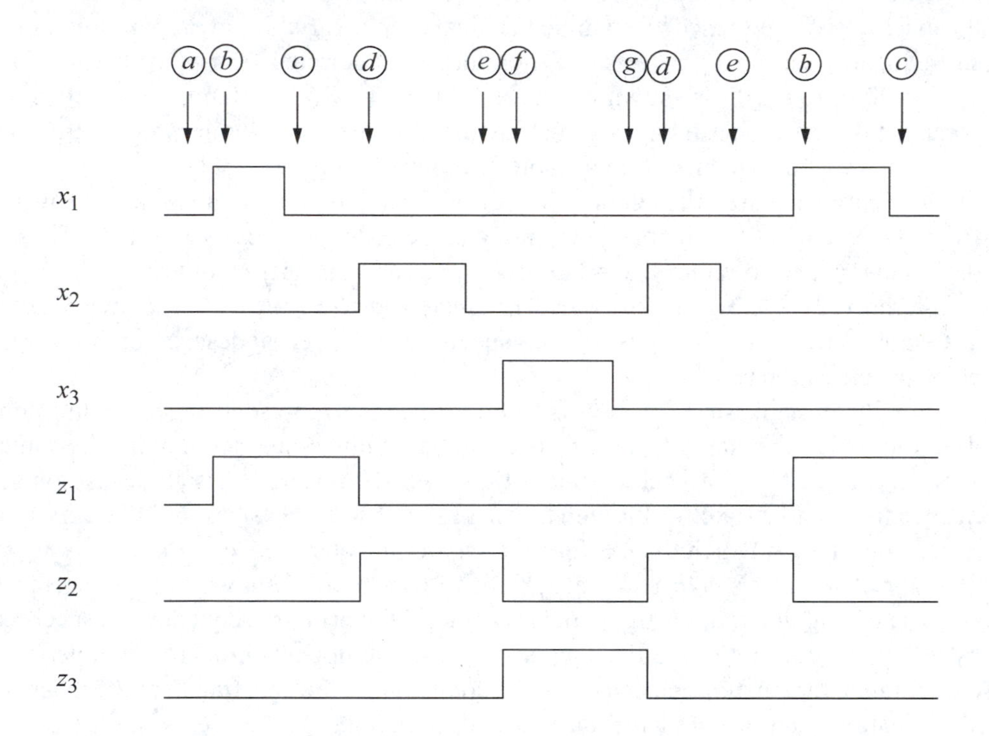

Figure 6.12 Typical input sequences for the asynchronous sequential machine of Example 6.3.

Since there are three input variables, eight columns are required in the primitive flow table, as shown in Figure 6.13. Stable state ⓐ represents a reset condition where $x_1x_2x_3z_1z_2z_3 = 000000$. This is shown in both the timing diagram and the primitive flow table.

There are only three possible paths, or vectors, from state ⓐ: $x_1x_2x_3 = 001, 010,$ or 100, since simultaneous input changes are not permitted. The dashes in row ⓐ indicate those sequences that do not adhere to the characteristics of a fundamental-mode machine.

Proceeding to the right from stable state ⓐ in the timing diagram, input x_1 becomes asserted and the machine sequences through transient state b to stable state ⓑ. The output vector in state ⓑ is $z_1z_2z_3 = 100$, which corresponds directly to input vector $x_1x_2x_3 = 100$. Since overlapping inputs are not permitted and since simultaneous input changes are a diametric contradiction to the fundamental-mode of operation, there is only one allowable input vector to affect a transition from state ⓑ: $x_1x_2x_3 = 000$. Thus, when x_1 returns to an inactive state, the machine proceeds through transient state c to stable state ⓒ, where the output vector remains at a value of $z_1z_2z_3 = 100$, in accordance with the machine specifications. The six dashes in row ⓑ represent invalid input sequences from stable state ⓑ.

$x_1x_2x_3$: 000	001	011	010	110	111	101	100	z_1	z_2	z_3
ⓐ	f	–	d	–	–	–	b	0	0	0
c	–	–	–	–	–	–	ⓑ	1	0	0
ⓒ	f	–	d	–	–	–	b	1	0	0
e	–	–	ⓓ	–	–	–	–	0	1	0
ⓔ	f	–	d	–	–	–	b	0	1	0
g	ⓕ	–	–	–	–	–	–	0	0	1
ⓖ	f	–	d	–	–	–	b	0	0	1

Figure 6.13 Primitive flow table for the asynchronous sequential machine Example 6.3.

In the timing diagram, the next change occurs when x_2 becomes active. This new input condition sequences the machine to state ⓓ, where output z_2 is asserted, output z_1 is deasserted, and z_3 remains inactive; that is, $z_1z_2z_3 = 010$. The path for this state transition sequence is from state ⓒ through transient state d to state ⓓ, as shown in the primitive flow table. In state ⓓ, if x_2 becomes inactive, resulting from an input vector of $x_1x_2x_3 = 000$, then the machine returns to column $x_1x_2x_3 = 000$, but must retain the output vector of $z_1z_2z_3 = 010$. Thus, the machine proceeds to state ⓔ where the output vector is $z_1z_2z_3 = 010$.

The next change in the timing diagram occurs when input x_3 becomes active. The machine proceeds from state ⓔ through transient state f to state ⓕ, where the output vector is set to $z_1z_2z_3 = 001$. The dashes in row ⓕ represent state transition sequences that are invalid with reference to the machine specifications or to the fundamental-mode of operation. When x_3 is deasserted, the machine returns to a stable state in column $x_1x_2x_3 = 000$ and maintains the output vector at $z_1z_2z_3 = 001$, as shown in stable state ⓖ.

Since the timing diagram portrays only a few of the possible variations in input sequences, the remaining squares of the primitive flow table are filled according to the verbal description of the machine specifications. The machine will proceed from any stable state in column $x_1x_2x_3 = 000$ to stable state ⓕ if the input vector changes from $x_1x_2x_3 = 000$ to 001. Stable state ⓕ is the only stable state in column $x_1x_2x_3 = 001$ to provide the requisite output vector of $z_1z_2z_3 = 001$. Likewise, the machine will traverse from any stable state in column $x_1x_2x_3 = 000$ to state ⓓ in column $x_1x_2x_3 = 010$ or to state ⓑ in column $x_1x_2x_3 = 100$ as a result of a change to the input vector from $x_1x_2x_3 = 000$ to 010 or to 100, respectively. Because of the constraints imposed by the machine specifications and the fundamental-mode of operation, stable states ⓑ, ⓓ, and ⓕ must proceed to a next state in column $x_1x_2x_3 = 000$ — all other input vectors from these stable states are invalid.

After analyzing the primitive flow table of Figure 6.13, it is evident that only four columns are required, since the verbal description specifies that only one input will be active at a time. Thus, only columns $x_1x_2x_3 = 000$, 001, 010, and 100 are necessary. The entries in all other columns are treated as "don't care" conditions. Retaining only those columns with state names, yields the compact primitive flow table shown in Figure 6.14. The table retains the essential information contained in the expanded version while omitting the invalid states. The compact primitive flow table still characterizes the complete operation of the machine.

$x_1x_2x_3$ 000	001	010	100	z_1	z_2	z_3
ⓐ	*f*	*d*	*b*	0	0	0
c	–	–	ⓑ	1	0	0
ⓒ	*f*	*d*	*b*	1	0	0
e	–	ⓓ	–	0	1	0
ⓔ	*f*	*d*	*b*	0	1	0
g	ⓕ	–	–	0	0	1
ⓖ	*f*	*d*	*b*	0	0	1

Figure 6.14 Compact primitive flow table for the asynchronous sequential machine of Example 6.3.

Example 6.4 As a final example in this section, consider an asynchronous sequential machine with two inputs x_1 and x_2 and one output z_1. Output z_1 is asserted at the rising edge of x_2, but only if x_1 is deasserted at that time. Output z_1 remains active for the duration of the x_2 pulse. If x_1 becomes asserted after the rising edge of x_2, then the duration of the output pulse is not affected — output z_1 is deasserted only when x_2 becomes deasserted.

Since a timing diagram is not always included as part of the machine specifications, the primitive flow table will be developed solely from the verbal description. Figure 6.15 illustrates a partial primitive flow table obtained after evaluating the more obvious primary sequences.

x_1x_2 00	01	11	10	z_1
ⓐ	*b*	–	*d*	0
	ⓑ	*c*	–	1
–		ⓒ	*d*	1
a	–		ⓓ	0

Figure 6.15 Partial primitive flow table for the asynchronous sequential machine of Example 6.4.

Let stable state ⓐ represent the initial state of the machine; that is, $x_1x_2z_1 = 000$. If input x_2 becomes asserted, then the machine proceeds through transient state b to stable state ⓑ, where output z_1 is asserted. In state ⓑ, if x_1 is asserted, then the machine sequences to state ⓒ, where z_1 maintains an active condition because x_2 is still asserted. Assume that x_2 is then deasserted. The machine enters state ⓓ and z_1 becomes inactive.

From a reset condition in stable state ⓐ, the assertion of x_1 advances the machine through transient state d to state ⓓ. Since x_2 is inactive in state ⓓ, output z_1 remains deasserted. If x_1 now becomes inactive, then the machine returns to the reset condition in state ⓐ. The dashes in rows ⓐ through ⓓ represent invalid input vectors that are inconsistent with the fundamental-mode of operation.

The remaining squares will now be filled with transient state names to complete the less obvious subsequences. Additional rows will be added if necessary to include all possible state transition sequences. Refer to the complete primitive flow table of Figure 6.16 for the discussion which follows. Row ⓐ is complete. In stable state ⓑ, if x_2 becomes inactive, then the machine returns to the initial condition of stable state ⓐ and z_1 is deasserted.

x_1x_2 00	01	11	10	z_1
ⓐ	b	–	d	0
a	ⓑ	c	–	1
–	b	ⓒ	d	1
a	–	e	ⓓ	0
–	f	ⓔ	d	0
a	ⓕ	e	–	0

Figure 6.16 Complete primitive flow table for the asynchronous sequential machine of Example 6.4.

From stable state ⓒ, the deassertion of x_1 takes the machine to a stable state in column $x_1x_2 = 01$. Because x_2 remains active, output z_1 must also remain active. Thus, the machine proceeds from state ⓒ to state ⓑ.

In stable state ⓓ, x_1 is asserted, whereas x_2 is deasserted, requiring that output z_1 be inactive. If x_2 becomes asserted in state ⓓ, then z_1 must remain inactive, since z_1 is asserted only if x_1 is inactive when x_2 is asserted. Thus, a change of input vector in stable state ⓓ from $x_1x_2 = 10$ to 11 must sequence the machine to a stable state in column $x_1x_2 = 11$, in which z_1 has a value of 0. Since stable state ⓒ has an output of $z_1 = 1$, a new row must be created to accommodate this new state transition sequence. The machine leaves stable state ⓓ and proceeds through transient state e to stable state ⓔ, where z_1 remains in an inactive state.

In stable state ⓔ, if x_2 changes from 1 to 0, then a transition occurs from state ⓔ to state ⓓ. Stable states ⓓ and ⓔ both represent conditions in which output z_1 is deasserted. The state transition sequence ⓓ $\rightarrow e \rightarrow$ ⓔ $\rightarrow d \rightarrow$ ⓓ ... is logically inconsequential to cause the assertion of output z_1, because the sequence represents a series of x_2 pulses that occurs after x_1 is already asserted, as shown in Figure 6.17. Since z_1 is asserted only if x_1 is inactive before the assertion of x_2, consequently, z_1 remains at a value of 0 for this sequence.

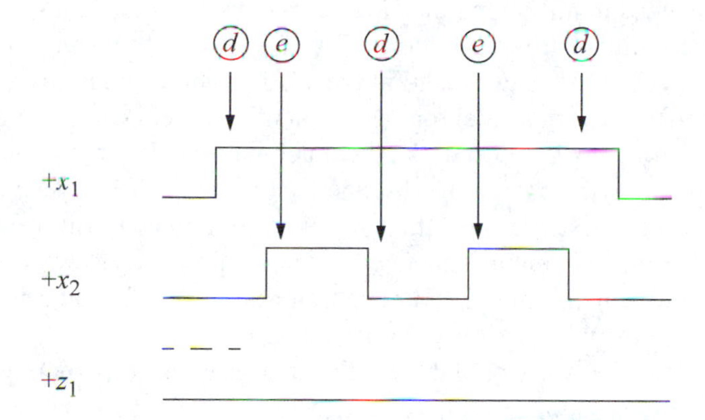

Figure 6.17 Timing diagram depicting the state transition sequence ⓓ → e → ⓔ → d → ⓓ. This is a sequence that is logically inconsequential to cause the assertion of output z_1, because x_1 is active before x_2 is asserted.

In state ⓔ, if x_1 changes from 1 to 0, then the machine enters a stable state in column $x_1x_2 = 01$. This destination stable state must maintain output z_1 in an inactive state to retain compatibility with state ⓔ. Stable state ⓑ does not meet this requirement. Therefore, a new row must be added in which a stable state is inserted in column $x_1x_2 = 01$ to provide an output value of $z_1 = 0$. The required state transition sequence is from state ⓔ through transient state f to state ⓕ, where z_1 retains a value of 0.

In state ⓕ, if x_2 changes from 1 to 0, then the machine will return to state ⓐ. A change in value of x_1 from 0 to 1 will result in a transition to a stable state in column $x_1x_2 = 11$. Since x_2 has not changed value in this transition, the destination stable state in column $x_1x_2 = 11$ must possess the same output characteristics as state ⓕ. Stable state ⓔ meets this requirement. Therefore, the state transition sequence is ⓕ → e → ⓔ. The primitive flow table is now complete and all major sequences and subsequences have been tabulated.

In the examples of this section, the output values in the primitive flow tables were associated with the stable states of their respective rows. The outputs are of interest primarily when the machine is in a stable state. The outputs associated with the unstable states can be either specified or left unspecified. For a Moore machine, the outputs of the unstable states are made equal to the outputs corresponding to the stable states of their respective rows. For a Mealy machine, the outputs of the unstable states are left unspecified until a later step. A greater degree of freedom exists if the outputs associated with the unstable states are left unspecified.

The choice of output values for unstable states is made during the construction of the output maps. The output values for the unstable, or transient, states must not generate an erroneous output during a state transition sequence. Also, the choice of output values can result in a machine with either a fast or a slow response to an applied input vector. If machine speed is unimportant, then the outputs for the transient states can be assigned values that minimize the output logic. The choice of output values for transient states will be expounded in greater detail in the section on output maps.

Unless otherwise specified, most asynchronous sequential machines must be strongly connected; that is, from any given state Y_j, the machine must be able to proceed to any other state Y_k by applying a prescribed sequence of inputs. This prevents a situation where a machine enters a stable state, then cycles continuously through two or more stable states for all applicable input sequences and never proceeds to any other state in the primitive flow table. This phenomenon can be easily detected in a merged flow table in which a row contains only stable states.

The examples in this section have illustrated the techniques used in constructing a primitive flow table from a given set of machine specifications. Each machine operates according to its prescribed rules and is characterized by a unique primitive flow table. The principle of fundamental-mode operation is of paramount importance in this step and underlies the construction of the table.

A primitive flow table with two inputs contains one square in each row in which a dash is inserted to comply with the fundamental-mode principle of disallowing simultaneous changes of two or more input signals. A table with three inputs contains four squares in each row in which dashes are inserted. An asynchronous sequential machine with four binary inputs requires a primitive flow table with 16 columns enumerated in the Gray code. There are 11 squares in each row in which two or more inputs change value simultaneously with respect to a single stable state in the row. A table of this magnitude becomes increasingly unwieldy when attempting to satisfy the operational requirements of every possible valid input sequence.

The importance of obtaining an accurate and comprehensive primitive flow table is cardinal in the synthesis of asynchronous sequential machines. The primitive flow table is the foundation upon which the entire synthesis procedure is predicated. Accuracy and completeness in this step will yield a machine that precisely operates according to the machine specifications.

6.2.3 Equivalent States

In the previous section, primitive flow tables were derived in which some states were left unspecified. This resulted from constraints imposed by the fundamental-mode characteristic or from certain invalid input combinations prescribed by the machine specifications; in some situations, both restrictions applied. Machines of this type are categorized as incompletely specified machines. This section describes methods to identify equivalent states and thus, to eliminate redundant states in the primitive flow tables of incompletely specified machines.

When constructing a primitive flow table for an asynchronous sequential machine, redundant states may be incognizantly inserted in the same column as a previously defined state. There may be no obvious indication that a new state is a duplicate of a previous state. This may be a result of haste in constructing the table or simply a precaution to assure that all possible input sequences are considered. It is possible to synthesize the machine directly from the primitive flow table, but redundant states may result in a design that is not minimum. The operation of the machine will be identical whether it is implemented from a primitive flow table or from a reduced primitive flow table.

Unless the redundant states are identified and eliminated, the number of feedback variables may be excessive. Since the number of storage elements is directly related to the number of feedback variables, the cost of the machine will increase accordingly; each feedback variable corresponds to a storage element and each row of a primitive flow table corresponds to a different combination of feedback variables. Reducing the number of rows in a primitive flow table may result in a machine with a fewer number of storage elements and thus, a minimal number of logic gates. Even if the number of storage elements remains unchanged, the δ next-state logic will be reduced, because the excitation equations will be less complex.

The state alphabet for an asynchronous sequential machine is a finite set of states such that, $Y = \{Y_0, Y_1, Y_2, \ldots, Y_{2^p-1}\}$. Two states Y_i and Y_j are equivalent if and only if, beginning from the same input conditions and applying every possible input sequence, the output sequences are indistinguishable, whether the initial state was Y_i or Y_j. Thus, no input sequence distinguishes state Y_i or state Y_j. The same rationale applies to three or more states. Therefore, if states Y_i, Y_j, and Y_k of the set Y are equivalent, then two rows in the primitive flow table can be eliminated. All references to states Y_j and Y_k are changed to Y_i to reflect this equivalence. The preceding statements can be generalized to any number of states, $Y_0, Y_1, Y_2, \ldots, Y_q$.

Eliminating redundant states from a primitive flow table will generate a reduced primitive flow table which is equivalent to the original table. A primitive flow table — unless already containing a minimal number of rows — can be reduced to provide a table that contains fewer rows than the original table and yet completely characterizes the operation of the machine.

State equivalence, denoted by the symbol (≡), can be considered an equivalence relation, which is characterized by the following properties:

Reflexive: $Y_i \equiv Y_i$ for every Y_i in the set Y.

Symmetric: If $Y_i \equiv Y_j$, then $Y_j \equiv Y_i$ for every Y_i and Y_j in the set Y.

Transitive: If $(Y_i \equiv Y_j) \bullet (Y_j \equiv Y_k)$, then $Y_i \equiv Y_k$ for every Y_i, Y_j, and Y_k in the set Y. Thus, $Y_i \equiv Y_j \equiv Y_k$.

The reflexive, symmetric, and transitive properties are intuitively obvious and partition the set Y into subsets of equivalent elements, or states. The partitioning provides nonempty, disjoint sets of equivalent states. As mentioned previously, if two or more stable states are equivalent, then only one state is retained in the set of equivalent states — all other states are eliminated.

The formal definition of general equivalence can be restated in simpler terms as applied to asynchronous sequential machines. Two stable states Y_i and Y_j in a primitive flow table are defined to be equivalent if and only if all of the following rules apply:

1. The stable states have the same input vector; that is, the states are in the same column.

2. The outputs associated with Y_i and Y_j have the same value; that is, $Z_r(Y_i) = Z_r(Y_j)$.

3. The next states for Y_i and Y_j are the same or equivalent for every column in the two rows of the primitive flow table. That is, for each input combination in the rows of the two stable states, the following is observed:

    (a) Identical or equivalent state names, or
    (b) Two dashes.

The definition of equivalence is recursive. Note that rule 3 (a) specifies that two states are equivalent (assuming that the requirements of the other rules are met) if the next states for Y_i and Y_j for a particular input vector are the same or have been previously found to be equivalent. Thus, the equivalence statement $Y_i \equiv Y_j$ if $Y_k \equiv Y_l$ implies that, if stable states Y_k and Y_l have already been shown to be equivalent, then Y_i and Y_j are equivalent. Thus, state pair (Y_i, Y_j) implies state pair (Y_k, Y_l). The task of determining equivalence between Y_i and Y_j is now replaced by ascertaining the equivalence of state pair (Y_k, Y_l).

Rule 3 (a) can be extended to include interdependence between states. Thus, the equivalence statements $Y_i \equiv Y_j$ if $Y_k \equiv Y_l$ and $Y_k \equiv Y_l$ if $Y_i \equiv Y_j$ infer that $Y_i \equiv Y_j$ and $Y_k \equiv Y_l$. That is, if state pair (Y_i, Y_j) implies state pair (Y_k, Y_l) and (Y_k, Y_l) implies (Y_i, Y_j), then Y_i and Y_j are equivalent and so also are Y_k and Y_l.

Examples will now be presented using primitive flow tables in which redundant states have been inserted. Using the three rules for equivalence, the tables will be analyzed and equivalent states indicated. The redundant states will then be eliminated to yield a reduced primitive flow table.

Example 6.5 Figure 6.18 illustrates a primitive flow table for an asynchronous sequential machine which contains redundant states. In order for two or more states to be equivalent, the conditions of all three equivalence rules must be relevant. Since stable state ⓐ is the only stable state in column $x_1x_2 = 00$, it cannot be equivalent to any other stable state (Rule 1). In column $x_1x_2 = 01$, stable states ⓑ and ⓚ cannot be equivalent, because the outputs are different (Rule 2).

Similarly, in column $x_1x_2 = 11$, stable state pairs (ⓓ, ⓕ), (ⓓ, ⓖ), (ⓕ, ⓘ), and (ⓖ, ⓘ) cannot be equivalent — each state pair has different outputs. However, state pairs (ⓓ, ⓘ) and (ⓕ, ⓖ) may be equivalent, because the conditions specified by rules 1 and 2 are satisfied.

x_1x_2 00	01	11	10	z_1
(a)	b	–	c	0
a	(b)	g	–	0
a	(k)	d	–	1
–	k	(d)	e	1
–	b	(f)	e	0
–	b	(g)	h	0
–	k	(i)	j	1
a	–	d	(c)	0
a	–	f	(e)	0
a	–	i	(h)	0
a	–	f	(j)	0

Figure 6.18 Primitive flow table for the asynchronous sequential machine of Example 6.5.

Consider stable states (d) and (i) for equivalence. Both states have identical entries in columns $x_1x_2 = 00$ and 01. In column $x_1x_2 = 10$, stable state (d) has a next state of (e) whereas, stable state (i) has a next state of (j). Thus, if states (e) and (j) are equivalent, then states (d) and (i) are also equivalent. In column $x_1x_2 = 10$, states (e) and (j) are examined for equivalence. Both states have the same output and both proceed to the same next state in columns $x_1x_2 = 00$, 01, and 11. Therefore, stable states (e) and (j) are equivalent and so also are stable states (d) and (i). Row (i) can be eliminated. Every reference to stable state (i) is replaced with an uncircled *d* entry.

Now consider stable states (f) and (g) for equivalence. Both states conform to the conditions of rules 1 and 2, because both are entered from the same input vector and both have identical outputs. With respect to rule 3, both have the same next states in columns $x_1x_2 = 00$ and 01. In column $x_1x_2 = 10$, however, state (f) proceeds to state

ⓔ, whereas, state ⓖ proceeds to state ⓗ. Therefore, the equivalence of stable states ⓕ and ⓖ depend on the equivalence of stable states ⓔ and ⓗ.

In column $x_1x_2 = 10$, states ⓔ and ⓗ have identical outputs and sequence to the same entries in columns $x_1x_2 = 00$ and 01. In column $x_1x_2 = 10$, however, if x_2 changes from 0 to 1, then state ⓔ sequences to state ⓕ while state ⓗ proceeds to state ⓘ. Examining states ⓕ and ⓘ in column $x_1x_2 = 11$, it is easily verified that stable states ⓕ and ⓘ are not equivalent, because their outputs are different. Therefore, stable states ⓕ and ⓖ are not equivalent. Thus far, the following equivalence relations have been determined:

ⓑ $\neg\equiv$ ⓚ, ⓓ $\neg\equiv$ ⓕ, ⓓ $\neg\equiv$ ⓖ, ⓕ $\neg\equiv$ ⓘ, ⓖ $\neg\equiv$ ⓘ, ⓕ $\neg\equiv$ ⓖ

ⓓ $\equiv$ ⓘ, ⓔ $\equiv$ ⓙ

Now examine the remaining state pairs in column $x_1x_2 = 10$ for equivalence. The possibility of equivalence exists between state pairs (ⓒ, ⓔ), (ⓒ, ⓗ), (ⓒ, ⓙ), (ⓔ, ⓗ), and (ⓗ, ⓙ). State pair (ⓒ, ⓔ) implies state pair (ⓓ, ⓕ). States ⓓ and ⓕ, however, have already been shown to be nonequivalent. Therefore, states ⓒ and ⓔ are not equivalent.

Now consider stable states ⓒ and ⓗ for equivalence. The conditions of all three equivalence rules are satisfied: Both states have the same input vector ($x_1x_2 = 10$); both have identical outputs; and both proceed to the same or equivalent next state (ⓓ $\equiv$ ⓘ). Therefore, states ⓒ and ⓗ are equivalent, and every reference to state ⓗ is replaced by an uncircled *c*. Similarly, it can be demonstrated that ⓒ $\neg\equiv$ ⓙ, ⓔ $\neg\equiv$ ⓗ, and ⓗ $\neg\equiv$ ⓙ.

The following state pairs have been shown to be equivalent: (ⓓ, ⓘ), (ⓔ, ⓙ), and (ⓒ, ⓗ). All uncircled references to a deleted row are replaced by an uncircled entry to the corresponding equivalent state. Thus, rows ⓘ, ⓙ, and ⓗ are eliminated and every reference to stable states ⓘ, ⓙ, and ⓗ is replaced by *d*, *e*, and *c*, respectively. The reduced primitive flow table is shown in Figure 6.19. If no additional row reduction is possible, then only three feedback variables are required in contrast to four for the original primitive flow table.

The procedure described above for identifying equivalent states is inherently more heuristic than algorithmic. A more systematic method is to partition the stable states into sets based on the conditions of the equivalence rules. All stable states in a column constitute a set of potentially equivalent states. Equivalence rule 1 partitions the sets as follows:

{ⓐ}, {ⓑ, ⓚ}, {ⓓ, ⓕ, ⓖ, ⓘ}, {ⓒ, ⓔ, ⓗ, ⓙ}

The stable states in each set possess the common attribute of having the same input vector.

Equivalence rule 2 further partitions each set into subsets in which the common characteristic is identical outputs. This partitioning yields the following sets:

{ⓐ}, {ⓑ}, {ⓚ}, {ⓓ, ⓘ}, {ⓕ, ⓖ}, {ⓒ, ⓔ, ⓗ, ⓙ}

x_1x_2 00	01	11	10	z_1
(a)	b	–	c	0
a	(b)	g	–	0
a	(k)	d	–	1
–	k	(d)	e	1
–	b	(f)	e	0
–	b	(g)	c	0
a	–	d	(c)	0
a	–	f	(e)	0

Figure 6.19 Reduced primitive flow table for Figure 6.18 for the asynchronous sequential machine of Example 6.5.

Equivalence rule 3 generates a final partitioning in which every next state of two potentially equivalent states is either the same or equivalent. This yields the following sets:

{ⓐ}, {ⓑ}, {ⓚ}, {ⓓ, ⓘ}, {ⓕ}, {ⓖ}, {ⓒ, ⓗ}, {ⓔ, ⓙ}

Stable states ⓔ and ⓙ are immediately equivalent, since both states proceed to the same next state for an applied input vector of x_1x_2 = 11 or 00. Using this equivalence relation, states ⓓ and ⓘ are then found to be equivalent, which in turn establishes the equivalence of states ⓒ and ⓗ.

An alternative method for identifying equivalent states is by means of the implication table described in Chapter 3. The implication table is a lower-left triangular matrix in which the rows are labeled by all states except the first and the columns are labeled by all states except the last. The intersection of a row and column identifies the stable states that are being tested for equivalence.

If two states are found to be nonequivalent, then the symbol × is entered at the intersection; if the states are determined to be equivalent, then the equivalence symbol ≡

is entered. The implication table for Example 6.5 is shown in Figure 6.20 and represents the first pass through the primitive flow table. During the first pass, any obvious nonequivalences and equivalences are entered in the implication table.

b	×									
c	×	×								
d	×	×	×							
e	×	×	*d,f*	×						
f	×	×	×	×	×					
g	×	×	×	×	×	*e,h*				
h	×	×	*d,i*	×	*f,i*	×	×			
i	×	×	×	*e,j*	×	×	×	×		
j	×	×	*d,f*	×	≡	×	×	*f,i*	×	
k	×	×	×	×	×	×	×	×	×	×
	a	*b*	*c*	*d*	*e*	*f*	*g*	*h*	*i*	*j*

Figure 6.20 Implication table for the asynchronous sequential machine of Example 6.5 after the first pass. The symbol × indicates nonequivalent states; the symbol ≡ indicates equivalent states.

As previously indicated, state ⓐ is not equivalent to any other state, because the conditions of equivalence rule 1 are not met. Thus, the symbol × is inserted at the intersection of all rows in column *a*. Similarly, the equivalence of state ⓑ can be excluded. Stable states ⓔ and ⓙ satisfy all the conditions for equivalence. Thus, the equivalence symbol is entered at their mutual intersection.

If equivalence between two states is implied by one or two state pairs, then the implied pairs are entered at the intersection. Successive passes are then made through the table to establish further equivalences.

State pairs (ⓒ, ⓔ) and (ⓒ, ⓙ) both imply (ⓓ, ⓕ). Since stable states ⓓ and ⓕ are not equivalent, therefore, state pairs (ⓒ, ⓔ) and (ⓒ, ⓙ) are not equivalent. Thus, the symbol × is inserted at the intersection of row *e*, column *c* and row *j*, column *c*, as shown in Figure 6.21, which represents the second pass through the implication table. The results of the second pass are shown in bold-lined squares. The second pass also establishes nonequivalence between the following state pairs: (ⓗ, ⓔ), (ⓖ, ⓕ), and (ⓙ, ⓗ).

The equivalence between states ⓔ and ⓙ is used to authenticate the equivalence between state pairs (ⓓ, ⓘ) and (ⓒ, ⓗ), as shown in Figure 6.21. The information provided by the implication table clearly indicates those states that are nonequivalent and those states that are equivalent. The nonequivalent states must be retained, whereas, for equivalent state pairs and mutually equivalent states, all but one stable state can be eliminated.

b	×									
c	×	×								
d	×	×	×							
e	×	×	×	×						
f	×	×	×	×	×					
g	×	×	×	×	×	×				
h	×	×	≡	×	×	×	×			
i	×	×	×	≡	×	×	×	×		
j	×	×	×	×	≡	×	×	×	×	
k	×	×	×	×	×	×	×	×	×	×
	a	*b*	*c*	*d*	*e*	*f*	*g*	*h*	*i*	*j*

Figure 6.21 Implication table for the asynchronous sequential machine of Example 6.5 after the second pass. The results of the second pass are shown in bold-lined squares.

Example 6.6 Identify equivalent states in the primitive flow table of Figure 6.22, then eliminate redundant states and obtain a reduced primitive flow table. Equivalence rule 1 specifies that possible equivalence exists between state pairs (ⓐ, ⓔ), (ⓐ, ⓘ), and (ⓔ, ⓘ), because all three stable states have the same inputs. Stable state ⓘ, however, cannot be equivalent to any other stable state in column $x_1x_2 = 00$, because the output vector in state ⓘ is different than the output vectors for states ⓐ and ⓔ — the requirement for equivalence rule 2 is not met. Therefore, ⓐ $\neg\equiv$ ⓘ and ⓔ $\neg\equiv$ ⓘ. Possible equivalence exists, however, between stable states ⓐ and ⓔ.

Similarly, equivalence rule 1 indicates that possible equivalence exists between stable state pairs (ⓑ, ⓕ), (ⓑ, ⓙ), and (ⓕ, ⓙ). Stable state ⓙ, however, generates different outputs than either states ⓑ or ⓕ. In states ⓑ and ⓕ, $z_1z_2 = 11$; in state ⓙ, $z_1z_2 = 10$. Thus, ⓑ $\neg\equiv$ ⓙ and ⓕ $\neg\equiv$ ⓙ. Possible equivalence exists, however, between states ⓑ and ⓕ. Stable state pairs (ⓒ, ⓖ) and (ⓓ, ⓗ) may also be equivalent. The following states have been shown to be nonequivalent:

ⓐ $\neg\equiv$ ⓘ, ⓔ $\neg\equiv$ ⓘ, ⓑ $\neg\equiv$ ⓙ, ⓕ $\neg\equiv$ ⓙ

x_1x_2 00	01	11	10	z_1	z_2
ⓐ	b	–	d	0	0
e	ⓑ	c	–	1	1
–	j	ⓒ	d	1	0
e	–	g	ⓓ	0	1
ⓔ	f	–	d	0	0
a	ⓕ	g	–	1	1
–	j	ⓖ	h	1	0
a	–	c	ⓗ	0	1
ⓘ	j	–	d	0	1
i	ⓙ	c	–	1	0

Figure 6.22 Primitive flow table for the asynchronous sequential machine of Example 6.6.

Thus far, the criteria specified by equivalence rules 1 and 2 have been utilized to separate the stable states into nonequivalent and equivalent sets. To determine the equivalence of state pairs (ⓐ, ⓔ), (ⓑ, ⓕ), (ⓒ, ⓖ), and (ⓓ, ⓗ), the state pairs must be examined with respect to equivalence rule 3. That is, for each state pair, the next states must be the same or equivalent for all possible valid input sequences. The equivalence relations between the remaining stable state pairs in which equivalence is possible are shown below.

ⓐ ≡ ⓔ if ⓑ ≡ ⓕ

ⓑ ≡ ⓕ if ⓐ ≡ ⓔ • ⓒ ≡ ⓖ

ⓒ ≡ ⓖ if ⓓ ≡ ⓗ

ⓓ ≡ ⓗ if ⓒ ≡ ⓖ • ⓐ ≡ ⓔ

The equivalence statements listed above indicate multiple interdependencies between states. Since interdependent state pairs are equivalent, the following state pairs are equivalent:

ⓐ ≡ ⓔ, ⓑ ≡ ⓕ

ⓒ ≡ ⓖ, ⓓ ≡ ⓗ

These equivalence relations can be easily verified. Beginning in stable state ⓐ, apply a unique sequence of valid inputs. The output sequence thus obtained, is identical to that observed if stable state ⓔ was the initial state and the same input sequence was applied to the machine.

For example, beginning in stable states ⓐ and ⓔ, apply the following sequence of input vectors to the machine: x_1x_2 = 01, 11, 10, 11, 01, 00, 10, 00. Figure 6.23 shows the outputs for each input vector when the initial stable state is either ⓐ or ⓔ. After the application of the input vectors, both initial stable states will have sequenced the machine to the same stable state (ⓙ). The same rationale is used to verify the equivalence of the remaining state pairs (ⓑ, ⓕ), (ⓒ, ⓖ), and (ⓓ, ⓗ).

x_1x_2 = 01 11 10 11 01 00 10 00

ⓐ → ⓑ → ⓒ → ⓓ → ⓖ → ⓙ → ⓘ → ⓓ → ⓔ
z_1z_2 = 11 10 01 10 10 01 01 00

ⓔ → ⓕ → ⓖ → ⓗ → ⓒ → ⓙ → ⓘ → ⓓ → ⓔ
z_1z_2 = 11 10 01 10 10 01 01 00

Figure 6.23 Output sequence for stable states ⓐ and ⓔ for a unique input sequence.

Thus, rows ⓔ, ⓕ, ⓖ, and ⓗ are redundant and can be eliminated. Every reference to stable states ⓔ, ⓕ, ⓖ, or ⓗ is replaced by an uncircled entry to stable states ⓐ, ⓑ, ⓒ, or ⓓ, respectively. The reduced primitive flow table is shown in Figure 6.24.

x_1x_2 00	01	11	10	z_1	z_2
ⓐ	b	–	d	0	0
a	ⓑ	c	–	1	1
–	j	ⓒ	d	1	0
a	–	c	ⓓ	0	1
ⓘ	j	–	d	0	1
i	ⓙ	c	–	1	0

Figure 6.24 Reduced primitive flow table for Figure 6.22 for the asynchronous sequential machine of Example 6.6.

Example 6.7 A primitive flow table for an asynchronous sequential machine is shown in Figure 6.25. If no reduction is possible, then four storage elements are required to provide four feedback variables. Since there are 12 rows in the primitive flow table, four additional rows are necessary to accommodate the $2^4 = 16$ combinations of four feedback variables. The entry in each square of the four extra rows is a dash, indicating a "don't care" condition, which can be used in a later step for minimization. However, state reduction is possible in this example. Using the rules for equivalence, redundant states will be identified and then eliminated.

The stable states will be partitioned into separate sets with common attributes. The stipulation of equivalence rule 1 provides the first partitioning. All stable states in the same column are potentially equivalent. Thus, the following sets are obtained, in which the stable states in each set are characterized by having the same input vector:

{ⓐ, ⓖ}, {ⓑ, ⓕ, ⓗ, ⓛ}, {ⓒ, ⓔ, ⓘ, ⓚ}, {ⓓ, ⓙ}

Now apply equivalence rule 2 to each set to provide further partitioning. Each set will now contain only those stable states that have the same inputs (rule 1) and the same outputs (rule 2). The following sets reflect this second level of partitioning:

{ⓐ, ⓖ}, {ⓑ, ⓕ, ⓗ}, {ⓛ}, {ⓒ, ⓘ}, {ⓔ}, {ⓚ}, {ⓓ, ⓙ}

x_1x_2 00	01	11	10	z_1	z_2
(a)	b	–	d	0	0
a	(b)	e	–	0	1
–	b	(c)	d	1	1
a	–	c	(d)	1	0
–	f	(e)	j	1	0
g	(f)	e	–	0	1
(g)	f	–	j	0	0
a	(h)	e	–	0	1
–	h	(i)	d	1	1
a	–	i	(j)	1	0
–	l	(k)	d	0	0
a	(l)	k	–	0	0

Figure 6.25 Primitive flow table for the asynchronous sequential machine of Example 6.7.

Only those sets with two or more stable states qualify for further possible partitioning. Thus, equivalence may exist between stable states in the four following sets:

{ⓐ, ⓖ}, {ⓑ, ⓕ, ⓗ}, {ⓒ, ⓘ}, {ⓓ, ⓙ}

Now apply equivalence rule 3 to the four sets of stable states. To be equivalent, two stable states in a set must proceed to the same or equivalent next state. The determination of equivalence using rule 3 is not always as immediately apparent as when assessing equivalence using rules 1 and 2.

First, look for obvious equivalences. Stable states ⓑ and ⓗ both proceed to states ⓐ and ⓔ for an applied input vector of $x_1x_2 = 00$ and 11, respectively. Therefore, states ⓑ and ⓗ are equivalent.

Now use the established equivalence between states ⓑ and ⓗ to find further equivalences. Consider column $x_1x_2 = 11$. The following equivalence relation exists: ⓒ ≡ ⓘ if ⓑ ≡ ⓗ. Since states ⓑ and ⓗ are equivalent, therefore, states ⓒ and ⓘ are equivalent.

By examining the primitive flow table, other equivalences can be found based upon previously established equivalences. For example, in column $x_1x_2 = 10$, the following equivalence relation is identified: ⓓ ≡ ⓙ if ⓒ ≡ ⓘ. Since states ⓒ and ⓘ have already been shown to be equivalent, therefore, states ⓓ and ⓙ are equivalent.

Continue to examine the remaining stable states of the primitive flow table for additional equivalences using states that have been previously found to be equivalent. In columns $x_1x_2 = 00$ and 01, the following equivalence relations exist:

ⓐ ≡ ⓖ if ⓑ ≡ ⓕ • ⓓ ≡ ⓙ

ⓑ ≡ ⓕ if ⓐ ≡ ⓖ

Since states ⓓ and ⓙ have been established as being equivalent, there is an interdependence relationship between stable state pairs (ⓐ, ⓖ) and (ⓑ, ⓕ). Therefore, states ⓐ and ⓖ are equivalent as are states ⓑ and ⓕ.

All stable states have now been examined for equivalence. The resulting equivalent states are as follows:

ⓐ ≡ ⓖ

ⓑ ≡ ⓗ ≡ ⓕ

ⓒ ≡ ⓘ

ⓓ ≡ ⓙ

The five rows ⓖ, ⓗ, ⓕ, ⓘ, and ⓙ are redundant and can, therefore, be eliminated. Every reference to stable state ⓖ is replaced by an uncircled *a*, which results in a transition to equivalent stable state ⓐ. Similarly, every denotation to states ⓗ or ⓕ is replaced by an uncircled *b*. References to either stable state ⓘ or ⓙ are replaced by transient states *c* and *d*, respectively. The reduced primitive flow table is shown in Figure 6.26.

Equivalent states occur in a primitive flow table because previously defined states have not been specified as next states when appropriate. The removal of redundant states immediately reduces the number of rows in a primitive flow table and hence, the number of feedback variables. When the reduced primitive flow table obtained by eliminating redundant states is independent of the order in which the equivalencies are established, the flow table is unique.

x_1x_2 00	01	11	10	z_1	z_2
ⓐ	b	–	d	0	0
a	ⓑ	e	–	0	1
–	b	ⓒ	d	1	1
a	–	c	ⓓ	1	0
–	b	ⓔ	d	1	0
–	l	ⓚ	d	0	0
a	ⓛ	k	–	0	0

Figure 6.26 Reduced primitive flow table for Figure 6.25 for the asynchronous sequential machine of Example 6.7.

Referring to the primitive flow table of Figure 6.25, it is observed that for any pair of stable states in any column, the permissible changes to the input vector are identical. The allowable changes differ for each dissimilar column, however, but this is irrelevant since only those stable states in the same column are tested for equivalence. Thus, the order in which stable states are examined for equivalence is immaterial.

For example, Figure 6.27 shows stable states ⓑ, ⓕ, and ⓗ excerpted from the primitive flow table of Figure 6.25. Once the equivalence of states ⓐ and ⓖ has been established, it makes no difference in the order in which states ⓑ, ⓕ, and ⓗ are examined for equivalence. The only allowable transitions for the three states are from $x_1x_2 = 01$ to 00 or 11. It is evident, therefore, that the three states are mutually equivalent.

6.2.4 Merger Diagram

Merging is a process of combining two or more rows of a reduced primitive flow table into a single row. The merging process reduces the number of rows in the flow table and thus, may reduce the number of feedback variables. A reduction of feedback variables decreases the number of storage elements in the machine. The stable states in the rows that are merged are entered in the same location in the single merged row. When merging, the outputs associated with each row are disregarded. Thus, two rows of a reduced primitive flow table can merge, regardless of the output values of the rows under consideration.

x_1x_2 00	01	11	10	z_1	z_2
a	(*b*)	*e*	–	0	1
g	(*f*)	*e*	–	0	1
a	(*h*)	*e*	–	0	1

Figure 6.27 Stable states (*b*), (*f*), and (*h*) excerpted from the primitive flow table of Figure 6.25.

Two rows can merge into a single row if the entries in the same column of each row satisfy the requirements of the following three merging rules:

1. Identical state entries, either stable or unstable
2. A state entry and a "don't care"
3. Two "don't care" entries

Merging rule 1 specifies that there must be no conflict in state name entries in the same column of the two rows under consideration. That is, two different states cannot both be active for the same input vector and the same combination of feedback variables. Three or more rows can merge into a single row if and only if all pairs of rows satisfy the conditions of the merging rules.

A stable state entry and an unstable state entry of the same name in the same column of two different rows are merged as the stable state entry, since the resultant state must be stable. Thus, in many cases, the merging process eliminates the transient unstable states. Two identical unstable states merge as an unstable state. Both stable and unstable states merge with a "don't care" entry as a stable and unstable state, respectively.

The merging process is facilitated by means of a merger diagram. The merger diagram depicts all rows of the reduced primitive flow table in a graphical representation. Each row of the table is portrayed as a vertex in the merger diagram. The rows in which merging is possible are connected by lines.

A set of rows in a reduced primitive flow table can merge into a single row if and only if the rows are strongly connected. That is, every row in the set must merge with all other rows in the set. For example, Figure 6.28 illustrates strongly connected sets of three, four, and five rows each. Each set in Figure 6.28 is a maximal compatible set, in which compatible pairs {ⓐ, ⓑ}, and {ⓑ, ⓒ) implies the compatibility of {ⓐ, ⓒ}. Thus, the transitive property applies to maximal compatible sets. Adding another state to a maximal compatible set negates the transitive property on the new set, unless the set remains strongly connected.

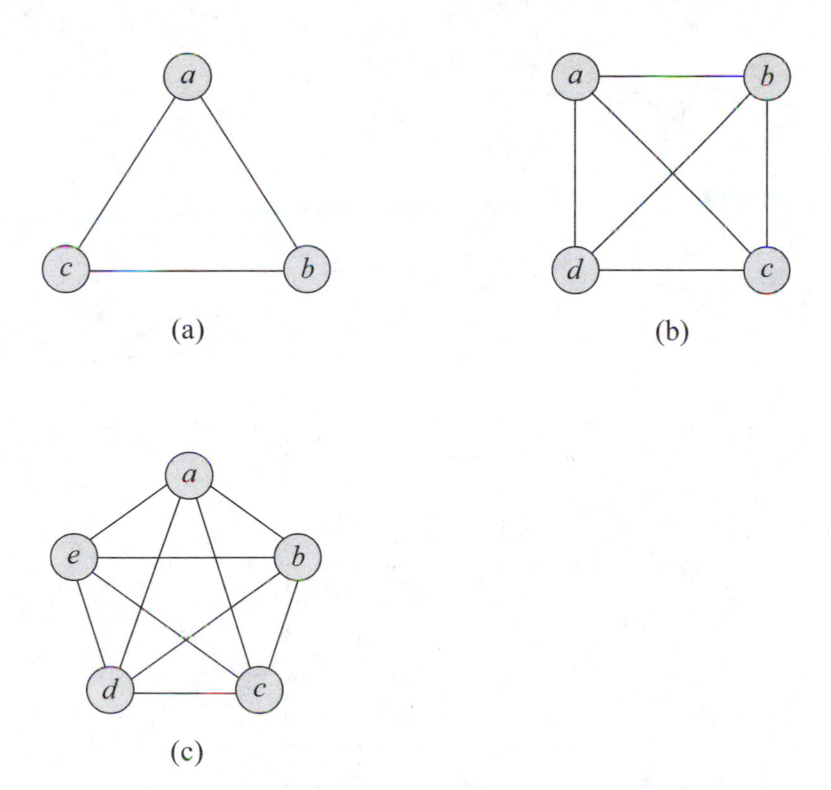

Figure 6.28 Strongly connected sets in which each row in the set can merge with all other rows in the set: (a) a 3-row set; (b) a 4-row set; and (c) a 5-row set.

The objective of merging is to combine the maximal number of rows into a single merged row while maintaining the fewest number of merged rows. All strongly connected sets can be combined into single individual rows, one set per row.

Example 6.8 Consider the reduced primitive flow table of Figure 6.29. List all rows in a circular format as shown in the merger diagram of Figure 6.30. Commence the merging process by attempting to merge row ⓐ with each subsequent row in turn, then row ⓑ with each succeeding row, then rows ⓒ through ⓖ with each successive row until all combinations of row pairs have been examined. That is, determine if the following pairs of rows can merge:

(ⓐ, ⓑ), (ⓐ, ⓒ), · · · , (ⓐ, ⓕ), (ⓐ, ⓖ), (ⓐ, ⓗ)

(ⓑ, ⓒ), (ⓑ, ⓓ), · · · , (ⓑ, ⓖ), (ⓑ, ⓗ)

(ⓒ, ⓓ), (ⓒ, ⓔ), · · · , (ⓒ, ⓗ)

.

.

.

(ⓕ, ⓖ), (ⓕ, ⓗ)

(ⓖ, ⓗ)

x_1x_2: 00	01	11	10	z_1
ⓐ	b	–	d	0
a	ⓑ	e	–	1
–	f	ⓒ	d	1
a	–	e	ⓓ	0
–	f	ⓔ	h	1
g	ⓕ	c	–	1
ⓖ	b	–	h	0
a	–	e	ⓗ	0

Figure 6.29 Reduced primitive flow table for the asynchronous sequential machine of Example 6.8.

Consider row ⓐ with respect to all other rows. Row ⓐ can merge with row ⓑ. In column $x_1x_2 = 00$, stable state ⓐ merges with unstable state a with no conflict (rule 1). The resulting entry in the merged row is ⓐ. Similarly, column $x_1x_2 = 01$ of rows ⓐ and ⓑ merge to yield an entry of ⓑ. Merging rule 3 applies to columns $x_1x_2 = 11$ and 10 of rows ⓐ and ⓑ. Both columns merge an unstable state with a "don't care" entry. The resulting merged row is shown in Figure 6.31. To indicate the merging compatibility of rows ⓐ and ⓑ, a line is drawn to connect the two rows, as shown in the partial merger diagram of Figure 6.32.

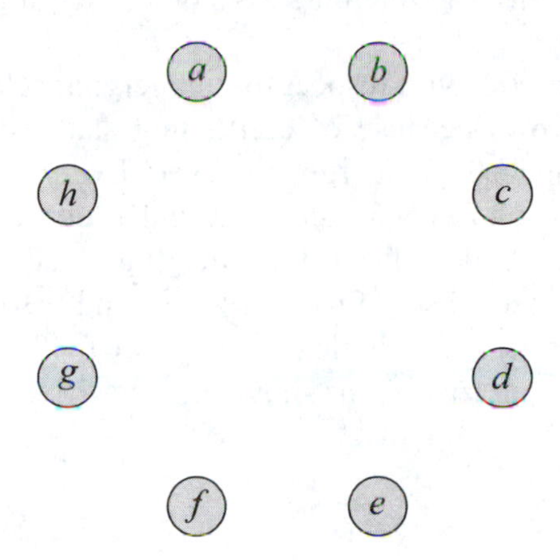

Figure 6.30 Merger diagram general format.

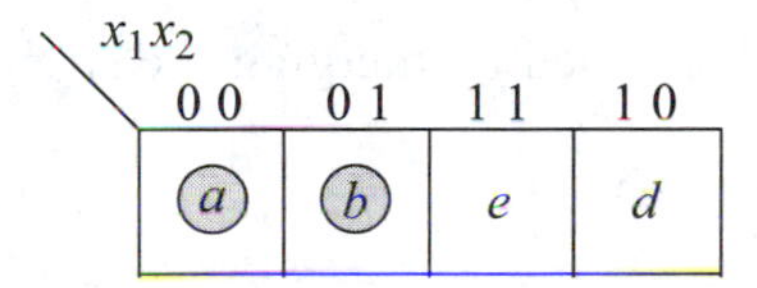

Figure 6.31 The resulting row obtained by merging rows ⓐ and ⓑ of Figure 6.29.

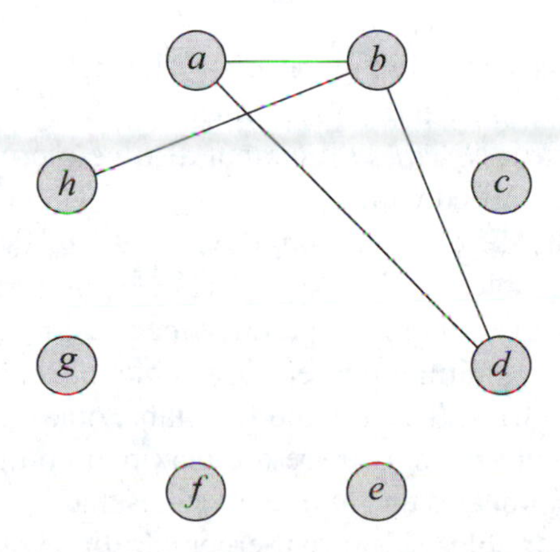

Figure 6.32 Partial merger diagram for Figure 6.29 depicting merger compatibility of rows ⓐ, ⓑ, and ⓓ into a single merged row.

Rows ⓐ and ⓒ cannot merge, because of the conflict in state names in column $x_1x_2 = 01$ — states *b* and *f* cannot merge into the same location. Rows ⓐ and ⓓ can merge into a single row, which is noted by the connecting line between rows ⓐ and ⓓ in the partial merger diagram of Figure 6.32. No other merging is possible with row ⓐ.

Now examine rows ⓑ and ⓒ with respect to the merging rules. No merger is possible between these two rows because of conflicting state entries in columns $x_1x_2 = 01$ and 11. Rows ⓑ and ⓓ can merge, however. Every column in rows ⓑ and ⓓ has either identical state names or a state name and a "don't care." Therefore, a line connects rows ⓑ and ⓓ to indicate this merging capability, as shown in Figure 6.32. The strongly connected set of rows ⓐ, ⓑ, and ⓓ is graphically depicted in Figure 6.32 and illustrated as a single row in Figure 6.33. The only remaining possible merger with row ⓑ is row ⓗ, as shown in the partial merger diagram of Figure 6.32.

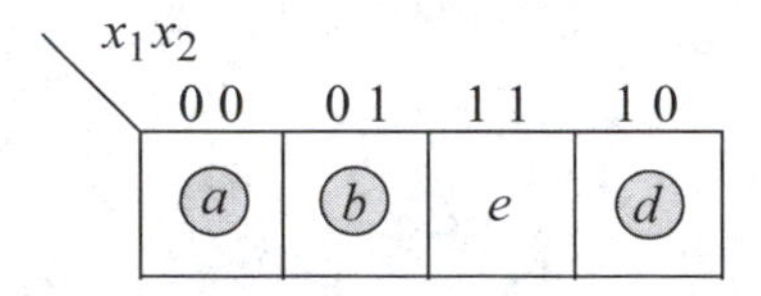

Figure 6.33 Merged row obtained by merging the strongly connected rows ⓐ, ⓑ, and ⓓ of Figure 6.29.

In a similar manner, row ⓒ is discerned to merge with row ⓕ only. Row ⓓ cannot merge with any succeeding row, because of dissimilar entries in columns $x_1x_2 = 00$, 11, and 10. Therefore, no line connects row ⓓ to any of the rows ⓔ through ⓗ. Continue to examine the remaining rows of the reduced primitive flow table in an analogous manner to obtain the complete enumeration of maximal compatible sets.

The final merger diagram is shown in Figure 6.34. The maximal compatibles are {ⓐ, ⓑ, ⓓ}, {ⓒ, ⓕ}, {ⓔ, ⓗ}, and {ⓖ}. Since row ⓑ is an integral part of the maximal compatible set {ⓐ, ⓑ, ⓓ}, it cannot also be merged with row ⓗ. Each stable state is uniquely identified with a particular row. That is, the same stable state cannot be characterized by different combinations of the feedback variables.

Another partition of mergeable sets is {ⓐ, ⓓ}, {ⓑ, ⓗ}, {ⓒ, ⓕ}, {ⓔ}, and {ⓖ}. This partition, however, requires three feedback variables, since five essential rows are required in the resulting merged flow table, plus an additional three rows containing "don't care" entries. The merged flow table obtained by the merging process contains at least as many rows as there are maximal compatible sets. The number of sets in the partition indicates the number of states that must be encoded and thus, the number of feedback variables. Using a merger partition with a small number of maximal compatible sets will yield a merged flow table with a minimal number of rows.

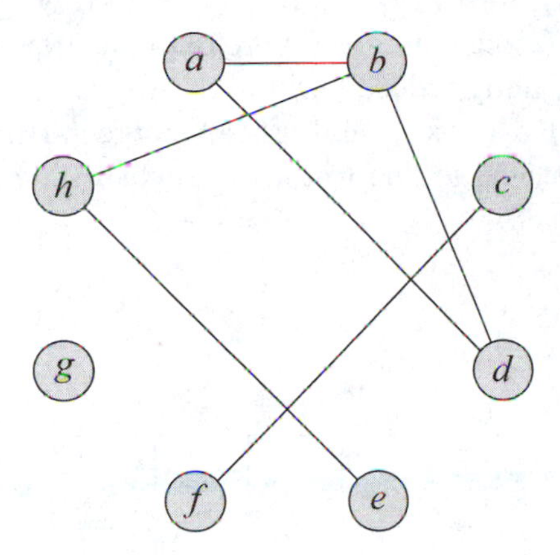

Figure 6.34 Merger diagram for the reduced primitive flow table of Figure 6.29 for Example 6.8.

There is no algorithmic method of identifying the optimal merger partition of strongly connected sets. In the merger diagram of Figure 6.34, the choice of strongly connected sets is unique. This is not always the case.

If the merger diagram indicates that multiple partition choices exist, where each partition produces the same number of rows, then a partition should be chosen which reduces the complexity of the λ output logic. By judiciously selecting a partition of strongly connected sets, the asserted output values of different states may form a pattern whereby, multiple minterm locations can be combined in the corresponding output map to minimize the λ output logic.

Whenever the merging process results in a merger diagram with a single set of strongly connected rows, then only one row is necessary in the merged flow table. In this case, no storage element is required and the circuit can be realized with combinational logic only. It must be remembered, however, that a single-line merged flow table represents the operation of a fundamental-mode machine. Thus, simultaneous changes of two or more input variables is not allowed.

The choice of selecting a partition from multiple partitions may also influence the speed of the machine. For example, if a state transition sequence causes the machine to proceed to a different row in a merged flow table, then the feedback variables must change values. However, if the partition chosen causes the change of state to proceed horizontally in the same row, then the feedback variables do not change, resulting in faster operational speed. If multiple partitions are characterized by the same number of strongly connected rows, then the partitions should be examined with respect to minimizing the λ output logic and increasing system response time.

The merging process is reflexive, since a row can merge with itself. The operation is also symmetric; that is, if row ⓘ can merge with row ⓙ, then row ⓙ can merge with row ⓘ. As mentioned previously, the merging process is not always transitive. Thus, the merger of row ⓘ with row ⓙ and row ⓙ with row ⓚ does not necessarily mean that rows ⓘ and ⓚ can merge. Figure 6.35 illustrates the nontransitive property of three rows in a partially reduced primitive flow table. Rows ⓘ and ⓙ can merge with no conflict. Also, rows ⓙ and ⓚ can merge into a single row. However, rows ⓘ and ⓚ cannot merge due to the conflict in state names in column $x_1x_2 = 10$.

x_1x_2 00	01	11	10	z_1
ⓘ	j	–	l	1
i	ⓙ	m	–	0
i	–	m	ⓚ	1

Figure 6.35 Partially reduced primitive flow table illustrating the nontransitive property of three rows.

The partitions of strongly connected sets represent sets that are mutually exclusive. Each set in the partition of disjoint sets contains unique rows that are not evident in any other set of the partitions.

Example 6.9 Consider the reduced primitive flow table of Figure 6.36. Since rows can be merged irrespective of the outputs of the rows under consideration, the output values for z_1 and z_2 can be ignored. Examine the table with respect to row ⓐ. Every entry in column $x_1x_2 = 00$ can merge with stable state ⓐ. In column $x_1x_2 = 01$, every row except rows ⓚ and ⓛ can merge with row ⓐ. In column $x_1x_2 = 11$, there is no conflict with any row, since row ⓐ contains an unspecified entry. There is also no conflict with any row in column $x_1x_2 = 10$, since the entries are either state d or unspecified. Thus, row ⓐ can merge with every row except rows ⓚ and ⓛ. This merging capability is depicted in Figure 6.37, in which a line is drawn from row ⓐ to rows ⓑ, ⓒ, ⓓ, and ⓔ, successively.

Row ⓑ can merge with row ⓔ but not with any other row due to a conflict of state names in column $x_1x_2 = 11$; row ⓒ can merge with row ⓓ but not with any other succeeding row for the same reason. That is, the attempted mergers in succeeding rows do not conform to the conditions specified in merging rule 1. Rows ⓚ and ⓛ represent the final merged pair. The complete merger diagram is illustrated in Figure 6.37.

x_1x_2 00	01	11	10	z_1	z_2
ⓐ	b	–	d	0	0
a	ⓑ	e	–	0	1
–	b	ⓒ	d	1	1
a	–	c	ⓓ	1	0
–	b	ⓔ	d	1	0
–	l	ⓚ	d	0	0
a	ⓛ	k	–	0	0

Figure 6.36 Reduced primitive flow table for the asynchronous sequential machine of Example 6.9.

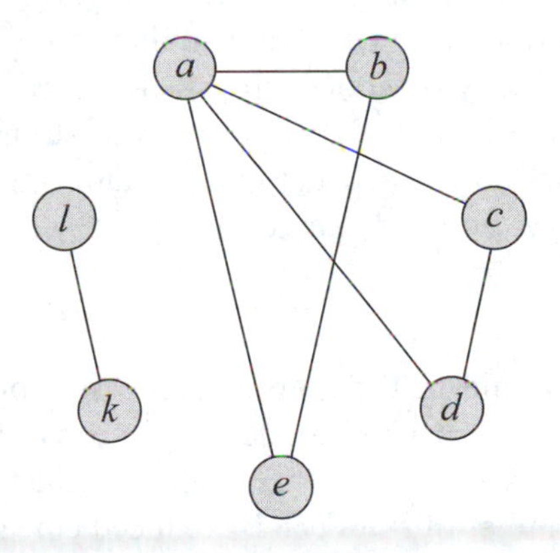

Figure 6.37 Merger diagram for the reduced primitive flow table of Figure 6.36 for Example 6.9.

The following partitions of disjoint sets are generated from the merger diagram:

{ⓐ, ⓒ, ⓓ}, {ⓑ, ⓔ}, {ⓚ, ⓛ}

{ⓐ, ⓑ, ⓔ}, {ⓒ, ⓓ}, {ⓚ, ⓛ}

The merged flow table can be derived from either partition, since both partitions represent sets of maximal compatibles. The merged flow table will contain four rows: one row for each set of maximal compatibles and one row of unspecified entries.

When choosing a particular partition from several partitions, it is imperative that the selected partition consists of maximal compatible disjoint sets. For example, assume that the following partitions were obtained from a merger diagram:

{ⓑ, ⓒ, ⓓ}, {ⓐ, ⓔ}, {ⓚ}, {ⓛ}

{ⓑ, ⓒ, ⓓ}, {ⓐ, ⓚ}, {ⓔ}, {ⓛ}

{ⓐ, ⓑ, ⓓ}, {ⓒ}, {ⓔ}, {ⓚ}, {ⓛ}

{ⓐ, ⓑ, ⓒ}, {ⓓ}, {ⓔ}, {ⓚ}, {ⓛ}

All partitions contain disjoint sets, but only the first two are maximal compatible disjoint sets. The first and second partitions necessitate only four rows in the merged flow table, whereas, the third and fourth partitions require five rows. Thus, the merged flow table should be derived from either the first or second partition.

Merging may eliminate all unspecified ("don't care") entries in a reduced primitive flow table. In many cases, merging is possible only because of the existence of unspecified entries. The merging process can assign to the "don't care" entries a stable or an unstable state name. The criteria specified in the merging rules then integrates the mergeable rows into a partition of maximal compatible sets.

The merger diagram graphically illustrates all possible partitions with which to generate the merged flow table. Each different partition of strongly connected sets provides a different merged flow table. Each merged flow table must then be analyzed with respect to machine minimization and operational speed.

Example 6.10 As a final example of the merging process, consider the reduced primitive flow table of Figure 6.38. Unless reduced by means of merging, the machine would require four feedback variables to accommodate the ten rows. Using the merging rules, however, the number of rows can be reduced considerably.

Proceeding carefully through the flow table and applying the merging rules methodically, maximal compatibility can be discerned in the four rows ⓒ, ⓓ, ⓔ, and ⓘ, as shown in the partial merger diagram of Figure 6.39 (a). Each row can merge with every row in the set, providing the strongly connected set {ⓒ, ⓓ, ⓔ, ⓘ}. Although rows ⓒ, ⓓ, ⓔ form a strongly connected set, as do rows ⓒ, ⓓ, ⓘ and rows ⓓ, ⓔ, ⓘ, strive always to produce a maximal compatible set. That is, attempt to combine as many rows as possible into one merged row. Utilizing a minimal number of rows reduces machine complexity and hardware cost.

Further observation discloses a maximal compatible set consisting of the three rows ⓐ, ⓕ, and ⓖ. The merging lines of Figure 6.39 (b) illustrate the compatibility of this strongly connected set.

x_1x_2 00	01	11	10	z_1	z_2
ⓐ	f	–	b	0	0
j	–	–	ⓑ	1	1
–	d	ⓒ	–	0	1
e	ⓓ	c	–	1	1
–	d	ⓗ	–	0	0
a	ⓕ	g	–	1	0
–	f	ⓖ	b	1	0
ⓔ	–	–	i	0	1
e	–	c	ⓘ	1	0
ⓙ	f	–	i	1	1

Figure 6.38 Reduced primitive flow table for the asynchronous sequential machine of Example 6.10.

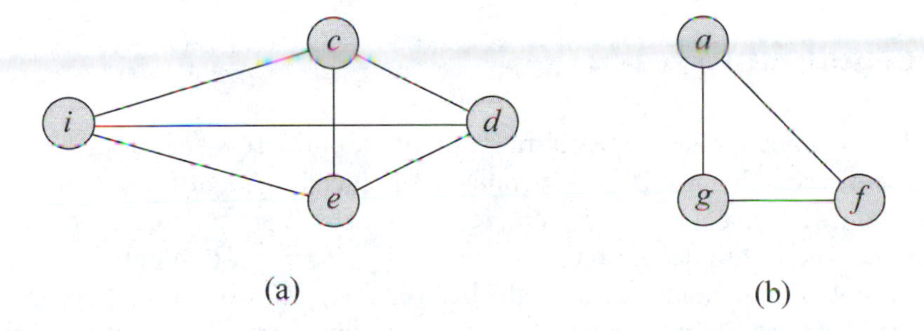

Figure 6.39 Partial merger diagram showing strongly connected sets for Example 6.10: (a) {ⓒ, ⓓ, ⓔ, ⓘ}; and (b) {ⓐ, ⓕ, ⓖ}.

Row ⓑ can merge with rows ⓒ, ⓖ, or ⓗ. However, since rows ⓒ and ⓖ have already been used to form other strongly connected sets, they cannot also be used as prospective merging rows with row ⓑ. Thus, row ⓑ is merged with row ⓗ to form a single merged row.

The remaining row ⓙ cannot merge with any other row due to the constraint imposed by merging rule 1, which specifies that there be no conflict in any column — columns $x_1x_2 = 00$, 01, and 10 have conflicting state names with respect to row ⓙ. Therefore, row ⓙ uniquely defines the fourth maximal compatible set.

Figure 6.40 shows the complete merger diagram, depicting all possible mergers. In this example, the merger diagram provides the following single partition of maximal compatible sets:

{ⓒ, ⓓ, ⓔ, ⓘ}, {ⓐ, ⓕ, ⓖ}, {ⓑ, ⓗ}, {ⓙ}

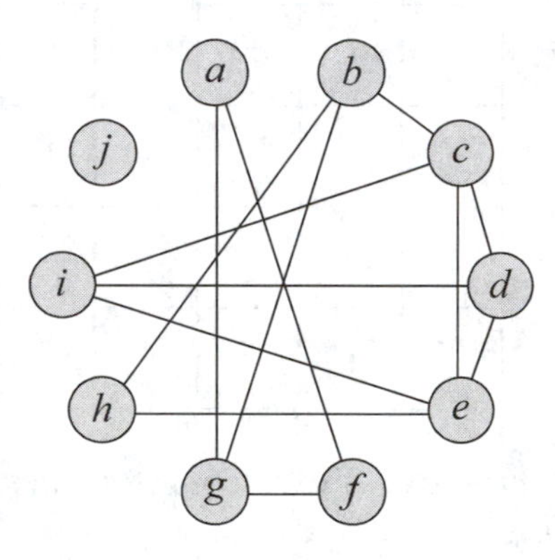

Figure 6.40 Merger diagram for the reduced primitive flow table of Figure 6.38 for Example 6.10.

6.2.5 Merged Flow Table

The merged flow table represents the culmination of the entire row-reduction process. The number of rows in the primitive flow table are reduced by identifying equivalent states using the rules for equivalence. The redundant states are then eliminated and every reference to an eliminated state is replaced by the corresponding equivalent state. Elimination of redundant states results in a reduced primitive flow table. Next, rows in the reduced primitive flow table are merged, where possible, using the rules for merging. The merging process yields a partition of maximal compatible sets, which are derived from a merger diagram. Each set contains strongly connected rows (or states), where each row in a set merges with all other rows in the set.

The next step in the synthesis procedure is the generation of a merged flow table. The merged flow table specifies the operational characteristics of the machine in a manner analogous to that of the primitive flow table and the reduced primitive flow table, but in a more compact form. Each row in a merged flow table represents a set of maximal compatible rows.

Most unspecified entries in the reduced primitive flow table are replaced with either a stable or an unstable state entry in the merged flow table. Since more than one stable state is usually present in each row of a merged flow table, many state transition sequences do not cause a change to the feedback variables. Thus, faster operational speed is realized.

The merged flow table is derived from the merger diagram in conjunction with the reduced primitive flow table. The partition of sets of maximal compatible rows obtained from the merger diagram dictates the minimal number of rows in the merged flow table.

If the number of sets in the optimal partition is not a power of two, then the merged flow table will contain the number of rows specified by the partition, plus additional rows such that, the total number of rows will be a power of two. The additional rows contain unspecified entries. Thus, if the partition contains four sets of strongly connected rows, then four rows are required in the merged flow table; if the partition contains three sets, then the merged flow table will contain one row for each set, plus an additional row of unspecified entries, for a total of four rows. Similarly, if the selected partition contains six sets, then the merged flow table will be constructed with eight rows, two of which contain unspecified entries.

One of the properties exhibited by logarithms is shown in Equation 6.3, which states that the variable p is equal to the logarithm to the base 2 of the variable r. Equation 6.3 can be rewritten as shown in Equation 6.4 and can be extended to include the form shown in Equation 6.5. In Equation 6.5, the notation $\lceil \log_2 r \rceil$ denotes the ceiling of $\log_2 r$ and signifies the smallest integer not less than $\log_2 r$.

Let p represent the number of storage elements and r the number of rows in a merged flow table. The property given in Equation 6.5 can be restated as shown in Equation 6.6, which states that the number of storage elements required for a specific merged flow table is greater than or equal to $\log_2 r$.

$$p = \log_2 r \qquad 6.3)$$

$$r = 2^p \qquad (6.4)$$

$$p = \lceil \log_2 r \rceil \qquad (6.5)$$

$$p \geq \log_2 r \qquad (6.6)$$

For example, using Equation 6.6, let the number of rows be $r = 8$. Therefore, the number of storage elements is $p \geq \log_2 8$, where $8 \leq 2^p$, requiring that the number of storage elements be $p = 3$ to accommodate eight rows in the merged flow table. Now, let the number of rows be $r = 5$. Therefore, the number of storage elements is $p \geq \log_2 5$, where $5 \leq 2^p$. The equation indicates that a value of $p = 3$ specifies the minimum number of storage elements necessary to represent five rows. Thus, the merged flow table will contain the five merged rows, plus three rows of unspecified entries.

Example 6.11 A reduced primitive flow table and a merger diagram are shown in Figure 6.41 and Figure 6.42, respectively. The merger diagram distinguishes the following single partition of sets containing maximal compatible rows:

{ⓑ, ⓒ, ⓔ}, {ⓓ, ⓕ}, {ⓐ}

Thus, the merged flow table is comprised of three rows—one row per set, plus a fourth row of unspecified entries.

x_1x_2 00	01	11	10	z_1
ⓐ	e	–	c	0
ⓑ	e	–	c	1
b	–	d	ⓒ	1
–	f	ⓓ	c	0
b	ⓔ	d	–	1
a	ⓕ	d	–	0

Figure 6.41 Reduced primitive flow table for the asynchronous sequential machine of Example 6.11.

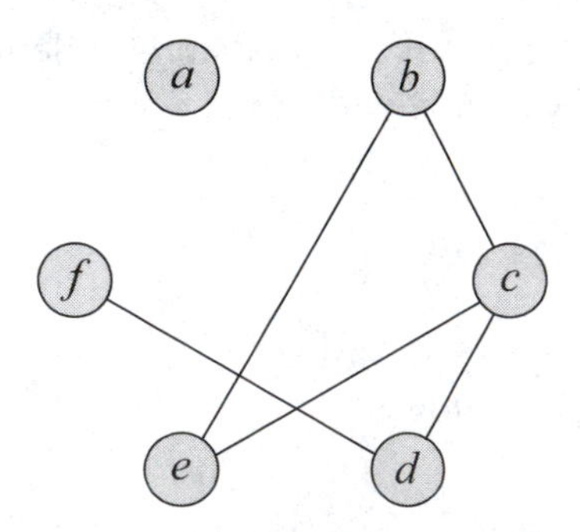

Figure 6.42 Merger diagram obtained from the reduced primitive flow table of Figure 6.41 for the asynchronous sequential machine of Example 6.11.

To merge rows ⓑ, ⓒ, and ⓔ into a single row, simply transcribe each row, one row at a time, from the reduced primitive flow table to the merged flow table. For example, row ⓑ transfers as shown in Figure 6.43 (a). Then transfer row ⓒ to the same row in the merged flow table, superimposing row ⓒ on row ⓑ, as shown in Figure 6.43 (b). Finally, transfer row ⓔ to the merged flow table, superimposing row ⓔ on previously transferred rows ⓑ and ⓒ, as shown in Figure 6.43 (c). Notice that no conflict occurs during the merging of rows ⓑ, ⓒ, and ⓔ into a single merged row. This is a necessary requirement and exemplifies the rationale for merging.

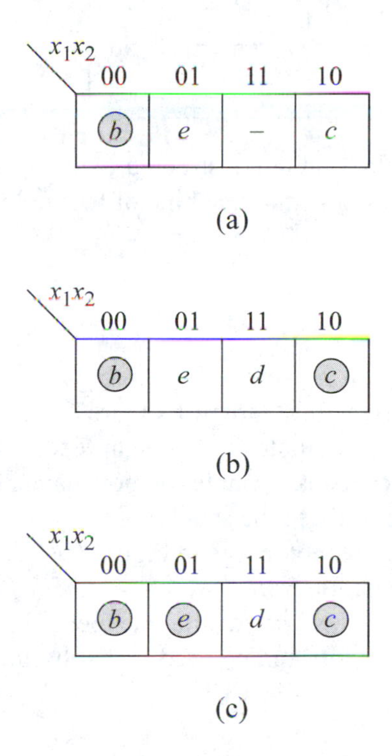

Figure 6.43 Top row of the merged flow table illustrating the transcribing of rows ⓑ, ⓒ, and ⓔ singly, from the reduced primitive flow table to the merged flow table: (a) row ⓑ transferred; (b) row ⓒ transferred and superimposed on row ⓑ; and (c) row ⓔ transferred and superimposed on rows ⓑ and ⓒ.

Similarly, rows ⓓ and ⓕ transfer from the reduced primitive flow table to the merged flow table. Since row ⓐ is not merged with any other row, no superimposing is necessary; thus, the third row of the merged flow table is represented by row ⓐ only. Figure 6.44 illustrates the complete merged flow table in which three rows are required for the maximal compatible sets, plus a fourth row of unspecified entries. Two feedback variables are required.

	x_1x_2 00	01	11	10
ⓑ, ⓒ, ⓔ	ⓑ	ⓔ	d	ⓒ
ⓓ, ⓕ	a	ⓕ	ⓓ	c
ⓐ	ⓐ	e	–	c
	–	–	–	–

Figure 6.44 Merged flow table constructed from the merger diagram of Figure 6.42 for the asynchronous sequential machine of Example 6.11.

In the reduced primitive flow table of Figure 6.41, a state transition sequence from state ⓑ to state ⓔ necessitates a change in feedback variables. In the merged flow table, however, the feedback variables do not change for this sequence, since the operation proceeds horizontally in the same row.

Since rows ⓐ and ⓑ cannot merge in Figure 6.41 due to the conflict in column $x_1x_2 = 00$, therefore, a transition from state ⓐ to state ⓔ results in a change in value for the feedback variables. This is evident in the merged flow table. The unspecified entries in the merged flow table can be used in a subsequent step to minimize the δ next-state logic.

Example 6.12 A merger diagram does not always produce a singular merged flow table. The sets of maximal compatible rows exhibited by the merger diagram may be combined in different configurations to yield two or more partitions. In this case, each partition uniquely defines sets of maximal compatible rows. For example, consider the reduced primitive flow table and merger diagram of Figure 6.45 and Figure 6.46, respectively.

The merger diagram indicates a maximal compatible set of three rows {ⓔ, ⓕ, ⓖ}. Since the objective of merging is to merge the greatest number of rows into a single row, the strongly connected set {ⓔ, ⓕ, ⓖ} should be left intact. The remaining rows ⓐ, ⓑ, ⓒ, ⓓ, and ⓗ can be configured in pairs to provide different combinations of strongly connected — or maximal compatible — sets. The merger diagram provides the following partitions of maximal compatible rows:

{ⓔ, ⓕ, ⓖ}, {ⓐ, ⓗ}, {ⓑ, ⓒ}, {ⓓ}
{ⓔ, ⓕ, ⓖ}, {ⓐ, ⓑ}, {ⓒ, ⓓ}, {ⓗ}
{ⓔ, ⓕ, ⓖ}, {ⓐ, ⓗ}, {ⓒ, ⓓ}, {ⓑ}

All three partitions produce merged flow tables consisting of four rows each, as shown in Figure 6.47. Rows ⓔ, ⓕ, and ⓖ are transcribed from the reduced primitive flow table to the merged flow table using the method described in Example 6.11. Each of the three rows is transferred, column by column, to the same row in the merged flow table. Row ⓔ is transferred first, then row ⓕ is superimposed on row ⓔ, and then row ⓖ is superimposed on previously transferred rows ⓔ and ⓕ. The remaining combinations of row pairs and the single row are then transferred to the merged flow table using the same process.

x_1x_2 00	01	11	10	z_1	z_2	z_3
ⓐ	b	–	h	0	1	0
a	ⓑ	c	–	1	0	0
–	b	ⓒ	d	1	1	0
e	–	c	ⓓ	0	1	1
ⓔ	f	–	h	1	1	1
e	ⓕ	g	–	1	0	1
–	f	ⓖ	h	0	1	1
a	–	g	ⓗ	1	0	0

Figure 6.45 Reduced primitive flow table for the asynchronous sequential machine of Example 6.12.

The relative position of the rows can be changed, if necessary, to eliminate race conditions and to provide an optimal pattern of 1s in the excitation maps in order to minimize the δ next-state logic. Excitation maps are covered in Section 6.2.6.

6.2.6 Excitation Maps and Equations

The merged flow table is the foundation from which the excitation maps are derived. The excitation maps directly formulate the equations necessary to implement the logic for the excitation variables.

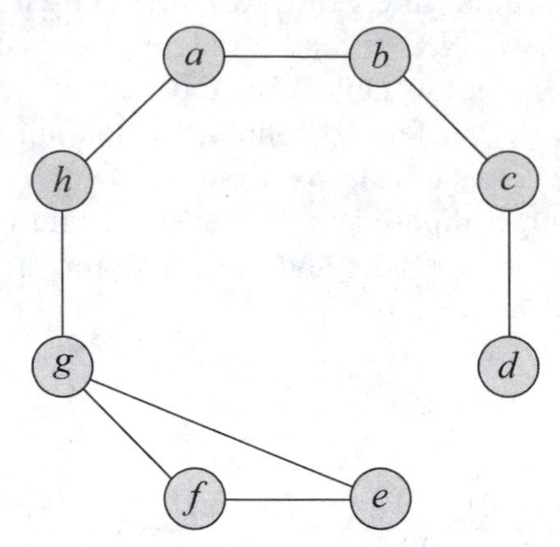

Figure 6.46 Merger diagram for the reduced primitive flow table of Figure 6.45 for Example 6.12.

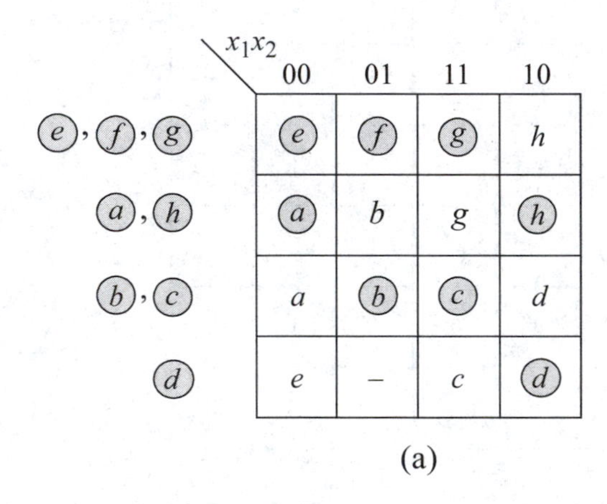

	x_1x_2 00	01	11	10
(e), (f), (g)	(e)	(f)	(g)	h
(a), (h)	(a)	b	g	(h)
(b), (c)	a	(b)	(c)	d
(d)	e	–	c	(d)

(a)

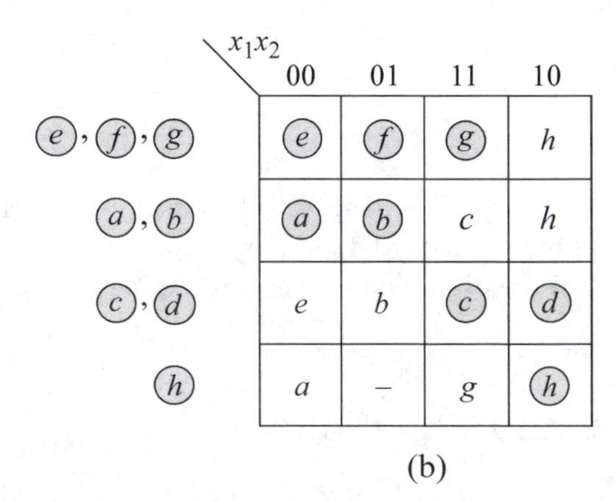

	x_1x_2 00	01	11	10
(e), (f), (g)	(e)	(f)	(g)	h
(a), (b)	(a)	(b)	c	h
(c), (d)	e	b	(c)	(d)
(h)	a	–	g	(h)

(b)

Figure 6.47 Merged flow table derived from the merger diagram of Figure 6.46: (a) partition 1; (b) partition 2; and (c) partition 3.

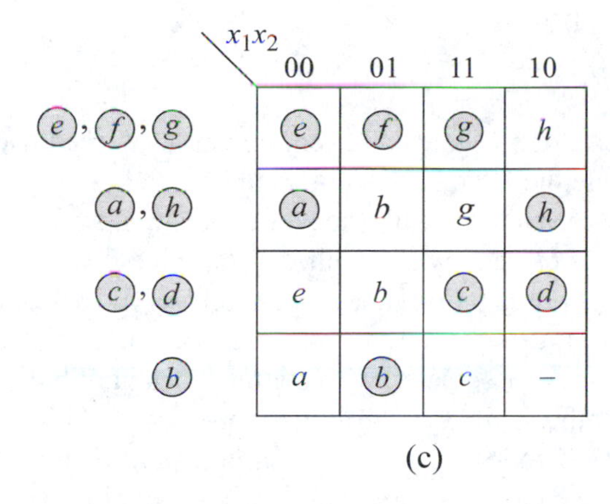

Figure 6.47 (Continued)

Each row of the merged flow table is assigned a unique combination of values for the feedback (or secondary) variables. The values of the feedback variables for each row then determine the values of the excitation variables for the stable state entries in the corresponding row of the excitation map.

Recall that a machine is stable in a particular state when the feedback variables are equal to the excitation variables. Thus, the values assigned to the excitation variables in each stable state of the excitation map are identical to those of the feedback variables for that row. The placement of each stable state of the excitation map represents a one-to-one mapping of the stable states in the corresponding locations of the merged flow table. The entries in the excitation map that correspond to unstable states in the merged flow table specify the next state to which the machine will sequence due to a change in the input vector.

The problems associated with simultaneous multiple changes to the binary input variables were nullified by requiring that the machine operate in fundamental mode only. Fundamental-mode operation precludes the possibility of simultaneous input changes by allowing only one input variable to change value at a time. The same problem arises, however, when a state transition sequence causes a simultaneous change to the excitation variables.

Two examples of this condition are:

$$\textcircled{00} \rightarrow 11 \rightarrow \textcircled{11} \text{ and}$$
$$\textcircled{01} \rightarrow 10 \rightarrow \textcircled{10}$$

To rectify this situation, the excitation values assigned to the unstable state should provide a change to only one variable. The resulting path will generate a cycle from the beginning stable state to the destination stable state, as shown below for the previous two conditions.

$$\text{(00)} \rightarrow 01 \rightarrow 11 \rightarrow \text{(11)} \text{ and}$$
$$\text{(01)} \rightarrow 11 \rightarrow 10 \rightarrow \text{(10)}$$

A longer path now exists between stable states and requires at least one unspecified entry in the cycle to accommodate the added transient state.

Since the propagation delays associated with the storage elements and other circuit components vary widely, it is extremely difficult to guarantee that the selected logic elements will have identical propagation delays. Therefore, the assignment of values for the feedback variables must assure that machine operation is predictable and correct even if different delays are associated with the feedback variables and other circuit components. In the synthesis of asynchronous sequential machines, one of the primary objectives in assigning codes for the state variables is the prevention of critical races. Minimization of the δ next-state logic and the λ output logic is also important, but to a lesser degree.

Methods will now be presented for assigning values to the feedback variables so that each state transition sequence will involve a change of only one excitation variable between logically contiguous rows in the cycle. Two changes of excitation variables will still be allowed, however, provided that the resulting race condition does not generate a critical race. Each transition in the merged flow table must be examined to ensure that the assigned feedback values differ by a change of only one variable between the row containing the beginning stable state and each successive pair of rows in the cycle, including the row containing the destination stable state. That is, the feedback variables must be assigned adjacent p-tuples for each successive row in the cycle.

Example 6.13 Consider the merged flow table of Figure 6.48 consisting of three rows labeled 1, 2, and 3. Inspection of column $x_1x_2 = 00$ indicates that the feedback values assigned to row 1 must be adjacent to those assigned to row 2, to provide a race-free cycle from state (*b*) or (*g*) through unstable state *a* to state (*a*). The same adjacency requirement is observed in column $x_1x_2 = 11$.

x_1x_2	00	01	11	10
1	(*a*)	*b*	(*e*)	*c*
2	*a*	(*b*)	*e*	(*g*)
3	(*f*)	*b*	(*d*)	(*c*)

Figure 6.48 Merged flow table for the asynchronous sequential machine of Example 6.13.

Examination of column $x_1x_2 = 10$ identifies an adjacency requirement between rows 1 and 3. This requirement accommodates a transition from either state ⓐ or ⓔ through unstable state c to state ⓒ.

Similarly, rows 2 and 3 must have adjacent feedback values to realize a transition from state ⓓ or ⓕ through unstable state b to state ⓑ. The observation of the above adjacency requirements for race-free operation are summarized as follows:

Column $x_1x_2 = 00$: Rows 1 and 2 must be adjacent.
Column $x_1x_2 = 01$: Rows 1 and 2 must be adjacent.
Rows 2 and 3 must be adjacent.
Column $x_1x_2 = 11$: Rows 1 and 2 must be adjacent.
Column $x_1x_2 = 10$: Rows 1 and 3 must be adjacent.

In column $x_1x_2 = 01$, only one stable state is specified. Therefore, a critical race condition is impossible. A noncritical race, however, may occur from row 1 or 3 to row 2, depending on the assigned feedback values. Since noncritical races do not present a problem in the deterministic operation of an asynchronous sequential machine, the assignment of values for the feedback variables is not crucial.

The preceding requirements listed for each column specify the adjacencies that are needed to establish race-free operation for the indicated transitions. The same information is portrayed graphically in the *transition diagram* of Figure 6.49. For a merged flow table containing three rows, the rows are listed in a triangular arrangement. Each row is represented by a vertex. Each pair of rows, for which adjacency is required, is connected by a line. The connecting line indicates a requisite transition between the pair of rows.

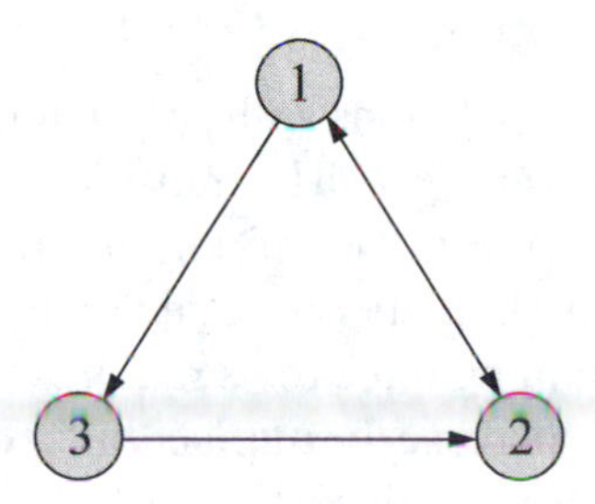

Figure 6.49 Transition diagram for the merged flow table of Figure 6.48 for Example 6.13.

There is no need to specify the direction of the lines, since adjacency is the only information that is relevant. However, directed lines in a transition diagram may be advantageous in visualizing the sequence of transitions.

The next step is to assign values to the vertices which will represent the values of the feedback variables. Each pair of logically adjacent rows must be assigned adjacent state codes. The transition diagram of Figure 6.49 illustrates that all three rows of the merged flow table must be adjacent. It is obviously not possible to assign 2-tuples to the three rows so that each row is adjacent to all other rows.

To illustrate this impracticability, observe the transition diagram of Figure 6.50. The codes assigned to the state variables in Figure 6.50 (a) produce a race condition for a transition between rows 2 and 3. Figure 6.50 (b) generates a critical race for a transition between rows 1 and 3. If the transition diagram contained four row vertices, then a 2-tuple Gray code assignment would realize a race-free operation.

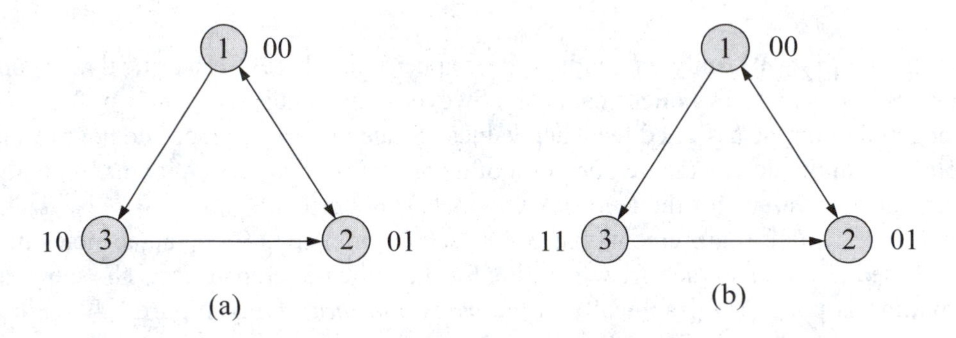

Figure 6.50 Transition diagrams illustrating state code assignments for a 3-row merged flow table containing critical races: (a) a race between rows 2 and 3; and (b) a race between rows 1 and 3.

The vertices of triangular or other polygons with an odd number of sides cannot be encoded with adjacent p-tuples for every row. The transition diagram must be altered so that triangular polygons do not appear. Modifying the transition diagram in this way may require more than a minimal number of state variables.

Since a merged flow table containing three rows requires two feedback variables, the addition of a fourth row will not increase the number of feedback variables. Recall that the number of rows in a table satisfies the expression of Equation 6.7, where r is the number of rows and p is the number of states, or feedback variables. Whether $r = 3$ or 4, the number of feedback variables remains the same. Therefore, a fourth row consisting of unspecified entries will be appended to the bottom row of the merged flow table. The additional row is not associated with any stable state. The unspecified entries will be used, where applicable, to establish intermediate unstable states which will direct the machine to the appropriate destination stable state. Two state variables are required to encode four rows.

$$r = 2^p \tag{6.7}$$

Figure 6.51 depicts the augmented merged flow table containing the original three rows plus a fourth row of unspecified entries. The transition diagram for the augmented flow table is shown in Figure 6.52. The transition from row 1 to row 3 is replaced by an equivalent sequence from row 1 to row 4 and then to row 3, as shown by the arrows in Figure 6.51 and Figure 6.52. This sequence represents a transition from stable state ⓐ or ⓔ through transient state c in column $x_1x_2 = 10$, then to the unspecified entry in row 4, then to state ⓒ in row 3. All other transitions involve a change of only one feedback variable.

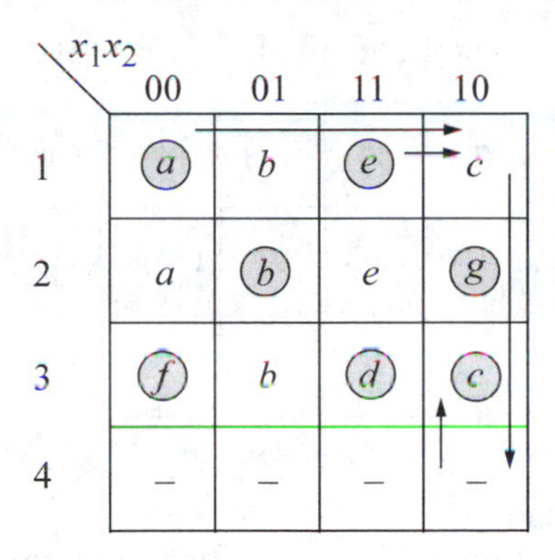

x_1x_2	00	01	11	10
1	ⓐ	b	ⓔ	c
2	a	ⓑ	e	ⓖ
3	ⓕ	b	ⓓ	ⓒ
4	–	–	–	–

Figure 6.51 Augmented merged flow table for the asynchronous sequential machine of Example 6.13.

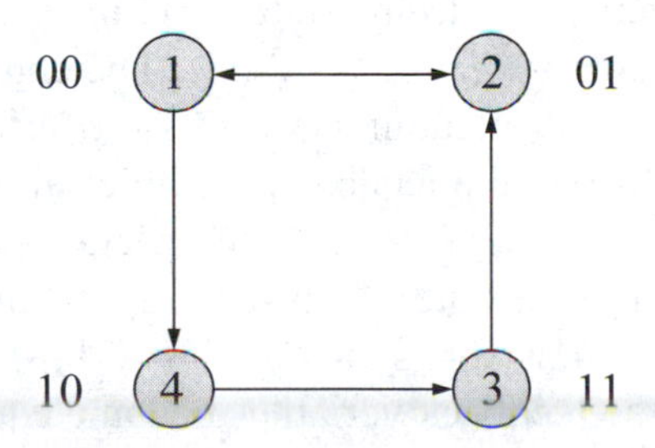

Figure 6.52 Transition diagram for the augmented merged flow table of Figure 6.51.

The codes next to each row vertex indicate the assigned values for the feedback variables. The choice of state codes is arbitrary in this context. Any assignment of sequential Gray code 2-tuples is a suitable choice. Each state name is now replaced by its corresponding assigned 2-tuple.

The excitation map for the augmented merged flow table is shown in Figure 6.53. This is a combined excitation map for excitation variables Y_{1e} and Y_{2e}. The feedback variables are $y_{1f}y_{2f}$. The stable states in each row are assigned excitation values that are equal to the feedback values of the corresponding row. The unstable states are assigned excitation values that direct the machine to the destination stable state.

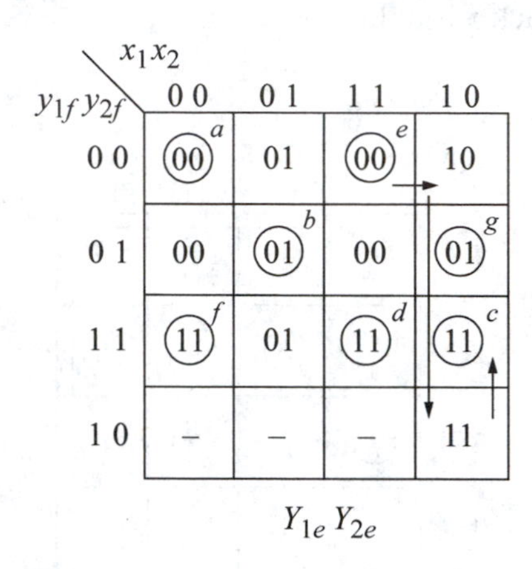

Figure 6.53 Combined excitation map for the augmented merged flow table of Figure 6.51.

For example, the transition from state ⓐ to state ⓑ is specified by ⓐ → b → ⓑ. Thus, $Y_{1e}Y_{2e} = \textcircled{00} \rightarrow 01 \rightarrow \textcircled{01}$. The entry $Y_{1e}Y_{2e} = 01$ for unstable state b in row $y_{1f}y_{2f} = 00$, column $x_1x_2 = 01$, specifies the excitation values which will become the values of the feedback variables after a delay of Δt, directing the machine to state ⓑ in row $y_{1f}y_{2f} = 01$, column $x_1x_2 = 01$.

Likewise, the transition from state ⓖ to state ⓔ requires excitation values of $Y_{1e}Y_{2e} = 00$ to be entered in unstable state e in row $y_{1f}y_{2f} = 01$, column $x_1x_2 = 11$. Thus, after a delay of Δt, the feedback variables become equal to the excitation variables and the machine enters state ⓔ. The "don't care" entries in row 4 can be used as intermediate transient states to introduce cycles which direct the machine to the desired stable state. Code assignments must be avoided that would cause the machine to cycle continuously between unstable states.

The transition from state ⓔ to state ⓒ necessitates the values for $Y_{1e}Y_{2e}$ as shown in Figure 6.54. The sequence from state ⓔ to state ⓒ is graphically illustrated by the arrows in Figure 6.53. A similar path is realized for a transition from state ⓐ to state ⓒ. Thus, the addition of the fourth row in the excitation map resolves the adjacency problem and thus, the possible critical race condition of the original merged flow table of Figure 6.48. Other arrangements are possible for the four rows of the augmented merged flow table of Figure 6.51.

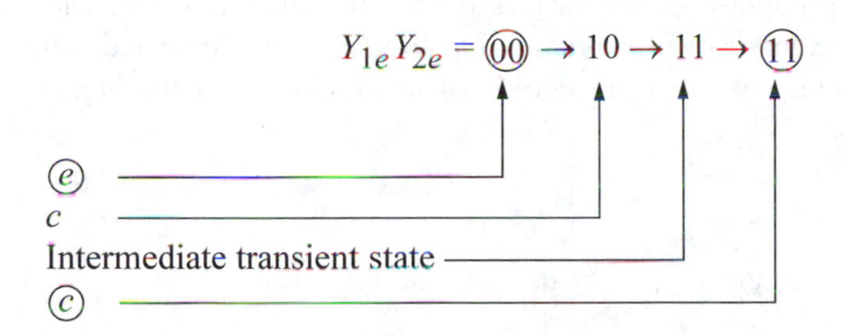

Figure 6.54 State transition sequence depicting the transition from stable state ⓔ to stable state ⓒ.

The combined excitation map of Figure 6.53 is now separated into its constituent parts to obtain individual excitation maps for Y_{1e} and Y_{2e}, as shown in Figure 6.55. To obtain the excitation map for Y_{1e}, simply transfer the values for Y_{1e} from the minterm locations in the combined map to the same squares in the individual map. Repeat the process to obtain the excitation map for Y_{2e}.

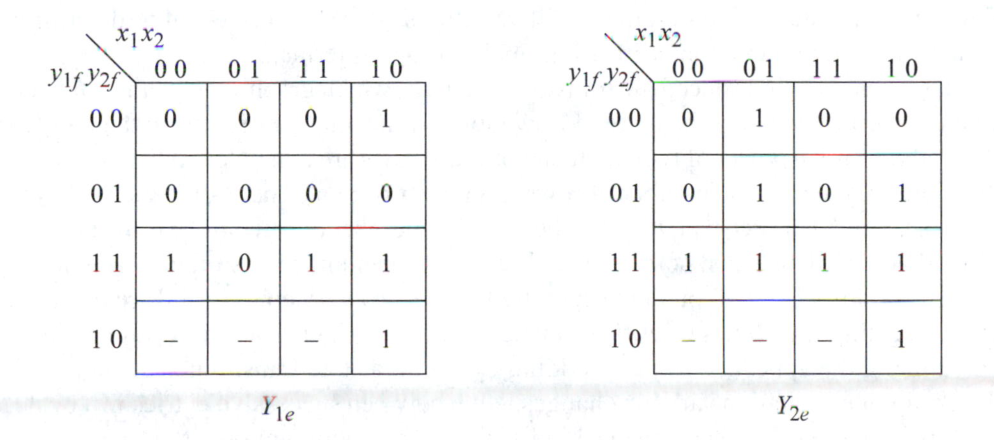

Figure 6.55 Individual excitation maps obtained from Figure 6.53 for the asynchronous sequential machine of Example 6.13.

The equations for the excitation variables are derived directly from the excitation maps. The equations can be specified in either a sum-of-products form or a product-of-sums form. Regardless of the form used, the boolean equations must be free of static-1 and static-0 hazards. The sum-of-products notation is shown in Equation 6.8 for

Y_{1e} and Y_{2e}; the product-of-sums form is shown in Equation 6.9 for each excitation variable. All equations are free of static hazards. In some instances, the equations may be manipulated — using the laws of boolean algebra — to obtain a network with fewer logic gates.

$$\begin{aligned} Y_{1e} &= y_{1f}x_1 + y_{1f}x_2' + y_{2f}'x_1x_2' \\ Y_{2e} &= y_{1f} + x_1'x_2 + y_{2f}x_1x_2' \end{aligned} \tag{6.8}$$

$$\begin{aligned} Y_{1e} &= (x_1 + x_2')(y_{1f} + y_{2f}')(y_{1f} + x_1)(y_{1f} + x_2') \\ Y_{2e} &= (y_{1f} + x_1 + x_2)(y_{1f} + x_1' + x_2')(y_{1f} + y_{2f} + x_1') \\ &\quad (y_{1f} + y_{2f} + x_2) \end{aligned} \tag{6.9}$$

Hazard cover

By rearranging the rows of the augmented merged flow table, other variations of the table are possible. There is no efficient method of determining the optimal configuration of state code assignment. All variations should be inspected to determine which form yields equations with a minimal number of terms.

Example 6.13 introduced the transition diagram which graphically depicts the adjacency requirements for the merged flow table. If the number of rows in the merged flow table is not a power of two, then one or more rows are appended to the bottom of the table so that the total number of rows is a power of two. That is, the merged flow table is augmented such that, $r = 2^p$, where r is the number of rows and p is the number of feedback variables that corresponds directly to the number of storage elements.

The additional rows contain unspecified entries in every column and are useful in circumventing the effects of critical race conditions. All critical races can be eliminated using the appended rows by adding cycles to a state transition sequence such that, only one excitation variable changes value between successive entries in a cycle. The creation of cycles alters the machine's internal behavior without affecting the machine specifications. The unspecified entries can also be used to minimize the excitation equations. Care must be taken when using the unspecified entries for this purpose, however. Entries must not be inserted that would generate a new stable state or cause the machine to cycle continuously between unstable states. For example, in row $y_{1f}y_{2f} = 10$ of the maps for Y_{1e} and Y_{2e}, values of $Y_{1e}Y_{2e} = 10$ would establish a new stable state that was not included in the machine specifications.

Example 6.14 A merged flow table for an asynchronous sequential machine is shown in Figure 6.56 with the accompanying transition diagram in Figure 6.57. The transition diagram indicates that all four rows must be adjacent. Since this is clearly impossible, adjacency can be achieved by redirecting some state transitions through rows containing unspecified entries. Appending four rows of unspecified entries to the bottom of the flow table would certainly resolve the problem of critical races.

x_1x_2	00	01	11	10
1	ⓐ	c	–	b
2	ⓓ	c	e	ⓑ
3	–	ⓒ	ⓕ	g
4	a	–	ⓔ	ⓖ

Figure 6.56 Merged flow table for the asynchronous sequential machine of Example 6.14.

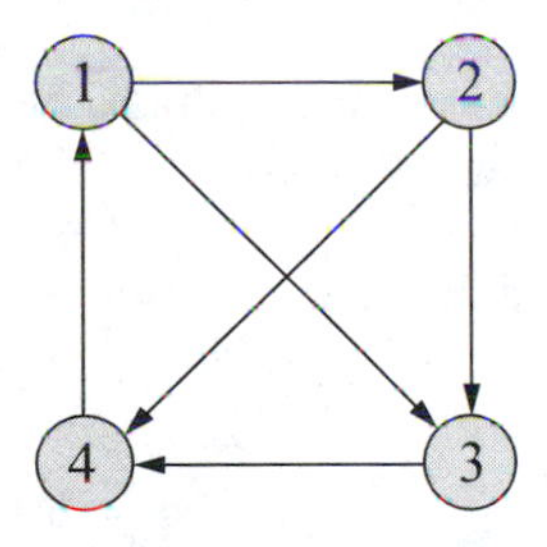

Figure 6.57 Transition diagram for the merged flow table of Figure 6.56 for Example 6.14, in which critical races are possible.

However, closer examination of the merged flow table and transition diagram reveals another possibility. Recall that polygons with an odd number of sides cannot be encoded with adjacent p-tuples for every row. If the transition from row 1 to row 3 could be rerouted through another row, then one triangle would be eliminated. The same rationale applies to the transition from row 2 to row 4.

Consider the transition from row 1 to row 3. In state ⓐ, if x_2 changes from 0 to 1, then the machine proceeds from state ⓐ in row 1 through unstable state c to state ⓒ in row 3. Since column $x_1x_2 = 01$ contains an unspecified entry in row 4, the sequence can be redirected from unstable state c in row 1 to an intermediate unstable state in row 4, then to state ⓒ in row 3. This new sequence would eliminate the triangle indicated by rows by rows 1, 3, and 4 in the transition diagram.

A similar procedure can be used to eliminate the triangle portrayed by rows 2, 3, and 4 in the transition diagram. In the merged flow table, the transition from state ⓑ to state ⓔ can be achieved with a different cycle by utilizing the unspecified entry in row 1, column $x_1x_2 = 11$. Thus, the new sequence will be from state ⓑ through unstable state *e*, through the intermediate unstable state in row 1, then to state ⓔ in row 4. The longer cycles are indicated by the arrows in the merged flow table of Figure 6.58. This is the same flow table as Figure 6.56, but illustrates the modified paths which utilize the two unspecified entries in columns $x_1x_2 = 01$ and 11.

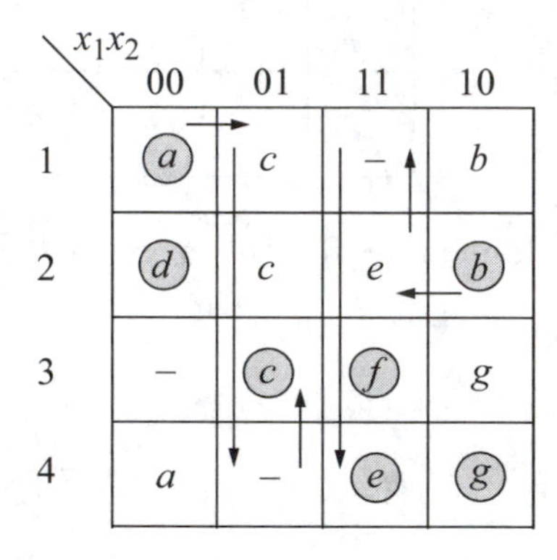

x_1x_2	00	01	11	10
1	ⓐ	*c*	–	*b*
2	ⓓ	*c*	*e*	ⓑ
3	–	ⓒ	ⓕ	*g*
4	*a*	–	ⓔ	ⓖ

Figure 6.58 Merged flow table of Figure 6.56 showing the modified transitions from row 1 to row 3 and from row 2 to row 4.

The new state transitions add one more transient state to each of the two cycles. The new transition diagram acquired from the modified cycles is shown in Figure 6.59. All triangular polygons have been removed, providing a 4-vertex transition diagram in which a Gray code sequence provides the necessary codes for the feedback variables. The codes assigned to the state variables preclude the possibility of races. The transition from row 1 to row 3 has been replaced by a transition from row 1 to row 4, then to row 3, as indicated by the arrows in Figure 6.58 and Figure 6.59. Similarly, the transition from row 2 to row 4 has been replaced by a transition from row 2 to row 1, then to row 4.

The combined excitation map is shown in Figure 6.60. The values for the feedback variables $y_{1f}y_{2f}$ are obtained from the transition diagram of Figure 6.59. The order in which the Gray code is assigned is not important at this point. The only requirement is that contiguous rows in a cycle be logically adjacent. The codes for the state variables could also be assigned as follows or any other combination such that, the appropriate rows differ by a change of only one feedback variable:

Row 1 = 00, row 4 = 01, row 3 = 11, row 2 = 10
Row 1 = 11, row 2 = 01, row 3 = 00, row 4 = 10

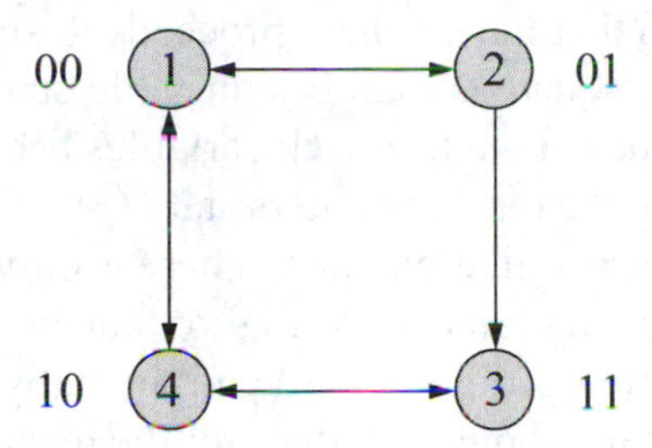

Figure 6.59 Transition diagram for the merged flow table of Figure 6.56 after rerouting two state transition sequences.

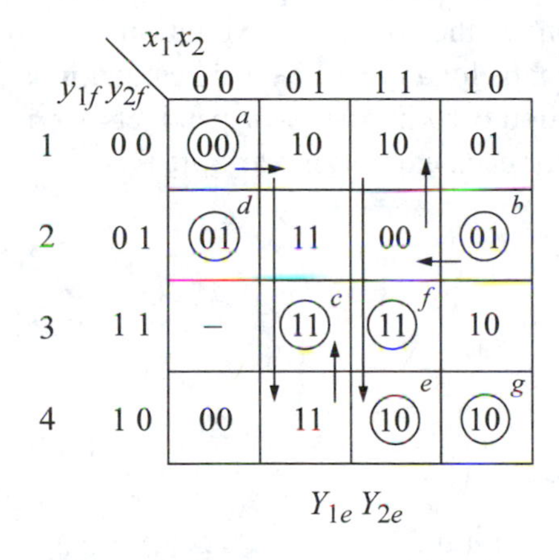

Figure 6.60 Combined excitation map for Y_{1e} and Y_{2e} obtained from the merged flow table of Figure 6.58 for Example 6.14 showing the lengthened cycles for transitions from ⓐ to ⓒ and from ⓑ to ⓔ.

The entries for the excitation variables in the stable states of the combined excitation map are identical to the values of the feedback variables for the corresponding row. Thus, state ⓐ contains the entry $Y_{1e}Y_{2e} = 00$; states ⓑ and ⓓ contain identical entries of $Y_{1e}Y_{2e} = 01$; states ⓒ and ⓕ are assigned values of $Y_{1e}Y_{2e} = 11$; and states ⓔ and ⓖ have excitation values of $Y_{1e}Y_{2e} = 10$.

The unstable states contain excitation values which direct the machine to the next row in a cycle after a delay of Δt. Thus, the entry for unstable state c in row $y_{1f}y_{2f} = 00$, column $x_1x_2 = 01$ is $Y_{1e}Y_{2e} = 10$, which causes a transition to row

$y_{1f}y_{2f} = 10$ in column $x_1x_2 = 01$ after a delay of Δt. The excitation values in this intermediate transient state are $Y_{1e}Y_{2e} = 11$, which effect a transition to state ⓒ after an appropriate delay.

Consider the transition from state ⓐ to state ⓑ. The excitation values for unstable state *b* must be such that the machine proceeds to state ⓑ after a delay of Δt. Therefore, the entry in the square containing unstable state *b* must be $Y_{1e}Y_{2e} = 01$. After a delay of Δt, the values of the feedback variables become equal to the values of the excitation variables and the machine enters state ⓑ. The entries for all other unstable states are obtained in a similar manner. There is only one unspecified entry remaining in the combined excitation map of Figure 6.60.

After further examination of Figure 6.60 and mentally separating the values for Y_{1e} and Y_{2e}, it appears that the choice of values for the feedback variables is adequate. The grouping of 1s for the excitation variables should yield boolean equations with a minimal number of terms. In some cases, however, it may be advantageous to assign different combinations of feedback variables, then rearrange the rows to obtain an optimal pattern of 1s and 0s.

The next step is to separate the combined excitation map into individual maps for Y_{1e} and Y_{2e}. These maps are shown in Figure 6.61. The values for Y_{1e} and Y_{2e} are transcribed separately to their respective maps. There is a one-to-one correspondence from the minterm locations in the combined excitation map to the same minterm locations in the respective maps for Y_{1e} and Y_{2e}. The equations for the excitation variables are shown in Equation 6.10 in a sum-of-products form. One hazard cover is required in the equation for each excitation variable.

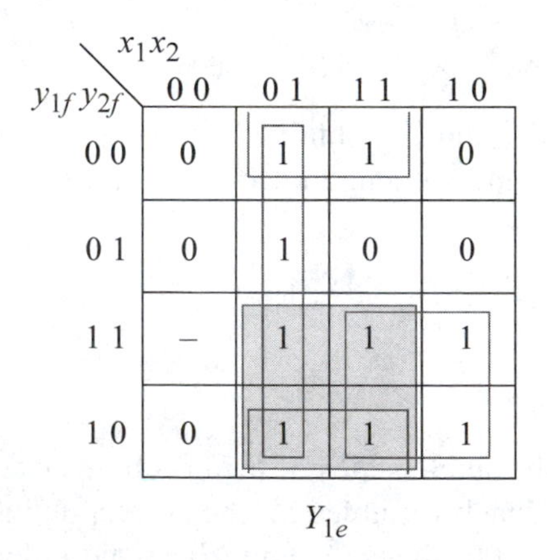

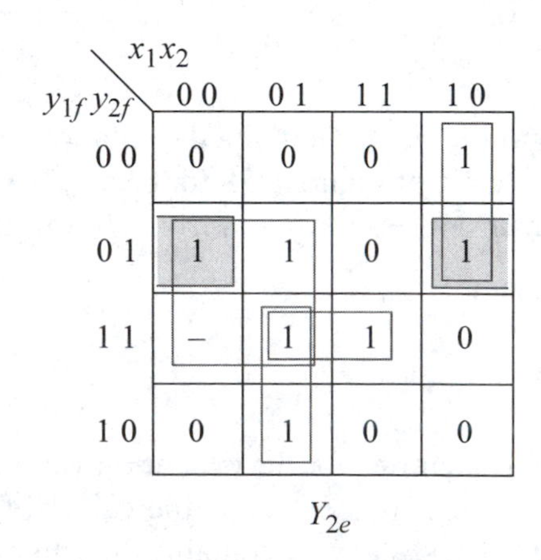

Figure 6.61 Individual excitation maps for Y_{1e} and Y_{2e} obtained from Figure 6.60. The maps are used to obtain a sum-of-products equation for the excitation variables.

$$Y_{1e} = x_1'x_2 + y_{1f}x_1 + y_{2f}'x_2 + y_{1f}x_2$$

Hazard cover (arrow to $y_{1f}x_2$)

$$Y_{2e} = y_{2f}x_1' + y_{1f}y_{2f}x_2 + y_{1f}x_1'x_2 + y_{1f}'x_1x_2' + y_{1f}'y_{2f}x_2' \quad (6.10)$$

Hazard cover (arrow to $y_{1f}'y_{2f}x_2'$)

To confirm the need for a hazard cover, consider the equation for Y_{1e} with the hazard cover $y_{1f}x_2$ removed. The logic necessary to implement Y_{1e} with no hazard cover is shown in Figure 6.62. Beginning in minterm location $y_{1f}y_{2f}x_1x_2 = 1111$ — which corresponds to state ⓕ — assume that input x_1 changes from 1 to 0. This input change causes a transition from state ⓕ to state ⓒ, as shown in Figure 6.62. Notice that Y_{1e} should not change value for this sequence. The output values for each gate are shown in Figure 6.62 as the machine proceeds through three gate delays for a transition from state ⓕ to state ⓒ. This state transition sequence may result in a static-1 hazard, as shown in the output for Y_{1e} in Figure 6.62. Thus, a hazard cover is required.

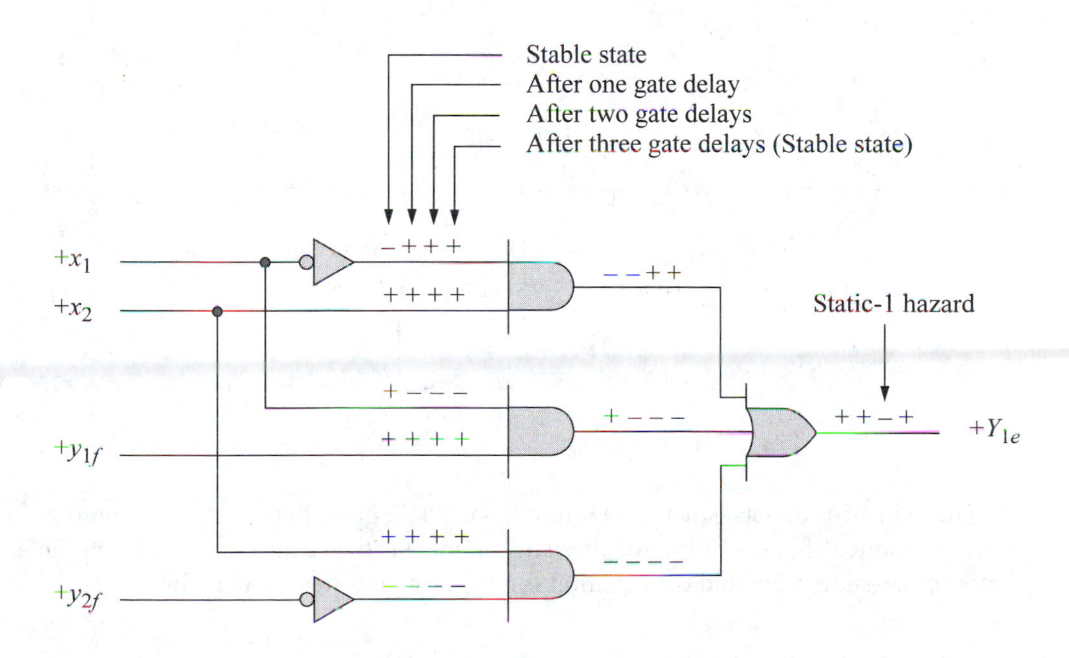

Figure 6.62 Logic diagram that depicts the implementation of excitation variable Y_{1e} illustrating a static-1 hazard when input x_1 changes from 1 to 0; that is, a change from $y_{1f}y_{2f}x_1x_2 = 1111$ to 1101.

Figure 6.61 is reproduced as Figure 6.63 to illustrate the derivation of the equations in a product-of-sums form. The equations are shown in Equation 6.11. Although no hazard cover is required in the equation for Y_{1e}, three hazard covers are necessary for the equation for Y_{2e}.

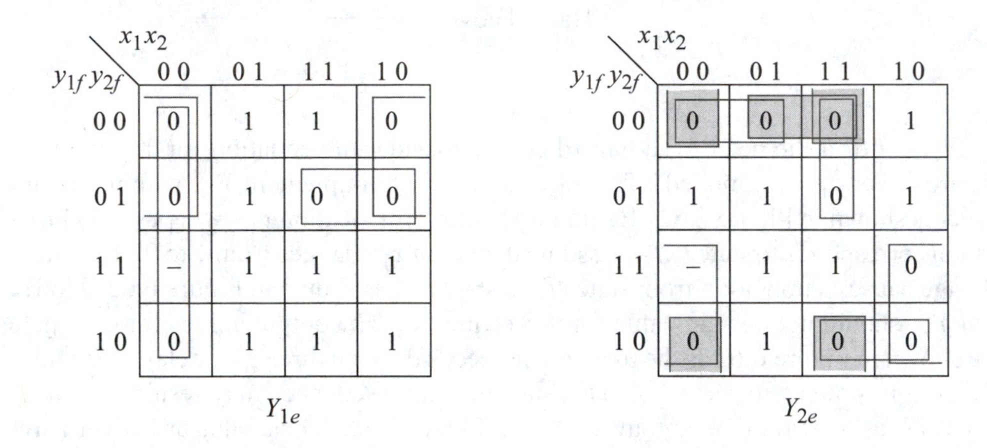

Figure 6.63 Individual excitation maps for Y_{1e} and Y_{2e} obtained from Figure 6.60. The maps are used to obtain a product-of-sums equation for the excitation variables.

$$
\begin{aligned}
Y_{1e} &= (x_1 + x_2)(y_{1f} + x_2)(y_{1f} + y_{2f}' + x_1') \\
Y_{2e} &= (y_{1f}' + x_2)(y_{1f}' + y_{2f} + x_1')(y_{1f} + y_{2f} + x_1)(y_{1f} + x_1' + x_2') \\
&\quad (y_{2f} + x_1 + x_2)(y_{1f} + y_{2f} + x_2')(y_{2f} + x_1' + x_2')
\end{aligned} \tag{6.11}
$$

Hazard covers (arrows pointing to $(y_{2f} + x_1 + x_2)$, $(y_{1f} + y_{2f} + x_2')$, $(y_{2f} + x_1' + x_2')$)

The sum-of-products equations require fewer logic gates in their implementation than the product-of-sums form. An alternative choice is to use the equation for Y_{1e} in a product-of-sums form and the equation for Y_{2e} in a sum-of-products form.

Example 6.15 Determining codes for the state variables in the synthesis of asynchronous sequential machines, such that critical races are eliminated, is not necessarily a trivial task. When modifying a merged flow table to avoid critical races, consideration must be given to machine cost and operating speed. Also, strict adherence to the functional specifications must be maintained.

In some cases, the structure of the merged flow table mandates the addition of another feedback variable. Consider the merged flow table of Figure 6.64. Four rows require two feedback variables. Examination of the flow table reveals that each row must be adjacent to every other row. Since there are no unspecified entries in the merged flow table, critical races cannot be eliminated by inserting intermediate unstable states in the merged flow table.

x_1x_2	00	01	11	10
1	ⓐ	ⓑ	*d*	*c*
2	*a*	*g*	ⓓ	ⓔ
3	*a*	ⓖ	*f*	ⓒ
4	*a*	*b*	ⓕ	*e*

Figure 6.64 Merged flow table for the asynchronous sequential machine of Example 6.15.

The transition diagram of Figure 6.65 graphically depicts the adjacency requirements for each row. Eliminating the triangles by rerouting certain transitions through other rows would eliminate any critical race conditions. Since column $x_1x_2 = 00$ contains the same state name in all rows, cycles can be created to remove races from row 2 to row 1 and from row 3 to row 1. However, adjacency from row 1 to row 2, specified by the sequence ⓑ $\rightarrow d \rightarrow$ ⓓ cannot be modified. Also, adjacency from row 1 to row 3 is still required, as indicated by the sequence ⓐ $\rightarrow c \rightarrow$ ⓒ. Although some of the directed lines can be removed from the transition diagram, this does not alter the general structure of the diagram. All triangles cannot be removed and adjacency is still required between every pair of rows. Therefore, a third feedback variable is necessary.

By adding a third feedback variable, the total number of rows will be $2^p = 8$, where p is the number of feedback variables. The augmented merged flow table must be functionally equivalent to the original merged flow table. Any additional intermediate unstable state entries must meet the requirements of the state transitions in the original flow table.

Figure 6.66 shows the augmented merged flow table. The top four rows are unchanged from the original table; the bottom four rows contain unspecified entries in every column and will be used to introduce intermediate transient states, where applicable. An intermediate state will be inserted whenever a transition causes two or more excitation variables to change state simultaneously. The assignment of feedback variables will be obtained from a transition diagram.

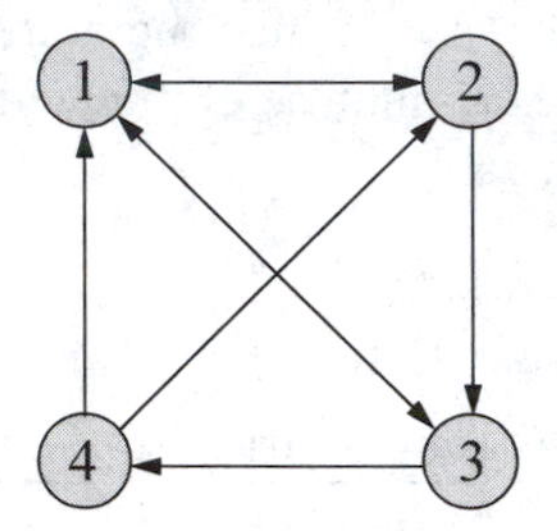

Figure 6.65 Transition diagram for the merged flow table of Figure 6.64.

x_1x_2	00	01	11	10
1	(a)	(b)	d	c
2	a	g	(d)	(e)
3	a	(g)	f	(c)
4	a	b	(f)	e
5	–	–	–	–
6	–	–	–	–
7	–	–	–	–
8	–	–	–	–

Figure 6.66 Augmented merged flow table obtained from Figure 6.64 for Example 6.15.

A convenient method of illustrating a transition diagram for eight rows is shown in Figure 6.67. To encode p feedback variables, the corresponding transition diagram is drawn as a p-dimensional cube. Figure 6.67 represents only one of several possible encoding schemes for the state variables. Notice that the top surface of the cube, represented by rows 1, 2, 6, and 4, is encoded in the Gray code 000, 001, 011, and 010, respectively. Similarly, the bottom surface, corresponding to rows 3, 5, 8, and 7, is encoded as 100, 101, 111, and 110, respectively.

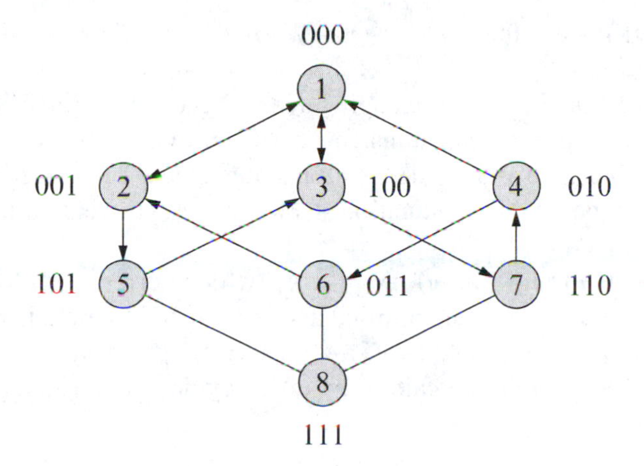

Figure 6.67 Transition diagram for the augmented merged flow table of Figure 6.66.

The encoding shown in Figure 6.67 is also consistent with the Hamming distance concept. Two code words are adjacent if they are separated by unit distance when plotted in a p-dimensional space. The distance between code words specifies the number of bits that are different in the code words under consideration. Thus, two code words are adjacent if they differ in only one bit position. In Figure 6.67, each vertex is adjacent to three other vertices, as indicated by the connecting lines. If a transition is required between two nonadjacent vertices (or rows), then a race-free sequence can be provided by directing the state transition sequence through intermediate vertices such that, each contiguous vertex involves a change of only one state variable. That is, the required adjacencies can be accommodated by creating cycles.

Transitions between rows 1 and 2 require no modification, since the two rows are adjacent. Thus, adjacency requirements are satisfied for the following sequences:

1 (000) → 2 (001) ;ⓑ → d → ⓓ
2 (001) → 1 (000) ;ⓔ → a → ⓐ

Similarly, rows 1 and 3 conform to the adjacency requirements for the following sequences:

1 (000) → 3 (100) ;ⓐ → c → ⓒ
3 (100) → 1 (000) ;ⓖ → a → ⓐ

3 (100) → 1 (000) ;ⓒ → *a* → ⓐ

Also, adjacency between rows 1 and 4 does not require the use of an appended row, since rows 1 and 4 differ by one state variable. This assignment accommodates the following sequence:

4 (010) → 1 (000) ;ⓕ → *b* → ⓑ

All remaining transitions in the augmented merged flow table of Figure 6.66 require the insertion of intermediate transient states to provide a race-free cycle between the beginning stable state and the destination stable state. The added unstable states provide an indirect path to the destination stable state and increase the duration of the state transition sequence.

For example, the transition ⓓ → *g* → ⓖ, which corresponds to row transition 2 → 3, requires that the cycle be modified and lengthened to include an intermediate transient state in row 5 such that, 2 → 5 → 3, as shown by the arrows in Figure 6.66. The transition ⓖ → *f* → ⓕ contains a modified cycle which changes the row transition from 3 → 4 to 3 → 7 → 4.

The transition ⓒ → *f* → ⓕ, specified by 3 → 4 , is also changed to include row 7, yielding 3 → 7 → 4. The final transition requiring the addition of an intermediate transient state is ⓕ → *e* → ⓔ, which changes 4 → 2 to 4 → 6 → 2. The modified row transitions are listed in Figure 6.68 together with their corresponding state transitions.

2 (001) → 5 (101) → 3 (100) ; ⓓ → *g* → ⓖ
3 (100) → 7 (110) → 4 (010) ; ⓖ → *f* → ⓕ
3 (100) → 7 (110) → 4 (010) ; ⓒ → *f* → ⓕ
4 (010) → 6 (011) → 2 (001) ; ⓕ → *e* → ⓔ

Figure 6.68 Modified row transitions to avoid races in the corresponding state transitions for Example 6.15.

The combined excitation map with all the requisite unstable states is shown in Figure 6.69. The codes for the feedback variables are listed in sequence using the Gray code format together with the corresponding row number. Although the machine takes longer to stabilize for some transitions, all noncritical and critical races have been eliminated.

The next step is to obtain the individual maps for the excitation variables Y_{1e}, Y_{2e}, and Y_{3e}. The procedure is identical to that of previous examples: Transfer the entries for Y_{1e} in each row of column $x_1x_2 = 00$ of the combined excitation map to the same locations in the individual excitation map for Y_{1e}. The process is repeated for the remaining columns of Y_{1e}, then likewise for Y_{2e} and Y_{3e}. The individual excitation maps are shown Figure 6.70 and the associated excitation equations in Equation 6.12.

	$y_{1f}y_{2f}y_{3f}$ \ x_1x_2	0 0	0 1	1 1	1 0
1	0 0 0	(000) a	(000) b	001	100
2	0 0 1	000	101	(001) d	(001) e
6	0 1 1	–	–	–	001
4	0 1 0	000	000	(010) f	011
7	1 1 0	–	–	010	–
8	1 1 1	–	–	–	–
5	1 0 1	–	100	–	–
3	1 0 0	000	(100) g	110	(100) c

$Y_{1e}Y_{2e}Y_{3e}$

Figure 6.69 Combined excitation map for the augmented merged flow table of Figure 6.66 with rows arranged in accordance with their respective state variable codes.

Race conditions have been identified and eliminated by altering the merged flow table to provide transitions through the unspecified entries in the appended rows. The remaining unspecified entries can be used to advantage to minimize the excitation equations and thus, reduce the cost of the machine. Care must be taken, however, to avoid assigning values to the excitation variables in the unspecified entries that are identical to the values of the feedback variables of the corresponding row. This would create an unwanted stable state which would be inconsistent with the machine specifications.

An alternative method of illustrating the transition sequences is by means of a Karnaugh map. The augmented merged flow table of Figure 6.66 requires three feedback variables which are coded in the transition diagram of Figure 6.67. The same information is portrayed in a different format using a Karnaugh map as shown in Figure 6.71. The feedback variables are represented by y_{1f}, y_{2f}, and y_{3f}. The entries in the minterm locations specify the rows of the flow table. The entries in parentheses represent the appended rows in the augmented flow table.

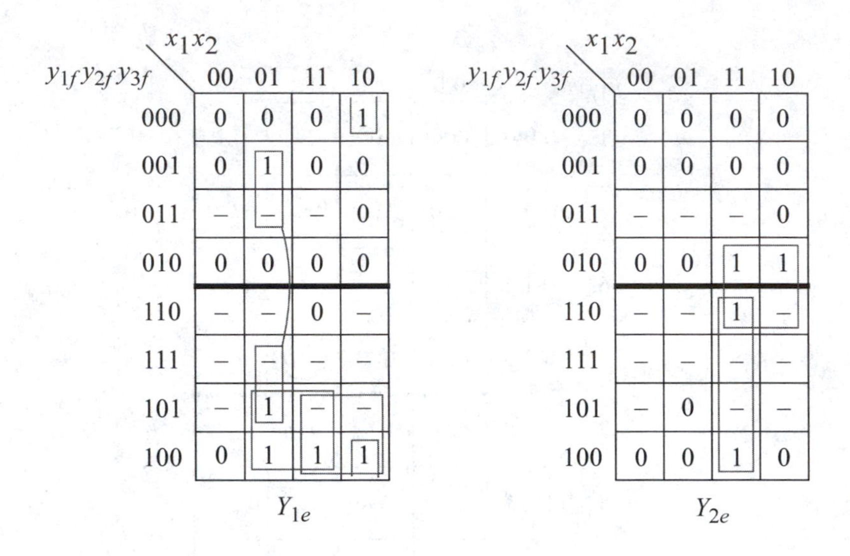

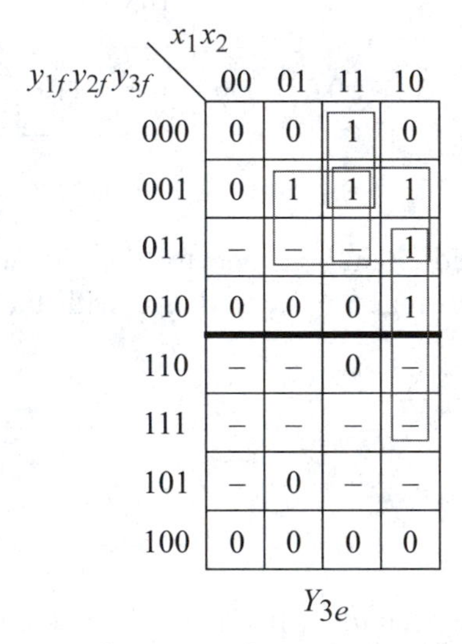

Figure 6.70 Individual excitation maps for Y_{1e}, Y_{2e}, and Y_{3e} obtained from Figure 6.69.

$$
\begin{aligned}
Y_{1e} &= y_{3f}x_1'x_2 + y_{2f}'y_{3f}'x_1x_2' + y_{1f}y_{2f}'x_2 + y_{1f}y_{2f}'x_1 \\
Y_{2e} &= y_{2f}y_{3f}'x_1 + y_{1f}x_1x_2 \\
Y_{3e} &= y_{2f}x_1x_2' + y_{1f}'y_{2f}'x_1x_2 + y_{1f}'y_{3f}x_1 + y_{1f}'y_{3f}x_2
\end{aligned}
\tag{6.12}
$$

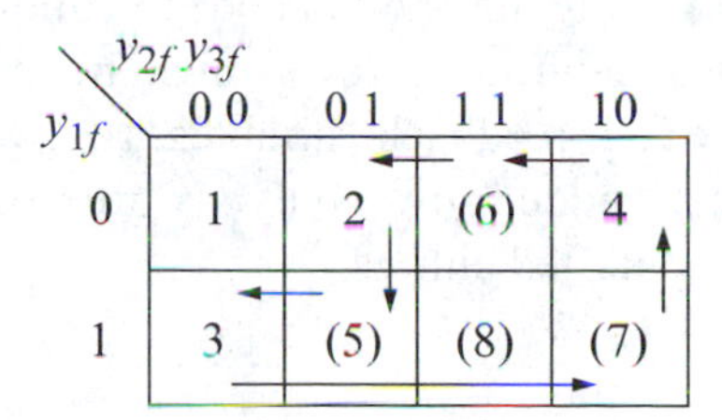

Figure 6.71 Alternative representation to illustrate state transition sequences for the asynchronous sequential machine of Example 6.15.

Since minterm locations that are physically adjacent are also logically adjacent (differ by only one state variable), a direct transition can be made from row 1 to row 2. This is evident in both the transition diagram and the Karnaugh map. Similar direct transitions can also be effected between rows 1 and 3 and between rows 1 and 4.

However, the transition from row 2 to row 3 is directed through row 5 to satisfy the adjacency requirement of allowing only one state variable to change value between contiguous rows in a cycle. This transition is depicted by the arrows in Figure 6.71. An alternative path for the same transition from row 2 to row 3 can be realized by utilizing row 1 as the intermediate transient state. This path also meets the adjacency requirements; however, there is no unspecified entry in row 1, column $x_1x_2 = 01$. Therefore, the only valid path from row 2 to row 3 is through row 5.

Similarly, the transition from row 3 to row 4 is directed through row 7 and the sequence from row 4 to row 2 passes through the intermediate transient state in row 6. All indirect transitions are indicated by arrows in Figure 6.71.

One-hot codes Using unspecified entries to provide intermediate unstable states may result in cycles with differing transition times. If equal cycle times are required for all state transitions, then this can be achieved by utilizing only one asserted feedback variable for each row of the merged flow table. Thus, the assignment of state codes is straightforward, thereby, eliminating the sometimes tedious task of finding an optimal set of intermediate unstable states. Since only one feedback variable is active in each row, the codes are classified as *one-hot codes*.

In general, the feedback variables are denoted as $y_{1f}y_{2f}y_{3f} \cdots y_{pf}$. In row i of the merged flow table, feedback variable $y_{if} = 1$, while all other variables are equal to 0. When a transition is required from row i to row j, y_{jf} is set to a value of 1 such that, $y_{if}y_{jf} = 11$. Then y_{if} is reset to 0. Thus, each transition between two rows passes through one intermediate transient state in an appended row. This allows the duration of all state transition sequences to be identical; that is, each transition from a stable state in row i to a stable state in row j requires two state transition durations of Δt_i and Δt_j.

Example 6.16 Consider the merged flow table of Figure 6.72. The feedback variables can be assigned values using the one-hot code technique. To achieve the desired equal state transition times, additional rows must be incorporated, as shown in Figure 6.73. Since there are four feedback variables, 16 rows are required for the augmented flow table. Of the 12 added rows, only six are necessary; the remaining six rows consist entirely of unspecified entries.

	$y_{1f}y_{2f}y_{3f}y_{4f}$ \ x_1x_2	00	01	11	10
1	1000	Ⓐ	Ⓑ	d	c
2	0100	a	g	Ⓓ	Ⓔ
3	0010	a	Ⓖ	f	Ⓒ
4	0001	a	b	Ⓕ	e

$Y_{1e}Y_{2e}Y_{3e}Y_{4e}$

Figure 6.72 Merged flow table of Figure 6.64 illustrating the one-hot state code assignment technique.

Every transition between stable states in different rows can now be accomplished without races and with the same number of intermediate transient states. The values of the feedback variables will be adjacent in contiguous rows of a cycle. For example, the transition from state Ⓐ to state Ⓒ is specified by the following sequence:

$$1\ (1000) \rightarrow 6\ (1010) \rightarrow 3\ (0010) \qquad ;Ⓐ \rightarrow c \rightarrow Ⓒ$$

The transition from state Ⓑ to state Ⓓ can be realized by the following sequence:

$$1\ (1000) \rightarrow 5\ (1100) \rightarrow 2\ (0100) \qquad ;Ⓑ \rightarrow d \rightarrow Ⓓ$$

The arrows in Figure 6.73 show four of the ten possible state transition sequences. The remaining row transitions are listed in Figure 6.74 together with their corresponding state transitions. All transitions in the original merged flow table are directed through one of the appended rows in the augmented merged flow table.

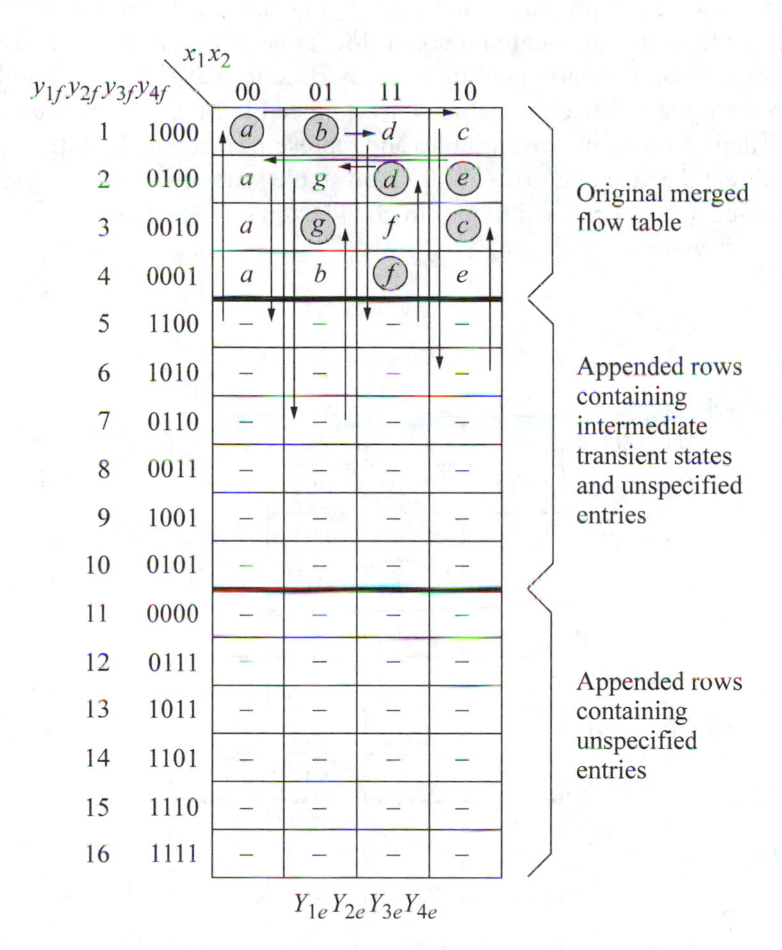

	$y_{1f}y_{2f}y_{3f}y_{4f}$ \ x_1x_2	00	01	11	10
1	1000	ⓐ	ⓑ	d	c
2	0100	a	g	ⓓ	ⓔ
3	0010	a	ⓖ	f	ⓒ
4	0001	a	b	ⓕ	e
5	1100	–	–	–	–
6	1010	–	–	–	–
7	0110	–	–	–	–
8	0011	–	–	–	–
9	1001	–	–	–	–
10	0101	–	–	–	–
11	0000	–	–	–	–
12	0111	–	–	–	–
13	1011	–	–	–	–
14	1101	–	–	–	–
15	1110	–	–	–	–
16	1111	–	–	–	–

Figure 6.73 Augmented merged flow table of Figure 6.72 using the one-hot state code assignment.

1 (1000) → 6 (1010) → 3 (0010) ;ⓐ → c → ⓒ
1 (1000) → 5 (1100) → 2 (0100) ;ⓑ → d → ⓓ
2 (0100) → 7 (0110) → 3 (0010) ;ⓓ → g → ⓖ
2 (0100) → 5 (1100) → 1 (1000) ;ⓔ → a → ⓐ
3 (0010) → 6 (1010) → 1 (1000) ;ⓖ → a → ⓐ
3 (0010) → 8 (0011) → 4 (0001) ;ⓖ → f → ⓕ
3 (0010) → 8 (0011) → 4 (0001) ;ⓒ → f → ⓕ
3 (0010) → 6 (1010) → 1 (1000) ;ⓒ → a → ⓐ
4 (0001) → 9 (1001) → 1 (1000) ;ⓕ → b → ⓑ
4 (0001) → 10 (0101) → 2 (0100) ;ⓕ → e → ⓔ

Figure 6.74 Modified row transitions to avoid races and to generate equal time durations for all state transition sequences.

The combined excitation map for Y_{1e}, Y_{2e}, Y_{3e}, and Y_{4e} is shown in Figure 6.75 and is derived from the augmented merged flow table of Figure 6.73. The values of the feedback variables are arranged in the Gray code format. The row numbers correspond to the same feedback variable codes as listed in Figure 6.73. The entries in the squares depict the excitation variables and represent either stable states or unstable states that direct the machine to the destination stable state. All remaining entries are left unspecified and can be used to minimize the excitation equations.

	$y_{1f}y_{2f}y_{3f}y_{4f}$ \ x_1x_2	0 0	0 1	1 1	1 0
11	0000	–	–	–	–
4	0001	–	1001	(0001) f	0101
8	0011	–	–	0001	–
3	0010	1010	(0010) g	0011	(0010) c
7	0110	–	0010	–	–
12	0111	–	–	–	–
10	0101	–	–	–	0100
2	0100	1100	0110	(0100) d	(0100) e
5	1100	1000	–	0100	–
14	1101	–	–	–	–
16	1111	–	–	–	–
15	1110	–	–	–	–
6	1010	1000	–	–	0010
13	1011	–	–	–	–
9	1001	–	1000	–	–
1	1000	(1000) a	(1000) b	1100	1010

$Y_{1e}Y_{2e}Y_{3e}Y_{4e}$

Figure 6.75 Combined excitation map for variables Y_{1e}, Y_{2e}, Y_{3e}, and Y_{4e} derived from the augmented merged flow table of Figure 6.73.

All transitions between two stable states in Figure 6.75 proceed through one unstable state. Thus, all state transition sequences consist of identical time durations. The state transitions listed in Figure 6.74 can be identified in the combined excitation map of Figure 6.75.

For example, in state ⓕ, the feedback values equal the excitation values; that is, $y_{1f}y_{2f}y_{3f}y_{4f} = Y_{1e}Y_{2e}Y_{3e}Y_{4e} = 0001$. In state ⓕ, if x_2 changes from 1 to 0, then the machine proceeds to state ⓔ. The transition from state ⓕ to state ⓔ proceeds from row 4 to the unstable state entry in row 4 where $y_{1f}y_{2f}y_{3f}y_{4f} = 0101$. After a delay of Δt, the values of the feedback variables become equal to the values of the excitation variables and the operation proceeds to row 10 ($y_{1f}y_{2f}y_{3f}y_{4f} = 0101$) where the excitation values in column $x_1x_2 = 10$ are 0100. After a further delay, the feedback values become 0100, and the machine enters state ⓔ in row 2, where $y_{1f}y_{2f}y_{3f}y_{4f} = Y_{1e}Y_{2e}Y_{3e}Y_{4e} = 0100$.

Similarly, observe the sequence from state ⓖ to state ⓐ. In state ⓖ, if x_2 changes from 1 to 0, then the machine enters unstable state $Y_{1e}Y_{2e}Y_{3e}Y_{4e} = 1010$ in column $x_1x_2 = 00$ of row 3. After a delay of Δt, the feedback variables become equal to the excitation variables and the operation proceeds to row 6, where $y_{1f}y_{2f}y_{3f}y_{4f} = 1010$. The entry 1000 in column $x_1x_2 = 00$ of row 6 is also unstable. Therefore, after an additional delay, the machine sequences to row 1, where the feedback values equal the excitation values ($y_{1f}y_{2f}y_{3f}y_{4f} = Y_{1e}Y_{2e}Y_{3e}Y_{4e} = 1000$) and the machine enters state ⓐ.

The combined excitation map is now separated into its constituent parts of four individual maps, one each for excitation variables Y_{1e}, Y_{2e}, Y_{3e}, and Y_{4e}. The individual excitation maps are shown in Figure 6.76. As in previous examples, the individual maps are obtained by transcribing the values for Y_{ie} in each column of the combined excitation map to the corresponding locations in the individual map for Y_{ie}. The equations for the four excitation variables are listed in Equation 6.13. Each equation must provide a hazard cover, where applicable.

The excitation equations using the one-hot code assignment are relatively easy to derive. The advantage is equal time delay for all state transition sequences. Transitions from one row in the merged flow table to another row are always achieved in a two-transition cycle by changing only one state variable at a time. Thus, all state transitions are free of races. The disadvantage of utilizing the one-hot code method is increased machine cost, since additional state variables must be employed.

The excitation equations can also be derived from the original merged flow table of Figure 6.72 without appending additional rows. The equations will be equivalent to those derived using the excitation maps, but may contain more terms. In Figure 6.72, consider the transition from state ⓐ to state ⓒ. Stable state ⓒ is in row 3. Therefore, excitation variable Y_{3e} will be set to 1 if $Y_{1e} = 1$ and the input vector changes from $x_1x_2 = 00$ to 10. Thus,

$$Y_{3e} = y_{1f}x_1x_2' + \cdots$$

The path for this sequence is ⓐ $\rightarrow c \rightarrow$ ⓒ, or $Y_{1e}Y_{2e}Y_{3e}Y_{4e} = 1000 \rightarrow 1010 \rightarrow 0010$. In order for the cycle to function properly, Y_{1e} must remain set until Y_{3e} becomes set, then Y_{1e} must be reset. Therefore, the equation for excitation variable Y_{1e} must contain the term $y_{1f}y_{3f}'$ as follows:

$$Y_{1e} = y_{1f}y_{3f}' + \cdots$$

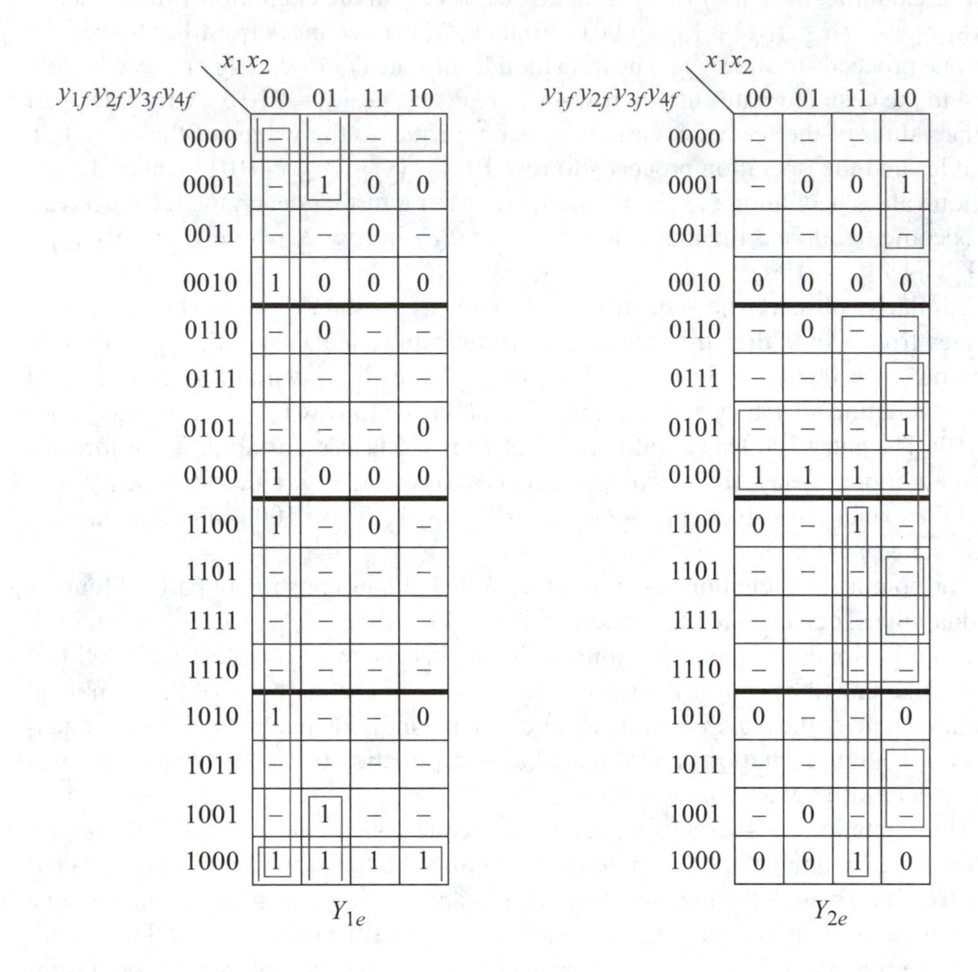

Figure 6.76 Individual maps for excitation variables Y_{1e}, Y_{2e}, Y_{3e}, and Y_{4e}.

Next, consider the sequence from state ⓑ to state ⓓ. Excitation variable Y_{2e} will be set to 1 if $Y_{1e} = 1$ and the input vector changes from $x_1x_2 = 01$ to 11. Thus,

$$Y_{2e} = y_{1f}x_1x_2 + \cdots$$

Since the path for this sequence is ⓑ $\rightarrow d \rightarrow$ ⓓ or $Y_{1e}Y_{2e}Y_{3e}Y_{4e} = 1000 \rightarrow 1100 \rightarrow 0100$, Y_{1e} must remain set until Y_{2e} is set, then Y_{1e} must be reset. Therefore, the equation for Y_{1e} must contain the contingent variable y_{2f}', which is appended to the existing term for Y_{1e} to yield the product term

$$Y_{1e} = y_{1f}y_{3f}'y_{2f}' + \cdots$$

Thus, Y_{1e} will remain set to a value of 1 only if $y_{2f}y_{3f} = 00$. If either y_{2f} of y_{3f} is equal to 1, then Y_{1e} will be reset to 0. This occurs in the destination stable state.

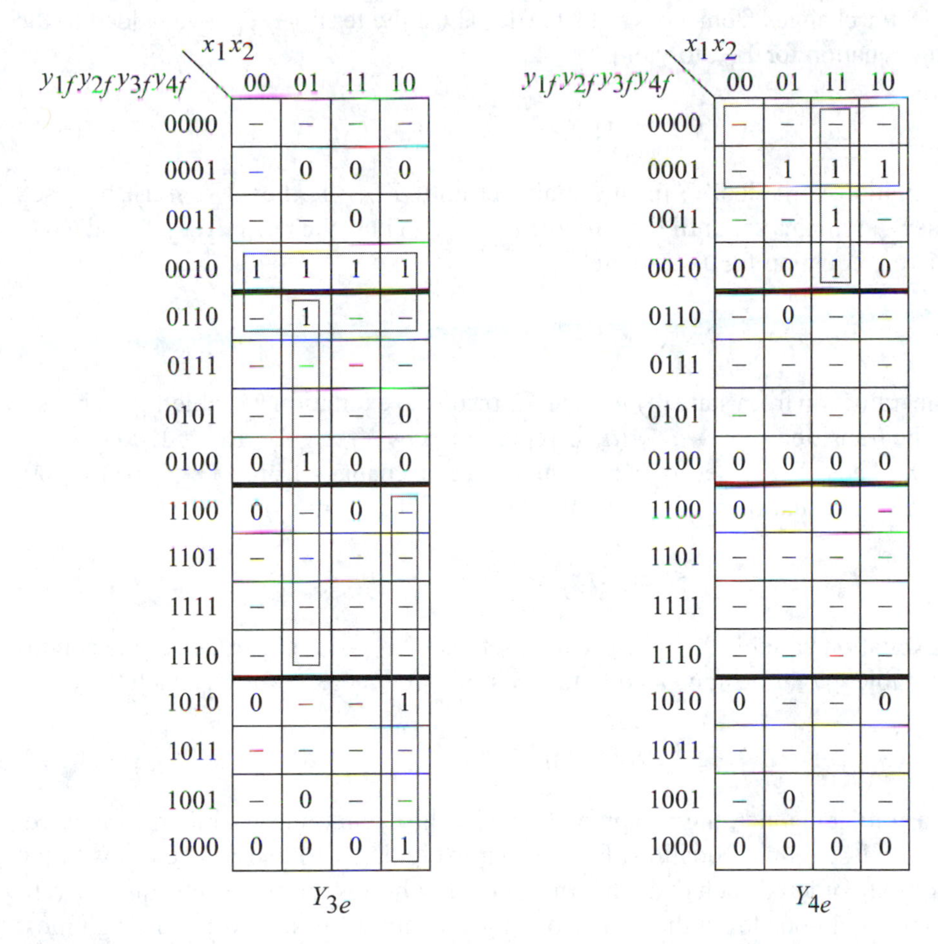

Figure 6.76 (Continued)

$$Y_{1e} = x_1'x_2' + y_{2f}'y_{3f}'x_1'x_2 + y_{2f}'y_{3f}'y_{4f}' \qquad \text{No hazards}$$

$$Y_{2e} = y_{1f}'y_{2f}y_{3f}' + y_{1f}x_1x_2 + y_{4f}x_1x_2' + y_{2f}x_1 + y_{1f}y_{4f}x_1$$

Hazard cover (arrows to $y_{2f}x_1$ and $y_{1f}y_{4f}x_1$)

$$Y_{3e} = y_{1f}'y_{3f}y_{4f}' + y_{2f}x_1'x_2 + y_{1f}x_1x_2' \qquad \text{No hazards}$$

$$Y_{4e} = y_{1f}'y_{2f}'y_{3f}' + y_{1f}'y_{2f}'x_1x_2 \qquad \text{No hazards} \qquad (6.13)$$

The transition from state ⓓ to state ⓖ causes excitation variable Y_{3e} to be set. The path for this sequence is ⓓ $\rightarrow g \rightarrow$ ⓖ or $Y_{1e}Y_{2e}Y_{3e}Y_{4e} = 0100 \rightarrow 0110 \rightarrow 0010$. Using the same rationale, the equation for Y_{3e} is set to 1 if Y_{2e} is set and the input vector changes from $x_1x_2 = 11$ to 01. Thus, the term $y_{2f}x_1'x_2$ is added to the existing equation for Y_{3e}, to yield

$$Y_{3e} = y_{1f}x_1x_2' + y_{2f}x_1'x_2 + \cdots$$

Excitation variable Y_{2e} must remain set until $Y_{3e} = 1$, then Y_{2e} must be reset. That is, Y_{2e} remains set until the variable $y_{3f}' = 1$. Thus, the term $y_{2f}y_{3f}'$ is added to the existing equation for Y_{2e}, to yield

$$Y_{2e} = y_{1f}x_1x_2 + y_{2f}y_{3f}' + \cdots$$

The transition from state ⓔ to state ⓐ results in excitation variable Y_{1e} being set to 1. The transition ⓔ $\rightarrow a \rightarrow$ ⓐ is represented by $Y_{1e}Y_{2e}Y_{3e}Y_{4e} = 0100 \rightarrow 1100 \rightarrow 1000$. Thus, Y_{1e} is set to 1 if the input vector changes from $x_1x_2 = 10$ to $= 00$. Therefore, the equation for Y_{1e} contains the additional term $y_{2f}x_1'x_2'$ such that,

$$Y_{1e} = y_{1f}y_{3f}'y_{2f}' + y_{2f}x_1'x_2' + \cdots$$

Also, excitation variable Y_{2e} must remain set until $Y_{1e} = 1$. This requires the contingent variable y_{1f}' to be appended to the existing product term for Y_{2e} such that,

$$Y_{2e} = y_{1f}x_1x_2 + y_{2f}y_{3f}'y_{1f}' + \cdots$$

In a similar manner, the remaining terms for the excitation variables are obtained. Equation 6.14 lists the equations for the four excitation variables. The terms in the equations are ordered such that, all terms but the rightmost term in each equation represent the conditions to set the corresponding excitation variable to 1. The rightmost term expresses the conditions which maintain the corresponding excitation variable in an active state until the one-hot variable in the destination stable state is set to a value of 1.

$$Y_{1e} = y_{2f}x_1'x_2' + y_{3f}x_1'x_2' + y_{4f}x_1'x_2 + y_{1f}y_{2f}'y_{3f}'$$

$$Y_{2e} = y_{1f}x_1x_2 + y_{4f}x_1x_2' + y_{1f}'y_{2f}y_{3f}'$$

$$Y_{3e} = y_{1f}x_1x_2' + y_{2f}x_1'x_2 + y_{1f}'y_{3f}y_{4f}'$$

$$Y_{4e} = y_{3f}x_1x_2 + y_{1f}'y_{2f}'y_{4f} \qquad (6.14)$$

The sum-of-products expressions in Equation 6.13 and Equation 6.14 are equivalent; however, the equations in Equation 6.14 do not provide hazard covers. The need for hazard covers is more readily apparent from the excitation maps of

Figure 6.76. While both methods for obtaining the excitation equations by utilizing the one-hot code technique yield equivalent equations, the additional logic required to implement the equations may be prohibitive. The equivalence can be verified by observing that all entries with a value of 1 in Figure 6.76 are covered by both sets of equations.

6.2.7 Output Maps and Equations

The next step in the synthesis procedure is to assign values to the output variables. A Karnaugh map — referred to as an output map — facilitates this process. This step utilizes the results of two previous operations: the merged flow table and the reduced primitive flow table.

The merged flow table indicates the location of all stable states. Since the outputs are associated with stable states, the merged flow table indicates the location of the stable state output variables in their respective output maps. The merged flow table defines the format of the output map, which is constructed directly from the merged flow table and the reduced primitive flow table. The reduced primitive flow table specifies the output values of the corresponding stable states in the output map. The merged flow table, the excitation map, and the output map have the same number of inputs — and thus, the same number of columns — and the same number of feedback variables, necessitating the same number of rows. The values assigned to the input variables and the feedback variables are the same in both the excitation map and the output map.

The speed of circuit operation can be established during this phase. If different output values are associated with the initial and destination stable states for a state transition with only one intermediate state, then the intermediate unstable state can be assigned a value that is equal to either the original or the final stable state output value.

By assigning appropriate values to the unstable states in the output map, the outputs can be made to change value as soon as possible or as late as possible for a particular state transition. If the output value of the original stable state is assigned to the intermediate state, then the change to the output is delayed until the machine enters the destination stable state. If, however, the output value of the destination stable state is assigned to the intermediate state, then the output value changes before the machine reaches the destination stable state.

Example 6.17 To illustrate the technique for establishing output response time for a particular state transition, consider the merged flow table and output map of Figure 6.77. The output map is derived directly from the reduced primitive flow table and the merged flow table, of which it is identical in format. Assume that the output values for z_1 corresponding to states ⓐ, ⓑ, ⓒ, ⓓ, and ⓔ are 1, 1, 0, 0, and 1, respectively.

For a transition from state ⓓ to state ⓐ, the machine proceeds through unstable state a ($y_{1f}x_1x_2 = 100$). Since the initial and destination stable states have different output values, the value assigned to location $y_{1f}x_1x_2 = 100$ can be left unspecified. If the output change for this transition is to be as fast as possible, then a value of 1 must be inserted. If, however, the machine specifications require that the output change is to be as slow as possible, then a 0 is inserted.

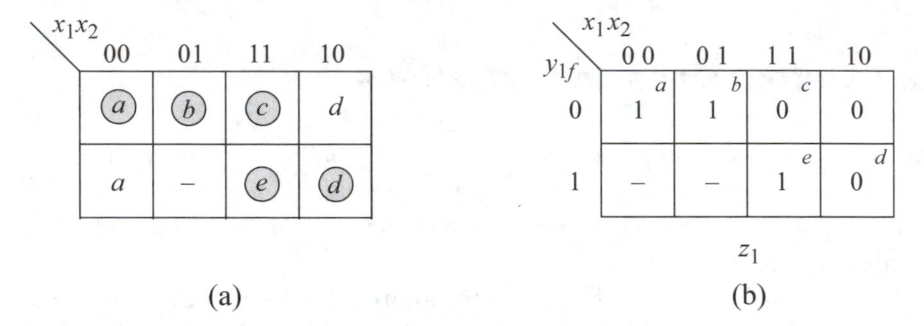

Figure 6.77 Merged flow table and output map for z_1: (a) merged flow table; and (b) output map.

Another important consideration is the avoidance of erroneous output signals, or glitches. If the original and final stable states in a state transition have the same output values, then all intermediate unstable states must contain that same value. That is, every unstable state through which the machine passes in a cycle must contain the same output value as the stable states, otherwise, a momentary false output will occur.

The value in each minterm location of the output map specifies the output value for the corresponding state. For example, in Figure 6.77, the transition ⓒ → d → ⓓ is possible. Since states ⓒ and ⓓ both specify output z_1 to be deasserted, therefore, the intermediate state $y_{1f}x_1x_2 = 010$ must also specify z_1 to be deasserted. The insertion of a 0 in unstable state $y_{1f}x_1x_2 = 010$ maintains the output in the deasserted state for the entire transition. If a value of 1 were placed in the unstable state, then a momentary 1 output would be generated. The output response would then be equivalent to a static-0 hazard.

A third consideration exists when two stable states with different output values proceed through the same unstable state to the same destination stable state. For a transition between stable states with different output values, the entry in the unstable transient state can be stipulated as an unspecified entry. However, the output value in the unstable state is predicated by the more stringent requirement imposed by transitions between stable states with identical outputs. It is this condition which ultimately determines the output value of the unstable state. In this situation, the transient state output value must be identical to that of the initial and destination stable states in order to avoid an erroneous signal on the output variable.

For example, again consider Figure 6.77. Examining the merged flow table, the following two transitions are evident:

ⓐ $\rightarrow d \rightarrow$ ⓓ
ⓒ $\rightarrow d \rightarrow$ ⓓ

Regarding the transition ⓐ $\rightarrow d \rightarrow$ ⓓ only, the value placed in the unspecified state $y_{1f}x_1x_2 = 010$ of the output map may be either a 0 or a 1 to effect a fast or slow change, respectively. However, the dominant transition is ⓒ $\rightarrow d \rightarrow$ ⓓ, which mandates that a value of 0 be inserted in state $y_{1f}x_1x_2 = 010$ in order to avoid an erroneous momentary output.

If the speed with which the outputs change value is inconsequential, then the unstable states in the output map may be considered as unspecified entries, where appropriate. The unspecified entries can then be used to minimize the λ output logic. The choice of output values for the unstable state entries is made in accordance with the following general guidelines:

(a) The output changes are to occur as soon as possible.
(b) The output changes are to occur as late as possible.
(c) The λ output logic is to be in minimal form.

In Figure 6.77, assume that output response time is immaterial and that the λ output logic is to be in a minimal sum-of-products form. Taking into consideration all state transitions and the necessity to avoid output glitches, the output map for z_1 is shown in Figure 6.78 with an entry of 1 in locations $y_{1f}x_1x_2 = 100$ and 101. The corresponding output equation is shown in Equation 6.15.

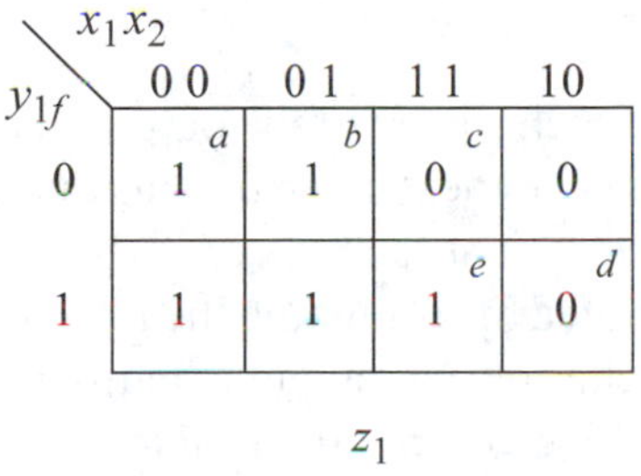

y_{1f} \ x_1x_2	00	01	11	10
0	1 (*a*)	1 (*b*)	0 (*c*)	0
1	1	1	1 (*e*)	0 (*d*)

z_1

Figure 6.78 Completed output map for Figure 6.77 (b)

$$z_1 = x_1' + y_{1f}x_2 \tag{6.15}$$

In the output map of Figure 6.78, a value of 0 was entered in location $y_{1f}x_1x_2 = 010$ to prevent a glitch on z_1 for the transition from state ⓒ to state ⓓ. Figure 6.79 (b) shows the same output map with an entry of 1 in location

$y_{1f}x_1x_2 = 010$. Figure 6.79 (a) illustrates the excitation map. The excitation and output equations are shown in Equation 6.16.

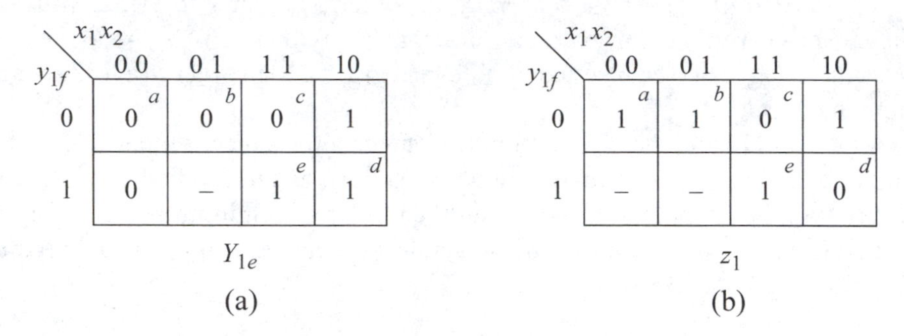

Figure 6.79 Excitation and output maps for Example 6.17: (a) excitation map; and (b) output map.

$$
\begin{aligned}
Y_{1e} &= x_1x_2' + y_{1f}x_1 \\
z_1 &= x_1' + y_{1f}x_2 + y_{1f}'x_2'
\end{aligned}
\tag{6.16}
$$

Because the value in each square of the output map specifies the value of the output variable for that state, a momentary erroneous output will be exhibited by z_1 as the machine passes through transient state $y_{1f}x_1x_2 = 010$. The occurrence of this static-0 hazard can be demonstrated by examining the logic diagram of Figure 6.80, which is synthesized from the excitation and output equations of Equation 6.16.

Assume that the machine is currently in state ⓒ and that input x_2 changes from 1 to 0. The machine will then proceed from state ⓒ through transient state $y_{1f}x_1x_2 = 010$ to state ⓓ. The transition from state ⓒ to state ⓓ is accomplished after a delay of Δt. A further delay takes place through the λ output logic that generates z_1. The effect of this transition is illustrated by the sequence of + and – symbols on inputs x_1 and x_2, on excitation variable Y_{1e}, and on output z_1. The waveform for z_1 contains two erroneous transitions which are similar to those encountered in a static-0 hazard.

When properly synthesized with no false outputs, the machine is implemented from the excitation equation of Equation 6.16 and the output equation of Equation 6.15. The logic diagram is shown in Figure 6.81. Consider the previous transition from state ⓒ to state ⓓ. The machine remains in state ⓒ, where $Y_{1e}y_{1f}z_1 = 000$, until an input variable changes value. When x_2 changes from 1 to 0, the excitation variable Y_{1e} changes from 0 to 1. After a delay of Δt, the feedback variable y_{1f} attains a value of 1. However, AND gate 4 is disabled coincident with the deassertion of input x_2. The deassertion of x_2 effectively maintains z_1 in an inactive state, providing a state transition that is free of output glitches.

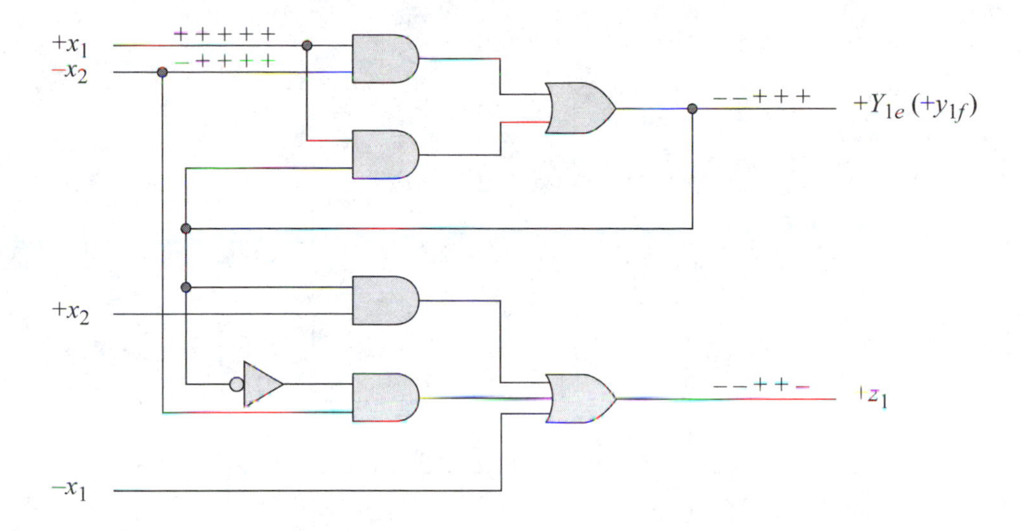

Figure 6.80 Logic diagram for Example 6.17 obtained from Equation 6.16 illustrating an erroneous output on z_1 when x_2 changes from 1 to 0.

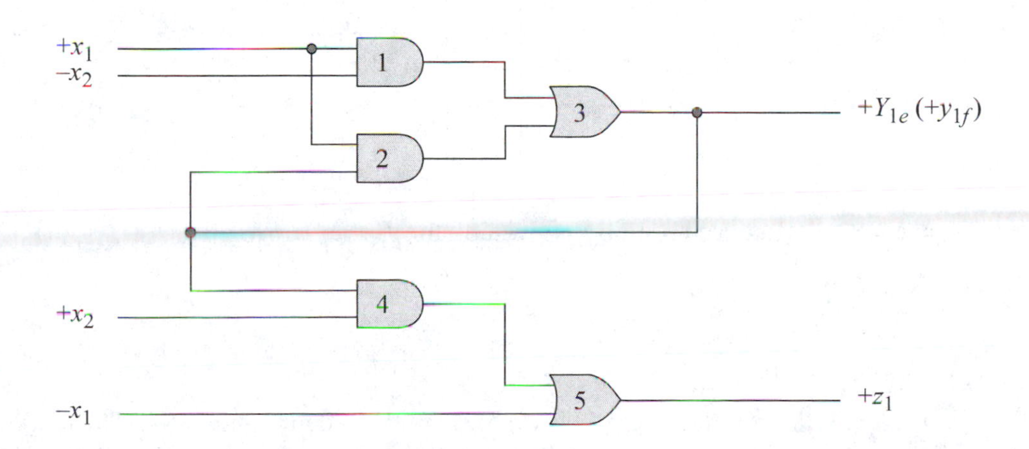

Figure 6.81 Correctly synthesized logic diagram for Example 6.17 obtained from Equation 6.16 for Y_{1e} and from Equation 6.15 for z_1, in which the state transition ⓒ $\rightarrow d \rightarrow$ ⓓ is free of output glitches.

Example 6.18 Figure 6.82 shows a reduced primitive flow table for an asynchronous sequential machine with three outputs. Using the rules for merging presented in Section 6.2.4, the merger diagram is constructed as shown in Figure 6.83. From the merger diagram, three possible partitions of maximal compatible sets can be obtained, as shown below:

1. {ⓔ, ⓕ, ⓖ}, {ⓐ, ⓗ}, {ⓑ, ⓒ}, {ⓓ}
2. {ⓔ, ⓕ, ⓖ}, {ⓐ, ⓑ}, {ⓒ, ⓓ}, {ⓗ}
3. {ⓔ, ⓕ, ⓖ}, {ⓐ, ⓗ}, {ⓒ, ⓓ}, {ⓑ}

x_1x_2 00	01	11	10	z_1	z_2	z_3
ⓐ	b	–	h	0	1	0
a	ⓑ	c	–	1	0	0
–	b	ⓒ	d	1	1	0
e	–	c	ⓓ	0	1	1
ⓔ	f	–	h	1	1	1
e	ⓕ	g	–	1	0	1
–	f	ⓖ	h	0	1	1
a	–	g	ⓗ	1	0	0

Figure 6.82 Reduced primitive flow table for Example 6.18.

Using partition 1, the merged flow table is presented in Figure 6.84. The merged flow table indicates the location of all stable states. The reduced primitive flow table specifies the values of the output variables z_1, z_2, and z_3 for all stable states. Therefore, the two tables will be used in conjunction to establish the output maps.

The transition diagram for the merged flow table is shown in Figure 6.85. Since the transition diagram consists of a polygon with an even number of sides, all state transitions will be free of race conditions. The arrangement of the four rows, therefore, can be left unchanged. The codes for the state variables can be realized by utilizing a Gray code for two variables.

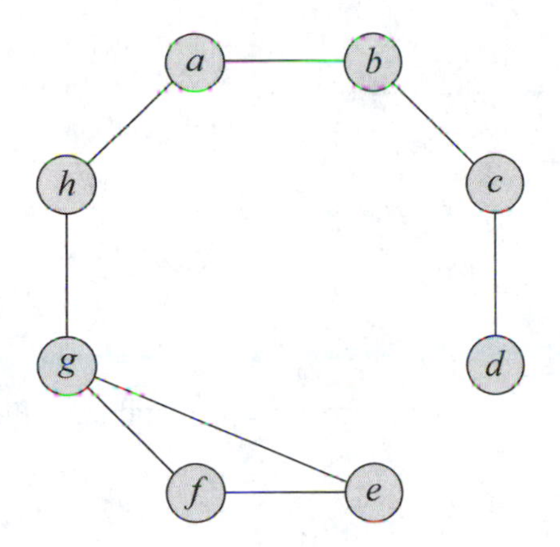

Figure 6.83 Merger diagram for the reduced primitive flow table of Figure 6.82.

		x_1x_2 00	01	11	10
1	ⓔ, ⓕ, ⓖ	ⓔ	ⓕ	ⓖ	h
2	ⓐ, ⓗ	ⓐ	b	g	ⓗ
3	ⓑ, ⓒ	a	ⓑ	ⓒ	d
4	ⓓ	e	–	c	ⓓ

Figure 6.84 One possible merged flow table derived from the merger diagram of Figure 6.83.

The output maps for z_1, z_2, and z_3 are presented in Figure 6.86. The locations of the stable states in each map are specified in accordance with the merged flow table. The values inserted in the stable state locations are obtained from the reduced primitive flow table. For example, stable states ⓐ, ⓑ, ⓒ, and ⓓ are placed in locations $y_{1f}y_{2f}x_1x_2$ = 0100, 1101, 1111, and 1010, respectively, in all output maps. There is a one-to-one mapping of the stable state locations in the merged flow table with the stable state locations in the output maps. The values inserted in the stable states are obtained directly from the corresponding stable state locations of the reduced primitive flow table for the respective outputs.

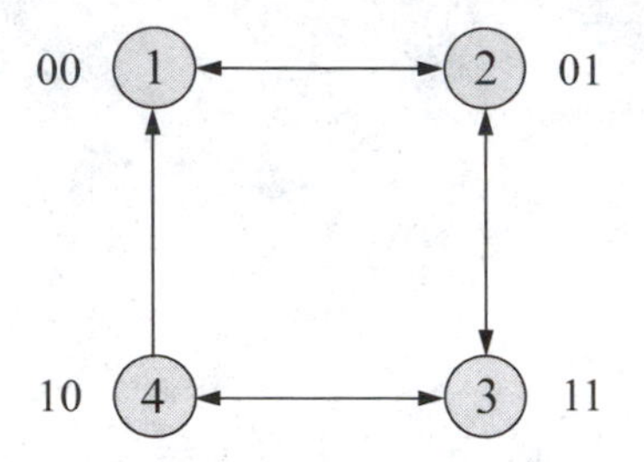

Figure 6.85 Transition diagram for the merged flow table of Figure 6.84.

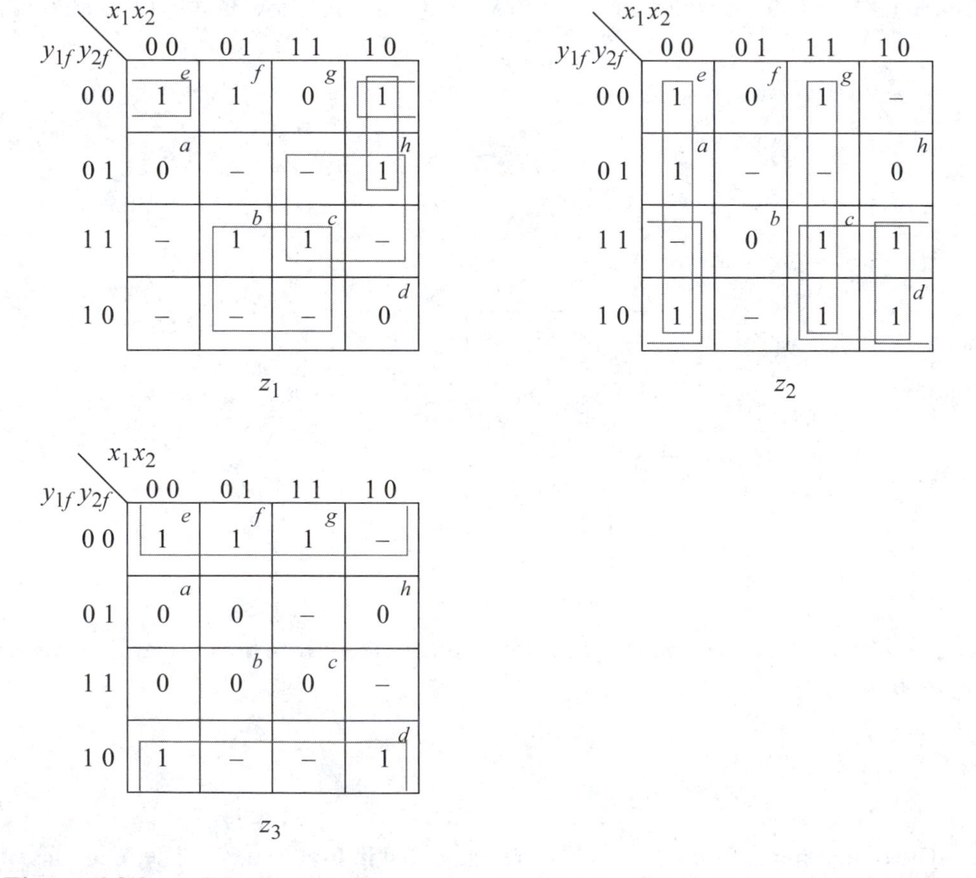

Figure 6.86 Output maps for z_1, z_2, and z_3 obtained from the merged flow table of Figure 6.84 and the reduced primitive flow table of Figure 6.82. The entries in the intermediate transient states were chosen to provide output equations in minimal form.

Consider state ⓐ in the three output maps. In state ⓐ of the reduced primitive flow table, the output values are defined as $z_1z_2z_3 = 010$. Therefore, location $y_{1f}y_{2f}x_1x_2 = 0100$ of each map contains the value 0, 1, and 0 for z_1, z_2, and z_3, respectively. Stable state ⓑ is in location $y_{1f}y_{2f}x_1x_2 = 1101$ of all output maps. In state ⓑ of the reduced primitive flow table, the output values are specified as $z_1z_2z_3 = 100$. Therefore, location $y_{1f}y_{2f}x_1x_2 = 1101$ of each map contains the value 1, 0, and 0 for z_1, z_2, and z_3, respectively. In a similar manner, the remaining stable states in each output map are filled in with appropriate values as specified by the output columns in the corresponding stable states of the reduced primitive flow table.

All other squares in the output maps represent transient states. Values are entered in the transient states according to the requirements of the state transition sequences. The avoidance of output glitches is of primary importance. The next main concern is machine performance — whether circuit response is to be fast or slow. When the conditions for false outputs and machine speed have been successively negotiated, consideration can be given to logic minimization.

Consider the output map for z_1 in Figure 6.86. Assume that output response time is immaterial and that the output equations are to be in minimal form. The reduced primitive flow table specifies that a transition will occur from state ⓔ to state ⓕ or ⓗ if the input vector changes from $x_1x_2 = 00$ to 01 or 10, respectively. The transition from state ⓔ to state ⓕ requires no change to the feedback variables and thus, does not proceed through an unstable state.

The transition from state ⓔ to state ⓗ, however, passes through unstable state $y_{1f}y_{2f}x_1x_2 = 0010$. Since both stable states assert output z_1, therefore, the intermediate unstable state must contain a value of 1 in order to avoid a momentary false 0 output on z_1.

In state ⓐ, the machine will proceed to state ⓑ or ⓗ if the input vector changes from $x_1x_2 = 00$ to 01 or 10, respectively. Since states ⓐ and ⓗ are in the same feedback row, only the sequence from state ⓐ to state ⓑ need be considered. Stable states ⓐ and ⓑ have different output values; therefore, the intermediate transient state $y_{1f}y_{2f}x_1x_2 = 0101$ can be left unspecified. In a similar manner, values are entered in the remaining squares of the output map for z_1.

The map for output z_2 in Figure 6.86 requires three mandatory values to prevent the occurrence of erroneous outputs. Locations $y_{1f}y_{2f}x_1x_2 = 1000$, 1011, and 1110 must contain a value of 1 to avoid a momentary false 0 output for state transitions ⓓ → ⓔ, ⓓ → ⓒ, and ⓒ → ⓓ, respectively. All other transient states can be left unspecified.

Similarly, the output map for z_3 requires compulsory values of 0, 1, and 0 in locations $y_{1f}y_{2f}x_1x_2 = 0101$, 1000, and 1100 to prevent momentary false outputs for state transitions ⓐ → ⓑ, ⓓ → ⓔ, and ⓑ → ⓐ, respectively. All remaining transient states are left unspecified.

The output maps, like the excitation maps, must be analyzed to determine if static, dynamic, or essential hazards exist. The output map for z_1 in Figure 6.86 contains a possible hazard for the transition ⓖ → ⓗ. The hazard can be negated by inclusion of the term $y_{1f}'x_1x_2'$ in the equation for z_1.

The output map for z_2 contains a static-1 hazard when x_2 changes from 1 to 0 for the transition ⓒ → ⓓ. Figure 6.87 shows the logic for output z_2 depicting the static-1 hazard. The hazard can be eliminated by including the redundant prime implicant

$y_{1f}x_1$ in the equation for z_2. The output equations are shown in Equation 6.17 in a minimal sum-of-products form in which all potential hazards have been eliminated.

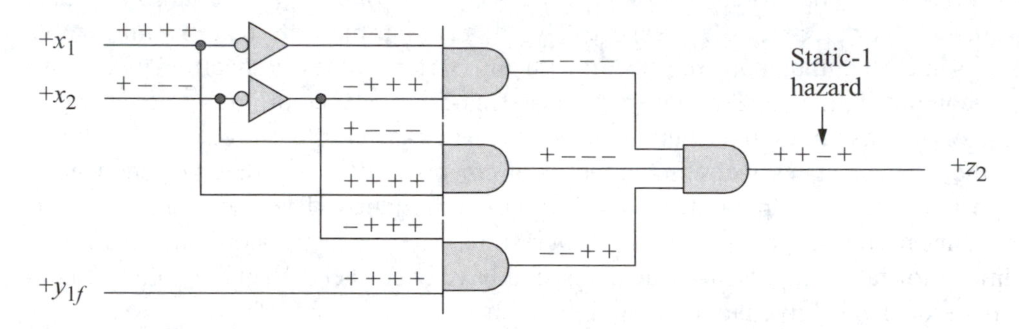

Figure 6.87 Logic diagram for output z_2, synthesized from Equation 6.17, depicting a static-1 hazard for the transition ⓒ → ⓓ. The hazard cover, using the redundant prime implicant $y_{1f}x_1$, has been omitted in order to illustrate the hazard.

Essential hazards may also be present. Recall that an essential hazard is the result of excessive propagation delay in certain network elements. The hazard can be eliminated by introducing appropriate delays in other components. Essential hazards can be detected by analyzing the merged flow table. An essential hazard is possible if, beginning in a stable state, input x_i changes value and sequences the machine to a stable state that is different than the stable state reached after three successive changes to x_i.

Thus, an asynchronous sequential machine contains a possible essential hazard if a single change to an input variable results in a transition from stable state S_j to stable state S_k, whereas, three consecutive changes to x_i sequences the machine to stable state S_l, where $S_k \neq S_l$.

$$\begin{aligned} z_1 &= y_{1f}x_2 + y_{2f}x_1 + y_{1f}'y_{2f}'x_2' + y_{1f}'x_1x_2' \\ z_2 &= x_1'x_2' + x_1x_2 + y_{1f}x_2' + y_{1f}x_1 \\ z_3 &= y_{2f}' \end{aligned} \tag{6.17}$$

The essential hazard can be eliminated by inserting delay elements in critical paths or by using a different merger partition of maximal compatible rows. All merger partitions should be examined to determine the optimal pattern of 1s and 0s in the output maps and thus, derive minimized output equations.

The equations of Equation 6.17 represent the λ output logic for the machine using combinational logic. In some situations, the output equations can be modified by using the axioms and theorems of boolean algebra to yield lower-cost circuits. In general, the output alphabet Z is described as a nonempty, finite set of outputs such that,

$$Z = \{Z_0, Z_1, Z_2, \cdots, Z_{2^m-1}\}$$

Since $m = 3$ in Example 6.17, each output vector Z_r is composed of three binary variables z_1, z_2, and z_3, as shown in Equation 6.18.

$$\begin{aligned}
Z_0 &= z_1'z_2'z_3' = 000 \\
Z_1 &= z_1'z_2'z_3 = 001 \\
Z_2 &= z_1'z_2 z_3' = 010 \\
&\vdots \\
Z_6 &= z_1 z_2 z_3' = 110 \\
Z_7 &= z_1 z_2 z_3 = 111
\end{aligned} \tag{6.18}$$

Figure 6.88 (a) and (b) show output maps for z_1 configured to permit fast and slow output changes, respectively. In Figure 6.88 (a), transient state $y_{1f}y_{2f}x_1x_2 = 0010$ must contain an entry of 1 to prevent a false output on z_1 for the transition ⓔ → ⓗ. All remaining unstable states contain values that permit fast output changes. The entries in these unstable states, therefore, must contain values that are identical to the output values of the destination stable state for each state transition sequence.

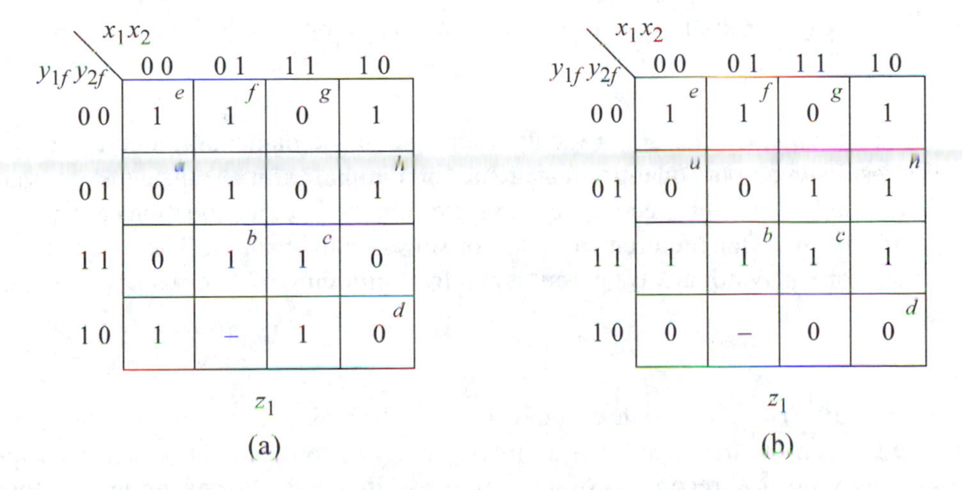

Figure 6.88 Output map for z_1 with values assigned to permit fast or slow output changes: (a) fast output change; and (b) slow output change.

For example, the transition ⓓ → ⓔ proceeds from state ⓓ, where $z_1 = 0$, through transient state $y_{1f}y_{2f}x_1x_2 = 1000$, where $z_1 = 1$, to state ⓔ, where $z_1 = 1$. This sequence results in a fast assertion for output z_1. Correspondingly, a value of 0 in location $y_{1f}y_{2f}x_1x_2 = 0111$ causes a fast deassertion for z_1 for the transition ⓗ → ⓖ. The equation for output z_1 that represents a fast output change is shown in Equation 6.19 in both a sum-of-products and a product-of-sums form utilizing hazard covers.

$$z_1 = x_1'x_2 + y_{1f}x_2 + y_{2f}'x_1' + y_{1f}'x_1x_2' + y_{1f}'y_{2f}'x_2'$$

$$z_1 = (y_{2f}' + x_1 + x_2)(y_{1f} + x_1' + x_2')(y_{1f}' + x_1' + x_2)(y_{1f}' + y_{2f}' + x_2) \quad (6.19)$$

Figure 6.88 (b) shows the output map for z_1 in which transient state values are chosen to generate slow output changes. As in Figure 6.88 (a), the entry in location $y_{1f}y_{2f}x_1x_2 = 0010$ prevents a momentary erroneous 0 output on z_1 for the transition ⓔ → ⓗ. The values in all remaining transient states are chosen to satisfy the requirement of slow output assertions and deassertions. Consider again the transition ⓓ → ⓔ. The entry in location $y_{1f}y_{2f}x_1x_2 = 1000$ maintains z_1 in a reset condition until the sequence terminates. The sequence ⓒ → ⓓ passes through unstable state $y_{1f}y_{2f}x_1x_2 = 1110$ in which $z_1 = 1$ to maintain the output in a set condition until the sequence is complete. The equation for z_1 that represents a slow output change is shown in Equation 6.20 in both a sum-of-products form and a product-of-sums form utilizing hazard covers.

$$z_1 = y_{1f}y_{2f} + y_{2f}x_1 + y_{1f}'y_{2f}'x_1' + y_{1f}'y_{2f}'x_2' + y_{1f}'x_1x_2'$$

$$z_1 = (y_{1f}' + y_{2f})(y_{1f} + y_{2f}' + x_1)(y_{2f} + x_1' + x_2') \quad (6.20)$$

The restriction of fast or slow output changes in fundamental-mode machines usually results in output equations that are not in minimal form. Transient states that would normally be left unspecified to provide minimized output equations must now be specified to meet the requirement of fast or slow output changes. The assigned entries, therefore, may not always permit a maximal grouping of 1s or 0s in the output maps.

Example 6.19 The appropriate output value must be maintained in every intermediate state for as many transient states as are necessary to complete the cycle. This applies to any sequence, regardless of the output requirements. For example, if output response time is not critical and different outputs are associated with the source and destination stable states, then all intermediate states contain unspecified entries. The unspecified entries can then be assigned suitable values to provide a minimal output equation.

Likewise, if the source and destination stable states have identical output values, then every intermediate state in the cycle must be assigned the same value as specified in the initial and final stable states. In this example, some state transitions proceed through more than one intermediate state before entering the destination stable state.

Figure 6.89 represents a reduced primitive flow table for an asynchronous sequential machine with two outputs. Using the rules for merging as described in Section 6.2.4, the merger diagram is obtained as shown in Figure 6.90. The merger diagram provides the following optimal partition of strongly connected sets:

$$\{ⓐ, ⓑ\}, \{ⓒ, ⓖ\}, \{ⓓ, ⓔ\}, \{ⓕ\}$$

Any other partition would generate more than four rows in the merged flow table.

x_1x_2 00	01	11	10	z_1	z_2
ⓐ	b	–	c	0	0
a	ⓑ	d	–	0	1
a	–	f	ⓒ	1	1
–	g	ⓓ	e	1	1
a	–	d	ⓔ	0	0
–	b	ⓕ	e	0	1
a	ⓖ	f	–	1	0

Figure 6.89 Reduced primitive flow table for Example 6.19.

Accordingly, the merged flow table is shown in Figure 6.91. Each row of the reduced primitive flow table is transferred to the assigned row in the merged flow table, superimposing a previously transferred row, if necessary. By identifying the state transition sequences in the four rows of the merged flow table, the transition diagram of Figure 6.92 is obtained. The transition diagram signifies that all four rows must be adjacent. That is, each row must be adjacent to every other row, where adjacency is specified as a unit distance between contiguous state code words of a state transition sequence.

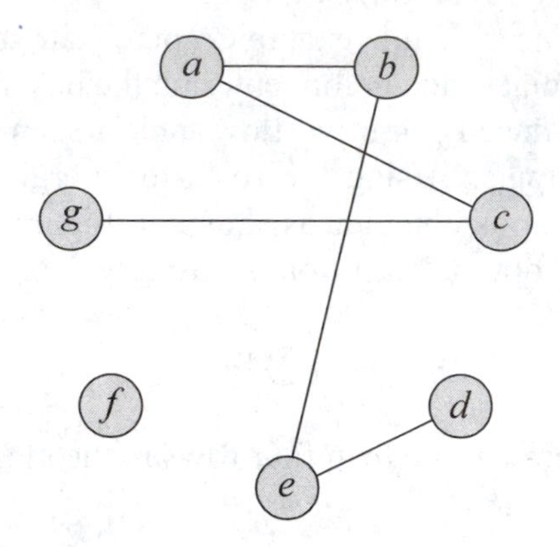

Figure 6.90 Merger diagram obtained from the reduced primitive flow table of Figure 6.89.

	x_1x_2	00	01	11	10
1	(a),(b)	(a)	(b)	d	c
2	(c),(g)	a	(g)	f	(c)
3	(d),(e)	a	g	(d)	(e)
4	(f)	–	b	(f)	e

Figure 6.91 Merged flow table generated from the merger diagram of Figure 6.90.

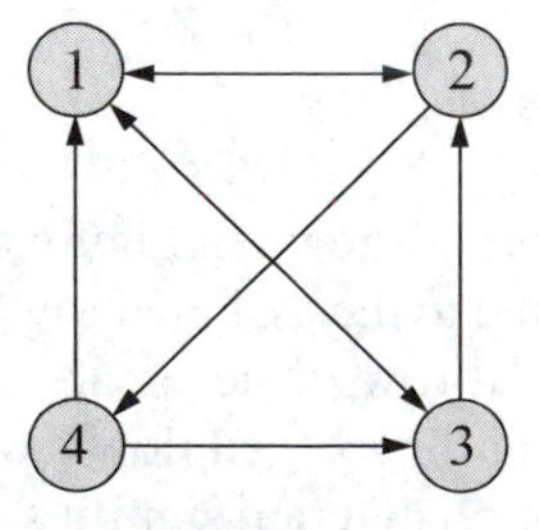

Figure 6.92 Transition diagram for the merged flow table of Figure 6.91.

Since adjacency is obviously impossible between each pair of rows in the transition diagram, a third feedback variable is required. The addition of a third variable makes possible the augmentation of the merged flow table by appending four additional rows to the table. The augmented merged flow table is shown in Figure 6.93 in which adjacency requirements for three state transitions are satisfied by multiple intermediate states, as identified by the arrows.

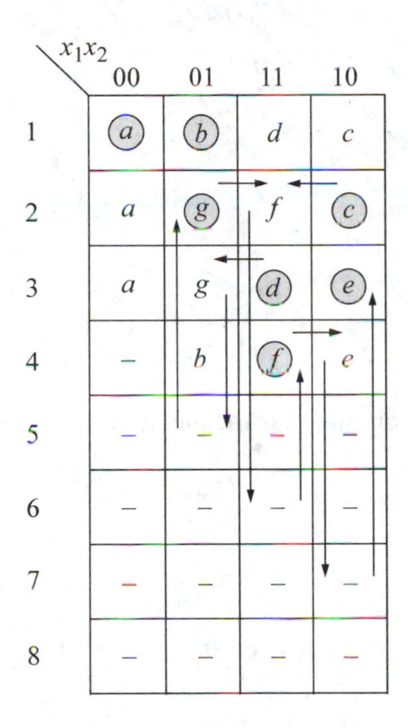

x_1x_2	00	01	11	10
1	ⓐ	ⓑ	*d*	*c*
2	*a*	ⓖ	*f*	ⓒ
3	*a*	*g*	ⓓ	ⓔ
4	–	*b*	ⓕ	*e*
5	–	–	–	–
6	–	–	–	–
7	–	–	–	–
8	–	–	–	–

Figure 6.93 Augmented merged flow table for Example 6.19.

The transition diagram for the augmented merged flow table is shown in Figure 6.94 as a 3-dimensional cube. The transition from row 2 to row 4 is accommodated by the row sequence $2 \rightarrow 6 \rightarrow 4$. This sequence is for state transitions ⓖ $\rightarrow f \rightarrow$ ⓕ and ⓒ $\rightarrow f \rightarrow$ ⓕ. In a similar manner, multiple intermediate states are necessary for row transitions $3 \rightarrow 5 \rightarrow 2$ and $4 \rightarrow 7 \rightarrow 3$, which correspond to state transitions ⓓ $\rightarrow g \rightarrow$ ⓖ and ⓕ $\rightarrow e \rightarrow$ ⓔ, respectively.

All requisite information is now available to construct the output maps for z_1 and z_2. The output map for z_1 is illustrated in Figure 6.95 and is derived from the reduced primitive flow table of Figure 6.89 and the augmented merged flow table of Figure 6.93. The merged flow table identifies the locations of all stable states in their respective rows. The reduced primitive flow table specifies the output values for z_1 for all stable states. The output map depicts the same row numbers for the stable states as specified in the augmented merged flow table, but correlates them to the Gray code assignment for feedback variables y_{1f}, y_{2f}, and y_{3f}. The state codes for the feedback variables — and thus the row numbers — are obtained directly from the transition diagram for the augmented merged flow table.

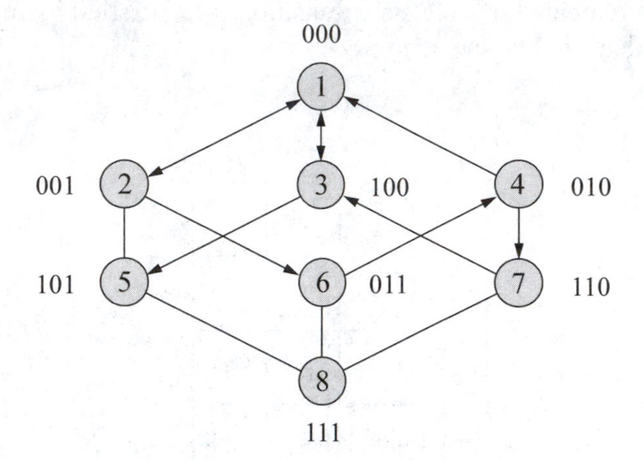

Figure 6.94 Transition diagram for the augmented merged flow table of Figure 6.93.

	$y_{1f}\,y_{2f}\,y_{3f}$ \ x_1x_2	00	01	11	10
1	000	0 (a)	0 (b)	–	–
2	001	–	1 (g)	–	1 (c)
6	011	–	–	–	–
4	010	–	0	0 (f)	0
7	110	–	–	–	0
8	111	–	–	–	–
5	101	–	1	–	–
3	100	0	1	1 (d)	0 (e)

z_1

Figure 6.95 Output map for z_1 derived from the reduced primitive flow table of Figure 6.89 and the augmented merged flow table of Figure 6.93.

All state transition sequences will now be examined to determine the values of the intermediate transient states. Assume that minimization of the λ output logic is of primary importance. The transitions ⓐ → ⓑ, ⓑ → ⓐ, ⓓ → ⓔ, and ⓔ → ⓓ require no change to the feedback variables and thus, preclude the necessity of an unstable state.

The transition sequence ⓐ ($z_1 = 0$) → ⓒ ($z_1 = 1$) passes through transient state $y_{1f}y_{2f}y_{3f}x_1x_2 = 00010$, which can be left unspecified, since the initial and final stable states have different output values. The same rationale can be applied to the sequence ⓑ ($z_1 = 0$) → ⓓ ($z_1 = 1$). The state transition sequences ⓖ ($z_1 = 1$) → ⓐ ($z_1 = 0$) and ⓒ ($z_1 = 1$) → ⓐ ($z_1 = 0$) both pass through the intermediate transient state $y_{1f}y_{2f}y_{3f}x_1x_2 = 00100$, which can remain unspecified due to different output values in the initial and final stable states.

Stable states ⓕ and ⓑ both manifest output z_1 in an inactive state. Thus, the sequence ⓕ ($z_1 = 0$) → ⓑ ($z_1 = 0$) must maintain z_1 in an inactive state for the entire state transition sequence. The unstable state at location $y_{1f}y_{2f}y_{3f}x_1x_2 = 01001$, therefore, must contain an entry of 0 for z_1. Similarly, the transition ⓔ ($z_1 = 0$) → ⓐ ($z_1 = 0$) requires that a value of $z_1 = 0$ be inserted in transient state $y_{1f}y_{2f}y_{3f}x_1x_2 = 10000$; otherwise an erroneous asserted output will be generated on z_1.

The remaining three state transitions are characterized by longer cycles, in which the machine passes through two intermediate transient states while proceeding to the destination stable state. The transition ⓖ ($z_1 = 1$) → ⓕ ($z_1 = 0$) passes through unstable states $y_{1f}y_{2f}y_{3f}x_1x_2 = 00111$ and 01111. Since the initial and final stable states have different output values, the intermediate states can be left unspecified. Care must be taken, however, to ensure that no spurious output occurs on z_1 by assigning improper output values to the intermediate states.

For example, the sequence ⓖ ($z_1 = 1$) → ⓕ ($z_1 = 0$) can be comprised of the following three unique sets of valid output values for z_1, which do not contribute to momentary erroneous outputs:

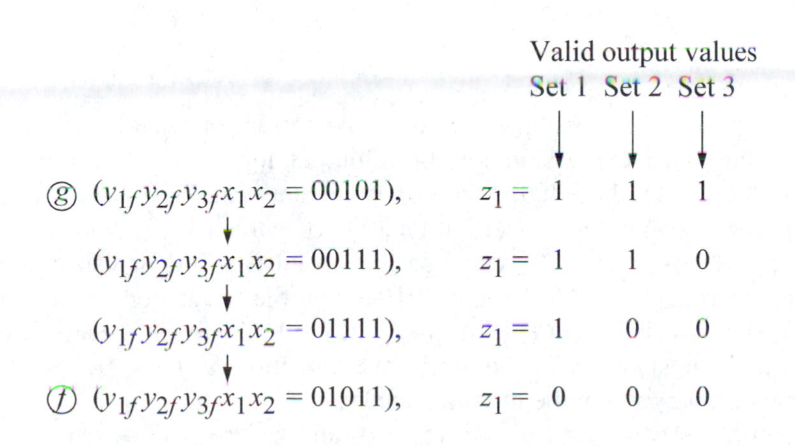

The following set of outputs, however, must not be used even though the set may yield an optimal pattern of 1s and 0s that generates a minimal boolean expression for the output variable:

$$
\begin{array}{lll}
\textcircled{g} & (y_{1f}y_{2f}y_{3f}x_1x_2 = 00101), & z_1 = 1 \\
& \downarrow & \\
& (y_{1f}y_{2f}y_{3f}x_1x_2 = 00111), & z_1 = 0 \\
& \downarrow & \\
& (y_{1f}y_{2f}y_{3f}x_1x_2 = 01111), & z_1 = 1 \\
& \downarrow & \\
\textcircled{f} & (y_{1f}y_{2f}y_{3f}x_1x_2 = 01011), & z_1 = 0
\end{array}
$$

The output value of $z_1 = 1$ in transient state $y_{1f}y_{2f}y_{3f}x_1x_2 = 01111$ generates an erroneous signal on the output. It must be emphasized that the values in the minterm locations of the output map become the prescribed values of the output variable as the machine passes through that state.

Consider the sequence Ⓕ → ⓔ in the output map for z_1 of Figure 6.95. The operation traverses through two unstable states ($y_{1f}y_{2f}y_{3f}x_1x_2 = 01010$ and 11010) before terminating in stable state ⓔ. Since the initial and final stable states contain identical output values of $z_1 = 0$, therefore, both intermediate states must provide output values of $z_1 = 0$.

The final transition to be considered is from state ⓓ to state ⓖ. This transition passes through two unstable states ($y_{1f}y_{2f}y_{3f}x_1x_2 = 10001$ and 10101) while sequencing to state ⓖ. In order to prevent a momentary false output of $z_1 = 0$, all intermediate states must contain a value of $z_1 = 1$. Thus, locations $y_{1f}y_{2f}y_{3f}x_1x_2 = 10001$ and 10101 are assigned values of $z_1 = 1$. The output map for z_1 contains no static hazards. The equation for z_1 is shown in Equation 6.21.

$$z_1 = y_{3f} + y_{1f}x_2 \tag{6.21}$$

The output map for z_2 is shown in Figure 6.96. Again, assume that the main consideration is a minimized expression for the λ output logic. For the transition ⓐ ($z_2 = 0$) → ⓒ ($z_2 = 1$), the intermediate transient state can be left unspecified. The unstable states $y_{1f}y_{2f}y_{3f}x_1x_2 = 01010$ and 11010, which are included in the transition Ⓕ ($z_2 = 1$) → ⓔ ($z_2 = 0$), can also remain unspecified, as can the transient states $y_{1f}y_{2f}y_{3f}x_1x_2 = 10001$ and 10101 for the transition ⓓ ($z_2 = 1$) → ⓖ ($z_2 = 0$). The transition ⓑ ($z_2 = 1$) → ⓓ ($z_2 = 1$) obviously necessitates a value of $z_2 = 1$ in the unstable state. Similarly, the transition ⓔ ($z_2 = 0$) → ⓐ ($z_2 = 0$) requires a value of $z_2 = 0$ in the intermediate state.

The two sequences Ⓕ ($z_2 = 1$) → ⓔ ($z_2 = 0$) and ⓓ ($z_2 = 1$) → ⓖ ($z_2 = 0$) both contain unspecified entries in the intermediate transient states. Therefore, the transient states can contain values of 0 or 1. However, once a value has been decided for z_2 in an intermediate state that is different than the value in the initial stable state, the value must not change for all subsequent intermediate states in the cycle. Thus, for the transition

$$ⓕ \rightarrow y_{1f}y_{2f}y_{3f}x_1x_2 = 01010 \rightarrow y_{1f}y_{2f}y_{3f}x_1x_2 = 11010 \rightarrow ⓔ$$

the following respective values for z_2 are valid: z_2 = 1110, 1100, or 1000. The sequence z_2 = 1010, however, is invalid, due to the erroneous output in state $y_{1f}y_{2f}y_{3f}x_1x_2$ = 11010.

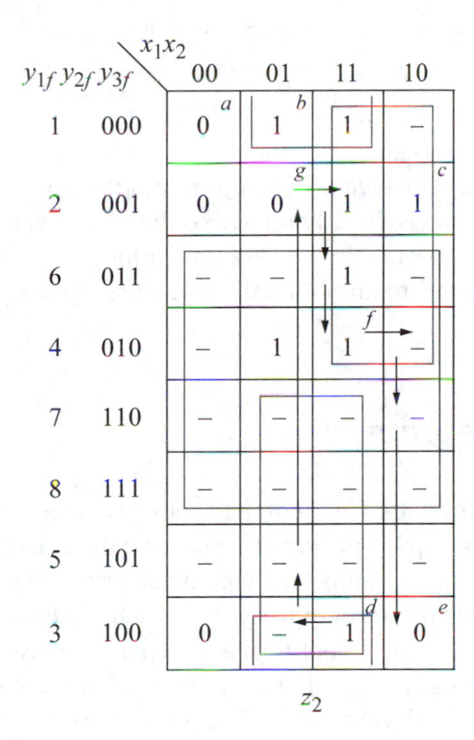

Figure 6.96 Output map for z_2 derived from the reduced primitive flow table of Figure 6.89 and the augmented merged flow table of Figure 6.93.

By the same rationale, the transition ⓓ (z_2 = 1) → ⓖ (z_2 = 0) can be correctly traversed by applying the same respective output values for z_2 as specified for the sequence ⓕ (z_2 = 1) → ⓔ (z_2 = 0). Thus, the transition

$$ⓓ \rightarrow y_{1f}y_{2f}y_{3f}x_1x_2 = 10001 \rightarrow y_{1f}y_{2f}y_{3f}x_1x_2 = 10101 \rightarrow ⓖ$$

must be assigned one of the following sets of output values in the corresponding states of the cycle: z_2 = 1110, 1100, or 1000. The set of output values z_2 = 1010 is invalid.

The transition ⓒ ($z_2 = 1$) → ⓐ ($z_2 = 0$) indicates that the intermediate state can remain unspecified. However, the transition ⓖ ($z_2 = 0$) → ⓐ ($z_2 = 0$) takes precedence, requiring that location $y_{1f}y_{2f}y_{3f}x_1x_2 = 00100$ be assigned a value of $z_2 = 0$ to prevent a glitch on z_2 as the machine passes through the unstable state.

The same reasoning applies to the two transitions ⓖ ($z_2 = 0$) → ⓕ ($z_2 = 1$) and ⓒ ($z_2 = 1$) → ⓕ ($z_2 = 1$). Since the transition ⓒ → ⓕ is the dominant transition with respect to the output values in the intermediate states of the cycle, all transient states must be assigned a value of $z_2 = 1$. The hazard-free equation for output z_2 is shown in Equation 6.22.

$$z_2 = y_{1f}'x_1 + y_{1f}x_2 + y_{2f} + y_{2f}'y_{3f}'x_2 \tag{6.22}$$

The output equations can be in either a sum-of-products form or in a product-of-sums form, depending on which notation provides the fewest number of logic gates. The output response time is directly related to the output values inserted in the intermediate transient states for the appropriate state transitions. Output values are not optional, however, when the initial and final stable states specify the same output values.

6.2.8 Logic Diagram

The logic diagram is the culmination of the synthesis procedure and represents the result of all the previous steps. The implementation can be in any desired form such as, sum-of-products, product-of-sums, or a canonical form. The machine is synthesized directly from the excitation and output equations, which should be in minimal boolean expressions. The logic family is optional and depends on available technology. Discrete logic gates are usually utilized in the implementation of small machines, whereas, programmable logic devices, including gate arrays, are ideally suited for the implementation of large asynchronous sequential machines.

When properly implemented in accordance with the synthesis procedure, the machine will be free of critical races, static hazards, and momentary false outputs. Identical logical terms can be shared between excitation variables. However, if error detection is incorporated in the design, then all logical terms should be assigned to their respective excitation variables only; otherwise, multiple errors can occur which may negate the detection capability of the error checking circuits.

Example 6.20 Consider the merged flow table of Figure 6.97. The procedures in previous sections are used to construct the diagrams and maps in this example. The state transition sequences are graphically depicted in the transition diagram of Figure 6.98. By rearranging the rows of the merged flow table in accordance with the transition diagram, all state transitions can be realized without regard for race conditions. Thus, all contiguous rows in a cycle are adjacent.

x_1x_2	00	01	11	10
1	(*a*)	*c*	(*b*)	*d*
2	*e*	(*f*)	*b*	(*d*)
3	*a*	(*c*)	*b*	(*g*)
4	(*e*)	*f*	–	*g*

Figure 6.97 Merged flow table for Example 6.20.

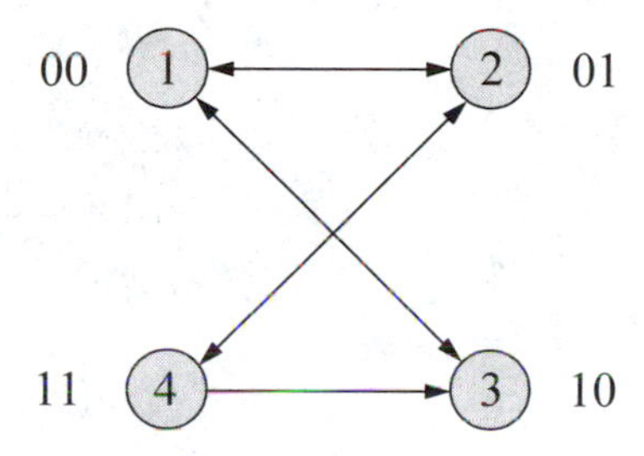

Figure 6.98 Transition diagram for Figure 6.97.

Figure 6.99 illustrates the combined excitation map in which the rows are reordered and assigned the appropriate state variable codes. The individual excitation maps are shown in Figure 6.100. The excitation equations are listed in Equation 6.23 in both a sum-of-products form and a product-of-sums form. Both forms are free of static-1 and static-0 hazards, respectively. The sum-of-products form will be used in the implementation of the machine, since the product-of-sums equation requires additional logic gates.

Assume that output z_1 has the following assigned values:

Stable state	Output z_1
(*a*)	0
(*b*)	1
(*c*)	0
(*d*)	1

Stable state	Output z_1
(*e*)	0
(*f*)	1
(*g*)	1

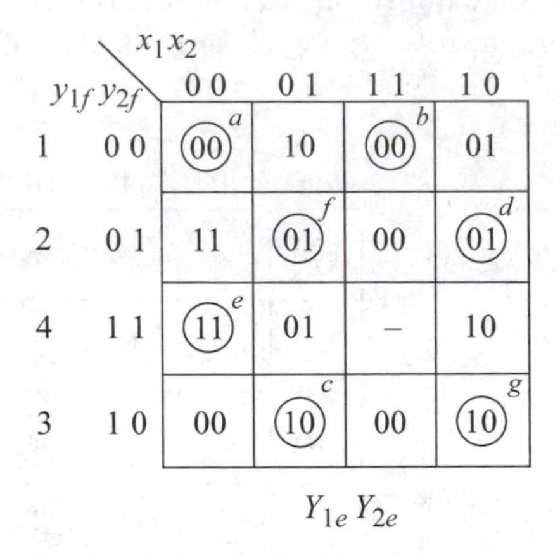

Figure 6.99 Combined excitation map for Example 6.20.

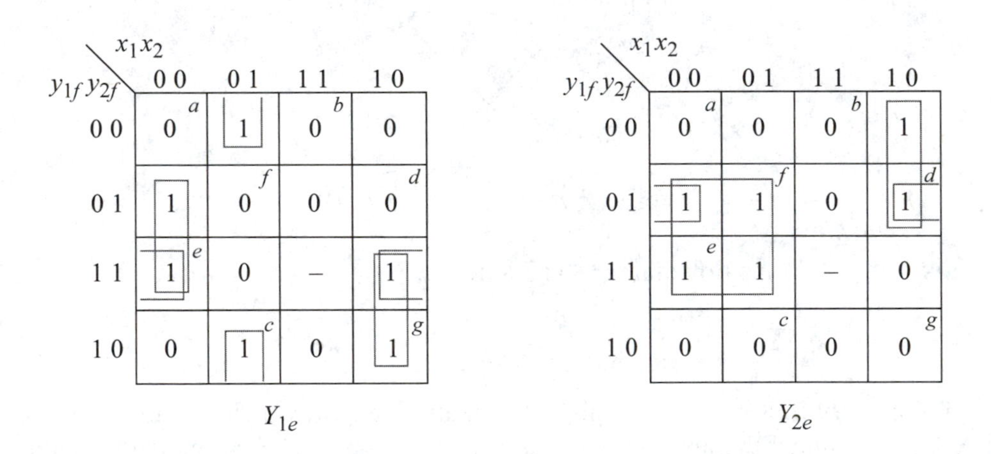

Figure 6.100 Individual excitation maps for Example 6.20.

$$Y_{1e} = y_{2f}x_1'x_2' + y_{2f}'x_1'x_2 + y_{1f}x_1x_2' + y_{1f}y_{2f}x_2'$$

$$Y_{1e} = (y_{1f} + x_1')(x_1' + x_2')(y_{2f}' + x_2')(y_{2f} + x_1 + x_2)(y_{1f} + y_{2f} + x_2)$$

$$Y_{2e} = y_{2f}x_1' + y_{1f}'x_1x_2' + y_{1f}'y_{2f}x_2'$$

$$Y_{2e} = (y_{1f}' + x_1')(y_{2f} + x_1)(x_1' + x_2')(y_{1f}' + y_{2f})(y_{2f} + x_2') \tag{6.23}$$

The output map is shown in Figure 6.101. All state transitions preclude the possibility of momentary false outputs. This is achieved by assigning appropriate output values to the intermediate transient states of a cycle, in which the initial and final stable states contain identical output values. Also, output response time is devised to be as fast as possible. Thus, if the initial and final stable states have different output values, then the output variable changes value in the first unstable state of a cycle. The output equation is shown in Equation 6.24.

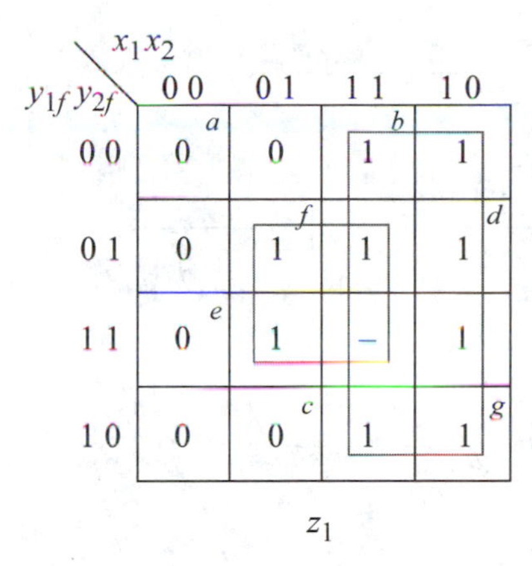

Figure 6.101 Output map for the asynchronous sequential machine of Example 6.20. Output values are assigned such that, there will be no momentary false outputs and output changes will be as fast as possible.

$$z_1 = x_1 + y_{2f}x_2 \tag{6.24}$$

The network is synthesized using the excitation equations of Equation 6.23 and the output equation of Equation 6.24. Implementation is realized using NOR logic with complementary outputs. The logic diagram for the asynchronous sequential machine of Example 6.20 is shown in Figure 6.102. To avoid clutter and aid in comprehending the operational characteristics of the machine, input signals are duplicated, where necessary. The + and – symbols preceding the signal name indicate the active level of the signal for both input and output variables.

Section 6.2 presented the necessary steps in the synthesis of an asynchronous sequential machine from a given set of machine specifications. The generation of a state diagram is an optional step and is usually omitted in the synthesis procedure. However, its inclusion adds completeness to the process. In most cases, the synthesis proceeds from the machine specifications directly to the primitive flow table.

The primitive flow table is the most critical step in the synthesis procedure. The table must accurately reflect the machine specifications for all possible input sequences. The construction of the primitive flow table is usually a two-pass process: The first pass lists all obvious stable states and transitions as specified in the timing diagram and/or verbal description of the operational characteristics; the second pass completes the table by filling in the remaining blank entries for those sequences that are more abstruse.

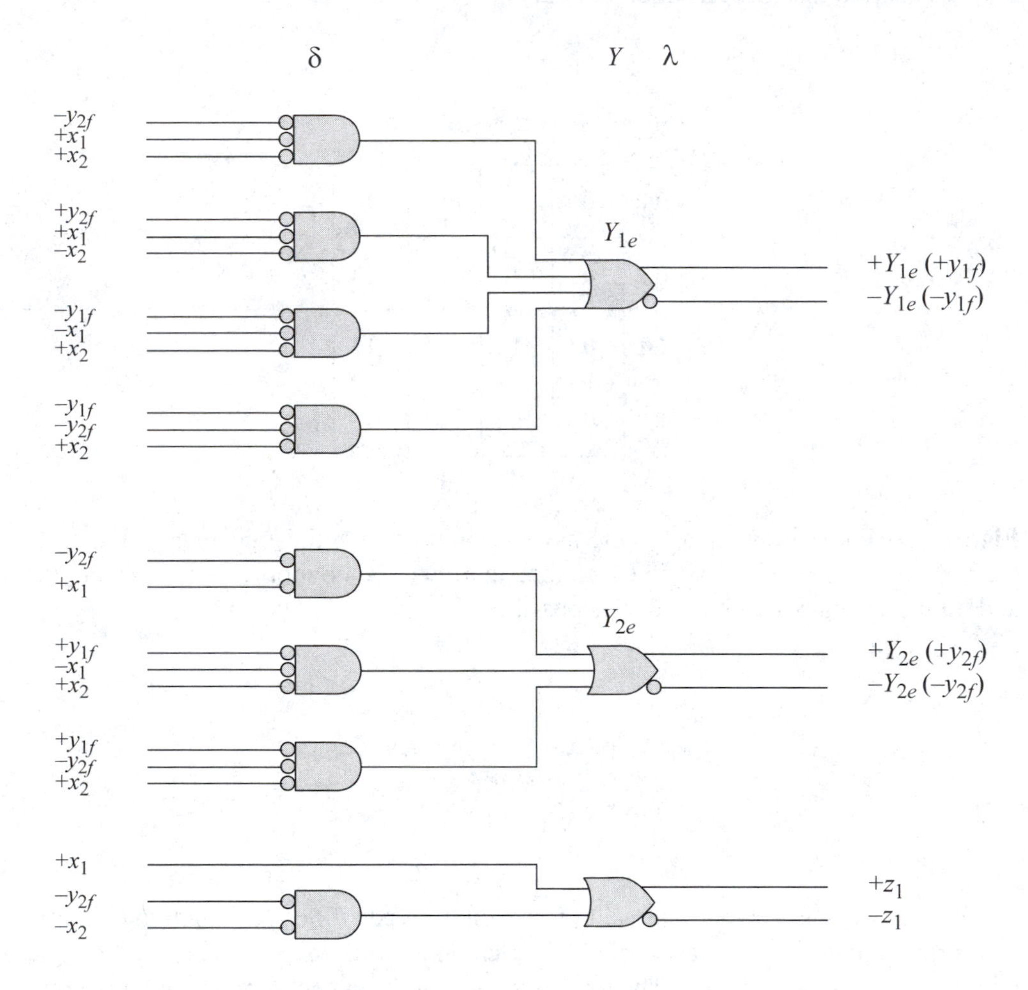

Figure 6.102 Logic diagram using NOR logic for the asynchronous sequential machine of Example 6.20.

Some state transitions are invalid due to fundamental-mode restrictions or machine characteristics. In these cases, unspecified entries are inserted in appropriate squares in the form of dashes. Thus, every location in the completed primitive flow table will contain a stable state, an unstable state, or an unspecified entry. The primitive flow table is a pivotal step, since all subsequent steps are contingent on the accuracy of the table, either directly or indirectly.

In most cases, the number of rows in the primitive flow table can be reduced by identifying equivalent states and then eliminating redundant states. This process yields a reduced primitive flow table containing a minimal number of rows. Fewer rows results in fewer feedback variables with a subsequent decrease in logic gates.

Further reduction in the number of rows can usually be achieved by merging two or more rows into a single row. This operation results in a merged flow table. In order to prevent race conditions between excitation variables when a single input changes value in the merged flow table, adjacency is required between contiguous rows in a state transition sequence. Thus, the state variable codes must provide a unit distance by allowing only one state variable to change value between each sequential row in a cycle.

Adjacency requirements can be determined by utilizing a transition diagram. The transition diagram specifies those rows in the merged flow table that must be adjacent in order to satisfy all required state transitions. In many instances, adjacency requirements can be realized within the existing flow table. If this is not possible, then an additional state variable must be provided, yielding an augmented merged flow table containing additional rows.

The excitation maps are derived directly from the merged flow table, which specifies the location of all stable states. The stable state locations in the excitation maps contain excitation values that are identical to those of the feedback variables in the corresponding row. One Karnaugh map is generated for each excitation variable. The equation for the excitation variable is derived from the excitation map in either a sum-of-products form or a product-of-sums form, with preference given to the form containing the fewest number of logic gates. All static-1 and static-0 hazards must be eliminated by inclusion of one or more redundant prime implicant terms in the excitation equation.

The output maps are derived from the reduced primitive flow table and the merged flow table. The reduced primitive flow table specifies the output values of each stable state; the merged flow table specifies the location of each stable state in the output maps. The output values assigned to the map entries in the intermediate transient states of a cycle must be such that no momentary false outputs are generated on the output variables.

Output response time is established during construction of the output map. If the initial and final stable states of a state transition sequence have different output values, then output response can be established as follows: For a fast output change, the first, and subsequent, transient states of a cycle must contain an output value that is the same as the output value of the final stable state; correspondingly, for a slow response, the first, and subsequent, transient states of a cycle must contain an output value that is the same as the output value of the initial stable state. Also, output hazards must be detected and eliminated in a manner analogous to that used for the excitation maps.

Section 6.2 developed the synthesis procedure as individual fragments of the complete process. Using this approach, each step in the procedure can be reviewed as a separate entity. This is useful when the details of a particular step have become obscured with time. To review the concepts of a specific step, simply go to the section in question and review the procedure by means of the documented examples. This fragmented approach negates the need to review the entire design process to locate a particular technique in a particular step.

6.3 Synthesis Examples

This section presents several examples illustrating the synthesis of asynchronous sequential machines in their entirety. The machine specifications for these examples will consist of a verbal description only or a verbal description in conjunction with a representative timing diagram.

If the verbal description is the only means of conveying the operational characteristics of the machine, then the description must be comprehensive and precise. Sufficient detail must be provided in the description so that no additional information is required to elucidate the machine specifications. If the machine specifications are delineated in sufficient detail, then a timing diagram can be created as a further aid, if necessary.

When defining the operational characteristics by means of a timing diagram with two or more input variables and at least one output variable, a comprehensive representation of machine characteristics is usually prohibitive. This is due, in part, to space limitations for the diagram in order to show the arrangement and relationship of all combinations of the input and output binary variables.

In the following examples, the solution may not be unique. If there is more than one possible solution, then all representations of the logic network will be presented. The synthesis specifications will stipulate whether the λ output logic is to be as fast as possible, as slow as possible, or in minimal form. In all examples, however, there must be no race conditions, no static-1 or static-0 hazards, and no momentary false outputs.

6.3.1 Mealy Machine with Two Inputs and One Output

Example 6.21 An asynchronous sequential machine has two inputs x_1 and x_2 and one output z_1. The machine is reset initially; that is, $x_1x_2z_1 = 000$. A specific condition of the operational characteristics is that input x_1 must envelope all occurrences of the x_2 pulse. Thus, the allowable input vectors are $x_1x_2 = 00$, 10, or 11; the input combination of $x_1x_2 = 01$ will never occur. Output z_1 is to be asserted coincident with the assertion of every second x_2 pulse and is to remain asserted until the deassertion of x_2. Output assertion is to be as fast as possible. A representative timing diagram is shown in Figure 6.103. Although the timing diagram illustrates a valid input sequence to generate an output, other variations are possible and must be considered to adequately represent the operation of the machine for all valid input sequences.

Primitive flow table As stated previously, a complete and accurate primitive flow table is of paramount importance in the synthesis of asynchronous sequential machines. The primitive flow table is constructed by plotting the sequence of input and output changes as indicated by the machine specifications, while permitting only one stable state per row. This allows the output variables for a row to be uniquely specified as a function of a particular stable state.

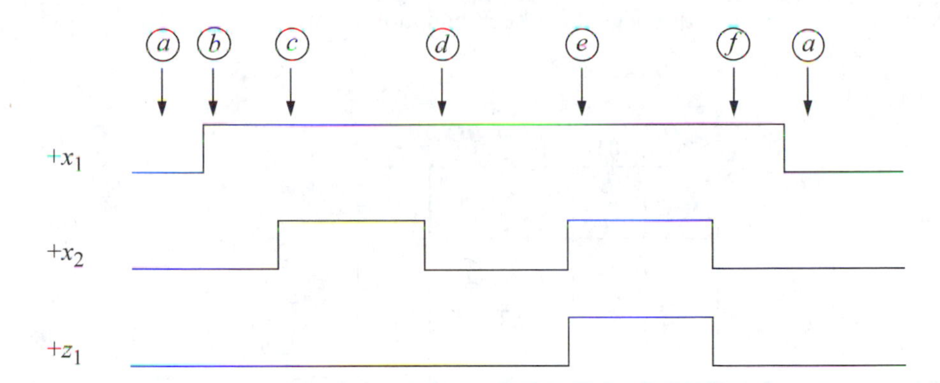

Figure 6.103 Timing diagram for the asynchronous sequential machine of Example 6.21.

Beginning at the leftmost section of the timing diagram where $x_1x_2z_1 = 000$, proceed left to right assigning a unique stable state name to each different combination of the input vector and the associated output z_1. Label location $x_1x_2z_1 = 000$ in the primitive flow table as stable state ⓐ, as shown in Figure 6.104. In state ⓐ, output $z_1 = 0$. The next change to the input vector occurs when x_1 changes from 0 to 1, providing a new stable state at location $x_1x_2 = 10$ in the primitive flow table, labeled state ⓑ. Thus, the inputs change from $x_1x_2 = 00$ to 10 and the operation proceeds from state ⓐ, through transient state b , to state ⓑ, where z_1 remains inactive.

Continuing in a left-to-right sequence, the next change occurs when input x_2 becomes asserted for the first time. This new stable state is designated ⓒ. Since this is the first occurrence of an active x_2 input, output z_1 remains deasserted.

When x_2 then becomes deasserted, the machine leaves state ⓒ and proceeds from $x_1x_2 = 11$ to 10, which represents a new stable state, designated ⓓ. The characteristics of state ⓓ are specified as $x_1x_2z_1 = 100$. Although state ⓑ was also specified as $x_1x_2z_1 = 100$, states ⓑ and ⓓ are not equivalent. Stable state ⓑ is the precursor of the first x_2 pulse, while stable state ⓓ immediately precedes the second x_2 pulse.

The second assertion of x_2 causes the machine to enter state ⓔ, where $z_1 = 1$. This state transition sequence is represented as a transition from state ⓓ, through transient state e, to state ⓔ, where $x_1x_2z_1 = 111$. The assertion of output z_1 conforms to the precise duration of the second assertion of input x_2. Thus, when the input vector next changes from $x_1x_2 = 11$ to 10, output z_1 becomes deasserted.

When the second x_2 pulse becomes deasserted, the machine proceeds to a stable state which precedes the first x_2 pulse. Stable state ⓑ satisfies this requirement. However, in order to review the method of identifying equivalent states, let the machine proceed to a new stable state designated ⓕ and identified by $x_1x_2z_1 = 100$. When input x_1 then becomes inactive, the machine returns to state ⓐ and the process repeats. This completes the first pass in constructing the primitive flow table. All the obvious state transition sequences have been accounted for.

x_1x_2 00	01	11	10	z_1
(a)	–	–	b	0
a	–	c	(b)	0
–	–	(c)	d	0
a	–	e	(d)	0
–	–	(e)	f	1
a	–	c	(f)	0

Figure 6.104 Primitive flow table for the asynchronous sequential machine of Example 6.20.

The remainder of the primitive flow table must now be completed. The fundamental-mode of operation specifies that two or more binary variables of the input vector must not change value simultaneously. Therefore, unspecified entries in the form of dashes are inserted in the appropriate locations in each row of the primitive flow table. For example, in row ⓐ, the transition from state ⓐ to location $x_1x_2 = 11$ is not allowed, necessitating the insertion of an unspecified entry in location $x_1x_2 = 11$. Similarly, in row ⓑ, a transition from state ⓑ to location $x_1x_2 = 01$ is not permitted. Thus, an unspecified entry is entered in location $x_1x_2 = 01$.

The fundamental-mode of operation prevents possible race conditions and associated timing problems that would otherwise be inherent in the operating characteristics of the machine. Once an input vector has been established, no further changes to the input variables will occur until the machine enters a stable state. These restrictions imposed by the fundamental-mode of operation permit the synthesized machine to be deterministic in nature. A unique characteristic of this machine is that the input combination of $x_1x_2 = 01$ will never occur. Therefore, the entire column $x_1x_2 = 01$ can be filled with unspecified entries, as shown in the primitive flow table of Figure 6.104.

The second pass establishes entries for any previously unspecified transient state and may result in the addition of one or more rows to the table. The remaining entries in the table represent subsequences that may not be expressly specified by the verbal description or the timing diagram. These subsequences must be carefully analyzed to ensure that strict adherence to the machine specifications is maintained.

In row ⓑ of the primitive flow table, a change of input vector from $x_1x_2 = 10$ to 00 will sequence the machine through transient state *a* to stable state ⓐ. Since output z_1 directly corresponds to the assertion of input x_2, state ⓐ represents a condition where all three binary variables are reset. This is the initial condition from which the machine begins checking for a new sequence. A similar situation occurs in state ⓓ when the inputs change from $x_1x_2 = 10$ to 00. Rows ⓐ, ⓑ, ⓒ, ⓓ, and ⓔ are now complete.

In stable state ⓕ, a change of input vector from $x_1x_2 = 10$ to 11 represents the assertion of the first x_2 pulse of a pair of x_2 pulses. The state that depicts this condition is state ⓒ. Thus, the transition is from state ⓕ, through transient state *c*, to state ⓒ. Finally, in stable state ⓕ, if the inputs change from $x_1x_2 = 10$ to 00, then the machine again enters the initial state, designated state ⓐ. The primitive flow table is now complete.

Equivalent states The next step is to identify all equivalent stable states and then to eliminate redundant states. In order for two stable states to be equivalent, they must have the same input vector, the same output value, and the same, or equivalent, next state for all valid input sequences. The only possible equivalences exist between stable state pairs {ⓒ, ⓔ}, {ⓑ, ⓓ}, {ⓑ, ⓕ}, and {ⓓ, ⓕ}.

Stable states ⓒ and ⓔ are not equivalent, because the output values are different. Next, states ⓑ and ⓓ are tested for equivalence. Both have the same input vector ($x_1x_2 = 10$) and both have the same output value ($z_1 = 0$). However, when the input vector changes from $x_1x_2 = 10$ to 11, the next state from state ⓑ is state ⓒ; whereas, the next state from state ⓓ is state ⓔ. Since states ⓒ and ⓔ have already been shown to be nonequivalent, therefore, states ⓑ and ⓓ are not equivalent. The same reasoning applies to stable state pair ⓓ and ⓕ, which are also not equivalent.

Stable state pair ⓑ and ⓕ, however, satisfy all equivalence requirements: Both are entered from the same input vector ($x_1x_2 = 10$); both have identical output values ($z_1 = 0$); and both proceed to the same next stable state ⓒ or ⓐ for an applied input vector of $x_1x_2 = 11$ or 00, respectively. Therefore, stable states ⓑ and ⓕ are equivalent. State ⓕ is redundant and can be eliminated from the primitive flow table. Every occurrence of state *f* is replaced by equivalent state *b*. The reduced primitive flow table is shown in Figure 6.105.

Merger diagram The number of rows in a reduced primitive flow table can usually be decreased by merging two or more rows into a single row. Recall the three requirements for merging two rows into a single merged row: Each column in the two rows under consideration must contain identical state names, either stable or unstable, or a state name and an unspecified entry, or two unspecified entries.

In the reduced primitive flow table of Figure 6.105, rows ⓐ and ⓑ can merge, because there is no conflict in any column of the two rows. This merging capability is indicated by a line connecting vertices ⓐ and ⓑ in the merger diagram of Figure 6.106. The only other row with which row ⓐ can merge is row ⓔ —all other rows have a conflict in at least one column. Rows ⓑ, ⓒ, and ⓓ cannot merge with any succeeding row due to conflicting state names in certain columns. The merger diagram of Figure 6.106 yields the following two partitions of maximal compatible sets:

1. {ⓐ, ⓑ}, {ⓒ}, {ⓓ}, {ⓔ}
2. {ⓐ, ⓔ}, {ⓑ}, {ⓒ}, {ⓓ}

x_1x_2	00	01	11	10	z_1
	ⓐ	–	–	b	0
	a	–	c	ⓑ	0
	–	–	ⓒ	d	0
	a	–	e	ⓓ	0
	–	–	ⓔ	b	1

Figure 6.105 Reduced primitive flow table obtained from the primitive flow table of Figure 6.104.

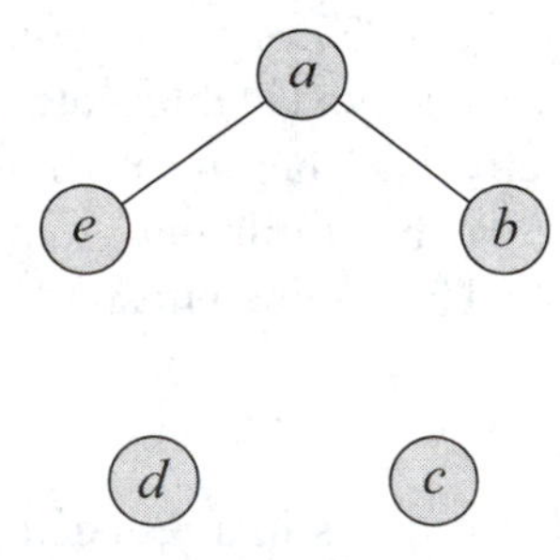

Figure 6.106 Merger diagram for the reduced primitive flow table of Figure 6.105.

Merged flow table Both partitions should be analyzed by means of a merged flow table to determine which partition yields the fewest number of logic gates. The first partition produces the merged flow table shown in Figure 6.107 (a). Each row of the merged flow table is generated by transferring the individual rows from the reduced primitive flow table to the merged flow table in accordance with the partition assignments.

After enumerating the rows of the merged flow table, all state transition sequences can be identified with reference to individual rows. The state transitions are illustrated in graphical form by means of a transition diagram. The transition diagram for the merged flow table of Figure 6.107 (a) is shown in Figure 6.108 (a). Row 1 proceeds to row 2 by the sequence ⓑ → c → ⓒ, as illustrated by the directed line from row 1 to row 2 in Figure 6.108 (a).

(a)

		x_1x_2 = 00	01	11	10
1	ⓐ, ⓑ	ⓐ	–	c	ⓑ
2	ⓒ	–	–	ⓒ	d
3	ⓓ	a	–	e	ⓓ
4	ⓔ	–	–	ⓔ	b

(b)

		x_1x_2 = 00	01	11	10
1	ⓐ, ⓔ	ⓐ	–	ⓔ	b
2	ⓑ	a	–	c	ⓑ
3	ⓒ	–	–	ⓒ	d
4	ⓓ	a	–	e	ⓓ

Figure 6.107 Merged flow tables obtained from the two partitions derived from the merger diagram of Figure 6.106: (a) partition 1: {ⓐ, ⓑ}, {ⓒ}, {ⓓ}, {ⓔ}; and (b) partition 2: {ⓐ, ⓔ}, {ⓑ}, {ⓒ}, {ⓓ}.

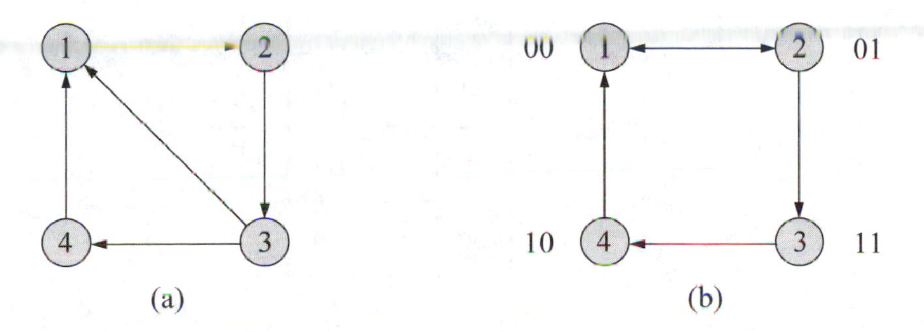

Figure 6.108 Transition diagram for the merged flow tables of Figure 6.107: (a) transition diagram for Figure 6.107 (a); and (b) transition diagram for Figure 6.107 (b).

Notice that row 3 can proceed to two different rows by the following sequences: ⓓ → a → ⓐ, which represents a transition from row 3 to row 1, and ⓓ → e → ⓔ, which represents a transition from row 3 to row 4. Thus, in state ⓓ, a change of input vector from x_1x_2 = 10 to 00 or 11 results in a transition from row 3 to row 1 or row 4, respectively.

The transition diagram of Figure 6.108 (a) contains a triangular polygon specified by rows 1 , 2 , and 3 or by rows 1 , 3 , and 4. Since three rows cannot all be adjacent, row 3 can proceed to row 1 through row 4, eliminating the need for a line from row 3 to row 1. Providing an additional intermediate state to the cycle from state ⓓ to state ⓐ does not alter the operational characteristics of the machine, but does generate a slightly slower transition.

The transition diagram for the merged flow table of Figure 6.107 (b) is depicted in Figure 6.108 (b). Since the transition diagram contains no polygons with an odd number of sides, the state transitions do not have to be altered. All transitions proceed through only one transient state. The merged flow table of Figure 6.107 (b) and the transition diagram of Figure 6.108 (b) will be used to generate the excitation and output equations.

Excitation maps and equations The combined excitation map for excitation variables Y_{1e} and Y_{2e} is shown in Figure 6.109. The stable states are assigned excitation values that are the same as the feedback values of the corresponding rows. It is important to not inadvertently assign excitation values to the "don't care" states that would generate a stable state. The individual excitation maps are shown in Figure 6.110 and the resulting hazard-free excitation equations in Equation 6.25 in a sum-of-products form. The rightmost term in each equation is the hazard cover. The excitation equations are shown in a product-of-sums notation in Equation 6.26.

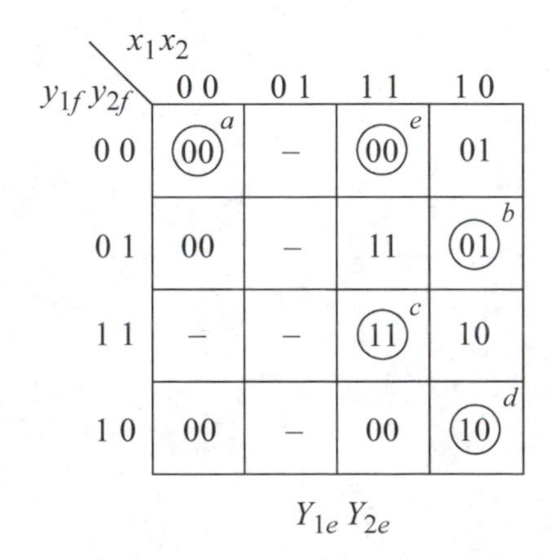

Figure 6.109 Combined excitation map for the merged flow table of Figure 6.107 (b).

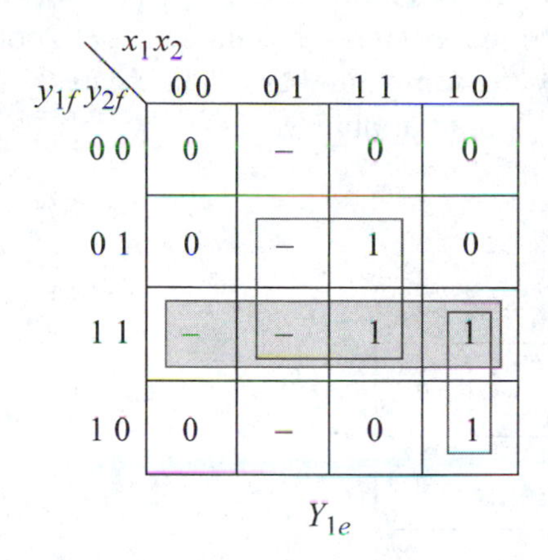

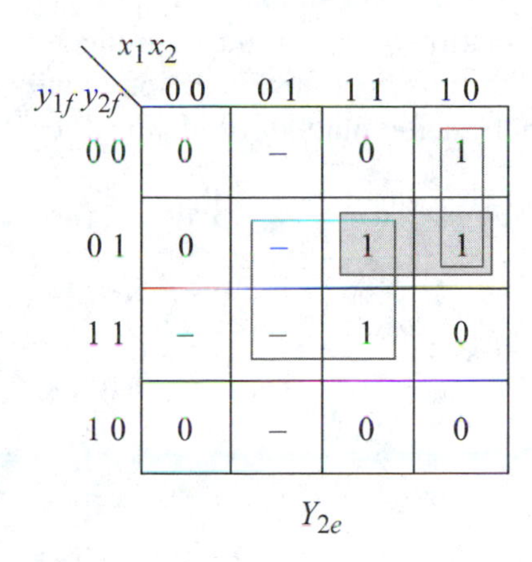

Figure 6.110 Individual excitation maps for Y_{1e} and Y_{2e} obtained from the combined excitation map of Figure 6.109.

$$
\begin{aligned}
Y_{1e} &= y_{2f}x_2 + y_{1f}x_1x_2' + y_{1f}y_{2f} \\
Y_{2e} &= y_{2f}x_2 + y_{1f}'x_1x_2' + y_{1f}'y_{2f}x_1
\end{aligned}
\tag{6.25}
$$

$$
\begin{aligned}
Y_{1e} &= (x_1)(y_{2f} + x_2')(y_{1f} + x_2) \\
Y_{2e} &= (x_1)(y_{2f} + x_2')(y_{1f}' + x_2)
\end{aligned}
\tag{6.26}
$$

Output map and equation The output map is constructed from the merged flow table of Figure 6.107 (b) and the reduced primitive flow table of Figure 6.105. The merged flow table indicates the location of the stable states and the reduced primitive flow table specifies the output values of the stable states. The output map is shown in Figure 6.111.

The transitions ⓐ → ⓑ, ⓑ → ⓐ, ⓑ → ⓒ, ⓒ → ⓓ, and ⓓ → ⓐ represent paths in which the determination of the output value is critical for the intermediate transient states. All transitions have the common characteristic of identical output values for both the initial and final stable states. Output glitches are avoided in all cycles by inserting output values in the transient states that are the same as the output values of the initial and final stable states.

The transition ⓓ → ⓔ contains an unspecified entry in the intermediate state, because stable states ⓓ and ⓔ have different output values. However, since the machine specifications require output changes to be as fast as possible, a value of $z_1 = 1$ is inserted in location $y_{1f}y_{2f}x_1x_2 = 1011$. The equation for output z_1 is shown in Equation 6.27 as a sum of products and as a product of sums. It is interesting to note that both forms of the equation require not only the same number of logic gates, but also the same number of identical feedback and input variables.

$y_{1f}y_{2f}$ \ x_1x_2	00	01	11	10
00	0 *a*	–	1 *e*	0
01	0	–	0	0 *b*
11	–	–	0 *c*	0
10	0	–	–	0 *d*

z_1

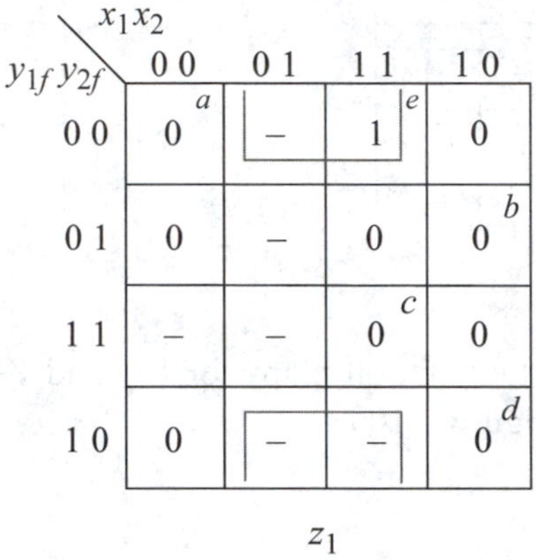

Figure 6.111 Output map for Example 6.21.

$$
\begin{aligned}
z_1 &= y_{2f}{}'x_2 \\
z_1 &= y_{2f}{}' + x_2
\end{aligned} \qquad (6.27)
$$

Logic diagram The logic diagram is shown in Figure 6.112 (a) using Equation 6.25 and in Figure 6.112 (b) using Equation 6.26. It is apparent that the product-of-sums form of Figure 6.112 (b) requires not only the fewest number of gates, but also the fewest number of inputs per gate.

Either of the logic diagrams shown in Figure 6.112 will operate according to the machine specifications. Machine operation can be verified by applying a specific sequence of inputs to the logic diagram and observing the output levels. For example, using the input sequence depicted in the timing diagram of Figure 6.103, the sequence of output levels will be precisely as shown in the timing diagram. The same results can be realized by utilizing the output map. Using the prescribed input sequence, the state transitions in the output map can be ascertained and the output values observed.

To understand the intricacies of internal machine operation for Example 6.21, the logic diagram of Figure 6.112 (a) will be analyzed using the timing diagram of Figure 6.103. The combined excitation map of Figure 6.109 is also useful in examining the inner relationships of the excitation, feedback, and input variables.

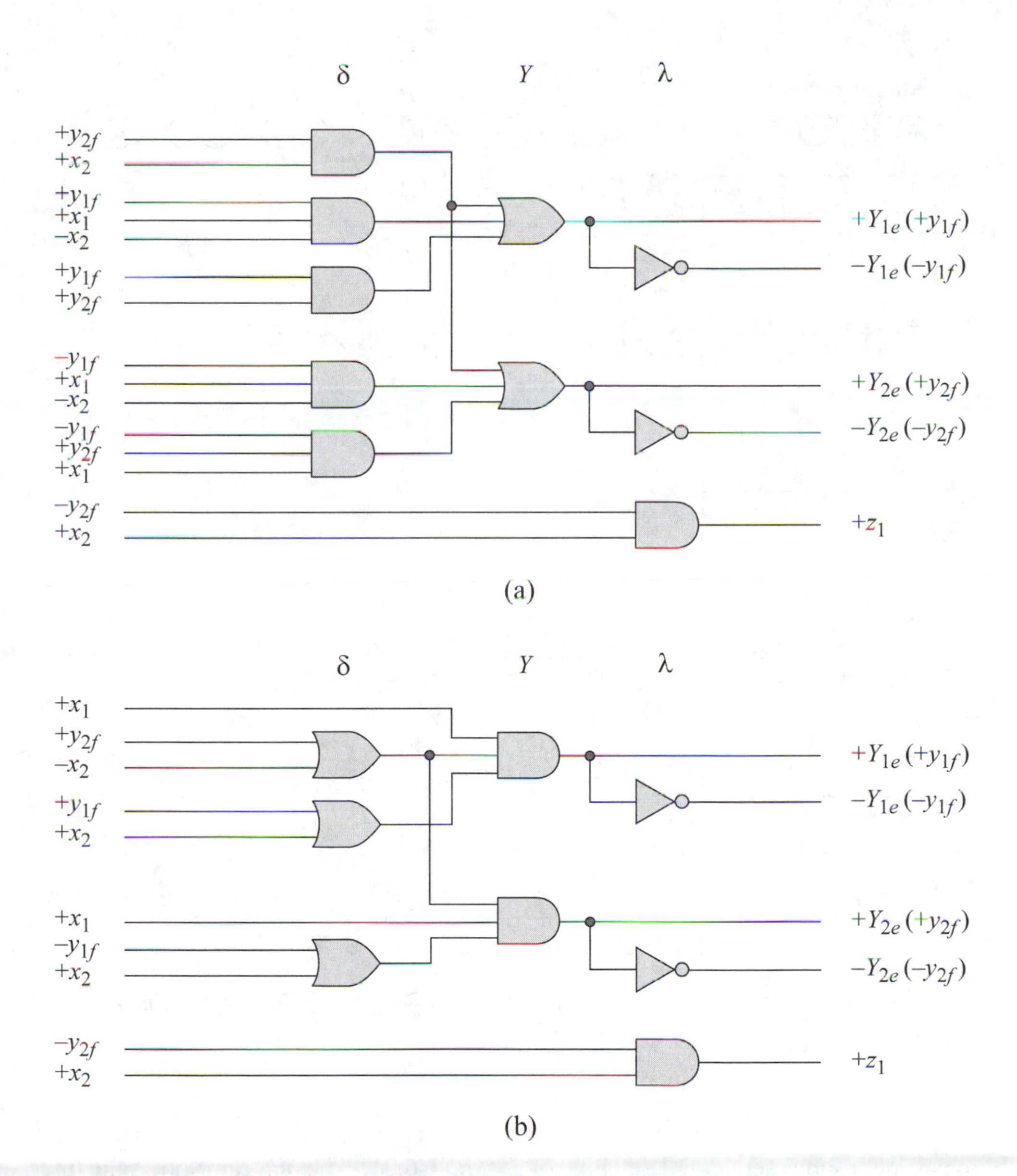

Figure 6.112 Logic diagram for Example 6.21: (a) sum-of-products form; and (b) product-of-sums form.

Figure 6.112 (a) is redrawn in Figure 6.113 in an expanded format so that the logic levels between gates can be clearly indicated. The logic levels of the variables and intergate signals are specified by the +/– symbols. The machine is reset to stable state ⓐ where all variables are deasserted. The reset condition is shown in the leftmost column designated as state ⓐ. The active-low reset signal (not shown) is connected to logic gates 2, 3, and 6.

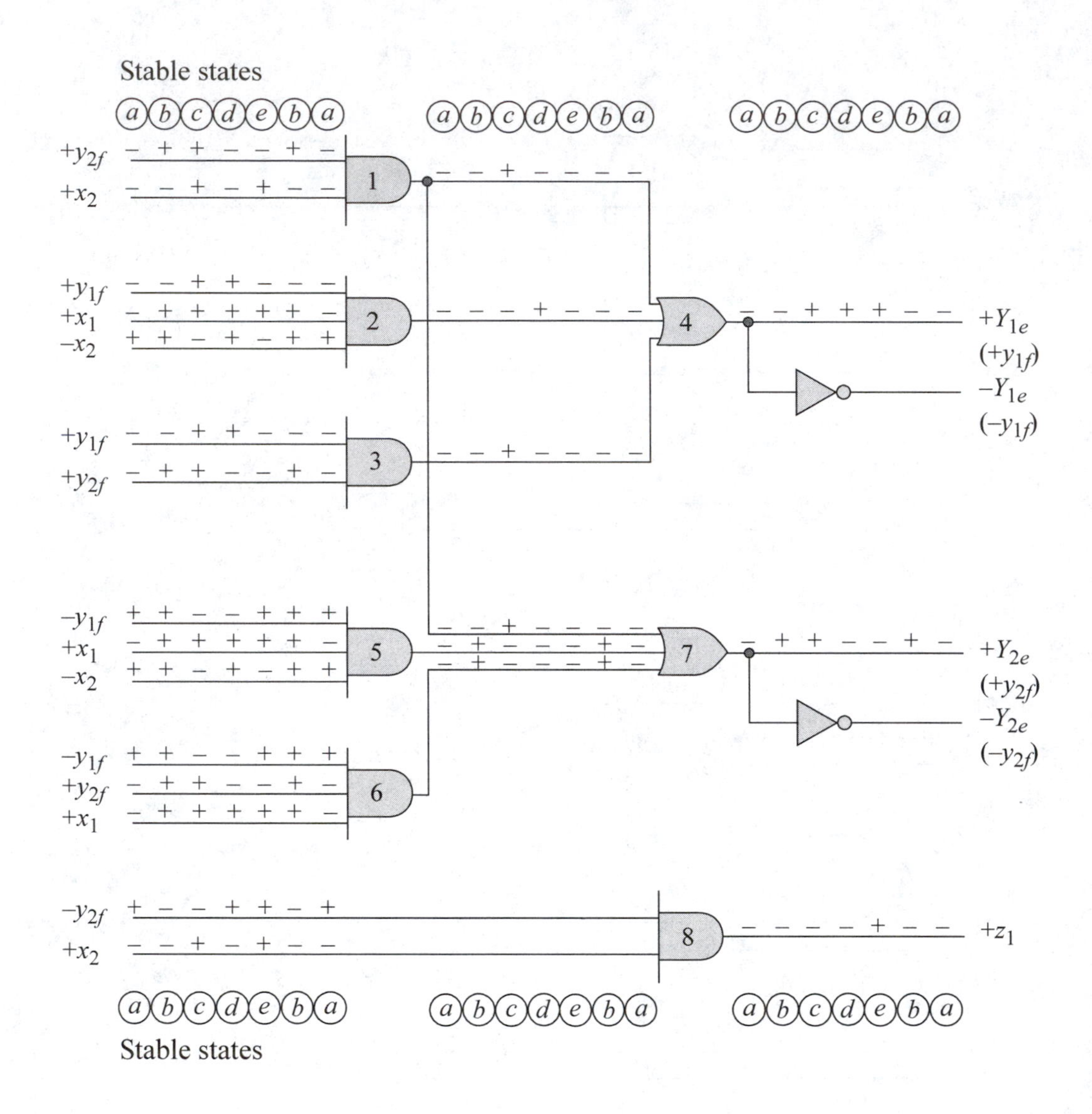

Figure 6.113 Logic diagram expanded from Figure 6.112 (a) used in the analysis of the asynchronous sequential machine of Example 6.21. The diagram shows the inter-gate signals for the input sequence shown in Figure 6.103.

By applying the input sequence specified by the timing diagram of Figure 6.103, the machine will proceed through the following stable states:

	ⓐ	ⓑ	ⓒ	ⓓ	ⓔ	ⓑ	ⓐ
$x_1x_2z_1 =$	000,	100,	110,	100,	111,	100,	000

The above input sequence indicates that output z_1 will be asserted in state ⓔ and coincident with the assertion of the second x_2 pulse. From state ⓐ ($x_1x_2z_1 = 000$), the machine proceeds to state ⓑ when x_1 changes from 0 to 1. Because of the fundamental-mode restriction, the input vector will not change again until the machine has stabilized. The effect of the transition from state ⓐ to state ⓑ is reflected by the +/– symbols in column ⓑ of Figure 6.113 for both the input variables and the logic gate outputs.

As the machine continues to sequence through the prescribed stable states, the output levels of the logic gates can be observed for each change of input vector. As indicated on the output of gate 8, z_1 is asserted in state ⓔ for the duration of x_2. The assertion of z_1 is the result of the deassertion of excitation variable Y_{2e} — and thus the deassertion of y_{2f} after a delay of Δt — and the assertion of x_2. When the input vector again changes, a state transition occurs from state ⓔ to state ⓑ, where both Y_{2e} and x_2 change value. That is, $Y_{2e}y_{2f}x_2 = 110$, providing two logic low voltage levels to the inputs of gate 8, thus deasserting output z_1, as required.

The asynchronous sequential machine of Example 6.21 was synthesized from a set of machine specifications — which included a verbal description and a timing diagram — and then analyzed by applying the same input sequence, as indicated by the timing diagram, to verify the functional operation of the machine.

6.3.2 Mealy Machine with Two Inputs and One Output Using a Programmable Logic Array (PLA)

Example 6.22 Consider an asynchronous sequential machine with two inputs x_1 and x_2 and one output z_1. The machine specifications state that z_1 is to be asserted coincident with the first assertion of x_2 if and only if x_1 is already asserted. Output z_1 is to be deasserted simultaneously with the deassertion of x_1. The machine will be implemented with a programmable logic array. A representative timing diagram depicting one possible sequence is shown in Figure 6.114.

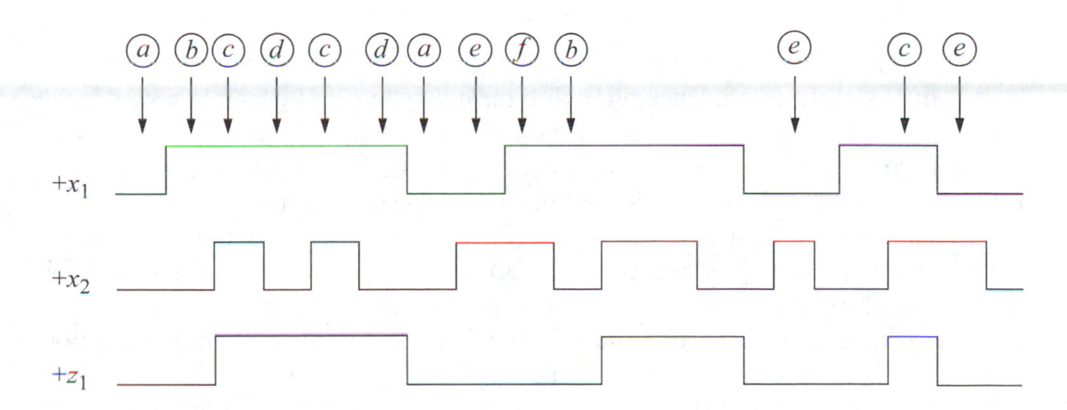

Figure 6.114 A representative timing diagram for the asynchronous sequential machine of Example 6.22.

Primitive flow table Assume that the machine is reset initially so that $x_1x_2z_1 = 000$. To construct the primitive flow table, begin at the leftmost section of the timing diagram and assign a stable state to every change of the input vector. The reset state is designated as ⓐ as shown in the primitive flow table of Figure 6.115. Following the operational characteristics of the machine specifications, proceed in a left-to-right manner in the timing diagram. The assertion of x_1 designates a new stable state labeled ⓑ. Stable state ⓑ is entered from state ⓐ, through transient state *b*, when the input vector changes from $x_1x_2 = 00$ to 10. Output z_1 remains inactive for states ⓐ and ⓑ. At this point, any assertion of x_2 will generate an output on z_1.

x_1x_2 00	01	11	10	z_1
(a)	e	–	b	0
a	–	c	(b)	0
–	e	(c)	d	1
a	–	c	(d)	1
a	(e)	f	–	0
–	e	(f)	b	0

Figure 6.115 Primitive flow table for the asynchronous sequential machine of Example 6.22.

The next change occurs when x_2 is asserted in state ⓑ. This new input vector causes the machine to sequence through transient state *c* and generate a new stable state labeled ⓒ. The conditions for the assertion of output z_1 have now been satisfied and z_1 becomes active coincident with the positive transition of x_2. Row ⓒ of the primitive flow table shows the result of this transition and the assertion of z_1.

In stable state ⓒ, if x_2 is deasserted, the operation proceeds through transient state *d* to stable state ⓓ, where z_1 remains asserted — output z_1 is deasserted only when x_1 is deasserted. When x_2 is asserted in state ⓓ, the machine returns to state ⓒ, where z_1 remains asserted. Output z_1 remains active through multiple transitions of ⓒ $\rightarrow d \rightarrow$ ⓓ $\rightarrow c \rightarrow$ ⓒ $\rightarrow \ldots$. Output z_1 becomes deasserted only when x_1 becomes deasserted, as shown in both the timing diagram and row ⓓ of the primitive flow table. In state ⓓ, if x_1 is deasserted, then the machine returns to state ⓐ, where z_1 is deasserted.

Now, in state ⓐ, if x_2 changes from 0 to 1, then the machine sequences to state ⓔ. In state ⓔ, the input vector is $x_1x_2 = 01$; therefore, z_1 remains inactive, since the stipulation that the assertion of x_1 must precede the assertion of x_2 has not been met. Thus, the third x_2 pulse will not produce an output. This situation is depicted by the sequence ⓔ → *f* → ⓕ in the primitive flow table, where $x_1x_2z_1 = 110$. From state ⓕ, the machine proceeds to state ⓑ when x_2 changes from 1 to 0. In state ⓑ, z_1 remains inactive until x_2 is again asserted — provided that x_1 does not become inactive before the assertion of x_2.

In state ⓑ, the next assertion of x_2 will generate an output on z_1, since x_1 is already asserted. This input change causes the machine to return to state ⓒ, where $x_1x_2z_1 = 111$. If x_1 now becomes inactive, then the negative transition causes the deassertion z_1. In a similar manner, the remainder of the timing diagram is examined for stable states and the primitive flow table is filled in with appropriate entries. The primitive flow table is complete when all possible valid input sequences have been examined and the necessary states inserted. The complete primitive flow table is shown in Figure 6.115.

Equivalent states The following pairs of stable states are potentially equivalent, since each pair is characterized by the same input vector: (ⓒ, ⓕ) and (ⓑ, ⓓ). Stable states ⓒ and ⓕ, however, are not equivalent, because the output values are different. The same is true for state pair ⓑ and ⓓ. Therefore, the primitive flow table contains no equivalent states and Figure 6.115 represents the reduced primitive flow table.

Merger diagram The number of rows in the primitive flow table may be reduced by the merging process. Rows ⓐ and ⓑ can merge into one row, since there is no conflict in state names in any column of the two rows. This merging capability is indicated by the line connecting vertices ⓐ and ⓑ in the merger diagram of Figure 6.116. Row ⓐ can also merge with rows ⓔ and ⓕ.

Row ⓑ cannot merge with any subsequent row due to conflicting state names in rows ⓒ and ⓓ, column $x_1x_2 = 10$ and rows ⓔ and ⓕ, column $x_1x_2 = 11$. Continued application of the merging rules for the remaining rows results in the merger diagram of Figure 6.116.

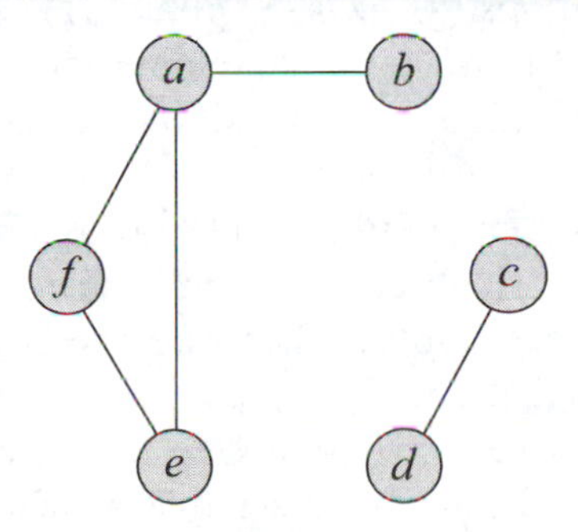

Figure 6.116 Merger diagram obtained from the reduced primitive flow table of Figure 6.115 for the asynchronous sequential machine of Example 6.22.

Merged flow table The merger diagram produces the following partition of maximal compatible sets:

{ⓐ, ⓔ, ⓕ}, {ⓑ}, {ⓒ, ⓓ}

The strongly connected set of {ⓐ, ⓔ, ⓕ} forms one merged row. The two remaining sets constitute two additional rows, providing the merged flow table of Figure 6.117. Since there are only three sets in the partition, the fourth row of the merged flow table contains unspecified entries.

	x_1x_2	00	01	11	10
1	ⓐ, ⓔ, ⓕ	ⓐ	ⓔ	ⓕ	b
2	ⓑ	a	–	c	ⓑ
3	ⓒ, ⓓ	a	e	ⓒ	ⓓ
4		–	–	–	–

Figure 6.117 Merged flow table for Example 6.22.

After enumerating the rows of the merged flow table, the state transitions can be identified with reference to individual rows. The transitions are graphically represented by means of the transition diagram of Figure 6.118 (a). The transition from row 1 to row 2 is the result of a change of input vector from $x_1x_2 = 00$ or 11 to $x_1x_2 = 10$, as indicated by the two transitions ⓐ $\rightarrow b \rightarrow$ ⓑ and ⓕ $\rightarrow b \rightarrow$ ⓑ. Two transitions emanate from row 2: row 2 to rows 1 and 3 delineated by ⓑ $\rightarrow a \rightarrow$ ⓐ and ⓑ $\rightarrow c \rightarrow$ ⓒ, respectively. Finally, two transitions occur from row 3 to row 1: ⓓ $\rightarrow a \rightarrow$ ⓐ and ⓒ $\rightarrow e \rightarrow$ ⓔ.

Since the transition diagram of Figure 6.118 (a) contains a triangle formed by vertices 1 , 2 , and 3 , no state variable code assignment can be realized in which all state transitions are free of race conditions. Therefore, the direct transition from row 3 to row 1 will be rerouted such that, row 3 proceeds first to row 4 and then to row 1. With the triangular polygon removed, a two-bit Gray code can be assigned as shown in Figure 6.118 (b) in which all transitions are free of races. Figure 6.119 shows the merged flow table with the transitions ⓓ $\rightarrow$ ⓐ and ⓒ $\rightarrow$ ⓔ routed through intermediate states in row 4.

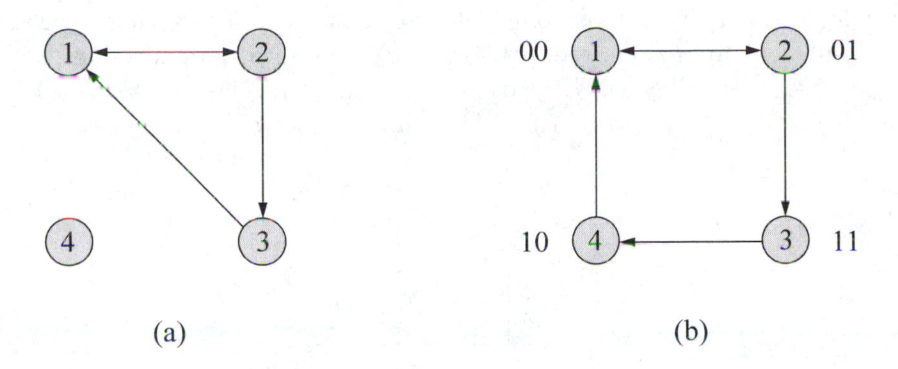

Figure 6.118 Transition diagram for the merged flow table of Figure 6.117: (a) transition diagram with races; and (b) transition diagram without races.

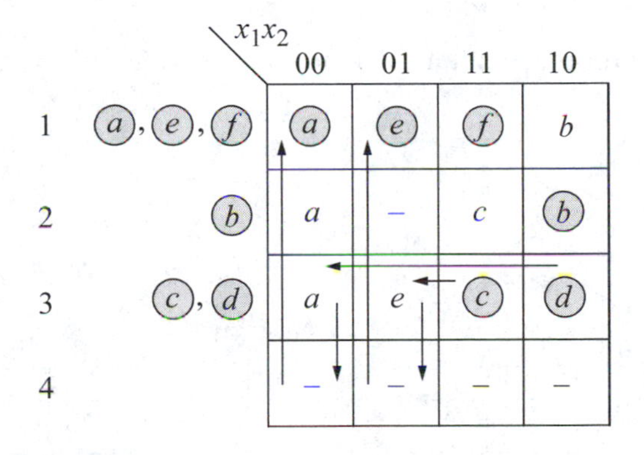

Figure 6.119 Merged flow table for Example 6.22 showing additional intermediate states for transitions ⓓ → ⓐ and ⓒ → ⓔ.

Excitation maps and equations The combined excitation map of Figure 6.120 is derived directly from the merged flow table of Figure 6.119 using the state variable code assignment of Figure 6.118 (b). The stable states of each row are assigned excitation values that are identical to the feedback values of the corresponding row The individual excitation maps are shown in Figure 6.121. The excitation variables in the individual maps are obtained from the combined map. For example, in state ⓑ of the combined map, $Y_{1e}Y_{2e} = 01$. Thus, in location $y_{1f}y_{2f}x_1x_2 = 0110$ of the individual maps, $Y_{1e} = 0$ and $Y_{2e} = 1$. Since both excitation maps contain overlapping prime implicants (groupings of 1s), there will be no static-1 hazards in the operation of the machine. The excitation equations are listed in Equation 6.28 in both a sum-of-products form and a product-of-sums form. As derived from the excitation maps, the sum-of-products form requires three logic gates for each excitation variable, whereas, the product-of-sums form requires only two gates. The sum-of-products form, however, can be reduced to two gates by using the distributive law such that, $Y_{1e} = y_{2f}$ $(y_{1f} + x_2)$ and $Y_{2e} = x_1$ $(y_{2f} + x_2')$, which is identical to the product-of-sums form.

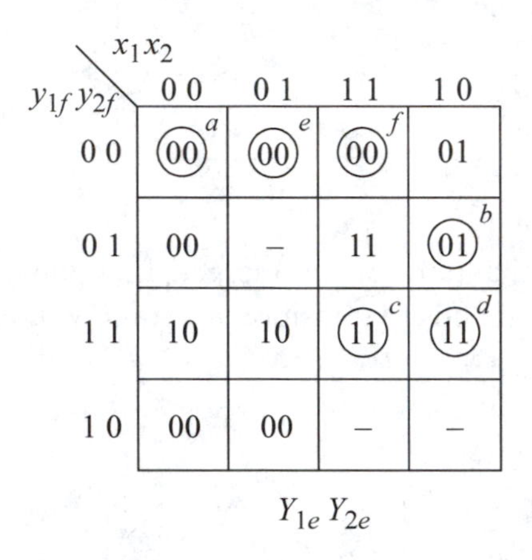

Figure 6.120 Combined excitation map for the merged flow table of Figure 6.119.

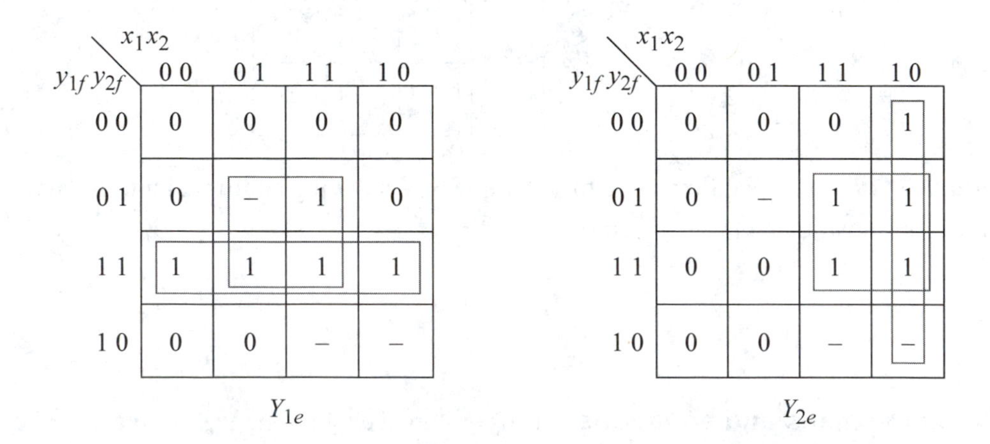

Figure 6.121 Individual excitation maps for Example 6.22.

$$Y_{1e} = y_{1f}y_{2f} + y_{2f}x_2$$
$$Y_{2e} = y_{2f}x_1 + x_1x_2'$$

$$Y_{1e} = (y_{2f})(y_{1f} + x_2)$$
$$Y_{2e} = (x_1)(y_{2f} + x_2') \tag{6.28}$$

Output map and equation The output map is derived from the merged flow table of Figure 6.119 and the reduced primitive flow table of Figure 6.115. The merged flow table specifies the location of the stable states in the output map and the reduced primitive flow table specifies the output values of the stable states. The output map is shown in Figure 6.122. The criterion for the output map of Figure 6.122 (a) is the least amount of hardware for z_1, whereas, Figure 6.122 (b) represents the fastest possible output change for z_1. Figure 6.122 (c) specifies entries that provide the slowest possible change for z_1. Note that the slowest change also represents the least amount of logic gates.

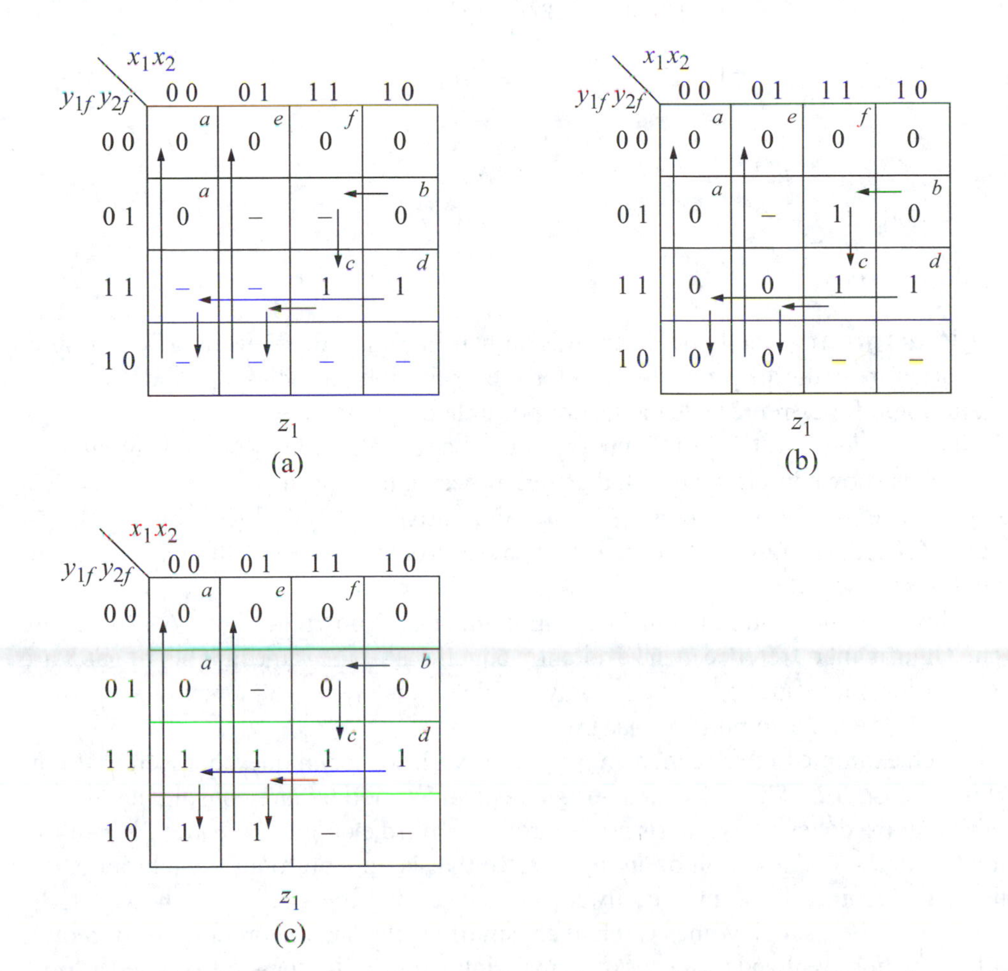

Figure 6.122 Output maps for Example 6.22 representing three criteria for z_1: (a) least amount of hardware; (b) fastest possible change; and (c) slowest possible change.

Each of the three output map configurations is free of hazards. The output equation is shown in Equation 6.29 in six formats: sum-of-products and product-of-sums notation for fewest logic gates, fastest operation, and slowest operation.

$$z_1 = y_{1f}$$
$$z_1 = y_{1f}x_1 + y_{2f}x_1x_2$$
$$z_1 = y_{1f}$$

$$z_1 = y_{1f}$$
$$z_1 = (x_1)(y_{2f})(y_{1f} + x_2)$$
$$z_1 = y_{1f} \tag{6.29}$$

Logic diagram The logic diagram is shown in Figure 6.123 using a PLA implementation in a sum-of-products form for the excitation variables Y_{1e} and Y_{2e}. The output logic for z_1 represents the fastest possible output changes.

The merged flow table of Figure 6.117 can be reconfigured by interchanging rows 2 and 4, as shown in Figure 6.124, then renumbering the rows in ascending numerical order. The transition diagram for the alternative merged flow table is shown Figure 6.125. The only difference is that row 3 proceeds to row 1 through row 2, instead of row 4.

The combined and individual excitation maps are shown in Figure 6.126. There is no significant difference in the excitation equations — the same number of gates are utilized for both equations: Excitation variable $Y_{1e} = y_{1f}x_1 + x_1x_2'$; $Y_{2e} = y_{1f}y_{2f} + y_{1f}x_2$. There are also no static hazards.

Each example in this section employs the synthesis techniques that were presented in Section 6.2. The individual design steps in Section 6.2 are consolidated in this section into a cohesive synthesis procedure. As a third example, consider the requirement to delay the assertion of an input pulse that is applied to an asynchronous sequential machine. This can be easily accomplished by using a delay line; however, the time delay of most delay lines is relatively small. Delays in the order of milliseconds or larger can be realized by means of a monostable multivibrator — also called a "one-shot" or "single-shot." The pulse duration is variable and can be adjusted for a specific duration by utilizing external components.

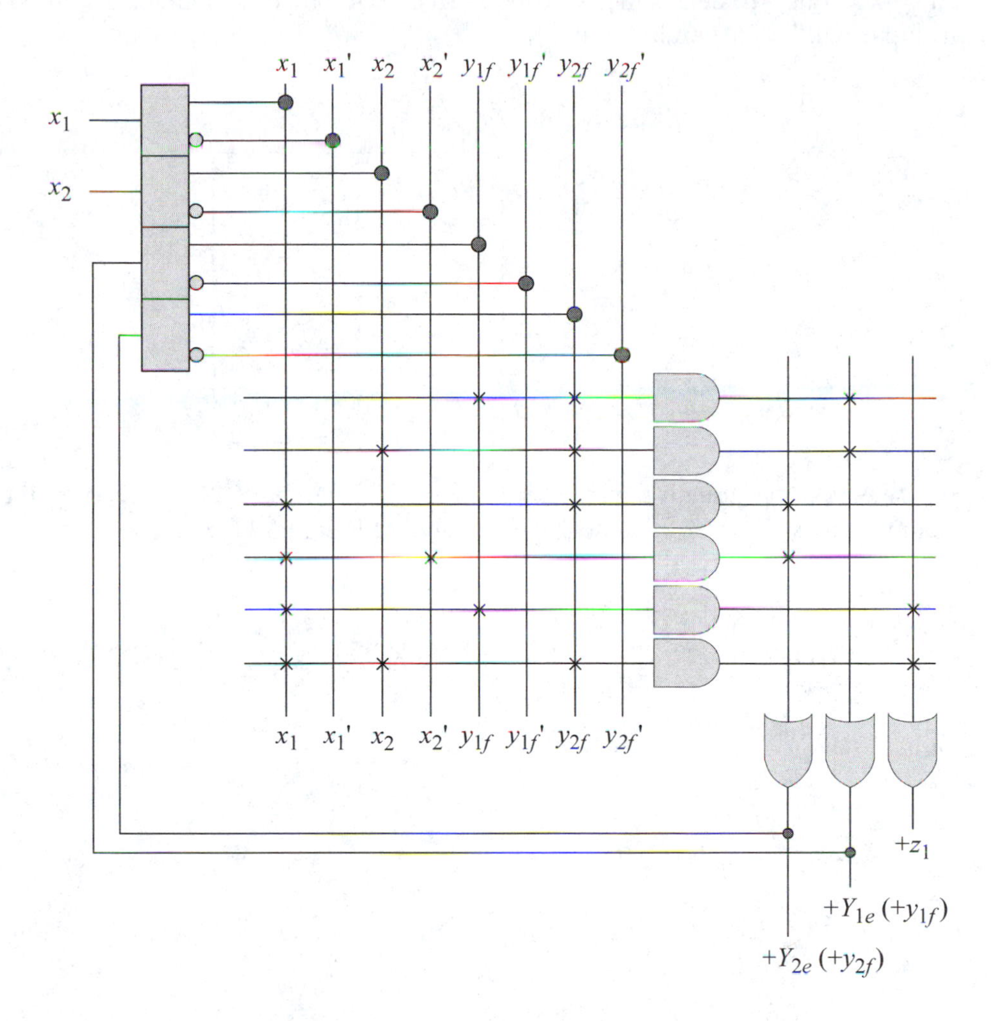

Figure 6.123 Logic diagram for Example 6.22 using a PLA implementation, which represents the fastest possible output changes. Excitation variable $Y_{1e} = y_{1f}y_{2f} + y_{2f}x_2$; $Y_{2e} = y_{2f}x_1 + x_1x_2'$; $z_1 = y_{1f}x_1 + y_{2f}x_1x_2$.

6.3.3 Moore Machine with One Input and One Output

Example 6.23 An asynchronous sequential machine has one input x_1 and one output z_1. The assertion of x_1 causes a monostable multivibrator to be triggered, generating a pulse of a predetermined duration. Input x_1 will remain asserted at least until the deassertion of the multivibrator pulse, at which time z_1 becomes asserted. Output z_1 is deasserted only when x_1 is deasserted. A timing diagram depicting this series of events is shown in Figure 6.127, where the multivibrator pulse is designated as x_2. A network of this type delays the assertion of an output for a time duration specified by the pulse width of the multivibrator output.

	x_1x_2	00	01	11	10
1	(a), (e), (f)	(a)	(e)	(f)	b
2		–	–	–	–
3	(c), (d)	a	e	(c)	(d)
4	(b)	a	–	c	(b)

Figure 6.124 Alternative merged flow table for Example 6.22, derived by interchanging rows 2 and 4 of the merged flow table of Figure 6.117.

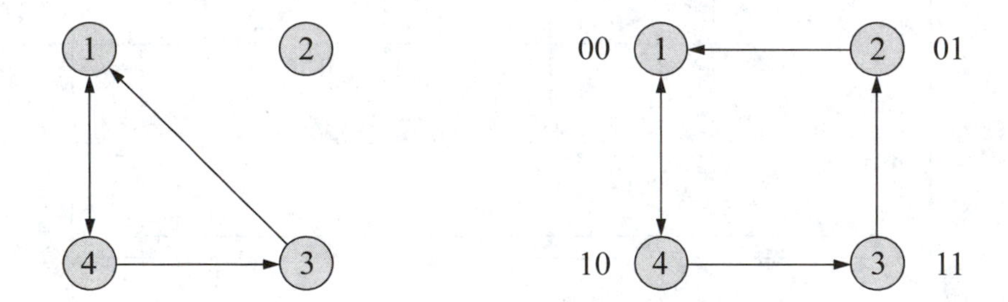

Figure 6.125 Transition diagram for the alternative merged flow table of Figure 6.124.

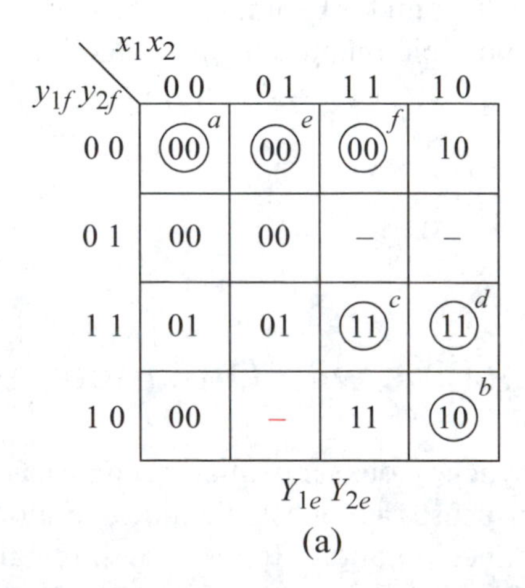

Figure 6.126 Excitation maps for the alternative merged flow table of Figure 6.124: (a) combined map for Y_{1e} and Y_{2e}; (b) individual map for Y_{1e}; and (c) individual map for Y_{2e}.

$y_{1f}\,y_{2f}$ \ x_1x_2	0 0	0 1	1 1	1 0
0 0	0	0	0	1
0 1	0	0	–	–
1 1	0	0	1	1
1 0	0	–	1	1

Y_{1e}

(b)

$y_{1f}\,y_{2f}$ \ x_1x_2	0 0	0 1	1 1	1 0
0 0	0	0	0	0
0 1	0	0	–	–
1 1	1	1	1	1
1 0	0	–	1	0

Y_{2e}

(c)

Figure 6.126 (Continued)

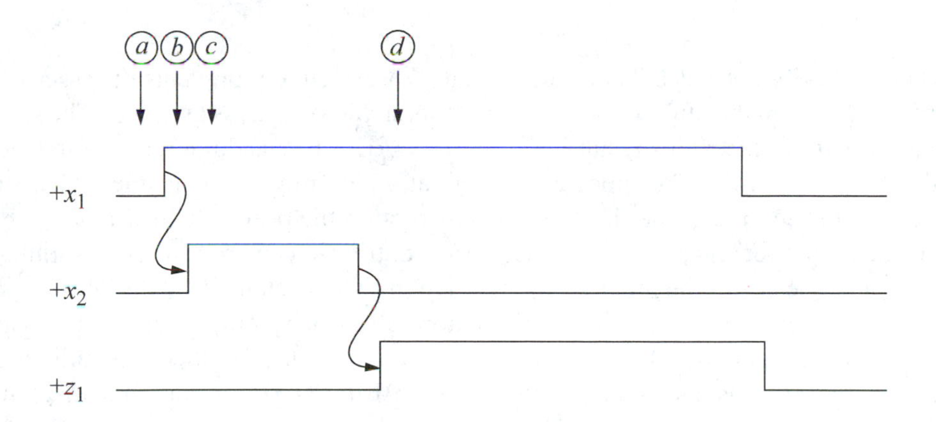

Figure 6.127 Timing diagram for the asynchronous sequential machine of Example 6.23.

Primitive flow table Although the circuit has only one external input, the output of the multivibrator can be considered as an input (designated as x_2) while generating the primitive flow table. Let the initial state of the machine be $x_1x_2z_1 = 000$. The initial condition is designated as stable state ⓐ in the primitive flow table of Figure 6.128. The first pass in the construction of the primitive flow table can now begin. The assertion of input x_1 triggers the multivibrator, which generates a pulse of the requisite duration. The state transition sequence to generate the multivibrator output is (a) $\rightarrow b \rightarrow$ (b) $\rightarrow c \rightarrow$ (c).

x_1x_2	00	01	11	10	z_1
	(a)	–	–	b	0
	–	–	c	(b)	0
	–	–	(c)	d	0
	a	–	–	(d)	1

Figure 6.128 Primitive flow table for the asynchronous sequential machine of Example 6.23.

The multivibrator pulse that occurs in state (c) effectively prevents the assertion of output z_1. When the multivibrator returns to its stable state, z_1 is asserted. The state transition that produces this result is (c) $\rightarrow d \rightarrow$ (d). The machine remains in state (d), with $z_1 = 1$, until x_1 becomes deasserted, at which time the machine returns to state (a). This completes the first pass in constructing the primitive flow table.

To begin the second pass, insert unspecified entries in each row where a simultaneous change occurs to the input vector relative to a stable state. In state (b), x_1 will not become inactive until after the multivibrator pulse has become inactive. Thus, an unspecified entry is inserted in column $x_1x_2 = 00$ in row (b). By the same rationale, an unspecified entry is inserted in column $x_1x_2 = 01$ in row (c), because the machine specifications state explicitly that a change of input vector from $x_1x_2 = 11$ to 01 cannot occur; that is, x_1 will remain active until the multivibrator output is deasserted. An unspecified entry is also inserted in column $x_1x_2 = 11$ of row (d), since the multivibrator will not be triggered again until the next assertion of input x_1. The primitive flow table is now complete.

Equivalent states Only states (b) and (d) are potentially equivalent, since both states are entered from the same input vector — $x_1x_2 = 10$. However, the two states have different outputs; in state (b), $z_1 = 0$; in state (d), z1 = 1. Thus, the primitive flow table contains no equivalent states and Figure 6.128 represents the reduced primitive flow table.

Merger diagram Rows ⓐ and ⓑ of Figure 6.128 can merge into a single row with no conflict. The columns in rows ⓐ and ⓑ contain either a state name coupled with an unspecified entry, two unspecified entries, or two identical state names. Row ⓐ cannot merge with rows ⓒ or ⓓ, due to conflicting state names in column $x_1x_2 = 10$.

Similarly, row ⓑ cannot merge with any succeeding row, due to conflicting state names in column $x_1x_2 = 10$. Finally, rows ⓒ and ⓓ can merge into a single row, because each column in the two rows contains either a state name and an unspecified entry, or two unspecified entries, or two identical state names. The merger diagram is shown in Figure 6.129. The merger diagram yields the following partition of maximal compatible sets: {ⓐ, ⓑ}, {ⓒ, ⓓ}.

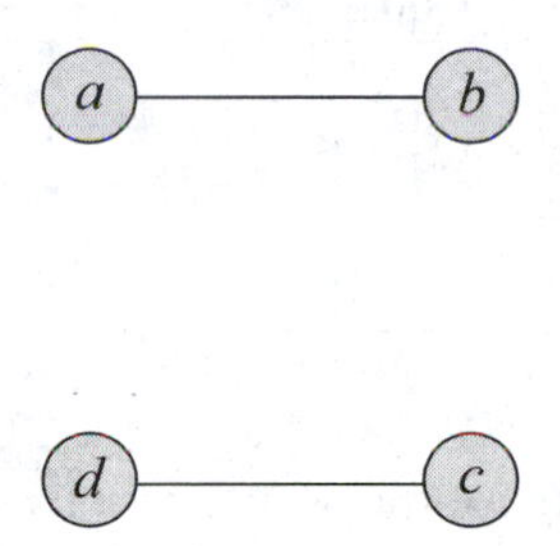

Figure 6.129 Merger diagram obtained from the reduced primitive flow table of Figure 6.128.

Merged flow table The merged flow table contains two rows, one for each maximal compatible set. Each row of the reduced primitive flow table of Figure 6.128 is transferred to the merged flow table in accordance with the partition assignment. The merged flow table is shown in Figure 6.130. Since there are only two rows in the merged flow table, the transition diagram can be omitted. Adjacency requirements for the two rows is easily established to provide race-free operation for all transitions.

	x_1x_2	00	01	11	10
1	ⓐ,ⓑ	ⓐ	–	c	ⓑ
2	ⓒ,ⓓ	a	–	ⓒ	ⓓ

Figure 6.130 Merged flow table obtained from the merger diagram of Figure 6.129 and the reduced primitive flow table of Figure 6.128 for the asynchronous sequential machine of Example 6.23.

Excitation map and equations The excitation map is constructed from the merged flow table, which indicates the location of the stable states, and from the reduced primitive flow table, which illustrates the state transition sequences. The excitation map for Y_{1e} is shown in Figure 6.131. For an excitation map with only two rows, the determination of the excitation values for the transient states is comparatively easy — each transient state must contain an excitation value of 0 or 1, whichever is the complement of the feedback value for the corresponding row. Otherwise, a stable state would be introduced. The excitation equation is shown in Equation 6.30 in both a sum-of-products form and a product-of-sums form.

y_{1f} \ x_1x_2	00	01	11	10
0	(0) *a*	–	1	(0) *b*
1	0	–	(1) *c*	(1) *d*

Y_{1e}

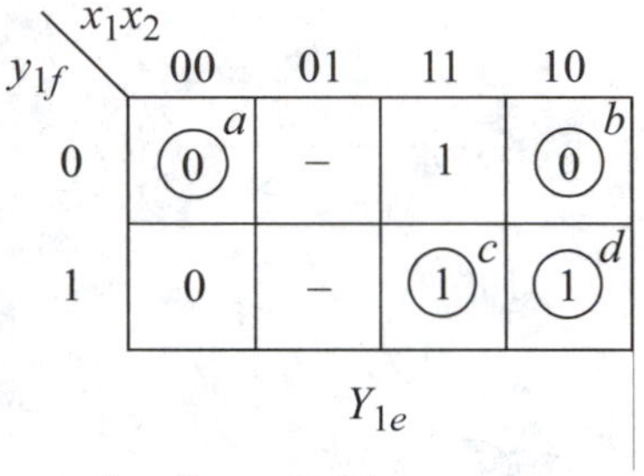

Figure 6.131 Excitation map for the asynchronous sequential machine of Example 6.23.

$$Y_{1e} = x_2 + y_{1f}x_1$$

$$Y_{1e} = (x_1)(y_{1f} + x_2) \tag{6.30}$$

Output map and equation The output map of Figure 6.132 is derived from the merged flow table, which designates the location of the stable states and from the reduced primitive flow table, which defines the output value for each stable state. The output entry of $z_1 = 0$ in location $y_{1f}x_1x_2 = 011$ is necessary to prevent an erroneous output for the transition ⓑ → ⓒ. The entry of $z_1 = 1$ in location $y_{1f}x_1x_2 = 100$ specifies a slow output change and also provides an output equation with the least amount of logic. The equation for z_1 is shown in Equation 6.31, which specifies z_1 in both a sum-of-products form and a product-of-sums form.

Logic diagram The logic diagram for the pulse delay circuit is shown in Figure 6.133 using the sum-of-products notation for the excitation and output variables. The storage element is drawn in the conventional form for an *SR* latch. The machine is self-resetting; that is, when input x_1 becomes deasserted, the latch is reset and z_1 is deasserted. The machine can also be used as a pulse shortener.

y_{1f} \ x_1x_2	00	01	11	10
0	0 *a*	–	0	0 *b*
1	1	–	0 *c*	1 *d*

z_1

Figure 6.132 Output map for the asynchronous sequential machine of Example 6.23.

$$z_1 = y_{1f} x_2'$$
$$z_1 = (y_{1f})\,(x_2') \tag{6.31}$$

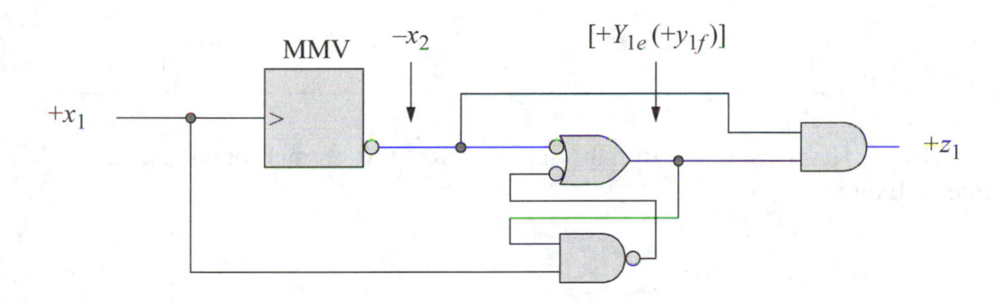

Figure 6.133 Logic diagram for the asynchronous sequential machine of Example 6.23.

Only by exposure to several worked examples and persistence in solving problems can the synthesis of asynchronous sequential machines be effectively learned. Proficiency of this relatively complex process is achieved only through dedicated perseverance. Therefore, additional examples will be presented detailing the synthesis procedure in a step-by-step process.

6.3.4 Mealy Machine with Two Inputs and Two Outputs

Example 6.24 An asynchronous sequential machine is controlled by applying a series of pulses on two inputs x_1 and x_2. The machine generates two outputs z_1 and z_2 in accordance with a prescribed input sequence. Output z_1 will be asserted coincident with every second x_2 pulse, but only if x_1 is asserted for the duration of the pair of x_2 pulses. Output z_2 is asserted for every second x_2 pulse, but only if x_1 is deasserted for the duration of the pair of x_2 pulses. Input x_1 will not change state while x_2 is asserted. The outputs will never be active simultaneously, because the outputs are asserted for different values of input x_1. A representative timing diagram is shown in Figure 6.134 illustrating the two sequences that assert outputs z_1 and z_2.

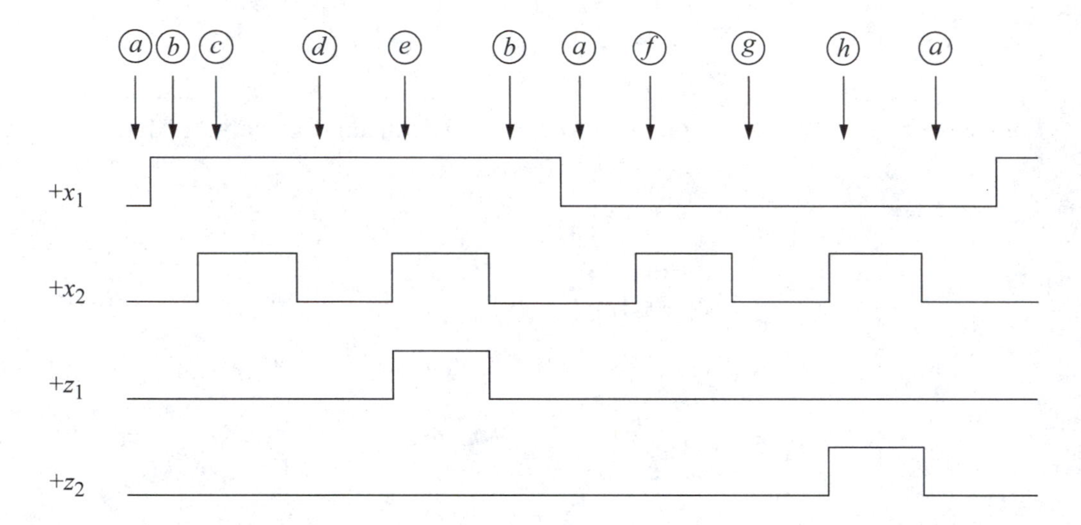

Figure 6.134 Representative timing diagram for the asynchronous sequential machine of Example 6.24.

Primitive flow table From an initial stable state of $x_1x_2z_1z_2$ = 0000, the timing diagram is examined in a left-to-right sequence assigning a unique stable state to every new combination of the input vector for a specific sequence. The first pass in constructing the primitive flow table will proceed through the two valid input sequences that generate the two outputs. The initial condition is designated as stable state ⓐ as shown in the primitive flow table of Figure 6.135.

The assertion of input x_1 causes the machine to proceed to state ⓑ and then to state ⓒ with the assertion of x_2. Since this is the first occurrence of x_2, output z_1 remains inactive. Input x_2 then changes from 1 to 0 and the machine proceeds to state ⓓ. Although states ⓑ and ⓓ are both defined by the same input vector x_1x_2 = 10, the two states are not equivalent — state ⓑ is the precursor for the first occurrence of x_2, whereas, state ⓓ anticipates the second x_2 pulse.

When x_2 next becomes active, the machine proceeds to state ⓔ and asserts output z_1. The input sequence to assert z_1 is represented in the primitive flow table as

ⓐ → b → ⓑ → c → ⓒ → d → ⓓ → e → ⓔ

Thus far, in all stable states except ⓔ, $z_1z_2 = 00$. The next deassertion of x_2 returns the machine to state ⓑ where the process repeats until x_1 is deasserted.

x_1x_2 00	01	11	10	z_1	z_2
ⓐ	f	–	b	0	0
a	–	c	ⓑ	0	0
–	–	ⓒ	d	0	0
a	–	e	ⓓ	0	0
–	–	ⓔ	b	1	0
g	ⓕ	–	–	0	0
ⓖ	h	–	b	0	0
a	ⓗ	–	–	0	1

Figure 6.135 Primitive flow table for the asynchronous sequential machine of Example 6.24.

Assume that the machine is again initialized to state ⓐ. With input x_1 in an inactive state, the assertion of x_2 causes the machine to proceed to state ⓕ, where $z_2 = 00$. This constitutes the beginning of a valid sequence to assert output z_2. When the input vector changes from $x_1x_2 = 01$ to 00, the machine enters state ⓖ. Since this is the first x_2 pulse of a pair of x_2 pulses, output z_2 remains inactive. The second x_2 pulse produces a sequence from state ⓖ to state ⓗ, where z_2 is asserted. The input sequence to assert z_2 is shown in the primitive flow table as

ⓐ $\rightarrow f \rightarrow$ ⓕ $\rightarrow g \rightarrow$ ⓖ $\rightarrow h \rightarrow$ ⓗ

The first pass in constructing the primitive flow table is now complete.

The remaining entries are established based on the fundamental mode of operation and the machine specifications. An unspecified entry is inserted in the row of each stable state where a change of input vector results in a simultaneous change to the input variables, initiated from the stable state. Row ⓐ is complete, since every possible transition has been realized. In row ⓑ, if x_1 becomes inactive before the assertion of x_2, then the machine returns to the initial condition in state ⓐ. Row ⓑ is now complete.

An unspecified entry is inserted in row ⓒ, column $x_1x_2 = 01$, because the verbal description specifies that x_1 will not become inactive while x_2 is asserted. Row ⓒ is now complete. In row ⓓ, if x_1 becomes inactive before the occurrence of the second x_2 pulse, then the machine returns to state ⓐ, completing the entries in row ⓓ. Row ⓔ, column $x_1x_2 = 01$ contains an unspecified entry for the same reason as row ⓒ — x_1 will not be deasserted while x_2 is active.

Similar reasoning applies to the empty columns in rows ⓕ, ⓖ, and ⓗ. In state ⓕ, input x_1 will not change value while x_2 is asserted; therefore, an unspecified entry is inserted in row ⓕ, column $x_1x_2 = 11$. In state ⓖ, if x_1 is asserted, then the machine proceeds to state ⓑ, which is identical to the process that occurs from the initial condition in state ⓐ. In this case, an odd number of x_2 pulses has occurred.

Finally, in state ⓗ, input x_1 will not change value while x_2 is asserted. Therefore, an unspecified entry is entered in row ⓗ, column $x_1x_2 = 11$. The remaining entry in row ⓗ is in column $x_1x_2 = 00$. In state ⓗ, if x_2 becomes inactive, then the machine returns to state ⓐ and the process repeats. The primitive flow table is now complete.

Equivalent states Equivalent states are now identified and then redundant states are eliminated. Two stable states are equivalent if they are entered from the same input vector and have the same outputs and proceed to the same, or equivalent, next state for all combinations of the inputs. Possible equivalence exists between stable state pairs {ⓐ, ⓖ}, {ⓕ, ⓗ}, {ⓒ, ⓔ}, and {ⓑ, ⓓ}.

Stable states ⓐ and ⓖ have the same inputs and the same outputs. For a change of input vector from $x_1x_2 = 00$ to 10, both stable states proceed to state ⓑ. For a change of input vector from $x_1x_2 = 00$ to 01, however, the sequences for the two stable states are ⓐ $\rightarrow f \rightarrow$ ⓕ and ⓖ $\rightarrow h \rightarrow$ ⓗ. Therefore, states ⓐ and ⓖ are equivalent if and only if states ⓕ and ⓗ are equivalent. Stable states ⓕ and ⓗ, however, are not equivalent, because the outputs for the two states are different; state ⓕ has outputs of $z_1z_2 = 00$, whereas, state ⓗ has outputs of $z_1z_2 = 01$. Therefore, stable states ⓐ and ⓖ are not equivalent.

Stable state pair ⓕ and ⓗ have already been shown to be nonequivalent, due to different outputs. Likewise, states ⓒ and ⓔ are not equivalent, because of different outputs.

The last pair of possible equivalent states is {ⓑ, ⓓ}. Both stable states are entered from the same input vector and both have the same output values — $z_1z_2 = 00$. Both proceed to the same next state (stable state ⓐ) when the inputs change from $x_1x_2 = 10$ to 00. However, when the inputs change from $x_1x_2 = 10$ to 11, the state transitions are ⓑ $\rightarrow c \rightarrow$ ⓒ and ⓓ $\rightarrow e \rightarrow$ ⓔ. It is has already been shown that states ⓒ and ⓔ are not equivalent; thus, states ⓑ and ⓓ are not equivalent. Therefore, since there are no equivalent states, the primitive flow table of Figure 6.135 is also the reduced primitive flow table.

Merger diagram The next step in the synthesis procedure is to attempt to reduce the number of rows in the reduced primitive flow table — and thus the number of feedback variables — by the merging process. Row ⓐ is tested for merging capability with each succeeding row ⓑ through ⓗ. Then row ⓑ is tested for merging capability with each succeeding row ⓒ through ⓗ. This process is continued until all rows have been checked for possible merging.

Rows ⓐ and ⓑ can merge into a single row, since there is no conflict in state names in any column of the two rows. Each column contains either identical state names (ⓐ, *a*) and (*b* , ⓑ) or a state name and an unspecified entry (*f*, –) and (–, *c*). The merging of rows ⓐ and ⓑ is indicated by the line connecting vertices ⓐ and ⓑ in the merger diagram of Figure 6.136. The only other row with which ⓐ can merge is row ⓔ, indicated by the line connecting vertices ⓐ and ⓔ. All remaining rows have conflicting state names in at least one column.

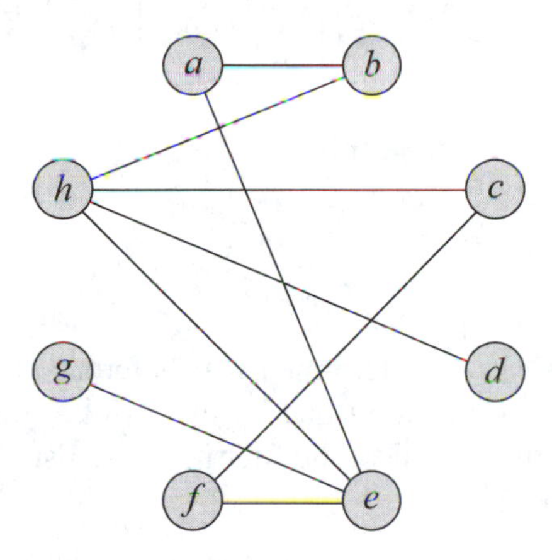

Figure 6.136 Merger diagram obtained from the reduced primitive flow table of Figure 6.135.

Row ⓑ can merge only with row ⓗ in the succeeding rows. Row ⓒ cannot merge with rows ⓓ or ⓔ due to conflicting state names in column $x_1x_2 = 11$ nor with row ⓖ due to conflicting state names in column $x_1x_2 = 10$. Rows ⓒ, ⓕ and ⓒ, ⓗ, however, can merge. In a similar manner, the remaining rows are checked for merging capabilities. The completed merger diagram is shown in Figure 6.136 and yields the following partition of maximal compatible sets:

{ⓐ, ⓑ}, {ⓒ, ⓕ}, {ⓓ, ⓗ}, {ⓔ, ⓖ}

Merged flow table Each row of the merged flow table of Figure 6.137 is obtained by transferring the individual rows from the reduced primitive flow table to the merged flow table in accordance with the partition assignment. Row ⓐ of the reduced primitive flow table transfers as ⓐ, *f*, –, *b*. Row ⓑ transfers as *a*, –, *c*, ⓑ and is superimposed on row ⓐ to yield the following merged row: ⓐ, *f*, *c*, ⓑ. In a similar manner, the remaining rows are transferred from the reduced primitive flow table to the merged flow table. The completed merged flow table consists of four rows as shown in Figure 6.137.

		x_1x_2 00	01	11	10
1	ⓐ, ⓑ	ⓐ	*f*	*c*	ⓑ
2	ⓒ, ⓕ	*g*	ⓕ	ⓒ	*d*
3	ⓓ, ⓗ	*a*	ⓗ	*e*	ⓓ
4	ⓔ, ⓖ	ⓖ	*h*	ⓔ	*b*

Figure 6.137 Merged flow table obtained from the merger diagram of Figure 6.136.

The state transitions can be determined with reference to individual rows in the merged flow table. This is accomplished by means of a transition diagram, which clearly identifies all race conditions. The four rows are listed as shown in the transition diagram of Figure 6.138.

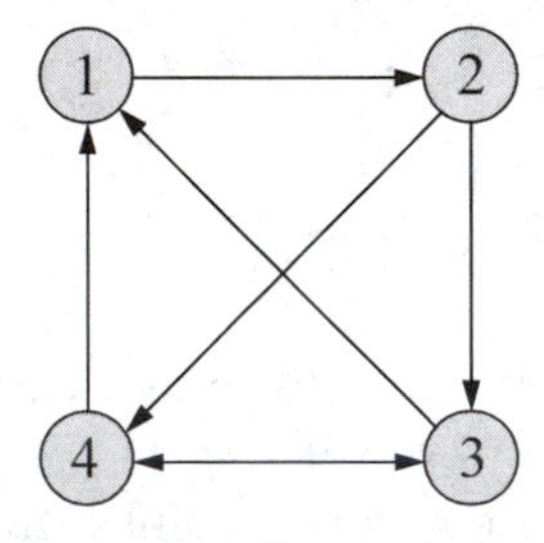

Figure 6.138 Transition diagram obtained from the merged flow table of Figure 6.137 for the asynchronous sequential machine of Example 6.24.

Row 1 proceeds to row 2 by the following two transitions: ⓐ → *f* → ⓕ and ⓑ → *c* → ⓒ. This is illustrated by the directed line connecting vertices 1 and 2 . A transition occurs from row 2 to rows 3 and 4 by state transitions ⓒ → *d* → ⓓ and ⓕ → *g* → ⓖ, respectively, as indicated by the lines connecting row 2 to rows 3 and 4.

A transition is realized between rows 3 and 1 from state ⓗ when the input vector changes from $x_1x_2 = 01$ to 00 or from state ⓓ when the input vector changes from $x_1x_2 = 10$ to 00. Also, a transition can emanate from row 3 to row 4 from stable states ⓗ or ⓓ when the inputs change from $x_1x_2 = 01$ to 11 or from $x_1x_2 = 10$ to 11, respectively. Finally, a transition can occur from row 4 to row 3 by the sequence ⓖ → h → ⓗ and from row 4 to row 1 by the sequence ⓔ → b → ⓑ.

Since it is impossible to assign two-tuple state variable codes so that all four rows are adjacent, it is necessary to introduce an additional state variable. Adding a state variable produces an augmented merged flow table containing eight rows as shown in Figure 6.139. The top four rows are unchanged from the original table. The bottom four rows contain unspecified entries in every column. The lower rows will be used to insert intermediate transient states, where applicable. An intermediate state will be inserted in the cycle whenever a state transition sequence causes two or more excitation variables to change state simultaneously. The values for the feedback variables will be obtained from a transition diagram.

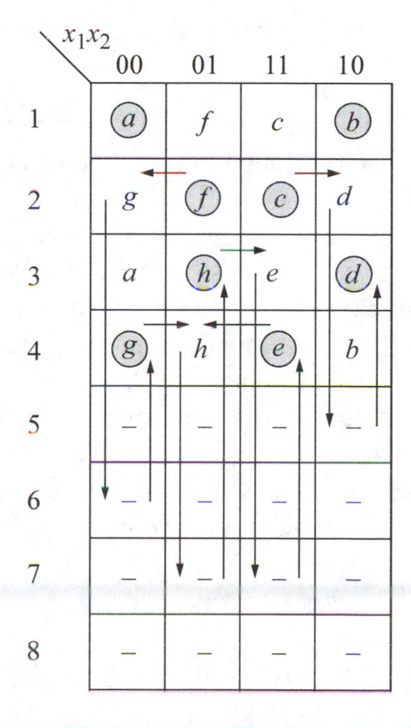

	x_1x_2 00	01	11	10
1	ⓐ	f	c	ⓑ
2	g	ⓕ	ⓒ	d
3	a	ⓗ	e	ⓓ
4	ⓖ	h	ⓔ	b
5	–	–	–	–
6	–	–	–	–
7	–	–	–	–
8	–	–	–	–

Figure 6.139 Augmented merged flow table for the asynchronous sequential machine of Example 6.24.

The transition diagram for eight rows is shown in Figure 6.140. This is a cube in which the top surface is assigned the three-tuple Gray code 000, 001, 011, 010 for row vertices 1, 2, 6, 4, respectively. The bottom surface is assigned Gray code values of 100, 101, 111, 110 for row vertices 3, 5, 8, 7, respectively.

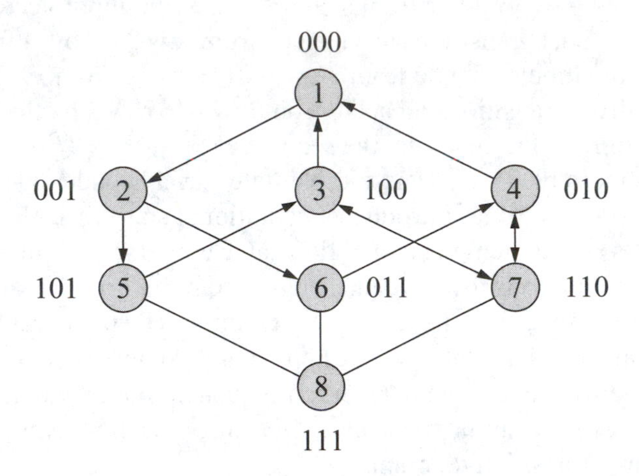

Figure 6.140 Transition diagram for the augmented merged flow table of Figure 6.139.

Any state transition that results in two or more state variables changing values simultaneously can be rerouted through unspecified entries in rows 5 through 8 such that, each individual transition in the cycle takes place between adjacent state variable codes. For example, the transition Ⓕ → *g* → Ⓖ from row 2 to row 4 can be directed first to row 6 and then to row 4. Each step of the cycle proceeds through contiguous vertices that involve a change of only one state variable. The transitions in the merged flow table of Figure 6.137 in which race conditions exist are shown redirected in the augmented merged flow table of Figure 6.139.

The transition from state Ⓕ (row 2) to state Ⓖ (row 4) passes through an intermediate state in row 6, as shown by the arrows. The transition from state Ⓒ (row 2) to state Ⓓ (row 3) causes two state variables to change (from 001 to 100), as shown in the transition diagram. To avoid this race condition, the transition from row 2 is directed first to row 5 and then to row 3. This modified state transition is

2 (001) → 5 (101) → 3 (100); Ⓒ → *d* → Ⓓ

In a similar manner, the remaining transitions are modified in which race conditions are present. The arrows in Figure 6.139 illustrate the redirected transitions, in which each contiguous vertex (row) in the cycle involves a change of only one state variable. The modified row transitions are listed in Figure 6.141 together with their corresponding state transitions.

The transition diagram of Figure 6.140 can also be represented as a Karnaugh map for three feedback variables y_{1f}, y_{2f}, and y_{3f}, as shown in Figure 6.142. The transitions shown by the arrows correspond to the transitions listed in Figure 6.141.

2 (001) → 6 (011) → 4 (010);	ⓕ → g → ⓖ
2 (001) → 5 (101) → 3 (100);	ⓒ → d → ⓓ
3 (100) → 7 (110) → 4 (010);	ⓗ → e → ⓔ
3 (100) → 7 (110) → 4 (010);	ⓓ → e → ⓔ
4 (010) → 7 (110) → 3 (100);	ⓖ → h → ⓗ
4 (010) → 7 (110) → 3 (100);	ⓔ → h → ⓗ

Figure 6.141 Modified row transitions to avoid races in the corresponding state transitions of the augmented merged flow table of Figure 6.139 for Example 6.24.

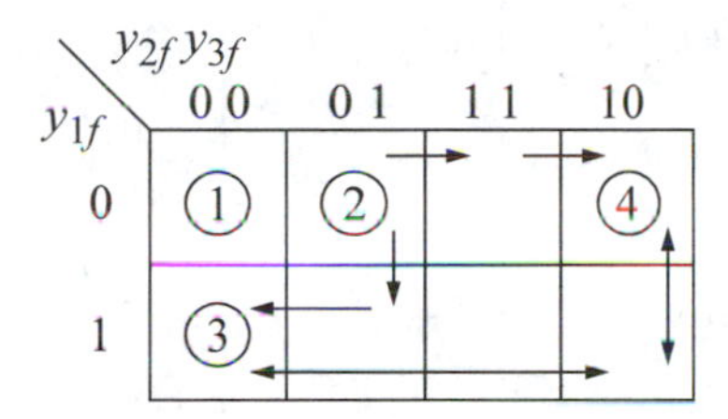

Figure 6.142 Alternative representation for the transition diagram of Figure 6.140.

Excitation maps and equations The next step in the synthesis procedure is to derive the excitation maps. The combined excitation map is derived directly from the augmented merged flow table of Figure 6.139 and is shown in Figure 6.143. The combined map shows the location of all stable states and the requisite unstable states as specified by the augmented merged flow table and the transition diagram. The codes for the feedback variables are listed in sequence using the Gray code format together with the corresponding row number.

Stable states ⓐ and ⓑ are in row 1 as depicted in the augmented merged flow table. Row 1 contains state code values of 000 as specified by the transition diagram; that is, $y_{1f}y_{2f}y_{3f} = 000$. Therefore, the values of the excitation variables assigned to states ⓐ and ⓑ are $Y_{1e}Y_{2e}Y_{3e} = 000$, since the machine is stable when the feedback values are equal to the excitation values. Stable states ⓕ and ⓒ are in row 2 which has assigned feedback values of $y_{1f}y_{2f}y_{3f} = 001$; therefore, the excitation values for states ⓕ and ⓒ are $Y_{1e}Y_{2e}Y_{3e} = 001$.

Stable states ⓗ and ⓓ are in row 3 of the augmented merged flow table. Row 3 is assigned a state variable code of $y_{1f}y_{2f}y_{3f} = 100$ in the transition diagram. Therefore, row 3 is the bottom row of the combined excitation map and states ⓗ and ⓓ are placed in the appropriate columns of row 3 and are assigned excitation values of $Y_{1e}Y_{2e}Y_{3e} = 100$ indicating a stable condition. Finally, stable states ⓖ and ⓔ are placed in row 4 in the combined excitation map and are assigned excitation values of $Y_{1e}Y_{2e}Y_{3e} = 010$ to coincide with the feedback values of $y_{1f}y_{2f}y_{3f} = 010$.

	$y_{1f}y_{2f}y_{3f}$ \ x_1x_2	00	01	11	10
1	000	(000) a	001	001	(000) b
2	001	011	(001) f	(001) c	101
6	011	010	–	–	–
4	010	(010) g	110	(010) e	000
7	110	–	100	010	–
8	111	–	–	–	–
5	101	–	–	–	100
3	100	000	(100) h	110	(100) d

$Y_{1e}Y_{2e}Y_{3e}$

Figure 6.143 Combined excitation map constructed from the augmented merged flow table of Figure 6.139 and the transition diagram of Figure 6.140.

The intermediate transient states are assigned excitation values that cause the machine to proceed through each state in a cycle such that, each transient state is adjacent to both the preceding state and the next state. The following state transition sequences require only one intermediate state:

$$ⓐ \rightarrow ⓕ, ⓑ \rightarrow ⓒ, ⓗ \rightarrow ⓐ, ⓓ \rightarrow ⓐ, ⓖ \rightarrow ⓑ, ⓔ \rightarrow ⓑ$$

For example, the sequence ⓐ $\rightarrow$ ⓕ is established by the path ⓐ ($Y_{1e}Y_{2e}Y_{3e}$ = 000) $\rightarrow Y_{1e}Y_{2e}Y_{3e} = 001 \rightarrow y_{1f}y_{2f}y_{3f} = 001 \rightarrow$ ⓕ ($Y_{1e}Y_{2e}Y_{3e} = 001$).

The remaining state transitions require two intermediate states. For example, the transition ⓕ $\rightarrow$ ⓖ is initiated by a change of input vector from $x_1x_2 = 01$ to 00. This results in a change of value for excitation variables Y_{2e} and Y_{3e} as follows: ⓕ ($Y_{1e}Y_{2e}Y_{3e} = 001$) $\rightarrow$ g ($Y_{1e}Y_{2e}Y_{3e} = 010$). The race condition resulting from the simultaneous change of excitation variables Y_{2e} and Y_{3e} is resolved by inserting an additional transient state in the cycle. The expanded transition that provides race-free operation is

ⓕ $(Y_{1e}Y_{2e}Y_{3e} = 001)$
$\rightarrow Y_{1e}Y_{2e}Y_{3e} = 011$
$\rightarrow y_{1f}y_{2f}y_{3f} = 011$
$\rightarrow Y_{1e}Y_{2e}Y_{3e} = 010$
$\rightarrow y_{1f}y_{2f}y_{3f} = 010$
ⓖ $(Y_{1e}Y_{2e}Y_{3e} = 010)$

Consider the transition ⓒ $(Y_{1e}Y_{2e}Y_{3e} = 001) \rightarrow$ ⓓ $(Y_{1e}Y_{2e}Y_{3e} = 100)$ when the input vector changes from $x_1x_2 = 11$ to 10. A race condition occurs for this sequence between excitation variables Y_{1e} and Y_{3e}. If the transition sequence is not modified, then a critical race will result. The critical race occurs if Y_{3e} resets before Y_{1e} sets. The destination will be state ⓑ, not the required destination of state ⓓ. By inserting an additional intermediate state, the race condition is eliminated. Thus, the sequence from state ⓒ to state ⓓ proceeds as follows:

ⓒ $(Y_{1e}Y_{2e}Y_{3e} = 001)$
$\rightarrow Y_{1e}Y_{2e}Y_{3e} = 101$
$\rightarrow y_{1f}y_{2f}y_{3f} = 101$
$\rightarrow Y_{1e}Y_{2e}Y_{3e} = 100$
$\rightarrow y_{1f}y_{2f}y_{3f} = 100$
ⓓ $(Y_{1e}Y_{2e}Y_{3e} = 100)$

The remaining transient states and unspecified entries are shown in Figure 6.143.

The individual excitation maps for Y_{1e}, Y_{2e}, and Y_{3e} are shown in Figure 6.144. The maps are obtained in the usual manner by transferring the values of Y_{1e} in the four columns $x_1x_2 = 00$, 01, 11, and 10 to the corresponding columns in the individual excitation map for Y_{1e}. The same procedure is used in constructing the excitation maps for Y_{2e} and Y_{3e}. The equations for the excitation variables are shown in Equation 6.32 in a hazard-free sum-of-products form and product-of-sums form.

Output maps and equations The output maps for z_1 and z_2 are shown in Figure 6.145. The maps are derived from the reduced primitive flow table of Figure 6.135 and the augmented merged flow table of Figure 6.139 in conjunction with the combined excitation map of Figure 6.143. The combined excitation map shows the relocated positions of the stable states relative to the row numbers and the feedback variable code assignments.

As indicated in the reduced primitive flow table, only stable states ⓔ and ⓗ have asserted outputs; in state ⓔ, $z_1 = 1$; in state ⓗ, $z_2 = 1$. The output values assigned to the transient states provide the fewest number of logic gates and insure that there will be no erroneous outputs. The output equations are shown in Equation 6.33 in both a sum-of-products form and a product-of-sums form.

$y_{1f}y_{2f}y_{3f}$ \ x_1x_2	00	01	11	10
000	0	0	0	0
001	0	0	0	1
011	0	–	–	–
010	0	1	0	0
110	–	1	0	–
111	–	–	–	–
101	–	–	–	1
100	0	1	1	1

Y_{1e}

$y_{1f}y_{2f}y_{3f}$ \ x_1x_2	00	01	11	10
000	0	0	0	0
001	1	0	0	0
011	1	–	–	–
010	1	1	1	0
110	–	0	1	–
111	–	–	–	–
101	–	–	–	0
100	0	0	1	0

Y_{2e}

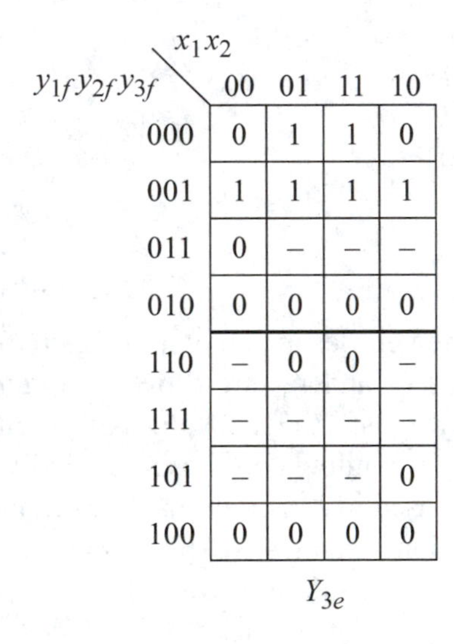

$y_{1f}y_{2f}y_{3f}$ \ x_1x_2	00	01	11	10
000	0	1	1	0
001	1	1	1	1
011	0	–	–	–
010	0	0	0	0
110	–	0	0	–
111	–	–	–	–
101	–	–	–	0
100	0	0	0	0

Y_{3e}

Figure 6.144 Individual excitation maps obtained from Figure 6.143 for the asynchronous sequential machine of Example 6.24.

$$Y_{1e} = y_{3f}x_1x_2' + y_{1f}y_{2f}'x_1 + y_{1f}y_{2f}'x_2 + y_{2f}x_1'x_2 + y_{1f}y_{3f}$$

$$Y_{2e} = y_{2f}x_1'x_2' + y_{3f}x_1'x_2' + y_{1f}x_1x_2 + y_{1f}'y_{2f}x_1' + y_{1f}'y_{2f}x_2 + y_{2f}x_1x_2$$

$$Y_{3e} = y_{1f}'y_{2f}'y_{3f} + y_{1f}'y_{2f}'x_2$$

$$
\begin{aligned}
Y_{1e} = {} & (x_1 + x_2)(y_{2f}' + x_1')(y_{1f} + y_{2f} + x_2')(y_{1f} + y_{2f} + y_{3f}) \\
& (y_{1f} + x_1' + x_2')(y_{2f}' + x_2)(y_{1f} + y_{2f} + x_1) \\
Y_{2e} = {} & (x_1' + x_2)(y_{1f}' + x_1)(y_{1f} + y_{2f} + y_{3f})(y_{1f} + y_{2f} + x_2')(y_{1f}' + x_2) \\
& (y_{1f} + y_{2f} + x_1') \\
Y_{3e} = {} & (y_{1f}')(y_{2f}')(y_{3f} + x_2)
\end{aligned}
\tag{6.32}
$$

z_1

$y_{1f}y_{2f}y_{3f}$ \ x_1x_2	00	01	11	10
000	0^a	0	0	0^b
001	0	0^f	0^c	0
011	0	–	–	–
010	0^g	0	1^e	0
110	–	0	–	–
111	–	–	–	–
101	–	–	–	0
100	0	0^h	–	0^d

z_2

$y_{1f}y_{2f}y_{3f}$ \ x_1x_2	00	01	11	10
000	0^a	0	0	0^b
001	0	0^f	0^c	0
011	0	–	–	–
010	0^g	–	0^e	0
110	–	–	0	–
111	–	–	–	–
101	–	–	–	0
100	0	1^h	0	0^d

Figure 6.145 Output maps for the asynchronous sequential machine of Example 6.24.

$$
\begin{aligned}
z_1 &= y_{2f}x_1x_2 \\
z_2 &= y_{1f}x_1'x_2
\end{aligned}
$$

$$
\begin{aligned}
z_1 &= (x_1)(y_{2f})(x_2) \\
z_2 &= (x_1')(y_{1f})(x_2)
\end{aligned}
\tag{6.33}
$$

Logic diagram The logic diagram is shown in Figure 6.146 using the product-of-sums forms of Equation 6.32 and Equation 6.33.

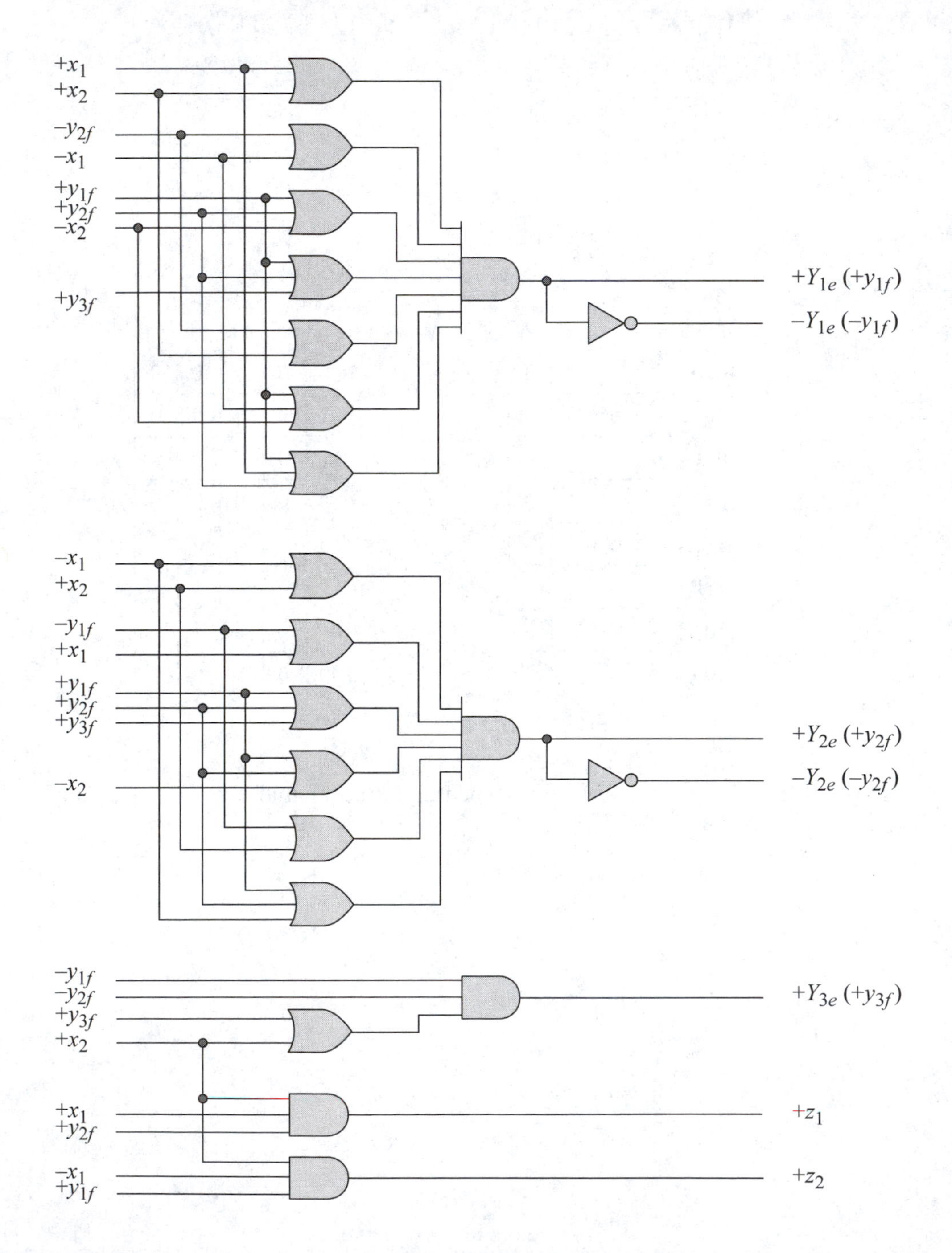

Figure 6.146 Logic diagram for the asynchronous sequential machine of Example 6.24.

6.3.5 Mealy Machine with Three Inputs and One Output

Example 6.25 Consider an asynchronous sequential machine with three inputs x_1, x_2, and x_3 and one output z_1. The machine operates according to the specifications defined below and the representative timing diagram of Figure 6.147.

The input signals are nonoverlapping, disjoint positive pulses of equal duration. Valid input vectors are $x_1x_2x_3$ = 000, 001, 010, and 100. Between each input vector in which a pulse x_i is asserted, a vector of $x_1x_2x_3$ = 000 is inserted, as shown in Figure 6.147.

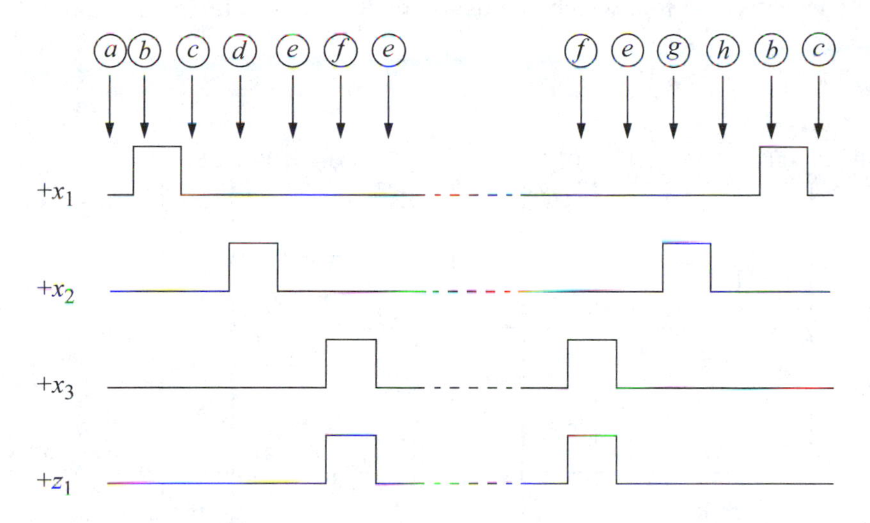

Figure 6.147 Representative timing diagram for the asynchronous sequential machine of Example 6.25.

The initial assertion of z_1 occurs coincident with the first assertion of x_3 if and only if x_3 is preceded by an input sequence of x_1x_2 = 10, 00, 01, 00. Thus, a valid sequence, which includes the assertion of z_1, is $x_1x_2x_3z_1$ = 0000, 1000, 0000, 0100, 0000, 0011, 0000, 0011, ... , 0000, 0011, 0000, 0100, 0000, 1000, 0000, ... , as shown in the timing diagram.

Therefore, once the initial output has been generated, every x_3 pulse will generate a z_1 pulse, provided that neither x_1 nor x_2 has been asserted during the x_3z_1 sequence. The duration of output z_1 is identical to the duration of input x_3. Input x_1 frames the $x_2x_3z_1$ sequence of pulses; input x_2 frames the x_3z_1 sequence of pulses. Simultaneous input changes will not occur.

Primitive flow table By careful analysis of the machine specifications in conjunction with the timing diagram, a primitive flow table can be derived in which there are no equivalent states. This is the preferred method, since the primitive flow table is less cumbersome than if the table included redundant states. It must be stressed, however, that every possible input sequence must be considered to exactly replicate the operational characteristics of the machine. Some redundant states have been purposely injected so that the procedure for identifying equivalent states can be reviewed.

Assume an initial state of $x_1x_2x_3z_1 = 0000$, as indicated by state ⓐ in both the timing diagram and the primitive flow table of Figure 6.148. First, a valid input sequence will be generated to assert z_1, then the inverse sequence, which includes the framing pulses x_2 and x_1. From state ⓐ, the machine proceeds to state ⓑ when x_1 becomes asserted. This is followed by a null vector of $x_1x_2x_3 = 000$, which causes a transition to state ⓒ. Then, when x_2 becomes asserted, the machine proceeds to state ⓓ. This sequence is also followed by a null vector, which results in a transition to state ⓔ.

$x_1x_2x_3$ 000	001	011	010	110	111	101	100	z_1
ⓐ	*i*	–	*j*	–	–	–	*b*	0
c	–	–	–	–	–	–	ⓑ	0
ⓒ	*k*	–	*d*	–	–	–	*b*	0
e	–	–	ⓓ	–	–	–	–	0
ⓔ	*f*	–	*g*	–	–	–	*b*	0
e	ⓕ	–	–	–	–	–	–	1
h	–	–	ⓖ	–	–	–	–	0
ⓗ	*i*	–	*j*	–	–	–	*b*	0
a	ⓘ	–	–	–	–	–	–	0
a	–	–	ⓙ	–	–	–	–	0
a	ⓚ	–	–	–	–	–	–	0

Figure 6.148 Primitive flow table for the asynchronous sequential machine of Example 6.25.

The next change in a valid sequence to assert output z_1 is the assertion of input x_3. This results in a transition from state ⓔ to state ⓕ. Output z_1 is activated coincident with the assertion of x_3 and remains active until x_3 is deasserted. When x_3 is deasserted, a null vector is generated and the machine returns to state ⓔ. The transitions ⓔ $\rightarrow f \rightarrow$ ⓕ $\rightarrow e \rightarrow$ ⓔ $\rightarrow f \rightarrow$ ⓕ $\rightarrow \cdots$ may then repeat with recurring sequences of $x_1x_2x_3z_1$ = 0011, 0000, 0011, 0000. Note that only state ⓕ will generate an output.

The input sequence described next restores the machine to its initial state. This is the inverse sequence to that which generated the assertion of z_1. From stable state ⓕ ($x_1x_2x_3z_1$ = 0011), the machine returns to state ⓔ ($x_1x_2x_3z_1$ = 0000). The next change in the reverse sequence occurs when x_2 becomes active, resulting in a transition to state ⓖ, as shown in the timing diagram. This is followed by a null vector, which in turn is followed by the assertion of x_1.

The assertion of x_1 indicates a new beginning for a possible valid sequence. This is the second assertion of x_1 in the timing diagram and is denoted as state ⓑ, which is the framing pulse for a possible new valid sequence. Assigning a stable state other than ⓑ, in this case, would produce a redundant state, since the new state would ultimately be shown to be equivalent to state ⓑ. When x_1 becomes deasserted, the machine enters state ⓒ and the process repeats. This completes the first pass in constructing the primitive flow table.

Since there are three input variables, the primitive flow table contains eight columns. Adjacency requirements between minterm locations are maintained by using the vertical centerline as a hinge. Thus, the left four columns and the right four columns fold together. Any squares that are then physically adjacent are also logically adjacent. For example, in row ⓐ, columns $x_1x_2x_3$ = 001 and 101 are adjacent, because they differ by only one variable.

There will also be more unspecified entries in each row than in a table with only two input variables, due to the restriction imposed by the fundamental mode of operation. Since simultaneous changes to the input variables are not allowed, the machine cannot proceed from state ⓐ ($x_1x_2x_3$ = 000) to columns $x_1x_2x_3$ = 011, 110, 111, or 101. Similar reasoning applies to the remaining rows of the primitive flow table.

Begin the second pass in stable state ⓐ. If x_3 becomes active in state ⓐ, then this represents an invalid sequence to assert output z_1. Therefore, the machine proceeds to a new stable state designated as ⓘ. The only possible transition from state ⓘ is to state ⓐ. This is due to the restrictions imposed by the fundamental mode of operation and the machine requirements, which specify that two or more inputs will not be asserted simultaneously.

A similar sequence occurs if x_2 becomes active in state ⓐ. This also represents an invalid sequence to assert output z_1 from the initial condition. In this case, the machine proceeds to state ⓙ. The only possible transition from state ⓙ is to state ⓐ, for the same reasons as stated previously for state ⓘ. Row ⓐ is now complete. Row ⓑ is also complete, since no new sequences are possible.

In row ⓒ, the machine proceeds to state ⓓ if the input vector changes from $x_1x_2x_3$ = 000 to 010. Two additional transitions are possible from state ⓒ. If x_3 becomes active in state ⓒ, then the machine will sequence to a new stable state labeled ⓚ, because the input sequence $x_1x_2x_3$ = 000, 100, 000, 001 does not represent a valid sequence to assert z_1; that is, input x_1 cannot be immediately followed by an x_3 pulse.

The remaining transition from state ⓒ is to state ⓑ if the input vector changes from $x_1x_2x_3 = 000$ to 100. This change represents the beginning of a new valid sequence. Row ⓒ is now complete. Row ⓓ is also complete, since the only valid transition is from state ⓓ to state ⓔ.

In row ⓔ, the transitions ⓔ → ⓕ and ⓔ → ⓖ have already been discussed. The remaining transition is from state ⓔ to state ⓑ, which represents the start of a valid sequence. The entries in row ⓔ are now complete. There are no more additional entries in rows ⓕ and ⓖ other than those already described, if the machine is to operate within the prescribed limitations. Therefore, rows ⓕ and ⓖ are complete.

In row ⓗ, the transition to state ⓑ has already been described. There are two remaining transitions from state ⓗ. If the input vector changes from $x_1x_2x_3 = 000$ to 001, then the machine proceeds to state ⓘ, from which only one transition is possible: ⓘ → ⓐ. In state ⓗ, if the input vector changes from $x_1x_2x_3 = 000$ to 010, then the machine proceeds to state ⓙ, from which only one transition is possible: ⓙ → ⓐ. Row ⓗ is now complete. Row ⓘ is also complete as described previously. The remaining two rows ⓙ and ⓚ have only one valid sequence each: ⓙ → ⓐ and ⓚ → ⓐ. The complete primitive flow table is shown in Figure 6.148.

Equivalent states Equivalent states will now be identified and redundant states eliminated. Recall that two or more stable states are equivalent if each of the following rules apply:

1. The stable states have the same input vector.
2. The outputs associated with the stable states have the same value.
3. Each stable state has the same or equivalent next state for all valid input sequences.

The row matching technique will be used to determine equivalence between states. Table 6.2 lists the possible equivalences that exist in columns $x_1x_2x_3 = 000$, 001, and 010. All other columns contain either unspecified entries only or a single stable state.

Table 6.2 Possible equivalences in the primitive flow table of Figure 6.148

Column $x_1x_2x_3 = 000$	Column $x_1x_2x_3 = 001$	Column $x_1x_2x_3 = 010$
ⓐ,ⓒ	ⓕ,ⓘ	ⓓ,ⓖ
ⓐ,ⓔ	ⓕ,ⓚ	ⓓ,ⓙ
ⓐ,ⓗ	ⓘ,ⓚ	ⓖ,ⓙ
ⓒ,ⓔ		
ⓒ,ⓗ		
ⓔ,ⓗ		

All stable states except ⓕ have an output value of $z_1 = 0$. Since stable states ⓐ and ⓒ have identical outputs, therefore, ⓐ ≡ ⓒ if ⓘ ≡ ⓚ • ⓙ ≡ ⓓ. Figure 6.149 portrays the series of equivalence expressions used to determine the equivalence of states ⓐ and ⓒ. Since states ⓘ and ⓕ are not equivalent due to different output values, therefore, this evaluative information reverses through the levels of equivalence expressions in Figure 6.149 to establish the following equivalence relations:

ⓘ ¬≡ ⓕ, ⓐ ¬≡ ⓔ, ⓙ ¬≡ ⓓ, ⓐ ¬≡ ⓒ

ⓐ ≡ ⓒ if ⓘ ≡ ⓚ • ⓙ ≡ ⓓ
ⓘ ≡ ⓚ
ⓙ ≡ ⓓ if ⓐ ≡ ⓔ
ⓐ ≡ ⓔ if ⓘ ≡ ⓕ • ⓙ ≡ ⓖ
ⓘ ¬≡ ⓕ

Figure 6.149 Series of equivalence expressions used to determine the equivalence of stable states ⓐ and ⓒ.

Stable states ⓐ and ⓗ are directly equivalent, because both states proceed to the same next state for the same input change and both states have the same output. The following equivalence relation exists between states ⓒ and ⓔ: ⓒ ≡ ⓔ if ⓚ ≡ ⓕ • ⓓ ≡ ⓖ. However, ⓚ ¬≡ ⓕ, because of different output values. Therefore, ⓒ ¬≡ ⓔ. Also, ⓓ ¬≡ ⓖ, because ⓔ ¬≡ ⓗ.

The following equivalence relation applies to states ⓒ and ⓗ: ⓒ ≡ ⓗ if ⓚ ≡ ⓘ • ⓓ ≡ ⓙ. It has been shown in Figure 6.149 that states ⓚ and ⓘ are directly equivalent, but states ⓓ and ⓙ are not equivalent, because ⓐ ¬≡ ⓔ. Therefore, ⓓ ¬≡ ⓙ and ⓒ ¬≡ ⓗ. Stable states ⓖ and ⓙ are equivalent, because ⓐ ≡ ⓗ. Finally, states ⓔ and ⓗ are tested for equivalence using the following equivalence expression: ⓔ ≡ ⓗ if ⓘ ≡ ⓕ • ⓙ ≡ ⓖ. Stable states ⓘ and ⓕ are not equivalent due to different output values. Therefore, states ⓔ and ⓗ are not equivalent.

Table 6.3 summarizes the equivalence relations between the stable states in columns $x_1x_2x_3$ = 000, 001, and 010 of the primitive flow table of Figure 6.148. Since ⓐ ≡ ⓗ, state ⓗ can be eliminated and every occurrence of unstable state h is replaced by unstable state a. Similarly, states ⓚ and ⓙ can be eliminated and every occurrence of unstable state k or j is replaced by unstable state i or g, respectively. The reduced primitive flow table is shown in Figure 6.150.

Table 6.3 Summary of equivalence relations for the primitive flow table of Figure 6.148

Column $x_1x_2x_3 = 000$	Column $x_1x_2x_3 = 001$	Column $x_1x_2x_3 = 010$
ⓐ $\neg\equiv$ ⓒ	ⓕ $\neg\equiv$ ⓘ	ⓓ $\neg\equiv$ ⓖ
ⓐ $\neg\equiv$ ⓔ	ⓕ $\neg\equiv$ ⓚ	ⓓ $\neg\equiv$ ⓙ
ⓐ $\equiv$ ⓗ	ⓘ $\equiv$ ⓚ	ⓖ $\equiv$ ⓙ
ⓒ $\neg\equiv$ ⓔ		
ⓒ $\neg\equiv$ ⓗ		
ⓔ $\neg\equiv$ ⓗ		

$x_1x_2x_3$ 000	001	011	010	110	111	101	100	z_1
ⓐ	i	–	g	–	–	–	b	0
c	–	–	–	–	–	–	ⓑ	0
ⓒ	i	–	d	–	–	–	b	0
e	–	–	ⓓ	–	–	–	–	0
ⓔ	f	–	g	–	–	–	b	0
e	ⓕ	–	–	–	–	–	–	1
a	–	–	ⓖ	–	–	–	–	0
a	ⓘ	–	–	–	–	–	–	0

Figure 6.150 Reduced primitive flow table for the asynchronous sequential machine of Example 6.25.

Merger diagram Recall that two or more rows in a reduced primitive flow table can merge into a single row if the entries in the same column of each row satisfy one of the following requirements:

1. Identical state names
2. A state name and an unspecified entry
3. Two unspecified entries

The eight rows of the reduced primitive flow table are arranged as shown in the merger diagram of Figure 6.151.

Using the rules for merging listed above, row ⓐ is analyzed for merging capability with all subsequent rows. Then row ⓑ is tested for merging capability with all following rows. This merging process repeats until all rows have been checked for possible mergers. In the merger diagram of Figure 6.151, a line connecting a pair of row vertices indicates that the two rows can merge with no conflict in state names. The merger diagram yields the following partitions of maximal compatible sets:

1. {ⓑ, ⓒ}, {ⓓ, ⓕ}, {ⓔ}, {ⓐ, ⓖ, ⓘ}
2. {ⓑ, ⓒ}, {ⓔ, ⓕ}, {ⓓ}, {ⓐ, ⓖ, ⓘ}

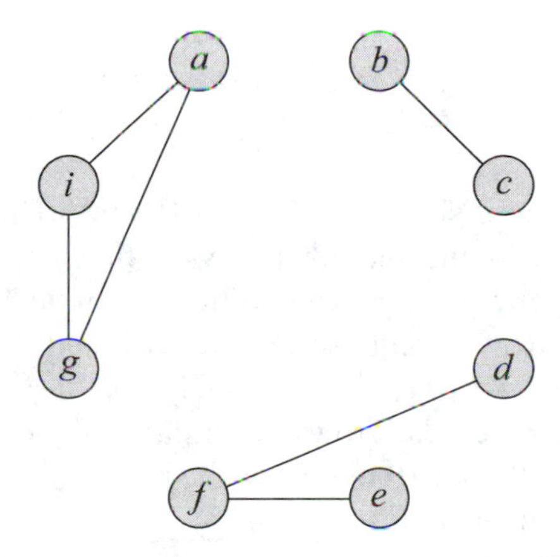

Figure 6.151 Merger diagram obtained from the reduced primitive flow table of Figure 6.150.

Merged flow table The merged flow table is constructed from the reduced primitive flow table using the assignments of partition 1. Rows ⓑ and ⓒ are transferred from the reduced primitive flow table to the first row of the merged flow table as shown in Figure 6.152. The three rows ⓐ, ⓖ, and ⓘ are shown below individually, together with the resulting merged row.

Row ⓐ =	ⓐ	*i*	–	*g*	–	–	–	*b*
Row ⓖ =	*a*	–	–	ⓖ	–	–	–	–
Row ⓘ =	*a*	ⓘ	–	–	–	–	–	–
Resulting merged row =	ⓐ	ⓘ	–	ⓖ	–	–	–	*b*

	$x_1x_2x_3$	000	001	011	010	110	111	101	100
1	ⓑ, ⓒ	ⓒ	i	–	d	–	–	–	ⓑ
2	ⓓ, ⓕ	e	ⓕ	–	ⓓ	–	–	–	–
3	ⓔ	ⓔ	f	–	g	–	–	–	b
4	ⓐ, ⓖ, ⓘ	ⓐ	ⓘ	–	ⓖ	–	–	–	b

Figure 6.152 Merged flow table for the asynchronous sequential machine of Example 6.25 obtained from the reduced primitive flow table of Figure 6.150 using partition 1.

The transition diagram is shown in Figure 6.153 (a). The state transitions can be determined with reference to the individual rows of the merged flow table. The transition diagram clearly portrays all race conditions as indicated by polygons with an odd number of sides. Row 1 sequences to rows 2 and 4 by the state transitions ⓒ $\rightarrow d \rightarrow$ ⓓ and ⓒ $\rightarrow i \rightarrow$ ⓘ, respectively, as depicted by the arrows in Figure 6.153 (a). Row 2 proceeds to row 3 from states ⓕ or ⓓ when the input vector changes from $x_1x_2x_3 = 001$ or 010 to 000. State transitions can occur from row 3 to rows 1, 2, and 4 by the sequences ⓔ $\rightarrow b \rightarrow$ ⓑ, ⓔ $\rightarrow f \rightarrow$ ⓕ, and ⓔ $\rightarrow g \rightarrow$ ⓖ, respectively. A transition occurs from row 4 to row 1 by the sequence ⓐ $\rightarrow b \rightarrow$ ⓑ.

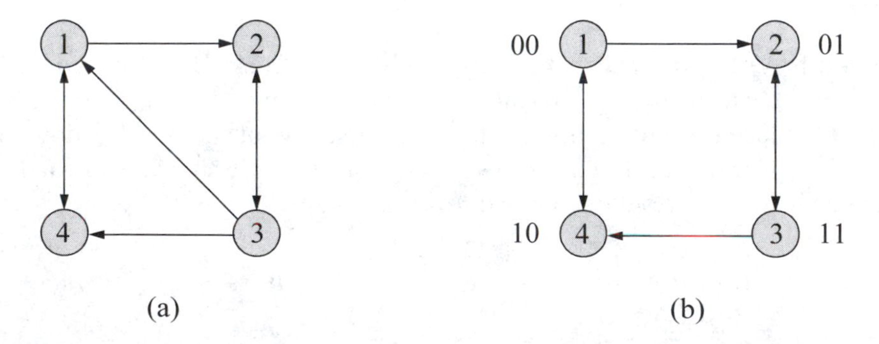

Figure 6.153 Transition diagram obtained from the merged flow table of Figure 6.152: (a) illustrates a race condition caused by triangular vertices 1, 3, and 4; and (b) the race condition is eliminated by modifying the sequence $3 \rightarrow 1$ to $3 \rightarrow 4 \rightarrow 1$.

The triangle depicted by row vertices 1, 3, and 4 in Figure 6.153 (a) indicates a race condition, since all three rows cannot be mutually adjacent. To eliminate the race condition, row 3 is directed first to row 4 and then to row 1, as shown in Figure 6.153 (b).

Excitation maps and equations The combined excitation map for Y_{1e} and Y_{2e} is shown in Figure 6.154. The map is constructed from the merged flow table of Figure 6.152 and the transition diagram of Figure 6.153 (b). Note the redirected transition from state ⓔ to state ⓑ. The sequence ⓔ → ⓑ is achieved by the following transition:

$$
\begin{aligned}
&ⓔ\ (Y_{1e}Y_{2e} = 11)\\
&\quad \rightarrow Y_{1e}Y_{2e} = 10\\
&\quad \rightarrow y_{1f}y_{2f} = 10\\
&\quad \rightarrow Y_{1e}Y_{2e} = 00\\
&\quad \rightarrow y_{1f}y_{2f} = 00\\
&ⓑ\ (Y_{1e}Y_{2e} = 00)
\end{aligned}
$$

The individual excitation maps are shown in Figure 6.155 and the corresponding excitation equations in Equation 6.34 in both a sum-of-products form and a product-of-sums form.

$y_{1f}y_{2f}$ \ $x_1x_2x_3$	000	001	011	010	110	111	101	100
0 0	(00) c	10	–	01	–	–	–	(00) b
0 1	11	(01) f	–	(01) d	–	–	–	–
1 1	(11) e	01	–	10	–	–	–	10
1 0	(10) a	(10) i	–	(10) g	–	–	–	00

Figure 6.154 Combined excitation map constructed from the merged flow table of Figure 6.152 and the transition diagram of Figure 6.153 (b).

Output map and equation The output map for z_1 is shown in Figure 6.156. The map is derived from the reduced primitive flow table and the merged flow table. The merged flow table shows the location of the stable states and the reduced primitive flow table specifies the value of z_1 for the corresponding stable states. The values assigned to z_1 for the intermediate transient states provide glitch-free operation for all state transition sequences.

$y_{1f}y_{2f}$ \ $x_1x_2x_3$	000	001	011	010	110	111	101	100
00	0	1	–	0	–	–	–	0
01	1	0	–	0	–	–	–	–
11	1	0	–	1	–	–	–	1
10	1	1	–	1	–	–	–	0

Y_{1e}

$y_{1f}y_{2f}$ \ $x_1x_2x_3$	000	001	011	010	110	111	101	100
00	0	0	–	1	–	–	–	0
01	1	1	–	1	–	–	–	–
11	1	1	–	0	–	–	–	0
10	0	0	–	0	–	–	–	0

Y_{2e}

Figure 6.155 Individual excitation maps for Y_{1e} and Y_{2e} obtained from the combined excitation map of Figure 6.154.

$$Y_{1e} = y_{2f}x_2'x_3' + y_{1f}y_{2f}'x_1' + y_{1f}x_2 + y_{2f}'x_1'x_3 + y_{1f}x_1'x_3'$$

$$Y_{2e} = y_{2f}\,x_1'x_2' + y_{1f}'x_2 + y_{1f}'y_{2f}$$

$$\begin{aligned} Y_{1e} &= (y_{1f} + x_2')\,(y_{2f} + x_1')\,(y_{2f}' + x_3')\,(y_{1f} + y_{2f} + x_2 + x_3) \\ Y_{2e} &= (y_{2f} + x_2)\,(y_{1f}' + x_2')\,(x_1') \end{aligned} \tag{6.34}$$

$y_{1f}y_{2f}$ \ $x_1x_2x_3$	000	001	011	010	110	111	101	100
00	0 (c)	0	–	0	–	–	–	0 (b)
01	0	1 (f)	–	0 (d)	–	–	–	–
11	0 (e)	–	–	0	–	–	–	0
10	0 (a)	0 (i)	–	0 (g)	–	–	–	0

z_1

Figure 6.156 Output map for z_1 for the asynchronous sequential machine of Example 6.25.

The state transition ⓕ $(z_1 = 1) \rightarrow$ ⓔ $(z_1 = 0)$ passes through transient state $y_{1f}y_{2f}x_1x_2x_3 = 01000$. Therefore, the value for z_1 in the transient state could be left unspecified to provide either a fast or slow circuit response. However, the dominant path to state ⓔ is ⓓ $(z_1 = 0) \rightarrow$ ⓔ $(z_1 = 0)$. Since both stable states contain a value of $z_1 = 0$, therefore, the intermediate state must also contain a value of $z_1 = 0$, otherwise, a momentary false output would be generated. The equation for output z_1 is shown in Equation 6.35.

$$z_1 = y_{2f}x_3 \tag{6.35}$$

Logic diagram The logic diagram is shown in Figure 6.157 using the product-of-sums form of Equation 6.34. The product-of-sums form yields the fewest number of logic gates. The machine is synthesized using the logic primitives of AND, OR, and NOT. The feedback variables y_{1f} and y_{2f} become equal to the excitation variables after a delay of Δt, at which time the machine has stabilized. This is indicated by the signals $\pm y_{1f}$ and $\pm y_{2f}$ in parentheses.

6.3.6 Mealy Machine with Two Inputs and Two Outputs

Example 6.26 As a final example, consider an asynchronous sequential machine which has two inputs x_1 and x_2 and two outputs z_1 and z_2. Output z_1 will be asserted coincident with the assertion of every second x_2 pulse, but only if x_1 is asserted for the duration of the first x_2 pulse and remains active at least until the assertion of the second x_2 pulse. Output z_1 will remain active until x_2 is deasserted.

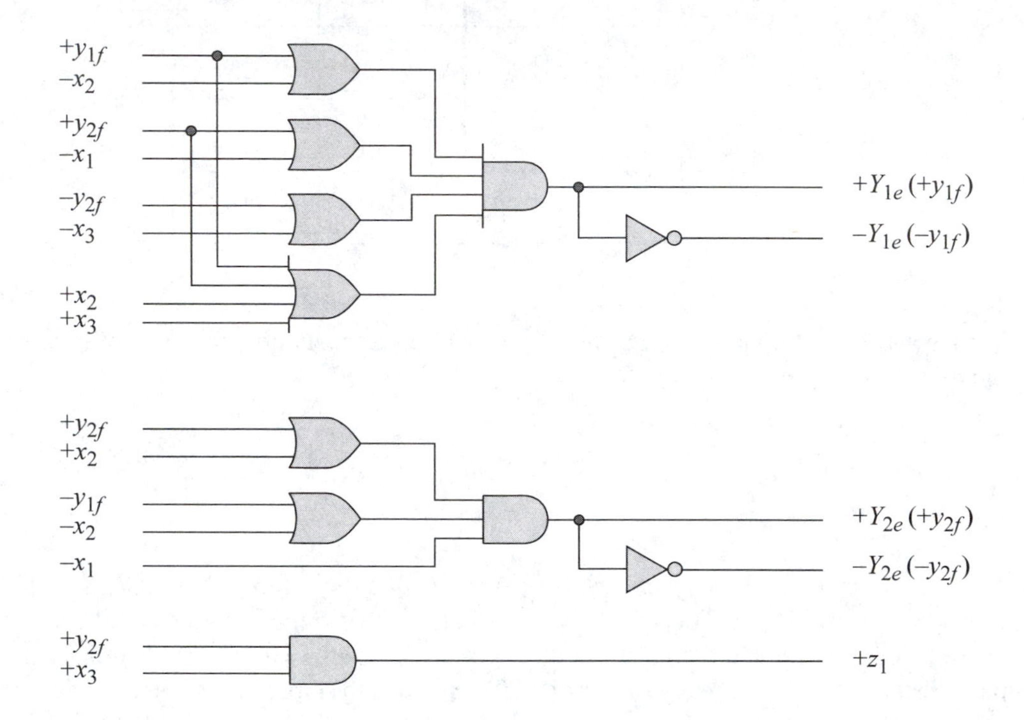

Figure 6.157 Logic diagram for the asynchronous sequential machine of Example 6.25.

Output z_2 will be asserted coincident with the assertion of every second x_2 pulse, but only if x_1 is deasserted for the duration of the first x_2 pulse and remains inactive at least until the assertion of the second x_2 pulse. Output z_2 will remain active until x_2 is deasserted. Input x_1 may change state at any time, but will adhere to the fundamental-mode of operation. The equations for excitation variable Y_{1e} and output z_1 only, will be derived. A representative timing diagram is shown in Figure 6.158.

Primitive flow table Starting from the left side of Figure 6.158 where all inputs and outputs are reset, the timing diagram is analyzed in a left-to-right sequence and all stable states are identified. Every change to the input vector will result in a stable state. The construction of the primitive flow table proceeds as described in previous examples. The stable states depicted in the timing diagram represent the first pass in constructing the primitive flow table. The complete primitive flow table is shown in

Figure 6.159. The stable states in the primitive flow table that do not appear in the timing diagram are established during the second pass.

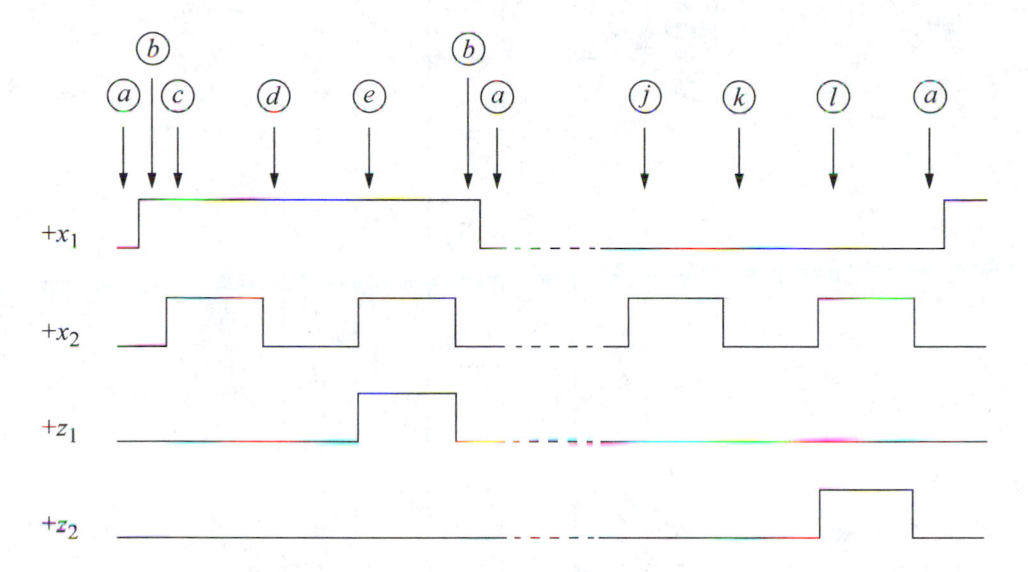

Figure 6.158 Representative timing diagram for the asynchronous sequential machine of Example 6.26.

Equivalent states Using the rules for equivalence as defined in Section 6.2.3, the following stable state pairs are equivalent: (ⓗ ≡ ⓟ), (ⓛ ≡ ⓝ), (ⓔ ≡ ⓖ), and (ⓘ ≡ ⓞ). Therefore, rows ⓟ, ⓝ, ⓖ, and ⓞ are redundant and can be eliminated. All occurrences of unstable states *p*, *n*, *g*, and *o* are replaced by unstable states *h*, *l*, *e*, and *i*, respectively. The reduced primitive flow table is shown in Figure 6.160.

Merger diagram Using the rules for merging as defined in Section 6.2.4, the merger diagram is generated as shown in Figure 6.161. The diagram depicts the following partitions of maximal compatible sets:

Partition 1: {ⓐ, ⓑ}, {ⓒ}, {ⓓ, ⓕ}, {ⓔ}, {ⓗ, ⓘ}, {ⓙ}, {ⓚ, ⓜ}, {ⓛ}

Partition 2: {ⓐ, ⓑ}, {ⓒ}, {ⓓ, ⓕ}, {ⓔ}, {ⓗ, ⓘ}, {ⓙ}, {ⓚ}, {ⓛ, ⓜ}

Partition 3: {ⓐ, ⓑ}, {ⓒ}, {ⓓ}, {ⓔ, ⓕ}, {ⓗ, ⓘ}, {ⓙ}, {ⓚ, ⓜ}, {ⓛ}

Partition 4: {ⓐ, ⓑ}, {ⓒ}, {ⓓ}, {ⓔ, ⓕ}, {ⓗ, ⓘ}, {ⓙ}, {ⓚ}, {ⓛ, ⓜ}

x_1x_2	00	01	11	10	z_1	z_2
	(a)	j	–	b	0	0
	a	–	c	(b)	0	0
	–	h	(c)	d	0	0
	a	–	e	(d)	0	0
	–	f	(e)	b	1	0
	a	(f)	g	–	1	0
	–	f	(g)	b	1	0
	a	(h)	i	–	0	0
	–	h	(i)	b	0	0
	k	(j)	o	–	0	0
	(k)	l	–	b	0	0
	a	(l)	m	–	0	1
	–	n	(m)	b	0	1
	a	(n)	m	–	0	1
	–	p	(o)	b	0	0
	a	(p)	o	–	0	0

Figure 6.159 Primitive flow table for the asynchronous sequential machine of Example 6.26.

Merged flow table Partition 1 will be used in constructing the merged flow table, which is shown in Figure 6.162. The rows are transferred from the reduced primitive flow table to the merged flow table and superimposed, where necessary. Each row of the merged flow table is assigned a number 1 through 8, which will represent vertices in the transition diagram. The following transitions occur between rows in the merged flow table:

$1 \rightarrow 2$ $1 \rightarrow 6$
$2 \rightarrow 3$ $2 \rightarrow 5$
$3 \rightarrow 1$ $3 \rightarrow 4$
$4 \rightarrow 1$ $4 \rightarrow 3$
$5 \rightarrow 1$
$6 \rightarrow 5$ $6 \rightarrow 7$
$7 \rightarrow 1$ $7 \rightarrow 8$
$8 \rightarrow 1$ $8 \rightarrow 7$

x_1x_2: 00	01	11	10	z_1	z_2
(*a*)	*j*	–	*b*	0	0
a	–	*c*	(*b*)	0	0
–	*h*	(*c*)	*d*	0	0
a	–	*e*	(*d*)	0	0
–	*f*	(*e*)	*b*	1	0
a	(*f*)	*e*	–	1	0
a	(*h*)	*i*	–	0	0
–	*h*	(*i*)	*b*	0	0
k	(*j*)	*i*	–	0	0
(*k*)	*l*	–	*b*	0	0
a	(*l*)	*m*	–	0	1
–	*l*	(*m*)	*b*	0	1

Figure 6.160 Reduced primitive flow table derived from the primitive flow table of Figure 6.159 for the asynchronous sequential machine of Example 6.26.

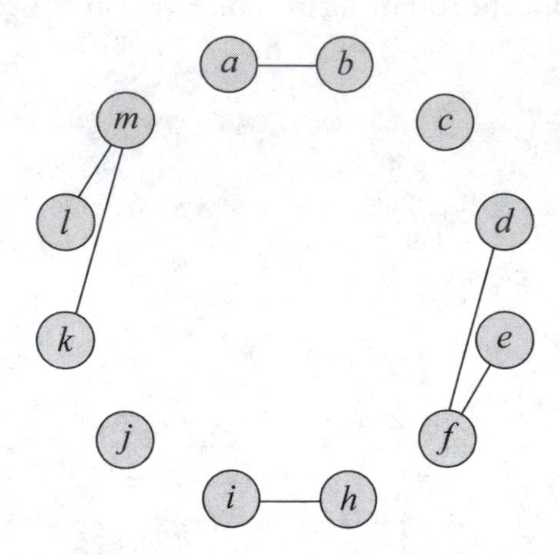

Figure 6.161 Merger diagram derived from the reduced primitive flow table of Figure 6.160 for the asynchronous sequential machine of Example 6.26.

		x_1x_2 00	01	11	10
1	(a), (b)	(a)	j	c	(b)
2	(c)	–	h	(c)	d
3	(d), (f)	a	(f)	e	(d)
4	(e)	–	f	(e)	b
5	(h), (i)	a	(h)	(i)	b
6	(j)	k	(j)	i	–
7	(k), (m)	(k)	l	(m)	b
8	(l)	a	(l)	m	–

Figure 6.162 Merged flow table using partition 1 for the asynchronous sequential machine of Example 6.26.

As seen from the row transitions, row 1 must be adjacent to rows 2, 3, 4, 5, 6, 7, and 8, which is clearly impossible, since there are no unspecified entries in columns $x_1x_2 = 01$ or 11 in which to add intermediate transient states. Since it is impossible to assign three-tuple state variable codes so that all rows are adjacent, one more state variable must be added. The fourth state variable will supply a sufficient number of unspecified entries to accommodate all row transitions such that, each pair of contiguous states in a cycle will be adjacent. The number of state variables is now four, which provides 16 rows as shown in the augmented merged flow table of Figure 6.163. The arrows indicate the transitions through transient states.

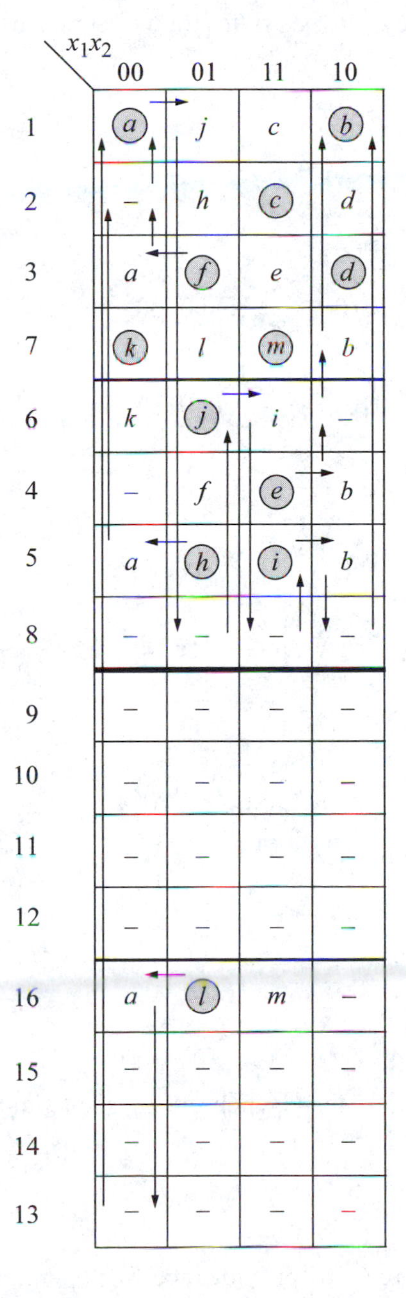

x_1x_2	00	01	11	10
1	*a*	*j*	*c*	*b*
2	–	*h*	*c*	*d*
3	*a*	*f*	*e*	*d*
7	*k*	*l*	*m*	*b*
6	*k*	*j*	*i*	–
4	–	*f*	*e*	*b*
5	*a*	*h*	*i*	*b*
8	–	–	–	–
9	–	–	–	–
10	–	–	–	–
11	–	–	–	–
12	–	–	–	–
16	*a*	*l*	*m*	–
15	–	–	–	–
14	–	–	–	–
13	–	–	–	–

Figure 6.163 Augmented merged flow table constructed from the merged flow table of Figure 6.162 and the transition diagram of Figure 6.164.

The transition diagram is shown in Figure 6.164, represented by two cubes. The transition diagram depicts the 16 rows, which are assigned a four-tuple Gray code. Each adjacent pair of row vertices is connected by a line. The state variable code assignment shown in the transition diagram is only one of several possible code assignments. The transition diagram graphically portrays all state transition sequences as specified by the merged flow table of Figure 6.162. Note that row 8 of the merged flow table, containing stable state ⓛ, is reassigned to row 16 in the augmented merged flow table to provide adjacency between states for the sequence ⓚ → ⓛ and ⓜ → ⓛ.

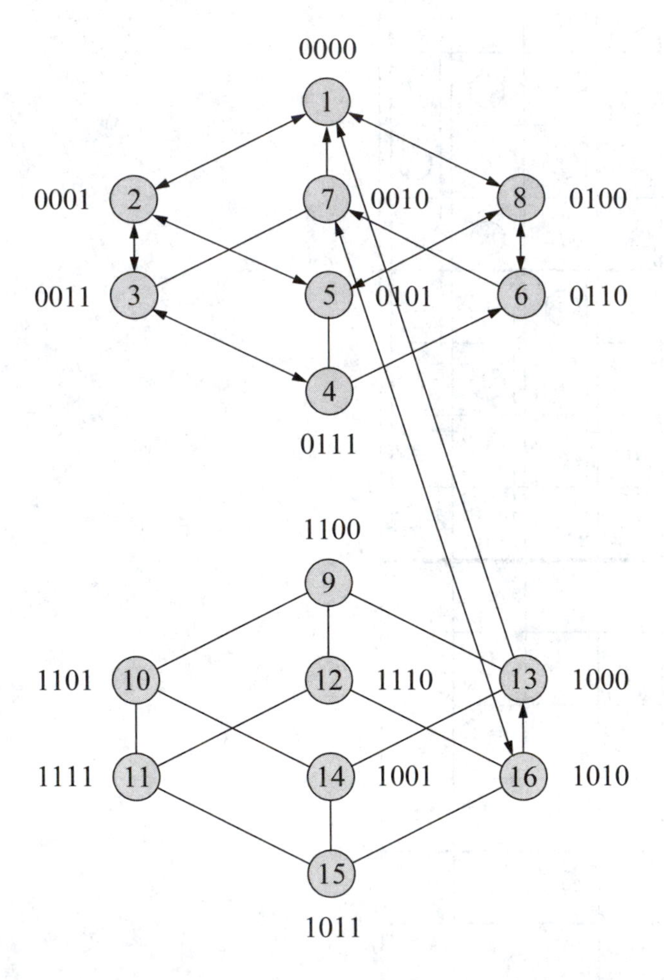

Figure 6.164 Transition diagram for the asynchronous sequential machine of Example 6.25.

The row transitions listed in Figure 6.165 provide race-free operation for all state transition sequences. Several transitions are directed through unspecified entries which are used as intermediate transient states to provide adjacency between contiguous states in a cycle.

1 (0000) → 2 (0001) ; ⓑ → ⓒ
1 (0000) → 8 (0100) → 6 (0110) ; ⓐ → ⓙ

2 (0001) → 3 (0011) ; ⓒ → ⓓ
2 (0001) → 5 (0101) ; ⓒ → ⓗ

3 (0011) → 2 (0001) → 1 (0000) ; ⓕ → ⓐ, ⓓ → ⓐ
3 (0011) → 4 (0111) ; ⓕ → ⓔ, ⓓ → ⓔ

4 (0111) → 6 (0110) → 7 (0010) → 1 (0000) ; ⓔ → ⓑ
4 (0111) → 3 (0011) ; ⓔ → ⓕ

5 (0101) → 2 (0001) → 1 (0000) ; ⓗ → ⓐ
5 (0101) → 8 (0100) → 1 (0000) ; ⓘ → ⓑ

6 (0110) → 8 (0100) → 5 (0101) ; ⓙ → ⓘ
6 (0110) → 7 (0010) ; ⓙ → ⓚ

7 (0010) → 1 (0000) ; ⓚ → ⓑ
7 (0010) → 16 (1010) ; ⓚ → ⓛ, ⓜ → ⓛ

16 (1010) → 13 (1000) → 1 (0000) ; ⓛ → ⓐ
16 (1010) → 7 (0010) ; ⓛ → ⓜ

Figure 6.165 Row transitions with no race conditions obtained from the transition diagram of Figure 6.164. See also, the combined excitation map of Figure 6.166.

Excitation maps and equations The combined excitation map is shown in Figure 6.166 and is generated from the augmented merged flow table and the transition diagram. The additional eight rows contain unspecified entries that are used to provide intermediate transient states such that, every state transition sequence is free of race conditions. The stable states are assigned excitation values that are identical to the values of the feedback variables of the corresponding row. The intermediate transient states are assigned excitation values that cause the machine to proceed to the row containing the same values for the feedback variables, after a delay of Δt.

For example, the sequences ⓔ → ⓑ and ⓛ → ⓐ are accomplished by the following cycles:

$$
\begin{array}{llllll}
\textcircled{e} & Y_{1e}Y_{2e}Y_{3e}Y_{4e} & = 0111 & \rightarrow & & Y_{1e}Y_{2e}Y_{3e}Y_{4e} = 0110 \rightarrow \\
 & y_{1f}y_{2f}y_{3f}y_{4f} & = 0110 & \rightarrow & & Y_{1e}Y_{2e}Y_{3e}Y_{4e} = 0010 \rightarrow \\
 & y_{1f}y_{2f}y_{3f}y_{4f} & = 0010 & \rightarrow & & Y_{1e}Y_{2e}Y_{3e}Y_{4e} = 0000 \rightarrow \\
 & y_{1f}y_{2f}y_{3f}y_{4f} & = 0000 & \rightarrow & \textcircled{b} & Y_{1e}Y_{2e}Y_{3e}Y_{4e} = 0000
\end{array}
$$

$$
\begin{array}{llllll}
\textcircled{l} & Y_{1e}Y_{2e}Y_{3e}Y_{4e} & = 1010 & \rightarrow & & Y_{1e}Y_{2e}Y_{3e}Y_{4e} = 1000 \rightarrow \\
 & y_{1f}y_{2f}y_{3f}y_{4f} & = 1000 & \rightarrow & & Y_{1e}Y_{2e}Y_{3e}Y_{4e} = 0000 \rightarrow \\
 & y_{1f}y_{2f}y_{3f}y_{4f} & = 0000 & \rightarrow & \textcircled{a} & Y_{1e}Y_{2e}Y_{3e}Y_{4e} = 0000
\end{array}
$$

	$y_{1f}y_{2f}y_{3f}y_{4f}$ \ x_1x_2	0 0	0 1	1 1	1 0
1	0000	(0000) ^{a}	0100	0001	(0000) ^{b}
2	0001	0000	0101	(0001) ^{c}	0011
3	0011	0001	(0011) ^{f}	0111	(0011) ^{d}
7	0010	(0010) ^{k}	1010	(0010) ^{m}	0000
6	0110	0010	(0110) ^{j}	0100	0010
4	0111	–	0011	(0111) ^{e}	0110
5	0101	0001	(0101) ^{h}	(0101) ^{i}	0100
8	0100	–	0110	0101	0000
9	1100	–	–	–	–
10	1101	–	–	–	–
11	1111	–	–	–	–
12	1110	–	–	–	–
16	1010	1000	(1010) ^{l}	0010	–
15	1011	–	–	–	–
14	1001	–	–	–	–
13	1000	0000	–	–	–

$Y_{1e}Y_{2e}Y_{3e}Y_{4e}$

Figure 6.166 Combined excitation map obtained from the augmented merged flow table of Figure 6.163 and the transition diagram of Figure 6.164 for the asynchronous sequential machine of Example 6.26.

The individual excitation map for Y_{1e} is shown in Figure 6.167 and is obtained from the combined excitation map by transferring the values for Y_{1e} to the corresponding locations in the individual map. The equation for Y_{1e} is shown in Equation 6.36.

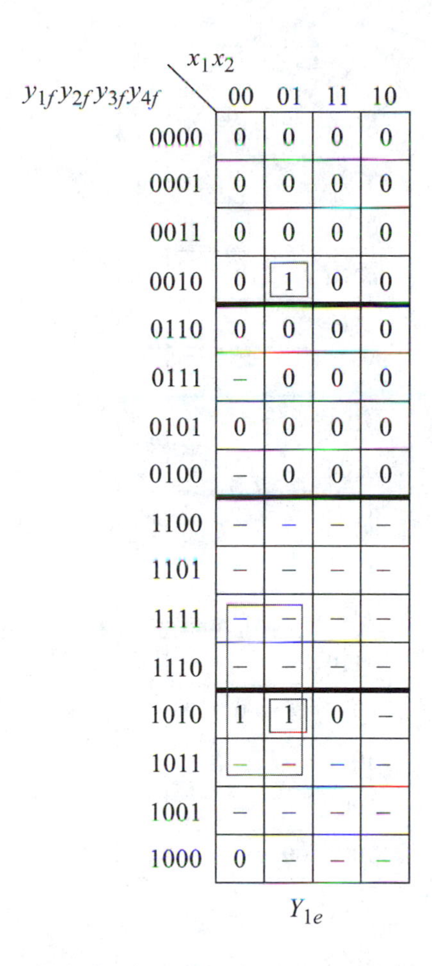

Figure 6.167 Individual excitation map for Y_{1e} obtained from the combined excitation map of Figure 6.166.

$$Y_{1e} = y_{2f}'y_{3f}y_{4f}'x_1'x_2 + y_{1f}y_{3f}x_1' \tag{6.36}$$

Output map and equation The output map for z_1 is shown in Figure 6.168 which is generated from the reduced primitive flow table and the augmented merged flow table. The location of the stable states are obtained from the augmented merged flow table of Figure 6.163; the values for output z_1 are obtained from the reduced primitive flow table of Figure 6.160. The equation for z_1 is shown in Equation 6.37.

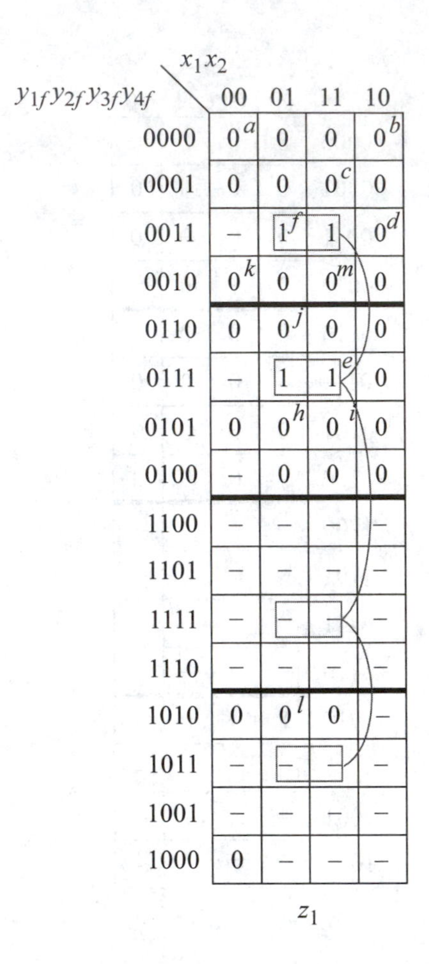

Figure 6.168 Output map for z_1 obtained from the reduced primitive flow table of Figure 6.160 and the augmented merged flow table of Figure 6.163.

$$z_1 = y_{3f} y_{4f} x_2 \tag{6.37}$$

The previous synthesis examples consolidated the distinct steps of the synthesis procedure into a cohesive process. When constructing the primitive flow table, it is advantageous to derive a reduced primitive flow table directly from the machine specifications, if possible. This reduces the amount of time required for the synthesis process, since there will be no equivalent states. As mentioned previously, however, the primary consideration is to provide a reduced primitive flow table that completely delineates the machine's inherent operating characteristics.

The merging process may further reduce the number of rows in the primitive flow table. The procedure attempts to merge two or more rows of the reduced primitive flow table into a single row. The number of rows in the flow table ultimately determines the number of feedback variables and thus, the size and cost of the machine. Fewer rows correlate to fewer logic gates.

When deriving equations from the excitation maps, consideration must be given to static hazards. All static-1 and static-0 hazards can be eliminated by combining adjacent groups of 1s or 0s with redundant prime implicants. Similarly, the equations obtained from the output maps must be free of static hazards.

When assigning values to unspecified entries in the output maps, the first consideration is to insert entries so that there will be no erroneous transient outputs. When this condition has been satisfied, then the remaining unspecified entries can be assigned values that provide either a slow or a fast output response, or a network that yields the fewest number of logic gates in the λ output logic.

6.4 Problems

6.1 Obtain the primitive flow table for an asynchronous sequential machine that has two inputs x_1 and x_2 and two outputs z_1 and z_2. If x_1 is asserted before x_2, then z_1 is asserted and remains active until z_2 is asserted. If x_2 is asserted before x_1, then z_2 is asserted and remains active until z_1 is asserted. Except for a reset condition, the outputs are mutually exclusive. Do not obtain a reduced primitive flow table. The timing diagram shown below represents some typical input sequences.

6.2 Obtain the primitive flow table for an asynchronous sequential machine that has two inputs x_1 and x_2 and one output z_1. Output z_1 reflects the value of x_1 when x_2 is asserted. The output will change only on the positive transition of x_2 and will remain in that state until the next assertion of x_2. Input x_1 will not change value while x_2 is active. A representative timing diagram is shown below illustrating some typical input sequences. Do not obtain a reduced primitive flow table.

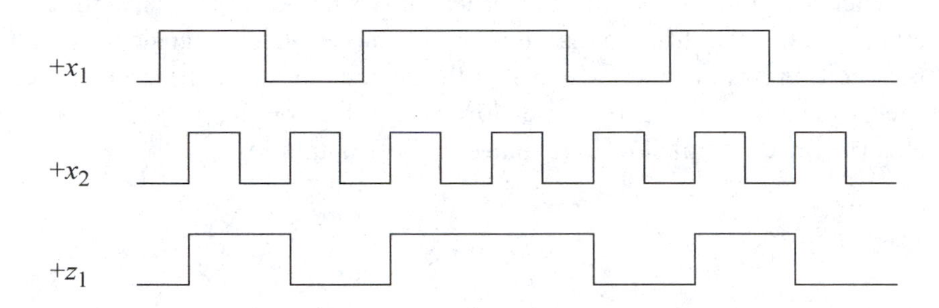

6.3 Obtain the primitive flow table for an asynchronous sequential machine that has two inputs x_1 and x_2 and one output z_1. Input x_1 is a periodic clock signal for a synchronous sequential machine under control of an asynchronous sequential machine. Input x_2 is used to control a single-step operation; that is, when x_2 is asserted, a single full width x_1 pulse — represented by output z_1 — is transferred to the synchronous sequential machine.

Input x_2 must be asserted before x_1 in order to generate a corresponding pulse on output z_1. If x_2 is deasserted before x_1 is deasserted, then output z_1 maintains its active state for the duration of x_1. A representative timing diagram is shown below.

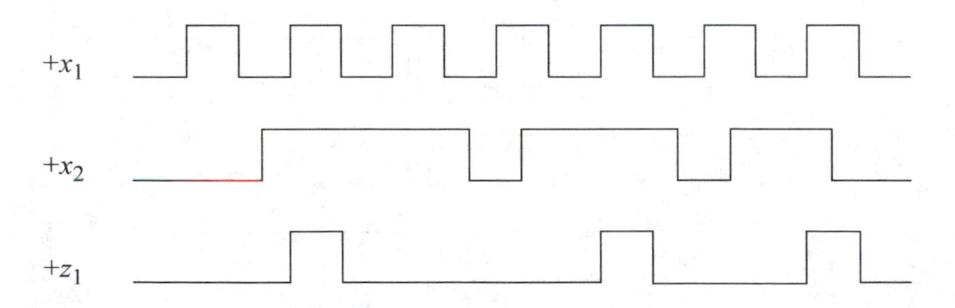

6.4 Obtain the primitive flow table for an asynchronous sequential machine that has three inputs x_1, x_2, and x_3 and one output z_1. The truth table shown below characterizes the operation of the machine. The inputs adhere to the fundamental-mode principle.

$x_1x_2x_3$	z_1
0 0 0	1
0 0 1	0
0 1 0	0
0 1 1	1
1 0 0	0
1 0 1	1
1 1 0	1
1 1 1	0

6.5 Identify equivalent states in the primitive flow table shown below. Then eliminate redundant states and obtain a reduced primitive flow table.

x_1x_2 00	01	11	10	z_1
(a)	b	–	c	0
a	(b)	d	–	1
f	–	e	(c)	1
–	h	(d)	g	0
–	h	(e)	g	0
(f)	b	–	c	0
a	–	d	(g)	0
f	(h)	e	–	0

6.6 Identify equivalent states in the primitive flow table shown below. Then eliminate redundant states and obtain a reduced primitive flow table.

x_1x_2 00	01	11	10	z_1
(*a*)	*b*	–	*d*	0
a	(*b*)	*c*	–	1
–	–	(*c*)	*d*	1
g	–	*e*	(*d*)	0
–	*f*	(*e*)	*d*	1
g	(*f*)	*e*	–	1
(*g*)	*b*	–	*h*	0
a	–	*c*	(*h*)	1

6.7 Derive the merger diagram from the reduced primitive flow table below.

x_1x_2 00	01	11	10	z_1	z_2
(*a*)	*b*	–	*d*	0	0
a	(*b*)	*c*	–	1	1
–	*j*	(*c*)	*d*	1	0
a	–	*c*	(*d*)	0	1
(*i*)	*j*	–	*d*	0	1
i	(*j*)	*c*	–	1	0

6.8 Identify equivalent states in the primitive flow table shown below. Then eliminate redundant states and obtain a reduced primitive flow table.

x_1x_2 00	01	11	10	z_1	z_2
(*a*)	*g*	–	*b*	1	0
a	–	*c*	(*b*)	0	1
–	*g*	(*c*)	*d*	0	0
e	–	*j*	(*d*)	0	1
(*e*)	*i*	–	*f*	0	0
a	–	*c*	(*f*)	0	1
a	(*g*)	*j*	–	1	1
(*h*)	*i*	–	*k*	1	0
a	(*i*)	*j*	–	1	1
–	*i*	(*j*)	*k*	1	0
h	–	*c*	(*k*)	0	1

6.9 Obtain the maximal compatible sets of merged rows from the merger diagram shown below.

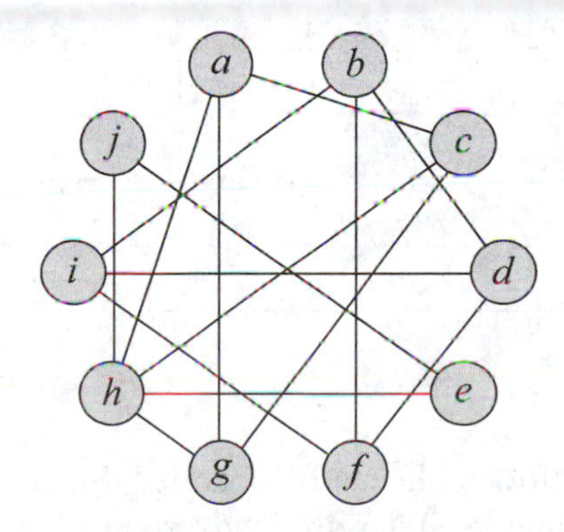

6.10 Identify equivalent states in the primitive flow table shown below. Then eliminate redundant states and obtain a reduced primitive flow table.

x_1x_2	00	01	11	10	z_1	z_2
	(a)	e	–	i	1	0
	(b)	e	–	j	1	0
	(c)	e	–	i	1	0
	(d)	f	–	j	1	0
	d	(e)	h	–	1	1
	c	(f)	g	–	0	0
	–	f	(g)	i	0	1
	–	e	(h)	i	0	1
	a	–	g	(i)	0	0
	b	–	h	(j)	1	1

6.11 Derive a reduced primitive flow table for an asynchronous sequential machine which operates according to the timing diagram shown below. There are two inputs x_1 and x_2 and one output z_1. When x_2 is active, output z_1 is set to the value of input x_1. When x_2 is inactive, any change to x_1 will not affect the output. Input x_1 will not change value while x_2 is asserted.

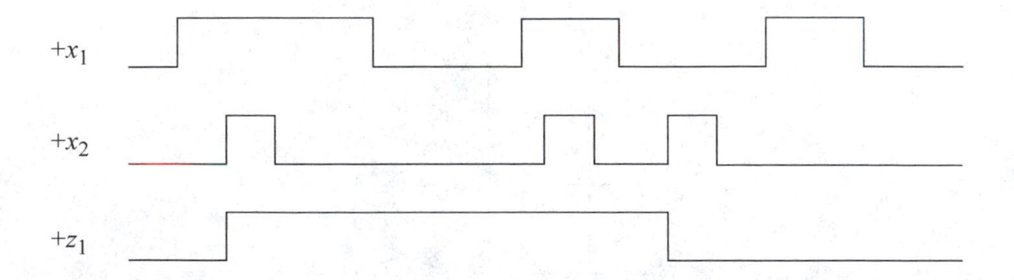

The reduced primitive flow table can be obtained directly from the machine specifications if no redundant states are inadvertently inserted.

6.12 Derive the reduced primitive flow table for an asynchronous sequential machine which operates according to the following specifications:

(a) The machine has two inputs x_1 and x_2 and one output z_1.
(b) Output z_1 is asserted when x_2 is asserted, but only if x_1 is already asserted.
(c) Output z_1 is deasserted only when x_2 becomes deasserted.

The reduced primitive flow table can be obtained directly from the machine specifications if no redundant states are inadvertently inserted.

6.13 Identify equivalent states in the primitive flow table shown below. Then eliminate redundant states and obtain a reduced primitive flow table.

x_1x_2 00	01	11	10	z_1
ⓐ	f	–	k	0
ⓒ	d	–	j	0
c	ⓓ	i	–	1
a	–	h	ⓚ	1
ⓑ	f	–	j	0
c	ⓕ	i	–	1
–	e	ⓖ	l	1
b	ⓔ	i	–	0
–	d	ⓗ	k	0
–	d	ⓘ	k	0
c	–	i	ⓙ	1
a	–	g	ⓛ	0

6.14 Identify equivalent states in the primitive flow table shown below. Then eliminate redundant states and obtain a reduced primitive flow table.

x_1x_2: 00	01	11	10	z_1	z_2
(*a*)	*g*	–	*d*	1	1
(*b*)	*e*	–	*d*	0	1
–	*g*	(*c*)	*k*	1	0
b	–	*c*	(*d*)	0	0
f	(*e*)	*i*	–	1	1
(*f*)	*g*	–	*k*	0	1
a	(*g*)	*n*	–	1	0
(*h*)	*l*	–	*d*	0	1
–	*g*	(*i*)	*m*	0	1
–	*g*	(*j*)	*d*	1	0
h	–	*j*	(*k*)	0	0
f	(*l*)	*i*	–	1	1
h	–	*n*	(*m*)	1	1
–	*l*	(*n*)	*k*	0	0

6.15 Identify equivalent states in the primitive flow table shown below. Then eliminate redundant states and obtain a reduced primitive flow table.

x_1x_2 00	01	11	10	z_1	z_2	z_3
(*a*)	*g*	–	*d*	1	1	0
(*b*)	*e*	–	*d*	0	1	1
(*f*)	*e*	–	*k*	0	1	1
(*h*)	*l*	–	*d*	0	1	1
f	(*e*)	*i*	–	1	1	0
a	(*g*)	*n*	–	1	0	0
f	(*l*)	*i*	–	1	1	0
–	*g*	(*c*)	*k*	1	0	1
–	*g*	(*i*)	*m*	0	1	0
–	*g*	(*j*)	*d*	1	0	1
–	*l*	(*n*)	*k*	0	0	1
b	–	*c*	(*d*)	0	0	1
h	–	*j*	(*k*)	0	0	1
h	–	*n*	(*m*)	1	1	1

6.16 Identify equivalent states in the primitive flow table shown below. Then eliminate redundant states and obtain a reduced primitive flow table.

x_1x_2: 00	01	11	10	z_1	z_2
(*a*)	*b*	–	*f*	0	0
a	(*b*)	*e*	–	1	0
(*c*)	*d*	–	*f*	0	0
a	(*d*)	*e*	–	1	0
–	*g*	(*e*)	*f*	0	1
c	–	*e*	(*f*)	1	1
c	(*g*)	*e*	–	1	1

6.17 Derive the merger diagram from the reduced primitive flow table shown below.

x_1x_2: 00	01	11	10	z_1	z_2
(*a*)	*b*	–	*d*	0	1
c	(*b*)	*e*	–	1	0
(*c*)	*b*	–	*d*	1	1
a	–	*e*	(*d*)	0	0
–	*f*	(*e*)	*d*	0	1
g	(*f*)	*e*	–	1	0
(*g*)	*b*	–	*d*	1	0

6.18 Identify equivalent states in the primitive flow table shown below. Then eliminate redundant states and obtain a reduced primitive flow table.

x_1x_2 00	01	11	10	z_1	z_2
(a)	b	–	c	0	0
a	(b)	g	–	0	1
a	–	d	(c)	0	1
–	k	(d)	e	1	1
a	–	f	(e)	0	1
–	b	(f)	e	0	0
–	b	(g)	h	0	0
a	–	i	(h)	0	1
–	k	(i)	j	1	1
a	–	f	(j)	0	1
a	(k)	d	–	1	0

6.19 Given the merger diagram shown below, obtain a partition which contains sets of maximal compatible rows.

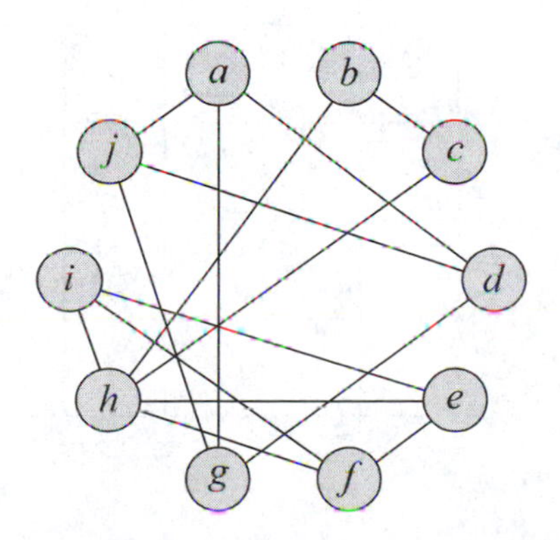

6.20 Derive the merger diagram from the reduced primitive flow table shown below.

x_1x_2 00	01	11	10	z_1	z_2
(a)	d	–	b	0	0
a	–	e	(b)	0	1
(c)	d	–	f	0	1
a	(d)	e	–	1	0
–	d	(e)	f	0	0
g	–	e	(f)	1	1
(g)	d	–	b	1	1

6.21 Derive the merger diagram from the reduced primitive flow table shown below.

x_1x_2 00	01	11	10	z_1
(a)	b	–	c	0
a	(b)	g	–	0
a	(k)	d	–	1
–	k	(d)	e	1
–	b	(f)	e	0
–	b	(g)	c	0
a	–	d	(c)	0
a	–	f	(e)	0

6.22 Derive the merger diagram from the reduced primitive flow table shown below.

x_1x_2 00	01	11	10	z_1
(a)	b	–	c	0
a	(b)	d	–	0
a	–	d	(c)	0
–	b	(d)	e	0
f	–	g	(e)	1
(f)	h	–	e	1
–	h	(g)	c	1
f	(h)	g	–	1

6.23 Derive the merger diagram from the reduced primitive flow table shown below.

x_1x_2 00	01	11	10	z_1	z_2
(a)	b	–	d	0	0
a	(b)	c	–	0	1
–	–	(c)	d	0	0
a	–	–	(d)	1	0
–	f	(e)	d	1	0
g	(f)	c	–	1	0
(g)	h	–	d	0	1
g	(h)	e	–	1	1

6.24 Given the primitive flow table shown below, obtain the reduced primitive flow table, the merger diagram, and all merged flow tables.

x_1x_2 00	01	11	10	z_1	z_2
(a)	b	–	j	0	0
e	(b)	d	–	1	0
g	–	h	(c)	1	1
–	b	(d)	c	0	0
(e)	f	–	j	1	0
e	(f)	h	–	1	0
(g)	f	–	c	0	0
–	f	(h)	c	0	0
(i)	f	–	j	0	0
i	–	d	(j)	1	1

6.25 Given the merger diagram shown below, obtain a partition which contains sets of maximal compatible rows.

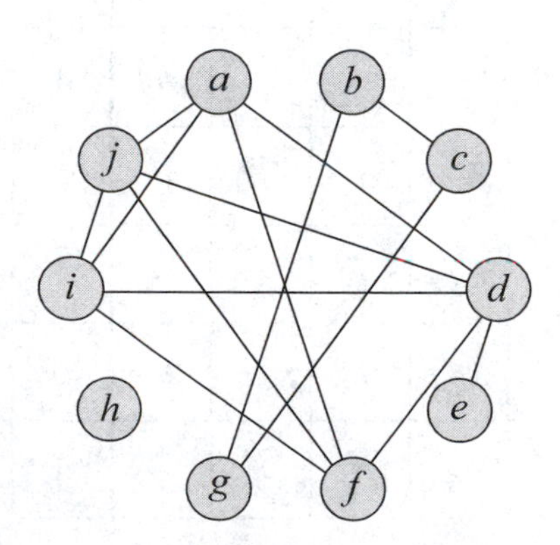

6.26 Given the reduced primitive flow table and the merger diagram shown below, obtain the merged flow table.

x_1x_2 00	01	11	10	z_1
(a)	b	–	c	0
a	(b)	d	–	0
a	–	d	(c)	0
–	b	(d)	e	0
f	–	g	(e)	1
(f)	h	–	e	1
–	h	(g)	c	1
f	(h)	g	–	1

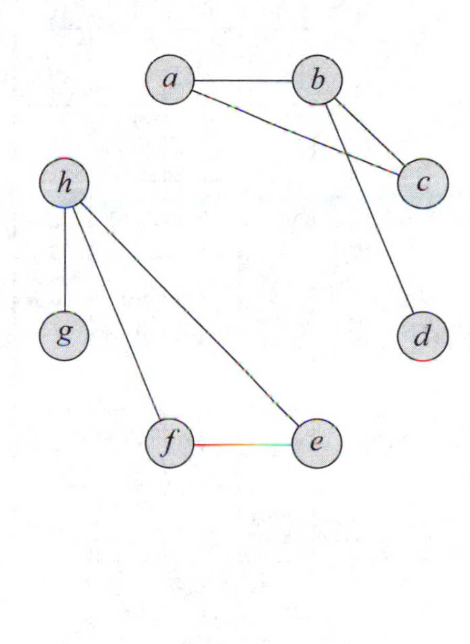

6.27 Given the merger diagram shown below, obtain a partition which contains sets of maximal compatible rows.

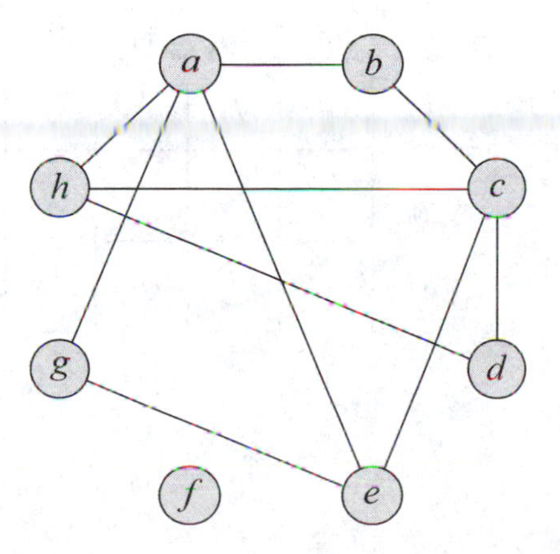

6.28 Given the reduced primitive flow table and the merger diagram shown below, obtain the merged flow table.

x_1x_2 00	01	11	10	z_1
(a)	b	–	d	0
a	(b)	c	–	0
–	b	(c)	d	0
a	–	e	(d)	0
–	f	(e)	d	1
g	(f)	c	–	1
(g)	f	–	d	1

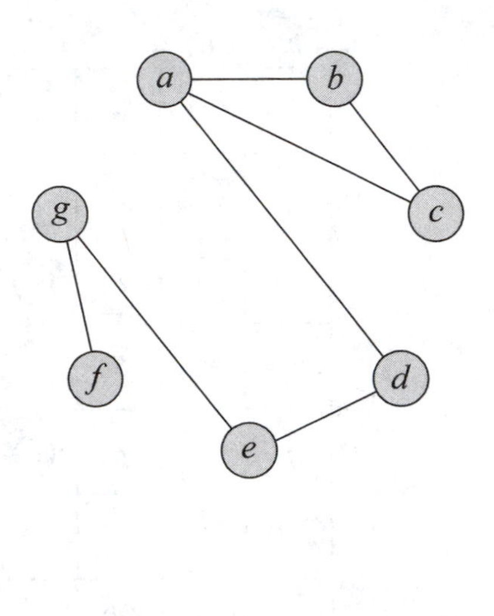

6.29 Derive the transition diagram for the merged flow table shown below. Resolve all adjacency conflicts, then obtain the excitation maps and equations.

	x_1x_2 00	01	11	10
1	(a)	f	(b)	e
2	(c)	(d)	b	(e)
3	c	(f)	b	–

6.30 Derive the transition diagram for the merged flow table shown below. Resolve all adjacency conflicts, then obtain the excitation maps and equations.

	00	01	11	10
1	(*a*)	(*b*)	*g*	*f*
2	–	*d*	(*c*)	*e*
3	*a*	(*d*)	*c*	(*e*)
4	*a*	*d*	(*g*)	(*f*)

6.31 Derive the transition diagram for the merged flow table shown below. Resolve all adjacency conflicts, then obtain the excitation maps and equations.

	00	01	11	10
1	(*a*)	*e*	*h*	(*b*)
2	*g*	(*c*)	(*d*)	*f*
3	*a*	(*e*)	*d*	(*f*)
4	(*g*)	*c*	(*h*)	*b*

6.32 Derive the transition diagram for the merged flow table shown below. Resolve all adjacency conflicts, then obtain the excitation maps and equations.

	00	01	11	10
1	(*a*)	(*b*)	*c*	(*g*)
2	–	*e*	(*c*)	*g*
3	*a*	–	*f*	(*d*)
4	*a*	(*e*)	(*f*)	*d*

6.33 Derive the transition diagram for the merged flow table shown below. Resolve all adjacency conflicts, then obtain the excitation maps and equations.

x_1x_2	00	01	11	10
1	(*a*)	(*b*)	–	*c*
2	(*f*)	*g*	–	*c*
3	*f*	(*e*)	*d*	(*c*)
4	*a*	(*g*)	(*d*)	*c*

6.34 Derive the transition diagram for the merged flow table shown below. Resolve all adjacency conflicts, then obtain the excitation maps and equations.

x_1x_2	00	01	11	10
1	(*a*)	*b*	–	*c*
2	(*d*)	*e*	–	*c*
3	*d*	(*b*)	*f*	–
4	*d*	(*e*)	(*f*)	(*c*)

6.35 Derive the transition diagram for the merged flow table shown below. Resolve all adjacency conflicts, then obtain the excitation maps and equations.

x_1x_2	00	01	11	10
1	(*a*)	*c*	(*e*)	*h*
2	*b*	(*c*)	*e*	(*g*)
3	(*b*)	*d*	(*f*)	*g*
4	*a*	(*d*)	*f*	(*h*)

6.36 Derive the transition diagram for the merged flow table shown below. Resolve all adjacency conflicts, then obtain the excitation maps and equations.

x_1x_2	00	01	11	10
1	ⓐ	b	ⓖ	c
2	f	ⓑ	g	ⓓ
3	ⓕ	ⓗ	e	ⓒ
4	a	h	ⓔ	d

6.37 Derive the transition diagram for the merged flow table shown below. Resolve all adjacency conflicts, then obtain the excitation maps and equations.

x_1x_2	00	01	11	10
1	ⓐ	f	c	ⓑ
2	g	ⓕ	ⓒ	d
3	a	ⓗ	e	ⓓ
4	ⓖ	h	ⓔ	b

6.38 Obtain the output map for the reduced primitive flow table shown below. The output equation for z_1 is to be in minimal form.

00	01	11	10	z_1
ⓐ	b	–	c	0
a	ⓑ	–	–	0
a	–	d	ⓒ	0
–	–	ⓓ	e	1
f	–	d	ⓔ	1
ⓕ	b	–	e	1

6.39 Given the reduced primitive flow table shown below, obtain the output map and equation for z_1 in minimized form. There must be no race conditions, no static hazards, and no momentary false outputs.

x_1x_2 00	01	11	10	z_1
(a)	–	–	b	0
a	–	c	(b)	0
–	–	(c)	d	0
a	–	e	(d)	0
–	–	(e)	b	1

6.40 Given the merged flow table shown below, obtain the output map for z_1. The output equation is to be in a minimal sum-of-products form. Output z_1 has the following assigned values:

Stable state	Output z_1
(a)	0
(b)	1
(c)	1
(d)	0
(e)	1
(f)	0
(g)	0
(h)	1

	x_1x_2 00	01	11	10
1	(a)	(b)	e	d
2	(c)	g	(f)	(d)
3	c	–	(e)	(h)
4	a	(g)	f	–

6.41 Use the synthesis procedure for asynchronous sequential machines to design an *SR* latch. Let $SR = x_1x_2$.

6.42 Obtain the output maps for the merged flow table shown below. The output equations are to be in a sum-of-products form. Outputs z_1 and z_2 have the following assigned values:

Stable state	Outputs z_1	z_2
(a)	0	0
(b)	0	1
(c)	1	1
(d)	1	0
(e)	1	0
(f)	1	1
(g)	0	1
(h)	0	0

x_1x_2	00	01	11	10
1	(a)	(b)	f	d
2	(c)	b	h	(d)
3	(e)	b	(f)	d
4	(g)	b	(h)	d

The output variables are to be free of false transient outputs and static hazards. Output response is to be as fast as possible.

6.43 Obtain the output maps and equations for the merged flow table shown below. Outputs z_1 and z_2 have the following assigned values:

Stable state	Outputs z_1	z_2
(a)	0	0
(b)	1	1
(c)	1	0
(d)	0	1
(e)	0	0
(f)	1	1
(g)	0	1
(h)	1	0

x_1x_2	00	01	11	10
1	(a)	c	(b)	h
2	e	(c)	b	(d)
3	(e)	g	(f)	d
4	a	(g)	f	(h)

Specify the equations in both a sum-of-products form and a product-of-sums form. The output variables are to be free of momentary erroneous outputs.

6.44 Given the reduced primitive flow table shown below, obtain the output maps and equations. The output equations are to be in a minimal sum-of-products and product-of-sums form. The outputs must be free of static hazards and transient false outputs.

x_1x_2 00	01	11	10	z_1	z_2	z_3
(a)	b	–	c	1	0	1
a	(b)	g	–	1	0	0
a	(h)	d	–	1	0	1
–	h	(d)	e	1	0	1
–	b	(f)	e	0	1	0
–	b	(g)	c	0	1	0
a	–	d	(c)	0	0	0
a	–	f	(e)	0	0	0

6.45 Given the reduced primitive flow table shown below, obtain the output map for z_1. Output glitches cannot be tolerated. The output map entries must produce the fastest possible output changes. Derive the output equations in a sum-of-products and a product-of-sums form.

x_1x_2 00	01	11	10	z_1
(a)	b	–	d	0
a	(b)	c	–	0
–	b	(c)	d	0
a	–	e	(d)	0
–	f	(e)	d	1
g	(f)	c	–	1
(g)	f	–	d	1

6.46 Given the merged flow table shown below, obtain the excitation and output equations in both a sum-of-products form and a product-of-sums form. Output z_1 has the following assigned values:

Stable state	Output z_1
(a)	0
(b)	0
(c)	0
(d)	1
(e)	0
(f)	1
(g)	1

x_1x_2	00	01	11	10
1	g	(a)	(b)	d
2	(c)	e	f	(d)
3	g	(e)	(f)	d
4	(g)	a	–	d

6.47 Obtain the excitation and output equations for the asynchronous sequential machine represented by the reduced primitive flow table shown below. Use all partitions of maximal compatible sets. Show the equations in both a sum-of-products form and a product-of-sums form. There are to be no static-1 or static-0 hazards. The λ output logic must have the fastest possible response to input changes.

x_1x_2: 00	01	11	10	z_1
(a)	b	–	c	0
a	(b)	g	–	0
a	–	d	(c)	0
–	h	(d)	e	1
a	–	f	(e)	0
–	b	(f)	e	0
–	b	(g)	c	0
a	(h)	d	–	1

6.48 Synthesize an asynchronous sequential machine which has two inputs x_1 and x_2 and one output z_1. Output z_1 is asserted for the duration of x_2 if and only if x_1 is already asserted. Assume that the initial state of the machine is $x_1x_2z_1 = 000$. A representative timing diagram is shown below. Obtain the excitation equations in both a sum-of-products and a product-of-sums form.

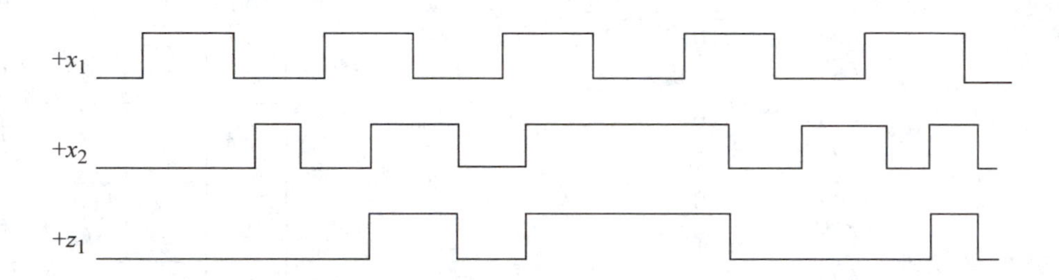

6.49 Synthesize an asynchronous sequential machine which has one input x_1 and one output z_1. The machine operates according to the timing diagram shown below. The assertion of x_1 toggles output z_1. Assign values to the transient states in the output map such that, the λ output logic will be minimized. Obtain the excitation equations in both a sun-of-products form and a product-of-sums form. Use NOR logic only.

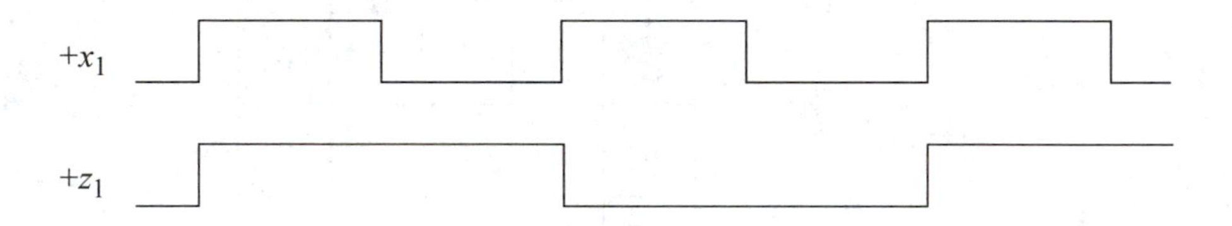

6.50 Synthesize an asynchronous sequential machine that has one input x_1 and one output z_1 which operates according to the timing diagram shown below. The output response is to be as fast as possible. Use NAND logic only.

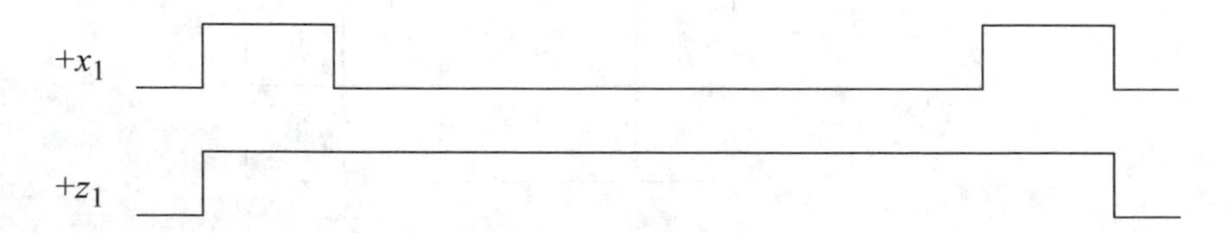

6.51 Obtain the excitation and output equations for an asynchronous sequential machine which has one input x_1 and two outputs z_1 and z_2. Output z_1 is toggled at the positive transition of x_1; output z_2 is toggled at the negative transition of x_1. Show two sets of output maps: one set for fast machine operation and one set for the fewest number of gates. Assume that the initial state of the machine is $x_1z_1z_2 = 000$.

6.52 Synthesize an asynchronous sequential machine which has two inputs x_1 and x_2 and one output z_1. Output z_1 will be asserted coincident with the assertion of x_2, but only if x_1 is already asserted. The deassertion of x_2 causes the deassertion of z_1. Input x_1 will not become deasserted while x_2 is asserted.

6.53 Obtain the excitation and output equations for an asynchronous sequential machine which has one input x_1 and two outputs z_1 and z_2. Output z_1 is asserted for the duration of every second x_1 pulse; output z_2 is asserted for the duration of every second z_1 pulse. The outputs are to respond as fast as possible to changes in the input vector. A representative timing diagram is shown below.

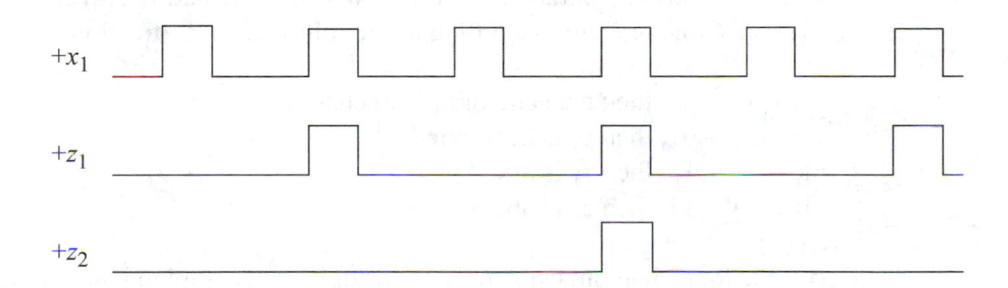

6.54 Synthesize an asynchronous sequential machine which has two inputs x_1 and x_2 and one output z_1. Output z_1 will be asserted coincident with the assertion of the first x_2 pulse and will remain active for the duration of the first x_2 pulse. The output will be asserted only if the assertion of x_1 precedes the assertion of x_2. Input x_1 will not become deasserted while x_2 is asserted. The λ output logic must have a minimal number of logic gates. A representative timing diagram is shown below.

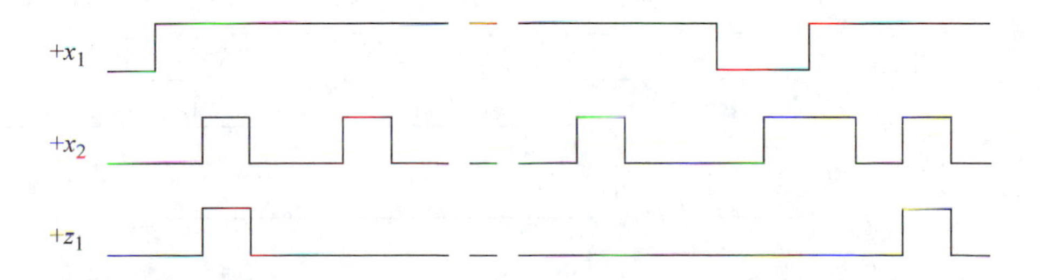

6.55 Synthesize an asynchronous sequential machine which has one input x_1 and two outputs z_1 and z_2. The machine functions as a two-output bistable multivibrator, whose operation is characterized by the timing diagram shown below. Output z_1 toggles on the positive transition of x_1 and output z_2 toggles on the negative transition of x_1. Obtain equations for z_1 and z_2 which produce the fastest possible output changes and equations which yield the least amount of logic.

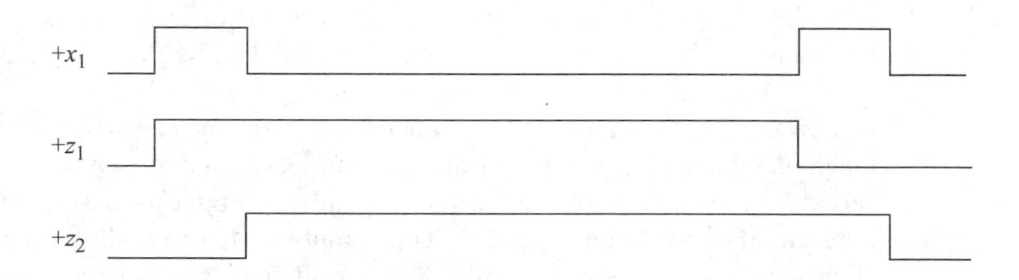

6.56 An asynchronous sequential machine has two inputs x_1 and x_2 and one output z_1. The machine operates according to the following specifications:

If $x_1x_2 = 00$, then the state of z_1 is unchanged.
If $x_1x_2 = 01$, then z_1 is deasserted.
If $x_1x_2 = 10$, then z_1 is asserted.
If $x_1x_2 = 11$, then z_1 changes state.

Derive the logic diagram using only NOR logic with complementary outputs. The inputs are available in both high and low assertion. Assume that the initial conditions are $x_1x_2z_1 = 000$.

6.57 Synthesize an asynchronous sequential machine which has two inputs x_1 and x_2 and one output z_1. The initial conditions are: $x_1x_2z_1 = 000$. Output z_1 is asserted whenever $x_1x_2 = 11$ if and only if the input sequence was $x_1x_2 =$ 00, 01, 11. The waveforms shown below depict some typical input sequences.

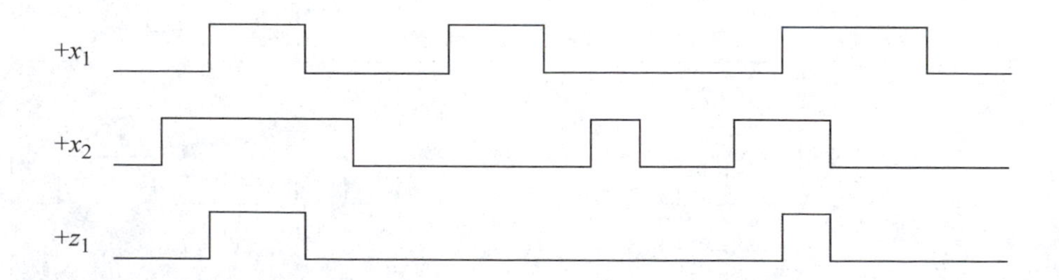

6.58 An asynchronous sequential machine has two inputs x_1 and x_2 and one output z_1. Input x_1 will always be asserted whenever x_2 is asserted; that is, there will never be a situation where x_1 is deasserted and x_2 is asserted. Output z_1 is asserted coincident with every third x_2 pulse and remains active for the duration of x_2. Obtain the excitation and output equations in a sum-of-products form and a product-of-sums form. A representative timing diagram is shown below.

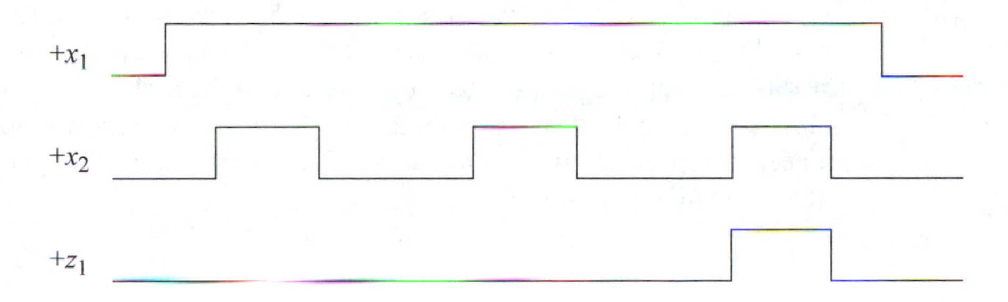

6.59 Obtain the excitation and output equations for an asynchronous sequential machine which has two nonoverlapping, disjoint positive inputs x_1 and x_2. Thus, the input vector $x_1x_2 = 11$ is invalid. There is one output z_1, which is asserted at the positive transition of input x_1 if the preceding sequence of input vectors was $x_1x_2 = 00, 10, 00, 01, 00$. Output z_1 is deasserted at the next positive transition of x_2. Thus, a valid sequence to assert and then deassert z_1 is $x_1x_2z_1 = 000, 100, 000, 010, 000, 101, 001, 010$.

The timing diagram shown below illustrates the necessary input vectors to assert and deassert output z_1. Assign output values to the intermediate states of the output map such that, the equation for z_1 will yield the fewest number of logic gates for the λ output logic. As in all asynchronous sequential machines, the fundamental-mode of operation specifies that simultaneous input changes are invalid.

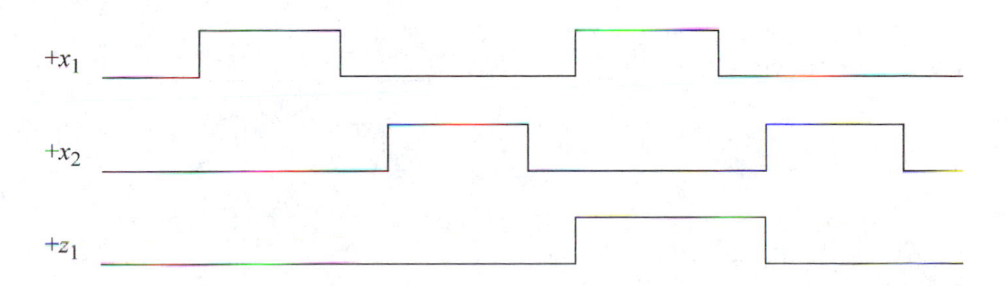

6.60 Synthesize an asynchronous sequential machine which has two inputs x_1 and x_2 and two outputs z_1 and z_2. The two inputs may overlap, but will not change state simultaneously. Only the following sequences are valid:

$$x_1x_2 = 00 \rightarrow 10 \rightarrow 11 \rightarrow 01 \rightarrow 00$$
$$x_1x_2 = 00 \rightarrow 10 \rightarrow 11 \rightarrow 10 \rightarrow 00$$
$$x_1x_2 = 00 \rightarrow 10 \rightarrow 00$$
$$x_1x_2 = 00 \rightarrow 01 \rightarrow 00$$

Output z_1 is asserted whenever x_1 is active and x_2 is asserted or when x_2 is active and x_1 is asserted. Output z_1 will be deasserted when either x_1 or x_2 is deasserted. Output z_2 is asserted coincident with the assertion of z_1 and remains active until the deassertion of the last active input of an overlapping sequence. A representative timing diagram is shown below. Use NOR logic with complementary outputs.

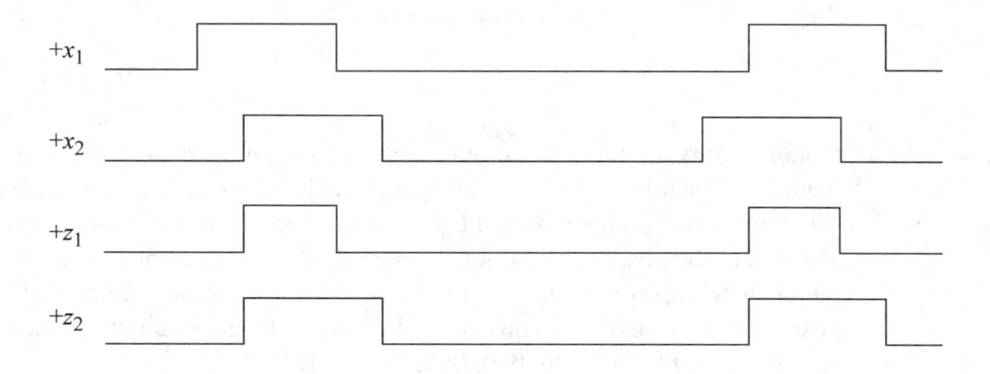

6.61 Obtain the excitation and output equations for Y_{1e} and z_1 only, for an asynchronous sequential machine which has two inputs x_1 and x_2 and two outputs z_1 and z_2. Output z_1 toggles at the negative transition of input x_1. When z_1 is active, z_2 reflects the state of input x_2, but only if z_1 is already asserted. Output z_2 will be asserted for the duration of input x_2. A representative timing diagram is shown below. Assume that the initial condition is $x_1x_2z_1z_2 = 0000$.

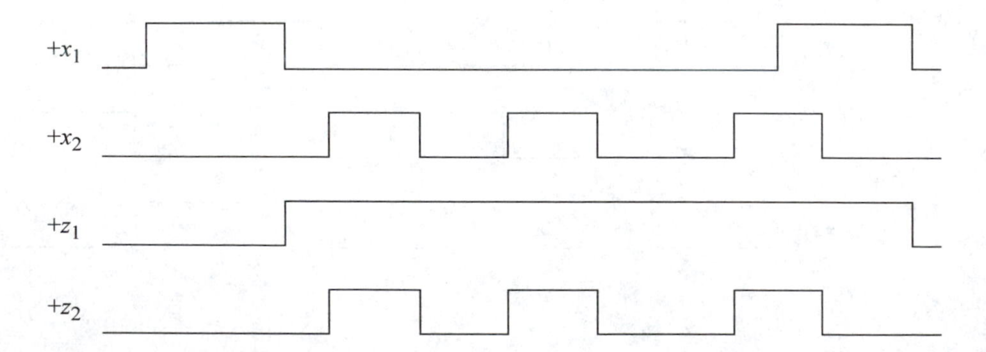

6.62 The merged flow table of Example 6.26 is redrawn below for convenience. Rearrange the rows so that adjacency requirements can be satisfied within the eight rows, where possible. Augment the merged flow table with eight additional rows, 9 through 16. The additional rows contain unspecified entries which will be used to provide intermediate transient states such that, contiguous rows in a cycle will be adjacent. That is, every state transition sequence will pass through one or more transient states, in which the values of the state variables are adjacent to both the preceding and the following state variables values. Obtain the augmented merged flow table, the transition diagram, and the combined excitation map. There is more than one correct answer.

		x_1x_2 00	01	11	10
1	(a), (b)	(a)	j	c	(b)
2	(c)	–	h	(c)	d
3	(d), (f)	a	(f)	e	(d)
4	(e)	–	f	(e)	b
5	(h), (i)	a	(h)	(i)	b
6	(j)	k	(j)	i	–
7	(k), (m)	(k)	l	(m)	b
8	(l)	a	(l)	m	–

7.1 Analysis Procedure
7.2 Synthesis Procedure
7.3 Problems

7

Pulse-Mode Asynchronous Sequential Machines

Many situations are encountered in digital engineering where the input signals occur as pulses and in which there is no periodic clock signal to synchronize the operation of the sequential machine. Typical examples which use the principles of *pulse-mode* techniques are vending machines, demand-access road intersections, and automatic toll booths.

In the presentation of synchronous sequential machines, the data input signals were asserted as voltage levels. A periodic clock input was also required such that, state changes occurred on the active clock transition. In the synthesis of asynchronous sequential machines, the input variables were also considered as voltage levels; however, there was no machine clock to synchronize state changes. The operation of pulse-mode machines is similar, in some respects, to both synchronous and asynchronous sequential machines. State changes occur on the application of input pulses which trigger the storage elements, rather than on a clock signal. The input pulses, however, occur randomly in an asynchronous manner.

In pulse-mode sequential machines, each variable of the input alphabet X is active in the form of a pulse. The duration of the pulse is less than the propagation delay of the storage elements and associated logic gates. Thus, an input pulse will initiate a state change, but the completion of the change will not take place until after the corresponding input has been deasserted. Multiple inputs cannot be active simultaneously. There is no separate clock input in pulse-mode machines.

Unlike a system clock, which has a specified frequency, the input pulses can occur randomly and more than one input pulse can generate an output. A typical example of a pulse-mode circuit is a vending machine in which coins of various denominations produce pulses that determine — together with a switch — the selection criteria for the product.

Figure 7.1 illustrates a general block diagram for a pulse-mode sequential machine. The machine is similar in structure to a Moore machine if $\lambda(Y)$ or to a Mealy machine if $\lambda(X,Y)$. The input variables are conditioned by the δ next-state combinational logic to provide inputs to the storage elements. The inputs may also be connected to the λ output logic, providing characteristics of a Mealy machine.

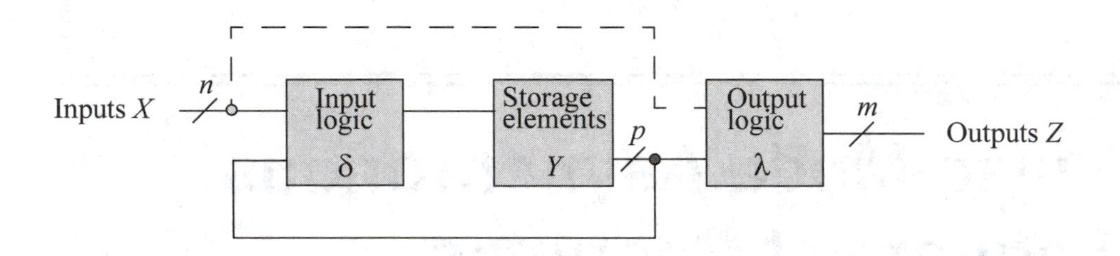

Figure 7.1 Block diagram of a pulse-mode sequential machine.

Since the storage element inputs are in the form of pulses, every term in the input equations will contain an input variable x_i. For example, the input equation to set a storage element labeled y_1 may be defined as shown in Equation 7.1, where x_1 and x_2 are pulsed inputs and y_1, y_2, and y_3 are storage elements. The outputs of a pulse-mode machine may be either voltage levels or pulses, representing Moore or Mealy outputs, respectively. A Moore output z_1 is represented as a boolean product of two or more storage elements, as shown in Equation 7.2. A Mealy output z_2 is also represented as a boolean product of storage elements with the inclusion of an input variable, as shown in Equation 7.3.

$$Sy_1 = x_2 + y_1'y_2y_3'x_1 \tag{7.1}$$

$$z_1 = y_1y_2'y_3 \tag{7.2}$$

$$z_2 = y_2'y_3x_1 \tag{7.3}$$

The storage elements in pulse-mode machines are usually level-sensitive rather than edge-triggered devices. Thus, *Set/Reset* (*SR*) latches using NAND or NOR logic are typically used in the implementation of pulse-mode machines. In order for the operation of the machine to be deterministic, some restrictions apply to the input pulses:

1. Input pulses must be of sufficient duration to trigger the storage elements.

2. The time duration of the pulses must be shorter than the minimal propagation delay through the combinational input logic and the storage elements, so that the pulses are deasserted before the storage elements can again change state.

3. The time duration between successive input pulses must be sufficient to allow the machine to stabilize before application of the next pulse.

4. Only one input pulse can be active at a time.

If the input pulse is of insufficient duration, then the storage elements may not be triggered and the machine will not sequence to the next state. If the pulse duration is too long, then the pulse will still be active when the machine changes from the present state $Y_{j(t)}$ to the next state $Y_{k(t+1)}$. The storage elements may then be triggered again and sequence the machine to an incorrect next state. If the time between consecutive pulses is too short, then the machine will be triggered while in an unstable condition, resulting in unpredictable behavior.

Since pulse inputs cannot occur simultaneously, a pulse-mode machine with n input signals can have only $n+1$ combinations of the input alphabet, instead of 2^n combinations as in synchronous sequential machines and asynchronous sequential machines that are not inherently characterized by pulse-mode operation. For example, for a pulse-mode machine with two inputs x_1 and x_2, three possible valid combinations can occur: $x_1x_2 = 00$, 10, or 01. However, since no changes are initiated by an input vector of $x_1x_2 = 00$, it is necessary to consider only the vectors $x_1x_2 = 10$ and 01 when analyzing or designing a pulse-mode asynchronous sequential machine.

Similarly, for a machine with three inputs x_1, x_2, and x_3, only the following input vectors need be considered: $x_1x_2x_3 = 100$, 010, and 001. The absence of a clock signal implies that state transitions occur only when an input is asserted.

The examples presented in this chapter will assume that pulse duration and duty cycle restrictions have been satisfied. A synthesis example will be presented in which D flip-flops are utilized to introduce delay characteristics in the output of the SR latches.

7.1 Analysis Procedure

Pulse-mode machines respond immediately to the assertion of an input signal without waiting for a clock signal. Analysis of pulse-mode sequential machines — as for synchronous sequential machines — consists of deriving a next-state table, input maps and equations, output maps and equations, and a state diagram for a given logic diagram. A timing diagram may also prove useful in analysis.

7.1.1 *SR* Latches as Storage Elements

Example 7.1 The analysis proceeds in a manner analogous to that described for synchronous sequential machines. The predominant differences are the absence of a clock signal and the input restrictions mentioned previously. A Moore pulse-mode asynchronous sequential machine is shown in Figure 7.2, and will be analyzed with respect to input and output maps and a state diagram.

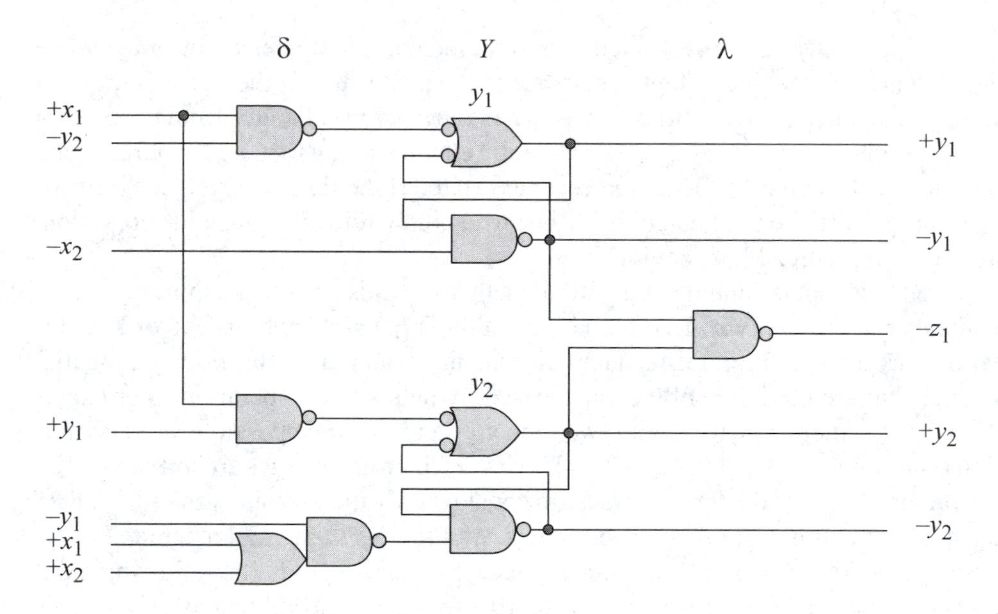

Figure 7.2 Pulse-mode Moore machine for Example 7.1.

Assume that the machine is reset initially such that, $y_1y_2 = 00$. Inputs x_1 and x_2 are assigned values of $x_1x_2 = 10$ and 01 for each specified state. The equations to set and reset y_1 and y_2 are derived from the logic diagram and are shown in Equation 7.4.

If x_1 is pulsed when the machine is in an initial reset condition, then latch y_1 will set. Input pulse x_1 must be deasserted, however, before latch y_1 stabilizes in a set condition, otherwise, the conditions to set latch y_2 would be established. If y_2 sets before x_1 is deasserted, then an erroneous next state will be generated, because the input pulse must be deasserted before the state change is received at the input logic. Assuming that the input restrictions are met, latch y_2 will remain in a reset state. Thus, from an initial condition of $y_1y_2 = 00$, a pulse on input x_1 causes the machine to proceed to state $y_1y_2 = 10$, as shown in the first row of Table 7.1. Output z_1 remains inactive. If input x_2 is pulsed in state $y_1y_2 = 00$, then both latches will receive a reset pulse coincident with the assertion of x_2; the pulse will be active for the duration of x_2. Therefore, latches y_1 and y_2 will remain reset, as shown in the second row of Table 7.1. Output z_1 remains inactive.

$$Sy_1 = y_2'x_1$$

$$Ry_1 = x_2$$

$$Sy_2 = y_1x_1$$

$$Ry_2 = y_1'(x_1 + x_2) \tag{7.4}$$

Table 7.1 Next-state table for the pulse-mode Moore machine of Figure 7.2

State name	Present state $y_1\ y_2$	Inputs $x_1\ x_2$	Next state $y_1\ y_2$	Output z_1
a	0 0	1 0	1 0	0
	0 0	0 1	0 0	0
b	1 0	1 0	1 1	0
	1 0	0 1	0 0	0
c	1 1	1 0	1 1	0
	1 1	0 1	0 1	0
d	0 1	1 0	0 0	1
	0 1	0 1	0 0	1

Assume that the machine is now in state $y_1y_2 = 10$ and that the input vectors $x_1x_2 = 10$ and 01 are applied in sequence. Using either Equation 7.4 or the logic diagram of Figure 7.2, it is evident that a pulse on x_1 will cause y_1 to remain set and y_2 to be set. If x_2 is pulsed, then y_1 will be reset and y_2 will remain reset.

Let the present state be $y_1y_2 = 11$ and the input vectors $x_1x_2 = 10$ and 01 be applied in sequence. When x_1 is pulsed, no change occurs to latch y_1, because $Sy_1 = y_2'x_1 = 01 = 0$. Thus, y_1 remains set. Latch y_2 will also remain set, because $Sy_2 = y_1x_1 = 11 = 1$. When x_2 is asserted, y_1 will be reset and y_2 will remain unchanged at $y_2 = 1$, because $Ry_2 = y_1'(x_1 + x_2) = 0(0 + 1) = 0$.

Finally, when the machine is in state $y_1y_2 = 01$, output z_1 is asserted unconditionally, due to the Moore characteristics of the output. If input x_1 is pulsed, then no change occurs to latch y_1, because $Sy_1 = y_2'x_1 = 01 = 0$. There is also no set pulse to latch y_2, because $Sy_2 = y_1x_1 = 01 = 0$. However, a reset pulse is generated for y_2 when x_1 is asserted, since $Ry_2 = y_1'(x_1 + x_2) = 1(1 + 0) = 1$. A pulse on x_2 provides a reset pulse to y_1; thus, y_1 remains reset. Latch y_2 is also reset by an x_2 pulse, because $Ry_2 = y_1'(x_1 + x_2) = 1(0 + 1) = 1$. Therefore, the assertion of x_1 or x_2 in state $y_1y_2 = 01$ returns the machine to the initial state of $y_1y_2 = 00$. Table 7.1 lists all possible states of two storage elements with the associated input vectors and the corresponding next states. Column z_1 lists the output values for the present state.

The input maps can be derived from either the set and reset equations or from the next-state table. The maps are constructed using one of two methods: separate maps for set and reset conditions or combined maps which specify both set and reset conditions in the same map.

In the first method, there are individual set and reset maps for each latch for each input variable, as shown in Figure 7.3. Thus, considering input x_1, the map representing the set condition for latch y_1 in Figure 7.3 (a) is designated as Sy_1. Both the input equations and the next-state table indicate that y_1 will be set if y_2 is in a reset state and x_1 is pulsed. Therefore, 1s are entered in minterm locations 0 and 2 in the map for Sy_1. Input x_1 is not a factor in the reset equation for y_1. Thus, the reset map for y_1, designated as Ry_1 in Figure 7.3 (a), contains a 0 in each minterm location.

Figure 7.3 (b) reflects the conditions that set or reset latch y_1 when x_2 is pulsed. Since input x_2 does not contribute to the set condition of y_1, the map for Sy_1 contains a 0 in each minterm location. The reset equation for latch y_1 specifies that y_1 will be reset when x_2 is pulsed; thus, the map for Ry_1 contains a 1 in each minterm location. Table 7.1 also indicates that an x_2 pulse will always reset y_1 regardless of the present state of the machine.

In a similar manner, the set and reset input maps for latch y_2 are derived, as shown in Figure 7.3 (c) and (d) relative to inputs x_1 and x_2, respectively. The maps that represent the set conditions for latch y_2 are labeled Sy_2; the maps that represent the reset conditions are labeled Ry_2. The set equation for y_2 is a function of y_1 and x_1. Therefore, 1s are entered in minterm locations 2 and 3 in the map of Figure 7.3 (c) labeled Sy_2. Since the reset condition for y_2 is a function of both x_1 and x_2, the input maps for Ry_2 contain 1s in the corresponding minterm locations that represent $y_1'x_1$ and $y_1'x_2$. Input x_2 is not a contributing variable in the conditions to set latch y_2; therefore, 0s are entered in each square of the map for Sy_2 in Figure 7.3 (d).

Consider now, the second method of constructing the input maps, in which each map contains both set and reset information with respect to an input variable. This method provides a more compact map, yet contains the same information as the separate set and reset maps. The combined maps will be used throughout the remainder of this chapter.

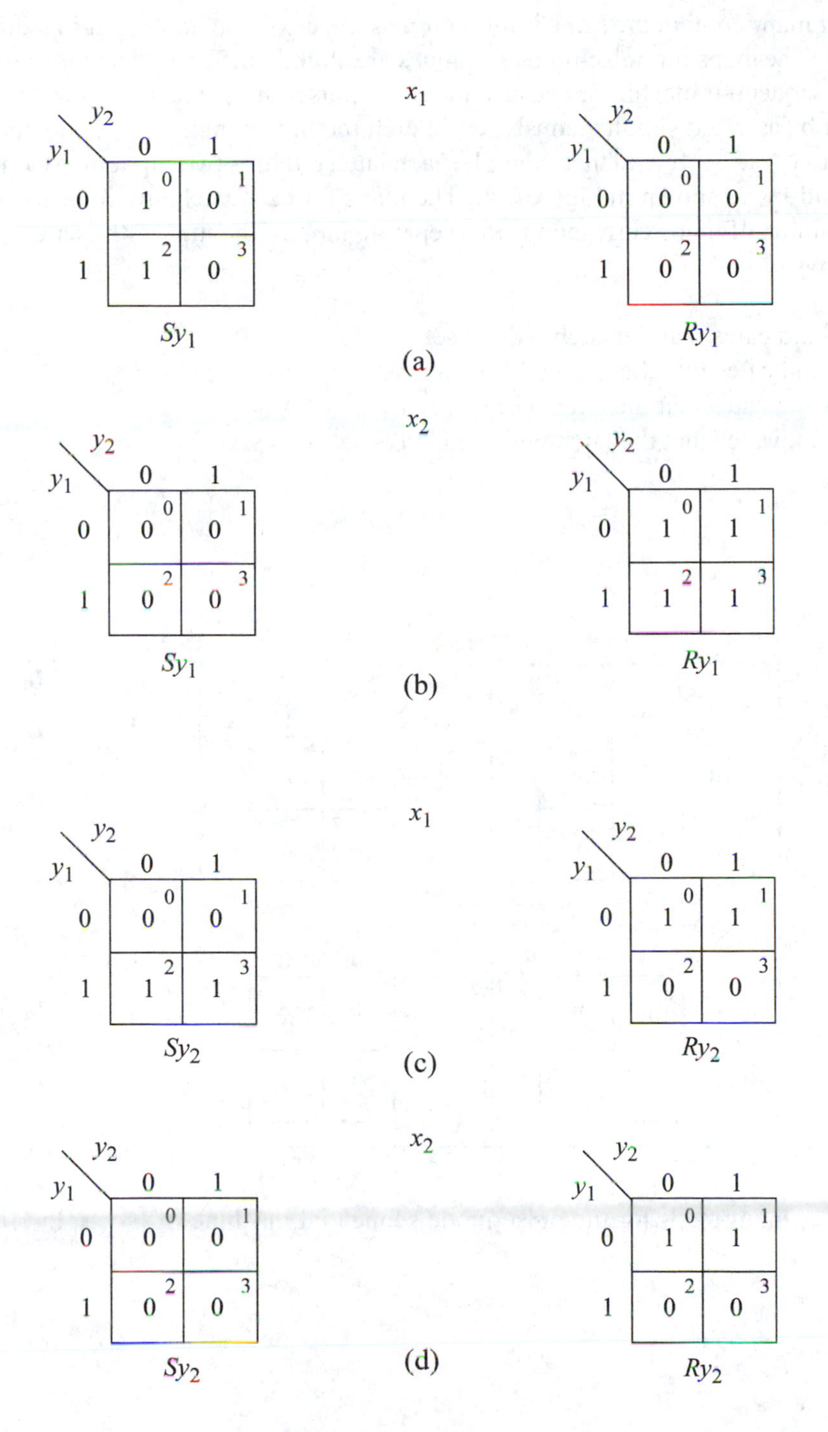

Figure 7.3 Set and reset input maps for latches y_1 and y_2 for the pulse-mode Moore machine of Example 7.1: (a) latch y_1 for input x_1; (b) latch y_1 for input x_2; (c) latch y_2 for input x_1; and (d) latch y_2 for input x_2.

The input maps contain the same information as the next-state table, but in a different format. The maps for pulse-mode machines are slightly different than those for synchronous sequential machines, because the input pulses are exclusive. Since the inputs cannot be asserted simultaneously, each latch requires n input maps, one map for each input $x_1, x_2, \cdots, x_n$. In this example, each latch requires two input maps, one each for x_1 and x_2, as shown in Figure 7.4. The maps for each latch are in the same row. Each column of maps corresponds to a separate input. The map entries are defined as follows:

S indicates that the latch will be set.
s indicates that the latch will remain set.
R indicates that the latch will be reset.
r indicates that the latch will remain reset.

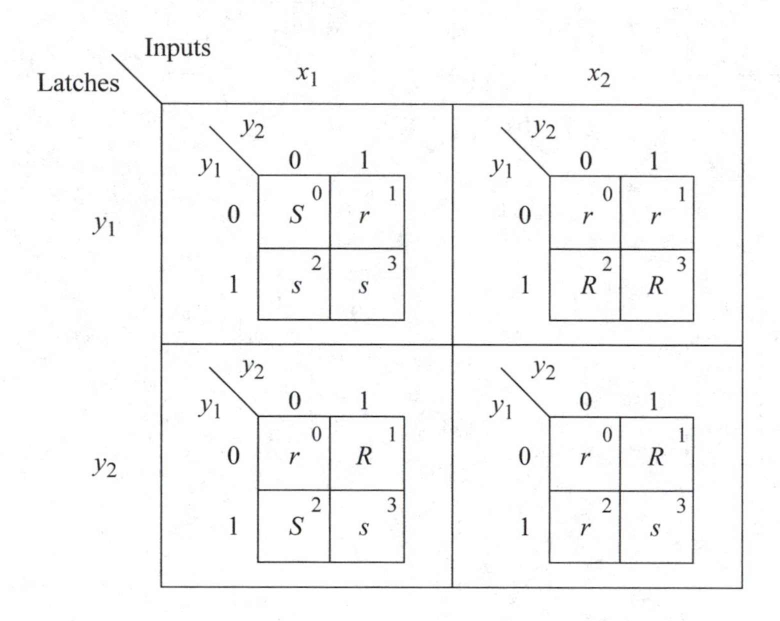

Figure 7.4 Input maps for the pulse-mode sequential machine of Example 7.1, Figure 7.2.

In the presentation which follows, the set and reset equations of Equation 7.4 in conjunction with the next-state table of Table. 7.1 will be utilized in deriving the entries for the input maps. The four maps will be constructed in parallel by considering each stable state in turn for each input vector.

Consider the input map in row y_1, column x_1 for an initial condition of $y_1y_2 = 00$ when x_1 is pulsed. The next state for latch y_1 will be 1 for an input vector of

$x_1x_2 = 10$. That is, y_1 will change from $y_1 = 0$ to 1, which represents a set condition. Thus, the letter S is inserted in minterm location 0 of the map in row y_1, column x_1. The map in row y_2, column x_1 specifies the next states for latch y_2 when x_1 is pulsed. In state $y_1y_2 = 00$, the set equation for latch y_2 is $Sy_2 = y_1x_1 = 01 = 0$. Thus, there will be no change to the state of y_2, which remains reset, as indicated by the letter r.

Consider the map for y_1 in row y_1, column x_2 for an initial condition of $y_1y_2 = 00$. The assertion of x_2 provides a reset pulse to latch y_1. Since y_1 remains in a reset state, the letter r is inserted in minterm location 0. Similarly, the letter r is entered in minterm location 0 of the map for y_2 in row y_2, column x_2, because $Ry_2 = y_1'(x_1 + x_2) = 1(0 + 1) = 1$.

Consider row y_1, column x_1, for a present state of $y_1y_2 = 01$. When x_1 is asserted, there will be no change to latch y_1, because $Sy_1 = y_2'x_1 = 01 = 0$; that is, x_1 will not produce a set pulse to y_1. Thus, y_1 remains in a reset state, as indicated by the letter r. There is also no set pulse for y_2 when x_1 is asserted, because $Sy_2 = y_1x_1 = 01 = 0$. Latch y_2, however, receives a reset pulse when x_1 is asserted, because $Ry_2 = y_1'(x_1 + x_2) = 1(1 + 0) = 1$. Since y_2 changes from $y_2 = 1$ to 0, therefore, the letter R is entered in minterm location 1.

For a present state of $y_1y_2 = 01$ in column x_2, the letters r and R are inserted in minterm location 1 for y_1 and y_2, respectively. Input x_2 provides a reset pulse to latch y_1, which remains reset. Input x_2 also provides a reset pulse for y_2 as specified by $Ry_2 = y_1'(x_1 + x_2) = 1(0 + 1) = 1$. Since the state of y_2 changes from 1 to 0, the letter R is entered in minterm location 1 of row y_2, column x_2.

In a similar manner, the remaining entries are derived. In column x_1, a pulse on input x_1 will provide either a set or reset pulse or leave the latches unchanged. In column x_2, a pulse on input x_2 will either reset the latches or leave them unchanged. Comparison of the entries in Table 7.1 with the entries in the input maps show a one-to-one correspondence for each state for identical input vectors.

When deriving the equations from the input maps, only the upper-case letters need be considered. The lower-case letters are treated as "don't care" entries and are used only if they aid in minimizing the equation. The equations derived from the input maps are identical to those shown in Equation 7.4. For example, consider the map in row y_1, column x_1. Minterms 0 and 2 can be combined, since both represent a set condition. Minterms 0 and 2 have common variables y_2' and x_1. Therefore, $Sy_1 = y_2'x_1$. The map in row y_1, column x_2, contains a reset entry in every location, indicating that x_2 will either reset latch y_1 or leave the latch in a reset state. Therefore, $Ry_1 = x_2$.

The map in row y_2, column x_1 contains the letters S and s in minterm locations 2 and 3, respectively, where s is considered a "don't care" entry. Since both squares posses the common variables y_1 and x_1, the set condition for latch y_2 is $Sy_2 = y_1x_1$. The map also contains entries of r and R in minterm locations 0 and 1, respectively, where r is considered a "don't care" entry. The same values exist in the same locations for the map in row y_2, column x_2. Therefore, the reset equation for latch y_2 is $Ry_2 = y_1'x_1 + y_1'x_2 = y_1'(x_1 + x_2)$.

Sequential machines that operate in pulse mode will not have race conditions — either noncritical or critical — because only one input is active at a time and the machine is stable even in the absence of input pulses. Each product term in the set and reset equations contains at least one input variable. Also, no input variable appears in a complemented form, because the complement of a variable corresponds to an inactive input signal.

The equation for output z_1 is obtained directly from the logic diagram and is shown in Equation 7.5. The output map is shown in Figure 7.5. If the outputs of a pulse-mode machine are required to be levels, then Moore-type outputs are generated in which the outputs are a function of the present state only. If the outputs are required to be pulses, then Mealy-type outputs are generated in which the outputs are a function of both the present state and a pulsed input.

$$z_1 = y_1'y_2 \tag{7.5}$$

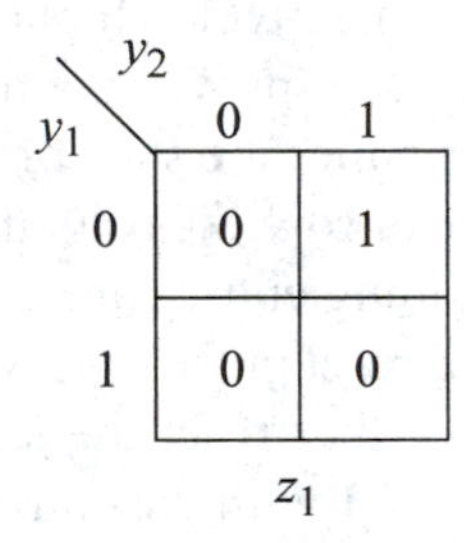

Figure 7.5 Output map for z_1 for the pulse-mode machine of Example 7.1.

The state diagram is generated from the next-state table directly. Assume that state *a* represents the initial condition of $y_1y_2 = 00$. The next state will be either *a* or *b* for an input pulse of x_2 or x_1, respectively, as shown in the next-state table of Table 7.1 and the state diagram of Figure 7.6. In state *b*, a pulse on x_2 causes the machine to return to state *a*, whereas, a pulse on x_1 sequences the machine to state *c*, as depicted by the directed lines emanating from state *b* in Figure 7.6. A transition occurs from state *c* to state *d* if x_2 is asserted; the machine remains in state *c* if x_1 is pulsed. In state *d*, the Moore-type output z_1 is asserted. The machine then proceeds to state *a* whether x_1 or x_2 is pulsed. Note that there are no complemented inputs in the state diagram, since a complemented input corresponds to the absence of that input.

A timing diagram may also be appropriate in the analysis of pulse-mode machines. An arbitrary sequence of input pulses can be applied to the machine and the corresponding output sequence generated. The state of the storage elements can also be included in the timing diagram. Observing the state and output sequences lends completeness to the analysis procedure.

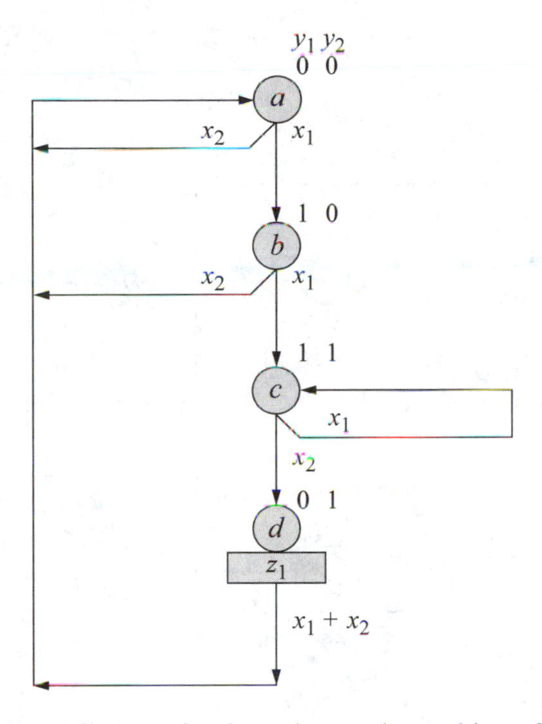

Figure 7.6 State diagram for the pulse-mode machine of Example 7.1 obtained from the next-state table of Table 7.1.

7.1.2 *SR* Latches with *D* Flip-Flops as Storage Elements

Example 7.2 A pulse-mode machine is shown in the logic diagram of Figure 7.7. The machine has two inputs x_1 and x_2 and one output z_1. The storage elements consist of *SR* latches and *D* flip-flops. The output of each latch connects to the *D* input of the associated flip-flop forming a master-slave relationship. Since the *D* flip-flops are clocked on the trailing edge of the positive input pulses, state changes are not fed back to the δ next-state logic until the active input has been deasserted. Clocking the flip-flops on the negative edge of the positive input pulses delays the next state from affecting the input logic while an input pulse is still active. Thus, the machine operates in a deterministic manner.

The purpose of analysis is to separate the machine into its constituent parts and to determine the output sequence relative to an input sequence. The machine will be analyzed with respect to its next-state table, input maps, input equations, output map, output equations, and state diagram. A representative timing diagram will also be generated. Although not shown in the logic diagram, each latch and flip-flop has a separate reset input. Assume that the machine is reset initially such that, $Ly_1 Ly_2 y_1 y_2 = 0000$.

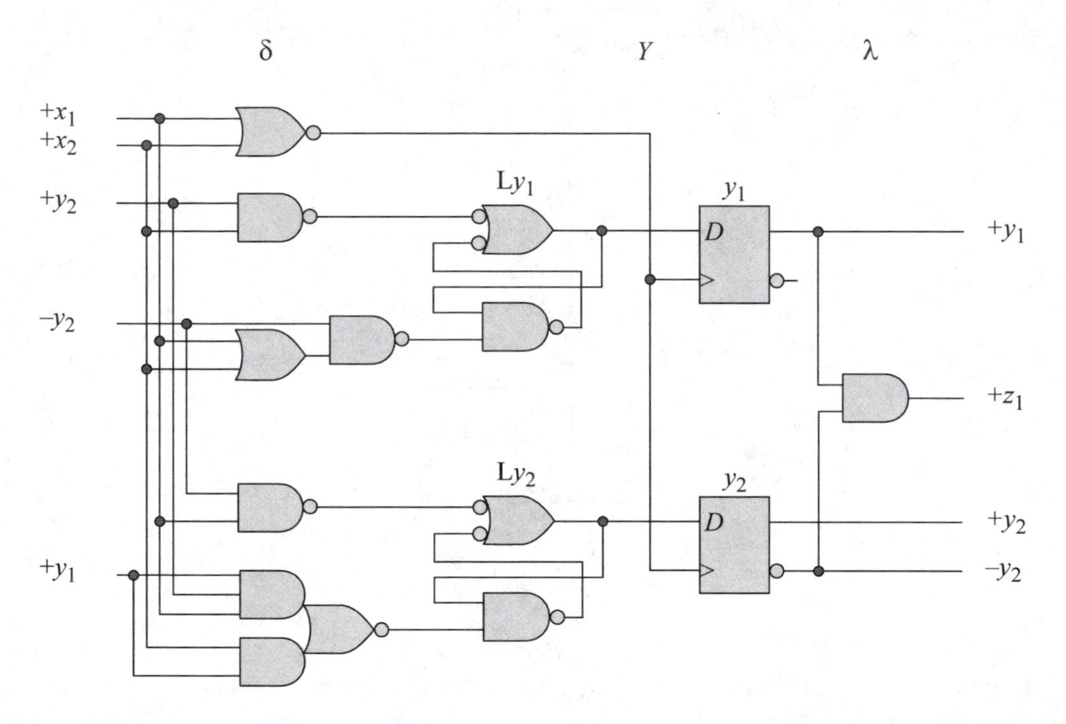

Figure 7.7 Logic diagram for the pulse-mode sequential machine of Example 7.2.

The input equations are obtained directly from the logic diagram and are shown in Equation 7.6. Latch Ly_1 will be set if flip-flop y_2 is set and x_2 is pulsed. Latch Ly_1 will be reset if y_2 is reset and either x_1 or x_2 is pulsed. Thus, the set and reset conditions for latch Ly_1 are $SLy_1 = y_2 x_2$ and $RLy_1 = y_2'(x_1 + x_2)$, respectively.

$$SLy_1 = y_2 x_2$$
$$RLy_1 = y_2' x_1 + y_2' x_2$$

$$SLy_2 = y_2' x_1$$
$$RLy_2 = y_1 y_2 x_1 + y_1 x_2 \qquad (7.6)$$

Similarly, the set and reset conditions for latch Ly_2 are obtained. Latch Ly_2 will be set if flip-flop y_2 is reset and x_1 is pulsed. Latch Ly_2 will be reset if flip-flops y_1 and y_2 are both set and x_1 is pulsed, or if flip-flop y_1 is set and x_2 is pulsed. These conditions yield the set and reset equations $SLy_2 = y_2'x_1$ and $RLy_2 = y_1y_2x_1 + y_1x_2$, respectively.

The input equations for Dy_1 and Dy_2 are not required, since the state of each latch is transferred to its associated flip-flop on the negative transition of the active input variable.

Using the equations of Equation 7.6, the next-state table can now be constructed. If the machine is in the initial reset state of $y_1y_2 = 00$ and input x_1 is pulsed, then latch Ly_1 remains reset and latch Ly_2 is set, as shown in the first row of the next-state table of Table 7.2. If x_2 is pulsed in the initial reset condition, then both latches remain reset. Output z_1 remains inactive.

Table 7.2 Next-state table for the pulse-mode Moore machine of Figure 7.7

State name	Present state $y_1\ y_2$	Inputs $x_1\ x_2$	Next state $y_1\ y_2$	Output z_1
a	0 0	1 0	0 1	0
	0 0	0 1	0 0	0
b	0 1	1 0	0 1	0
	0 1	0 1	1 1	0
c	1 1	1 0	1 0	0
	1 1	0 1	1 0	0
d	1 0	1 0	0 1	1
	1 0	0 1	0 0	1

In state $y_1y_2 = 01$, if x_1 is asserted, then latches Ly_1 and Ly_2 will remain reset and set, respectively, and the machine will not change state. If x_2 is asserted, however, then latch Ly_1 will set and latch Ly_2 will remain set, because $SLy_1 = y_2x_2 = 11 = 1$ and $Ry_2 = y_1y_2x_1 + y_1x_2 = 010 + 01 = 0$, resulting in a transition from state $y_1y_2 = 01$ to state $y_1y_2 = 11$. Output z_1 remains inactive.

In state $y_1y_2 = 11$, if x_1 is pulsed, then latch Ly_1 remains set while latch Ly_2 resets. The same set and reset conditions apply if x_2 is pulsed. Thus, a state transition will occur from $y_1y_2 = 11$ to 10 if either x_1 or x_2 is pulsed.

Finally, in state $y_1y_2 = 10$, output z_1 is asserted unconditionally, due to the Moore characteristics of the output variable. If x_1 is pulsed in state $y_1y_2 = 10$, then latch Ly_1 is reset, whereas, latch Ly_2 is set and the machine proceeds to state $y_1y_2 = 01$. If x_2 is pulsed, then latch Ly_1 is reset and latch Ly_2 remains reset, causing a transition to the initial state of $y_1y_2 = 00$.

The input maps can be derived directly from the next-state table or from the input equations. The input maps are shown in Figure 7.8, where the letters *S* and *s* indicate that the latch will be set or remain set, respectively and the letters *R* and *r* indicate that the latch will be reset or remain reset, respectively.

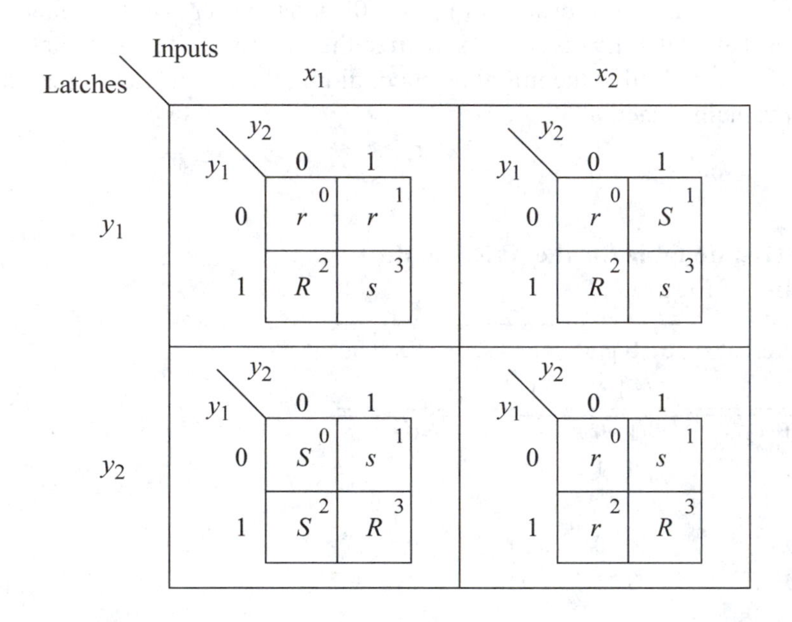

Figure 7.8 Input maps for the pulse-mode sequential machine of Example 7.2, Figure 7.7.

The map entries correlate directly to the entries in the next-state table. For example, in state $y_1y_2 = 10$, latch Ly_1 will be reset if x_1 is pulsed, as indicated by the letter *R* in minterm location 2 of the map in row y_1, column x_1. Also, in state $y_1y_2 = 10$, latch Ly_2 will be set if x_1 is pulsed, as indicated by the letter *S* in minterm location 2 of the map in row y_2, column x_1.

Now consider the effect when input x_2 is asserted in state $y_1y_2 = 10$. When x_2 is activated, a reset pulse is applied to latch Ly_1, as shown in the logic diagram and the next-state table. Since latch Ly_1 was set, the letter *R* is entered in minterm location 2 of the map for latch Ly_1 in row y_1, column x_2. The assertion of x_2 will also generate a reset pulse to latch Ly_2. However, since latch Ly_2 is already reset, an x_2 pulse causes latch Ly_2 to remain in a reset state, as specified by the letter *r* in minterm location 2 of the map for latch Ly_2 in row y_2, column x_2.

Since the pulse-mode machine of Example 7.2 is a Moore machine, the output is a function of the present state only. Thus, output z_1 is asserted if flip-flops y_1 and y_2 are set and reset, respectively, yielding the equation of Equation 7.7. The output map is shown in Figure 7.9.

$$z_1 = y_1 y_2' \tag{7.7}$$

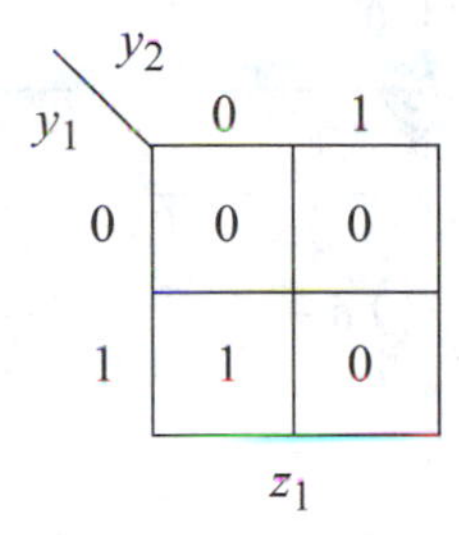

Figure 7.9 Output map for the pulse-mode sequential machine of Example 7.2.

The state diagram for the Moore pulse-mode asynchronous sequential machine of Example 7.2 is readily constructed from the next-state table. From an initial condition of $y_1y_2 = 00$ in state *a*, inputs x_1 and x_2 are applied in sequence and cause a transition to state $y_1y_2 = 01$ or 00, respectively, as shown in both the next-state table and the state diagram of Figure 7.10. In state *b* ($y_1y_2 = 01$), an x_1 pulse results in no change of state, whereas, an x_2 pulse produces a state transition to state *c* ($y_1y_2 = 11$).

In state *c* ($y_1y_2 = 11$), a pulse on either x_1 or x_2 results in a change of state to state *d* ($y_1y_2 = 10$) where output z_1 is asserted. In state *d*, an x_1 pulse causes a transition back to state *b* ($y_1y_2 = 01$) and an x_2 pulse produces a transition to state *a* ($y_1y_2 = 00$).

The state diagram provides a flowchart representation of the operation of the machine for all possible valid input vectors. The state transitions are clearly delineated, resulting in a comprehensive analysis of the operational characteristics of the machine.

Although race conditions are not possible in pulse-mode sequential machines, output glitches can occur. In Moore-type outputs, where the output signals are a function of the *D* flip-flops only, a state transition which causes two or more flip-flops to change state simultaneously, may produce an erroneous signal on the output variable. The output glitch is the result of differing propagation delays through the internal circuitry of the flip-flops.

A timing diagram completes the analysis of the pulse-mode Moore machine of Figure 7.7. The timing diagram of Figure 7.11 illustrates a representative input sequence and a corresponding output sequence. As depicted in the state diagram of

Figure 7.10, an input vector sequence of $x_1x_2 = 10, 01$ from the initial condition of state a will cause a state transition from state a to state b and then to state c. This sequence is also shown in the timing diagram.

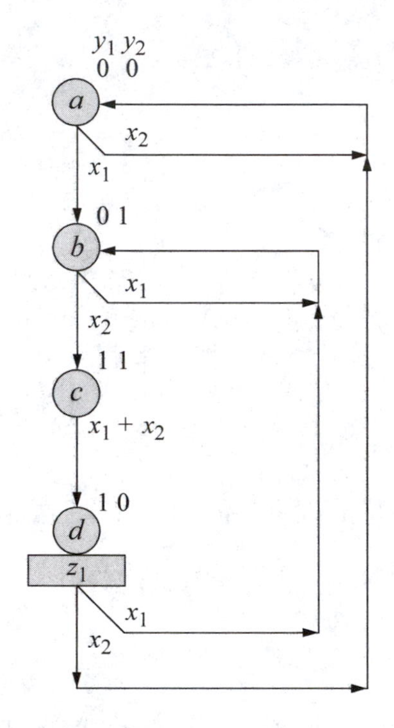

Figure 7.10 State diagram for the pulse-mode sequential machine of Example 7.2, Figure 7.7.

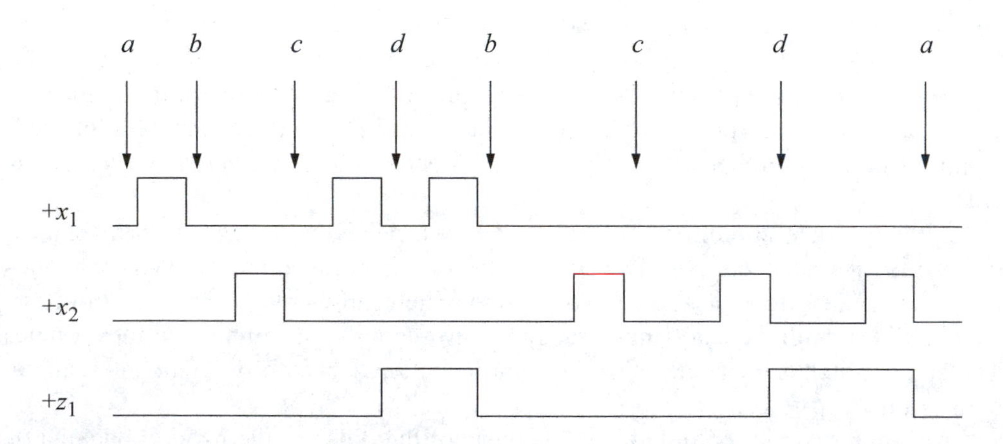

Figure 7.11 Representative timing diagram for the pulse-mode sequential machine of Figure 7.7.

In state *c*, any input pulse will produce a transition to state *d*, where z_1 is asserted. The output is asserted coincident with the trailing edge of any input pulse if and only if the preceding input sequence was an x_1 pulse followed by an x_2 pulse. Output z_1 will remain asserted in state *d* until either x_1 or x_2 is pulsed. At the negative transition of the next x_1 or x_2 input pulse in state *d*, output z_1 is deasserted and the machine proceeds to state *b* or state *a*.

Since z_1 is a Moore-type output, the assertion of z_1 is generated as a level as shown in the timing diagram of Figure 7.11, because the output is a function of the present state only. If z_1 is specified as a Mealy-type output, then the output is a function of both the present state and an input variable. The output signal, therefore, would be asserted as a pulse coincident with and for the same duration as the corresponding input pulse.

7.2 Synthesis Procedure

Due to the stringent requirements of input pulse characteristics, pulse-mode machines are less preferred than synchronous sequential machines, especially when the machine is implemented with subnanosecond logic. The synthesis concepts, however, are of sufficient importance to warrant a detailed presentation of the synthesis procedure. Reliability of pulse-mode machines can be increased by inserting delay circuits of an appropriate duration in the output networks of the storage elements. The aggregate delay of the storage elements and the delay circuit must be of sufficient duration so that the input pulse will be deasserted before the storage element output signals arrive at the δ next-state logic.

Three techniques are commonly used to insert delays in the storage element outputs: An even number of inverters are connected in series with each latch output; a linear delay circuit is connected in series with each latch output; or an edge-triggered *D* flip-flop is connected in series with each latch output. The flip-flops are set to the state of the latches, but are triggered on the trailing edge of the input pulses. Thus, the flip-flop outputs — and therefore the state of the machine as represented by the *SR* latch outputs — are received at the δ next-state logic only when the active input pulse has been deasserted. The *SR* latches and the *D* flip-flops constitute a master-slave relationship.

The synthesis procedure will be described using several different types of storage elements: *SR* latches; *SR* latches configured as *T* flip-flops; *SR* latches configured as *T* flip-flops with separate set and reset inputs; and *SR* latches with *D* flip-flops arranged in a master-slave configuration.

7.2.1 *SR* Latches as Storage Elements

Example 7.3 Since *SR* latches are used in the implementation of this machine, the *SR* latch characteristics will be reviewed. Table 7.3 lists all possible state transitions of a latch with corresponding *SR* values. Referring to rows 1 and 3, if the present state is $Y_{j(t)} = 0$ and input S is inactive, then the next state $Y_{k(t+1)}$ will also be 0. Thus, a 0-to-0 transition requires that input S remain deasserted, but input R can be either asserted or not asserted.

Table 7.3 *SR* latch characteristics

Row	Present state $Y_{j(t)}$	Inputs S R	Next state $Y_{k(t+1)}$	State transition sequence
1	0	0 0	0	$0 \rightarrow 0$
2	1	0 0	1	$1 \rightarrow 1$
3	0	0 1	0	$0 \rightarrow 0$
4	1	0 1	0	$1 \rightarrow 0$
5	0	1 0	1	$0 \rightarrow 1$
6	1	1 0	1	$1 \rightarrow 1$
7	–	1 1	–	Invalid sequence
8	–	1 1	–	Invalid sequence

Likewise, in rows 2 and 6, the latch will remain set if the present state is $Y_{j(t)} = 1$ and input R is inactive. Thus, a 1-to-1 transition requires that input R remain deasserted, but input S can be either asserted or not asserted. The requirements for a 1-to-0 transition (line 4) specify that inputs S and R must be deasserted and asserted, respectively. The final sequence is a 0-to-1 transition (line 5), which specifies that inputs S and R must be asserted and deasserted, respectively. There are no state values for lines 7 and 8, because $SR = 11$ is an invalid condition.

Using SR latches, a Mealy machine will be synthesized that operates according to the following specifications: The machine has two inputs x_1 and x_2 and one output z_1. The inputs are pulses and will never be active concurrently. Output z_1 is also a pulse and is asserted coincident with x_2 whenever x_2 immediately follows exactly two x_1 pulses, as shown in Figure 7.12. No output will be generated for three or more consecutive x_1 pulses. For this occurrence, the machine will be reinitialized by the next x_2 pulse. Assume that timing restrictions for pulse width and duty cycle have been satisfied.

The operation of the machine is graphically depicted by the state diagram of Figure 7.13. Since the inputs will not be asserted simultaneously, the state transition sequence depends on the occurrence of only a single pulse, either x_1 or x_2. The machine is initialized to state a ($y_1y_2 = 00$). Exactly two consecutive x_1 pulses will sequence the machine to state c ($y_1y_2 = 11$). If an x_2 pulse immediately follows the two consecutive x_1 pulses, then output z_1 is asserted coincident with the x_2 pulse and the machine proceeds to state *a* where the process repeats. If, however, there are three consecutive x_1 pulses, then the machine proceeds to state *d* and remains in state *d* until an x_2 input pulse sequences the machine to state *a* where the process repeats.

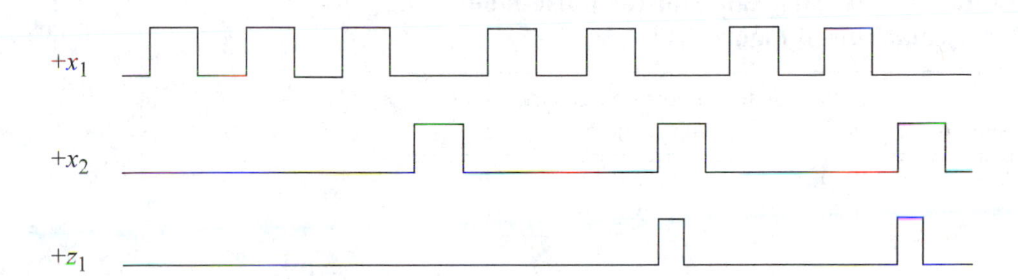

Figure 7.12 Representative timing diagram for the pulse-mode Mealy machine of Example 7.3.

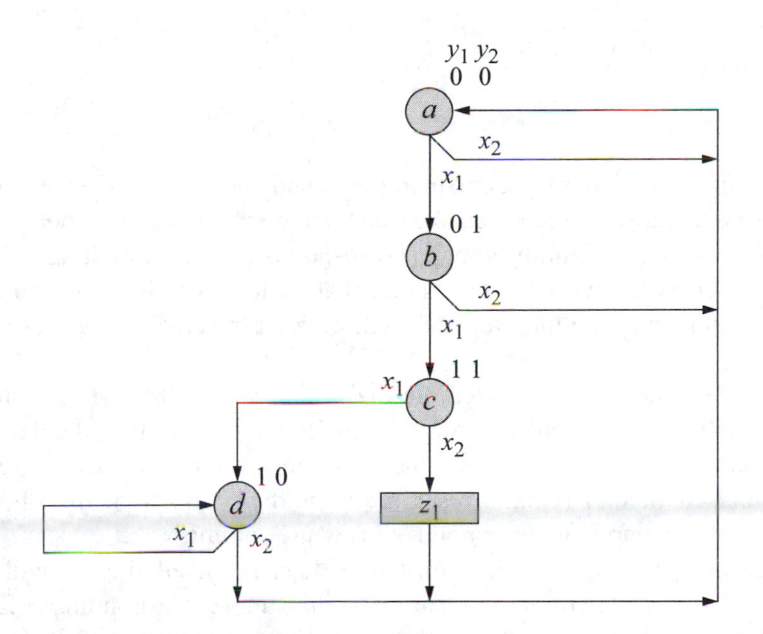

Figure 7.13 State diagram for the pulse-mode Mealy machine of Example 7.3.

Table 7.4 presents the same information as the state diagram, but in a tabular representation. Although there are two input variables, only one input can be asserted at a time; therefore, only two combinations are listed for x_1 and x_2: x_1x_2 = 10 and 01.

The remaining two combinations, $x_1x_2 = 00$ and 11, are not used in pulse-mode synthesis. If $x_1x_2 = 00$, then the machine does not change state; if $x_1x_2 = 11$, then this represents an invalid combination, since input pulses cannot occur simultaneously.

Table 7.4 Next-state table for the pulse-mode Mealy machine of Figure 7.13

State name	Present state $y_1\ y_2$	Inputs $x_1\ x_2$	Next state $y_1\ y_2$	Output z_1
a	0 0	1 0	0 1	0
	0 0	0 1	0 0	0
b	0 1	1 0	1 1	0
	0 1	0 1	0 0	0
c	1 1	1 0	1 0	0
	1 1	0 1	0 0	1
d	1 0	1 0	1 0	0
	1 0	0 1	0 0	0

Each latch requires two input maps, one map for x_1 and one map for x_2, as shown in Figure 7.14. The maps are arranged such that, the maps corresponding to each latch are in the same row, and each column of maps corresponds to a separate input. The map entries are defined as follows: *S* and *s* indicate that the latch will be set or remain set, respectively; *R* and *r* indicate that the latch will be reset or remain reset, respectively.

Refer to Table 7.4 and Figure 7.14 for the discussion which follows. In state *a* ($y_1y_2 = 00$) for latch y_1, if x_1 is pulsed, then y_1 will remain reset. Thus, the letter *r* is inserted in minterm location 0 in the map that corresponds to row y_1, column x_1. Likewise, if x_2 is pulsed, then y_1 remains reset, requiring the letter *r* to be inserted in minterm location 0 of the map that corresponds to row y_1, column x_2.

Next, consider the operation of latch y_2 in state *a*. If x_1 is pulsed, then y_2 will be set. Thus, the letter *S* is placed in minterm location 0 of the map corresponding to row y_2, column x_1. If input x_2 is pulsed in state *a*, then latch y_2 remains reset and the letter *r* is inserted in minterm location 0 of the input map in row y_1, column x_2.

Referring to state *b* ($y_1y_2 = 01$) for latch y_1, if x_1 is pulsed, then latch y_1 sequences from 0 to 1. Thus, y_1 is set and the letter *S* is placed in minterm location 1 of the map in row y_1, column x_1. If x_2 is pulsed in state *b*, then y_1 remains reset, requiring the letter *r* to be inserted in minterm location 1 of the map in row y_1, column x_2.

Next, consider the transition for y_2 in state *b*. If x_1 is pulsed, then latch y_2 executes a 1-to-1 transition and remains set. This sequence necessitates that the letter *s* be inserted in minterm location 1 of the map in row y_2, column x_1. If x_2 is pulsed, then

y_2 changes from 1 to 0; that is, y_2 is reset. Therefore, the letter R is placed in minterm location 1 of the map in row y_2, column x_2.

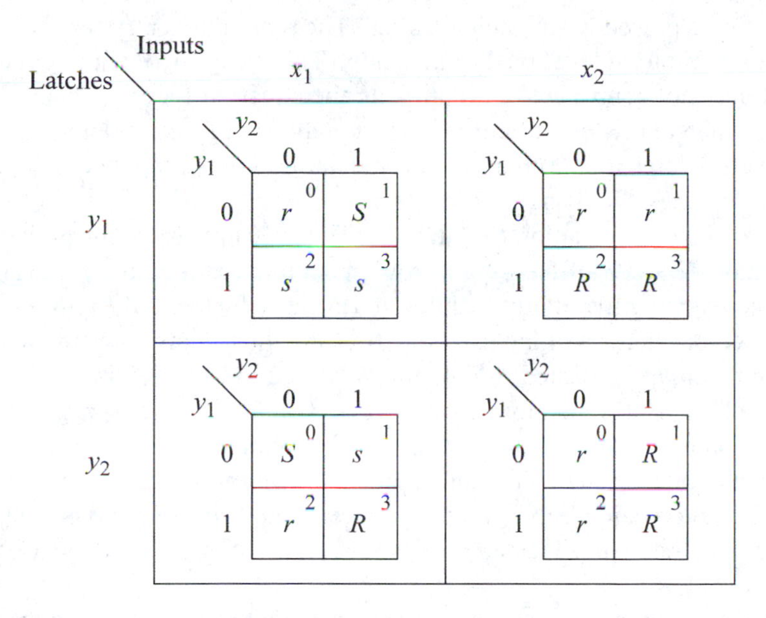

Figure 7.14 Input maps for the pulse-mode Mealy machine of Figure 7.13.

In a similar manner, the map entries are obtained for states c ($y_1y_2 = 11$) and d ($y_1y_2 = 10$). In state c, an x_1 pulse will cause y_1 to remain set (s) and y_2 to be reset (R); an x_2 pulse will reset (R) both y_1 and y_2. Output z_1 will be asserted for a duration which is the shorter of the two propagation delays to reset y_1 and y_2 plus the delay through the AND gate. In state d, x_1 will cause y_1 to remain set (s) and y_2 to remain reset (r). Finally, in state d, if x_2 is pulsed, then y_1 is reset (R) and y_2 remains reset (r).

When obtaining the equations for the latches from the input maps, all uppercase letters must be considered, whereas, lowercase letters are used only if they facilitate minimizing the corresponding set or reset equation. The input equations are shown in Equation 7.8.

$$Sy_1 = y_2x_1$$
$$Ry_1 = x_2$$

$$Sy_2 = y_1'x_1$$
$$Ry_2 = y_1x_1 + x_2 \tag{7.8}$$

Consider the input maps for latch y_1. The only entry that must be considered for setting the latch is the *S* entry in minterm location 1 of the map in row y_1, column x_1. Minterm locations 1 and 3 of this map combine to yield $Sy_1 = y_2x_1$, because both minterms contain a set function and y_2 is common to both terms. That is, latch y_1 will be set (*S*) if input x_1 is pulsed when latch y_2 is set. The remaining entry for setting y_1 is the lowercase *s* in minterm location 2 of this map. This entry can be ignored, however, because it does not contribute to minimizing the equation for Sy_1. There is no uppercase *S* in the map in row y_1, column x_2. The conditions that cause latch y_1 to remain set are not considered as a primary factor in determining the input equation, because the state of y_1 does not change.

Consider now, the reset equation for latch y_1. The only uppercase *R*s are located in minterm locations 2 and 3 of the map in row y_1, column x_2. Minterm locations 2 (*R*) and 3 (*R*) combine with minterm locations 0 (*r*) and 1 (*r*) to yield $Ry_1 = x_2$; that is, every square contains a form of the reset function and thus, contributes to the reset equation. Since this map is located in column x_2 of latch y_1, therefore, y_1 is reset whenever x_2 is pulsed, as is evident in Table 7.4.

The set equation for latch y_2 is derived in a similar manner. The only uppercase *S* is located in minterm location 0 of the map in row y_2, column x_1. Both minterm locations 0 (*S*) and 1 (*s*) contain a form of the set function, and since $y_1 = 0$ is common to both minterms and the map is located in column x_1, therefore, the set equation for y_2 is $Sy_2 = y_1'x_1$.

The reset equation for latch y_2 contains two terms, both resulting from an uppercase *R* in the two input maps for latch y_2. Using the map in row y_2, column x_1, minterm locations 2 (*r*) and 3 (*R*) combine to yield the reset term y_1x_1. The second term in the reset equation for latch y_2 results from combining all four minterms in the input map in row y_2, column x_2, yielding the reset term x_2. Thus, the complete reset equation for latch y_2 is $Ry_2 = y_1x_1 + x_2$.

Figure 7.15 shows the output map for z_1. Since z_1 is a function not only of the present state but also of the present input x_2, the equation for z_1 includes the x_2 input variable, as shown in Equation 7.9. Thus, output z_1 is asserted when the machine is in state *c* ($y_1y_2 = 11$) and input x_2 is pulsed. The duration of z_1 will not exceed the duration of the x_2 pulse.

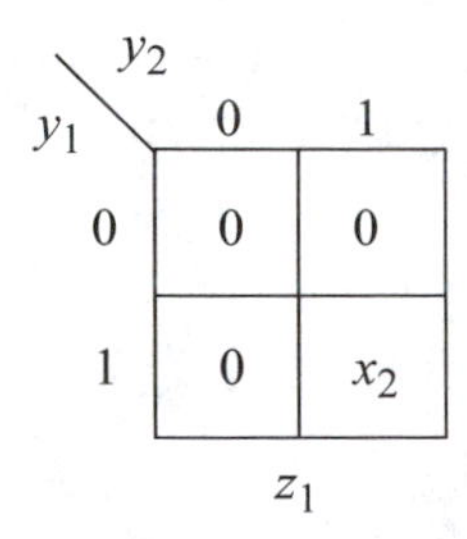

Figure 7.15 Output map for z_1 for the pulse-mode Mealy machine of Figure 7.13.

$$z_1 = y_1y_2x_2 \tag{7.9}$$

Figure 7.16 illustrates the logic diagram for this Mealy machine. The machine is initialized to state a (y_1y_2 = 00). By pulsing inputs x_1 and x_2 in the appropriate sequence, the machine proceeds through the various states depicted in the state diagram of Figure 7.13. It is important to remember that the input pulses must be deasserted before the latch output signals arrive at the δ next-state logic. This is an extremely significant consideration to ensure correct operation of the machine. If the duration of the x_1 pulse is too long in state a, then latch y_1 will be set, resulting in an invalid state change.

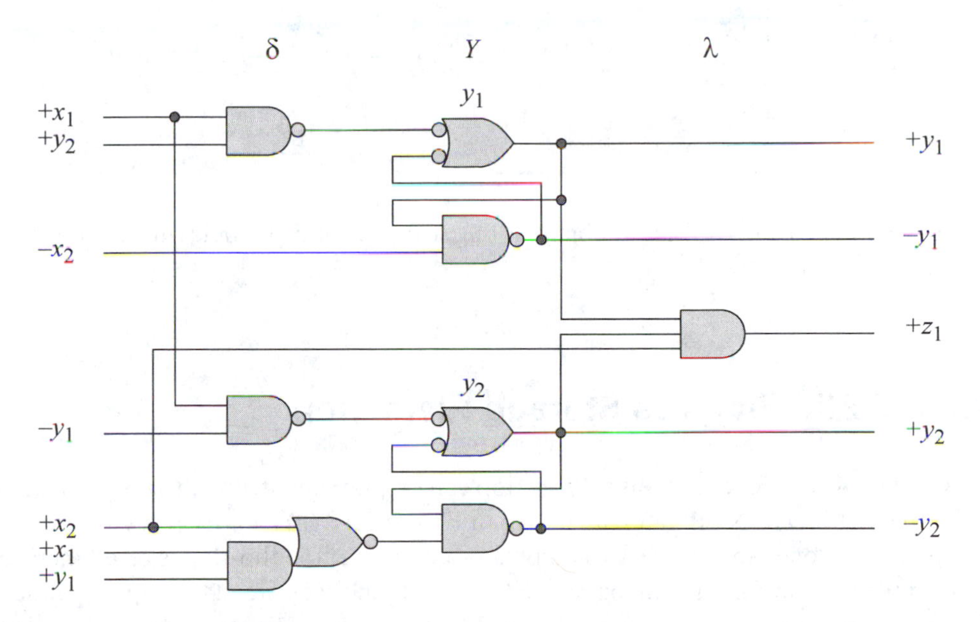

Figure 7.16 Logic diagram for the pulse-mode Mealy machine of Figure 7.13.

In state a (y_1y_2 = 00), both latches are reset. If input x_1 is asserted, then y_1 will remain reset, but y_2 will be set. Input x_1 must be deasserted before the output of y_2 conditions the NAND gate that sets y_1. Latch y_2 will then function autonomously and be set due to the feedback connections in the latch, even though x_1 has been deasserted.

The machine is now in state b (y_1y_2 = 01). If x_1 is again pulsed, the output of latch y_1 will be asserted high. The x_1 pulse must now be deasserted so that latch y_2 will not be reset due to the y_1x_1 term in the reset logic. With the x_1 set pulse deasserted, latch y_1 operates without further control and enters the latched, or set, state.

The machine is now in state c (y_1y_2 = 11). If x_2 is now asserted, output z_1 will be asserted coincident with the x_2 pulse and both latches will begin to reset. When x_2 is deasserted, output z_1 will be deasserted and both latches will be reset, returning the machine to state a ($y_1y_2 = 00$). To obtain a longer pulse for z_1, the delay circuit shown in Figure 7.17 can be inserted in the reset path for the flip-flops. As stated previously, the critical factor in the synthesis of pulse-mode sequential machines is controlling the pulse duration of the input signals. If the pulse duration is less than the minimal width to trigger a latch or greater than the maximal width that ensures only one state change, then the machine will not function reliably in a deterministic manner.

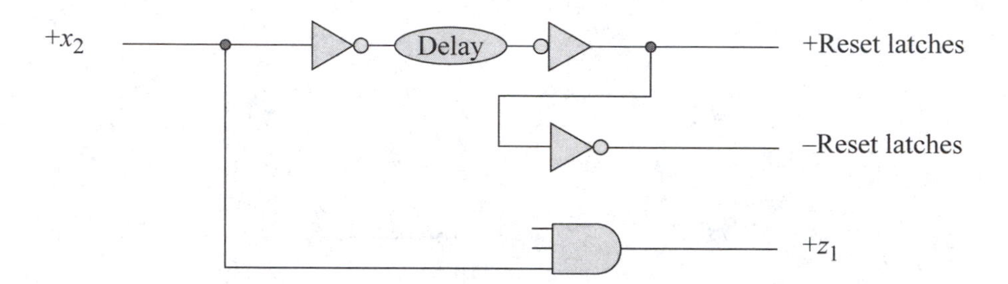

Figure 7.17 Delay circuit to obtain a longer pusle for z_1 by delaying the reset of the latches.

7.2.2 *T* Flip-Flops as Storage Elements

Example 7.4 A level-sensitive *T* flip-flop can be implemented with an *SR* latch using either NAND or NOR gates, as shown in Figure 7.18 (a) and (b), respectively. A *T* flip-flop has one input *T*, and two outputs y_1 and y_1'. If the flip-flop is reset, then an active pulse on the *T* input will toggle the flip-flop to the set state; if the flip-flop is set, then a pulse on the *T* input will toggle the flip-flop to the reset state. The *T* flip-flop characteristics are shown in Table 7.5.

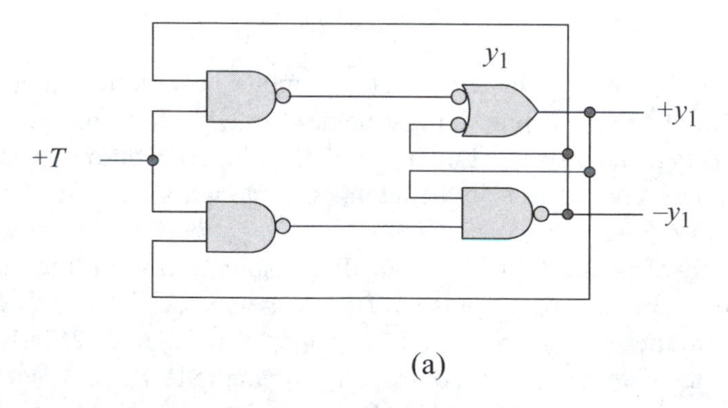

Figure 7.18 Implementation of a *T* flip-flop using an *SR* latch: (a) using NAND gates; and (b) using NOR gates.

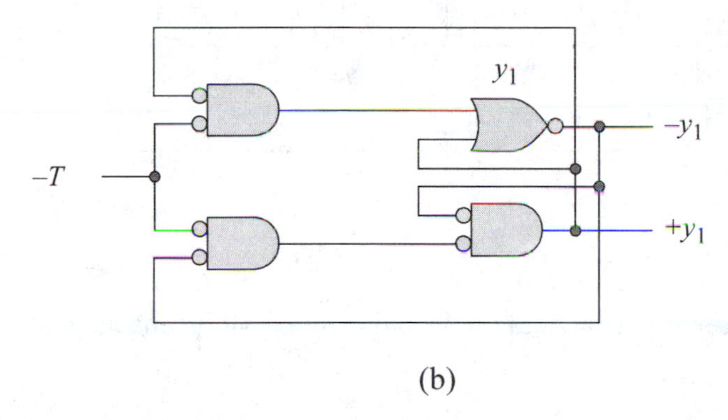

(b)

Figure 7.18 (Continued)

In this example, T flip-flops will be used as the storage elements for a Moore machine which operates according to the following specifications: A pulse on input x_2 will assert output z_1. Output z_1 will be deasserted by the second x_1 pulse in a sequence of consecutive x_1 pulses, but only if the two x_1 pulses are preceded by an x_2 pulse, as shown in the timing diagram of Figure 7.19.

Table 7.5 ***T* flip-flop characteristics**

Present state $Y_{j(t)}$	Input T	Next state $Y_{k(t+1)}$	State transition sequence
0	0	0	$0 \rightarrow 0$
1	0	1	$1 \rightarrow 1$
0	1	1	$0 \rightarrow 1$
1	1	0	$1 \rightarrow 0$

Figure 7.20 depicts the state diagram for this machine. Since this is a Moore machine, the choice of state codes is critical to ensure that output z_1 will not glitch. The machine is initially reset to state a ($y_1 y_2 = 00$). Pulses on input x_1 maintain the machine in state a, whereas an x_2 pulse sequences the machine to state b, where output z_1 is asserted as a level. The continued assertion of x_2 pulses in state b causes the machine to remain in state b, according to the machine specifications. The next x_1 pulse sequences the machine to state c, where z_1 remains asserted. It is only after the second x_1 pulse that the machine proceeds to state a, where z_1 is deasserted and the process repeats.

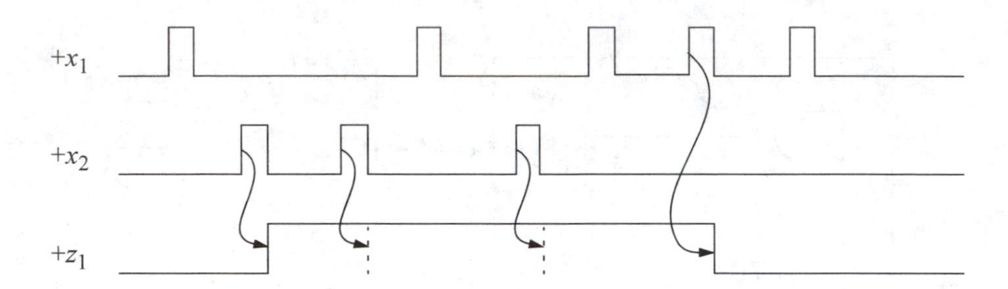

Figure 7.19 Representative timing diagram for the pulse-mode Moore machine of Example 7.4.

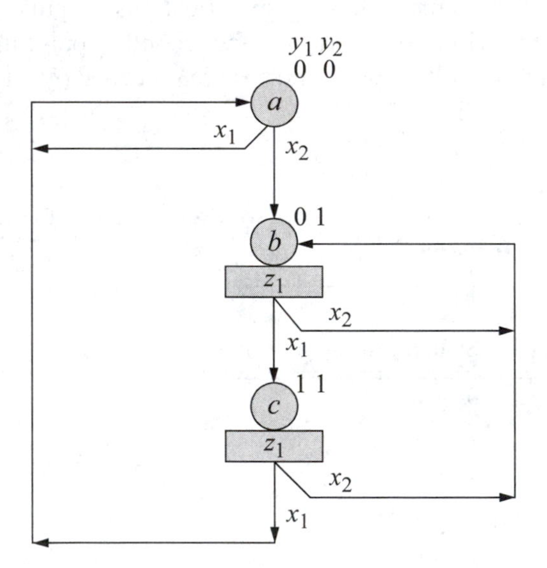

Figure 7.20 State diagram for the pulse-mode Moore machine of Example 7.4. There is one unused state: $y_1y_2 = 10$.

Figure 7.21 shows the same timing diagram as Figure 7.19, but illustrates the pulse duration in relationship with the assertion and deassertion of flip-flops y_1 and y_2. Note that input pulses x_1 and x_2 have both become inactive before either flip-flop y_1 or y_2 has changed state; that is, the combined propagation delay of the δ next-state logic and the flip-flops is longer than the duration of the input pulse. It is only under these conditions that the machine will operate reliably.

The next-state table is shown in Table 7.6. As in the previous example, only the exclusive disjunctive values of inputs x_1 and x_2 are listed; that is, $x_1 x_2 = 10$ or 01, since inputs cannot be active concurrently. Output z_1 is asserted in states b and c. There is one unused state: $y_1 y_2 = 10$.

The input maps are illustrated in Figure 7.22 using the same format as in the previous example. The maps representing flip-flop y_1 are in the row corresponding to y_1; the maps representing flip-flop y_2 are in row y_2. The input variables x_1 and x_2 denote the column headings. The following different types of map entries are shown in Figure 7.22 and are obtained directly from the state diagram using the attributes of the *T* flip-flop:

> *T* indicates that the flip-flop toggles from 0 to 1 of from 1 to 0.
> *s* indicates that the flip-flop remains set.
> *r* indicates that the flip-flop remains reset.

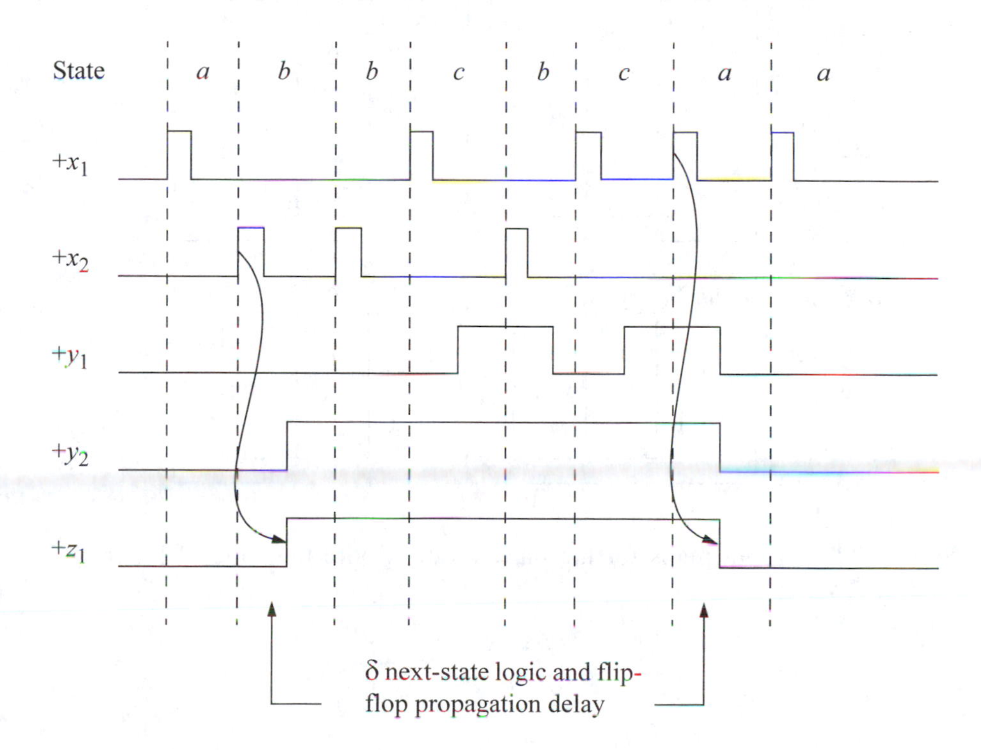

Figure 7.21 Representative timing diagram for the pulse-mode Moore machine of Figure 7.20 showing input pulse duration relative to the δ next-state logic and flip-flop propagation delay.

Table 7.6 Next-state table for the pulse-mode Moore machine of Figure 7.20

State name	Present state $y_1\ y_2$	Inputs $x_1\ x_2$	Next state $y_1\ y_2$	Output z_1
a	0 0	1 0	0 0	0
	0 0	0 1	0 1	0
b	0 1	1 0	1 1	1
	0 1	0 1	0 1	1
c	1 1	1 0	0 0	1
	1 1	0 1	0 1	1

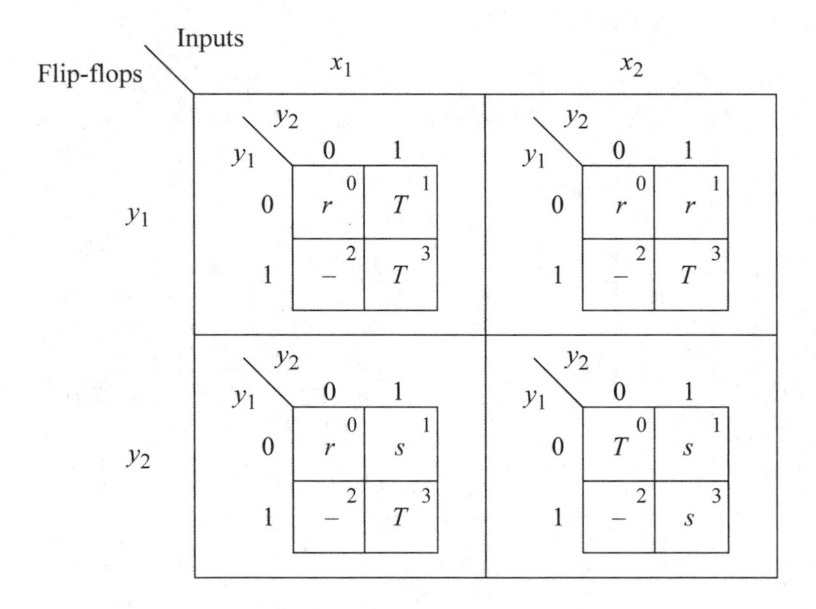

Figure 7.22 Input maps for the pulse-mode Moore machine of Figure 7.20 for Example 7.4.

In state a ($y_1y_2 = 00$), a pulse on input x_1 maintains state a as the present state; that is, both y_1 and y_2 remain reset. This is indicated by the letter r in minterm location 0 of the maps in column x_1. If input x_2 is pulsed in state a, then y_1 remains reset and y_2 toggles from 0 to 1, as indicated by the entries r and T in minterm location 0 for flip-flops y_1 and y_2, respectively.

In state b ($y_1 y_2 = 01$), output z_1 is asserted unconditionally. An x_1 pulse will toggle flip-flop y_1 from 0 to 1 while flip-flop y_2 remains set, as indicated by the entries T and s, respectively, in the maps of column x_1. In state b, an x_2 pulse causes the machine to remain in state b, where z_1 maintains its active state. Thus, the letters r and s are placed in minterm location 1 of the maps in column x_2, for y_1 and y_2, respectively. The machine then proceeds to state c, where z_1 remains asserted. Output z_1 will not be deasserted in the transition between states.

In state c ($y_1 y_2 = 11$), an x_1 pulse causes the machine to proceed to state a ($y_1 y_2 = 00$), where both y_1 and y_2 toggle from 1 to 0. Thus, minterm location 3 of the maps in column x_1 contain the entry T for flip-flops y_1 and y_2. If input x_2 is pulsed in state c, then the machine returns to state b, where flip-flop y_1 toggles from 1 to 0 and y_2 remains set, as indicated by the entries T and s in minterm location 3 of the maps in column x_2 for y_1 and y_2, respectively.

To derive the input equations, refer to the input maps of Figure 7.22. The entries that must be considered are the T entries, since these are the only entries for a T flip-flop that result in a change of state for y_1 and y_2. The unused state $y_1 y_2 = 10$ can be used for minimization. The s and r entries cannot combine with the T in the minimization process, since these entries maintain a constant flip-flop state, whereas a T will change the state of the corresponding flip-flop.

The equations for toggling flip-flops y_1 and y_2 consist of the terms shown in Equation 7.10. Refer to the input map in row y_1, column x_1, state $y_1 y_2 = 01$. Flip-flop y_1 will toggle from 0 to 1 if input x_1 is pulsed and the machine is in state $y_1 y_2 = 01$ or from 1 to 0 if x_1 is pulsed in state $y_1 y_2 = 11$. The T entries in minterm locations 1 and 3 can combine, resulting in the term $y_2 x_1$. The other occurrence where y_1 will toggle (from 1 to 0) is: if x_2 is pulsed in state $y_1 y_2 = 11$. These conditions, when combined with the unused state $y_1 y_2 = 10$, yield two toggle terms, $y_2 x_1$ and $y_1 x_2$. Thus, the complete toggle equation for flip-flop y_1 is $Ty_1 = y_2 x_1 + y_1 x_2$.

The toggle equation for flip-flop y_2 is obtained in a similar manner. There are two situations in which y_2 will toggle. In state $y_1 y_2 = 11$, flip-flop y_2 will toggle from 1 to 0 if x_1 is pulsed. Using the unused state in combination with the T entry in minterm location 3 of the map in row y_2, column x_1, the toggle term $y_1 x_1$ is obtained. The second term is obtained by combining the T entry in minterm location 0 of the map in row y_2, column x_2 with the unused state to yield $y_2' x_2$. The complete equation for flip-flop y_2 is $Ty_2 = y_1 x_1 + y_2' x_2$.

$$
\begin{aligned}
Ty_1 &= y_2 x_1 + y_1 x_2 \\
Ty_2 &= y_1 x_1 + y_2' x_2
\end{aligned}
\tag{7.10}
$$

Since both flip-flops change state as the machine sequences from state c to state a, transient states $y_1 y_2 = 01$ or 10 may be entered. State $y_1 y_2 = 01$ refers to state b where z_1 is asserted. This transient state presents no hazard — z_1 simply remains asserted for a slightly longer duration. Transient state $y_1 y_2 = 10$ is an unused state.

The output map for z_1 is shown in Figure 7.23. Output z_1 will be asserted whenever flip-flop y_2 is set. The equation for z_1 is shown in Equation 7.11 and is synthesized from the input and output equations of Equation 7.10 and Equation 7.11. The storage elements are *T* flip-flops synthesized with NAND gate latches of the type shown in Figure 7.18 (a). Correct operation of the machine can be verified by applying an appropriate sequence of x_1 and x_2 pulses and observing that the machine functions in accordance with the machine specifications as depicted by the state diagram of Figure 7.20.

7.2.3 *SR-T* Flip-Flops as Storage Elements

Example 7.5 Another flip-flop that is occasionally used in pulse-mode machines is the *SR-T* flip-flop. The *SR-T* flip-flop possesses the combined operational characteristics of both the *SR* latch and the *T* flip-flop. The *SR-T* flip-flop contains three inputs, set (*S*), reset (*R*), and toggle (*T*), as shown in Figure 7.25 using NAND logic. If the flip-flop is reset ($y_1 = 0$), then a pulse on either the *S* input, the *T* input, or both will set the flip-flop. A pulse on the *R* input will have no effect. If the flip-flop is set ($y_1 = 1$), then a pulse on either the *R* input, the *T* input, or both will reset the flip-flop. A pulse on the *S* input will cause no change.

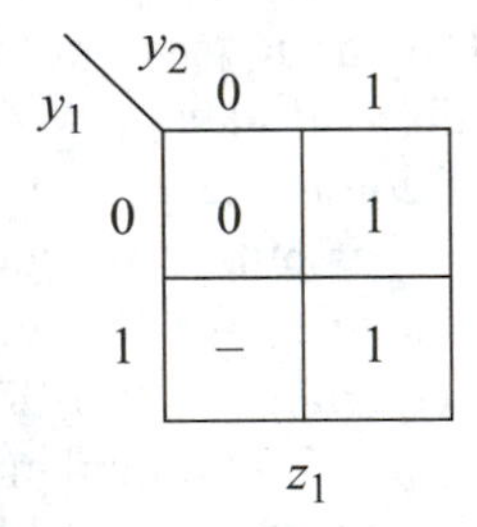

Figure 7.23 Output map for z_1 for the pulse-mode Moore machine of Figure 7.20.

$$z_1 = y_2 \tag{7.11}$$

Some timing restrictions apply to the input signals. Simultaneous pulses on both the *S* and *R* inputs is an invalid combination of input signals, because the resulting state of the flip-flop is indeterminate. When the flip-flop is set, simultaneous pulses on the *S* and *T* inputs is invalid — the flip-flop would attempt to set to 1 and toggle from 1 to 0 at the same time. This situation is negated, however, by the inclusion of a NAND gate in the set logic which allows the set pulse to be propagated only if the flip-flop is reset. Also, simultaneous pulses on the *R* and *T* inputs is invalid when the flip-flop is reset — the flip-flop would attempt to reset to 0 and toggle from 0 to 1 at the same time. The *R* input, however, is propagated only if the flip-flop is set. The operating characteristics are summarized in Table 7.7.

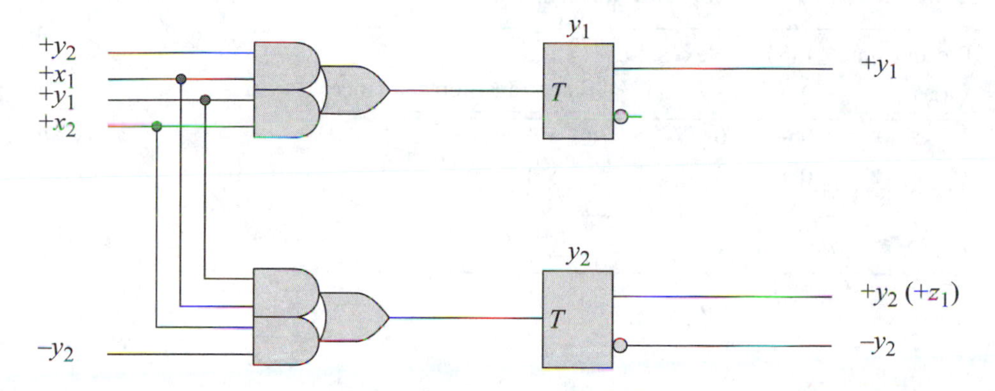

Figure 7.24 Logic diagram for the pulse-mode Moore machine of Figure 7.20 using positive-input T flip-flops for the storage elements.

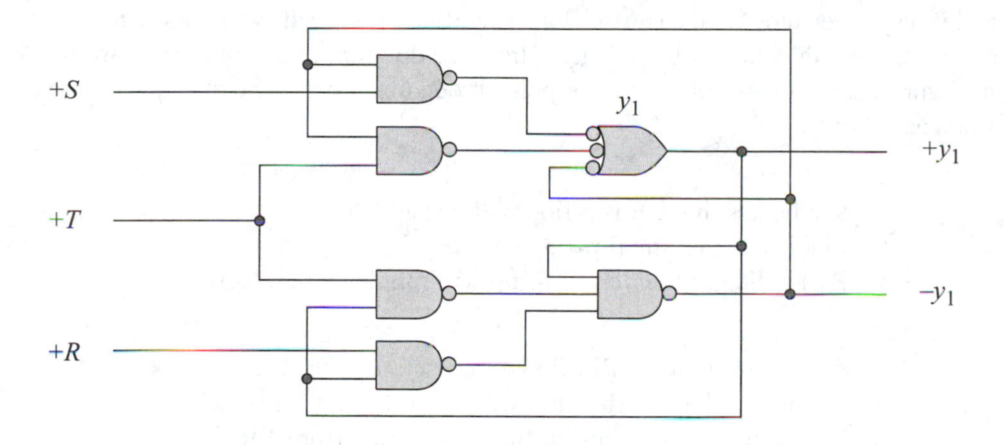

Figure 7.25 The SR-T flip-flop using NAND logic.

Table 7.7 *SR-T* flip-flop characteristics

Present state $Y_{j(t)}$	Inputs $S\,R\,T$	Next state $Y_{k(t+1)}$	State transition sequence
0	0 – 0	0	$0 \rightarrow 0$
0	1 0 –	1	$0 \rightarrow 1$
0	– 0 1	1	$0 \rightarrow 1$

Continued on next page

Table 7.7 ***SR-T*** **flip-flop characteristics**

Present state $Y_{j(t)}$	Inputs $S\ R\ T$	Next state $Y_{k(t+1)}$	State transition sequence
1	0 – 1	0	$1 \rightarrow 0$
1	0 1 –	0	$1 \rightarrow 0$
1	– 0 0	1	$1 \rightarrow 1$

The Mealy machine of Example 7.3 in Section 7.2.1 will be redesigned using *SR-T* flip-flops. For convenience, the timing diagram is redrawn in Figure 7.26. Recall that output z_1 is asserted as a function of input x_2 if x_2 is preceded by exactly two x_1 pulses. Output z_1 is asserted coincident with the assertion of x_2. The operation of the machine is graphically depicted by the state diagram of Figure 7.27 in which the state codes have been redefined. The machine specifications are also tabulated in the next-state table of Table 7.8 as obtained from the state diagram. The input maps are shown in Figure 7.28 in the usual format for pulse-mode machines. The map entries are defined as follows:

S indicates that the flip-flop will be set.
s indicates that the flip-flop will remain set.
T(*S*) indicates that the flip-flop will toggle from 0 to 1.

R indicates that the flip-flop will reset.
r indicates that the flip-flop will remain set.
T(*R*) indicates that the flip-flop will toggle from 1 to 0.

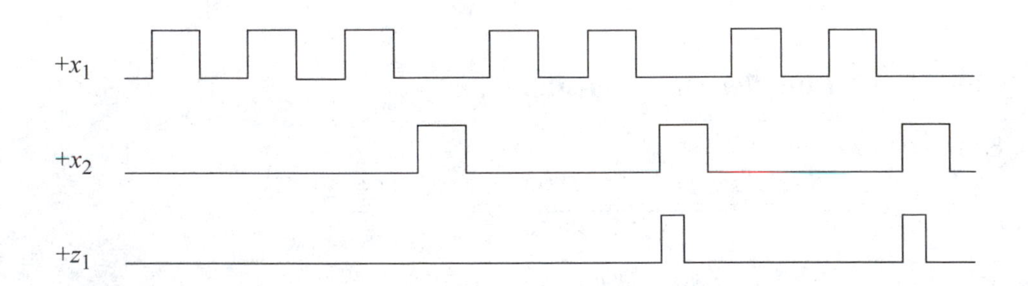

Figure 7.26 Representative timing diagram for the pulse-mode Mealy machine of Example 7.5.

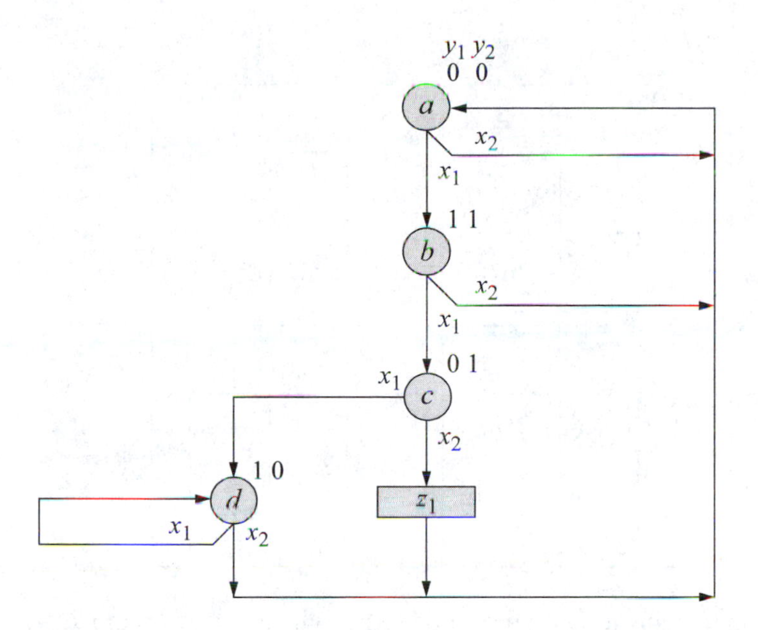

Figure 7.27 State diagram for the pulse-mode Mealy machine of Example 7.5.

Table 7.8 Next-state table for the pulse-mode Mealy machine of Example 7.5

State name	Present state $y_1\ y_2$	Inputs $x_1\ x_2$	Next state $y_1\ y_2$	Output z_1
a	0 0	1 0	1 1	0
	0 0	0 1	0 0	0
b	1 1	1 0	0 1	0
	1 1	0 1	0 0	0
c	0 1	1 0	1 0	0
	0 1	0 1	0 0	1
d	1 0	1 0	1 0	0
	1 0	0 1	0 0	0

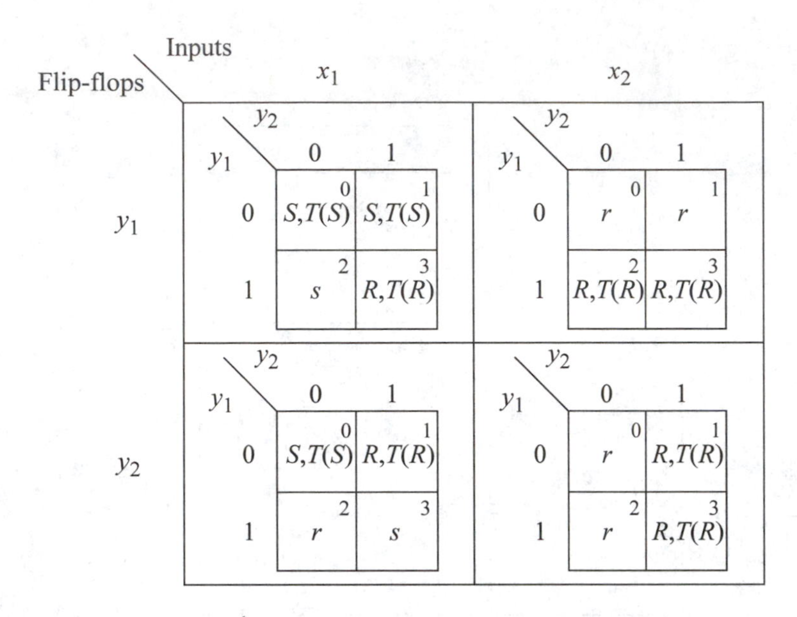

Figure 7.28 Input maps for the pulse-mode Mealy machine of Figure 7.27.

Refer to the state diagram or the next-state table for the derivation of the input maps. In state a ($y_1y_2 = 00$), if x_1 is pulsed, then the machine proceeds to state b, where both y_1 and y_2 are set to a value of 1. Since the previous state was $y_1y_2 = 00$, both flip-flops toggle. Therefore, two entries are placed in minterm location 0 for the maps in column x_1: S which specifies a set condition and $T(S)$ which specifies a toggle condition from 0 to 1. Although the two entries culminate in the same result, a choice is available for minimization. If x_2 is pulsed in state a, then the machine remains in state a and the letter r is placed in minterm location 0 of the two maps in column x_2.

In state b ($y_1y_2 = 11$), if x_1 is pulsed, then the machine sequences to state c, where y_1 is reset and y_2 remains set. Therefore, the entries $R,T(R)$ and s are placed in minterm location 3 of the maps in column x_1 for y_1 and y_2, respectively. If x_2 is pulsed in state b, then the machine returns to state a, where both flip-flops are reset. Thus, the entries $R,T(R)$ are placed in minterm location 3 of the maps in column x_2 for y_1 and y_2.

In state c ($y_1y_2 = 01$), an x_1 pulse sequences the machine to state d, where both flip-flops change state. Therefore, $S,T(S)$ and $R,T(R)$ are placed in minterm location 1 of the maps in column x_1 for flip-flops y_1 and y_2, respectively. An x_2 pulse in state c will assert output z_1 and sequence the machine to state a. Output z_1 is asserted coincident with the x_2 pulse with a duration equal to the propagation delay to reset y_2 plus the delay through the AND gate which decodes z_1.

Finally, in state d ($y_1y_2 = 10$), an x_1 pulse causes the machine to remain in state d; therefore, the entries s and r are placed in minterm location 2 of the maps in column x_1 for y_1 and y_2, respectively. An x_2 pulse sequences the machine to state a, where flip-flop y_1 is reset and y_2 remains reset. Thus, $R,T(R)$ and r are entered in minterm location 2 of the maps in column x_2 for y_1 and y_2, respectively.

When deriving the equations from the input maps, occasionally more than one equation will satisfy the transition requirements. The equation which sets flip-flop y_1, for example, is obtained from the *S* and *T*(*S*) entries in the map of row y_1, column x_1. There are two methods for deriving the equation which will result in y_1 acquiring a value of 1. Combine the *S* entries of minterm locations 0 and 1 or the *S* and *s* entries of minterm locations 0 and 2, yielding the following equations:

$$Sy_1 = y_1'x_1 \text{ or}$$
$$Sy_1 = y_2'x_1$$

The *T*(*S*) entries in minterm locations 0 and 1 may also be combined. The resulting equation, however, would be redundant. A more encompassing toggle equation results from the combination of entries in minterm locations 1 [*T*(*S*)] and 3 [*T*(*R*)] of the map in row y_1, column x_1. Refer to the entry in minterm location 1. If $y_1 = 0$, then $y_1'y_2x_1$ will toggle y_1. Since $y_1 = 0$, only variables x_1 and y_2 are necessary to toggle flip-flop y_1 from 0 to 1. Similarly, if $y_1 = 1$, then $y_1y_2x_1$ will toggle y_1. Since $y_1 = 1$, only variables x_1 and y_2 are necessary to toggle flip-flop y_1 from 1 to 0. In both cases, therefore, y_1 will toggle for the following condition:

$$Ty_1 = y_2x_1$$

Thus, the equation to set y_1 to a value of 1 is

$$Sy_1 = y_1'x_1$$

The equation which resets flip-flop y_1 is obtained from the *R*,*T*(*R*) entries in minterm location 3 of the map in row y_1, column x_1, yielding $y_1y_2x_1$. This term can be ignored, however, since it is accounted for in the equation for Ty_1. All of the minterm entries in the map of row y_1, column x_2 combine to establish the reset equation for flip-flop y_1. That is, the entries *r*, *r*, *R*, and *R* of minterm locations 0 through 3, respectively, yield the following reset equation for y_1:

$$Ry_1 = x_2$$

The toggle equation, $Ty_1 = y_1x_2$, obtained from minterm locations 2 [*T*(*R*)] and 3 [*T*(*R*)] is redundant, since it is included in the more general reset equation of $Ry_1 = x_2$.

The equations which cause flip-flop y_2 to acquire a value of 1 are derived in a similar manner. Only one entry need be considered — minterm location 0 of the map in row y_2, column x_1. This results in the following equation:

$$Sy_2 = y_1'y_2'x_1$$

The letter s in minterm location 3 can be disregarded, because it does not change the state of flip-flop y_2 and cannot be used to minimize the equation, since it is not logically adjacent to minterm 0. There is a second equation that will result in y_2 obtaining a value of 1. Refer to the map in row y_2, column x_1. If $y_2 = 0$, then $y_1'y_2'x_1$ will toggle [$T(S)$] flip-flop y_1. Since $y_2 = 0$, only variables x_1 and y_1' are necessary to toggle y_2 from 0 to 1. The remaining toggle entry in this map is in minterm location 1 [$T(R)$]. Thus, if $y_2 = 1$, then the term $y_1'y_2x_1$ will toggle y_2. Since $y_2 = 1$, only the variables x_1 and y_1' are needed, resulting in the following toggle equation for flip-flop y_2:

$$Ty_2 = y_1'x_1$$

The reset equation for flip-flop y_2 is derived from the maps in row y_2. The $T(R)$ entry in minterm location 1 of the map in row y_2, column x_1 is contained in the toggle equation $Ty_2 = y_1'x_1$ and can be disregarded. Likewise, the R entry can be ignored, since the reset function of minterm 1 has already been evaluated. The letter r in minterm location 2 cannot be used in minimizing the reset equation for y_2, because it is not logically adjacent to minterm 1.

The last consideration for resetting y_2 is in the map in row y_2, column x_2. All minterms combine to yield the following reset equation:

$$Ry_2 = x_2$$

The $T(R)$ entries in minterm locations 1 and 3 can be disregarded, because they are included in the more general reset equation of $Ry_2 = x_2$. The complete set of equations for controlling the state transition sequence of flip-flops y_1 and y_2 are listed in Equation 7.12.

$$\begin{aligned} Sy_1 &= y_1'x_1 \\ Ty_1 &= y_2x_1 \\ Ry_1 &= x_2 \\ \\ Sy_2 &= y_1'y_2'x_1 \\ Ty_2 &= y_1'x_1 \\ Ry_2 &= x_2 \end{aligned} \tag{7.12}$$

The output map of Figure 7.29 is generated from either the state diagram or the next-state table. The output equation is shown in Equation 7.13.

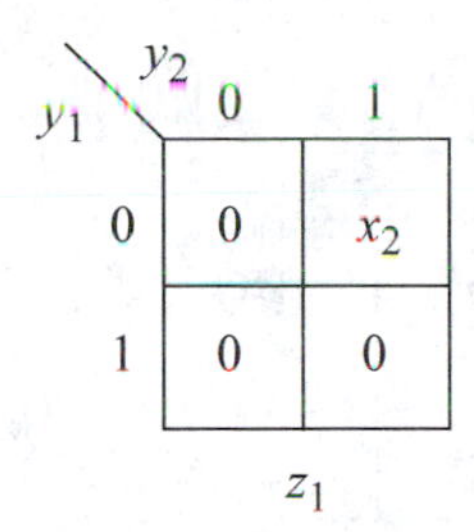

Figure 7.29 Output map for z_1 for the pulse-mode Mealy machine of Figure 7.27.

$$z_1 = y_1 ' y_2 x_2 \tag{7.13}$$

The logic diagram is shown in Figure 7.30 and is synthesized using the input and output equations of Equation 7.12 and Equation 7.13, respectively. The machine is initially reset to state a ($y_1 y_2 = 00$). By applying input pulses in an appropriate sequence and by observing the restrictions on input pulse duration, the machine will proceed through the correct state transitions according to the machine specifications.

In state a ($y_1 y_2 = 00$), both flip-flops are reset. A pulse on input x_1 will set y_1 and toggle y_2 from 0 to 1. This sequence results from equations $Sy_1 = y_1 ' x_1$ and $Ty_2 = y_1 ' x_1$. The x_1 pulse must be deasserted before the new states of y_1 and y_2 arrive at the δ next-state logic, otherwise, flip-flop y_1 will toggle if x_1 is still active when y_2 is set, resulting in an incorrect state transition sequence.

In state b ($y_1 y_2 = 11$), an x_1 pulse will toggle y_1 ($Ty_1 = y_2 x_1$) from 1 to 0, but will have no effect on flip-flop y_2 — y_2 remains set. An x_2 pulse will reset both flip-flops, returning the machine to state a.

In state c ($y_1 y_2 = 01$), an x_1 pulse will set y_1 ($Sy_1 = y_1 ' x_1$) and toggle y_2 ($Ty_2 = y_1 ' x_1$), sequencing the machine to state d ($y_1 y_2 = 10$). An x_2 pulse occurring in state c ($y_1 y_2 = 01$) will assert output z_1 ($z_1 = y_1 ' y_2 x_2$) coincident with input x_2. Since z_1 is a function of the present state and the present input x_2 for this Mealy machine, z_1 is deasserted when the present state changes to the next state. The x_2 pulse will reset both y_1 and y_2, returning the machine to state a, where the process repeats.

In state d ($y_1 y_2 = 10$), an x_1 pulse will not be propagated through the δ next-state AND gates, because y_1 is set and y_2 is reset. Therefore, the machine remains in state d. An x_2 pulse, however, will directly reset both flip-flops returning the machine to state a, where the process repeats. As stated previously, the duration of the input pulses in relation to the combined propagation delay of the δ next-state logic and the flip-flop transitions is critical for correct operation of pulse-mode sequential machines.

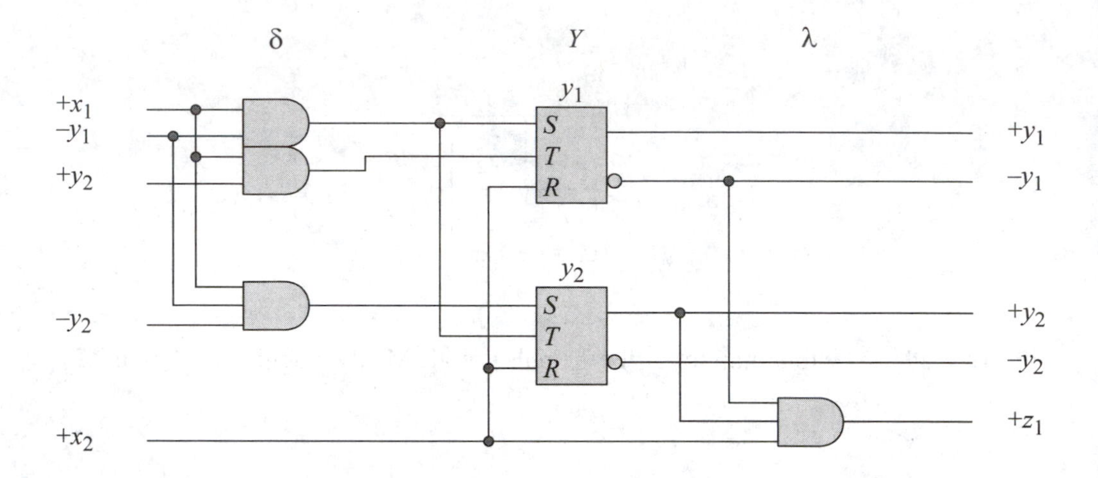

Figure 7.30 Logic diagram for the pulse-mode Mealy machine of Figure 7.27 using *SR-T* flip-flops as the storage elements.

7.2.4 *SR* Latches with *D* Flip-Flops as Storage Elements

The pulse width restrictions that are dominant in pulse-mode sequential machines can be eliminated by including *D* flip-flops in the feedback path from the *SR* latches to the δ next-state logic. Providing edge-triggered *D* flip-flops as a constituent part of the implementation negates the requirement of precisely-controlled input pulse durations. This is by far the most reliable means of synthesizing pulse-mode machines. The *SR* latches, in conjunction with the *D* flip-flops, form a master-slave configuration as shown in Figure 7.31.

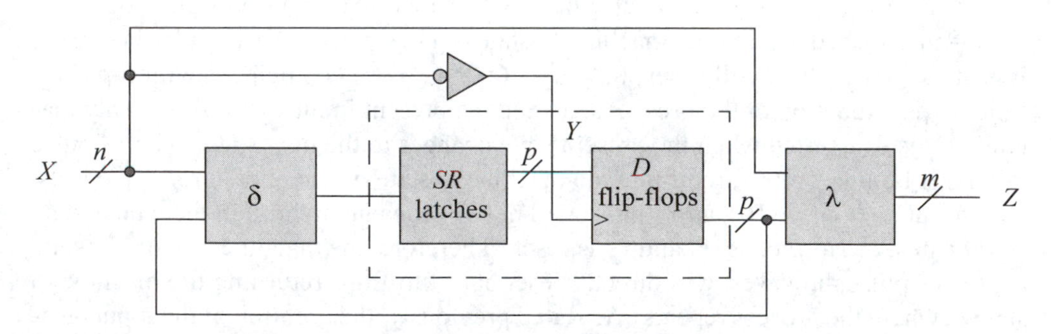

Figure 7.31 General block diagram of a pulse-mode sequential machine using *SR* latches and *D* flip-flops in a master-slave configuration.

The output of each latch connects to the D input of its associated flip-flop which in turn connects to the δ next-state logic. The flip-flops are clocked on the complemented trailing edge of the active input variable. For example, if the input pulses are active high, then the D flip-flops are triggered on the inverted negative transition of the active input pulse; that is, the flip-flops are triggered on a positive transition, as shown in Figure 7.31. JK flip-flops may also be used as the slave storage elements. The output alphabet Z of pulse-mode machines can be generated as either levels for Moore machines or as pulses for Mealy machines.

The set and reset equations for each latch will be of the form shown in Equation 7.14. The set equation SLy_j for latch Ly_j is a function f of the present input variables $X_{i(t)}$ and the present state $Y_{j(t)}$. The reset equation RLy_j is a function g of $X_{i(t)}$ and $Y_{j(t)}$. Each term of the input equations must contain a pulsed input variable, since pulses trigger the latches. Since the latches are level-sensitive devices and connect directly to the D input of their corresponding slave flip-flops, the set and reset equations also serve as the input equations for the flip-flops.

$$
\begin{aligned}
SLy_j &= f(X_{i(t)}, Y_{j(t)}) \\
RLy_j &= g(X_{i(t)}, Y_{j(t)}) \\
Z_r &= h(Y_{j(t)}) \qquad \text{Moore} \\
Z_r &= h'(X_{i(t)}, Y_{j(t)}) \ \text{Mealy}
\end{aligned}
\tag{7.14}
$$

Equation 7.14 also lists the λ output equations for the output vector Z_r. The output equations will be a function h of the present state for Moore-type outputs, as indicated by $Z_r = h(Y_{j(t)})$. Since Moore-type outputs are not a function of the input vector $X_{i(t)}$, the outputs will be represented as levels. The outputs for a Mealy machine, however, will be a function h' of the input vector and the present state, as indicated by the equation $Z_r = h'(X_{i(t)}, Y_{j(t)})$. Mealy-type outputs, therefore, will be generated as pulses.

Example 7.6 A Mealy machine will be synthesized which has three pulse input variables x_1, x_2, and x_3 and one output z_1 that is asserted coincident with x_3 whenever the sequence $x_1x_2x_3 = 100, 010, 001$ occurs. The storage elements will consist of SR latches and positive-edge-triggered D flip-flops.

A representative timing diagram displaying valid input sequences and corresponding outputs is shown in Figure 7.32. The state diagram is shown in Figure 7.33. State code assignment is arbitrary, since input pulses trigger all state transitions and the machine does not begin to sequence to the next state until the input pulse, which initiated the transition, has been deasserted. Thus, output z_1 will not glitch.

A tabular representation of the state diagram is shown in Table 7.9. Since only one input variable can be active at a time, only three combinations are listed: $x_1x_2x_3 = 100, 010$, and 001. With the exception of $x_1x_2x_3 = 000$, all other combinations of the inputs are invalid.

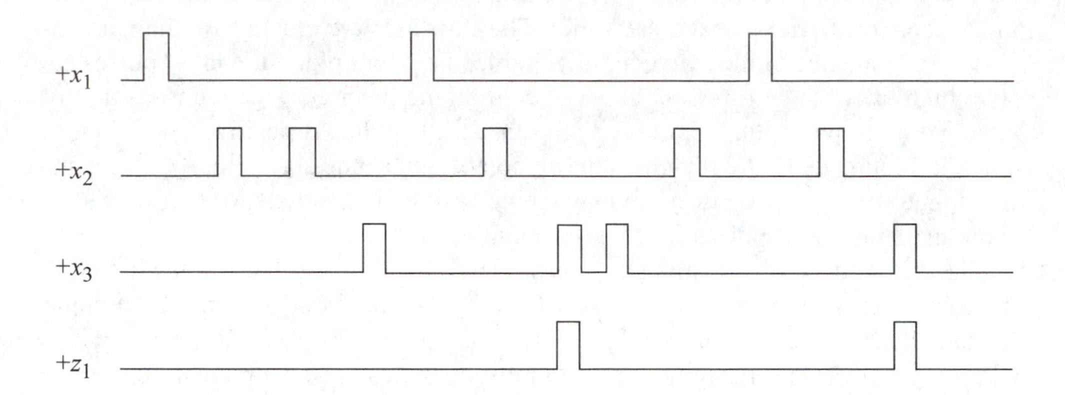

Figure 7.32 Representative timing diagram for the pulse-mode Mealy machine of Example 7.6.

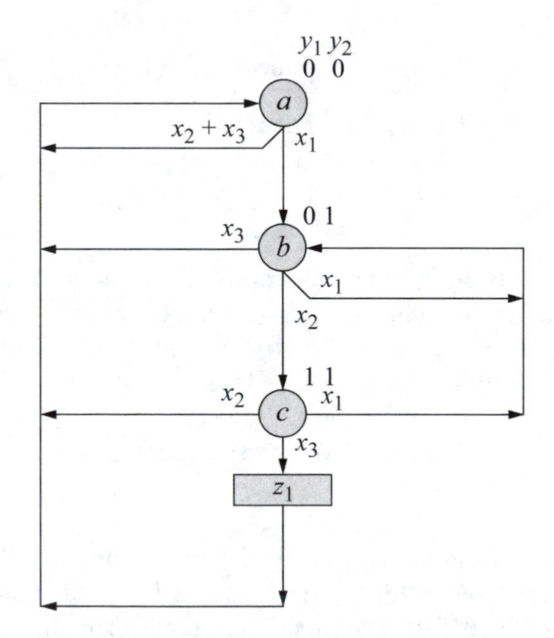

Figure 7.33 State diagram for the pulse-mode Mealy machine of Example 7.6.

Each latch requires three input maps, one each for x_1, x_2, and x_3, as shown in Figure 7.34. As in previous examples, the maps are arranged such that, the maps corresponding to each latch are in the same row, and each column of maps corresponds to a unique input. The map entries are defined as follows:

S indicates that the latch will be set.
s indicates that the latch will remain set.
R indicates that the latch will be reset.
r indicates that the latch will remain reset.

Table 7.9 Next-state table for the pulse-mode Mealy machine of Figure 7.33

State name	Present state $y_1\ y_2$	Inputs $x_1\ x_2\ x_3$	Next state $y_1\ y_2$	Output z_1
a	0 0	1 0 0	0 1	0
	0 0	0 1 0	0 0	0
	0 0	0 0 1	0 0	0
b	0 1	1 0 0	0 1	0
	0 1	0 1 0	1 1	0
	0 1	0 0 1	0 0	0
c	1 1	1 0 0	0 1	0
	1 1	0 1 0	0 0	0
	1 1	0 0 1	0 0	1

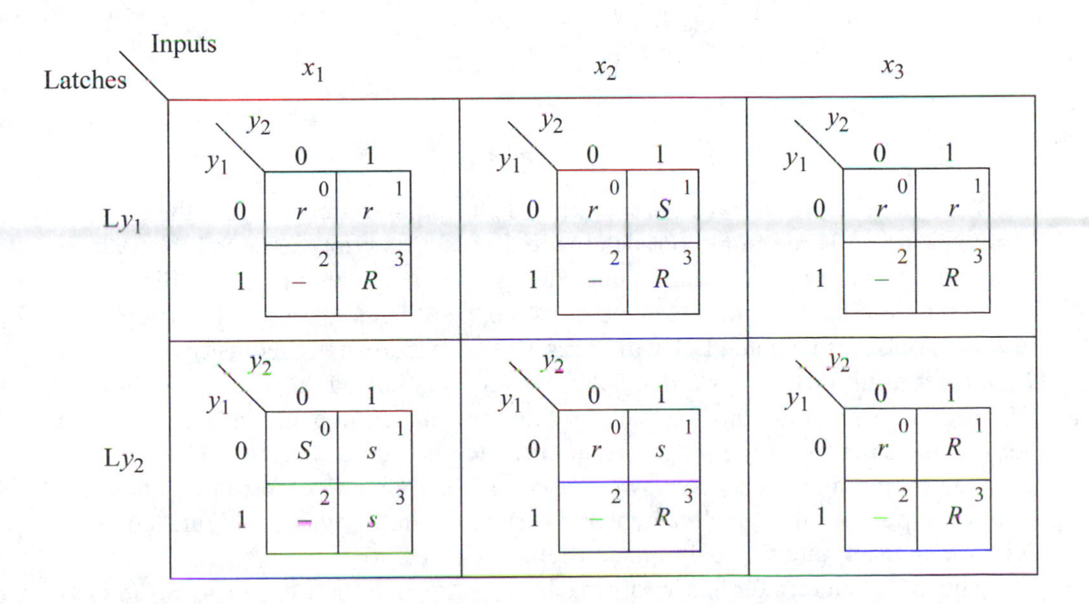

Figure 7.34 Input maps for the pulse-mode Mealy machine of Figure 7.33.

The map entries are obtained as in previous examples. Refer to the state diagram and minterm location 0 of the map in row Ly_1, column x1. In state *a* ($y_1 y_2 = 00$), if x_1 is pulsed, then the machine sequences to state *b* ($y_1 y_2 = 01$) where flip-flop y_1 remains reset. Thus, the letter *r* is inserted in minterm location 0. In the same map, minterm location 3 contains the entry *R*. That is, in state *c* ($y_1 y_2 = 11$), flip-flop y_1 is reset if x_1 is pulsed.

The map in row Ly_2, column x_1 contains the letter *s* in minterm location 1. In state *b* ($y_1 y_2 = 01$), if x_1 is pulsed, then the machine remains in state *b* and flip-flop y_2 remains set. In minterm location 0 of the same map, the letter *S* is entered. In state *a* ($y_1 y_2 = 00$), if x_1 is pulsed, then the machine proceeds to state *b* ($y_1 y_2 = 01$) where flip-flop y_2 is set. In a similar manner, the remaining input maps are derived.

When obtaining the equations for the latches from the input maps, only the uppercase letters must be considered. The lowercase letters and the unused states are used only if they contribute to a minimized equation. The set and reset input equations are listed in Equation 7.15, where *S*Ly_1, *R*Ly_1 and *S*Ly_2, *R*Ly_2 are the set and reset equations for latches Ly_1 and Ly_2, respectively. Note that all equations contain an input variable x_i, since the machine is triggered by input pulses.

$$SLy_1 = y_1{}'y_2x_2$$
$$RLy_1 = x_1 + y_1x_2 + x_3$$

$$SLy_2 = x_1$$
$$RLy_2 = y_1x_2 + x_3 \qquad (7.15)$$

Latch Ly_1 is set only if $y_1y_2x_2 = 011$. This can be observed from the state diagram in state *b*. Three conditions determine the reset function for latch Ly_1. Latch y_1 will always be in a reset state after input x_1 is pulsed, as shown in the state diagram for the following state transition sequences: $a \rightarrow b$, $b \rightarrow b$, and $c \rightarrow b$. Also, if Ly_1 is set and x_2 is pulsed, then latch Ly_1 will be reset. This occurs in the sequence $c \rightarrow a$. The final reset term for Ly_1 is x_3; that is, whenever x_3 is pulsed, latch Ly_1 will be reset. The set and reset equations for latch Ly_2 can be verified in a similar manner by observing the state transition sequences in the state diagram of Figure 7.33.

The output map for z_1 is shown in Figure 7.35. Since z_1 is asserted coincident with x_3, input x_3 is used as a map-entered variable in state *c* ($y_1 y_2 = 11$) and combines with the unused state to yield Equation 7.16.

The logic diagram for this Mealy machine is shown in Figure 7.36, using *SR* latches and *D* flip-flops in a master-slave relationship. The logic is synthesized from the input and output equations of Equation 7.15 and Equation 7.16, respectively. Each of the three mutually exclusive input pulses is inverted through the three-input NOR gate. When the pulses are active, a low voltage level is applied to the clock inputs of D flip-flops y_1 and y_2. The active level of the pulses also sets or resets latches Ly_1 and Ly_2, depending on the present state and the present input. The latches stabilize to their respective next states while the input pulse is still active and provide the next-state values to the D inputs of flip-flops y_1 and y_2. When the input pulse is deasserted, a positive transition is applied to the clock inputs of flip-flops y_1 and y_2, which then sequence the machine to the next state.

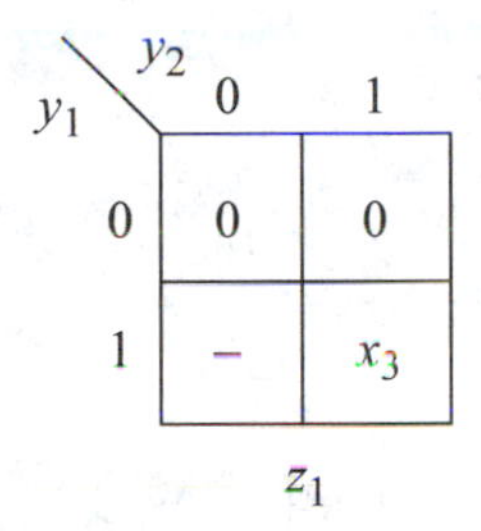

Figure 7.35 Output map for z_1 for the pulse-mode Mealy machine of Figure 7.33.

$$z_1 = y_1 x_3 \tag{7.16}$$

The machine is reset initially to state a. In state a ($y_1y_2 = 00$), input x_1 will directly reset latch Ly_1 and set latch Ly_2 sequencing the machine to state b ($y_1y_2 = 01$). Input x_2 has no effect on either latch, whereas x_3 will reset both latches.

In state $b(y_1y_2 = 01)$, x_1 will not affect the state of latches Ly_1 or Ly_2 — input x_1 maintains Ly_1 and Ly_2 in a reset and set state, respectively. Input x_2 will set latch Ly_1, but will have no effect on latch Ly_2, because flip-flop y_1 is reset. Thus, the machine proceeds from state b ($y_1y_2 = 01$) to state c ($y_1y_2 = 11$). Input x_3 will directly reset both latches. The machine sequences to state a when x_3 is deasserted.

In state c ($y_1y_2 = 11$), x_1 will directly reset latch Ly_1 and set latch Ly_2, sequencing the machine to state b. Input x_2 will reset both latches, because flip-flop y_1 is set. The machine then proceeds to state a when x_2 becomes deasserted. Input x_3 will directly reset both latches and assert output z_1, because flip-flops y_1 and y_2 remain set until x_3 is deasserted.

The concept of pulse-mode machines is frequently encountered in a variety of common applications, such as vending machines and traffic control for demand-access intersections. Of the four methods of pulse-mode sequential machines presented in this section, only the SR latch with an edge-triggered D flip-flop offers a high degree of reliability. The timing constraints of the other three methods demand an extremely restrictive tolerance on input pulse duration.

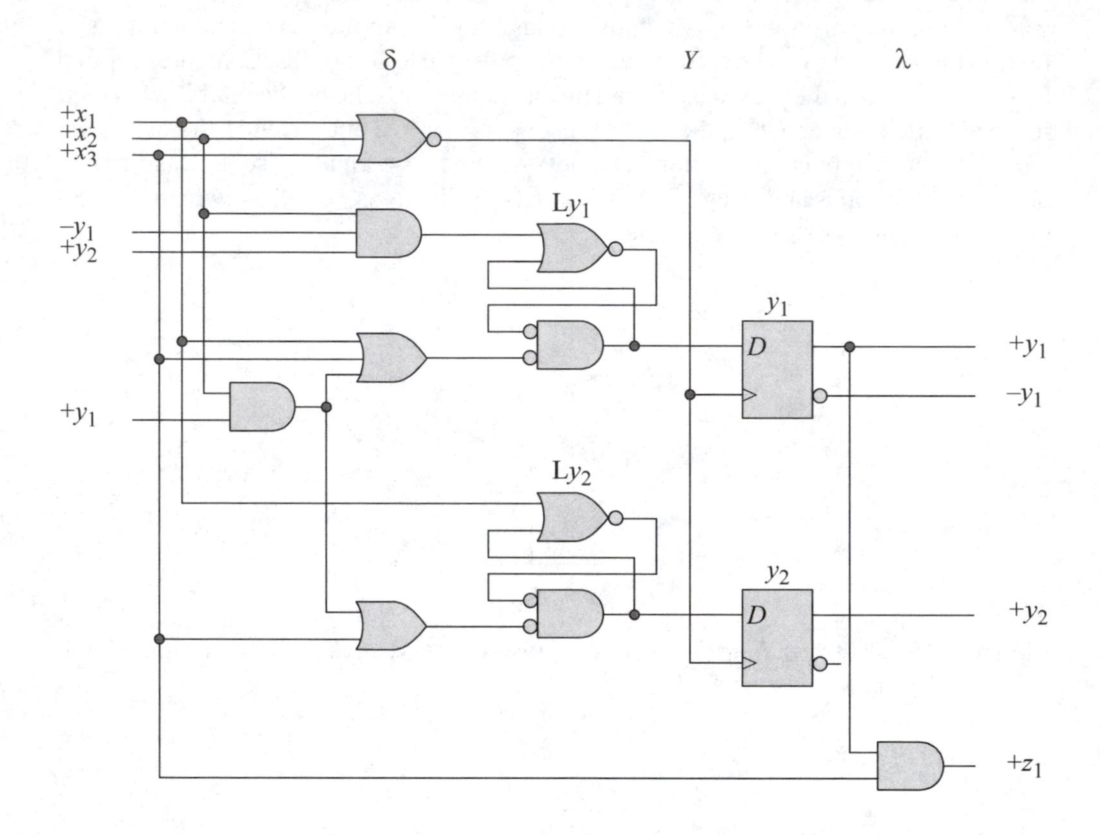

Figure 7.36 Logic diagram for the pulse-mode Mealy machine of Figure 7.33, using *SR* latches and *D* flip-flops in a master-slave configuration.

7.3 Problems

7.1 Synthesize a Mealy pulse-mode sequential machine which has two inputs x_1 and x_2 and one output z_1. Output z_1 is asserted coincident with every second x_2 pulse, if and only if the pair of x_2 pulses is immediately preceded by an x_1 pulse. Use only NAND logic, inverters, and *SR* latches as the storage elements.

7.2 Synthesize the Mealy pulse-mode sequential machine of Problem 7.1 using NAND logic and *T* flip-flops as the storage elements.

7.3 Analyze the Mealy pulse-mode sequential machine shown below. Obtain the next-state table, the input maps and equations, the output map and equations, and the state diagram.

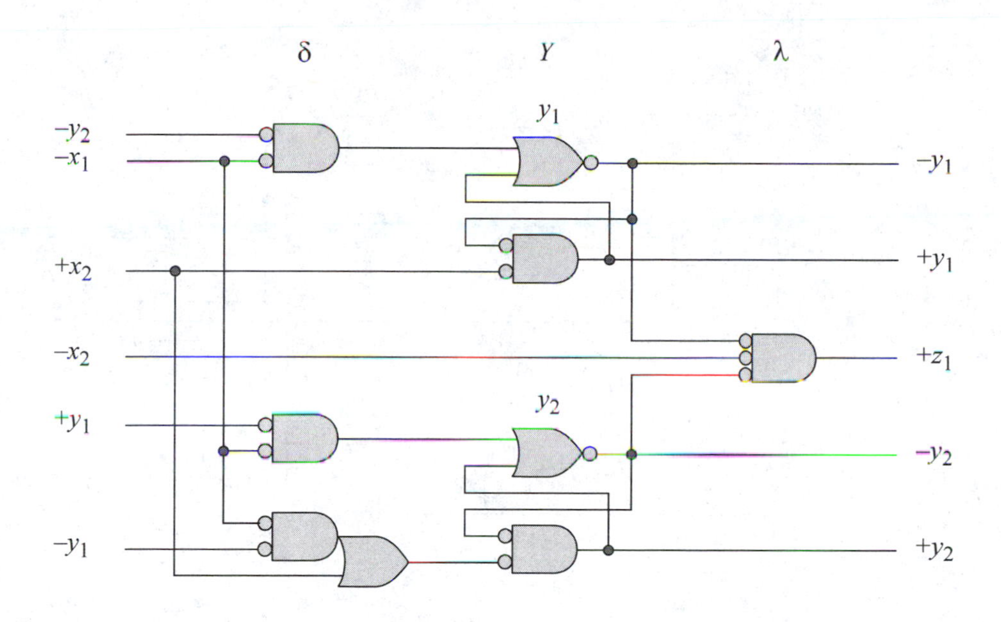

7.4 Analyze the Mealy pulse-mode sequential machine shown below. Obtain the next-state table and the state diagram.

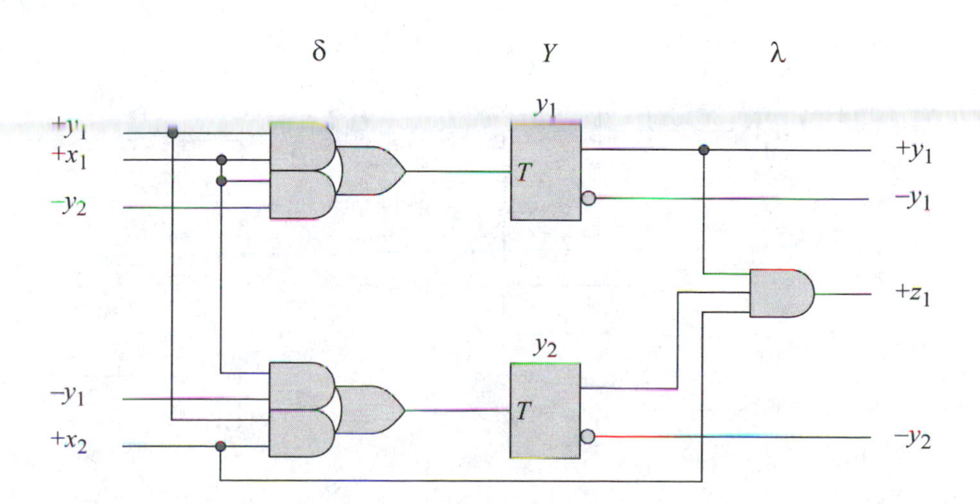

7.5 Analyze the Mealy pulse-mode sequential machine shown below. Obtain the next-state table, the input maps and equations, the output map and equation, the state diagram, and a representative timing diagram.

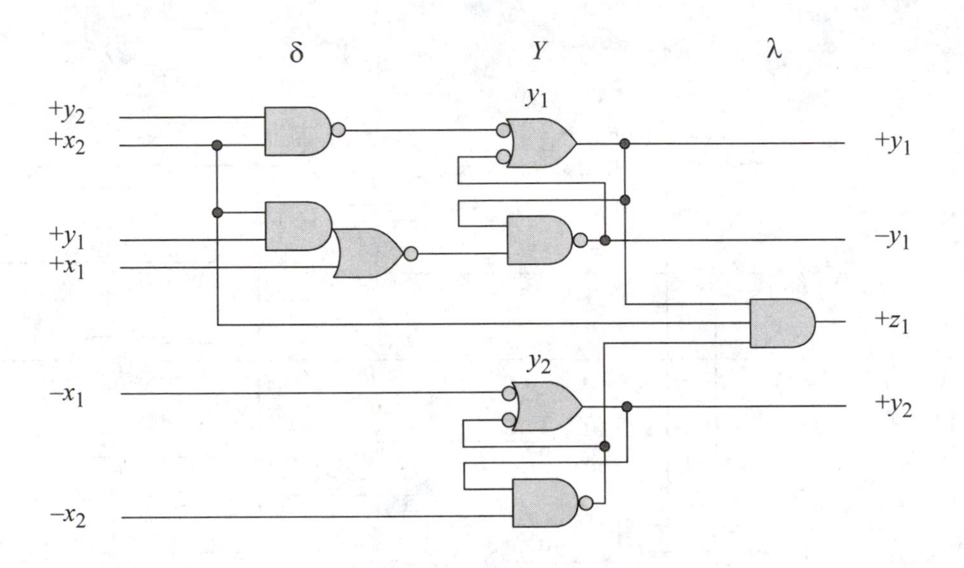

7.6 Analyze the Mealy pulse-mode sequential machine shown below. Obtain the next-state table, the input maps and equations, the output map and equation, and the state diagram.

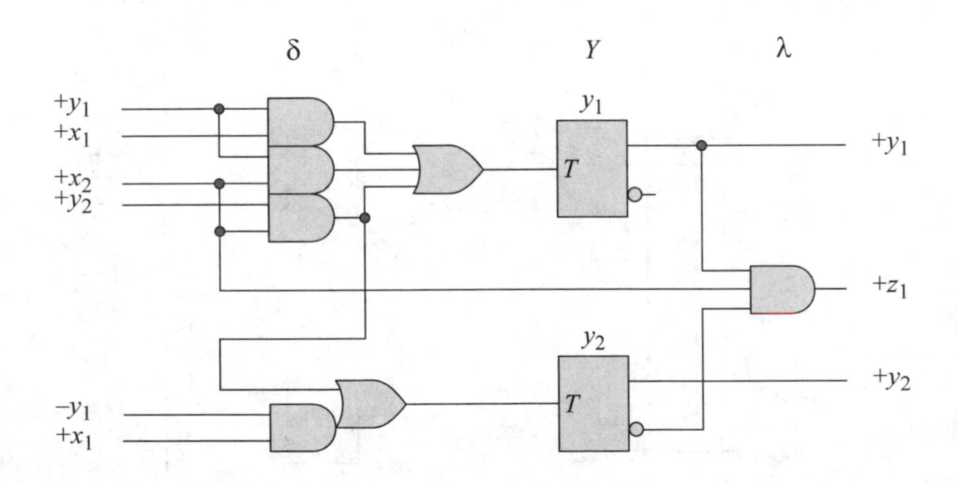

7.7 Analyze the Mealy pulse-mode sequential machine shown below. Obtain the next-state table, the input maps and equations, the output map and equation, and the state diagram.

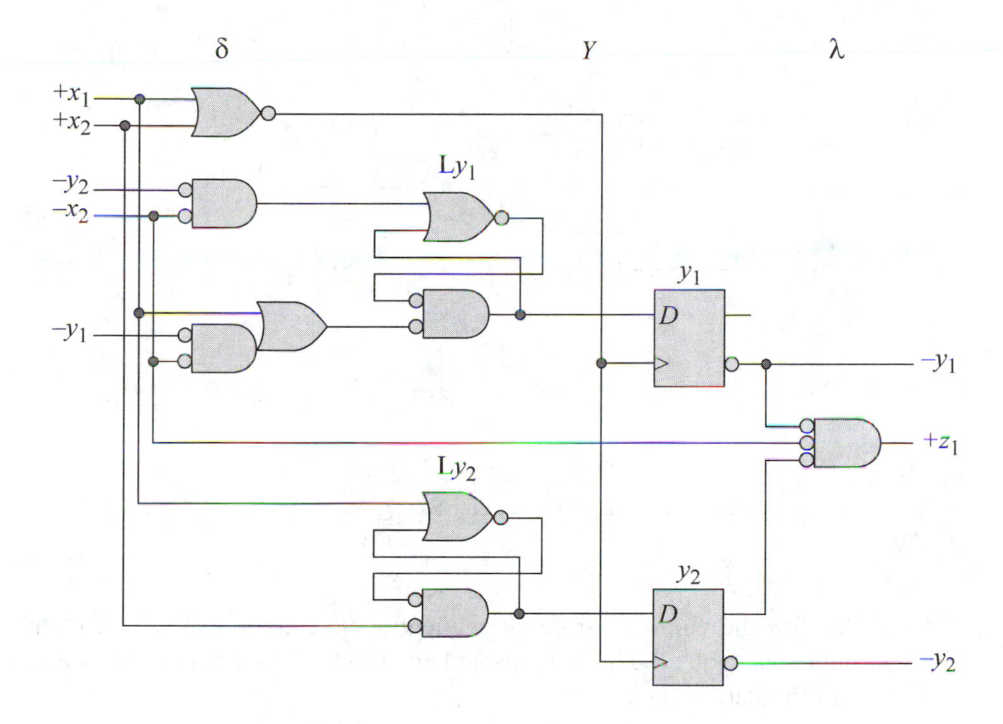

7.8 Analyze the Moore pulse-mode sequential machine shown below. Obtain the next-state table, the input maps and equations, the output map and equation, and the state diagram.

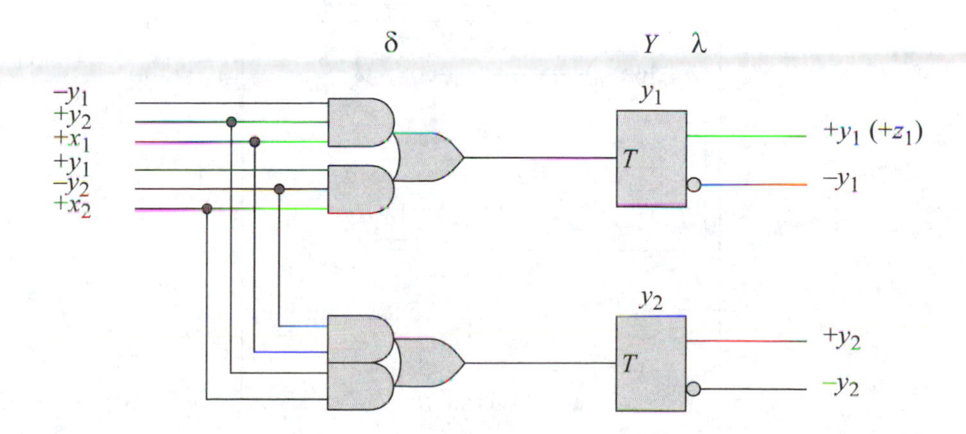

7.9 Analyze the Moore pulse-mode sequential machine shown below. Obtain the next-state table, the input maps and equations, the output map and equation, and the state diagram.

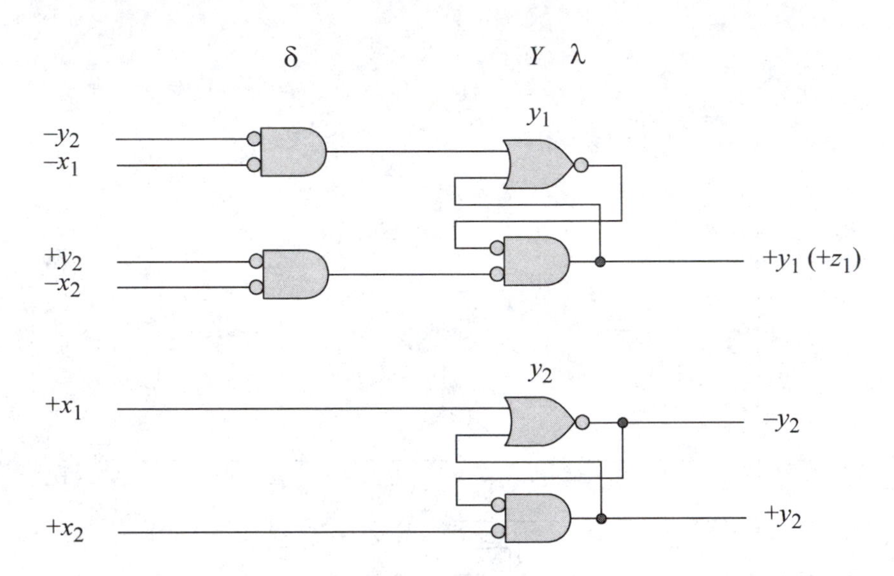

7.10 Analyze the Moore pulse-mode sequential machine shown below. Obtain the next-state table, the input maps and equations, the output map and equation, and the state diagram.

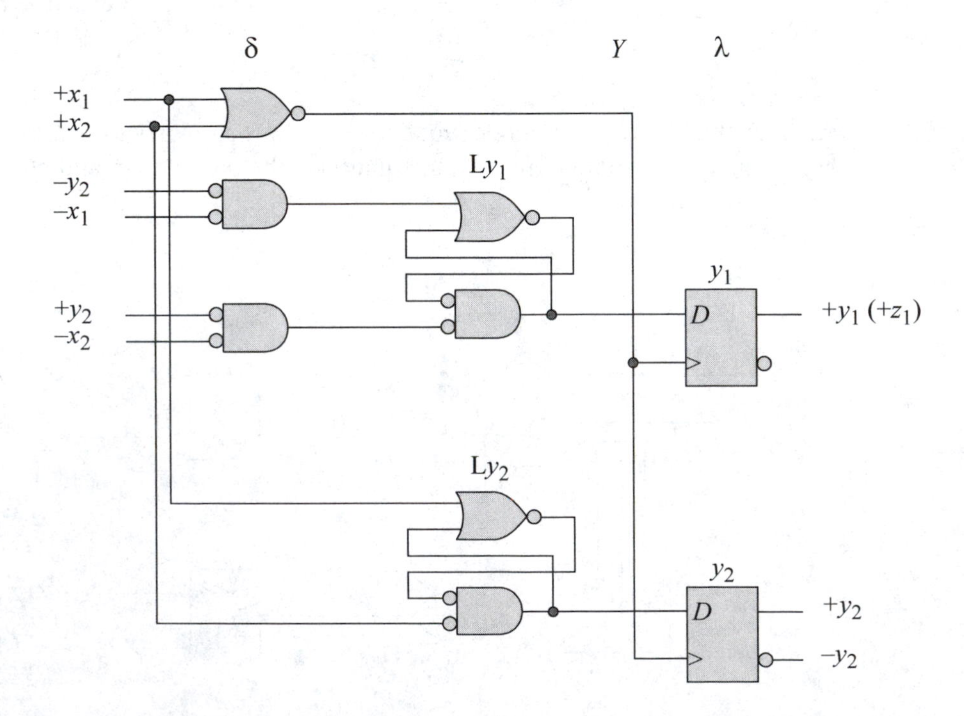

7.11 Analyze the Moore pulse-mode sequential machine shown below. Obtain the next-state table, the input maps and equations, the output map and equation, and the state diagram.

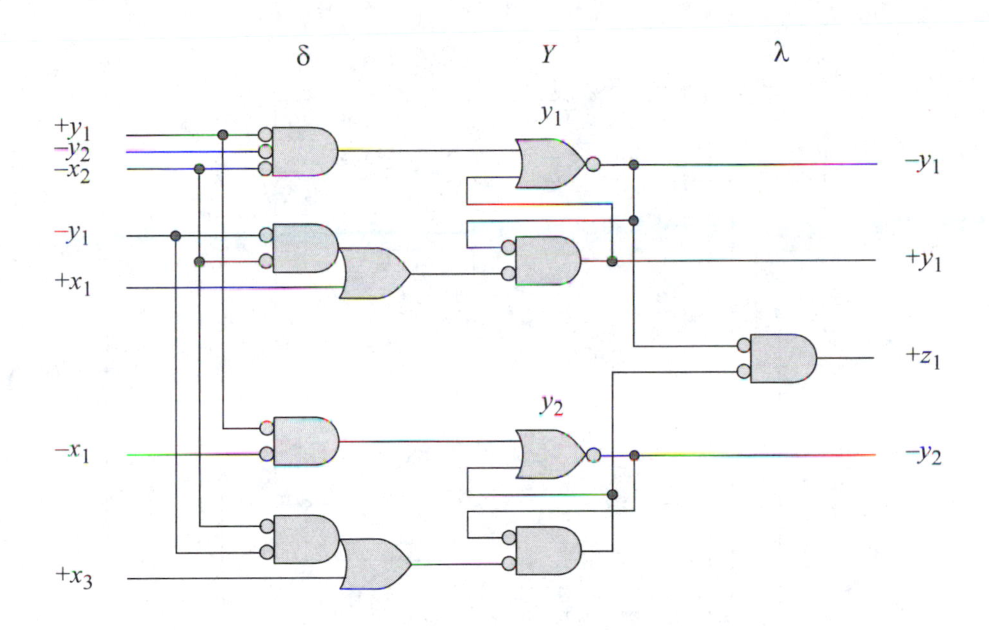

7.12 Analyze the Moore pulse-mode sequential machine shown below. Obtain the next-state table, the input maps and equations, the output map and equation, and the state diagram.

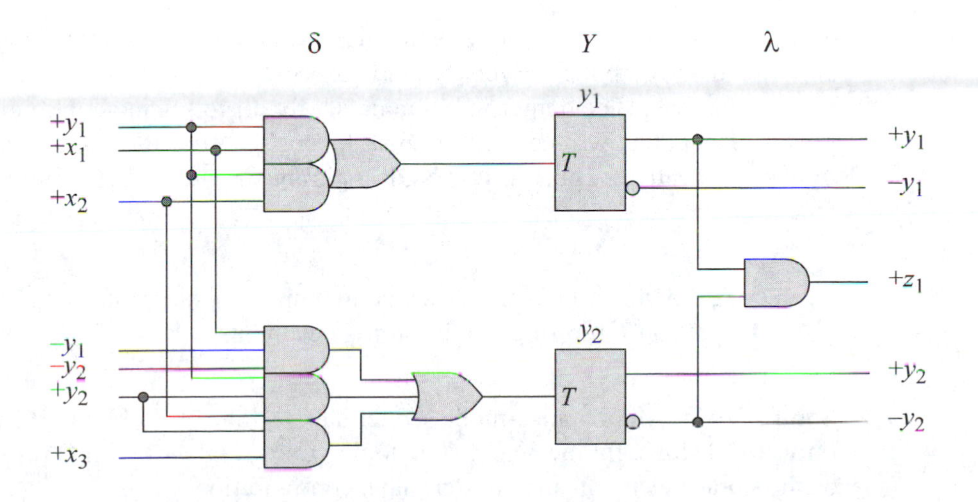

7.13 Analyze the Moore pulse-mode sequential machine shown below. Obtain the next-state table, the input maps and equations, the output map and equation, and the state diagram.

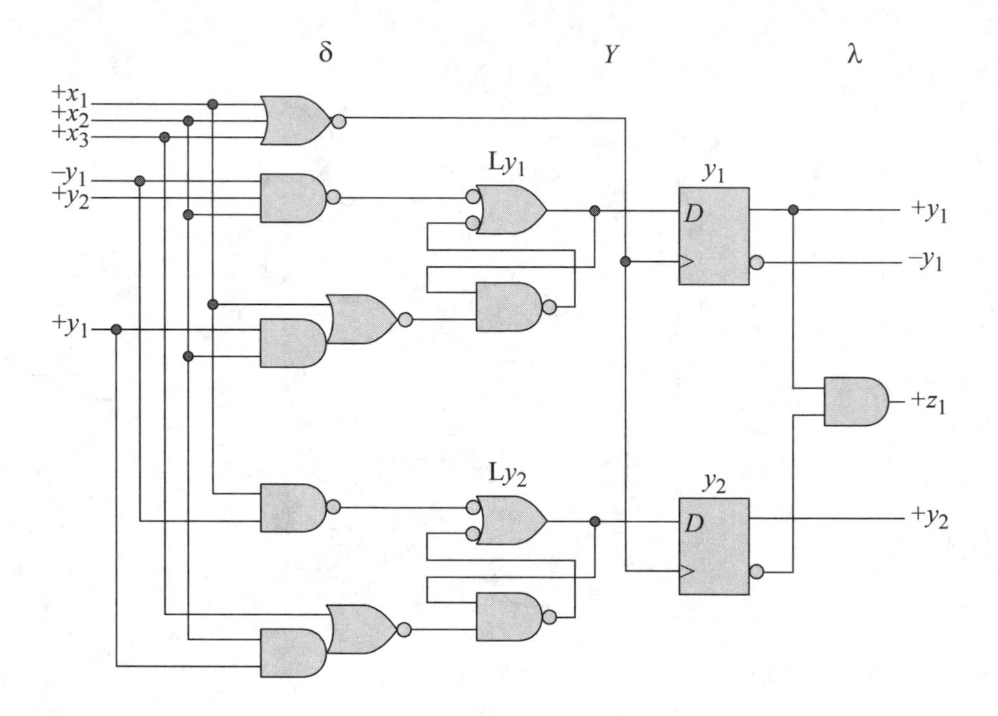

7.14 Synthesize the Mealy pulse-mode sequential machine of Problem 7.1 using *SR* latches and *D* flip-flops in a master-slave configuration. Use only NOR logic and inverters for the logic primitives.

7.15 Synthesize a Moore pulse-mode sequential machine which has two inputs x_1 and x_2 and one output z_1. Every second consecutive x_1 pulse will assert output z_1 as a level. The output will remain set for all following contiguous x_1 pulses. The output will be deasserted at the positive transition of the second of two consecutive x_2 pulses. Use NOR logic and *SR* latches as the storage elements.

7.16 Synthesize the Moore pulse-mode sequential machine of Problem 7.15 using NAND logic and *T* flip-flops as the storage elements.

7.17 Synthesize the Moore pulse-mode sequential machine of Problem 7.15 using only NAND logic for the δ next-state logic. Use *SR* latches with *D* flip-flops as the storage elements in a master-slave configuration.

7.18 Given the state diagram shown below for a pulse-mode Moore machine, synthesize the machine using *SR* latches and *D* flip-flops in a master-slave configuration. Obtain the next-state table, the input maps and equations, and the logic diagram.

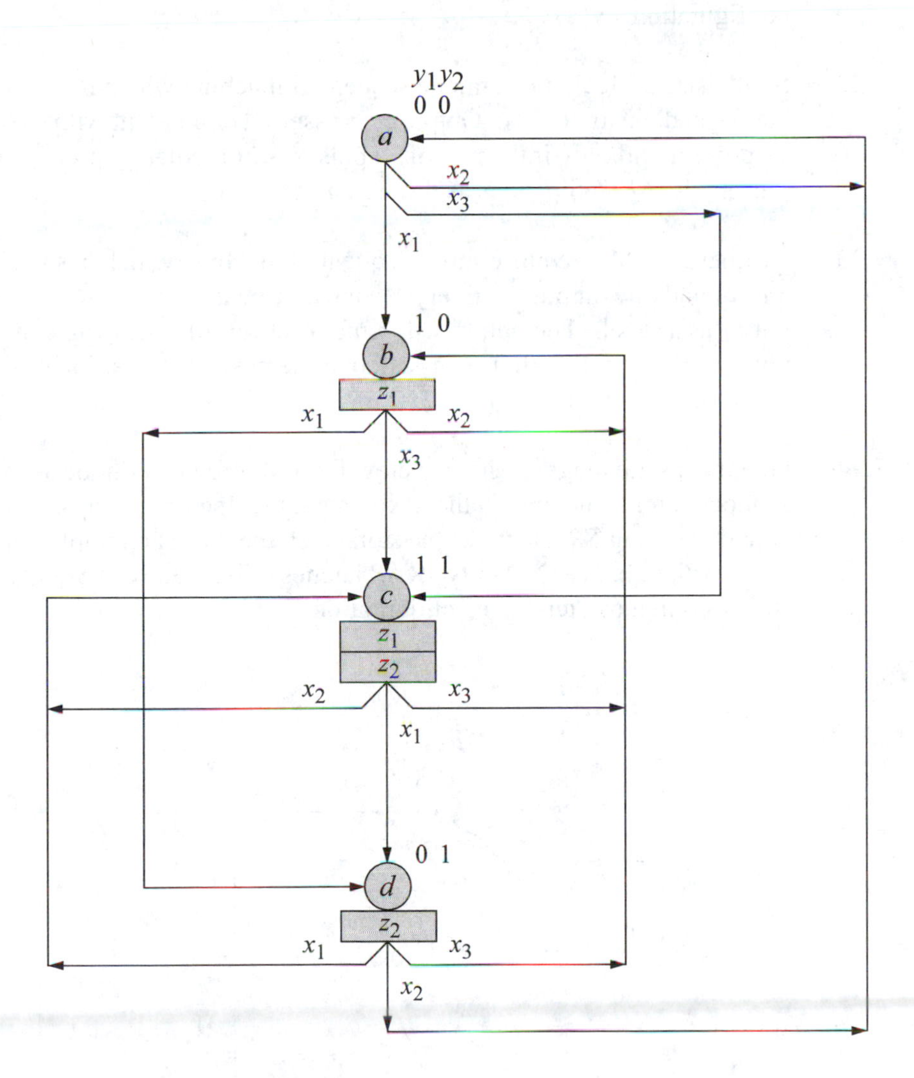

7.19 Synthesize a Moore pulse-mode sequential machine which has three inputs x_1, x_2, and x_3 and one output z_1. Output z_1 will be asserted coincident with the assertion of the x_3 pulse if and only if the x_3 pulse was preceded by an x_1 pulse followed by an x_2 pulse. That is, the input vector must be $x_1 x_2 x_3 = 100$, 000, 010, 000, 001 to assert z_1. Output z_1 will be deasserted at the next active x_1 pulse or x_2 pulse. Use *SR* latches as the storage elements.

7.20 Synthesize the Moore pulse-mode sequential machine of Problem 7.19 using NAND logic and *T* flip-flops as the storage elements.

7.21 Synthesize the Moore pulse-mode sequential machine of Problem 7.19 using NOR *SR* latches with *D* flip-flops as the storage elements in a master-slave configuration.

7.22 Synthesize a Mealy pulse-mode sequential machine which has two inputs x_1 and x_2 and one output z_1. Output z_1 is asserted coincident with every second x_2 pulse if and only if the pair of x_2 pulses is immediately preceded by an x_1 pulse. Use *SR-T* flip-flops as the storage elements.

7.23 Synthesize a Moore pulse-mode sequential machine which has two inputs x_1 and x_2 and one output z_1. Every second consecutive x_1 pulse will assert output z_1 as a level. The output will remain set for all following contiguous x_1 pulses. The output will be deasserted at the positive transition of the second of two consecutive x_2 pulses. Use *SR-T* flip-flops as the storage elements.

7.24 Given the state diagram shown below for a Moore pulse-mode asynchronous sequential machine, obtain the next-state table, the input maps, and the input equations using *SR* latches as the storage elements. Then implement the machine with either NAND or NOR *SR* latches only, then with *SR* latches and *D* flip-flops in a master-slave configuration.

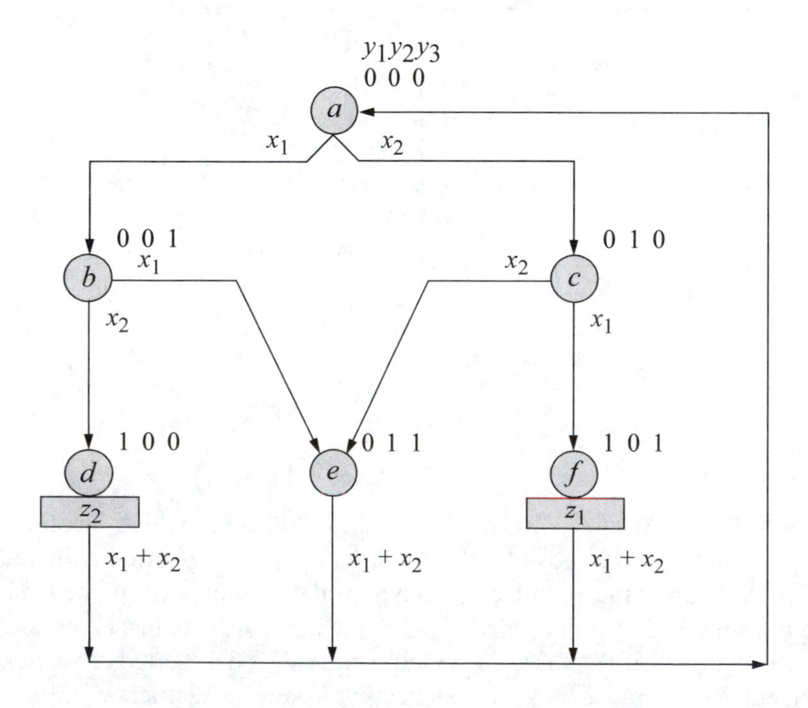

7.25 A vending machine operates as a Mealy pulse-mode asynchronous sequential machine. The product costs 35 cents. The machine accepts nickels, dimes, and quarters in any combination. Coins are entered until the amount is equal to or greater than the price of the product — no change is returned. A push button then selects the desired product. Let the pulse inputs and the output be defined as follows:

x_1 = Nickel
x_2 = Dime
x_3 = Quarter
x_4 = Selection push button
z_1 = Product

The selection push button can generate an output only when the requisite amount has been entered. When the product has been obtained, the machine returns to the initial state.

Generate the state diagram, the next-state table, the input maps and equations, and the output map and equation using *SR* latches as the storage elements. Then implement the machine with either NAND or NOR *SR* latches only, then with *SR* latches and *D* flip-flops in a master-slave configuration.

Appendix

Answers to Selected Problems

Chapter 1

1.4

(a)	Decimal integer	Radix 4	(b)	Decimal integer	Radix 12
	0	00		0	00
	1	01		1	01
	2	02		2	02
	3	03		3	03
	4	10		4	04
	5	11		5	05
	6	12		6	06
	7	13		7	07
	8	20		8	08
	9	21		9	09
	10	22		10	(10)
	11	23		11	(11)
	12	30		12	(01)(00)
	13	31		13	(01)(01)
	14	32		14	(01)(02)
	15	33		15	(01)(03)

For decimal 15, radix 12: $(01) \times 12^1 + (03) \times 12^0 = 12 + 3 = 15$

1.7 10_{10} in radix 3 = 101_3 (Divide by 3 repeatedly)

10_{10} in radix 7 = 13_7 (Divide by 7 repeatedly)

111_4 in radix 3 = 210_3 (Convert to radix 10, then divide by 3 repeatedly)

111_4 in radix 7 = 30_7 (Convert to radix 10, then divide by 7 repeatedly)

1.9

2s complement	Sign magnitude	
0111 1111	0111 1111	+117
1000 0001	1111 1111	–127
0000 1111	0000 1111	+15
1111 0001	1000 1111	–15
1111 0000	1001 0000	–16

1.14 $x_1 + x_1'x_2 = (x_1 + x_1')(x_1 + x_2)$ Distributive law
$= 1(x_1 + x_2)$ Complementation law
$= x_1 + x_2$ Identity law

1.19 $x_3(x_1'x_2' + x_1x_2) \neq x_3$ $\quad x_1'x_2' + x_1x_2 \neq 1$
$(x_1x_2)' + x_1x_2 = 1$

1.23 $z_1 = x_1'x_4' + x_1'x_2x_3' + x_3x_4' + x_2'x_3$
$z_1 = (x_1' + x_3)(x_2' + x_3' + x_4')(x_2 + x_3 + x_4')$

1.27 $z_1 = x_1'x_3'\text{A} + x_1'x_2x_3' + x_1x_3\text{A}' + x_1x_2'\text{A}' + x_1x_2'x_3$

1.29 $z_1 = x_3x_4' + x_2'x_4' + x_1'x_2x_3 + x_1x_2x_3'x_4$

1.39 $\text{Maj}_1 = x_1x_2 + x_1x_3 + x_2x_3$

$\text{Maj}_2 = x_1'x_2' + x_1'x_3' + x_2'x_3'$

$\text{Maj}_3 = \text{Maj}_1\,\text{Maj}_2 + \text{Maj}_1\,x_1x_2' + \text{Maj}_2\,x_1x_2'$

$z_1 = \text{Maj}_3 = 0 + x_1x_2'x_3 + x_1x_2'x_3'$

$z_1 = x_1x_2'$

Chapter 2

2.13

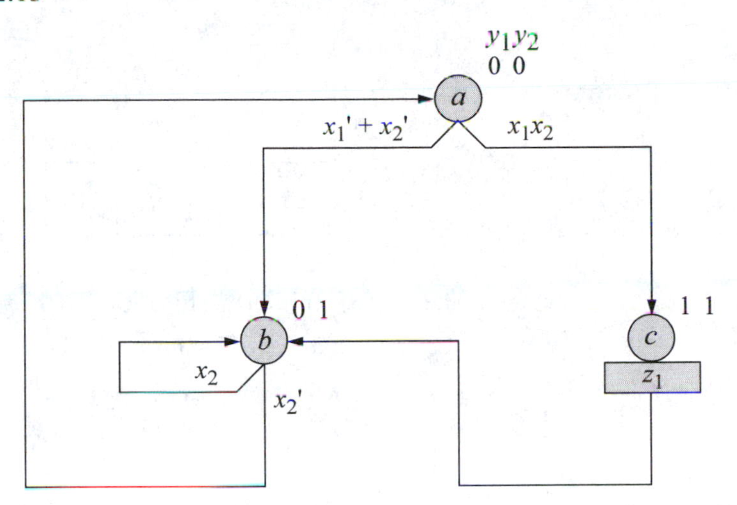

2.15

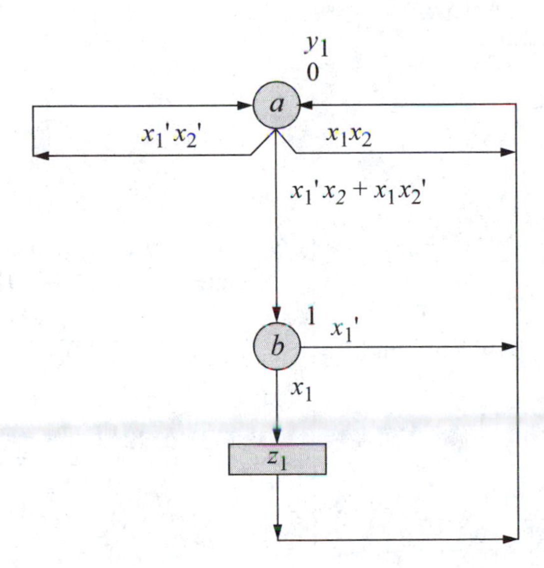

2.18 The function of z_1 is sum. The function of z_2 is carry.

2.20

Next-state map

$y_1 \backslash y_2y_3$	00	01	11	10
0	1 (0)	1 (1)	1 (3)	– (2)
1	– (4)	* (5)	– (7)	1 (6)

y_1

* Minterm location 101 $= x_1'x_3' + x_1x_2x_3' + x_1x_2'x_3$

Minimize minterm location 101 by a Karnaugh map

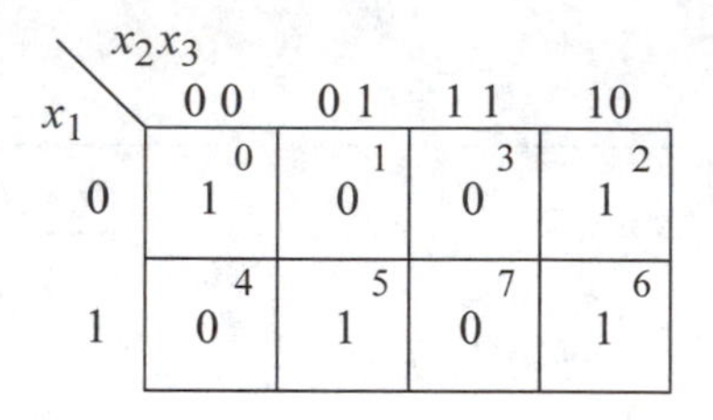

Therefore, minterm location 101 $= x_1'x_3' + x_2x_3' + x_1x_2'x_3$

2.23

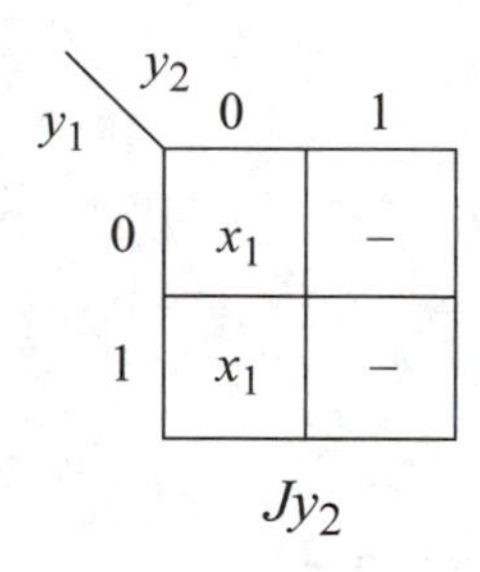

$y_1 \backslash y_2$	0	1
0	–	x_1'
1	–	x_1'

Ky_2

2.25

In state 101, $y_3 = 1$, if $x_1'x_2x_3 + x_1x_2 + x_1x_2'x_3 + x_1x_2'x_3'$

Minimize the equation by a Karnaugh map.

$x_1 \backslash x_2x_3$	00	01	11	10
0	0 (0)	0 (1)	1 (3)	0 (2)
1	1 (4)	1 (5)	1 (7)	1 (6)

Flip-flop has a next state of 1 if, $y_3 = x_1 = x_2x_3$

Now construct a partial next-state table for state a ($y_1y_2y_3 = 101$)

Present state $y_1y_2y_3$	Inputs $x_1x_2x_3$	Next state y_3	Jy_3	Ky_3
1 0 1	0 0 0	0	–	1
	0 0 1	0	–	1
	0 1 0	0	–	1
	0 1 1	1	–	0
	1 0 0	1	–	0
	1 0 1	1	–	0
	1 1 0	1	–	0
	1 1 1	1	–	0

Therefore, $Jy_3 = -$

$Ky_3 = x_1'x_3' + x_1'x_2'$

2.28

2.30 $z_1 = \{[x_1' + (x_1x_2)''] [x_2' + (x_1x_2)'']\}'$
$= (x_1' + x_1x_2)' + (x_2' + x_1x_2)'$
$= x_1(x_1x_2)' + x_2(x_1x_2)'$
$= x_1(x_1' + x_2') + x_2(x_1' + x_2')$
$= x_1x_2' + x_1'x_2$

Chapter 3

3.4 $a \equiv e, \quad b \equiv h \equiv d \equiv f, \quad c \equiv g$

3.8 (a) $y_1y_2 = 00, 11, 01, 10, 00, \ldots$
(b) $y_1y_2 = 00, 10, 01, 11, 00, \ldots$

3.14 $Jy_1 = y_3'$ $\quad Ky_1 = y_3$
$Jy_2 = y_1$ $\quad Ky_2 = y_1'$
$Jy_3 = y_2$ $\quad Ky_3 = y_2'$

3.23

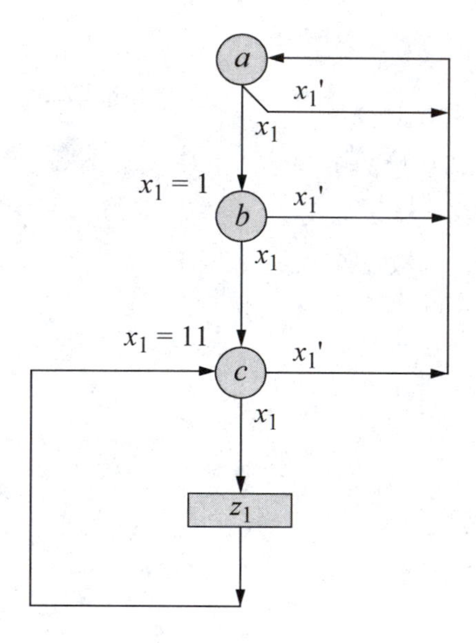

3.28 (d) = 110, (e) = 100

3.31 (c) = 011, (e) = 001

3.35 $z_1 \uparrow t_1 \downarrow$No

3.38 (a) $z_1 = \uparrow t_1 \downarrow t_2$
(b) $z_1 = \uparrow t_1 \downarrow t_2$ or $z_1 = \uparrow t_3 \downarrow t_4$
(c) $z_1 = \uparrow t_1 \downarrow t_3$ or $z_1 = \uparrow t_4 \downarrow$Next t_4

Chapter 4

4.1 $s_0 = y_3$ $s_1 = y_2$ $s_2 = y_1$
$d_0 = 0$ $d_1 = 0$ $d_2 = 1$ $d_3 = 1$
$d_4 = x_1 x_2$ $d_5 = x_3$ $d_6 = x_1$ $d_7 = x_1 + x_2$

4.6 $s_0 = y_3$ $s_1 = y_2$
$d_0 = x_1$ $d_1 = y_1'$ $d_2 = 1$ $d_3 = 1$

4.10 $s_0 = y_2$ $s_1 = y_1$
$d_0 = x_1' + x_2'$ $d_1 = x_1 x_2$ $d_2 = 1$ $d_3 = 1$

4.13 Mux y_1: $s_0 = y_3$ $s_1 = y_1$
$d_0 = y_2' x_1'$ $d_1 = 0$ $d_2 = 0$ $d_3 = (x_2' + x_3) y_2'$

Mux y_2: $s_0 = y_3$ $s_1 = y_1$
$d_0 = 0$ $d_1 = x_2$ $d_2 = 0$ $d_3 = y_2' x_2'$

Mux y_3: $s_0 = y_3$ $s_1 = y_1$
$d_0 = y_2'$ $d_1 = 0$ $d_2 = 0$ $d_3 = y_2' x_2' x_3'$

z_1 = decoder output 2 z_2 = decoder output 4 z_3 = decoder output 7

4.20 $Dy_1 = y_4'$ $Dy_2 = y_1 y_4'$ $Dy_3 = y_2 y_4'$ $Dy_4 = y_3 y_4'$
$z_1 = y_4 \bullet$ Clock delayed

4.26 (a) Address input 0 = $+A_{12}$ Enable high input = +I/O/M'
Address input 1 = $+A_{14}$ Enable low input = –Write
Address input 2 = $+A_{15}$ Enable low input = $+A_{13}$

Decoder output 4 is active $A_{15}\ A_{14}\ A_{13}\ A_{12} = 1000$

(b) Address input 0 = $+A_{13}$ Enable high input = +I/O/M'
Address input 1 = $+A_{14}$ Enable low input = –Read
Address input 2 = $+A_{15}$ Enable low input = $-A_{12}$

Decoder output 6 is active $A_{15}\ A_{14}\ A_{13}\ A_{12} = 1101$

4.28 (a) States f, g, h, i, j, k, l and m cannot be used if states a, b, c, d, or e is asserted.

(b) Decoder address input 0 = Outport 5
Decoder address input 1 = Outport 6
Decoder address input 2 = Outport 7

Enable high input = 1
Enable low input = 0
Enable low input = State a + State b + State c + State d + State e

Decoder output 0 = State f
Decoder output 1 = State g
Decoder output 2 = State h
Decoder output 3 = State i
Decoder output 4 = State j
Decoder output 5 = State k
Decoder output 6 = State l
Decoder output 7 = State m

4.37 $Jy_1 = y_2 y_3' y_4'$ $\quad Ky_1 = y_2' y_3' y_4'$
$Jy_2 = y_1' y_3 y_4'$ $\quad Ky_2 = y_1 y_3 y_4'$
$Jy_3 = y_4(y_1 \oplus y_2)'$ $\quad Ky_3 = y_4(y_1 \oplus y_2)$
$Jy_4 = (y_1 \oplus y_2 \oplus y_3)'$ $\quad Ky_4 = y_1 \oplus y_2 \oplus y_3$

The parity flip-flop is set during initialization.
$Jy_p = 1$ $\quad Ky_p = 1$

Error = $(y_1 \oplus y_2 \oplus y_3 \oplus y_4 \oplus y_5 \oplus y_p)'$

4.38 $Dy_1 = y_4'$ $\quad Dy_2 = y_1$ $\quad Dy_3 = y_2$ $\quad Dy_4 = y_3$
y_p is wired in toggle mode.
Error = $(y_1 \oplus y_2 \oplus y_3 \oplus y_4 \oplus y_p)'$

4.39

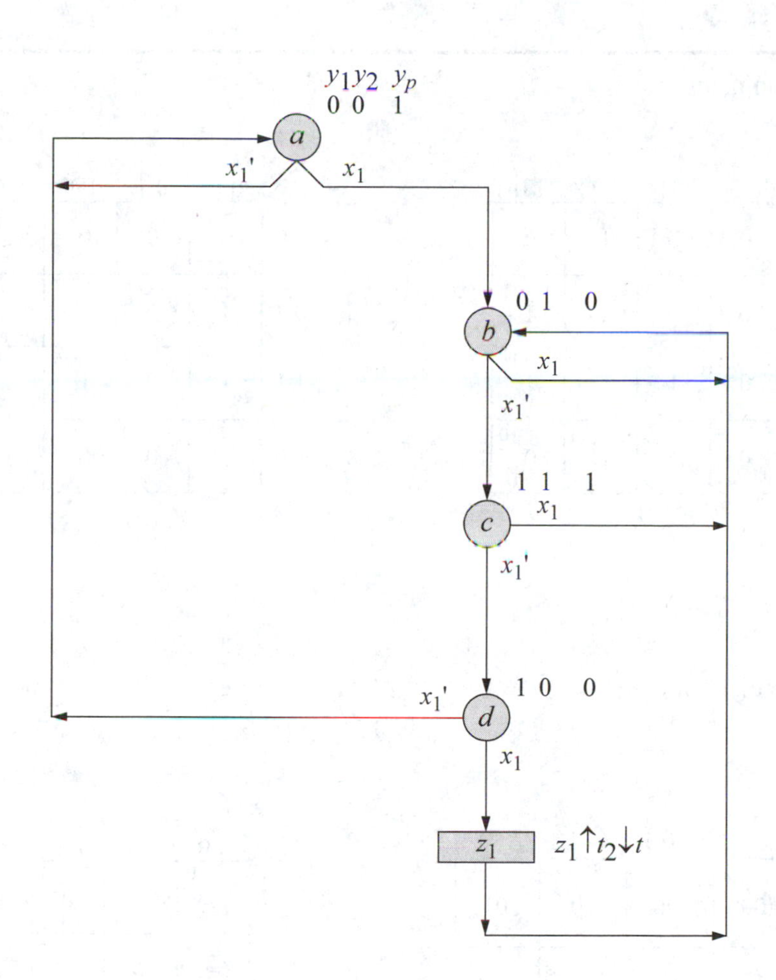

(a)
$$Dy_1 = y_2 x_1'$$
$$Dy_2 = x_1 + y_1' y_2$$
$$Dy_p = x_1' y_1' + x_1' y_2'$$
$$= x_1' (y_1' + y_2')$$

(b)

$Jy_1 = y_2 x_1'$	$Ky_1 = x_1 + y_2'$
$Jy_2 = x_1$	$Ky_2 = y_1 x_1'$
$Jy_p = x_1'$	$Ky_p = x_1 + y_1 y_2$

$$z_1 = y_1 y_2' x_1 \text{clk}'$$

Chapter 5

5.3 Excitation maps

$y_{1f}y_{2f}$ \ x_1x_2	00	01	11	10
00	0 (0)	0 (1)	1 (3)	0 (2)
01	0 (4)	0 (5)	1 (7)	1 (6)
11	0 (12)	1 (13)	1 (15)	1 (14)
10	0 (8)	1 (9)	1 (11)	0 (10)

Y_{1e}

$y_{1f}y_{2f}$ \ x_1x_2	00	01	11	10
00	0 (0)	0 (1)	0 (3)	1 (2)
01	1 (4)	0 (5)	1 (7)	1 (6)
11	1 (12)	0 (13)	1 (15)	1 (14)
10	0 (8)	0 (9)	0 (11)	1 (10)

Y_{2e}

Combined excitation map

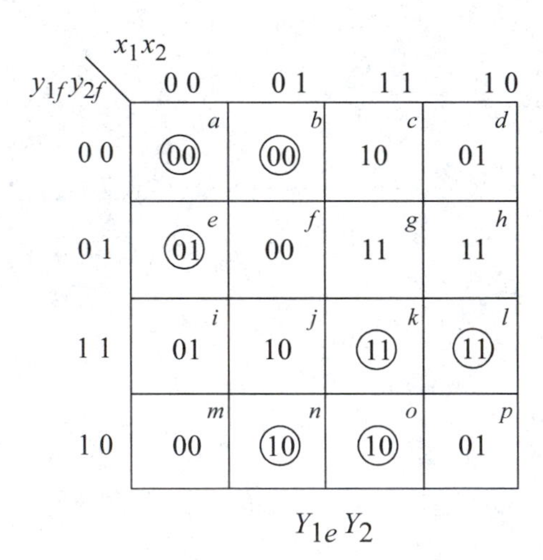

$y_{1f}y_{2f}$ \ x_1x_2	00	01	11	10
00	*a* (00)	*b* (00)	*c* 10	*d* 01
01	*e* (01)	*f* 00	*g* 11	*h* 11
11	*i* 01	*j* 10	*k* (11)	*l* (11)
10	*m* 00	*n* (10)	*o* (10)	*p* 01

$Y_{1e}Y_2$

Flow table

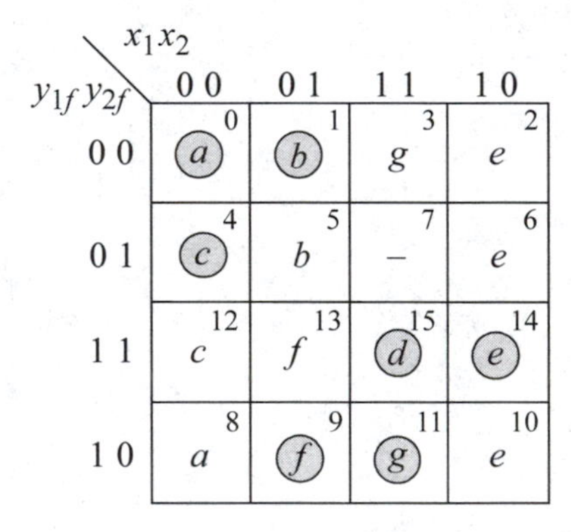

$y_{1f}y_{2f}$ \ x_1x_2	00	01	11	10
00	(*a*) 0	(*b*) 1	*g* 3	*e* 2
01	(*c*) 4	*b* 5	– 7	*e* 6
11	*c* 12	*f* 13	(*d*) 15	(*e*) 14
10	*a* 8	(*f*) 9	(*g*) 11	*e* 10

5.14

x_1x_2 \ x_3x_4	0 0	01	1 1	1 0
0 0	[0] 1	[1] 1	[3] 0	[2] 0
0 1	[4] 1	[5] 1	[7] 1	[6] 0
1 1	[12] 0	[13] 0	[15] 1	[14] 1
1 0	[8] 0	[9] 0	[11] 0	[10] 1

z_1

Sum-of-products implementation

$x_1x_2x_3x_4$ = 0101 → 0111 produces no static hazard.
$x_1x_2x_3x_4$ = 0111 → 0101 produces a static-1 hazard.
$x_1x_2x_3x_4$ = 1111 → 1110 produces no static hazard.
$x_1x_2x_3x_4$ = 1110 → 1111 produces no static hazard.

Product-of-sums implementation

$x_1x_2x_3x_4$ = 0011 → 0010 produces no static hazard.
$x_1x_2x_3x_4$ = 0010 → 0011 produces a static-0 hazard.
$x_1x_2x_3x_4$ = 1001 → 1011 produces a static-0 hazard.
$x_1x_2x_3x_4$ = 1011 → 1001 produces no static hazard.

5.16 $x_1x_2x_3x_4x_5$ = 11111 → 11011 produces a static-1 hazard.
The hazard cover is $x_1x_2x_4x_5$

$x_1x_2x_3x_4x_5$ = 11011 → 11010 produces a static-1 hazard.
The hazard cover is $x_1x_2x_3'x_4$

5.19 (a) and (b) contain no static-1 hazards.

5.20 $x_1'x_3'x_4 + x_2x_3x_4 + x_1x_2'x_3x_4' + x_1'x_2x_4$ ← Hazard cover.

5.28 Oscillations are: ⓕ $\rightarrow g \leftrightarrow c$

ⓙ $\rightarrow k \rightarrow g \leftrightarrow c$

5.30 Oscillations are: ⓔ $\rightarrow f \leftrightarrow b$
ⓔ $\rightarrow h \leftrightarrow l$

ⓚ $\rightarrow l \leftrightarrow h$
ⓚ $\rightarrow j \rightarrow f \leftrightarrow b$

ⓜ $\rightarrow p \rightarrow l \leftrightarrow h$
ⓜ $\rightarrow n \rightarrow j \rightarrow f \leftrightarrow b$

5.33 Noncritical race: ⓦ $\rightarrow g$
$y_{1f}y_{2f}x_1x_2x_3 = 01111 \rightarrow 01110$
The machine will always terminate in state ⓞ

Noncritical race: ⓓ $\rightarrow c$
$y_{1f}y_{2f}x_1x_2x_3 = 00100 \rightarrow 00110$
The machine will always terminate in state ⓞ

Noncritical race: ⓗ $\rightarrow g$
$y_{1f}y_{2f}x_1x_2x_3 = 01100 \rightarrow 01110$
The machine will always terminate in state ⓞ

Critical race: (bb) $\rightarrow l$
$y_{1f}y_{2f}x_1x_2x_3 = 11101 \rightarrow 11100$
The machine will terminate in either state ⓓ or ⓗ

Critical race: ⓜ $\rightarrow cc$
$y_{1f}y_{2f}x_1x_2x_3 = 10000 \rightarrow 10001$
The machine will terminate in state ⓠ, ⓤ, or ⓨ

5.36 There are no noncritical races and no critical races. The following oscillations exist:

ⓕ $\rightarrow g \leftrightarrow c$
ⓙ $\rightarrow k \rightarrow g \leftrightarrow c$

Chapter 6

6.2

x_1x_2 00	01	11	10	z_1
(a)	f	–	b	0
a	–	c	(b)	0
–	–	(c)	d	1
a	–	c	(d)	1
(e)	f	–	d	1
a	(f)	–	–	0

6.5 Equivalent states are: (a) ≡ (f), (d) ≡ (e)

x_1x_2 00	01	11	10	z_1
(a)	b	–	c	0
a	(b)	d	–	1
a	–	d	(c)	1
–	h	(d)	g	0
a	–	d	(g)	0
a	(h)	d	–	0

6.16 The following states are equivalent:
ⓑ ≡ ⓓ, ⓐ ≡ ⓒ. Delete rows ⓒ and ⓓ.

x_1x_2 00	01	11	10	z_1	z_2
ⓐ	b	–	f	0	0
a	ⓑ	e	–	1	0
–	g	ⓔ	f	0	1
a	–	e	ⓕ	1	1
a	ⓖ	e	–	1	1

6.17

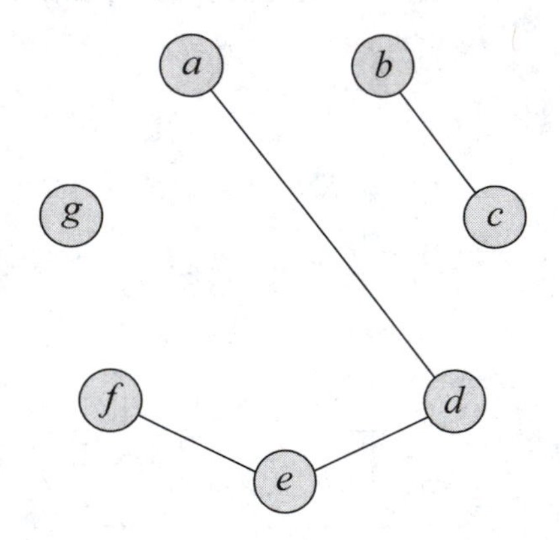

6.18

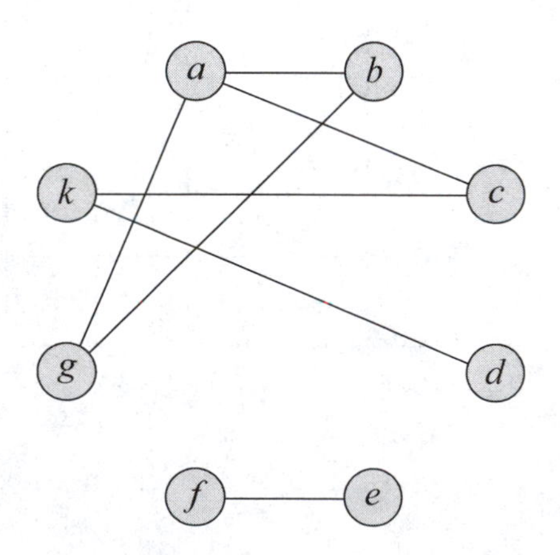

6.26

	x_1x_2 00	01	11	10
(a), (b), (c)	(a)	(b)	d	(c)
(e), (f), (h)	(f)	(h)	g	(e)
(d)	–	b	(d)	e
(g)	–	h	(g)	c

6.31 Transition diagram

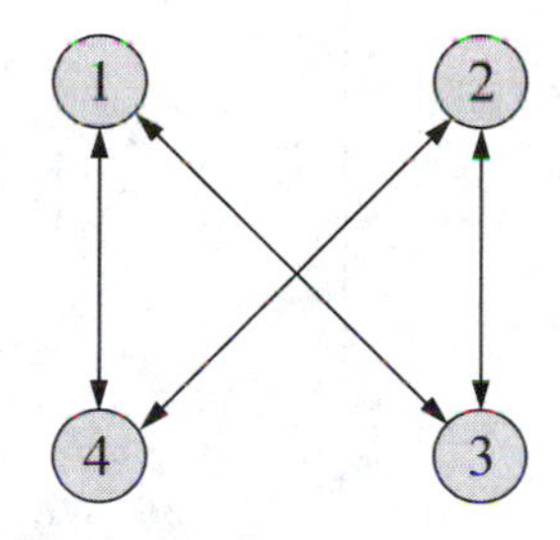

Excitation maps

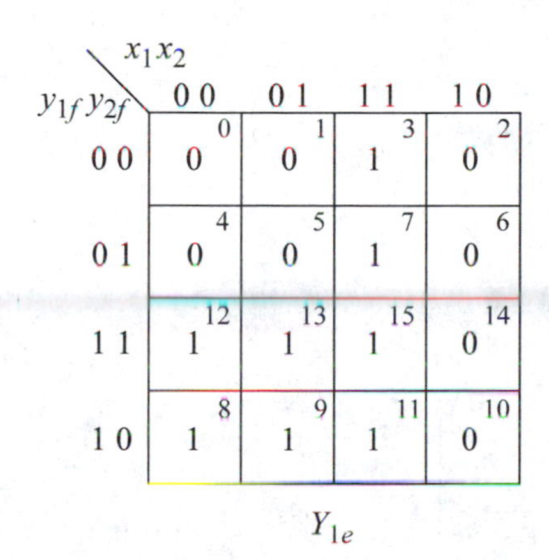

$y_{1f}y_{2f}$ \ x_1x_2	0 0	0 1	1 1	1 0
0 0	^{0} 0	^{1} 1	^{3} 0	^{2} 0
0 1	^{4} 0	^{5} 1	^{7} 1	^{6} 1
1 1	^{12} 0	^{13} 1	^{15} 1	^{14} 1
1 0	^{8} 0	^{9} 1	^{11} 0	^{10} 0

Y_{2e}

$Y_{1e} = y_{1f}x_1' + x_1x_2 + y_{1f}x_2$

Hazard cover → $y_{1f}x_2$

$Y_{2e} = x_1'x_2 + y_{2f}x_1 + y_{2f}x_2$

Hazard cover → $y_{2f}x_2$

6.39 Merger diagram

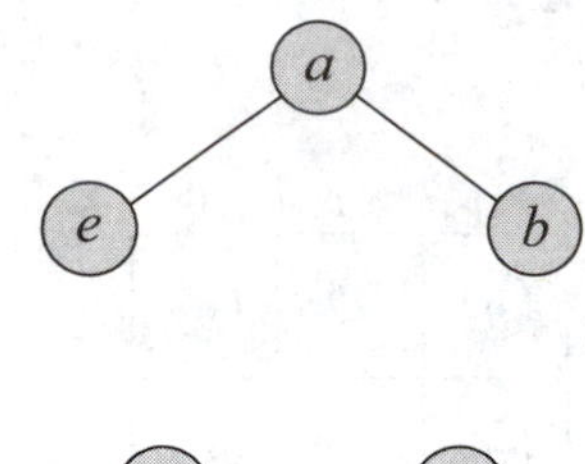

Merged flow table

		x_1x_2 00	01	11	10
1	(a), (b)	(a)	–	c	(b)
2	(c)	–	–	(c)	d
3	(d)	a	–	e	(d)
4	(e)	–	–	(e)	b

Transition diagram

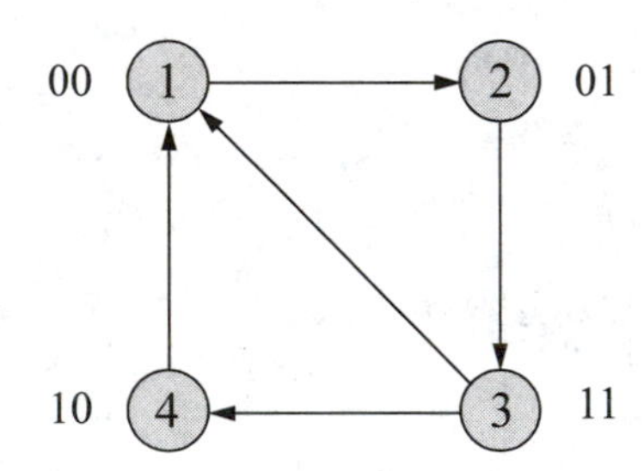

Output map

$y_{1f}y_{2f}$ \ x_1x_2	0 0	0 1	1 1	1 0
0 0	[0] 0	[1] –	[3] 0	[2] 0
0 1	[4] –	[5] –	[7] 0	[6] 0
1 1	[12] 0	[13] –	[15] –	[14] 0
1 0	[8] 0	[9] –	[11] 1	[10] –

$z_1 = y_{1f}x_2$

6.49

Primitive flow table

	$x_1 = 0$	$x_1 = 1$	z_1
	(a)	b	0
	c	(b)	1
	(c)	d	1
	a	(d)	0

Combined excitation map

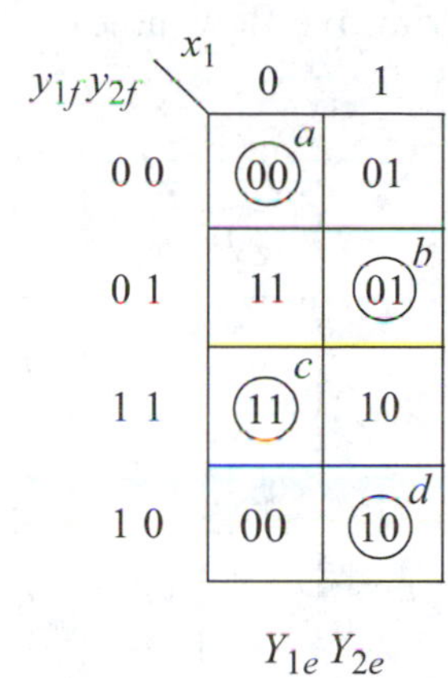

Individual excitation maps

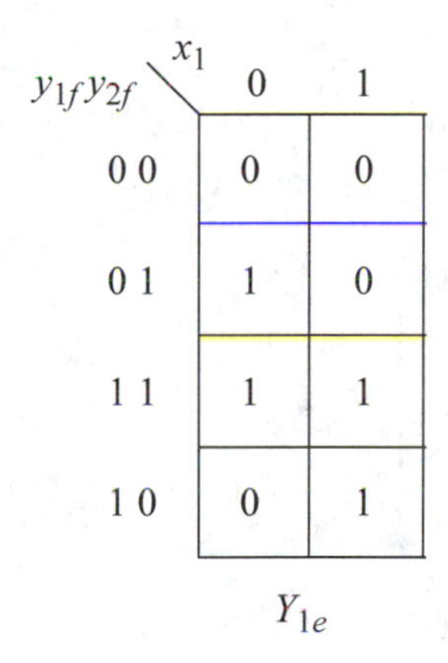

$y_{1f}y_{2f}$ \ x_1	0	1
0 0	0	1
0 1	1	1
1 1	1	0
1 0	0	0

Y_{2e}

Output map

$y_{1f}y_{2f}$ \ x_1	0	1
0 0	0 (a)	–
0 1	1	1 (b)
1 1	1 (c)	–
1 0	0	0 (d)

$$Y_{1e} = y_{1f}x_1 + y_{2f}x_1' + y_{1f}y_{2f}$$

$$Y_{2e} = y_{1f}'x_1 + y_{2f}x_1' + y_{1f}'y_{2f}$$

$$Y_{1e} = (y_{1f} + x_1')(y_{2f} + x_1)(y_{1f} + y_{2f})$$

$$Y_{2e} = (y_{1f}' + x_1')(y_{2f} + x_1)(y_{1f}' + y_{2f})$$

$$z_1 = y_{2f}$$

6.56

Primitive flow table

x_1x_2	00	01	11	10	z_1
	(a)	d	–	b	0
	c	–	f	(b)	1
	(c)	d	–	b	1
	a	(d)	e	–	0
	–	d	(e)	b	1
	–	d	(f)	b	0

Merger diagram

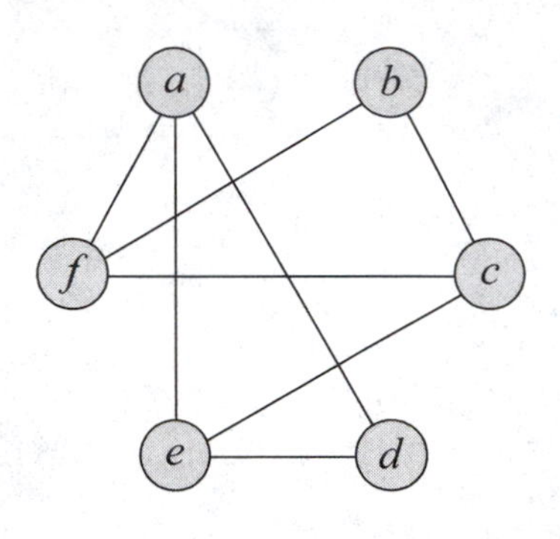

Merged flow table

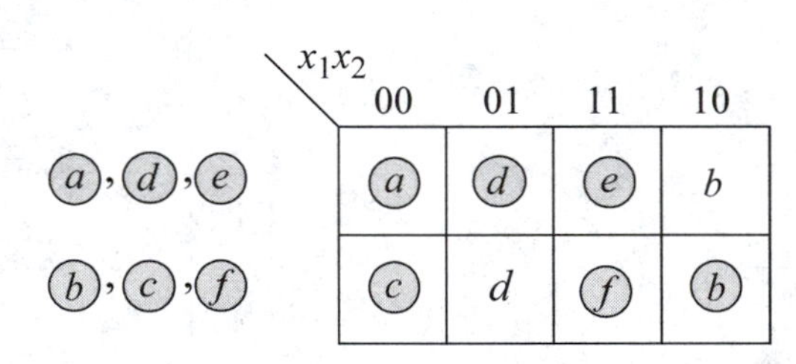

	x_1x_2 00	01	11	10
(a), (d), (e)	(a)	(d)	(e)	b
(b), (c), (f)	(c)	d	(f)	(b)

Excitation map

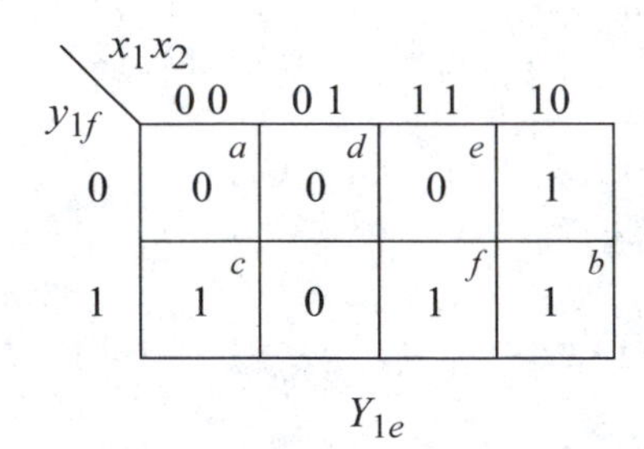

y_{1f} \ x_1x_2	00	01	11	10
0	0 (a)	0 (d)	0 (e)	1
1	1 (c)	0	1 (f)	1 (b)

Y_{1e}

Output map

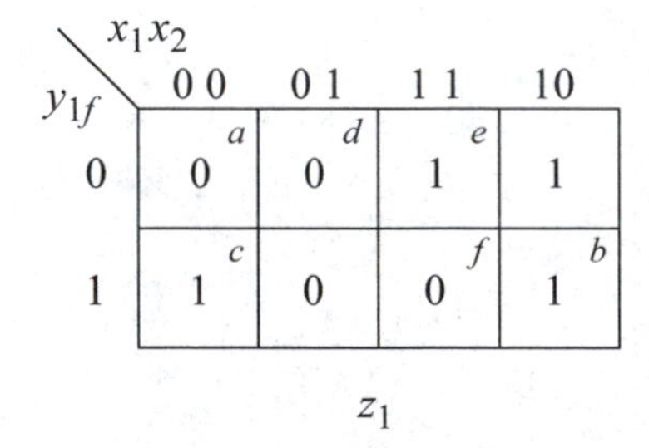

y_{1f} \ x_1x_2	00	01	11	10
0	0 (a)	0 (d)	1 (e)	1
1	1 (c)	0	0 (f)	1 (b)

z_1

$$Y_{1e} = x_1x_2' + y_{1f}x_1 + y_{1f}x_2'$$

$$z_1 = y_{1f}'x_1 + y_{1f}x_2' + x_1x_2'$$

6.58

Primitive flow table

x_1x_2	00	01	11	10	z_1
	(a)	–	–	b	0
	a	–	c	(b)	0
	–	–	(c)	d	0
	a	–	e	(d)	0
	–	–	(e)	f	0
	a	–	g	(f)	0
	–	–	(g)	b	1

Merger diagram

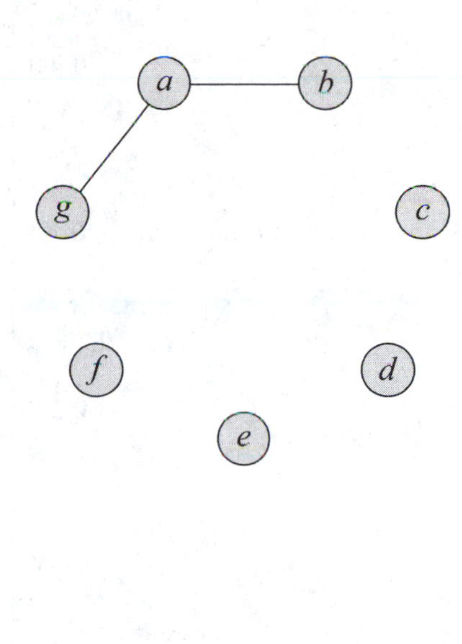

Merged flow table

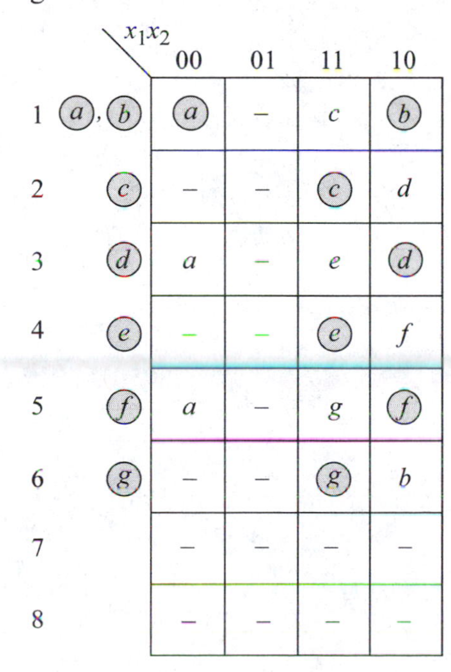

	x_1x_2	00	01	11	10
1	(a), (b)	(a)	–	c	(b)
2	(c)	–	–	(c)	d
3	(d)	a	–	e	(d)
4	(e)	–	–	(e)	f
5	(f)	a	–	g	(f)
6	(g)	–	–	(g)	b
7		–	–	–	–
8		–	–	–	–

Transition diagram

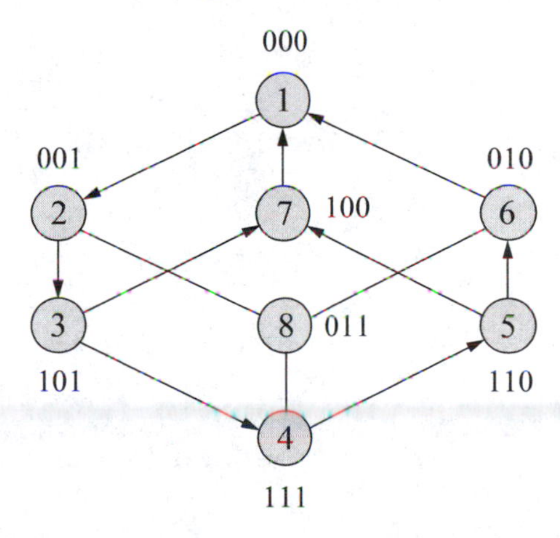

Combined excitation map

	$y_{1f}y_{2f}y_{3f}$ \ x_1x_2	00	01	11	10
1	000	(000) a	–	001	(000) b
2	001	–	–	(001) c	101
8	011	–	–	–	–
6	010	–	–	(010) g	000
5	110	100	–	010	(110) f
4	111	–	–	(111) e	110
3	101	100	–	111	(101) d
7	100	000	–	–	–

$Y_{1e}Y_{2e}Y_{3e}$

$$Y_{1e} = y_{3f}x_2' + y_{1f}y_{3f} + y_{1f}y_{2f}x_2'$$

$$Y_{1e} = (y_{2f} + y_{3f})(y_{1f} + y_{2f}')(y_{1f} + x_2')(y_{2f}' + y_{3f} + x_2')$$

$$Y_{2e} = y_{1f}x_2 + y_{2f}x_2 + y_{1f}y_{2f}x_1$$

$$Y_{2e} = (x_1)(y_{1f} + y_{2f})(y_{2f} + x_2)(y_{1f} + x_2)$$

$$Y_{3e} = y_{2f}'x_2 + y_{3f}x_2 + y_{2f}'y_{3f}x_1$$

$$Y_{3e} = (x_1)(y_{2f}' + y_{3f})(y_{2f}' + x_2)(y_{3f} + x_2)$$

$$z_1 = y_{1f}'y_{2f}$$

$$z_1 = (y_{1f}')(y_{2f})$$

Chapter 7

7.1 State diagram

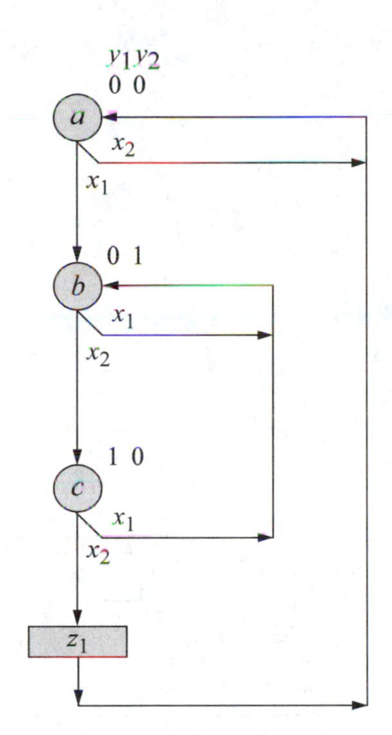

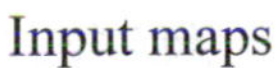
Input maps

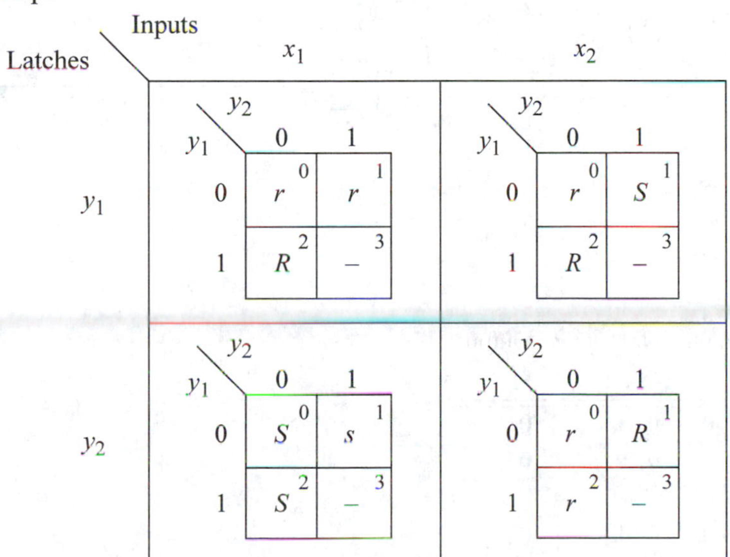

$Sy_1 = y_2x_2$ $Sy_2 = x_1$

$Ry_1 = x_1 + y_1x_2$ $Ry_2 = x_2$

Output map

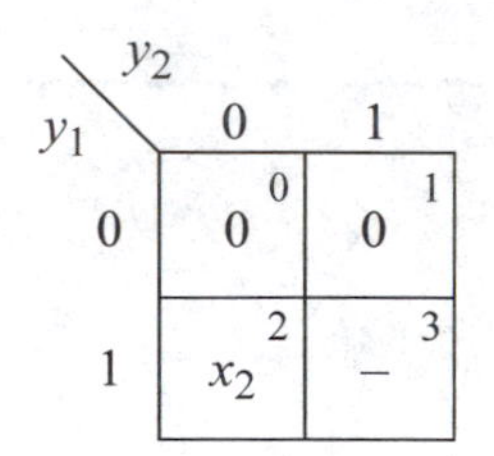

$z_1 = y_1 x_2$

Logic diagram

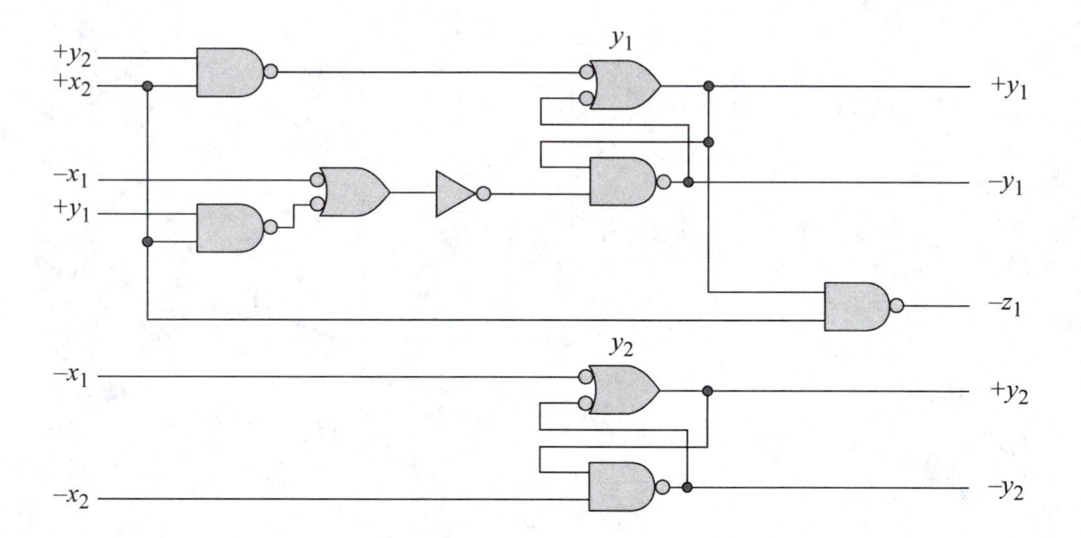

7.5

Present state $y_1\ y_2$	Inputs $x_1\ x_2$	Next state $y_1\ y_2$	Output z_1
0 0	1 0	0 1	0
0 0	0 1	0 0	0
0 1	1 0	0 1	0
0 1	0 1	1 0	0
1 0	1 0	0 1	0
1 0	0 1	0 0	1

Input maps

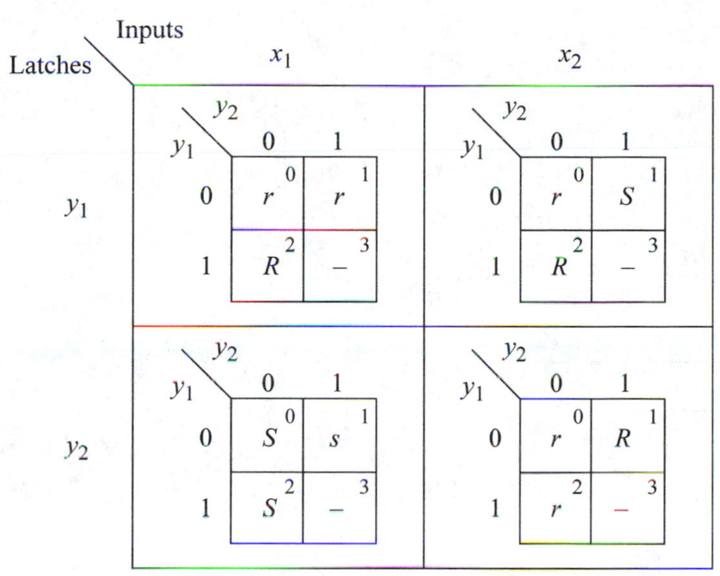

$Sy_1 = y_2 x_2$ $\quad$ $Sy_2 = x_1$

$Ry_1 = x_1 + y_1 x_2$ $\quad$ $Ry_2 = x_2$

Output map

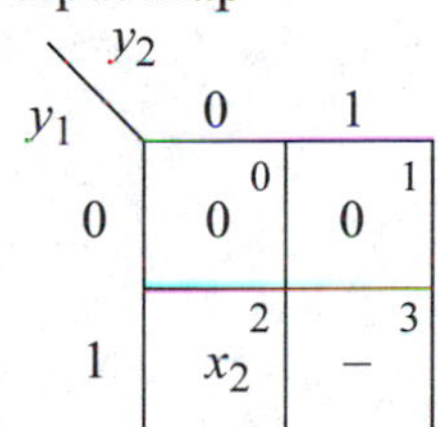

$z_1 = y_1 y_2' x_2$

$= y_1 x_2$ (Minimized)

State diagram

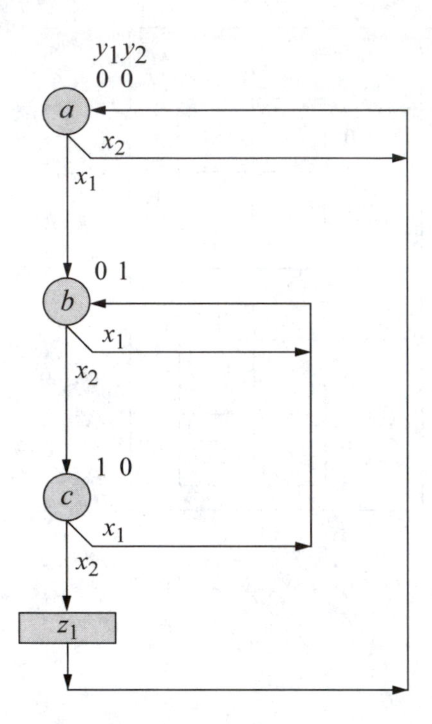

7.8

Present state $y_1\ y_2$	Inputs $x_1\ x_2$	Next state $y_1\ y_2$	Output z_1
0 0	1 0	0 1	0
0 0	0 1	0 0	0
0 1	1 0	1 1	0
0 1	0 1	0 0	0
1 1	1 0	1 1	1
1 1	0 1	1 0	1
1 0	1 0	1 1	1
1 0	0 1	0 0	1

Input maps

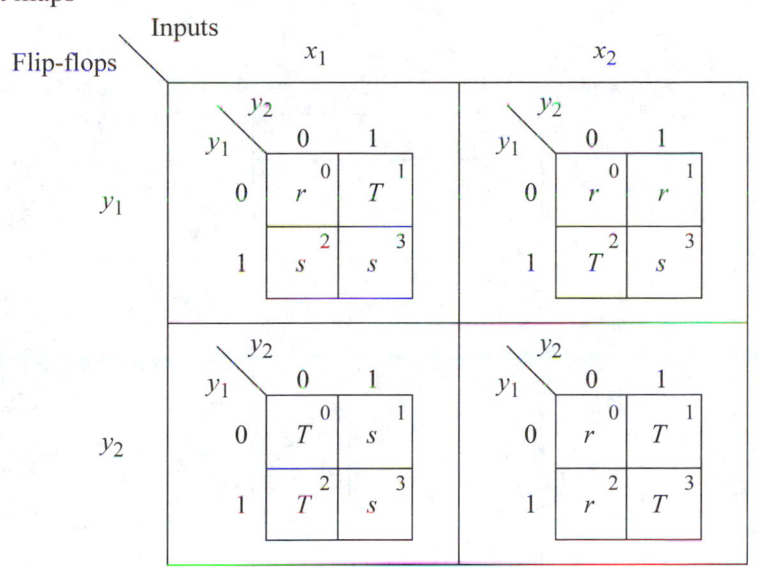

$Ty_1 = y_1'y_2x_1 + y_1y_2'x_2$ $\qquad$ $Ty_2 = y_2'x_1 + y_2x_2$

Output map

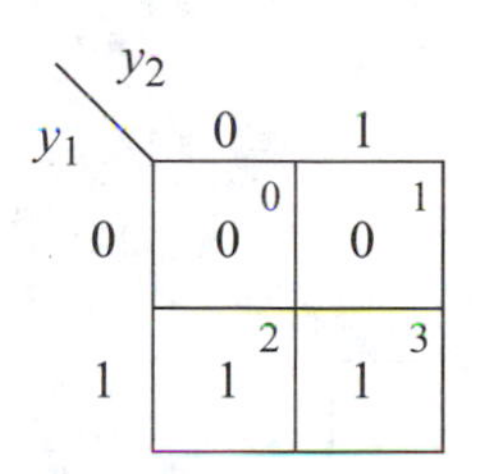

$z_1 = y_1$

State diagram

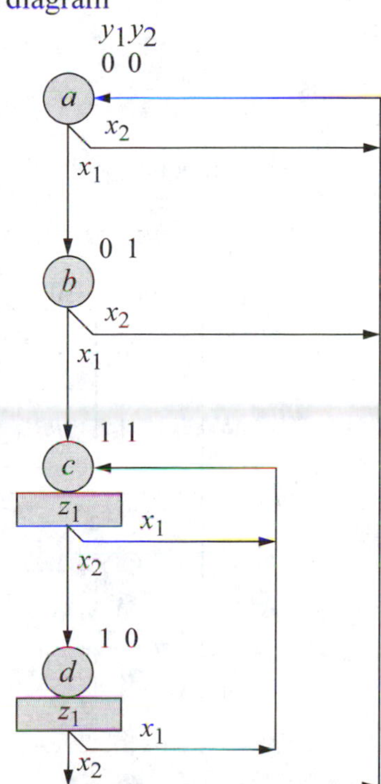

7.14 State diagram

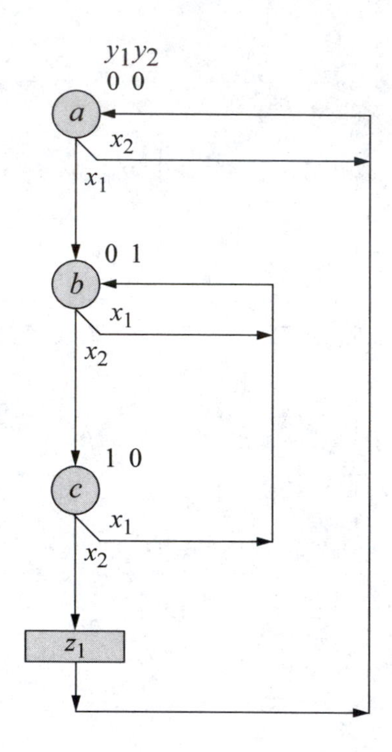

Input maps

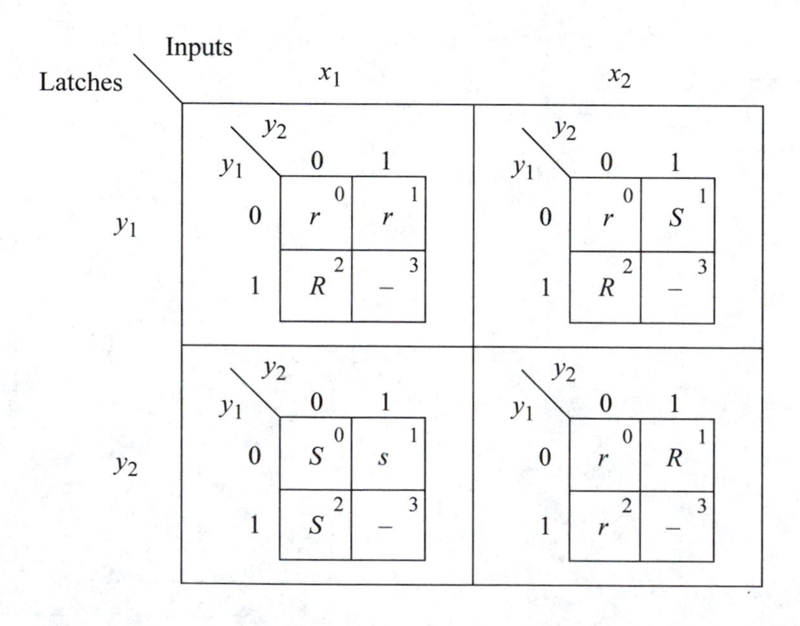

Logic diagram

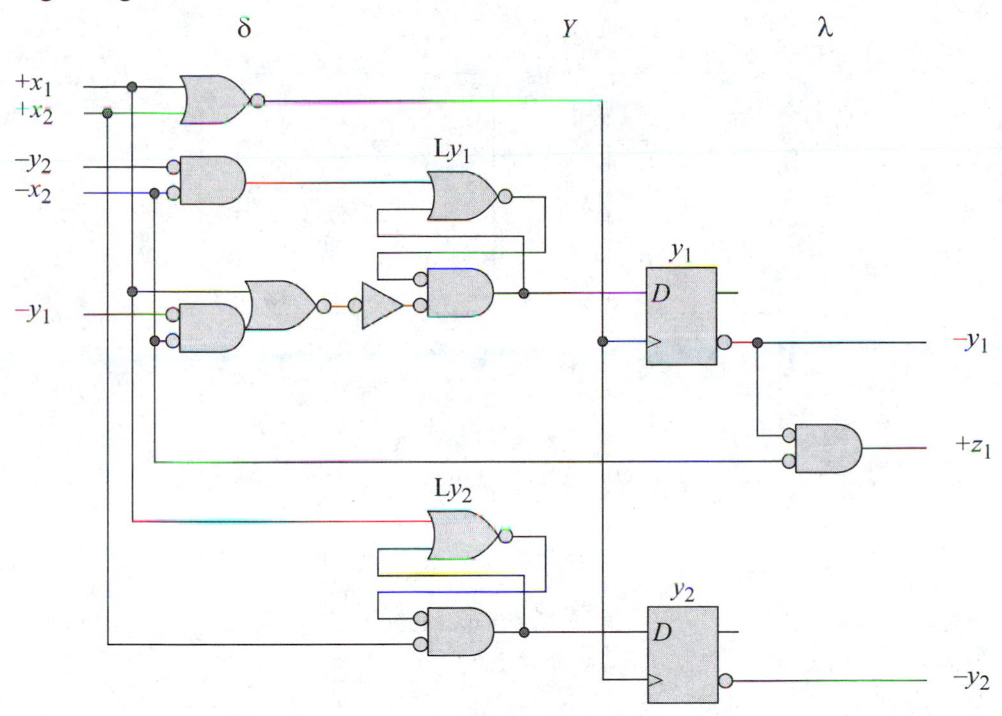

7.22

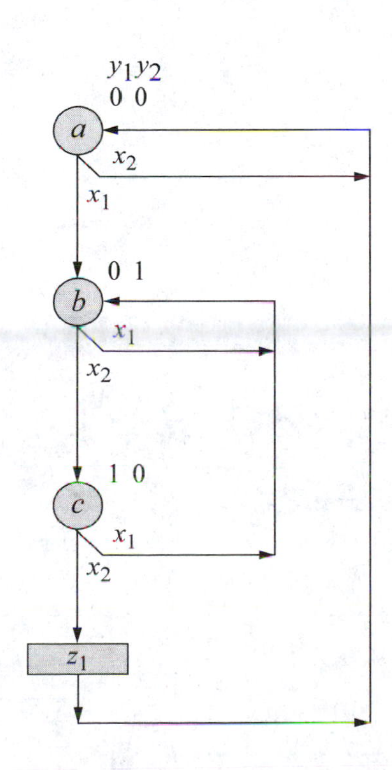

Input maps

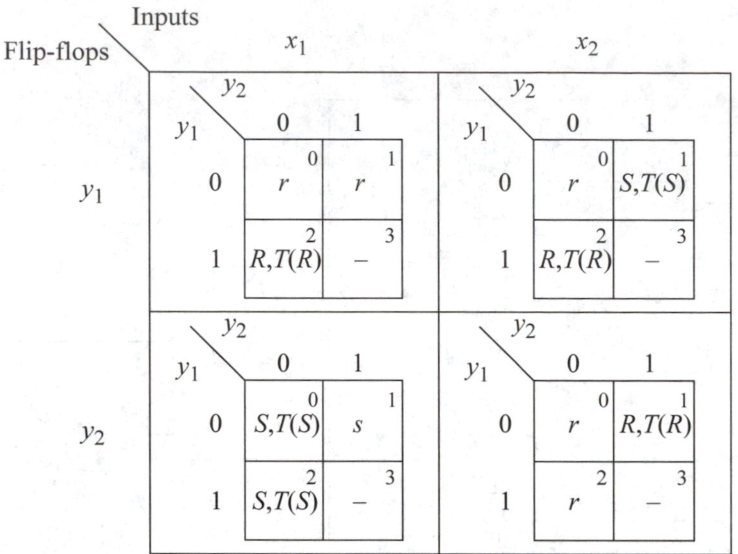

$Sy_1 = y_2x_2$
$Ty_1 = y_1x_1$
$Ry_1 = x_1 + y_1x_2$

$Sy_2 = x_1$
Ty_2 = Included in Sy_2 and Ry_2
$Ry_2 = x_2$

Output map

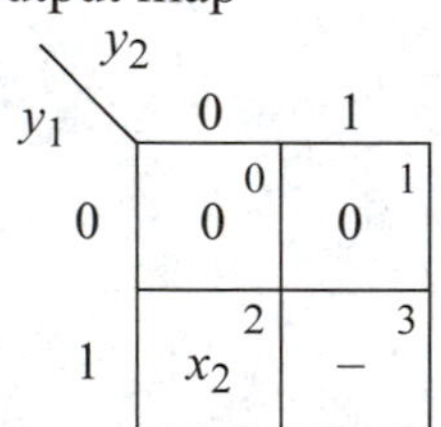

$z_1 = y_1x_2$

Logic diagram

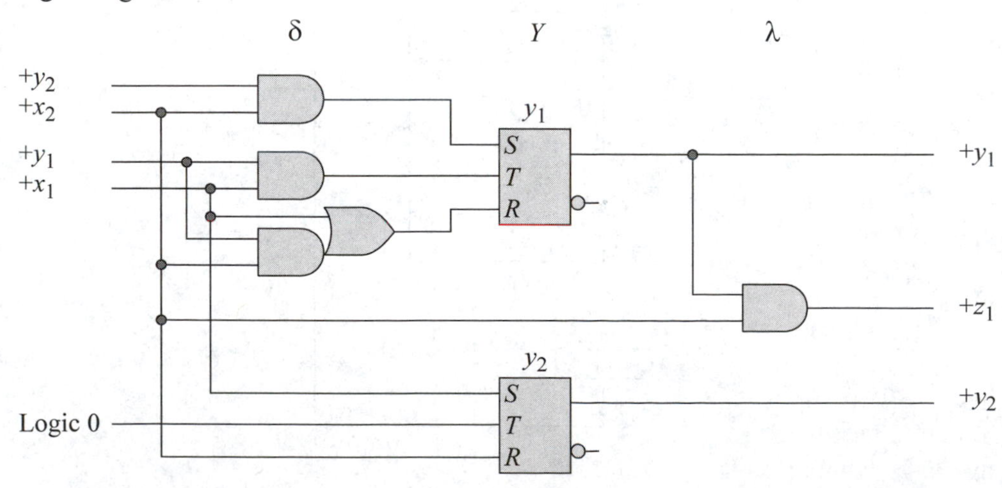

INDEX

Symbols

(r – 1) complement, 9

Numerics

0 and 1 associated with a variable, 14
0 and 1 complement, 14

A

absorption law 1, 15
absorption law 2, 15
active voltage levels, 43
adjacency rules, 258
algebraic manipulation, 19
algebraic structure, 126
analysis of combinational logic, 37
analysis procedure, 809–823
- *SR* latches, 810
- *SR* latches with *D* flip-flops, 817

AND gate, 36
- logical truth table, 39
- physical truth table, 39

AND operator, 12
ANSI/IEEE Std. 91-1984, 35
assertion levels, 37
assertion/deassertion times, 145
associative laws, 13
asynchronous counter, 99, 104
asynchronous sequential machine, 79, 121, 519
- fundamental-mode, 522
- general configuration, 121
- hazards, 553
- methods of analysis, 526
- oscillations, 578
- races, 582

augmented merged flow table, 667
autonomous machine, 105
axiom 1: boolean set definition, 13
axiom 2: closure laws, 13
axiom 3: identity laws, 13
axiom 4: commutative laws, 13
axiom 5: associative laws, 13
axiom 6: distributive laws, 13
axiom 7: complementation laws, 14

B

Binary-coded decimal (BCD) counter, 234
binary counter, 223
binary number system, 3
binary polynomial, 212
binary-coded decimal, 5
binary-coded hexadecimal, 6
binary-coded octal, 4
binary-to-Gray code converter, 248
boolean algebra, 12–17
- 0 and 1 associated with a variable, 14
- 0 and 1 complement, 14
- absorption law 1, 15
- absorption law 2, 15
- associative laws, 13
- axiom, 12
- axiom 3: identity laws, 13
- boolean set definition, 13
- closure laws, 13
- commutative laws, 13
- complementation laws, 14
- DeMorgan's laws, 15
- distributive laws, 13
- dual, 13
- idempotent laws, 14
- involution law, 15
- summary of axioms and theorems, 17

boolean product, 13
burst mode, 210

C

canonical product of sums, 17
canonical sum of products, 16
Cartesian product, 82
check symbol, 30
classes of sequential machines, 89
- additional definitions, 123
- asynchronous sequential machines, 121
- counters, 98
- Mealy machines, 110
- Moore machines, 105
- registers, 93

clock, 77
clock pulses, 79
code coverter
  binary to Gray, 248
combinational logic, 1, 89
  analysis, 37
  synthesis, 42
combinational shifter, 218
  shift left algebraic, 218
  shift left logical, 218
  shift right algebraic, 218
  shift right logical, 218
combined excitation map, 668, 681
commutative laws, 13
comparators
  connected in cascade, 55
  equality function, 56
compendium of output glitches, 339
complementation laws, 14
complete synchronous sequential machine, 161
conjunctive normal form, 17
counters, 98–105
  asynchronous, 99, 104
  count-down, 99
  count-up, 99
  Gray code, 100
  Johnson, 100
  modulo-8, 102
  parity checked, 493
  ripple, 99
  synchronous, 99, 102
critical race, 586, 588
cycles, 586
cyclic redundancy check (CRC), 212

D

D flip-flop, 57
  characteristic table, 58
  edge-triggered device, 57
  excitation equation, 58
decimal number system, 4
  binary-coded decimal, 5
decoders, 47–52, 400
  2:4 decoder, 49
  3:8 decoder, 50
  4:10 decoder, 51
  4:16 decoder, 51
  definition, 400
  demultiplexer, 401
  minterm, 47
decoders for output logic, 400–412
delta next-state function
  definition, 86
DeMorgan's laws, 15
deterministic synchronous
  sequential machines, 254
diminished-radix complement, 9
disjunctive normal form, 16
distributive laws, 13
divide-by-two operation, 197
domain, 126
don't care, 24, 32
duality, 13
dynamic hazards, 568

E

enable, 96
encoders, 53–54
  8:3 encoder, 53
  10:4 encoder, 53
  priority encoder, 53
equality function, 56
equation equation, 661
equivalence relation, 184, 636
  reflexive, 184
  symmetric, 184
  transitive, 184
equivalent machines, 132
equivalent sequential machine, 485
equivalent states 183, 184, 633, 719, 729, 738, 744, 758, 767
  definition, 633
  interdependence, 634
  reflexive, 633
  symmetric, 633
  transitive, 633
error detection, 489–500
  block diagram, 489
  check generation, 490
  duplicate machine, 490
  input regeneration, 490
  parity checking, 493
essential hazards, 572, 700

essential prime implicant, 31
excitation equation, 722, 731, 740, 749, 763, 773
excitation map, 526, 612, 661, 722, 731, 740, 749, 763, 773
excitation variable, 122, 523, 526
exclusive-NOR function, 36
exclusive-OR function, 36

F

feedback path, 520
feedback variable, 122, 523, 526
field-programmable gate array (FPGA), 61, 437
    definition, 438
    organization, 440
finite-state machine, 78
flow table, 533
full adder, 119
    carry equation, 119
    sum equation, 119
functionally complete gates, 42
fundamental-mode, 524, 609, 619, 632
fundamental-mode model, 522

G

generator polynomial, 212
George Boole, 12
glitch, 263, 307, 451
glitch elimination, 312–333
    using complemented clock, 323
    using delayed clock, 326
    using state code assignment, 312
    using storage elements, 319
glitches and output maps, 333
Gray code, 20, 249, 252
Gray code counter, 100

H

hazards, 553–578
    dynamic, 568
    essential, 572
    multiple-order, 577
    static, 553
hexadecimal number system, 5
    binary-coded hexadecimal, 6
hierarchical method, 182
homomorphic image, 127
homomorphism, 126
horizontal organization, 464

I

idempotent laws, 14
identity elements, 13
identity laws, 13
implicant, 22
implication method, 196
implication symbol, 489
implication table, 187, 637
    implication method, 196
    implied pair, 189
    interdependence, 196
    self-implied pair, 190
implied pair, 189
incomplete synchronous sequential machine, 161, 162
individual excitation map, 676, 682
input alphabet, 79, 608
input map, 138
input port, 450
interdependence, 196, 197, 634
into mapping, 84
inverse machine, 490
involution law, 15
isomorphism, 101, 130
iterative network, 98, 481
    block diagram 482
    cascade, 98
    functionally equivalent sequential machine, 485
    parallel, 98
    typical cell, 483

J

JK flip-flop, 59
    characteristic table, 59, 228
    edge-triggered device, 59
    excitation equation, 60
    excitation table, 59, 230
Johnson counter, 100, 245

K

Karnaugh maps, 20–28

alternative map for five variables, 21
five variables, 21
four variables, 21
grouping the 0s, 22
grouping the 1s, 22
logically adjacent squares, 20
map-entered variables, 25
physically adjacent squares, 20
three variables, 21
two variables, 21
unspecified entries, 24

L

lambda output function
definition, 86
linear feedback shift register, 212, 213
linear-select multiplexer, 363
block diagram, 363
one-to-one mapping, 363
logic diagram, 724, 734, 740, 754, 765
logic primitives, 36
logic symbols, 35
assertion levels, 37
compact form, 37
logic gate, 37
polarity symbol, 35
qualifying function symbols, 35

M

machine alphabets, 79
input alphabet, 79
output alphabet, 81
state alphabet, 80
summary, 88
machine homomorphism, 126
machine isomorphism, 130
machine state, 77
machine state augmentation, 461
majority circuit, 75
map-entered variables, 25
mapping
into, 84
one-to-one, 84
onto, 84
maxterm, 16
maxterm expansion, 17
Mealy machine, 110, 277, 453, 608
definition, 111
general configuration, 110
next state, 112
output alphabet, 112
summary of equations, 121
merged flow table, 611, 656, 720, 730, 739, 745, 761, 768
merger diagram, 611, 719, 729, 739, 745, 760, 767
merging, 645
merging process, 646, 647
merging rules, 646
metastable, 88, 141, 266
methods of analysis, 136–160, 526–553
analysis examples, 146
input map, 138
next-state map, 137
next-state table, 136
output map, 140
present-state map, 137
state diagram, 143
timing diagram, 141
microprocessor-controlled sequential machines, 448–480
microprocessors, 448
advantages, 448
general considerations, 449
glitches, 451
inport, 467
input port, 450
interrupt structure, 480
machine state augmentation, 461
Mealy machine, 453
Moore and Mealy outputs, 466
multiple machines, 477
output, 467
output port, 450
state codes, 454
state generation, 464
state register, 466
system architecture, 467, 470
minimization techniques, 18–34
algebraic minimization, 19
Karnaugh maps, 20

Quine-McCluskey algorithm, 28
minterm, 16, 31
minterm expansion, 16
modulo-10 counter, 234
equivalent states, 234
input maps, 236
logic diagram, 239
next-state table, 234
output maps, 239
state code assignment, 234
state diagram, 234
modulo-2 addition, 252
modulo-8 counter, 223
equivalent states, 225
erroneous output, 233
input maps, 225, 230
logic diagram, 227, 232
next-state table, 225, 230
output maps, 227, 232
state code assignment, 225
state diagram, 223
Moore machine, 98, 105, 254, 608
definition, 105
general configuration, 105
next state, 106
output alphabet, 106
summary of equations, 121
Moore-Mealy equivalence, 298
Moore-to-Mealy transformation, 298, 307
multigate network
compact format, 41
multiple oscillations, 580
multiple-order hazards, 577
second-order, 577
third-order, 577
multiplexer, 45
2:1 multiplexer, 47
4:1 multiplexer, 47
8:1 multiplexer, 47
16:1 multiplexer, 47
data input selection, 45
definition, 361
linear-select, 363
nonlinear-select, 377
select inputs, 45
multiplexers for next-state logic, 361–400
multiply-by-two operation, 197

N

NAND gate, 36
logical truth table, 40
physical truth table, 40
next state, 82, 85
next-state function, 84
next-state map, 137
next-state table, 136
noncritical race, 584
noncritical races, 583
nonlinear-select multiplexer
block diagram, 379
NOR gate, 36
NOT (inverter), 36
notation, 82
states, 82
vectors, 82
number representations, 8–11
diminished-radix
complement, 9
overflow, 11
radix complement, 10
sign magnitude, 8
number systems, 2–8
positional number system, 2

O

octal number system, 4
binary-coded octal, 4
one-hot codes, 683
one-to-one mapping, 84, 96
onto mapping, 84
OR gate, 36
logical truth table, 39
physical truth table, 39
ordered n-tuple, 82
ordered pairs, 82
ordered quadruple, 82
ordered triple, 82
oscillations, 578–582
output alphabet, 81, 489, 608
output equation, 723, 733, 740, 751, 763, 775
output function, 84
output glitches, 152, 307–343
output map, 140, 612, 691, 723, 733, 740, 751, 763, 775

fast response, 699
minimization, 699
output glitches, 699
slow response, 699
output port, 450
output symbol, 144
overflow, 11

P

parallel-in, parallel-out registers, 95, 197
parallel-in, serial-out registers, 201
parity checking, 493
parity prediction, 492
Petrick algorithm, 33
polarity symbol, 35
positional number systems, 2
binary number system, 3
decimal number system, 4
hexadecimal number system, 5
octal number system, 4
postulate, 12
present input vector, 82
present output vector, 82
present state, 82, 85
present-state map, 137, 372
prime implicant, 22, 23, 30
prime implicant chart, 31
primitive flow table, 611, 616, 716, 728, 737,
742, 755, 766
priority encoder, 53
product of maxterms, 17
product of sums, 17, 22, 23
product term, 16
programmable array logic (PAL), 61, 64, 421
definition, 421
organization, 421
programmable logic array (PLA), 61, 67, 432
definition, 432
organization, 433
programmable logic devices, 61–68, 412–447
AND array, 61
OR array, 61
programmable array logic, 64
programmable logic array, 67
programmable read-only memory (PROM), 61, 413
floating input, 63
fused connection, 62
general block diagram, 416
hard-wired connection, 62
organization, 414
programming, 64
proper subset, 124, 489
pulse mode
analysis procedure, 809
input maps, 814, 820, 827, 834, 840
input pulse restrictions, 809
logic diagram, 810, 818, 829, 850
state diagram, 817, 822
synthesis procedure, 823

Q

qualifying function symbols, 35
Quine-McCluskey algorithm, 28
check symbol, 30
don't care, 32
essential prime implicant, 31
Petrick algorithm, 33
prime implicant chart, 31
secondary essential prime implicant, 31
quotient polynomial, 212

R

race condition, 523
races, 582
critical, 586
cycles, 586
noncritical, 583
radix complement, 10
radix point, 3
range, 126
reduced primitive flow table, 611
reflexive property, 184
registers, 93, 197
parallel-in, parallel-out, 94, 95
parallel-in, serial-out, 94
serial-in, parallel-out, 94

serial-in, serial-out, 94, 96
relations, 82
remainder polynomial, 212
residue checking, 492
right shift count, 219
ripple counter, 99
row-matching, 184

S

second-order hazard, 577
secondary essential prime implicant, 31
secondary variables, 523
self-implied pair, 190
sequential circuits, 78–88
machine alphabet, 79
sequential iterative machines, 481–489
sequential logic circuit, 78
sequential machines
classes, 89
serial adder, 117, 120
serial-in, parallel-out registers (SIPO), 205
serial-in, serial-out registers (SISO), 208
set/reset (*SR*) latch, 57, 609
characteristic table, 57
complementary outputs, 57
negative feedback, 57
setup time, 88, 266
shift left algebraic (SLA), 218, 220
shift left logical (SLL), 220
shift registers, 94
shift right algebraic (SRA), 218, 221
shift right logical (SRL), 218, 220
shifting left, 197
shifting right, 197
sign magnitude, 8
stable state, 523, 531, 612, 617
definition, 524
standard product of sums, 17
standard sum of products, 16
state, 77
state alphabet, 80, 608
state code, 77, 80, 144, 454
adjacency, 258
state diagram, 143, 153, 183, 533
state generation, 464
horizontal organization, 464
vertical organization, 464
state machine, 123
general model, 123
state names, 612
state symbol, 143
static hazards, 553
static-0 hazard, 554
static-1 hazard, 554
storage elements, 56–61
D flip-flop, 57
JK flip-flop, 59
SR latch, 57
T flip-flop, 60
strongly connected machine, 125
strongly connected sets, 703
submachine, 124
subset, 124, 489
sum of minterms, 16
sum of products, 17, 22
sum term, 16
summary of equations, 121
symbol, 79
symmetric property, 184
synchronous counters, 99, 102, 223
BCD counter, 234
binary-to-Gray, 248
Johnson counter, 245
modulo-8, 223
modulo-10, 234
synchronous registers, 197–222
error correction, 212
error detection, 212
parallel in, serial out, 201
parallel-in, parallel-out, 197
serial-in, parallel-out, 205
serial-in, serial-out, 208
synchronous sequential machines, 79, 82, 123, 181
additional definitions, 123
complete, 161
definition, 84
deterministic, 85
equivalent machines, 132
fault, 489
general model, 86
incomplete, 161, 162
machine homomorphism, 126

machine isomorphism, 130
parity checking, 497
state machine, 123
strongly connected, 125
summary, 134
terminal state, 125
synthesis examples, 716–777
equivalent states, 719, 729, 738, 744, 758, 767
excitation equation, 731, 740, 749, 763, 773
excitation map, 722, 731, 740, 749, 763, 773
logic diagram, 724, 734, 740, 754, 765
merged flow table, 720, 730, 739, 745, 761, 768
merger diagram, 719, 729, 739, 745, 760, 767
output equation, 723, 733, 740, 751, 763, 775
output map, 723, 733, 740, 763, 775
primitive flow table, 716, 728, 737, 742, 755, 766
synthesis of combinational logic, 42–45
synthesis procedure, 43, 182, 610–715, 823–850
equivalent states, 184, 611, 632
excitation equation, 612, 661
excitation map, 612, 661
implication table, 187
logic diagram, 612, 710
merged flow table, 611, 656
merger diagram, 611, 645
next-state table, 186
output equation, 612, 691
output map, 612, 691
primitive flow table, 611, 616
reduced primitive flow table, 611
row matching, 184
SR latches, 823
SR latches with *D* flip-flops, 844
SR-T flip-flops, 836
state diagram, 183, 185, 611, 612
T flip-flops, 830
sythesis examples
output map, 751

T

T flip-flop, 60, 830
characteristic table, 60
excitation equation, 61
tabular method, 28
terminal state, 125, 143, 162
theorem 1: 0 and 1 associated with a variable, 14
theorem 2: 0 and 1 complement, 14
theorem 3: idempotent laws, 14
theorem 4: involution law, 15
theorem 5: absorption laws, 1 15
theorem 6: absorption laws, 2 15
theorem 7: DeMorgans laws, 15
third-order hazard, 577
timing diagram, 141
top-down approach, 182
transient state, 612, 617
transition diagram 665, 670, 748
transitive property, 184
triple redundancy, 492
two-level network, 91

U

unspecified entries, 24, 659, 670
unstable state
definition, 524
output values, 693

V

vector, 79
vertical organization, 464